Biological Inorganic Chemistry

Structure and Reactivity

Biological Inorganic Chemistry

Structure and Reactivity

Ivano Bertini
University of Florence

Harry B. Gray
California Institute of Technology

Edward I. Stiefel
Princeton University

Joan Selverstone Valentine
UCLA

UNIVERSITY SCIENCE BOOKS
Sausalito, California

University Science Books
www.uscibooks.com

Production Manager: Mark Ong
Manuscript Editor: Jeannette Stiefel
Design: Mark Ong
Cover Design: George Kelvin
Illustrator: Lineworks
Compositor: Asco Typesetters
Printer & Binder: Maple-Vail Book Manufacturing Group

This book is printed on acid-free paper.

ISBN 10: 1-891389-43-2
ISBN 13: 978-1-891389-43-6

Library of Congress Cataloging-in-Publication Data

Biological inorganic chemistry : structure and reactivity / edited by Ivano Bertini . . . [et al.].
p. cm.
Includes bibliographic references (p.).
ISBN 1-891389-43-2 (alk. paper)
1. Bioinorganic chemistry. I. Bertini, Ivano.
QP531.B547 2006
612'.01524—dc22

2006044712

Printed in the United States of America
10 9 8 7 6 5 4 3 2 1

Dedicated to Jeannette and Ed Stiefel,
with admiration and great affection

Contents in Brief

Contents

List of Contributors

Philip Aisen, Department of Physiology and Biophysics, Albert Einstein College of Medicine, Bronx, New York 10461

Michael W. W. Adams, Department of Biochemistry and Molecular Biology and Center for Metalloenzyme Studies, University of Georgia, Athens, Georgia 30602

Bruce A. Averill, Department of Chemistry, University of Toledo, Toledo, Ohio 43606

Gerald T. Babcock, Department of Chemistry, Michigan State University, East Lansing, Michigan 48828

Lucia Banci, Magnetic Resonance Center and Department of Chemistry, University of Florence, Sesto Fiorentino, Italy 50019

Helmut Beinert, Institute for Enzyme Research, University of Wisconsin, Madison, Wisconsin 53726

Ivano Bertini, Magnetic Resonance Center and Department of Chemistry, University of Florence, Sesto Fiorentino, Italy 50019

Joan B. Broderick, Department of Chemistry and Biochemistry, Montana State University, Bozeman, Montana 59717

Alison Butler, Department of Chemistry and Biochemistry, University of California, Santa Barbara, Santa Barbara, California 93106

Stefano Ciurli, Laboratory of Bioinorganic Chemistry, Department of Agro-Environmental Science and Technology, University of Bologna, I-40127, Bologna, Italy

J. A. Cowan, Chemistry, Ohio State University, Columbus, Ohio 43210

Valeria Culotta, Environmental Health Sciences, Johns Hopkins University School of Public Health, Baltimore, Maryland 21205

David M. Dooley, Department of Chemistry and Biochemistry, Montana State University, Bozeman, Montana 59717

Torbjörn Drakenberg, Department of Biophysical Chemistry, Lund University, SE-22100 Lund, Sweden

David J. Eide, Department of Nutritional Sciences, University of Wisconsin, Madison, Wisconsin 53706

Shelagh Ferguson-Miller, Biochemistry and Molecular Biology, Michigan State University, East Lansing, Michigan 48824

Bryan Finn, IT Department, Swedish University of Agricultural Sciences, SE-23053 Alnarp, Sweden

Marc Fontecave, Université Joseph Fourier, CNRS–CEA, CEA–Grenoble, 38054 Grenoble, France

Sture Forsén, Department of Biophysical Chemistry, Lund University, SE-22100 Lund, Sweden

C. David Garner, The School of Chemistry, The University of Nottingham, Nottingham NG7 2RD, United Kingdom

Edith B. Gralla, Department of Chemistry and Biochemistry, UCLA, Los Angeles, California 90095

Harry B. Gray, Beckman Institute, California Institute of Technology, Pasadena, California 91125

Hans-Juergen Hartmann, Anorganische Biochemie Physiologisch Chemisches Institut, University of Tübingen, Tübingen, Germany

James A. Ibers, Department of Chemistry, Northwestern University, Evanston, Illinois 60208

Geoffrey B. Jameson, Centre for Structural Biology, Institute of Fundamental Sciences, Chemistry, Massey University, Palmerston North, New Zealand

M. Claire Kennedy, Department of Chemistry, Gannon University, Erie, Pennsylvania 16561

Judith Klinman, Departments of Chemistry and of Molecular and Cell Biology, University of California, Berkeley, Berkeley, California 94720

Jean LeGall, Instituto de Tecnologia Química e Biológica, Universidade Nova de Lisboa, Oeiras, Portugal

Peter F. Lindley, Instituto de Tecnologia Química e Biológica, Universidade Nova de Lisboa, Oeiras, Portugal

Yi Lu, Department of Chemistry, University of Illinois at Urbana-Champaign, Urbana, Illinois 61801

Claudio Luchinat, Magnetic Resonance Center and Department of Agricultural Biotechnology, University of Florence, Sesto Fiorentino, Italy 50019

Thomas J. Lyons, Department of Chemistry, University of Florida, Gainesville, Florida 32611

John S. Magyar, Beckman Institute, California Institute of Technology, Pasadena, California 91125

Stephen Mann, School of Chemistry, University of Bristol, Bristol BS8 1TS, United Kingdom

Michael J. Maroney, Department of Chemistry, University of Massachusetts, Amherst, Amherst, Massachusetts 01003

Jonathan McMaster, The School of Chemistry, The University of Nottingham, Nottingham NG7 2RD, United Kingdom

Christopher Muncie, School of Chemistry, University of Edinburgh, Edinburgh, United Kingdom

Aram Nersissian, Chemistry Department, Occidental College, Los Angeles, California 90041

William E. Newton, Department of Biochemistry, The Virginia Polytechnic Institute and State University, Blacksburg, Virginia 24061

Thomas V. O'Halloran, Chemistry Department, Northwestern University, Evanston, IL 60208

Thomas L. Poulos, Departments of Molecular Biology and Biochemistry, Chemistry, and Physiology and Biophysics, University of California, Irvine, Irvine, California 92617

Lawrence Que, Jr., Department of Chemistry and Center for Metals in Biocatalysis, University of Minnesota, Minneapolis, Minnesota 55455

Stephen W. Ragsdale, Department of Biochemistry, University of Nebraska, Lincoln, Nebraska 68588

Keith Rickert, Department of Cancer Research WP26-462, Merck & Co., P. O. Box 4, West Point, Pennsylvania 19486

James A. Roe, Department of Chemistry and Biochemistry, Loyola Marymount University, Los Angeles, California 90045

Roopali Roy, Department of Biochemistry and Molecular Biology and Center for Metalloenzyme Studies, University of Georgia, Athens, Georgia 30602

Peter J. Sadler, School of Chemistry, University of Edinburgh, Edinburgh, United Kingdom

Bibudhendra Sarkar, Structural Biology and Biochemistry, The Hospital for Sick Children and the University of Toronto, Toronto, Ontario M5G1X8 Canada

Bryan F. Shaw, Department of Chemistry and Biochemistry, UCLA, Los Angeles, California 90095

Michelle A. Shipman, School of Chemistry, University of Edinburgh, Edinburgh, United Kingdom

Edward I. Stiefel, Department of Chemistry, Princeton University, Princeton, New Jersey 08544

JoAnne Stubbe, Departments of Chemistry and Biology, Massachusetts Institute of Technology, Cambridge, Massachusetts 02139

Elizabeth C. Theil, Children's Hospital Oakland Research Institute and the University of California, Berkeley, Oakland, California 94609

Paola Turano, Magnetic Resonance Center and Department of Chemistry, University of Florence, Sesto Fiorentino, Italy 50019

Joan Selverstone Valentine, Department of Chemistry and Biochemistry, UCLA, Los Angeles, California 90095

Ulrich Weser, Anorganische Biochemie Physiologisch Chemisches Institut, University of Tübingen, Tübingen, Germany

James W. Whittaker, Environmental and Biomolecular Systems, Oregon Health and Science University, Beaverton, Oregon 97006

Dennis R. Winge, Departments of Medicine and Biochemistry, University of Utah Health Sciences Center, Salt Lake City, Utah 84132

Jay R. Winkler, Beckman Institute, California Institute of Technology, Pasadena, California 91125

António V. Xavier, Instituto de Tecnologia Química e Biológica, Universidade Nova de Lisboa, Oeiras, Portugal

Charles Yocum, Chemistry and MCD Biology, University of Michigan, Ann Arbor, Michigan 48109

Shinya Yoshikawa, Department of Life Science, University of Hyogo, Kamigohri Akoh, Hyogo 678-1297, Japan

Preface

Life depends on the proper functioning of proteins and nucleic acids that very often are in combinations with metal ions. Elucidation of the structures and reactivities of metalloproteins and other metallobiomolecules is the central goal of biological inorganic chemistry.

One of the grand challenges of the 21st century is to deduce how a specific gene sequence codes for a metalloprotein. Such knowledge of genomic maps will contribute to the goal of understanding the molecular mechanisms of life. Specific annotations to a sequence often allude to the requirement of metals for protein function, but it is not yet possible to read that information from sequence alone. Work in biological inorganic chemistry is critically important in this context.

Our goal at the outset was to capture the full vibrancy of the field in a textbook. Our book is divided into Part A, "Overviews of Biological Inorganic Chemistry," which sets forth the unifying principles of the field, and Part B, "Metal Ion Containing Biological Systems," which treats specific systems in detail. Tutorials are included for those who wish to review the basics of biology and inorganic chemistry; and the Appendices provide useful information, as does "*Physical Methods in Bioinorganic Chemistry*" (see Appendix III), which we highly recommend.

Biological inorganic chemistry is a very hot area. It has been our good fortune to work with many exceptionally talented contributors in putting together a volume that we believe will be a valuable resource both for young investigators and for more senior scholars in the field.

—The Editors

Acknowledgments

Working with so many gifted authors has been a real treat for us. The project also has presented many challenges. We would not have made it to the finish line without the able assistance of many colleagues. First and foremost, the brilliant editorial hand of Jeannette Stiefel made the manuscript a real book rather than just a random collection of vignettes. We cannot thank Jeannette enough for her contributions to the final product. In Florence, Paola Turano kept everyone in line; she was simply fantastic! At Caltech, John Magyar helped immensely in reading all the proof sheets and offering many suggestions for improvements. Both John and Paola played a leading role in the most critical stages of the project.

We are greatly in debt to Larry Que for his contributions; in addition to numerous helpful suggestions over the course of the project, Larry worked very closely with us in all aspects of writing and editing Chapter 11. We only have ourselves to blame if the final product does not meet his very high standards.

Edith Gralla, Aram Nersissian, and Bryan Shaw at UCLA, and Jim Roe at Loyola Marymount put together tutorials that have greatly enhanced the pedagogical value of the book. The book was class tested at Princeton and UCLA. We thank all the students who made helpful comments.

We lost three coauthors during the course of the project. Jerry Babcock, Jean LeGall, and Antonio Xavier were great scientists and dear friends. We miss them very much.

Our publisher, Bruce Armbruster, and his team at University Science Books cheered us on through what seemed to some of us to be an eternity. We especially thank Kathy Armbruster for her patience and unwavering support, Jane Ellis for her persistence and good humor, and Mark Ong for putting all the pieces together to bring the project to a successful conclusion. We acknowledge six other colleagues: Catherine May and Rick Jackson at Caltech; Margaret Williams and Rhea Rever at UCLA; Ingrid Hughes at Princeton; and Simona Fedi at CERM (Florence) with thanks for their dedication to our cause.

Ivano Bertini
Harry B. Gray
Edward I. Stiefel
Joan Selverstone Valentine

CHAPTER I

Introduction and Text Overview

Ivano Bertini
Magnetic Resonance Center and Department of Chemistry
University of Florence
Sesto Fiorentino, Italy 50019

Harry B. Gray
Beckman Institute
California Institute of Technology
Pasadena, CA 91125

Edward I. Stiefel
Department of Chemistry
Princeton University
Princeton, NJ 08544

Joan Selverstone Valentine
Department of Chemistry and Biochemistry, UCLA,
Los Angeles, CA 90095

Contents

I.1. The Elements of Life

Eleven elements of the periodic table are required for *all* forms of life and some thirteen additional elements are essential components of most living species. An additional seven or eight elements are used by *some* organisms on our planet. The periodic table of biological elements, coded by their relative importance for life, is shown in Fig. I.1. This table highlights the >30 elements required for the diversity of life on Earth.

The familiar elements, carbon, hydrogen, nitrogen, oxygen, phosphorus, and sulfur (CHNOPS; the big six, which are well covered in biochemistry texts) provide the building blocks for major cellular components including proteins, nucleic acids, lipids–membranes, polysaccharides, and metabolites. Yet, despite the tremendous organic diversity that is exhibited, life cannot survive with only these principal elements. It is now clear that >20 additional elements are essential for most species to function. Conduction of nerve impulses, hydrolysis and formation of adenosine triphosphate (ATP), regulation of gene expression, control of cellular processes and signaling, and catalysis of many key reactions of metabolism require elements beyond the big six. Understanding the roles that metallic and nonmetallic elements play in biological systems is the goal of biological inorganic (bioinorganic) chemistry.

I.2. Functional Roles of Biological Inorganic Elements

Inorganic elements are involved in all life processes including:

- Charge Balance and Electrolytic Conductivity: Na, K, Cl
- Structure and Templating: Ca, Zn, Si, S
- Signaling: Ca, B, NO,
- Brønsted Acid–Base Buffering: P, Si, C

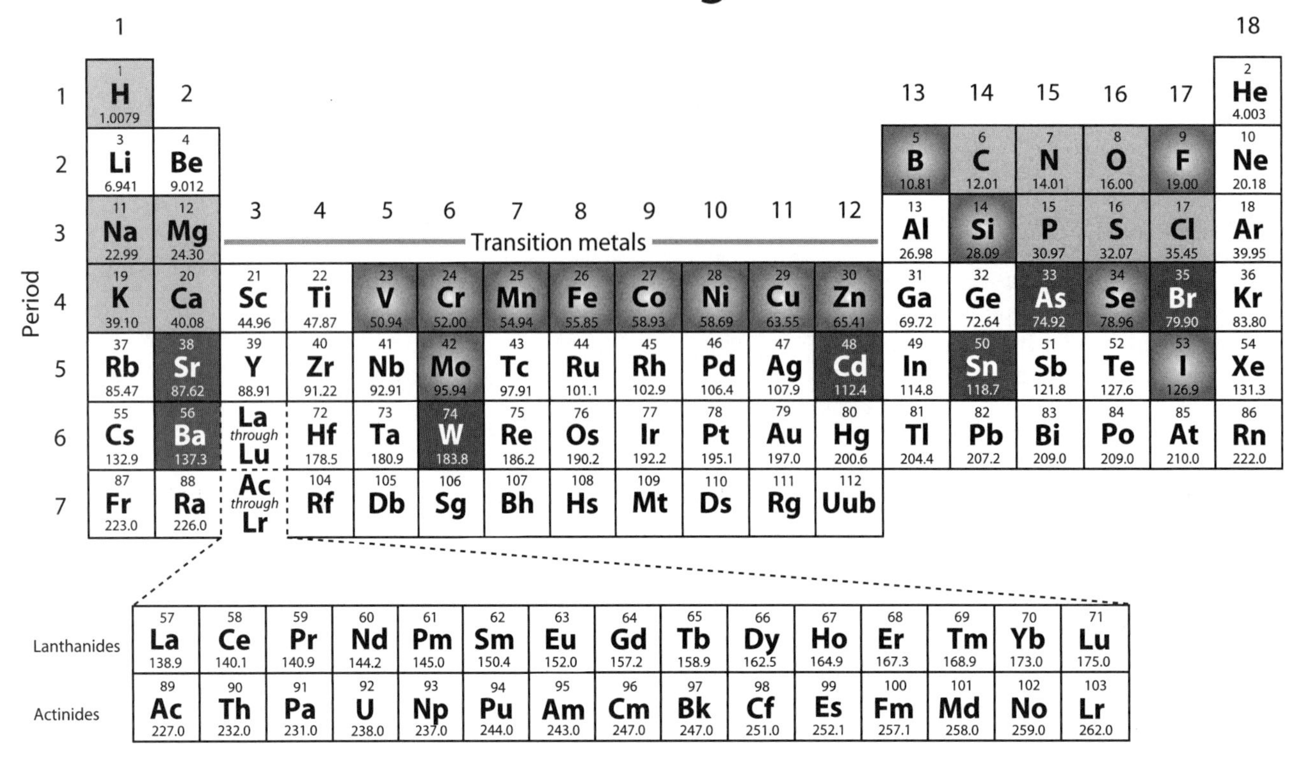

Fig. I.1.
The periodic table of the elements highlighting those elements used by living organisms. The degree of utilization by life on Earth is indicated by the shading of the elements (as defined in the key). In time, additional elements may be discovered that are required by humans or by other organisms. Such elements have not yet been identified, either because the organism that uses them has not been thoroughly studied, or the elements are required in such trace amounts that we have not yet recognized their essentiality.

- Lewis Acid–Base Catalysis: Zn, Fe, Ni, Mn
- Electron Transfer: Fe, Cu, Mo
- Group Transfer (e.g., CH_3, O, S): V, Fe, Co, Ni, Cu, Mo, W
- Redox Catalysis: V, Mn, Fe, Co, Ni, Cu, W, S, Se
- Energy Storage: H, P, S, Na, K, Fe
- Biomineralization: Ca, Mg, Fe, Si, Sr, Cu, P

Owing to the great advances in research in biological inorganic chemistry, we now know the structures of many components of the systems that biology has adapted through evolution to perform these essential functions. Moreover, many relationships between structure and function have been elucidated and the electronic structural, mechanistic, and genetic underpinnings of these relationships are being revealed for the first time. What is more, biological inorganic chemistry has profoundly impacted both environmental science (Chapter II) and medicine (Chapter VII).

I.3. A Guide to This Text

A cornerstone of bioinorganic chemistry is the biochemistry of the major organic constituents of life. Readers should feel free to consult Tutorial I on "Cell Biology, Biochemistry, and Evolution" before tackling those parts of the text that require a strong background in biology and biochemistry. A second major foundation for bioinorganic chemistry is the coordination chemistry of the transition metal elements. Tutorial II on "Coordination Chemistry" should prove helpful to those readers who would like to review this area of inorganic chemistry.

The book is divided into two main parts: Part A contains general overviews of broad areas of biological inorganic chemistry as well as introductory material that is necessary for understanding the material in part B. Specifically, Chapters III and IV on "Metal Ions in Proteins" and "Special Cofactors and Metal Clusters", respectively, are clearly prerequisites for the study of much of the material in the text. Chapters II, V, VI, and VII provide overviews of large subareas of bioinorganic chemistry. The chapters in Part A are generally longer, more pedagogically oriented, and contain bibliographies to allow the reader to access the material in further depth.

Part B contains more specific treatments of metal ions in biological systems. Here we strive for detailed coverage with general and specific cited references. The goal in Part B is to bring the reader to the state of the art in each subarea of the field.

Appendix I lists the abbreviations used in this text. Appendix II is a glossary of the terms that are common in biological inorganic chemistry.

In recent years, a great many three-dimensional (3D) metalloprotein structures have been determined using X-ray crystallography and/or nuclear magnetic resonance (NMR) spectroscopy. You can examine these structures in full detail, thanks to the Protein Data Bank (PDB). We strongly recommend that you view protein and other structures in color. Accordingly, for every protein where a 3D structure is available, we have given the four-letter PDB ID for that structure. By simply accessing the PDB on the Internet, you can view and manipulate the structure on your own computer. Thus, in using this text we expect that you, the reader, will go to the PDB web site (www.rcsb.org/pdb) and view a protein in full detail. Appendix IV provides an introduction to the PDB, which should allow you to render the PDB structures using your own computer and view, rotate, zoom, and color the structure to your own specification. Examining these images in conjunction with the text will give you a feel for any given structure and will help in understanding the material in each chapter. We expect that the use of the PDB will be an integral part of any course based on this text.

PART A

OVERVIEWS OF BIOLOGICAL INORGANIC CHEMISTRY

CHAPTER II

Bioinorganic Chemistry and the Biogeochemical Cycles

Edward I. Stiefel
Department of Chemistry
Princeton University
Princeton, NJ 08544

Contents

II.1. Introduction

The surface of the Earth is shaped by life. Our planet is unique in the Solar System and, quite possibly, in the Universe. Earth's unusual atmosphere and biosphere, which exist and *sustain themselves* far from chemical equilibrium, are both truly unusual. The atmospheric composition of Earth has evolved over geological time by the combination of geological and biological processes. Organisms on Earth execute diverse chemical transformations in conjunction with their metabolic requirements for survival. The resultant conversions and byproducts have transformed Earth and now maintain niches within the planet for myriad extant life forms. We understand, at least in qualitative form, the biogeochemical cycles of the elements, which maintain the global, regional, and local environments that facilitate the existence, continuance, and diversity of life on Earth.

Each of the major biological elements [C, H, N, O, P, and S (CHNOPS)] has a well-defined cycle. Key "trace" elements, including the transition metals V, Mn, Fe, Co, Ni, Cu, Zn, Mo, and W, play major roles in controlling the major element cycles through their essential roles in metalloenzymes. The second half of the twentieth century has seen the elucidation of many of the central reaction steps and enzymatic catalysts of the biogeochemical cycles. The quantitative delineation of these cycles and the understanding of their evolution, integration, and future development are key challenges for the twenty-first century.

Biological inorganic chemistry may be organized in many ways. One may systematize the field in terms of particular elements that define classes of enzymes (those containing Fe, Cu, Zn, etc.), or, one can organize the subject in terms of reaction types (oxygenases, ligases, proteases, etc.). Alternatively, a biomedical or an agronomic approach can be used to categorize the relevant processes in terms of their physiological roles in particular biological systems.

In this chapter, the bioinorganic systems that are portrayed in molecular detail in this text are placed in the context of the biogeochemical cycles that define Earth and its varied habitats. Such an organization demands not only inspection of present systems, but also requires scrutiny of the past. In contrast to pure chemistry or physics, biology and geology are historical sciences. The evolution of life on Earth and the continued differentiation of its life forms are the result of untold numbers of evolutionary machinations, whose details have led us to the present state. Therefore, to appreciate *where we are and how we got there,* it is important to think in terms of the history and evolutionary pressures that shaped our current biosphere and its inhabitant organisms. Inclusion of such a perspective adds an additional powerful approach to attempts to sort out the *raison d'être* and comprehend the continuing and future development of molecular bioinorganic systems and the biogeochemical cycles that they control.

II.2. The Origin and Abundance of the Chemical Elements

II.2.1. Overview and Astrobiological Connection

The chemical elements were originally produced through nucleosynthesis, first in the "Big Bang" and then in stars such as our Sun and in supernova explosions. These processes produced the O, Mg, Al, Si, Fe, and other elements that eventually led to the formation of rocky planets such as Earth. The same processes provided the C, N, O, P, S, and the other elements necessary for life.

Current cosmological models invoke the Big Bang as the seminal event in which fundamental constituents of matter (protons, neutrons, and electrons) first formed. During early moments of the Big Bang, as the nascent Universe cooled, the temperature and density remained high enough to produce some helium and tiny amounts of deuterium, lithium, beryllium, and boron. However, as the expansion of the Big Bang continued, the Universe cooled and its density became too low to sustain additional nucleosynthesis. In the absence of other processes, the Universe was mostly hydrogen, with helium constituting most of the remainder. The dearth of light and heavy elements in the Universe at that point would have made life (and this book!) highly improbable. However, in the course of time other processes emerged.

Gravitational attraction led to the aggregation of galaxies and stars, in which nucleosynthetic processes continued. Hydrogen fusion to form helium (called hydrogen burning) is the major process in first generation stars. After a certain proportion of the hydrogen is converted to helium, the star contracts, its internal temperature rises, and so-called helium burning is initiated. This process produces many of the elements with atomic mass up to 40, in small amounts. Exhaustion of the hydrogen and helium fuel results in further gravitational collapse which, depending on the size of the star, may culminate in a nova or supernova explosion. The ferocity of the latter is such that virtually all elements in the periodic table are produced to some extent, with iron, whose isotope ^{56}Fe is the most stable nuclear species, produced in significant amounts. The explosion of a supernova also spews the elements into interstellar space where they become raw materials for the next generation of star and planet formation.

Second or later generation stars, such as our Sun, are formed from the hydrogen and helium, which are still the dominant elements; but these stars also contain small amounts of nitrogen and carbon that serve as catalysts for hydrogen-burning reactions. Thus, the elements produced in nova and supernova explosions alter the character of the later generation stars (including our Sun) and allow the formation of rocky planets in the solar systems that form around them.

The elements essential for the formation of Earth-like planets and the >25 elements used in the chemistry of life are the legacy of astronomical events that occurred billions of years ago, including the Big Bang, the evolution of stars, and their violent explosions. The raw material of biological inorganic chemistry is stardust.

II.2.2. Abundance of Elements and Molecules in the Solar System

The legacy of elements from which the Solar System condensed allowed formation of rocky bodies in the inner solar system and moons around the giant planets. The abundance of elements in the Solar System, as shown in Fig. II.1, provided the raw materials for the evolution of life on Earth and possibly on other bodies such as Mars and Europa. For example, the presence of oxygen and hydrogen permitted such planets or moons to retain water. Moreover, moderate surface or subsurface temperature fosters the maintenance of water in its liquid state, either at present or at some time in the past. Liquid water is considered a *sine qua non* for the existence of life, as we know it.

II.2.3. Earth as a Planetary System: Comparison to Venus and Mars

The comparison of Earth's atmosphere with that of its nearest planetary neighbors, Mars and Venus, is illuminating (Fig. II.2). Venus has a surface temperature of ~500°C and an atmospheric density almost 100 times that of Earth's (9.321 MPa for Venus vs. 0.101 MPa for Earth). The major constituent of the Venusian atmosphere is CO_2 (95.6%) with N_2 (3.2%) as the second most common gas (Fig. II.2). Mars, whose surface temperature averages −63°C, has an atmospheric pressure (at 700–900 Pa) < 1/100th that of Earth. Thus, Mars and Venus differ by a factor of $>10^4$ in the density of their atmospheres. Yet, the atmospheric *composition* of Mars and Venus is almost identical, with Mars also having CO_2 (95.0%) and N_2 (2.7%) as its most common gases (Fig. II.2).

When the three terrestrial planets are viewed from deep space, Venus and Mars, respectively, the nearest and farthest of these planets from the Sun, have atmospheres

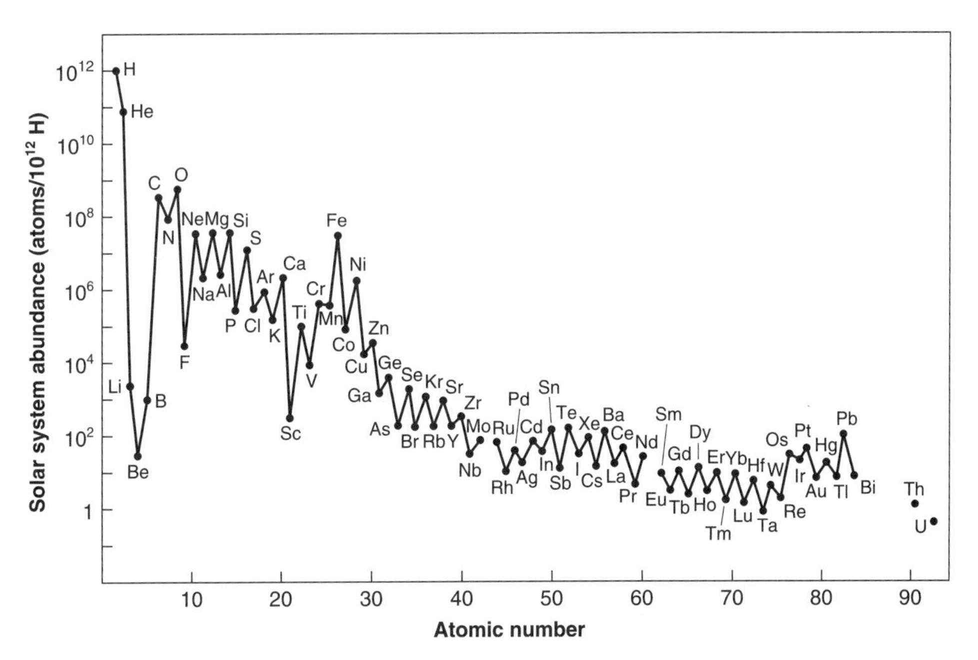

Fig. II.1.
The abundance of elements in the Solar System. A logarithmic plot of relative abundance in parts per billion (ppb) on the vertical axis versus periodic position on the horizontal axis. Hydrogen and helium are far and away the most common elements. [Adapted from Cox, 1989.]

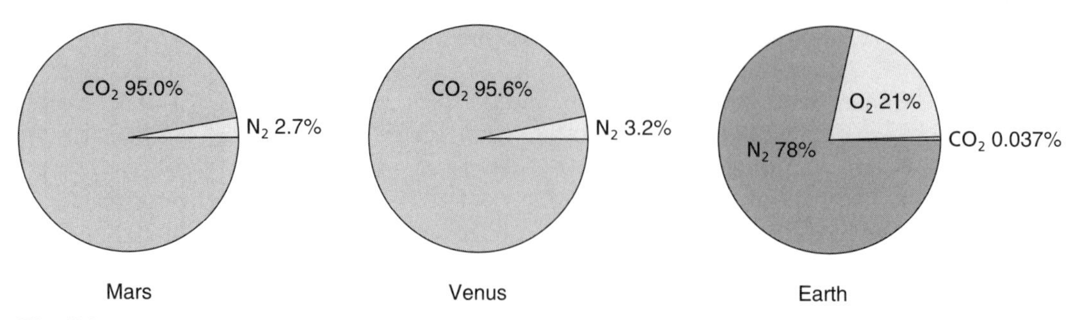

Fig. II.2.
Atmospheric compositions of Mars, Venus, and Earth.

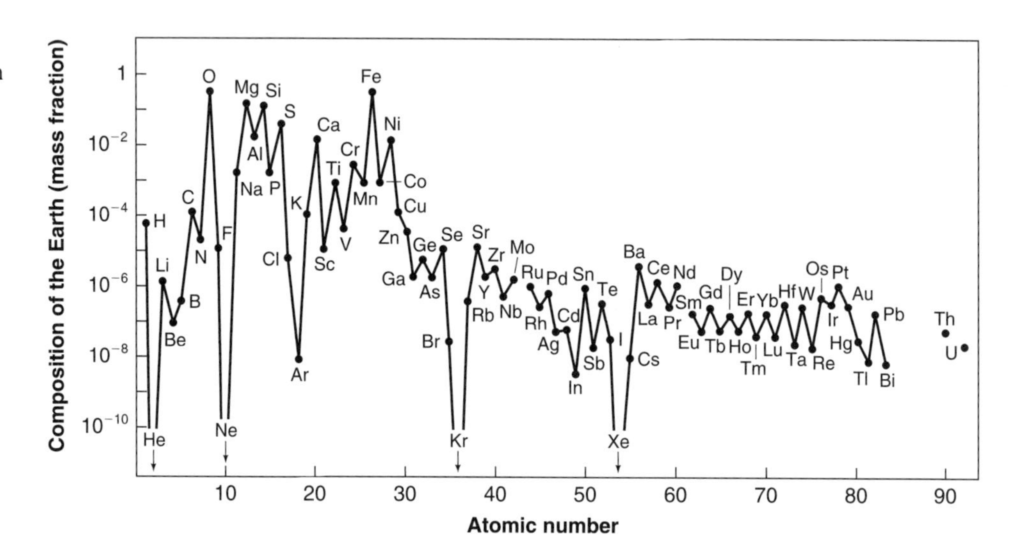

Fig. II.3.
The abundance of elements in the Earth. Comparison with Fig. II.1 shows the dearth of hydrogen and the relative abundance of the first-row transition elements. [Adapted from Cox, 1989.]

dominated by CO_2. Earth, lying at the intermediate distance and temperature, is clearly anomalous, with its present atmosphere dominated by N_2 (78%) and O_2 (21%) and containing only ~0.03% CO_2. Molecular oxygen is a rare and unlikely gas in our Solar System and probably in any planetary system. High oxygen levels are particularly anomalous, since there are reducing materials of both (geo)chemical and biological origin in the atmosphere, hydrosphere, and upper lithosphere, which should react with oxygen, depleting it over geological time scales. Yet, oxygen is present in large amounts in the atmosphere and oceans and in many lakes and rivers on Earth. The presence of oxygen is entirely the result of one biological process—oxygen-evolving photosynthesis executed by higher plants, algae, and cyanobacteria.

II.2.4. Abundance of Elements on Earth

The abundance of elements in the whole Earth and in the uppermost layer of the solid Earth, the crust, are shown in Figs. II.3 and II.4, respectively. These percentages are often quoted as a measure of elemental availability for biological systems. However, whole Earth and crustal abundances alone are not a sufficient criterion for the utilization of an element by life. While a minimum abundance is clearly required, of equal importance is the chemical *availability* of the element in a solid or soluble form, from which organisms can effectively acquire the element. In this respect, the abundance of elements in the aqueous phases of the planet is particularly relevant to the biological use of elements on Earth.

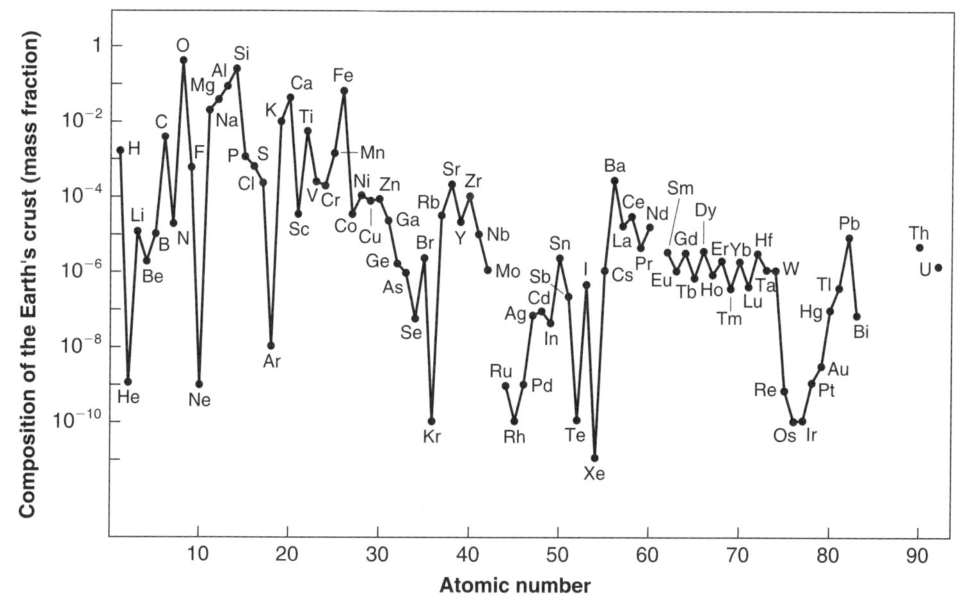

Fig. II.4.
The abundance of elements in the Earth's crust shown on a logarithmic scale. Comparison with Figs. II.1 and II.3 reveals differences attributable to segregation of elements between the core and mantle versus the crust. [Adapted from Cox, 1989.]

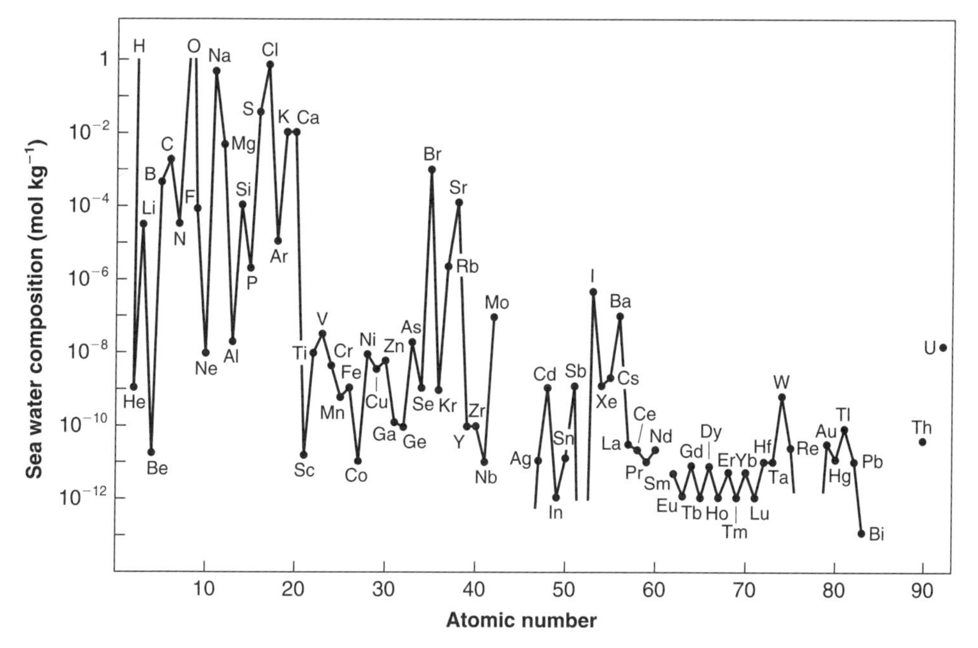

Fig. II.5.
The abundance of elements in the oceans of Earth. Note the enormous differences with Fig. II.4 for many elements (e.g., Fe, Mo, Cl, Br, Al) showing that the crucial aqueous phases do not reflect the overall composition of the planet. [Adapted from Cox, 1989.]

II.2.5. Abundance of Elements in the Oceans

The abundance of elements in the oceans of modern Earth is shown in Fig. II.5. The concentrations of the transition metals are particularly relevant to this text. The first-row transition elements have low, but not insignificant, abundances although in most cases it is not yet clear what physical or chemical form these elements take.

For the second and third transition periods, molybdenum and tungsten, respectively, are the most abundant metals in the oceans. This correlates with the fact that molybdenum and tungsten are the only elements of their respective periods that have known biological functions. Indeed, molybdenum is the most abundant transition metal in the oceans. To what extent this abundance is responsible for the extensive usage of Mo in biology, as opposed to its unique chemistry, is not entirely clear. Prior to the development of an aerobic hydrosphere and atmosphere, molybdenum was not as available, since it was likely trapped in the form of MoS_2. Molybdenum and tungsten are the only transition metals that have anionic forms at pH 8.3, the value of the modern oceans. The highly soluble molybdate (MoO_4^{2-}) and tungstate (WO_4^{2-}) ions, while clearly available, must be transported into cells by methods that differ dramatically from those used to acquire the largely cationic (or organic-bound) di- or trivalent first-row transition species present in the marine environment.

II.2.6. The Suitability of the Chemical Elements

In addition to being sufficiently abundant and potentially available, a chemical element must be exploitable by organisms for their metabolic function and survival. Thus, in response to selective pressure, specific elements are adopted and adapted through Darwinian evolution (natural selection) to provide particular chemical attributes within functional biological molecules. Such a process has been called *The Natural Selection of the Chemical Elements* by Williams and Frausto da Silva.

II.3. The Carbon/Oxygen/Hydrogen Cycles

II.3.1. Carbon and Oxygen Cycles

The carbon cycle has received great public attention owing to the increasing level of carbon dioxide in Earth's atmosphere and the potential radiative forcing this provides for global warming through an enhanced "greenhouse effect". Indeed, atmospheric CO_2 has increased from the preindustrial value of ~280 ppmV (parts per million by volume = ppmV) to current levels of ~370 ppmV (in the year 2003). While this small value is far from the ~960,000 ppmV on Mars and Venus (96%), and much lower than that found in earlier epochs on Earth, the rapidity of the current increase and its potential for significantly enhanced greenhouse warming are cause for concern. Extant discussions focus on reducing CO_2 emissions or increasing CO_2 uptake through increased photosynthetic productivity or other methods of sequestration. Issues central to the carbon cycle involve, at their core, biochemistry controlled by metalloenzymes.

At least on a global scale, the carbon and oxygen cycles cannot be considered independently. A simplified picture of the intertwined nature of these cycles is presented in Fig. II.6. On early Earth, oxygen was absent and CO_2 levels were far higher than present values, perhaps as high, proportionately, as on present day Mars and Venus. Early organisms used solar photons to drive the light reactions of photosynthesis (phototrophy) or the inorganic chemical reactions (chemolithotrophy). These reactions provided energy necessary to produce the low-potential electrons required to reduce the abundant CO_2 or bicarbonate, HCO_3^-, the dominant forms of inorganic carbon. The source of the reducing equivalents (electron donor) was likely different on early Earth than it is now.

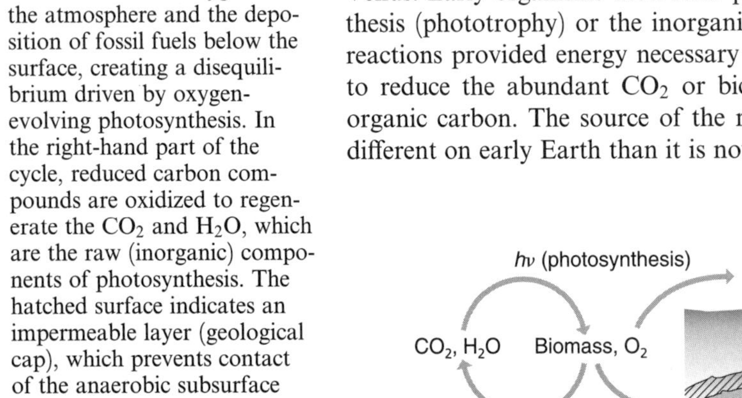

Fig. II.6.
A simplified schematic of the biological part of the global biogeochemical cycle of carbon. The left half of the cycle represents the accumulation of molecular oxygen in the atmosphere and the deposition of fossil fuels below the surface, creating a disequilibrium driven by oxygen-evolving photosynthesis. In the right-hand part of the cycle, reduced carbon compounds are oxidized to regenerate the CO_2 and H_2O, which are the raw (inorganic) components of photosynthesis. The hatched surface indicates an impermeable layer (geological cap), which prevents contact of the anaerobic subsurface phases with the aerobic strata and atmosphere above.

At present, in the quantitatively most significant form of photosynthesis on Earth, the electrons ultimately come from water. A complex coupled system involving two photosystems, PSI and PSII, provides the reducing potential, which through chemical mediators and catalysts leads to reduction of CO_2 to organic compounds and the oxidation of H_2O to O_2 (Sections X.3; X.4; Tutorial I). Oxygenic photosynthesis is found in a wide variety of eukaryotic plants and in a specific group of bacteria, the cyanobacteria (formerly called blue-green algae).

A simpler form of photosynthesis, present today in certain photosynthetic bacteria, was likely operative on early Earth. This form of photosynthesis, found exclusively in the realm of bacteria, uses reduced compounds of sulfur (H_2S, S, or thiosulfate), H_2, or organic acids (acetate, succinate) as electron donors, and, generally, uses only a single photosystem. Because the reducing agents necessary for this reaction are in limited supply, non-oxygen-evolving photosynthesis is capable of carrying out only a small amount of carbon dioxide reduction compared to oxygen-evolving photosynthesis. On the other hand, once the evolutionary innovation of oxygen-evolving photosynthesis was achieved, with ubiquitous H_2O serving as the electron donor, an increase in planetary biomass by as much as two or three orders of magnitude became sustainable. The oxidative part of the photosynthetic complex is ultimately anchored by the manganese- and calcium-containing oxygen-evolving complex (OEC), which serves as the water-oxidizing center coupled to PSII (Section X.4). This center produces O_2 as a byproduct of water oxidation.

The triumph and spread of oxygen-evolving photosynthesis ultimately resulted in the presence of oxygen in the atmosphere at 21% (210,000 ppmV), essentially all of which comes from this biological process. The same process also produces equivalent amounts of reduced carbon compounds in the form of the cellular materials of biomass. Thus, photosynthesis produces a disequilibrium in which the solar photons have driven the CO_2 and H_2O to the higher energy state of biomass and O_2. What is the fate of the reduced carbon and the atmospheric oxygen?

At first, oxygen was a new and surprising component of the atmosphere and waters. It can be viewed as a noxious and poisonous gas that may have led to the extinction of many inhabitants of early Earth while driving others deep into the soil or water, where they could continue their anaerobic life style. On the other hand, some organisms adapted to this newly available powerful oxidant. Among them are our distant ancestors.

The right side of the cycle in Fig. II.6 shows that the processes of respiration and combustion are able to collapse the disequilibrium that is produced by photosynthesis. These processes regenerate the CO_2 and H_2O that complete the cycle by serving as the inorganic raw materials of photosynthesis. Many organisms, including ourselves, use aerobic respiration reactions to generate the energy required to perform the functions of life—metabolism, motility, reproduction, autopoiesis, and isolation from the environment. The oxidative part of the cycle is consummated by an array of aerobic bacteria (the decomposers) and higher organisms that are able to exploit the large free energy change provided by the reaction of free oxygen with reduced carbon compounds. This chemistry required the evolution of heme and non-heme iron- and copper-containing enzymes to catalyze and control the reactions of oxygen discussed in Chapter XI.

If the simple cycle in Fig. II.6 were quantitatively complete, no oxygen would ever accumulate in the atmosphere, as mass balance would require the consumption of as much oxygen on the right side of the cycle as is produced on the left. However, over geological time scales a new factor comes into play. Some of the reduced carbon in the form of sediments and biomass becomes buried, mostly in the form of plants on land and algae in shallow marine basins and in lakes. This buried organic matter rapidly becomes anaerobic and begins a remarkable transformation that ends with the formation of "fossil fuels" sequestered in anaerobic formations. Thus, each year a small amount of the carbon from photosynthesis is buried and gradually becomes unavailable for reoxidation. While constituting only a small percentage of the cycle each year, over geological time scales of hundreds of millions to billions of years, this buried reduced organic matter accumulates and is, in part, transformed into the coal, shale, oil, and natural gas that constitute the majority of our fossil fuel.

At the same time that this reduced carbon is buried, the oxygen that would have been used to complete the cycle can begin to accumulate in the atmosphere. Thus, over the time scale of many hundreds of millions of years, the same process that created our fossil fuel reserve produced the molecular oxygen in the atmosphere and drew down the high levels of CO_2 (which likely made early Earth compositionally more like Mars and Venus). On early Earth, the global accumulation of O_2 was slow, because vast amounts of reduced (ferrous) iron and sulfide were present in the oceans. This iron was oxidized by the O_2, yielding banded ferric iron formations, which are found as major iron-ore deposits throughout the world. Once the reduced iron (and, some reduced sulfur) was consumed, molecular oxygen began to accumulate in the atmosphere reaching present day levels somewhat <1 billion years ago. Although there have been excursions (e.g., during the carboniferous period, where atmospheric O_2 likely reached 35% contemporaneous with the massive burial of land plants, leading to the formation of enormous deposits of coal) the oxygen level has stayed in the vicinity of 21% over the last half billion years.

By recovering and burning the ancient legacy of fossil fuels through combustion, we are now beginning to collapse the long existing disequilibrium of dioxygen and reduced carbon on a very short time scale. The combustion reactions are obviously the major factor, but each year some of the oil or gas reaches the surface through natural or human activities. That fraction that is not combusted can biodegrade at various rates and also represents a completion of the carbon cycle of Fig. II.6 with a very short time scale. While the reactions of fossil fuels with O_2 run little risk of significantly decreasing planetary oxygen levels (which are nonetheless declining), increasing CO_2 levels are proportionately far more significant and remain of great concern with respect to global climate change.

II.3.2. Anaerobic Aspects and the Hydrogen Cycle

The aerobic cycle displayed in Fig. II.6 is clearly important on a global scale. However, particular locales may manifest either aerobic or anaerobic aspects of the carbon cycle. For example, anaerobic reactions of the carbon cycle occur in the sediments of marine or lacustrine (lake) environments and in the lower layers of the soil. In these strata, reduced carbon compounds formed by photosynthesis undergo a fermentative (decay) reaction (essentially a disproportionation) facilitated by bacteria and fungi to produce, among other products, reactive small molecules including formate, acetate, CO, CO_2, and H_2.

A simplified illustration of a fresh water environment is shown in Fig. II.7. The inorganic compounds produced in the lower anaerobic zone of such an environment derive from photosynthetic detritus and serve as the raw material for fermentative bacteria. The bacterial reaction products provide raw materials that ultimately allow archaea to produce methane (methanogens = methane formers) and bacteria to produce acetate (acetogens = acetate formers). These reactions are discussed in Chapter XII, where it is seen that Fe-, Co-, Ni-, and Mo-based metalloenzymes play key roles in the requisite one-carbon chemistry.

The methane (CH_4) produced by the methanogens is a volatile gas; and, unless the lake is very cold and very deep, it wends its way upward. However, in many cases the CH_4 never reaches the surface. At some point, molecular oxygen diffusing down

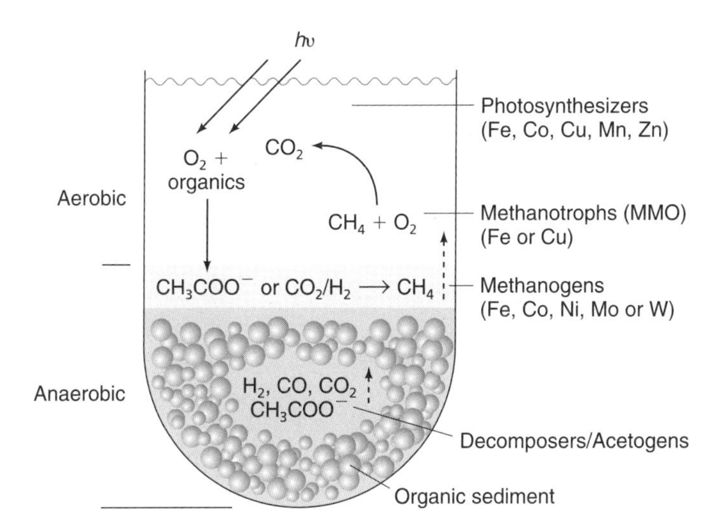

Fig. II.7.
A schematic diagram of a lake in which the lower sediments are anaerobic and produce methane from the photosynthetically produced detritus that reaches the bottom. The methane rises from the sediments whereupon it meets oxygen descending from the phototrophic layer. Here the methanotrophs, using the methane monooxygenase (MMO) enzyme based on copper or iron, are active in oxidizing the methane.

from the phototrophic layer meets methane bubbling up from the anaerobic sediments. At this conjunction, the thermodynamically highly favorable reaction of methane and oxygen is exploited by the methanotrophs (methane eaters) to provide energy and carbon intermediates required for their metabolism. The key first enzyme for methanotrophic organisms is methane monooxygenase (MMO), an iron- or copper-dependent system that catalyzes the formation of methanol, the key first step in the oxidative metabolism of methane (Section XII.4).

If the lake is very shallow, or if it is really no longer a lake, such as in a bog, or a rice paddy, or the rumen of grazing animals such as sheep and cows, then the methane may escape into the atmosphere. There, it is a potent greenhouse gas, but has a limited lifetime ($\tau_{1/2} \sim 12$ years) because of its reaction with •OH radicals in the atmosphere. If the basin is very deep and/or a sediment overlayer is present, massive gas hydrates of methane can form just below the surface layer. These gas hydrates may constitute the largest fossil fuel reservoir on this planet.

In the anaerobic region (bottom of Fig. II.7) myriad other reactions can take place. In this anoxic world, molecular hydrogen (H_2) plays a dramatic role (Fig. II.8). Fermentative organisms, such as clostridia, can take organic compounds such as sugars and oxidize them to pyruvate. The further conversion of pyruvate to acetate allows these organisms to produce adenosine triphosphate (ATP) (substrate level phosphorylation) and generates reducing equivalents, which are taken up by oxidized ferredoxins. Indeed, it is in such organisms (and in certain photosynthetic bacteria) that the ubiquitous ferredoxins were first recognized, almost 40 years ago, and found to contain iron and acid-labile sulfide. The detailed structures and crucial roles of the ferredoxins in redox chemistry are discussed in Chapter IV and Section X.1.

In the fermentative process, ferredoxin must be reoxidized in order to be available for the next cycle of pyruvate oxidation. In clostridia and other fermentative organisms, there are no available or usable strong oxidants (such as O_2 or nitrate) and the electron acceptor may be the proton, with H_2 generated in the process. The free energy for this process is made favorable, with a little help from other resident organisms (see below) that keep the product H_2 level low. The fermentative organisms clearly use the proton as their oxidant–electron acceptor and produce H_2. The enzyme responsible for the H_2 production reaction is the *all*-iron hydrogenase discussed in Section XII.1.

The H_2 generated fermentatively can be used by other organisms in the anaerobic milieu as shown in Fig. II.8. These organisms employ hydrogenases that specialize in

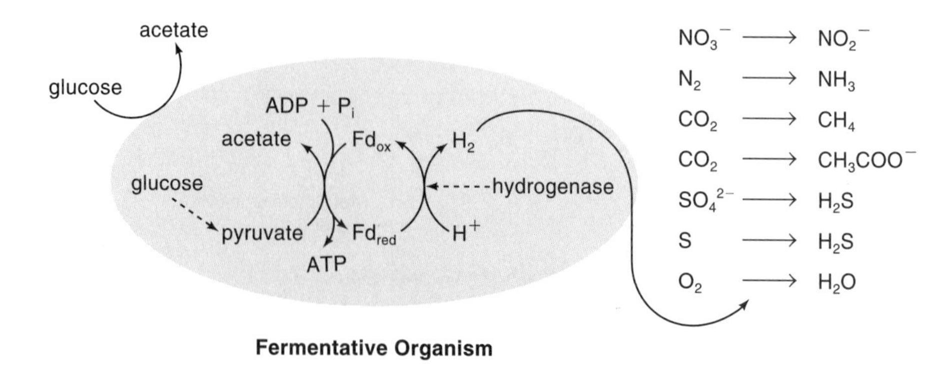

Fig. II.8.
Hydrogen cycling in a largely anaerobic environment. Sugars or other organic matter, usually ultimately derived from photosynthesis, end up in sediments or soils that become anaerobic. Fermentative organisms, such as *Clostridium pasteurianum,* can oxidize glucose to pyruvate, and then oxidize the pyruvate to acetate in the course of substrate level adenosine triphosphate (ATP) synthesis. The electron acceptor is oxidized ferredoxin (Fd_{ox}), which is converted to reduced ferredoxin (Fd_{red}). The Fd_{red} serves as the electron donor to protons to form H_2, catalyzed by the enzyme hydrogenase.

The right of the diagram shows that the H_2 produced can serve as a reductant for (from the top down): nitrate reduction (by denitrifiers); nitrogen fixation; methane formation (by methanogens); acetate formation (by acetogens); sulfate reduction (by sulfate reducing bacteria); sulfur reduction (by photosynthetic bacteria); and oxygen reduction (in the Knallgas reaction). Thus, hydrogen is a key carrier of reducing equivalents, largely in the anaerobic world.

the uptake of H_2, using the reducing equivalents generated to carry out diverse reaction chemistry. Clearly, H_2 serves as a redox currency, with its reducing equivalents harnessed in the reduction of nitrate, sulfate, ferric iron, dinitrogen, and/or sulfur in particular habitats. If H_2 reaches the aerobic region, hydrogen-oxidizing aerobes use the "Knallgas" reaction to combine H_2 and O_2 to produce water. Such processes keep the H_2 level sufficiently low that formation of H_2 by the fermentative organisms remains thermodynamically favorable.

Virtually all of the organisms that use H_2 in this manner employ a nickel–iron hydrogenase. Remarkably, as discussed in Section XII.1, both types of hydrogenases, those used in the evolution of H_2 and those specializing in its uptake, have CO and CN^- ligands, bound to their Fe sites. Ligation by these "strong-field" diatomic ligands, which are generally toxic to most life, alters the coordination chemistry to favor activation and/or evolution of hydrogen gas.

Clearly, the cycles of carbon, oxygen, and hydrogen are intimately intertwined. Photosynthesis provides the reduced carbon that organisms use in their biosynthetic pathways, while oxidative transformation and decomposition of the organic compounds leads to a variety of organic or inorganic products. Other organisms use these compounds in their metabolism, allowing them to occupy a variety of ecological–biochemical niches. Moreover, photosynthesis provides molecular oxygen and reduced carbon, which play a major role in the remaining cycles, as discussed below.

II.4. The Nitrogen Cycle

Nitrogen is an essential element in proteins; nucleic acids; the tetrapyrroles of heme, chlorophyll, siroheme, and B_{12}; the metal-binding pyranopterin dithiolene (mpt) of molybdenum and tungsten enzymes; all B vitamins (cofactors/coenzymes); and many other cellular constituents. In all of these compounds, nitrogen is found in its most reduced form, that is, in the same state of oxidation as ammonia (NH_3) or the ammonium ion (NH_4^+). However, the principal form of nitrogen on Earth is molecular nitrogen (N_2) in the atmosphere, with a second far smaller reservoir consisting mostly of dissolved nitrate in aqueous phases. Therefore, to satisfy the needs of living systems, nitrogen must first be reduced to the ammonia level so that it becomes available for incorporation into the myriad essential nitrogen compounds.

From the perspective of animal life, the nitrogen cycle can be viewed as the set of enzymatically catalyzed chemical reactions that leads to an accessible supply of reduced nitrogen. Importantly, each individual reaction of the cycle takes place with a negative free energy change. Thus, as in the nonphotosynthetic parts of the carbon cycle, specific organisms have evolved to exploit available free energy changes and occupy the corresponding chemically defined ecological niches. While these organisms propel the cyclic process, they exploit the available free energy for their own growth and metabolism. As we shall see, in contrast to the carbon cycle, where plants and animals play a quantitatively major role, the interconversion of the various inorganic forms of nitrogen is carried out almost entirely by microorganisms.

There are four main processes in the inorganic nitrogen cycle: nitrogen fixation, nitrification, nitrate assimilation, and denitrification. A simplified schematic diagram of the cycle is shown in Fig. II.9.

II.4.1. Nitrogen Fixation

The dinitrogen molecule is the most abundant constituent of Earth's atmosphere. And yet, the availability of "fixed" nitrogen is very often growth limiting, in both terrestrial and marine environments. This paradoxical scarcity in the presence of seeming abundance is due to the exceedingly unreactive nature of the N_2 molecule. Indeed, chemists often work "under nitrogen" when they desire an inert atmosphere. Although small amounts of dinitrogen may react with oxygen under extreme conditions, such as in lightning discharges and in high-temperature combustion, these processes produce only a small proportion of the nitrogen needs of the biosphere. The remainder comes from the process of nitrogen fixation.

The biological reduction of nitrogen to ammonia is carried out by a range of prokaryotes, including both bacteria and archaea (see Section XII.2). The enzyme system for nitrogen fixation must function anaerobically, and oxygen rapidly and irreversibly destroys the enzyme components *in vitro*. Nevertheless, nitrogen fixation is found in a phylogenetically broad range of organisms including aerobes (such as azotobacter), facultatively aerobic organisms (e.g., klebsiella), and photosynthetic bacteria including the oxygen-evolving cyanobacteria (such as trichodesmium), as well as in strict anaerobes (e.g., clostridia).

Those organisms that function aerobically and fix nitrogen have devised a variety of strategies to avoid contact with molecular oxygen. Moreover, a significant number of important symbioses and associations have evolved between higher plants and nitrogen-fixing bacteria (e.g., rhizobia). The bacteria live in environments, such as root nodules, that have controlled and/or low oxygen tension. In these symbioses and associations, the plant provides the bacterium with reduced carbon compounds from photosynthesis, while the bacterium provides fixed nitrogen to the plant.

Thermodynamically, a reduction potential of about $-300\,mV$ is required for biological nitrogen fixation. This potential requires reducing agents of moderate strength, which are readily accessed by biological systems (Section X.1). Reduced ferredoxins or flavin-containing electron-transfer proteins can deliver electrons at the required reducing potential. The reducing equivalents can usually be traced back to organic molecules produced during photosynthesis.

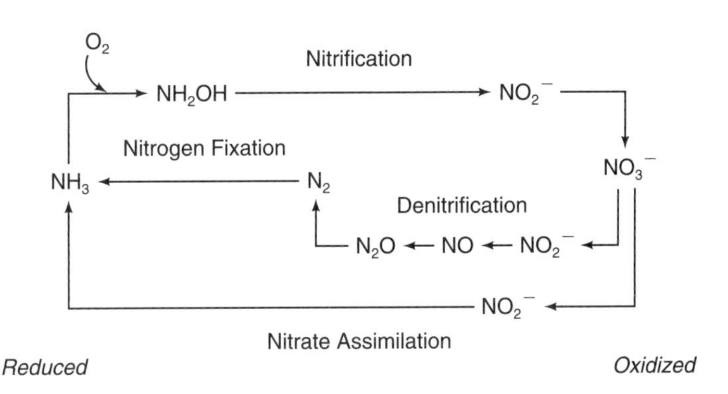

Fig. II.9.
The nitrogen cycle. The four major inorganic steps in the nitrogen cycle are nitrogen fixation, nitrification, nitrate assimilation, and denitrification. See text and Sections XII.3 and XII.4 for a full discussion.

While nitrogen fixation is a facile reaction thermodynamically, such is not the case kinetically. There is a large kinetic barrier to reducing N_2, apparently caused by the need to form bound intermediates (possibly at the diazene or hydrazine levels) in the conversion of dinitrogen to ammonia. Organisms must therefore expend a great deal of metabolic energy in the form of ATP hydrolysis [for which $\Delta G \approx -8\,\text{kcal mol}^{-1}$ for conversion to adenosine diphosphate (ADP) and inorganic phosphate (P_i), see Tutorial I] to overcome the activation energy required to produce the key intermediates in the nitrogen-fixation process. The reduction of N_2, which consumes 16 equiv of ATP for each equivalent of N_2 fixed, is clearly a costly process for any organism. Given the opportunity, most organisms capable of nitrogen fixation will use available fixed nitrogen (ammonia, nitrate, or nitrite) and fully repress the synthesis of the elaborate nitrogen-fixation system (Section XII.3).

The biological process of nitrogen fixation is dependent on a two-protein system where the major transition metal ions are Mo and Fe. Details of the process and the proteins are presented in Section XII.3. The iron–molybdenum cofactor, FeMoco, is the key active-site structure that appears to be responsible for the binding and reduction of dinitrogen. This $MoFe_7S_9X$ metallocluster, where X is the newly found central light atom (either N, or less likely C, or O), is also held together by sulfur bridges and is linked to the protein by one cysteine, bound to iron, and one histidine, bound to molybdenum, which also carries a homocitrate ligand. Despite the great attention that this protein has attracted, the manner in which the FeMoco center activates the dinitrogen molecule remains obscure.

In addition to those enzymes that contain the iron–molybdenum cofactor, some organisms can use alternative nitrogenases, which are expressed only when molybdenum is unavailable. These alternative systems use vanadium or iron, in the form of FeVco or FeFeco, respectively (Section XII.3). The alternative enzymes are less efficient than the FeMoco version and are only present as back-up systems for organisms that also possess the primary system. The presence of three genetically and biochemically distinct, albeit related, nitrogenases is testimony to the importance of nitrogen fixation for these organisms.

Once the nitrogen is in the reduced form, organisms use a variety of mechanisms to incorporate the nitrogen into organic molecules so that it can be made available for the biosynthetic processes of the cell.

The industrial fixation of nitrogen (usually called ammonia synthesis) involves the Haber–Bosch process, which provides a technological alternative and significant augmentation to the total nitrogen available to life on Earth. The impact of this additional input is discussed in Section II.4.5.

II.4.2. Nitrification

The processes of biological nitrogen fixation and synthetic fertilizer production each provide ammonia, which is incorporated into the organonitrogen compounds of biomass. When biomass decays, nitrogen is released in the form of ammonia (or ammonium depending on the conditions). The organisms that oxidatively convert this ammonia to nitrite and/or nitrate are called *nitrifiers* and the process is called *nitrification* (Section XII.3).

The conversion of nitrogen to the nitrate level, that is, to nitrogen in its most oxidized form, is crucial to most life on Earth. While nitrogen fixation is limited to a range of prokaryotic organisms, nitrate reduction is part of the nitrogen assimilation pathway common to prokaryotes, many fungi, and virtually all green plants (see below and Fig. II.9). Indeed, green plants in agricultural settings, unless they harbor symbiotic nitrogen-fixing bacteria, generally use nitrogen in the form of nitrate. The small group of prokaryotic nitrifiers thus performs the crucial task of providing nitrate for the growth of many natural and agricultural flora. A major portion of the biomass on Earth gets its nitrogen directly from the nitrate formed through nitrification.

The oxidation of NH_3 by molecular oxygen to nitrite or nitrate is a very favorable process thermodynamically. The significant negative free energy change is exploited by organisms that occupy the ecological niche defined by the aerobic oxidation of ammonia to nitrogen oxyanions. These *aerobic* organisms use ammonia monooxygenase (AMO), which is related to the copper-containing MMO, to first convert ammonia to hydroxylamine. Then, hydroxylamine is oxidized to nitrite by a multiheme enzyme that conserves the reducing equivalents, which, in turn, act as donors to the AMO (which, like all monooxygenases, requires two electrons in addition to oxygen) and to a cytochrome oxidase (to generate energy-storing cross-membrane proton gradients). These enzymes are discussed in Section XII.3.

Nitrifiers can be autotrophic, which means that the energy harnessed through ammonia oxidation must be sufficient to provide all the driving force and substrate reducing power required by the organism for this mode of life. Organisms, generally distinct from those that form nitrite, complete the nitrification step by converting nitrite to nitrate. The required nitrite oxidase is a molybdenum enzyme that is an interesting counterpart to the better known nitrate reductase, which catalyzes the reverse reaction (Section XII.4).

II.4.3. Nitrate Assimilation

The nitrate assimilation pathway occurs in bacteria, fungi, and plants. It consists of the two steps shown in Fig. II.9. The first step involves nitrate reductase, which converts nitrate to nitrite. The assimilatory nitrate reductase is a molybdenum cofactor enzyme discussed in Section XII.6. This enzyme contains 1 equiv of the metal-binding pyranopterin dithiolene that, together with sulfur from a cysteine side chain, provides three sulfur atoms that bind to one active-site molybdenum atom.

The nitrite produced by the molybdenum enzyme is then reduced by nitrite reductase in a six-electron step to yield ammonia. The nitrogen-containing intermediates remain bound to the enzyme during the redox process. This assimilatory nitrite reductase strongly resembles the assimilatory sulfite reductase (see below and Section XII.5), which also effects a six-electron reduction without dissociated intermediates. Both the nitrite reductase and the sulfite reductase have an unusual active site grouping of a siroheme (Chapter IV) and a Fe_4S_4 cluster (Chapter IV) bridged by a cysteine side chain (Section XII.5). The siroheme iron apparently provides the binding site for nitrite (through its nitrogen) and allows the multielectron reductive process to occur.

Once the reduction is complete, the ammonia can be enzymatically incorporated into a variety of biosynthetic intermediates. The reverse process, involving degradation–ammonification, regenerates ammonia from cellular components; organisms can scavenge the ammonia for new biosynthesis or, in aerobic situations, use the ammonia in the nitrification pathway.

II.4.4. Denitrification

There is an important alternative fate for the nitrate that is produced in the nitrification process (Fig. II.9). This process is known as denitrification. The ultimate product of the complete denitrification pathway is N_2, so denitrification, which takes NO_3^- to N_2, is *not the reverse of nitrification,* which takes NH_3 to NO_3^- (although there are a few organisms that produce NH_3 from nitrite to gain energy). The denitrification pathway utilizes the strong oxidizing potential of nitrate to generate energy through the oxidation of organic compounds.

At first glance, there would appear to be overlap between the nitrification and nitrate assimilation processes, at least in their respective first steps, the conversion of nitrate to nitrite. However, the nitrate reductase enzymes responsible for this step are dramatically different for the two pathways. Although both are molybdenum enzymes (Section XII.6), their active sites and protein structures are distinct and the enzymes appear to be genetically unrelated. The dissimilatory nitrate reductases contain 2 equiv of the metal-binding pyranopterin dithiolene per molybdenum (cf. 1 equiv for the assimilatory enzyme).

In denitrification, nitrate serves as an alternative electron acceptor to O_2 and is a sufficiently strong oxidant (remember nitric acid!) to provide a large free energy change and a niche for a significant number of bacteria and some fungi. Denitrifiers are often facultative anaerobes. In other words, while they can live aerobically if O_2 is present, they can also live anaerobically using nitrate as their oxidant for organic substrates. The enzymes in this pathway are coupled to the transmembrane bioenergetic system of the cell and thus most are membrane bound or membrane associated.

As discussed in Section XII.4, the four enzymes in the pathway (dissimilatory nitrate reductase, nitrite reductase, NO reductase, and N_2O reductase) produce, respectively, nitrite, nitric oxide, nitrous oxide, and N_2 (Fig. II.9). These enzymes depend on molybdenum, copper, and heme iron and are discussed extensively in Sections XII.4 and XII.6. The denitrifiers are important because they complete the nitrogen cycle. Moreover, they are currently of great concern, because, under certain conditions, especially when there is limiting organic reductant or limiting copper, nitrous oxide is produced as the final product. Since N_2O is both a potent greenhouse gas and, once it reaches the stratosphere, an ozone depleting catalyst, the production of N_2O, especially in agricultural settings, is of significant concern.

II.4.5. Technological Impacts

The impact of the expanding human population and modern technology on the nitrogen cycle is enormous. It is estimated that one-third to one-half of all nitrogen "fixed" on Earth occurs through the technologically highly developed Haber–Bosch process. The added nitrogen enters the nitrogen cycle and adds flux to the nitrogen-fixation segment of the existing natural cycle. The additional nitrogen flux can lead to eutrophication or nitrite contamination of ground water from agricultural runoff. Worldwide, there is no obvious let-up in the application of fertilizer to crop plants. Clearly, augmentation of the flow through a specific segment of the nitrogen cycle (nitrogen fixation) potently affects the entire cycle. Moreover, this technological perturbation to the biogeochemical nitrogen cycle is, as a percentage of the annual flux through the cycle, much larger than the human perturbation of the carbon cycle.

II.5. The Sulfur Cycle

Sulfur is essential to all life forms. In addition to its presence in the amino acids cysteine and methionine, sulfur is found in many key active-site structures including the inorganic sulfide in Fe–S proteins (Chapter IV) organic structures such as biotin, thiamine, and lipoic acid; and the molybdenum- and tungsten-binding pyranopterin dithiolate ligand (Section XII.6). Moreover, many organisms obtain energy for their metabolism by the oxidation or reduction of inorganic compounds of sulfur. These organisms occupy important niches in the sulfur cycle.

The sulfur cycle is quite complex and less completely defined than the nitrogen cycle. The extremes of biological sulfur chemistry are found in sulfate, the most oxidized form and in H_2S or its ionized forms (HS^- or sulfide), the most reduced forms.

Major and well-studied classes of organisms carry out the reactions of Fig. II.10. The sulfate reducing bacteria (SRB) reduce sulfate to sulfide under anaerobic conditions, oxidizing organic compounds in the process. The sulfide-oxidizing organisms are generally, but not always, aerobic and use O_2 to oxidize sulfide (or disulfide, polysulfide, elemental sulfur, or thiosulfate) to sulfate. Figure II.10 is an incomplete version of the sulfur cycle, only detailing some of the intermediate species. The sulfur cycle is described in more detail in Section XII.5.

II.5.1. The Sulfate Reducers

The SRBs, use sulfate as their terminal electron acceptor and generate sulfide under anaerobic conditions. These strictly anaerobic organisms are prevalent in anoxic marine environments, and in other niches where both sulfate and reduced organic matter are present (e.g., in the rumen of sheep and cattle). The SRBs are discussed in Section XII.5, and have importance in sulfide ore formation, biocorrosion, the souring of petroleum under anaerobic conditions, the Cu–Mo antagonism in ruminants, and many other physiological, ecological, and biogeochemical contexts.

Sulfate reduction can be divided into two main reactions. First, the relatively unreactive sulfate must be activated, which is achieved through the expenditure of ATP to forming adenosine phosphosulfate (APS) and phosphoadenosine phosphosulfate (PAPS) (Fig. II.10). These "active-sulfate" intermediates can be used in the sulfation of organic molecules, for example, in the formation of chondroitin sulfate and other polysaccharide sulfates, or can be reduced to sulfite by APS reductase. Once at the sulfite level, the conversion of sulfite to sulfide is catalyzed by the enzyme sulfite reductase, which carries out the six-electron conversion without the release of free intermediates. Similar to the nitrite reductase discussed above, sulfite reductase contains a siroheme (Chapter IV; Section XII.5) and an Fe_4S_4 center (Chapter IV) bridged by a cysteine ligand.

II.5.2. Sulfide Oxidation Reactions

The oxidative part of the sulfur cycle is the province of a large number of organisms, mostly bacteria, that gain energy from various interconversions of the type shown in

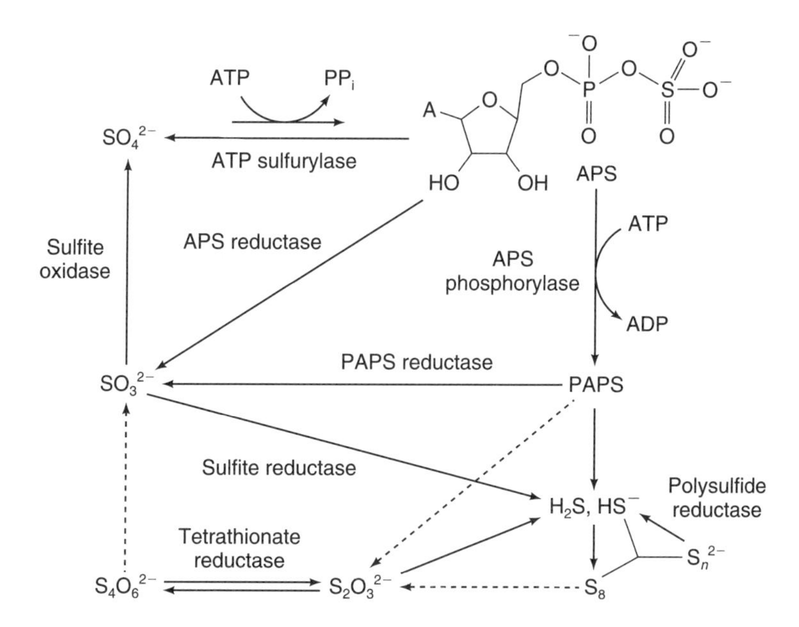

Fig. II.10.
The sulfur cycle. Anaerobes are involved in the reduction of sulfate that is activated by the formation of adenosine phosphosulfate (APS) and its subsequent reduction to sulfite, and then in a six-electron process to H_2S. Sulfide oxidizers use oxygen or occasionally nitrate (see Fig. II.9), to oxidize the sulfide to sulfate. See Section XII.5 and text for discussion.

Fig. II.10. Thiobacillus is a well-studied example of an aerobic sulfide-oxidizing organism. Some thiobacillus species oxidize the sulfide in ores, for example, iron sulfides. The oxidation of sulfide to sulfate produces an acidic environment in which some species of thiobacillus thrive. The sulfide oxidizers severely alter the pH to produce acidic conditions favorable to their own metabolic needs. For example, in acid mine drainage, water can have a microbially produced pH as low as 1.5! These organisms have been harnessed in the mobilization of metals from sulfide ores. For example, *Thiobacillus ferrooxidans* not only oxidizes the sulfur in iron sulfide deposits but also oxidizes the ferrous iron to ferric iron, which is solubilized in the acidic mine water solutions. Thiobacilli are autotrophs and thus, like the ammonia oxidizing nitrifiers, they use the energy obtained from sulfide oxidation for all their cellular needs, including the fixation of carbon through the Calvin cycle (see Tutorial I).

As in the nitrification pathway, most sulfide oxidizers use molecular oxygen to oxidize the reduced sulfur species from sulfide to sulfite, with sulfite oxidation being the terminal step in the process. However, in the lower portions of the oxidation range there are conversions that occur in the anaerobic regime. Some of the reactions that occur are shown in Fig. II.10. The enzymes use iron, nickel, cobalt, and molybdenum to effect the conversions of the reduced sulfur species.

II.5.3. Deep Sea Hydrothermal Vents

There are many special habitats on Earth where sulfur-rich waters elicit extraordinary manifestations of the sulfur cycle. For example, at the deep sea hydrothermal vents, entire communities of organisms live off the oxidation of sulfur compounds. It is often stated that this is an example of life that is independent of photosynthesis. This claim is untrue. All of the *macroscopic* organisms that live in the vicinity of the vents are *aerobic* in nature. These organisms, or more precisely, symbiotic bacteria living within the organisms, use molecular oxygen as their oxidant. This molecular oxygen comes from photosynthesis. Thus, while the autotrophic vent organisms do perform chemosynthesis of organic compounds (and do not require the reduced products of photosynthesis), they are fully dependent on the oxidized product of the photosynthetic reaction, namely, molecular oxygen.

Many extraordinary reactions are found in the organisms of the vents. One organism where a biological inorganic perspective is particularly edifying is the giant tube worm, riftia. The retractable red "heads" of riftia can emerge from the tubes that anchor the tube worms to the sea floor near the vents. Riftia are filter feeders and their heads are red because of the presence of hemoglobin. This hemoglobin is remarkable in that it not only carries O_2 and CO_2, as other hemoglobins and myoglobins do, but it is also capable of binding sulfide! The hemoglobin transports the oxygen, carbon dioxide, and sulfide into special organs called trophosomes, where symbiotic bacteria use the oxygen to oxidize the sulfide to sulfate. The reducing power and energy generated in the process is used to fix CO_2 by the Calvin cycle. The overall process is made possible by the availability of molecular oxygen, a product of oxygenic photosynthesis. Macroscopic life at the deep sea vents is thus dependent on past photosynthesis for its energy and metabolic capability.

In contrast, to the aerobic nature of the vent animals, there exist microniches in the vicinity of the black smoker vents, where there is no molecular oxygen; and thriving bacterial communities live using strictly anaerobic reactions. It is this community of organisms, the microorganisms, many of which are thermophilic or hyperthermophilic archaea and bacteria, that indeed represent life without any input from photosynthesis. These microorganisms, not the large visible organisms at the vents, may have implications for the origins and early evolution of life on Earth.

II.5.4. Volatile Organic Sulfur: Dimethyl Sulfide

Except for H_2S there are no volatile components of the inorganic sulfur cycle discussed above. However, H_2S does not have a long residence time in the atmosphere as it is rapidly oxidized through chemical and photochemical reactions. Therefore, in contrast to the carbon, nitrogen, and oxygen cycles, the natural *inorganic* conversions of sulfur do not have significant global effects. Nevertheless, these sulfur interconversions are critical in local or regional contexts, as any one who has ever smelled swamp gas can attest. However, there is one volatile organic compound of sulfur that may have important global implications and this is dimethyl sulfide (DMS).

Dimethyl sulfide is produced by the decomposition of dimethylsulfoniopropionate (DMSP), a compound produced by algae, which possibly functions as: a flotation aid; an agent of osmotic balance; a chemical deterrent to potential grazers; an antioxidant; and/or a storage compound for sulfur or carbon. Dimethyl sulfide is produced in DMSP decomposition upon the death of the algae. The DMS produced can be oxidized chemically or biologically to dimethyl sulfoxide (DMSO), which in turn serves as a terminal electron acceptor for microorganisms, in the process being reduced to the volatile DMS (Fig. II.11). The DMS is relatively insoluble in the oceans and enters the atmosphere where it is responsible for the smell of the sea and serves to attract sea birds to areas of algal productivity. The molybdenum enzymes, DMSO reductase and DMS dehydrogenase, are involved in the DMS–DMSO interconversion (Section XII.6).

From a global perspective, the key reactions occur when DMS enters the atmosphere and is chemically and/or photochemically converted to methyl sulfonates and sulfates, which serve as cloud condensation nuclei. The production of DMS and its atmospheric oxidation provide one of the few mechanisms for cloud formation in open ocean regions where there is a dearth of dust (which facilitates cloud formation in other areas). High temperatures enhance algal dying, which leads to release of DMS into the atmosphere, giving a cooling effect, as the DMS generated clouds increase Earth's albedo (its ability to reflect light), and therefore contribute to cooling. It has been suggested that the production of DMS is part of a negative feedback loop that may help regulate temperature, possibly with global implications.

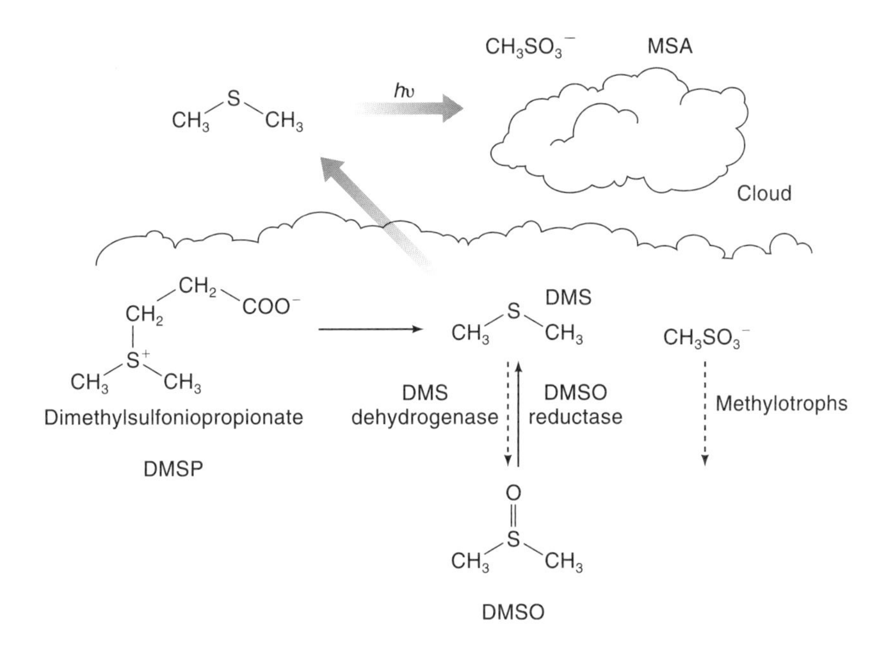

Fig. II.11.
The formation of DMS by the decomposition of algal DMSP. The interconversion of DMS and DMSO is catalyzed by the molybdenum enzymes DMSO reductase and DMS dehydrogenase. Oxidation products of DMS, such as methane sulfonic acid (MSA), that enter the atmosphere serve as cloud condensation nuclei. See discussion in text.

II.6. The Interaction and Integration of the Cycles

The biogeochemical cycles have each been described, more or less, with emphasis on the specific element that gives the cycle its name. However, it is clearly not possible to talk of any of the principal elements in the absence of the others.

II.6.1. Direct Interface of the Nitrogen and Sulfur Cycle: Thioploca

There are many local examples that illustrate the intimate interaction of the cycles. For example, in marine environments, there are bacteria called thioploca that juxtapose aspects of the nitrogen and sulfur cycles (Fig. II.12). Thioploca live at a level in sediments where H_2S, produced by the SRBs, is moving upward and comes into contact with soluble nitrate moving downward into the sediment. Thioploca has the remarkable ability to accumulate nitrate in vacuoles, where its concentration can be as high as 0.1 *M* compared to 10^{-7} *M* in the surrounding waters. Thioploca oxidizes the sulfide to sulfate, but it does not use oxygen to accomplish this reaction. Rather, it uses nitrate as the oxidizing agent. Thioploca first oxidizes the sulfide to elemental sulfur, which it stores in cellular bodies. Then, when H_2S is not immediately available, the stored sulfur is oxidized using the stored nitrate to produce sulfate and N_2. Thus, thioploca is a sulfide oxidizing bacterium, but also is a denitrifying organism (Fig. II.12). Thioploca and other organisms define a unique ecological niche in which they thrive by combining major features of the nitrogen and sulfur cycles in a single organism.

II.6.2. The Oxygen Cycle: A Reprise

The production of molecular oxygen has been discussed as a product of photosynthesis. In any discussion of the carbon cycle, it is unavoidable to include the major role played by water and oxygen. However, in the nitrogen and sulfur cycles oxygen also plays a major role. The nitrification pathway of the nitrogen cycle and the sulfur

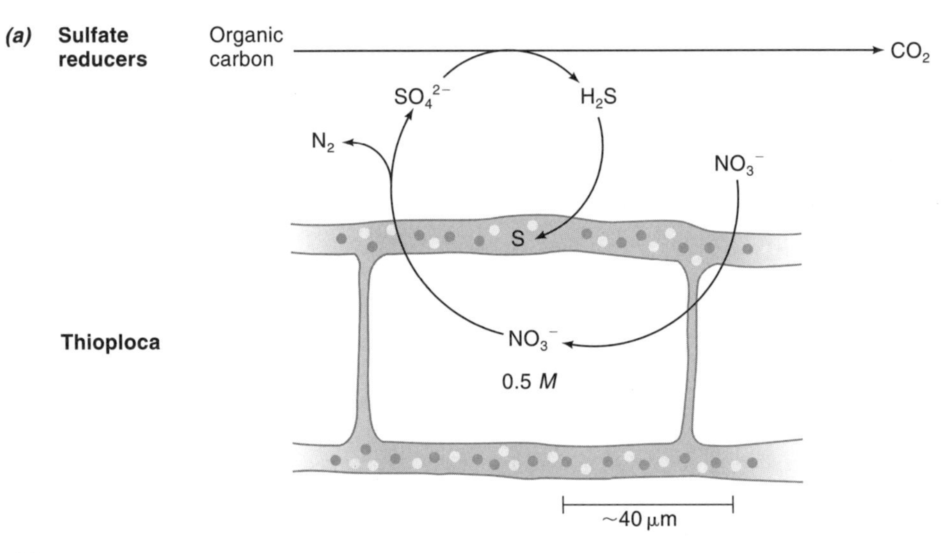

(b)

$$2\,NO_3^- + 5\,HS^- + 7\,H^+ \longrightarrow N_2 + 5/8\,S_8 + 6\,H_2O$$
$$6\,NO_3^- + 5/8\,S_8 + 2\,H_2O \longrightarrow 3\,N_2 + 5\,SO_4^{2-} + 4\,H^+$$
$$8\,NO_3^- + 5\,HS^- + 3\,H^+ \longrightarrow 4\,N_2 + 5\,SO_4^{2-} + 4\,H_2O$$

Fig. II.12.
Thioploca is an example of the overlap of the nitrogen and sulfur cycles. (*a*) The thioploca organism can accumulate nitrate and elemental sulfur to high levels and then, as required, oxidize the sulfur to sulfate to gain energy, while reducing the nitrate to N_2. (*b*) The equations representing the combination of the nitrogen (denitrification) and sulfur cycle (sulfide oxidation) that is executed in thioploca metabolism.

oxidation pathway of the sulfur cycle each depend on oxygen as their oxidizing agent. Figure II.13 places the emphasis on the oxygen cycle. Special aspects of this cycle involve the formation of ozone in the troposphere (where it is a devastating pollutant) and in the stratosphere where ozone is the critical ultraviolet (UV) shield that protects surface life from the powerful solar UV radiation. Other aspects of oxygen reactivity, both biological and technological, illustrate the crucial role of oxygen reactivity on Earth.

Many key activities in Fig. II.13 (the oxidation of carbon compounds of biomass through respiration or combustion; the oxidation of ammonia to nitrite and nitrate; and the oxidation of hydrogen sulfide or sulfur to sulfate or sulfuric acid) have already been discussed in considering the carbon, nitrogen, and sulfur cycles, respectively.

II.6.3. The Effects of Human Activities on the Cycles

Figure II.13 includes aspects of human intrusions on the biogeochemical cycles. Thus, for the carbon cycle, the use of fossil fuels as a source of reduced carbon is responsible for much of the increase in atmospheric CO_2 and its consequences as a greenhouse gas. In the nitrogen cycle, the ammonia synthesized by the Haber–Bosch process and the products produced from ammonia represent a major perturbation on the natural nitrogen cycle. Almost one-half of all nitrogen fixed on Earth occurs through technological and agricultural rather than natural biological reduction. As shown in Fig. II.13, nitrogen oxides are produced unintentionally in internal combustion engines where they contribute to photochemical smog and acid rain. Nitrogen oxides and nitrogen oxyanions are produced intentionally through the Ostwald process, which uses oxygen to convert synthetic ammonia to nitrate salts for use in fertilizer. The nitrate produced in the process also enters the nitrogen cycle and, sometimes, the natural cycles and reservoirs are inadequate sinks for the excess input. Under such conditions, nitrate or nitrite may end up as undesirable components of ground water.

The combustion of sulfur-rich fossil fuels (especially coal used for power generation) leads to sulfur oxides in the atmosphere, which contribute to acid rain in the region down wind from the source. Photosynthetically derived oxygen is the oxidant in these combustion processes, and, here again, the human activity combines with the natural cycle to produce untoward consequences, in this case on the regional level.

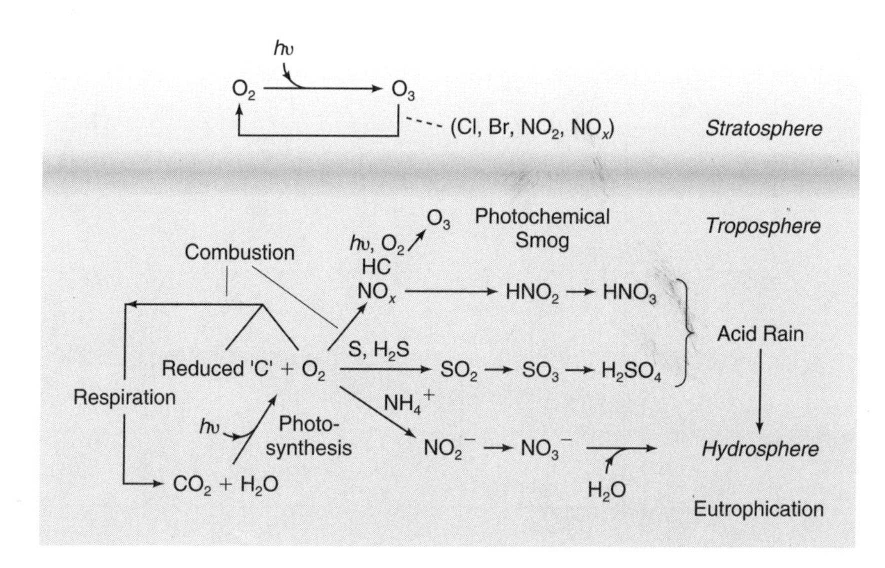

Fig. II.13.
The oxygen cycle showing aspects of the sulfur, nitrogen, and carbon cycles from the perspective of oxygen. All of the O_2 is produced by photosynthesis and then serves as the oxidant for the reduced compounds of carbon, nitrogen, and sulfur by both biological and human (nonbiological) activities (industry, agriculture, transportation, heating, etc.). In addition, there are special aspects of the oxygen cycle, such as the production of the protective ozone in the stratosphere and the formation of photochemical smog (HC = hydrocarbon derived), including ozone, in the troposhere.

It is incumbent upon us as scientists (and citizens) to understand the natural cycles and to apprehend the significant effects of technologies, which through anthropogenic technology and its byproduct chemistry and biology, impinge upon and quantitatively imbalance the natural cycles.

II.6.4. A Summary of Key Bioenergetic Aspects

As we look at the overall flow of materials through the natural biogeochemical cycles we are struck by the number of key reactions that are catalyzed by metalloenzyme systems. From photosynthesis and nitrogen fixation to respiration, denitrification, and sulfate reduction, virtually all of the key steps are catalyzed by metalloenzyme systems with Mn, Fe, Co, Ni, Cu, Zn, and Mo playing key roles at the active sites of Nature's crucial catalysts.

As we view the fluxes through the various cycles, we recognize different organisms or groups of organisms as responsible for particular pathways within the cycles. In each case, the individual organisms (except for us) are oblivious to the grand scheme. The roles played by particular classes of organisms are not parsed out or planned in advance. Rather, from the point of view of the organism, a negative free energy change is available and the organism exploits that free energy change to live. The reactions of the biogeochemical cycles associated with these free energy changes are redox reactions. The electrochemical scale in Fig. II.14 displays the redox potentials for biologically important chemical species. The strongest oxidizing agents are found on the upper left of the vertical line, while the strongest reductants are found on the bottom right of the line. Clearly, the most free energy is available from the oxidation of organic compounds with molecular oxygen. Ultimately, this energy is available because of the disequilibrium created by photosynthesis.

However, if O_2 is not available, it is clear from Fig. II.14 that other oxidizing agents, including nitrate, ferric iron, manganic ion, sulfate, and even CO_2 (for methanogens that use H_2) and protons (for fermentative organisms) can act as oxidants,

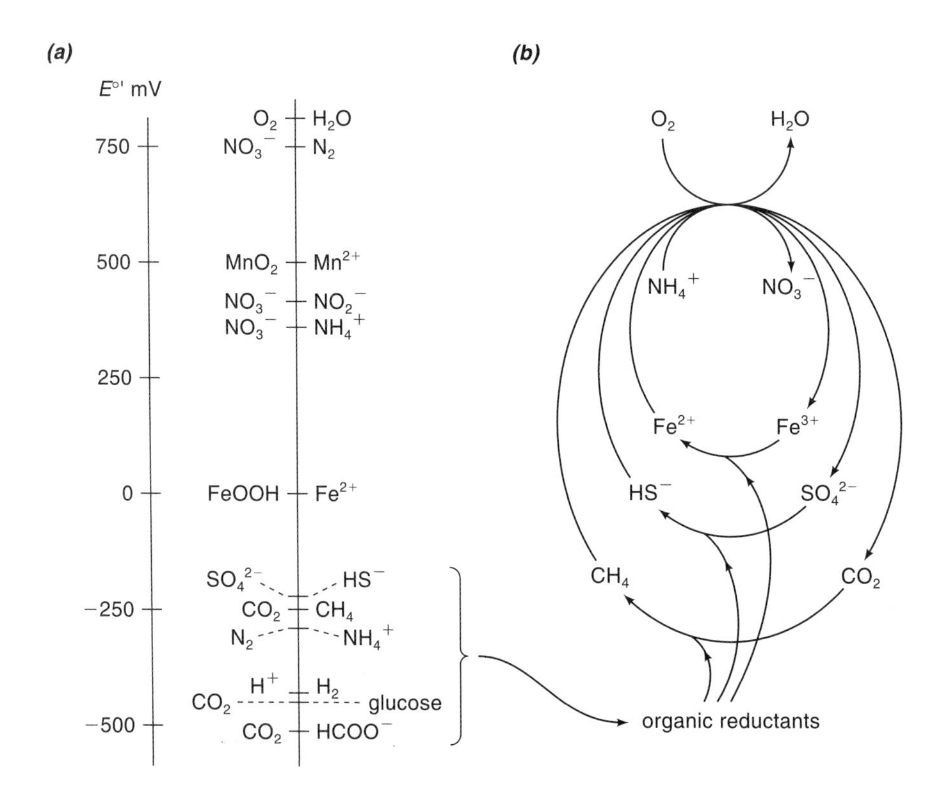

Fig. II.14.
Redox seriatim of the oxidants and reductants available to biological systems. (*a*) Simple scale showing the strongest oxidants on top and the most powerful reductants on bottom. (*b*) The same seriatim with the addition of parts of the biogeochemical cycle to illustrate the crucial role that O_2 plays in generating the most powerful available oxidants including nitrate, ferric iron, and sulfate.

producing smaller but sufficient amounts of energy to drive metabolism. However, all of the larger free energy changes within this group are also ultimately dependent on oxygen, and hence on oxygenic photosynthesis. As shown in Fig. II.14*b*, the presence of nitrate, ferric iron, manganic ion, and sulfate are each dependent on the reaction of reduced versions of these compounds with molecular oxygen. Thus, a world without oxygen (e.g., early Earth or a deep subsurface environment) would not only lack O_2 as the terminal electron acceptor, but also nitrate, sulfate and, possibly, ferric iron may not be available. Organisms in such habitats would be restricted to the lower part of the electrochemical series, and hence have much smaller amounts of free energy available for metabolic purposes and other more creative activities such as the development of shells, locomotion, and reproductive variety. Nevertheless, bacteria and archaea survive and thrive in these anaerobic habitats. While these microbial ecosystems do not garner the attention of the global biogeochemical systems, they are clearly ancient and ubiquitous, and, if we are not wise, they may survive us and the other macroorganisms that depend on the oxygenic environment of Earth.

II.6.5. The Key Role of Prokaryotes

Except for the carbon cycle, the biogeochemical cycles described here are largely executed by groups of microorganisms in which bacteria and archaea (prokaryotes) play inordinately large roles. The carbon and oxygen cycles involve complex (sometimes called "higher") plants and animals and other eukaryotes as well as prokaryotes in particular niches. In contrast, the cycles of hydrogen, nitrogen, and sulfur are largely the dominion of the prokaryotes. Only plant nitrate and nitrite reductases are major exceptions; here, eukaryotic enzymes play an important role in one step of the nitrogen cycle. The chemical conversions of the hydrogen and sulfur cycles lie almost exclusively in the realm of the prokaryotes.

Green plants contribute to the production of oxygen that we have seen to be the dominant and ultimate oxidant on the surface of Earth. Large animals conduct portions of the oxidative degradation of the reduced carbon compounds. However, microorganisms carry out virtually all of the reactions of the nitrogen and sulfur cycles and a significant number of the reactions in the degradation of fixed carbon compounds. No eukaryote can digest cellulose, and therefore even though animals may be said to eat grass, it is the symbiotic bacteria living in their rumens that in fact digest the cellulose. No plant can fix nitrogen; it is symbiotic bacteria that are responsible for "nitrogen-fixing crops".

The biodiversity of Earth is now coming into dramatic focus driven by the power of molecular genetic analysis. As one inspects this diversity, as shown in Fig. II.15, one is struck by the magnitude of the genetic diversity that lies outside the realm of plants, animals, and fungi. The latter familiar and visible organisms occupy but a small part of genetic space and indeed carry out some fairly specialized albeit quantitatively massive reactions. However, the great genetic diversity and the great metabolic diversity is present in the lower eukaryotes and, especially, the prokaryotes, which occupy the vast majority of chemically defined ecological niches on this planet (see Tutorial I for more detail). Indeed, without the continual contributions of prokaryotes to the biogeochemical cycles on Earth, life on this planet would totally cease.

Despite the critical importance of prokaryotes, we are only vaguely aware of the full extent of their genetic and metabolic capabilities. In probing environmental samples by genetic means, it is common to find that >95% of the organisms have not been cultured, and therefore have not been studied in the laboratory. Clearly, the microbial diversity responsible for much of the chemistry of Earth is just now being apprehended.

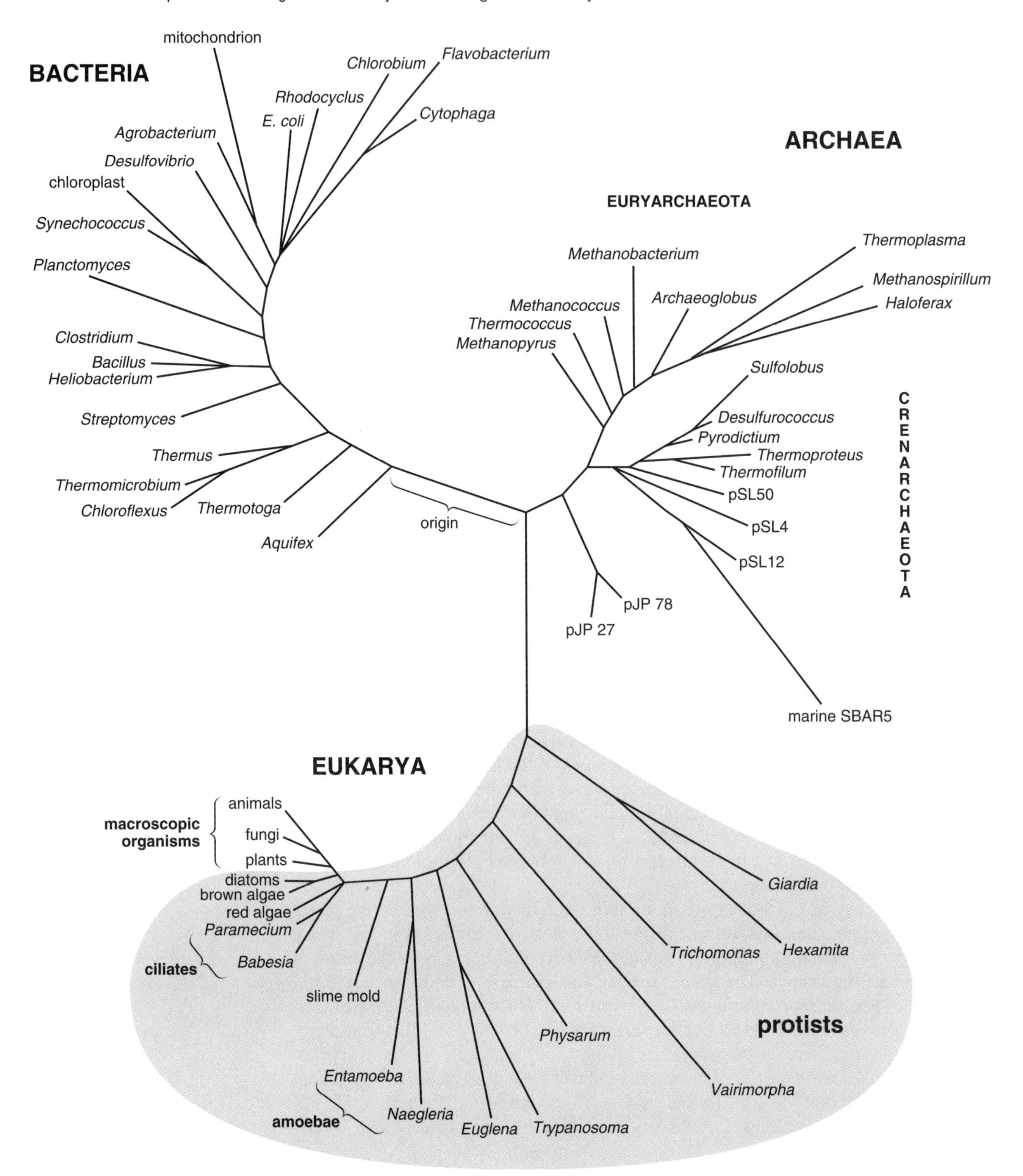

Fig. II.15.
Simplified version of the tree of life showing the three domains of life: The Eubacteria, the Archaea, and the Eukarya, based on the work of Olsen, Bames, Sogin, and their co-workers. [Adapted from Banfield and Nealson, 1997.]

II.7. Conclusions

Increasingly, it is found that the metabolic conversions of the biogeochemical cycles involve unique (biological) inorganic chemistry, both in the reactions that are catalyzed and in the metalloactive sites of the enzyme catalysts. We must appreciate and value our (mostly micro-) organism partners as we continue to evolve a sustainable relationship with planet Earth.

Bibliography

The Formation and Abundances of the Chemical Elements

Greenwood, N. N. and Earnshaw, A., *The Chemistry of the Elements,* Butterworth-Heinemann, Oxford, 2nd ed., 1997, pp. 1–15.

Pagel, B. J., *Nucleosynthesis and Chemical Evolution of Galaxies,* Cambridge University Press, Cambridge, UK, 1997, pp. 1–378.

Trimble, V. L., Origin and Evolution of the Chemical Elements, in *Origin and Evolution of the Universe,* Zuckerman, B. and Malkan, M. A., Eds., Jones and Bartlett Publishers, Boston, 1996, pp. 89–108.

Cox, P. A., *The Elements on Earth: Inorganic Chemistry in the Environment,* Oxford University Press, Oxford, 1995, pp. 1–287.

Lewis, J. S., *Physics and Chemistry of the Solar System,* Academic Press, San Diego, 1995, pp. 8–43.

Cox, P. A., *The Elements: Their Origin, Abundance, and Distribution,* Oxford University Press, Oxford, 1989, pp. 1–202.

Origins of Life and Astrobiology

Gilmour, I., Sephton, M. A., Conway, A., Jones, B. W., Rothery, D. A., and Zarnecki, J. C., *Introduction to Astrobiology,* Cambridge University Press, Cambridge, UK, 2004, pp. 1–358.

Bennett, J., Shostak, S., and Jakosky, B., *Life in the Universe,* Addison Wesley, Boston, 2003, pp. 1–346.

Darling, D., *Life Everywhere,* Basic Books, New York, 2001, pp. 1–224.

Bakich, M. E., *The Cambridge Planetary Handbook,* Cambridge University Press, Cambridge, UK, 2000, pp. 1–336.

Ward, P. D. and Brownlee, D., *Rare Earth: Why Complex Life is Uncommon in the Universe,* Copernicus/Springer Verlag, New York, 2000, pp. 1–333.

Mason, S. F., *Chemical Evolution: Origin of The Elements, Molecules, and Living Systems,* Clarendon Press, Oxford, 1992, pp. 1–317.

Photosynthesis

Falkowski, P. G. and Raven, J. A., *Aquatic Photosynthesis,* Blackwell Science, Oxford, 1997, pp. 1–375.

Hall, D. O. and Rao, K. K., *Photosynthesis,* 5th ed., Cambridge University Press, Cambridge, UK, 1994, pp. 1–211.

Suitability of the Elements

Williams, R. J. P. and Frausto da Silva, J. J. R., *The Natural Selection of the Chemical Elements,* Oxford University Press, Oxford, 1996, pp. 1–646.

Biogeochemical Cycles and Earth History

Anbar, A. D. and Knoll, A. H., "Proterozoic Ocean Chemistry and Evolution: A Bioinorganic Bridge?" *Science,* **297,** 1337–1342 (2002).

Jacobson, M. C., Charlson, R. J., Rodhe, H., and Orians, G. H., Eds., *Earth System Science: From Biogeochemical Cycles to Global Change,* Academic Press, New York, 2000, pp. 1–523.

Kaplan, I. R. and Bartley, J. K., "Global Biogeochemical Cycles: Carbon, Sulfur and Nitrogen", in *Earth Systems: Processes and Issues,* Ernst, W. G., Ed., Cambridge University Press, Cambridge, UK, 2000, pp. 278–196.

Schlesinger, W. H., *Biogeochemistry: An Analysis of Global Change,* 2nd ed., Academic Press, New York, 1997, pp. 1–588.

Berner, E. K. and Berner, R. A., *Global Environment: Water, Air, and Geochemical Cycles,* Prentice-Hall, Upper Saddle River, NJ, 1996, pp. 1–376.

Ecological Stoichiometry

Sterner, R. W. and Elser, J. J., *Ecological Stoichiometry: The Biological Elements from Molecules to the Biosphere,* Princeton University Press, Princeton, New Jersey, 2002, pp. 1–439.

Carbon/Oxygen/Hydrogen Cycle

Adams, M. W. W. and Stiefel, E. I., "Organometallic Iron—Key to Biological Hydrogen Metabolism", *Curr. Opin. Chem. Biol.,* **4,** 214–220 (2000).

Holmén, K., "The Global Carbon Cycle", in *Earth System Science: From Biogeochemical Cycles to Global Change,* Jacobson, M. C., Charlson, R. J., Rodhe, H., Orians, G. H., Eds., Academic Press, New York, 2000, pp. 282–321.

Wigley, T. M. L. and Schimel, D. S., Eds., *The Carbon Cycle,* Cambridge University Press, Cambridge, UK, 2000, pp. 1–292.

Adams, M. W. W. and Stiefel, E. I., "Biological Hydrogen Production: Not So Elementary", *Science,* 1842–1843 (1998).

Keeling, R. F., Najjar, R. P., Bender, M. L., and Tans, P. P., "What atmospheric oxygen measurements can tell us about the global carbon cycle", *Global Biogeochem. Cycles,* **7,** 37–67 (1993).

Nitrogen Cycle

Jaffe, D. A., "The Nitrogen Cycle", in *Earth System Science: From Biogeochemical Cycles to Global Change,* Jacobson, M. C., Charlson, R. J., Rodhe, H., Orians, G. H., Eds., Academic Press, New York, 2000, pp. 322–342.

Postgate, J., *Nitrogen Fixation,* 3rd ed., Cambridge University Press, Cambridge, UK, 1998, pp. 1–100.

Vitousek, P. M., et al., "Human Alterations of the Global Nitrogen Cycle: Sources and Consequences" *Iss. Ecol.,* **7,** 737–750 (1997).

Zumft, W. G., "Cell Biology and the Molecular Basis of Denitrification", *Microbiol. Mol. Biol. Rev.,* **61,** 533–616 (1997).

Sulfur Cycle

Charlson, R. J., Anderson, T. L., and McDuff, R. E., "The Sulfur Cycle", in *Earth System Science: From Biogeochemical Cycles to Global Change,* Jacobson, M. C., Charlson, R. J., Rodhe, H., and Orians, G. H., Eds., Academic Press, New York, 2000, pp. 343–359.

Barton, L. L., Ed., *Sulfate-Reducing Bacteria,* Kluwer Academic/Plenum, New York, 1995, pp. 1–352.

Kelly, D. P., Shergill, J. K., Lu, W.-P., and Wood, A. P., "Oxidative metabolism of inorganic sulfur compounds by bacteria", *Antonie Van Leeuwenhoek,* **71,** 95–107 (1997).

Stiefel, E. I., "Molybdenum Bolsters the Bioinorganic Brigade", *Science,* **272,** 5999–6000 (1996).

Microbiology

Oren, A., "Prokaryotic Diversity and Taxonomy: Current Status and Future Challenges", *Philos. Trans. R. Soc. London. Ser. B-Biol. Sci.,* **359,** 623–638 (2004).

Hugenholtz, P., Goebel, B. M., and Pace, N. R., "Impact of Culture-Independent Studies on the Emerging View of Bacterial Diversity", *J. Bacteriol.,* **180,** 4765–4774 (1998).

Banfield, J. F. and Nealson, K. H., Eds., *Reviews in Mineralogy,* Vol. 35, The Mineralogy Society of America, Washington, DC, 1997, p. 37.

Madigan, M. T., Martinko, J. M., and Parker, J., *Brock Biology of Microorganisms,* 8th ed., Prentice Hall, Upper Saddle River, NJ, 1997.

Microbial Biodiversity

Microbial biodiversity and, in particular the genomes of the prokaryotes that have been sequenced, are accessible through the following ***WebSites***:

http://www.ucmp.berkeley.edu/exhibit/phylogeny.html

http://www.cbs.dtu.dk/services/GenomeAtlas

http://genome.jgi-psf.org/mic_cur1.html

CHAPTER III

Metal Ions and Proteins: Binding, Stability, and Folding

Ivano Bertini and Paola Turano
Magnetic Resonance Center and Department of Chemistry
University of Florence
Sesto Fiorentino,
Italy 50019

Contents

III.1. Introduction

This chapter deals with the interactions between metal ions and proteins. It provides the definitions of "metalloprotein" and "metalcofactor" and describes the properties of the amino acids that act as ligands and their possible binding modes with metals. The formation of the metal–ligands bond profoundly affects properties of both the metals and the amino acid ligand. Hints on how to browse genomes with the aim of identifying potential metalloproteins are provided in Section III.4. Some concepts about the possible role of metal ions in the overall structural stability and correct folding of proteins are described in Section III.5. Section III.6 introduces the idea of kinetic control of metal delivery, which is gaining more importance with the identification of complex intracellular networks of metal trafficking pathways and has allowed the estimation of free metal content within the cell.

III.2. The Metal Cofactor

Life has coevolved with the minerals of the Earth's crust and the ions in Earth's waters (see Chapter II). Therefore, it is not surprising that living beings have evolved the capability to use inorganic elements for key biological processes and to defend themselves from poisoning by other elements. Some metal ions, when associated with polypeptides, can help catalyze unique chemical reactions and perform specific physiological functions. We call such metal ions "metal cofactors".

Amino acids and proteins alone are not sufficient to perform all the reactions needed for life. For example, the Fe^{3+}/Fe^{2+} and Cu^{2+}/Cu^{+} redox couples play critical roles as cofactors for electron-transfer reactions and in the catalysis of redox reactions. The Fe^{2+} ion can reversibly bind dioxygen (O_2) if a coordination site is available. The Zn^{2+} ion can act as a Lewis acid to produce a bound hydroxyl ion nucleophilic Zn–OH^- moiety from water around neutral pH. Much of this book deals with these and other metal cofactors and their reactions.

In Chapter I, Figure I.1 presents a periodic table in which metal ions essential for life are highlighted. Some of these (like Fe, Cu, and Zn) are strongly associated with proteins and form the so-called metalloproteins. The affinity between the metal ion and the protein can be expressed as

$$\mathrm{M} + \mathrm{P} \leftrightarrows \mathrm{MP}$$

$$K = [\mathrm{MP}]/[\mathrm{M}][\mathrm{P}]$$

where K is the equilibrium binding constant between the metal ion, M, and the polypeptide chain, P, to form the metalloprotein, MP. The thermodynamic stability of such adducts is often very high and K can have values greater than $\sim 10^8\ M^{-1}$. The non-transition metal ions, such as Na^+, K^+, Mg^{2+} (and Ca^{2+} in selected systems) form complexes with proteins with a significantly smaller affinity constant. These latter systems are properly considered as metal–protein complexes. Manganese(II) sometimes behaves like magnesium and sometimes is as strongly associated as the metal ions in true metalloproteins. Proteins that function as biological catalysts are referred to as enzymes, and consistently, metalloprotein catalysts are termed metalloenzymes.

In any cellular compartment, the concentration of "free" transition metal ions (i.e., aqua ions) is usually very low (even near zero). In contrast, there can be *relatively* high concentrations of the non-transition metal ions sodium, potassium, calcium, and magnesium in living tissues. In multicellular organisms, sodium and calcium are largely extracellular, while potassium and magnesium are largely intracellular. Calcium and magnesium are often metal activators in proteins to which they bind with relatively low affinity. Under appropriate circumstances these metal ions induce conformational changes in the protein upon binding and in doing so they may transmit a signal (e.g., along the axon of a neuron). The Na^+ and K^+ ions are able to bind to sites containing oxygen donors within preorganized cavities, such as in ATP-ases (i.e., those proteins whose function is coupled to ATP hydrolysis). Such "weakly" bound metal ions are being found more and more frequently in specific locations in protein X-ray structures, implicating a likely role in electrostatic ordering, which may be relevant to the proper functioning of the protein.

Some metal ions are found deeply buried within proteins. Such metal ions are often "structural" in function. Their interaction with the protein helps insure the optimal protein structure and contributes to the stability and appropriate acid–base behavior necessary for the physiological function. The occurrence of structural metal ions is very common. For example, there are four calcium ions in thermolysin, a thermophilic protein with peptidase (i.e., peptide bond hydrolysis) activity, and at least two of them have been shown to be important for the high-temperature stability. Peroxidases (see Section XI.3) contain several Ca ions that keep the nonheme bound (distal) histidine in the right position and with the right pK_a. The Zn^{2+} ions in Zn fingers (see Section XIV.2), which are transcription factors (see Tutorial I) are critical for the adoption of the proper shape of the protein, which allows it to interact with DNA.

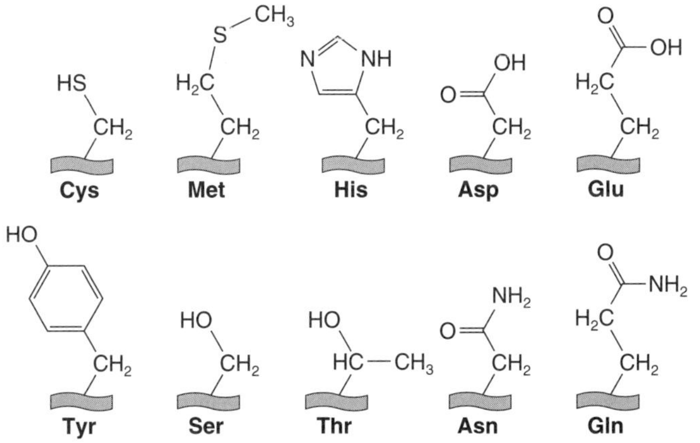

Fig. III.1.
Amino acids containing donor atoms in their side chains. The protein backbone is schematically represented as a gray ribbon.

III.3. Protein Residues as Ligands for Metal Ions

The common metal-donor atoms in proteins are sulfur atoms of cysteines and methionines, nitrogen atoms of histidines, and oxygen atoms of glutamates, aspartates, and tyrosinates. Oxygen atoms of peptide carbonyl moieties, and of threonines and serines, and nitrogen atoms of (deprotonated) backbone amides, and of lysine side chains are also potential donor atoms (see Fig. III.1 and Table III.1). Histidine can bind the metal ion either through the imidazole Nε or Nδ. Glutamate and aspartate can behave as monodentate, bidentate, or bridging ligands.

Examples of the coordination modes of side chains of amino acids in proteins of known structure are listed below and summarized in Fig. III.2. This compilation is not intended as an exhaustive description of all coordination spheres observed in metalloproteins, but rather serves to illustrate some general features of the coordination chemistry of residue side chains. The metal ion preference for a certain amino acid side chain follows the general rules of coordination chemistry and often can be simply interpreted using the theory of "hard and soft" acids and bases (see Tutorial II). "Soft" donor atoms favor binding to soft metal ions, while "hard" donor atoms favor "hard" ions. For example, sulfur is a soft donor atom that binds well to Cd(II) and Cu(I), whereas oxygen is a hard donor atom that binds well to hard cations such as Ca(II) and Mn(II). Zn(II) is intermediate in hardness–softness and binds both oxygen and sulfur as well as nitrogen donors, the latter being intermediate in terms of hardness between oxygen and sulfur. The hard–soft character of a metal ion depends on oxidation state; for example, Fe(III) is harder than Fe(II), and Cu(II) is harder than Cu(I).

What follows is a description of each of the principal ligand donor types that occur in metalloproteins:

Table III.1
The pK_a Values for Free Amino Acids Containing Protonated Groups in Their Side Chains

Free Amino Acid	pK_a
Cys	8.3
Met[a]	
His^+/His	6.0
His/His^-	14.0
Glu	4.3
Asp	3.9
Tyr	10.1
Ser	13.0
Thr	13.0
Asn[a]	
Gln[a]	
Lys^+/Lys	10.5
Arg^+/Arg	12.5

[a] These residues do not display pK_a values in biologically accessible pH regions.

Cysteine (Cys): The thiolate sulfur atom of Cys residues is often involved in the coordination of Cu, as well as of Zn, Fe, Ni, and Mo. A few examples are known of a neutral cysteine serving as a ligand in ferrous heme iron, but cysteine binds most frequently in its deprotonated, cysteinate form. For example, ferric heme iron in cytochrome P450 binds Cys in its deprotonated form (Fig. III.3). A nice example of Cys coordination to nonheme iron is provided by the iron site of nitrile hydratase. In this enzyme, the ferric iron is bound to a conserved stretch of residues composed of the side chains of Cys109, Cys112, Cys114 and the amide nitrogens of Ser113 and Cys114. Blue copper proteins represent a well-known example of Cu–Cys binding (Fig. III.4; Chapter IV). The Cys residue binds to iron in all

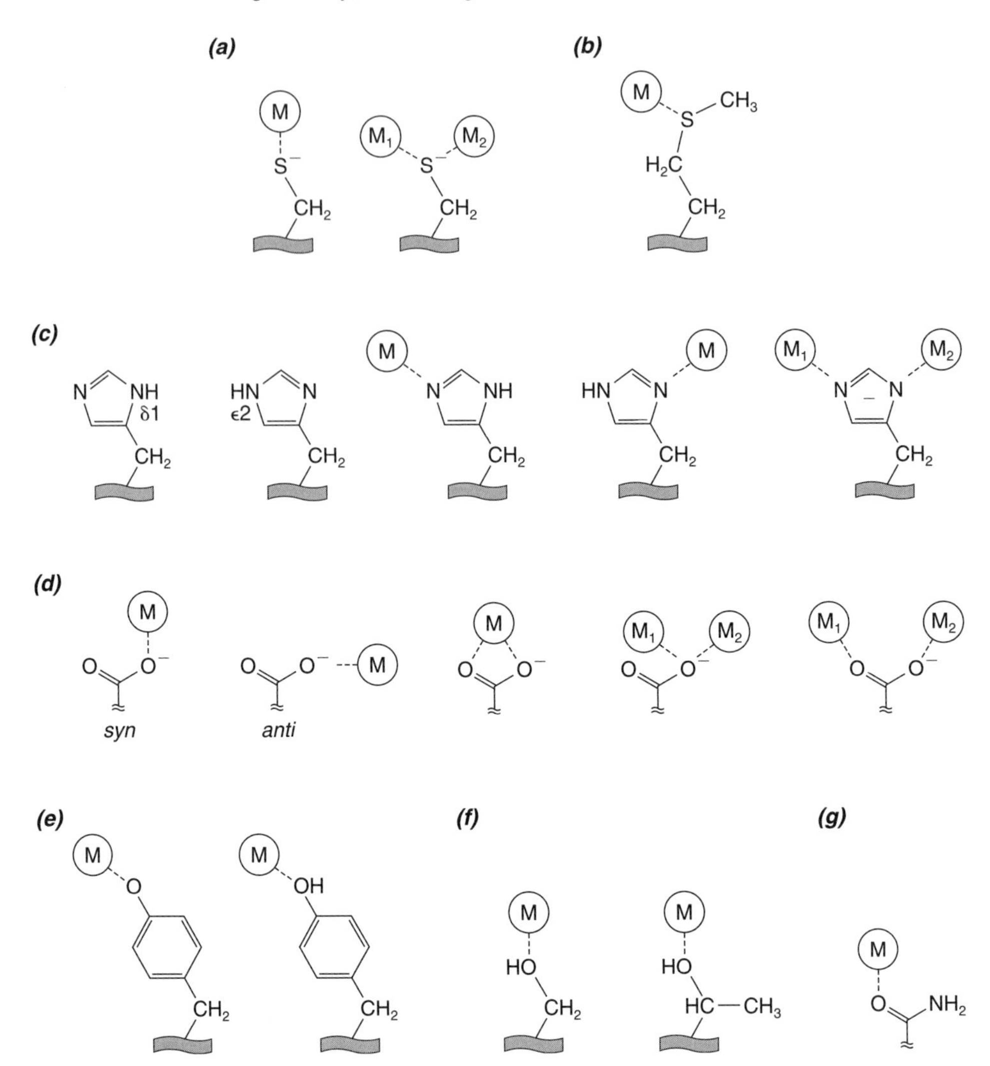

Fig. III.2.
Schematic representation of the coordination modes of amino acid side chains. In all the examples, the metal ion is shown as a "M" in a circle, and the coordination bonds as dashed lines. (*a*) The depronated form of Cys can bind a single metal ion or act as a bridging ligand between two metal ions; (*b*) Met binds through the sulfur lone pair; (*c*) The two drawings an the left represent the tautomeric equilibrium of His, while the other three show the coordination of His through the Nε2, the Nδ1 and as bridging ligand; (*d*) the common terminal carboxylate of Glu and Asp can bind a metal ion as monodentate ligand in syn or anti conformation, can act as bidentate, or as a bridging ligand using one or both oxygen atoms; (*e*) binding of the deprotonated and protonated Tyr side chain; (*f*) Ser and Thr behave like monodentate neutral ligands; (*g*) binding of Asn and Gln through their side chain oxygen.

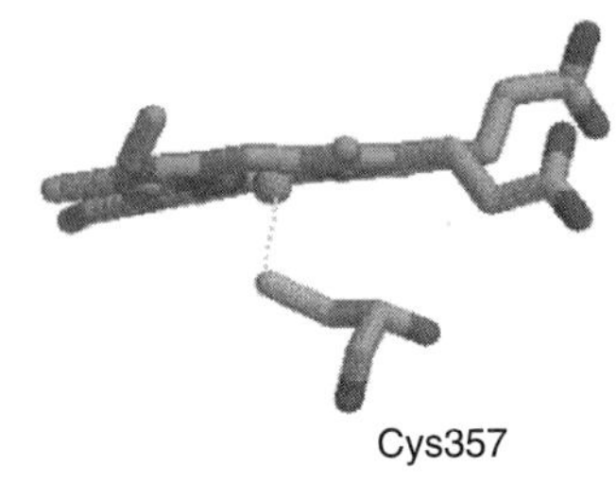

Fig. III.3.
The coordination environment of the ferric P450cam from *Pseudomonas putida* showing Cys357 in the axial coordination position of Fe(III) (PDB code: 1DZ4). The metal ion is shown as a sphere.

types of iron–sulfur proteins, from the mononuclear rubredoxin to all the very complex clusters that are discussed in Chapter IV. The Cys residue binds nickel in [NiFe]–hydrogenase. It has also been found that a Cys sulfur can act as bridging ligand between two metal ions (see Sections XII.1 and XII.6). Cysteine is a ligand of zinc in alcohol dehydrogenase and other zinc enzymes (transferases, ligases) as well as in the very common regulatory zinc finger domains (Fig. III.5 and Sections XIV.1 and XIV.2). A unique distribution of Cys residues also characterizes the metal-binding metallothioneins, which provide soft sulfur donor atoms to bind Zn(II) and Cu(I) as well as Cd(II) and other heavy metal ions.

Methionine (Met): Methionine ligation as well as Cys ligation is often found for Cu(I). As an example, Met ligands are present in the coordination environment of blue copper proteins (Fig. III.4; Chapter IV) (where, however, the copper–methionine bond is very long), blue oxidases (Section XI.7), in the copper A center in cytochrome *c* oxidase (Section XI.5A), and in nitrite reductase (Section XII.4).

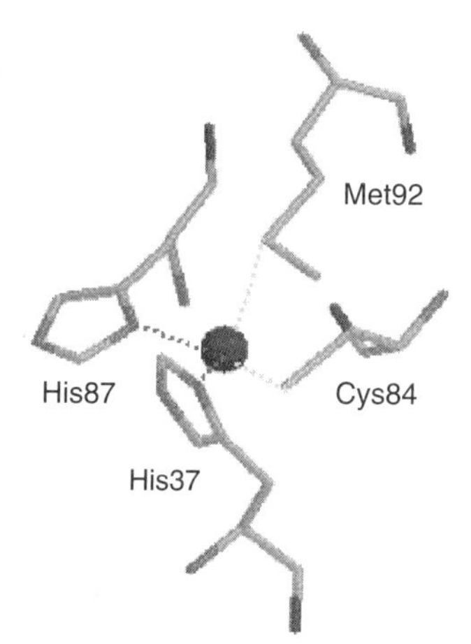

Fig. III.4.
The coordination environment of Cu in plastocyanin (PDB code: 1IUZ). The metal ion is shown as a sphere.

The neutral Met is a much weaker donor than the anionic cysteinate ligand, and the affinity of Met S for the metal ion is modulated by the redox state of the latter. For example, in the electron-transfer *c*-type cytochromes the iron–sulfur bond is weaker for the ferric form compared to the ferrous form (see Chapter IV).

Histidine (His): Histidines are often encountered in the coordination spheres of Fe, Cu, Zn, and Ni. Deprotonated, neutral, imidazole nitrogen atoms coordinate metal ions using a lone pair on a nitrogen atom. In protic solvents, the NH protons of free His participate in a tautomeric equilibrium, the Nε–H tautomer being the predominant form (Fig. III.2). The presence of metal ions may induce preferential binding for one of the two positions (Nδ1 or Nε2). Coordination through either of the two imidazole nitrogen atoms is almost equally frequent in metalloproteins and the preference for one of the two donor atoms is often determined only by their different stereochemical requirements. Indeed, mixed coordination is commonly encountered in the coordination sphere of the same metal ion, for example, in Cu,Zn-superoxide dismutase (SOD) (Fig. III.6), and in zinc carbonic anhydrase. The oxidized form of the former protein represents a nice example of the ability of His to act as a bridging ligand between two metal ions (Cu^{2+} and Zn^{2+}). This bridge is broken upon Cu reduction, providing an example of His-based flexibility. Generally, His imidazole (Im) ligands provide a relatively rigid framework. Imidazole coordination to a metal ion can lower the histidine–histidinate transition pK_a (by ~2 units).

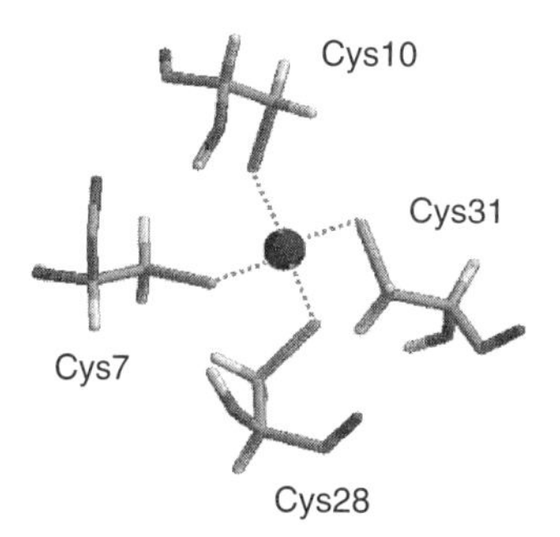

Fig. III.5.
The tetrahedral coordination of Zn in the zinc finger domain of the erythroid transcription factor GATA-1 (PDB code: 1GAT). The metal ion is shown as a sphere.

Second-coordination sphere effects may also significantly influence the metal site. In some proteins, second-sphere hydrogen bonding of His ligands with carboxylate residues is present. This hydrogen bonding allows the His ligands to attain partial anionic character. Indeed, the formation of a strong hydrogen bond between the hydrogen atom bound to this nitrogen and an oxygen of a nearby carboxylate can be viewed as a partial deprotonation of the imidazole ring, as depicted in Fig. III.7. The ligand properties of the His are affected by such hydrogen-bond formation and, through this interaction, both redox properties and reactivity of the bound metal ion can be effectively tuned.

Glutamate (Glu) and Aspartate (Asp): These two amino acids can be treated together because their side chains differ only in length. The chemical properties of their terminal carboxylate donors are essentially identical (pK_a, coordination properties of the carboxylate oxygens). The donor atom(s) being oxygen implies a tendency to bind to relatively hard metal ions. Indeed, these carboxylate side chains are present in the coordination environment of Ca(II) (EF-hand proteins), of Fe(III) in nonheme, non-iron–sulfur iron proteins (e.g., transferrin; 2,3-dihydroxybiphenyl dioxygenase; 4-hydroxyphenylpyruvate dioxygenase; extradiol dioxygenase), and of Zn(II) in zinc hydrolases. Both Glu and Asp residues also bind to K^+ (pyruvate kinase; tryptophanase), Na^+, Mg^{2+} (isocitrate dehydrogenase), where they act as monodentate ligands. The carboxylate group can act as a bidentate ligand as in the case of Asp362 in the mononuclear site of naphthalene dioxygenase or in the case of Glu72 in carboxypeptidase (Fig. III.8). An example of carboxylate acting as a bridging ligand is found in the active site of arginase (Fig. III.9), where Asp124 binds to both Mn(II) ions bridging a binuclear

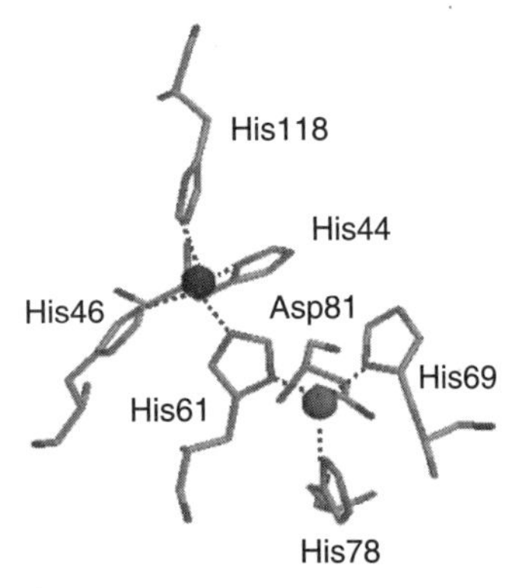

Fig. III.6.
The coordination environment of the Cu(II),Zn(II) center in Cu,ZnSOD (PDB code: 1E9Q). Note the mixed Nδ1 (His44, His69, His78) and Nε2 (His46, His118) coordination of the histidines, and the presence of His61 acting as a bridging ligand. The two metal ions are shown as dark gray and light gray spheres, respectively.

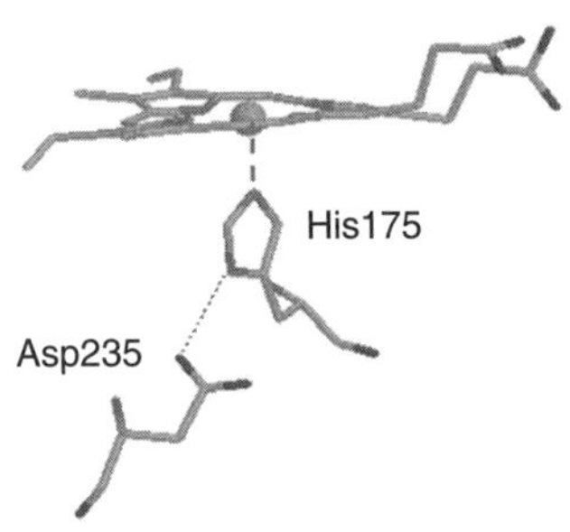

Fig. III.7.
The hydrogen bond between Asp235 and the metal-binding His175 in cytochrome *c* peroxidase (PDB code: 2CYP) provides an example of second-coordination sphere effects. The metal ion is shown as a sphere.

metal center, as do Asp106 and Glu58 in the di-iron site of hemerythrin. Upon redox changes or substrate binding, carboxylate ligands can easily vary their mode of coordination, contributing to the flexibility of the metal environment. As shown in Fig. III.2, metal ions display syn or anti stereochemistry as they interact with the carboxylate anion, but the syn-carboxylate–metal ion stereochemistry is far more common.

Tyrosine (Tyr): The deprotonated form of tyrosine binds Fe(III) in intradiol dioxygenases, in transferrin (Fig. III.10), and in the binuclear center of mammalian and plant purple acid phosphatase. Protonation of Tyr ligands can modulate their metal–oxygen bond length and even cause the bond to dissociate. This effect represents a source of flexibility in the active center of nonheme, non-sulfur iron proteins.

Serine (Ser): An example of Ser binding is found in the coordination site of K^+ in dialkylglycine decarboxylase (Fig. III.11). The metal ion coordinates to five protein ligands and a water molecule. One of these ligands is Ser80, providing its side-chain hydroxyl oxygen as a donor atom. The coordination site of K^+ in this protein nicely exemplifies the strong preference of monovalent alkaline metal ions for oxygen donor atoms. Indeed, the other amino acid ligands are one of the two oxygens of the carboxylate of an Asp and three main chain peptide carbonyl groups.

Threonine (Thr): The hydroxyl oxygen of this amino acid side chain has binding properties that closely resemble those of the hydroxyl oxygen of Ser. Threonine OH coordination is found in the coordination environment of K^+ in the cation-binding site of ascorbate peroxidase (Fig. III.12).

Asparagine (Asn) and Glutamine (Gln): The side-chain oxygen of an asparagine is bound to Mn^{2+} in the active site of phosphatase. The Gln330 residue is coordinated to Mn^{2+} in Mn-substituted isopenicillin-*N* synthase, although it does not bind to the iron of the native enzyme. The side-chain oxygen atoms of Gln and Asn residues bind to Zn in the binuclear center of plant purple acid phosphatase (Fig. III.13).

Post-translational modifications at the metal-binding site occur in a number of proteins. In these cases, the nascent apoprotein undergoes a reaction that induces a covalent change in a specific amino acid, whose modified side chain provides the donor atom(s). Examples are the reactions that transform Glu, Asp, Asn, and Ser to γ-carboxy-Glu, β-hydroxy-Asp, β-hydroxy-Asn, and phospho-Ser, respectively. A nice example of a further kind of post-translational modification is provided by galactose oxidase (Chapter XII). This protein contains a Tyr side chain linked by a thioether bond to a Cys residue. The modified tyrosinate binds Cu(I) and becomes oxidized to a tyrosyl radical as the metal is oxidized to Cu(II). Many examples of such modifications are reviewed by Kuchar and Hausinger (2003).

The case of galactose oxidase also introduces the important topic of radical binding to metal ions, which is discussed in Chapter XII. As considered in the relevant chapters, in some cases the active site of a metal ion that formally would appear as M^{n+}, is more properly described as $M^{(n-1)+}$ bound to a cation radical ligand. The radical may be delocalized on a binding amino acid side chain as in galactose oxidase, or be associated with one of the metal-binding entities present in the special metal cofactors, which are described in Chapter IV.

The ability of the amino acid side chains described above to complete the metal coordination environment in the case of special metal cofactors is described in detail in Chapter IV.

III.4. Genome Browsing

In general, similarities in the coordination geometry and nature of ligand amino acids in the coordination sphere of a given metal ion and oxidation state provide a so-called "consensus sequence", where ligand side chains belong to amino acids in corresponding positions along the primary sequence of a set of proteins. This finding opens an important bioinorganic application of genome browsing, which provides a new way to identify metal-binding proteins from the analysis of genomes of different organisms.

Genome browsing in bioinorganic chemistry is not limited to the analysis of amino acid identity or similarity in primary sequence, but also takes into account the nature and spacing of amino acids that act as potential metal ligands. (Figure III.14 shows key examples of consensus sequences used to identify specific proteins in gene data banks.) The nature of other key residues in the metal-binding site could also be considered. These considerations illustrate the great relevance of databases and their interrogation for consensus sequences of metalloproteins using specific bioinformatic tools for the search. At present, partial information on metal-binding sites is available at the Metalloprotein Database at The Scripps Research Institute (see references at the end of this chapter for more details).

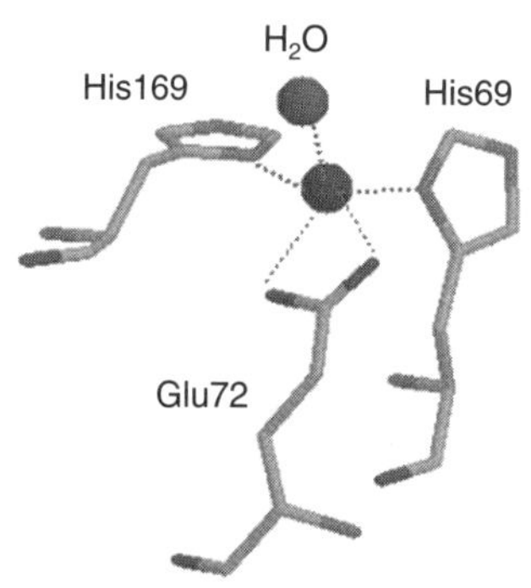

Fig. III.8.
The coordination environment of zinc in carboxypeptidase (PDB code: 5CPA). Note the presence of Glu72 acting as bidentate. The oxygen atom of a bound water molecule is represented as a light gray sphere. The metal ion is shown as a dark sphere.

III.5. Folding and Stability of Metalloproteins

Most of the proteins that have been structurally characterized to date are globular proteins. In order to function, these proteins assume a unique, "native", three-dimensional (3D) structure, usually designated as the folded state of the protein. It is

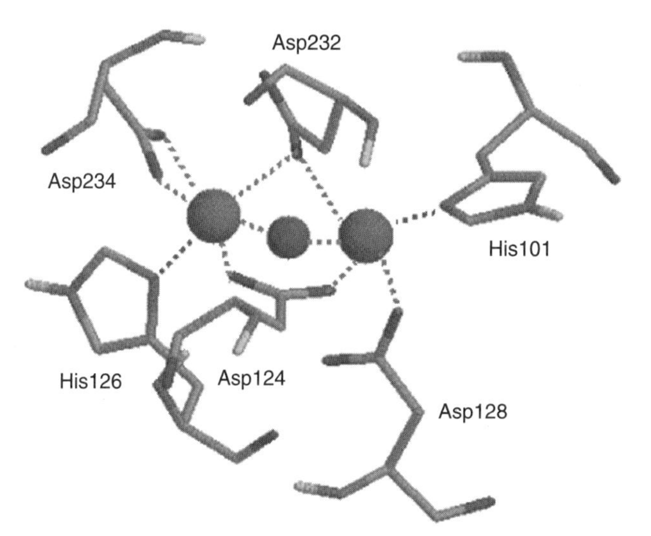

Fig. III.9.
The coordination environment of arginase (PDB code: 1RLA) represents an example of the different coordination modes of carboxylate groups. In this protein Asp124 binds both Mn(II) ions constituting the binuclear metal center (represented as large spheres) using the two oxygen atoms of the carboxylate, while Asp232 bridges the two metal ions using a single oxygen. Asp234 binds one of the manganese ions as bidentate, while Asp128 behaves as monodentate. The small sphere represents the oxygen of a bridging OH group.

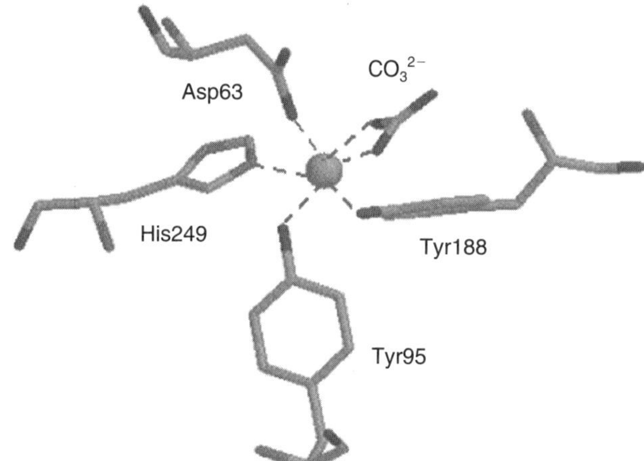

Fig. III.10.
The coordination environment of the ferric iron in the CO_3^{2-} complex of transferrin, N-terminal domain (PDB code: 1TFD). The metal ion is shown as a sphere.

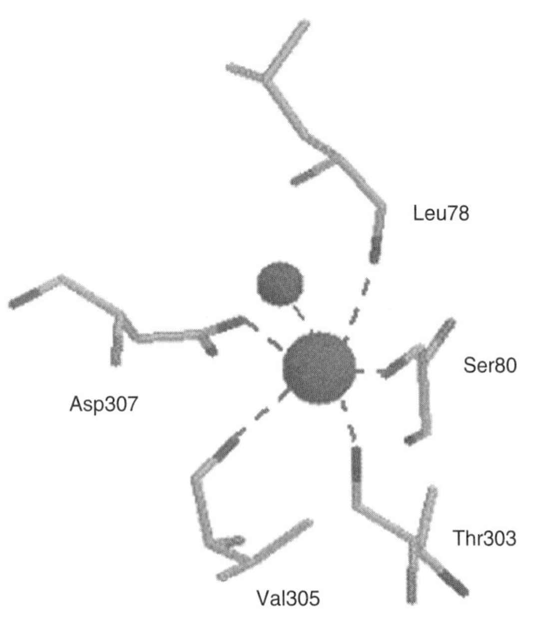

Fig. III.11.
The coordination environment of K^+ in dialkylglycine decarboxylase (PDB code: 1DKA). The potassium ion is shown as a large sphere. The donor atoms are all oxygen atoms, provided by the side chains of a Ser and of a monodentate Asp, as well as by the carbonyl oxygen atoms of a Thr, Leu, and Val. The oxygen atom of a bound water molecule is represented as a small sphere.

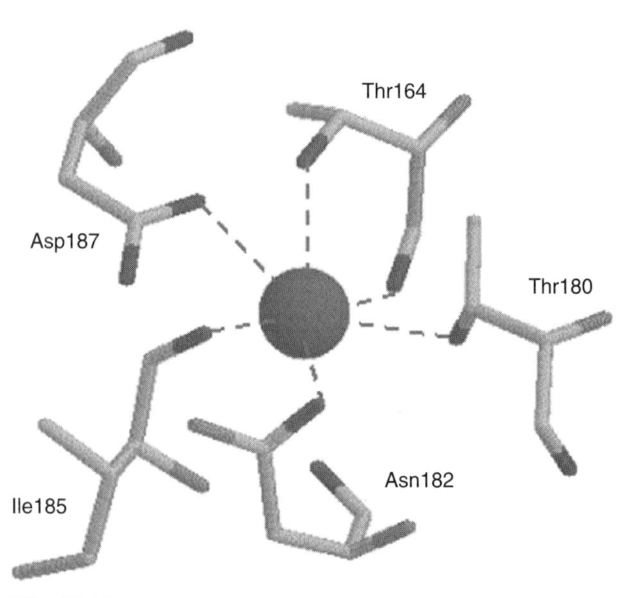

Fig. III.12.
The coordination environment of K^+ (large sphere) in the cation binding site of ascorbate peroxidase (PDB code: 1APX) is formed by the side chain oxygen atoms of Thr180 and Thr164, of Asn182 and monodentate Asp187. The backbone carbonyl oxygen atoms of Thr164 and Ile185 complete the metal ion coordination.

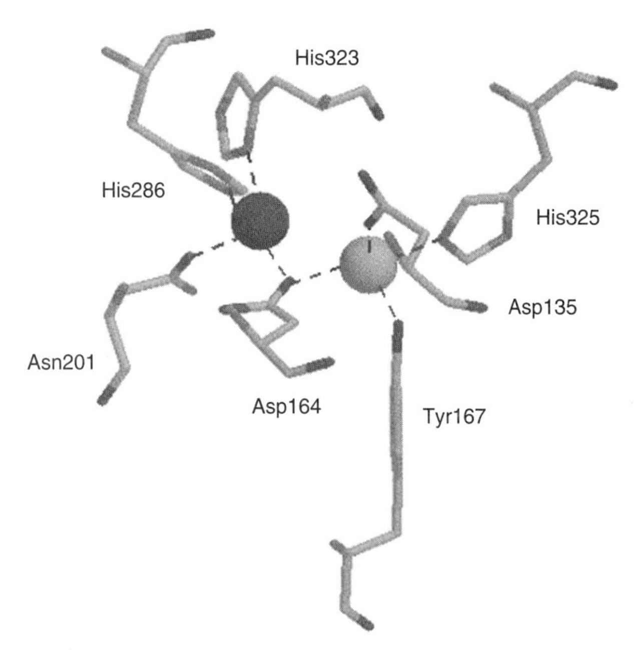

Fig. III.13.
The side chain oxygen atoms of an Asn residue bind Zn in the binuclear center of plant purple acid phosphatase (PDB code: 1KBP). The zinc and iron metal ions are shown as dark and light gray spheres, respectively. Only amino acid ligands are shown in this figure to improve readability.

– Search criteria for metallochaperones

Input sequence	Atx1 from *S. cerevisiae*
Metal binding motif	x'-x''-Cys-x'''-x''''-Cys (recurring residues at positions x)

– Search criteria for metal transporting ATPases

Input sequence	Ccc2a from *S. cerevisiae*
Metal binding motif	x'-x''-Cys-x'''-x''''-Cys (recurring residues at positions x)

Fig. III.14.
Consensus sequences derived from *Saccharomyces cerevisiae*, are used to find potential metallochaperones in gene data banks. Recurring residues are found at *x*-type positions in corresponding proteins from different organisms.

now emerging that such a state does not constitute a universal feature of functional proteins. For example, intrinsically unstructured proteins are responsible for cellular signaling and regulation, lending an increased importance to the role of flexibility in biomolecular recognition. What is the role of metal ions in stabilizing either type of protein?

The compact, folded state of a globular protein is usually characterized by the presence of a well-defined hydrophobic core with restricted fluctuations about bond torsion angles. In contrast, the residues outside this core, and in particular surface residues, may be highly disordered. The well-defined conformational state is the result of a closely packed structure (the proportion of space occupied by the component atoms being $\sim 75\%$, if conventional van der Waals radii are used), with well-defined, optimized interactions among amino acid side chains.

In general, a close balance exists between the enthalpic and entropic contributions to the stability of the protein, resulting in a small overall free-energy change for the protein folding reaction. However, the small net folding free energy of a protein is the result of the compensation between various large contributions: the enthalpic content associated with the sum of all the nonbonding interactions between atoms in the folded conformation; the formation of disulfide bonds; and the coordination of protein ligands to metal ions. The entropy change is due to the loss in degrees of freedom for the polypeptide backbone and side chains during the folding process. Very importantly, the difference in solvation enthalpy and entropy between folded and unfolded states also makes a significant contribution. The enthalpy–entropy compensation leads to the quite surprising result that the functional state of a protein is stabilized by only a few kilojoules per mole. Consequently, the favored status of a folded protein occurs under a narrow range of physical and chemical conditions.

How does the presence of a metal ion influence the thermodynamic stability of a metalloprotein? Clearly, the formation of metal–protein chemical bonds contributes to the protein folding. The affinity constant value discussed in Section III.2 is a measure of the free energy stabilization upon metal binding. A favorable enthalpic change is associated with the formation of the coordination bonds. At the same time, when the metal ion enters the protein cavity, it is denuded of water molecules and waters in the cavity are also set free. This water release provides a significant contribution to the entropy and thus to the free energy of stabilization of the metal–protein complex (see Tutorial II). However, metal binding decreases the degree of freedom of bound amino acid side chains, which is also accompanied by reduced flexibility of the protein backbone, in particular when at least one of the ligands belongs to unstructured parts of the protein.

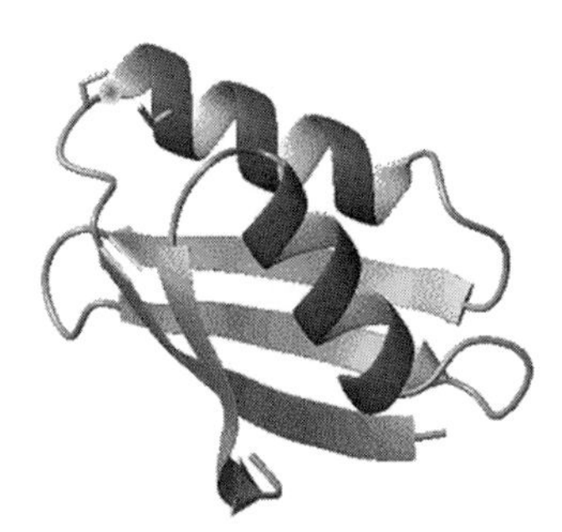

Fig. III.15.
Ribbon representation of the 3D structure of the Cu transport protein Ccc2 (PDB code: 1FVS).

At this point one may ask: What is the structure, if any, of the metalloprotein without the metal ion? The answer differs on a case by case basis. If the protein has elements of secondary structure organized into a tertiary structure, with strong hydrophobic interactions, the removal of the metal ion may introduce some disorder at its binding site, but the structure as a whole is essentially maintained. This, for example, is the case in the Cu transport protein Ccc2, whose structure is of $\alpha\beta\alpha\beta\alpha\beta$ type (where α stands for α-helix and β for β-structure) with a well-defined hydrophobic core and with the copper-binding domain in an external loop (Fig. III.15). The same situation occurs in the blue copper proteins, which essentially have an all-β-structure and the metal-binding site is in a loop. In the case of Cu,Zn SOD, which has a predominantly β-barrel secondary structure, the native structure is greatly affected upon removal of metal ions, as shown from the structural study of the apoprotein (Fig. III.16). In this case, the presence of metal ions has a key role in maintaining the correct arrangement of the two β-sheets that constitute the protein and imposes constraints on the metal-binding loops that rigidify the overall structure.

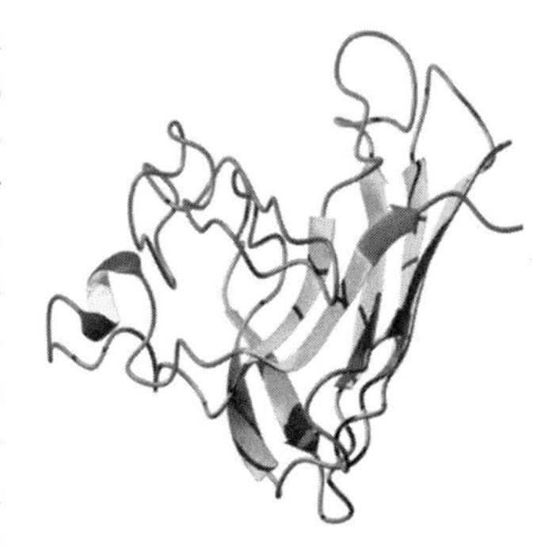

Fig. III.16.
Ribbon representation of the 3D structure of the apo-form of monomeric superoxide dismutase (PDB code: 1RK7).

From a practical point of view, the metal ion of a metalloprotein can be removed by chemical manipulation. If the metalloprotein is dialyzed against a large concentration of a ligand that is specific for the metal, then the metal ion is distributed between the protein matrix and the ligand and, by changing the dialyzing solution, eventually the metal ion is removed from the protein.

The presence of a metal ion is also relevant with respect to protein stability in the presence of unfolding agents. Compounds like guanidinium chloride (GdmCl) and urea form strong hydrogen bonds with the lone pairs of protein amino acids, thereby disrupting the existing network of hydrogen bonds and eventually unfolding the protein. The attack of these chaotropes (i.e., substances able to induce chaos in the protein structure) in general starts from loops and peripheral hydrogen bonds. The extra stability provided by the metal ion increases the resistance to such unfolding agents.

An instructive example is provided by the behavior of the monomeric mutant of Cu,Zn SOD. Upon addition of GdmCl, only minor changes in structure and dynamics are detected for the loops and some of the β-strands, but the overall structure of the protein is maintained up to 3.5 *M* GdmCl. At this concentration of the chaotropic agent, metals are released from the native binding sites and the protein largely unfolds. The coordination bonds are clearly the bulwark against the collapse of the structure and native protein folding exists only as long as there are coordination bonds. Additional examples will be discussed in Chapter IV, for proteins containing "special metal cofactors".

Last, but not least, a trigger role for the metal ion during the folding process can be considered, where kinetic control of the collapse of the random coil structure is due to the formation of native and/or non-native coordination bonds. The random-coil state represents the complete lack of secondary structure, where the conformation of each single amino acid in the polypeptide chain is completely independent of that of the others, and free rotation around chemical bonds is possible.

An emerging view of proteins that are unstructured, or only partially structured under physiological conditions, may provide new insight in the investigation of metalloproteins. For example, the calcium-based signaling proteins are in general constituted by multiple domains connected by flexible loops. The presence of Ca makes the connecting loops less flexible, and thereby has an important role in controlling the relative orientation of the domains.

III.6. Kinetic Control of Metal Ion Delivery

From the picture developed above, a fundamental need emerges in living organisms: The proper metal must encounter the proper protein in such a way that the latter assumes the proper structure and the metalloprotein executes its proper function. The protein must encounter the correct metal, acquire, and retain it. An emerging concept in biology is that a selective coordination environment may not be sufficient, at least for the first two of these steps. Evolution has not achieved selectivity solely through metal-binding preferences. *In vitro* high affinity of a certain protein toward a certain metal ion is not proof that this protein binds the same metal ion *in vivo*.

A prerequisite for metal binding *in vivo* is that the metal ion is accessible to the protein. However, free metal ions are often present in very low, or effectively zero, concentration within the cell. The way to make the metal accessible to the target protein may involve a cascade of accessory proteins, which accompany the metal ion from the external membrane to the final target. Metal trafficking factors control the kinetics of metal ion exchange between proteins, providing preferential pathways within the cell compartments and between them, thereby assisting metal delivery to the proper protein target(s) (i.e., chaperoning). So, a specific pool of dedicated proteins is necessary for the homeostasis of a particular metal ion, giving rise to discriminatory metal sensing, transport, and trafficking. Moreover, selectivity in metal binding is also achieved through selectivity in protein–protein recognition. The interactions of pools of proteins involved in the trafficking of specific metal ions are beginning to be unraveled, which represents an exciting new frontier in bioinorganic research (see Chapters V and VIII).

Bibliography

Metals and Proteins

Bertini, I. and Rosato, A., "Bioinorganic chemistry in the postgenomic era", *Proc. Natl. Acad. Sci. U.S.A.*, **100,** 3601–3604 (2003).

Cavet, J. S., Borrelly, G. P., and Robinson, N. J., "Zn, Cu and Co in cyanobacteria: selective control of metal availability", *FEMS Microbiol. Rev.*, **27,** 165–181 (2003).

Finney, L. A. and O'Halloran, T. V., "Transition metal speciation in the cell: insights from the chemistry of metal ion receptors", *Science,* **300,** 931–936 (2003).

Williams, R. J. P., "The fundamental nature of life as a chemical system: the part played by inorganic elements", *J. Inorg. Biochem.*, **88,** 241–250 (2002).

Williams, R. J. P., "Chemical selection of elements by cells", *Coord. Chem. Rev.*, **216–217,** 583–595 (2001).

Genome Browsing

Andreini, C., Banci, L., Bertini, I., and Rosato, A., "Counting the Zinc-Proteins Encoded in the Human Genome", *J. Proteome Res.*, **5,** 196–201 (2006).

Protein Folding/Unfolding

Inorg. Chem., Special Forum on Metalloprotein Folding, **43,** 7893–7960 (2004).

Assfalg, M., Banci, L., Bertini, I., Turano, P., and Vasos, P. R., "Superoxide dismutase folding/unfolding pathway: role of the metal ions in modulating structural and dynamical features", *J. Mol. Biol.*, **330,** 145–158 (2003).

Dobson, C. M., "Protein-misfolding diseases: Getting out of shape", *Nature* (London), **418,** 729–730 (2002).

Dyson, H. J. and Wright, P. E., "Insights into the structure and dynamics of unfolded proteins from nuclear magnetic resonance", *Adv. Protein Chem.*, **62,** 311–340 (2002).

Acc. Chem. Res., Protein Folding Special Issue, **31,** 697–780 (1998).

Miranker, A. D. and Dobson, C. M., "Collapse and cooperativity in protein folding", *Curr. Opin. Struct. Biol.*, **6,** 31–42 (1996).

Metal Trafficking

Banci, L. and Rosato, A., "Structural genomics of proteins involved in copper homeostasis", *Acc. Chem. Res.*, **36,** 215–221 (2003).

Luk, E., Jensen, L. T., and Culotta, V. C., "The many highways for intracellular trafficking of metals", *J. Biol. Inorg. Chem.*, **8,** 803–809 (2003).

Outten, C. E. and O'Halloran, T. V., "Femtomolar sensitivity of metalloregulatory proteins controlling zinc homeostasis", *Science,* **292,** 2488–2492 (2001).

Web Sites

http://metallo.scripps.edu is a metalloprotein database and browser; containing quantitative information on geometrical parameters of metal-binding sites in metalloprotein structures deposited in the Protein Data Bank (PDB), as well as statistical information about the recurrence of consensus sequences (called patterns) within the PDB.

http://metallo.scripps.edu/PROMISE is a database that collects structural and functional information (including bibliography) on metalloproteins, with emphasis on the properties of the metal site. The proteins have been categorized on the basis of the metal cofactor bound. Unfortunately, the database is no longer updated.

http://www.postgenomicnmr.net contains links to a database of metalloprotein modelled structures. The metalloproteins of unknown structure have been identified through a genome browsing approach of the type described in the text, and their structures have been modeled using as a template a protein whose structure is available in the Protein Data Bank (i.e., homology modeling).

CHAPTER IV

Special Cofactors and Metal Clusters

Lucia Banci, Ivano Bertini, Claudio Luchinat, Paola Turano
Magnetic Resonance Center and Department of Chemistry
University of Florence
Sesto Fiorentino, Italy 50019

Contents

IV.1. Why Special Metal Cofactors?

A cofactor is a relatively small chemical entity that is needed in association with biological macromolecules [i.e., protein, ribonucleic acid (RNA) and perhaps deoxyribonucleic acid (DNA)], to allow the combination to perform a biological function. Generally, neither the small molecule nor the macromolecule can perform the function by itself. The small molecule cofactor is sometimes called a "prosthetic group" indicating its ability to help the macromolecule carry out its function. Typically, metal ions are cofactors that are used to catalyze reactions or to help provide the required geometry of a protein. Apparently, however, the versatile behavior of metal ions when coordinated to various donor atoms from protein residues is not sufficient for the biological demand. Nitrogen, oxygen, and sulfur donors are provided in different flavors by amino acid side chains (see Chapter III), but these donors do not cover the full range of ligand-field strengths, π-donor–acceptor properties, and chelating abilities, which coordination chemists have thoroughly described in the second half of the twentieth century.

Therefore, besides directly incorporating a variety of metal ions into proteins, Nature, through evolution, has also selected other inorganic or organic ligands for the metal center. The two most common constructs of non-protein origin are iron–sulfur clusters and tetrapyrroles (including hemes, chlorophylls, and corrins). These prosthetic groups are largely used in electron-transfer (ET) proteins as their redox potentials can conveniently be tuned over large ranges. This capability is evident in the catalytic properties of the heme in metabolic redox reactions. Note that once Nature has learned how to produce a special cofactor, it tries to use it for as many purposes essential to life as possible. Therefore, hemes are found in dioxygen (O_2) transport and activation, and in many oxidative reactions where high-valent iron can be formed as an intermediate. Analogously, iron–sulfur clusters are found in regulatory proteins that interact with DNA or RNA, in redox sensing, and in dehydration reactions. The above mentioned functions are just the most common. There are others (see Section IV.2) and many more may be discovered in the future.

Other less common cofactors are the Co corrin, which is present in vitamin B_{12} and will be covered in Section XIII.3 and the Mo and W cofactors family treated in Sections XII.6 and XII.7. Additional more specific cofactors, such as the Ni cofactor, the Fe–Mo cofactor, and the H center of hydrogenase, will be discussed within their specific contexts. Assemblies of more than one metal ion that are characteristic of specific systems, like the Cu_A construct in cytochrome *c* oxidase and nitrite reductase, the dicopper center of type 3 copper, the dinickel center in urease, and so on, also will be described in the appropriate chapters. Of course, several wholely organic cofactors including flavins, adenosine triphosphate (ATP), adenosine diphosphate (ADP), and nicotinamide adenine dinucleotide, oxidized form (NAD^+) (Scheme IV.1) are used in biological processes, but as these do not contain metal ions, they are hence outside the scope of this chapter.

In this chapter, we discuss the iron–sulfur and heme-iron cofactors. Other important cofactors are depicted in Figs. IV.1–IV.3 and discussed elsewhere in this book.

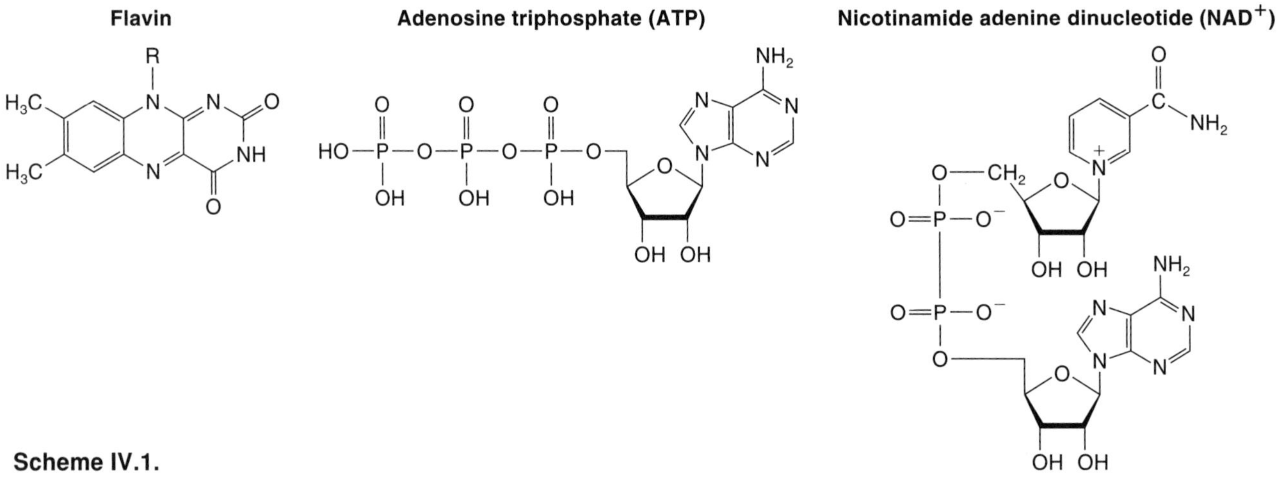

Flavin **Adenosine triphosphate (ATP)** **Nicotinamide adenine dinucleotide (NAD^+)**

Scheme IV.1.

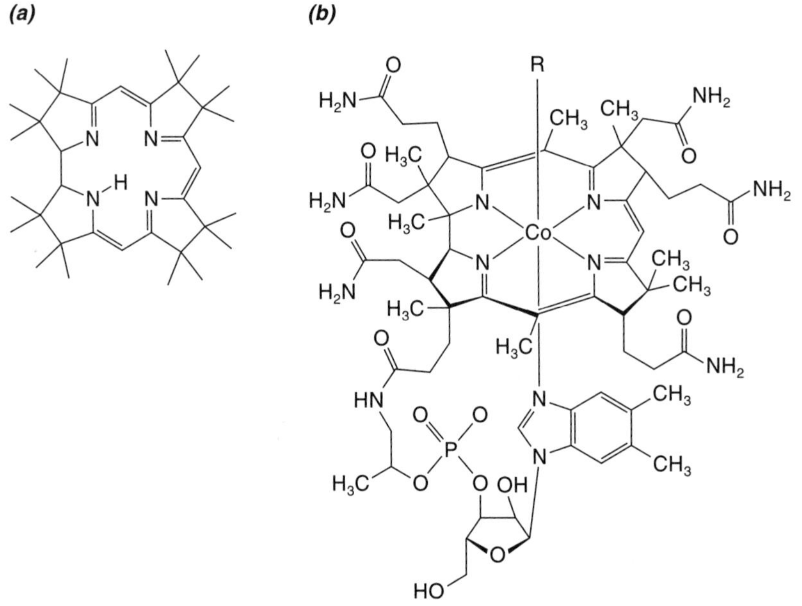

Fig. IV.1.
(*a*) The schematic representation of a corrin ring, an essential component of the cobalamin structure reported in (*b*). This metal complex constitutes vitamin B_{12}, when R = CN, and coenzyme B_{12}, when R = adenosyl.

(a) ***(b)***

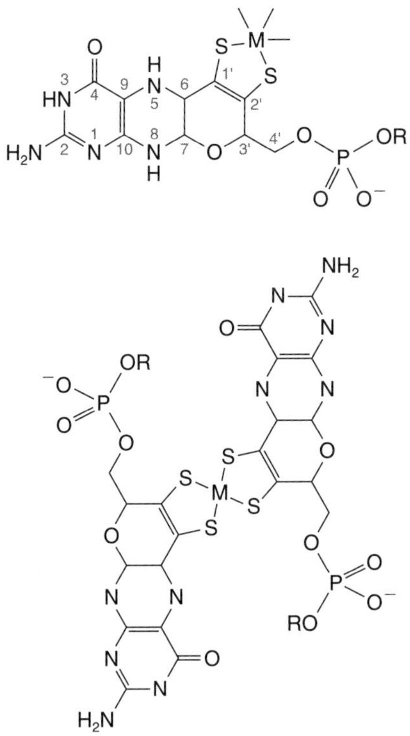

Fig. IV.2.
The two types of Mo and W cofactors. Here M indicates the metal ion (Mo or W). For the metal cofactor reported in the top panel, R can be either hydrogen or adenosine. For the metal cofactors in the lower panel, R can be hydrogen, adenosine, cytosine, guanosine, or hypoxanthine. The metal ion completes its coordination geometry by forming two or three more bonds with protein ligands and/or O, S, or Se exogenous atoms.

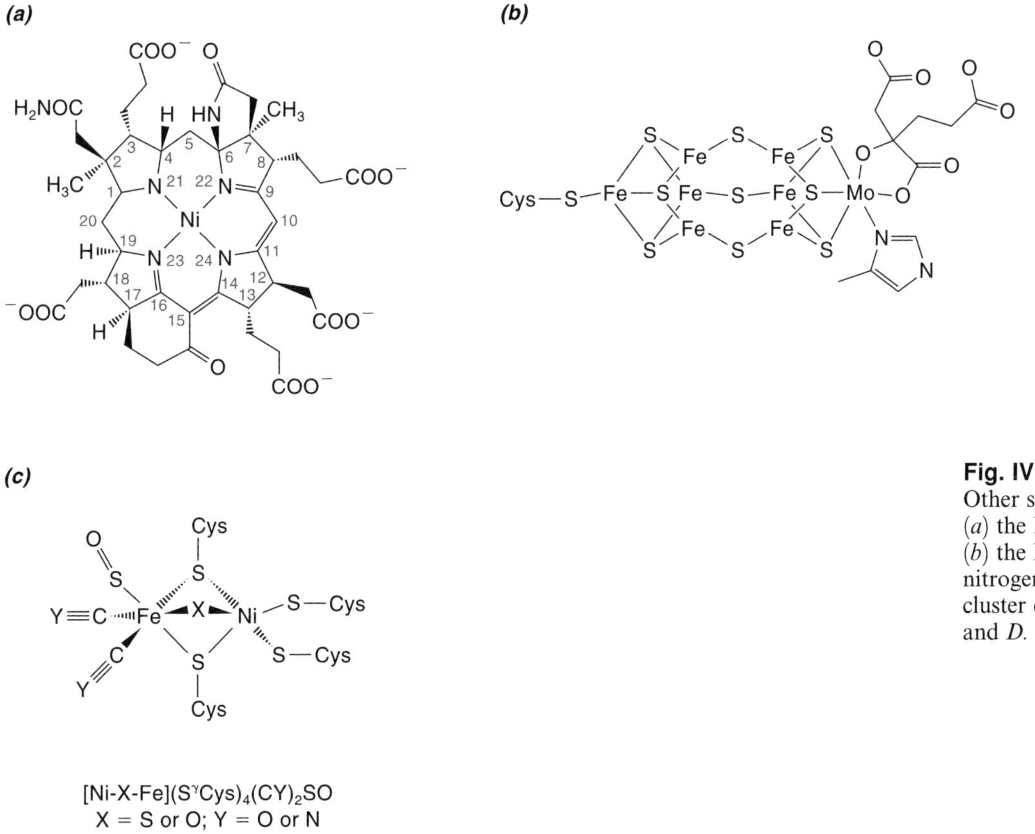

(a)

(b)

(c)

$[Ni\text{-}X\text{-}Fe](S^{\gamma}Cys)_4(CY)_2SO$
X = S or O; Y = O or N

Fig. IV.3.
Other specific cofactors: (*a*) the F430 nickel cofactor, (*b*) the Fe–Mo cofactor of nitrogenase, (*c*) the [NiFe] cluster of reduced *D. vulgaris* and *D. baculatus* hydrogenases.

IV.2. Types of Cofactors, Structural Features, and Occurrence

IV.2.1. Iron–Sulfur Cofactors

Here, we present the general structural properties of iron–sulfur clusters. Many variations on this theme occur in individual protein systems, which are described in the appropriate sections. The most common types of iron–sulfur cofactors are illustrated in Fig. IV.4. Figure IV.4**A** does not really represent an iron–sulfur cluster, but shows an iron ion coordinated to four cysteine residues from a protein. Proteins of this class, called rubredoxins, are often treated together with iron–sulfur proteins, and the properties of their metal cofactor constitute a reference to understand those of iron–sulfur clusters. Figure IV.4**B** represents the simplest type of iron–sulfur cluster. Here, four cysteines from the protein coordinate a so-called "diamond", which consists of two iron ions and two sulfide ions. Figure IV.4**C** shows a variation on the same theme, encountered in a class of iron–sulfur proteins called Rieske proteins, where the protein ligands to one of the iron ions are histidines rather than cysteines. Another very common iron–sulfur cluster is the so-called "cubane" structure, shown in Fig. IV.4**E**. The protein ligands are still four cysteines, as in **A** and **B**, but here they host a cluster formed by four iron and four sulfur ions. The cluster shown in Fig. IV.4**D** is also common. It can be described as a cubane lacking one of the iron ions and the corresponding Cys.

Structural features of these clusters are summarized in Table IV.1. In all cases, each iron ion is coordinated by four donors in an approximately tetrahedral environment. In the proteins containing $[Fe_2S_2]$ clusters, each metal is bound to two protein donors (Cys sulfur or His nitrogen atoms) and to two sulfide ions that in turn bridge the two metals. In the proteins containing the cubane $[Fe_4S_4]$ clusters, each iron is coordinated by one protein donor (most often a Cys) and by three sulfide ions that in turn bridge three metals. In both classes of iron–sulfur proteins, the sulfur–metal–sulfur

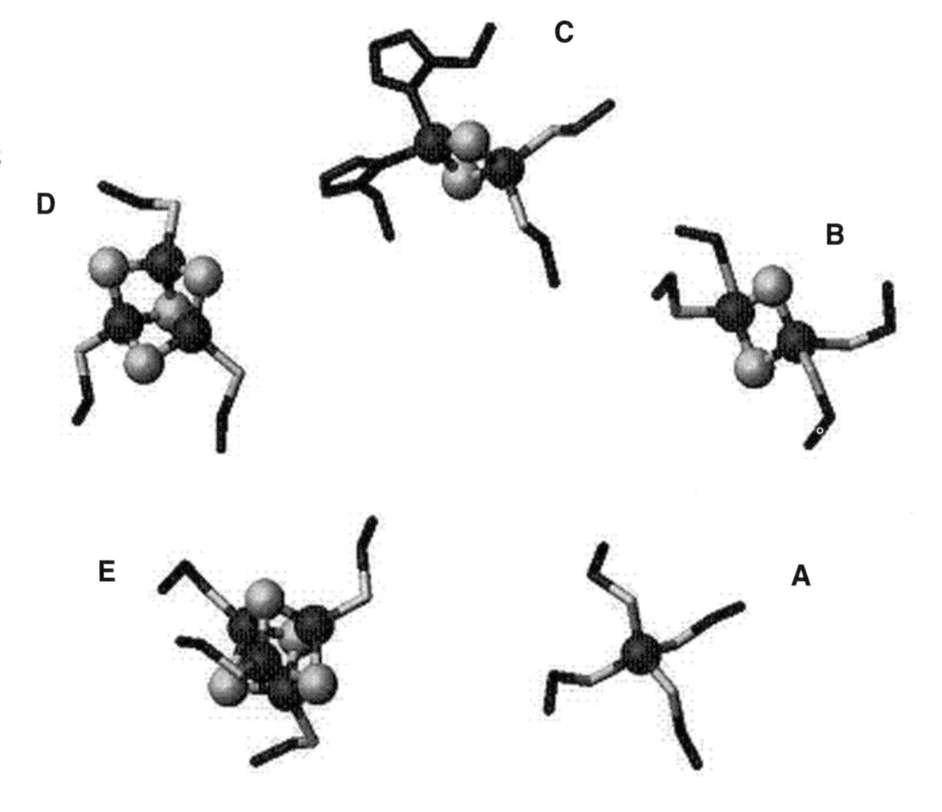

Fig. IV.4.
Ligand geometries and arrangement of iron and sulfide ions in clusters of: **A** rubredoxins; **B** plant-type ferredoxins; **C** Rieske proteins; **D** 3Fe–4S and 7Fe–8S ferredoxins; and **E** 4Fe–4S, 7Fe–8S, 8Fe–8S ferredoxins and high potential iron–sulfur proteins (HiPIPs).

Table IV.1.
Structural Features of Iron–Sulfur Clusters[a]

Geometric Factors[b]	Rubredoxins (ox: 1BRF; red: 1RB9)[c]	Model Compounds
Fe–SR Distances[c]	2.26–2.31 (ox) 2.24–2.29 (red)	2.252–2.278 (ox) 2.324–2.378 (red)
RS–Fe–SR Angles	102.5–114.4 (ox) 104.3–114.9 (red)	106.67–112.20 (ox) 103.5–114.9 (ox)
	$[Fe_2S_2]^{2+}$ (1AWD)[c]	Model Compounds
Fe–SR Distances	2.27–2.36	2.303–2.306
Fe–S* Distances	2.16–2.23	2.185–2.232
S*–Fe–S* Angles	104.8–101.4	104.7
Fe–S*–Fe Angles	76.3–77.2	75.3
	$[Fe_3S_4]^{0/+}$ (6FDR)[c,d]	Model Compounds
Fe–SR Distances	2.26–2.30	2.310–2.327
Fe–S* Distances	2.23–2.33	2.242–2.275
Fe–S*3 Distances	2.20–2.32	2.273–2.333
S*–Fe–S* Angles	100.2–113.9	101.63–106.06; 111.5–113.8
Fe–S*–S Angles	69.4–75.6	70.05–72.70; 71.71–74.42
	$[Fe_4S_4]^{2+}$ in Ferredoxins (1FCA)[c]	Model Compounds
Fe–SR Distances	2.29–2.32	2.291–2.299
Fe–S* Distances	2.29–2.34	2.337–2.368; 2.269–2.299
S*–Fe–S* Angles	101.3–106.7	103.1–106.1
Fe–S*–Fe Angles	71.3–72.7	71.98–73.59
	$[Fe_4S_4]^{2+}$ in HiPIPs (1CKU)[c]	Model Compounds
Fe–SR Distances	2.25–2.30	2.291–2.299
Fe–S* Distances	2.23–2.33	2.337–2.368; 2.269–2.299
S*–Fe–S* Angles	103.5–105.7	103.1–106.1
Fe–S*–Fe Angles	72.1–73.7	71.98–73.59
	$[Fe_4S_4]^{3+}$ in HiPIPs (1ISU)[c]	Model Compounds
Fe–SR Distances	2.13–2.31	
Fe–S* Distances	2.16–2.34	
S*–Fe–S* Angles	100.8–108.7	
Fe–S*–Fe Angles	70.5–76.1	

[a] Distances are given in angstroms (Å).

[b] SR indicates the Cys sulfur, S* is the inorganic sulfur, S*3 is the inorganic sulfur bridging three irons in the Fe_3S_4 cluster.

[c] The PDB code is reported in parentheses.

[d] The structures of the $[Fe_3S_4]^0$ and $[Fe_3S_4]^+$ clusters in reduced and oxidized *Azotobacter vinelandii* ferredoxin I, respectively, are the same within the experimental error of X-ray crystallography at 1.4-Å resolution.

$$[Fe(SR)_4]^{2-} \rightleftharpoons [Fe(SR)_4]^{-}$$

$$[Fe_2S_2(SR)_4]^{4-} \rightleftharpoons [Fe_2S_2(SR)_4]^{3-} \rightleftharpoons [Fe_2S_2(SR)_4]^{2-}$$

$$[Fe_3S_4(SR)_3]^{5-} \rightleftharpoons [Fe_3S_4(SR)_3]^{4-} \rightleftharpoons [Fe_3S_4(SR)_3]^{3-} \rightleftharpoons [Fe_3S_4(SR)_3]^{2-}$$

$$[Fe_4S_4(SR)_4]^{4-} \rightleftharpoons [Fe_4S_4(SR)_4]^{3-} \rightleftharpoons [Fe_4S_4(SR)_4]^{2-} \rightleftharpoons [Fe_4S_4(SR)_4]^{-} \rightleftharpoons [Fe_4S_4(SR)_4]^{0}$$

Scheme IV.2.

Table IV.2.
Biological Functions of Iron–Sulfur Clusters[a]

Function	Cluster	Protein
Electron transfer (Sections III.3, X.1, XII.1, XII.3, XII.6)	Cys_4Fe^-	Rubredoxin, desulforedoxin
	2Fe–2S	Rieske proteins
	2Fe–2S and/or 3Fe–4S and/or 4Fe–4S	Ferredoxins Iron-only hydrogenase Subunit B of fumarate reductase
	4Fe–4S	High-potential iron–sulfur protein Nitrogenase iron protein Trimethylamine dehydrogenase Pyruvate:ferredoxin oxidoreductase
Catalysis of a nonredox reaction (Sections VIII.2, IX.4)	4Fe–4S	Aconitase
Catalysis of redox reactions (Sections XII.1, XII.5)	H-cluster (see Fig. IV.3, structure **C**)	Fe-only hydrogenase
	4Fe–4S + siroheme (see Fig. IV.6)	Sulfite reductase hemoprotein
Stabilization of protein structure for DNA repair (Section IX.2)	4Fe–4S	Endonuclease III, MutY
Sensing and regulation:		
(i) Oxygen sensors: loss of original cluster and of activity (Sections II.4, IX)	4Fe–4S	Glutamine PRPP amidotransferase
	4Fe–4S/2Fe–2S	FNR protein
	4Fe–4S/3Fe–4S	Aconitase
(ii) Sensor of O_2^- and NO: redox-regulated control of transcription (Chapter XIV)	2Fe–2S	SoxR protein
(iii) Iron sensor: post-transcriptional regulation (Section IX.2)	4Fe–4S	Iron regulatory protein/aconitase
Redox-mediated generation of free radicals (Chapter XIII)	4Fe–4S	Anaerobic ribonucleotide reductase, pyruvate formate-lyase activating enzyme Biotin synthase
Stabilization of an intermediate in disulfide reduction	4Fe–4S	Ferredoxin: thioredoxin reductase

[a] Reference is made to chapters where these systems are further treated.

angles are approximately tetrahedral and, as a corollary, the metal–sulfur–metal angles are acute.

Many of the biological functions of these clusters stem from their redox properties. The functionally relevant redox states of iron in rubredoxin are only the 2+ ($[Fe(SR)_4]^{2-}$ in Scheme IV.2) and the 3+ ($[Fe(SR)_4]^{-}$) states. This situation contrasts with hemes, where the extended aromatic porphyrin and axial ligands may stabilize higher oxidation states (see Section IV.2.2). However, the variability is regained by the presence of more than one Fe center. The oxidation states that are formally available in all the clusters depicted in Fig. IV.4 are shown in Scheme IV.2. Those states not known to be functionally relevant are shown in gray. Even if more than two redox states are available, usually only two are functionally relevant for the same protein. Different proteins may utilize different redox pairs among those accessible.

The biological functions of proteins containing iron–sulfur clusters are summarized in Table IV.2. In this table, reference is made to the sections where more information on each system can be obtained.

Much is known about the electronic structure of iron–sulfur clusters in their various oxidation states. For example, in $[Fe_2S_2]$ systems, addition of one electron to the diferric cluster results in the localized reduction of one of the two iron ions, the other one remaining ferric. On the other hand, reduction of the all-ferric $[Fe_3S_4]^+$ cluster ($[Fe_3S_4(SR)_3]^{2-}$) results in the extra electron being delocalized onto a pair of iron ions having a 2.5+ oxidation state (the so-called mixed-valence pair), while the third iron ion remains ferric. Similarly, $[Fe_4S_4]^{3+}$ clusters consist of a mixed-valence pair and a ferric pair, $[Fe_4S_4]^{2+}$ clusters contain all mixed-valence irons, and $[Fe_4S_4]^+$ clusters have a mixed valence and a ferrous pair. This variation in the distribution of oxidation number is the result of specific ferro- and antiferromagnetic coupling between the Fe ions. When the Fe ions are all oxidized, only antiferromagnetic coupling occurs. By considering a pair of Fe(III) ions, the five unpaired electrons with an $S = \frac{5}{2}$ ground state on each ion sense those on the neighbor Fe through the sulfur bridges in such a way as to provide a total ground state with $S' = 0$ (i.e., with all the electron spins paired). In the case of different oxidation states, the five electrons of Fe(III) with a total spin $S = \frac{5}{2}$ and the six electrons of Fe(II) with a total spin $S = 2$, provide a total ground state $S' = \frac{5}{2} - 2 = \frac{1}{2}$ in the case of antiferromagnetic coupling. However, if the Fe(III)–Fe(II) pair interacts strongly, the electron can be considered fully delocalized between the two metal centers and the oxidation state for each Fe is 2.5+. In this case, the ground state has nine unpaired electrons. The result is the same as that obtained when there is ferromagnetic coupling and the two S values of $\frac{5}{2}$ and 2 sum to $S' = \frac{9}{2}$. In this case, we say that a mixed-valence pair occurs. In polymetallic systems, a mixed-valence pair with $S = \frac{9}{2}$ may couple in an antiferromagnetic fashion with either a single Fe ion or another pair of Fe ions. In the resulting cluster, the spin takes the smallest possible value. The situation for iron–sulfur clusters is summarized in Fig. IV.5, and the resulting global and local spin states are summarized in Table IV.3.

The spectra of iron–sulfur proteins in the visible and near-ultraviolet (UV) range show a broad absorption envelope resulting from several overlapping absorption bands deriving from transitions with predominant $S \rightarrow Fe^{III}$ charge-transfer character, where S can be either the Cys or the inorganic sulfur. Circular dichroism (CD) and magnetic circular dichroism (MCD) spectra partially resolve this complex envelope thanks to the different selection rules and the appearance of both positive and negative bands. As indicated in Table IV.2, iron–sulfur clusters contain high-spin Fe(III) and Fe(II): while the former has no spin-allowed d–d transitions, high-spin Fe(II) has spin-allowed transitions of the type $d_{z^2} \rightarrow d_{xz}$ and $d_{z^2} \rightarrow d_{yz}$ ($e \rightarrow t_2$ in a tetrahedral environment). Bands attributable to these d–d transitions have been detected in CD spectra. Meaningful differences are detected in the electronic and

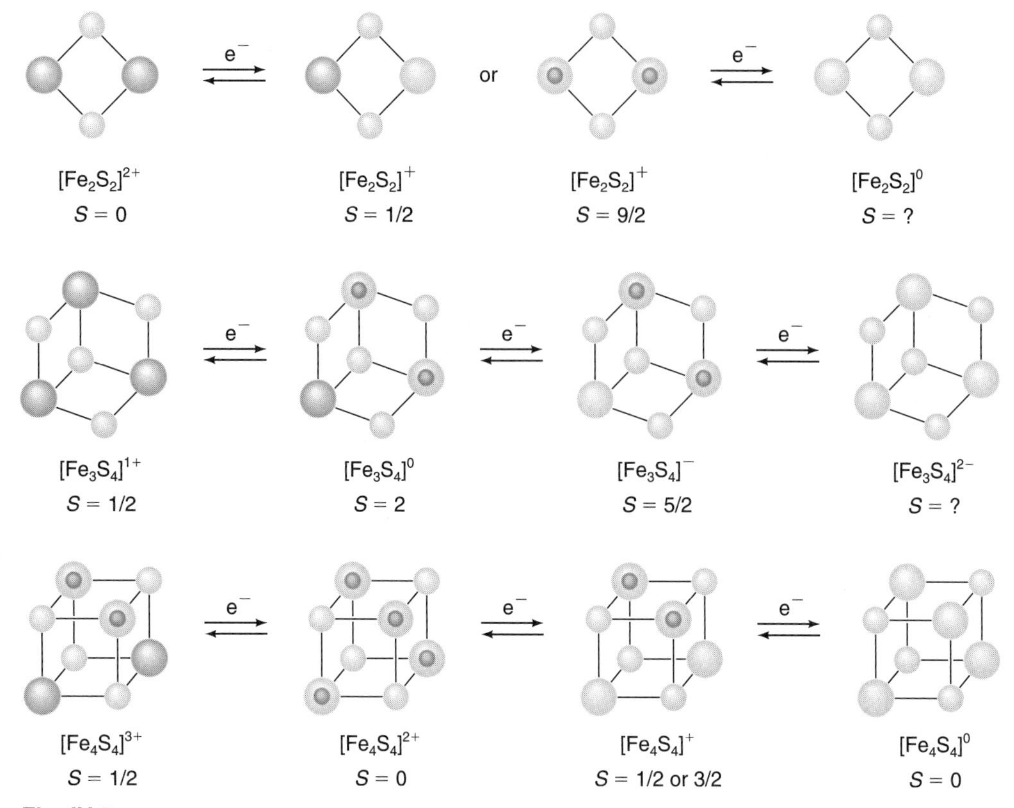

Fig. IV.5.
Total spin of the cluster and individual redox states of the iron ions in various iron–sulfur clusters as a function of the cluster oxidation state. Dark gray circles indicate Fe(III) ions, light gray circles Fe(II) ions, and dual color circles iron ions with a 2.5+ oxidation state.

Table IV.3.
Spin Distribution within Iron–Sulfur Clusters

Cluster	Individual Spins	Subspins[a]	Total S
$[Fe_2S_2]^{2+}$	5/2, 5/2	5/2, 5/2	0
$[Fe_2S_2]^{+}$	2, 5/2	2, 5/2	1/2
$[Fe_3S_4]^{0}$	2, 5/2, 5/2	9/2*, 5/2	2
$[Fe_3S_4]^{+}$	5/2, 5/2, 5/2	2, 5/2 (3, 5/2)	1/2
$[Fe_4S_4]^{0}$	2, 2, 2, 2	0, 0 (?)	0
$[Fe_4S_4]^{+}$	2, 2, 2, 5/2	9/2*, 4 (?)	1/2
$[Fe_4S_4]^{2+}$	2, 2, 5/2, 5/2	9/2*, 9/2* (?)	0
$[Fe_4S_4]^{3+}$	2, 5/2, 5/2, 5/2	9/2*, 4 (7/2*, 3) (9/2*, 4 + 7/2*, 4)	1/2

[a] Subspin related to mixed-valence pairs within the polymetallic center are marked with an asterisk. Alternate ground states are reported in parentheses.

CD spectra of Rieske centers with respect to those of proteins with four Cys sulfurs coordinated to the Fe_2S_2 cluster. In CD spectroscopy, systems belonging to high-symmetry point groups in general show less intense spectra than systems belonging to lower symmetry point groups. Therefore, Rieske clusters display particularly intense CD spectra. This spectroscopic signature can be used to identify the Rieske cluster.

IV.2.2. Hemes

The iron-porphyrins present in heme proteins are heme *a*, heme *b*, heme *c*, heme *d*, heme d_1, heme *o*, heme P460, and siroheme. Their structures show a common skeleton constituted by the tetrapyrrole ring, but differ in the peripheral substituents (Fig. IV.6). Although extensive electronic delocalization of the porphyrin favors planarity, heme nevertheless possesses a relatively high degree of plasticity. Significant deviations from planarity are often encountered in proteins.

Heme *b* (or protoporphyrin IX) is the simplest representative. This heme has methyl groups at positions 1, 3, 5, and 8, two propionate groups at positions 6 and 7, and two vinyl groups at positions 2 and 4. Protons are present at the so-called α, β, γ, and δ-meso positions. In the free heme, the conformation of the vinyl groups coplanar with the heme is energetically favored, by the interaction between π orbitals of the vinyl group and the aromatic heme plane. However, in proteins, vinyl group orientation is governed by the steric requirements of the protein site and hydrophobic interactions with protein amino acids. Many heme proteins contain noncovalently bound hemes.

Hemes *a* and *o* both carry a farnesylhydroxyethyl side chain, but heme *a* differs from heme *o* by an additional formyl group at position 8 of the tetrapyrrole ring. In heme *d*, the C ring of the tetrapyrrole is saturated. Heme d_1 and siroheme have unconjugated A and B rings. Different extents of conjugation in the various hemes impart different electronic delocalizations on the heme plane, while different substituents are designed to optimize heme–protein interactions. Hemes *c* and some modified hemes *b* are covalently linked to the protein. In *c*-type heme proteins, a heme *c* is covalently bound to the protein by two thioether bonds involving sulfhydryl groups of Cys residues, in effect through a Markovnikov addition of the sulfhydryl groups of cysteines to the heme *b* vinyl groups. A few *c*-type cytochromes lack one of the conserved cysteines involved in heme linkage. In addition, mutant proteins have been prepared where one of the vinyl groups of a heme *b* has been substituted with a thioether bond to a protein cysteine.

The heme P460 contained in the N-terminal domain of hydroxylamine oxidoreductase (see Section XII.3) is attached to the protein matrix through two thioether bonds analogous to those of *c*-type cytochromes; here an additional covalent bond exists between the heme α-meso carbon and the Cε of a Tyr on an adjacent subunit. At variance with plant and fungal peroxidases, which contain heme *b*, lactoperoxidase and myeloperoxidase contain a modified heme *b* with $-CH_2R$, and $-CH_2R'$ groups substituting for the $-CH_3$ usually present at positions 1 and 5. The groups R and R′ are a Glu and an Asp residue, respectively, and covalent linkages exist with the protein through the side-chain carboxylate groups of the two amino acids. In myeloperoxidase, an additional covalent bond has been proposed to exist at position 2 between the heme and the protein through a thioether bridge, which involves a Met residue forming a sulfonium cation.

Heme iron in proteins can be five or six coordinate. Axial ligation appears strongly related to protein function (see Table IV.4). Pentacoordination is usually found in globins and in heme enzymes. In globins, the imidazole ring of the "proximal" His residue provides the fifth heme-iron ligand; the other axial heme-iron position remains essentially free for O_2 coordination. A similar situation is encountered in nitrophorin, the NO transport protein. In heme enzymes, the sixth coordination position is available for substrate binding. The nature of the fifth (or proximal) ligand modulates the redox potential of the heme iron, while distal site features are

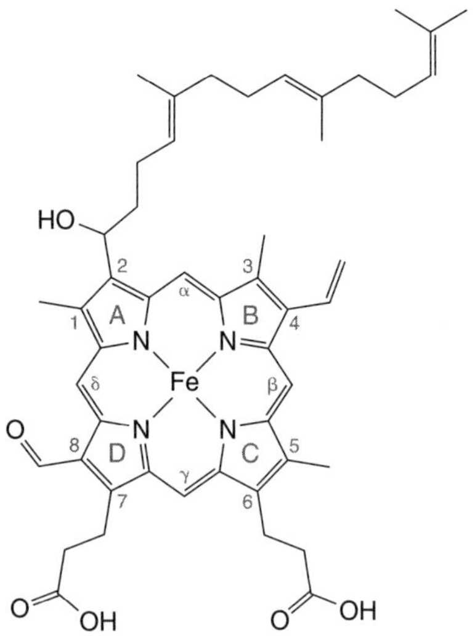

Heme *a*

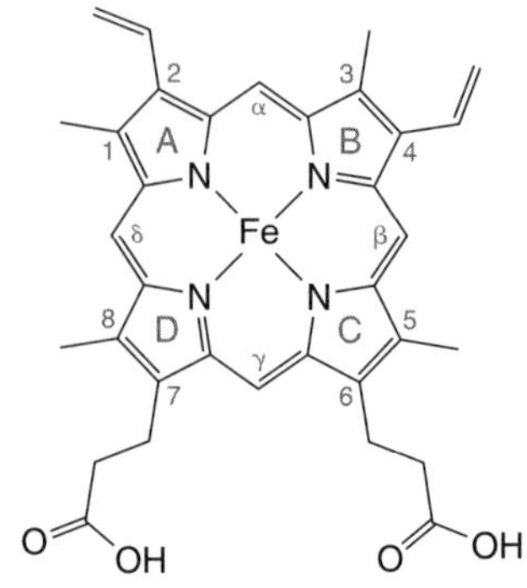

Heme *b*

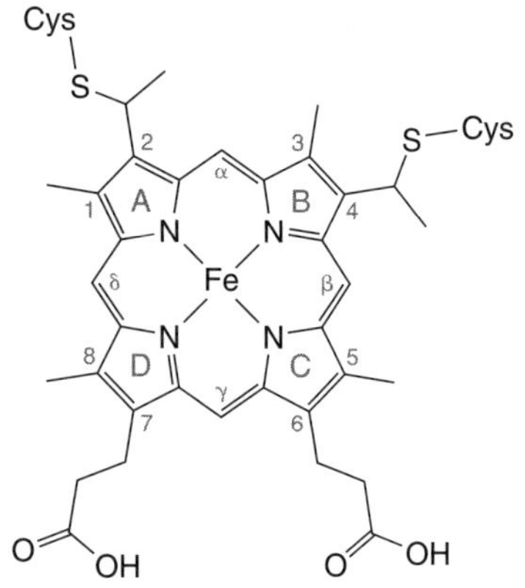

Heme *c*

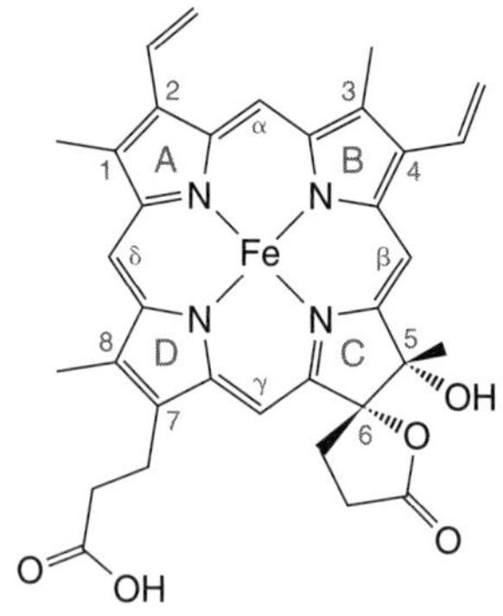

Heme *d*

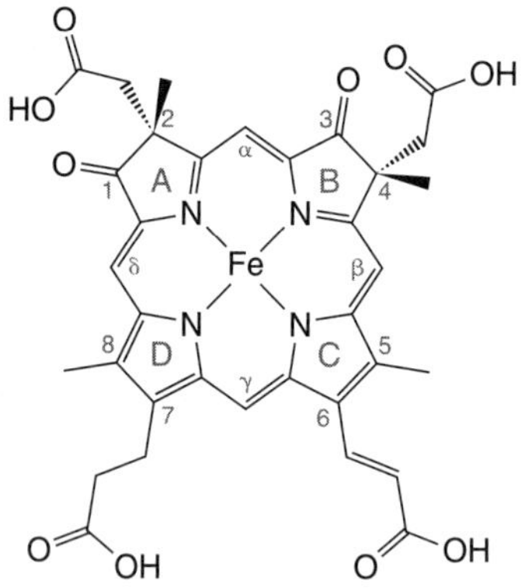

Heme d_1

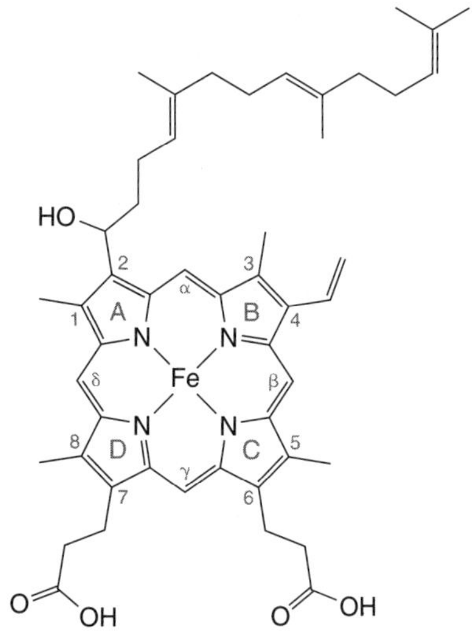

Heme *o*

Fig. IV.6.
Naturally occurring iron-porphyrins.

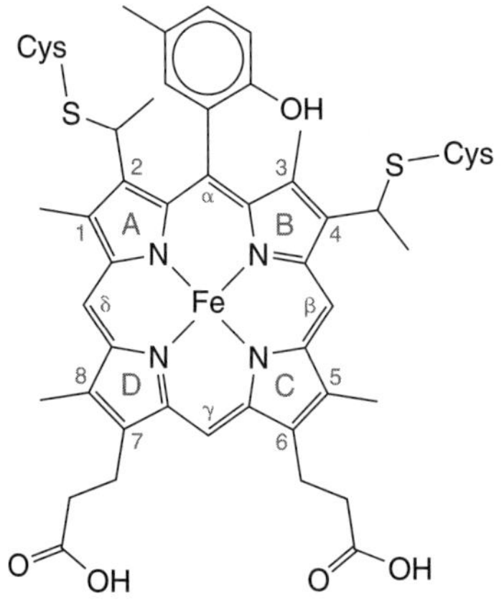

Heme P460

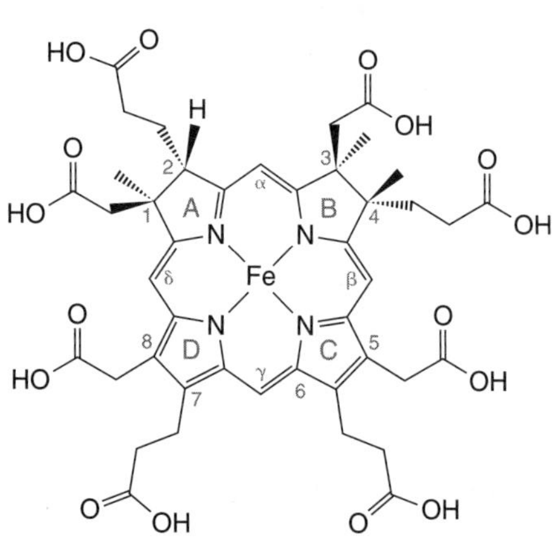

Siroheme

Table IV.4.
Biological Functions of Redox-Active Heme Proteins[a]

Protein Class/Family	Function	Heme Type	Axial Ligation	Formal Iron Oxidation/ Spin States
Globins (Section XI.4)	O_2 transport and storage	*b*	His Nε	Fe(II) ($S = 2$)
Nitrophorin (Section XIV.3)	Nitric oxide transport	*b*	His Nε	
Catalases (Section XI.3)	$H_2O_2 + H_2O_2 \rightarrow 2\ H_2O + O_2$	*b*	Tyr Oη	Fe(III) ($S = 5/2$)
Peroxidases (Section XI.3)	$A\text{–}H_2 + H_2O_2 \rightarrow A + 2\ H_2O$	*b*[b]	His Nε ($+H_2O$)	Fe(III) ($S = 5/2$)
Chloroperoxidases (Section XI.3)	$AH + X^- + H^+ + H_2O_2 \rightarrow AX + 2\ H_2O$	*b*	Cys Sγ	Fe(III) ($S = 5/2$)
Nitric oxide synthases (Section XIV.3)	L-Arg $+ 2\ O_2 + \frac{3}{2}$ NADPH $\rightarrow$ citrulline + nitric oxide $+ \frac{3}{2}\ NADP^+$	*b*	Cys Sγ	Fe(III) ($S = 5/2$)
P450 proteins (Sections VI.3, IX.3, XI.5)	$RH + O_2 + 2\ H^+ + 2\ e^- \rightarrow ROH + H_2O$	*b*	Cys Sγ ($+H_2O$ or OH^-)	Fe(III) ($S = 5/2$)
Cytochrome *c* oxidase (heme a_3) (Section XI.6)	$O_2 + 4\ e^- + 4\ H^+ \rightarrow 2\ H_2O$	a_3	His Nε	Fe(III) ($S = 5/2$) Spin coupled to Cu_B
Cytochrome *bo* quinol oxidase (heme *o*)	$O_2 + 4\ e^- + 4\ H^+ \rightarrow 2\ H_2O$	*o*	His Nε	Fe(III) ($S = 5/2$) ?
Cyt *c*554 (heme 2) (Sections IV, XII.3)	oxidation of ammonia to nitrite	*c*	His Nε	Fe(III) ($S = 5/2$)
Nitrite reductases (Section XII.4)	$NO_2^- + 8\ H^+ + 6\ e^- \rightarrow NH_4^+ + 2\ H_2O$	Siroheme	Cys Sγ	Fe(II) ($S = 1$ or $S = 2$) Spin coupled to $[Fe_4S_4]^+$
Hydroxylamine oxidoreductase (heme P460) (Section XII.3)	$NH_2OH + H_2O \rightarrow NO_2^- + 4\ e^- + 5\ H^+$	P460	His Nε	Fe(III) ($S = 5/2$ or $S = 3/2$) Spin coupled to heme 6
Cytochrome cd_1 nitrite reductase (Section XII.4)	$NO_2^- + 2\ H^+ + e^- \rightarrow NO + 2\ H_2O$ and $O_2 + 4\ H^+ + 4\ e^- \rightarrow 2\ H_2O$	d_1	His Nε or His Nε + Tyr Oη	Fe(III) ($S = 1/2$)
Sulfite reductases (Section XII.5)	$HSO_3^- + 6\ H^+ + 6\ e^- \rightarrow HS^- + 3\ H_2O$ $3\ HSO_3^- + 3\ H^+ + 2\ e^- \rightarrow S_3O_6^{2-} + 3\ H_2O$ $2\ HSO_3^- + 4\ H^+ + 4\ e^- \rightarrow S_2O_3^{2-} + 3\ H_2O$	Siroheme	Cys Sγ	Fe(II) ($S = 1$ or $S = 2$) Spin coupled to $[Fe_4S_4]^+$

[a] Reference is made to chapters where these systems are further treated. For ET cytochromes see Chapter XI. The heme type and axial ligation are also reported, as well as the resting state oxidation and spin state for the heme iron.

[b] As mentioned in the text, the only exception is constituted by lactoperoxidase and myeloperoxidase, which contain modified hemes covalently bound to protein amino acids.

designed to facilitate the catalytic reaction. The redox states of the heme iron in the resting states of redox-active heme proteins are reported in Table IV.4.

The nature of the axial ligands also influences the spin state of the heme iron. Penta- or hexacoordinated iron with a water molecule as the sixth ligand is high spin. Binding of CO, O_2, NO, and CN^- to the sixth (axial) position produces a low-spin complex. His/Met, bis(His), or bis(Met) coordination always corresponds to a low-spin state for the iron atom. In cytochromes, there is no need to accommodate exogenous ligands for biological function. The heme iron must simply switch between the Fe(III) and Fe(II) oxidation states during the electron-transfer reaction. The axial ligands of the heme iron are usually two histidines or one histidine and one methionine. Cytochrome *f* has a His and a backbone amide nitrogen as axial ligands. The only pentacoordinate heme that probably has an electron-transfer function is that of cytochrome *c′*, where the fifth ligand is a His residue. In all of these cases, the donor atom for the His ligand is the Nε of the imidazole ring. To date, a single exception has been reported for the recently resolved structure of cytochrome c_{554}, where one of the four heme irons has a six-coordinate Nε His-Nδ His axial ligation. Structural features of cytochromes will be discussed in detail in Section X.1.

It is noteworthy to mention that, in heme enzymes involving oxygen, iron oxidation states higher than 3+ are usually encountered during the catalytic cycle. The features of Fe(IV) and formal Fe(V) heme will be discussed in detail in Chapter XI, which deals with these enzymes.

The electronic spectra of heme proteins are quite characteristic and are dominated by internal ligand bands. The main spectroscopic feature is the presence of a strong band ($\varepsilon = 1\text{–}2 \times 10^5\ M^{-1}\,cm^{-1}$) at ~400 nm, called the Soret band, which is due to heme π–π* transitions. Two other bands of lower intensity ($\varepsilon = 1.0\text{–}1.5 \times 10^4\ M^{-1}\,cm^{-1}$), called α and β, are usually detected in the range 500–600 nm. A charge-transfer band with $\varepsilon < 10^4\ M^{-1}\,cm^{-1}$ can be present at 600–800 nm when sulfur ligation is present. When detected by MCD spectroscopy, these bands are indicative of the coordination number, nature of the axial ligands, and spin state of the heme iron.

IV.3. Cofactor Biosynthesis

The selection of iron–sulfur clusters and hemes as cofactors had to rely on the availability of suitable ingredients, so it is not surprising that these cofactors were mainly based on iron, and that the first to be developed, when all life was still anaerobic, used only sulfide ions as the second ingredient. Only later did the biosynthesis of porphyrins and corrins became viable, and cofactors based on these macrocyclic ligands were developed. For example, the genetic code for heme proteins contains instructions not only to express the protein that hosts the heme, but also to express an array of enzymes devoted to the biosynthesis of the heme itself, and chaperone proteins to help its incorporation in the protein matrix (see below).

Inorganic chemistry has taught us that iron and sulfide ions assemble spontaneously under reducing conditions, in solutions containing thiolates, to form iron–sulfur clusters. Similar processes may have occurred at the origin or early stages of life. Theories have been developed where metal sulfides could either act as catalysts to form organic compounds from carbon monoxide and hydrogen sulfide, or even act as a carbon fixation source through oxidation to pyrite. It is thus conceivable that some of the early proteins containing cysteines might have spontaneously wrapped around Fe ions and formed iron–sulfur clusters in the presence of sulfide ions (see also Section X.1). These constructs revealed themselves to be much more versatile than single protein-coordinated Fe ions as far as redox properties were concerned, and possibly also for other catalytic purposes.

Interestingly, iron–sulfur proteins have been found to be involved in the regulation of genes responsible for iron uptake by cells (see Chapter V). Furthermore, iron–sulfur proteins are spread throughout evolution from archaea to humans, and are particularly well represented in the former group of organisms. This finding suggests an evolutionary picture where (1) iron was among the first metallic elements to be involved in life, and iron–sulfur clusters could have been employed from the very beginning as redox centers and, possibly, hydrolytic or condensation catalysts; (2)

later in evolution, aerobic organisms developed a need for protection against overload; (3) overload could manifest itself in the form of loading a particular iron–sulfur protein with the full complement of iron; (4) this fully loaded protein could be involved in a regulation mechanism for iron uptake. Studies are in progress to identify individual proteins involved in *in vivo* iron–sulfur cluster assembly, which includes the delivery of iron and sulfur constituents.

Heme is a necessary component in a variety of oxygen-binding proteins, electron-transfer proteins, and enzymes. The porphyrin macrocycle binds Fe with high affinity and upon insertion in the protein matrix, the metal ion redox potential can be fine tuned, making Fe oxidation states accessible that would be difficult to obtain in solution with simple Fe complexes.

Heme biosynthesis is a multistep process that involves many enzymes. With only rare exceptions, organisms that utilize heme possess the entire biosynthetic pathway to produce this tetrapyrrole compound. The current body of experimental data for heme-synthesizing organisms suggests that all proceed via the same pathway once the first compound (5-aminolevulinate; $^{-}OOC{-}CH_2{-}CH_2{-}CO{-}CH_2{-}NH_2$) is formed. The formation of this compound is known to occur via two distinct means: in plants and most bacteria there is the so-called five-carbon pathway, where 5-aminolevulinate is formed from glutamate; in animals and other bacteria this step involves the condensation of glycine with succinylcoenzyme A and is called the four-carbon pathway. [$-OOC{-}CH_2{-}CH_2{-}CO{-}CoA$; the formula of coenzyme A (CoA) is reported in Scheme IV.3]. From this precursor to the formation of protoporphyrin IX the reaction proceeds through the same intermediates, although the enzyme involved can be different from one organism to another. In particular, there are some steps where there is a difference between aerobic and anaerobic conversion.

Once protoporphyrin IX is obtained, ferrochelatases are needed to catalyze the insertion of the metal ion. It has been proposed that these enzymes first bind ferrous iron and then the porphyrin substrate: metalation occurs when the macrocycle is bent, allowing metal insertion concomitant with proton release from the porphyrin. Once the heme is formed it becomes planar and is released as product. The heme *b* so obtained is not only incorporated directly into *b*-type heme proteins, but also serves as a precursor for the formation of the hemes *a, c, d,* and *o*. Heme d_1 and siroheme on the other hand are formed in pathways that branch-off from the above pathway after the first three common steps following formation of 5-aminolevulinate.

Noncovalent binding of hemes to their proteins does not appear to require the assistance of any other protein. In contrast, a key step in *c*-type heme protein expression is the covalent ligation of the heme to the protein matrix. From genetic and biochemical studies, it is clear that three distinct systems have evolved in Nature to assemble these proteins. In Gram-negative bacteria and plant/protozoa mitochondria, there is a system with eight genes (called *ccm*-genes) involved in the cytochrome

CoA

Scheme IV.3.

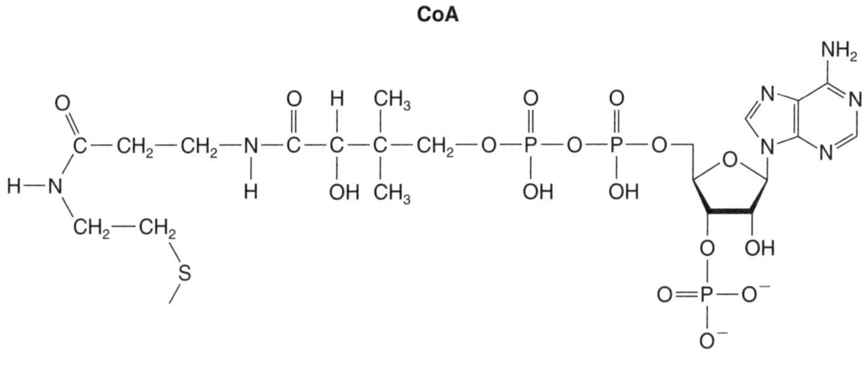

c maturation. In Gram-positive bacteria and chloroplasts, a simpler system exists involving diffusion of heme through a transmembrane protein. The third known system has evolved in mitochondria of fungi, invertebrates, and vertebrates; A fundamental role in this system is played by an enzyme called cytochrome *c* lyase, which attaches the heme group to the protein matrix.

Bibliography

Iron–Sulfur Proteins

Luchinat, C., Capozzi, F., and Bentrop, D., *Iron–Sulfur Proteins,* in *Handbook on Metalloproteins,* Bertini, I., Sigel, A., and Sigel, H., Eds., Marcel Dekker, New York, pp. 357–460, 2001.

Physical Methods in Bioinorganic Chemistry: Spectroscopy and Magnetism, L. Que, Ed., University Science Books, Sausalito, CA, 2000, pp. 1–556.

Beinert, H. and Kiley, P. J., "Fe–S Proteins in Sensing and Regulatory Functions", *Curr. Opin. Chem. Biol.*, **3,** 152–157 (1999).

Link, T. A., "The Structures of Rieske and Rieske-Type Proteins", *Adv. Inorg. Chem.*, **47,** 83–157 (1999).

Capozzi, F., Ciurli, S., and Luchinat, C., "Coordination Sphere versus Protein Environment as Determinants of Electronic and Functional Properties of Iron–Sulfur Proteins", *Struct. Bonding,* **90,** 127–160 (1998).

Johnson, M. K., "Iron–Sulfur Proteins: New Roles for Old Clusters", *Curr. Opin. Chem. Biol.*, **2,** 173–181 (1998).

Beinert, H., Holm, R. H., and Münck, E., "Iron–Sulfur Clusters: Nature's Modular Multipurpose Structures", *Science,* **277,** 653–659 (1997).

Beinert, H., Kennedy, M. C., and Stout, C. D., "Aconitase as Iron–Sulfur Protein, Enzyme, and Iron-Regulatory Protein", *Chem. Rev.*, **96,** 2335–2373 (1996).

Flint, D. H. and Allen, R. M., "Iron–Sulfur Proteins with Nonredox Functions", *Chem. Rev.*, **96,** 2315–2334 (1996).

Bertini, I., Ciurli, S., and Luchinat, C., "The Electronic Structure of FeS Centers in Proteins and Models. A Contribution to the Understanding of their Electron Transfer Properties", *Struct. Bonding,* **83,** 1–54 (1995).

Advances in Inorganic Chemistry. Iron Sulfur Proteins, Vol. 47, Cammack, R. and Sykes, A. G., Eds., Academic Press, San Diego, 1992, pp. 1–514.

Johnson, M. K., Robinson, A. E., and Thomson, A. J., Low-temperature Magnetic Circular Dichroism Studies of Iron–sulfur Proteins, in *Iron–sulfur proteins,* Spiro, T. G., Ed., Wiley-Interscience, New York, pp. 367–406, 1982.

Heme-Proteins

Turano, P. and Lu, Y., "Iron in Heme and Related Proteins", in *Handbook on Metalloproteins,* Bertini, I., Sigel, A., and Sigel, H., Eds., Marcel Dekker, New York, pp. 269–356, 2001.

Physical Methods in Bioinorganic Chemistry: Spectroscopy and Magnetism, Que, L., Ed., University Science Books, Sausalito, CA, 2000.

Banci, L., Bertini, I., Luchinat, C., and Turano, P., *The Porphyrin Handbook;* Academic Press, Burlington, MA, 1999, pp. 323–350.

Dailey, H. A., "Enzymes of Heme Biosynthesis", *JBIC,* **2,** 411–417, 1997.

Gadsby, P. M. A. and Thomson, A. J., "Assignment of the Axial Ligands of Ferric Ion in Low-Spin Hemoproteins by Near-Infrared Magnetic Circular Dichroism and Electron Paramagnetic Resonance", *J. Am. Chem. Soc.*, **112,** 5003–5011, 1990.

Makinen, M. W. and Churg, A. K., "Structural and Analytical Aspects of the Electronic Spectra of Hemeproteins", in *Iron Porphyrins,* Part One; Lever, A. B. P. and Gray, H. B., Eds., Addison-Wesley, Reading, MA, pp. 141–235, 1983.

Scheidt, W. R. and Gouterman, M., "Ligands, Spin State, and Geometry in Hemes and Related Metalloporphyrins", in *Iron Porphyrins,* Part One; Lever, A. B. P. and Gray, H. B., Eds., Addison-Wesley, Reading, pp. 89–139, 1983.

Cofactor Biosynthesis

Kuchar, J. and Hausinger, R. P., "Biosynthesis of Metal Sites", *Chem. Rev.*, **104,** 509–525 (2004).

Allere, J. W. A., Daltrop, O., Stevens, J. H., and Ferguson, S. J., "c-Type cytochromes: diverse structures and broadness pose evolutionary problems", *Philos. Trans. R. Soc. London B,* **358,** 255–266 (2003).

Dailey, H. A., Dailey, T. A., Wu, C. K., Medlock, A. E., Wang, K. F., Rose, J. P., and Wang, B. C., "Ferrochelatase at the Millennium: Structures, Mechanisms and [2Fe–2S] Clusters", *Cell Mol. Life Sci.*, **57,** 1909–26 (2000).

Thoeny-Meyer, L., "Biogenesis of Respiratory Cytochromes in Bacteria", *Microbiol. Rev.*, **61,** 337–376 (1997).

Dailey, H. A., "Ferrochelatase", in *Mechanisms of Metallocenter Assembly,* Hausinger, R. P., Eichorn, G. L., and Marzilli, L. G., Eds., VCH, New York, 1996, pp. 77–98.

Mansy, S. S. and Cowan, J. A., "Iron–sulfur cluster Biosynthesis: Toward an understanding of cellular machinery and molecular mechanism", *Acc. Chem. Res.*, **37,** 719–725 (2004).

CHAPTER V

Transport and Storage of Metal Ions in Biology

Thomas J. Lyons
Department of Chemistry
University of Florida
Gainesville, FL 32611

David J. Eide
Department of Nutritional Sciences
University of Wisconsin
Madison, WI 53706

Contents

V.1. Introduction

Metal ions have unique chemical properties that allow them to play diverse roles in cellular biochemistry. Whether it is the capability of Cu to catalyze oxidation–reduction (redox) chemistry or of Zn to act as a Lewis acid in hydrolytic enzymes, these properties have rendered certain metal ions indispensable for living organisms. Some metals, like Zn, have become so biologically abundant that it is difficult to imagine a living organism being able to adapt to life without them. In addition to its enzymatic role, Zn is a structural cofactor for thousands of proteins that mediate protein–protein, protein–nucleic acid, and protein–lipid interactions. Perhaps the most commonly recognized of these motifs is the ubiquitous *zinc finger* domain first identified in transcription factor IIIA (TFIIIA). It has been estimated that as many as 1% of proteins encoded by the human genome contain zinc-binding domains of this type.

The essentiality of metal ions in biology is unquestionable. Yet, despite the relative abundance of inorganic minerals on Earth, many formidable hurdles impede the acquisition of metal ions by living organisms, thus making metal nutrient sufficiency a perpetual problem. The exigent nature of this need requires organisms to go to great lengths to scrounge enough nutrition from the environment to survive. To recognize this problem, one needs only the image of a deer rooting around in and drinking from a brackish mud seep in an attempt to acquire necessary salts, like sodium, that it cannot get from its diet of vegetation.

Of course, the acquisition of metal ions is important for humans as well. Many genetic diseases in humans are caused by mutations that alter metal ion metabolism (see http://www3.ncbi.nlm.nih.gov/disease/Transporters.html). Furthermore, as many as 2 billion people worldwide suffer from malnutrition owing to deficiencies of micronutrients such as iron (see http://www.who.int/nut/). These deficiencies are due in part to inadequate food supplies in many parts of the world. However, poor nutritional quality of some regional crop plants and nutrient-depleted soil also play major roles in this problem.

Mineral nutrient resources are frequently so limited that organisms often wage war to obtain them. For example, the success of microbial pathogens in causing human disease depends critically on their ability to obtain metal ions from the blood stream and host tissues. In fact, the concerted effort to limit the availability of metal ions to pathogens is a major part of human immune defense against infection. Invading microbes respond reciprocally by up-regulating their own metal ion scavenging mechanisms, precipitating a veritable tug-of-war. Both pathogen and host subsequently produce molecules known as cytolytic agents that punch holes in the cellular membranes of the opposing side allowing stockpiles of nutrients to leak out.

An understanding of metal ion transport is crucial for improving human health. However, a proper understanding of metal ion transport processes requires a comprehension of those factors that essentially limit the ability of living cells to distribute these ions to the right place at the right time. After all, the uniquely useful properties of metal ions are worthless unless they can be effectively harnessed by the cellular machinery.

What are the major obstacles preventing the acquisition of metal ions? First and foremost, one must look at the *bioavailability* of the elements. This term implies more than just the incidence of an element on Earth and includes its prevalence in environments where life is found. There may be plenty of nickel in the Earth's core, but life certainly does not persist there. This supply is of no use for an organism that needs nickel. Another aspect of bioavailability is the form in which an element is commonly found. Zinc sulfide minerals are certainly common enough in the biosphere, but in this form Zn is not very usable. Few organisms possess the ability to mobilize Zn from such a source. Lastly, inherent in the term bioavailability is the presence of chemical competitors that impede acquisition of a desired nutrient. Molybdenum may be the most abundant transition metal ion in the ocean and it certainly possesses many desirable chemical properties. However, it is mainly found as the oxyanion, molybdate, a species that, while highly soluble and amenable for uptake, is very similar to the much more common sulfate and phosphate oxyanions. Competition from these other oxyanions may seriously hinder the ability to transport molybdate into the cell.

Other chemical properties that influence uptake include the redox chemistry, hydrolysis, solubility, chelation of the free ion, and ligand exchange rates of metal ion chelates. As we will see, for the widely coveted nutrient, iron, the aqueous chemistry of a metal ion in the ecological niches where life persists governs the mechanism by which the nutrient is acquired.

Once an organism has found a bioavailable source of minerals another problem arises: Cellular membranes are effective permeability barriers that block passive diffusion of charged molecules such as metal ions. Thus an organism may find and ingest the correct amount of a mineral, but it still needs to absorb it. To deal with this obstacle, exquisitely efficient uptake mechanisms have evolved. These uptake systems use molecules embedded in cellular membranes, which we refer to generically as *transporters*, to facilitate the selective movement of inorganic ions across the barrier. Not only do these transporters facilitate the absorption of nutrients from the environment, they are also responsible for the proper distribution of metal ions within whole organisms and individual cells.

Another critical problem in the delivery of metal ions to their ultimate targets is transport within the cytoplasm or the space between membranes. The cytoplasm is an environment filled with metal ion chelators such as soluble proteins, peptides (e.g., glutathione), and organic metabolites (e.g., citrate). These molecules may impede the ability of metalloproteins to acquire their metal cofactor by acting as competitive chelators. In the case of Cu, specific soluble transport proteins, called copper chaperones, have evolved to facilitate the safe transfer of Cu from the plasma membrane to various copper-containing proteins (see Section VIII.6). Similar proteins may exist for other metal ions.

The level of a particular metal ion available to an organism in its diet or environment can change drastically over time. Often, when food is plentiful, organisms are exposed to quantities of metal ions that exceed their requirement. The ability to store nutrients in a usable form during these times of plenty represents a sensible means of ensuring adequate nutrition during times of starvation. Indeed, storage mechanisms have evolved for a variety of nutrients. Storage mechanisms also provide a means by which excess metal ions can be detoxified. Overaccumulation can lead to a variety of toxic effects. As the intracellular level of a metal ion becomes excessive, the metal can inhibit critical processes, for example, by competing with other metal ions for enzyme active sites and other biologically important ligands. Excess Fe and Cu can also generate reactive oxygen species that damage DNA, lipids, and proteins.

Should the optimal range of metal ions be exceeded, various mechanisms limit the deleterious effects of the metal. These mechanisms include: *exclusion* of metals at the level of uptake, *extrusion* of accumulated metals out of the cytoplasm, and *detoxification* of metals by transformation into a harmless adduct or incorporation into an inert state. These mechanisms are also necessary to survive exposure to metal ions, such as Pb, that are not biologically essential, yet common enough to pose a serious biochemical threat.

Since metal ions can be both essential and toxic, a delicate balance, or *homeostasis,* must be maintained to keep the intracellular metal ion levels within an optimal range. Metal ion homeostasis is generally regulated tightly by sensors that govern the activity of transporters, storage molecules, and detoxifying enzymes (see Section XV.1). If metal ion deficiency is sensed, net ion uptake is increased and stored metals are mobilized. If metal ion excess is sensed, then net uptake is curtailed and storage capability is expanded. The two processes are intimately linked.

V.2. Metal Ion Bioavailability

The abundance of different metals in the Earth's crust can vary over several orders of magnitude (Chapter II; Table V.1). Iron, for example, is the fourth most abundant element in the Earth's crust and is 100-fold more abundant than Cu (geochemical considerations are discussed in greater detail in Chapter II). Despite its abundance in terrestrial environments, the level of iron in some surface sea water can be as low as 50 pM (5×10^{-5} μM). Could this extremely low level of iron restrict growth of aquatic organisms? This hypothesis was proven correct in experiments where iron supplementation at the surface of specific areas in the Pacific and Southern Oceans led to increased phytoplankton populations as measured by biomass levels. Thus, in many aquatic environments, as well as in some terrestrial environments, the abundance of metal ions can be growth limiting.

For mammalian cells, the source of metal ions is the blood plasma. Plasma levels of metal ions are controlled within a narrow range of concentrations by regulating absorption and/or excretion of metal ions from the body. Cells that obtain their metal ions from plasma accumulate most of these elements to much higher levels than the plasma itself (Table V.1). The observation that cells—mammalian and otherwise—must often acquire metal ions against a concentration gradient demonstrates the importance of efficient metal ion uptake systems. Notice in Table V.1 the

Table V.1.
Average Relative Abundance of Selected Elements in the Earth's Crust, Sea Water, Mammalian Blood Plasma, and in Mammalian Cells or Tissue

Element	Crust (ppm)	Sea Water (μM)	Blood Plasma (μM)	Cell/Tissue[a] (μM)
Ca	4×10^4	1×10^4	2×10^3	1×10^3
Cd	0.2	1×10^{-3}		
Co	25	2×10^{-5}	2.5×10^{-5}	
Cu	55	4×10^{-3}	8–24	~68
Fe	5×10^4	1×10^{-3}	22	0.001–10
K	3×10^4	1×10^4	4×10^3	1.5×10^5
Mg	2×10^4	5×10^4	500	9×10^3
Mn	950	5×10^{-4}	0.1	180
Mo	1.5	0.1		5×10^{-3}
Na	3×10^4	5×10^5	1×10^5	1×10^4
Ni	75	8×10^{-3}	0.04	2
V	135	0.03	0.07	0.5–30
W	1.5	5×10^{-3}		
Zn	70	0.01	17	180

[a] Approximate values based on total content rather than labile concentration.

contrasting behavior of Ca and Na. These elements are found in greater quantities in plasma than in cells. In these cases, it is equally difficult for cells to extrude the ions against a concentration gradient.

The chemical properties of metal ions can greatly affect their availability to organisms. The properties of Fe in particular have a major impact on its bioavailability. In general, the chemical properties of other metal ions are less of an obstacle to their uptake and simpler transport systems may be required. In a sense, the discussion of Fe chemistry will serve as an extreme example of the problems cells must overcome to obtain metal ions.

V.2.1. Iron: A Case Study

Approximately one-third of our planet's mass is iron (see Chapter VI). This great abundance would suggest that Fe is readily available to organisms for their use. However, this is far from the case because of iron's chemical properties. First, in aerobic environments, Fe is most prevalent in the Fe^{3+} oxidation state. Hydrated Fe^{3+} (ferric) ions are only stable in strongly acidic solutions. At higher pH values, hydrated ferric ions readily undergo hydrolysis to form iron hydroxide [ferrihydrite, $Fe(OH)_3$, Eq. 1]. The solubility product of iron hydroxide is 10^{-38} (Eq. 2). By using Eqs. 3 and 4 we can calculate the concentration of hydrated ferric ions at pH 7 to be $\sim 10^{-17}$ M. Clearly, the free ferric ion is all but insoluble at physiological pH in aqueous solutions and as such would be growth limiting.

$$[Fe(H_2O)_6]^{3+} \rightarrow Fe(OH)_3 + 3\ H^+ + 3\ H_2O \tag{1}$$

$$K_{sp} = [Fe^{3+}][OH^-]^3 \approx 10^{-38}\ M \tag{2}$$

$$[Fe^{3+}] = 10^{-38}/[OH^-]^3 \tag{3}$$

$$\text{At pH 7.0, } [Fe^{3+}] = 10^{-38}/(10^{-7})^3 = 10^{-17}\ M \tag{4}$$

Soluble Fe^{3+} can exist at much higher concentrations in an aqueous solution of neutral pH if it is bound to chelators such as citrate, ethylenediaminetetraacetic acid (EDTA), and so on. Such binding explains why there is radically more iron dissolved in the ocean than would be predicted by Eqs. 1–4 (see Table V.1). However, while these chelators allow for a higher level of soluble Fe^{3+}, they present additional problems to the process of cellular iron accumulation. Once bound, the metal ion may no longer be available for uptake. An important determinant in the bioavailability of chelated metal ions is their ligand exchange rates. The rate of ligand exchange for ferric ion chelates is generally very slow. Consider the following equation (Eq. 5):

$$[FeL_6]^{n+} + L' \rightarrow FeL_5L'^{n+} + L \qquad (5)$$

where L and L′ are different ligands in an octahedral complex. The exchange rate for water bound to Fe^{3+} in such a reaction is on the order of $3 \times 10^3\ s^{-1}$ as compared to $3 \times 10^6\ s^{-1}$ for Fe^{2+} (ferrous iron). This reaction is representative of other ligand exchange rates, indicating that dissociation of Fe^{3+} from a chelator complex is much slower than for Fe^{2+}. Ligand exchange reactions are an important consideration in understanding transport of metal ions because the transport process can be thought of as principally a ligand-exchange reaction between the chelator and the transporter (Eq. 6).

$$[FeL_6]^{3+} + \text{Transporter} \rightarrow Fe^{3+}\text{-Transporter} + 6\ L \rightarrow \text{Transport} \qquad (6)$$

Clearly, given its slow rate of ligand-exchange reactions, this reaction would probably not be an effective means of obtaining Fe^{3+}. Moreover, Fe is usually bound in multidentate chelate complexes such that there are additional entropic forces limiting the release of the metal (i.e., the chelate effect, see Tutorial II).

As described in more detail later in this chapter, given the solubility problems and slow ligand-exchange reactions of Fe^{3+}, a commonly used strategy of organisms to obtain iron is to reduce Fe^{3+} to Fe^{2+} prior to its mobilization. This strategy holds several advantages; Fe^{2+} is many orders of magnitude more soluble than Fe^{3+} at physiological pH values. The solubility product constant (K_{sp}) for $Fe^{2+}(OH)_2$ is 10^{-15}, thus $[Fe^{2+}]$ at pH 7 is 0.08 M. Moreover, Fe^{2+} is bound much less tightly by most organic acid chelators. As an example of this latter effect, the affinity of EDTA for Fe^{3+} is $\sim 10^8$-fold higher than its affinity for Fe^{2+}.

Another factor affecting the accumulation of metal ions in general, and Fe in particular, is the presence of high levels of other cations. For example, in humans, high levels of dietary Cu or Zn can inhibit Fe accumulation from the diet. This inhibition is probably due to competition of these different metal ions for a particular transporter (see Section V.3).

V.3. General Properties of Transport Systems

Before we discuss some of the specific systems responsible for metal ion transport, some general concepts of biological transport must be introduced. These concepts are critical to our understanding of the movement of solutes in and out of living cells and their intracellular compartments. Transport of an atom or molecule across a lipid bilayer can occur either by simple diffusion (i.e., nonmediated transport) or by transport mediated by molecules embedded in the membrane. Nonmediated transport occurs efficiently when the transported molecule is itself hydrophobic (like O_2) and can pass across the membrane unaided. Being charged entities, free metal ions require protein- or ionophore-mediated transport to cross a lipid bilayer.

Ionophores are a diverse class of organic molecules that increase the permeability of membranes to particular ions. Many ionophores are antibiotics produced by bacteria to inhibit the growth of other organisms by discharging the concentration gradients of important ions. One such example is gramicidin, an oligopeptide comprised of 15 alternating L- and D-amino acids. When inserted in the membrane of a Gram-positive bacterium, gramicidin forms a channel that is permeable to a variety of cations (Fig. V.1). Several non-peptidic ionophores function not as pores, but as carriers that dissipate ion gradients by physically shuttling ions across membranes. One such ionophore—the aromatic organic chelator pyrithione—is very effective at dissipating Zn^{2+} gradients. While ionophores are important for medical inorganic biochemistry, we focus here on protein-mediated transport because such processes are required for physiological metal ion transport in all cells.

The nomenclature of the protein-mediated transport field can be confusing, but there are essentially only three types of transport proteins: *channels, carriers,* and *pumps* (Fig. V.2). All of these proteins share the common feature of having several regions that span the membrane—designated transmembrane domains. These domains can be either α-helices or β-strands depending on the specific transporter. For our purposes, carriers are also referred to as *permeases*.

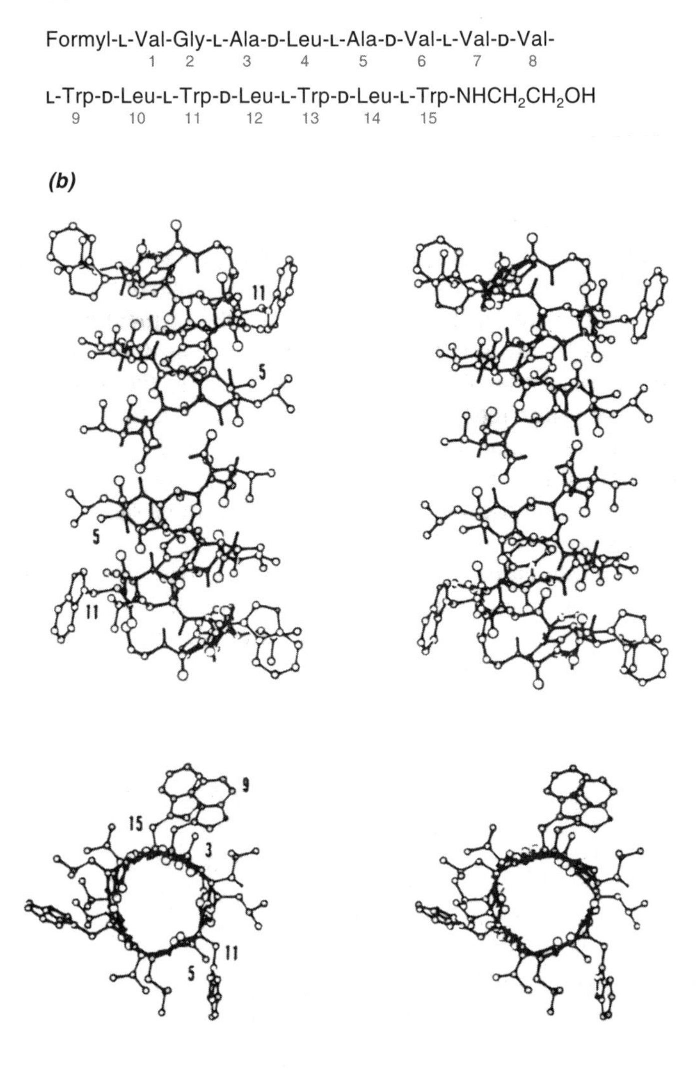

Fig. V.1.
The gramicidin channel.
(*a*) The chemical structure of gramicidin, a polypeptide with alternating L- and D-amino acids. (*b*) A stereoprojection model for the structure of the channel, based on physical measurements. The upper half of the figure shows a side view, the lower half an end-on view through the channel down the axis of the molecule. (Figure from Stein, **1990** with permission.)

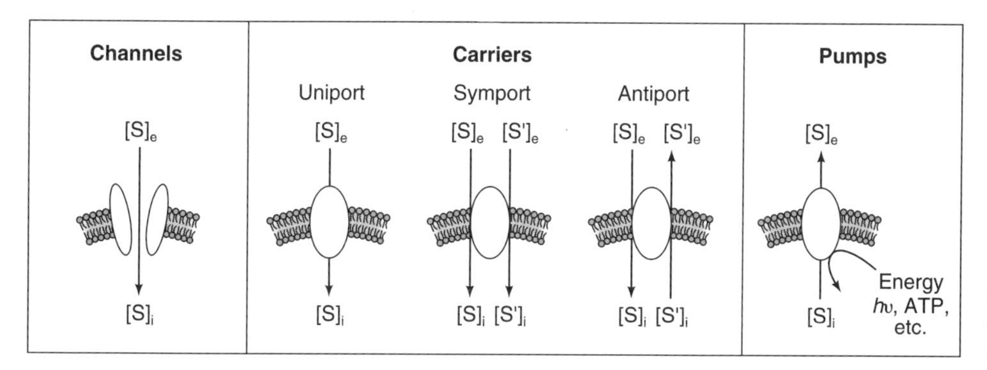

Fig. V.2.
Types of transport proteins. Three types of systems can mediate metal ion transport. Channel proteins utilize the chemical or electrical potential of the membrane to drive the transport of the substrate [S] through a pore in the membrane. The substrate is denoted by $[S]_e$ on the external side of the membrane and $[S]_i$ denotes substrate on the internal side of the membrane. Carrier proteins undergo a conformational change to facilitate movement of ions across membranes using the concentration gradient of [S] or the gradient of a cosubstrate [S′]. The cosubstrate can be transported in the same direction as the substrate (symport) or in the opposite direction (antiport). Pumps utilize energy, usually provided by adenosine triphosphate (ATP) hydrolysis (directly) to drive the transport of substrate ($h\nu$ = energy).

One crucial aspect of a transport system is the force that determines the direction of substrate movement. For proteins that simply facilitate the diffusion of cations, the concentration gradient or difference in *chemical potential* of the substrate is the major *driving force* such that the substrate will diffuse only in the direction from its high to its low concentration. For charged substrates such as metal ions, the cell membrane's *electrical potential* can also be an important factor. Membrane electrical potentials in living cells can be measured using microelectrodes and often range from -50 to $-100\,\text{mV}$ (inside negative). This electrical potential will favor uptake of cations and oppose their efflux from the cell.

V.3.1. Channels

Channels are proteins that form *pores* in the membrane that allow the movement of substrate across the membrane by diffusion. These pores can be gated such that they only open in response to signals such as changes in membrane electrical potential or ligand binding. In general, channels allow rapid transport of large quantities of substrate across membranes, thus dissipating a concentration gradient. An example of a channel protein is the ligand-gated *N*-methyl-D-aspartate (NMDA) receptor on the postsynaptic membrane of neuronal cells. This protein senses glutamate, an excitatory neurotransmitter, and responds by opening a Ca^{2+}-permeable channel that allows rapid influx of Ca^{2+} from the synaptic cleft, thereby transmitting a signal from one neuron to another.

V.3.2. Carriers

Carriers bind their substrate(s) on one side of a membrane, undergo a conformational change, and then release the substrate on the opposite side of the membrane. Some carrier proteins are *uniporters,* that is, proteins that transport only one type of substrate atom or molecule. Uniporters carry out a process known as *facilitated diffusion* and generally only allow transport along the concentration gradient. An example of such a system is the glucose transporter on the plasma membrane of human erythrocytes. A site on the external side of the membrane has a high affinity for glucose. Upon binding glucose, the conformation of the carrier protein changes in such a way that the site moves to the internal side of the membrane, where, if the concentration of glucose is low, glucose is released. For example, there may be a high glucose flux through the glycolytic pathway due to a high demand for ATP and a consequent rapid rate of glucose oxidation.

Alternatively, carriers can be *cotransporters* that transport two or more substrates simultaneously. Cotransport of different substrates by a single transporter can occur in the same direction, called *symport,* or in opposite directions called *antiport*. As will be discussed in Section V.4, the Dmt1 transporter is responsible for the uptake of Fe into mammalian cells. This transporter is an Fe^{2+}–H^+ symporter that uses a concentration gradient of protons and the electrical potential of the membrane to drive the uptake of Fe^{2+}. Because the driving force for transport of one substrate can be provided by the transport of the other, cotransporters allow for the accumulation of ions against a concentration gradient and are examples of *secondary active transport*.

V.3.3. Pumps

Finally, *pumps* use energy derived directly from the hydrolysis of ATP or other energy sources (e.g., light) to provide the energy for transport. Pumps are *primary active transport* systems and allow for accumulation of ions against a chemical gradient. An example of this type of primary active transport is the Ca^{2+}-ATPase on the membrane of the sarcoplasmic reticulum (SR). This transporter is responsible for the maintenance of low cytoplasmic calcium after muscle contraction by vesicular sequestration of cytoplasmic Ca^{2+}. The pump accomplishes this task by having two conformations (Fig. V.3). The E_1 conformation has two Ca^{2+}-binding sites on the cytoplasmic side of the membrane with a high affinity for Ca^{2+}. The Ca^{2+}- and ATP-dependent phosphorylation drives the conversion of the enzyme to the E_2 state. This conformational change has two main effects: (1) the Ca^{2+}-binding sites move to the luminal (inside) surface of the SR membrane, where (2) their relative affinity for Ca^{2+} decreases. Dissociation of Ca^{2+} followed by dephosphorylation prompt a return to the E_1 state.

Protein-mediated transport systems normally display a relationship between rates of substrate movement versus concentration that fits the Michaelis–Menten equation describing the rates of enzyme reactions (Eq. 7):

$$V = \frac{V_{max}[S]}{K_m + [S]} \tag{7}$$

where V is the rate of transport, [S] is the substrate concentration, V_{max} is the maximal rate of transport, and K_m is the concentration of substrate that gives a rate equal to one-half V_{max}. This relationship gives rise to a hyperbolic curve with saturation occurring at high substrate concentrations (Fig. V.4). This saturation effect is a characteristic of all protein-mediated transport systems. The parameter V_{max} values are

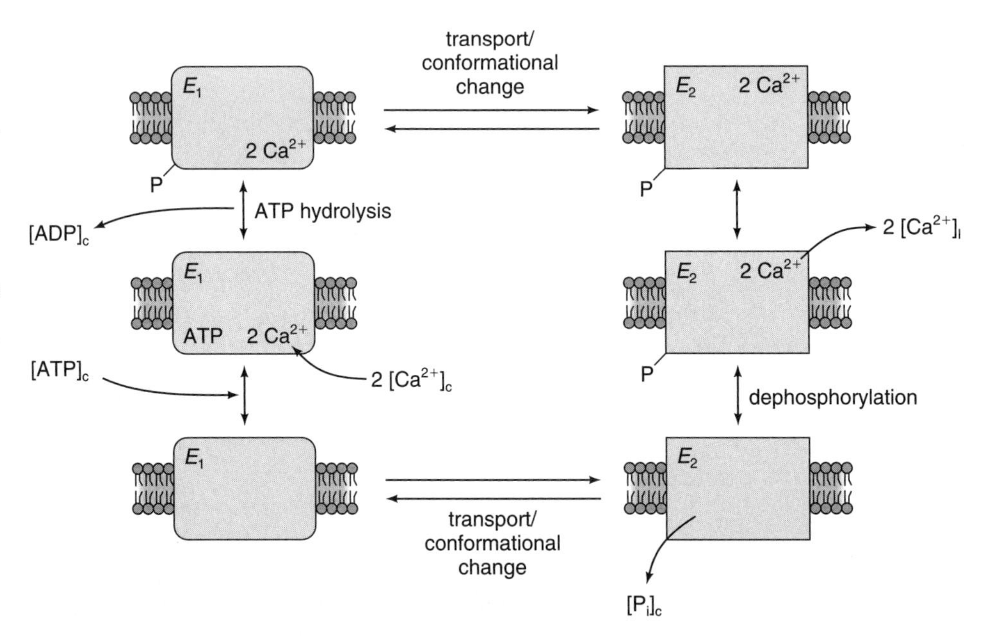

Fig. V.3.
Model of the Ca^{2+}-ATPase of the sarcoplasmic reticulum (SR). Low cytosolic calcium levels are maintained in the muscle cells via the Ca^{2+}-ATPase on the membrane of the SR. This protein transports Ca^{2+} into the SR using ATP hydrolysis as the driving force. This is achieved by a two-state mechanism (E_1 and E_2). In the E_1 state, the Ca^{2+}-binding sites are cytoplasmic and have a high affinity for Ca^{2+}. In the alternative conformational state, E_2, the Ca^{2+}-binding sites move to the lumenal side of the SR and have a much lower affinity for Ca^{2+}. Conversion from one state to the other is mediated by ATP hydrolysis and protein phosphorylation–dephosphorylation. Subscripts denote substrates on specific sides of the membrane; $[S]_c$ = cytoplasmic, $[S]_l$ = lumenal, P = phosphate, ADP = adenosine diphosphate.

indicative of either the prevalence of transporters on the membrane or the capacity of individual transporters, if their numbers can be measured. The K_m value of a transporter gives an estimate of the relative affinity of the protein for its substrate. The K_m values of metal ion transporters range from the low nanomolar (nM) range for carrier proteins to the millimolar (mM) range for most channel proteins.

Organisms often employ multiple transporters with widely varying affinities for the same substrate depending on its availability. For example, the yeast *S. cerevisiae* possesses 20 different hexose transporters (HXT), more than any other organism. Many of these HXTs function in the uptake of glucose from the medium and have widely varying K_m values for glucose transport. This allows the yeast to transport glucose (and therefore grow) at an optimal rate over a wide range of extracellular glucose concentrations (from μM to M).

Finally, what determines the substrate specificity of metal ion transporters? At this time, we know very little about how metal ion transporters distinguish between different metal ions. Presumably, this specificity is due to the same factors that affect metal ion binding by other metalloproteins (i.e., ionic radii, coordination geometries, and ligand preferences; see Chapter III and Tutorial II). Metal ion transporters have the added complexity of providing both specific and high-affinity metal binding along with a high degree of lability such that substrates are transported and efficiently

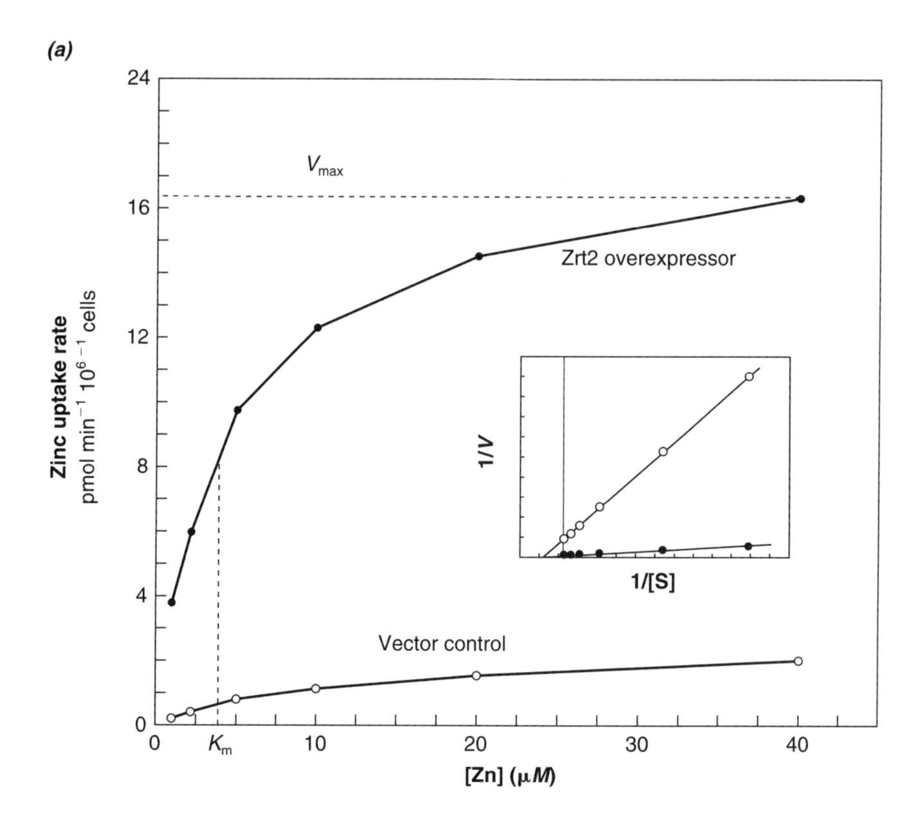

(b)

Strain	K_m	V_{max}
Zrt2 Overexpressor	3.6 ± 0.1	17 ± 2
Vector control	10 ± 1	2.0 ± 0.1

Fig. V.4.
Enzyme kinetics for the Zrt2 low-affinity zinc transporter from *Saccharomyces cerevisiae*. The Zrt2 transporter is overexpressed in a yeast strain that is deficient in high-affinity Zn uptake. The graph demonstrates that transporter activity is governed by classic Michaelis–Menten kinetics. Saturation of uptake, V_{max}, occurs at high concentrations of substrate; K_m, a measure of relative substrate affinity, is the substrate concentration at half-maximal uptake velocity. Inset shows a Lineweaver–Burk plot used to calculate kinetic parameters.

released following movement across the membrane. Elucidation of the way in which this combination of specificity, high affinity, and lability occurs represents an exciting area of research in bioinorganic chemistry.

V.4. Iron Illustrates the Problems of Metal Ion Transport

Iron is an essential nutrient to almost all organisms. But, as was discussed earlier, it can often be very scarce in the environment. While unusual in the natural world, one ingenious strategy used by some organisms to deal with the challenges posed by iron scarcity is simply to live without it. One organism that apparently does not require Fe is *Borrelia burgdorferi,* the bacterial pathogen that causes Lyme disease. This pathogen is an obligate parasite that survives Fe limitation in its human host by using manganese instead of iron in some metalloproteins and doing without other enzymes altogether. The lack of several of these important enzymes also explains why this bacterium is an obligate parasite and cannot grow outside of its host.

For those organisms that do require Fe for growth, several different pathways of Fe accumulation can be present in the same cell to ensure an adequate supply under a variety of conditions. For example, the baker's yeast *S. cerevisiae* has at least seven different Fe uptake systems (Fig. V.5). Gram-negative bacteria such as *Escherichia coli* have a similarly high number of uptake pathways available to them. Clearly, when it comes to obtaining Fe for growth, organisms leave little to chance.

Prior to transport, organisms must mobilize Fe. There are three general means by which mobilization is accomplished: *chelation, reduction,* and *acidification.* Each strategy serves to maintain Fe in a soluble form. Two general systems are then used to facilitate the transport of Fe across cellular membranes. Interestingly, mammalian Fe uptake systems combine all of these strategies and systems.

V.4.1. Chelation

Bacteria, fungi, and some plants use chelating agents called *siderophores* to obtain iron. Siderophores are small organic molecules that bind ferric iron with high affinity. Their structure, chemistry, and uptake mechanisms are discussed in Section VIII.3. These molecules are synthesized by bacteria, some plants, and some fungi and are secreted directly into the extracellular environment. Upon binding Fe, the extracellular iron–siderophore complex can be bound by receptors at the cell surface that also function as transporters for the complex. One such receptor–transporter is the FhuA protein of *E. coli,* discussed at length in Section VIII.3. Alternatively, the Fe^{3+} can be dissociated from the siderophore complex on the external surface and subsequently taken up by other transporters.

Siderophores have remarkably high affinity for iron due to their multidentate ligand character. Moreover, in addition to their high affinity for Fe^{3+}, these compounds also show extremely high specificity for ferric iron over other metals such that siderophores are very effective agents for specifically mobilizing extracellular iron, even when other metal ions might be present in much higher concentration.

Siderophores are effective devices in the competition between microorganisms for available Fe in the environment. Once bound by the siderophore, the Fe is no longer available to other microbes that are unable to use that particular iron–siderophore complex. Some organisms can actually take up Fe^{3+} complexes composed of siderophores produced by other organisms. One example is *S. cerevisiae,* an organism that does not produce its own siderophores, but has transport systems for a number of different iron–siderophore complexes (Fig. V.5). This clever strategy vividly illustrates the cut-throat world of microbial competition for scarce resources.

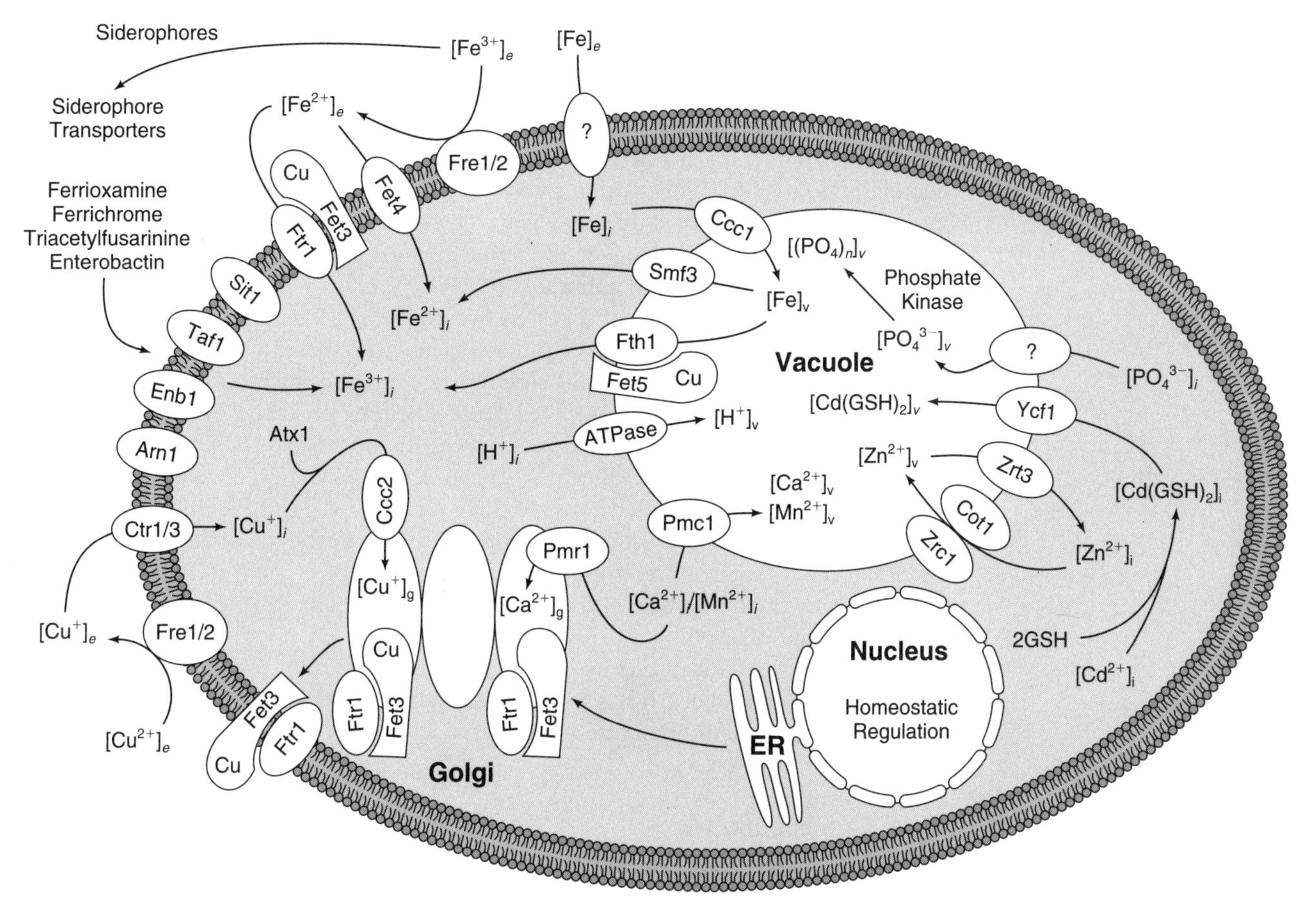

Fig. V.5.
Overview of Fe and Cu uptake systems and vacuolar cation storage in *S. cerevisiae*. Yeast possess at least seven different systems for acquiring Fe from the external milieu. Strains that lack all known Fe uptake systems are still viable suggesting that other routes of entry for iron are yet to be identified. The Fet3/Ftr1 high-affinity Fe uptake system requires Cu as a cofactor that is supplied by the action of Ctr1, Ctr3, Atx1, and Ccc2. Ion sensors in the nucleus and elsewhere regulate the activity of these systems and maintain ion homeostasis. The yeast vacuole is vital for metal ion homeostasis. Via its ability to accumulate large amounts of metals, the vacuole is critical for resistance to excess quantities of metal ions like Zn, Cu, Fe, Mg, Ca, and Mn. Vacuolar acidification and polyphosphate synthesis play crucial roles in this capacity. Mechanisms for mobilization of Fe and Zn from this storage pool are beginning to be understood (see text for details). Subscripts denote substrates in specific subcellular compartments; $[S]_e$ = external, $[S]_i$ = internal, $[S]_v$ = vacuolar, and $[S]_g$ = Golgi; GSH = glutathione.

V.4.2. Reduction

A second strategy of Fe uptake is to reduce extracellular Fe^{3+} to Fe^{2+} prior to uptake. This reduction dramatically increases the solubility of particulate Fe and can even liberate Fe from ores such as magnetite. The Fe^{2+} product is then transported into the cell by Fe^{2+}-specific transporters. This mechanism of uptake, typified by the major pathways of iron accumulation in *S. cerevisiae* (Fig. V.5), is common among other fungi, most plants, and mammals. The benefits of this strategy should be clear from the chemistry of iron. As discussed, Fe^{2+} is significantly more soluble, more kinetically labile, and binds with lower affinity to most chelators compared with Fe^{3+}. Thus, reduction increases the bioavailability of extracellular iron. In an atmosphere that has a significant partial pressure of dioxygen (pO_2), the newly generated Fe^{2+} oxidizes readily. Therefore, it is necessary to reduce the Fe close to the site of transport or in excess amounts. However, as we discuss later, many environments are anaerobic or microaerobic and organisms that live under these conditions do not necessarily encounter the Fe solubility problems created by aerobiosis.

Reduction of Fe^{3+} is mediated by a class of enzymes found in the plasma membrane called ferrireductases (or ferric-chelate reductases). These reductases are often cytochromes that transfer electrons donated by intracellular reductants [e.g., reduced nicotinamide adenine dinucleotide phosphate (NADPH)] across the plasma membrane of the cell. The available evidence suggests that two heme groups embedded in the transmembrane domains of the reductases are responsible for the transfer of electrons across the plasma membrane.

For ferrireductases to be useful in iron accumulation, they must be capable of reducing Fe^{3+} bound to a variety of different chelators that are found in the extracellular environment. This finding is clearly true for many such enzymes. In *S. cerevisiae,* for example, the Fre1 and Fre2 ferrireductases are capable of reducing Fe bound to chelators as structurally diverse as citrate, nitrilotriacetic acid, EDTA, and the siderophore ferrioxamine. Since *S. cerevisiae* possesses seven putative ferric reductase genes, it is possible that some of them have evolved to function in the uptake of Fe bound to specific chelators or siderophores. Obviously, the ability to reduce extracellular microbial iron–siderophore complexes must be of great utility to this yeast when it is growing in the wild and competing with other microbes for what is often a limiting amount of iron.

A reductive mechanism of Fe absorption is also important for the uptake of nonheme iron in human diets. It has been shown that cell-surface ferrireductases line the cells of the mammalian small intestine and reduce dietary Fe^{3+} prior to its uptake by Fe^{2+}-specific transporters.

V.4.3. Acidification

The third strategy for solubilizing Fe^{3+}, often used in conjunction with reduction-dependent pathways, is acidification of the extracellular environment. For example, iron deficiency in plants causes an increased extrusion of protons from the roots into the soil. Activation of plasma membrane H^+-pumping ATPases in response to iron deficiency is responsible for this acidification. The rate of proton release can be quite fast, reducing the pH of the surrounding soil to values of 3 or even lower. This lowered pH serves to increase the soluble concentration of Fe^{3+} by inhibiting formation of hydrolysis products. For example, at pH 3, the soluble maximum concentration of free Fe^{3+} is $10\,\mu M$ (see Eq. 3). This level of iron is sufficient to sustain cell growth. The lowered pH also stabilizes ferrous iron relative to ferric ion, thus disfavoring reoxidation by O_2. Moreover, the increased $[H^+]$ competitively inhibits binding of both Fe^{3+} and Fe^{2+} by extracellular chelators.

V.4.4. Iron Transport Systems: Fe^{2+} Transporters

The molecular identification of Fe^{2+} transport systems began with the cloning of the *FeoB* gene of *E. coli.* The FeoB gene encodes an inner-membrane protein that transports Fe^{2+} into the cytoplasm of the cell. Among eukaryotes, several Fe^{2+} transporters have now been identified. In *S. cerevisiae,* the Fet4 protein is responsible for Fe^{2+} transport (Fig. V.5). In plants, the Irt1 protein mediates uptake of Fe^{2+} from the soil into the roots, while in mammals, the Dmt1 transporter mediates uptake of Fe^{2+} in the intestinal lumen into the enterocytes lining the small intestine. In each case, these transporters are integral membrane proteins with multiple transmembrane domains that transport Fe^{2+} directly across the membrane.

All Fe^{2+} transporters described so far function in environments that have very low oxygen tension or are anaerobic, for example, the intestine (FeoB, Dmt1) or the rhizosphere (Irt1). Under oxygen-poor conditions, ferrous iron will predominate and it is energetically more favorable to simply transport this species. Yeast are also capable of growing anaerobically. It is no surprise then that transcription of the gene that encodes the Fet4 ferrous iron transporter is induced 1000-fold by a shift from aerobic to anaerobic growth.

V.4.5. Iron Transport Systems: Ferrous Oxidase-Permease Linked Transport

Other Fe^{2+} transport systems use an ingenious mechanism of ion transport that takes advantage of the redox chemistry of Fe to drive the transport process. In *S. cerevisiae,* when iron is abundant in the medium, the Fet4 transporter is responsible for iron uptake. This protein has a relatively low affinity for its substrate with a K_m of 600-n*M* free Fe^{2+}. When iron becomes limiting, a second system is induced that has a 200-fold higher affinity for iron.

One of the surprising observations regarding this high-affinity Fe uptake system was its dependence on copper: Copper-deficient cells are also Fe deficient because of a defect in the high-affinity Fe^{2+} uptake system. This mystery was solved by the molecular and biochemical characterization of the protein subunits that make up the high-affinity transporters, Ftr1 and Fet3. The Ftr1 transporter is an integral membrane protein with multiple transmembrane domains that probably serves as the iron transporter. The Fet3 transporter is also an integral plasma membrane protein with a single transmembrane domain at its carboxy-terminus. The amino-terminus of the protein is remarkably similar to a family of enzymes known as multicopper oxidases. The properties and mechanism of multicopper oxidases will be discussed in greater detail in Section XII.7. These proteins share the ability to catalyze four single-electron oxidations of substrate with the concomitant four-electron reduction of O_2 to H_2O. In the case of oxidase-permease based Fe uptake systems, the substrate is Fe^{2+}. In the yeast system, the Fe^{2+} produced by ferrireductases (or present in the environment through the action of extracellular reductants) is oxidized to Fe^{3+} by the Fet3 multicopper oxidase on the external surface of the cell. The Fe^{3+} product is transferred directly from Fet3 to an Fe^{3+}-binding site on the Ftr1 permease. Then, following a conformational change, the Fe is passed across the membrane into the cell.

V.4.6. Fe^{2+} Transport versus Oxidase-Permease Mediated Transport: A Comparison of Strategies

A confusing aspect of oxidase-permease mediated iron transport systems is understanding why cells go to the energy-demanding trouble of reducing Fe^{3+} to Fe^{2+} and then reoxidizing it back to Fe^{3+} while it is still on the extracellular surface of the plasma membrane. One possible explanation for the paradox is that extracellular reoxidation may confer greater substrate selectivity on the high-affinity uptake system. To be transported by this system, the Fe substrate must go through several steps (i.e., binding to the oxidase, oxidation, transfer to the permease, and transport), each with some degree of specificity for Fe^{2+} or Fe^{3+}. The combined effect of this multistep pathway is complete specificity for iron over other metal ions. This specificity is clearly manifest in the yeast high-affinity system, which cannot transport any metal ions other than iron.

In contrast, all of the simpler Fe^{2+} transporters studied to date, including Fet4, Irt1, and Dmt1, can transport several metal ions other than Fe^{2+} including Zn^{2+}, Mn^{2+}, and Cd^{2+}. Apparently, the means by which the Fe^{2+} transporters identify and transport Fe cannot completely distinguish between Fe^{2+} and many other metal ions as transport substrates. This fact is important, given that other metal ions are usually found in much greater concentrations than iron and as such will be efficient competitors for transport. This promiscuity definitely poses a problem for iron-deficient plants, which upregulate the iron transporter, Irt1, to scavenge Fe from the soil. These plants also hyperaccumulate Zn^{2+}, Mn^{2+}, and Cd^{2+} in the process.

If the hypothesis is correct that ferrous-oxidase mediated transporters exist to confer specificity for Fe over other metal ions, then one must logically ask, why do Fe^{2+} transporters exist at all? One answer may be as simple as where the transporters are meant to function. Oxidase-permease based uptake depends on the availability of oxygen as a cosubstrate. In microaerobic or anaerobic environments such as those discussed above, the oxidase would not be functional. Thus, under these conditions, Fe^{2+} transporters may be the only reliable means to obtain iron, despite their apparent substrate promiscuity.

V.4.7. Mammalian Iron Transport: A Combination of Strategies

Uptake of dietary nonheme iron from the lumen of the mammalian intestine is mediated by an Fe^{2+} uptake system conceptually similar to that described above for low-affinity yeast iron uptake. Dietary iron that is reduced either by cell surface ferrireductases or by reductants (e.g., ascorbate) in the diet itself is taken up into cells called enterocytes that line the small intestine. The transporter responsible for this uptake is the previously described Dmt1 protein. Once inside the enterocyte cytoplasm, the iron is subsequently exported across the membrane of the enterocyte into the blood stream by ferroportin/IREG1. Before its release into the serum, the iron is oxidized by hephaestin, and then bound as Fe^{3+} by transferrin, an ~700 amino acid serum protein. Transferrin binds Fe with high affinity ($K_d = 10^{-22}$ M at pH 7.0) and plays the principal role in delivering Fe^{3+} to the cells of most tissues in the body. Transferrin is discussed in greater detail in Section VIII.1.

Uptake of Fe from transferrin into other cell types combines all three of the strategies of iron mobilization described above (i.e., chelation, reduction, and acidification). Transferrin plays an analogous role to the siderophores used for Fe uptake by bacteria, fungi, and plants; by binding Fe^{3+} with high affinity, transferrin maintains the Fe in serum in a soluble form that can be used by cells.

Analogously to siderophores, iron-loaded transferrin is delivered to receptors on the surface of cells for uptake, where it undergoes a process called *receptor-mediated endocytosis* (discussed in Section VIII.1). Once inside the endocytic compartment, Fe^{3+} is released from the protein by acidification of the endosome, reduced by a ferrireductase, and transported into the cytoplasm by the same transporter, Dmt1, that was responsible for intestinal Fe uptake. Thus, in order for iron to reach its intended cellular targets in mammalian cells, a combination of all of the different iron solubilization and transport strategies must be used.

V.5. Transport of Metal Ions Other Than Iron

The chemistry of copper also poses many problems for its uptake and distribution. Like iron the solubility and stability of copper are highly dependent on environment, oxidation state, and chelators. Fortunately, Cu^{2+} is stable and soluble in an aerobic atmosphere. However, the divalent form of ionic Cu does not seem to be the preferred species for transport. In the case of yeast, the Ctr1 and Ctr3 high-affinity transporters mediate copper uptake. Uptake by these systems is similar to iron uptake in that Cu^{2+} is first reduced to Cu^{+} by plasma membrane cuprireductases. In fact, the same reductases that reduce Fe^{3+} are responsible for Cu^{2+} reduction (Fig. V.5). The Cu^{+} ion is then the substrate for Ctr1 and Ctr3. This system seems potentially inefficient, as free Cu^{+} undergoes rapid disproportionation to Cu^{2+} and Cu^{0}

in aqueous solution and readily oxidizes in air. However, despite this inherent instability, Cu^+ has faster ligand exchange rates and is bound by many organic chelators, although with lesser affinity than Cu^{2+}. The Cu^+ ion also has more stringent electronic requirements regarding the geometrical distribution and chemical properties of its preferred ligand set, so reduction of Cu^{2+} to Cu^+ prior to transport is likely to allow for a more specific transport system. Thus, reduction of Cu^{2+} to Cu^+ during copper uptake provides some of the same advantages as the reduction of Fe^{3+} does in iron uptake.

While many gene products responsible for the uptake and distribution of metal ions are known, the transport mechanisms are still poorly understood. Some hydrated metal ions, such as alkali metals (Na^+, K^+) and alkaline earth metals (Mg^{2+}, Ca^{2+}), exist only in a single oxidation state. These ions are also common, highly soluble, and form complexes with very fast exchange rates. Thus, transport of these ions is less complicated and probably proceeds in a straightforward manner. Zinc (Zn^{2+}) is very similar in that it has only one available oxidation state. However, Zn is much less abundant and is insoluble at higher pH values, so acidification followed by high-affinity transport represents an efficient means for its acquisition.

Other transition metals such as Mn, Co, and Ni can exist in multiple oxidation states, yet exploitation of their redox chemistry during transport is not suspected since the divalent species (Mn^{2+}, Co^{2+}, Ni^{2+}) of each ion is most stable in aqueous environments. Again, acidification is the primary means for mobilization as their solubilities and stabilities are enhanced at low pH values. Cobalt is unique in that it is often acquired as part of Vitamin B_{12} and as such may be transported in this complex. These metal ions are inserted into proteins, where their oxidation state may change. It is unclear which oxidation state is utilized during the intracellular transport and insertion processes.

Finally, transition metals like V, Mo, and W are most often found in oxygen-rich environments as oxyanions (VO_4^{3-}, MoO_4^{2-}, WO_4^{2-}). In alkaline conditions, these inorganic anions are structurally similar to phosphates and sulfates and are probably transported by proteins that resemble phosphate and sulfate permeases. However, acidic conditions tend to complicate their chemistry and promote formation of polyanionic species containing multiple metal ions or cationic species, depending on the concentration. Therefore, the transport of these metals as oxyanions may not be as simple as is commonly suspected.

Toxic metal ions, such as cadmium and silver, probably enter the cell via transporter proteins involved in taking up essential elements. For example, the Dmt1 Fe^{2+} transporter can also transport Cd^{2+} and Pb^{2+}. Thus, Dmt1 probably represents a significant entry point into the body for these toxic elements that can sometimes contaminate our diets.

V.6. Mechanisms of Metal Ion Storage and Resistance

The levels of essential metal ion nutrients available to an organism in its diet or surrounding environment can vary widely. Therefore, in addition to possessing efficient uptake mechanisms for metal ion accumulation, organisms also have the means to store essential metal ions when they are abundant for use at later times when those elements become scarce. These storage systems have the added benefit of allowing the accumulation of high intracellular levels of metal ions without the toxic consequences that such accumulation might otherwise entail. It should come as no surprise that many of the mechanisms involved in storing essential metal ion nutrients also play a role in detoxifying toxic, non-nutrient metals. Therefore, these two topics are considered together in this section.

Cells employ two main strategies for metal ion storage that also serve to prevent toxicity due to overload of essential and non-nutrient metal ions. First, the metal ion may be bound by cytoplasmic proteins or macromolecules thereby keeping the level of free metal ions low. Certainly, the best understood systems of metal ion storage are the ferritin and metallothionein proteins (see Sections VIII.2 and VIII.4). Alternatively, metal ions can be transported into membrane-bound compartments within the cell (e.g., the plant and fungal vacuole), where the metal is less damaging. An additional strategy for detoxification of toxic metal ions, such as Cd^{2+}, is to pump them out of the cell so they will be diluted in the extracellular environment. Metal ion efflux does not just play a role in detoxification of harmful metals; it is also a major mechanism regulating levels of nutrient accumulation in mammals.

A recurring theme for metal ion storage and detoxification systems is that they are induced by metal ion exposure. This regulation links the storage and detoxification capacity of the cell to the level of exposure. As you will see in Chapter VIII, this control can be exerted at transcriptional, translational, or even post-translational levels.

V.6.1. Ferritin

Found in vertebrates, plants, some fungi, and bacteria, ferritin is the primary site of Fe storage in most organisms. The structure, chemistry, and biology of ferritin are discussed in greater detail in Section VIII.3. In brief, ferritin is a spherical molecule with an outer coat of protein and an inner core of hydrous ferric oxide $[Fe_2O_3(H_2O)_n]$. As many as 4500 atoms of Fe can be stored in a single ferritin molecule. Thus, when fully loaded with iron, soluble Fe concentrations as high as 0.25 M can be easily achieved *in vitro*. Comparing this value to the solubility of Fe^{3+} in solution at pH 7 (10^{-17} M) demonstrates just how useful this system is to the cell for storing iron in a usable form. Ferritin levels increase in cells treated with high levels of iron due to a post-transcriptional control mechanism that regulates translation of the messenger ribonucleic acid (mRNA) (Section VIII.2). However, while a great deal is known about the structure, metal-binding properties, and regulation of ferritin protein synthesis, we do not yet fully comprehend the mechanisms that control Fe loading during abundance and mobilization during scarcity.

V.6.2. Metallothionein

Another cytoplasmic metal-binding protein involved in metal ion storage and detoxification is metallothionein. Metallothioneins are small, cysteine-rich proteins that bind Zn^{2+}, Cd^{2+}, Cu^{2+}, and other metal ions by virtue of their Cys ligands. As many as seven Zn^{2+} or Cd^{2+} ions can be accommodated in the metallothionein structure. Widespread in Nature, metallothioneins are found in cyanobacteria, fungi, plants, insects, and vertebrates. A more detailed discussion of these proteins is found in Section VIII.4.

Metallothioneins bind metal ions with high affinity. For example, the stability constant for the Zn_7MT II complex is 3.2×10^{13} M^{-1}. Despite this high affinity of binding, at least some of the metal ions bound to metallothionein are kinetically very labile and can be readily donated to other ligands, such as Zn^{2+}-binding proteins, *in vitro*. Therefore, metallothioneins are thought to be an *in vivo* store of labile metal ions, particularly Zn^{2+}. Synthesis of some metallothionein isozymes is induced at the level of gene transcription by metal treatment. As with ferritin, this induction links the metal storage and detoxification capacity to the availability of the ion.

Mice lacking metallothionein show reproductive deficits during dietary zinc deficiency. These same mice are also more sensitive to the toxic effects of Cd. Thus, metallothioneins seem to play a dual role in Zn^{2+} storage and Cd^{2+} detoxification.

V.6.3. Other Intracellular Chelators

Some organisms use intracellular chelators in metal ion storage. In the fungus *Neurospora crassa,* the hydroxamate siderophore ferricrocin is used as an Fe storage pool in spores. Alternatively, small polypeptides called phytochelatins (PCs) can also play a role in metal ion storage and detoxification in plants and fungi. These PCs have the structure (γ-Glu-Cys)$_n$-Gly, where $n > 1$. While PCs are polypeptides, they are generated by enzymatic synthesis rather than by translation. The structure of phytochelatin indicates that it is related to glutathione ($n = 1$) and, indeed, glutathione is used as a precursor by the enzyme, phytochelatin synthase, that produces this compound (Fig. V.6).

As was the case with ferritin and metallothionein, the synthesis of PCs is also regulated by metal ions. This control occurs at a third level of regulation, post-translational control of enzyme activity. As depicted in Fig. V.6, a number of metal ions stimulate the activity of PC synthase *in vitro*. Synthesis continues until the metal ion is bound by the accumulated PC (or by adding a chelator such as EDTA) at which point the reaction stops. *In vivo* this mechanism provides an elegantly simple method to regulate PC synthesis in response to metal ion levels in the cell.

V.6.4. Intracellular Transport in Metal Ion Storage and/or Resistance

It is clear that macromolecules can act as internal stores for essential metals and as inert sinks for toxic metals. Metal ions can also be stored within membrane-bound organelles inside the cell, thereby eliminating their harmful effects in the cytoplasm. For example, the vacuole of fungi and plants appears to play a storage role for many metal ions including Fe^{2+}, Ca^{2+}, and Zn^{2+}. Several vacuolar functions have been shown to be essential for resistance to high levels of these same ions. One of these functions is the vacuolar H^+-ATPase, which is responsible for the acidification of the compartment. Another function is the accumulation of polyphosphate anions, which may provide a counterion to balance the accumulated positive charge as well as perhaps binding the metals in an inert form (Fig. V.5).

Other organelles seem to play a role in metal ion sequestration. The Ca^{2+} ion accumulates in mitochondria and secretory vesicles where it is stored for signaling purposes. Labile Zn^{2+} has been detected in membrane-bound organelles in mammalian cells and these may also play a storage or signaling function.

If these organelles (vacuole, secretory vesicle, mitochondria) are storing metal ions for later use, how are these metal ions mobilized? Recent advances in this field have come from the study of zinc and iron accumulation in the yeast vacuole. While two yeast proteins of overlapping function, Zrc1 and Cot1, have been shown to transport zinc into the vacuole, another protein, Zrt3, has been shown to be involved in the mobilization of this pool during zinc deficiency. In a similar fashion, iron is stored in the yeast vacuole via a protein called Ccc1. An oxidase-permease system comprised of the Fet5 and Fth1 proteins (homologous to Fet3 and Ftr1) has been postulated to mobilize stored iron pools. Another protein, called Smf3, is also suspected of

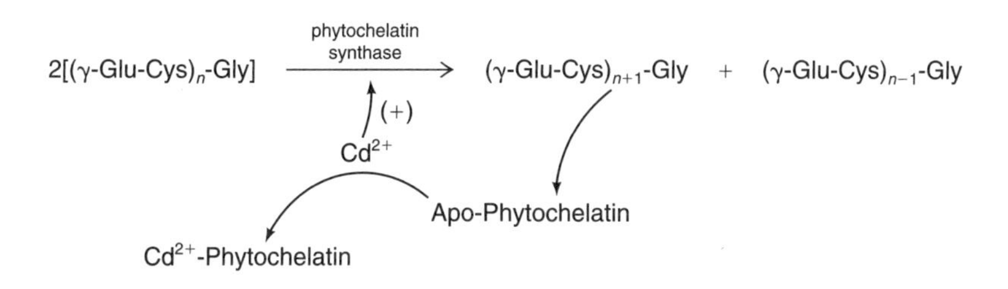

Fig. V.6.
Mechanism of phytochelatin synthase. Phytochelatins are synthesized from glutathione monomers. In response to heavy metals (e.g., Cd^{2+}), the enzyme phytochelatin synthase is activated via metal binding to thiol residues. The enzyme synthesizes phytochelatin by polymerizing γ-glutamylcysteine units (from glutathione or other phytochelatins). The newly synthesized phytochelatins then bind heavy metal ions, thus inhibiting the activity of phytochelatin synthase.

mobilizing vacuolar iron stores. Not surprisingly, the gene that encodes Smf3 is induced by anaerobiosis so that iron can still be exported under anoxic conditions when the Fet5/Fth1 oxidase-permease is nonfunctional. Mobilization of metals from cytosolic pools like ferritin and metallothionein are equally challenging issues, such that the exact nature and mobilization of these stored pools of metal ions remains an exciting line of research.

V.6.5. Exclusion

When it comes to the toxic effects of metal accumulation, *exclusion* is often one of the best strategies. Many cell types simply exclude the offending metal ion from the internal milieu by removing its routes of entry. One such method is the post-translational regulation of the activity or localization plasma membrane transporters. An example of this approach is the zinc-induced endocytosis and degradation of the high- and low-affinity zinc uptake systems in *S. cerevisiae*. Another elegant example of exclusion is the secretion of sulfide by yeasts in order to form insoluble extracellular complexes with metals like Cu and Cd.

V.6.6. Detoxification

Exclusion, however, is not always the best option for preventing metal toxicity. Sometimes, *detoxification* of a metal ion is more advantageous. One strategy for *detoxification* is to render metals harmless by internal chelation. Metallothionein and phytochelatin play this role for Cd^{2+}. A classic, yet certainly unusual example of metal ion detoxification is the *mer* operon in bacteria. This system involves the deliberate uptake of toxic Hg^{2+} ions and their conversion to volatile Hg^0 vapor thus rendering them diffusable and harmless to the organism. An ingenious use of this mercuric reductase system is in the process of phytoremediation. In principle, a plant that overexpresses a bacterial mercuric reductase could detoxify Hg^{2+} contaminated soil. This was indeed shown to be possible with the development of poplar trees that were transgenic for the mercuric-reductase gene. These transgenic plants volatilized Hg^0 at 10 times the rate of untransformed plants.

V.6.7. Extrusion

One of the simplest mechanisms to detoxify metals is simple *extrusion* from the cytoplasm. In *E. coli,* for example, excessive Zn accumulation activates transcription of a pump known as ZntA, a Zn^{2+} translocating P-type ATPase that expels excess Zn from the cell. Similar transporters are responsible for the efflux and detoxification of Cd, Cu, and Ag in a variety of bacterial species. *Extrusion* is not limited to expulsion from the cell entirely. Some resistance mechanisms transport toxic metal ions into storage compartments rendering the metals inert. Yeast compartmentalize excess Ca^{2+} and Mn^{2+} into both the vacuolar and Golgi compartments using the Pmc1 and Pmr1 ATPases, respectively. In an interesting combination of strategies, Cd^{2+} is detoxified first by conjugation to glutathione and then by transport into the vacuole by an ABC-type transporter called Ycf1.

V.7. Intracellular Metal Ion Transport and Trafficking

V.7.1. Trafficking

Metal ion transport into intracellular compartments for storage and detoxification requires that specific transporter proteins be present in the membranes of these organelles to facilitate this movement. Moreover, metal ions need to be transported into cellular organelles to function as cofactors of compartmentalized metal-dependent enzymes. For example, Fe, Cu, Mn, and Zn are required in the matrix of the mitochondrion. Iron is required in the matrix for the formation of heme and Fe–S clusters. Copper is required for the function of cytochrome *c* oxidase in the electron-transport chain. Manganese is a cofactor of the mitochondrial isozyme of superoxide dismutase, while Zn is required for the activity of proteins that proteolytically process matrix proteins and also for function of the mitochondrial RNA polymerase. For each of these metal ions, transporter proteins are required in the mitochondrial membranes to allow their movement into the organelle. Similarly, metal ions are required for various functions in the organelles of the secretory pathway, and transporters are needed to get them there as well.

In general, we know little about the transport mechanisms that are responsible for organellar metal ion trafficking. However, one excellent example of intracellular transport where the participants are well known is the transport of Cu into the secretory pathway. This transport is required for the activation of secreted or cell surface copper-dependent proteins such as lysyl oxidase, ceruloplasmin, ethylene receptors, and the yeast multicopper ferrous oxidase, Fet3. Moreover, in the mammalian liver, regulated Cu transport into the secretory pathway is the major site of control for Cu levels in the body. Excess Cu is transported into the bile caniculi for excretion.

The transport of Cu^{+} into the secretory pathway, specifically organelles of the Golgi apparatus, is mediated by eukaryotic members of the P-type ATPase transporter family (i.e., the same type of proteins responsible for metal ion efflux in bacteria). These proteins are discussed in Section VIII.5. A distinguishing feature of these proteins is the presence of a structural domain containing the heavy metal-binding motif MXCXXC. In yeast, the copper transporting ATPase of the secretory pathway is known as Ccc2 (Fig. V.5). In mammals, closely related transporters are called Mnk and WD because mutations in these genes cause the disorders of copper homeostasis known as Menkes and Wilson's disease.

V.7.2. Metallochaperones

Recent advances in our understanding of the intracellular trafficking of copper highlight the potency of research at the interface of biology and chemistry. Given the potential toxicity of Cu and its tendency to bind to adventitious (undesirable) sites, it is unlikely that the free ion is normally found at appreciable levels in the cytoplasm. In fact, it was recently demonstrated that there is no detectable free Cu in the cytoplasm. This surprisingly low level of Cu available for newly synthesized Cu enzymes presents a remarkable challenge to the cell, that is, How can a desired ligand compete with all the other potential copper-binding sites in the cell for miniscule amounts of available copper? We now know that the cell uses specific targeting proteins, called metallochaperones, to deliver Cu to intracellular targets. These proteins are discussed in Section VIII.6.

Copper metallochaperones are soluble, cytoplasmic copper-binding proteins. The proteins bind Cu after it enters the cell and deliver it to their corresponding recipient proteins. In yeast, for example, Atx1 is the copper metallochaperone that delivers Cu to the Ccc2 P-type ATPase (Fig. V.5) and Lys7 delivers copper to Cu/Zn superoxide dismutase. Most often, copper chaperones possess significant structural similarity to their target proteins and they often possess the heavy metal associated MXCXXC motif found in metal ion translocating ATPases. An important question is, Do metallochaperones exist for other metal ions? While we do not yet know the answer, it is entirely possible. Like Cu, Fe is very toxic to cells and homeostatic control mechanisms probably maintain very low levels of labile iron. Thus, the same challenges that lead to the evolution of copper chaperones also exist for iron.

V.8. Summary

This chapter has introduced many of the concepts underlying biological metal ion transport and storage. Subsequent chapters will consider, in much greater detail, several aspects presented briefly here. The chemistry of metal ions can create serious challenges to their accumulation by cells and this is especially true for iron. Organisms have evolved three general strategies to mobilize Fe from the extracellular environment: chelation, reduction, and acidification. The process of transferrin-dependent iron uptake in mammals is a good example of how all three of these strategies can be combined into a single iron transport pathway. Transport of mobilized metal ions depends heavily on the environmental conditions and the chemical properties of the metal. Many distinct pathways of Fe transport exist depending on external iron concentration, the presence of chelators, and the level of oxygen in the environment. Similar considerations may influence the uptake strategies for other metals as well. The uptake of metal ion nutrients is a highly regulated process governed by their intracellular availability.

Intracellular transporters are involved in moving metal ions in and out of organelles and across the plasma membrane. Metallochaperones shuttle their wards across the perilous cytoplasm full of potent metal chelators and sensitive components. Sensors regulate the activities of all of these proteins so that a balance, or homeostasis, is achieved and an optimal range of metal ion concentration is maintained.

When this balance is overwhelmed, or when an organism encounters a nonessential metal ion, storage and resistance mechanisms become necessary. Two basic strategies exist for metal ion storage. The first is binding by intracellular macromolecules and the second is their transport into intracellular organelles.

Three strategies exist whereby the toxic effects of metal ions can be abrogated. Metals can be excluded from the cell by regulation of transporter activity or by external sequestration. They can be detoxified within the cell by sequestration in macromolecules or by conversion to inert forms. Finally, they can be extruded from the cytoplasm either into storage compartments or into the extracellular space.

Bibliography

Environmental Chemistry

Jonnalagadda, S. B. and Rao, P. V., "Toxicity, bioavailability and metal speciation", *Comp. Biochem. Physiol., C.*, **106**(3), 585–595 (1993).

Cox, P. A., *The Elements,* Oxford University Press, 1989.

General Transporters

Hille, B., *Ion Channels and Excitable Membranes,* Sinauer Associates, Inc., Sunderland, MA, 2001.

Stein, W. D., *Channels, Carriers, and Pumps: An introduction to membrane transport,* Academic Press, New York, 1990.

Metal Ion Transporters

Forbes, J. R. and Gros, P., "Divalent-metal ion transport by NRAMP proteins at the interface of host–pathogen interactions", *TRENDS Microbiol.*, **9**(8), 397 (2001).

Culotta, V. C., "Manganese transport in microorganisms", *Metal Ions Biol. Systems,* **37,** 35 (2000).

Guerinot, M. L., "ZIP family of metal transporters", *Biochim. Biophys. Acta,* **1465**(1–2), 190 (2000).

Moncrief, M. B. and Maguire, M. E., "Magnesium transport in prokaryotes", *J. Biol. Inorg. Chem.*, **4**(5), 523 (1999).

Butler, A., "Acquisition and utilization of transition metal ions by marine organisms", *Science,* **281**(5374), 207 (1998).

McMahon, R. J. and Cousins, R. J., "Mammalian zinc transporters", *J. Nutrition,* **128**(4), 667 (1998).

Paulsen, I. T. and Saier, M. H., Jr., "A novel family of ubiquitous heavy metal ion transport proteins", *J. Membrane Biol.*, **156**(2), 99 (1997).

Cunningham, K. W. and Fink, G. R., "Ca^{2+} transport in *Saccharomyces cerevisiae*", *J. Exp. Biol.*, **196,** 157 (1994).

Metals and Infection

Posey, J. E. and Gherardini, F. C., "Lack of a role for iron in the Lyme disease pathogen", *Science,* **288**(5471), 1651 (2000).

Braun, V. and Focareta, T., "Pore-forming bacterial protein hemolysins (cytolysins)", *Crit. Rev. Microbiol.*, **18**(2), 115 (1999).

Docampo, R. and Moreno, S. N., "Acidocalcisome: A novel Ca^{2+} storage compartment in trypanosomatids and apicomplexan parasites", *Parasitol. Today,* **15**(11), 443–448 (1999).

Rolfs, A. and Hediger, M. A., "Metal ion transporters in mammals: structure, function and pathological implications", *J. Physiol.*, **518**(1), 1 (1999).

Supek, F., Supekova, L., Nelson, H., and Nelson, N., "Function of metal–ion homeostasis in the cell division cycle, mitochondrial protein processing, sensitivity to mycobacterial infection and brain function", *J. Exp. Biol.*, **200**(2), 321 (1997).

Metal Metabolism Reviews

Aisen, P., Enns, C., and Wessling-Resnick, M., "Chemistry and biology of eukaryotic iron metabolism", *Int. J. Biochem. Cell Biol.*, **33**(10), 940 (2001).

De Luca, N. G. and Wood, P. M., "Iron uptake by fungi: contrasted mechanisms with internal or external reduction", *Adv. Microbial. Physiol.*, **43,** 39 (2000).

Eide, D. J., "Metal ion transport in eukaryotic microorganisms: insights from *Saccharomyces cerevisiae*", *Adv. Microbial. Physiol.*, **43,** 1 (2000).

Labbe, S. and Thiele, D. J., "Pipes and wiring: the regulation of copper uptake and distribution in yeast", *Trends Microbiol.*, **7**(12), 500 (1999).

Linder, M. C. et al., "Copper transport in mammals", *Adv. Exper. Med. Biol.*, **448,** 1 (1999).

Cuajungco, M. P. and Lees, G. J., "Zinc metabolism in the brain: relevance to human neurodegenerative disorders", *Neurobiol. Disease,* **4**(3–4), 137–169 (1997).

Grunden, A. M. and Shanmugam, K. T., "Molybdate transport and regulation in bacteria", *Arch. Microbiol.*, **168**(5), 345 (1997).

Metallochaperones

Finney, L. A. and O'Halloran, T. V., "Transition metal speciation in the cell: insights from the chemistry of metal ion receptors", *Science,* **300**(5621), 931–936 (2003).

O'Halloran, T. V. and Culotta, V. C., "Metallochaperones, an intracellular shuttle service for metal ions", *J. Biol. Chem.*, **275**(33), 25057 (2000).

Web Sites

http://www3.ncbi.nlm.nih.gov/disease/Transporters.html

http://www.who.int/nut/

CHAPTER VI

Biominerals and Biomineralization

Stephen Mann
School of Chemistry
University of Bristol
Bristol BS8 1TS, UK

Contents

VI.1. Introduction

The biological process that gives rise to the inorganic-based solid-state structures of life, such as bones, shells, and teeth, is called *biomineralization*. Over the last two decades, the study of biomineralization has shifted toward a more chemical perspective, and in so doing has become established as a new branch of bioinorganic chemistry that represents an increase in the length scale of the interplay between biological processes and inorganic chemistry. Whereas biocoordination chemistry principally focuses on interactions between metal atoms and ligands at the level of the coordination sphere, biomineralization addresses the chemistry between collections of inorganic atoms (nucleation clusters, crystal faces, etc.) and multiple arrays of ligands arranged across organic surfaces (insoluble proteins, lipid membranes, etc.). The research aims of biomineralization in the context of bioinorganic chemistry include the following: structural and compositional characterization of biominerals; understanding the functional properties of biominerals; and elucidation of the processes through which organic macromolecules and organic structures control the synthesis, construction, and organization of inorganic mineral-based materials.

In this chapter, we describe some of the main biominerals, their functions, and the general principles that govern the formation of these fascinating materials.

VI.2. Biominerals: Types and Functions

Table VI.1 is a summary of the main types of biominerals and their biological locations and functions. Of the >25 essential elements required by living organisms, H, C, O, Mg, Si, P, S, Ca, Mn, and Fe are common constituents of >60 different

Table VI.1.
The Types and Functions of the Main Inorganic Solids Produced by Controlled Biomineralization

Mineral	Formula	Organism/Function
CALCIUM CARBONATE		
Calcite	$CaCO_3$[a]	Algae/exoskeletons Trilobites/eye lens
Aragonite	$CaCO_3$	Fish/gravity device Mollusks/exoskeleton
Vaterite	$CaCO_3$	Ascidians/spicules
Amorphous	$CaCO_3.nH_2O$	Plants/Ca store
CALCIUM PHOSPHATE		
Hydroxyapatite	$Ca_{10}(PO_4)_6(OH)_2$	Vertebrates/endoskeletons teeth, Ca store
Octacalcium phosphate	$Ca_8H_2(PO_4)_6$	Vertebrates/precursor phase in bone
Amorphous	?	Mussels/Ca store Vertebrates/precursor phases in bone
CALCIUM OXALATE		
Whewellite	$CaC_2O_4.H_2O$	Plants/Ca store
Weddellite	$CaC_2O_4.2H_2O$	Plants/Ca store
GROUP 2A (IIA) METAL SULFATES		
Gypsum	$CaSO_4$	Jellyfish larvae/gravity device
Barite	$BaSO_4$	Algae/gravity device
Celestite	$SrSO_4$	Acantharia/cellular support
SILICON DIOXIDE		
Silica	$SiO_2.nH_2O$	Algae/exoskeletons
IRON OXIDES		
Magnetite	Fe_3O_4	Bacteria/magnetotaxis Chitons/teeth
Geothite	α-FeOOH	Limpets/teeth
Lepidocrocite	γ-FeOOH	Chitons (Mollusca) teeth
Ferrihydrite	$5Fe_2O_3.9H_2O$	Animals and plants Fe storage proteins
IRON SULFIDES		
Greigite	Fe_3S_4	Bacteria/magnetotaxis

[a] Magnesium-substituted calcites are also formed.

biological minerals. Other essential elements such as N, F, Na, K, Cu, and Zn are also constituents of biominerals, but are less widespread. Nonessential elements, such as Ag, Au, Pb, and U, are found in association with the external cell walls of bacteria, and Sr and Ba are accumulated and deposited as intracellular sulfate minerals in certain algae.

VI.2.1. Calcium Biominerals

Shells and mineralized tissues, such as bone and teeth, are composed of calcium carbonate or calcium phosphate minerals, respectively, in combination with a complex organic macromolecular matrix of proteins, polysaccharides, and lipids. These minerals have high lattice energies and low solubilities, are thermodynamically stable in biological environments, and display several polymorphic structures (same compositions, but different unit cell arrangements). The predominance of Ca biominerals over other Group 2A (IIA) metals can be explained by the low solubility products of the carbonates, phosphates, pyrophosphates, sulfates, and oxalates, and the relatively high levels of Ca^{2+} in extracellular fluids (10^{-3} M). However, a few examples of $BaSO_4$ and $SrSO_4$ biominerals are documented. In contrast, magnesium salts are generally more soluble and no simple Mg biominerals are known although Mg^{2+} has an important role in influencing the structure of both carbonate and phosphate biominerals through lattice and surface substitution reactions. For example, Mg^{2+} ions are readily accommodated in the calcite lattice and magnesium-containing calcites are common biominerals.

Calcium carbonate biominerals, such as calcite and aragonite, and to a lesser extent vaterite, are often used for structural support (e.g., in seashells). The mother-of-pearl (nacre) layer of seashells is a laminate of 0.5-μm thick calcium carbonate (aragonite) polygonal tablets sandwiched between thin (30-nm) sheets of a protein-polysaccharide organic matrix (Fig. VI.1). The matrix plays a key spatial role in limiting the thickness of the crystals and is structurally important in the mechanical "design" of the shell. The organic matrix reduces the number of voids in the shell wall and so inhibits crack propagation by dissipating the energy associated with an expanding defect along the organic layers and not through the inorganic crystals.

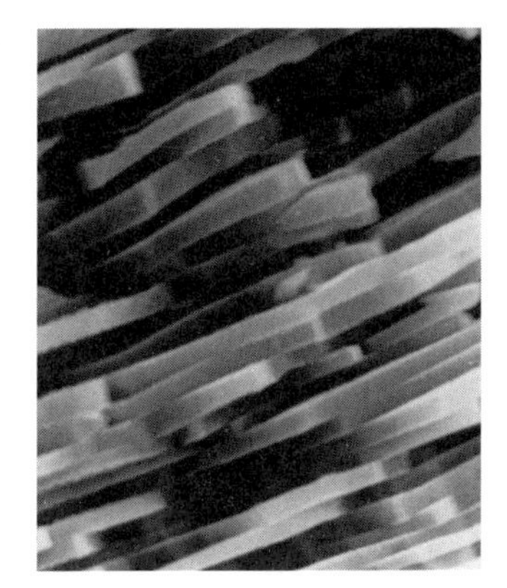

Fig. VI.1.
Cross-section of a mollusk shell showing the brick wall arrangement of oriented tabular aragonite ($CaCO_3$) crystals. Thin sheets of protein are sandwiched between adjacent layers of the crystals. The crystals are ~0.5 μm in thickness.

Biominerals also have some unusual functions. For example, crystals of calcite are used as gravity sensors in a wide range of animals. The optical properties of calcite are exploited in the lenses of the compound eyes of extinct creatures called trilobites, which are preserved as fossils. Vaterite is the least thermodynamically stable of the three nonhydrated crystalline polymorphs of calcium carbonate and rapidly transforms to calcite or aragonite in aqueous solution. Nevertheless, vaterite occurs as spicules in a few marine sponges (the majority of calcareous sponges have magnesium-rich calcite spines), where it possibly acts as a structural support or as a deterrent against predators. *Amorphous* calcium carbonate is deposited in the leaves of many plants where it acts as a calcium store. Although this material is exceedingly unstable in inorganic systems due to rapid phase transformation in aqueous solution, the non-crystalline biogenic mineral appears to be stabilized through the adsorption of biological macromolecules such as polysaccharides at the solid surface.

Bone and teeth are made from calcium phosphate in the form of a mineral, hydroxyapatite (HAP), along with a large number of proteins. The structural chemistry of biological hydroxyapatite is very complex because the mineral is not compositionally pure, often being Ca deficient and enriched in CO_3^{2-}, which replaces PO_4^{3-} ions in various lattice sites. Several other calcium phosphate polymorphs (Table VI.1), as well as an amorphous phase, have been identified as intermediates in bone mineralization. The structure and mechanical properties of bone are derived from the organized mineralization of hydroxyapatite (often as carbonated apatite) within a fibrous matrix of a structural protein, collagen, along with proteoglycans (proteins with sugar side chains). The distinction between an inorganic and a bioinorganic mineral is clearly seen in bone, which is close to being described as a "living mineral", since it undergoes continual growth, dissolution, and remodeling.

The structure and organization of tooth enamel, like bone, derives from a highly complex system designed to withstand specific types of mechanical stress. However, enamel is much less tough because it is close to 95 wt% calcium phosphate (bone on average is ~65 wt% mineral), but gains some structural resistance by interweaving long ribbon-like crystals into an inorganic fabric. Interestingly, enamel starts out with a high proportion of proteins (amelogenins and enamelin), which are progressively removed as the biomineral matures to produce the high mineral volume fraction. Dentine, on the other hand, which resides within the central regions of the tooth, is more similar in structure and composition to bone.

Although not as widespread as the carbonates or phosphates, calcium oxalate minerals are deposited in relatively large amounts in certain plants.

VI.2.2. Silica

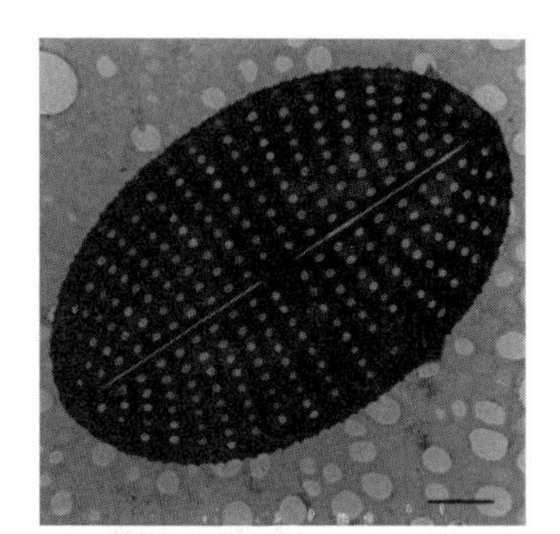

Fig. VI.2.
Silica shell of a unicellular diatom. Note the exquisite patterning of the perforated shell. Scale bar = 2 μm.

Although most biominerals are ionic salts, the stability of Si–O–Si units in water and the variability of the Si–O–Si bond angle gives rise to inorganic polymers in the form of amorphous silica with various degrees of hydroxylation ($[SiO_{n/2}(OH)_{4-n}]_m$ $n = 1-4$). A wide variety of elaborate mineral shapes is produced by unicellular organisms, such as diatoms and radiolaria (Fig. VI.2). Why some organisms utilize amorphous silica rather than a crystalline mineral like calcium carbonate as a structural material is unknown. One possibility is the absence of fracture and cleavage planes, which are inherent to crystalline structures. The amorphous biomineral can subsequently be moulded without loss of strength into a wide variety of complex architectures. In plants, silica spines and nodules in the cell walls act as a deterrent against predators.

VI.2.3. Iron Oxides

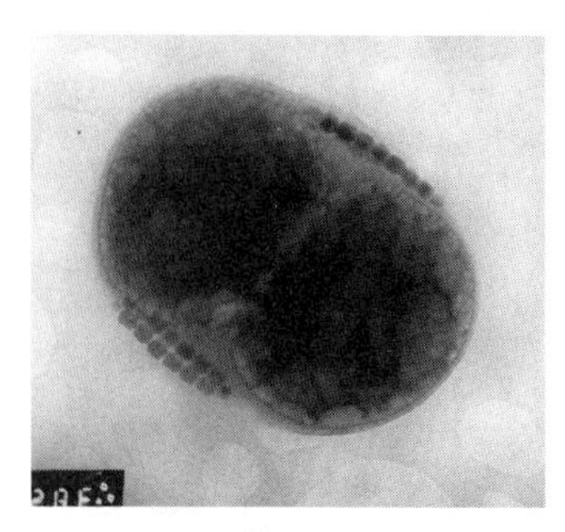

Fig. VI.3.
Image of a magnetotactic bacterial cell containing chains of magnetite (Fe_3O_4) crystals. Each crystal is ~100 nm in length.

Bioinorganic iron oxides are widespread and serve several different functions (Table VI.1). These solids have important inorganic counterparts that are extensively used in catalytic and magnetic devices. A mixed-valence compound, magnetite (Fe_3O_4), is of special biological relevance. Magnetite is synthesized in the form of discrete crystallographically oriented inclusions in a wide range of magnetotactic bacteria (Fig. VI.3). The crystals have dimensions compatible with that of a single magnetic domain, and are aligned in chains so that they function efficiently as biomagnetic compasses. These organisms are aligned in the Earth's magnetic field such that in the northern hemisphere they swim downward (north-seeking) toward the oxygen-depleted zone at the sediment–water interface of fresh water and marine environments.

An important and widespread iron oxide is a hydrated mineral called ferrihydrite. This red-brown gelatinous precipitate is readily formed in a test tube by the addition of sodium hydroxide to an Fe(III) solution, the so-called amorphous "ferric hydroxide". An iron storage protein, ferritin, contains a 5–7-nm size central core of this mineral wrapped in a protein coat. Encapsulation of the mineral in this way protects the organism from "rusting", as well as providing a means of cellular protection from the deleterious effects of labile iron. Ferritin iron may also be utilized in biochemical processes such as hemoglobin synthesis. Ferrihydrite is a disordered material with relatively high solubility, which is consistent with the dynamic behavior of this iron store.

Other iron oxides such as goethite (α-FeOOH) and lepidocrocite (β-FeOOH) are deposited in the teeth of certain mollusks. For example, the common limpet is armed with sabre-like rust-colored goethite teeth. During feeding, these hardened structures are rasped across rocks encrusted with algae. In certain species called chitons, the teeth are composed of both lepidocrocite and magnetite and are therefore magnetic!

VI.2.4. Sulfides

Many iron sulfide minerals are formed in association with sulfate-reducing bacteria, but they show none of the characteristics of biologically controlled mineralization. Most of these minerals are adventitious and arise from the reaction of metabolic products such as H_2S with Fe(III) species in the surrounding environment. However, recent studies have shown that certain types of magnetotactic bacteria present in sulfide-rich environments synthesize and organize crystals of a ferrimagnetic mineral, greigite (Fe_3S_4). The inclusions are discrete single crystals of narrow size distribution. They have species-specific morphologies and appear to be crystallographically aligned within chains, just like the magnetite crystals described above.

Elements such as Cu, Zn, and Pb are deposited on the external walls of bacterial cells also in the form of metal sulfides. Interestingly, some yeasts mineralize quantum dot semiconductors! Nanometer-size intracellular CdS particles are nucleated within short chelating peptides consisting of repeats of carboxylate (glutamate) and thiol (Cys = cysteine) amino acids. As the number of CdS units per particle is small (~ 85), these peptide–mineral complexes can be considered as large nuclearity clusters capped by cysteinyl thiolate ligands, analogous with proteins such as metallothionein that sequester smaller (three- and four-metal ion) clusters.

VI.3. General Principles of Biomineralization

VI.3.1. Biologically Induced and Biologically Controlled Mineralization

Although there is a continuum of biological involvement in biomineralization, two fundamental processes have been described. In biologically induced biomineralization, inorganic minerals are deposited as secondary events originating from interactions between metabolic activity and the surrounding environment. For example, calcification can be induced from saturated calcium bicarbonate solutions by metabolic removal of carbon dioxide for photosynthesis, according to the chemical equilibrium:

$$Ca^{2+}(aq) + 2\ HCO_3^-(aq) \rightleftharpoons CaCO_3 + CO_2 + H_2O$$

Photosynthesis drives this equilibrium to the right by removing CO_2 (Le Chatelier's principle). In certain algae, this results in the deposition of calcium carbonate between the cells (intercellular mineralization) such that the colony is entombed within the inorganic structure. The scale of such processes—as illustrated by the formation of coral reefs—can be enormous.

Biologically induced biomineralization is common in bacteria where the cell walls are involved in the extrusion of many different types of metabolic products, such as ions, gases, polypeptides, and electrons. These can coprecipitate with extraneous metal ions along the surface of the cell (epicellular mineralization) to produce a wide range of minerals, including oxides, phosphates, sulfides, and ice. Because there is minimal cellular control, the crystals are deposited adventitiously and are therefore heterogeneous in crystal size, shape, and organization. In many cases, the biominerals look very similar to their inorganic counterparts formed by simple inorganic precipitation reactions in the laboratory.

In contrast, biologically controlled biomineralization is a highly regulated process that produces materials such as bones, shells, and teeth, which clearly have specific biological functions. The biominerals have well-defined structures, shapes, particle

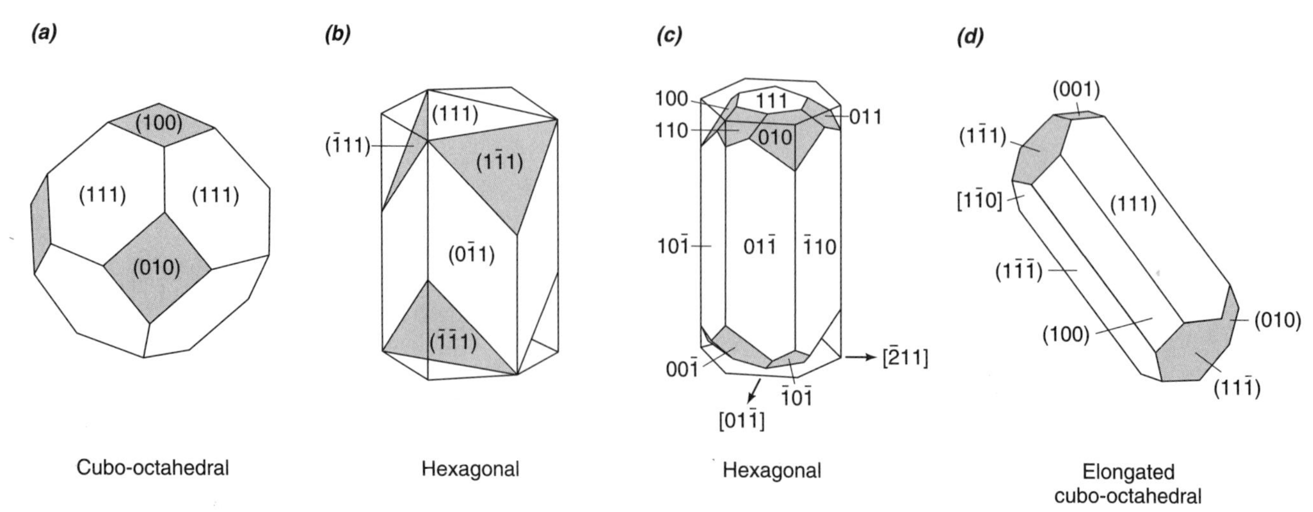

Fig. VI.4.
Crystal morphologies of bacterial magnetite are well defined and species specific; (*a*) cubo-octahedral, (*b*) and (*c*) are hexagonal prisms, and (*d*) is elongated cubo-octahedral.

sizes, and are highly organized. Although only a few examples of this process are known in bacteria—most notable are the magnetite crystals of magnetobacteria (Fig. VI.4)—many of the biominerals formed in unicellular organisms such as unicellular algae and protozoa are precisely controlled. In these organisms, mineral formation occurs within specific microenvironments or compartments within the cells (intracellular mineralization).

In multicellular organisms, such as higher plants and animals, biologically controlled mineralization often occurs in well-defined noncellular spaces (extracellular mineralization). For example, the formation of bones, shells, and teeth is precisely regulated through the activity of specialized cells that secrete biopolymers, such as collagen and chitin, into nearby extracellular spaces. There is a strong genetic and hormonal regulation of these processes, and when they go wrong, serious pathologies such as kidney stones and bone demineralization (osteoporosis) occur.

VI.3.2. Boundary-Organized Biomineralization

Biologically controlled biomineralization involves the precise regulation of four physicochemical properties: solubility (and solubility product), supersaturation, nucleation, and crystal growth. The solubility of the inorganic mineral is a critical factor in determining the thermodynamic conditions for precipitation, and the extent that a solution is out of equilibrium is given by the level of supersaturation, which in turn influences rates of nucleation and growth. Biological control over supersaturation levels is generally achieved by partitioning discrete regions of the intracellular or extracellular environment so that these enclosed spaces can be chemically separated from the background activity of the cells. The boundaries of these spatially delineated sites also act as physical constraints that regulate the size, volume, and shape of the mineralization environment.

Intracellular mineralization occurs within sealed membrane-enclosed compartments (vesicles). The individual vesicles are assembled from lipids and have selective permeabilities to certain metal ions and molecules. After mineralization, the vesicles and their mineral products may remain inside the cell or be translocated out of the cell and further processed on the cell surface or in the extracellular space. For example, in the shells and spines of adult sea urchins, the vesicles are produced from individual cells and fuse together to form an extensive organic framework that serves as the underlying structural foundation for inorganic precipitation.

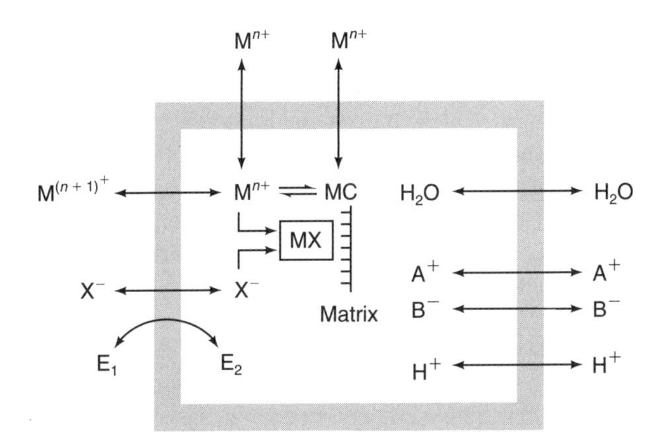

Fig. VI.5.
Generalized strategies for controlling supersaturation in biomineralization. The mechanisms can be either direct (membrane pumps, complexation, and enzymatic regulation) or indirect (H_2O, H^+, and ion fluxes).

Supersaturation levels in the vesicles are controlled by a variety of membrane transport processes, which along with several examples are listed below (Fig. VI.5).

1. Control of ionic concentrations: For example, the concentration of Ca^{2+} ions can be controlled by cation-gated systems via membrane-bound pumps and molecular channels. For iron, the transport process may be facilitated by redox changes at the membrane surface [e.g., reduction of Fe(III) to Fe(II) may facilitate transport into a vesicle space (see Chapter V)].
2. Inhibition and release: For example, ion-complexation of transported cations with ligands such as citrate or pyrophosphate may be followed by controlled release of the ion through the destabilization of the complex. Alternatively, the stabilization and destabilization of mineralized precursors, such as amorphous phases, can generate controlled pulses of supersaturation.
3. Enzymic regulation of ion concentrations: For example, depletion of H_2CO_3 by carbonic anhydrase can be coupled to calcification processes, according to the equilibrium:

$$Ca^{2+}(aq) + 2\,HCO_3^-(aq) \rightleftharpoons CaCO_3 + H_2CO_3$$

4. Increases in activity products by changes in the background ionic strength of the enclosed space: For example, selective transport of ions such as Na^+ and Cl^- can increase or decrease the ionic strength.
5. Increases in the activity products and concentrations of ions by water extrusion from localized volumes: For example, silica condensation is induced in the leaves of plants by transpiration.
6. Changes in pH by proton pumping or cellular metabolism: Changes in pH will influence the acid–base equilibria of oxyanions involved in biomineralization, such as HCO_3^-/CO_3^{2-} and HPO_4^{2-}/PO_4^{3-}, as well as the hydrolysis of metal ions like Fe(III). For example, magnetite (Fe_3O_4) formation is highly sensitive to changes in redox and pH,

$$Fe_2O_3.nH_2O(s) + FeOH^+(aq) \rightleftharpoons Fe_3O_4(s) + H^+ + nH_2O$$

In extracellular environments, the mineralization compartments are produced within a biopolymeric structure formed from insoluble organic macromolecules. The inorganic crystals nucleate and grow preferentially with confined spaces delineated by the organized framework. For example, hydroxyapatite crystals are located within regular gaps that occur between collagen fibers in bone, whereas aragonite crystals in the seashell grow between an ordered array of insoluble protein sheets. Supersaturation levels in these environments are controlled by ion transport from neighboring cells or controlled dissolution of precursor mineral phases, such as amorphous calcium phosphate, that are initially deposited on the organic matrix.

VI.3.3. Organic Matrix-Mediated Biomineralization

The widescale involvement of insoluble organic structures in controlled biomineralization is often referred to as organic matrix-mediated biomineralization. The organic matrix is specifically synthesized under genetic regulation, before and sometimes during the biomineralization process. The matrix is usually a polymeric framework that consists of a complex assemblage of macromolecules (e.g., proteins, polysaccharides).

In extended structures, such as bone and shells, insoluble polymeric structures are constructed before the onset of inorganic precipitation. In bone, the primary building unit is based on the nucleation of calcium phosphate in nanospaces organized within a supramolecular assembly of collagen fibrils. The structure of collagen is based on a helical arrangement of three noncoaxial helical polypeptides and is stabilized by interchain hydrogen bonding. This arrangement is considered to be the basic subunit for the higher order assembly of collagen fibrils that can take place by a variety of permutations depending on the chemical composition of the surrounding medium. Type I collagen, which is produced by bone cells (osteoblasts), has a specific fibril structure—called the revised quarter-stagger model—in which the triple helical macromolecules are staggered by 64 nm along their long axis (Fig. VI.6). Each macromolecule is divided into five zones, the first four of equal length (64 nm) and the fifth only 25 nm. Adjacent molecules are transposed by 64 nm along their length by strong intermolecular cross-linking and this leaves a 40×5-nm space (the hole zone) between the ends of each molecule. In three-dimensions (3D), adjacent hole zones overlap to produce grooves that are organized in parallel rows at the same height in the fibrils. These sites are considered to be the loci for the specific nucleation and growth of oriented plate-like crystals of hydroxyapatite only $45 \times 20 \times 3$ nm in dimension.

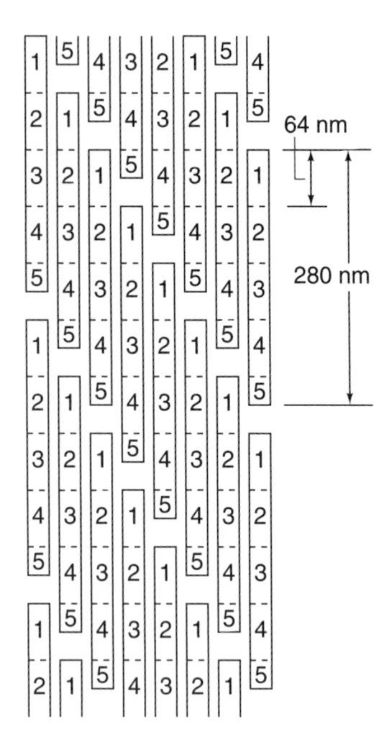

Fig. VI.6.
The revised quarter-stagger model for collagen.

Many shells and teeth are constructed within frameworks that may be lamellar, columnar, or reticular. In each case, there appear to be two key features in the assembly. First, a relatively inert hydrophobic structural frame is built from insoluble proteins and/or polysaccharides (chitin). The proteins have an antiparallel β-pleated sheet silk-like structure that consists of polypeptide chains high (~85%) in glycine (Gly), alanine (Ala), and serine (Ser) in the form of corrugated sheets linked by strong interchain hydrogen bonds. Second, acidic macromolecules, rich in carboxylate (typically 30% aspartic and 17% glutamic acids) or phosphate residues derived from the post-translational phosphorylation of the amino acids Ser and threonine (Thr), are assembled on the hydrophobic scaffold, often in association with sulfated polysaccharides. In terms of biomineral processing, the shell matrix is thought to control the polymorph—aragonite rather than calcite in the nacreous layer—as well as the crystallographic orientation of the aragonitic tablets. The antiparallel arrangement of adjacent polypeptide chains results in a periodic array of amino acid side chains that are potential nucleation sites for inorganic mineralization. This ordered array implies that the surface of the protein sheets contains amino acid groups that interact with specific crystal faces of the aragonite lattice.

The nature of the interface between the crystal surface of the mineral phase and the underlying surface of the organic matrix is considered to be of pivotal importance, particularly in the nucleation of biominerals.

VI.3.4. Matrix-Mediated Nucleation

Many biominerals are crystallographically aligned with regard to the underlying organic matrix. For example, the aragonitic tablets in the nacreous layers of the shell are preferentially nucleated so that the *c* axis of the unit cell is perpendicular to the plane of the organic sheets. The rate of nucleation is related to the corresponding activation energy required to form the solid–liquid interface of the nuclei present in a supersaturated solution. In general, the role of the organic matrix is to lower this activation energy and therefore increase the rate of nucleation of the preferentially oriented mineral phase.

We can consider the role of an organic matrix in inorganic nucleation to be analogous to that of an enzyme in solution, with the incipient inorganic nucleus taking the place of the corresponding substrate molecule. In both cases, the lowering of the activation energy depends on electronic and stereochemical factors between the two components acting over molecular distances. But in nucleation, longer range interactions must also explicitly be considered. In particular, the electrostatic forces and space symmetries of ionic surfaces indicate that factors such as lattice geometry, spatial charge distribution, hydration, defect states, and surface relaxation also need to be considered at the inorganic–organic interface.

Figure VI.7 shows diagrammatically how an organic matrix can be used to influence the activation energy and control structural aspects of nucleation in biomineralization. Consider a mineral nucleated in the absence (state 1) or presence (state 2) of an organic surface. The activation energy corresponding to state 2 is always lower than state 1 because the organic surface promotes nucleation. The inorganic phase also has two possible structural states, A and B, where A is the more kinetically favored in the absence of the organic matrix. For example, A and B could be two different structures (e.g., calcite and aragonite polymorphs) or two different crystal faces (for the case of oriented nucleation). Then, depending on the relative changes in the activation energies of nucleation for A and B in the presence of the organic matrix, there are three possibilities:

1. Promotion of nonspecific nucleation in which both polymorphs (or crystal faces) have reduced activation energies because of the presence of the matrix surface, but there is no change in the outcome of mineralization.
2. Promotion of structure-specific nucleation of polymorph (or crystal face) B due to more favorable crystallographic recognition at the matrix surface.
3. Promotion of a sequence of structurally nonspecific-to-highly specific nucleation depending on how the degree of recognition of nuclei A and B and the reproducibility of matrix structure change with external factors such as genetic, metabolic, and environmental processes.

The outcome of the processes illustrated in Fig. VI.7 very much depends on the nature and extent of molecular recognition at the inorganic–organic interface. The specific lowering of the activation energy of nucleation in the presence of the organic matrix implies charge, structural, and stereochemical complementarity between ions in the mineral phase and functional groups on the protein surface. The simplest form of interfacial complementarity involves similarities in charge matching and ion coordination environments between ions in the mineral phase and ions bound to organic ligands exposed at the organic surface. However, because of the long-range ordering of crystal surfaces, interfacial recognition based on ion binding and association must occur over distances larger than the first coordination sphere. Clustering of binding sites is therefore required in order to produce localized regions of high spatial

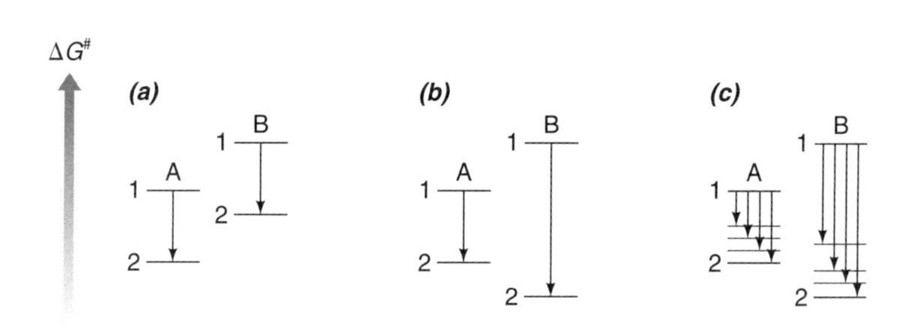

Fig. VI.7. Diagrammatic representation of the activation energies, $\Delta G^{\ddagger}$, of nucleation, and of inorganic minerals in the absence (state 1) and presence (state 2) of an organic surface involved in biomineralization. See text for details.

(surface) charge density that can electrostatically accumulate and stabilize the number of ions that are sufficient for nucleation. This is sometimes termed an "ionotropic" mechanism.

The structural rearrangement of the dispersed ionic cluster into the more ordered structure of the condensed phase is dependent on the dynamical freedom imposed by the matrix interactions. For example, high-affinity binding at the matrix surface might favor an amorphous nucleus, whereas high capacity, low-affinity interactions could facilitate the movement of ions into periodic lattice sites of a crystalline polymorph. Clearly, the shape of the nucleation site can influence the spatial charge distribution of functional groups on the organic surface, and hence their ability to stabilize inorganic clusters. For example, whereas concave sites can give rise to 3D clustering, convex surfaces spread out and dissipate the charge density. Planar substrates, on the other hand, have the additional important property that they can generate periodic arrangements of functional groups and bound ions that can be used as a template for the oriented nucleation of a specific crystal face.

In this case, the fundamental idea is that there is geometric matching (epitaxy) between the lattice spacings of ions in certain crystal faces and functional groups that are periodically predisposed across the organic surface (Fig. VI.8). So far, the best understood system is that based on X-ray and electron diffraction studies of thin flakes of nacre from the mollusk shell. The results show that the *a* and *b* axes of the antiparallel β-pleated sheet of the matrix are aligned with the *a* and *b* crystallographic directions in the (001) face of aragonite. If we now compare the distances between Ca^{2+} ions in this crystal face with the matrix periodicity, there is good matching along the *a* axes (0.496 and 0.47 nm, respectively). The lattice match along the corresponding *b* axes is not as high (0.797 and 0.69 nm, respectively), but the periodicities are essentially commensurate over longer distances such as seven Ca^{2+} ions (4.8 nm). Because the shell proteins are enriched in aspartic (Asp) acid, the model proposes that alignment of these residues along the matrix in the form of repeated domains of [Asp-X] (where X = neutral amino acid) provides a strong correlation between the carboxylate binding sites and the lattice arrangement of Ca^{2+} ions required for nucleation of the (001) surface of aragonite. Moreover, because calcium binding to carboxylate groups is generally cooperative involving at least two or three ligands, there needs to be enough flexibility in the geometric arrangement to accommodate these stereochemical requirements for oriented nucleation.

VI.3.5. Growth of Biominerals

In pure solutions, crystal growth of inorganic minerals occurs by addition of ions and clusters to active sites, such as steps and kinks, on the crystal surface, and terminates when the supersaturation level falls to the equilibrium solubility limit. Clearly, biological fluids contain countless components and many of these can interfere with the growth of biominerals. The major influence of extraneous ions and molecules on the rate of crystal growth is inhibition through binding of the additives to the active sur-

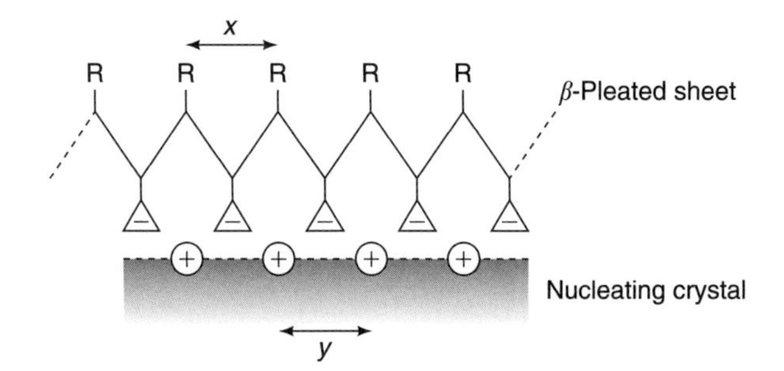

Fig. VI.8.
Geometric matching (epitaxy) in biomineralization. Cation–cation distances in one specific crystal face and polymorph structure are commensurate with the spacing of periodic-binding sites on the organic surface.

face sites. Thus, many soluble proteins and polysaccharides involved with biomineralization will slow down the rate of growth of analogous minerals grown under laboratory conditions.

Another possibility is that these soluble macromolecules can influence the course of crystal growth by interference with phase transformations that involve the structural modification of precursor phases such as amorphous or hydrated phases, or other crystalline polymorphs with reduced thermodynamic stability. For example, the growth of hydroxyapatite (HAP) crystals in biology is often preceded by the nucleation of amorphous calcium phosphate (ACP) that subsequently transforms to octacalcium phosphate (OCP), and finally with time to HAP. How far the phase transformations proceed along a pathway of intermediates depends on the solubilities of the amorphous precursor and polymorphic intermediates and the free energies of activation for their interconversions. The latter can be significantly influenced by organic macromolecules in the surrounding environment.

There are many known biological inhibitors of HAP formation that act by stopping the transformation of ACP to crystalline intermediates or, in a few cases, OCP to HAP. Their inhibitory action can be overcome in the case of phosphometabolites and proteins by the addition of enzymes for which these molecules act as natural substrates. This suggests that "promotion" of various polymorphs does not occur in the literal sense, but that the mineralization pathway is controlled through intermittent release of a system under chemical repression. (In fact, most biological fluids are supersaturated with respect to inorganic minerals.) The crystallographic selection between minerals such as vaterite, aragonite and calcite, or polymorphs of calcium phosphate, then becomes dependent on the extent of the sequence of chemical inhibition and release within the biomineralization site (Fig. VI.9). Modulation of the sequence in time and biological location allows development of alternative crystallographic structures during the stages of growth and changes in mineral polymorphs in different regions of the large-scale structures such as the seashell.

One important consequence of this multiphase pathway for biomineralization is that the selection of the final mineral structure can arise either from the chemical regulation of phase-transformation processes by soluble macromolecules or structural control involving similar molecules immobilized on an organic matrix, or both. Indeed, there is growing evidence that amorphous granules containing high levels of inorganic and organic components are prevalent in the early stages of many biomineralization systems. These structures are often formed at some distance from the mineralization site and subsequently transported to the organic matrix where they aggregate in large numbers. They are then destabilized (dissolved) to produce a pulse of supersaturation that initiates the phase-transformation pathway. Depending on the system, this may be a single event or an episodic process at the mineralization front.

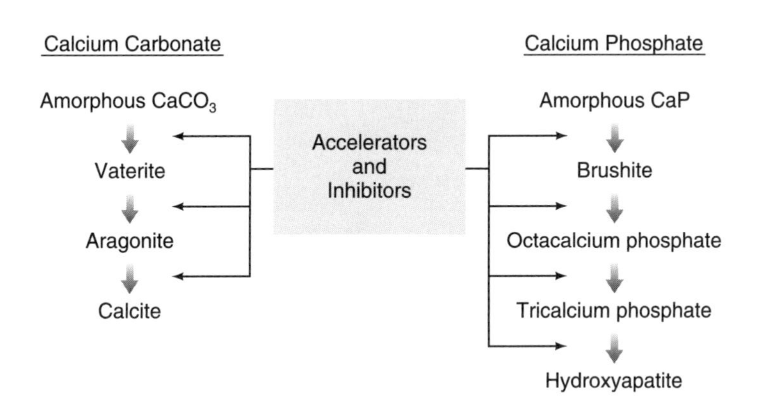

Fig. VI.9. Control of polymorph structure in biomineralization by controlled release of molecules that act as accelerators or inhibitors of crystal growth and phase transformation.

VI.3.6. Pattern and Form in Biomineralization (Morphogenesis)

The elaborate shapes of biominerals arise from patterned organic assemblies of vesicles and frameworks that originate from biosynthetic pathways and are subsequently "turned to stone" through the chemical medusa of biomineralization. Often, the inorganic mineral becomes a direct replica of a patterning process that has terminated, and is therefore analogous to a cast produced in a mold. Alternatively, mineralization and vesicle shaping proceed in concert, with the mineralization front remaining some distance behind the developing organic structure. Under these circumstances, synergistic interactions between the mineral and vesicle induce changes in the patterning process through coupling of the inorganic and organic assembly processes. For example, as the mineral begins to dominate the replicated morphology, there is no longer any requirement for the vesicle to be held in place by associated biological structures in the cell.

The general features of pattern formation in biomineralization are illustrated in Fig. VI.10 in which the dynamic shaping of vesicles takes place by anchoring the lipid membrane through the use of directing agents, such as microtubules (protein cables) to an underlying scaffold such as the cell wall. This phenomenon is widespread both among crystalline and amorphous biominerals. The intracellular space is criss-crossed with microskeletal networks and associated stress fields, such that the equilibrium spherical shape of a vesicle membrane is readily distorted by mechanical and structural forces operating locally and at a distance. Empirically, it appears that the shaping of a vesicle can be directed by two perturbing force fields acting either tangentially along the surface of the cell wall or an internal organelle, or radially along structural filaments based on a protein, tubulin. Independently, or in combination, these force fields can generate a wide range of mineralized patterns.

As an example, the diversity of patterns seen in the porous silica shells of diatoms can be explained by geometrical deviations in the close packing of vesicles against the curved surface of the cellular membrane. The vesicles are arranged into a thin polygonal foam with organized interstitial spaces that become mineralized. The exquisite patterning of the microskeletons can then be rationalized using a simple geometric model in which the silica framework is a direct replica of the spaces around and between the vesicles, and the pores are the voids created by removal of the vesicles after biomineralization. The process is not controlled by surface tension, as originally postulated, but is the consequence of programmed cellular organization within the interstitial spaces. In the diatom *Coscinodiscus*, for example, tubular vesicles are secreted and assembled along with microtubules in the gaps between the large areolar

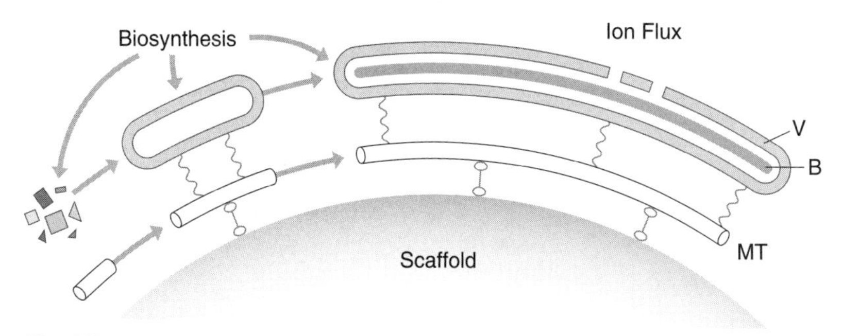

Fig. VI.10.
Cell walls, intracellular organelles, and cellular assemblages can act as scaffolds for the assembly of microtubules (MT), which in turn are used as directing agents for the patterning of vesicles (V) involved in biomineralization (B).

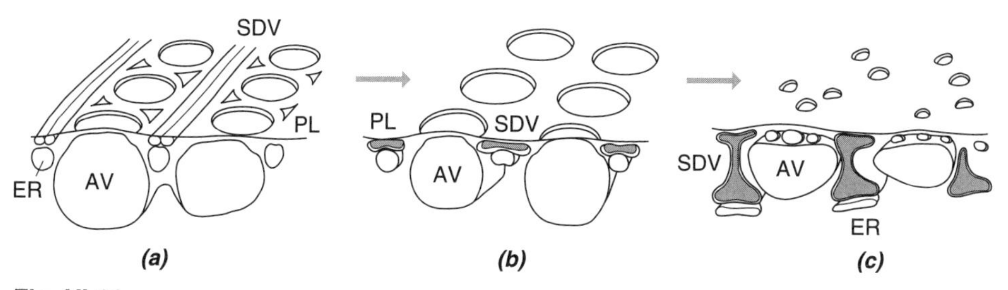

Fig. VI.11.
Illustration of the key stages in the formation of the siliceous diatom exoskeleton. (*a*) Silica deposition vesicles (SDV) are preorganized with microtubules around the boundary spaces of large areolar vesicles (AV) attached to the plasmalemma (PL). (*b*) The SDVs are mineralized with amorphous silica to give a patterned porous wall. (*c*) The mineralized wall is thickened by extension of each SDV in association with the endoplasmic reticulum (ER). In some diatoms, detachment and retraction of the areolar vesicles from the plasmalemma results in infiltration with new SDVs and further mineralization of the pore spaces.

vesicles prior to mineralization (Fig. VI.11). Silica deposition is then confined tangentially to this tubular system such that an open geometric mesh of mesopores is established. The vesicles then detach and withdraw from the plasmalemma. The resulting space is infiltrated with smaller vesicles that produce a thin-patterned shell of silica across the top of the mesopores.

VI.3.7. Higher Order Assemblies

Many biomineral structures are associated with a variety of constructional mechanisms involving the cellular processing of shaped and patterned biominerals into higher order assemblies with micro- or macroscopically organized architecture. Construction of these higher order architectures occurs within or outside the cellular space, or on the cell surface by sequential and concerted processes of organization.

As an example, the formation of magnetite crystals in the vesicles of bacteria does not by itself result in magnetotaxis because individual crystallites are randomly ordered and a compass needle requires north and south polarity. If the vesicles are disorganized, then the magnetic fields from different crystals cancel out and there is no permanent dipole moment. The bacteria solve this problem by organizing the crystals into relatively rigid chains by arranging the vesicles accordingly along the cell wall. In this sequential process, the vesicles are synthesized and added to the ends of the growing chain. In other single-celled organisms called *choanoflagellates*, curved silica rods are first formed within intracellular vesicles. The rods are then transported sideways through the cell membrane and out into the extracellular space, where they act as building blocks for the construction of an open-ended basket-like framework. Approximately 150 silica girders are finally locked into place by a "glue" of silica and organic material during construction.

Finally, macroscopic architectures, such as bone, shells, and teeth, are associated with extended extracellular matrices and associated tissues containing patterns of specialized cells such as osteoblasts (bone formation) and osteoclasts (bone resorption) in bone matrix, layers of epithelial cells in shell mantle, and ameloblasts in enamel. The intimate association with these cells produces complex higher order composite materials that have hierarchical structures often capable of being remodeled by active cellular processing. The structural hierarchy of bone does not stop with the supramolecular structure of the collagen fibrils, but is operational at all levels throughout the system (Fig. VI.12). In fact, the collagen fibril only constitutes the

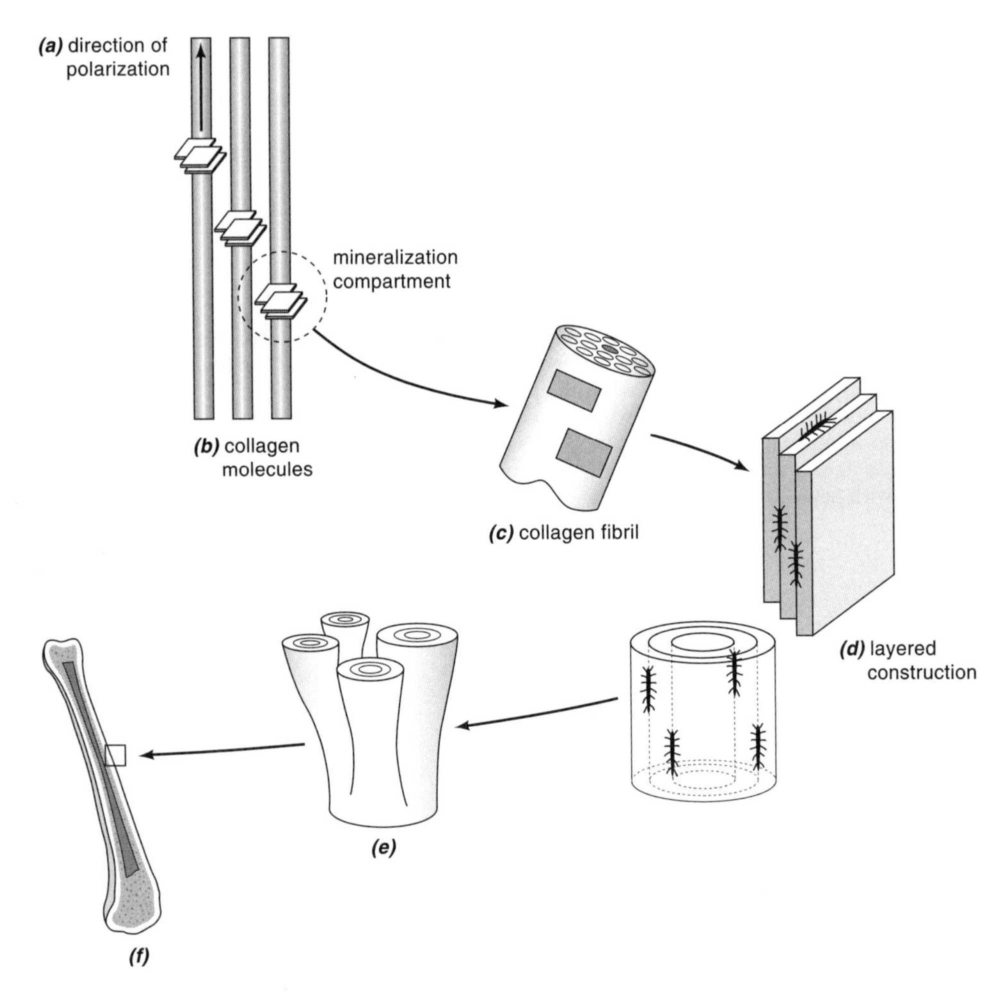

Fig. VI.12.
The hierarchical structure of bone. [Reprinted with permission from VCH-Wiley Publishers.]

second of five levels of embedded architecture in the mature bone. The collagen fibrils are themselves organized into sheets that are concentrically layered into a structure called the osteon. Each layer of the osteon has its constituent fibrils oriented in alternate directions, like plywood. On a longer scale, the osteons are grouped to give various microstructures (woven bone, lamellar bone, haversian bone, etc.) each of which has the same underlying building blocks, but different spatial organizations according to structure–function relationships within the whole bone.

The macroscopic shape of the whole bone is determined by cellular differentiation during formation of the embryo. For example, the long bone of the leg is shaped prior to mineralization in the form of a soft biodegradable model of the structure composed of cartilage—a highly hydrated nonmineralized matrix of cells and macromolecules. The cells that first deposit the preformer cartilage then assemble around it and differentiate into osteoblasts that secrete collagen around the model. Bone mineralization occurs in the collagen to give a mineralized collar around the cartilage. The bone collar starves the cells inside the matrix causing degradation, which forms a hollow cavity filled with marrow. The newly formed bone is then remodeled into different microstructures. Remodeling is influenced by gravitational and mechanical force fields placed on the new bone, under the guidance of bone cells that reside within the mineralized structure where they appear to act as an interconnected network of biological "strain gauges".

VI.4. Conclusions

We have described how biominerals, such as calcium carbonate, calcium phosphate, amorphous silica, and iron oxides, are deposited as functional materials in a wide range of organisms. Many of these biominerals have remarkable levels of complexity, and we have reviewed some of the general principles that help us to understand how such exquisite structures can be produced. Several processes—*boundary-organized biomineralization, organic matrix-mediated nucleation and growth, morphogenesis, and higher order assembly*—have been described, each of which is underpinned by chemical control. Bioinorganic chemists therefore continue to play a crucial role in extending our knowledge of biomineralization, and in related fields such as biomimetic or bioinspired materials research.

Bibliography

Textbooks

Mann, S., *Biomineralization: Principles and Concepts in Bioinorganic Materials Chemistry,* Oxford Chemistry Masters, 5, Oxford University Press, New York, 2001.

General Research Books

Bäuerlein, E., *Biomineralization: From Biology to Biotechnology and Medical Application.* Wiley-VCH, Weinheim, 2000.

Lowenstam, H. A. and Weiner, S., *On Biomineralization,* Oxford University Press, New York, 1989.

Mann, S., Webb, J. and Williams, R. J. P., Eds. *Biomineralization: Chemical and Biochemical Perspectives,* VCH Verlagsgesellschaft, Weinheim, 1989.

Simkiss, K. and Wilbur, K., *Biomineralization: Cell Biology and Mineral Deposition,* Academic Press, San Diego, 1989.

Miller, A., Phillips, D., and Williams, R. J. P., "Mineral Phases in Biology", *Philos. Trans. R. Soc. London Ser. B,* **304,** 409–588, Royal Society, London, 1984.

Principles and Concepts

Mann, S., "Biomineralization: The Form(id)able Part of Bioinorganic Chemistry!", *J. Chem. Soc. Dalton Trans.,* 3953–3961 (1997).

Berman, A., Hanson, J., Leiserowitz, L., Koetzle, T. F., Weiner, S., and Addadi, L., "Biological Control of Crystal Texture: A Widespread Strategy for Adapting Crystal Properties to Function", *Science,* **259,** 776–779 (1993).

Mann, S., "Molecular Tectonics in Biomineralization and Biomimetic Materials Chemistry", *Nature* (*London*), **365,** 499–505 (1993).

Addadi, L. and Weiner, S., "Control and Design Principles in Biological Mineralization", *Angew. Chem. Int. Ed. Engl.,* **31,** 153–169 (1992).

Mann, S., "Molecular Recognition in Biomineralization", *Nature* (*London*), **332,** 119–124 (1988).

Weiner, S., "Organization of Extracellularly Mineralized Tissues: A Comparative Study of Biological Crystal Growth", *Crit. Rev. Biochem.,* **20,** 365–408 (1986).

Lowenstam, H. A., "Minerals Formed by Organisms", *Science,* **211,** 1126–1131 (1981).

Calcium Biominerals

Kono, M., Hayashi, N., and Samata, T., "Molecular Mechanism on the Nacreous Layer Formation in *Pinctada maxima*", *Biochem. Biophys. Res. Commun.,* **269,** 213–218 (2000).

Young, J. R., Davis, S. A., Bown, P. R., and Mann, S., "Coccolith Ultrastructure and Biomineralisation", *J. Struct. Biol.,* **12,** 195–215 (1999).

Belcher, A. M., Wu, X. H., Christensen, R. J., Hansma, P. K., Stucky, G. D., and Morse, D. E., "Control of Crystal Phase Switching and Orientation by Soluble Mollusk-Shell Proteins", *Nature* (*London*), **381,** 56–58 (1996).

Aizenberg, J., Hanson, J., Koetzle, T. F., Leiserowitz, L., Weiner, S., and Addadi, L., "Biologically Induced Reduction in Symmetry: A Study of Crystal Texture of Calcitic Sponge Spicules", *Chem. Eur. J.,* **7,** 414–422 (1995).

Krampitz, G. and Graser, G., "Molecular Mechanisms of Biomineralization in the Formation of Calcified Shells", *Angew. Chem. Int. Ed. Engl.,* **27,** 1145–1156 (1988).

Weiner, S. and Traub, W., "Organization of Hydroxyapatite Crystals within Collagen Fibrils", *FEBS Letts.,* **206,** 262–266 (1986).

Currey, J., *The Mechanical Adaptations of Bones,* Princeton University Press, New Jersey, 1984.

Miller, A., "Collagen: The Organic Matrix of Bone", *Philos. R. Soc. London Ser. B,* **304,** 455–477 (1984).

Watabe, N., "Crystal Growth of Calcium Carbonate in the Invertebrates", *Prog. Cryst. Growth Charact.,* **4,** 99–147 (1981).

Weiner, S. and Traub, W., "X-Ray Diffraction Study of the Insoluble Organic Matrix of Mollusk Shells", *FEBS Letts.,* **111,** 311–316 (1980).

Silica Biominerals

Perry, C. C., Keeling-Tucker, T., "Biosilicification: The Role of the Organic Matrix in Structure Control", *J. Biol. Inorg. Chem.*, **5**, 537–550 (2000).

Kröger, N., Deutzmann, R., and Sumper, M., "Polycationic Peptides from Diatom Biosilica that Direct Silica Nanosphere Formation", *Science,* **286,** 1129–1132 (1999).

Leadbeater, B. S. C., "Silica Deposition and Lorica Assembly in Choanoflagellates", in *Biomineralization in lower plants and animals,* Leadbeater, B. S. C. and Riding, R. Eds., Systematics Association, Vol. 30, Oxford University Press, Oxford, 1986, pp. 345–359.

Volcani, B. E. and Simpson, T. L., "*Silicon and Siliceous Structures in Biological Systems*", Springer Verlag, Berlin, 1982.

Iron Oxides

Frankel, R. B. and Blakemore, R. P., Iron Biominerals, Plenum Press, New York, 1991.

Harrison, P. M., Andrews, S. C., Artymiuk, P. J., Ford, G. C., Guest, J. R., Hirzmann, J., Lawson, D. M., Livingstone, J. C., Smith, J. M. A., Treffry, A., and Yewdall, S. J., "Probing Structure–Function Relations in Ferritin and Bacterioferritin", *Adv. Inorg. Chem.*, **36,** 449–486, (1991).

Ford, G. C., Harrison, P. M., Rice, D. W., Smith, J. M. A., Treffry, A., White, J. L., and Yariv, J., "Ferritin: Design and Formation of an Iron-Storage Molecule", *Philos. Trans. R. Soc. London Ser. B,* **304,** 551–565, (1984).

Frankel, R. B., Papaefthymiou, G. C., Blakemore, R. P., and O'Brien, W. D., "Fe_3O_4 Precipitation in Magnetotactic Bacteria", *Biochim. Biophys. Acta,* **763,** 147–159 (1983).

CHAPTER VII

Metals in Medicine

Peter J. Sadler, Christopher Muncie, and Michelle A. Shipman
School of Chemistry
University of Edinburgh, UK

Contents

VII.1. Introduction

Inorganic compounds have been used in medicine for many centuries, but often only in an empirical fashion with little attempt to design the compounds used and with little or no understanding of the molecular basis of their mechanisms of action. This chapter is concerned largely with compounds containing metals. The aim is to show that there is enormous scope for the design of novel therapeutic and diagnostic metal compounds. Compounds that are in clinical use and others that show promise are described. Where possible, proposals are made for the molecular basis of their mechanism of action by relating their chemistry to biological and pharmacological activity.

As indicated in Fig. VII.1, key areas in the design of active compounds are the control of toxicity (side effects) and targeting of the metal to specific tissues, organs, or cells where activity is needed. The toxicity of an element will depend on the element itself, its oxidation state, and the nature and number of coordinated ligands, as well as on the dose, mode of administration, and biochemical status of the host. Moreover, the effect of one element may depend on the presence or availability of another.

It is seldom useful to describe elements as "toxic" or "nontoxic". Even so-called toxic compounds can usually be tolerated in low doses, and may exhibit therapeutic effects within narrow concentration ranges, and biochemically essential elements can become toxic at high doses. The *Bertrand Diagram* (Fig. VII.2) schematically summarizes this situation.

At least 24 elements are currently thought to be essential for mammalian life: H, C, N, O, F, Na, Mg, Si, P, S, Cl, K, Ca, V, Mn, Fe, Co, Ni, Cu, Zn, Se, Mo, Sn, and I. As implied, this list may not be complete. For example, B and Cr may

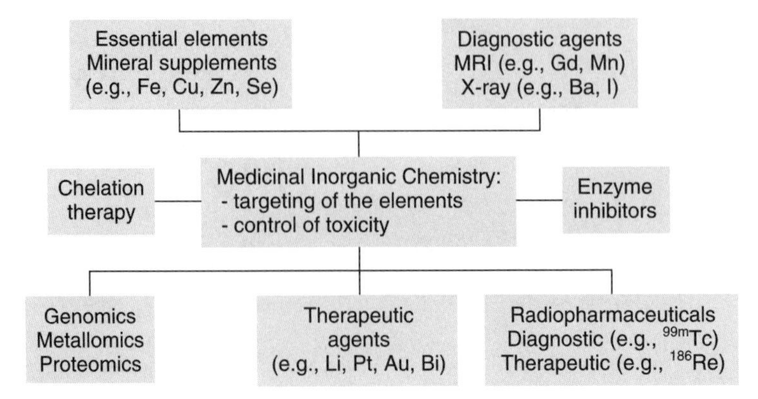

Fig. VII.1.
Some of the areas of medicinal inorganic chemistry.

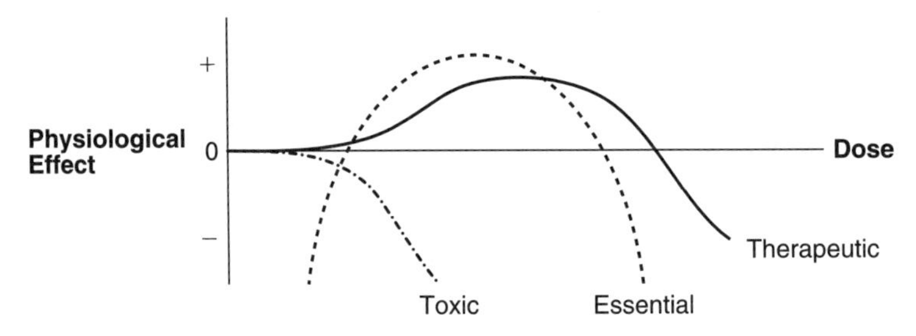

Fig. VII.2.
A Bertrand diagram, showing that physiological and toxic effects are a continuum. [Adapted from Bertrand, G. (1912) *8th Int. Congr. Appl. Chem.* **28,** 30.]

turn out to be essential, and it has been suggested that Si is essential only to prevent Al from being toxic. The mammalian biochemistry of several essential elements is poorly understood (e.g., V, Ni, and Sn). For Co only one specific compound, the coenzyme vitamin B_{12}, appears to be essential. Selenium provides a good example of the importance of speciation (the nature of the molecule or ion that contains the element). Selenium is an essential element and yet some of its compounds are highly toxic (e.g., H_2Se).

Both essential and nonessential metals can be used in therapy and diagnosis and examples of compounds in current clinical use are listed in Table VII.1. It is important to ask which parts of the compound are essential for activity: the metal itself, the ligands, or the intact complex of metal plus at least some of the ligands? Many metallodrugs are "prodrugs": They undergo ligand substitution and/or redox reactions before they reach the target site.

VII.2. Metallotherapeutics

VII.2.1. Anticancer Therapeutic Agents

Cancer is one of the top three killers worldwide and is a difficult disease to treat. It is hard (if not impossible) to find drugs that are both effective and have low toxicity to the human body as a whole.

Table VII.1
Some Metal Compounds in Clinical Use

Compound Example (Brand Name)	Function	Comment
ACTIVE COMPLEXES[a]		
cis-$[Pt^{II}Cl_2(NH_3)_2]$ (Cisplatin)	Anticancer	Trans isomer is inactive
$[Gd^{III}(DTPA)(H_2O)]^{2-}$ (Magnevist)	Extracellular MRI[b] contrast agent	Low toxicity
$[^{99m}Tc^{I}(CNCH_2C(CH_3)_2OCH_3)_6]^+$ (Cardiolite)	Myocardial imaging	Positively charged complex taken up by the heart
Vitamin B_{12}	Coenzyme	Deficiency causes pernicious anemia
ACTIVE METALS		
Li_2CO_3	Prophylaxis bipolar disorders	Li(I) forms weak complexes, labile
Au^{I}(thiomalate) (Myocrisin)	Antirheumatoid arthritic	Facile thiol exchange on Au(I)
Ammonum potassium Bi^{III} citrate (De-Nol)	Antibacterial, antiulcer	Strong binding of Bi(III) to thiols, facile exchange
$Na_2[Fe^{II}(CN)_5NO]\cdot 2H_2O$ (Nipride)	Hypotensive	Releases NO, relaxes vascular muscle
Bleomycin	Anticancer	Requires Fe for DNA[c] attack
p-Xylyl-bicyclam·8HCl (AMD3100)	Anti-HIV[d]	May bind metals *in vivo*
$CaCO_3$, $Mg(OH)_2$	Antacids	Slow release of alkali
$La_2(CO_3)_3$ (Fosnol)	Chronic renal failure	Reduces phosphate absorption ($LaPO_4$ insol)

[a] Diethylenetriaminepentaacetate = DTPA.

[b] Magnetic resonance imaging = MRI.

[c] Deoxyribonucleic acid = DNA.

[d] Human immunodeficiency virus = HIV.

Discovery of Cisplatin. The Pt(II) complex cisplatin (Chart VII.1) is one of the most widely used anticancer drugs, particularly for the treatment of testicular cancer and ovarian carcinoma. Cisplatin was first approved for the clinical treatment of genitourinary tumors in 1978 and is often used in combination with one, two, three, or even

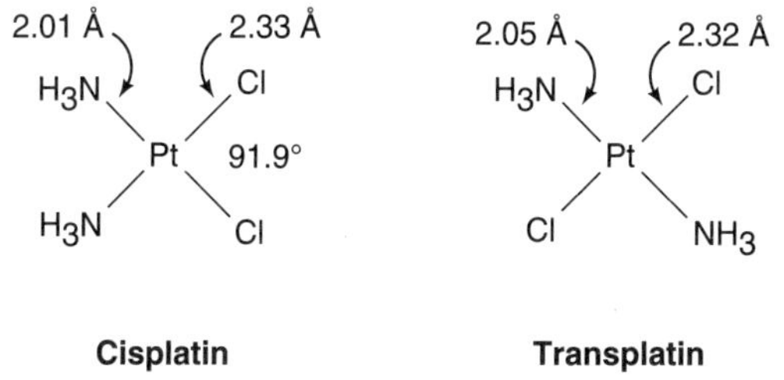

2.01 Å 2.33 Å
H_3N Cl
Pt 91.9°
H_3N Cl
Cisplatin

2.05 Å 2.32 Å
H_3N Cl
Pt
Cl NH_3
Transplatin

Chart VII.1.
Structural differences between cisplatin and transplatin.

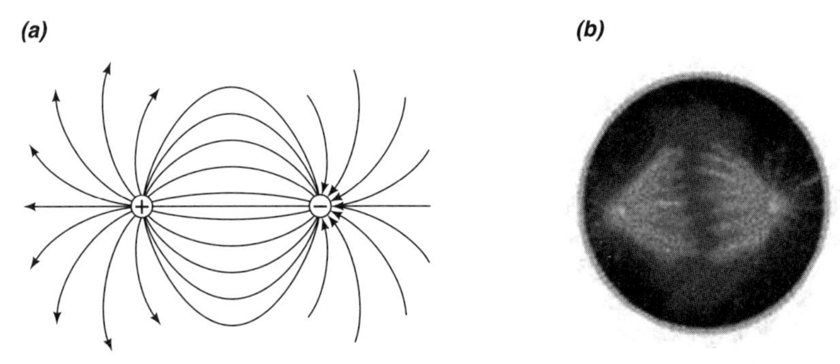

Fig. VII.3.
(*a*) The electric field lines for two equal and opposite point charges (an electric dipole)
(*b*) Mitotic spindle formation during division of a eukaryotic cell. (Reprinted by permission from *Nature Cell Biology,* Vol. 3, No. 1, pp. E28–34. Copyright © 2001 Macmillan Magazines Ltd.)

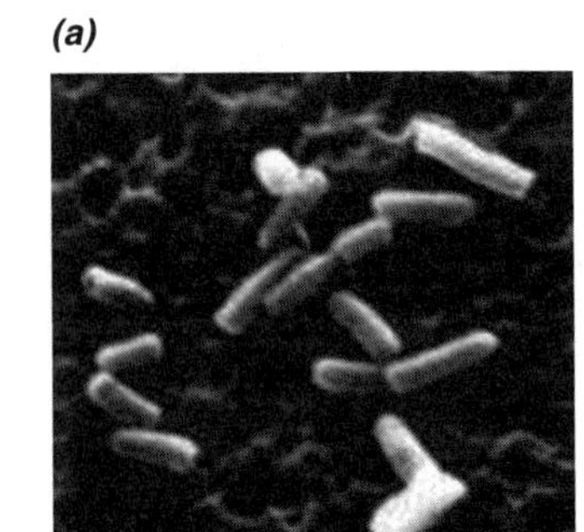

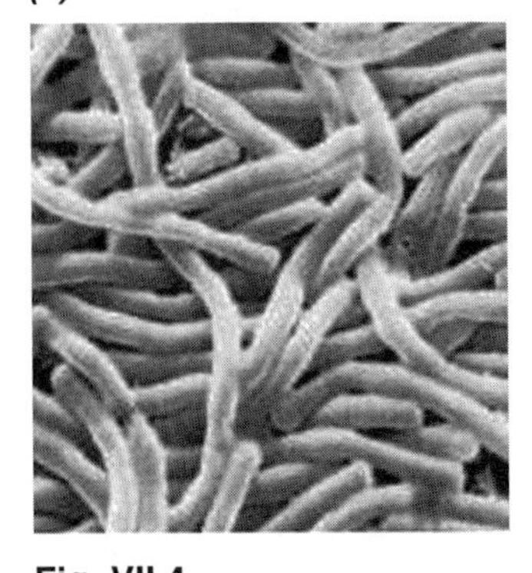

Fig. VII.4.
(*a*) *E. coli* bacteria, and (*b*) the filamentous growth of the same bacteria in the presence of cisplatin. (Pictures courtesy of D. Beck, Bowling Green State University.)

four organic anticancer drugs, such as 5-fluorouracil, cyclophosphamide, or gemcitabine. Testicular cancer in particular has become largely curable since the introduction of cisplatin. The anticancer activity of cisplatin was discovered serendipitously by Barnett Rosenberg, a biophysicist working at Michigan State University in 1965. Rosenberg devised an experiment to investigate the effect of electric fields on cell division, because he thought that spindle formation by eukaryotic cells seen by microscopy during cell division resembled electric field lines between two equal and opposite point charges, Fig. VII.3. Rosenberg and his colleague Loretta VanCamp passed an alternating electric current through two Pt electrodes immersed in a beaker containing cells in a cell growth medium (in fact, they used *Escherichia coli* bacteria, which do not show spindle formation!). Instead of dividing normally into more rod-like bacteria, the cells formed filaments (spaghetti growth). Eventually, with the help of inorganic chemists, they tracked down the cause of this unusual observation to electrolysis and subsequent reaction of Pt ions with the growth medium:

$$\text{Pt(electrode)} \xrightarrow{\text{electrolysis}} [\text{Pt}^{\text{IV}}\text{Cl}_6]^{2-} \xrightarrow[h\nu]{\text{NH}_4{}^+} [\text{Pt}^{\text{IV}}\text{Cl}_4(\text{NH}_3)_2]$$

However, the initial product, $[PtCl_6]^{2-}$, is bacteriostatic and does not promote filamentous growth. Substitution reactions of Pt(IV), which is kinetically inert, low-spin $5d^6$, are usually very slow, but the beaker was in sunlight, which probably catalyzed reactions with ammonia. An initial step in the preparation of $[Pt^{IV}Cl_4(NH_3)_2]$ involved the synthesis of $[Pt^{II}Cl_2(NH_3)_2]$. When they tested the cis and trans isomers of this Pt(II) complex, only the cis isomer induced filamentous growth, Fig. VII.4. Rosenberg subsequently showed that only *cis*-$[Pt^{II}Cl_2(NH_3)_2]$ and not *trans*-$[Pt^{II}Cl_2(NH_3)_2]$ could prevent the growth of cancer cells *in vivo*. Typically, cisplatin kills cancer cells at micromolar doses.

Chemistry of Cisplatin. Cisplatin is unstable in water (half-life 2.5 h at 310 K) and, for administration as an anticancer drug, it is formulated in saline solution to prevent aquation (hydrolysis), Scheme VII.1. The aquated species are much more reactive than cisplatin, and are more damaging to the kidneys. Aqua (H_2O) ligands bound to Pt(II) are *acidic* with pK_a values in the range of 5–8. The extent and rate of aquation of platinum am(m)ine complexes and the pK_a values of aqua ligands can be determined from proton nuclear magnetic resonance (^{1}H NMR) titration curves, or from ^{15}N NMR titration curves if the complex is labeled with ^{15}N.

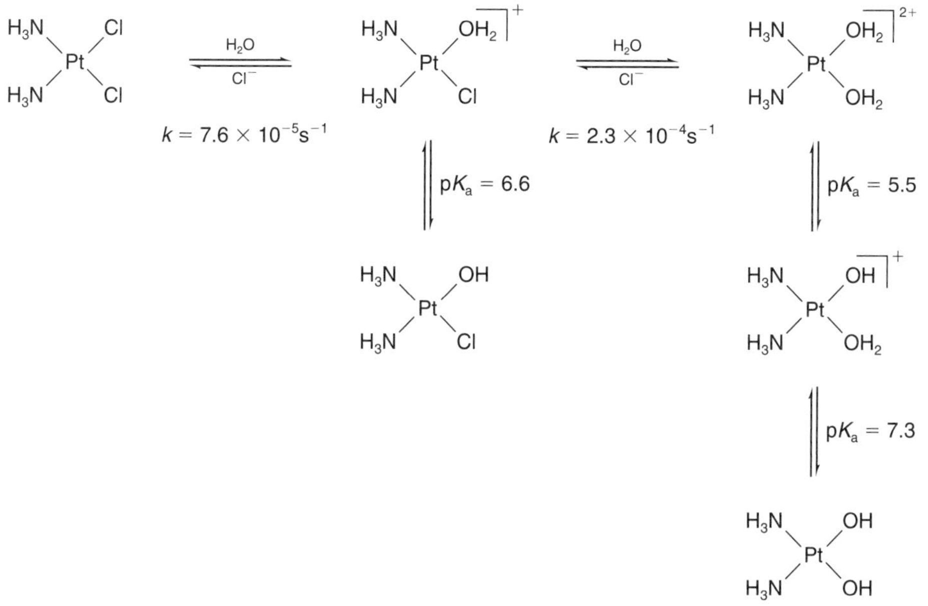

Scheme VII.1.

$$\mathrm{cis\text{-}[Pt(NH_3)_2Cl_2]} \underset{Cl^-}{\overset{H_2O}{\rightleftharpoons}} \mathrm{[Pt(NH_3)_2(OH_2)Cl]^+} \underset{Cl^-}{\overset{H_2O}{\rightleftharpoons}} \mathrm{[Pt(NH_3)_2(OH_2)_2]^{2+}}$$

$k = 7.6 \times 10^{-5} s^{-1}$; $k = 2.3 \times 10^{-4} s^{-1}$

$$\mathrm{[Pt(NH_3)_2(OH_2)Cl]^+} \rightleftharpoons \mathrm{[Pt(NH_3)_2(OH)Cl]} \quad pK_a = 6.6$$

$$\mathrm{[Pt(NH_3)_2(OH_2)_2]^{2+}} \rightleftharpoons \mathrm{[Pt(NH_3)_2(OH)(OH_2)]^+} \quad pK_a = 5.5$$

$$\mathrm{[Pt(NH_3)_2(OH)(OH_2)]^+} \rightleftharpoons \mathrm{[Pt(NH_3)_2(OH)_2]} \quad pK_a = 7.3$$

Hydroxo ligands bound to Pt(II) are much *less reactive* than aqua ligands and are good bridging ligands, so that hydroxo-bridged dimers and trimers readily form:

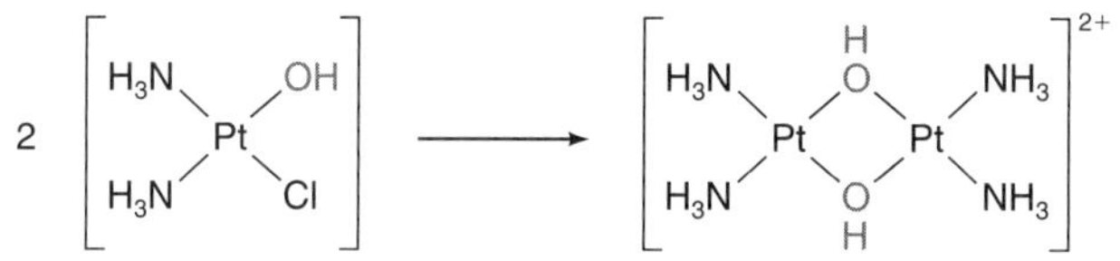

$$2\,\mathrm{[(H_3N)_2Pt(OH)Cl]} \longrightarrow \mathrm{[(H_3N)_2Pt(\mu\text{-}OH)_2Pt(NH_3)_2]^{2+}}$$

In blood plasma at pH 7.4, where the chloride concentration is ~104 m*M*, dichloro and chloro–hydroxo complexes are the dominant species for cisplatin, whereas in the cell nucleus, where the chloride concentration is lower (~4–20 m*M*), there are likely to be higher concentrations of aqua species, Table VII.2. *Aquation of cisplatin is therefore more likely to occur inside cells,* where reactive species are generated that can attack DNA.

Table VII.2.
Mole Fraction of Chloro, Aqua, and Hydroxo Adducts of Cisplatin in Media with Low and High Chloride Concentrations

[Cl] (m*M*)	Cl, Cl	Cl, OH	Cl, H_2O	H_2O, H_2O	OH, H_2O	OH, OH
104	0.67	0.26	0.04	0.0001	0.009	0.012
4	0.03	0.30	0.05	0.003	0.28	0.35

STRUCTURE–ACTIVITY RELATIONSHIPS

A very large number of active Pt antitumor complexes exhibit the following features:

1. Electrostatically neutral.
2. Square-planar Pt(II) or octahedral Pt(IV), with two cis monodentate N ligands or one chelated diamine, and at least one H on each N (ammine, primary, or secondary amines).

3. Two monodentate leaving groups or one bidentate leaving group, for example, Cl^- or carboxylate [Pt(II)–ligand bonds must not be too reactive (e.g., OH_2, SO_4^{2-}) or too unreactive (e.g., S ligands)].

How does cisplatin differ from transplatin in chemistry, and might this help to explain the dramatic difference in efficacy? The two Cl^- ligands (i.e., the reactive sites) are further apart in transplatin (4.64 Å) compared to cisplatin (3.29 Å), which affects the way Pt(II) can cross-link sites on DNA. Although the first aquation step for transplatin ($t_{1/2}$ 2 h) is as rapid as that for cisplatin, the *second aquation step is very slow* for transplatin because of the stabilizing effect of the new trans oxygen ligand. Therefore, the differences between the cis and trans isomers are both thermodynamic and kinetic.

New Generation Platinum Drugs. There is a need to develop new Pt anticancer drugs because:

1. Cisplatin is a very toxic compound and can have severe side effects, for example, nephrotoxicity (kidney poisoning), ototoxicity (loss of high frequency hearing), and peripheral neuropathy, although some side effects can be controlled.
2. Activity is required against a wider range of cancer types such as lung, breast, and colon cancers.
3. Cancer cells can become resistant to cisplatin after repeated treatment.

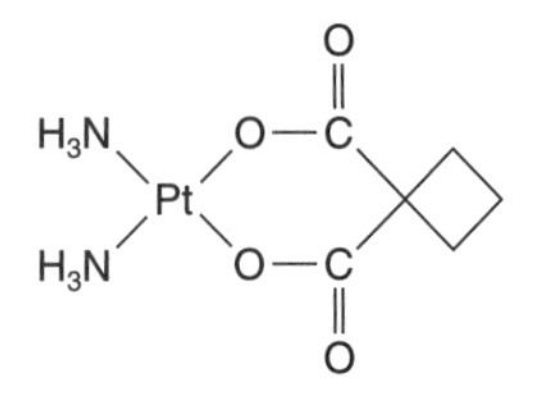

Carboplatin

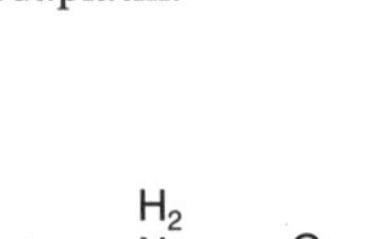

Nedaplatin

Chart VII.2.
The new generation anticancer drugs carboplatin and nedaplatin.

Carboplatin (Chart VII.2) was the second Pt anticancer drug to be approved for clinical use. It is less toxic than cisplatin. Reactions of carboplatin with H_2O and Cl^- ions lead to chelate ring opening and activation of carboplatin. However, these reactions are very slow (half-life in $H_2O > 4$ years, in saline solution ~ 10 days, at 310 K). The mechanism of activation of carboplatin is not fully understood.

The third complex to be approved, nedaplatin (Chart VII.2), contains a chelated glycolate ligand. Cisplatin, carboplatin, and nedaplatin are administered via intravenous injection. Typical weekly doses are 25–100 mg m^{-2} for cisplatin, and 300–400 mg m^{-2} for carboplatin (i.e., based on body area). There has also been much interest in 1,2-diaminocyclohexane (dach) complexes because they are often active against cisplatin-resistant cancer cell lines. An example is oxaliplatin (Chart VII.3). The diamine ligand has geometric (cis, trans: axial/equatorial, equatorial/equatorial) and optical isomers [*trans-d*-(1*S*,2*S*) and *trans-l*-(1*R*,2*R*)]. Platinum(II) complexes containing different isomers of dach differ in their biological activity. Oxaliplatin (trade name Eloxatin) was approved for clinical use in August, 2002 for use in the treatment of metastatic carcinoma of the colon or rectum as a combination infusion with 5-fluorouracil and leucovorin (5-FU/LV).

Rule Breakers

Active Pt complexes are now known that do not obey the normal structure–activity rules, and examples are shown in Chart VII.4.

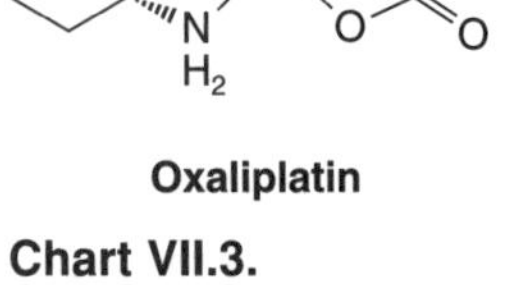

Oxaliplatin

Chart VII.3.
An example of an active dach complex, the drug oxaliplatin [Pt(1*R*,2*R*-dach)(oxalate)] (Eloxatin).

Ligand Substitution in Square-Planar Pt(II) Complexes. The drug AMD473 reacts *much more slowly* than cisplatin. Aquation of the Cl ligand cis to the 2-methylpyridine (2-picoline) ligand is about five times slower than for cisplatin. This rate difference is due to the steric effect of the methyl group, which hinders ligand attack and raises the energy of the five-coordinate intermediate in associative substitution reactions.

The mechanism of substitution reactions in square-planar Pt(II) complexes (and indeed all associative square-planar substitution reactions) is illustrated in Fig. VII.5 for the substitution reaction Pt–X + Y → Pt–Y + X.

Chart VII.4.
Examples of active Pt complexes that deviate from the commonly accepted structure–activity rules.

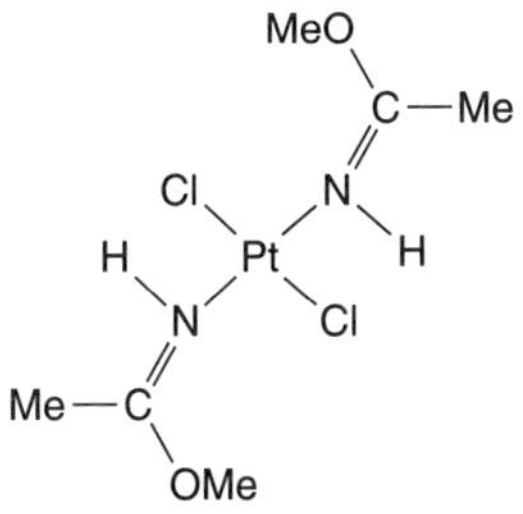

Trans-EE complex with iminoether ligands (more active than *cis* isomer)

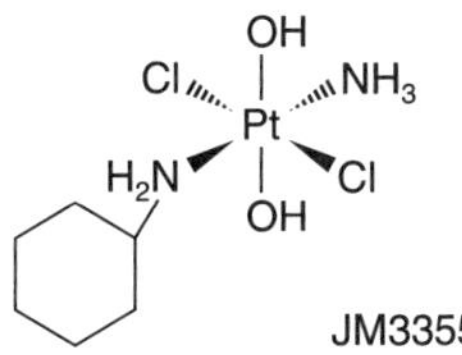

JM3355

The Pt(II) analogue of this Pt(IV) complex, without axial OH^- ligands, is inactive

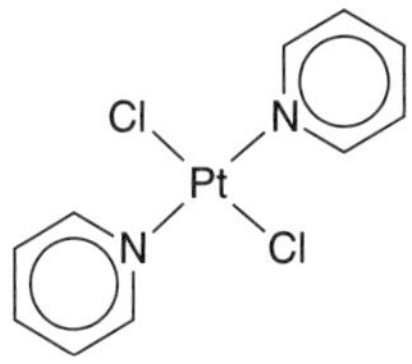

Trans complex with pyridine ligands

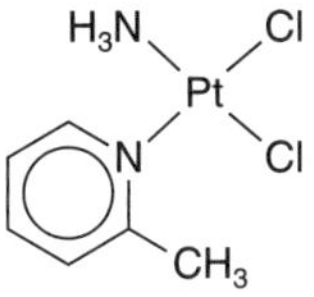

Two cis-N ligands, but only one has NH group

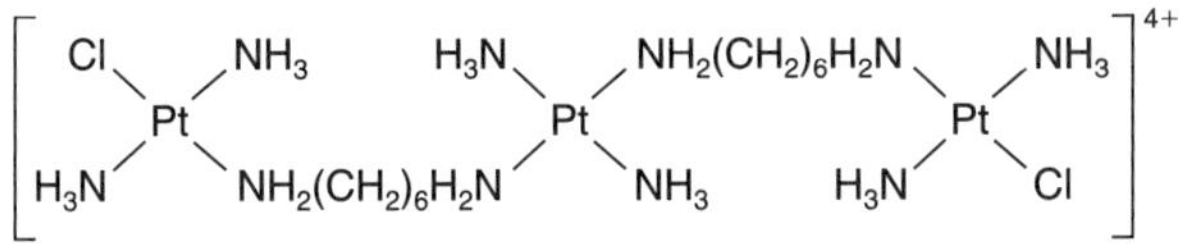

BBR3464

High positive charge, only one leaving group on each terminal Pt

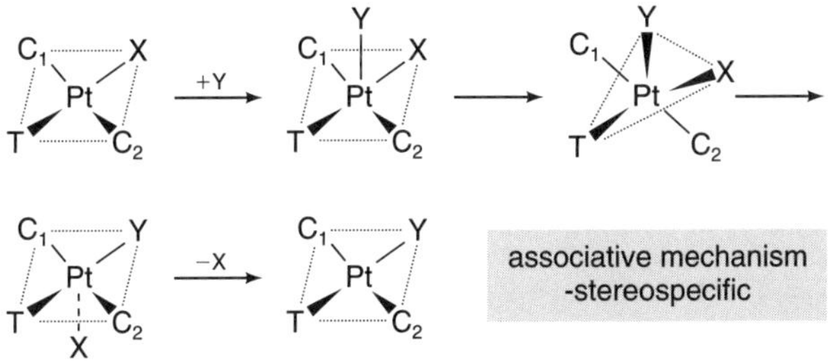

Fig. VII.5.
Mechanism for ligand substitution in square-planar Pt(II) complexes.

associative mechanism -stereospecific

$$\text{Rate law} \qquad \frac{-d[\text{PtX}]}{dt} = k_1[\text{PtX}] + k_2[\text{PtX}][\text{Y}]$$

The first term (k_1) is a solvent-assisted pathway.

Trans Effect

The *trans effect* is the effect of a coordinated ligand upon the rate of substitution of ligands opposite to it. It is a *kinetic* effect. For Pt(II) complexes, the order of the trans effects of ligands is generally

$$CN^-, C_2H_4, CO, NO > R_3P, H^- > SC(NH_2)_2 > CH_3^- > C_6H_5^- > SCN^- >$$
$$NO_2^- > I^- > Br^- > Cl^- > NH_3 > OH^- > OH_2$$

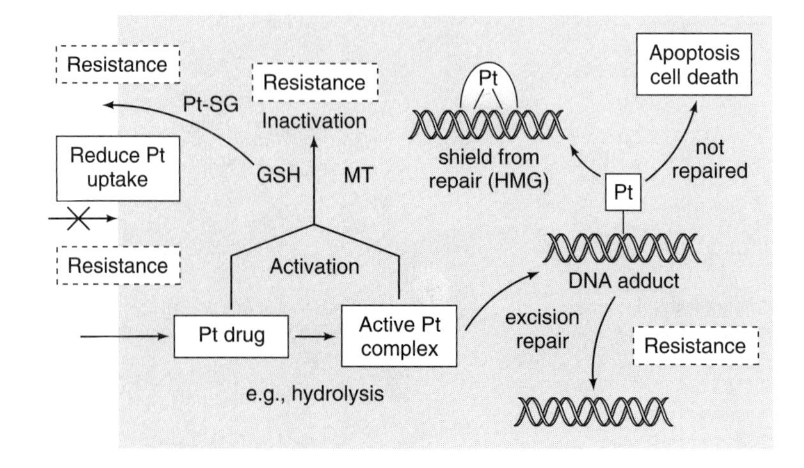

Fig. VII.6.
Some important sites of interaction of cisplatin with cells and resistance mechanisms. High mobility group = HMG, GSH = glutathione, MT = metallothionein.

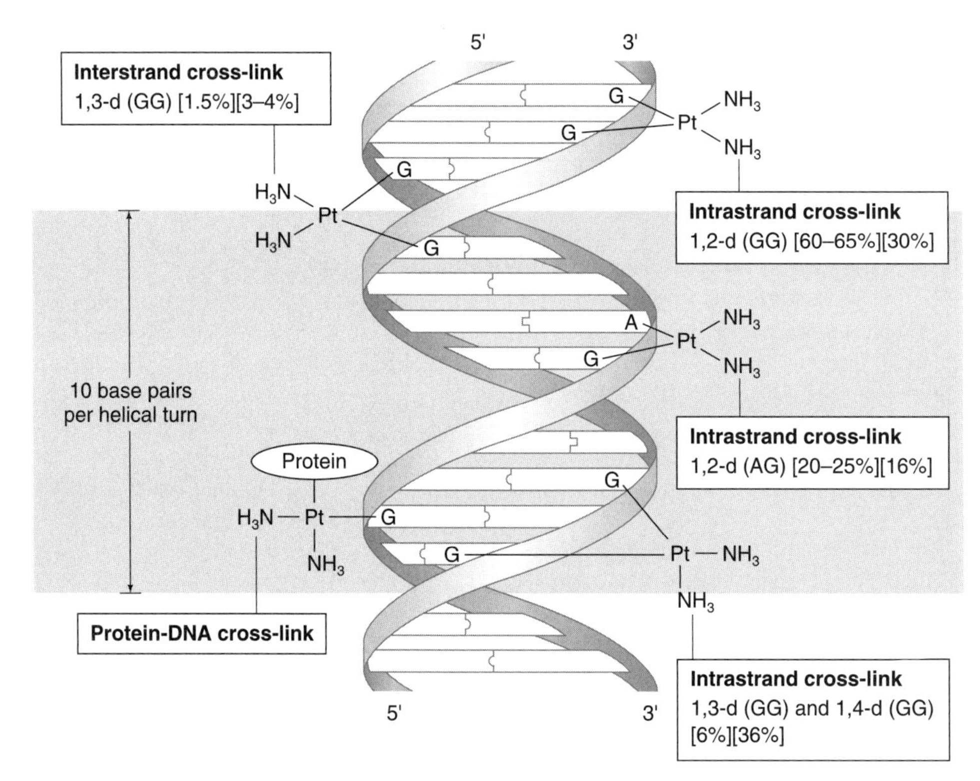

Fig. VII.7.
Platinum adducts with DNA. Typical populations after reaction with cisplatin are shown in the first [ ] and for carboplatin in the second [ ] brackets.

The rates of substitution of ligands trans to the π-acceptor ligands CO, CN^-, and C_2H_4 are several orders of magnitude faster than for ligands at the opposite end of the series. Due to the trans effect, cisplatin is the product of the reaction of $[PtCl_4]^{2-}$ with NH_3, whereas transplatin results from the reaction of $[Pt(NH_3)_4]^{2+}$ with Cl^-. Reactions of Pt anticancer complexes with S ligands, which have high trans effects, can lead to the facile release of coordinated am(m)ines.

The *trans influence* is the thermodynamic effect of one ligand on the strength of the bond between the metal and the ligand trans to it. The trans influence can be seen in ground-state properties such as bond lengths, vibrational frequencies, and NMR coupling constants. For example, the high trans influence of sulfur can be seen from the following values of $^1J(^{195}Pt{-}^{15}N)$ coupling constants:

$^1J(H_3{}^{15}N{-}^{195}Pt{-}X)$	(Hz)	390	310	285	265
	X =	H_2O	Cl	NH_3	R_1R_2S

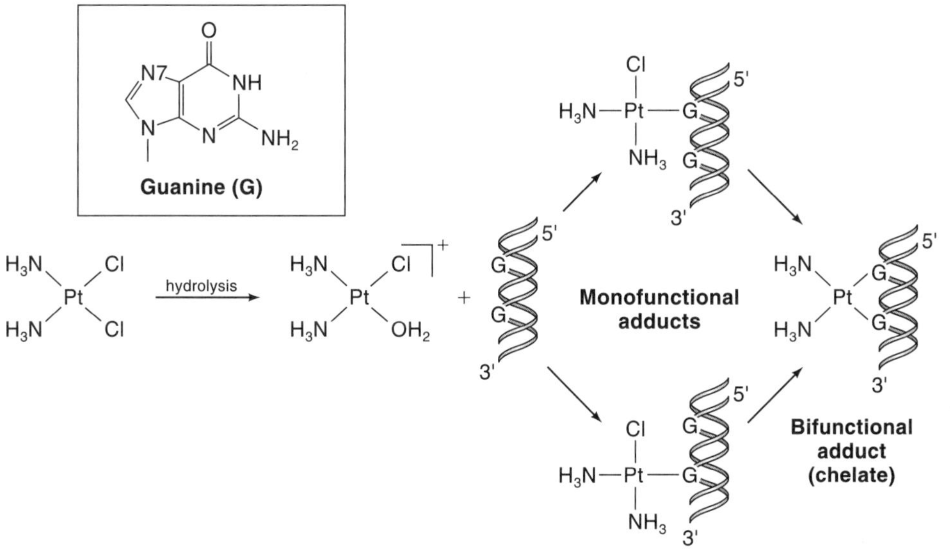

Fig. VII.8.
Pathways for GG intrastrand cross-linking of DNA by cisplatin. The insert shows the structure of guanine and the position of N7, the major Pt-binding site.

Mechanism of Action of Cisplatin. A simplified overview of the effect of Pt drugs on cells is shown in Fig. VII.6. The target site for cisplatin is DNA. The major binding site is N7 of guanine (G), the most electron-rich site on DNA. The N7 atom of G is readily accessible to Pt in the major groove of B-DNA. Other possible binding sites for Pt(II) include N7 of adenine, and N3 of cytosine (although this is usually less accessible). The most prevalent platinated adducts on DNA from cancer cells treated with cisplatin are 1,2 intrastrand cross-links involving G and sometimes adenine (A): GG $>$ AG $\gg$ GA (Fig. VII.7). A pathway for GG intrastrand cross-linking of DNA by cisplatin is shown in Fig. VII.8: aquation is followed by monofunctional adduct formation, and then ring closure to give the bifunctional GG macrochelate. In the GG chelate, the two G bases are in the head-to-head conformation with a dihedral angle of $\sim 26°$ between them. Hydrogen bonding between Pt–NH and backbone phosphate or C6 carbonyl groups can play a role in stabilizing such cross-links. The GG platination causes B-DNA to bend (kink) by ~ 35–$40°$. This bent DNA is recognized by proteins such as high mobility group HMG proteins. These proteins bind strongly to bent platinated DNA, shown in Fig. VII.9, and may shield these adducts from repair and be hijacked from their normal function as transcription factors. Platinated DNA adducts that are not repaired in cells trigger a series of events called apoptosis (programmed cell death), which results ultimately in the digestion of DNA by endonuclease enzymes.

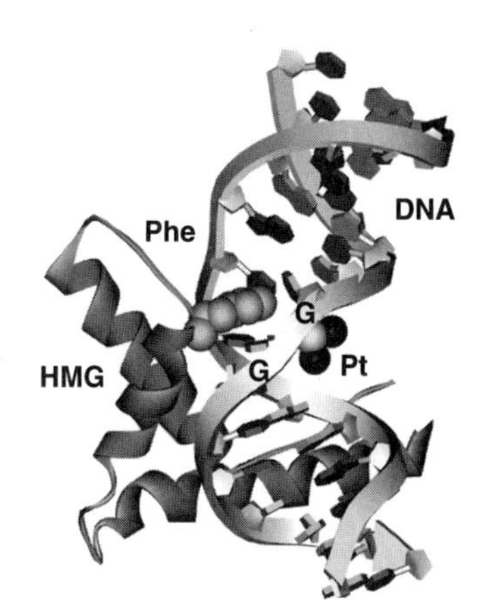

Fig. VII.9.
Intercalation of a phenylalanine side chain of HMG protein into a hydrophobic notch at the platinum cross-linked d(GpG) site of a platinated DNA 16-mer. (From PDB code: 1CKT.)

Interstrand DNA cross-links also cause major structural changes in DNA. These are rarer, but could be important for cytotoxicity. For example, the cytosine (C) residues opposite the platinated Gs are extrahelical in the double helix d[CATAGCTATG]$_2$ [where d = deoxyribose sugars in the molecular backbone, in contrast to the ribose sugars in ribonucleic acid (RNA)], which is interstrand cross-linked by *cis*-$\{Pt(NH_3)_2\}^{2+}$. Why are normal cells not damaged by cisplatin? They are, but the repair enzymes in normal cells can probably excise the Pt and correct the damage. Interstrand GG cross-links are important for the trinuclear complex BBR3464 (Chart VII.4), which can cross-link G bases on opposite strands, up to six base pairs apart. Phase II clinical trials of BBR3464 were completed in 2003, but these yielded results similar to cisplatin and other platinum agents currently on the market, probably due to binding and degradation of BBR3464 by human plasma proteins. Other methods of delivering these interesting tri- and dinuclear Pt complexes to cells, including the use of liposomes and antibodies, are now being explored.

The Pt(IV) complex JM216, *cis,trans,cis*-$[PtCl_2(Ac)_2(NH_3)(C_6H_5NH_2)]$, is active by oral administration. In general octahedral Pt(IV) complexes (low-spin d^6) are much more kinetically inert than square-planar Pt(II) complexes. The Pt(IV) complexes can be readily reduced *in vivo* to Pt(II) by reductants such as ascorbate or thiols [e.g., Cys, glutathione (GSH)]. The reduction potentials [cyclic voltammetry E_p values in mV vs. normal hydrogen electrode (NHE)] for Pt(IV) complexes correlate with the rate of reduction of analogous Pt(IV) anticancer complexes that have undergone clinical trials: *trans*-$[PtCl_4(NH_3)_2]$ > *cis,trans,cis*-$[PtCl_2(Ac)_2(NH_3) \cdot (C_6H_5NH_2)]$ > *cis,trans,cis*-$[PtCl_2(OH)_2(i\text{-}PrNH_2)_2]$, where Ac = acetate. Some E_p values for Pt(IV) ethylenediamine (en) complexes are given below.

cis,trans-$[Pt(en)Cl_2X_2]$	X	=	Cl	CH_3CO_2	OH
	E_p (mV)	=	−4	−326	−664

Platinum(IV) is highly polarizing and ammine ligands coordinated to it readily deprotonate at physiological pH values (i.e., near pH 7).

Resistance

The development of cellular resistance after prolonged treatment with cisplatin can be a problem in the clinic. The major resistance mechanisms are

1. Reduced transport across the cell membrane (Pt pumped out via transmembrane P-glycoprotein).
2. Formation of unreactive adducts with thiol ligands of the tripeptide GSH and the protein metallothionein MT.
3. Repair of DNA damage by excision-repair enzymes.

Platinum(II) forms strong bonds to S-containing ligands. Pt–s(thiolate) bonds tend to form irreversibly, whereas Pt–S(thioether) bonds form reversibly. Platinum drugs are inactivated in cells by reactions with the thiolate S of Cys in the tripeptide GSH (γ-L-Glu-L-Cys-Gly), which is highly abundant (millimolar concentrations in cells), and by reaction with the inducible protein MT that contains 20 Cys groups, and usually transports and stores Zn(II) and Cu(I). Besides Cys and Met, histidine (His) N can also be a target site for Pt in proteins.

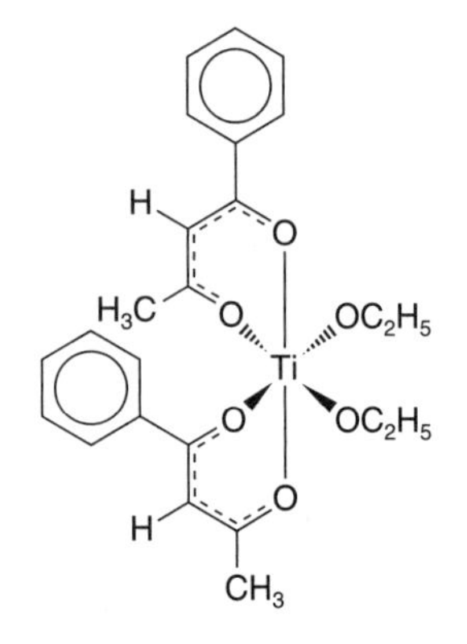

Budotitane

Titanocene dichloride

Chart VII.5.
Titanium agents containing reactive cis ligands.

Palladium Analogues. Palladium(II) complexes are usually isostructural with those of Pt(II). However, substitution reactions are much more rapid than those of Pt(II) ($\sim \times 10^4$–10^5 times) and so, with some exceptions, Pd(II) complexes tend to undergo side reactions before reaching the DNA target.

Titanium and Metallocenes. The success of cisplatin prompted a search for analgous transition metal complexes containing reactive cis chloride ligands. The octahedral Ti(IV) complex budotitane (Chart VII.5) showed promising activity against colorectal carcinomas and entered Phase I clinical trials in 1986. The cis, cis, cis isomer is predominant in solution. For the sarcoma 180 ascitic tumor, activity follows the order: Ti ~ Zr > Hf > Mo > Sn > Ge (inactive). There is little dependence of activity on the leaving group (e.g., OEt, Cl), but aromatic substitiuents on the β-diketonato ligands are important. Budotitane undergoes rapid aquation reactions in water (half-life ~20 s) with the eventual formation of TiO_2. It was formulated in micelles in an attempt to prevent hydrolysis. However, these difficulties have led to the abandonment of clinical trials until new formulations are found.

Clinical trials of the tetrahedral organometallic Ti(IV) complex titanocene dichloride [$TiCp_2Cl_2$] (Cp = cyclopentadienyl) (Chart VII.5) began in 1995. Hepato and gastrointestinal (GI) toxicity appear to be dose limiting. The anticancer activity of metallocene complexes [MCp_2Cl_2] is dependent on the metal ion: M = Ti, V, Nb, and Mo are active, but M = Ta and W show marginal activity, and M = Zr and Hf are inactive. Titanocene dichloride and vanadocene dichloride exhibit the best activity (e.g., against lung, breast, and GI cancers in mice), although this behavior has as yet not been confirmed in the clinic.

The [MCp_2Cl_2] complexes undergo rapid aquation reactions in water. Loss of the first Cl^- ligand from [$TiCp_2Cl_2$] is too rapid to measure and the second Cl^- has a half-life of ~45 min. The ion $[TiCp_2(H_2O)_2]^{2+}$ is acidic (pK_{a_1} 3.5, pK_{a_2} 4.4). It is not clear whether DNA is the primary target for Ti(IV). Binding to N sites on DNA bases appears to be weak at pH 7, although stronger at acidic pH values. Phosphate groups are stronger binding sites, and binding to proteins may also be important. The Ti(IV) ion can bind to the Fe(III) transport protein transferrin and may interfere with iron biochemistry.

Ruthenium. The octahedral Ru(III) complexes *cis*-$[RuCl_2(NH_3)_4]^+$ and *fac*-$[RuCl_3(NH_3)_3]$ (Chart VII.6) were shown early on to exhibit excellent anticancer activity, but they are too insoluble for pharmacological use. Solubility can be increased by increasing the number of Cl^- ligands. Trans complexes of the type

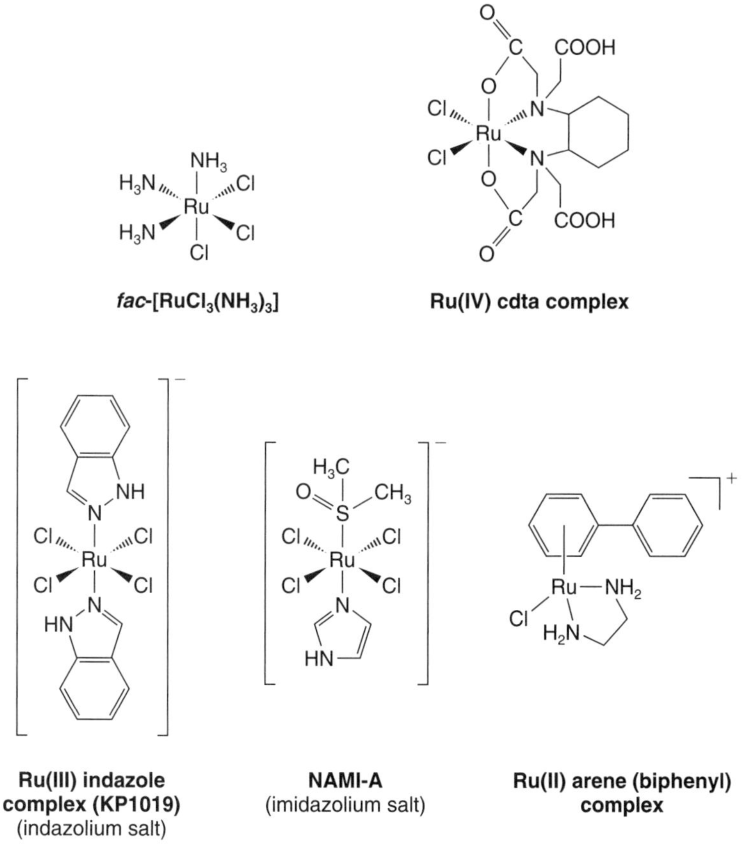

***fac*-$[RuCl_3(NH_3)_3]$**

Ru(IV) cdta complex

Ru(III) indazole complex (KP1019)
(indazolium salt)

NAMI-A
(imidazolium salt)

Ru(II) arene (biphenyl) complex

Chart VII.6.
Examples of ruthenium-containing anticancer agents. (cdta = 1,2-cyclohexanediaminetetraacetato).

$(LH)[Ru^{III}Cl_4L_2]$, where L = imidazole (Im) or indazole, exhibit excellent activity against a range of cancers. Octahedral Ru(III) and even Ru(IV) polyaminocarboxylate complexes are highly water soluble and are active anticancer agents. Examples are $[Ru^{IV}(cdta)Cl_2]$ and *cis*-$[Ru^{III}(ptda)Cl_2]$, where cdta = 1,2-cyclohexanediaminetetraacetate and pdta = 1,2-propylenediaminetetraacetate. Reduction of Ru(IV) to Ru(III) and Ru(II) may occur *in vivo:* E° values: Ru(IV/III) 0.78 V, Ru(III/II) −0.01 V. Dimethyl sulfoxide (dmso) complexes of Ru(II) and Ru(III) are relatively nontoxic, for example, *cis*- and *trans*-$[Ru^{II}Cl_2((CH_3)_2SO)_4]$.

The octahedral imidazole (Im) Ru^{III} dmso complex NAMI-A *trans*-$[Ru^{III}Cl_4((CH_3)_2SO)(Im)](ImH)$ (Chart VII.6) entered clinical trials in the year 2000 as an antimetastatic agent, the first Ru complex to enter the clinic. NAMI-A is not very toxic to primary cancer cells; it is hoped that it will prevent the spread of cancer (metastasis). In December 2003, the complex indazolinium [tetrachlorobisindazole)Ru^{III}] (KP1019; Chart VII.6) became the second Ru complex to enter clinical trials. This complex is active against colon carcinomas and their metastases.

As for cisplatin, aquation is thought to be an important activation mechanism for Ru(III) complexes.

$$trans\text{-}[RuCl_4(Im)_2]^- + H_2O \underset{310\,K}{\overset{t_{1/2}\ 3\,h}{\rightleftharpoons}} trans\text{-}[RuCl_3(H_2O)(Im)_2] + Cl^-$$

A second important activation mechanism for Ru(III) is thought to be reduction to Ru(II). Tumors are often hypoxic (low in O_2) and contain reducing agents such as thiols (e.g., GSH, $E^{\circ\prime} = -0.24\,V$). Reduction of Ru(III) to Ru(II) weakens bonds to π-donor ligands and increases ligand substitution rates; π acceptors such as dmso can raise the redox potential.

$$trans\text{-}[Ru^{III}Cl_4(Im)_2]^- \qquad trans\text{-}[Ru^{III}Cl_3(dmso)(Im)_2]$$
$$E^{\circ\prime} = -0.24\,V \qquad E^{\circ\prime} = 0.235\,V$$

Both Ru(II) and Ru(III) bind strongly to DNA bases with a strong preference for G-N7. Protein binding may also play an important part in the mechanism of action of

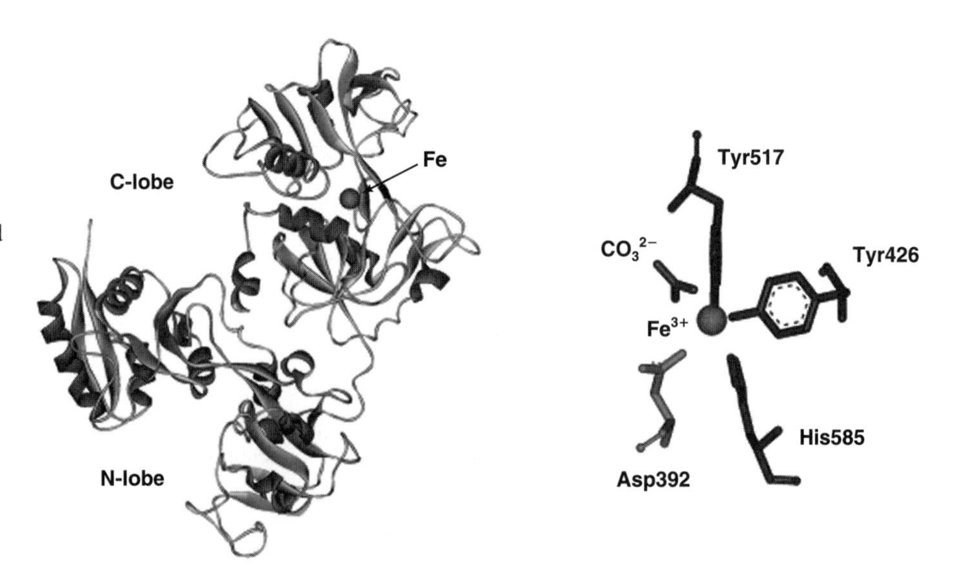

Fig. VII.10.
X-ray crystal structure of serum transferrin. There are two metal-binding sites one in the N and one in the C lobe. Only the C-lobe site is occupied in this picture (sphere). (Coordinates supplied by H. Zuccola.)

Ru complexes. The uptake of Ru(III) by cells appears to be mediated by transferrin. Transferrin (Fig. VII.10) is an 80-kDa blood plasma glycoprotein with two specific binding sites for Fe(III). Ruthenium(III) can bind reversibly at these sites, and Ru–transferrin complexes themselves exhibit anticancer activity. Cancer cells have a high density of transferrin receptors on their cell surfaces, and rapid cell division requires a high rate of Fe uptake. Ruthenium, a periodic congener of Fe, may therefore exhibit part of its activity by interfering with Fe biochemistry. Enzyme targets for Ru may also include topoisomerase and matrix metalloproteinases.

Ruthenium(II) is stabilized in organometallic Ru(II) arene complexes $[(\eta^6\text{-arene})Ru(en)Cl]PF_6$ (where en = ethylenediamine), which have the characteristic "piano-stool" structure (Chart VII.6) and exhibit anticancer activity. These complexes contain a reactive Ru–Cl bond and undergo hydrolysis in aqueous solution. They bind strongly to G through N7 coordination together with en NH···OC6 G hydrogen bonding and with arene-purine base stacking when the arene is large enough (e.g., biphenyl or tetrahydroanthracene). Such coordinated arenes can therefore intercalate into DNA.

Gallium, Tin, and Arsenic. Intravenous Ga(III) nitrate is effective for treating some types of cancer, hypercalcemia, and Paget's bone disease. It is likely that Ga(III) interferes strongly with Fe biochemistry since much of the Ga(III) in blood is taken up by transferrin. A promising new agent is the octahedral complex $[Ga(III)(maltolate)_3]$, Chart VII.7. Oral absorption into the plasma is rapid, followed by almost complete transfer of Ga(III) onto transferrin. A related compound $[Ga(8\text{-hydroxyquinoline})_3]$ (KP46) entered clinical trials as an oral anticancer drug in 2003.

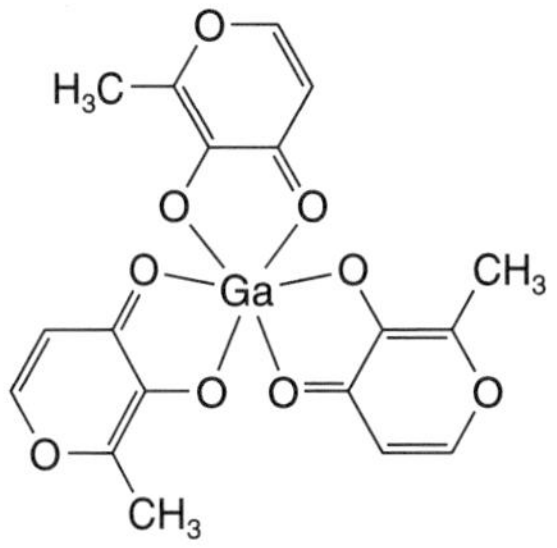

Chart VII.7. $[Ga^{III}(maltolate)_3]$, a potential anticancer drug.

A large number of tin compounds exhibit activity against P388 leukemia, but are inactive in most other screens. In the organo Sn(IV) series, the triorganotin complexes are usually the most toxic, with diorganotin and monoorganotin successively less toxic:

$$R_3Sn > R_2Sn > RSn$$

Tetraorganotin R_4Sn is dealkylated to R_3Sn *in vivo*. The toxicity is usually due to inhibition of mitochondrial oxidative phosphorylation. The most promising tin anticancer complexes are diorganotin complexes such as dibutyl tin(IV) glycylglycinate. The mechanism of action of Sn complexes is not understood.

Arsenic compounds are often associated only with toxicity, but in parts of China, As_2O_3 is the therapeutic agent of choice for treating acute promyelocytic leukemia giving "complete remission in most patients". In September 2000, arsenic trioxide (under the trade name Trisenox) was approved by the FDA for use in the United States. Solutions of As_2O_3 can be infused to a level of 1–2 μM in blood plasma without giving hematopoietic toxicity. Cell treatment leads to collapse of mitochondrial membrane potentials, release of cytochrome *c* into the cytosol, and apoptosis.

In aqueous solution, at physiological pH, As_2O_3 should exist as $As(OH)_3$ since it has pK_a values of 9.29, 13.5, and 14.0. It is a much weaker acid than $P(OH)_3$ (phosphorous acid). Arsenic(III) readily undergoes oxidative methylation in cells:

$$As^{III}(OH)_3 \xrightarrow{CH_3^+} CH_3As^{V}O(OH)_2 \xrightarrow{2e^-} CH_3As^{III}(OH)_2 \xrightarrow{CH_3^+} \text{etc.}$$

Bacterial genes encode for As(III) (usually called arsenite) efflux proteins (ArsB and ArsAB). Arsenic(III) is also transported as a GSH conjugate, and a reductase (ArsC) converts arsenate, $As^{V}O_4^{3-}$, to As(III).

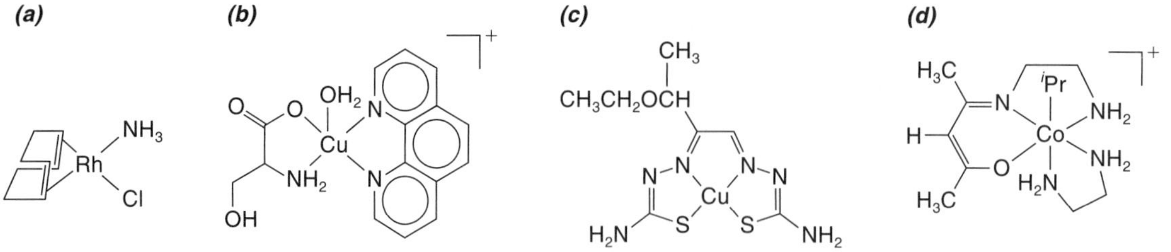

Chart VII.8.
Anticancer agents containing various metal centers. (*a*) A Rh(I) cyclooctadiene derivative, (*b*) a Cu(II) complex with phenanthroline and Ser ligands, (*c*) a Cu(II) thiosemicarbazone complex, and (*d*) an example of an alkyl Co(III) derivative, a delivery agent for alkyl radicals.

Other Anticancer Complexes. Complexes of several other metals exhibit anticancer activity, but have yet to enter the clinic, usually because they are not active against a wide enough range of cancers, or have severe toxic side effects.

The activity of Rh(II) carboxylate dimers is dependent on the length of the side-chain R, being optimum with five carbons. These dimers readily react with DNA, especially adenine. Square-planar Rh(I) complexes such as *a* in Chart VII.8 are also active, as are Rh(II) analogues of Ru(II) anticancer complexes. A number of Cu(I) and Cu(II) complexes exhibit anticancer activity, including mixed-ligand phenanthroline Cu(II) amino acid complexes such as *b*, and thiosemicarbazone complexes such as *c*.

The use of alkylcobalt(III) complexes as precursors for alkyl radicals, which can damage DNA, is potentially a novel anticancer strategy. The pH value inside solid tumors is typically 0.2–0.5 units lower than in normal tissues, and the decomposition of compound *d* in Chart VII.8 via homolytic fission yields alkyl radicals more rapidly at pH 6.5 than 7.5.

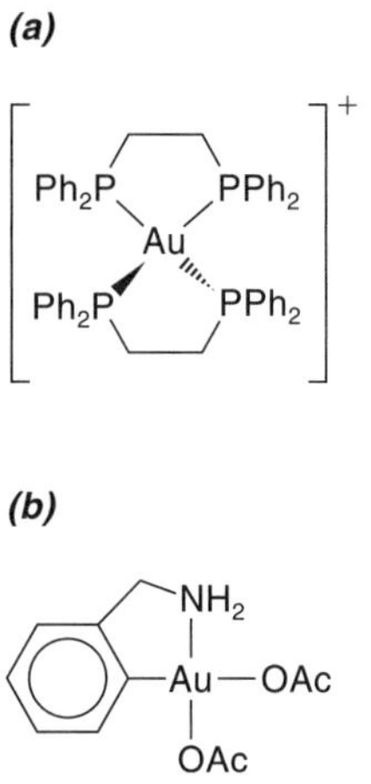

Gold(I) bis(diphosphine) complexes such as $[Au(dppe)_2]^+$ (dppe = 1,2-diphenylphosphinoethane), illustrated in Chart VII.9, are active against various types of cancer and kill cells via damage to mitochondria. Heart toxicity has so far prevented their clinical use, but it may be possible to circumvent this problem by careful choice of substituents on the phosphine and by tuning the lipophilicity of the cation.

Chart VII.9.
The Au(I) bis(diphosphine) anticancer complex, $[Au(dppe)_2]^+$, and an example of an active organoAu(III) complex.

Gold(III) complexes are isoelectronic ($5d^8$) and isostructural (square planar) with Pt(II), but gold analogues of cisplatin are too reactive and are readily reduced (faster ligand substitution rates, higher reduction potentials). However, some organogold(III) complexes (Chart VII.9) are stabilized toward reduction and exhibit anticancer activity.

Radiosensitizers. Radiotherapy using X-rays or γ-rays (^{60}Co) can be very successful for killing cancer cells, but relies on free radical damage propagated by O_2. In *hypoxic tumors* (lacking O_2), this treatment is less successful. *Electrophilic* molecules that mimic oxygen can act as radiosensitizers. Cisplatin itself can act as a radiosensitizer, and Pt(II) and Ru(II) complexes incorporating N-heterocyclic ligands with reducible nitro substituents are often effective. Cobalt(III) complexes, such as that shown in Chart VII.10, exhibit selective cytotoxicity toward hypoxic tumor cells. Inside the cell, they are reduced to the more labile Co(II) with release of coordinated nitrogen mustard ligands.

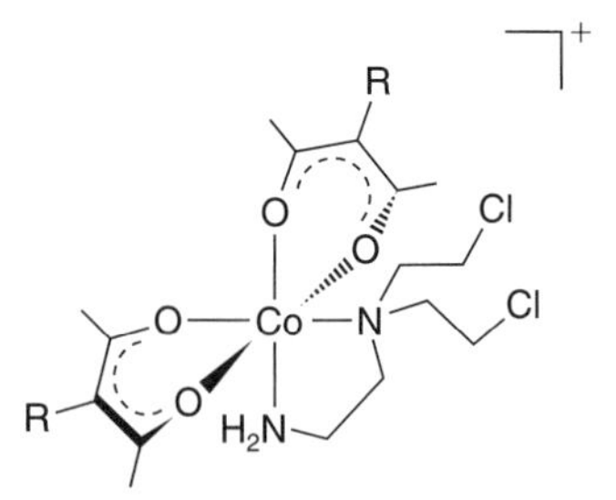

Chart VII.10.
An example of a cobalt-containing radiosensitizer, with nitrogen mustard ligands.

The Gd(III) complex of the texaphyrin (expanded porphyrin) derivative motexafrin (see Chart VII.11) is in clinical development and enhances the radiation response of tumors by a mechanism that involves the catalytic oxidation of intracellular reducing metabolites (e.g., GSH and ascorbate).

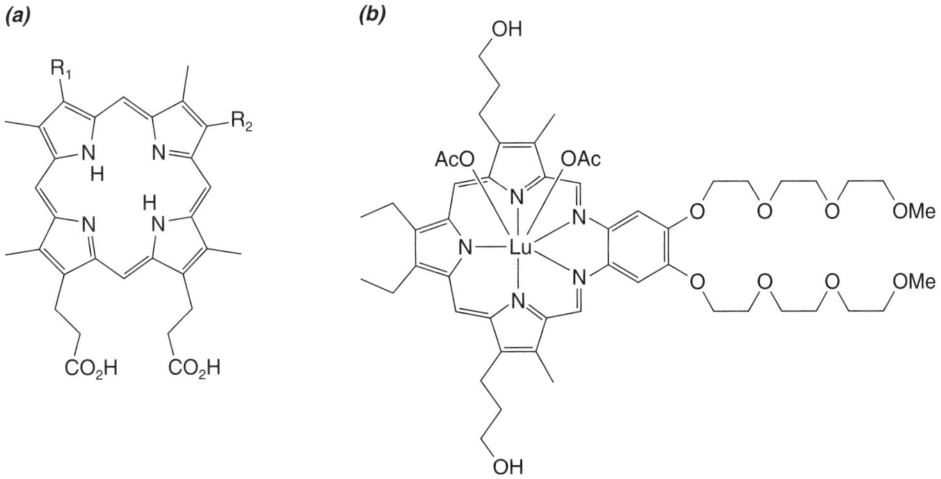

Chart VII.11.
(*a*) The hematoporphyrin Photofrin and (*b*) an expanded metalloporphyrin: the Lu(III) complex of the texaphyrin derivative motexafrin (LuTex). The Gd(III) complex of motexafrin (Xcytrin, Gd-Tex) is in clinical development as an adjuvant for radiation therapy of brain metastases.

Photosensitizers. In photodynamic therapy, a light-absorbing molecule is introduced into the cell and activated with laser light. The sensitizer should preferably absorb red light (>650 nm ε 10^3–10^5 M^{-1} cm^{-1}) strongly, since longer wavelength light penetrates tissues more effectively than more energetic, shorter wavelength light. However, light can be delivered internally through catheters, so photodynamic therapy is not limited to the skin. The sensitizer is excited to a triplet state and transfers energy to an organic substrate or O_2. The latter yields the reactive cytotoxic species, singlet oxygen, $^1O_2$ (half-life in water ~6 μs).

The only clinically approved sensitizer is hematoporphyrin (Photofrin, compound *a* in Chart VII.11—a mixture of derivatives). Injected porphyrins tend to localize in tumor tissue. A variety of related metal porphyrins, expanded metalloporphyrins (e.g., *b* in Chart VII.11) and phthalocyanines can also act as photosensitizers and may become second generation drugs. Phthalocyanine complexes with diamagnetic ions [e.g., Zn(II), Al(III), and Ga(III)] give the highest triplet yields and longest lifetimes.

Neutron Capture Therapy. The nuclei of ^{10}B and ^{157}Gd have high cross-sections for neutron capture. If compounds containing these nuclei are delivered to cancer cells, and then the cells are bombarded with neutrons, the subsequent fission generates destructive α-particles:

$$^{10}_{5}B + ^{1}_{0}n \rightarrow ^{7}_{3}Li + ^{4}_{2}He + 2.7\,MeV$$

Clinical facilities for neutron capture therapy are being constructed in various parts of the world, especially for the treatment of glioma (tumors originating from the brain or spinal cord) and cutaneous melanoma. High doses of the precursor nucleus must reach the cell (~10^9–10^{10}B atoms per tumor cell). Hence, the use of polyhedral boranes, such as $[B_{12}H_{11}SH]_2^-$ offers an effective way of achieving high B concentrations. Delivery should be selective for tumor cells so as to minimize damage to surrounding tissues.

VII.2.2. Gold Antiarthritic Drugs

Injectable Au(I) thiolate drugs and one oral Au(I) phosphine drug are widely used in clinics today for the treatment of difficult cases of rheumatoid arthritis. There is also interest in the potential use of Au compounds for treating asthma, the autoimmune disease pemphigus (a chronic, sometimes fatal, skin disease), malaria, cancer, and HIV.

For thousands of years Au has been considered to be of medical, therapeutic value. Even in the nineteenth century Au was considered to be a "cure-all" for diseases. The alchemists knew that metallic Au dissolves in aqua regia and can be reduced back to metallic gold in the form of a stable colloid (gold sol), the color of which depends on the size of the particles and ranges from blue (large particles) to dark purple (small particles; for example, "Purple of Cassius" first prepared by Andreas Cassius in 1685). Neutralized solutions ("aurum potabile", drinkable gold) were widely used in therapy during the Middle Ages, but their true value is in doubt. The greatest practitioner of gold therapy in medical history is said to have been Leslie Keeley. By the 1890s every state in the United States, and nearly every country, had a Keeley Institute, typically advertising:

> "Gold cure for opium habit $10
> Gold cure for drunkenness $9
> Gold cure for neurasthenia $8
> Remedies sold only in pairs"

The patients seemed to be pleased with their treatment and formed the "*Bichloride of Gold Club*" (but note that $[Au^{I}Cl_2]^-$ is not very stable!). The chemical nature of Keeley's gold preparations is uncertain. Keeley died in 1900, but his Institutes continued until 1960.

The rational use of Au in medicine began in the early twentieth century when the bacteriologist Robert Koch discovered that $K[Au(CN)_2]$ could kill the bacteria that cause tuberculosis. In common with many Au(I) complexes, $[Au^{I}(CN)_2]^-$ contains linear, two-coordinate Au(I) with carbon-bound cyanide: $[NC{-}Au{-}CN]^-$. Three coordinate Au(I) and four-coordinate tetrahedral Au(I) complexes are known, but are less common. Gold(I) is a class "b" or "soft" metal ion and the most stable complexes contain heavier ligands, for example, P in preference to N, S in preference to O. The aqua ion $[Au^{I}(H_2O)_x]^+$ is unknown. Gold(I) is stabilized by π-acceptor ligands. In biochemical systems, the most strongly preferred ligand for Au(I) is thiolate sulfur (from Cys in proteins). The affinity of Au(I) for DNA is very low.

It is common to find weak Au(I)–Au(I) interactions in the X-ray structures of Au(I) complexes, perpendicular to the axis of linear coordination. The Au–Au distances of 2.5–3.3 Å are much shorter than the van der Waals contact distance of 3.6 Å. This attraction of Au ions for each other is termed "*aurophilicity*", and can be attributed to the strong influence of relativistic effects on gold chemistry (inner-shell electrons moving with velocities approaching the speed of light (heavy electrons) cause shell contractions). In contrast to Cu, which is also in group 11(IB), gold in oxidation state +II, Au(II), is very unstable. Square-planar Au(III) complexes are readily prepared, but tend to be reduced in biological media [to Au(I) and/or Au(0)].

During the early twentieth century, Au treatment for tuberculosis switched from $K[Au(CN)_2]$ to the less toxic Au(I) thiolate complexes. In the early 1930s, the French physician Forestier was the first to use these thiolate complexes to treat rheumatoid arthritis, a condition that he believed to be related to tuberculosis. The use of Au for the treatment of rheumatoid arthritis has continued ever since. The major injectable antiarthritic Au drugs are the water-soluble complexes aurothiomalate (Myocrisin), aurothioglucose (Solgonal), sodium bis(thiosulfate) Au(I), and sodium thiopropanolsulfonate-*S*-gold(I).

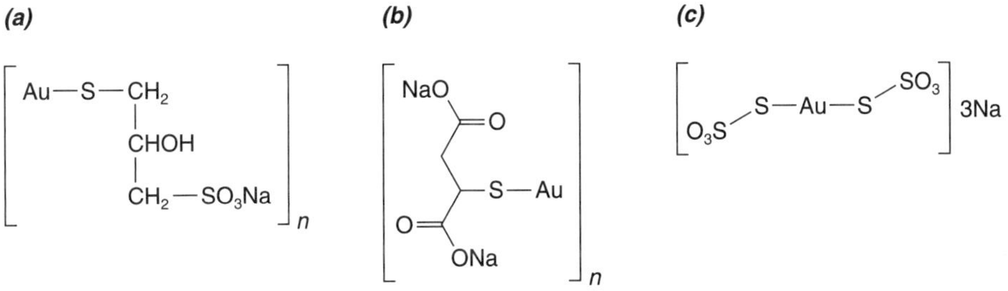

Chart VII.12.
Examples of water-soluble antiarthritic Au drugs. (*a*) Sodium thiopropanolsufonate-*S*-gold(I), (*b*) Myocrisin, and (*c*) sodium bis(thiosulfate)-gold(I). The thiolate complexes are polymers, see Fig. VII.9.

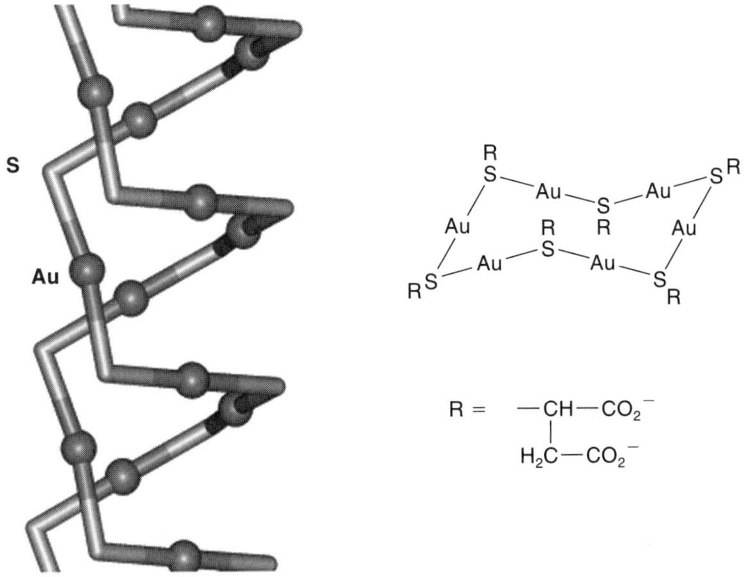

Fig. VII.11.
X-ray crystal structure of an aurothiomalate complex similar to the antiarthritic drug Myocrisin, showing the double helical chains, and a schematic drawing of a ring structure for a 1:1 Au(I) thiolate complex.

The antiarthritic Au(I) thiolate complexes (Chart VII.12) are formulated as approximate 1:1 complexes, but their structures in solution are complicated. Gold(I) must be at least two coordinate and thiolate sulfur acts as a bridge between Au(I) ions: –*S*–Au–*S*–Au–*S*–Au–. Chains and cyclic structures are possible. The X-ray crystal structure of $[CsNa_2HAu_2\{SCH(CO_2)CH_2(CO_2)\}_2]$, which has a formula closely related to the drug Myocrisin, contains intertwined helices (Fig. VII.11). The right-handed helices contain exclusively (*R*)-thiomalate, and the left-handed helices (*S*)-thiomalate.

The oral drug auranofin (Ridaura) is a monomer and contains linear two-coordinate Au(I) (Chart VII.13). Auranofin is a stable white crystalline complex, sparingly soluble in water, but soluble in organic solvents. It is remarkable to find this first example of the use of a trialkylphosphine in a drug. Trialkylphosphines alone are often associated with high reactivity, toxicity, and pungent smells!

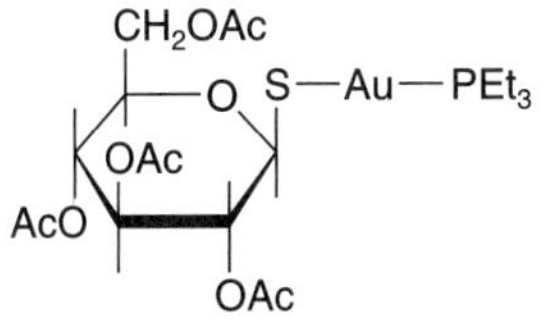

Chart VII.13.
The oral Au(I) antiarthritic complex, Auranofin.

Gold drugs are "*pro-drugs*" because ligand substitution reactions are relatively facile on Au(I). They have low activation energies and proceed via three-coordinate intermediates. Thiol exchange reactions are important *in vivo*. The initial ligands on the gold drugs are displaced (substitution of thiols, displacement, and oxidation of PEt_3 to $OPEt_3$). In the blood, most of the Au(I) is carried by the thiol of Cys34 of albumin.

Gold concentrations in blood can rise to ~20–40 μ*M* after injection of Au drugs. The half-life for Au excretion is ~5–31 days, but gold can remain in the body for many years. A major deposit site is in lysosomes (*aurosomes*), the membrane-bound intracellular compartments that house destructive enzymes. The inhibition of enzymes that destroy joint tissue may be the key to the antiarthritic activity of Au, although the cause of rheumatoid arthritis itself is unknown.

Patients who smoke and are treated with gold drugs attain much higher concentrations of Au in their red blood cells than nonsmokers. Inhaled tobacco smoke contains up to 1700 ppm HCN, and Au has a very high affinity for cyanide (log β_2 36.6). Cyanide reacts with the administered gold drug to form $[Au(CN)_2]^-$, which readily passes through cell membranes. Traces of cyanide appear to be present naturally in the body (formed from SCN^-), and $[Au(CN)_2]^-$ is a metabolite of gold drugs even in patients who are not smokers, reaching levels of 5–560 n*M* in their urine.

The formation of Au(III) may be responsible for some of the toxic side effects of gold drugs. Although most of the Au *in vivo* will be present as Au(I), powerful oxidants, such as hypochlorous acid (HOCl), which can oxidize Au(I) to Au(III), are generated at sites of inflammation, and white blood cells from patients treated with gold drugs become sensitive to Au(III). Further understanding of the redox cycling of Au may lead to a better understanding of these side effects.

VII.2.3. Lithium: Control of Bipolar Affective Disorders

Lithium, usually administered in tablet form as Li_2CO_3, in doses of up to 2 g day^{-1}, is a safe drug currently received by between 0.5 and 1 million patients worldwide for the (mainly prophylactic) treatment of bipolar affective disorders such as manic depression. The psychiatric activity of Li was first discovered by the Australian psychiatrist John Cade in 1949. He was interested in the effects of uric acid and chose to work with lithium urate because it was the most soluble salt of uric acid. Unexpectedly, it was the Li that turned out to have a pharmacological effect in his experiments and not urate!

The Li(I) ion is very small (ionic radius 0.60 Å) and mobile (rapid ligand exchange: residence time of coordinated water ~1 ns). Lithium(I) binds weakly to ligands, and is strongly hydrated in water with a hydrated radius of ~3.4 Å. After administration, Li is widely distributed in body tissues and blood plasma, with concentrations typically in the range 0.4–0.8 m*M*. There are no useful radioisotopes of Li, but the stable, naturally occurring isotopes ^6Li (7.4% abundance, nuclear spin quantum number $I = 1$) and ^7Li (92.6% $I = \frac{3}{2}$) are useful NMR nuclei. In particular, ^7Li NMR has been used to measure the rate and extent of Li uptake into biological cells. The Li(I) transport across red blood cell membranes occurs via Li(I)/Na(I) exchange; Li(I)/$CO_3{}^{2-}$ cotransport; Li(I) replacement of K(I) in Na, K-ATPase; the Cl-dependent Na/K transport system; and a leak mechanism.

The chemistry of Li(I) is similar to that of Mg(II) (diagonal relationship in the periodic table). Lithium(I) is thought to bind to the second Mg(II) site in the enzyme inositol monophosphatase and inhibits this enzyme at therapeutic doses of Li(I). In this way, Li(I) could interfere with Ca(II) metabolism since inositol phosphates control the mobilization of Ca(II) inside cells. Calcium(II) is responsible for triggering a number of signaling processes in cells.

There are potential new uses for Li in therapy other than in psychiatry. Lithium(I) succinate is licensed for the treatment of seborrhoeic dermatitis and Li(I) compounds inhibit the replication of some DNA viruses. A chemical challenge is to design compounds that could release Li(I) slowly in the body.

VII.2.4. Bismuth Antiulcer Drugs

Bismuth compounds have been used in medicine for >200 years to treat a wide variety of conditions, including GI disorders and syphilis. Current interest centers on their antiulcer activity, in particular antimicrobial activity against *Helicobacter pylori*, a bacterium that can prevent ulcers from healing.

Bismuth, atomic number 83, is the heaviest stable element in the periodic table and occurs as a single isotope in Nature: ^{209}Bi [Xe] $4f^{14}5d^{10}6s^{2}6p^{3}$. The oxidation state of interest in medicine is Bi(III). Bismuth(V) is known, but tends to be a strong oxidant. Bismuth(III), with an ionic radius of ~1.03 Å, is similar in size to Ca(II), and adopts variable coordination numbers from 3 to 10 with a wide range of geometries. The $6s^2$ lone pair of electrons sometimes exhibits a stereochemical effect, the "*inert-pair effect*". Bismuth(III) is a highly acidic metal ion. The first deprotonation of the aqua ion has a pK_a of 1.5:

$$[Bi(H_2O)_9]^{3+} \rightleftharpoons [Bi(H_2O)_8(OH)]^{2+} + H^+ \qquad pK_a = 1.5$$

Further deprotonation to give coordinated hydroxide and oxide is facile, and oxygen-bridged clusters such as $[Bi_6O_5(OH)_3]^{5+}$ and $[Bi_6O_4(OH)_4)]^{6+}$ readily form in aqueous solution. It is common for Bi(III) complexes to contain oxide (bismuthyl ion, BiO^+) and hydroxide in addition to other ligands, and they are usually referred to as "*basic*", "*oxy*", or "*sub*" salts.

The most widely used bismuth compounds for treating GI disorders are bismuth subsalicylate (BSS, e.g., Pepto-Bismol), colloidal bismuth subcitrate (CBS, e.g., De-Nol), and ranitidine bismuth citrate (RBC, Pylorid). The chemical nature of the bismuth compounds in these preparations is not fully understood. Bismuth(III) citrate [Bi(Hcit)] is insoluble but can be solubilized with alkali (including ammonia and amines such as ranitidine—itself an antiulcer drug). Citric acid, Chart VII.14, with pK_a values of 2.9, 4.3, and 5.6, exists as a trianion at pH 7. In addition, metal ions, such as Al^{3+}, Fe^{3+}, and Ga^{3+}, as well as Bi^{3+}, can displace the proton from the central hydroxyl group. The Bi(III) citrate complexes have complicated structures, which are often based on the dimeric unit $[(cit)BiBi(cit)]^{2-}$, Chart VII.15, where cit is tetra-deprotonated citric acid, containing tridentate citrate, together with one terminal carboxylate bridging to the neighboring Bi(III). The Bi(III)–O(alkoxide) bond is very short (2.2 Å) and strong, being part of a five-membered chelate ring. The Bi(III) citrate dimers can associate to give chain and sheet structures via further bridging and hydrogen bonding. Such polymers may be deposited on the surface of ulcers. At pH values < 3.5 in dilute HCl, BiOCl precipitates.

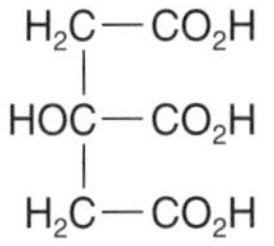

H_2C-CO_2H
HOC$-CO_2H$
H_2C-CO_2H

Chart VII.14.
Citric acid.

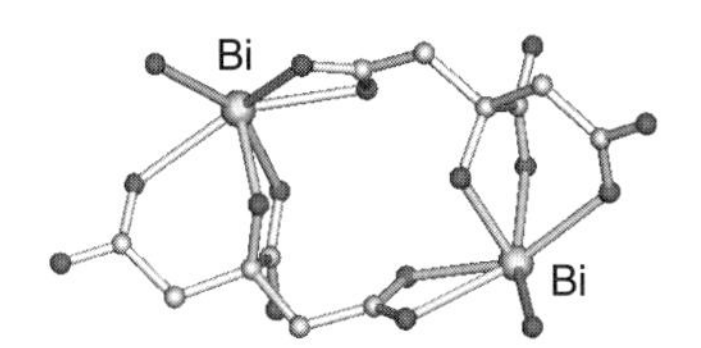

Chart VII.15.
The X-ray crystal structure of the Bi(III) citrate dimer, $[Bi_2(cit)_2]^{2-}$, a possible constituent of Bi antiulcer agents. The additional O attached to each Bi is from a neighboring dimer so forming a chain structure.

Bismuth(III) citrates react readily with thiols such as the tripeptide GSH, with formation of $[Bi(SG)_3]$, in which Bi(III) is bound to the thiolate S. Even though $[Bi(SG)_3]$ is a highly stable complex (log $K = 29.6$), the thiolate ligands are kinetically labile and exchange with free thiol on a millisecond time scale. Therefore Bi(III) may be a highly mobile ion inside cells.

The bacterium *H. pylori* lives under highly acidic conditions in the stomach and uses the Ni enzyme urease to make ammonia to neutralize the acid, and thereby to survive:

$$H^+ + 2\,H_2O + H_2NCONH_2 \rightleftharpoons HCO_3^- + 2\,NH_4^+$$

Inhibition of urease by Bi(III) thiolate complexes may play a role in the antibacterial activity of Bi(III).

In general, Bi(III) compounds are relatively nontoxic. Cells are probably protected against Bi(III) by the thiol-rich protein MT (see Section VIII.4). Bismuth(III) can induce the synthesis of MT and pretreatment with Bi(III) is an effective mechanism for minimizing the toxicity of Pt drugs. Curiously, Bi is deposited in membrane-bound vesicles in the nuclei of cells as "bismuth inclusion bodies", but the chemical nature of these deposits is unknown. The most serious side effects of bismuth drugs were encountered in France and Australia in the 1960s and 1970s when outbreaks of *encephalopathy* were reported. The chelating agent 2,3-dimercapto-1-propanesulfonic acid (DMPS) is an effective antidote for acute Bi intoxification. The Bi(III) may interfere with the biochemistry of Fe(III). The binding constants of Bi(III) for a range of O- and N-donor ligands correlate with those of Fe(III). Bismuth(III) binds tightly to the blood serum protein transferrin, which transports Fe(III) and has receptors in the brain.

VII.3. Imaging and Diagnosis

VII.3.1. Radiodiagnostic and Radiopharmaceutical Agents

Radionuclides are used for both imaging and therapy. For diagnostic radioimaging, the half-life of the radionuclide must be long enough to allow the synthesis of the radiopharmaceutical complex; long enough to allow accumulation in target tissue in the patient and clearance through nontarget organs; and yet, short enough to minimize the radiation dose to the patient. Cost, availability, and the energy of the radiation are also important considerations.

Two types of radioimaging are used in the clinic.

1. Single-photon emission computed tomography (SPECT). This technique requires a pharmaceutical labeled with a γ-emitting radionuclide with an optimum energy in the range 100–250 keV. γ-Rays are highly penetrating high-energy photons ($\lambda < 1$ pm), which have low ionization effects and are not deflected by magnetic and electric fields. γ-Emission usually accompanies other decay processes. Useful γ-emitting nuclides (half-lives in parentheses) include

   ^{67}Ga (3.3 day), ^{111}In (2.8 day), ^{99m}Tc (6.0 h), ^{201}Tl (3.0 day)

   ^{99m}Tc is used in >85% of all diagnostic scans in hospitals because of its ideal properties.
2. Positron emission tomography (PET). This technique requires a radiopharmaceutical labeled with a positron β^+-emitting radionuclide. The emitted positron is antimatter and reacts virtually instantaneously with a nearby electron anihilating both with the production of two 511-keV photons 180° apart which, with multiple annihilations, can pinpoint the site of emission.

Useful positron (β^+) emitters (half-lives in parentheses) include

^{55}Co (17.5 h), ^{64}Cu (12.7 h), ^{66}Ga (9.5 h), ^{68}Ga (1.1 h), ^{82}Rb (79 s), ^{86}Y (14.7 h)

Therapeutic radiopharmaceuticals should deliver localized cytotoxic doses of ionizing radiation. The radionuclides used emit β^- particles (electrons) or α particles (helium-4 nuclei, $^4_2He^{2+}$). A major aim is often to treat secondary or metastatic cancer sites. Most radiotherapeutic nuclides used in the clinic are β^- emitters. Examples are (half-lives in parentheses):

^{32}P (14.3 day), ^{47}Sc (3.3 day), ^{64}Cu (0.5 day), ^{67}Cu (2.6 day)
^{89}Sr (50.5 day), ^{90}Y (2.7 day), ^{105}Rh (1.5 day), ^{111}Ag (7.5 day)
^{117m}Sn (13.6 h), ^{131}I (8.0 day), ^{149}Pm (2.2 day), ^{153}Sm (1.9 day)
^{166}Ho (1.1 day), ^{177}Lu (6.8 day), ^{186}Re (3.8 day), ^{188}Re (0.7 day)

The choice of isotope is determined by the ease of production, cost, and intensity and duration of the radiation delivered. Iodine(131) is highly effective for the treatment of hyperthyroidism and thyroid cancer.

HEDP EDTMP

Chart VII.16.
The phosphonate ligands HEDP, and EDTMP, which are used in radiopharmaceuticals of ^{153}Sm and ^{186}Re.

Since 1942, radiopharmaceuticals have been used to relieve pain from skeletal metastases. A major constituent of bone is the mineral hydroxyapatite. This mineral can be targeted with ^{32}P-orthophosphate, ^{89}Sr-strontium chloride [similar chemistry of Ca(II) and Sr(II)] and with phosphonate complexes of radionuclides such as ^{153}Sm and ^{186}Re. Typical phosphonate ligands, hydroxyethylidene-1,1-diphosphonate (HEDP) and ethylenediaminetetramethylenephosphonate (EDTMP) are shown in Chart VII.16. An example of skeletal imaging is shown in Fig. VII.12.

Radiation from radionuclides emitting α particles extends only a few cell diameters (40–100 μm). These high-energy He nuclei are highly cytotoxic and effective in the treatment of tumors with small diameters, while causing little damage to normal tissues. Most attention has focused on the α-emitters:

$$^{211}At(t_{1/2}\ 7.2\,h) \quad \text{and} \quad ^{212}Bi(t_{1/2}\ 1\,h)$$

The half-life of ^{212}Bi is short ($t_{1/2}$ 1 h), but ^{212}Pb ($t_{1/2}$ 10.6 h) can be used as an *in vivo* generator of ^{212}Bi. Attempts are being made to use monoclonal antibodies to target these isotopes. However, the labeling reactions must be rapid, and the resultant conjugates must be stable enough to allow them to reach the target site (see Receptor Targeting of Radiopharmaceuticals).

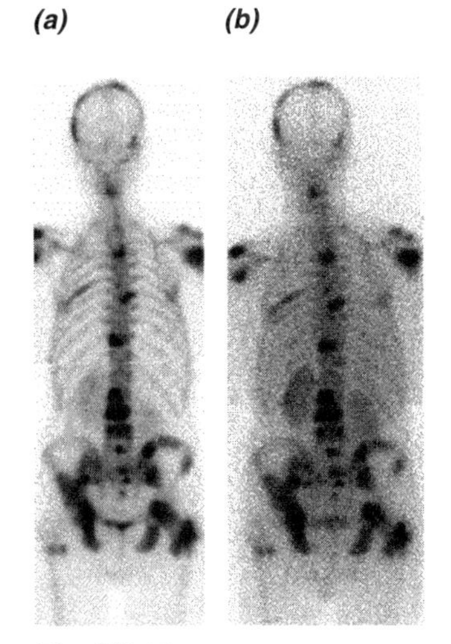

Fig. VII.12.
Scintigrams of a patient with prostatic cancer. (*a*) After dosing with ^{99m}Tc-hydroxymethylenediphosphonate (imaging), and (*b*) after dosing with ^{186}Re-hydroxyethylidene-1,1-diphosphonate (therapy). The images are identical and show the localization of the metastases. (Reproduced with permission from A. van Dijk et al., *Technetium and Rhenium in Chemistry and Nuclear Medicine: 4*, M. Nicolini, G. Bandoli, U. Mazzi, Eds., SG Editoriali, Padova, 1995, p. 529.)

^{99m}Tc Imaging Agents. Technetium, Tc, [Kr] $4d^5\ 5s^2$, atomic number 43, does not occur naturally on Earth, except for traces from spontaneous fission of ^{238}U. Technetium was first made by Perrier and Segrè in 1937. In accordance with Mattauch's rule, Tc does not have stable, non-radioactive isotopes. All 22 isotopes (including 9 isomers) are radioactive with half-lives of 0.3 s–4.2 million years. The ^{99}Tc isotope can be obtained in large enough quantities for preparative chemistry, is only a weak β-emitter, has a long half-life (213,000 years), and can be handled safely in laboratory glassware with elementary precautions. Technetium-99m is a short-lived metastable isomer of ^{99}Tc, and is a strong γ-emitter with a half-life of 6 h.

Stable oxidation states of Tc range from -1 to $+7$ with coordination numbers of 4–9. The $[Tc^{VII}O_4]^-$ ion is only a weak oxidizing agent, in contrast to $[MnO_4]^-$. Technetium(II) forms strong complexes, but does not exist as a simple aquated ion, in contrast to $[Mn(H_2O)_6]^{2+}$. Technetium complexes are often isostructural with those of Re. Radiopharmaceuticals containing ^{99m}Tc are used at such low doses (concentrations of 10^{-6}–10^{-8} *M*) that pharmacological effects do not need to be considered.

Over 80% of radiopharmaceuticals used worldwide are labeled with ^{99m}Tc because of

1. Optimal nuclear properties of ^{99m}Tc [142 keV γ-rays, effectively detected by NaI (doped with Tl) scintillation counters].
2. Convenience of production from the long-lived isotope ^{99}Mo.
3. Low cost of commercial generator columns and facile separation by chromatography.
4. Short half-life, low-radiation exposure to patient—negligible tissue damage.

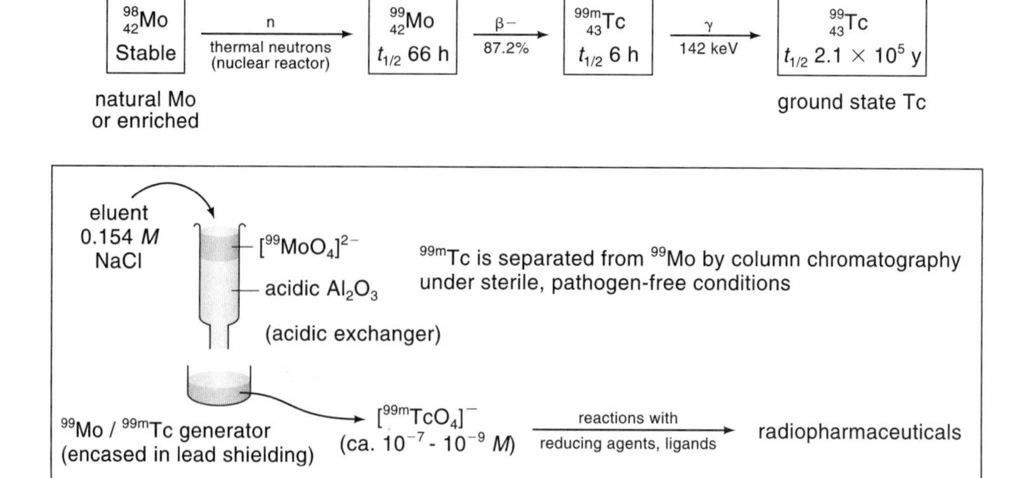

Fig. VII.13.
A procedure for the production of ^{99m}Tc from ^{99}Mo that is effective for clinical use.

The procedure used for the generation of ^{99m}Tc in the clinic is illustrated in Fig. VII.13.

Selective imaging of specific tissues and organs in the body can be achieved by a suitable choice of the oxidation state of ^{99m}Tc and the nature of the coordinated ligands. The initial product from the $^{99}Mo/^{99m}Tc$ generator is a dilute solution ($\sim 10^{-8}$ *M*) of the pertechnetate ion $[^{99m}Tc^{VII}O_4]^-$ in 0.154 *M* saline. Pertechnetate can be used to image the thyroid, gastric mucosa, and blood pool, but for other applications new ^{99m}Tc coordination complexes must be synthesized rapidly, in as few steps as possible, and in high yield. However, localization of ^{99m}Tc in the target tissue does not have to be complete. Uptake of a few percent of the administered dose can often be sufficient to produce useful images.

The complex $[^{99m}Tc^V$-*d,l*-HM-PAO] (HM-PAO = hexamethyl-propyleneamine oxine) was the first agent to be approved for human *cerebral perfusion imaging*. Maximum count rate is achieved in the brain 30–40 s after injection. The complex containing the meso ligand diffuses out of the brain more rapidly than that of the racemic *d,l* ligand. The latter complex reacts more rapidly with GSH and the ^{99m}Tc becomes trapped in the brain.

Positively charged ^{99m}Tc complexes such as hexakis isonitrile $^{99m}Tc(I)$ complexes are readily prepared in a rapid single step by heating $[^{99m}TcO_4]^-$ with excess isonitrile [e.g., as a sterile preformed Cu(I) complex] and a reducing agent (e.g., $SnCl_2$). Octahedral complexes such as the 2-methylether-*n*-propylisonitrile $^{99m}Tc(I)$ complex Tc-Sestamibi (Cardiolite) are water and air stable. Hundreds of isonitrile ligands were investigated to obtain the optimum balance between uptake and clearance by the heart compared to other organs (blood, liver, lungs). The six methoxy ligands are sequentially metabolized to hydroxyl groups in the liver. This transformation turns the complex into increasingly hydrophilic species, which are not retained in myocardial tissues. Other examples of myocardial perfusion agents are the phosphine complex $[^{99m}Tc(V)(tetrofosmin)_2]^+$ (Myoview), which contains a O=Tc=O core, and $[^{99m}Tc(III)(furifosmin)]^+$ (Tc-Q12), containing a tetradentate Schiff base in the equatorial plane and tertiary phosphines as axial ligands (Chart VII.17). The latter has rapid blood, lung, and liver clearance.

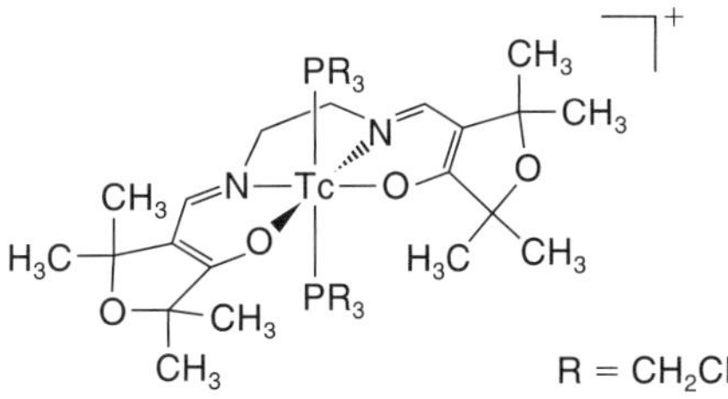

Tc-Q12

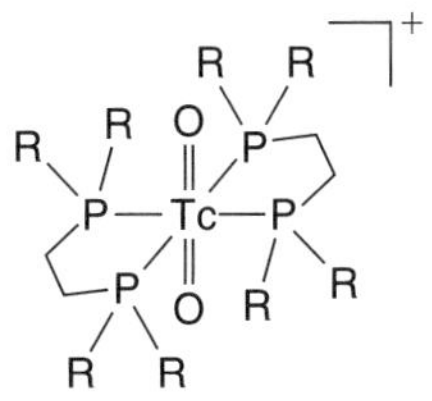

$R = CH_2CH_2OCH_3$

Myoview

$R = CH_2CH_2OCH_2CH_3$

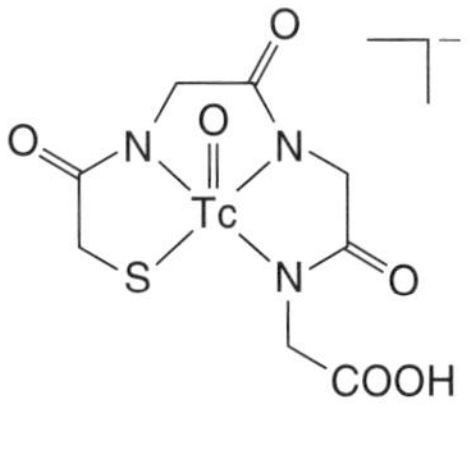

$[^{99m}TcO(MAG_3)]^-$

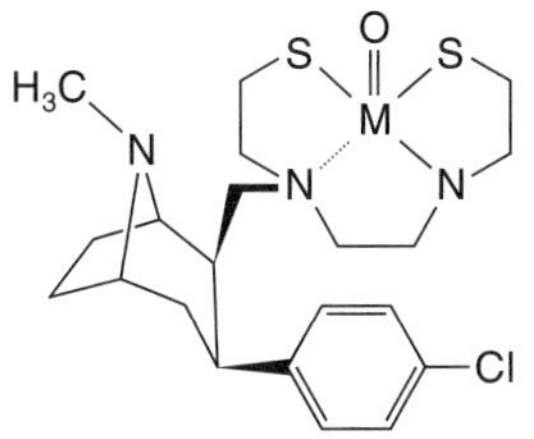

M-TRODAT-1
(2): M = ^{99m}Tc
(3): M = Re

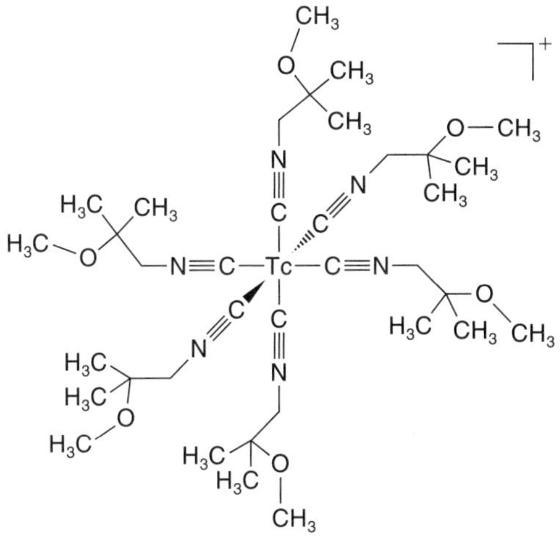

Cardiolite

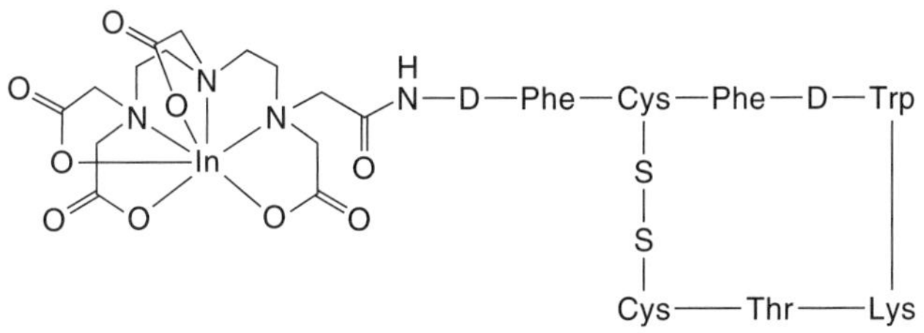

H
N—D—Phe—Cys—Phe—D—Trp
S
S
Cys—Thr—Lys

Somatostatin analogue

Chart VII.17.
Examples of ^{99m}Tc and ^{111}In imaging agents.

Agents used clinically for imaging the *kidney* include ^{99m}Tc complexes with DTPA, dimercaptosuccinate (DMSA), and mercaptoacetylglycine (MAG). The latter complex $[^{99m}TcO(MAG_3)]^-$ is almost all protein-bound and is transported to the proximal renal tubes in the kidney. This complex is chiral even though the ligand itself does not have a chiral center, Chart VII.17. The phosphonate complexes of ^{99m}Tc are widely used for *skeletal imaging,* Fig. VII.12. Nanoparticles containing ^{99m}Tc including sulfur and protein colloids are also used for imaging the *liver, spleen,* and *bone marrow.*

Other Radiodiagnostic Isotopes. The $^{201}Tl^+$ ion is successful for *heart imaging* because it has similar chemistry to K^+ and is readily taken up by myocytes (heart cells). Such selective uptake is also the case for ^{67}Ga(III)-citrate, which is widely used for the clinical diagnosis of Hodgkin's disease, lung cancer, malignant melanoma, and leukemia. In the blood, Ga(III) is readily taken up by transferrin ($\log K = 19$), and by lactoferrin in activated leukocytes. There are both transferrin and non-transferrin mediated ^{67}Ga uptake mechanisms for tumor cells.

Indium-111 is also a cyclotron-produced γ-emitter used for labeling white blood cells as a lipophilic 8-hydroxyquinoline (oxine) complex. Oxine is displaced inside cells and ^{111}In becomes trapped.

In contrast to ^{67}Ga and ^{111}In, which are cyclotron produced, the β^+ (positron) emitter ^{68}Ge is available via a commercial ^{68}Ge/^{68}Ga generator kit with a long lifetime (1–2 years since ^{68}Ge has $t_{1/2} = 271$ d).

Receptor Targeting of Radiopharmaceuticals. To improve the selectivity of the agent for the target, a complex (a conjugate) is usually constructed of the type radiometal–chelate–biomolecule. The general principles described here are applicable to a range of radionuclides as well as to paramagnetic metal complexes used for MRI contrast agents (Section VII.3.2). However, selective *binding* to receptors is difficult to achieve in MRI because of the high concentrations of metal required. Targeting strategies often involve labeling proteins with metal complexes by the following procedures.

1. Direct labeling: Binding metal to a protein side chain (e.g., Cys).
2. Prelabeling: Covalent coupling of metal chelate to protein (e.g., via Lys, Cys).
3. Postlabeling: Covalent coupling of chelating ligand to protein, followed by binding the metal.

Pretargeting involves binding of monoclonal antibody (mAb) itself labeled with a binding site (e.g., via conjugation to streptavidin, SA) for a rapidly cleared small molecule (e.g., biotin), clearing excess antibody from circulation, followed by addition of the radiolabeled small molecule (e.g., biotin-M*) to bind to the antibody on the cell.

$$\text{mAb–SA} \xrightarrow[\substack{\text{excess}\\ \text{mAb–SA}}]{\text{clear}} \text{Cell–mAb–SA} \xrightarrow{\text{biotin–M}^*} \text{mAb–SA–biotin–M}^*$$

Neuroreceptors are important targets. For example, the dopamine transporter can be targeted using tropane derivatives of antagonists (e.g., ^{99m}Tc-TRODAT-1, Chart VII.17).

Steroid-derivatized ^{99m}Tc chelates can target steroid receptors. Estrogen and progesterone receptors are present in ~65% of human breast cancers. Hormone receptors can be targeted with somatostatin analogues, Chart VII.17. Somatostatin is a 14-amino acid peptide involved in the regulation and release of hormones with receptors in the GI tract, central nervous system, exocrine, and endocrine pancreas. Stable analogues can be labeled with radionuclides. The compound ^{111}In-DTPA-Octreotide

(OctreoScan) has been approved for clinical diagnostic imaging of neuroendocrine tumors.

The Tc(I) ([Kr] $4d^6$) ion is kinetically inert, and the stable, water-soluble reagent $[^{99m}Tc^{I}(H_2O)_3(CO)_3]^+$ is useful for labeling biomolecules. The analogous Re(I) complex is also used.

VII.3.2. Magnetic Resonance Imaging Contrast Agents

In MRI (MRI = NMR imaging), the spatial distribution of protons (largely from H_2O) is detected in thin slices through the body. The technique is noninvasive and the applied magnetic field (1–1.5 T) and radiofrequency (40–70 MHz) are harmless. Scans only take a few minutes and achieve the same spatial resolution (~1 mm) as X-ray computer tomography. Contrast agents are used to reveal features of images that are not apparent in normal maps. Heavy atom compounds were used as contrast agents in X-ray imaging as early as 1895, typical examples being organoiodide compounds and $BaSO_4$. They are administered in gram quantities, and therefore should have a low toxicity. The first MRI contrast agent, the gadolinium complex $[Gd^{III}(DTPA)(H_2O)]^{2-}$, was used in 1988, and since then 30,000 kg of Gd have been administered to patients worldwide, and contrast agents are used in ~30% of all MRI examinations (e.g., in ~8 million of the 25 million MRI scans carried out in 1999).

Magnetic resonance imaging scans not only map the intensity of ^1H nuclear spins, but also their associated spin–lattice (longitudinal) and spin–spin (transverse) relaxation times, T_1 and T_2, respectively. The spin–lattice relaxation time T_1 is associated with the return of the induced proton magnetization along the applied magnetic field axis (z), whereas the spin–spin relaxation time T_2 is related to the width of the NMR peak at half-height ($\Delta\nu_{1/2}$): $\Delta\nu_{1/2} = 1/\pi T_2$ (decay of magnetization in the x, y plane).

If an agent that produces local magnetic fields fluctuating at about the ^1H Larmor frequency is introduced into specific regions of the body, then shortening of T_1 and T_2 in these regions produces contrast in the image. The most effective MRI contrast agents are the high-spin transition metal ions Cr^{III}: [Ar] $3d^3$, Mn^{II}: [Ar] $3d^5$, and Fe^{III}: [Ar] $3d^5$, and the lanthanide ion Gd^{III}: [Xe] $4f^7$.

The effectiveness of a contrast agent in relaxing H_2O is determined by its relaxivity (R_1 or R_2):

$$\left(\frac{1}{T_{1(\text{or }2)}}\right)_{\text{obs}} = \left(\frac{1}{T_{1(2)}}\right)_{\text{d}} + \left(\frac{1}{T_{1(2)}}\right)_{\text{p}}$$

$$(R_{1(2)})_{\text{obs}} = (R_{1(2)})_{\text{d}} + (R_{1(2)})_{\text{p}}$$

where "d" refers to the diamagnetic and "p" to the paramagnetic contribution to the water relaxation rate. The relaxivity of the paramagnetic complex is measured from the slope of the (usually linear) plot of $(1/T_{1(2)})_{\text{obs}}$ versus complex concentration (in units of $mM^{-1}\,s^{-1}$). In order to achieve contrast, the concentration of the paramagnetic complex at the target site must be $>\sim 50\,\mu M$. Contrast agents are therefore usually administered in gram doses.

Various factors contribute to the relaxivity of a complex, as illustrated in Fig. VII.14. For example, the following factors increase the relaxivity:

1. Increase in *hydration number* of the metal ion [q, up to 6 for Mn(II), 9 for Gd(III)]. However, too many coordinated H_2O ligands may make the metal ion too reactive and toxic. Hydation numbers for the lanthanide ions Yb(III) and Eu(III), which have a similar chemistry to Gd(III), can be determined from fluorescence quenching experiments in H_2O and D_2O.
2. Shortening of the *metal–water proton distance* ($1/r^6$ dependence) [r is typically 2.5 Å for Gd(III), 2.2 Å for Mn(II)].

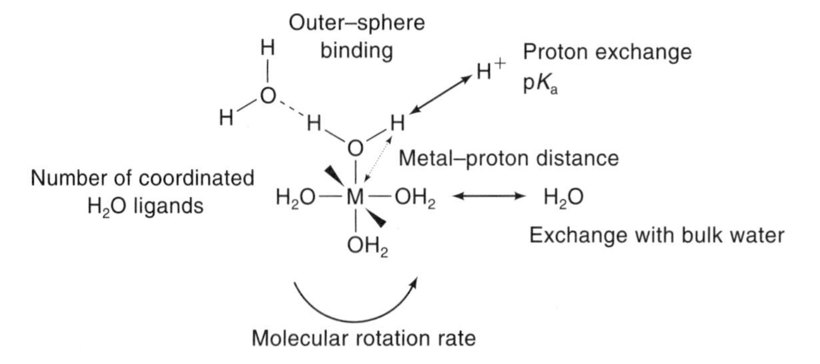

Fig. VII.14.
Factors that affect the relaxivity of a paramagnetic complex.

The rate of exchange of water bound to the paramagnetic metal ion with bulk water (k_{ex}) must be fast enough that the bulk water experiences relaxation enhancement. It is possible to vary the lifetime of water in the inner-coordination sphere of Gd(III) from nanoseconds to milliseconds by variation of the other ligands. For example, introduction of the chelating ligand DTPA greatly decreases the water exchange rate:

$[Gd^{III}(H_2O)_9]^{3+}$	$k_{ex} = 8 \times 10^8\ s^{-1}$	(associative mechanism)
$[Gd^{III}(DTPA)(H_2O)]^{2-}$	$k_{ex} = 4 \times 10^6\ s^{-1}$	(dissociative mechanism)

A major limitation for Cr(III) is the slow exchange of coordinated H_2O ($k_{ex} \sim 10^{-6}\ s^{-1}$).

Slow molecular rotation can increase the relaxivity of complexes. The rotational correlation time (time for rotation through 1 rad) can be estimated from the Stokes–Einstein equation $\tau = 4\pi a^3\eta/3RT$, where a = radius of molecule and η = solvent viscosity. There is a linear correlation between the relaxivity per Gd(III) and molecular weight of mono- and multi-Gd(III) chelates. The relationship between these factors and relaxivity is very complicated. Optimization of the relaxivity of a given field strength requires an understanding of the field dependence of these parameters (nuclear magnetic relaxation dispersion, NMRD).

The most effective contrast agents are those that give EPR signals at room temperature, that is, they have relatively *long electron spin relaxation times:* Gd(III), Mn(II). Six Gd(III) complexes have been approved for clinical use as MRI contrast agents. They are all extracellular agents present in blood plasma and interstitial spaces. The $[Gd(bopta)(H_2O)]^{2-}$ complex is lipophilic and can also be used for liver imaging. Table VII.3 lists Gd(III) complexes approved for clinical use as MRI contrast agents. These extracellular agents are not only highly stable thermodynamically, but Gd(III) macrocycle complexes are also kinetically inert [very slow Gd(III) release]. The possibility of Gd(III) displacement by, for example, Ca(II), Zn(II), and Cu(II) *in vivo* must also be considered, but adverse effects in humans are rare.

Extracellular agents are excreted by the kidney with an elimination half-life of ~1.5 h and are typically used for imaging brain tumors, Fig. VII.15. Most of the extracellular Gd(III) complexes are nine coordinate in the solid state and contain one bound H_2O. Examples of structures are shown in Chart VII.18. All have spin–lattice relaxivities, R_1, of $\sim 4\ mM^{-1}\ s^{-1}$.

Some of the goals of future work on MRI contrast agents are

1. To optimize the relaxivity of complexes, for example, by increasing the size (protein conjugates, dendrimers: Gd_{170}(dendrimer) has $R_1 = 5800\ mM^{-1}\ s^{-1}$).
2. To improve targeting, for example, by coupling to antibodies (needs high loading with Gd, use dendrimers).
3. To introduce agents that are sensitive to local biochemistry (functional agents), for example, activated by enzyme cleavage or by Ca(II) binding.

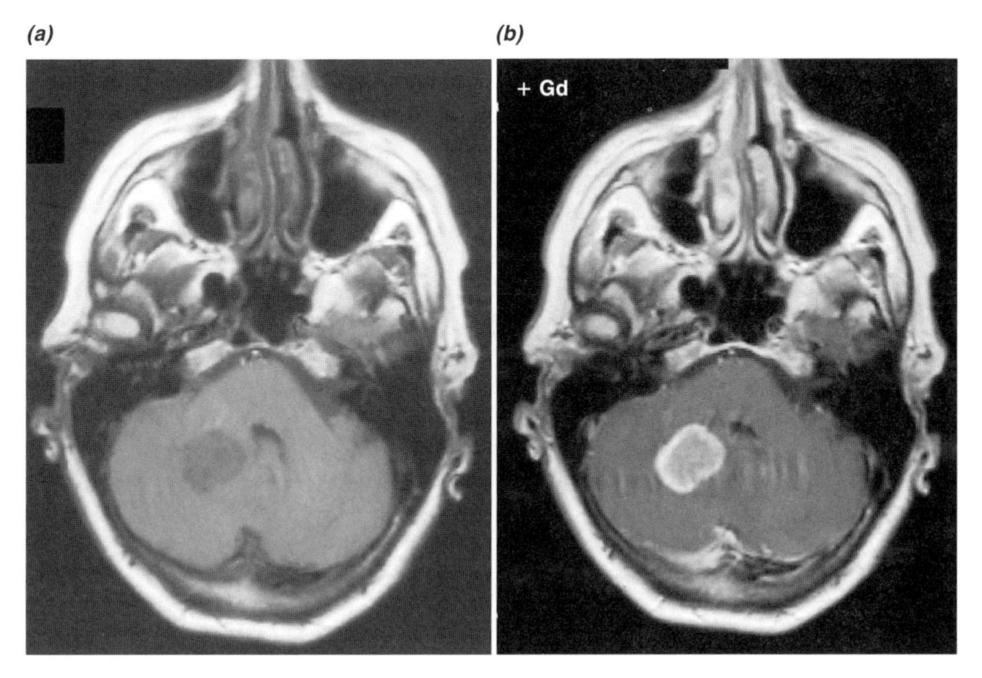

Fig. VII.15. Axial MRI images of the brain showing a tumor in the cerebellum. (*a*) Preinjection and (*b*), postinjection of a gadolinium chelate. The gadolinium contrast agent crosses the blood–brain barrier causing enhancement of the ^{1}H resonance intensity (which, in the particular pulse sequence used, occurs by shortening of ^{1}H T_1). This increases contrast in the region of the tumor, making it easier to see. (Pictures courtesy of D. Collie, SHEFC Brain Imaging Research Centre for Scotland.)

Table VII.3.
Some Gd(III) Complexes Approved for Clinical Use as MRI Contrast Agents

Complex[a]	Brand Name	log K^b	log K^{*c}
$[Gd(DTPA)(H_2O)]^{2-}$	Magnevist	22.5	17.7
$[Gd(DOTA)(H_2O)]^{-}$	Dotarem	25.3	18.3
$[Gd(DTPA\text{-}BMA)(H_2O)]$	Omniscan	16.8	14.9
$[Gd(HP\text{-}DO3A)(H_2O)]$	ProHance	23.8	17.2
$[Gd(bopta)(H_2O)]^{2-}$	MultiHance	22.5	
$[Gd(DO3A\text{-}butrol)(H_2O)]$	Gadovist		

[a] 1,4,7,10-Tetraazacyclododecane-1,4,7,10-tetraacetate = DOTA, DTPA–BMA = bis(methylamide) derivative of DTPA, HP-DO3A = hydroxypropyl derivative of DOTA, bopta = (9*R*,*S*)-2,5,8-tris(carboxymethyl)-12-phenyl-11-oxa-2,5,8-triazadodecane-1,9-dicarboxylate (a DTPA derivative), DO3A-butrol = 1,4,7-tris(carboxymethyl)-10-(1-(hydroxymethyl)-2,3-dihydroxypropyl)-1,4,7,10-tetraazacyclododecane (a DOTA derivative).

[b] Stability constant = K.

[c] Conditional stability constant (pH 7.4) = K^*.

The distorted octahedral Mn(II) complex Teslascan (mangafodipir trisodium salt, Chart VII.18) is in clinical use for enhancing contrast of the liver. Superparamagnetic nanoparticles consisting of iron oxide coated with dextran are also used as MRI contrast agents. The distribution of particles in the body depends on the particle size. Particles of 30-nm diameter are useful for blood pool imaging, 150 nm for liver imaging, and 300 nm for GI tract imaging after oral administration. Zeolites with Gd(III) incorporated into the pores are also used for GI tract imaging.

Chart VII.18.
The X-ray crystal structures of $[Gd(dota)]^-$ and $[Gd(bopta)]^{2-}$, MRI contrast agents for detection of extracellular (systemic) and liver abnormalities, respectively. The aqua ligand bound to Gd(III) is labeled O(W). Also shown is another agent for enhancing contrast in the liver, the Mn(II) complex of dipyridoxyl diphosphate (mangafodipir).

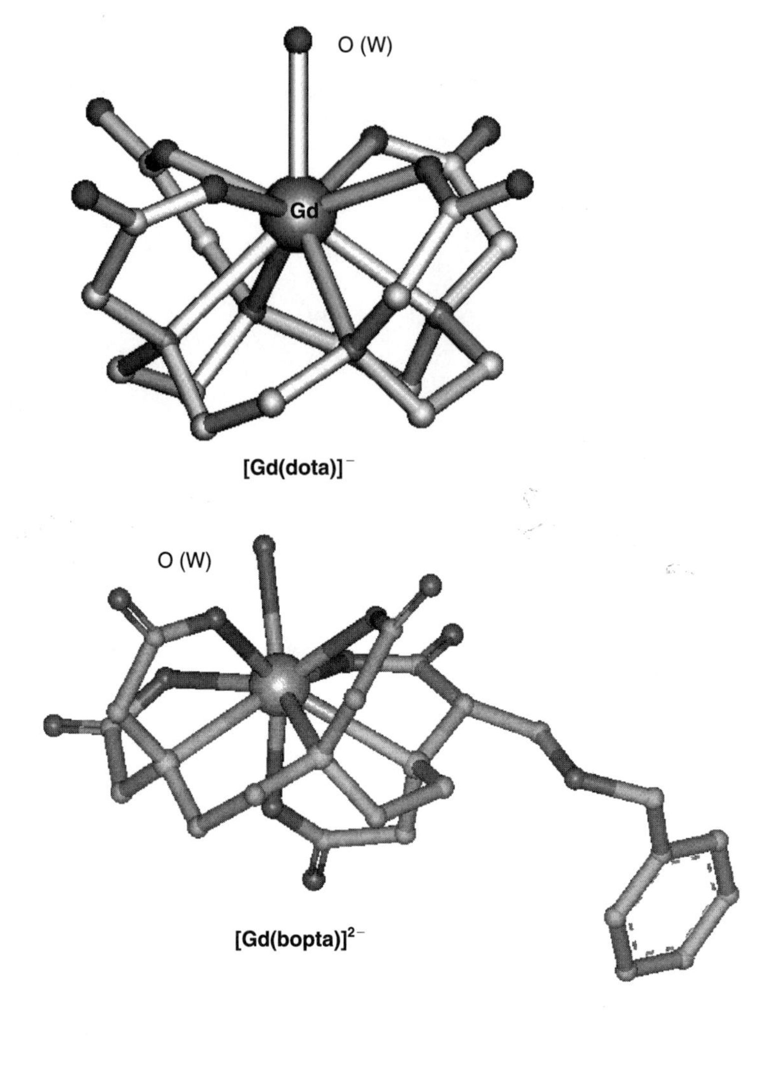

Teslascan

VII.4. Molecular Targets

VII.4.1. Metalloenzymes

Table VII.4 lists some metalloenzymes that are potential targets for novel therapeutic agents.

Matrix Metalloproteinases. Matrix metalloproteinases, matrixins (MMPs), are a family of at least 20 zinc-dependent enzymes that mediate the breakdown of connec-

Table VII.4.
Some Metalloenzymes that Are Targets for Therapeutic Agents

Enzyme	Metal	Inhibitor	Treatment
Matrix metalloproteinases	Zn (Ca)	Hydroxamate peptide mimics Drugs: Marimastat, AG3340, BAY 12-9566	Inflammatory, malignant, and degenerative diseases
Carbonic anhydrase	Zn	Sulfonamides	Hypertension
Angiotensin-converting enzyme (ACE)	Zn	Captopril	Hypertension
Enkephalinase	Zn	Thiorphan, phosphoryl-Leu-Phe	Analgesia
Metallo-beta-lactamases	Zn	*p*-Phenyl substituted dibenzyl succinate	Antibiotics
Cytochrome P450	Fe	Benzodioxoles, halothane, amphetamines, erythromycin	Drug metabolism
Heme oxygenase	Fe	Sn mesoporphyrin	Hyperbilirubinemia
Ribonucleotide reductase	Fe	Hydroxyureas	Cancer
Lipoxygenase	Fe	Acetohydroxamates	Inflammation, hypersensitivty
Prolyl 4-hydroxylase	Fe	Carboxypyridines	Fibroproliferation
Dopamine β-hydroxylase	Cu	Benzyl-imidazole-2-thiol	Neurodisorders

tive tissue. They include collagenases, gelatinases, and stromelysins, and degrade collagens, fibronectin, and gelatins. These enzymes can facilitate tumor growth by making connective tissue, including blood vessels, more fluid.

The improper expression of MMPs leads to a variety of diseases and illnesses, including the destruction of cartilage and bone in rheumatoid and osteoarthritis; tissue breakdown and remodeling during invasive tumor growth, and tumor angiogenesis; degradation of myelin-based protein in the blood–brain barrier following brain injury; loss of aortic wall strength in aneurisms; tissue degradation in gastric ulceration; and the breakdown of connective tissue in periodontal disease. Angiogenesis is the process of vascularization of a tissue involving the development of new capilliary blood vessels.

Inhibitors of MMPs possess: zinc-binding groups such as hydroxamate, thiolate, or carboxylate; at least one functional group that hydrogen bonds to the enzyme backbone; and one or more side chains that undergo van der Waals interactions with enzyme subsites. Hydroxamic acid peptides, which chelate the active-site Zn(II) and have a backbone complementary to the enzyme cleavage site are particularly effective.

The generalized structure of a peptide inhibitor is shown in Fig. VII.16. The substitutent R^1 increases activity against collagenase and can be modified to provide oral bioavailability; R^2 is the major determinant of activity and selectivity; R^3 and R^4 can encompass a wide range of functional groups, with aromatic and bulky groups preferred. Variation of these substituents affects the MMP-inhibitory characteristics of the formulated agent.

Chart VII.19 shows Batimastat [BB-94, 4-(*N*-hydroxyamino)-2(*R*)-isobutyl-3(*S*)-(thienylthiomethyl) succinyl]-L-phenylalanine-*N*-methylamide)] a broad spectrum inhibitor of matrixins including human neutrophil collagenase and human stromelysin.

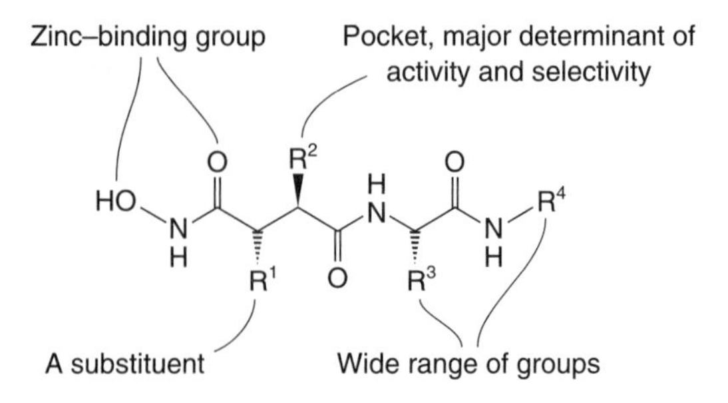

Fig. VII.16.
Backbone of matrix metalloproteinase inhibitors.

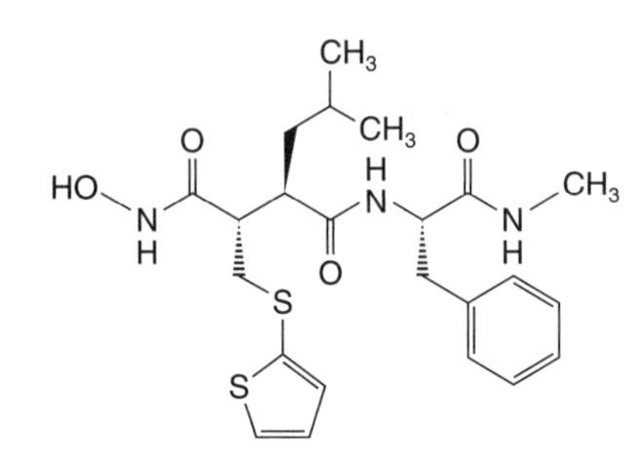

Chart VII.19.
Batimastat (BB-94), a potent inhibitor of human neutrophil collagenase and human stromelysin. The hydroxamic acid group binds to the catalytic zinc in the enzyme (see Fig. VII.17).

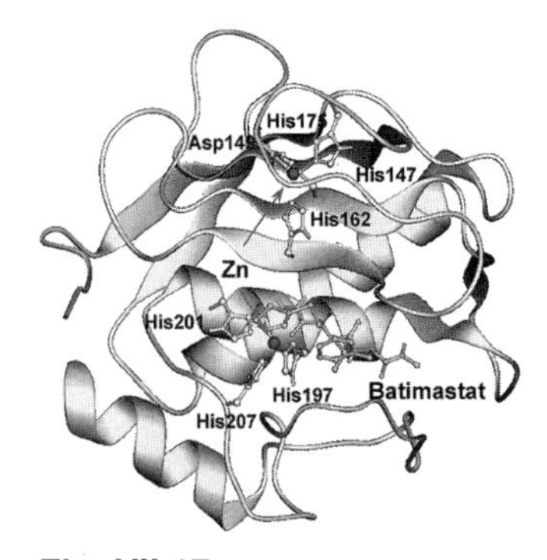

Fig. VII.17.
X-ray crystal structure of the batimastat–human neutrophil collagenase complex, showing the inhibitor coordinated to the catalytic Zn(II) center (PDB code: 1MMR).

It inhibits MMPs at nanomolar concentrations. X-ray studies show that Batimastat coordinates to the catalytic Zn(II) of human neutraphil collagenase in a bidentate manner via the NO^- and carbonyl oxygens of the hydroxamate group, as shown in Fig. VII.17. Other groups form hydrogen bonds to the enzyme and cause conformational changes around the active site. Batimastat has been in clinical trials for the treatment of breast and ovarian cancers. There also is interest in the use of such inhibitors for the treatment of other conditions.

VII.4.2. Insulin Mimetics

The usual treatment for insulin-dependent diabetes is daily injections of insulin. Insulin mimetics with a suitably low toxicity may be useful for treating this form of diabetes (type I). Vanadium and chromium complexes, in various oxidation states, mimic some of the effects of insulin. Vanadium compounds are perhaps better called *insulin-enhancing* drugs since they do not work in the absence of insulin.

Vanadium, V(V), as vanadate and V(IV) as vanadyl stimulate glucose uptake and oxidation, and glycogen synthesis. Vanadate has been shown to be effective in diabetic rats, although too toxic for humans. Vanadyl, as $VOSO_4$, is also unsuitable since high doses are needed on account of its poor oral absorption. However, when conjugated to organic ligands, the species formed have proved to be less toxic and can have improved aqueous solubility and lipophilicity. The square-pyramidal compound, bis(maltolato)oxovanadium(IV) (BMOV), illustrated in Chart VII.20, is three times more effective than $V^{IV}OSO_4$ as an orally active insulin-mimetic agent. The complex is rapidly oxidized to the dioxovanadium(V) species *in vivo*. The ligand, maltol, is itself an approved food additive and BMOV is now also a common constituent of nutritional supplements (a Web search is revealing!).

It is not clear whether V(IV) or V(V) is the active insulin-mimetic redox state of vanadium. Vanadates act as phosphate analogues, and there is evidence for potent inhibition of phosphotyrosine phosphatases. Insulin receptors are membrane-spanning tyrosine-specific protein kinases activated by insulin on the extracellular side to catalyze intracellular protein tyrosine phosphorylation. Peroxovanadium compounds, for example, *b* in Chart VII.20, can induce autophosphorylation at Tyr residues and inhibit the insulin-receptor-associated phosphotyrosine phosphatase. They show insulin-mimetic properties both *in vivo* and *in vitro*. Reducing agents such as GSH and ascorbic acid may reduce V(V) to V(IV).

The major accumulation of vanadium in the body is in the bones and kidneys. Vanadium accumulated in the bone is subsequently released slowly to other organs via the blood stream, which may give rise to long-term activity, perhaps negating the need to take the drug as often.

A chromium complex ("termed glucose tolerance factor") containing GSH and nicotinamide ligands and isolated from Baker's yeast was once thought to be an insulin mimetic. Evidence for this now is not strong. The lipophilic complex [Cr(III)(picolinate)$_3$], chromium picolinate, (marketed as a "nutritional" supplement under the same name and widely used as a weight-loss supplement), is said to affect metabolic parameters regulated by insulin. Chromium picolinate is extremely stable, and it remains to be seen whether (as suggested) it can take part in redox chemistry inside cells and become involved in the generation of hydroxyl radicals and potentiate DNA and lipid damage. Chromate, $[Cr^{VI}O_4]^{2-}$, has been shown to have insulin-like action on glucose transport. The toxicity of both Cr(VI) and Cr(III) may be due to redox reactions in cells that generate Cr(V).

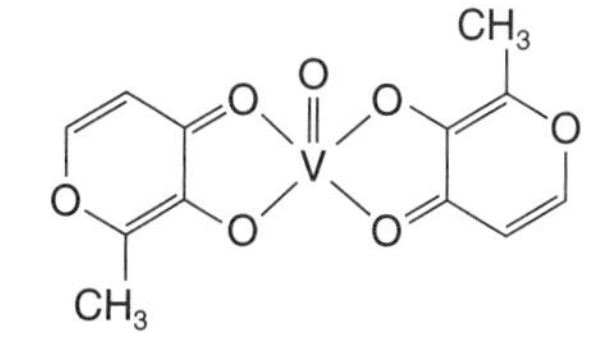

A naturally occurring oligopeptide of 10 amino acids, so-called *Low-Molecular-Weight Chromium-binding substance* (LMWCr, ~1500 Da) *isolated from yeast,* may function as part of a novel insulin-signaling autoamplification mechanism by stimulating insulin receptor protein kinase activity. Its activity is directly proportional to the Cr content of the oligopeptide [with maximum activity at four Cr(III) per oligopeptide]. The trinuclear chromium complex $[Cr^{III}{}_3O(O_2CCH_2CH_3)_6(H_2O)_3]^+$, Chart VII.21, can activate protein tyrosine kinase activity of the insulin receptor. When propionate is replaced by acetate, the complex inhibits both the membrane phosphatase and kinase activity, rather than activating it.

Molybdenum, which like chromium is a member of group 6 (VI.13) of the periodic table, is an essential element and is present in important redox enzymes such as xanthine oxidase. Ammonium tetrathiomolybdate is a copper-lowering drug, and is currently used in medicine for the treatment of Wilson's disease (copper overload, Section VIII.5.2), and as an inhibitor of angiogenesis. The $[MoS_4]^{2-}$ complex may be useful for the treatment of cancer, diseases of inflammation, and fibrosis. Ammonium tetrathiomolybdate can remove Cu from metallothionein, and is also the drug of choice for treating chronic copper poisoning in sheep. Interestingly, orally administered sodium tungstate has recently been reported to have antidiabetic properties.

(a)

(b)

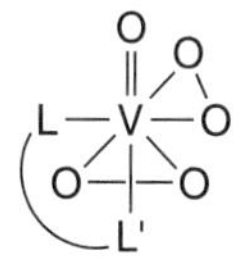

L-L' = phen, oxalate, or picolinate

Chart VII.20.
Orally active insulin-mimetic agents. Bis(maltolato)-oxovanadium(IV) (*a*), and a peroxovanadate complex (*b*).

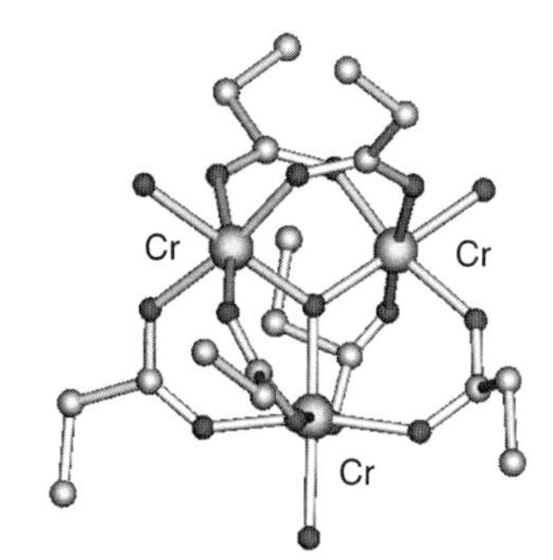

Chart VII.21.
X-ray crystal structure of the trinuclear Cr complex $[Cr_3O(O_2CCH_2CH_3)_6(H_2O)_3]^+$, a potential insulin mimetic.

VII.4.3. Antimicrobial and Antiviral Agents

Many silver(I) compounds are potent antibacterial agents although their mechanism of action is not understood. They are widely used for the treatment of burn wounds, which are readily infected. A typical agent is the insoluble polymeric compound *silver sulfadiazine,* shown in Chart VII.22, which slowly releases Ag(I) ions in the wound. Impregnation of *surgical materials* with Ag(I) improves their sterility. When suspended in solution, AgCl/TiO_2 composites (e.g., 20:80) formulated with sulfosuccinate salts maintain an equilibrium of ppm–ppb levels of cytotoxic Ag(I) ions. Recent work suggests that bacterial resistance to Ag(I) involves the induction of a His-rich protein.

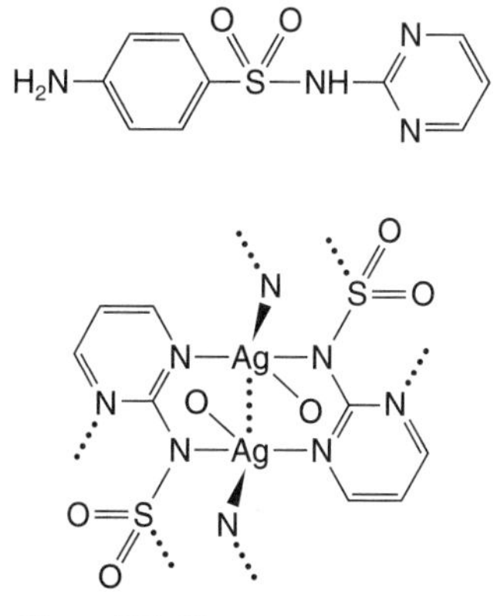

Some mercury compounds are effective antimicrobial agents [e.g., aryl Hg(II) complexes]. *Bacterial resistance* to Hg(II) is well understood and involves the conversion of arylHg^{II} to Hg^{II} by lyase enzymes, and conversion of Hg(II) to volatile Hg(0) by the enzyme mercuric reductase. Resistant bacteria synthesize these enzymes and a series of other proteins, which carefully capture the Hg at the cell membrane and transfer it to the reductase. A DNA-binding Hg sensor protein switches from being a repressor to a transcriptional activator on binding to Hg(II).

Chart VII.22.
Sulfadiazine and the polymeric antibacterial agent silver sulfadiazine.

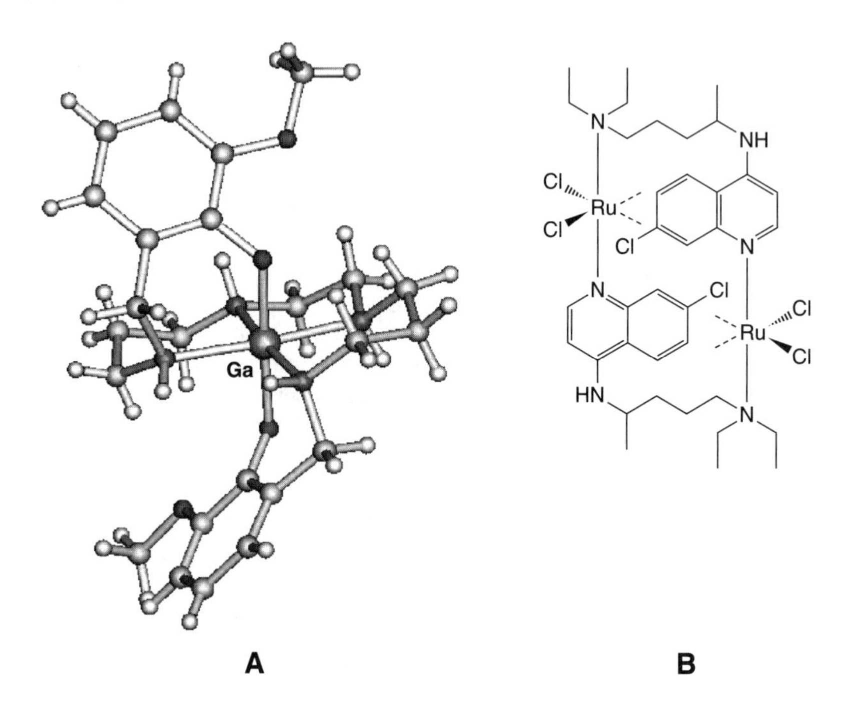

Chart VII.23. X-ray crystal structure of $[Ga(madd)]^{+}$ (**A**), and a potential antimalarial Ru(II) chlorotrimazole complex (**B**).

Antimony compounds such a *N*-methylglucamine antimonite and sodium stibogluconate have been widely used for the treatment of the tropical parasitic disease *leishmaniasis,* which infects ~10–15 million people worldwide. The target for these Sb(V) compounds appears to be the white blood cells called macrophages, where the parasites are destroyed. *In vivo,* reduction of Sb(V) to Sb(III) may occur. Antimony(III) compounds are usually more toxic than Sb(V) compounds. An example is tartar emetic $[Sb_2(tartrate)_2]^{2-}$, well-known in the Middle Ages as the byproduct produced by storing white wine in antimony vessels! Recently, promising *antileishmanial* activity has been found for Rh(III) and Pt(II) complexes, but a wider search for other active metal compounds is needed.

Malaria is caused by a protozoan parasite of the genus *Plasmodium* and is responsible for ~2 million deaths per year. The most common drug for prophylaxis and treatment of malaria is chloroquine, which accumulates within the parasite and interferes with the function of its digestive vacuole (where the host hemoglobin is digested). The Al(III), Ga(III), and Fe(III) complexes of the ligand madd [madd = 1,12-bis(2-hydroxy-5-methoxybenzyl)-1,5,8,12-tetraazadodecane], Chart VII.23, are highly active against chloroquine-resistant *Plasmodium falciparum* and inhibit heme polymerization in the digestive vacuole of the parasite. However, these new complexes have not yet reached the clinic. Metal complexation can increase the potency of organic antiparasitic agents. For example, the Au(I) complex of chloroquine is *more potent than chloroquine itself.* The Ru(II)-chlorotrimazole in Chart VII.23 causes a 90% inhibition of the proliferation of *Trypanosoma cruzi* at a concentration of $10\,\mu M$. This parasite is responsible for *Chagas' disease,* which affects millions of people in Latin America and is currently incurable.

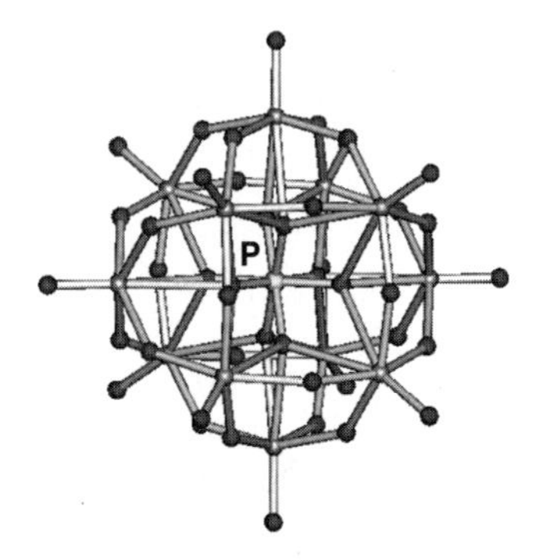

Chart VII.24. Crystal structure of $[PMo_{4.27}W_{7.73}O_{40}]^{6-}$, a Keggin-type polyoxometallate antiviral ion.

Early work in 1971 showed that polyoxometallates can exhibit antiviral activity. Polyoxotungstates of the Keggin and Wells-Dawson type, Chart VII.24, bind to the viral envelope and prevent virus adsorption. However, $[NaW_{21}Sb_9O_{86}](NH_4)_{17}(Na)$ (HPA-23) proved to be too toxic in clinical trials to allow prolonged use. Perhaps this problem can be avoided with other active compounds such as $K_{12}H_2[P_2W_{12}O_{48}]\cdot 24H_2O$.

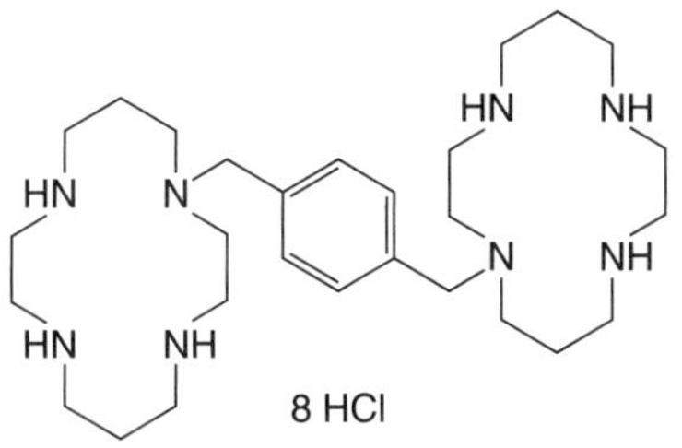

Chart VII.25.
The antiviral macrocyclic bicyclam AMD3100. Metal ions such as Ni(II), Cu(II), and Zn(II) can bind strongly to each ring.

Drugs to fight acquired immune deficiency syndrome (AIDS), caused by human immunodeficiency virus (HIV), are being intensely pursued. One of the *most potent inhibitors of the HIV* is the macrocyclic bicyclam AMD3100, Chart VII.25, which targets the early stages of the retrovirus replicative cycle and blocks HIV-1 cell entry by interacting with the membrane coreceptor protein CXCR4. The Zn_2(II)–bicyclam complex is equally potent. Clinical trials of AMD3100 for the treatment of AIDS have been halted by adverse side effects, but there is continued interest in this drug on account of its ability to mobilize stem cells (in clinical development under the trade name Mozobil). Aurocyanide $[Au(CN)_2]^-$, a natural metabolite of gold antiarthritic drugs, inhibits proliferation of *HIV* in cultured white blood cells at levels as low as 20 n*M* and might be useful for treating AIDS in combination with other drugs. This concentration of $[Au(CN)_2]^-$ is similar to that found in the blood of patients treated with gold antiarthritic drugs.

The radiation enhancing complex Gd-Tex (see the section on Radiosensitizers and Chart VII.11) selectively induces apoptosis in HIV-1-infected T cells and has potential as an anti-HIV agent.

VII.4.4. Superoxide Dismutase Mimics

The radical superoxide (O_2^-) is overproduced in the body after ischemia followed by reperfusion, exposure to radiation, or activation of white blood cells in autoimmune conditions such as arthritis. Protonated superoxide $HO_2^\bullet$ (the perhydroxyl radical) can initiate the autoxidation of lipid membranes and cause selective damage on DNA by abstracting the 5′-hydrogen from the deoxyribose ring of nucleotide units.

Superoxide is removed naturally by superoxide dismutase (SOD) enzymes that are Cu(I/II)/Zn(II) or Mn(II/III) enzymes in eukaryotic cells.

$$O_2^{\bullet-} + M^{n+1} \rightarrow O_2 + M^{n+}$$

$$HO_2^\bullet + H^+ + M^{n+} \rightarrow H_2O_2 + M^{n+1}$$

When these enzymes cannot cope with O_2^- overproduction, the use of a mimic (a synzyme) may be useful in therapy.

Complexes of Cu, Mn, Fe, and Ni can catalyze the dismutation of O_2^- to H_2O_2 and O_2. Aquated Cu(II) ions are nearly as active as Cu/Zn-SOD, but for toxicity and other reasons current emphasis is on the clinical potential of chelated Mn complexes. The Mn(III)–bis(salicylidene)ethylenediamine (salen) complexes such as EUK-8 have protective activity in models of oxidative stress, including neuroprotection. They mimic SOD. The seven-coordinate Mn(II) complex (SC-52608, Chart VII.26) containing 1,4,7,10,13-pentaazacyclopentadecane and trans-dichloro ligands is an efficient catalyst for O_2^- dismutation ($k = 4 \times 10^7\ M^{-1}\ s^{-1}$ at pH 7.4), has a reasonable thermodynamic stability ($\log K = 10.7$ at pH 7.4), and is kinetically inert. Its wide spectrum of biological activity includes inhibition of neutrophil-mediated killing of human aortic endothelial cells; attenuation of inflammation; protection

(a)

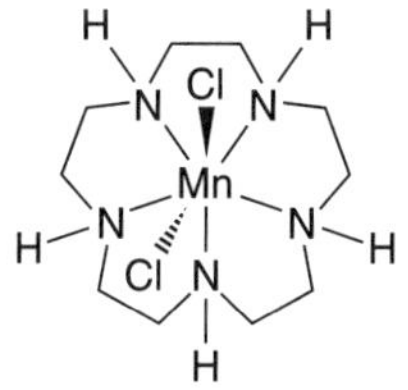

SC-52608

(b)

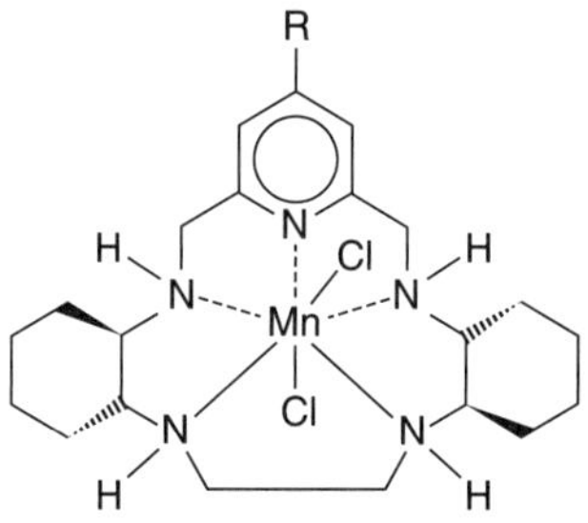

M40403, R = H
M40470, R = $SCH_2CH_2NH_2$

Chart VII.26.
The Mn(II) SOD mimics SC 52608 (*a*), M40403 (*b*), and M40470.

against myocardial ischemia and reperfusion injury; and inhibition of coronary tissue injury. The catalytic efficiency and stability of this complex can be improved by modifying the substituents on the macrocyclic ring. Complex M40403 shown in Chart VII.26 has an activity close to the enzyme itself, and is stable to high-temperature sterilization. Both M40403 and M40419 (Chart VII.26) are currently (2006) in Phase I–III human clinical trials for the treatment of pain including synergistic pain relief with opioids. The complex M40403 also has potential to reduce the erosion of cartilage and bone, as well as chronic inflammation, in rheumatoid arthritis. These SOD mimics can also provide protective surface treatments when covalently linked to implanted medical devices made of, for example, polyethylene, tantalum, platinum, and stainless steel.

VII.4.5. Nitric Oxide

The free radical nitric oxide NO is a muscle relaxant, vasodilator, and neurotransmitter involved in a variety of physiological processes such as regulation of cardiovascular function, signaling in the nervous system, and mediation of host defense against microorganisms and tumor cells. Overproduction or deficiency of NO can cause or contribute to several diseases.

Deficiency in NO can be overcome by sodium nitroprusside $Na_2[Fe^{II}(CN)_5NO]{\cdot}2H_2O$ (Chart VII.27), which is used clinically to lower blood pressure. Its hypotensive effect is rapid, controlling blood pressure within a few minutes. The complex releases NO, which relaxes vascular muscle.

Chart VII.27.
The nitric oxide delivery agent, sodium nitroprusside (*a*). The complex Ru(III) edta (*b*), a nitric oxide scavenger. A Mn(III) porphyrin (*c*) capable of catalytically decomposing peroxynitrite.

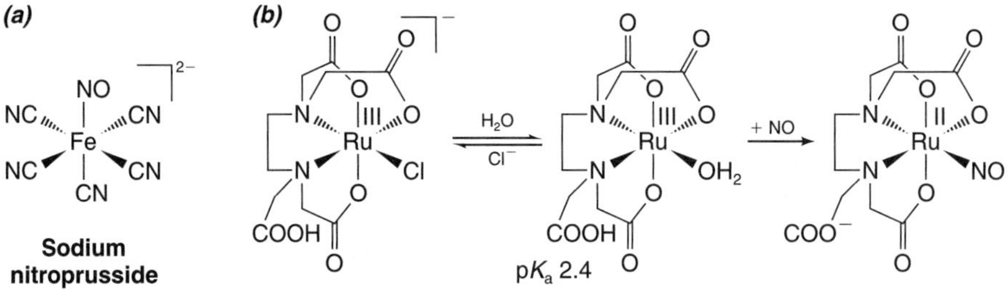

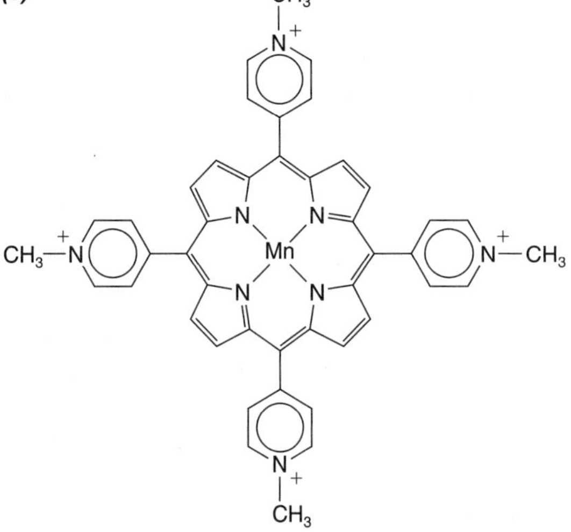

(a) **Sodium nitroprusside**

(b) pK_a 2.4

(c)

Excess levels of NO are thought to be involved in diseases such as inflammatory bowel disease, septic shock, arthritis, stroke, and psoriasis. Nitric oxide scavengers, such as the Ru^{III}–edta complex (edta = ethylenediaminetetraacetato) in Chart VII.26, are therefore potential therapeutic agents. Both the chloro complex and its aqua analogue adduct bind to NO rapidly (rate constant $> 10^8\ M^{-1}\ s^{-1}$) and strongly ($K > 10^8\ M^{-1}$) to form a Ru^{II}–NO complex. The aqua complex can reverse the poor response of arteries to vasoconstrictor drugs, a major clinical problem for patients with septic shock.

Peroxynitrite (ONO_2^-), formed from NO and O_2^-, is highly reactive, being capable of nitrating Tyr and oxidizing metal ions, DNA, lipids, and thiol, and thioether groups in proteins. It is formed as part of the macrophage immune response, for example, during ischemia reperfusion. Peroxynitrite has a short lifetime (~1 s at pH 7.4) but can rapidly diffuse across membranes. Metal complexes capable of catalytically decomposing ONO_2^- are potential drugs. Examples include the Mn(III) complex in Chart VII.27, which can prolong the survival of mice lacking Mn–SOD.

VII.5. Metal Metabolism as a Therapeutic Target

VII.5.1. Mineral Supplements

About 24 elements are essential for the human body. The recommended daily intake of several metals is shown in Table VII.5. Mineral supplements are used to correct deficiencies due to inadequate uptake or storage. Deficiency of essential elements can result in deleterious effects, but excesses can cause toxic effects, Table VII.5.

Iron deficiency causes anemia and leads to unusual tiredness, shortness of breath, a decrease in physical performance, and learning problems in children and adults. It may also increase the risk of infection. There is a need for more research into the optimization of the design of Fe compounds that promote iron absorption. For example, Fe(II) compounds (e.g., ferrous succinate, ferrous fumarate) and not Fe(III) compounds are used as oral iron supplements, but heme iron is well absorbed, as is finely divided Fe metal [so-called carbonyl iron, very fine Fe particles of high purity and uniform size obtained by removing CO from $Fe(CO)_5$], which has been used as a food additive.

Table VII.5.
Intake of Some Metals and Their Effects

Metal	Recommended Daily Dose (U.S.)	Result of Deficiency	Toxic Level	Toxic Effects
Ca	1 g	Bone deterioration	$>2.5\ g\ day^{-1}$	Magnesium deficiency
Cr	5–200 μg	May regulate insulin levels	>70 mg [Cr(III)]	Irregular heartbeat
Fe	10–15 mg	Anemia	$>60\ mg\ kg^{-1}$	Liver cirrhosis, vascular congestion
Cu	~2 mg	Brain disease, anemia, heart disease	7.5 g (death)	Hemolytic anemia
Zn	15 mg	Growth retardation, skin changes	$>500\ mg\ day^{-1}$	Heavy vomiting

Therapy with Fe(II) salts, especially in combination with ascorbate, can lead to artificial poisoning via the generation of reactive oxygen species (Fenton chemistry). It may therefore be safer to administer only Fe(III) compounds with reduction potentials lower than −324 mV at pH 7. These are not readily reduced *in vivo*. Examples are iron dextrin, iron dextran, and iron glucose.

Intake of excess iron can lead to anorexia, oliguria (low excretion of urine relative to fluid intake), diarrhea, hypothermia, diphasic shock, metabolic acidosis, and death. In the United States, iron poisoning is the leading cause of death caused by toxicological agents in children under 6 years of age. The toxic effects of Fe are mediated by the following chemical–biochemical disturbances: direct mucosal necrosis; inhibition of the Krebs cycle (affect on anaerobic metabolism); impairment of capillary permeability; vasodilation; and uncoupling of oxidative phosphorylation.

Deficiencies and excesses of other trace elements are also known. Zinc deficiency can result in growth retardation and underdevelopment of sex organs. Over 200 Zn enzymes are known and zinc-binding proteins regulate the activity of steroid hormones. Excessive intake of Zn, on the other hand, can result in severe vomiting and even death. Excess calcium intake interferes with iron absorption and can also lead to magnesium deficiency, causing hardening of the arteries. Chromium supplements are said to have some therapeutic value in treating diabetes, hypoglycemia, and heart disease although the essentiality of chromium has yet to be conclusively proven. Copper deficiency is an important factor in the development of hyperthroidism and also leads to anemia.

VII.5.2. Copper: Wilson's and Menkes' Diseases

Copper is essential for many cellular functions in the body, but excess Cu is toxic. Menkes' and Wilson's diseases are genetic disorders of copper metabolism in humans. Wilson's disease is the overaccumulation of Cu with subsequent toxic effects, while Menkes' disease is caused by a defect in intracellular Cu transport and subsequent deficit of biochemically available copper. The genes responsible for both diseases code for copper transporting ATPases (see Section VIII.5), transmembrane proteins of ~1400 amino acids, which hydrolyze ATP and have six copper-binding motifs on the cytoplasmic side. The Menkes' Cu-ATPase is in the intestinal mucosal cell membrane (used for Cu absorption), and the Wilson's protein is in the trans-Golgi network (providing Cu for proteins such as ceruloplasmin).

Wilson's disease is an autosomal recessive disorder of Cu transport. Copper accumulates in the cytosol of liver cells, leading to hepatic necrosis and the release of large amounts of Cu into the blood stream. This release causes damage to red cell membranes, leading to hemolytic anemia. Eventually, copper accumulates in other organs, especially the brain, kidneys, and cornea. Accumulation may be due to decreased efflux of copper (biliary excretion) from the body rather than to increased influx. New therapeutic approaches are needed for Wilson's disease. Current treatments include the use of chelating agents to increase Cu excretion (D-penicillamine, 2,3-dimercaptopropanol, triethylenetetramine), and agents that block intestinal absorption of Cu (e.g., Zn salts via induction of metallothionein, and $[MoS_4]^{2-}$), Chart VII.28. Tetrathiomolybdate ($[MoS_4]^{2-}$) has cytokine-inhibiting effects and is also being investigated for possible applications as a copper-lowering drug in the treatment of cancer, fibrosis, and inflammation.

Menkes' disease is a fatal genetic disorder characterized by inefficient intracellular copper transport. More than 9 out of 10 sufferers die by the age of 3. Treatment with copper-histidine, shown in Chart VII.29, is effective in preventing severe neurodegenerative problems, but tissue disorders can still remain, perhaps because Cu is not incorporated into the enzyme lysyl oxidase.

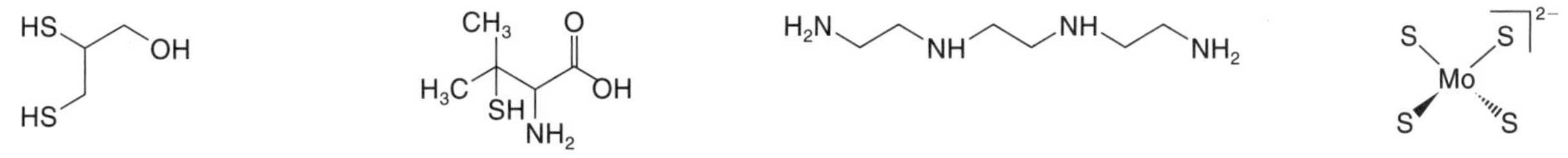

(a) 2,3-Dimercaptopropanol

(b) D-Penicillamine

(c) Triethylenetetramine

(d) Tetrathiomolybdate

Chart VII.28.
Examples of Cu chelating agents used in the treatment of Wilson's disease.

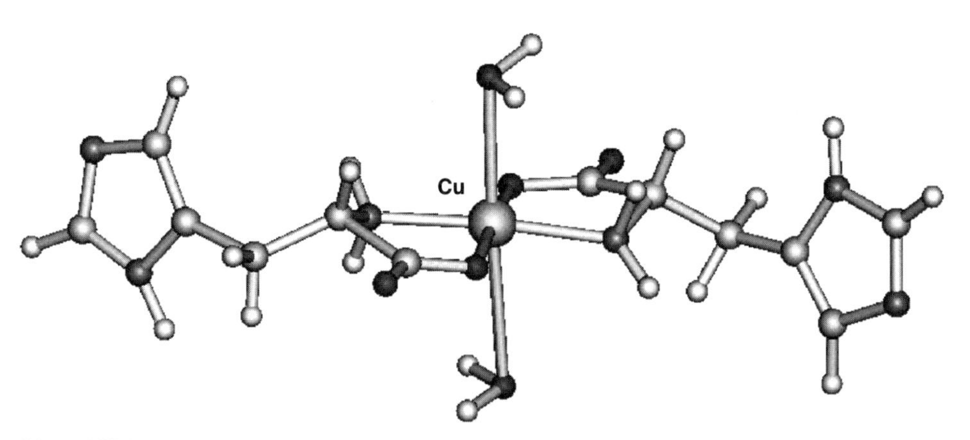

Chart VII.29.
X-ray crystal structure of $[Cu(His)(HHis)(H_2O)_2]^+$, a complex used for treating Menkes' disease.

VII.5.3. Thalassemia

Thalassemia is a genetic form of anemia in which there is an abnormality in the globin portion of hemoglobin. Affected individuals produce small, pale, short-lived red blood cells. People with Thalassemia minor carry the recessive gene, but are healthy. They are unaware that they carry the gene, but their children may inherit Thalassemia major (Cooley's anemia). Treatment for thalassemia usually involves blood transfusions, bone marrow transplants, and/or chelation therapy.

Patients treated intravenously (or subcutaneously) with the Fe(III) chelator Desferal (Desferrioxamine B) combined with blood transfusions are able to lead a normal life. Desferal contains three bidentate hydroxamate chelating groups and forms a very stable six-coordinate Fe(III) complex. However, this drug is slow acting, is rapidly excreted, and is not orally active.

The design criteria for new Fe chelators include

- High affinity for Fe(III), and selectivity over Ca, Mg, Cu, and Zn.
- Low toxicity and stability *in vivo*.
- Low synthetic cost and oral activity.
- Rapid excretion as the Fe complex.

Some new chelating agents that might become useful for iron chelation therapy are shown in Chart VII.30.

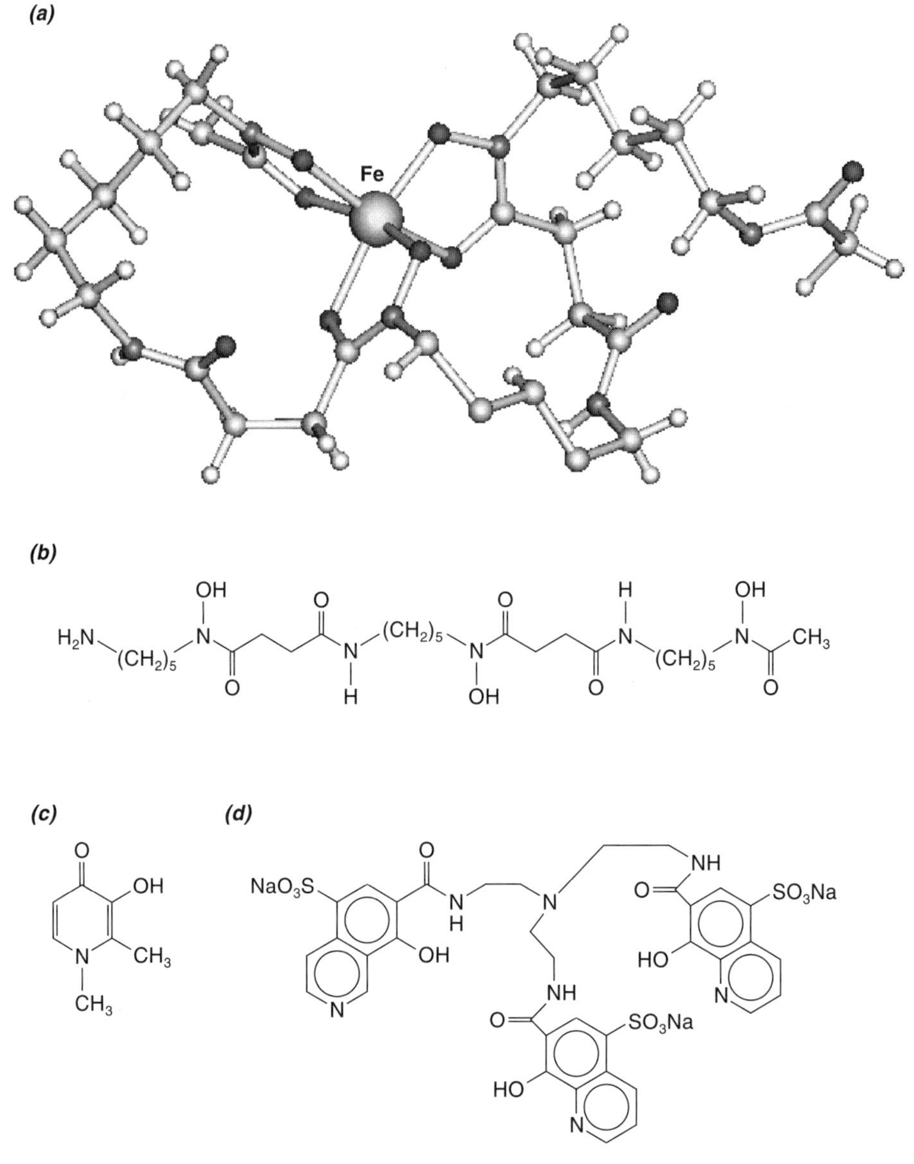

Chart VII.30.
X-ray crystal structure of Fe(III) desferrioxamine (*a*), desferrioxamine (*b*). The orally active Fe chelator 1,2-dimethyl-3-hydroxypyridin-4-one (*c*). The water soluble Fe chelator O-Trensox (*d*).

(a)

Fe

(b)

H_2N $(CH_2)_5$ N OH O O N H $(CH_2)_5$ N OH O O N H $(CH_2)_5$ N OH CH_3 O

(c)

O OH N CH_3 CH_3

(d)

NaO_3S O N H OH N N NH O SO_3Na HO N NH O SO_3Na HO N

VII.6. Conclusions

Inorganic chemistry is beginning to have a major impact on medicine. Although metal compounds in particular have been used in therapy for many centuries, there has been little molecular understanding of their mechanisms of action. This is now changing. The success of cisplatin, currently the leading anticancer drug, has established that structure–activity relationships can be constructed for metal compounds, and that they can be designed to achieve selectivity in biological activity. Organic drugs also often interact with metal ions, and metals in the active sites of metalloenzymes can be target sites for organic drugs.

In general, the activity of a metal complex will depend not only on the metal itself, but also on its oxidation state, on the number and types of bound ligands, and the coordination geometry of the complex. Metal drugs will often be "pro-drugs", which undergo ligand substitution and redox reactions before they reach the target site. It is important to learn how to control such processes and to devise novel drug delivery procedures for metal complexes. Control can involve both thermodynamics and kinetics. The success of Gd(III) complexes as MRI contrast agents has demonstrated how the toxicity and targeting of metals can be finely controlled by the appropriate choice of coordinated ligands. Similarly, radioactive metal complexes can be targeted both for diagnostic imaging and therapy.

Knowledge of genome sequences is likely to have a major impact on our understanding of essential elements. The 3.2×10^9 bases in the human genome code for $\sim$30,000 proteins. These play a major role in controlling the uptake, transport, and utilization of the elements of the periodic table. Although we already know much about the proteins that handle Fe and Cu, much less is known about the genes that control V, Mn, and Sn, for example. Microorganisms can bypass host defense systems by changing their elemental requirements. For example, the Lyme disease pathogen *Borrelia burgdorferi* has eliminated the need for iron, eliminating genes coding for Fe-proteins from its genome sequence. Such information will be very powerful for drug design.

Both essential and nonessential metals can be used in therapy. Nonessential elements should not merely be thought of as toxic. Aluminum, for example, can certainly have toxic effects (e.g., in cases of kidney failure), but for $>$70 years aluminum salts, such as alum, aluminum hydroxide, and aluminum phosphate have been used to add "oomph" to many vaccines, as a helper chemical (an adjuvant) to help the body's immune system react more effectively to antigens. Arsenic too is usually associated with toxicity, but arsenic trioxide has recently been approved for the treatment of certain types of leukemia. Improved understanding of the biological chemistry of these and many other elements will eventually require new techniques and new methods of study.

The exploration of the periodic table for diagnostic and therapeutic purposes in still in its infancy, but we can anticipate that the application of the principles of inorganic chemistry, in combination with modern molecular biology and biotechnology, will have a major impact on medicine in the future.

Acknowledgments

We thank Professor D. Beck (Bowling Green State University), Professor A. van Dijk (Utrecht), Dr. D. Collie (Edinburgh) for the provision of Figures VII.4*b*, VII.12, and VII.15, respectively, and many colleagues for helpful discussion, especially from the EC COST Actions D8 and D20.

Bibliography

Introduction to Metals in Medicine

Abrams, M. and Murrer, B. A., "Metal compounds in therapy and diagnosis", *Science*, **261,** 725–730 (1993).

Thompson, K. H. and Orvig, C., "Boon and bane of metal ions in medicine", *Science*, **300,** 936–939 (2003).

Farrell, N. P. et al., "Biomedical uses and applications of inorganic chemistry", *Coord. Chem. Rev.*, **232,** 1–230 (2002).

Farver, O., "Metals in medicine: inorganic medicinal chemistry", in *Textbook of Drug Design and Discovery*, 3rd ed., Taylor & Francis Ltd., London, UK, 2002.

Krogsgaard-Larsen, P., Liljefors, T., and Madsen, U., Eds., *Textbook of Drug Design and Discovery*, 2002, pp. 364–409.

Sun, H., "Metallodrugs", in *Encyclopedia of Nuclear Magnetic Resonance*, John Wiley & Sons Ltd., Chichester, UK, Vol. 9, 2002, pp. 413–427.

Guo, Z. and Sadler, P. J., "Medicinal Inorganic Chemistry", *Adv. Inorg. Chem.*, **49,** 183–306 (2000).

Clarke, M. J. and Sadler, P. J., *Topics in Biological Inorganic Chemistry*, Vols. 1–2, Metallopharmaceuticals I and II, Springer-Verlag, Berlin, 1999.

Farrell, N. P., Ed., *Uses of Inorganic Chemistry in Medicine*, Royal Society of Chemistry, Cambridge, UK, 1999.

Guo, Z. and Sadler, P. J., "Metals in Medicine", *Angew. Chem. Int. Ed. Engl.*, **38,** 1512–1531 (1999).

Orvig, C. and Abrams, M. J., "Medicinal Inorganic Chemistry", *Chem. Rev.,* **99,** 2201–2204 (1999).

Preusch, P. C., Ed., "Metals in Medicine: Targets, Diagnostics, and Therapeutics". http://www.nigms.nih.gov/news/meetings/metals.html

Sadler, P. J., Ed., "Chemistry of Metals in Medicine—The Industrial Perspective", *Met.-Based Drugs,* **4,** 1–73 (1997).

Berners-Price, S. J. and Sadler, P. J., "Coordination chemistry of metallodrugs: insights into biological speciation from NMR spectroscopy", *Coord. Chem. Rev.,* **151,** 1–40 (1996).

Berthon, G., Ed., *Handbook of Metal–Ligand Interactions in Biological Fluids—Bioinorganic Chemistry,* Marcel Dekker, New York, 1995.

Taylor, D. M. and Williams, D. R., "Trace Element Medicine and Chelation Therapy", Royal Society of Chemistry, Cambridge, UK, 1995.

Sadler, P. J., "Inorganic Chemistry and Drug Design", *Adv. Inorg. Chem.,* **36,** 1–48 (1992).

Sadler, P. J., in *Lectures in Bioinorganic Chemistry,* Nicolini, M. and Sindellari, L., Eds., Cortina Int./Raven, Verona/New York, pp. 1–24, 1991.

Farrell, N., *Transition Metal Complexes as Drugs and Chemotherapeutic Agents,* Kluwer Academic, Boston, 1989.

Howard-Lock, H. E. and Lock, C. J. L., in Wilkinson, G., Gillard, R. D., and McCleverty, J. A., Eds., *Comprehensive Coordination Chemistry,* Pergamon, Oxford, Vol. 6, Chap. 62.2, 1987, pp. 755–778.

Sigel, A. and Sigel, H., Eds., "Metal Ions and Their Complexes in Medication", *Met. Ions Biol. Syst.,* **41** (2004).

Sigel, A. and Sigel, H., Eds., "Metal Complexes in Tumor Diagnosis and as Anticancer Agents", *Met. Ions Biol. Syst.,* **42** (2004).

Gielen, M. and Tiekink, E. R. T., Eds., "Metallotherapeutic Drugs and Metal-Based Diagnostic Agents: The Use of Metals in Medicine", John Wiley & Sons, Inc. Hoboken, N.G., 2005.

Wang, D. and Lippard, S. J., "Cellular processing of platinum anticancer drugs", *Nature Rev. Drug Discov.,* **4,** 307–320 (2005).

Metallotherapeutics

Ali, H. and van Lier, J. E., "Metal Complexes as Photo- and Radiosensitizers", *Chem. Rev.,* **99,** 2379–2450 (1999).

Birch, N. J., "Inorganic Pharmacology of Lithium", *Chem. Rev.,* **99,** 2659–2682 (1999).

Briand, G. G. and Burford, N., "Bismuth Compounds and Preparations with Biological or Medicinal Relevance", *Chem. Rev.,* **99,** 2601–2657 (1999).

Clarke, M. J., Zhu, F., and Frasca, D. R., "Non-Platinum Chemotherapeutic Metallopharmaceuticals", *Chem. Rev.,* **99,** 2511–2533 (1999).

Gelasco, A. and Lippard, S. J., "Anticancer Activity of Cisplatin and Related Complexes", in *Topics in Biological Inorganic Chemistry,* Clarke, M. J. and Sadler, P. J., Eds., Vol. 1, Springer-Verlag, Berlin, 1999, pp. 1–43.

Lippert, B., Ed., *Cisplatin—Chemistry and Biochemistry of a Leading Anticancer Drug,* Wiley-VCH, Weinheim, 1999.

Sava, G., Alessio, E., Bergamo, A., and Mestroni, G., "Sulfoxide Ruthenium Complexes: Non-Toxic Tools for the Selective Treatment of Solid Tumour Metastases", in *Topics in Biological Inorganic Chemistry,* Clarke, M. J. and Sadler, P. J., Eds., Vol. 1, Springer-Verlag, Berlin, pp. 143–169, 1999.

Shaw, C. F. III, "Gold-Based Therapeutic Agents", *Chem. Rev.,* **99,** 2589–2600 (1999).

Wong, E. and Giandomenico, C. M., "Current Status of Platinum-Based Antitumor Drugs", *Chem. Rev.,* **99,** 2451–2466 (1999).

Sun, H., Li, H., and Sadler, P. J., "The Biological and Medicinal Chemistry of Bismuth", *Chem. Ber./Recueil,* **130,** 669–681 (1997).

Best, S. L. and Sadler, P. J., "Gold drugs: mechanism of action and toxicity", *Gold Bull.,* **29,** 87–93 (1996).

Pineto, H. M. and Schornagel, J. H., Eds., *Platinum and Other Metal Coordination Compounds in Cancer Chemotherapy,* Plenum, New York, 1996.

Reedijk, J., "Improved Understanding in Platinum Antitumour Chemistry", *Chem. Commun.,* 801–806 (1996).

Sadler, P. J. and Sue, R. E., "Gold Drugs", in *Handbook of Metal–Ligand Interactions in Biological Fluids, Bioinorganic Chemistry,* Berthon, G., Ed., Vol. 2, Marcel Dekker, New York, pp. 1039–1051, 1995.

Fricker, S. P., Ed., *Metal Complexes in Cancer Therapy,* Chapman & Hall, London, 1994.

Keppler, B. K., Ed., *Metal Complexes in Cancer Chemotherapy,* VCH, Weinheim, 1993.

Pineto, H. M. and Schornagel, J. H., Eds., *Platinum and Other Metal Coordination Compounds in Cancer Chemotherapy 2,* Plenum, New York, 1993.

Birch, N. J., Ed., *Lithium and the Cell,* Academic Press, London, 1991.

Clarke, M. J., Ed., *Ruthenium and Other Metal Complexes in Cancer Chemotherapy,* Springer-Verlag, Heidelberg, 1989.

Zhang, C. X. and Lippard, S. J., "New metal complexes as potential therapeutics", *Curr. Opin. Chem. Biol.,* **7,** 481–489 (2003).

Imaging and Diagnosis

Merbach, A. E. and Toth, E., Eds., *The Chemistry of Contrast Agents in Medicinal Magnetic Resonance Imaging,* John Wiley & Sons Ltd., Chichester, UK, 2001.

Anderson, C. J. and Welch, M. J., "Radiometal-Labeled Agents (Non-Technetium) for Diagnostic Imaging", *Chem. Rev.,* **99,** 2219–2234 (1999).

Caravan, P., Ellison, J. J., McMurry, T. J., and Lauffer, R. B., "Gadolinium(III) Chelates as MRI Contrast Agents: Structure, Dynamics, and Applications", *Chem. Rev.,* **99,** 2293–2352 (1999).

Jurisson, S. S. and Lydon, J. D., "Potential Technetium Small Molecule Radiopharmaceuticals", *Chem. Rev.*, **99,** 2205–2218 (1999).

Liu, S. and Edwards, D. S., "^{99m}Tc-Labeled Small Peptides as Diagnostic Radiopharmaceuticals", *Chem. Rev.*, **99,** 2235–2268 (1999).

Nicolini, M., Bandoli, G., and Mazzi, U., Eds., *Technetium and Rhenium and other Metals in Chemistry and Nuclear Medicine,* Vol. 5, SGEditorali, Padova, 1999.

Packard, A. B., Kronauge, J. F., and Brechbiel, M. W. "Metalloradiopharmaceuticals", in *Topics in Biological Inorganic Chemistry,* Clarke, M. J. and Sadler, P. J., Eds., Vol. 2, Springer-Verlag, Berlin, pp. 45–115, 1999.

Tweedle, M. F. and Kumar, K., "Magnetic Resonance Imaging (MRI) Contrast Agents", in *Topics in Biological Inorganic Chemistry,* Clarke, M. J. and Sadler, P. J., Eds., Vol. 2, Springer-Verlag, Berlin, pp. 1–43, 1999.

Volkert, W. A. and Hoffman, T. J., "Therapeutic Radiopharmaceuticals", *Chem. Rev.*, **99,** 2269–2292 (1999).

Nicolini, M., Bandoli, G., and Mazzi, U., Eds., *Technetium and Rhenium in Chemistry and Nuclear Medicine,* Vol. 4, SGEditorali, Padova, 1995.

Nicolini, M., Bandoli, G., and Mazzi, U., Eds., *Technetium and Rhenium in Chemistry and Nuclear Medicine,* Vol. 3, Cortina International, Verona, 1989.

Molecular Targets

Riley, D., Udipi, K., and Ornberg, R., "Radical Alternatives", *Chem. Br.*, **36,** 43–44 (2000).

Riley, D. P., "Functional Mimics of Superoxide Dismutase Enzymes as Therapeutic Agents", *Chem. Rev.*, **99,** 2573–2587 (1999).

Thompson, K. H., McNeill, and Orvig, C., "Vanadium Compounds as Insulin Mimics", *Chem. Rev.*, **99,** 2561–2571 (1999).

Whittaker, M., Floyd, C. D., Brown, P., and Gearing, A. J. H., "Design and Therapeutic Application of Matrix Metalloproteinase Inhibitors", *Chem. Rev.*, **99,** 2735–2776 (1999).

Metal Metabolism as a Therapeutic Target

Andersen, O., "Principles and Recent Developments in Chelation Treatment of Metal Intoxification", *Chem. Rev.*, **99,** 2683–2710 (1999).

Sarkar, B., "Treatment of Wilson and Menkes Diseases", *Chem. Rev.*, **99,** 2535–2544 (1999).

PART B

METAL ION CONTAINING BIOLOGICAL SYSTEMS

CHAPTER VIII

Metal Ion Transport and Storage

VIII.1. Transferrin

Philip Aisen
Department of Physiology and Biophysics
Albert Einstein College of Medicine
Bronx, New York 10461

Contents

VIII.1.1. Introduction: Iron Metabolism and the Aqueous Chemistry of Iron

VIII.1.1.1. Uses and Hazards of Iron in Biological Systems

Few elements are put to the range of uses Nature has found for iron in living organisms. Iron is essential for the transport of dioxygen (O_2) by hemoglobin and hemerythrin; it is indispensable in the energy-transducing pathways of electron transport; it plays a pivotal role in the formation of deoxyribonucleotides by ribonucleotide reductase; it participates in a multitude of oxidation and oxygenation reactions as well as in the elimination of noxious metabolites of O_2; it is crucial for the fixation of nitrogen and hydrogen; it functions in a variety of hydratases and hydrolases; it is a magnetosensor or geosensor for organisms as diverse as bacteria, honey bees, and homing pigeons; and it is even put to use in the mineralization of teeth by invertebrates of the limpet family.

Iron in many of its complexes shuttles more or less easily between its two principal oxidation states, the ferrous (Fe^{2+}) and ferric (Fe^{3+}) states, losing or gaining an electron. This one-electron transfer is the reaction underlying most of the biological functions of iron proteins and complexes. However, the acceptor or supplier of that electron, often O_2 or one of its derivatives, may then be converted to a reactive free radical, capable of attacking and damaging biological molecules. Such reactive free radicals are believed responsible for much of the cellular injury resulting from acute and chronic iron overload (see Section XI.1).

To survive in an aerobic world, an organism must be able to store iron and transport it from one location to another while maintaining it in soluble and bioavailable form and preventing it from undergoing uncontrolled redox chemistry. Two proteins that serve these requirements well are ferritin, the iron storage protein in a wide range of organisms (see Section VIII.1.2) and transferrin, the iron-transport protein in multicellular organisms from cockroach to human.

VIII.1.2. Transferrin: The Iron Transporting Protein of Complex Organisms

VIII.1.2.1. Nomenclature and Overall Structure

Three major classes of transferrins have been distinguished: (1) serum transferrin or serotransferrin, functioning to transport iron in blood to iron-dependent cells; (2) lactoferrin or lactotransferrin, found in many extracellular fluids and in the specific granules of polymorphonuclear leukocytes; and (3) ovotransferrin (formerly called conalbumin) of avian egg white. The latter two proteins are thought to provide a bacteriostatic milieu by sequestering iron in physiological fluids; a basic sequence (lactoferricin) in lactoferrin may also have direct microbicidal activity.

The transferrin molecule is a single-chain bilobal structure of molecular weight 75,000–80,000; human transferrin is a glycoprotein of 679 amino acid residues and two carbohydrate chains.[1] The lobes, representing the N- and C-terminal halves of the molecule, are joined by a short connecting peptide strand and each lobe bears a single iron-binding site. The two lobes have $\sim$50% sequence identity and similar overall structure, and the exon–intron organization of the human gene shows identical patterns for the lobes. Duplication and fusion of a precursor gene that specified a single-sited single-lobe structure must have given rise to the modern transferrin molecule, but no verified single-sited "half-transferrin" has yet been identified.

Each lobe is comprised of two domains, which form a hydrophilic cleft in which the iron-binding site of that lobe is embedded; the domain arrangement is of particular relevance in understanding the binding and release of iron by transferrins. The first domain in each lobe, designated N1 or C1 depending on the lobe in which it is found, is half completed by the first 90 or so amino acid residues of the peptide chain in that lobe. The chain then crosses to the second domain, N2 or C2, which is fully formed by the next $\sim$160 residues, before returning to complete the first domain.[2] Domains are thus joined by two antiparallel connecting strands, each of which contributes a ligand to the iron-binding site, with each domain also contributing a ligand. The two domains of each lobe are brought to close apposition as iron is bound,

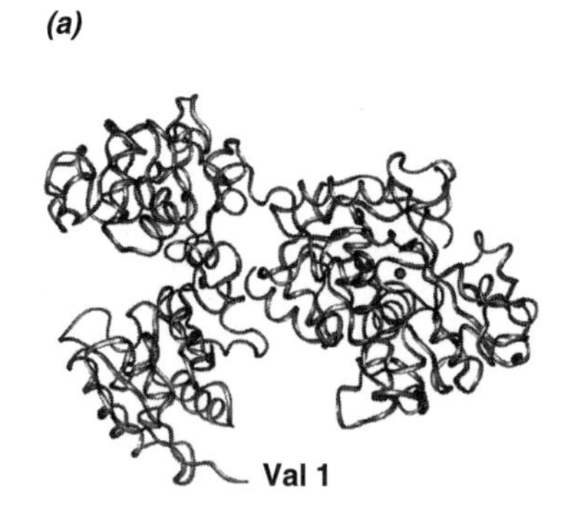

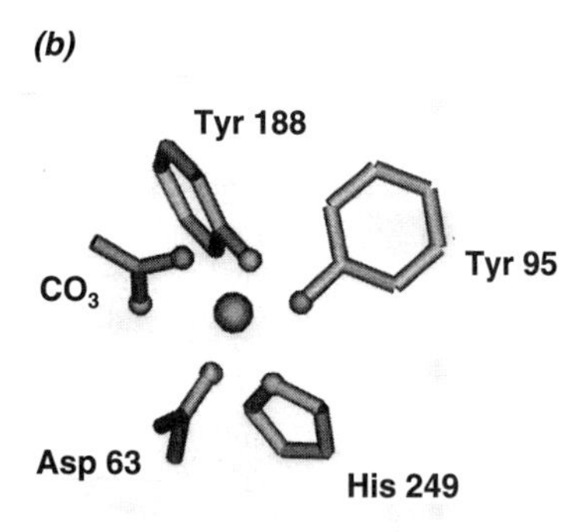

Fig. VIII.1.1. (*a*) Ribbon structure of monoferric human transferrin showing the unoccupied N-lobe in the open configuration and the iron-bearing C-lobe in the closed conformation. The Fe atom is shown as a ball. (Coordinates kindly provided by Dr. Harmon Zuccola.) (*b*) The distorted octahedral arrangement of Fe ligands in the N-lobe of human transferrin.[3]

but release of iron from transferrin is accompanied by a rigid rotation of the domains from the closed state to an open configuration, permitting access to a new incoming iron atom (Fig. VIII.1.1*a*). Little is known of the events initiating the switch from closed-to-open, or open-to-closed configurations, but the switch helps account for binding of iron that is sufficiently strong to resist hydrolysis, yet is physiologically reversible. Reversibility of iron binding enables each transferrin molecule in the human circulation to experience some 100–200 cycles of iron binding, transport, and release during its lifetime.

VIII.1.2.2. Anion-Binding Requirement of Transferrins

The binding of iron and other metals depends on the concomitant binding of a carbonate or other "synergistic" anion, which acts as a bridging ligand between protein and metal ion. Ordinarily, carbonate serves as the synergistic anion, but, in its absence, a variety of other bifunctional anions may take its place. Just as the sites differ in their affinities for metal ions, so do they differ in their affinities for synergistic anions, with carbonate the most tightly bound. Carbonate is anchored to transferrin by a complex network of electrostatic and hydrogen bonds involving the positively charged side chain of arginine (arg) and several peptide nitrogens. The degree of cooperativity between metal and anion binding is remarkable. Neither is bound strongly in the absence of the other. *Anion-dependent metal binding is unique to the transferrins and may be taken as the defining feature of the proteins.*

VIII.1.2.3. Iron Ligands and Iron Binding

The binding ligands at the specific sites of transferrin are provided by two tyrosine residues (Tyr 95 and Tyr 188), one aspartate (Asp 63), a histidine (His 249), and the synergistic carbonate in bidentate coordination to iron (Fig. VIII.1.1*b*; designations refer to the N-terminal lobe of the human protein). Binding ligands are widely dispersed in the linear sequences of each lobe, and therefore are not encoded by a single exon. In ovotransferrin, substitution of nitrilotriacetate for carbonate displaces two of the protein ligands, leaving only the two tyrosines coordinated and an open conformation despite persistence of bound iron.[4] Interlobe, as well as interdomain, interactions influence the iron-binding properties of the two sites and probably contribute to the differences in thermodynamic, kinetic, and spectroscopic properties of the binding sites despite their identical ligands. Site directed mutagenesis has been helpful in revealing remote effects of protein structure on iron-binding and iron-releasing properties of the lobes, particularly when correlated with X-ray structures.[5–8]

One remarkable feature distinguishing the N-lobe and helping to account for its weaker binding in the physiological pH range, as well as its greater susceptibility to attack by protons, is the presence of a dilysine pair formed by residues on each domain.[9] When both lysines are protonated, and hence positively charged, the resulting strong electrostatic repulsion drives the domains apart[10] and thus facilitates release of iron from the protein to the iron-dependent cell in which transferrin is localized to a low pH compartment where release takes place (see Section VIII.1.3). Thus, the protein fabric, apart from the ligating residues, exerts important and often distinctive effects on the iron-binding properties of the two lobes.

VIII.1.3. Iron-Donating Function of Transferrin

Transferrin is a major source of iron for all cells and the only physiological source of iron for most. The iron-donating interaction of transferrin with cells begins with its binding to the transferrin receptor on the plasma membrane, which selectively recognizes iron-bearing transferrin at pH 7.4. The complex of transferrin and receptor invaginates to form a pit that becomes coated with clathrin. Budding of the pit produces a clathrin-coated vesicle or endosome, which rapidly sorts for protection against lysosomal degradation. Upon maturation, the endosome develops proton-pumping competence in an ATP-dependent process that lowers endosomal pH to 5.3–6.0, depending on cell type.[11] At the lowered pH, the receptor favors iron-free apotransferrin, binding it more strongly than iron-bearing transferrin. The lowered pH facilitates iron release from transferrin by direct effect on the protein and via the action of receptor.[12] Released iron is probably reduced to make it available to DMT-1 (originally designated Nramp2), a divalent metal ion transporter,[13–15] and it is then exported from the endosome to the cytoplasm. Whether reduction occurs as part of the release mechanism, or subsequent to release, is still unclear. The reduction potential of Fe^{3+} bound to free transferrin is too low, less than -500 mV, for physiological reduction; but binding to receptor raises the potential to -285 mV, so that Fe^{3+} in the complex of transferrin and its receptor is accessible to reduction driven by pyridine nucleotides.[16] The Fe^{2+} ion is bound to transferrin much more

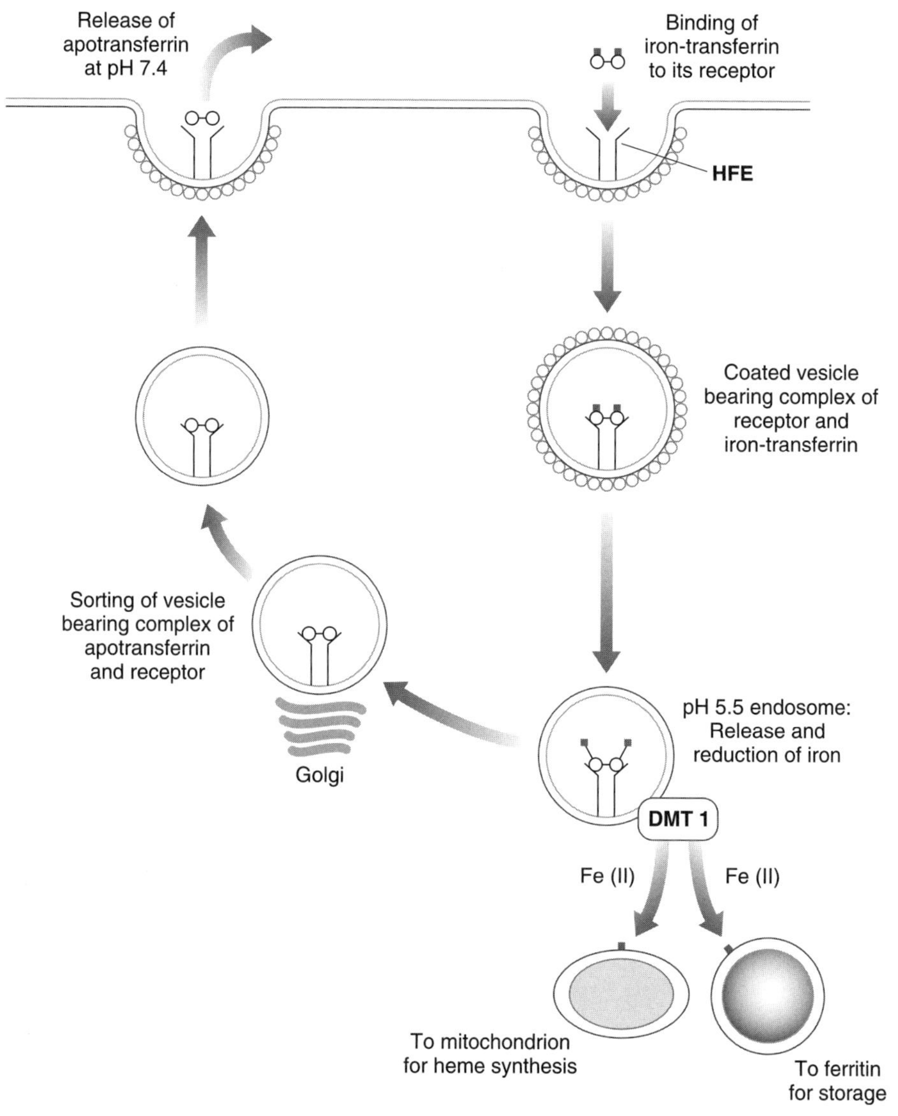

Fig. VIII.1.2.
The transferrin-to-cell cycle in iron metabolism.

weakly than Fe^{3+}, raising the possibility that reductive release of Fe from transferrin within the cell occurs.

Released iron then finds its way to storage by ferritin, incorporation into heme by ferrochelatase, utilization for synthesis of iron-dependent enzymes, or export to the circulation. Iron export is managed by a membrane exporter, variously named IREG1,[17] MTP1,[18] and ferroportin1,[19] which, like DMT1, accepts only Fe^{2+}. A specific iron reductase, Dcytb, has been identified in iron-absorbing duodenal cells,[20] and may function in intracellular iron transport as well. Since iron exported to the circulation is destined for transferrin, an oxidative event must accompany export. A multicopper oxidase, with similarities to ceruloplasmin, has been implicated in such an event.[21]

Meanwhile, the low-pH endosome, with its iron-depleted transferrin, is sorted and directed to the cell surface, where it fuses with the plasma membrane and reexposes transferrin to pH 7.4, thereby freeing the protein for a new cycle of iron binding, transport, and delivery to cells (Fig. VIII.1.2). In reticulocytes, the entire cycle is completed within 2 min; longer times are required by cells with less vigorous mechanisms of iron uptake.

VIII.1.4. Interaction of Transferrin with HFE

Hemochromatosis, a disorder leading to generalized iron overload and, if untreated, to death, is among the most prevalent of genetic diseases with a gene frequency $> 6\%$ in the population of the western world. After years of search by many groups, the genetic abnormality in hemochromatosis has recently been identified.[22] The defect is in a protein designated HFE (a protein that, when mutated, leads to hemochromatosis and generalized iron overload). An HFE molecule closely resembles those of the major histocompatibility complex (MHC). In its normal state, HFE associates with the transferrin receptor at the cell surface (Fig. VIII.1.2) and acts to depress iron uptake from transferrin.[23–25] Loss of such association and its effect on cellular iron uptake helps explain, at least in part, the deranged iron metabolism and increased iron absorption in hemochromatosis, although much remains unknown about the underlying molecular mechanisms.

References

General Reference

1. Aisen, P., "Transferrin, its receptor, and the uptake of iron by cells", in *Metal Ions in Biological Systems,* Vol. 35. Sigel, A. and Sigel, H., Eds., Marcel Dekker, New York, 1998, pp. 585–631.

Specific References

2. Baker, E. N. and Lindley, P. F., "New perspectives on the structure and function of transferrins", *J. Inorg. Biochem.,* **47,** 147–160 (1992).
3. MacGillivray et al., "Two high-resolution crystal structures of the recombinant N-lobe of human transferrin reveal a structural change implicated in iron release", *Biochemistry,* **37,** 7919–7928 (1998).
4. Mizutani, K., Yamashita, H., Kurokawa, H., Mikami, B., and Hirose, M., "Alternative structural state of transferrin—The crystallographic analysis of iron-loaded but domain-opened ovotransferrin N-lobe", *J. Biol. Chem.,* **274,** 10190–10194 (1999).
5. He, Q. Y., Mason, A. B., Woodworth, R. C., Tam, B. M., MacGillivray, R. T., Drady, J. K., and Chasteen, N. D., "Mutations at nonliganding residues Tyr-85 and Glu-83 in the N-lobe of human serum transferrin—Functional second shell effects", *J. Biol. Chem.,* **273,** 17018–17024 (1998).
6. Beatty, E. J., Cox, M. C., Frenkiel, T. A., Tam, B. M., Mason, A. B., MacGillivray, R. T., Sadler, P. J., and Woodworth, R. C., "Trp128Tyr mutation in the N-lobe of recombinant human serum transferrin: ^{1}H- and ^{15}N-NMR and metal binding studies", *Protein Eng.,* **10(5),** 583–591 (1997).
7. Zak, O., Tam, B., MacGillivray, R. T. A., and Aisen, P., "A kinetically active site in the C-lobe of human transferrin", *Biochemistry,* **36,** 11036–11043 (1997).
8. Halbrooks, P. J., He, Q. Y., Briggs, S. K., Everse, S. J., Smith, V. C., MacGillivray, R. T., and Mason, A. B., "Investigation of the mechanism of iron release from the C-lobe of human serum transferrin: Mutational analysis of the role of a pH sensitive triad", *Biochemistry,* **42,** 3701–3707 (2003).

9. Dewan, J., Mikami, B., Hirose, M., and Sacchettini, J. C., "Structural evidence for a pH-sensitive dilysine trigger in the hen ovotransferrin N-lobe: implications for transferrin iron release", *Biochemistry,* **32,** 11963–11968 (1993).
10. Steinlein, L. M., Ligman, C. M., Kessler, S., and Ikeda, R. A., "Iron release is reduced by mutations of lysines 206 and 296 in recombinant N-terminal half-transferrin", *Biochemistry,* **37,** 13696–13703 (1998).
11. Sipe, D. M., Jesurum, A., and Murphy, R. F., "Absence of Na^{+},K^{+}-ATPase regulation of endosomal acidification in K562 erythroleukemia cells. Analysis via inhibition of transferrin recycling by low temperatures", *J. Biol. Chem.,* **266,** 3469–3474 (1991).
12. Bali, P. K., Zak, O., and Aisen, P., "A new role for the transferrin receptor in the release of iron from transferrin", *Biochemistry,* **30,** 324–328 (1991).
13. Fleming, M. D., Trenor, C. C., III, Su, M. A., Foernzler, D., Beier, D. R., Dietrich, W. F., and Andrews, N. C., "Microcytic anaemia mice have a mutation in *Nramp2,* a candidate iron transporter gene", *Nat. Genet.,* **16(4),** 383–386 (1997).
14. Gunshin, H., Mackenzie, B., Berger, U. V., Gunshin, Y., Romero, M. F., Boron, W. F., and Nussberger, S., "Cloning and characterization of a mammalian proton-coupled metal-ion transporter", *Nature* (*London*), **388,** 482–488 (1997).
15. Fleming, M. D., Romano, M. A., Su, M. A., Garrick, L. M., Garrick, M. D., and Andrews, N. C., "*Nramp2* is mutated in the anemic Belgrade (*b*) rat: Evidence of a role for Nramp2 in endosomal iron transport", *Proc. Natl. Acad. Sci. U.S.A.,* **95,** 1148–1153 (1998).
16. Dhungana, S., Taboy, C. H., Zak, O., Larvie, M., Crumbliss, A. L., and Aisen, P., "Redox properties of human transferrin bound to its receptor", *Biochemistry,* **43,** 205–209 (2004).
17. McKie, A. T. et al., "A novel duodenal iron-regulated transporter, IREG1, implicated in the basolateral transfer of iron to the circulation", *Mol. Cell,* **5(2),** 299–309 (2000).
18. Abboud, S. and Haile, D. J., "A novel mammalian iron-regulated protein involved in intracellular iron metabolism", *J. Biol. Chem.,* **275,** 19906–19912 (2000).
19. Donovan, A. et al., "Positional cloning of zebrafish ferroportin1 identifies a conserved vertebrate iron exporter", *Nature* (*London*), **403,** 776–781 (2000).
20. McKie, A. T. et al., "An iron-regulated ferric reductase associated with the absorption of dietary iron", *Science,* **291,** 1755–1759 (2001).
21. Vulpe, C. D., Kuo, Y. M., Murphy, T. L., Cowley, L., Askwith, C., Libina, N., Gitschier, J., and Anderson, G. J., "Hephaestin, a ceruloplasmin homologue implicated in intestinal iron transport, is defective in the *sla* mouse", *Nat. Genet.,* **21,** 195–199 (1999).
22. Feder, J. N. et al., "A novel MHC class I-like gene is mutated in patients with hereditary haemochromatosis", *Nat. Genet.,* **13,** 399–408 (1996).
23. Feder, J. N., Penny, D. M., Irrinki, A., Lee, V. K., Lebron, J. A., Watson, N., Tsuchihashi, Z., Sigal, E., Bjorkman, P. J., and Schatzman, R. C., "The hemochromatosis gene product complexes with the transferrin receptor and lowers its affinity for ligand binding", *Proc. Natl. Acad. Sci. U.S.A.,* **95,** 1472–1477 (1998).
24. Lebrön, J. A. and Bjorkman, P. J., "The transferrin receptor binding site on HFE, the class I MHC-related protein mutated in hereditary hemochromatosis", *J. Mol. Biol.,* **289,** 1109–1118 (1999).
25. Salter-Cid, L., Brunmark, A., Li, Y. H., Leturcq, D., Peterson, P. A., Jackson, M. R., and Yang, Y., "Transferrin receptor is negatively modulated by the hemochromatosis protein HFE: Implications for cellular iron homeostasis", *Proc. Natl. Acad. Sci. U.S.A.,* **96(10),** 5434–5439 (1999).

VIII.2. Ferritin

Elizabeth C. Theil
Children's Hospital
Oakland Research Institute
and The University of
California-Berkeley
Oakland, CA 94609

Contents

VIII.2.1. Introduction: The Need for Ferritins

The chemistry of iron and dioxygen collide at the neutral values of pH in biology making Fe insoluble or highly toxic. Ferrous ions, Fe^{2+}, and metallic iron, Fe^{0}, oxidize (corrode) rapidly in the presence of dioxygen (O_2) at pH 7. The products are biologically toxic reactive oxygen species (ROS, see Chapter XI.1), hydrated H^+, and ferric ion, Fe^{3+}. Water coordinated to Fe^{3+} at neutral pH ionizes, Fe^{3+} multimers bridged by O are formed, and rust (hydrated ferric oxide) precipitates. Air-breathing organisms need complex strategies to control Fe availability and prevent toxicity due to ROS and rust. The solution Nature has found for the problem of achieving soluble ferric iron at concentrations needed for life is the ferritin family of proteins.[1]

VIII.2.2. Ferritin: Nature's Nanoreactor for Iron and Oxygen

Ferritin is a very soluble protein found in plants, animals, and bacteria that concentrates ferric iron to levels equivalent to millimolar, by directing iron and oxygen reactions to the cavity or "nanoreactor" ($V = 256\,\text{nm}^3$) in the protein center where a solid, ferric mineral forms.[1,2] Deletion of a ferritin gene is lethal and uses both DNA and mRNA regulation, which emphasizes the biological importance of ferritin.[3,4] Nonlethal mutations in ferritin genes cause a number of diseases in the central nervous system of humans. The nanoreactor cavity of ferritin can be filled with solvent, or partly or completely with the iron oxide mineral. When the protein has no mineral it is often called apoferritin as if the iron were a cofactor; but this is a misnomer in the case of almost all types of ferritin, since little or none of the iron is bound to the protein, and most of the iron in ferritin is in the mineralized solid, stabilized by the coating of protein on the mineral surface.

The ferritin protein is comprised of 24 polypeptide subunits ($\sim$20 kDa each, with an assembled molecular weight of $\sim$480,000 kDa) that spontaneously fold into bundles of α-helices. The polypeptide helix bundles self-assemble into the hollow sphere (Fig. VIII.2.1). Each ferritin protein sphere of 24 subunits has symmetry around the two-, three-, and fourfold axes, with eight pores (ion channels) that are created at the junction of three subunits (Fig. VIII.2.2).[2,5,6]

Subunit designations in ferritin are H, for subunits with ferroxidase (F_{ox}) sites (Fig. VIII.2.3) that catalyze the oxidation and coupling of two Fe^{2+} ions with oxygen to create mineral precursors, and L, for the rare, noncatalytic subunit, found only in animals. The ferritin subunit designations are historical and refer to identification of H in heart and L in liver. The H subunits self-assemble, and in animal tissue, where

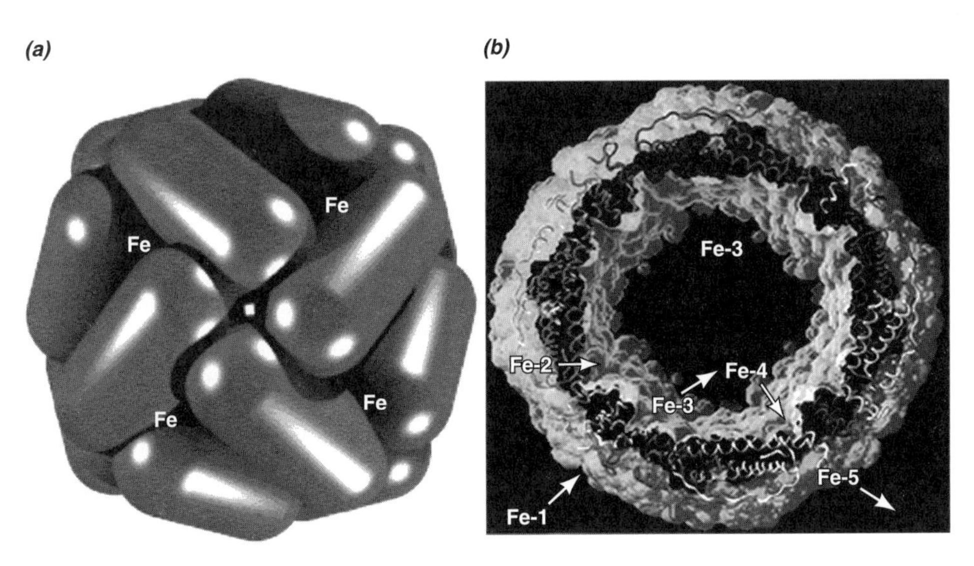

Fig. VIII.2.1.
Ferritin protein assembled (*a*) and as a cross-section through the center of the protein sphere (*b*). The volume of the protein is $\sim$864 nm^3 and the volume of the nanoreactor cavity, where the ferric iron mineral (hydrated ferric oxide) forms, is $\sim$256 nm^3. Amino acid side chains are deleted over the subunit polypeptide backbones; the drawings are modified from Ref. 5. [See color insert.]

Sites Fe-2 and Fe-5 are better studied than Fe-1, Fe-3, and Fe-4.
Sites for movement of Fe(II) and Fe(II) mineral precursors have not yet been identified
Fe-1 = Ferritin F_{ox} site
Fe-2 = Ferroxidase (F_{ox}) site
Fe-3 = Mineral nucleation site
Fe-4 = Nanomineral
Fe-5 = Fe Exit Pore

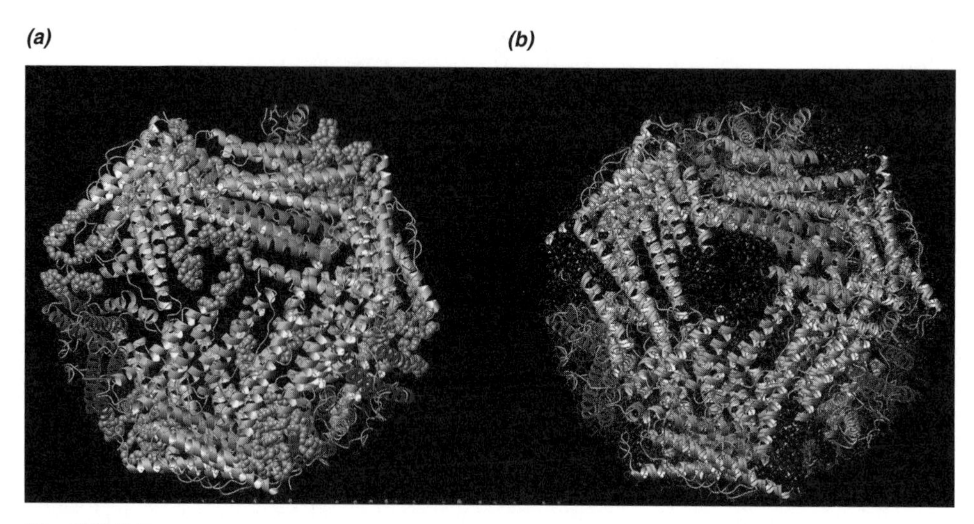

Fig. VIII.2.2.
Gated ferritin pores. Pore helices, composed of three pairs of helices, as in ferritin (one set from each subunit), assemble around the threefold axis of 24-subunit ferritins, and are exquisitely sensitive to heat, low concentrations (m*M*) of chaotropes, such as urea or guanidine, and mutation of conserved residues (pores). The gates open when the pore helices unfold differentially, with the global protein structure intact, to accelerate iron removal through increased access between the iron mineral, reductants, and chelators. (*a*) Closed gates with pore helices; (*b*) Open gates, where pore helices are unfolded and so disordered that they appear structureless in protein crystals, although the polypeptide chains are intact. [Modified by X. Liu from Ref. 6, Issue cover.] [See color insert.]

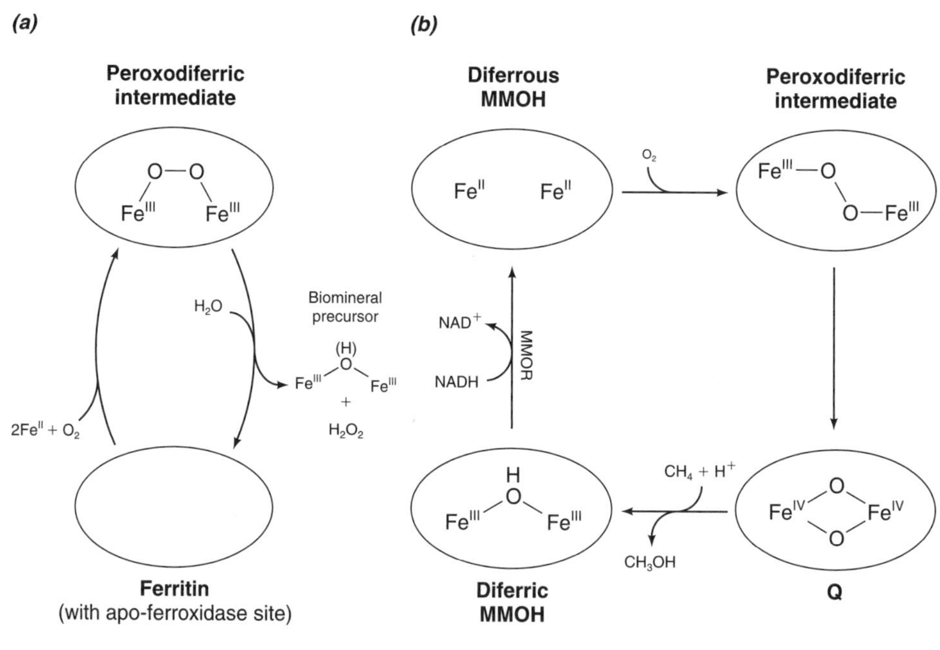

Fig. VIII.2.3.
Diferric peroxo intermediates in Fe(II) substrate or Fe(II) cofactor oxidation. A structure for the ferritin F_{ox} site is shown, based on cocrystals of recombinant ferritin (no iron) with Mg-ferritin cocrystals (from Ref. 5), where the metal–metal distance is 3.3 Å. In solution, based on extended X-ray absorption fine structure (EXAFS) studies, the Fe–Fe distance in the diferric peroxo intermediate is 2.5 Å[14,15] and for the diferric oxo/hydroxo mineral precursor is 3.0 Å.[11] Ligand differences at the di-iron sites in cofactor proteins and the F_{ox} site include, substitution at one of the iron sites of Glu for His and Asp for Glu, which may account for the weaker metal–protein interactions at the di-iron substrate site in ferritin. The F_{ox} site structure and Fe(II) oxidation pathway are shown from left and middle right, with the pathway for the di-iron cofactors on the right.

the transcription and translation produce a different ratio of H:L subunits in each type of differentiated cells, they coassemble with L subunits. Ferritin with different H:L subunit ratios appears to have F_{ox} activity matched to the metabolism of each cell type. Since each H ferritin subunit contains one F_{ox} site, there can be as many as 24 F_{ox} sites/ferritin protein, but it is not known how many of these sites can function in parallel.

Bacteria have three types of ferritin structure: the 24 subunit ferritin similar to that in higher plants and animals; used for managing cellular iron; A 24 subunit ferritin called bacterioferritin which contains up to 12 heme groups; and a mini-ferritin with the same spherical cavity structure created by 12 subunits that is used to manage oxygen and peroxide. All these ferritin types use the protein cavity as a nanoreactor for aqueous oxygen and iron chemistry that forms the iron mineral (rust). Before the structure of the bacterial mini-ferritins were known, many were called dps proteins (DNA Protection during Starvation) for their ability to increase bacterial cell and DNA resistance to hydrogen peroxide formed during stress and infection. The many human pathogens with dps proteins make the mini-ferritins potential drug targets. *Escherichia coli* has four different ferritin genes, encoding either 24-subunit ferritins or 12-subunit mini-ferritins to facilitate function in all the different oxygen and iron environments in which it grows.

VIII.2.2.1. Ferritin Mineralization: Iron Entry

The overall chemical reactions involved in ferritin mineralization are shown in Eqs. 1–3:

$$\begin{aligned}&\sim 36\text{–}2200 \text{ Hydrated } Fe^{2+} + \sim 18\text{–}1100\ O_2\\&\quad \rightarrow \sim 18\text{–}1100 \text{ hydrated } [Fe^{III}\text{–}O\text{–}O\text{–}Fe^{III}] \qquad (1)\end{aligned}$$

$$\begin{aligned}&\sim 18\text{–}1100 \text{ Hydrated}[Fe^{III}\text{–}O\text{–}O\text{–}Fe^{III}]\\&\quad \rightarrow \sim 36\text{–}2200 \text{ hydrated } Fe^{3+} + \sim 18\text{–}1100\ H_2O_2 \qquad (2)\end{aligned}$$

$$\begin{aligned}&\sim 36\text{–}2200 \text{ Hydrated } Fe^{3+} \rightarrow [Fe_2O_3.(H_2O)_x]_{\sim 18\text{–}1100}\\&\quad + \sim 54\text{–}3300 \text{ hydrated } H^+ \qquad (3)\end{aligned}$$

In the first step of ferritin mineralization (Eq. 1), a pair of Fe^{2+} atoms and O_2 are substrates at the F_{ox} site and are rapidly converted in milliseconds (ms) to a diferric peroxo intermediate (Fig. VIII.2.3).[1,7–14] The F_{ox} site in ferritin,[4,5] has structural similarity to diferric carboxylate catalysts such as methane monoxygenase hydroxylase (MMOH) (see Section 11.5), ribonucleotide reductase (RNR), and some fatty acid desaturases, based on ferritin-metal crystals of Ca^{2+} and Mg^{2+}. Because Fe is usually a substrate in ferritin, in contrast to a cofactor, cocrystallization with Fe is not feasible. In ferritin cocrystallized with Ca^{2+} and Mg^{2+}, the F_{ox} site model is likely closer to the Fe^{2+} substrate rather than the diferric peroxo intermediate or the diferric oxo/hydroxo product.

The iron ligands in ferritin F_{ox} sites diverge from the di-iron carboxylate protein cofactor sites of MMOH, RNR, and fatty acid desaturase by substitution of glutamine (Gln or Q), with an amide side chain, for glutamate (Glu or E), with a long carboxylate side chain, while aspartate (Asp or D), with a short carboxylate side chain) substitutes for histidine (His or H), with an aromatic amine side chain. The conserved di-iron cofactor sites are E,EXXH/E,EXXH, compared to E,EXXH/E, QXXD. The ligand differences between the di-iron cofactor sites in the organic oxidases and the iron substrate-binding sites in ferritin F_{ox} sites, account for retention of the di-iron cofactors or release of the diferric products. Since the DNA codons for Gln/Glu and Asp/His each differ by only one nucleotide, it is easy to envision the evolution of the major functional change from cofactor to substrate binding by the relatively minor genetic change of two nucleotide substitutions.

In some bacterial ferritins, such as the dps proteins, the ferroxidase reaction (Eq. 4) can use iron and hydrogen peroxide as substrates.

$$(2\ Fe^{2+})(H_2O)_6 + H_2O_2 \rightarrow (Fe_2O_3)(H_2O)_3 + 2\ H_2O + 4\ H^+ \qquad (4)$$

The putative oxidation site is on the inner surface of the protein at the mineral cavity, rather than in the center of each subunit, as in eukaryotic ferritins, where oxygen concentrates and detoxifies iron. Both types of ferritins make iron minerals.

The blue ($\lambda_{max} = 650$ nm) diferric peroxide (DFP) complex intermediate is the first detectable ferric species during oxidation at the F_{ox} site and has been characterized by Mössbauer,[11,12] resonance Raman,[13] and EXAFS spectroscopies.[14] The unusually short Fe–Fe distance in the ferritin DFP (2.54 Å),[14] has also been observed in the DFP of a ribonucleotide reductase variant R2 protein.[15] The ligands for the diferric peroxo intermediate have recently been determined,[9] using kinetic studies of chimeric proteins, since the formation and decay rates are too rapid to permit trapping in crystal structures. The minimal conserved ligands for site A are Glu, Glu, His, and are identical to those in other DFP forming proteins, while the minimal ligands for site B, which controls DFP kinetics, are Glu, Gln, Asp; natural variations of Asp influence DFP kinetics. Mammalian mitochondrial ferritin and frog H ferritin are examples of proteins with the Asp → Ser substitution.[9,10]

$$2\ Fe^{2+} + O_2 + 2\ H^+ \rightarrow Fe^{3+}\text{–O–}Fe^{3+} + H_2O_2 \qquad (5)$$

The F_{ox} reaction (Eq. 5) oxidizes and couples ferrous ions to form diferric products in the form of diferric oxo/hydroxo mineral precursors and hydrogen peroxide. Translocation of the hydrated Fe^{3+} ferritin mineral precursor species across the protein shell to the cavity and into the mineral requires minutes to hours, while oxidation and decay of the diferric peroxo species occurs in milliseconds. The migration path of the mineral precursors from the F_{ox} site to the nanocavity is a distance of 10–50 Å, which depends on the final location of the iron in the mineral (see sites Fe-2 and Fe-4 in Fig. VIII.2.1). Why migration of the mineral precursors is so slow is not understood.

During ferritin mineralization, protons diffuse from the protein during the hydrolysis of ferric ions. For a ferritin mineral with 2000 Fe atoms, ~5000 protons are produced (2.5 H^+/Fe). Proton diffusion from ferritin can be detected by titration (or if the buffer capacity of the solution is too low, by the white precipitate of acid-denatured ferritin). If the protons released by mineralizing 2000 Fe atoms in ferritin were retained in the cavity, the pH in the cavity would drop to ~0.3, low enough to hydrolyze the peptide bonds in the ferritin protein shell. The lag between formation of the diferric oxo mineral precursors (ms) and transfer to the mineral (hours), may relate to the rate of proton exit and the maintenance of protein stability. Hydrogen peroxide is also released during F_{ox} site activity (Fig. VIII.2.3), and can be detected by enzymes outside the protein.[16,17] In animals, where ferritin is composed of both F_{ox} active (H) and inactive (L) subunits, the subunit composition may relate to cellular activity of catalases or peroxidases or to whether the hydrogen peroxide from the diferric peroxo decay flows out of the protein into the cell, or into the protein to react with the ferric mineral.

VIII.2.2.2. Ferritin Demineralization and Ferritin Gated Pores

Ferritin iron can be released rapidly upon demand, although the mineral can be stable for long periods of time in the absence of biological signals. Examples of removing iron from ferritin are blood loss (synthesis of red blood cells), seed germination (synthesis of plant iron and heme proteins), embryonic development (rapid synthesis of fetal/adult red blood cells to replace the embryonic type), chloroplast maturation (synthesis of ferredoxin), and nitrogen fixation (synthesis of nitrogenase and leghemoglobin by nodules of leguminous plants).

Dissolving a mineral inside a protein is not a common reaction in biology, the mineralized iron core of ferritin and bone and tooth minerals being the primary characterized examples (see Chapter VI). It is no surprise, then, that the mechanism has been difficult to elucidate. The overall chemistry at first appears straightforward in that it is simply the reverse of mineralization, that is, ferric oxide hydration and reduction of Fe^{3+} to Fe^{2+}. The complexity of the process becomes apparent when we realize that the rates and locations of these reactions involve a phase transition from solid to solution. Moreover, the reaction rates must be carefully controlled to respond to specific biological signals and to avoid the havoc of the uncontrolled flow of electrons and protons.

The two current hypotheses for dissolving the ferritin mineral are not mutually exclusive and may occur in parallel or under different biological conditions. The first hypothesis, "destroy and dissolve", proposes a compartmentalization of ferritin inside an acidic cytoplasmic vesicle, the lysosome, followed by digestion of the ferritin protein, dissolution of the mineral, and export of the iron ions.[18]

The second hypothesis is regulated unfolding/opening the ferritin pores (Fig. VIII.2.2). The gates of the ferritin pores are a set of conserved amino acids that form bridges between the six helices of the pores.[1,6,19,20] When the gates are "open" or unfolded, reductants such as NADH (reduced nicotinamide adenine dinucleotide), FMN (flavin mononucleotide), and chelators have increased access to the ferritin iron mineral leading to rates of removal of iron from the minerals that are increased as much as 30-fold.[19,20] Unfolding of the ferritin pores is observed in crystals of recombinant protein with an amino acid substitution at a gate residue,[19] by circular dichroism (CD) analysis of helix content, and by the effects of temperature or very low concentrations (mM) of unfolding reagents such as urea or guanidine. Rates of mineral reduction and chelation increase without changes in global structure or function.[6]

The selective advantage of the ferritin pores may involve protection of the ferritin mineral from the excess of reductants found inside cells until the signal that iron is needed induces opening of the gated pores. Gated pores, composed of multiple polypeptide helices, as in ferritin, also occur in the ion channels of cells and organelles that manage the transfer of ions through the lipid bilayer. In ferritin, the pores are embedded in two layers of polypeptide helices stabilized by hydrophobic interactions to form the protein sphere in structural analogy to the lipid bilayer of cell and organelle membranes. The short helices of the gated pores in ferritin are functionally equivalent to the longer helices of gated pore–ion channel proteins that are embedded in membranes. Studies on gated pore–ion channel proteins are a source of information about iron transport into and out of the gated ferritin pores.

VIII.2.2.3. Ferritin Genes

All multicellular organisms and most single-celled organisms have ferritin and complex mechanisms to regulate ferritin gene expression. Iron and dioxygen, hormones, growth factors, and inflammation all regulate ferritin genes.[3,4] One part of the control system uses DNA as the target to regulate transcription [i.e., messenger ribonucleic acid (mRNA) synthesis], but a second part uses a mRNA target to regulate translation (i.e., protein synthesis), thereby increasing the range of signal response by the ferritin genes. The ferritin mRNA regulatory sequence called the iron-responsive element (IRE) has a metal-binding pocket related to selectivity of repressor protein binding. One of the protein repressors is related to aconitase, a metal-binding protein (see Section IX.4). A number of genes encoding proteins for iron and oxygen metabolism also encode mRNAs with IRE structures forming a natural combinatorial array, members of which have quantitatively different or hierarchal responses to the iron and oxygen signals.[3] To live with dioxygen and iron for the last 2 billion years on Earth has clearly required the evolution of complex and sophisticated protein and genetic machinery, which is dramatically illustrated by the ferritin family.

References

General References

1. Theil, E. C., "Ferritin", in *Handbook of Metalloproteins*, Messerschmidt, A., Huber, R., Poulos, T., and Wieghardt, K., Eds., John Wiley & Sons, Inc., Chichester, 2001, pp. 771–781.
2. Chasteen, N. D. and Harrison, P. M., "Mineralization in ferritin: an efficient means of iron storage", *J. Struct. Biol.*, **126,** 182–194 (1999).
3. Theil, E. C. and Eisenstein, R. S., "Combinatorial mRNA Regulation", *J. Biol. Chem.*, **275,** 40659–40662 (2000).
4. Torti, F. M., and Torti, S. V., "Regulation of ferritin genes and protein", *Blood,* **99,** 3505–3516 (2002).

Specific References

5. Trikha, J., Waldo, G. S., Lewandowski, F. A., Theil, E. C., Weber, P. C., and Allewell, N. M., "Crystallization and Structural Analysis of Bullfrog Red Cell L-Subunit Ferritins", *Proteins,* **18,** 107–118 (1994).
6. Liu, X., Jin, W., and Theil, E. C., "Opening protein pores with chaotropes enhances Fe reduction and chelation of Fe from the ferritin biomineral", *Proc. Natl. Acad. Sci. U.S.A.*, **100,** 3653–3658 (2003).
7. Hempstead, P. D., Yewdall, S. J., Fernie, A. R., Lawson, D. M., Artymiuk, P. J., Rice, D. W., Ford, G. C., and Harrison, P. M., "Comparison of the Three-dimensional Structures of Recombinant Human H and Horse L Ferritins at High Resolution", *J. Mol. Biol.*, **268,** 424–448 (1997).
8. Ha, Y., Shi, D., Small, G.W., Theil, E.C., and Allewell, N. M., "Crystal structure of bullfrog M ferritin at 2.8 Å resolution: analysis of subunit interactions and the binuclear metal center", *J. Biol. Inorg. Chem.*, **4,** 243–256 (1999).
9. Periera, A. S., Small, G. W., Krebs, C., Tavares, P., Edmondson, D. E., Theil, E. C., and Huynh, Boi-Hanh, "Direct Spectroscopic and Kinetic Evidence for the Involvement of a Peroxodiferric Intermediate during the Ferroxidase Reaction in Fast Ferritin Mineralization", *Biochemistry,* **37,** 9871–9876 (1998).
10. Bou-Abdallah, F., Papaefthymiou, G. C., Scheswohl, D. M., Stanga, S. D., Arosio, P., and Chasteen, N. D., "μ-1,2-Peroxobridged di-iron(III) dimer formation in human H-chain ferritin", *Biochem. J.*, **364,** 57–63 (2002).
11. Liu, X. and Theil, E. C., "Ferritin reactions: Direct Identification of the site for the diferric peroxide intermediate", *Proc. Natl. Acad. Sci. U.S.A.*, **101,** 8557–8562 (2004).
12. d'Estaintot, B., Samtambrogio, P., Granier, T., Gallois, B., Chevalier, J. M., Precigoux, G., Levi, S., and Arosio, P., "Crystal structure and biochemical proerties of the human mitochondrial ferritin and its mutant Ser 144A", *J. Mol. Biol.*, **340,** 277–293 (2004).
13. Moenne-Loccoz, P., Krebs, C., Herlihy, K., Edmondson, D. E., Theil, E. C., Huynh, B. H., and Loehr, T. M., "The ferroxidase reaction of ferritin reveals a diferric μ-1,2 bridging peroxide intermediate in common with other O_2-activating non-heme diiron proteins", *Biochemistry,* **38,** 5290–5295 (1999).
14. Hwang, J., Krebs, K., Huynh, B.-H., Edmondson, D. E., Theil, E. C., and Penner-Hahn, J. E., "An Unusually Short Fe-Fe Distance for the Peroxodiferric Intermediate in Iron Biomineralization by Ferritin", *Science,* **287,** 122–125 (2000).
15. Baldwin, J., Krebs, C., Saleh, L., Stelling, M., Huynh, B. H., Bollinger, J. M. Jr., and Riggs-Gelasco, P., "Structural characterization of the peroxodiiron(III) intermediate generated during oxygen activation by the W48A/D84E variant of ribonucleotide reductase protein R2 from *Escherichia coli*", *Biochemistry,* **42,** 13269–13279 (2003).
16. Jameson, G. N., Jin, W., Krebs, C., Perreira, A. S., Tavares, P., Liu, X., Theil, E. C., and Huynh, B. H., "Stoichiometric production of hydrogen peroxide and parallel formation of ferric multimers through decay of the diferric-peroxo complex, the first detectable intermediate in ferritin mineralization", *Biochemistry,* **41,** 13435–13443 (2002).
17. Zhao, G., Bou-Abdallah, F., Arosio, P., Levi, S., Janus-Chandler, C., and Chasteen, N. D., "Multiple pathways for mineral core formation in mammalian apoferritin. The role of hydrogen peroxide", *Biochemistry,* **42,** 3142–3153 (2003).
18. Askwith, C. and Kaplan, J., "Iron and copper transport in yeast and its relevance to human disease", *Trends Biochem. Sci.*, **23,** 135–138 (1998).
19. Takagi, H., Shi, D., Ha, Y., Allewell, N. M., and Theil, E. C., "Localized Unfolding at the Junction of Three Ferritin Subunits", *J. Biol. Chem.*, **273,** 18685–18688 (1998).
20. Jin, W., Takagi, H., Pancorbo, B., and Theil, E. C., "Opening" the ferritin pore for iron release by mutation of conserved amino acids at interhelix and loop sites", *Biochemistry,* **40,** 7525–7532 (2001).

VIII.3. Siderophores

Alison Butler
Department of Chemistry and Biochemistry
University of California
Santa Barbara, CA 93106

Contents

VIII.3.1. Introduction: The Need for Siderophores

The vast majority of all bacteria require iron for growth.[1,2] The importance of iron is magnified by its insolubility, K_{sp} for $Fe(OH)_3$ is 10^{-39}, at the near neutral pH conditions in which most bacteria grow. Microorganisms have evolved an elaborate mechanism to acquire Fe, which is tightly regulated and highly selective. Under aerobic conditions, bacteria and other microorganisms produce siderophores to solubilize and sequester Fe(III). Siderophores are low molecular weight compounds that coordinate Fe(III) with high affinity. (The name siderophore derives from the Greek for "iron-carriers.") In addition to siderophore production under low iron conditions, outer-membrane receptor proteins are produced that facilitate the transfer of iron into the bacterium. When sufficient levels of Fe have been acquired by microorganisms, the biosynthesis of siderophores and their outer-membrane receptor proteins is repressed.

VIII.3.2. Siderophore Structures

Hundreds of siderophore structures are known.[1,2] Within these structures, the two most common Fe(III)-binding moieties are catechols and hydroxamic acids. However, many other iron-binding functional groups are also found in siderophores.

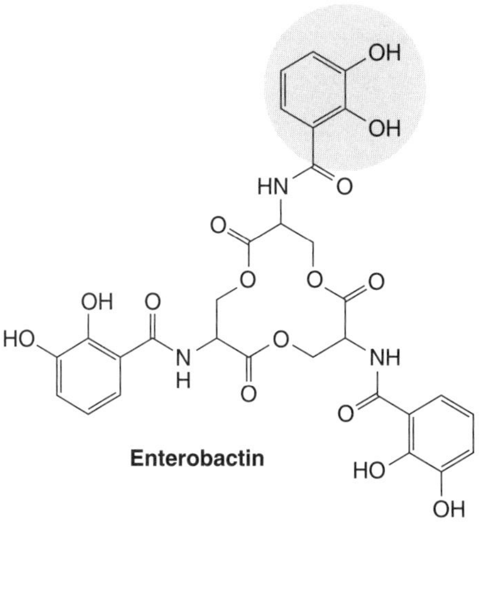

Enterobactin

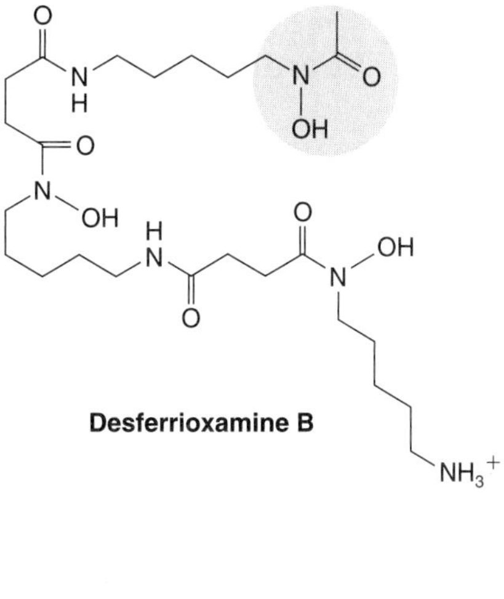

Desferrioxamine B

Enterobactin, produced by several species of *Enterobacteriaceae,* such as *Escherichia coli,* is an example of a triscatecholate-containing siderophore (see circled portion of the structure) and desferrioxamine B, produced by *Streptomyces* species, is an example of a tris(hydroxamate) siderophore (see circled portion of the structure). Although each of these siderophores only contains a single type of iron(III)-binding functional group, many siderophores coordinate iron(III) through a combination of ligand types.

Enterobactin is the cyclic triester of 2,3-dihydroxy-*N*-benzoyl-(*S*)-serine, which forms an octahedral complex with ferric ion. The ferric complex is optically active and can form Λ or Δ configurations, as a result of the (*S*)-serine configuration that are diastereomers. By comparison to model complexes, it has been found that ferric enterobactin forms the Δ configuration exclusively. A favored metal center configuration has also been observed for other siderophores such as ferrichrome (a tris(hydroxamate)-containing siderophore) and rhodotorulic acid (a dihydroxamate-containing siderophore; see below) and has been implicated as a chiral recognition factor in the transport of Fe(III).

While most siderophores form 1:1 complexes with ferric ion, some siderophores are only tetradentate, such as rhodotorulic acid, alcaligin, putrebactin and bisucaberin, among others. These siderophores form Fe_2(siderophore)$_3$ complexes to achieve octahedrally coordinated Fe(III). The crystal structure of the Fe_2(alcaligin)$_3$ complex shows that only one alcaligin acts to bridge the Fe(III) ions, as opposed to a triple-bridged helicate structure.

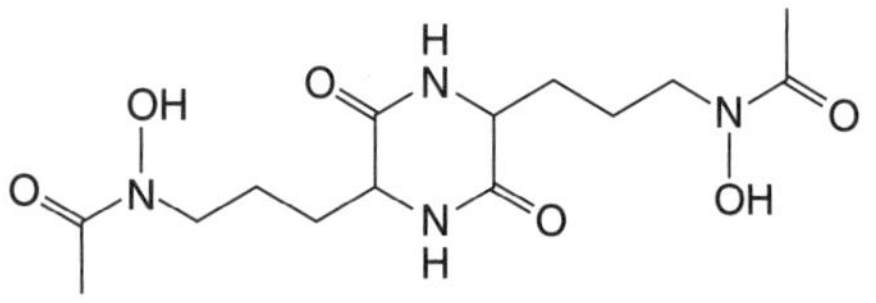

Rhodotorulic acid 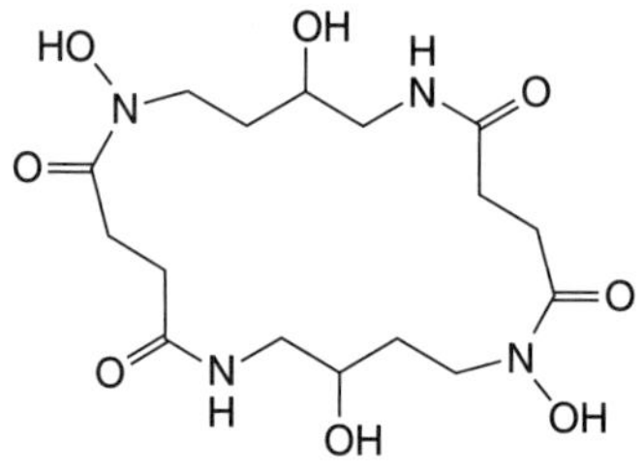 **Alcaligin** 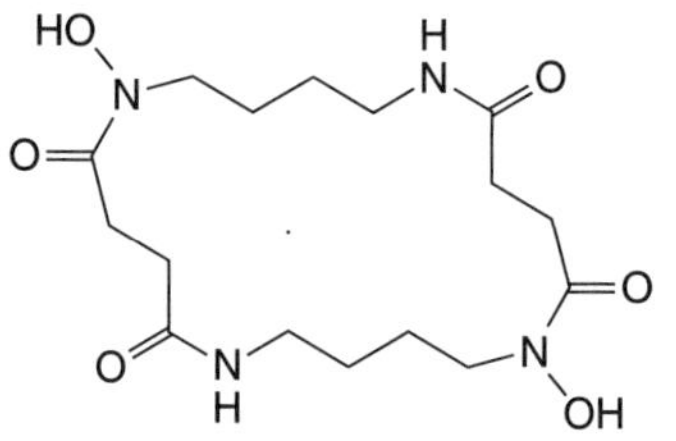 **Putrebactin**

While the biological role of siderophores is to facilitate iron acquisition during microbial growth, siderophores also have medicinal applications, such as the existing treatment for iron overload disease in which desferrioxamine B (Desferal) is used, as well as other potential applications such as contrast agents for diagnostic imaging and siderophore–antibiotic conjugates for treatment of drug-resistant microbes.

VIII.3.3. Thermodynamics of Ferric Ion Coordination by Siderophores

Siderophores complex ferric ion with particularly high-affinity constants. The stability constant is often reported as the complex formation constant, β_{FeLH} [FeLH indicates the ratio of iron to siderophore (L) to protons] between hydrated ferric ion, Fe^{3+}(aq), and the fully deprotonated siderophore ligand (Sid^{n-}):

$$Fe^{3+}(aq) + Sid^{n-} \overset{\beta_{110}^{siderophore}}{=} FeSid^{(3-n)}$$

$$\beta_{110}^{siderophore} = \frac{[FeSid^{(3-n)}]}{[Fe^{3+}(aq)][Sid^{n-}]}$$

Table VIII.3.1.
The β_{110} and pM Values for Selected Siderophores

Siderophore	log β_{110}	pM	Type of Siderophore
Enterobactin	49	35.5	Tris(catecholate)
Desferrioxamine B	30.6	26.6	Tris(hydroxamate)
Ferrichrome	29.07	25.2	Tris(hydroxamate)
Aerobactin	22.5	23.3	Bis(hydroxamate) and α-hydroxycarboxylate
Rhodotorulic acid	21.55 log β_{230} 62.2	21.8	Bis(hydroxamate)
Alcaligin	23.5 log β_{230} 64.66	23.0	Bis(hydroxamate)

The values of the formation constants span a wide range, from $10^{22.5}$ for ferric aerobactin to 10^{49} for the ferric enterobactin complex. Determination of these proton-independent formation constants requires knowledge of the pK_a values of each coordinating group.

Because little fully deprotonated siderophore would be present in solution at physiological pH, and to avoid problems associated with hydrolysis of ferric ion, the stability constant is often determined by competing siderophore against another ligand with well-defined thermodynamics of interaction with ferric ion, such as ethylenediaminetetraacetato (edta):

$$\mathrm{Fe(edta)^{1-} + Sid^{n-} \overset{K_{overall}}{\rightleftharpoons} FeSid^{(3-n)} + edta^{4-}}$$

$$K_{\text{overall}} = \frac{[\mathrm{FeSid}^{(3-n)}][\mathrm{edta}^{4-}]}{[\mathrm{Fe(edta)}^{1-}][\mathrm{Sid}^{n-}]}$$

$$= \left(\frac{[\mathrm{FeSid}^{(3-n)}]}{[\mathrm{Fe}^{3+}(\mathrm{aq})][\mathrm{Sid}^{n-}]}\right)\left(\frac{[\mathrm{edta}^{4-}][\mathrm{Fe}^{3+}(\mathrm{aq})]}{[\mathrm{Fe(edta)}^{1-}]}\right) = \frac{\beta_{110}^{\text{siderophore}}}{\beta_{110}^{\text{edta}}}$$

β values are comparable only for ligands of the same denticity.

Another measure of relative complexation behavior of a siderophore is the amount of free or uncomplexed Fe^{3+}(aq) left in solution. This concentration is represented as the pM value, which is the negative log of the Fe^{3+}(aq) concentration (p$M = -\log[Fe^{3+}(aq)]$) under the specific set of conditions at pH 7.4 in which the total siderophore concentration is $10\,\mu M$ and the total Fe concentration is $1\,\mu M$. The log β and pM values are given in Table VIII.3.1. The larger the pM value for a particular siderophore, the less free ferric ion in solution and the larger the formation constant. Thus the formation constants and the pM values are a measure of the selectivity of the siderophore for Fe(III) over competing metal ions, as well as a measure of the relative tendency to solubilize ferric ion.

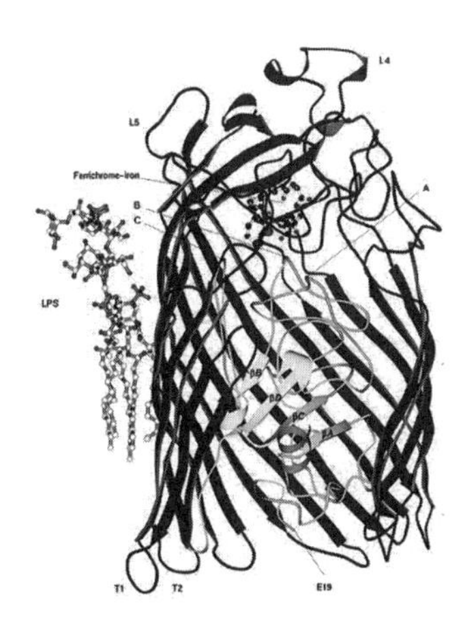

Fig. VIII.3.1.
Crystal structure of the FhuA–ferrichrome-iron complex showing a single molecule of lipopolysaccharides LPS non-covalently associated with the outer membrane (OM) protein complex. [From Ref. 3.] [See color insert.]

VIII.3.4. Outer-Membrane Receptor Proteins for Ferric Siderophores

Bacteria have evolved a class of high-affinity outer-membrane receptor proteins that recognize specific Fe(III)–siderophore complexes and are involved in an energy-dependent, active transport of the ferric siderophore complex across the outer membrane. Three receptors from *E. coli* whose crystal structures have recently been determined are FhuA (which stands for ferric hydroxamate uptake) (see Fig. VIII.3.1) [PDB codes 2FCP and 1FCP for FhuA and the FhuA–ferrichrome-Fe(III) complex, respectively; Ref. 3], FepA (which stand for ferric enterobactin permease; PDB code 1FEP; Ref. 4), and FecA [which stands for ferric citrate protein A; PDB codes 1KMO and 1KMP for FecA and the FecA–(Fe-citrate)$_2$ complex; Ref. 5]. FhuA is the receptor protein for ferric ferrichrome siderophores in *E. coli*. FepA is the receptor for ferric enterobactin. FecA is the receptor protein for dinuclear (ferric citrate)$_2$ complex.

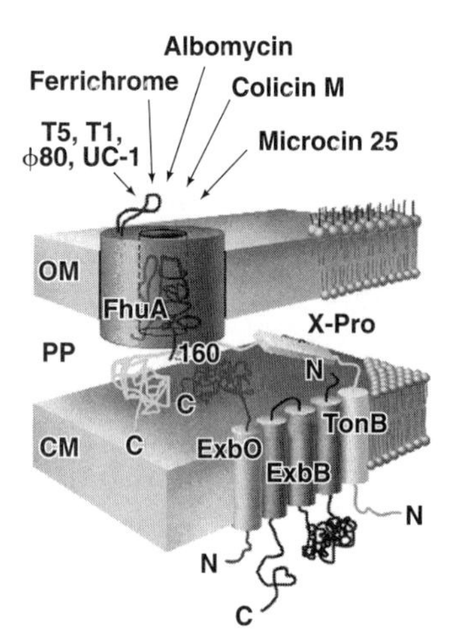

Fig. VIII.3.2.
Model of the FhuA-catalyzed transport of ferrichrome across the outer membrane (OM) of *E. coli*. (This figure also depicts the FhuA-mediated infection by the phages T5, T1, ϕ80, and UC-1.) [From Ref. 6.]

The structural core of these siderophore receptors is a 22-strand, membrane-spanning, antiparallel β-barrel similar in overall structure to other membrane porins, although these siderophore receptors have a segment of residues at the N-terminus that folds inside the β-barrel, effectively corking the barrel from the periplasmic side of the outer membrane. When the ferric siderophore complex binds to the receptor, FhuA, FepA, and FecA undergo conformational changes, which, through interactions with the TonB (transport of iron) protein, lead to the transport and release of the ferric siderophore complex into the periplasmic space. The TonB protein is a cytoplasmic membrane protein that spans the periplasmic space binding the siderophore receptor proteins and also binding the cytoplasmic membrane proteins, ExbB and ExbD. The TonB/ExbB/ExbD ternary complex mediates the signal transduction of the electrochemical potential of the cytoplasmic membrane to the outer membrane, allowing transport and release of the ferric siderophore complex (see Fig. VIII.3.2).

Once the ferric siderophore is released to the periplasmic space, it is bound by a high-affinity periplasmic-binding protein (e.g., FhuD), preventing the reverse transport of the iron–siderophore complex across the outer membrane.

The binding and transport of a ferric siderophore is usually highly specific. The kinetics of ferric siderophore uptake show saturation behavior with an apparent K_d value in the range of 0.1–100 n*M*. For example, FepA binds the $[Fe^{III}(enterobactin)]^{3-}$ complex with a K_d of <0.1 n*M*. Other complexes, such as $[Rh^{III}(catecholate)_3]^{3-}$ (a kinetically inert complex with respect to ligand substitution), do not affect the rate of uptake of $[Fe^{III}(enterobactin)]^{3-}$; however, Rh^{III}(*N*,*N*-dimethyl-2,3-dihydroxybenzamide)$_3$ does block uptake of $[Fe^{III}(enterobactin)]^{3-}$ at ~100 μ*M*. Thus, while the trilactone backbone of enterobactin is not required for recognition and uptake, the catecholamide moiety is an essential feature of enterobactin for Fe uptake to occur.

Several types of systems exist for the transport of Fe across the cytoplasmic membrane. These include the translocation of ionic ferric iron, ferrous iron, and the ferric siderophore complexes. Transport of the ferric ion and ferric siderophore complexes typically requires use of an ABC (ATP Binding Cassette) transporter system in an energy-dependent process.

VIII.3.5. Marine Siderophores

While Fe is abundant in most terrestrial environments, in much of the world's oceans, Fe is present at very low concentration (0.02–1 n*M*).[7] Over 99% of the dissolved ferric ion in the ocean is complexed by an organic ligand (or ligands), although at this point the structure is not known. These low levels of Fe have been shown to limit the growth of bacteria and other microorganisms. Many oceanic bacteria have been shown to produce siderophores, and the conditional stability constants of these siderophores are at least that of the natural ligand(s) complexing Fe(III) in the ocean, establishing that they can function in the Fe acquisition process. While relatively few marine siderophores have been structurally characterized, those that have been include both a few known terrestrial siderophores (e.g., aerobactin, desferrioxamine), as well as siderophores with unusual structures and properties. A new class of self-assembling amphiphilic peptide siderophores has been identified in distinct genera of marine bacteria (Fig. VIII.3.3).[8,9] Three families of siderophores,

Marinobactins

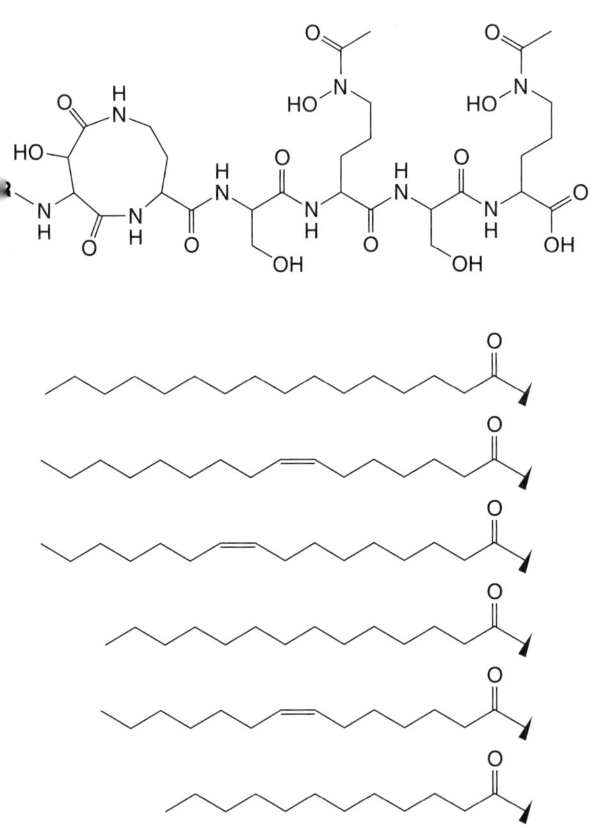

Aquachelins

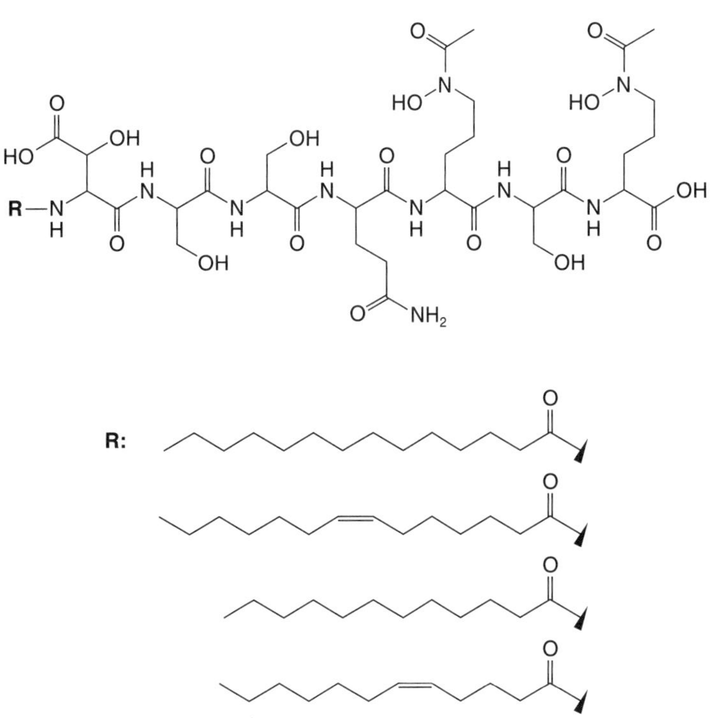

Amphibactins

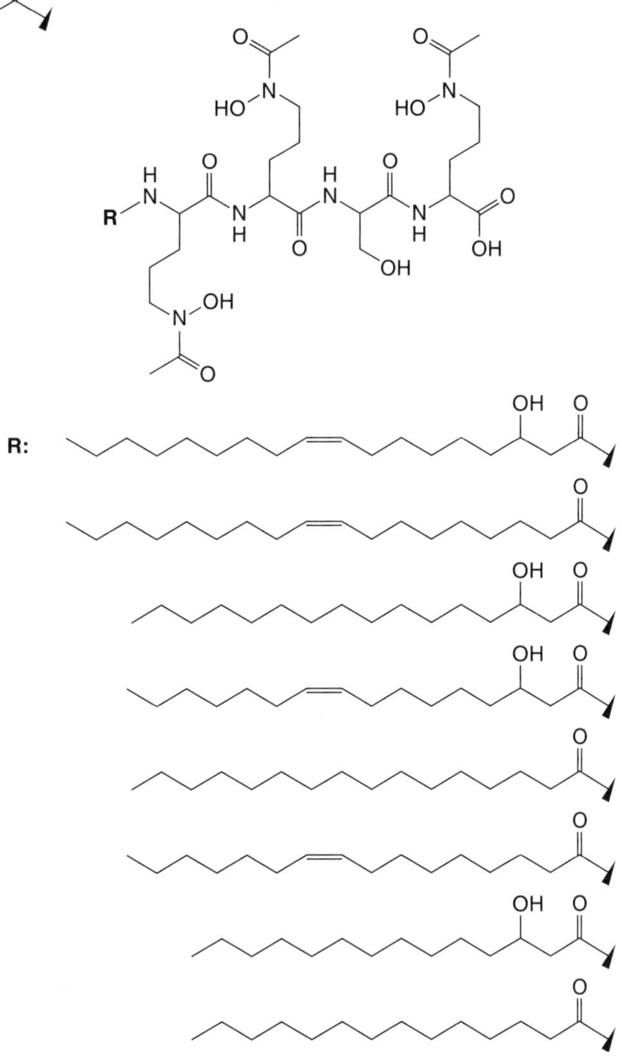

Fig. VIII.3.3.
Amphiphilic marine siderophores.[8,9]

represented below as the marinobactins, aquachelins, and amphibactins each contain a unique peptidic head group that coordinates Fe(III), and one of a series of fatty acid tails. The distinctive feature of these siderophores with their polar peptidic head groups and hydrophobic fatty acid tails is their amphiphilic, surface-active nature, which leads to the formation of self-assembled structures (micelles and vesicles). In addition, Fe(III) coordinated to the α-hydroxy acid groups (i.e., citrate and β-hydroxy aspartate) in these siderophores is photoreactive, leading to oxidation of the siderophore and reduction of Fe(III) to Fe(II).[10] The functional or physiological significance of this photoreactivity is not known yet and is an active area of research.

References

General References

1. *Iron Transport in Bacteria,* Crosa, J. H., Mey, A. R., and Payne, S. M., Eds., American Society for Microbiology (ASM) Press, Washington, DC, 2004, p. 384 (and references therein).
2. *Molecular and Cellular Iron Transport,* Templeton, D. M., Ed., Marcel Dekker, New York, 2002, p. 827.

Specific References

3. Ferguson, A. D., Hofmann, E., Coulton, J. W., Diederichs, K., and Welte, W., "Siderophore-mediated iron transport: Crystal structure of FhuA with bound lipopolysaccharide", *Science,* **282,** 2215–2220 (1998).
4. Buchanan, S. K., Smith, B. S., Venkatramani, L., Xia, D., Esser, L., Palnitakar, M., Chakraborty, R., Van Der Helm, D., and Deisenhofer, J., "Crystal structure of the outer membrane active transporter FepA from *Escherichia coli*", *Nat. Struct. Biol.,* **6,** 56–63 (1999).
5. Ferguson, A. D., Chakraborty, R., Smith, B. S., Esser, L., van der Helm, D., and Deisenhofer, J., "Structural Basis of Gating by the Outer Membrane Transporter FecA", *Science,* **295,** 1715–1719 (2002).
6. Braun, V., "Pumping Iron through Cell Membranes", *Science,* **282,** 2202–2205 (2000).
7. Butler, A., "Acquisition and Utilization of Transition Metal Ions by Marine Organisms", *Science,* **281,** 207–210 (1998).
8. Martinez, J. S., Zhang, G. P., Holt, P. D., Jung, H.-T., Carrano, C. J., Haygood, M. G., and Butler, A., "Self-Assembling Amphiphilic Siderophores from Marine Bacteria", *Science,* **287,** 1245–1247 (2000).
9. Martinez, J. S., Carter-Franklin, J. N., Mann, E. L., Martin, J. D., Haygood, M. G., and Butler, A., "Structure and Dynamics of a New Suite of Amphiphilic Siderophores Produced by a Marine Bacterium", *Proc. Natl. Acad. Sci., U.S.A.,* **100,** 3754–3759 (2003).
10. Barbeau, K., Rue, E. L., Bruland, K. W., and Butler, A., "Photochemical Cycling of Iron in the Surface Ocean Mediated by Microbial Iron(III)-Binding Ligands", *Nature* (*London*), **413,** 409–413 (2001).

VIII.4. Metallothioneins

Hans-Juergen Hartmann
Anorganische Biochemie
Physiologisch Chemisches
Institut, University of
Tübingen, Tübingen,
Germany

Ulrich Weser
Anorganische Biochemie
Physiologisch Chemisches
Institut, Universität of
Tübingen, Tübingen,
Germany

Contents

VIII.4.1. Introduction

Metallothioneins (MTs) were discovered in 1957 by Margoshes and Vallee in the course of their search for the distribution of Cd in horse kidney tissue. These low molecular mass, Cys-rich proteins are widely distributed in biological systems.[1–7] The MTs, which are able to bind a series of d^{10} metal ions including Zn(II), Cu(I), Cd(II), Ag(I), Au(I), Hg(II), Pt(II), and Bi(III) in the form of metal-thiolate clusters, have been the subject of extensive studies over the past 40 years. Of special importance is the coordination of Zn and Cu ions and the metabolic control of these essential biologically active metals.

VIII.4.2. Classes of Metallothioneins

Metallothioneins are ubiquitously occurring proteins that have been identified in vertebrates, invertebrates, plants, and both eukaryotic and some prokaryotic microorganisms. They are divided into three classes:

Class I. This class includes mammalian MTs with a relative molecular mass of 6–7 kDa. They usually contain 61–62 amino acid residues, 20 of which are cysteines. In all the mammalian proteins sequenced to date, the positions of cysteines are highly conserved. All Cys sulfurs are involved in metal coordination, many of which form bridges. Usually seven divalent metal ions are bound to four thiolate sulfurs. Copper-containing MTs, with Cu in its Cu(I) state, have different metal/protein ratios. Closely related class I proteins are found in the 18-Cys MTs of crustaceans and the short-chain seven-Cys MTs from the fungi *Neurospora crassa* and *Aspergillus bisporus* with only 26 and 25 amino acid residues, respectively. The seven cysteines are located at exactly the same positions as found in the 25-residue sequence of the N-terminal domain of mammalian proteins. A further member of class I MTs is the brain-specific 68 amino acid Cu,Zn-MT-3 that is identical to the so-called growth inhibitory factor (GIF). This protein contains two amino acid insertions relative to all other mammalian MTs.

Class II. Polypeptides that belong to this group display no or only a distant evolutionary relationship to mammalian MTs, and include proteins from yeast (*Saccharomyces cerevisiae*), sea urchin, and pea, for example. The common features between class I and II MTs are that both consist of small, single-chain proteins containing large quantities of heavy metals and Cys residues linked by metal–thiolate bonding.

Class III. This group consists of enzymatically synthesized peptides composed of atypical poly(γ-glutamyl-cysteinyl)glycines known as phytochelatins and cadystins. Cadmium-containing class III MTs often include inorganic sulfide. They form a metal–cysteinyl thiolate cluster composed of multiple peptides heterogeneous in length and metal ions. In some microorganisms they are components of CdS crystallites.

VIII.4.3. Induction and Isolation

Metallothioneins are induced by a variety of agents and conditions. Biosynthesis is considerably enhanced, both *in vivo* and in cultured cells, in the presence of a salt of Cd, Zn, Cu, Hg, Au, Ag, Co, Ni, or Bi. Likewise, certain hormones, cytokines, growth factors, tumor promoters, X-rays, and many other chemicals exert a stimulating effect. Metallothionein induction is also observed in the livers of animals after physical and chemical stress.

Induction mechanisms for the synthesis of MT have been studied mainly in mammalian species and are known in some respects for classes I and II. The regulation of biosynthesis occurs on the level of transcription initiation. Both class I and II MTs display genetic variability (polymorphism). Mammalian tissues usually contain two major fractions. For example, MT-1 and MT-2 are present in mouse and rat liver and can be separated by anion-exchange chromatography. They often vary by only a single amino acid residue, leading to the difference of one negative charge at neutral pH. Some mammalian species including primates generate multiple subisoforms of MT-1 and MT-2. In humans at least 10 iso-MT genes are expressed, some of which are tissue specific. This polymorphism makes the construction of an evolutionary pedigree quite difficult. Nevertheless, there is evidence for divergent evolution from an ancient precursor gene.

Mammalian MTs are richly abundant in soft (parenchymatous) tissue, the highest concentrations being found in kidney and liver. However, their occurrence and biosynthesis have been documented also in many other tissues and cell types. The absolute amounts present in different species and tissues are highly variable, reflecting differences in age, state of development, and dietary factors. In human and equine kidney and liver, which are natural sources of MTs, their concentration can vary by a factor of 10 or more. The biosynthesis can be increased by heavy metal administration to laboratory animals, such as rat and rabbit. Cadmium is the most effective inducer, followed by Zn and Cu. Hepatic tissue is most commonly used because the MT isoforms are well characterized and yields of up to $10\,mg\,g^{-1}$ wet weight are obtained.

Special isolation procedures are required to purify the oxygen-sensitive Cu/Zn-MTs that are naturally found in the livers of pre- and neonatal vertebrates (fetal MT), the hepatic canine MT, the brain-specific MT-3, the Cu-MTs from *N. crassa* and *A. bisporus,* and the class II yeast Cu-MT. The microbial MTs are usually induced by adding Cu salts to the growth medium of the respective organism. A common feature of all copper-containing MTs is their orange-red luminescence emission upon ultraviolet (UV)-irradiation at 300 nm, which is derived from buried, and thus water-shielded, intact Cu(I)-thiolate chromophores. This phenomenon may be used to identify fractions of these proteins during isolation.

VIII.4.4. Structural and Spectroscopic Properties

The primary structure of >50 MTs has been documented. A common structural motif of all class I and II forms is the recurrence of CysXCys tripeptide sequences (where X represents an amino acid residue other than cysteine). Dipeptide CysCys sequences are also observed (Fig. VIII.4.1).

There is a very close correspondence in the alignment of the cysteines along the protein chain. Fifty-six percent of all amino acid residues are conserved during evolution, including all 20 cysteines and nearly all lysines and arginines in the known mammalian MTs. The N-terminal Met is usually acetylated. Aromatic amino acids are not present, resulting in rather featureless electronic absorption spectra for the apoprotein. Metal binding leads to characteristic absorbances in the near-UV wavelength region, which are assigned to charge-transfer transitions. The first spectroscopic measurements, which were obtained using mammalian Cd/Zn-MTs, demonstrated features characteristic of tetrahedral metal–thiolate complexes. Optical data for copper- and mercury-containing proteins showed the same properties in that the absorbances between 220 and 450 nm were assigned to the metal–thiolate

Class I

	1	11	21	31	41	51	61
Human MT-2	MDPNCSCAAG	DSCTCAGSCK	CKECKCTSCK	KSCCSCCPVG	CAKCAQGCIC	KGASDKCSCC	A
N.crassa	MGDCGCSGAS	SCNCGSGCSC	SNCGSK				

Class II

	1	11	21	31	41	51
Yeast *(S.cerevisiae)*	QNEGHECQCQ	CGSCKNNEQC	QKSCSCPIGC	NSDDKCPCGN	KSEETKKSCC	SGK

Fig. VIII.4.1.
Amino acid sequences of typical class I and class II MTs.

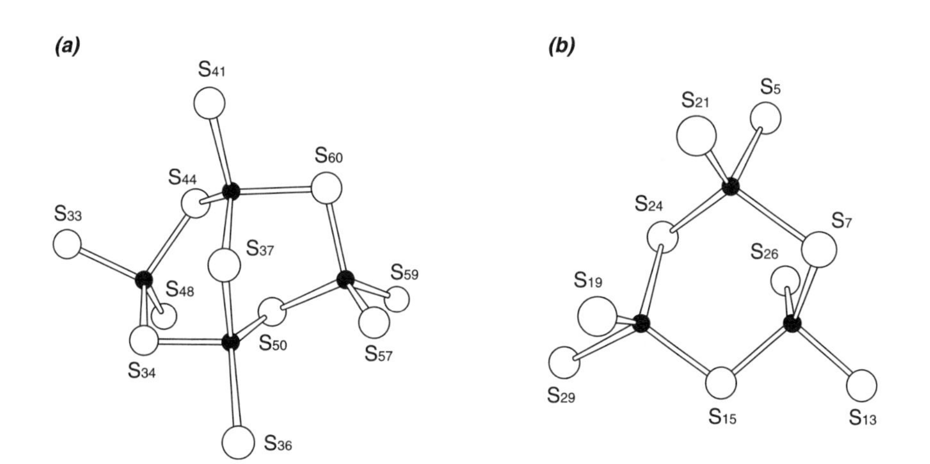

Fig. VIII.4.2.
Schematic drawing of the metal clusters in MT. (*a*) Cluster A, $(M)_4S_{11}$; (*b*) Cluster B, $(M)_3S_9$. Bivalent metal ions (Cd,Zn) are in black. Cysteine sulfur atoms (white) bear the number of the respective amino acid residue allocated in the polypeptide chain. [Adapted from entries made to the Brookhaven protein data bank (PDB) (PDB code: 4MT2, 1MRT, and 2MRT).]

bonding. Circular dichroism proved even more informative. The metal-binding sites exhibit specific chirality, which represents a marker for formation of clusters involving several metal–cysteinyl thiolate groups with both bridging and terminal thiolates present.

Detailed structural information was obtained from ^{113}Cd NMR (nuclear magnetic resonance) spectroscopy of rabbit liver Cd_5,Zn_2-MT-2 and Cd_7-MT-2. From these studies, it was concluded that MT is composed of two approximately equally sized globular domains. The protein binds 7 cadmium ions in two distinct clusters, one with 3 cadmium ions and 9 cysteines, consistent with a cyclohexane-like structure, and the other with 4 cadmium ions and 11 cysteines, consistent with a bicyclo(3.3.1)-nonane-like structure (Fig. VIII.4.2). The isolation of the discrete domains demonstrated that each end of the polypeptide chain generated one of the clusters. The α-domain (cluster A; 4 Cd, 11 Cys) is formed by the C-terminal part and the β-domain (cluster B; 3 Cd, 9 Cys) consists of the N-terminus of the protein.

One of the unique features of mammalian MT, relative to many other metalloproteins, is the major role played by coordinated metal ions in determining the structure of the folded metalloprotein. The metal-free protein is a random coil polypeptide that adapts the form of the holoprotein only in the presence of the appropriate metal ions. The spatial structure was determined from two-dimensional (2D) heteronuclear ^{1}H^{113}Cd NMR correlation spectroscopy of $^{113}Cd_7$-MT in aqueous solution, which allowed the assignment of the detailed connectivities of the

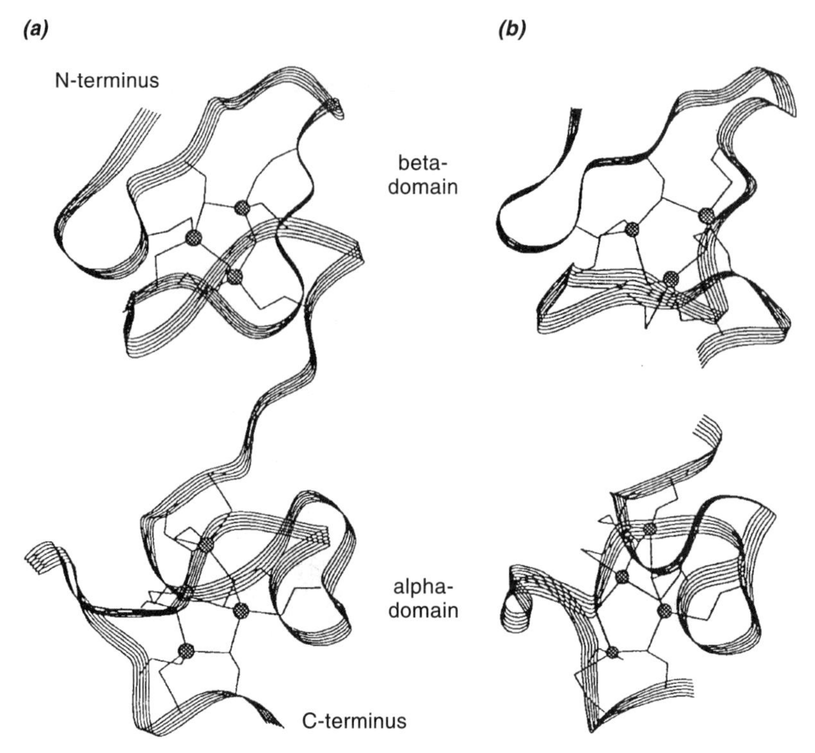

Fig. VIII.4.3.
Ribbon drawing of the rat MT three-dimensional (3D) structure as determined (*a*) by X-ray crystallography and (*b*) by NMR in aqueous solution. Metals are shown as shaded spheres connected to the protein backbone by Cys thiolate ligands. [Adapted from entries made to the Brookhaven data bank (PDB codes: 4MT2, 1MRT, and 2MRT).]

cadmium–thiolate bonds. It was found that the spatial structures of human MT-2, rabbit MT-2a, and rat MT-2 are closely similar, although the agreement in the course of the polypeptide backbone of the latter two MTs differs in >25% of their non-Cys amino acids. This result indicates that the conservative arrangement of the Cys residues dictates the organization of the metal–thiolate clusters and the conformation of the polypeptide chain. The X-ray crystal structure determination of native rat liver Cd_5,Zn_2-MT-2 at 2.0-Å resolution essentially confirmed the data obtained by 2D NMR spectroscopic studies (Fig. VIII.4.3).

Besides Zn, Cu is the second most important metal that binds to MT both *in vivo* and *in vitro*. Because of its pronounced oxidation–reduction properties, Cu needs special chaperoning. In this context, metallothioneins are good candidates to sequester the metal in its Cu(I) state in the form of Cu(I)-thiolates. Unlike the Cd,Zn-MTs, no 3D structure is known for mammalian Cu-MTs. The different copper/protein ratios of the Zn and Cu species are expected to affect the metal–thiolate cluster structure considerably. Nevertheless, it is assumed that a two-domain formation similar to that of the Cd,Zn-protein exists in class I Cu-MT. From spectroscopic measurements, it was deduced that 12 copper atoms are coordinated to the apoprotein. Six Cu(I) ions in each of the two domains were bound exclusively to 11 and 9 cysteinyl thiolate sulfurs, respectively. Mammalian Cu- and Cu,Zn-MTs are very sensitive to oxidation. Thus, the isolation and *in vitro* experiments must be carried out anaerobically and/or in the presence of reducing agents. Under reducing conditions, the Cu^{I}-thiolate chromophore is thermodynamically more stable than that of zinc.

Yeast Cu-MT (class II MT) is a 53 amino acid polypeptide containing 12 Cys residues and, in contrast to class I MTs, one His. From 2D NMR studies of the native protein and of the ^{109}Ag-substituted derivative, 7 metal ions were found to coordinate to 10 cysteinyl thiolates in a single cluster. Histidine has been excluded from metal binding. The 2 Cys residues at positions 49 and 50 in the polypeptide chain are not involved in metal complexation according to these studies. Nevertheless, the structure of the Cu polymetallic center may well be different from that of the Ag center because of differences in the coordination preferences of the 2 metal ions. A more resolved (^{1}H NMR) structure of the protein moiety using the isolated protein and a truncated form were compared to examine the survival of the Cu-thiolate cluster. In both species the oligomeric Cu–thiolate center remained unchanged (Fig. VIII.4.4).[8,9] At last, the long known controversy regarding the exact number of coordinated Cu(I) ions was finally resolved. In a recent crystal structure analysis, 6 trigonally coordinated coppers and 2 somewhat loosely bound digonally arranged Cu(I) ions were detected, supporting earlier spectroscopic conclusions.[3] This is the first example of an intact Cu_8–MT cluster.[10]

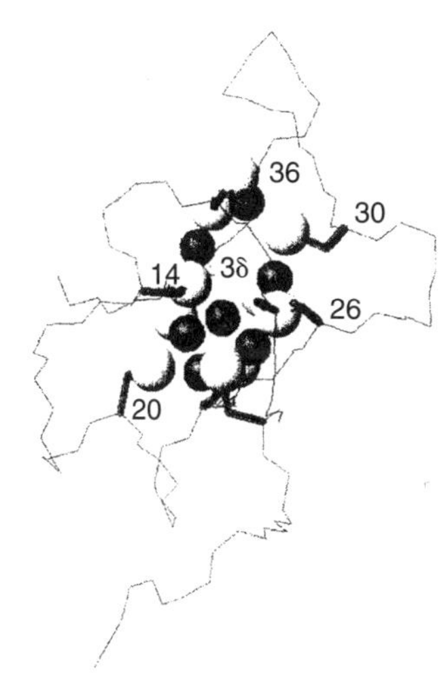

Fig. VIII.4.4.
One of five possible copper–sulfur arrangements in yeast (*S. cerevisiae*) Cu^I–MT consistent with 2D ^{1}H NMR data. Copper and sulfur atoms are shown as dark gray and white spheres, respectively. The numbers indicate the positions of the Cys residues within the polypeptide chain. (From Ref. 9, with permission of authors). In this structure, one of the loosely bound digonal Cu(I) ions was lost during the preparation.

VIII.4.5. Reactivity and Function

In vitro investigations have revealed an unusual reactivity of the MTs. Zinc ions are readily removed by a variety of competing ligands including polyaminocarboxylates, polyamines, nitrilotriacetate, ethylenediaminetetraacetato (edta), and others. In this context, Zn–MT has been suggested to act as an intermediate metal donor to apo-Zn-proteins. In contrast to this reactivity, the metal-free MT readily removes Zn ions from zinc finger transcription factors, thereby abrogating their transcription activation competence. Copper ion transfer is also observed with Cu–MT and apo-Cu-proteins. This process may occur as a direct Cu(I) incorporation into vacant Cu(I)-binding sites or as Cu(II) after controlled oxidation of the Cu^I–thiolate moieties. Apart from this reaction, both intracellular copper transport and insertion of the metal into copper enzymes is generally accomplished via specific Cu chaperone proteins (see Section VIII.6).

Metal exchange reactions are further notable features of the MTs. Both Cd and Cu(I) ions displace Zn very rapidly in the mammalian proteins, indicating unobstructed access to the metal-bound thiolates of each cluster. Alkylating and oxidizing agents are able to release metal ions from the protein moiety often leading to disulfide or to irreversible oxidized sulfur species, including RSO_2^- and RSO_3^-. Metal binding is prevented under these conditions. Demetalated MTs, which have been employed for many experiments, including metal substitution, are usually prepared by acidification of the protein followed by separation of the released metal ions. Special care is needed for protection of the acid-stable apoproteins as these compounds are highly sensitive to oxidation. Apart from the generation of oxidized thiolate groups that are unable to bind metals, formation of inter- and intramolecular disulfide bridging leads to uncontrolled denaturation processes in the protein. It is important to realize that metal binding to the apoprotein provides resistance to proteinase digestion. At the same time, the oxidoreductive sulfhydryl groups are protected and are available for maintaining a particular redox potential.

Another remarkable reactivity of the MTs may be seen in radical scavenging. These proteins provide cells with radioresistance. Hydroxyl- and superoxide radicals react with Cd–MT or Zn–MT with rate constants of 10^{12} and $5 \times 10^5\ M^{-1}\,s^{-1}$, respectively, suggesting that the cysteinyl thiolate groups are the primary attack targets of the radical species. The Cu–MTs exhibit even higher superoxide dismutase activity resembling that of a Fe-superoxide dismutase (see Section XI.2). The rate constant for the class II yeast Cu–MT is $k_2 = 0.8 \times 10^7 \, M^{-1} \, s^{-1}$. In the same protein, the generation of thiyl radicals is observed in the presence of superoxide. No Cu(II) is formed indicating that the redox process occurs primarily at the thiolate sulfur. The Cu^I–thiolate chromophores remain fully intact during this reversible thiolate redox cycle.[11]

A variety of possible functions for the MTs *in vivo* have been assumed during the last 40 years. Because of their high metal/protein ratio compared to other metalloproteins, their unusual structure, and their remarkable kinetic lability, they are believed to function primarily in rapid metal transfer and metalloregulatory processes. Initially, the cadmium-binding property of these proteins suggested their function to be detoxification of this metal. Later, it became apparent that MTs appear as prominent binding sites for zinc and copper. Thus, its participation in intestinal Zn and Cu absorption has been established. In general, MTs are highly suitable for sequestration and storage of these biometals in a way similar to that of the ferritins for iron storage (see Section VIII.2).

The phenomenon of a high transient Cu–MT level in fetal liver suggests that, in this developmental state, this Cu is provided for the biosynthesis of the many copper enzymes that are required for respiratory processes. By contrast, unbalanced metabolism of Cu is implicated in Wilson's and Menkes' diseases, and it has been suggested that MT may be involved in the biochemistry of these genetically determined disorders (see Section VIII.5).

Attention has also turned recently to the reactivity of the numerous thiolate groups toward (toxic) electrophiles and oxidizing agents. The nucleophilicity of the thiols may be of importance regarding the pharmacology of alkylating agents and metallodrugs. Other putative roles have been suggested, including the above-mentioned radical scavenging, the stress response, and the possible participation in cell differentiation and/or proliferation in relation to the localization of MT in the nucleus. These many diversified functions suggest strongly that MTs play many diverse and important roles in biological systems.

References

General References

1. Kägi, J. H. R. and Nordberg, M., Eds., *Metallothionein, Experientia Supplementum 34,* Birkhäuser, Basel, 1978.
2. Kägi, J. H. R. and Kojima, Y., Eds., *Metallothionein II, Experientia Supplementum 52,* Birkhäuser, Basel, 1987.
3. Riordan, J. F. and Vallee, B. L., Eds., Methods in Enzymology, *Metallobiochemistry Part B: Metallothionein and Related Molecules,* Vol. 205, Academic Press, San Diego, 1991.
4. Stillman, M. J., Shaw, G. F., III, and Suzuki, K. T., Eds., *Metallothioneins,* VCH Publishers, New York, 1992.
5. Suzuki, K. T., Imura, N., and Kimura, M., Eds., *Metallothionein III: Biological Roles and Medical Implications,* Birkhäuser, Basel, 1993.
6. Pountney, D. L., Kägi, J. H. R., and Vasák, M., in *Handbook of Metal–Ligand Interactions* in Biological Fluids; Bioinorganic Chemistry, Berthon, G., Ed., Vol. 1, pp. 431–442. Marcel Dekker, New York, 1995.
7. Klaassen, C. D., Ed., *Metallothionein IV,* Birkhäuser, Basel, 1999.

Specific References

8. Luchinat, C., Dolderer, B., Del Bianco, C., Echner, H., Hartmann, H.-J., Voelter, W., and Weser, U., "The $Cu(I)_7$ cluster in yeast thionein survives major shortening of the polypeptide backbone as deduced from electronic absorption, circular dichroism, luminescence and ^1H NMR", *J. Biol. Inorg. Chem.* **8,** 353–359 (2003).
9. Bertini, I., Hartmann, H.-J., Klein, T., Liu, G., Luchinat, C., and Weser, U., "High resolution solution structure of the protein part of Cu^7 metallothionein", *Eur. J. Biochem.* **267,** 1008–1018 (2000).
10. Calderone, V., Dolderer, B., Hartmann, H.-J., Echner, H., Luchinat, C., Del Bianco, C., Mangani, S., and Weser, U., "The crystal structure of yeast copper thionein: the solution of a long lasting enigma", *Proc. Natl. Acad. Sci. USA* **102,** 51–56 (2005).
11. Hartmann, H.-J., Sievers, Ch., and Weser, U., in *Metal Ions in Biological Systems; Interrelations Between Free Radicals and Metal Ions in Life Processes,* Sigel, A. and Sigel, H., Eds., Vol. 36, Marcel Dekker, New York, 1999, pp. 389–413.

VIII.5. Copper-Transporting ATPases

Contents

Bibudhendra Sarkar
Structural Biology and Biochemistry
The Hospital for Sick Children and the University of Toronto
Toronto, Ontario M5G1X8
Canada

VIII.5.1. Introduction: Wilson and Menkes Diseases

Wilson and Menkes diseases are genetic disorders of copper transport.[1] Wilson disease is an autosomal recessive disorder causing toxic accumulation of Cu in the liver and brain leading to progressive hepatic and neurological damage. Menkes disease is a fatal X-linked genetic disorder causing copper deficiency that leads to progressive neurodegeneration and death in children. The Wilson disease gene is expressed predominantly in the liver, kidney, and placenta, whereas the Menkes disease gene is expressed in the intestine and all tissues with the exception of the liver. Both of these proteins are localized to the membrane of the Golgi apparatus. Mutations in the Wilson disease gene cause toxic accumulation of Cu in those cells where the gene is expressed. Mutations in the Menkes disease gene cause copper deficiency and the lack of Cu incorporation in important enzymes. The genes for these two diseases encode copper-transporting ATPases, which belong to a subfamily of cation-transporting ATPases known as P-type ATPases. These ATPases are responsible for the translocation of ions across biological membranes and are critical to the maintenance of ion gradients, proper intracellular ion composition, and ion resistance. Membrane-bound cation transporting pumps are found in fungi, bacteria, plants and animals, and function by using the energy obtained from the hydrolysis of ATP to move ions across the membrane.

VIII.5.2. Structure and Function

The copper-transporting ATPases are the first heavy metal transporting ATPases identified in humans. They are part of a larger family of heavy metal transporters responsible for the transport of a diverse set of metals, including Cu, Cd, Zn, Hg, and As. All members of this family have between 6 and 8 transmembrane segments with only one pair of transmembrane segments on the carboxy terminal side of the cytoplasmic ATP-binding domain. This family also contains a highly hydrophobic region immediately preceding the TGEA conserved motif, which probably gives rise to two additional membrane segments in this region. However, the most striking and interesting feature of this type of transporter is the presence of a large N-terminal cytosolic domain that contains between one and six copies of a metal binding repeat, GMTCXXC. The variable number of these repeats leads to a great diversity of molecular weights within this class of ATPase.

The Menkes and Wilson disease copper-transporting ATPases share a high degree of homology with each other. The overall sequence identity between the two proteins is 57%, however, this figure rises to 79% and higher in the phosphatase, transduction–phosphorylation, adenosine triphosphate (ATP), and metal-binding domains. In spite of their similarities, the Menkes and Wilson disease ATPases do

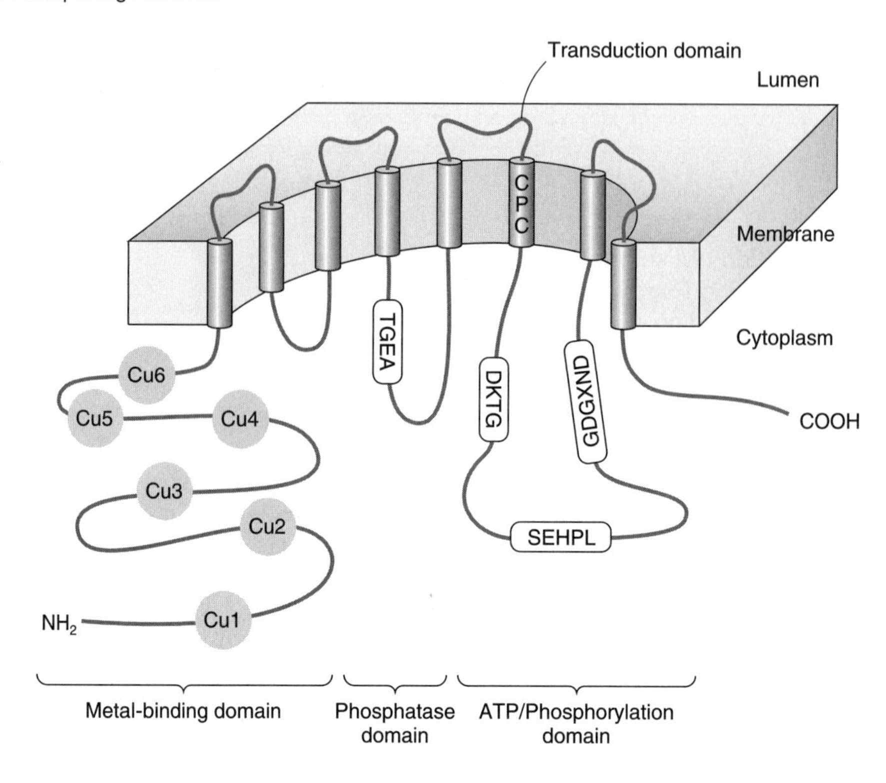

Fig. VIII.5.1.
Proposed structure for the Menkes and Wilson disease copper-transporting ATPases based on sequence comparison with other cation-transporting ATPases.

Fig. VIII.5.2.
Alignment of metal-binding domains (WDCu1 to WDCu6) from Wilson disease copper-transporting ATPase with other heavy metal transporters: Atx1 copper chaperone from *Saccharomyces cerevisiae*, ATOX1 human orthologue of Atx1, CopA copper ATPase from *E. hirae*, MerP mercury transport protein from *Pseudomonas aeruginosa*, CadA cadmium ATPase from *Staphylococcus aureus*. Conserved Cys residues are shown in bold.

WDCu1	ILGMT**C**QS**C**VKSIEDRISNLKGIISMKVSL
WDCu2	VEGMT**C**QS**C**VSSIEGKVRKLQGVVRVKVSL
WDCu3	IDGMH**C**KS**C**VLNIEENIGQLLGVQSIQVSL
WDCu4	IAGMT**C**AS**C**VHSIEGMISQLEGVQQISVSL
WDCu5	IKGMT**C**AS**C**VSNIERNLQKEAGVLSELVAL
WDCu6	ITGMT**C**AS**C**VHNIESKLTRTNGITYASVAL
Atx1	VV-MT**C**SG**C**SGAVNKVLTKLEPDVSKIDIS
ATOX1	VD-MT**C**GG**C**AEAVSRVLNKL--GGVKYDID
CopA	ITGMT**C**AN**C**SARIEKELNEQPGVMSATVNL
MerP	ITGMT**C**DS**C**AVHVKDALEKVPGVQSADVCT
CadA	VQGFT**C**AN**C**AGKFEKNVKKIPGVQDAKVNF

have some distinct differences. The greatest of these is a 78 amino acid deletion in the Wilson disease ATPase between the first and second metal-binding domains.

The structures of Wilson and Menkes disease ATPases contain many conserved motifs (Fig. VIII.5.1). Among these are the TGEA, CPC, DKTG, SEHPL, and GDGXND motifs. The TGEA motif is located in the cytoplasmic loop between transmembrane segments 4 and 5, which is implicated in energy transduction. The CPC motif in transmembrane segment 6 is characteristic of all heavy metal transporting ATPases. The DKTG, SEHPL, and GDGXND motifs are located in the large cytoplasmic loop between transmembrane segments 6 and 7. The DKTG motif is critical to the function of the ATPase, since it contains the conserved aspartic acid residue that is phosphorylated in the transport cycles of the proteins. The SEHPL motif is only present in heavy metal transporting P-type ATPases. The His in this motif is thought to be critical for the proper functioning of this protein since mutation of this residue to Ser in the Wilson disease ATPase leads to the Wilson disease phenotype. The GDGXND motif is located before the transmembrane segment 7 enters the membrane and is predicted to be highly alpha-helical, forming part of the "stalk" of the ATPase. The large cytosolic N-terminal domain of the Menkes and Wilson disease ATPase contains six tandemly repeated metal-binding motifs of $\sim$30 amino acids, each containing a GMTCXXC motif. A similar copper-binding motif is found in other heavy metal transporters. An alignment of the metal-binding domain from the Wilson disease ATPase with other heavy metal transporters is shown in Fig. VIII.5.2.

VIII.5.3. Metal Ion Binding and Conformational Changes

The N-terminal copper-binding domains of both the Wilson and Menkes disease ATPases have been cloned, expressed, purified, and shown to bind one Cu per metal-binding motif.[2–4] The metal-binding domain is able to bind a variety of transition metals and has a high affinity for Zn(II). However, several metals are able to successfully compete with Zn(II) for binding to the domain. In particular, Cd(II), Au(III), and Hg(II) seem to have the highest affinities for the domain relative to Zn(II), whereas Mn(II) and Ni(II) have little or no affinity relative to Zn(II). Results with Cu are very dramatic and unique compared to the other metals. At low concentration, Cu is able to decrease Zn(II)-binding substantially. However, as the concentration of Cu is raised, the affinity for Cu seems to increase rapidly. The behavior of copper binding is similar to that observed for the cooperative binding of ligands to proteins indicating a possible cooperative-binding mechanism for Cu that is not observed for other transition metals.

As increasing concentrations of Cu are added to the copper-binding domain of the Wilson disease ATPase, an X-ray absorption near-edge structure (XANES) feature is seen at 8983.5 eV,[3] consistent with $1s \rightarrow 4p$ transition of a Cu(I) center.[3] The intensity of the resolved peak is somewhat weaker than those of digonal Cu(I) thiolate compounds, but stronger than those corresponding to trigonal complexes. The intermediate nature of the XANES feature suggests that the Cu(I) site is distorted from the linear geometry of a two coordinate Cu(I) center with the S–Cu–S angle between 120 and 180°. The XANES feature also seems to be independent of Cu stoichiometry, which would imply that all the Cu(I) sites of the protein are quite similar. Extended X-ray absorption fine structure (EXAFS) spectra with incremental addition of Cu give rise to a very similar spectrum after each stoichiometric addition of copper. Therefore, the geometry of the Cu(I) sites is most likely not perturbed with different amounts of copper. All Cu atoms appear to be ligated by two sulfurs from Cys residues in a distorted linear arrangement with a Cu–S distance of 2.17–2.19 Å. A Cu–S distance of 2.16 Å has also been found in the related Menkes disease ATPase,[5] which has the same copper-binding sequence motif as the Wilson disease ATPase suggesting a similar Cu(I) coordination environment for both the ATPases.

Upon binding of Cu to the N-terminal domain of the Wilson disease ATPase, there is a conformational change in the protein involving both secondary and tertiary structures, as shown by CD spectral analysis.[3] Addition of incremental amounts of Cu results in a progressive increase in secondary structure, greatest between the apoprotein and the 1:2 complex and to a lesser extent from the 1:4 complex to the 1:6 complex. The secondary structure changes are paralleled by drastic changes in the tertiary structure of the protein. These structural changes are not centered around the copper-binding sites since the EXAFS results show that the coordination environment around the Cu atoms does not change upon the binding of successive amounts of copper. Large changes are observed in the aromatic region. There is also evidence that the disulfide bonds are affected upon increasing amounts of copper binding to the N-terminal domain of Wilson disease ATPase. As Cu binds, the Cys residues are involved in Cu ligation and therefore become unavailable for disulfide bond formation, resulting in a conformational change.

These conformational changes may play a critical role in the *in vivo* regulation of Wilson and Menkes disease ATPases. It is possible that the copper binding to the N-terminal domain triggers a series of conformational changes that could cause this domain to interact with other cytosolic loops of the copper-transporting ATPase, leading to its copper-transporting activity. A fuller understanding of the mechanism of this process must await further structural and functional studies of various domains of the Wilson and Menkes disease ATPases.

References

General Reference

1. Sarkar, B., "Treatment of Wilson and Menkes Diseases", *Chem. Rev.*, **99,** 2535–2544 (1999).

Specific References

2. DiDonato, M., Narindrasorasak, S., Forbes, J. R., Cox, D. W., and Sarkar, B., "Expression, purification, and metal binding properties of the N-terminal domain from the Wilson disease putative copper-transporting ATPase (ATP7B)", *J. Biol. Chem.*, **272,** 33279–33282 (1997).
3. DiDonato, M., Hsu, H. F., Narindrasorasak, S., Que, L., Jr., and Sarkar, B., "Copper-induced conformational changes in the N-terminal domain of the Wilson disease copper-transporting ATPase", *Biochemistry,* **39,** 1890–1896 (2000).
4. Lutsenko, S., Petrukhin, K., Cooper, M. J., Gilliam, C. T., and Kaplan, J. H., "N-terminal domains of human copper-transporting adenosine triphosphatases (the Wilson's and Menkes disease proteins) bind copper selectively *in vivo* and *in vitro* with stoichiometry of one copper per metal-binding repeat", *J. Biol. Chem.*, **272,** 18939–18944 (1997).
5. Ralle, M., Cooper, M. J., Lutsenko, S., and Blackburn, N. J., "The Menkes Disease Protein Binds Copper via Novel 2-Coordinate Cu(I)-Cysteinates in the N-Terminal Domain", *J. Am. Chem. Soc.*, **120,** 13525–13526 (1998).

VIII.6. Metallochaperones

Thomas V. O'Halloran
Chemistry Department
Northwestern University
Evanston, IL 60208

Valeria Culotta
Environmental Health Sciences
Johns Hopkins University
School of Public Health
Baltimore, MD 21205

Contents

VIII.6.1. Introduction

Metalloproteins are found at diverse locations inside the eukaryotic cell: within membranes; on the cell surface; or exported into the extracellular environment. In spite of low levels of available metals in the circulating plasma of a mammal or in the typical environment of a microorganism, most cells accumulate significant amounts of Fe, Zn, and Cu, among other metals; and the metalloproteins acquire the correct cofactor from this mixture. How are the metal ions brought into the cell and then distributed among diverse compartments for specific incorporation in the correct metalloenzymes? The problem becomes paramount when one considers the low available concentrations of these precious nutrients, the requirement for specificity in metal ion insertion into the correct metalloprotein, and the added challenge of avoiding catalysis of deleterious oxidation reactions of cell components by redox-active metal ions.

How does the cell, once it gathers these different metal ions, know which one to put into which protein? Until recently, the traditional perspective has been that metal selectivity is due to specific chelating properties of the individual apoproteins. In this scenario, apoproteins are thought to orient ligating side chains in such a fashion that they match the precise ionic radius and electronic preferences of the functional metal ions (e.g., Zn^{2+} in carbonic anhydrase) and to discriminate against all others (e.g., Cu^{2+} or Fe^{2+}). The apo form of a metalloprotein can then be viewed as a remarkably specific chelating agent, with finely tuned kinetic and thermodynamic properties, selected through evolution to bind only one type of transition metal ion. In this model, each apoprotein, after it is produced within the cell, simply selects the correct metal ion from the mixture of metal ions available to it in the cytoplasm.

There are several problems with this perspective, especially when one considers the similarities of the ionic radii and coordination preferences of many metal ions as well as the difficulty of inserting the correct metal ion into large enzymes with deeply buried active sites. It has recently become apparent that this traditional view, that a given apoprotein acquires the correct metal cofactor by virtue of its selective metal ion chelating properties, is not in general correct. There is another solution to the problem, using "metallochaperones".[1]

VIII.6.2. The Need for Metallochaperones

The notion of metallochaperones is relatively new; prior to 1997, there was no indication that such molecules were necessary to assist in metal ion insertion into apometalloproteins. In a test tube setting, most Cu enzymes easily acquire their metal without a helper protein. As an example, the copper-containing superoxide dismutase 1 (SOD1) binds copper ions with an extremely high affinity (K_d in the femtomolar range). However, in a living cell, where total copper can accumulate in the micromolar range, SOD1 is highly dependent on a helper protein (known as CCS, for copper chaperone for SOD) for acquiring the metal. This puzzle can be best explained in terms of the apparent absence of "free copper" ions in the cell under normal growth conditions. Copper is not only an essential nutrient, but is also quite toxic to living cells. All organisms have devised elaborate mechanisms for detoxifying the metal and preventing its accumulation in the dangerous free ionic form. Based on studies of *Saccharomyces cerevisiae* SOD1 and using the yeast cell as a model, it has been estimated that the total concentration of intracellular free Cu is less than attomolar, which corresponds to less than one atom of Cu per cell.[2] In spite of this apparent vacuum of free copper, the metallochaperones are capable of acquiring the metal, protecting the cofactor from intracellular copper chelators, and donating Cu to enzymes that need it. Thus the metallochaperones indeed function in a "chaperone"-like manner to guide Cu ions and protect the metal from the overchelation capacity of the cell.

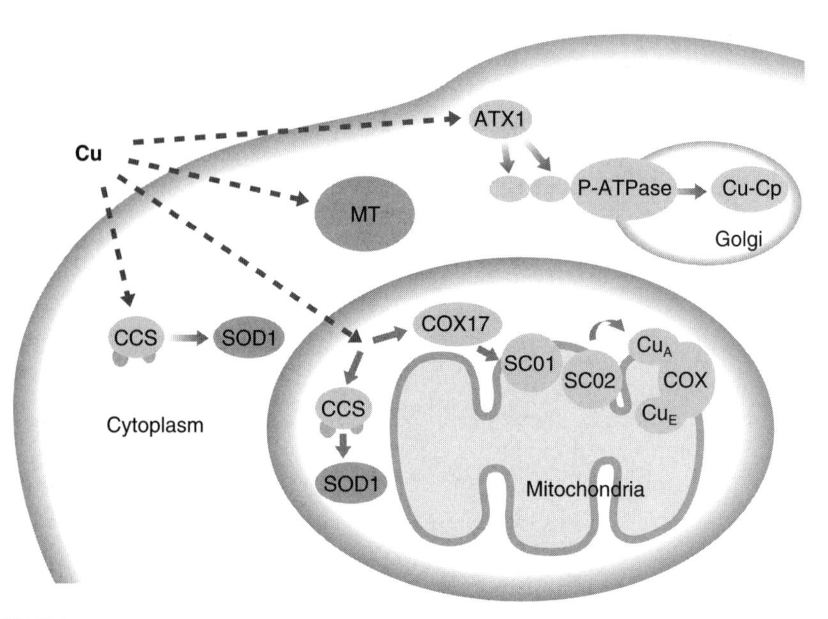

Fig. VIII.6.1.
A cartoon is shown illustrating distinct copper trafficking pathways. An intracellular Cu ion is managed in one of several pathways indicated by solid arrows. Excess Cu(I) is readily sequestered by detoxification factors such as metallothioneins (MT). Alternatively, the metal ion can enter one of the trafficking pathways that leads to the ultimate insertion of the ion into a copper metalloprotein. The pathway involving the copper chaperone ATX1 delivers copper to a copper-transporting P-type ATPase (P-ATPase) in the Golgi. This copper transporter then delivers the metal to the lumen of the Golgi where the ion can be incorporated into copper enzymes in the secretory pathway, such as the mammalian protein ceruloplasm (Cp see Section XI.7). In a second pathway of copper trafficking, the metal can be taken up by the CCS copper chaperone in the cytoplasm and then inserted into the soluble SOD1. Copper can also enter the intermembrane space (IMS) of the mitochondria by unknown mechanisms, where it can bind the mitochondrial forms of CCS and SOD1 or be delivered to the COX and SCO family of proteins. The COX17 protein in the IMS is believed to deposit the Cu onto the inner-membrane proteins SCO1 or SCO2, which in turn may facilitate insertion of the metal into the Cu_A site of cytochrome oxidase (COX). (Dashed arrows indicate multiple steps or unknown mechanisms.)

Thus far, three classes of copper chaperone-like molecules have been identified in eukaryotes (Fig. VIII.6.1).[3–8] Two of these have been demonstrated to transfer Cu directly to a specific intracellular target: the ATX1 protein directs Cu loading into vesicles in the secretory pathway and the copper chaperone for SOD1 (CCS) transfers copper to superoxide dismutase. In the case of ATX1, the partners are membrane-bound cation transporters in the CPx family of P-type ATPases. In the case of CCS, a partnership with the apo form of SOD1 was demonstrated both *in vivo* and *in vitro*. Another pathway exists for delivery of Cu to the mitochondrial enzyme cytochrome oxidase. This pathway involves the putative copper chaperone, COX17 (see Section VIII.6.3), and the molecules designated as "cochaperones", SCO1 and its homologue SCO2. In the current model, COX17 appears to donate Cu to SCO1 (or SCO2), which in turn transfers the metal to one of the subunits of cytochrome oxidase.[9,10] All of these various Cu carrying molecules (ATX1, CCS, and the COX17/SCO proteins) were originally discovered through genetic studies in the baker's yeast, *S. cerevisiae*, and subsequently, close homologues have been identified in numerous eukaryotes including plants, insects, and humans.

VIII.6.3. COX17

The *COX17* gene was discovered by Glerum et al. in 1996[11] in their search for yeast genes involved in the assembly of cytochrome oxidase (hence the name COX17). Cytochrome oxidase requires a total of three Cu ions to be inserted into two copper-binding subunits for full activity: the binuclear Cu_A site of subunit 2 and the single copper-binding site (Cu_B) of subunit 1 (see Section XI.6). Copper insertion into the two subunits appears to involve distinct pathways. Assembly of the Cu_B site involves the COX11 protein, whereas insertion of Cu into the binuclear Cu_A site requires at least two proteins: COX17, and in the inner membrane of the mitochondria, SCO1 (or its homologue, SCO2).[9,10] For the purposes of this section, we shall focus on the pathway involving COX17.

The COX17 polypeptide is 8.0 kDa in molecular weight and contains six cysteines that are conserved between the yeast and human proteins. While a variety of metal-bound states have been reported, spectroscopic analyses of one form of yeast COX17 isolated from bacterial overexpression systems is consistent with a "binuclear cuprous-thiolate cluster". In this model, several Cu ions are bound to the sulfurs of three cysteines, forming two clusters that bind the metal with an affinity that is relatively low compared to other copper proteins, such as metallothioneins.[12] Studies on the yeast protein have shown that COX17 localizes to both the cytoplasm and the intermembrane space of the mitochondria.[13] Winge and co-workers,[14] however, have shown that only the mitochondrial form of the protein is necessary to activate cytochrome oxidase with copper. Thus the rationale for a cytosolic form of the protein remains elusive. In the case of mitochondrial COX17, a likely target of Cu delivery includes the SCO proteins.

SCO1 and its close homologue SCO2 reside in the inner membrane of the mitochondria where they appear closely juxtaposed to the cytochrome oxidase target.[15] Interestingly, SCO1 and subunit 2 of cytochrome oxidase both have a CXXXC motif, which in the latter case is part of the Cu_A-binding site. Structural studies reveal that the soluble domain of SCO1 has a thioredoxin-like structure and that the conserved cysteines in the CXXXCP motif bind Cu(I).[16] In an attractive model, this region of SCO1 or SCO2 may directly transfer copper to cytochrome oxidase, whereas COX17 may act to shuttle Cu ions to the SCO molecules (Fig. VIII.1.5.1).[9,10] Alternative roles for these cytochrome oxidase assembly proteins, such as thioredoxin-like disulfide exchange, cannot yet be ruled out. Further studies are needed to directly test the sequential transfer of Cu from protein to protein.

VIII.6.4. ATX1

Saccharomyces cerevisiae ATX1 was originally identified in 1995 as a gene that provided some protection against oxidative damage when expressed at very high levels in yeast, hence the name "ATX1" for anti-oxidant.[17] However, this apparent antioxidant activity was later found not to be the primary function of ATX1. Instead, this protein acts to shuttle Cu(I) specifically to a Cu transporting ATPase located in the Golgi compartment of the secretory pathway (see cartoon of Fig. VIII.6.2).[1,18] This Cu transporter in turn delivers Cu into the lumen of the Golgi for insertion of the metal into copper enzymes destined for the cell surface or extracellular milieu. The human homologue of yeast ATX1 is denoted HAH1[19] or ATOX1, and functional homologues have also been described for mice, worms, and plants.

The protein that accepts Cu from ATX1 is a member of a large family of cation transporters known as "P-type transporting ATPases" that use energy from ATP

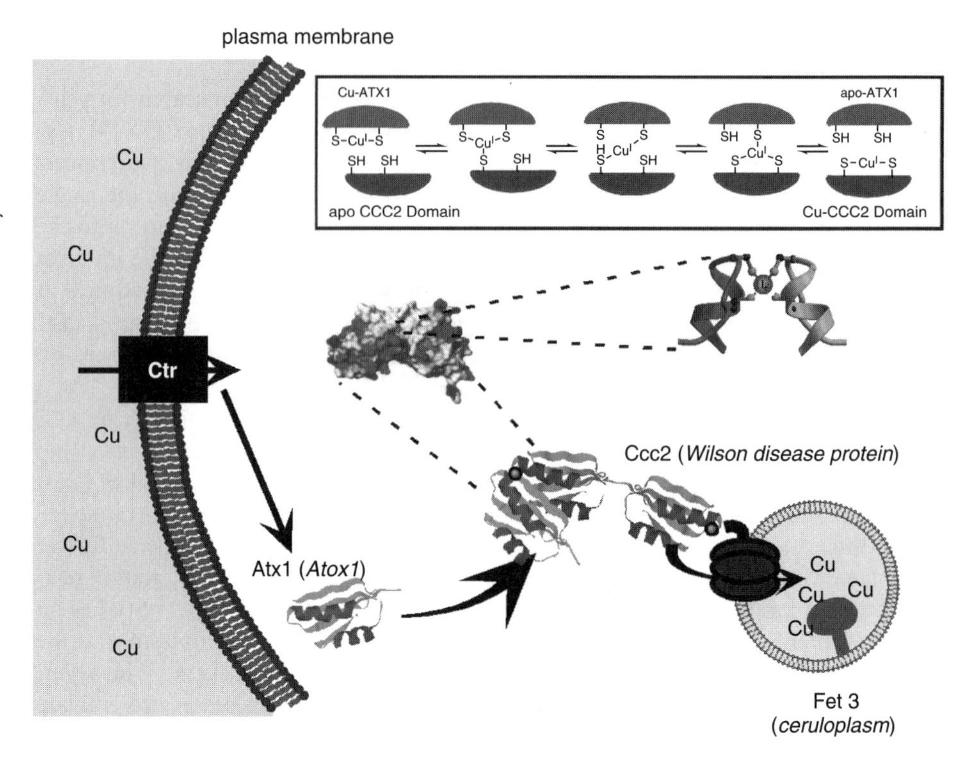

Fig. VIII.6.2.
Inorganic chemistry and structural biology of the ATX1 copper chaperone trafficking pathway (as summarized in Ref. 8). The Cu(I) ions are brought into the cell via unknown mechanisms, some of which involve the Ctr1 family of membrane-bound transporter proteins. Once loaded, the Cu–Atx1 complex can dock with a structurally homologous domain of a partner protein, the P-type ATPase Ccc2. Electrostatic complementarity between the Atx1 chaperone and its Ccc2 (see docked structures) provides a basis for partner recognition and for correct orientation of donor and acceptor Cys residues. A mechanism of copper transfer involving a series of two- and three-coordinate Cu(I)-thiolate intermediates (inset) is corroborated by spectroscopic, thermodynamic, and structural studies. Similar mechanisms are proposed for the mammalian forms (names in italics) of these yeast proteins. The PDB locators are Cu-Hah1: 1FEE; Atx1: 1FD8; Cu-Ccc2a: 1FVS. [See color insert.]

hydrolysis to drive the transport of ions across biological membranes. The subfamily of ATPases that transport Cu specifically harbor multiple copies of a copper-binding protein domain at the amino terminus of the polypeptide. Humans express two types of this copper transporter, known as the Wilson and Menkes proteins (see Section VIII.5). The yeast homologue to these copper transporters is known as CCC2 and is responsible for activating a copper protein involved in iron uptake (FET3, a homologue of human ceruloplasmin).[20] In all species, the copper transporter is presumed to rely heavily on its corresponding copper chaperone partner (i.e., ATX1) for acquiring the metal, particularly under copper starvation conditions.

The ATX1 copper chaperone is known to bind a single Cu(I) ion via two cysteine residues present as a MXCXXC motif [X = any amino acid, C = cysteine (Cys), M = methionine (Met)]. A third as-of-yet unidentified atom is typically observed by EXAFS near the Cu(I) ion in what is best described as a secondary bonding interaction. This bis-cysteine copper-binding site is unique to the family of proteins or domains that are homologous to ATX1.[1] The low coordination number provides tight binding of Cu(I), but at the same time allows for attack of an incoming nucleophilic ligand, for example, the Cys side chain of a partner protein, as a prelude to the rapid transfer of copper from one tight binding site to another (see Fig. VIII.6.2 inset).

Distinct subfamilies of copper proteins containing this motif are emerging from a series of recent structural studies including (1) the small (~70 amino acid) copper chaperone homologues to yeast ATX1 itself (identified in mammals, invertebrates, and plants); (2) the first domain of the CCS copper chaperone (see below); (3) the CopZ copper protein of bacteria; and (4) the N-terminal domains of the copper transporting ATPases including bacterial CopA, yeast CCC2, and mammalian Menkes and Wilson proteins. In the case of yeast ATX1, CCC2a, and metal-binding domain 4 of the Menkes disease protein (Mbd4), the solution NMR or X-ray crystal structure is known with Hg(II), Cu(I) or Ag(I) bound (reviewed in Ref. 21). In each instance, the protein is a monomer and the metal ion is bound to two cysteines in the first loop of the β_ββ_β-fold of the polypeptide. This fold is widely employed in biology and is typically found in small proteins and/or as a discrete domain of larger proteins. In most of those cases, the first loop and helix are employed to bind inorganic ions such as Cu(I) and Hg(II) as with the ATX1-like domains and the mercury resistance protein MerP, but can also bind iron–sulfur clusters in the case of ferredoxin II, or phosphate anion in the case of acylphosphatase (reviewed in Ref. 21). A "structural genomics" approach to understanding this fold suggests that additional functions and metal substrates will be found.[7,22]

How does the ATX1 copper chaperone specifically recognize the copper transporter? Studies with the yeast and human metallochaperone demonstrate that ATX1/HAH1 physically interacts with the ATX1-like domains of the copper transporter and that this interaction requires Cu ions and the CXXC copper site.[1] Docking of ATX1 with the copper transporter also appears to require electrostatic interactions involving a positively charged face on the copper chaperone and a negatively charged patch on the corresponding portion of the copper transporter domains (Fig. VIII.6.2).[23,24]

Once the metallochaperone has docked with its target, direct Cu transfer occurs in a facile manner. Analysis of the energetics of partner recognition and metal transfer have provided additional mechanistic insights. The overall equilibrium constant for metal transfer from Cu-ATX1 to CCC2a is small (~ 1.5) indicating a shallow thermodynamic gradient for copper transfer between these partner proteins. The overall driving force for copper transfer from ATX1 into the Golgi vesicle is most likely provided by the CCC2 copper transporter in the subsequent ATP hydrolysis step: The transporter is envisioned to undergo an ATP-driven conformational change that displaces the Cu(I) from the cytoplasmic face of CCC2, moving it across the vesicular membrane into a separate thermodynamic compartment of the Golgi lumen. The transfer step is rapid on the time scale of the assay ($t_{1/2} < 1$ min). Thus while these ATX1-like domains bind Cu tightly, they also allow for rapid equilibration of Cu(I) between physiologically related sites. In this manner, they guide the metal down a reaction pathway by controlling the kinetics of metal transfer to the copper-transporting partner.[25] Given the overcapacity for copper chelation in the cytoplasm, as discussed above, the copper chaperones are most likely to kinetically discriminate against Cu(I) transfer to nonpartner sites and in this manner provide vectorial delivery of Cu to specific targets.

In general, the mechanism by which ATX1 recognizes its target and delivers its Cu cargo is the best understood of all the eukaryotic copper chaperones. By comparison, the mechanism of action of the CCS copper chaperone (described below) is far more complex.

VIII.6.5. Copper Chaperone for SOD1

The CCS copper chaperone is responsible for inserting Cu into the Cu and Zn requiring SOD1. The SOD1 acts to scavenge toxic superoxide anion radicals through a reaction involving the bound Cu ion, thereby guarding cells against oxidative damage. The *in vivo* insertion of Cu into SOD1 requires a copper chaperone that was first identified in yeast as the *LYS7* gene.[26] Yeast cells containing a mutation in the *LYS7* gene express a form of SOD1 that is essentially apo for Cu, but still harbors a single atom of Zn per dimer enzyme.[27] The identification of the yeast copper chaperone for SOD1 quickly led to the cloning of the human homologue and both yeast and human molecules have been denoted as CCS.[4]

Cell biology studies have shown that CCS largely colocalizes with SOD1. The vast majority of SOD1 is in the soluble cytoplasm, as is its copper chaperone. However, a small fraction of SOD1 also enters the IMS of mitochondria where the enzyme is believed to scavenge superoxide derived from the respiratory chain. And, as expected, a fraction of CCS is also found in the IMS to facilitate the conversion of inactive mitochondrial SOD1 into a fully active copper-containing enzyme[28] (see cartoon of Fig. VIII.5.1).

The CCS proteins are the largest copper chaperones known to date (with molecular masses of 30–32 kDa versus 7–8 kDa for ATX1 and COX17). While ATX1 and COX17 are single-domain proteins, CCS is represented by three functionally distinct protein domains.[4]

The domain closest to the CCS amino terminus (Domain I) bears striking homology to ATX1, including the CXXC copper-binding site. Based on this homology, it was initially assumed that this domain was critical for the copper-transfer reaction. However, a CCS molecule lacking this domain can still insert Cu into SOD1 *in vivo,* but only when Cu is plentiful. It therefore appears that this ATX1-like domain works to maximize CCS activity specifically under Cu starvation conditions.

Domain II in the center of CCS bears remarkable similarity to the target of Cu delivery, SOD1. In the case of human CCS, this noted homology is so strong that a single mutation in the Cu site of Domain II activates a superoxide scavenging activity of CCS. Domain II does not appear to directly participate in copper transfer, but physically interacts with SOD1 to secure the enzyme during Cu insertion. The SOD1 normally exists as a homodimer of two identical subunits and it has been proposed that CCS docks with SOD1 to form either a transient heterodimer or heterotetramer as a prerequisite to copper transfer. In the case of the yeast proteins, a CCS–SOD1 heterodimer complex has been resolved by X-ray crystallographic analysis.[29] As predicted, the complex is stabilized by interactions at the dimer interface of SOD1 and the corresponding homologous region of CCS Domain II.

The carboxy terminal domain of CCS (Domain III) is relatively small ($\sim$30 amino acids), yet it plays a critical role in the *in vivo* activation of SOD1. Domain III shows high conservation among CCS molecules from diverse species, including an invariant CXC motif that has the potential to bind copper. It has been proposed that the C-terminal domain of CCS in concert with the N-terminal Domain I, acts to insert Cu into the SOD1 active site.[4] Interestingly, in the crystallographic analysis of the SOD1–CCS heterodimer, one of these cysteines forms an intermolecular disulfide with an essential cysteine in SOD1, perhaps representing an intermediate in the copper-transfer process.[29] The precise mechanism of copper transfer is not completely understood and several intriguing questions remain. For example, how does Cu move from an all-sulfur coordination environment (in CCS) to the all-nitrogen copper site in SOD1? Additionally, how does SOD1 acquire its Zn atom? A combination of biochemical, spectroscopic, and biological approaches are now in motion to address these questions.

VIII.6.6. Metallochaperones for Other Metals?

Although the bulk of current knowledge on metallochaperones is restricted to Cu, it is conceivable that analogous molecules exist to "chaperone" other metal ions within the cell.[8] Heavy metals such as Fe, Mn, Ni, and Zn also serve as cofactors for metalloenzymes, and like Cu, the free form of these metals is not expected to accumulate in living cells. As other features of metal trafficking emerge, the future is bound to bring new insight into the escorting and shuffling of metals within living cells.

VIII.6.7. Concluding Remarks

In summary, studies with Cu have resulted in a new concept for the trafficking of metal ions within the cell, a paradigm that involves the action of small soluble metallochaperones. These molecules not only carry the ion to its proper destination, but directly facilitate insertion of the Cu ion cofactor into the metalloenzyme target. Thus far, the bulk of mechanistic understanding of copper transfer has emerged from studies on the ATX1 and CCS proteins, and to a lesser degree on the COX17/SCO proteins. Although the targets for these copper site assembly factors are quite distinct, they share several interesting commonalities. First, all of these proteins bind the reduced Cu(I) form of the metal. Second, the various molecules are "look-alikes" of their target: ATX1 bears structural and sequence homology with the metal-binding domains of the copper transporter; the central Domain II of CCS shares strong homology with SOD1; and SCO1 is homologous to its target copper site in cytochrome *c* oxidase. The striking resemblance between copper chaperones and cognate targets has presumably evolved to facilitate target recognition and to maintain specificity for the copper-transfer reaction. Finally, it should be emphasized that the copper chaperones do not act to protect the cell from Cu toxicity. Rather these molecules have evolved to protect the precious Cu cargo from the metal ion scavengers in the cell (e.g., metallothioneins see Section VIII.4; Fig. VIII.6.1). The copper chelation capacity of the cell is indeed strong and prevents the toxic metal from accumulating in the deleterious free ionic form. In spite of this potent "vacuum" for Cu, the metallochaperones ensure the safe and accurate delivery of the metal to the enzymes that need it.

References

General References

1. Pufahl, R., Singer, C., Peariso, K. L., Lin, S. J., Schmidt, P., Fahrni, C., Culotta, V. C., Penner-Hahn, J. E., and O'Halloran, T. V., "Metal ion chaperone function of the soluble Cu(I) receptor Atx1", *Science*, **278,** 853–856 (1997).
2. Rae, T. D., Schmidt, P. J., Pufahl, R. A., Culotta, V. C., and O'Halloran, T. V., "Undetectable intracellular free copper: the requirement of a copper chaperone for superoxide dismutase", *Science*, **284,** 805–808 (1999).
3. Valentine, J. S. and Gralla, E. B., "Delivering copper inside yeast and human cells", *Science*, **278,** 817–818 (1997).
4. O'Halloran, T. V. and Culotta, V. C., "Metallochaperones: an intracellular shuttle service for metal ions", *J. Biol. Chem.*, **275,** 25057–25060 (2000).
5. Harrison, M. D., Jones, C. E., Solioz, M., and Dameron, C. T., "Intracellular copper routing: the role of copper chaperones", *TIBS*, **25,** 29–32 (2000).
6. Huffman, D. L. and O'Halloran, T. V., "Function, structure, and mechanism of intracellular copper trafficking proteins", *Annu. Rev. Biochem.*, **70,** 677–701 (2001).
7. Bertini, I. and Rosato, A., "Bioinorganic Chemistry Special Feature: Bioinorganic chemistry in the postgenomic era", *Proc. Natl. Acad. Sci. U.S.A.*, **100,** 3601–3604 (2003).
8. Finney, L. A. and O'Halloran, T. V., "Transition metal speciation in the cell: insights from the chemistry of metal ion receptors", *Science*, **300,** 931–936 (2003).
9. Winge, D. R., "Let's Sco1 Oxidase, Let's Sco!" *Structure*, **11,** 1313–1314 (2003).
10. Carr, H. S. and Winge, D. R., "Assembly of cytochrome *c* oxidase within the mitochondria", *Acc. Chem. Res.*, **36,** 309–316 (2003).

Specific References

11. Glerum, D. M., Shtanko, A., and Tzagoloff, A., "Characterization of *COX17*, a yeast gene involved in copper metabolism and assembly of cytochrome oxidase", *J. Biol. Chem.*, **271,** 14504–14509 (1996).
12. Heaton, D. N., George, G. N., Garrison, G., and Winge, D. R., "The mitochondrial copper metallochaperone Cox17 exists as an oligomeric, polycopper complex", *Biochemistry*, **40,** 743–751 (2001).
13. Beers, J., Glerum, D. M., and Tzagoloff, A., "Purification, Characterization, and Localization of Yeast Cox17p, a Mitochondrial Copper Shuttle", *J. Biol. Chem.*, **272,** 33191–33196 (1997).
14. Maxfield, A. B., Heaton, D. N., and Winge, D. R., "Cox17 is functional when tethered to the mitochondrial inner membrane", *J. Biol. Chem.*, **279**(7), 5072–5080 (2004).

15. Glerum, D. M., Shtanko, A., and Tzagoloff, A., "SCO1 and SCO2 act as high copy suppressors of a mitochondrial copper recruitment defect in Saccharomyces cerevisiae", *J. Biol. Chem.*, **271**, 20531–20535 (1996).
16. Balatri, E., Banci, L., Bertini, I., Cantini, F., and Ciofi-Baffoni, S., "Solution structure of Sco1: A thioredoxin-like protein involved in cytochrome *c* oxidase assembly", *Structure*, **11**, 1431–1443 (2003).
17. Lin, S. and Culotta, V. C., "The *ATX1* gene of *Saccharomyces cerevisiae* encodes a small metal homeostasis factor that protects cells against reactive oxygen toxicity", *Proc. Natl. Acad. Sci. U.S.A.*, **92**, 3784–3788 (1995).
18. Lin, S. J., Pufahl, R., Dancis, A., O'Halloran, T. V., and Culotta, V. C., "A role for the *Saccharomyces cerevisiae ATX1* gene in copper trafficking and iron transport", *J. Biol. Chem.*, **272**, 9215–9220 (1997).
19. Klomp, L. W. J., Lin, S. J., Yuan, D., Klausner, R. D., Culotta, V. C., and Gitlin, J. D., "Identification and functional expression of HAH1, a novel human gene involved in copper homeostasis", *J. Biol. Chem.*, **272**, 9221–9226 (1997).
20. Yuan, D. S., Stearman, R., Dancis, A., Dunn, T., Beeler, T., and Klausner, R. D., "The Menkes/Wilson disease gene homologue in yeast provides copper to a ceruloplasmin-like oxidase required for iron uptake", *Proc. Natl. Acad. Sci., U.S.A.*, **92**, 2632–2636 (1995).
21. Rosenzweig, A. C. and O'Halloran, T. V., "Structure and Chemistry of the Copper Chaperone Proteins", *Curr. Op. Chem. Biol.*, **4**, 140–147 (2000).
22. Arnesano, F., Banci, L., Bertini, I., Ciofi-Baffoni, S., Molteni, E., Huffman, D. L., and O'Halloran, T. V., "Metallochaperones and metal-transporting ATPases: a comparative analysis of sequences and structures", *Genome Res.*, **12**, 255–271 (2002).
23. Arnesano, F., Banci, L., Bertini, I., Cantini, F., Ciofi-Baffoni, S., Huffman, D. L., and O'Halloran, T. V., "Characterization of the binding interface between the copper chaperone Atx1 and the first cytosolic domain of Ccc2 ATPase", *J. Biol. Chem.*, **276**, 41365–41376 (2001).
24. Rosenzweig, A. C., "Copper delivery by metallochaperone proteins", *Acc. Chem. Res.*, **34**, 119–128 (2001).
25. Huffman, D. and O'Halloran, T., "Energetics of Copper Trafficking between the Atx1 Metallochaperone and the Intracellular Copper Transporter, Ccc2", *J. Biol. Chem.*, **275**, 18611–18614 (2000).
26. Culotta, V. C., Klomp, L., Strain, J., Casareno, R., Krems, B., and Gitlin, J. D., "The copper chaperone for superoxide dismutase", *J. Biol. Chem.*, **272**, 23469–23472 (1997).
27. Lyons, T. J., Nersissian, A., Goto, J. J., Zhu, H., Gralla, E. B., and Valentine, J. S., "Metal ion reconstitution studies of yeast copper-zinc superoxide dismutase: the "phantom" subunit and the possible role of Lys7p", *JBIC*, **3**, 650–662 (1998).
28. Sturtz, L. A., Diekert, K., Jensen, L. T., Lill, R., and Culotta, V. C., "A Fraction of Yeast Cu,Zn-Superoxide Dismutase and Its Metallochaperone, CCS, Localize to the Intermembrane Space of Mitochondria", *J. Biol. Chem.*, **276**, 38084–38089 (2001).
29. Lamb, A. L., Torres, A. S., O'Halloran, T. V., and Rosenzweig, A. C., "Heterodimeric structure of supeoxide dismutase in complex with its metallochaperone", *Nat. Struct. Biol.*, **8**, 751–755 (2001).

CHAPTER IX

Hydrolytic Chemistry

IX.1. Metal-Dependent Lyase and Hydrolase Enzymes. (I) General Metabolism

J. A. Cowan
Chemistry
Ohio State University
Columbus, OH 43210

Contents

IX.1.1. Introduction

The goal of Sections IX.1 and IX.2 is to review the chemistry of a selection of metal-dependent hydrolase and lyase enzymes (see Table IX.1.1), first describing metalloenzymes that function in general metabolic pathways and then focusing on enzymes that function in nucleic acid biochemistry. For most of the metalloenzymes that we describe here, the metal ion cofactor acts as a Lewis acid, serving to activate a bound nucleophile by promoting ionization to a more reactive anionic form (e.g., water to hydroxide), or to stabilize a bound intermediate. The protein-bound metal ion also provides a template that brings the reactant species into close proximity.

The identity of a metal ion cofactor in a metalloenzyme depends on several criteria, including availability and suitability for the task.[1] In the case of metal-dependent hydrolase and lyase enzymes, a metal cation with a high degree of Lewis acidity is essential to the function of the enzyme, allowing a large number of hydrolysis and condensation reactions to proceed under physiological conditions that would otherwise require extremes of pH. Non-redox metal ions are also preferred for these types of enzymes in order to protect sensitive functional groups in proteins and nucleic acids from oxidative damage. For these reasons, Zn^{2+} as well as Mg^{2+} (and occasionally Ca^{2+}) are commonly used as Lewis acids in the enzymatic catalysis of hydrolysis reactions. Of these, zinc is the stronger Lewis acid (reflected in the relative

Table IX.1.1.
Classification of Enzymes

Enzyme Class	Enzyme Function
Oxidoreductase	Oxidation–reduction
Transferase	Transfer of a group from one compound to another
Hydrolase	Hydrolysis
Lyase	Nonhydrolytic addition or removal of groups
Isomerase	Conversion of a substance into an isomeric form
Ligase	Synthesis of a large molecule from two smaller ones

pK_a values for bound H_2O: Zn–OH_2 $\rightarrow$ Zn–OH^- + H^+, p$K_a = 8.8$; Mg–OH_2 $\rightarrow$ Mg–OH^- + H^+, p$K_a = 11.4$) and is typically used in the hydrolysis of carbonyl functionality (esters and especially amides by protease enzymes).

It is interesting in this regard that the most common ligand geometry at the active sites of Zn^{2+} enzymes is tetrahedral, that is, relatively low coordination number, since Zn^{2+} in a tetrahedral ligand environment is typically the strongest Lewis acid among first-row transition metal ions. Zinc ion is normally tethered to enzymes through coordination to ligands that are intermediate in hard–soft character with extensive coordination to histidine (His) and occasionally cysteine (Cys) thiol or carboxylate bearing residues. The harder Mg^{2+} ion is typically coordinated to enzymes by carboxylate residues or water molecules, and is more frequently associated with phosphate ester hydrolysis and phosphoryl transfer, or stabilization of enolate anions and related species. The latter functional groups are common to the substrates of enzymes discussed in this section.

IX.1.2. Magnesium

Magnesium ion plays a role in enzymatic reactions in two general ways. First, an enzyme may bind the magnesium–substrate complex. In this case, the enzyme interacts principally with the substrate and shows little, or at best weak interaction with Mg^{2+} (e.g., Mg isocitrate in isocitrate lyase). Alternatively, Mg^{2+} binds directly to the enzyme and alters its structure and/or serves a catalytic role. Although other divalent metal ions may also activate these enzymes, this is frequently accompanied by a reduction of enzyme efficiency and/or substrate specificity. Magnesium binds weakly to proteins and enzymes, with association constants, $K_{a'} \leq 10^5\ M^{-1}$; magnesium-activated enzymes are therefore not necessarily isolated in the metal-bound form. Magnesium must be added to the enzyme solution for *in vitro* reactions, while background Mg^{2+} concentrations *in vivo* are of the order of several millimolar and therefore sufficient.

Many metabolic cycles in higher organisms (citric acid cycle, glycolytic cycle, etc.) are mediated by magnesium-dependent enzymes and follow a general mechanistic theme for involvement of Mg^{2+} ion, which is presented in Fig. IX.1.1.[2] In spite of the apparent complexity of the structural and kinetic models that have been developed, several common themes do emerge. First, most enzymes on these pathways require at least two metal-binding sites: an allosteric regulatory site modulating either structure or binding, while the metal ion binding to a second site typically serves a catalytic role. Also, in these cases, the substrate binds to the metal-activated enzyme and the principal substrate molecules (xylose, pyruvate, etc.) do not themselves have a high affinity for Mg^{2+}. Other enzymes tend to show isolated high-affinity metal-binding sites where the metal cofactor serves to stabilize an important reaction intermediate along the catalytic pathway.

Crystallographic evidence and supporting mutagenesis studies suggest that substrate binding occurs at only one of the metal sites. In some cases, the catalytic metal is delivered as a chelate complex of the substrate (e.g., Mg^{2+}-isocitrate with isocitrate lyase). However, in others, the enzyme makes a genuine complex with the Mg^{2+} cofactor; in these examples, the coordinating ligands are typically carboxylates. Occasionally, either an amide carbonyl from the backbone or a side-chain bind to Mg^{2+}, but the residual ligands that complete the octahedral coordination set are most often water molecules, some of which may be displaced by substrate binding.

IX.1.2.1. Xylose Isomerase and Isocitrate Lyase

Kinetic data for Mg^{2+}-activated xylose isomerase, for example, suggests that the first step in catalysis (Fig. IX.1.2) is the binding of Mg^{2+} to site 1 (having the higher binding affinity), which facilitates subsequent substrate binding and proper orientation and stabilizes reaction intermediates.[3] The second cation then binds, and turnover is initiated. A Mg^{2+}-bound water molecule is likely to assist in proton transfer between the hydroxyl and carbonyl oxygens. In the case of xylose isomerase, the second ion is essential for the isomerization reaction (i.e., for H^+ transfer from C1 to C2, and H^+

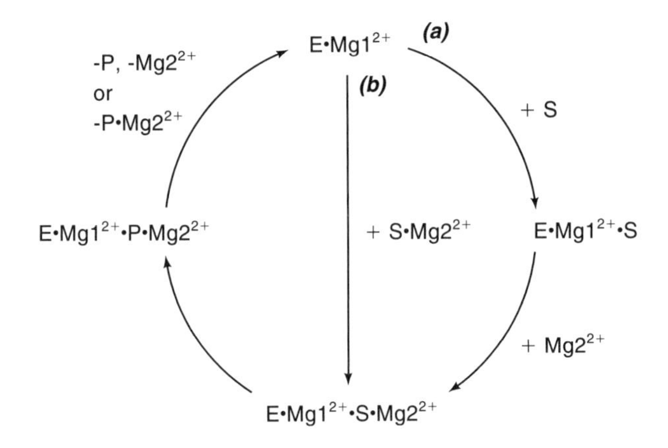

Fig. IX.1.1.
General mechanistic features for magnesium-catalyzed transformations of enzymes on the glycolytic pathway. Pathways (*a*) and (*b*) correspond to cases where the substrate does not, or does chelate the second Mg^{2+} cofactor, respectively.

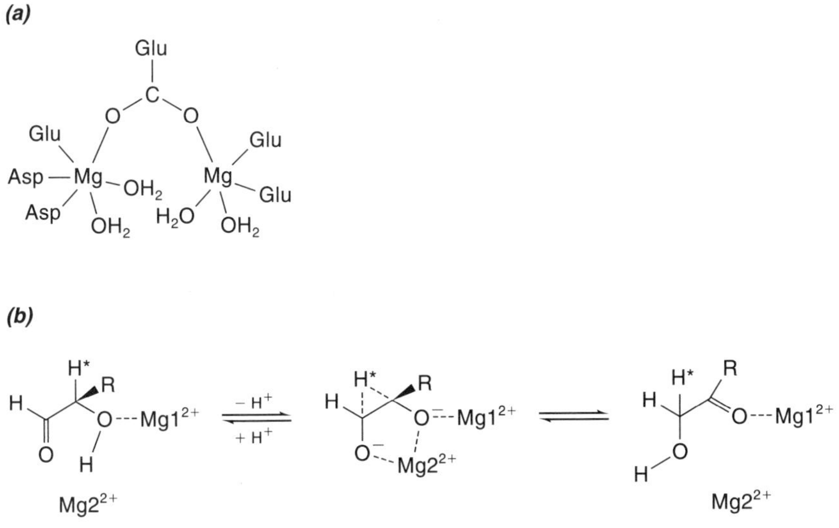

Fig. IX.1.2.
(*a*) A schematic illustration of the two metal ions of xyolse isomerase (PDB code: 1MUW). (*b*) A proposed mechanism for the metal-catalyzed 1,2-hydride shift in *Arthrobacter* xylose isomerase. The labile hydrogen is starred. Divalent Mg ions maintain octahedral coordination with extensive coordination to carboxylate residues. Note that $Mg2^{2+}$ is thought to be the catalytic ion, while the other most likely plays a structural role.[3]

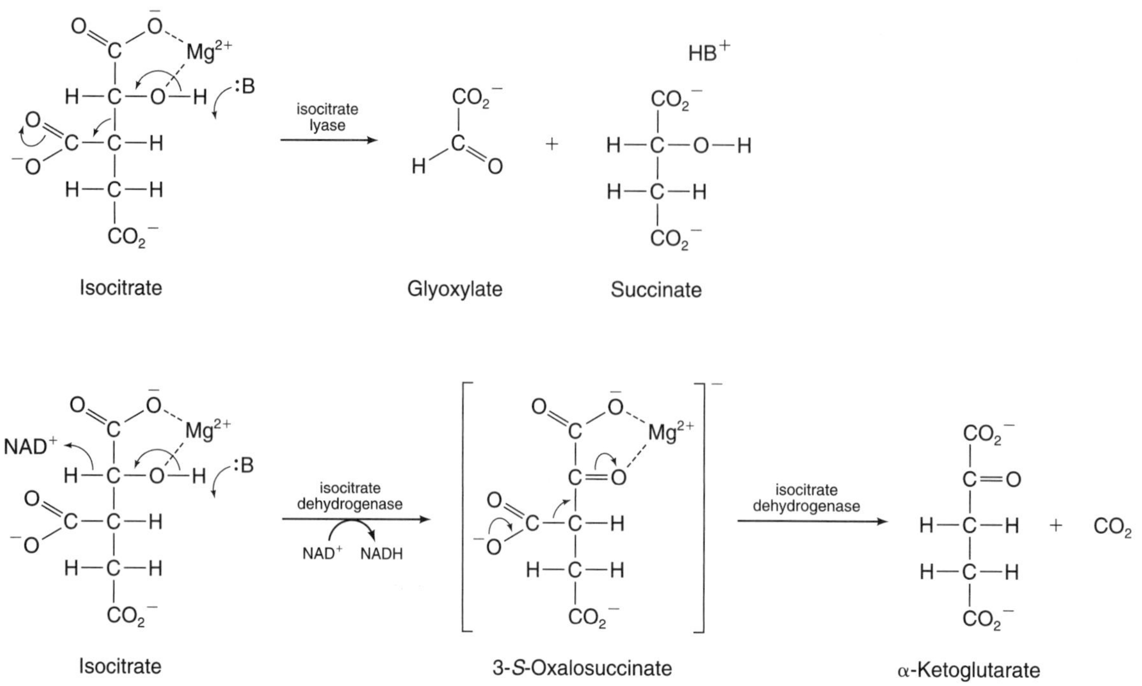

Fig. IX.1.3.
A proposed mechanism for isocitrate lyase illustrating the role of the catalytic magnesium cofactors (PDB code: 1DQU).[4]

transfer from O2 to O1); this Mg^{2+} most probably stabilizes the anionic substrate and polarizes the C–O bonds (Fig. IX.1.2).

Isocitrate lyase catalyzes the reversible cleavage of isocitrate to glyoxylate and succinate (Fig. IX.1.3) and shows an absolute requirement for divalent Mg to promote enzyme activity. The enzyme is tetrameric, with four identical subunits, each of which possesses an active site.[5] The substrate is the Mg^{2+}–isocitrate complex, although a second, but noncatalytic high-affinity Mg^{2+}-binding site (dissociation constant, $K_{d'} \sim 200\,\mu M$) has been identified, which activates the enzyme by inducing a conformational change, presumably optimizing the alignment of key catalytic residues in the active site. The Mg^{2+}–isocitrate complex binds tightly to the active site ($K_d \sim 40\,\mu M$) although free Mg^{2+} can also bind weakly ($K_d \sim 6\,mM$) and presumably accounts for the inhibition observed at higher $[Mg^{2+}]$. (The enzymes described in this section are inhibited at elevated $[Mg^{2+}]$, which typically only arises in those cases where excess free Mg^{2+} can compete with a Mg^{2+}–substrate complex.) Inhibition is also observed for xylose isomerase, but not in the case of L-aspartase.

IX.1.2.2. Glutamine Synthetase and L-Aspartase

Figure IX.1.4 summarizes the chemistry of glutamine synthetase, a dodecamer that catalyzes the formation of glutamine from glutamate and ammonia with accompanying hydrolysis of ATP.[6] Crystallographic and kinetic studies have identified two metal-binding sites (of high and low affinity), with an internuclear distance of 5.8 Å. The high-affinity site corresponds to the catalytic cofactor, while the weakly bound ion has been implicated with binding of an MgATP chelate.[7,8] Again, the catalytic Mg^{2+} ion provides a template for the reaction, electrostatic stabilization during the transition state, and Lewis acid catalysis.

L-Aspartase catalyzes the reversible deamination of L-aspartic acid to form fumaric acid and ammonia.[9] Unlike the aforementioned synthetase, there is no need for an MgATP cosubstrate, and so only one metal cofactor is required. At low pH, the strict cation requirement disappears, and residual activity occurs without the metal

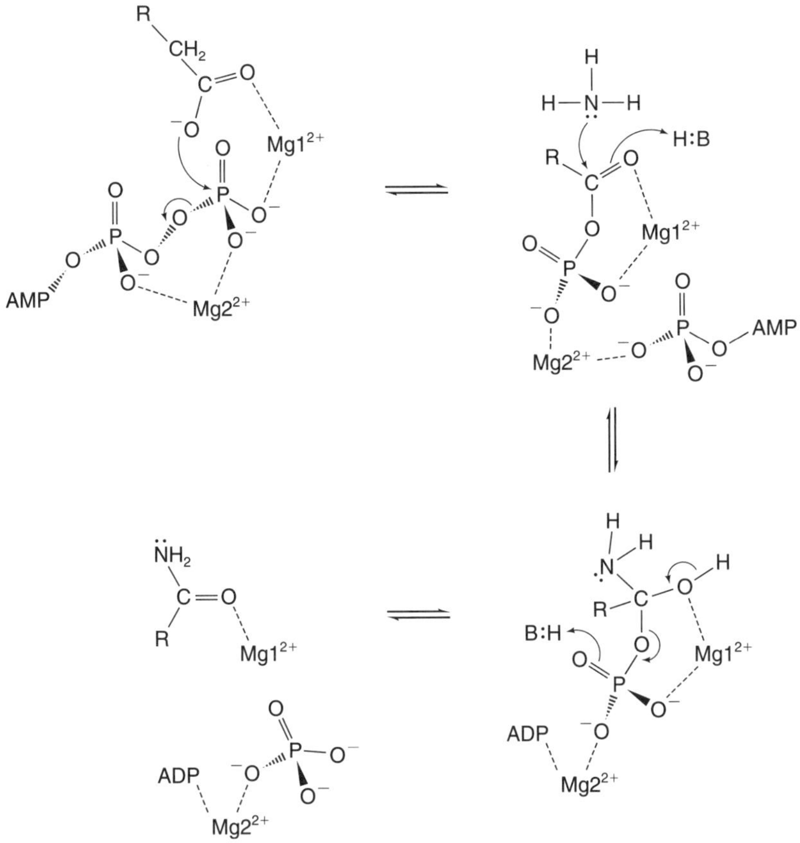

Fig. IX.1.4.
Structural details and a proposed mechanism for *E. coli* glutamine synthetase amination.[4] Two basic residues are thought to be involved in the reaction, supplying protons to the carboxyl and the phosphoryl groups. Divalent Mg ions maintain octahedral coordination with extensive coordination to carboxylate residues (PDB code: 1GDO).

cofactor. The metal cofactor stabilizes the active site conformation of the enzyme, but it does not coordinate the substrate (PDB code: 1JSW).

IX.1.3. Zinc

While the pK_a for Zn^{2+}(aq) is ~ 10, it can drop to ~ 7 in an enzyme environment where the more hydrophobic environment favors the deprotonation and consequent reduction in overall charge. In the case of Mg^{2+}, the pK_a is typically higher than physiological pH and a bound hydroxide is not usually formed. By contrast, Zn^{2+} readily activates bound water, and $[Zn–OH]^+$ with a bound HO^- can be regarded as a Lewis base. The Zn^{2+} ion also forms kinetically inert bonds to His residues that disfavor dissociation from the protein (Fig. IX.1.5), and so, again in contrast to Mg^{2+}, Zn^{2+} is usually found bound to the isolated protein. While Zn^{2+}–nitrogen coordination is fairly inert, water and hydroxide bound to Zn^{2+} are relatively labile ligands, a requirement for hydrolysis reactions. Interestingly, Zn^{2+}, Mn^{2+}, and Mg^{2+} each lack ligand-field stabilization contributions that might favor octahedral coordination. Expansion of coordination from four to five after binding substrate is therefore possible, with no preference for trigonal-bipyramidal or square-pyramidal geometries.[10]

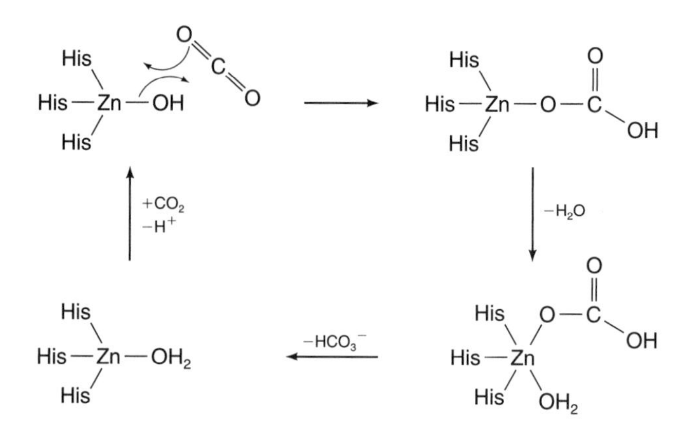

Fig. IX.1.5.
Structural details and a proposed mechanism for carbonic anhydrase (PDB code: 1ZNC).[10]

IX.1.3.1. Carbonic Anhydrase

Carbonic anhydrase (CA) catalyzes the hydration of CO_2, with acceleration of $\sim 10^7$ relative to the uncatalyzed reaction.

$$CO_2 + H_2O \rightarrow HCO_3^- + H^+ \tag{1}$$

The Zn^{2+} ion is coordinated in a distorted tetrahedron, with three His ligands and one labile H_2O (Fig. IX.1.5). The enzyme is prevalent in red blood cells and is important in respiration and pH buffering. The active site possesses an array of water molecules that play important structural and functional roles.[11] In particular, the waters mediate H^+ transfer—a key aspect of the hydrolysis reaction (Eq. 1). A zinc-bound hydroxide (formed by deprotonation via the water network via His64) attacks CO_2. The Zn^{2+} ion also serves to polarize the O=C=O and stabilize the developing negative charge. Subsequent release of HCO_3^- and uptake of a H_2O is followed by proton transfer to yield a new $[Zn^{2+}–OH]^+$ center.

IX.1.3.2. Carboxypeptidase A

Carboxypeptidase A (CPA) was one of the first Zn enzymes to be discovered and has been extensively investigated by kinetic, structural,[12,13] and spectroscopic methods. As a result, it is one of the best understood hydrolytic enzymes, providing a useful reference for related zinc enzymes.

Carboxypeptidase A is a digestive enzyme that favors C-terminal cleavage of residues with large aromatic side chains. It is an exopeptidase of molecular weight 34.5 kDa and thus is similar in size to carbonic anhydrase. A related protein, carboxypeptidase B, is much like CPA but favors N-terminal basic residues (Scheme IX.1.1).

Protein (or peptide)

Exopeptidase (e.g., CPB) → H_2N—————CO_2H ← Exopeptidase (e.g., CPA)

Endopeptidase (e.g., thermolysin)

Similar to carbonic anhydrase, the zinc cofactor in CPA is tetrahedral in geometry and located at the bottom of a cavity. Instead of the three His ligands and a H_2O

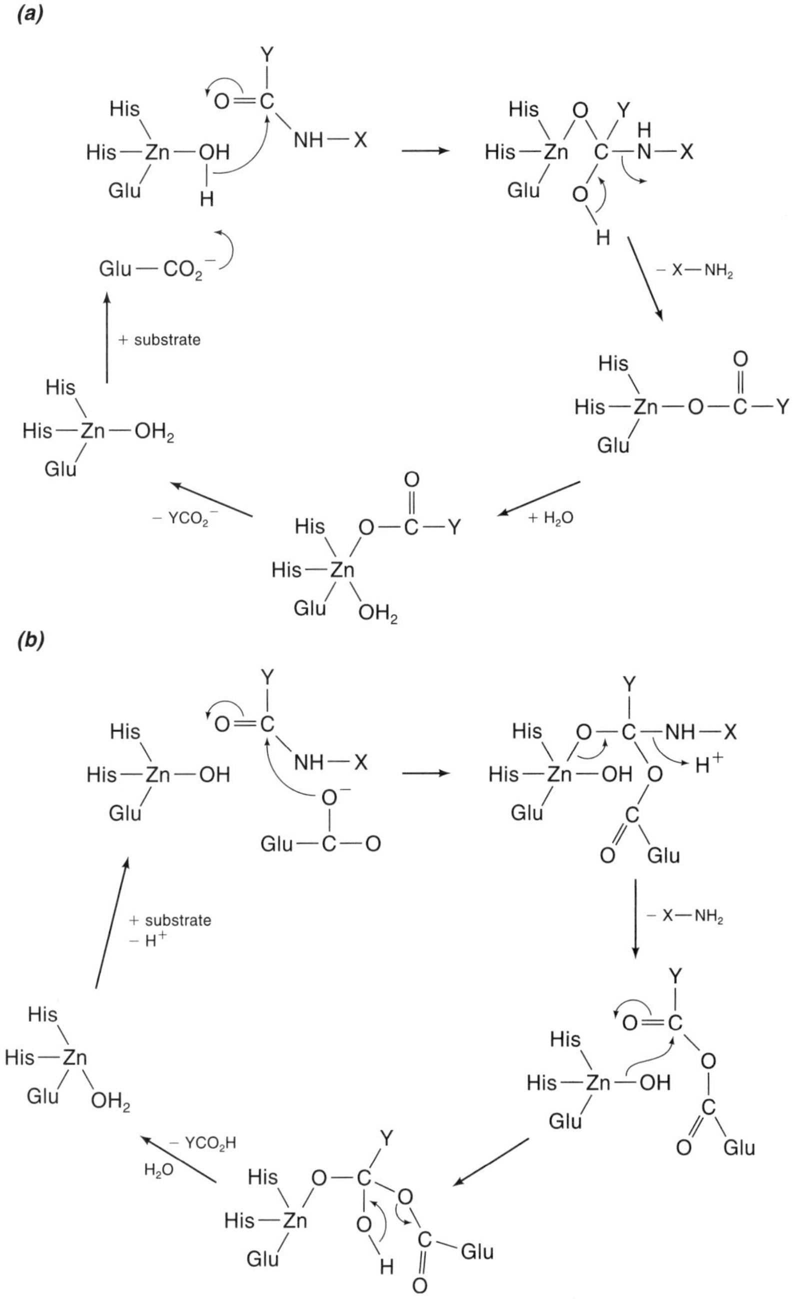

(a)

(b)

Fig. IX.1.6.
Structural details and a proposed mechanism for carboxypeptidase A (PDB code: 1YME); for carboxyprotein B (PDB code: 1CPB).[10] (*a*) Reaction via a nucleophilic water. (*b*) Reaction via an anhydride.

found in CA, the zinc in CPA is liganded by two His ligands, a bidentate glutamate (Glu), and one H_2O, which is hydrogen bonded to Ser197 and Glu270. In the presence of substrate, the bound carboxylate shifts to unidentate coordination. The enzyme can also hydrolyze esters, and other metal cofactors can promote the peptidase and esterase activities.

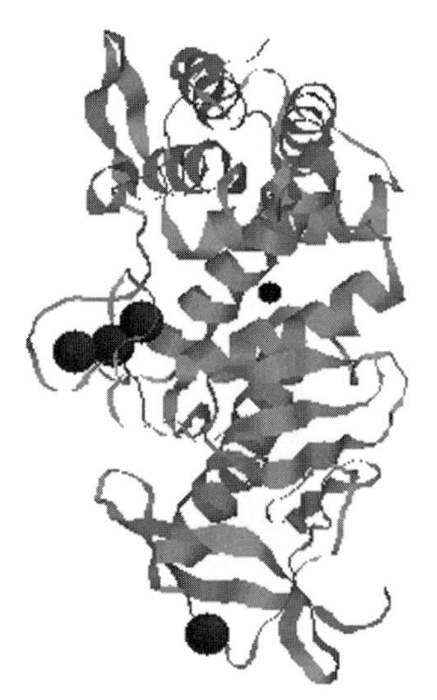

Fig. IX.1.7.
X-ray structure of the endopeptidase, thermolysin (PDB code: 1LND). The structurally more flexible loop regions are stabilized by bound Ca^{2+} ions (large circles). Zinc ion (small circle) is coordinated by His residues from the adjacent α-helices.

Two mechanistic hypotheses have been put forward for the CPA hydrolysis mechanism (Fig. IX.1.6). The first mechanism favors a promoted-water pathway where nucleophilic water is deprotonated by Glu270 (Fig. IX.1.6.*a*). A combination of the Zn^{2+} ion and neighboring positively charged residues lower the pK_a of the bound water ($pK_a \sim 7$). Note that there is no direct coordination of the carbonyl oxygen to Zn^{2+} prior to addition of water. The role of the Zn^{2+} is therefore to stabilize negatively charged intermediates formed during hydrolysis rather than to polarize a bound carbonyl, which would have the adverse effect of increasing the pK_a of bound water. Instead, the carbonyl interacts with a neighboring arginine (Arg) 127, which also stabilizes the negatively charged intermediates. Second, the zinc-bound hydroxide attacks the amide, with the Zn^{2+} providing electrostatic stabilization of the transition state. Neighboring positively charged Arg residues also stabilize the transition state. The second mechanism (Fig. IX.1.6.*b*) is similar, but in this case it is proposed that nucleophilic attack by Glu270 on a Zn^{2+}-bound amide yields an anhydride that is subsequently hydrolyzed.

Other peptidases include thermolysin (Fig. IX.1.7), which also binds four structural Ca^{2+} ions that stabilize structurally more accessible loop regions of the enzyme toward autoproteolysis. This enzyme is an endopeptidase that cleaves in the central part of the peptide chain, many residues from the termini.

IX.1.3.3. Liver Alcohol Dehydrogenase

While formally an oxidative conversion of an alcohol to an aldehyde (Eq. 2), we shall consider the activity of liver alcohol dehydrogenase (LADH) in this section since it utilizes many of the same features that we have just described for CA and CPA.

$$RCH_2OH + NAD^+ \rightarrow RCHO + NADH + H^+ \quad (2)$$

Figure IX.1.8 summarizes how Zn^{2+} ion is again coordinated in a tetrahedral fashion by two Cys and one His residue, with a bound hydroxide. The presence of two softer thiolate ligands is significant. Again, Zn^{2+} must expand its coordination num-

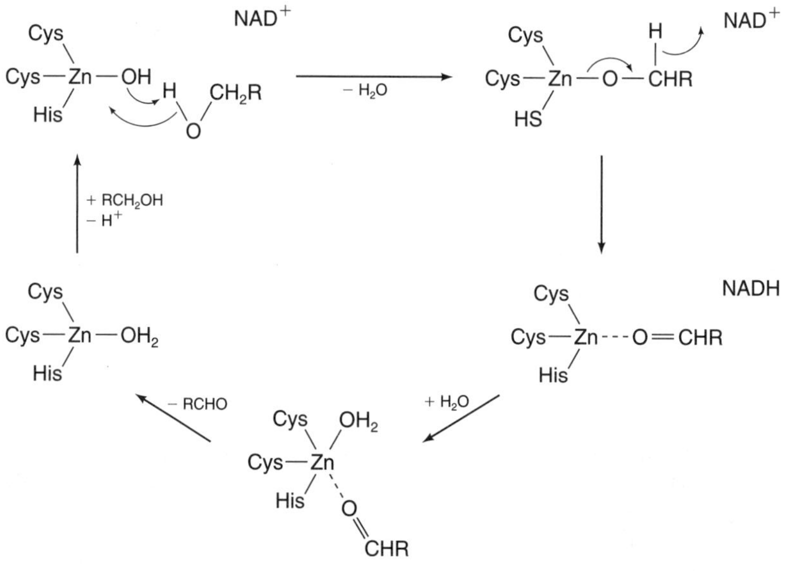

Fig. IX.1.8.
Structural details and a proposed mechanism for LADH (PDB code: 1DEH).[10]

ber to 5 after coordinating the alcohol. The bound hydroxide deprotonates the alcohol and the divalent ion stabilizes the bound alkoxide. However, the softer thiolate ligands do not favor coordination by a hard alkoxide ligand, and so loss of hydride to NAD^+ with formation of a softer carbonyl ligand results. Coordination by water results in release of the product.

IX.1.4. Manganese

IX.1.4.1. Arginase

Manganese is used less frequently than magnesium as a cofactor in biology,[14] reflecting its low abundance relative to Mg^{2+}, and Mg^{2+} serves just as well as Mn^{2+} in most of its functional roles; arginase is an exception. Liver arginase catalyzes the physiologically important final step in the urea cycle, a pathway that facilitates removal of nitrogenous waste during protein catabolism. Mammalian excretion of urea provides a nitrogen source for bacteria, fungi, and plants. In other tissues, arginase regulates the concentration of arginine and ornithine concentrations that are important mediators of the secondary messenger nitric oxide. (see Section XIV.3) Arginase is also the first structurally characterized enzyme with a physiological requirement for a binuclear Mn^{2+} center for activity. An interesting comparison can also be made with the enzyme urease (see Section IX.3), which contains a binuclear Ni^{2+} (PDB code: 1FWJ) site and displays a remarkable similarity in the disposition of catalytically important carboxylate and His residues. The Mn^{2+} ion produces low levels of activity in urease, but Ni^{2+} produces no activity in arginase.

Figure IX.1.9 summarizes the structural environment of the manganese-binding sites in each monomer subunit of the arginase trimer.[15] Each binuclear center is located at the bottom of a 15-Å deep substrate binding cleft. The Mn^{2+} ion that is more deeply embedded shows square-pyramidal geometry, while the other exhibits distorted-octahedral symmetry. Spin coupling of the two Mn^{2+} sites has previously been observed and is mediated by three bridging ligands: a bidentate bridging carboxylate (Asp124); a monodentate bridging carboxylate (Asp232); and a bridging hydroxide (Fig. IX.1.9). These yield a Mn^{2+}–Mn^{2+} separation of ~ 3.3 Å. Two other

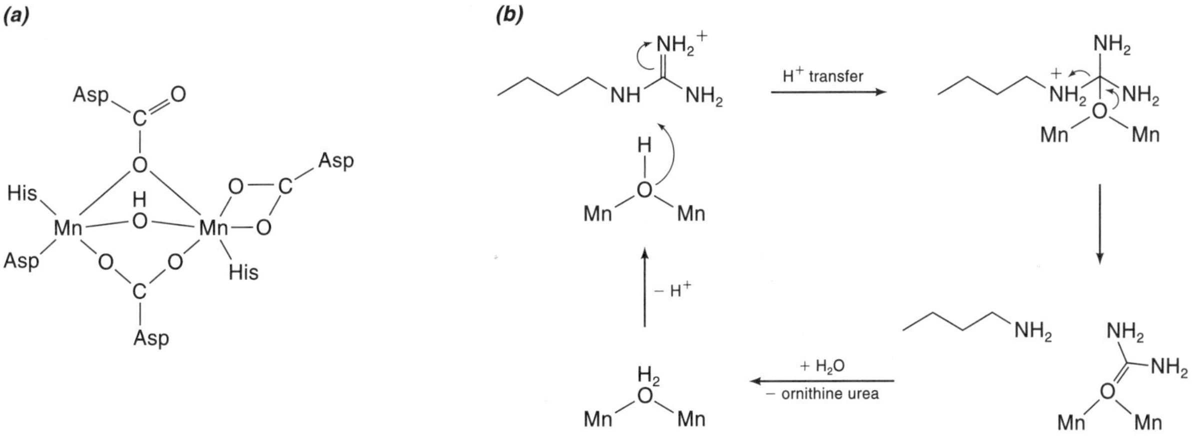

Fig. IX.1.9.
Structural details and a proposed mechanism for arginase (PDB code: PQ3).[15] (*a*) Coordination at each Mn site. (*b*) Mechanistic details excluding most Mn ligands for clarity.

polar residues lie in close proximity to the active metal center (Glu277 and His141) and both are most likely involved in catalysis. The metal centers do not appear to be directly involved in substrate binding, but more likely activate a nucleophilic water molecule, which has a pK_a of 7.9 (derived from turnover measurements) that is significantly lower than would be expected from coordination to one of the two Mn^{2+} sites alone, thereby tuning it to a physiologically accessible value.

The selection of Mn^{2+} rather than the more physiologically abundant Mg^{2+} center is also worthy of comment. As a result of the small ionic radius and divalent charge of the Mg^{2+} ion, close bridging contacts of two Mg^{2+} centers is not favored by the significant electrostatic repulsion due to the positive charge density that resides on each cation. Furthermore, hydroxide is a poor bridging ligand for two Mg^{2+} centers and does not sufficiently neutralize the electrostatic charge repulsion. By contrast, the larger manganous ion, with a lower charge density and more favorable bonding possibilities with hydroxide (as a result of *d*-orbital participation), more readily accommodates the bridged binuclear center configuration. It is also important to realize that the enzyme is isolated with the Mn^{2+}–Mn^{2+} center in place. This finding indicates a high affinity for the Mn^{2+} ions—a prerequisite if Mn^{2+} is to be used as the natural cofactor, given the low natural abundance (and so availability) of this metal cofactor ($<10^{-7}$ M).

Finally, it is noteworthy that the substrate does not bind directly to the metal cofactors. That is, the metal does not act as a classical Lewis acid—a feature that is becoming more commonly recognized in a variety of hydrolytic enzymes, such as carboxypeptidase A, and ribonuclease H (PDB code: 1RDD).

References

1. Cowan, J. A., *Inorganic Biochemistry. An Introduction,* 2nd ed., Chapts. 1 and 4., Wiley-VCH, New York 1997, pp. 1–63 and 167–202.
2. Cowan, J. A., Ed., *The Biological Chemistry of Magnesium,* VCH, New York, 1995.
3. van Bastelaere, P. B. M., Kersters-Hilderson, H. L. M., and Lambeir, A.-M., "Wild-type and mutant D-xylose isomerase from *Actinoplanes missouriensis*: metal-ion dissociation constants, kinetic parameters of deuterated and non-deuterated substrates and solvent-isotope effects", *Biochem. J.,* **307,** 135–142 (1995).
4. Black, C. B., Huang, H.-W., and Cowan, J. A., "Biological Coordination Chemistry of Magnesium, Sodium, and Potassium Ions. Protein and Nucleotide Binding Domains", *Coord. Chem. Rev.,* **135/136,** 165–202 (1994).
5. Britton, K. L., Langridge, S. J., Baker, P. J., Weeradechapon, K., Sedelnikova, S. E., De Lucas, J. R., Rice, D. W., and Turner, G., "The crystal structure and active site location of isocitrate lyase from the fungus *Aspergillus nidulans*", *Structure,* **8,** 349–362 (2000).
6. Liaw, S. H. and Eisenberg, D., "Structural model for the reaction mechanism of glutamine synthetase, based on five crystal structures of enzyme-substrate complexes", *Biochemistry,* **33,** 675–681 (1994).
7. Almassy, R. J., Janson, C. A., Hamlin, R., Xuong, N. H., and Eisenberg, D., "Novel subunit-subunit interactions in the structure of glutamine synthetase", *Nature* (*London*), **323,** 304–309 (1986).
8. Yamashita, M. M., Almassy, R. J., Janson, C. A., Cascio, D., Eisenberg, D., "Refined atomic model of glutamine synthetase at 3.5 Å resolution", *J. Biol. Chem.,* **264,** 17681–17690 (1989).
9. Schindler, J. F. and Viola, R. E., "Mechanism-Based Inactivation of L-Aspartase from Escherichia coli", *Biochemistry,* **33,** 9365–9370 (1994).
10. Bertini, I., Luchinat, C., and Monnani, R., "*Zinc enzymes*", *J. Chem. Edu.,* **62,** 924–927 (1985).
11. Liljas, A., Haakansson, K., Jonsson, B. H., and Xue, Y., "Inhibition and catalysis of carbonic anhydrase. Recent crystallographic analyses", *Eur. J. Biochem.,* **219,** 1–10 (1994).
12. Christianson, D. W. and Lipscomb, W. N., "Carboxypeptidase A", *Acc. Chem. Res.,* **22,** 62–69 (1989).
13. Rees, D. C. and Lipscomb, W. N., "Refined crystal structure of the potato inhibitor complex of carboxypeptidase A at 2.5 Å resolution", *J. Mol. Biol.,* **160,** 475–498 (1982).
14. Pecoraro, V. L., Ed., *Manganese Redox Enzymes,* VCH, New York, 1992.
15. Kanyo, Z. F., Scolnick, L. R., Ash, D. E., and Christianson, D. W., "Structure of a unique binuclear manganese cluster in arginase", *Nature* (*London*), **383,** 554–557 (1996).

IX.2. Metal-Dependent Lyase and Hydrolase Enzymes. (II) Nucleic Acid Biochemistry

J. A. Cowan
Chemistry
Ohio State University
Columbus, OH 43210

Contents

IX.2.1. Introduction

In this section, the focus of attention is on enzymes that mediate hydrolysis or condensation reactions involving nucleic acids. A smaller subclass of enzymes that catalyze phosphorylation and dephosphorylation chemistry is also discussed here. Almost all enzymes that participate in the biochemistry of nucleic acids require divalent metal ion cofactors to promote activity, and divalent magnesium is typically the metal ion of choice. Calcium and zinc ions are also used in specific cases and the distinct features of these reactions are also considered.

IX.2.2. Magnesium-Dependent Enzymes

IX.2.2.1. Characteristics of Magnesium-Promoted Activity in Nucleic Acid Biochemistry[1,2]

As discussed in Section IX.1, almost all Mg^{2+} enzymes possess a labile cofactor that is weakly ($K_a \sim 10^3$–$10^4\ M^{-1}$) coordinated to protein side chains. As such, this class is best described as metal dependent enzymes rather than as metalloenzymes (where the metal cofactor typically remains bound to the enzyme during isolation and purification). This chapter considers those enzymes that use free divalent Mg as a cofactor or a Mg–NTP complex as substrate (where NTP = nucleotide triphosphate). First, we consider pertinent facts relating to the suitability of Mg^{2+} for its role as an activator of enzymes that act on nucleic acid substrates, the required metal ion stoichiometry, and mechanistic details of function. This provides an essential background for a subsequent review of specific enzyme families.

The small ionic radius and resulting high charge density of the divalent magnesium ion results in a tendency to bind water molecules rather than bulkier ligands in its inner-coordination shell, leading to appropriate hydration states and slow solvent exchange rates.[3] These facts are also reflected in the hydration numbers for crystalline salts of Mg^{2+} relative to other alkaline earth metal ions (Table IX.2.1), which demonstrate the difficulty of coordinating large counterions or ligands to the rather small Mg^{2+} cation. Outer-sphere mediated hydrolytic pathways are followed by a large section of magnesium-dependent enzymes that carry out reactions on nucleic acid substrates (Fig. IX.2.1).[3] Such pathways are favored by the hydrogen-bonding network that can form between hydrated Mg^{2+} and the nucleic acid substrate (Fig. IX.2.2). According to this mechanism, the metal cofactor serves principally to

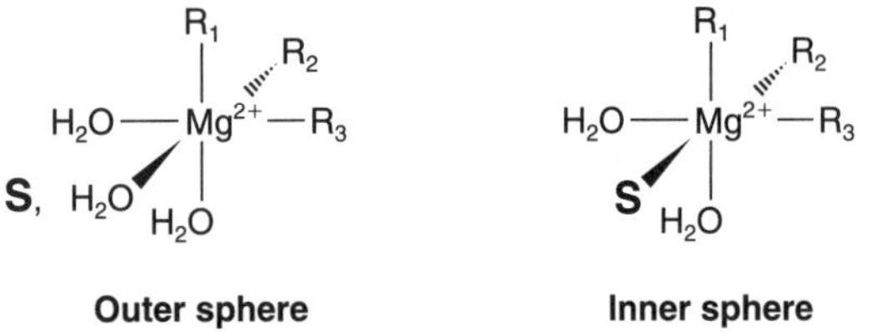

Fig. IX.2.1.
Comparison of inner- and outer-sphere modes of activation. A substrate **S** interacts with a catalytic Mg^{2+} ion that is tethered to an enzyme through ligation to residues R_1, R_2, and R_3, for example. (Note that the number of inner-sphere contacts to protein ligands varies considerably, depending on the enzyme.) From 0–5 contacts have been observed.) Both activation modes are observed for magnesium-promoted reactions with the latter more typical of nuclease activity.

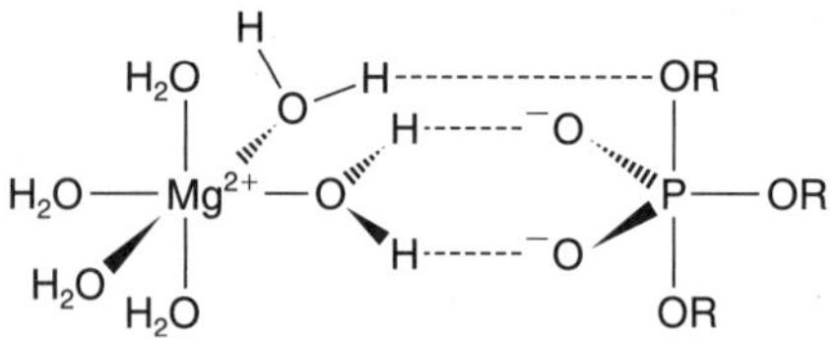

Fig. IX.2.2.
Illustration of transition state stabilization by outer-sphere complex formation to hydrated Mg ion. Both hydrogen bonding and electrostatics contribute to the stabilization of the increased negative charge in the transition state.

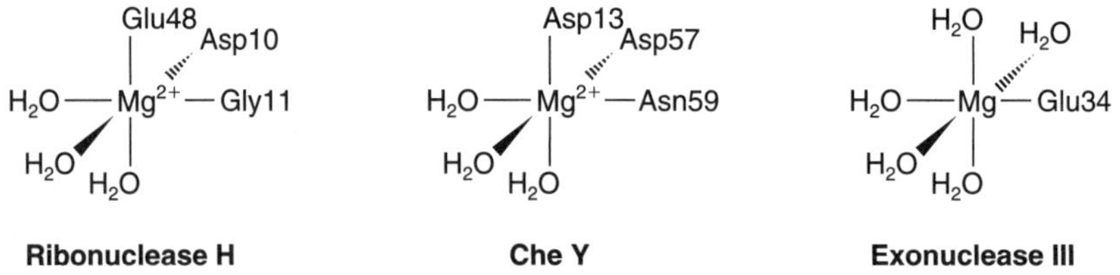

Fig. IX.2.3.
Comparison of Mg coordination spheres for selected magnesium proteins: (the metal binding domains of *Escherichia coli* ribonuclease H (PDB code: 1RDD); the chemotaxal protein CheY (PDB code: 1CHN), and the DNA repair enzyme exonuclease III (PDB code: 1AKC)).

Table IX.2.1.
Hydration States of Crystalline Salts of Alkaline Earth Metal Ions

Mg^{2+}	$MgSO_4 \cdot 7H_2O$, $MgCl_2 \cdot 6H_2O$, $Mg(CH_3CO_2)_2 \cdot 6H_2O$, $Mg(ClO_4)_2 \cdot 6H_2O$
Ca^{2+}	$CaSO_4 \cdot 2H_2O$, $CaCl_2 \cdot 2H_2O$, $Ca(CH_3CO_2)_2 \cdot H_2O$, $Ca(ClO_4)_2 \cdot 4H_2O$
Ba^{2+}	$BaSO_4$, $BaCl_2 \cdot 2H_2O$, $Ba(CH_3CO_2)$, $Ba(ClO_4)_2 \cdot 3H_2O$

stabilize the transition state, either electrostatically, and/or through hydrogen bonding from the metal-bound waters. The latter appears to be the larger contribution in the few cases that have been examined.[3,4] The solvation state of the metal cofactor, which is defined by the number of protein ligand contacts, is of critical importance for the proper functioning of these enzymes. The metal-binding pockets of metal-dependent nucleases have evolved to allow a large variation in the number of protein ligands (and thereby solvent water), while optimizing the binding affinity to physiological requirements. For example, Fig. IX.2.3 illustrates a selection of magnesium-binding modes to proteins with varying extents of solvation. Nevertheless, the K_a values are comparable and are tuned to physiological availability.

A mechanistic model that has been put forward for metal-mediated hydrolysis of the phosphodiester backbone of nucleic acids involves two metal ions working in concert. This two-metal-ion model is perhaps best defined as the requirement for two-metal ions, located in close proximity (<4 Å), that are bridged by a common

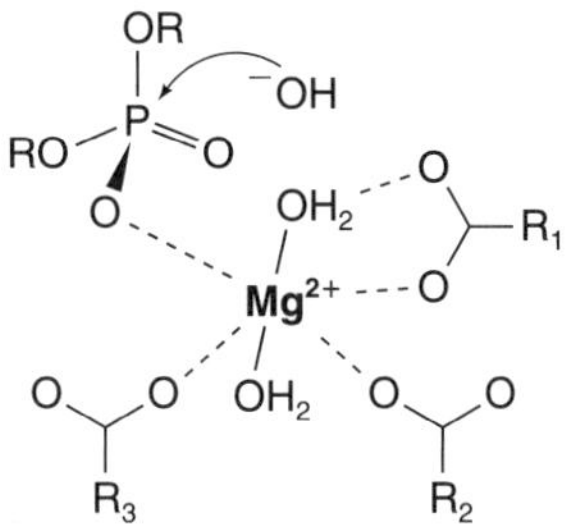

Fig. IX.2.4.
(*a*) Schematic illustration of the metal binding sites identified from crystallographic studies of the Mn^{2+} derivative of the DNA polymerase I Klenow fragment (PDB code: 1KFD). Site B is weakly populated in the presence of substrate or a substrate analogue. Site A is the principal site of both metal coordination and catalytic chemistry. (*b*) A minimalist scheme for Klenow activity showing one essential metal cofactor and an activated nucleophilic solvent water.[3]

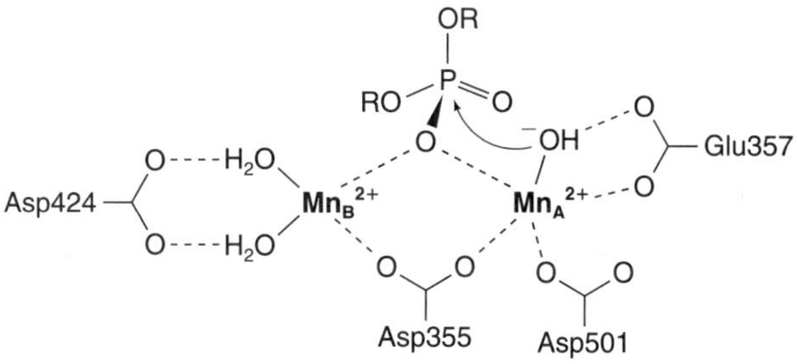

substrate. This allows a clear distinction from other enzymes that may bind two metal cofactors in the same catalytic domain, but do not function as a coherent catalytic unit. Examples of the latter include T4 RNase H (PDB code: 1TFR) in which two Mg^{2+} sites are separated by 7 Å, and T5 5′-exonuclease, in which crystallographic analysis shows two Mn^{2+} sites separated by 8.1 Å [10 Å in the case of Taq 5′-exonuclease (PDB code: 1EXN)]. An example of the two-metal-ion model is the mechanism proposed for the Klenow fragment (defined below), illustrated in Fig. IX.2.4*a*, in which one divalent ion activates a bound water and the other provides a template effect, drawing together the substrate and the nucleophile.

The two-metal-ion model has some drawbacks, however, in the case of Mg^{2+} enzymes. The first is that the high charge density of Mg^{2+} creates a high electrostatic barrier to such close placement of the two metal ions. The second is that Mg^{2+}-bound H_2O has a higher pK_a than Zn^{2+}-bound H_2O making metal activation of H_2O less favorable and thus less likely for divalent magnesium than for divalent zinc. An insightful theoretical analysis of the Klenow fragment from DNA polymerase I identifies electrostatic stabilization of a free hydroxide nucleophile as the principal role of the metal cofactor(s) (Fig. IX.2.3*b*).[3] That is, metal-bound H_2O is not necessarily required as a nucleophile, but rather a free solvent molecule may serve as the hydrolytic agent, perhaps with base catalysis from an acidic side chain. Some of the problems of interpretation in this area arise from inconsistencies in data from crystallographic and solution studies. In part, these problems arise from the use of metal analogues for Mg, as described next.

IX.2.2.2. Magnesium Analogues in Mechanistic Studies

The divalent Mg^{2+} ion has such a low electron density that it is difficult to distinguish in most spectroscopic and crystallographic experiments, and thus it is essentially spectroscopically invisible. For this reason, X-ray structures of proteins with labile Mg^{2+} cofactors are usually obtained by doping pre-existing crystals of enzyme with very high concentrations of other metal cofactors (relative to physiological concentrations). Several caveats must be borne in mind when using such high metal ion concentration in doping experiments: the coordination preferences for Mg^{2+} are often different from those of the analogues; and weak or physiologically irrelevant metal-binding sites may be occupied. With this in mind, the high-affinity "structural site" populated by Zn^{2+} or Mn^{2+} in Klenow in Fig. IX.2.4*a*, showing 4- or 5-coordination, depending on whether substrate is bound, would not be expected to be populated by Mg^{2+} at normal physiological concentrations (Table IX.2.2). Moreover, neither Zn^{2+} nor Mn^{2+} bind at this site at normal physiological concentrations.

The chemistry of magnesium-dependent enzymes has also been studied in solution with the use of transition metal probes and analogues.[3] Again the underlying assumption that Mg^{2+} analogues (especially Mn^{2+}) show similar chemistry to that displayed by Mg^{2+} is not a good general assumption. In fact, the stoichiometry and coordination mode of other metals frequently differ from Mg^{2+}. It is usually observed that Mn^{2+} (or other transition metals such as Co^{2+}, Fe^{2+}, or Zn^{2+}) confers higher levels of activity than the native enzyme, since these ions typically coordinate more tightly than the natural Mg^{2+} cofactor. For example, the plant-stimulated nuclease from *Fusarium solani* is stimulated by $Mn^{2+} > Ca^{2+} > Co^{2+} > Mg^{2+}$, but is inhibited strongly by Zn^{2+}.[3] These transition metal ions may also show changes in coordination geometry (Table IX.2.2), or induce conformational perturbations in active site residues that influence substrate specificity or form the basis for metal ion mutagenicity and nucleotide selectivity in the replication of DNA by magnesium-dependent enzymes. While the use of analogues can provide insight on function, there are differences in coordination chemistry that can lead to changes in substrate selectivity and metal ion stoichiometry. Such problems need to be recognized and accounted for.

IX.2.2.3. Mechanistic Probes

Clearly, the question of whether a magnesium cofactor is acting as an inner- or an outer-sphere complex is important and nontrivial inasmuch as the chemical roles for these two coordination modes are quite distinct. However, the relatively fast exchange kinetics of Mg^{2+} poses a rather significant problem for the elucidation of the appropriate coordination mode in solution. A major difficulty in distinguishing

Table IX.2.2.
Comparison of Coordination Properties of Mg^{2+} versus Analogues

Ion	Coordination Numbers	Geometry[a]	Radius (Å)	Ligand Preference
Mg^{2+}[b]	6	oct	0.65	O; (H_2O)
Mn^{2+}	5, 6	dist oct, sq pyr	0.85	O, N,
Co^{2+}	4, 5, 6	dist oct, tet, sq pyr	0.81	N, S
Zn^{2+}	4, 5	tet, sq pyr	0.79	O, N, S

[a] Octahedral = oct, distorted octahedral = dist oct, square pyramidal = sq pyr, tetrahedral = tet.

[b] Maguire, M. E. and Cowan, J. A., "Magnesium Chemistry and Biochemistry", *Biometals* **15,** 203–210 (2002).

inner- and outer-sphere pathways for magnesium-promoted reactions stems from the kinetic lability of this metal ion. One approach that has been successfully applied is the use of cobalt and chromium complexes that are substitutionally inert within the timeframe of the enzyme-catalyzed reactions, for example, $[Cr(NH_3)_x(H_2O)_{6-x}]^{2+}$, where x varies from 0 to 6.[4] If such complexes promote a magnesium-dependent reaction, this is evidence in favor of an outer-sphere pathway. Of course, the absence of activity promoted by such complexes does not necessarily preclude such a pathway. Although these substitutionally inert complexes are unable to coordinate directly to protein residues, it has been shown that the metal-binding pockets of surface Mg^{2+} sites typically show poor selectivity and can accommodate a large variation in metal size. The nature of the interaction between the metal cofactor and the nucleotide substrate is the critical issue. The metal-binding protein residues simply serve to tether a metal cofactor to the catalytic pocket on the enzyme surface.

Complexes of NTPs with Co^{3+} and Cr^{3+} ions have also been used to investigate the chemistry of MgNTP-dependent enzymes.[5] Since the former two are exchange inert, the various isomeric forms of the NTP derivatives have been isolated and used as substrates or inhibitors to test the stereochemical course of enzyme reaction pathways. As a paramagnetic ion, NTP complexes of Cr^{3+} have afforded structural insight on enzyme-bound complexes through measurement of relaxation enhancements of ^{1}H, ^{13}C, and ^{31}P NMR resonances. Since these complexes are chemically inert, an important intermediate or substrate analogue can often be trapped in the enzyme-bound form.

IX.2.2.4. Illustrative Examples of Metalloenzymes in Nucleic Acid Biochemistry

Restriction Endonucleases. Few magnesium-dependent nuclease enzymes have been studied in detail with regard to the role of the essential metal cofactor. Noteworthy exceptions are the restriction enzymes EcoRI, and especially EcoRV. A recent review summarizes what is known of the recognition and cleavage chemistry of type-II endonucleases,[6] the family to which both of the aforementioned enzymes belong. EcoRI is also a type-II endonuclease and shows a high degree of structural homology with the active site of EcoRV (Fig. IX.2.5), although there is no general sequence homology. A mechanistic model has been suggested for EcoRI and EcoRV (Fig. IX.2.5). Some uncertainty remains over the metal cofactor stoichiometry. However, an inner-sphere pathway appears likely (probes of outer-sphere paths do not promote

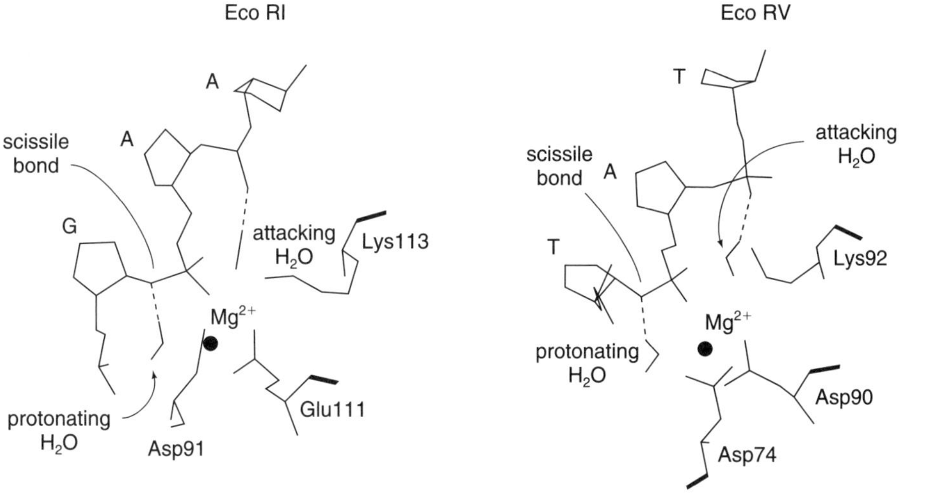

Fig. IX.2.5. A comparison of the metal cofactor binding domains, and other features of the catalytic pockets of Eco RI (PDB code: 1QCD) and Eco RV (PDB code: 1RVE), illustrating the high degree of local homology for these two restriction endonucleases and proposed mechanistic details.[3]

activity), possibly reflecting the stricter substrate recognition requirements of such enzymes.

Exonuclease. Exonuclease activity can be defined as the removal of nucleotide fragments from either the 3′- or 5′-end of a strand of nucleic acid (usually one of the strands in a double-stranded substrate) by hydrolysis of the terminal phosphodiester linkage. By contrast, endonucleases cleave phosphodiester linkages that may lie many bases from the ends of the nucleic acid chain. Most restriction enzymes are endonucleases. A well-defined exonuclease activity is that associated with the Klenow fragment, a proteolysis product of DNA polymerase I, which contains both the 3′-5′ exonuclease and polymerase domains. The 3′-5′ exonuclease activity indicates that the enzyme begins digestion from the 3′-end of a nucleic acid strand, and degrades the strand by removal of terminal nucleotides, one base at a time. Models for the catalytic behavior of the metal cofactor were described earlier and are illustrated in Fig. IX.2.4. In contrast, the *E. coli* exonuclease III is a monomeric DNA repair enzyme. The enzyme has several functions, including a 3′-5′ exonuclease activity. In this case, the enzyme appears to follow an outer-sphere pathway as defined earlier.[3]

Polymerases. A large number of enzymes utilize Mg^{2+} chelated by nucleotidyl di- or triphosphates (especially ATP and ADP), where the metal cofactor serves as a mediator of phosphoryl or nucleotidyl transfer chemistry. The functions of the former were considered in Section IX.1. Nucleotidyl transferases are the subject of greater focus here since this activity is relevant to the chemistry of RNA and DNA polymerases.

Polymerases are enzymes that catalyze the replication and synthesis of strands of DNA or RNA from a single-strand or double-strand template polynucleotide.[3] Some of these enzymes are multifunctional and contain other exonuclease or ribonuclease H activities that are required for their overall operation. These catalytic activities are carried out at different sites on the enzyme. A common feature of the polymerase active site is possession of a labile metal-binding site in addition to the Mg^{2+} that is carried in as a chelate with the NTP substrate (Fig. IX.2.6). Acidic active site residues can interact with this metal–phosphate center and contribute to active-site chemistry. Divalent Mg^{2+} binds to ATP^{4-} as a β,γ-chelate. This binding serves to promote nucleophilic attack at the γ-phosphate during phosphoryl transfer reactions (discussed below). Similarly, the chelate serves to stabilize the pyrophosphate leaving group following nucleophilic attack at the α-phosphate during nucleotidyl transfer

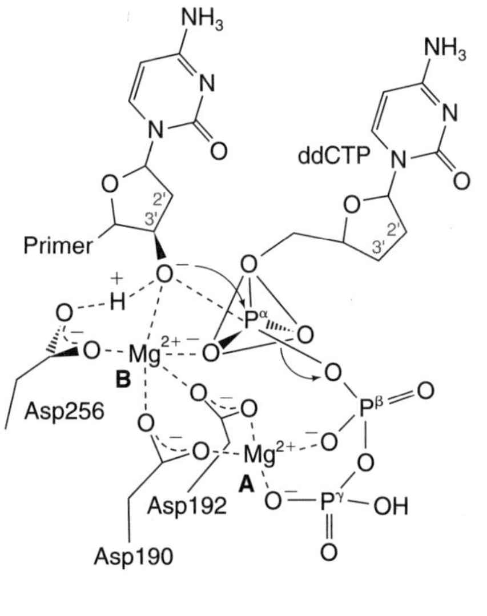

Fig. IX.2.6.
Stereoview of the polymerase active site of rat DNA polymerase β (PDB code: 2BPG), and a schematic model of the reaction chemistry. [Adapted from Ref. 7.]

reactions of polymerases (Fig. IX.2.6). The presence of the Mg^{2+} ion also tends to direct any H^+ delivered during catalysis to the β-phosphate (with release of Mg^{2+} to the more negatively charged terminal phosphate), which further enhances pyrophosphate as a leaving group.

Phosphoryl Transfer. Enzymes that catalyze phosphorylation of substrates (usually Ser, Thr, or Tyr residues on protein targets) typically use Mg^{2+} chelates of ATP as a cosubstrate. As detailed earlier, the bound Mg^{2+} serves to facilitate nucleophilic attack at the γ-phosphate of the highly negative ATP substrate (Fig. IX.2.6). Such enzymatic reactions are therefore mediated by Mg^{2+}. However, note that the hydrolysis of simple phosphate esters is often catalyzed by enzymes containing transition metal cofactors, for example, alkaline phosphatase and purple acid phosphatase, which contain binuclear zinc and iron centers, respectively. The terms alkaline and acid phosphatases indicate the pH optima for activity and reflect the relative pK_a values for bound water to the zinc and iron cofactors, respectively. Presumably, substrates for the latter are less reactive and require more potent Lewis acid catalysts. In passing, note that a structural role is played by the Mg^{2+} ion cofactor for the dinuclear Zn^{2+} *E. coli* enzyme, alkaline phosphatase.

Phosphorylation and Dephosphorylation. Phosphate and phosphoryl transfer reactions are ubiquitous in cellular biochemistry, and require the Mg^{2+} ion as an essential cofactor.[3] Distinctions between enzymes in this class can be related to the function of magnesium. Generally, Mg^{2+} will either make its principal bonding contacts with the phosphate moiety of a nucleotide di- or triphosphate, or with side chains in the enzyme. The catalytic role for Mg^{2+} in promoting NTP hydrolysis is well established, with activation of the β-phosphate toward protonation and formation of a "pyrophosphate" leaving group, and/or relief of electrostatic repulsion between an incoming nucleophile and the terminal phosphate.[8,9]

In the case of regulatory proteins, hydrolysis chemistry must be coupled to structural change. For example, the family of *Ras genes* code for regulatory proteins that bind guanine nucleotides (G-proteins). Ha-Ras p21 binds a divalent magnesium–GTP complex and has been examined by time-resolved crystallographic characterization. A change in metal ion coordination triggers substantial changes in structure during hydrolysis. Although the details are still unclear, it is now known that upon hydrolysis of the γ-phosphate, Mg^{2+} coordinates to the carbonyl group of Asp57, leaving only one phosphate ligand (Fig. IX.2.7). In addition to an added Asp contact, the Thr35 contact is lost. These differences in the coordination sphere around

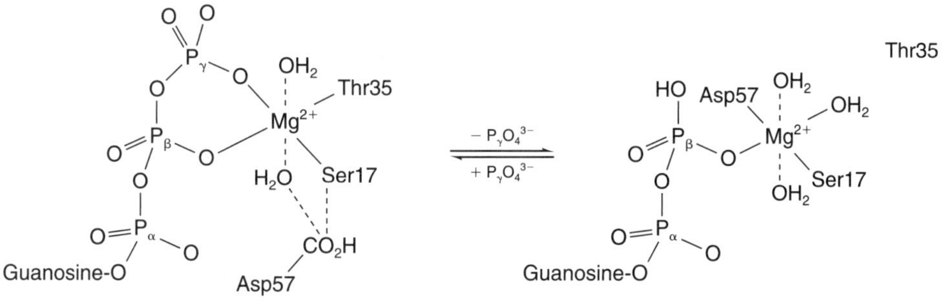

Fig. IX.2.7.
The magnesium binding site in Ha-Ras p21 (PDB code: 1Q21) before and after γ-phosphate transfer. Note the retention of two inner-sphere water molecules. Also, Asp57 replaces the γ-phosphate after transfer. The residue Thr35, which is part of the effector loop, moves to ~4 Å distance from Mg^{2+} after transfer. The uptake and release of protein residues results in structural change of the protein.

Mg^{2+} lead to large conformational changes in an "effector loop" of approximately seven residues in length. This loop is believed to be responsible for the binding of GAP (GTPase activating protein), which activates Ha-Ras p21 for catalysis. From this structural data, and previous mechanistic evidence, either a protein-bound water situated directly opposite the leaving group, or the β-phosphate oxygen may be the attacking nucleophile. After loss of the nucleophilic water molecule, a five-coordinate phosphate intermediate is formed. It is likely that Mg^{2+} increases the electrophilicity of the γ-phosphate center through coordination (Fig. IX.2.7) and stabilizes the product after transfer.

IX.2.3. Calcium

Calcium-dependent enzymes such as staphylococcal nuclease and deoxyribonuclease I bind Ca^{2+} tightly ($K_a \sim 10^6$–$10^7\ M^{-1}$) and are isolated with the metal *in situ*. Enzymes that have an absolute requirement for Ca^{2+} are typically extracellular digestive enzymes. The extracellular Ca concentration is millimolar, while intracellular levels of free Ca approach nanomolar. A selection mechanism for Ca^{2+} vis à vis Mg^{2+} can be made on the basis of coordination geometry (Table IX.2.2). This selectivity for Ca^{2+} serves as a safety mechanism to prevent premature activation by Mg^{2+} after synthesis in the cell. We shall now take a closer look at two of these calcium-dependent digestive enzymes.

IX.2.3.1. Staphylococcal Nuclease and DNase I

The extracellular 5′-phosphodiesterase from the yeast *Staphylococcus aureus* catalyzes the hydrolysis of DNA or RNA to give 3′-mononucleotide and dinucleotide products. The enzyme has been crystallographically characterized as the ternary enzyme–Ca^{2+} inhibitor complex (Fig. IX.2.8). The inhibitor (thymidine-3′,5′-bisphosphate) models substrate binding at the active site. Calcium binds to the phosphodiester group and stabilizes the leaving group after hydrolysis by a calcium-bound water molecule.

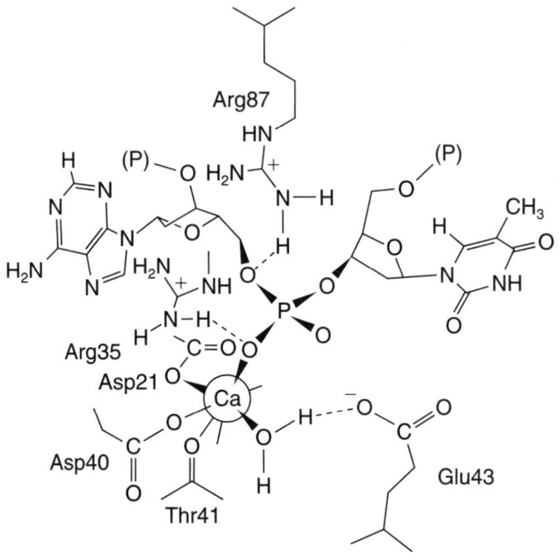

Fig. IX.2.8.
Structure of staphylococcal nuclease (PDB code: 1STN) showing a ternary complex with Ca^{2+} and an inhibitor thymidine-3′,5′-bisphosphate. The Ca^{2+} ion coordinates the phosphate and activates a bound H_2O, demonstrating both Lewis acid catalysis and the template effect referred to in Section IX.2. [Reprinted with permission from Ref. 10.]

By way of contrast, DNase I is structurally distinct from staphylococcal nuclease and shows both exo- and endo-nucleolytic behavior. The reaction pathway is now a paradigm for other nuclease reactions, showing an Asp-His-H_2O triad (Fig. IX.2.9). Similar to thermolysin in Chapter IX.1, in addition to the unique catalytic Ca^{2+}, additional structural calcium ions are found. Unlike the catalytic metal, there are no water ligands associated with these sites. It is likely that DNase I is actually a magnesium-dependent enzyme since the catalytic site can also be populated by Mg^{2+} at physiological concentrations and promotes higher activity levels than Ca^{2+}.

IX.2.3.2. Phospholipase A_2

This calcium-dependent enzyme is located in the outer membrane and catalyzes the hydrolysis of the 2-acyl ester bond in 1,2-diacylglycero-3-phospholipids, releasing free fatty acids. In mammals, fatty acids are transported through the blood stream to the liver for further digestion in the mitochondrial energy cycle. Snake venom phospholipase preferentially hydrolyzes neutral lecithin molecules (Fig. IX.2.10), whereas the mammalian pancreatic enzyme prefers negatively charged phospholipids such as phosphatidyl glycerol and phosphatidic acid. Two positively charged lysine residues in the binding domain for the pancreatic enzyme assist in the selectivity for negatively charged substrates. The calcium-binding site is heptacoordinate and does not readily bind Mg^{2+}, which prefers a regular octahedral geometry. Again, Ca^{2+} binds to the phosphate group and stabilizes the increased negative charge that arises in the transition state.

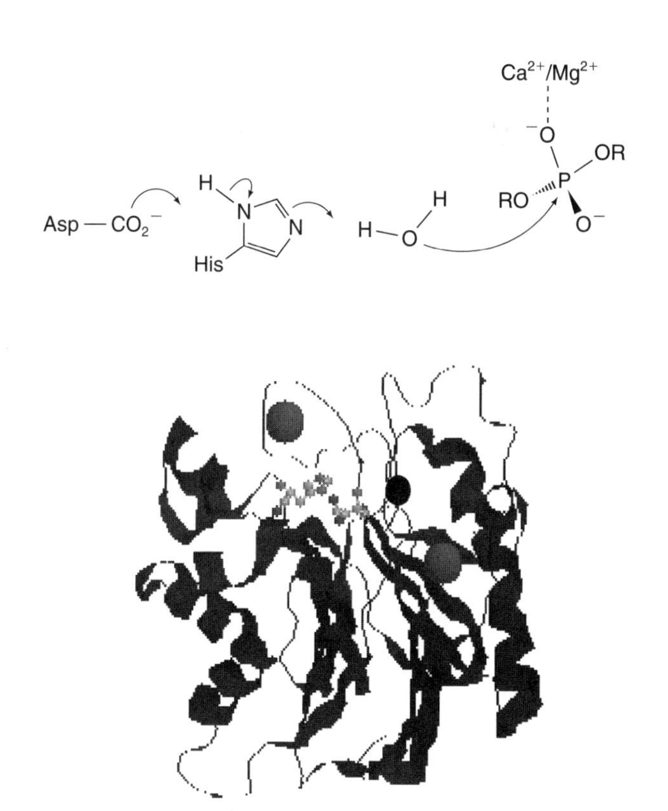

Fig. IX.2.9.
Structure of bovine DNase I (PDB code: 3DNI), highlighting the catalytic Ca^{2+}/Mg^{2+} ion (small sphere) and the two structural Ca^{2+} ions (large spheres). On the left is shown the proposed Asp-His-H_2O catalytic triad that executes hydrolytic attack at the phosphate diester.

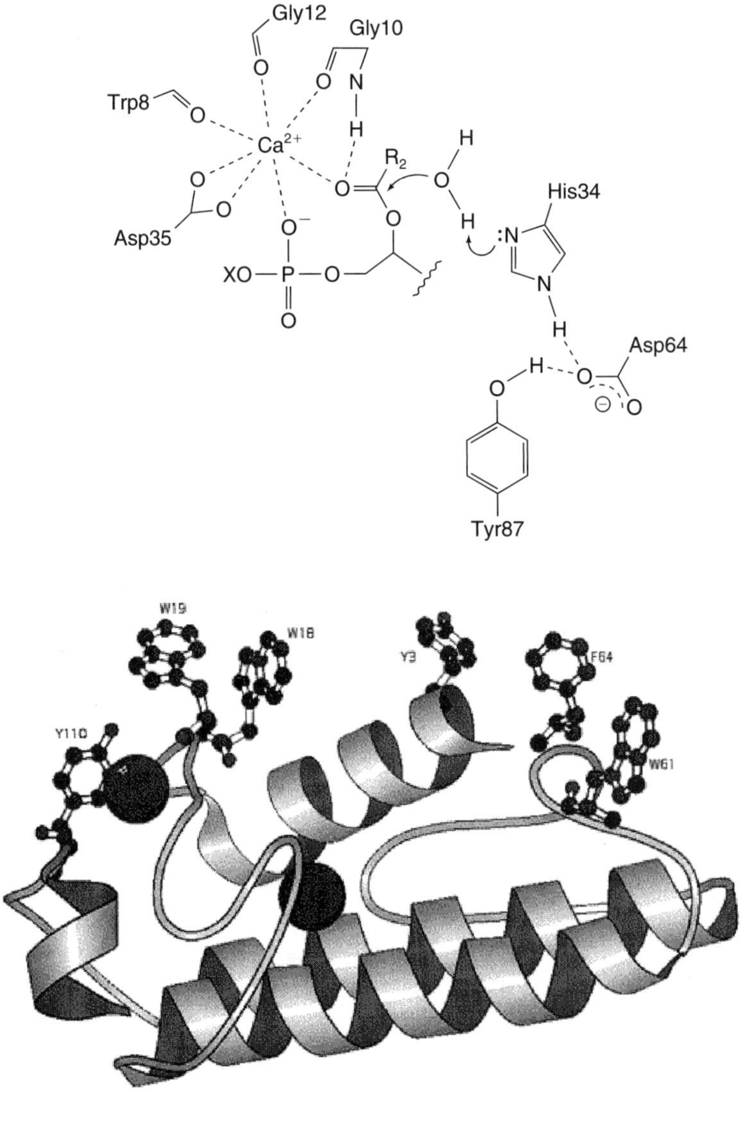

Fig. IX.2.10.
Structure of the active site of phospholipase A2 (PDB code: 1JIA) and a proposed mechanism. The tertiary structure and catalytic and structural Ca^{2+} cofactors are shown below. The hydrophobic residues that interface with a lipid surface are also highlighted. [Reprinted with permission from Refs. 11, 12.]

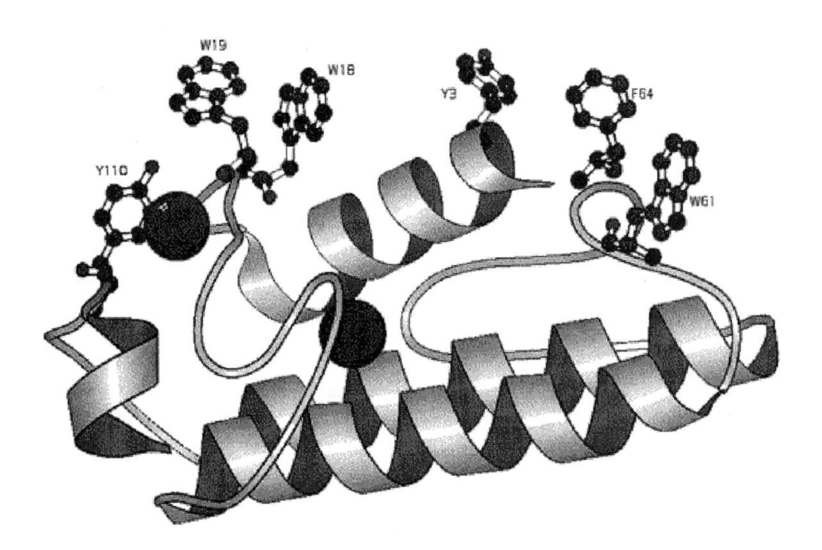

IX.2.4. Zinc

IX.2.4.1. Alkaline Phosphatase

Escherichia coli alkaline phosphatase is a dimer of 94-kDa subunits that hydrolyzes a variety of phosphate esters. The enzyme has optimal activity ~pH 8, hence the name. The basic reaction scheme is shown in Fig. IX.2.11 and is seen to proceed by way of a covalently phosphorylated enzyme intermediate (E-PO_3^-). Each subunit contains two Zn^{2+} ions (separated by a distance of 4 Å) and one Mg^{2+} that is located 5–7 Å from the binuclear zinc site. The two Zn^{2+} ions form the catalytic site. The role of Mg^{2+} is most likely structural, but appears unrelated to catalysis. The two Zn^{2+} ions are bound by His and Asp. One Zn^{2+} is located close to the Ser hydroxyl. This positioning decreases the pK_a of the hydroxyl and facilitates nucleophilic attack at the phosphate ester. The other Zn^{2+} ion stabilizes the developing negative charge on the leaving alkoxide group. The five-coordinate trigonal-bipyramidal intermediate is bound and stabilized by both Zn^{2+} and a neighboring positively charged side chain

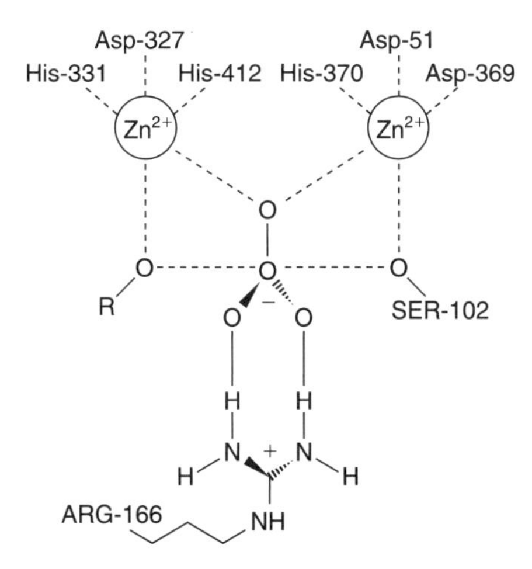

Fig. IX.2.11.
Structural details and a proposed hydrolytic mechanism for *E. coli* alkaline phosphatase (PDB code: 1ED). Zn1 interacts with imidazole nitrogen's of His412 and His331, one of the phosphate oxygens, and the carboxylate oxygens of Asp327. The Zn2 interacts with the imidazole nitrogen of His370, one of the phosphate oxygens, and the carboxylate oxygens of Asp51 and Asp369. Ser102, which is phosphorylated during the reaction, and Arg166, which interacts with the phosphate are also shown. [Reprinted with permission from Ref. 13.]

(an arginine guanidinium group) and the resulting serine phosphate ester remains bound to Zn^{2+}. The pK_a of the aquated Zn(II) ion is 8.8, and so the coordination of H_2O to Zn(II) can produce an OH^- nucleophile to displace the phosphoryl group from the phosphoserine intermediate to yield a binuclear Zn^{2+} site bridged by the product phosphate. The pH dependence of the hydrolase activity provides evidence for the existence of such a zinc hydroxide species.

IX.2.4.2. Purple Acid Phosphatase

Acid phosphatases, so-called because their pH optimum normally lies in the range 4.9–6.0, hydrolyze orthophosphate monoesters and are widespread in Nature. They possess a coupled binuclear metal center that exhibits an intense purple color from a tyrosinate to Fe(III) charge transfer, and so the name purple acid phosphatases. Typical mammalian phosphatases possess a binuclear ion center [Fe^{3+}–Fe^{2+}] while [Fe^{3+}–Zn^{2+}] centers have been identified in plant enzymes (Fig. IX.2.12). The different pH optima for the zinc-binding alkaline phosphatase and the iron-binding acid phosphatase arise in part from the distinct pK_a values of the metal-bound H_2O [Fe^{3+} p$K_a = 2.2$; Zn^{2+} p$K_a = 8.8$]. The enzyme-bound ferric iron is ligated by carboxylate and is unlikely to have a pK_a that is quite as low as the value quoted here for $[Fe(H_2O)_6]^{3+}$. Both reaction pathways proceed by way of metal–OH^- species and possess a similar catalytic apparatus that includes a binuclear metal site to bind the substrate and enzyme-bound intermediates prior to a final metal-activated hydrolysis of the enzyme-phosphate ester to yield a bridging phosphate (Fig. IX.2.12). The enzyme is active in the reduced form [Fe^{3+}–Fe^{3+} (oxidized, inactive); Fe^{3+}–Fe^{2+} (reduced, active)]. A plausible reaction scheme is shown in Fig. IX.2.12.

IX.2.4.3. Trinuclear Zinc Enzymes

A number of enzymes have recently been characterized that contain three Zn^{2+} ions in their active sites. These include: phospholipase C, which removes phosphate ester fragments from phospholipids; nuclease P1, which catalyzes hydrolysis of

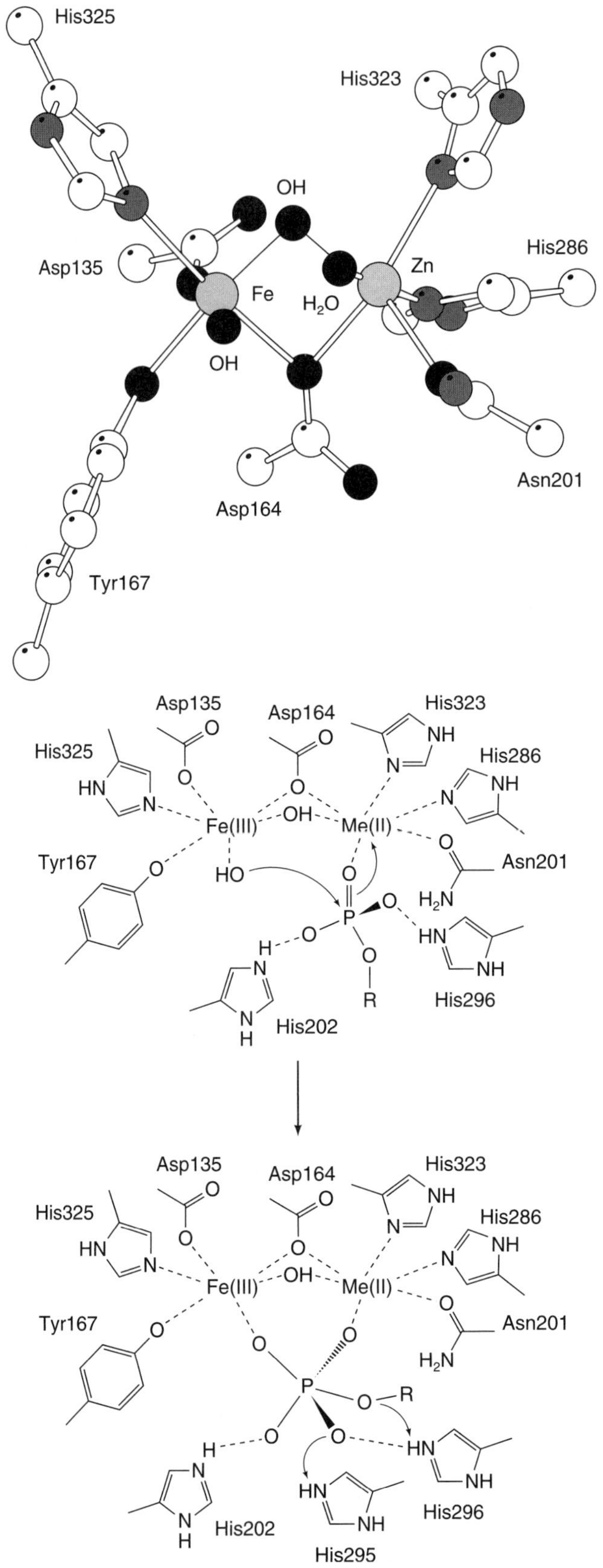

Fig. IX.2.12.
Structural details and a proposed mechanism for the $Fe^{III}Zn^{II}$ kidney bean (purple) acid phosphatase (PDB code: 1UTE). [Reprinted with permission from Ref. 14.]

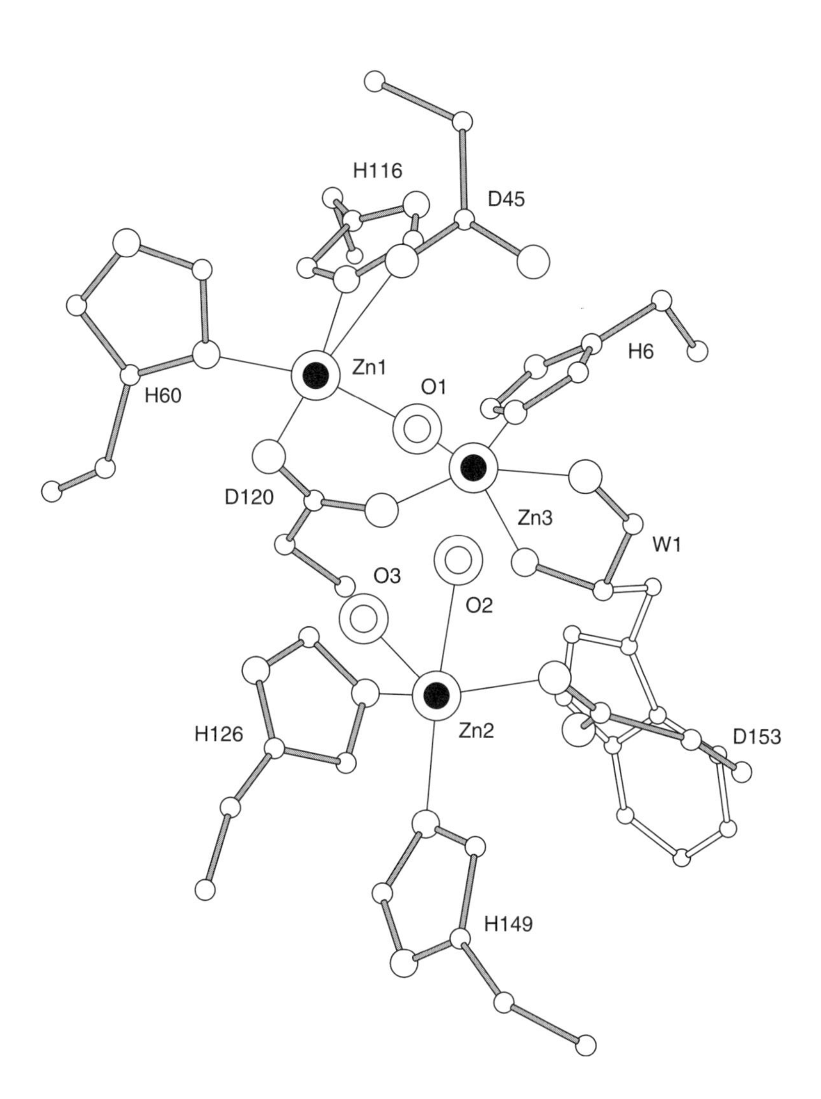

Fig. IX.2.13.
Structural details in the active site of the trinuclear zinc *P. citrinum* P1 nuclease (PDB code: 1AKO). [Reprinted with permission from Ref. 15.]

single-strand ribonucleic acid (RNA) and deoxyribonucleic acid (DNA), and also removal of 5′-phosphate groups; and organophosphate triesterases, an important family of enzymes that detoxify herbicides and nerve gas agents carrying phosphate triesters as their key functional group (PDB code: 1EYW). The general structural motif for the Zn centers is summarized in Fig. IX.2.13. A pair of carboxylate bridged Zn^{2+} ions and a neighboring isolated Zn^{2+} ion each show five coordination. A bound water is released when the phosphate center binds to the isolated Zn^{2+} ion. The binuclear Zn^{2+} center appears to serve a structural role in the case of the latter two enzymes, but bound phosphate bridges the binuclear site in phospholipase C.

References

General References

1. Cowan, J. A., *Inorganic Biochemistry. An Introduction,* 2nd ed., Chapts. 1 and 4, Wiley-VCH, New York, 1997, pp. 1–63 and 167–202.
2. Cowan, J. A., Ed., *The Biological Chemistry of Magnesium,* VCH, New York, 1995.

Specific References

3. Cowan, J. A., "Metal Activation of Enzymes in Nucleic Acid Biochemistry", *Chem. Rev.*, **98,** 1067–1087 (1999).
4. Black, C. B., Foster, M., and Cowan, J. A., "Evaluation of the Free Energy Profile for Ribonuclease H Metal-Promoted Catalysis of Phosphate Ester Hydrolysis. Use of Inert Chromium Complexes to Evaluate Hydrogen

Bonding and Electrostatic Contributions to Transition State Stabilization", *J. Biol. Inorg. Chem.*, **1,** 500–506 (1996).
5. Cleland, W. W. and Mildvan, A. S., "Chromium(III) and cobalt(III) nucleotides as biological probes", in *Advances in Inorganic Biochemistry,* Eichorn, G. L. and Marzilli, L., Eds., Vol. 1, Elsevier, New York, 1979, pp. 163–191.
6. Pingoud, A. and Jeltsch, A., "Recognition and cleavage of DNA by type-II restriction endonucleases", *Eur. J. Biochem.*, **246,** 1–22 (1997).
7. Pelletier, H., Sawaya, M. R., Kumar, A., Wilson, S. H., and Kraut, J., "Structures of ternary complexes of rat DNA polymerase β, a DNA template-primer, and ddCTP", *Science,* **264,** 1891–1903 (1994).
8. Cowan, J. A., "Metallobiochemistry of Magnesium. Coordination Complexes with Biological Substrates: Site Specificity, the Kinetics and Thermodynamics of Binding, and Implications for Activity", *Inorg. Chem.*, **30,** 2740–2747 (1991).
9. Sigel, H., "Metal-assisted stacking interactions and the facilitated hydrolysis of nucleoside 5′-triphosphates", *Pure Appl. Chem.*, **70,** 969–976 (1998).
10. Weber, D. J., Libson, A. J., Lebowitz, M. S., and Mildvan, A. S., "NMR docking of a substrate into the X-ray structure of the Asp21/Glu mutant of staphylococcal nuclease", *Biochemistry,* **33,** 8017–8028 (1994).
11. Annand, R. R., Kontoyianni, M., Penzotti, J. E., Dudler, T., Lybrand, T. P., and Gelb, M. H., "Active site of bee venom phospholipase A2: the role of histidine-34, aspartate-64 and tyrosine-87", *Biochemistry,* **35,** 4591–4601 (1996).
12. Sumandea, M., Das, S., Sumandea, C., and Cho, W., "Roles of Aromatic Residues in High Interfacial Activity of *Naja naja atra* Phospholipase A2", *Biochemistry,* **38,** 16290–16297 (1999).
13. Kim, E. E. and Wykoff, H. W., "Reaction mechanism of alkaline phosphatase based on crystal structures. Two-metal ion catalysis", *J. Mol. Biol.*, **218,** 449–464 (1991).
14. Strater, N., Klabunde, T., Tucker, P., Witzel, H., and Krebs, B., "Mechanism of Fe(III)-Zn(II) purple acid phosphatase based on crystal structures", *J. Mol. Biol.*, **259,** 737–748 (1996).
15. Volbeda, A., Lahm, A., Sakiyama, F., and Suck, D., "Crystal structure of *Penicillium citrinum* P1 nuclease at 2.8 Å resolution", *EMBO J.*, **10,** 1607–1618 (1991).

IX.3. Urease

Stefano Ciurli
Laboratory of Bioinorganic Chemistry
Department of Agro-Environmental Science and Technology
University of Bologna
1-40127 Bologna, Italy

Contents

IX.3.1. Introduction

Urea is the catabolic product of nitrogen-containing compounds and is hydrolyzed by urease (urea aminohydrolase E.C. 3.5.1.5) to ammonia and carbamate, which spontaneously decomposes to give a second molecule of ammonia and bicarbonate:[1,2]

$$H_2N{-}C({=}O){-}NH_2 \xrightarrow[\text{urease}]{H_2O} H_2N{-}C({=}O){-}O^- + NH_4^+$$

$$H_2N{-}C({=}O){-}O^- \xrightarrow{H_3O^+} HCO_3^- + NH_4^+$$

Urease was the first enzyme shown to contain Ni ions in its active site.[3] The highly conserved amino acid sequences of all known ureases and the constant presence of two Ni ions and their ligands in the active sites imply a conserved catalytic mechanism.

Jack bean urease (JBU) was the first enzyme to be crystallized, >70 years ago,[4] but the crystal structures of urease from *Klebsiella aerogenes* (KAU; PDB code: 1FWJ),[5] *Bacillus pasteurii* (BPU; PDB code: 2UBP),[6] and *Helicobacter pylori* (PDB code: 1E9Z)[7] were only recently reported. The structures of these microbial ureases are largely equivalent, and the following discussion is based on the structure of BPU, except for cases in which differences are observed.

IX.3.2. The Structure of Native Urease

Bacterial urease is a large heteropolymeric metalloprotein characterized by the presence of a dinuclear Ni^{2+} center in three independent active sites. In each active site (Fig. IX.3.1), the two Ni ions are closely spaced (3.7 Å) and are bridged by the carboxylate group of a carbamylated $Lys^{\alpha220^*}$. The Ni(1) atom is further coordinated by $His^{\alpha249}$ Nδ and $His^{\alpha275}$ Nε, while Ni2 is bound to $His^{\alpha137}$ Nε, $His^{\alpha139}$ Nε, and $Asp^{\alpha363}$ Oδ1. Three water–hydroxide molecules are directly bound to the metal ions. One of these (W_B) symmetrically bridges the two Ni ions, whereas the other two (W_1 and W_2) complete the coordination sphere of Ni1 and Ni2, respectively, forming hydrogen bonds with $His^{\alpha222}$ Nε and $Ala^{\alpha170}$ O, respectively. A fourth water

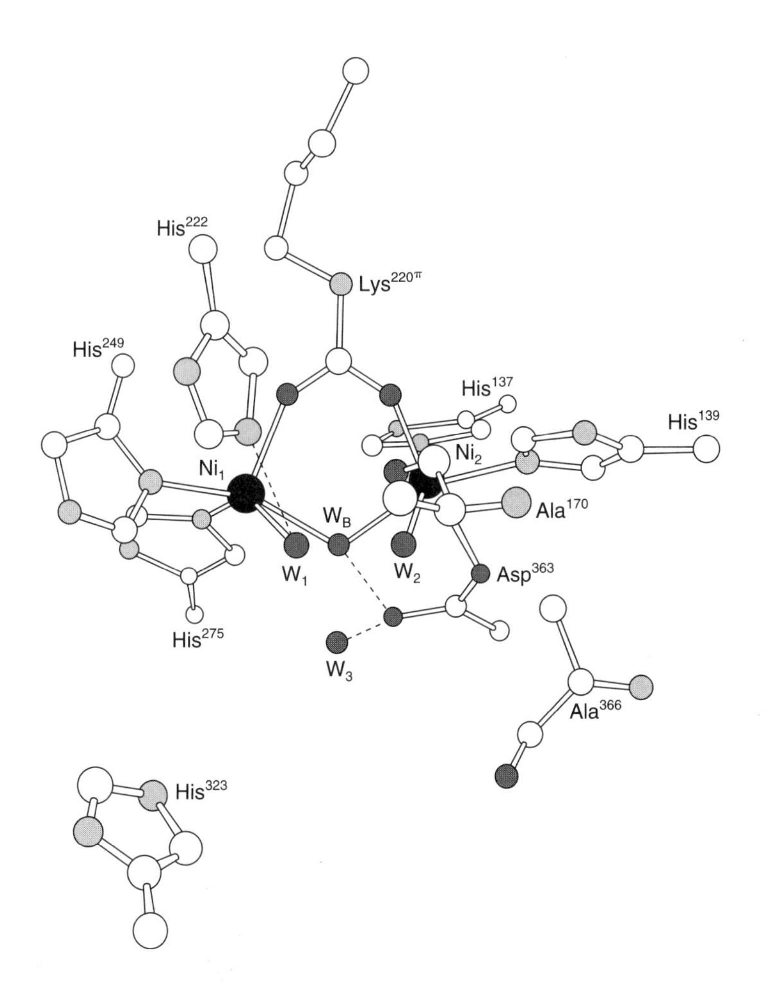

Fig. IX.3.1.
Model of the active site of native BPU, showing the positions of selected relevant residues. Atoms filled with pattern by element (C: hollow; N: light gray; O: dark gray). Hydrogen bonds are shown as dashed lines. All residues are sized according to their relative depths.

molecule, W_3, interacts with W_B, W_1, and W_2 through hydrogen bonds, while $Asp^{\alpha 363}$ Oδ2 is at hydrogen-bonding distance from both W_B and W_3.[6] Several structures of inhibitor complexes with *B. pasteurii* urease (BPU) are available, all of them indicating a substantial lability of the terminal water molecules and the bridging hydroxide, while the protein framework involving the ligands to the metal ions is rigidly maintained.

IX.3.3. The Structure of Urease Complexed with Transition State and Substrate Analogues

The structure of BPU crystallized in the presence of phenylphosphorodiamidate $(NH_2)_2P(O)OPh$ (PPD) (PDB code: 3UBP)[6] reveals the presence of a molecule of diamidophosphoric acid $(NH_2)_2P(O)OH$ (DAP) (Fig. IX.3.2) produced by the enzymatic hydrolysis of PPD along with phenol. The DAP is an analogue of the tetrahedral transition state of urea hydrolysis, and the determination of its binding mode to the active site of the enzyme has profound implications for understanding the catalytic mechanism. The structure reveals that DAP replaces the four water–hydroxide molecules found in native BPU, and is bound to Ni1 and to Ni2 using three of the four potential donor atoms. An oxygen atom symmetrically bridges the two Ni ions, while Ni1 and Ni2 are bound to one oxygen and one nitrogen atom, respectively. A nitrogen atom of DAP points away from the Ni center, toward the cavity opening. The structure is in agreement with extended X-ray absorption fine structure (EXAFS) studies.[8] An extensive hydrogen-bonding pattern stabilizes the inhibitor and directs its orientation in the cavity: (1) the Ni1-bound oxygen atom receives a hydrogen bond from $His^{\alpha 222}$ Nε; (2) the Ni2-bound DAP amido nitrogen forms two hydrogen bonds donated to the carbonyl oxygen atoms of $Ala^{\alpha 170}$ and $Ala^{\alpha 366}$, with the latter changing its conformation with respect to the structure of the native enzyme; (3) the $Asp^{\alpha 363}$ Oδ2 atom is at a hydrogen-bonding distance from both the Ni-bridging DAP

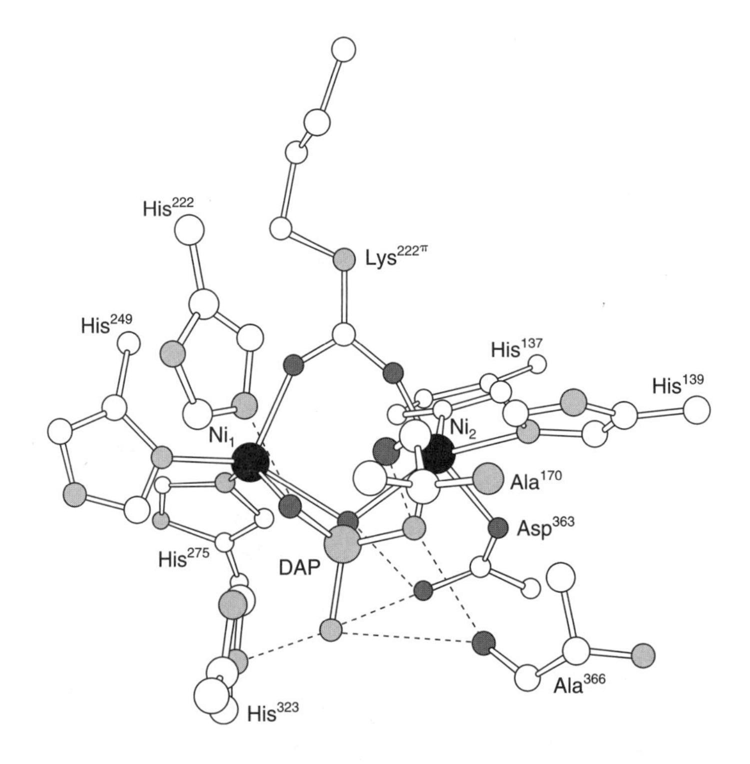

Fig. IX.3.2. Model of the active site of BPU inhibited with DAP, showing the positions of selected relevant residues. Atoms filled with pattern by element (C: hollow; N: light gray; O: dark gray; P: gray). Hydrogen bonds are shown as dashed lines. All residues are sized according to their relative depths.

oxygen and the distal DAP nitrogen; and (4) the distal DAP nitrogen donates a hydrogen bond to $Ala^{\alpha 366}$ and to $His^{\alpha 323}$ $N\varepsilon$. This latter residue is located on a helix–loop–helix motif that flanks the active site channel. This flap is found to be flexible and capable of assuming two different conformations: open, as in the native enzyme; or closed, as in the DAP-inhibited form.[6]

The structure of a complex between BPU and boric acid, $B(OH)_3$ (PDB code: 1S3T)[9] reveals the bridging binding mode of this neutral trigonal-planar molecule to the Ni ions in the active site (Fig. IX.3.3). Boric acid can be considered a substrate analogue, revealing the molecular details of the substrate binding step. Boric acid is symmetrically positioned between the Ni ions, replacing the three water molecules present in the cavity, and leaving in place the bridging hydroxide. The Ni–Ni distance is unperturbed with respect to the native enzyme. The Ni1-bound oxygen atom of $B(OH)_3$ receives a hydrogen bond from $His^{\alpha 222}$ $N\varepsilon$, the latter being protonated because of the interaction of $His^{\alpha 222}$ $N\delta$ with the peptide NH group of $Asp^{\alpha 224}$. The Ni2-bound O atom forms a hydrogen bond with $Ala^{\alpha 170}$ O, while the distal B–OH group is hydrogen bonded to a water molecule involved in a hydrogen-bonding network with an additional solvent molecule and $Ala^{\alpha 366}$ O. This hydrogen-bonding network highlights the role of neighboring residues in stabilizing an unprecedented binding mode of $B(OH)_3$ to a dinuclear center. The lack of reactivity of $B(OH)_3$ with the bridging hydroxide (placed at 2.1 Å from the B atom, in a direction almost perpendicular to the plane of the molecule, Fig. IX.3.3) could be due to unfavorable symmetry and energy of the highest occupied molecular orbital (HOMO) on the nucleophile and the lowest unoccupied molecular orbital (LUMO) on the boric acid.

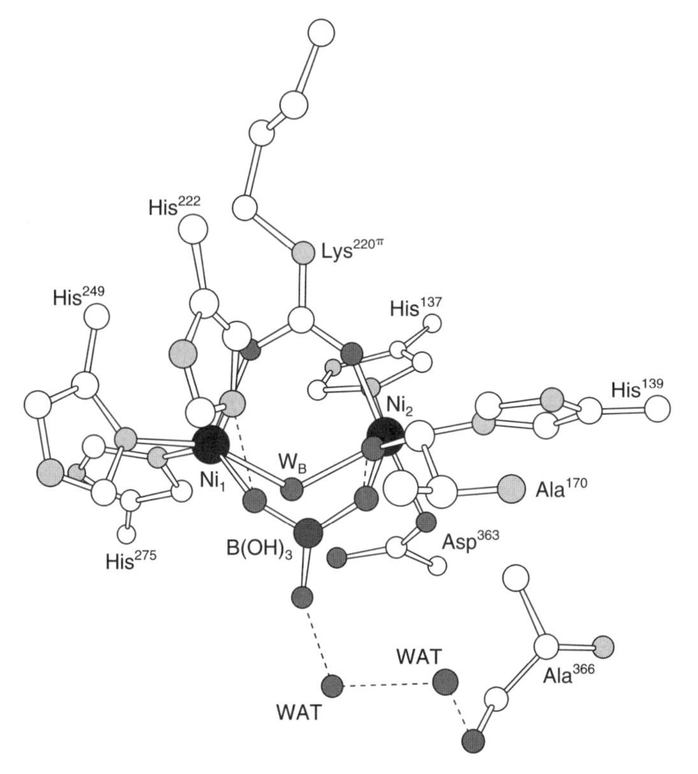

Fig. IX.3.3.
Model of the active site of BPU inhibited with $B(OH)_3$, showing the position of selected relevant residues. Atoms filled with pattern by element (C: hollow; N: light gray; O: dark gray; P: gray). Hydrogen bonds are shown as dashed lines. All residues are sized according to their relative depths.

IX.3.4. The Structure-Based Mechanism

The structure of DAP-inhibited BPU suggests a mechanism for urease that involves a direct role of both Ni^{2+} ions in binding and activating the substrate, and of the Ni bridging hydroxide as the nucleophile in the process of urea hydrolysis (Fig. IX.3.4*a*).[9,10] According to this mechanism, urea enters the active site cavity replacing W_1, W_2, and W_3, located in positions matching its molecular shape and dimensions, and binds Ni1 and Ni2 with the carbonyl oxygen and amide nitrogen, respectively. The position and orientation of the substrate are induced by the steric constraints as well as the asymmetric structural features of the active site, poised to donate hydrogen bonds in the vicinity of Ni1 and receive hydrogen bonds in the vicinity of Ni2. Urea is a poor chelating ligand because of the low Lewis base character of its NH_2 groups; however, the formation of strong hydrogen bonds with the nearby carbonyl oxygens could enhance the basicity of the NH_2 group and facilitate the interaction of the amide nitrogen with Ni2. This binding mode brings the urea carbon in close proximity to the Ni-bridging ligand, proposed here to be the nucleophile in the form of a hydroxide anion.

The structure of the tetrahedral transition state in the active site, generated upon bond formation between the bridging hydroxide and the urea carbon, is mimicked by the structure of DAP-inhibited BPU. The proton needed by the distal urea NH_2 group in order to form ammonia by cleavage of the C–N bond could easily be provided, once the HO–C_{urea} bond is formed, by the hydroxide nucleophile itself, which would then also act as a general acid, through a proton transfer assisted by $Asp^{\alpha363}$ Oδ2. The open-to-closed conformational change of the active site flap could be necessary to move $His^{\alpha323}$ in close proximity to the transition state so that it would act as a general base by stabilizing, in its deprotonated form, the positive charge that develops on the distal nitrogen atom during formation of nascent ammonia. This mechanistic model is in full agreement with the pH dependence of urease activity.[11] Its

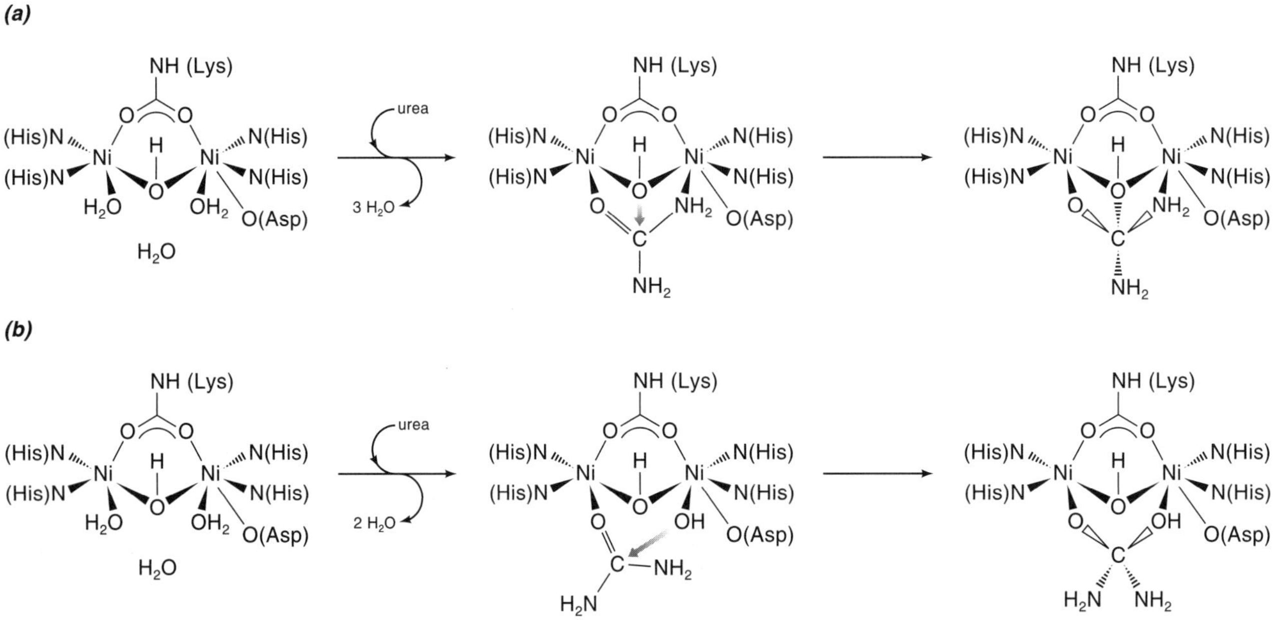

Fig. IX.3.4.
Scheme of the urease mechanism involving the bridging hydroxide (*a*) or the terminal hydroxide (*b*) as the nucleophile in urea hydrolysis.

defining feature is the nucleophilic role of the metal-bridging hydroxide, a peculiarity that may well constitute a general rule for all hydrolytic enzymes containing bimetallic catalytic sites.[12]

The fact that $B(OH)_3$ is a competitive inhibitor of urease, the isoelectronic structure of $B(OH)_3$ and urea, the bridging-chelating binding mode of boric acid, the topology of the hydrogen-bonding network, the replacement of the labile water molecules in the active site by a neutral trigonal molecule, and the presence of the nonsubstituted Ni-bridging hydroxide in the complex of urease with $B(OH)_3$, all provide evidence that gives further support to the proposed mechanism described above.

An alternative mechanism involves different roles played by the two Ni ions: Ni1 binding and activating urea; Ni2 binding and activating the nucleophile hydroxide (Fig. IX.3.4*b*).[5,13] However, this mechanism raises two problems: one involves the missing general base that would deprotonate the Ni2-bound water molecule at the optimum pH for enzyme activity (pH $\sim$ 8); and the second is the role of $His^{\alpha323}$ as a general acid, which must be protonated at pH 8 even though it has a pK_a of $\sim$6.5.[14] Therefore, according to this mechanism, only 0.3% of all urease molecules would be in the optimal protonation state for catalysis, an inconsistency justified by the "reverse protonation hypothesis".[13]

Kinetic data on the KAU inhibition by fluoride[15] as well as theoretical studies[16] provided consensus for the role of the OH^- ion bridging the two active site Ni ions acting as the nucleophile on the Ni-bound urea during enzymatic hydrolysis. Other steps of the mechanism, such as the substrate binding mode and the identity of the groups involved in the proton transfer, still require further investigation. In particular,

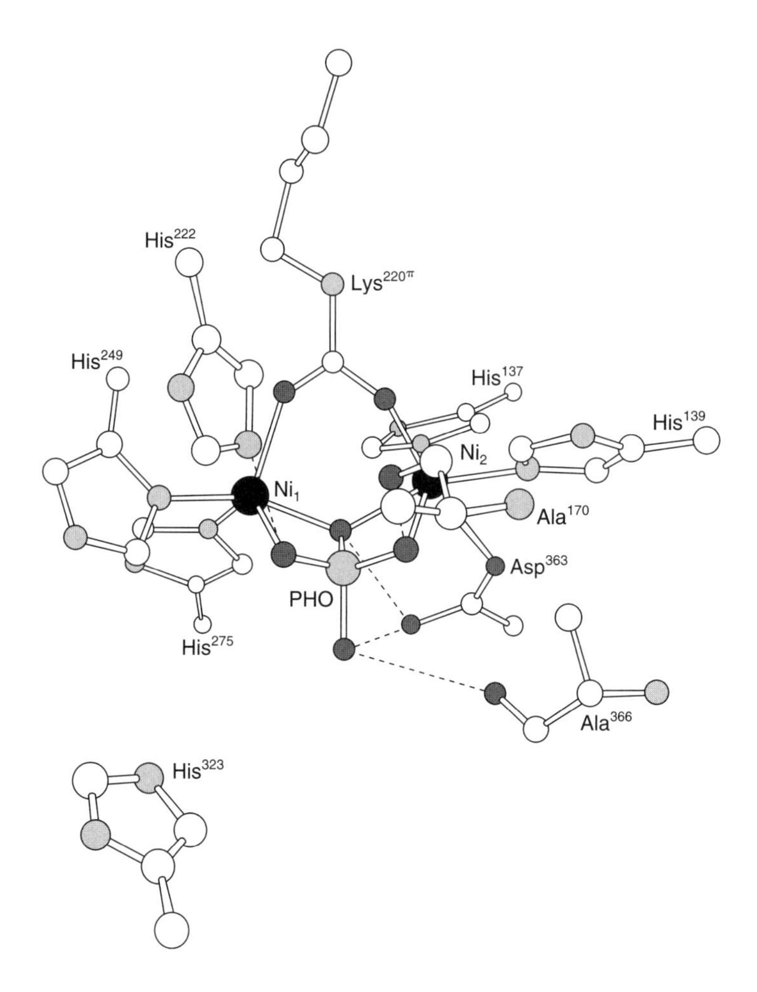

Fig. IX.3.5.
Model of the active site of BPU inhibited with PHO, showing the positions of selected relevant residues. Atoms filled with pattern by element (C: hollow; N: light gray; O: dark gray; P: gray). Hydrogen bonds are shown as dashed lines. All residues are sized according to their relative depths.

the role of His$^{\alpha323}$ acting as acceptor of a proton coming from the bridging HO–C$_{urea}$ (Fig. IX.3.4*a*), or as a proton donor shuttling it from water (Fig. IX.3.4*b*) is a matter of debate. The structure of BPU complexed with phosphate (PHO)[17] certainly helps in discriminating between these two possibilities (Fig. IX.3.5). The structure features a tetrahedral molecule of phosphate bound to the bimetallic center. The Ni-bridging phosphate oxygen atom is symmetrically placed between the two metal ions. The O_B atom of phosphate replaces the bridging hydroxide molecule (W_B) present in native BPU. Two of the other phosphate oxygen atoms coordinate Ni1 and Ni2, while the fourth (distal) phosphate oxygen points toward the cavity opening, away from the two Ni^{2+} ions. The hydrogen-bonding network between the inhibitor atoms and the active site residues provides a clear definition of the protonation state of the bound phosphate as the mono-anion $H_2PO_4^-$ and reveals that His$^{\alpha323}$ is deprotonated at pH 6.3, used for crystallization, rendering His$^{\alpha323}$ less likely to act as a proton donor during catalysis.

IX.3.5. The Structure of Urease Complexed with Competitive Inhibitors

Urease inhibitors could be used to overcome negative side effects caused by the abrupt overall pH increase during the enzymatic hydrolysis of urea.[18] Several classes of molecules (diphenols, quinones, hydroxamic acids, phosphoramides, and thiols) have been discovered through large-scale screenings. A rational structure-based molecular design of new and efficient urease inhibitors relies upon the elucidation of the structure of urease complexed with known inhibitors.

The structure of BPU inhibited with β-mercaptoethanol (BME) (PDB code: 1UBP) shows that the sulfur atom of BME symmetrically bridges the binuclear Ni center (Ni–Ni = 3.1 Å) while the inhibitor chelates Ni1 using its terminal OH group (Fig. IX.3.6).[19] Both Ni ions are pentacoordinated: Ni1 is distorted square pyramidal, while Ni2 is distorted trigonal bipyramidal. A second molecule of BME is involved in a mixed-disulfide bond with Cys$^{\alpha322}$, situated on the flap lining the active site channel, and is further involved in a hydrogen bond between its α-hydroxyl group and the carbonyl oxygen atom of Ala$^{\alpha366}$, positioned on a neighboring loop (Fig. IX.3.6). This interaction reduces the flexibility of the flap, and the resulting network seals the entrance to the active site by steric hindrance. In summary, inhibition with BME occurs by targeting enzyme sites that both directly (the metal centers) and indirectly (the cysteine side chain) participate in substrate positioning and activation.

The most studied urease inhibitor belonging to the hydroxamic acid derivatives is acetohydroxamic acid (AHA). The structure of BPU inhibited with AHA (PDB code 4UBP) has been reported recently.[20] One AHA molecule is complexed to the Ni ions in the active site (Fig. IX.3.7), and provides a symmetric bridge (2.0 Å) between Ni1 and Ni2 through the hydroxamate oxygen (O_B). A second AHA oxygen (O_T) is bound to Ni1. The coordination geometry of both Ni ions is trigonal bipyramidal. The Ni1–Ni2 distance is 3.5 Å. The Asp$^{\alpha363}$ carboxylate group is rotated about the Cβ–Cγ bond by 35° with respect to its conformation in native BPU, to receive a hydrogen bond from the AHA–NH group. Therefore, the AHA molecule behaves as a bridging–chelating ligand, in analogy with the binding mode of BME. The structure of KAU mutated close to the active site and complexed with AHA (PDB code: 1FWE)[21] shows a molecule of the inhibitor bound to the Ni ions in a manner similar to that observed for AHA-inhibited BPU. But some differences are significant: (1) a pronounced asymmetry is observed for the Ni-bridging atom (Ni1–O_B = 2.6 Å; Ni2–O_B = 1.8 Å); and (2) the AHA–NH is not involved in a hydrogen bond with Asp$^{\alpha363}$ Oδ2, but rather with the carbonyl group of Ala$^{\alpha363}$.

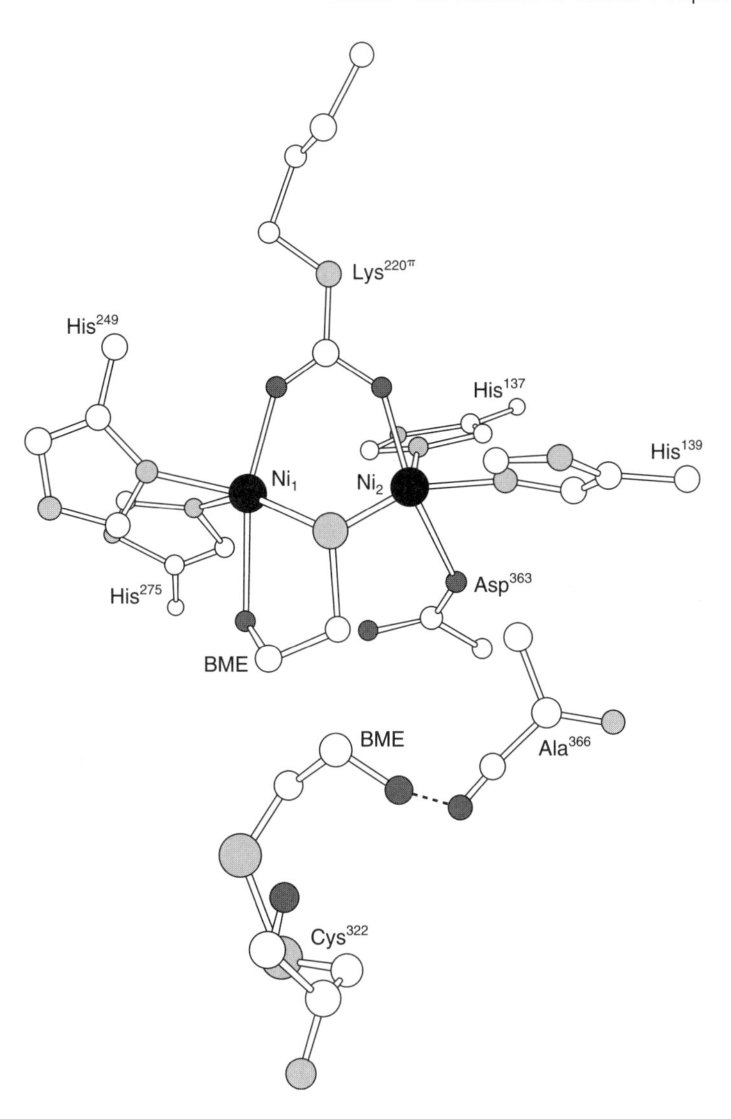

Fig. IX.3.6.
Model of the active site of BPU inhibited with BME, showing the position of selected relevant residues. Atoms filled with pattern by element (C: hollow; N: light gray; O: dark gray; S: gray). Hydrogen bonds are shown as dashed lines. All residues are sized according to their relative depths.

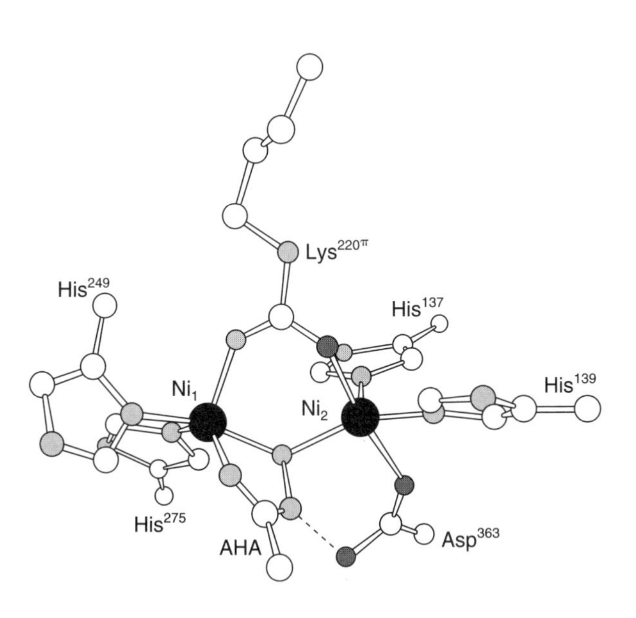

Fig. IX.3.7.
Model of the active site of BPU inhibited with AHA, showing the position of selected relevant residues. Atoms filled with pattern by element (C: hollow; N: light gray; O: dark gray). Hydrogen bonds are shown as dashed lines. All residues are sized according to their relative depths.

IX.3.6. The Molecular Basis for *in vivo* Urease Activation and Nickel Trafficking

The new frontier in the chemistry of urease is represented by the *in vivo* assembly of the protein, and in particular by the molecular mechanisms by which the Ni-containing active site is built. The biology of the transport and insertion of Ni^{2+} into the active site of urease has been extensively studied for *K. aerogenes*. The key players in this process are four accessory proteins,[18] named *Ka*UreD (~30 kDa), *Ka*UreF (~25 kDa), *Ka*UreG (~22 kDa), and *Ka*UreE (~18 kDa), which are the biosynthetic products of four corresponding genes clustered in the urease operon. This operon also includes the *ureA, ureB,* and *ureC* genes, encoding the three structural subunits of the $\alpha_3\beta_3\gamma_3$ heteropolymeric urease enzyme. The specific functions of these four accessory proteins are not fully understood. However, all evidence points to a temporally ordered mechanism as the basis of the *in vivo* urease activation, in which each urease chaperone plays a precise role.

Apo-urease can be partially activated *in vitro* by addition of Ni^{2+} in the presence of CO_2,[22] consistent with the presence of this metal ion and a Ni-bridging carbamylated Lys residue in the active site of the enzyme. The role of *Ka*UreD appears to be that of maintaining a proper conformation of apo-urease by forming a specific complex.[22–24] The protein *Ka*UreF binds to the *Ka*UreD–apo-urease complex[25] and is required to facilitate carbamylation of the Ni-bridging Lys residue in the active site by preventing Ni^{2+} ions from binding to the active site until the Lys has been carbamylated.[26] The protein *Ka*UreG exhibits clear sequence similarity to nucleotide triphosphate-binding proteins, featuring a so-called P-loop motif. This motif suggests a possible role in an energy-dependent step during *in vivo* urease assembly.[27–29] The protein *Ka*UreE acts as a metal ion carrier and donor metallochaperone, by delivering Ni^{2+} ions to the *Ka*UreDFG–apo-urease complex.[30] Gene knock-out experiments have shown that UreD, UreF, and UreG are absolutely necessary for urease active site assembly, while UreE is required to facilitate this process.[18]

The crystal structure of UreE from two different species, *B. pasteurii* (*Bp*UreE; PDB codes: 1EB0 and 1EAR, resolution 1.85 and 1.70 Å, respectively)[31] and *K. aerogenes* (*Ka*UreE; PDB codes: 1GMU, 1GMV, and 1GMW, resolution 1.50, 2.80, and 1.50 Å, respectively),[32] were reported recently (Fig. IX.3.8). The proteins *Bp*UreE and *Ka*UreE reveal a largely conserved and unique tertiary structure, made up of two distinct domains, separated by a short flexible linker. The N-terminal is composed of two three-stranded mixed parallel and anti-parallel β-sheets stacked upon each other in a nearly perpendicular fashion, with a short helical region between the two sheets. The C-terminal domain is organized in a βαββαβ-fold. The functional dimers of both *Bp*UreE and *Ka*UreE are built by a head-to-head interaction, involving the hydrophobic face of an amphiphilic helix in the C-terminal domain. A single Zn^{2+} ion is bound on the surface of the *Bp*UreE dimer through the coordination of two His100 Nε (one from each monomer), while a Cu^{2+} ion is bound to the homologous His96 in *Ka*UreE. The Zn^{2+} ion in *Bp*UreE can be replaced by Ni^{2+}, as revealed by paramagnetic nuclear magnetic resonance (NMR)[33] and X-ray anomalous diffraction difference maps.[31] In the crystal structure of *Ka*UreE, a second metal-binding site constituted by His110 and His112 (*Ka*UreE consensus sequence) is found on each monomer, resulting in the presence of two additional Cu^{2+} ions. In *Bp*UreE, this second binding site is absent because of the substitution of His110 and His112 by Tyr and Lys, respectively, suggesting that this additional binding site is not functionally relevant. In the structure of *Bp*UreE, the last few residues located at the C-terminal tail of each monomer are not visible because of disorder, probably caused by the dimerization of the functional dimer in the solid state. In the structure of *Ka*UreE, this region is altered by a truncation of

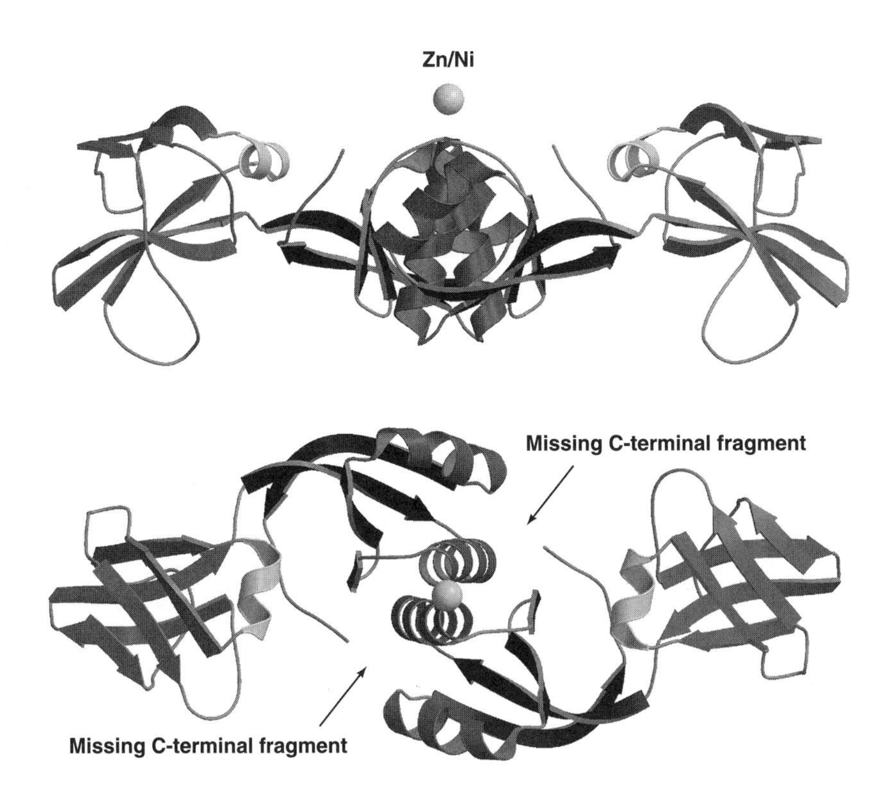

Fig. IX.3.8. Ribbon model of *Bp*UreE showing the position of the metal ion. The protein is shown with the metal-binding site toward the viewer (*a*) and rotated by 90° about the horizontal axis (*b*). The position is indicated for the C-terminal fragment not visible in the crystal structure due to disorder.

the protein sequence, artificially performed in order to eliminate a protein segment containing 15 residues, 10 of which are represented by histidines. Further studies are necessary to determine the role of this terminal region in the functional binding of the metal ion to UreE.

The study of the molecular basis for the *in vivo* assembly of the nickel-containing active site of urease is continuing[34] and will certainly provide interesting insights into the metal-trafficking modes of nickel in the cell.

References

General References

1. Ciurli, S. and Mangani, S., "Nickel-containing enzymes. In Handbook on Metalloproteins", Bertini, I., Sigel, A., and Sigel, H., Eds., Marcel Dekker, New York, 2001, pp. 669–708.
2. Hausinger, R. P. and Karplus, P. A., "Urease", in *Handbook of Metalloproteins,* Messerschmidt, A., Huber, R., Poulos, T., and Wieghardt, K., Eds., John Wiley & Sons, Inc., New York, 2001, pp. 867–879.

Specific References

3. Dixon, N. E., Gazzola, C., Blakeley, R., and Zerner, B., "Jack bean urease (EC 3.5.1.5). A metalloenzyme. A simple biological role for nickel?", *J. Am. Chem. Soc.,* **97,** 4131–4132 (1975).
4. Sumner, J. B., "The isolation and crystallization of the enzyme urease", *J. Biol. Chem.,* **69,** 435–441 (1926).
5. Jabri, E., Carr, M. B., Hausinger, R. P., and Karplus, P. A., "The crystal structure of urease from *Klebsiella aerogenes*", *Science,* **268,** 998–1004 (1995).
6. Benini, S., Rypniewski, W. R., Wilson, K. S., Miletti, S., Ciurli, S., and Mangani, S., "A new proposal for urease mechanism based on the crystal structures of the native and inhibited enzyme from *Bacillus pasteurii:* why urea hydrolysis costs two nickels", *Structure Fold. Des.,* **7,** 205–216 (1999).
7. Ha, N.-C., Oh, S.-T., Sung, J. Y., Cha, K. A., Lee, M. H., and Oh, B.-H., "Supramolecular assembly and acid resistance of *Helicobacter pylori* urease", *Nat. Struct. Biol.,* **8,** 505–509 (2001).
8. Benini, S., Ciurli, S., Nolting, H. F., and Mangani, S., "X-ray absorption spectroscopy study of native and phenylphosphorodiamidate-inhibited *Bacillus pasteurii* urease", *Eur. J. Biochem.,* **239,** 61–66 (1996).

9. Benini, S., Rypniewski, W. R., Wilson, K. S., Mangani, S., and Ciurli, S., "Molecular details of urease inhibition by boric acid: insights into the catalytic mechanism", *J. Am. Chem. Soc.*, **126,** 3714–3715 (2004).
10. Ciurli, S., Benini, S., Rypniewski, W. R., Wilson, K. S., Miletti, S., and Mangani, S., "Structural properties of the nickel ions in urease: novel insights into the catalytic and inhibition mechanisms", *Coord. Chem. Rev.*, **190–192,** 331–355 (1999).
11. Todd, M. J. and Hausinger, R. P., "Purification and characterization of the nickel-containing multicomponent urease from *Klebsiella aerogenes*", *J. Biol. Chem.*, **262,** 5963–5967 (1987).
12. Wilcox, D. E., "Binuclear metallohydrolases", *Chem. Rev.*, **96,** 2435–2458 (1996).
13. Karplus, P. A., Pearson, M. A., and Hausinger, R. P., "70 years of crystalline urease: what have we learned?", *Acc. Chem. Res.*, **30,** 330–337 (1997).
14. Park, I.-S. and Hausinger, R. P., "Diethylpyrocarbonate reactivity of *Klebsiella aerogenes* urease: effect of pH and active site ligands on the rate of inactivation", *J. Protein Chem.*, **12,** 51–56 (1993).
15. Todd, M. J. and Hausinger, R. P., "Fluoride inhibition of *Klebsiella aerogenes* urease: mechanistic implications of a pseudo-uncompetitive, slow-binding inhibitor", *Biochemistry,* **39,** 5389–5396 (2000).
16. Musiani, F., Arnofi, E., Casadio, R., and Ciurli, S., "Structure-based computational study of the catalytic and inhibition mechanism of urease", *J. Biol. Inorg. Chem.*, **3,** 300–314 (2001).
17. Benini, S., Rypniewski, W. R., Wilson, K. S., Ciurli, S., and Mangani, S., "Structure-based rationalization of urease inhibition by phosphate: novel insights into the enzyme mechanism", *J. Biol. Inorg. Chem.*, **6,** 778–790 (2001).
18. Mobley, H. L. T., Island, M. D., and Hausinger, R. P., "Molecular biology of microbial ureases", *Microbiol. Rev.*, **59,** 451–480 (1995).
19. Benini, S., Rypniewski, W. R., Wilson, K. S., Ciurli, S., and Mangani, S., "The complex of *Bacillus pasteurii* urease with β-mercaptoethanol from X-ray data at 1.65 Å resolution", *J. Biol. Inorg. Chem.*, **3,** 268–273 (1998).
20. Benini, S., Rypniewski, W. R., Wilson, K. S., Miletti, S., Ciurli, S., and Mangani, S., "The complex of *Bacillus pasteurii* urease with acetohydroxamate anion from X-ray data at 1.55 Å resolution", *J. Biol. Inorg. Chem.*, **5,** 110–118 (2000).
21. Pearson, M. A., Overbye, M. L., Hausinger, R. P., and Karplus, P. A., "Structures of Cys319 variants and acetohydroxamate-inhibited *Klebsiella aerogenes* urease", *Biochemistry,* **36,** 8164–8172 (1997).
22. Park, I.-S. and Hausinger, R. P., "Requirement of carbon dioxide for in vitro assembly of the urease nickel metallocenter", *Science,* **267,** 1156–1158 (1995).
23. Park, I.-S., Carr, M. B., and Hausinger, R. P., "*In vitro* activation of urease apoprotein and role of UreD as a chaperone required for nickel metallocenter assembly", *Proc. Natl. Acad. Sci., U.S.A.*, **91,** 3233–3237 (1994).
24. Park, I.-S. and Hausinger, R. P., "Metal ion interactions with urease and UreD-urease apoproteins", *Biochemistry,* **35,** 5345–5352 (1996).
25. Park, I.-S. and Hausinger, R. P., "Evidence for the presence of urease apoprotein complexes containing UreD, UreF, and UreG in cells that are competent for in vivo enzyme activation", *J. Bacteriol.*, **177,** 1947–1951 (1995).
26. Moncrief, M. C. and Hausinger, R. P., "Purification and activation properties of UreD-UreF-urease apoprotein complexes", *J. Bacteriol.*, **178,** 5417–5421 (1996).
27. Moncrief, M. C. and Hausinger, R. P., "Characterization of UreG, identification of a UreD-UreF-UreG complex, and evidence suggesting that a nucleotide-binding site in UreG is required for in vivo metallocenter assembly of *Klebsiella aerogenes* urease", *J. Bacteriol.*, **179,** 4081–4086 (1997).
28. Soriano, A. and Hausinger, R. P., "GTP-dependent activation of urease apoprotein in complex with the UreD, UreF, and UreG accessory proteins", *Proc. Natl. Acad. Sci. U.S.A.*, **96,** 11140–11144 (1999).
29. Soriano, A., Colpas, G. J., and Hausinger, R. P., "UreE stimulation of GTP-dependent urease activation in the UreD-UreF-UreG-urease apoprotein complex", *Biochemistry,* **39,** 12435–12440 (2000).
30. Lee, M. Y., Pankratz, H. S., Wang, S., Scott, R. A., Finnegan, M. G., Johnson, M. K., Ippolito, J. A., Christianson, D. W., and Hausinger, R. P., "Purification and characterization of *Klebsiella aerogenes* UreE protein: a nickel binding protein that functions in urease metallocenter assembly", *Protein Sci.*, **2,** 1042–1052 (1993).
31. Remaut, H., Safarov, N., Ciurli, S., and Van Beeumen, J. J., "Structural basis for Ni transport and assembly of the urease active site by the metallo-chaperone UreE from *Bacillus pasteurii*", *J. Biol. Chem.*, **276,** 49365–49370 (2001).
32. Song, H.-K., Mulrooney, S. B., Huber, R., and Hausinger, R. P., "Crystal structure of *Klebsiella aerogenes* UreE, a nickel-binding metallochaperone for urease activation", *J. Biol. Chem.*, **276,** 49359–49364 (2001).
33. Ciurli, S., Safarov, N., Miletti, S., Dikiy, A., Christensen, S. K., Kornetzky, K., Bryant, D. A., Vandenberghe, I., Devreese, B., Samyn, B., Remaut, H., and Van Beeumen, J. J., "Molecular characterization of *Bacillus pasteurii* UreE, a metal-binding chaperone for the assembly of the urease active site", *J. Biol. Inorg. Chem.*, **7,** 623–631 (2002).
34. Musiani, F., Zambelli, B., Stola, M., and Ciurli, S., "Nickel trafficking: insights into the fold and function of UreE, a urease metallochaperone", *J. Inorg. Biochem.*, **98,** 803–813 (2004).

IX.4. Aconitase

M. Claire Kennedy and Helmut Beinert
Department of Chemistry
Gannon University
Erie, PA 16564
and
Institute for Enzyme Research, University of Wisconsin, Madison, WI 53726

Contents

IX.4.1. Introduction

Aconitase, citrate(isocitrate)hydrolyase (EC4.2.1.3) is an important enzyme widely distributed in Nature that catalyzes the stereospecific isomerization of citrate to isocitrate via consecutive dehydration–hydration reactions, as shown in Eq. 1.[1,2]

$$\underset{\textbf{Citrate}}{\text{HO}-\text{C}(\text{COO}^-)_3} \;\underset{+\text{H}_2\text{O}}{\overset{-\text{H}_2\text{O}}{\rightleftharpoons}}\; \underset{\textbf{\textit{cis}-Aconitate}}{(\alpha)\,\text{COO}^-,\ (\beta)\,\text{COO}^-,\ (\gamma)\,\text{COO}^-} \;\underset{-\text{H}_2\text{O}}{\overset{+\text{H}_2\text{O}}{\rightleftharpoons}}\; \underset{\textbf{Isocitrate}}{\text{HO}-\text{COO}^-,\ \text{COO}^-,\ \text{COO}^-} \qquad (1)$$

The enzyme was discovered in mammalian systems >50 years ago and since has been isolated from a variety of species including an archeal aconitase from the thermoacidophile, *Sulfolobus acidocaldarus*. Early on it was shown that the enzyme lost activity during isolation and that the activity could be restored upon incubation with iron under reducing conditions. It is now well established that the Fe present in the enzyme and in many of the other Fe-dependent dehydratases, is present in the form of an iron–sulfur cluster that is essential for activity. As such, aconitase was among the first Fe–S proteins characterized that did not function in electron transfer. Although aconitase and some other Fe–S dehydratases (e.g., isopropylmalate isomerase) also function as isomerases, this finding is not true for all members of the family (e.g., *Escherichia coli* fumarase A and B).

Aconitase proteins can be divided into three phylogenetic categories: mitochondrial aconitases (mAcn); a group containing the cytosolic aconitases (cAcn), the Fe regulatory proteins (IRP1 and IRP2), and the bacterial aconitases A (AcnA); and, lastly, a family of bacterial only aconitases B (AcnB). The second group contains the bifunctional cytosolic aconitase of higher organisms, which, in its apo-form, functions as an iron regulatory protein (IRP1) by binding to specific RNA sequences called Fe responsive elements, thus affecting the translation (expression) of a number of proteins. The bacterial AcnAs and AcnBs share this same bifunctional property and under Fe replete conditions the proteins contain Fe–S clusters and function as enzymes. However, under oxidative stress or during Fe starvation, cluster disassembly may occur, converting the enzyme into an Fe regulatory protein. The non-enzymatic role of these proteins in Fe regulation will not be discussed further here.

The mitochondrial isoform of aconitase is a key enzyme in metabolism as a component of the tricarboxylic acid (TCA) and glyoxylate cycles. Extensive kinetic, chemical, and spectroscopic studies of mammalian mitochondrial aconitase, and the determination of three-dimensional (3D) structures by X-ray crystallography in a variety of states, have provided a wealth of information allowing a detailed understanding of the reaction mechanism and elucidation of the role of the Fe–S cluster in catalysis. For this reason, the following discussion is based primarily on what has been learned about this isoform of the enzyme.

IX.4.2. Stereochemistry of the Citrate–Isocitrate Isomerase Reaction

Early metabolic studies of the TCA cycle using radiolabeled substrates established that the loss of water from citrate, which is a symmetric molecule, occurred only from that portion of the molecule derived from oxaloacetate in the citrate synthase reaction, labeled α and β in Fig. IX.4.1. Following the trans elimination of the elements of water, the obligatory intermediate *cis*-aconitate thus formed must rotate 180° around an axis perpendicular to the C_α–C_β bond, that is, "flip" before hydration and formation of the correct stereoisomer, (2*R*,3*S*), of isocitrate can form. The active site of aconitase is thus able to accommodate four different structures: citrate, isocitrate, and *cis*-aconitate in either the citrate or isocitrate mode. During the reaction, the hydroxyl lost from the substrate equilibrates rapidly with solvent, whereas the proton removed can be transferred back to another substrate molecule (labeled H* in Fig. IX.4.1). Inhibitor studies utilizing the tight-binding nitro analogues of citrate and isocitrate established that the reaction occurs by a carbanion mechanism. To fully explain these observations, it is necessary to discuss properties of the enzyme learned from the spectroscopic and structural studies of the purified protein.

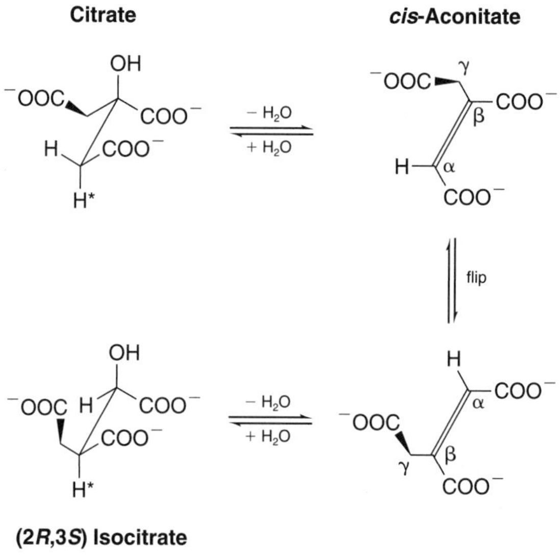

Citrate | ***cis*-Aconitate** | **(2*R*,3*S*) Isocitrate**

Fig. IX.4.1.
Stereochemistry of the aconitase reaction. Loss of the elements of water occurs by a trans elimination of H^+ and OH^- with the formation of *cis*-aconitate. To form the correct diastereoisomer of isocitrate, H^+ and OH^- must be added to the faces of *cis*-aconitate opposite to where they were removed. This requires a rotation of 180° about an axis perpendicular to the double bond (flip) for *cis*-aconitate to be bound to the enzyme in the isocitrate mode. Experiments with labeled substrates show that the OH^- lost equilibrates rapidly with solvent, whereas the proton (H*) removed by the enzyme may be added back to the substrate. The labeling of the carbons refers to those portions of the molecule derived from the citrate synthase reaction of the TCA cycle; α and β for oxaloacetate, and γ for acetyl coenzyme A (acetylCoA).

IX.4.3. Characterization and Function of the Fe–S Cluster

The enzyme, as isolated in its inactive state, is a monomeric protein (as are all aconitases identified thus far) of 754 amino acid residues, $M_r \sim 84$ kDa, that contains Fe and inorganic sulfide in a ratio of 3:4. This state of the protein exhibits a nearly isotropic electron paramagnetic resonance (EPR) signal at $g = 2.016$ that is now well established to be the signature for cuboidal $[3Fe–4S]^+$ clusters. Upon reaction with iron and reductant, a cubane $[4Fe–4S]^{2+}$, $S = 0$, cluster is formed (Fig. IX.4.2) and activity is restored to the enzyme. This form of the protein is EPR silent, but can be reduced to the $S = \frac{1}{2}$, 1+ state to yield a protein that retains activity. Although this latter species is not the native form of the active enzyme, it has served as an important spectroscopic tool in the study of the protein. The dramatic change in the EPR spectrum of the substrate-free reduced enzyme, $g_{1,2,3} = 2.06, 1.93, 1.86$ (Fig. IX.4.3*a*) upon substrate addition, $g_{1,2,3} = 2.04, 1.85, 1.78$ (Fig. IX.4.3*b*), or upon the addition of a variety of inhibitors, for example, *trans*-aconitate, $g_{1,2,3} = 2.01, 1.88, 1.80$ (Fig. IX.4.3*c*), was among the first pieces of evidence to suggest that the cluster was

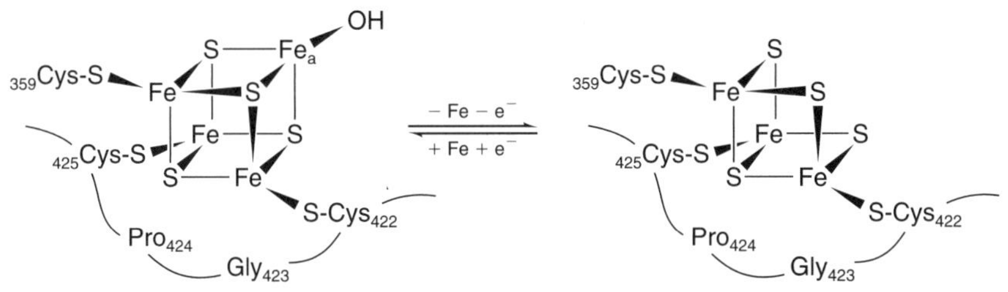

Fig. IX.4.2.
Interconversion and structure of the 4Fe and 3Fe clusters of aconitase. Inactivation of the enzyme occurs by the one-electron oxidation of the $[4Fe–4S]^{2+}$ cluster and loss of Fe_a. The enzyme can be reactivated by reduction of the $[3Fe–4S]^{1+}$ cluster followed by the rapid addition of Fe^{2+} to the vacant site.

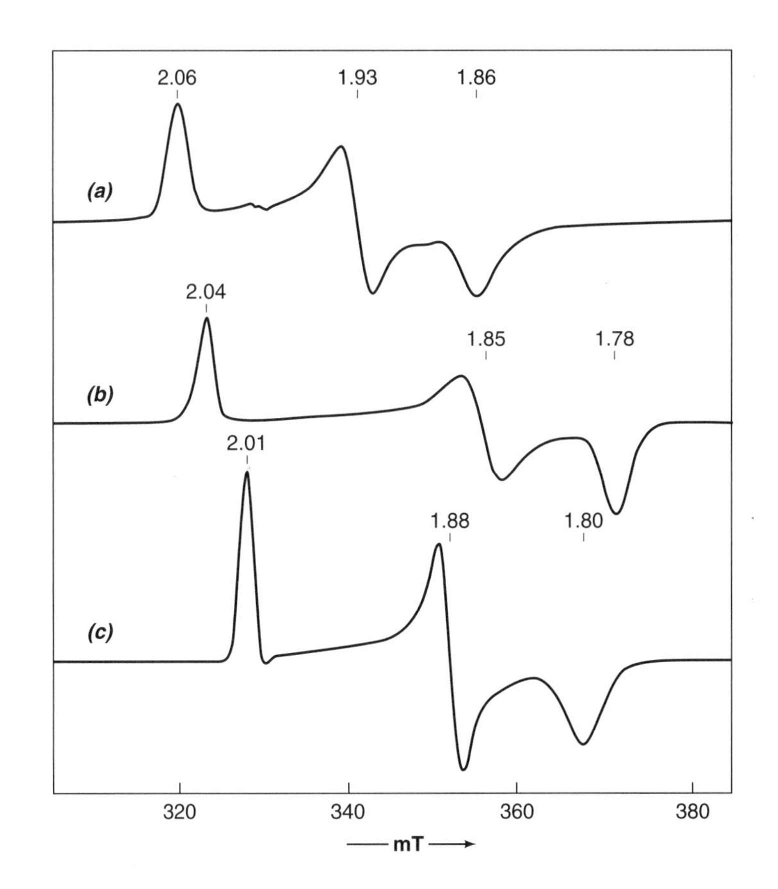

Fig. IX.4.3.
The EPR spectra of reduced active aconitase, $[4Fe–4S]^+$, (*a*) in the absence of substrate, (*b*) in the presence of substrate, and (*c*) in the presence of the inhibitor *trans*-aconitate.

directly involved in catalysis. A K_d of 1 μM for substrate was determined by titration of enzyme with citrate as monitored by EPR. It was ^{57}Fe Mössbauer spectroscopy, however, that was essential in establishing that the $g = 2.016$ EPR signal was in fact due to a 3 Fe cluster and that activation of the enzyme involved a 3 Fe to 4 Fe cluster interconversion (Fig. IX.4.2). The Fe added to the vacant site of the 3 Fe cluster, designated Fe_a, is the same Fe that is lost on inactivation of the enzyme during the isolation procedure or upon oxidation by chemical reagents. Note that the Fe is lost as Fe(II), which is of particular importance when this reaction occurs intracellularly. For example, the inactivation of aconitase by superoxide anion, $O_2^{\bullet -}$, with the formation of H_2O_2 can result in the production of damaging hydroxyl radicals as a result of the reaction between Fe(II) and H_2O_2. Mössbauer experiments in which the cluster was labeled with ^{57}Fe in the Fe_a-site, [$^{57}Fe_a$–3^{56}Fe–4S], resulted in spectra that changed significantly upon the addition of substrate, as was also observed by EPR. In addition to showing direct interaction of substrate and Fe_a of the cluster, these results indicated that the coordination geometry of Fe_a expanded from tetrahedral to octahedral. Results from both Mössbauer and EPR spectroscopy, and from experiments using radiolabeled substrates, demonstrate that substrate is very tightly bound to the protein, remaining bound even after passage through a desalting column. This observation is consistent with the findings from kinetic studies indicating that the rate-determining step of the catalytic reaction is the displacement of one substrate molecule from the active site by another.

Details of the interaction of substrate and solvent with the cluster were obtained from extensive electron nuclear double resonance spectroscopy (ENDOR) experiments by observing the hyperfine interactions of ^{57}Fe, ^{33}S, ^{17}O , ^{2}H, and ^{13}C labeled cluster, solvent, substrates, or inhibitors with reduced enzyme. An important discovery stemming from these experiments is that, whereas the ligation to Fe in most 4 Fe clusters consists of three cluster sulfides and a Cys from protein, in aconitase a solvent hydroxyl is bound to Fe_a in lieu of a Cys residue (Fig. IX.4.2). This unique arrangement imparts special characteristics to the cluster permitting it to participate in enzyme catalysis. These same ENDOR studies also demonstrated that upon binding the substrate to Fe_a, the ligated hydroxyl is protonated to a water molecule and a five-membered ring involving Fe_a, carboxylate and hydroxyl of substrate is formed. This result is in agreement with the findings of Mössbauer spectroscopy where a change in coordination number from 4 to 6 is observed for Fe_a upon the binding of substrate.

IX.4.4. Active Site Amino Acid Residues and the Reaction Mechanism

Although the spectroscopic studies of aconitase give precise information of the interaction of substrate and cluster, details of the binding of substrate to protein and of the catalytic mechanism, in particular the base involved in the dehydration reaction, came only from knowledge of the 3D structure as determined by X-ray crystallography. Data have been obtained for a number of structures, including the inactive 3 Fe (PDB code: 1ACO) form, as well as active enzyme with the substrate, isocitrate, (PDB code: 7ACN)[3] and inhibitors nitroisocitrate (PDB code: 8ACN)[3] and nitrocitrate (PDB code: 1NIS) bound. The protein folds into four domains, the first three of which are tightly associated. The fourth large C-terminal domain is attached to domain 3 by a flexible 25 amino acid chain and forms, with the other domains, a solvent-accessible cleft leading to the Fe–S cluster. This exposure of the cluster accounts for the ease of oxidative inactivation of the enzyme during isolation. The

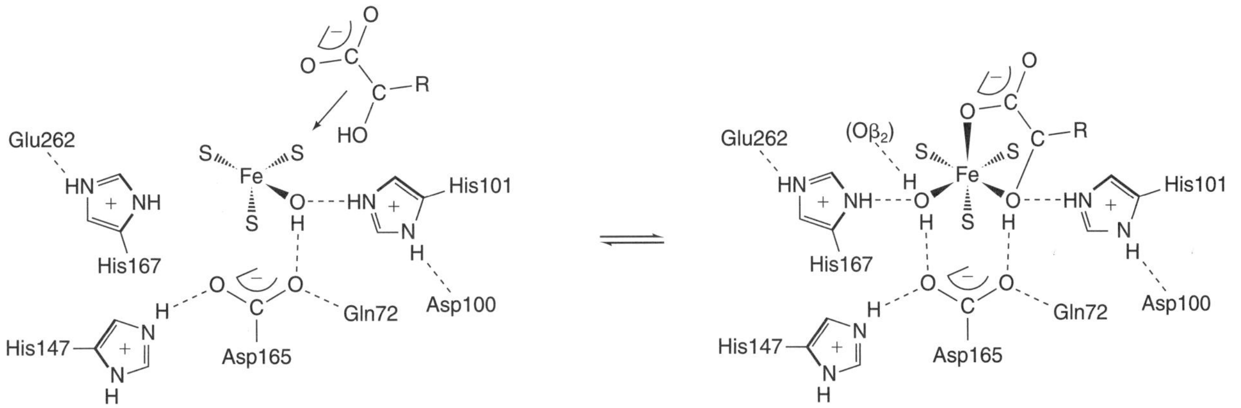

Fig. IX.4.4.
Schematic representation of the transition from substrate-free aconitase to enzyme with isocitrate bound. The solvent-derived hydroxyl of the four-coordinate Fe_a in the substrate-free state is hydrogen bonded to both His101 and Asp165. Upon binding of isocitrate, Fe_a becomes six-coordinate and the solvent derived hydroxyl is converted to a water molecule that is hydrogen bonded to Asp165, His167, and the β-carboxyl of substrate. The His101 residue is now hydrogen bonded to the –OH of substrate coordinated to Fe_a.

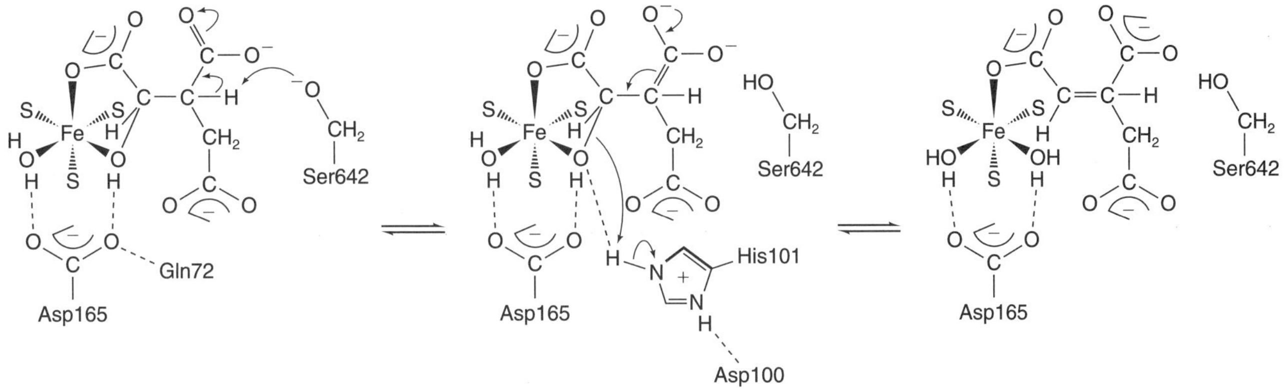

Fig. IX.4.5.
Schematic representation of the mechanism of the conversion of isocitrate to *cis*-aconitate, on the assumption that Ser642 of the enzyme is deprotonated to the alkoxide upon binding of substrate. The binding of the substrate –COOH and –OH to Fe_a activates the $C_β$–H bond for deprotonation by Ser642. Cleavage of the C–OH bond on collapse of the *aci*-acid intermediate to form *cis*-aconitate is assisted by the binding of substrate hydroxyl to Fe_a and by hydrogen bonding to His101.

cluster is bound to the protein by three Cys ligands, two of which are present in a -Cys-X-X-Cys- motif, common to many Fe–S proteins. The mode of binding of substrate to the Fe–S cluster was as deduced by spectroscopy (above), that is, formation of a five-membered ring of Fe_a with a hydroxyl and carboxyl (α- for isocitrate and β- for citrate) from substrate (Fig. IX.4.4). A solvent molecule bound to Fe_a was also observed. Chief among the binding residues is Arg580, which anchors the γ-carboxyl of all substrate forms to the protein. Identification of the catalytic base as a Ser residue was a surprise, but explains the slow exchange of the proton abstracted from substrate with solvent. It had always been assumed that the Fe of aconitase functioned as a Lewis acid in the elimination of hydroxyl from substrate. However, exactly how this occurred and the intrinsic role played by protein residues in this process was not clear. The 3D structure reveals that there are a number of His and acidic residues located in proximity to Fe_a (Fig. IX.4.4). Elimination of the substrate hydroxyl is facilitated through protonation by His101 and by hydrogen bonding of Asp165 to both the hydroxyl of substrate and the water on Fe_a (Fig. IX.4.5). The critical importance of Asp165 in the catalytic mechanism is reflected in the decrease in V_{max} of over five orders of magnitude in an Asp165Ser mutant of the enzyme. This decrease is of the same order of magnitude as that observed for a mutant where the catalytic base, Ser642, is changed to an alanine.[4]

Confirmation of the proposed mechanism was achieved when the crystal structure was solved for aconitase with the tight-binding inhibitor, hydroxy-*trans*-aconitate, bound (PDB code: 1FGH). Hydroxy-*trans*-aconitate is a mechanism-based inhibitor formed from the reaction of aconitase and the highly toxic (−)-*erythro*-2-fluorocitrate, which can be produced intracellularly from the rodenticide, fluoroacetate, also known as Compound 1080. The 2-fluorocitrate is first converted to fluoro-*cis*-aconitate, which then binds in the isocitrate mode after a 180° rotation about an axis perpendicular to the double bond (see "flip", Fig. IX.4.1), which is followed by the addition of hydroxyl with loss of fluoride and formation of hydroxy-*trans*-aconitate. The very tight binding of the inhibitor can be attributed to four short hydrogen bonds, 2.7 Å, involving inhibitor, water bound to the cluster, Asp165 and His167. Although not covalently bound to the protein, the inhibitor can only be displaced by a very large excess of substrate (10^4-fold) or by denaturation. This extremely tight binding of the hydroxy-*trans*-aconitate to aconitase effectively shuts down the TCA cycle and may result in death of the organism.

IX.4.5. Cluster Reactivity and Cellular Function

It has long been known that there are differences in the oxygen sensitivity of the 4 Fe clusters of the different isoforms of aconitase. The fact that cAcn is less susceptible to aerobic inactivation than mAcn led many to the use of cAcn in kinetic studies of the enzyme. The recent discovery of the dual role of cAcn and the bacterial aconitases as both enzymes and RNA-binding proteins (when in their apo forms), suggests that the differences in cluster stability to reactive oxygen and nitrogen species (e.g., nitric oxide) may indeed be related to the individual intracellular functions of the different proteins. It has been proposed that the inactivation of mAcn and Acn B of *E. coli* by $O_2^{\bullet -}$ under conditions of oxidative stress may serve as a regulatory mechanism during aerobic metabolism.[5] Removal of the stress results in a rapid reconversion to the active 4 Fe cluster. In addition, decreased activity of mAcn has been shown to be related to a number of mitochondrial diseases and to the aging process.[6,7] The rapid inactivation of mAcn contrasts with the slower inactivation of cAcn and with the intracellular interconversion of cAcn and IRP1, which occurs on a timescale of hours. There are subtle but significant differences between the EPR spectra of the mammalian forms of mAcn and cAcn for both the 3 Fe and reduced 4 Fe clusters, indicating differences in their individual electronic structures. Sequence comparisons of mammalian mAcn and cAcn reveal an identity of ~30%. However, the 23 residues assigned to the active site of mAcn that are involved in the binding of cluster and substrate are essentially conserved in cAcn and in most aconitases identified thus far. At present, there are no 3D structures for cAcn/IRP1, and therefore it is difficult to assess the role that the protein environment may play in altering the chemical reactivity of the Fe–S clusters and the cellular functions of these two isoforms of aconitase.

References

General References

1. Beinert, H., Kennedy, M. C., and Stout, C. D., "Aconitase as Iron–Sulfur Protein, Enzyme, and Iron-Regulatory protein", *Chem. Rev.*, **96,** 2335–2373 (1996).
2. Flint, D. H. and Allen, R. M., "Iron-Sulfur Proteins with Nonredox Functions", *Chem. Rev.*, **96,** 2315–2334 (1996).

Specific References

3. Lauble, H., Kennedy, M. C., Beinert, H., and Stout, C. D., "Crystal Structures of Aconitase with Isocitrate and Nitroisocitrate Bound", *Biochemistry*, **31,** 2735–2748 (1992).
4. Zheng, l., Kennedy, M. C., Beinert, H., and Zalkin, H., "Mutational analysis of active site residues in pig heart aconitase", *J. Biol. Chem.*, **267,** 7895–7903 (1992).
5. Walden, W. E., "From bacteria to mitochondria: Aconitase yields surprises", *Proc. Natl. Acad. Sci. U.S.A.*, **89,** 41380–4140 (2002).
6. Rotig, A., de Lonlay, P., Chretien, D., Foury, F., Koenig, M., Sidi, D., Munnich, A., and Rustin, P., "Aconitase and iron–sulphur protein deficiency in Friedreich ataxia", *Nat. Genet.*, **17,** 215–217 (1997).
7. Yan, L. J., Levine R. l., and Sohal, R. S., "Oxidative damage during aging targets mitochondrial aconitase", *Proc. Natl. Acad. Sci. U.S.A.*, **94,** 11168–11172 (1997).

IX.5. Catalytic Nucleic Acids

Yi Lu
Department of Chemistry
University of Illinois at
Urbana-Champaign
Urbana, IL 61801

Contents

IX.5.1. Introduction and Discovery of Catalytic Nucleic Acids

Catalytic nucleic acids are a new class of metalloenzymes.[1–10] Study of these enzymes has become a new frontier in bioinorganic chemistry. Historically, the central dogma in biology consists of transcription of DNA into RNA, followed by translation of RNA into proteins. However, enzymes are required for the synthesis of DNA, and these enzymes were believed to be proteins. Therefore, a typical "chicken-or-egg" question arose as to which one, DNA, RNA, or protein, came first. In the early 1980s, some RNA molecules were discovered to have enzymatic activities (and were therefore called catalytic RNA, ribozymes, RNA enzymes, or RNAzymes).[11,12] Those early studies and others since then support the idea of a prebiotic RNA world.[4,5] The recent crystal structure of the ribosome revealed that RNA may play a dominant role in protein synthesis.[13,14]

Since the difference between DNA and RNA is mainly the presence of a 2′-OH of the sugar moiety in the RNA (Fig. IX.5.1), a natural question is whether DNA can be as efficient a catalyst as RNA. This question was answered in 1994 by isolation of single-stranded DNA molecules that can catalyze RNA cleavage.[15] Therefore, catalytic DNA (also called deoxyribozymes, DNA enzymes or DNAzymes) became the newest member of the enzyme family after proteins and RNA.

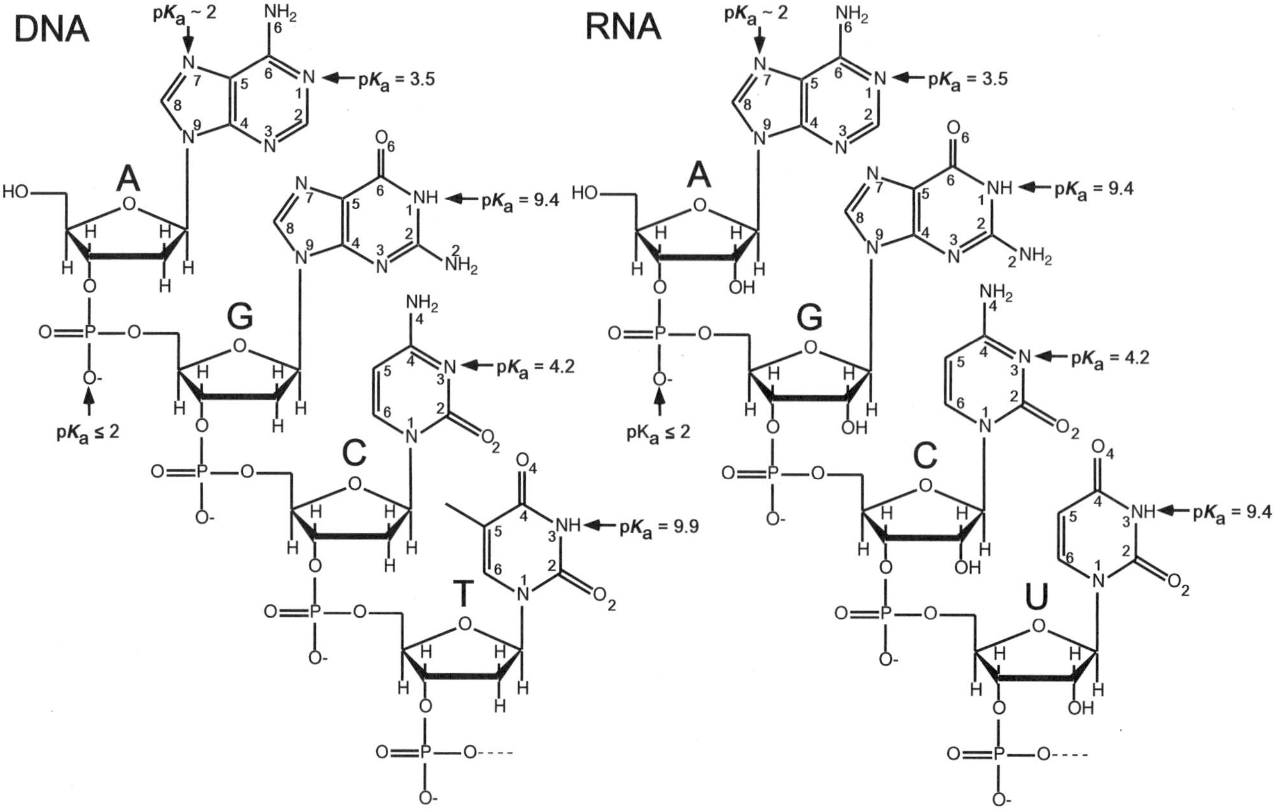

Fig. IX.5.1.
The molecules of DNA, RNA and nucleobases therein. Approximate pK_a values of important functional groups that can bind metal ions are also shown.

IX.5.2. Scope and Efficiency of Catalytic Nucleic Acids

Natural catalytic RNA molecules known to date catalyze several reactions such as transesterification, hydrolysis of phosphodiester bonds, or peptidyl transfer. In the early 1990s, a powerful method called *in vitro* selection[16–19] was developed to determine whether catalytic RNA could perform other functions. In this method, a small population of RNA with the desired functions is selected and amplified from a large pool of RNA molecules with random sequences, typically on the order of 10^{15} molecules. The selected catalytic RNA molecules are then subjected to further rounds of selection and amplification, often with more stringent conditions. In this way, a growing number of catalytic RNA molecules with activities previously known only for protein enzymes have been obtained (Table IX.5.1).[20] Furthermore, the same *in vitro* selection approach has been applied to the discovery of new catalytic DNA (Table IX.5.1).[6,21,22] The lower cost and higher stability of catalytic DNA makes it more attractive in biochemical, biotechnological, and pharmaceutical applications.

The enzymatic efficiency of the catalytic nucleic acids is often similar to those of protein enzymes. For example, the rate constant of self-cleaving hepatitis delta virus ribozyme (10^2–$10^4\ s^{-1}$) is close to the maximum cleavage rate of the protein enzyme RNase A ($1.4 \times 10^3\ s^{-1}$). Similarly, the catalytic efficiency of $10^9\ M^{-1}\ min^{-1}$ observed for the "10–23" deoxyribozyme[23] rivals the protein enzyme ribonuclease.

Table IX.5.1.
Reactions Catalyzed by Catalytic Nucleic Acids[a]

	Catalytic Activity			
Reaction[b]	Enzyme[c]	k_{cat} (min^{-1})	K_m (μM)	k_{cat}/k_{uncat}
Phosphoester transfer	R-nat	0.1	1×10^{-3}	10^{11}
	R-lab	0.3	0.02	10^{13}
Phosphoester cleavage	R-nat	1	0.05	10^6
	R-lab	0.1	0.03	10^5
	D-lab	3	8×10^{-4}	10^6
Polynucleotide ligation	R-nat	4	3	10^6
	R-lab	100	9	10^9
	D-lab	0.04	100	10^4
Polynucleotide phosphorylation	R-lab	0.3	40	$>10^5$
Mononucleotide aminoacylation	R-lab	0.3	5×10^3	$>10^7$
Polynucleotide aminoacylation	R-lab	1	9×10^3	10^6
Aminoacyl ester hydrolysis	R-lab	0.02	0.5	10
Aminoacyl transfer	R-lab	0.2	0.05	10^3
Amide bond cleavage	R-lab			10^2
Amide bond formation	R-lab	0.04	2	10^5
Peptide bond formation	R-lab	0.05	200	10^6
N-Alkylation	R-lab	0.6	1×10^3	10^7
S-Alkylation	R-lab			10^3
Oxidative DNA cleavage	R-lab			$>10^6$
Biphenyl rotation	R-lab	3×10^{-5}	500	10^2
Porphyrin metallation	R-lab	0.9	10	10^3
	D-lab	0.2	3×10^3	10^3
Diels–Alder cycloaddition	R-lab	>0.1	>500	10^3

[a] From Ref. (20).

[b] One example is listed for each class of reaction and each type of enzyme.

[c] (R-nat): catalytic RNA derived from naturally occurring sources; (R-lab): catalytic RNA obtained by *in vitro* selection; (D-lab): catalytic DNA obtained by *in vitro* selection.

IX.5.3. Classification of Catalytic Nucleic Acids with Hydrolytic Activity

Catalytic nucleic acids with hydrolytic activity can be classified according to mechanistic considerations (Fig. IX.5.2).[7] Examples for each class of ribozymes are shown in Figs. IX.5.3 and IX.5.4.[1–5] Selected examples of new deoxyribozymes (catalytic DNA) are shown in Fig. IX.5.5.[6,15,21–31] Class A activates an internal 2′-OH for nucleophilic attack on the phosphorus of an adjacent phosphodiester bond (Fig. IX.5.2*a*). The products of this reaction have 2′,3′-cyclic phosphate and 5′-OH termini. This class of catalytic nucleic acids includes the hammerhead, hairpin, hepatitis delta virus (HDV), *Neurospora* VS, and Mn^{2+}-dependent ribozymes as well as "10–23" deoxyribozymes (Figs. IX.5.3 and IX.5.5). However, a Pb^{2+}-dependent catalytic RNA called leadzyme and a Pb^{2+}-dependent catalytic DNA called "8–17"

(a)

Class A

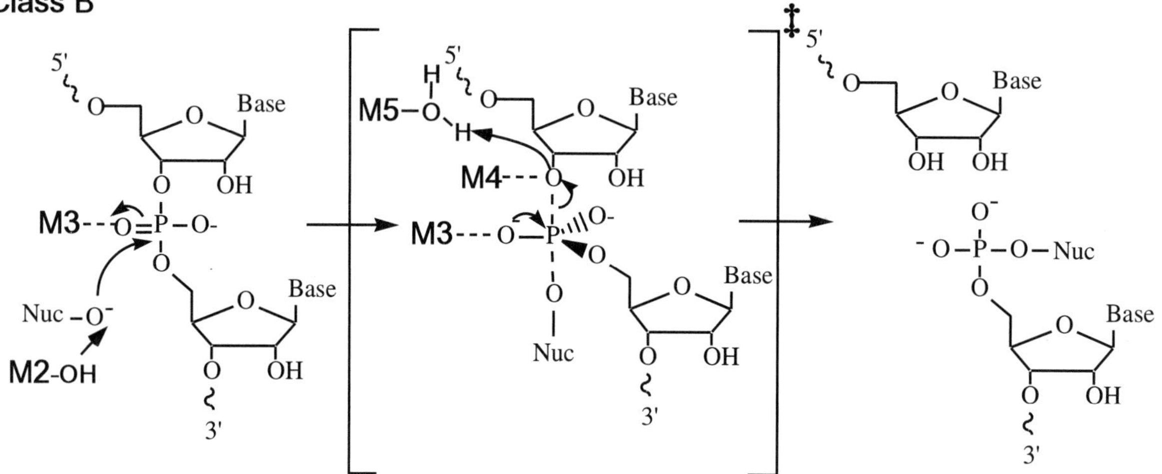

(b)

Class B

Fig. IX.5.2.
Two classes of catalytic nucleic acids with hydrolytic activities and the roles of metal ions in catalysis. [Adapted from Ref. 7.]

deoxyribozyme follow an interesting two-step mechanism by generating products containing 2′,3′-cyclic phosphate and 5′-OH groups, followed by further hydrolysis of the 2′,3′-cyclic phosphate to yield 3′-monophosphate (Fig. IX.5.2*a*).[32,33] In addition to the similarity of their reaction products, these catalytic nucleic acids are typically small ($<$200 bases in length).[34]

Class B of catalytic nucleic acids are larger ($>$400 bases) and include the RNA moiety of ribonuclease P and the group I and group II intron ribozymes (Fig. IX.5.4). These enzymes all employ an external nucleophile, such as an activated nucleotide or water, to carry out attack on the adjacent scissile phosphodiester, resulting in products containing 3′-OH and 5′-phosphate termini (Fig. IX.5.2*b*).

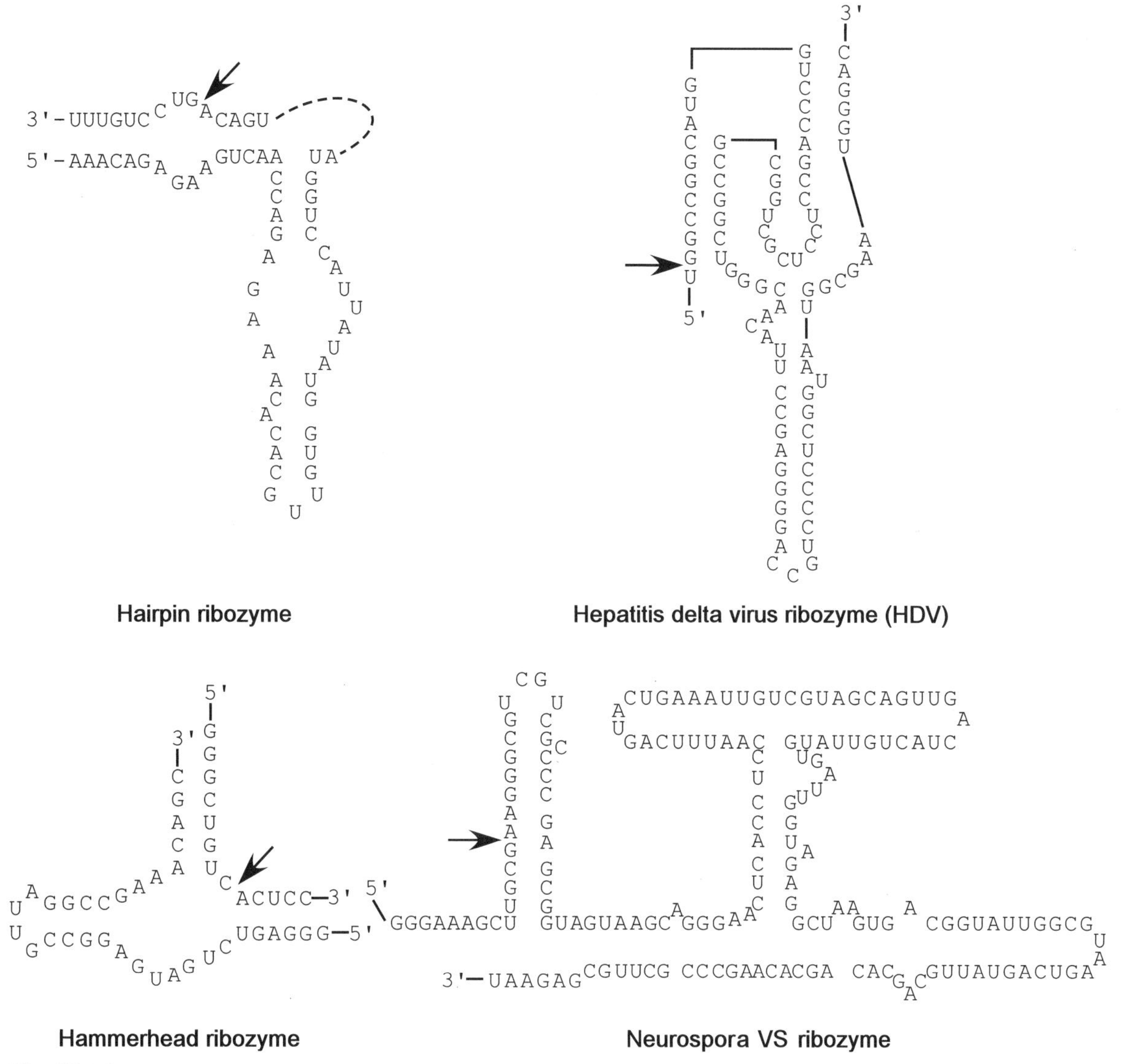

Fig. IX.5.3.
Sequences and secondary structures of the class A natural ribozymes.

IX.5.4. Metal Ions as Important Cofactors in Catalytic Nucleic Acids

In contrast to protein enzymes constructed from the 20 natural amino acids, structural repertoires of catalytic nucleic acids are limited because of the availability of only four different nucleotide building blocks for either DNA or RNA. The lack of a 2′-OH functional group in catalytic DNA in comparison to catalytic RNA makes catalytic DNA even more limited. In biology, one common solution to the lack of efficiency and diversity is the employment of cofactors, such as NADH, porphyrins, and especially metal ions. Protein enzymes are known to recruit these cofactors to broaden the scope of reactions, to increase catalytic efficiency, and to fine-tune the

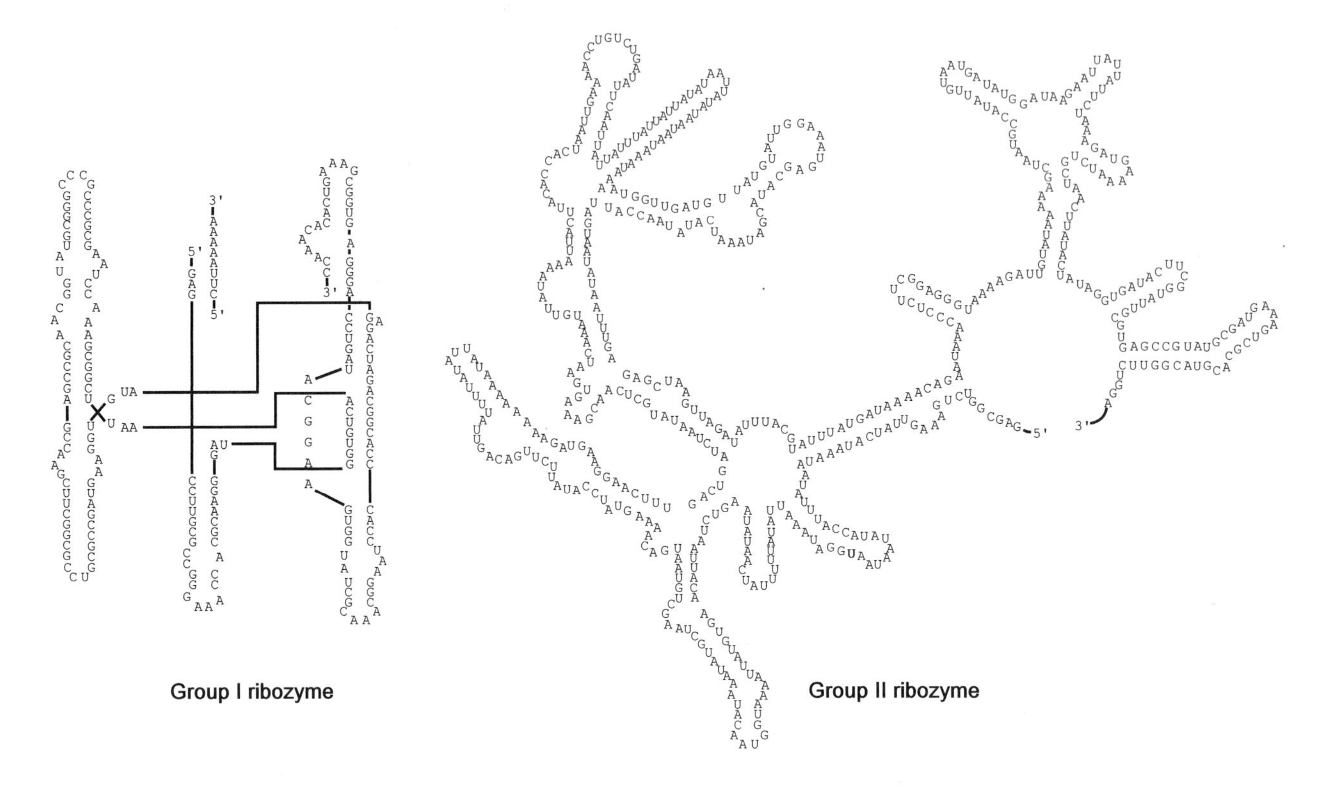

Fig. IX.5.4.
Sequences and secondary structures of the class B natural ribozymes.

reactivity for difficult reactions (see Chapter III).[35–38] The need for cofactors is even greater for catalytic DNA molecules, which lack the 2′-OH group. Indeed, despite the recent reports that some catalytic nucleic acids are active in the presence of monovalent ions[39] or amino acid cofactors,[40] divalent metal ions such as Mg^{2+}, Mn^{2+}, or Ca^{2+}, are essential for the catalytic function of the majority of catalytic nucleic acids under physiological conditions (Table IX.5.2).[7–10,41–43]

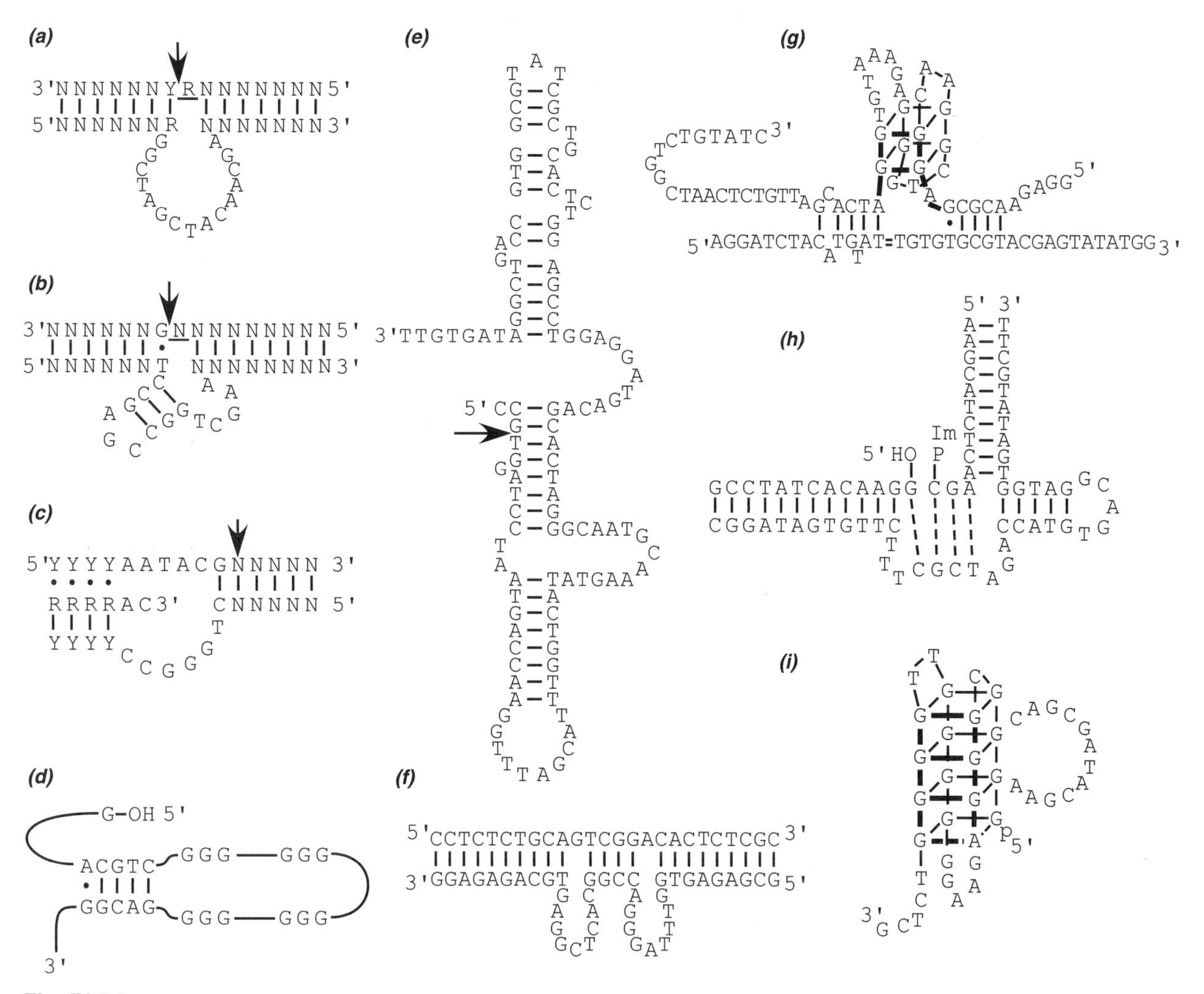

Fig. IX.5.5.
Examples of catalytic DNA molecules. (*a*) "10–23" deoxyribozyme with RNA nuclease activity;[23] (*b*) "8–17" deoxyribozyme with RNA nuclease activity;[23,28] (*c*) A catalytic DNA with DNA nuclease activity;[25] (*d*) A catalytic DNA with 3′-5′-phosphoramidate bond cleavage activity;[26] (*e*) A catalytic DNA with kinase activity;[27] (*f*) A catalytic DNA with RNA nuclease activity producing altered regio- or enantioselectivity;[30] (*g*) A catalytic DNA with repairing of thymine dimers in DNA;[31] (*h*) A catalytic DNA with ligase activity;[24] (*i*) A catalytic DNA with DNA capping activity.[29]

IX.5.5. Interactions between Metal Ions and Catalytic Nucleic Acids

The interactions between metal ions and nucleic acids have been classified into three types:[10,44] diffuse binding, site-bound outer-sphere binding, and site-bound inner-sphere binding. In diffuse binding, a mobile charged layer of hydrated metal ions is associated with the catalytic nucleic acids. The positively charged layer provides charge screening to overcome electrostatic repulsion of the negatively charged nucleic acid backbone consisting of phosphodiesters. This type of binding can play an important role in the folding of catalytic nucleic acids into complex globular shapes, like proteins. In site-bound outer-sphere binding, metal ions bind to specific sites of nucleic acids without forming direct contacts with the nucleic acid functional groups. They interact through the metal-bound water and the interaction can often be

Table IX.5.2.
Metal Ion Specificity of Various Catalytic Nucleic Acids[a]

Catalytic Nucleic Acids	Functional	Nonfunctional
Hammerhead[b]	Mg^{2+}, Mn^{2+}, Ca^{2+}, Cd^{2+}, Co^{2+}	Ba^{2+}, Sr^{2+}, $[Cr(NH_3)_6]^{3+}$, Pb^{2+}, Zn^{2+}, Tb^{2+}, Eu^{2+}
Hairpin[b]	All tested, including $[Cr(NH_3)_6]^{3+}$	
Hepatitis δ virus[b]	Mg^{2+}, Mn^{2+}, Ca^{2+}, Sr^{2+}	Cd^{2+}, Ba^{2+}, Co^{2+}, Pb^{2+}, Zn^{2+}
Neurospora VS[b]	Mg^{2+}, Mn^{2+}, Ca^{2+}	
RNase P[b]	Mg^{2+}, Mn^{2+}, Ca^{2+}	Sr^{2+}, Ba^{2+}, Zn^{2+}, Co^{2+}, Cu^{2+}, Fe^{2+}, Ni^{2+}
Tetrahymena Group I[b]	Mg^{2+}, Mn^{2+}	Ca^{2+}, Sr^{2+}, Ba^{2+}, Zn^{2+}, Co^{2+}, Cu^{2+}
Tetrahymena Group II[b]	Mg^{2+}	Ca^{2+}, Mn^{2+}
10–23[c]	Mg^{2+}, Mn^{2+}, Pb^{2+}, Ca^{2+}, Cd^{2+}, Sr^{2+}, Ba^{2+}	
8–17[c]	Pb^{2+}, Zn^{2+}, Co^{2+}, Mn^{2+}, Ca^{2+}, Mg^{2+}	Tb^{2+}, Eu^{2+}

[a] Adapted from Ref. (8).

[b] Ribozymes.

[c] Deoxyribozymes.

replaced by exchange-inert complexes, such as $[Co(NH_3)_3]^{3+}$. In Fig. IX.5.6, a site-bound outer-sphere binding of Mg^{2+} in an internal loop E of a 5S rRNA is shown.[45] Finally, in site-bound inner-sphere binding, at least one metal-bound water is replaced by a ligand from the nucleic acids, such as phosphoryl oxygens, N7 of purines (i.e., adenosines and guanosines), base keto groups (e.g., O6 of guanosine and O4 of uracil), or in the case of RNA, the ribose 2′-OH. For example, two Mg^{2+} ions coordinate directly to the phosphate oxygens of the A-rich bulge in the P4–P6 domain of the group I intron ribozyme (Fig. IX.5.7).[46] Generally, both outer- and inner-sphere binding can be observed, such as in the leadzyme (Fig. IX.5.8).[47,48]

IX.5.6. The Role of Metal Ions in Catalytic Nucleic Acids

As with protein enzymes, metal ions in catalytic nucleic acids can play structural or catalytic roles. Structural metal ions facilitate folding of the nucleic acids into stable tertiary structures by charge neutralization of the anionic phosphodiester backbone and promotion of secondary and tertiary interactions between different parts of the molecule via site-bound interactions.[49] Catalytic metal ions participate directly in the chemical reaction (either in ground or transition state) in the following ways (Fig. IX.5.2):[7]

M1: A metal ion may coordinate directly to the attacking nucleophile (internal 2′-OH for class A catalytic nucleic acids or the external nucleophile for class B catalytic nucleic acids) to increase its nucleophilicity.

M2: A metal-coordinated hydroxyl group may act as a general base by deprotonating the attacking nucleophile and thereby increasing its nucleophilicity.

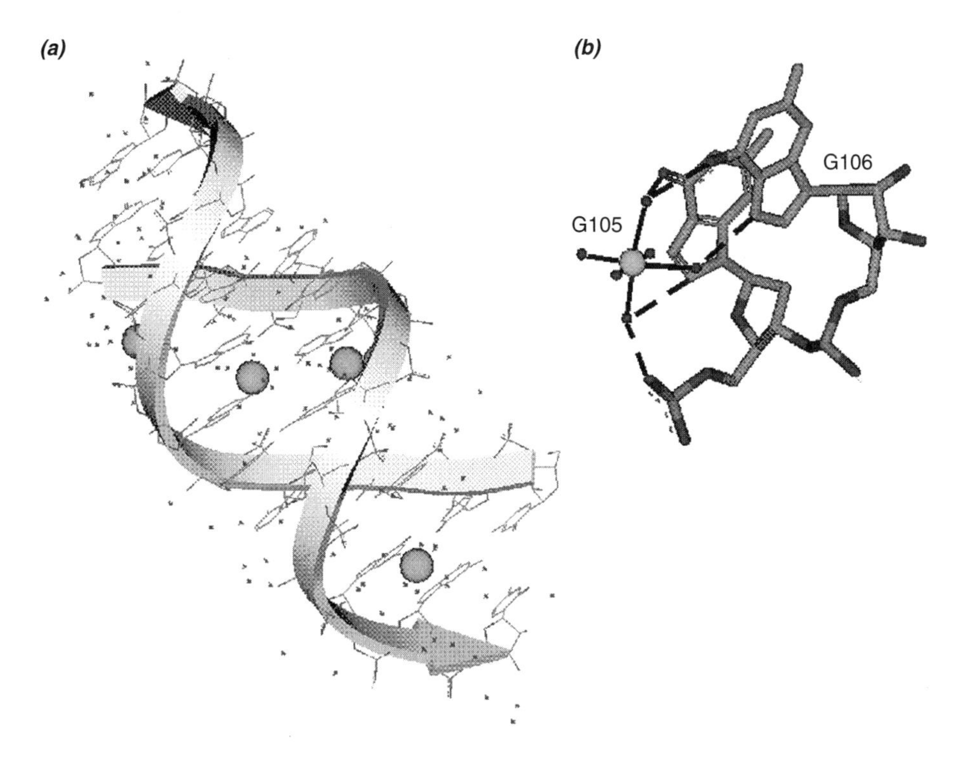

Fig. IX.5.6.
(*a*) The crystal structure of an internal loop E of a 5S rRNA with Mg^{2+} (PDB code: 354D).[45] (*b*) Outer-sphere binding of Mg^{2+} (large sphere) involving six water molecules (small spheres) observed in the crystal structure. [See color insert.]

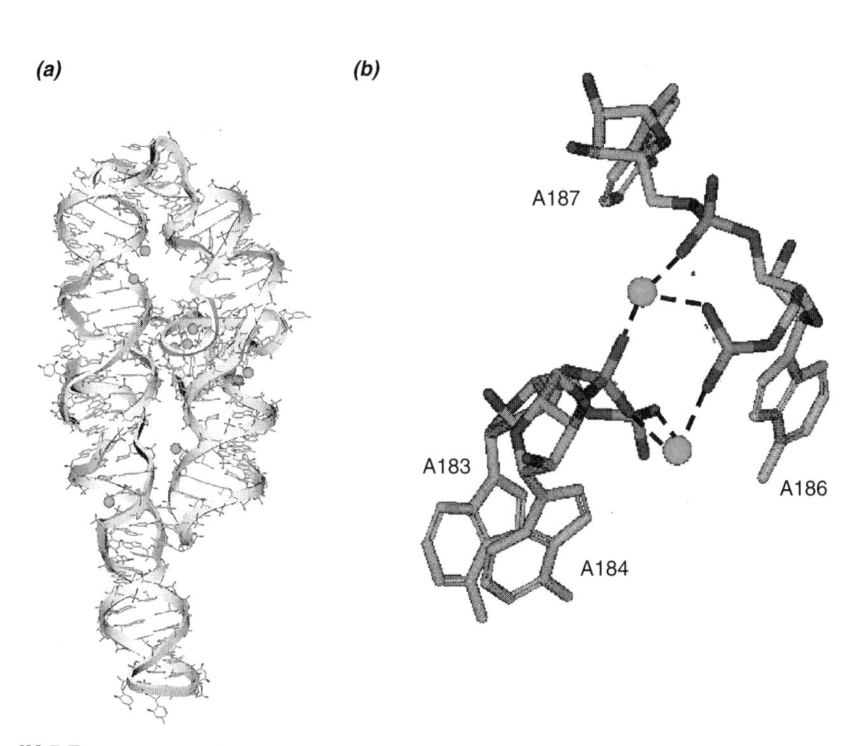

Fig. IX.5.7.
(*a*) Crystal structure of the P4–P6 domain of the group I ribozyme with Mg^{2+} (PDB code: 1GID).[46] (*b*) Structure of the A-rich bulge with two Mg^{2+} ions (large spheres) that coordinate to the phosphate of the A-rich bulge of the P4–P6 domain of the group I ribozyme. Both Mg^{2+} ions bind directly to the phosphates (inner-sphere binding). The average phosphate-oxygen-Mg^{2+} distance is 2.2 Å. [See color insert.]

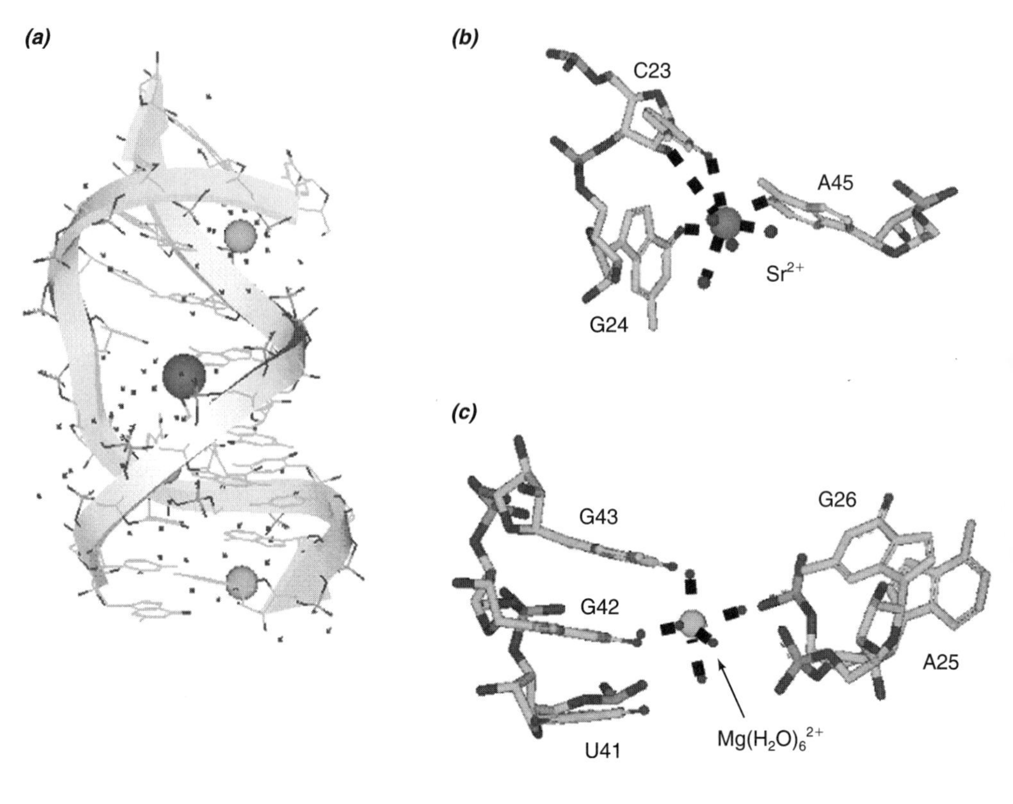

Fig. IX.5.8.
Crystal structure of the leadzyme ribozyme with Mg^{2+} and Sr^{2+} (*a*) (PDB code: 1NUV).[48] Inner-sphere binding of Sr^{2+} (*b*). Outer-sphere binding of Mg^{2+} (*c*). [See color insert.]

Fig. IX.5.9.
A proposed transition state for phosphodiester cleavage at the active site of the Tetrahymena ribozyme. Three metal ions (M_A, M_B and M_C) have been identified that participate in the catalysis. [Adapted from Ref. 51.]

M3: A metal ion may coordinate to the nonbridging phosphodiester oxygen to either render the phosphorus center more susceptible to nucleophilic attack, or help stabilize the negative charge of the oxyanion in the proposed trigonal-bipyramidal transition state.

M4: A metal ion may coordinate to the phosphodiester oxygen of the leaving group and stabilize the negative charge in the transition state.

M5: A hydrated metal ion may act as a general acid and donate a proton to the phosphodiester oxygen of the leaving group, thus facilitating cleavage.

Even though metal ions can participate in the catalysis through the above five binding modes, catalytic nucleic acids do not necessarily require five metal ions. One metal ion may fulfill multiple tasks. For example, three metal ions have been proposed to play critical roles in RNA cleavage catalyzed by the *Tetrahymena* group I intron (Fig. IX.5.9).[50–53] The metal ion M_C in Fig. IX.5.9 coordinates directly to the attacking nucleophile, the 3′-O of the external guanine, and therefore plays the role of M1. The metal ion M_B in Fig. IX.5.9 binds to the nonbridging phosphodiester oxygen and plays the role of M3. The metal ion M_A in Fig. IX.5.9, on the other hand, binds both the nonbridging phosphodiester oxygen and the phosphodiester oxygen of the leaving group and thus plays the roles of both M3 and M4.

IX.5.7. Expanding the Repertoire of Catalytic Nucleic Acids with Transition Metal Ions

Catalytic nucleic acids found to date cannot match protein enzymes in terms of either the variety of metal ions they employ or the specificity of their metal-binding sites. For example, while protein enzymes use metal ions from almost all groups of metals, including even the second- and third-row transition metal ions such as Mo and W, catalytic nucleic acids only utilize a limited number of metal ions such as Mg^{2+}, Ca^{2+}, and Mn^{2+}. Furthermore, it is well known that protein enzymes possess remarkable metal-binding affinity and specificity, and they are commonly classified by the metal ions they specifically bind (e.g., Cu proteins or Zn proteins). In contrast, catalytic nucleic acids tend to work with several metal ions with equal efficiency.[7–10,41–43] For example, hammerhead ribozymes are known to be active not only with Mg^{2+}, but also with Mn^{2+} or Co^{2+}.[54] Furthermore, the metal-binding affinity of catalytic nucleic acids is generally much weaker than that of protein enzymes, with dissociation constants typically in the micromolar to millimolar range for catalytic nucleic acids compared to nanomolar or stronger for proteins.

By using the *in vitro* selection technique mentioned earlier, nucleic acids that are specific for a variety of metal ions have been obtained in the laboratory. For example, new Zn(II)-binding RNA have been obtained.[55–57] *In vitro* selected variants of the group I intron[58] and the RNase P ribozyme[59] have shown greatly improved activity with Ca^{2+}, which is not an active metal ion cofactor for the native ribozyme. The Mg^{2+} concentration required for optimal hammerhead ribozyme activity has been lowered using *in vitro* selection to improve the enzyme's performance under physiological conditions.[60,61] Similarly, catalytic nucleic acids that are highly specific for Pb^{2+},[15,28,32] Cu^{2+},[24,25] Zn^{2+},[62] Co^{2+},[63–65] and porphyrins[66,67] have been obtained.

IX.5.8. Application of Catalytic Nucleic Acids

In addition to their significance in the debate about the origin of life, catalytic nucleic acids show promise in biochemical, biotechnological, and pharmaceutical applications.[68] They have been used as site-specific nucleases in cleaving genomic RNA, or as antiviral pharmaceutical agents against diseases such as acquired immune deficiency syndrome (AIDS) and leukemia.[69] This application can be accomplished by design of a catalytic nucleic acid that binds specifically to the genomic or viral RNA of interest. Upon binding, a unique 3D structure would be formed (as in protein enzymes) and substrate cleavage would occur. The combination of the specificity of an antisense oligonucleotide with hydrolytic cleavage and the catalytic turnover of an enzyme makes this method an attractive approach.

Catalytic nucleic acids have also been used as computational tools or as biosensors for metal ions and other analytes.[9] In the former application, a set of logic gates capable of generating any Boolean function have been obtained based on catalytic DNA such as the "8–17" deoxyribozyme.[70] In the latter application, the catalytic nucleic acids have been designed into fluorescent or colorimetric metal sensors by attaching either fluorophores or nanoparticles to the catalytic nucleic acids that are specific for metal ions.[71,72] The presence of metal ions results in cleavage of the nucleic acids, which in turn causes either an increase of fluorescent signals of the fluorophores or a change in color of the nanoparticle aggregates.

IX.5.9. From Metalloproteins to Metallocatalytic Nucleic Acids

Through many years of research, different classes of metal-specific proteins, such as heme, copper, or zinc proteins, are now known, including a comprehensive understanding of the sequence, structural, and functional features specific to each class of metalloproteins.[35–38] Similar information about metal-specific catalytic nucleic acids is virtually unknown. However, important progress has been made recently on biochemical and spectroscopic studies of metal-binding sites in Mg^{2+} and Ca^{2+}-dependent catalytic nucleic acids.[7–10,41–43] In addition, the implementation of *in vitro* selection now allows us to obtain different catalytic nucleic acids with high specificity and affinity for a metal ion of choice and thus lays a foundation for a detailed and systematic study of metal-binding sites in catalytic nucleic acids. Further biochemical and spectroscopic studies of these and other catalytic nucleic acids[73–79] will enrich our knowledge of metal-binding sites in catalytic nucleic acids so that a similar level of understanding as in protein enzymes can be achieved, thus possibly introducing a new paradigm into chemistry and biology.[80]

References

General References

1. Eckstein, F. and Lilley, D. M. J., *Catalytic RNA*. Springer-Verlag, Berlin, Germany, 1996.
2. Krupp, G. and Gaur, R. K., *Ribozyme: Biochemistry and Biotechnology*. Eaton Publishing Co., Natick, MA, 2000.
3. Doudna, J. A. and Cech, T. R., "The chemical repertoire of natural ribozymes", *Nature (London)*, **418**, 222–228 (2002).
4. Gesteland, R. F., Cech, T. R., and Atkins, J. F., *The RNA World, 2nd ed.*, Cold Spring Harbor Laboratory Press, Cold Spring Harbor, New York, 1999.
5. Joyce, G. F., "The antiquity of RNA-based evolution", *Nature (London)*, **418**, 214–221 (2002).
6. Li, Y. and Breaker, R. R., "Deoxyribozymes: new players in the ancient game of biocatalysis", *Curr. Opin. Struct. Biol.*, **9**, 315–323 (1999).
7. Pyle, A. M., "Ribozymes: a distinct class of metalloenzymes", *Science*, **261**, 709–714 (1993).
8. Feig, A. L. and Uhlenbeck, O. C., "The role of metal ions in RNA biochemistry", in *The RNA World*, Vol. 37, *Cold Spring Harbor Monogr. Ser.*, Gesteland, R. F., Cech, T. R., and Atkins, J. F., eds., Cold Spring Harbor Laboratory Press, Cold Spring Harbor, New York, 1999, pp. 287–319.
9. Lu, Y., "New Transition Metal-Dependent DNAzymes as Efficient Endonucleases and as Selective Metal Biosensors", *Chem. Eur. J.*, **8**, 4588–4596 (2002).
10. DeRose, V. J., Burns, S., Kim, N. K., and Vogt, M., "DNA and RNA as ligands", *Comp. Coord. Chem. II*, **8**, 787–813 (2004).

Specific References

11. Kruger, K., Grabowski, P. J., Zaug, A. J., Sands, J., Gottschling, D. E., and Cech, T. R., "Self-splicing RNA: autoexcision and autocyclization of the ribosomal RNA intervening sequence of Tetrahymena", *Cell*, **31**, 147–157 (1982).
12. Guerrier-Takada, C., Gardiner, K., Marsh, T., Pace, N., and Altman, S., "The RNA moiety of ribonuclease P is the catalytic subunit of the enzyme", *Cell*, **35**, 849–857 (1983).
13. Nissen, P., Hansen, J., Ban, N., Moore, P. B., and Steitz, T. A., "The structural basis of ribosome activity in peptide bond synthesis", *Science*, **289**, 920–930 (2000).
14. Moore, P. B. and Steitz, T. A., "The involvement of RNA in ribosome function", *Nature (London)*, **418**, 229–235 (2002).
15. Breaker, R. R. and Joyce, G. F., "A DNA enzyme that cleaves RNA", *Chem. Biol.*, **1**, 223–229 (1994).

16. Chapman, K. B. and Szostak, J. W., "*In vitro* selection of catalytic RNAs", *Curr. Opin. Struct. Biol.,* **4,** 618–622 (1994).
17. Joyce, G. F., "*In vitro* evolution of nucleic acids", *Curr. Opin. Struct. Biol.,* **4,** 331–336 (1994).
18. Osborne, S. E. and Ellington, A. D., "Nucleic Acid Selection and the Challenge of Combinatorial Chemistry", *Chem. Rev.,* **97,** 349–370 (1997).
19. Breaker, R. R., "DNA aptamers and DNA enzymes", *Curr. Opin. Chem. Biol.,* **1,** 26–31 (1997).
20. Joyce, G. F., "Reactions Catalyzed by RNA and DNA Enzymes", in *The RNA World,* Vol. 37, Gesteland, R. F., Cech, T. R., and Atkins, J. F., Eds., Cold Spring Harbor Laboratory Press, Cold Spring Harbor, New York, 1999, pp. 687–689.
21. Breaker, R. R., "DNA enzymes", *Nat. Biotechnol.,* **15,** 427–431 (1997).
22. Sen, D. and Geyer, C. R., "DNA enzymes", *Curr. Opin. Chem. Biol.,* **2,** 680–687 (1998).
23. Santoro, S. W. and Joyce, G. F., "A general purpose RNA-cleaving DNA enzyme", *Proc. Natl. Acad. Sci. U. S. A.,* **94,** 4262–4266 (1997).
24. Li, J., Zheng, W., Kwon, A. H., and Lu, Y., "*In vitro* selection and characterization of a highly efficient Zn(II)-dependent RNA-cleaving deoxyribozyme", *Nucleic Acids Res.,* **28,** 481–488 (2000).
25. Carmi, N., Shultz, L. A., and Breaker, R. R., "In vitro selection of self-cleaving DNAs", *Chem. Biol.,* **3,** 1039–1046 (1996).
26. Burmeister, J., Von Kiedrowski, G., and Ellington, A. D., "Cofactor-assisted self-cleavage in DNA libraries with a 3′-5′-phosphoramidate bond", *Angew. Chem., Int. Ed. Engl.,* **36,** 1321–1324 (1997).
27. Li, Y. and Breaker, R. R., "Phosphorylating DNA with DNA", *Proc. Natl. Acad. Sci. U. S. A.,* **96,** 2746–2751 (1999).
28. Ordoukhanian, P. and Joyce, G. F., "RNA-cleaving DNA enzymes with altered regio- or enantioselectivity", *J. Am. Chem. Soc.,* **124,** 12499–12506 (2002).
29. Chinnapen, D. J. F. and Sen, D., "A deoxyribozyme that harnesses light to repair thymine dimers in DNA", *Proc. Natl. Acad. Sci. U. S. A.,* **101,** 65–69 (2004).
30. Cuenoud, B. and Szostak, J. W., "A DNA metalloenzyme with DNA ligase activity", *Nature (London),* **375,** 611–614 (1995).
31. Li, Y., Liu, Y., and Breaker, R. R., "Capping DNA with DNA", *Biochemistry,* **39,** 3106–3114 (2000).
32. Pan, T. and Uhlenbeck, O. C., "A small metalloribozyme with a two-step mechanism", *Nature (London),* **358,** 560–563 (1992).
33. Brown, A. K., Li, J., Pavot, C. M. B., and Lu, Y., "A Lead-Dependent DNAzyme with a Two-Step Mechanism", *Biochemistry,* **42,** 7152–7161 (2003).
34. Mckay, D. B. and Wedekind, J. E., "Small ribozymes", in *The RNA World,* Vol. 37, Gesteland, R. F., Cech, T. R., and Atkins, J. F., Eds., Cold Spring Harbor Laboratory Press, Cold Spring Harbor, New York, 1999, pp. 265–286.
35. Bertini, I., Gray, H. B., Lippard, S. J., and Valentine, J. S., *Bioinorganic Chemistry,* University Science Books, Sausalito, CA, 1994.
36. Lippard, S. J. and Berg, J. M., *Principles of Bioinorganic Chemistry,* University Science Books, Mill Valley, CA, 1994.
37. Cowan, J. A., *Inorganic Biochemistry, 2nd ed.,* VCH, Weinheim, Germany, 1996.
38. Holm, R. H., Kennepohl, P., and Solomon, E. I., "Structural and functional aspects of metal sites in biology", *Chem. Rev.,* **96,** 2239–2314 (1996).
39. Geyer, C. R. and Sen, D., "Evidence for the metal-cofactor independence of an RNA phosphodiester-cleaving DNA enzyme", *Chem. Biol.,* **4,** 579–593 (1997).
40. Roth, A. and Breaker, R. R., "An amino acid as a cofactor for a catalytic polynucleotide", *Proc. Natl. Acad. Sci. U. S. A.,* **95,** 6027–6031 (1998).
41. Yarus, M., "How many catalytic RNAs? Ions and the Cheshire cat conjecture", *FASEB J.,* **7,** 31–39 (1993).
42. Pan, T., Long, D. M., and Uhlenbeck, O. C., "Divalent metal ions in RNA folding and catalysis", in *The RNA World,* Gesteland, R. F. and Atkins, J. F., Eds., Cold Spring Harbor Laboratory Press, Cold Spring Harbor, New York, 1993, pp. 271–302.
43. DeRose, V. J., "Metal ion binding to catalytic RNA molecules", *Curr. Opin. Struct. Biol.,* **13,** 317–324 (2003).
44. Pyle, A. M., "Metal ions in the structure and function of RNA", *J. Biol. Inorg. Chem.,* **7,** 679–690 (2002).
45. Correll, C. C., Freeborn, B., Moore, P. B., and Steitz, T. A., "Metals, motifs, and recognition in the crystal structure of a 5S rRNA domain", *Cell,* **91,** 705–712 (1997).
46. Cate, J. H., Gooding, A. R., Podell, E., Zhou, K., Golden, B. L., Kundrot, C. E., Cech, T. R., and Doudna, J. A., "Crystal structure of a group I ribozyme domain: Principles of RNA packing", *Science,* **273,** 1678–1685 (1996).
47. Wedekind, J. E. and Mckay, D. B., "Crystal structure of a lead-dependent ribozyme revealing metal binding sites revelant to catalysis", *Nat. Struct. Biol.,* **6,** 261–268 (1999).
48. Wedekind, J. E. and Mckay, D. B., "Crystal Structure of the Leadzyme at 1.8 Å Resolution: Metal Ion Binding and the Implications for Catalytic Mechanism and Allo Site Ion Regulation", *Biochemistry,* **42,** 9554–9563 (2003).
49. Doudna, J. A. and Doherty, E. A., "Emerging themes in RNA folding", *Folding Des.,* **2,** R65–R70 (1997).
50. Piccirilli, J. A., Vyle, J. S., Caruthers, M. H., and Cech, T. R., "Metal ion catalysis in the *Tetrahymena* ribozyme reaction", *Nature (London),* **361,** 85–88 (1993).
51. Shan, S.-O., Yoshida, A., Sun, S., Piccirilli, J. A., and Herschlag, D., "Three metal ions at the active site of the *Tetrahymena* group I ribozyme", *Proc. Natl. Acad. Sci. U. S. A.,* **96,** 12299–12304 (1999).

52. Shan, S.-O., Kravchuk, A. V., Piccirilli, J. A., and Herschlag, D., "Defining the Catalytic Metal Ion Interactions in the *Tetrahymena* Ribozyme Reaction", *Biochemistry,* **40,** 5161–5171 (2001).
53. Szewczak, A. A., Kosek, A. B., Piccirilli, J. A., and Strobel, S. A., "Identification of an Active Site Ligand for a Group I Ribozyme Catalytic Metal Ion", *Biochemistry,* **41,** 2516–2525 (2002).
54. Dahm, S. C. and Uhlenbeck, O. C., "Role of divalent metal ions in the hammerhead RNA cleavage reaction", *Biochemistry,* **30,** 9464–9469 (1991).
55. Ciesiolka, J., Gorski, J., and Yarus, M., "Selection of an RNA domain that binds Zn^{2+}", *RNA,* **1,** 538–550 (1995).
56. Ciesiolka, J. and Yarus, M., "Small RNA-divalent domains", *RNA,* **2,** 785–793 (1996).
57. Kawakami, J., Imanaka, H., Yokota, Y., and Sugimoto, N., "In vitro selection of aptamers that act with Zn^{2+}", *J. Inorg. Biochem.,* **82,** 197–206 (2000).
58. Lehman, N. and Joyce, G. F., "Evolution *in vitro* of an RNA enzyme with altered metal dependence", *Nature* (*London*), **361,** 182–185 (1993).
59. Frank, D. N. and Pace, N. R., "*In vitro* selection for altered divalent metal specificity in the RNase P RNA", *Proc. Natl. Acad. Sci. U. S. A.,* **94,** 14355–14360 (1997).
60. Conaty, J., Hendry, P., and Lockett, T., "Selected classes of minimised hammerhead ribozyme have very high cleavage rates at low Mg^{2+} concentration", *Nucleic Acids Res.,* **27,** 2400–2407 (1999).
61. Zillmann, M., Limauro, S. E., and Goodchild, J., "In vitro optimization of truncated stem–loop II variants of the hammerhead ribozyme for cleavage in low concentrations of magnesium under non-turnover conditions", *RNA,* **3,** 734–747 (1997).
62. Santoro, S. W., Joyce, G. F., Sakthivel, K., Gramatikova, S., and Barbas, C. F., III., "RNA Cleavage by a DNA Enzyme with Extended Chemical Functionality", *J. Am. Chem. Soc.,* **122,** 2433–2439 (2000).
63. Seetharaman, S., Zivarts, M., Sudarsan, N., and Breaker, R. R., "Immobilized RNA switches for the analysis of complex chemical and biological mixtures", *Nat. Biotechnol.,* **19,** 336–341 (2001).
64. Bruesehoff, P. J., Li, J., Augustine, I. A. J., and Lu, Y., "Improving metal ion specificity during in vitro selection of catalytic DNA", *Combinat. Chem. High Throughput Screening,* **5,** 327–335 (2002).
65. Mei, S. H. J., Liu, Z., Brennan, J. D., and Li, Y., "An Efficient RNA-Cleaving DNA Enzyme that Synchronizes Catalysis with Fluorescence Signaling", *J. Am. Chem. Soc.,* **125,** 412–420 (2003).
66. Conn, M. M., Prudent, J. R., and Schultz, P. G., "Porphyrin Metalation Catalyzed by a Small RNA Molecule", *J. Am. Chem. Soc.,* **118,** 7012–7013 (1996).
67. Li, Y. and Sen, D., "A catalytic DNA for porphyrin metallation", *Nat. Struct. Biol.,* **3,** 743–747 (1996).
68. Sun, L. Q., Cairns, M. J., Saravolac, E. G., Baker, A., and Gerlach, W. L., "Catalytic nucleic acids: From lab to applications", *Pharmacol. Rev.,* **52,** 325–347 (2000).
69. Sullenger, B. A. and Gilboa, E., "Emerging clinical applications of RNA", *Nature* (*London*), **418,** 252–258 (2002).
70. Stojanovic, M. N., Mitchell, T. E., and Stefanovic, D., "Deoxyribozyme-Based Logic Gates", *J. Am. Chem. Soc.,* **124,** 3555–3561 (2002).
71. Li, J. and Lu, Y., "A Highly Sensitive and Selective Catalytic DNA Biosensor for Lead Ions", *J. Am. Chem. Soc.,* **122,** 10466–10467 (2000).
72. Liu, J. and Lu, Y., "A Colorimetric Lead Biosensor Using DNAzyme-Directed Assembly of Gold Nanoparticles", *J. Am. Chem. Soc.,* **125,** 6642–6643 (2003).
73. Legault, P., Farmer, B. T., Ii, Mueller, L., and Pardi, A., "Through-Bond Correlation of Adenine Protons in a 13C-Labeled Ribozyme", *J. Am. Chem. Soc.,* **116,** 2203–2204 (1994).
74. Hoogstrate, C. G., Legault, P., and Pardi, A., "NMR solution structure of the lead-dependent ribozyme: evidence for dynamics in RNA catalysis", *J. Mol. Biol.,* **284,** 337–350 (1998).
75. Cunningham, L. A., Li, J., and Lu, Y., "Spectroscopic evidence for inner-sphere coordination of metal ions to the active site of a hammerhead ribozyme", *J. Am. Chem. Soc.,* **120,** 4518–4519 (1998).
76. Morrissey, S. R., Horton, T. E., Grant, C. V., Hoogstraten, C. G., Britt, R. D., and DeRose, V. J., "Mn^{2+}-Nitrogen Interactions in RNA Probed by Electron Spin-Echo Envelope Modeulation Spectroscopy: Application to the Hammerhead Ribozyme", *J. Am. Chem. Soc.,* **121,** 9215–9218 (1999).
77. Morrissey, S. R., Horton, T. E., and DeRose, V. J., "Mn^{2+} Sites in the Hammerhead Ribozyme Investigated by EPR and Continuous-Wave Q-band ENDOR Spectroscopies", *J. Am. Chem. Soc.,* **122,** 3473–3481 (2000).
78. Maderia, M., Hunsicker, L. M., and DeRose, V. J., "Metal-Phosphate Interactions in the Hammerhead Ribozyme Observed by 31P NMR and Phosphorothioate Substitutions", *Biochemistry,* **39,** 12113–12120 (2000).
79. Hoogstraten, C. G., Grant, C. V., Horton, T. E., DeRose, V. J., and Britt, R. D., "Structural Anaysis of Metal Ion Ligation to Nucleotides and Nucleic Acids Using Pulsed EPR Spectroscopy", *J. Am. Chem. Soc.,* **124,** 834–842 (2002).
80. I wish to thank Dr. Daiske Miyoshi for help in making the figures for Section IX.5 and Ms. Debapriya Maeumdar for proofreading the manuscript.

CHAPTER X

Electron Transfer, Respiration, and Photosynthesis

X.1. Electron-Transfer Proteins

Lucia Banci
CERM and
Department of Chemistry
University of Florence
Florence, Italy

Ivano Bertini
CERM and
Department of Chemistry
University of Florence
Florence, Italy

Claudio Luchinat
CERM and Department of
Agricultural Biotechnology
University of Florence
Florence, Italy

Paola Turano
CERM and Department of
Chemistry
University of Florence
Florence, Italy

Contents

X.1.1. Introduction

Electron transfer in biology uses specialized protein systems. Efficient electron transfer has several requirements.

1. One electron at a time is normally required, which implies that the electron-transfer protein must be capable of one-electron oxidation–reduction. Many redox active organic molecules, with a few exceptions, do not meet this requirement. On the other hand, metal-containing proteins solve the problem quite naturally because redox active metals often have accessible redox states differing by only one electron.
2. Electron transfer is faster when the free energy change for the electron to be transferred from the donor to the acceptor is negative, albeit not too much so. Therefore, the reduction potential must be fine-tuned to be in between those of the donor (the electron-transfer protein must be able to accept the electron) and the acceptor, when the same protein (in an electron-transfer chain) donates and accepts the electron. Although the biologically relevant redox metals are relatively few, they are used in a large variety of biological systems because their reduction potentials can be regulated over a wide range by their coordination chemistry and by the electrostatics of the metal ion environment.

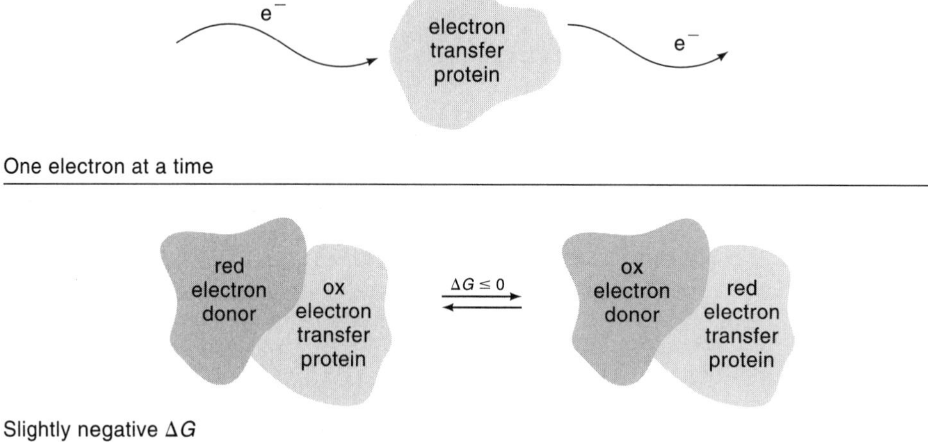

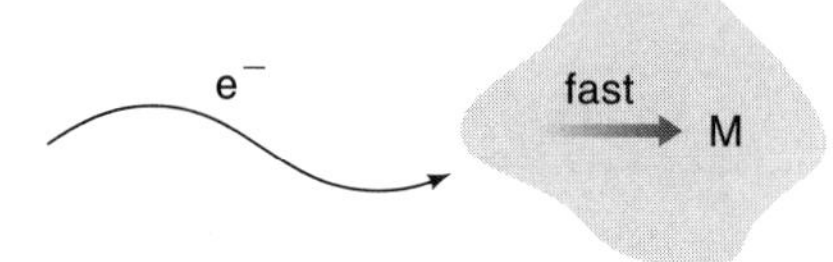

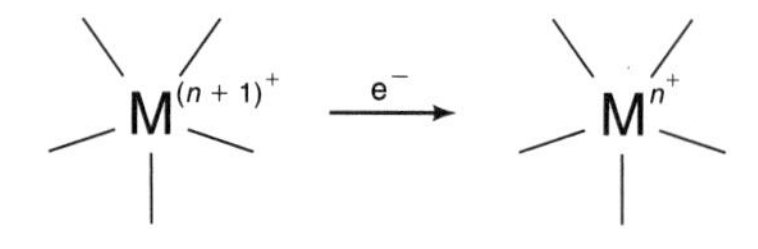

Fig. X.1.1.
Criteria for a good electron-transfer protein.

3. There must be efficient electron-transfer pathways that connect the donor–acceptor interaction site(s) with the redox site. These pathways are often shortened by the fact that the metal is either ligated by a large "conducting" cofactor like a heme or is part of a "conducting" metal cluster.
4. The reorganization energy on passing from one redox state to another should be small. For the metal, this implies minimizing the geometric changes that accompany the change in oxidation state. This minimization may be accomplished by special coordination geometries that are in between those preferred in the two oxidation states (or possibly by distributing the electron over several metal ions, as in the case of metal clusters, see Chapter IV). Rules for a good electron-transfer protein are summarized in Fig. X.1.1.

X.1.2. Determinants of Reduction Potentials

Nature has very often chosen metalloproteins for electron transfer, because the reduction potentials of metal ions are more easily tunable than those of organic compounds. The metals of choice have been iron and copper. Iron-containing electron-transfer proteins belong to two main classes: heme proteins and iron–sulfur proteins.

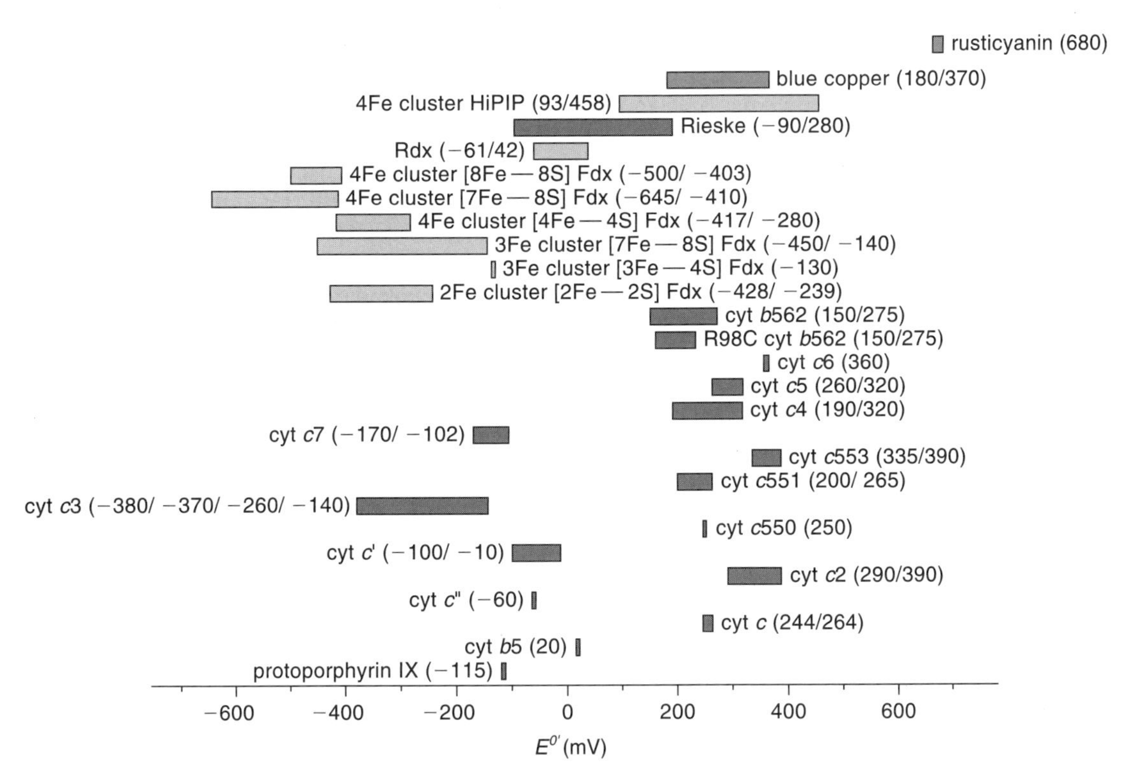

Fig. X.1.2.
Ranges of reduction potentials, $E^{\circ\prime}$ (mV), covered by the different classes of iron–sulfur proteins, heme proteins, and copper proteins. (HiPIP = High potential iron–sulfur proteins, Rdx = rubredoxins, Fdx = ferredoxins, Cyt = cytochromes.)

These classes are further subdivided according to the axial donors (hemes) and number of iron and sulfur atoms clustered together (iron–sulfur proteins). Copper-containing electron-transfer proteins belong to the class of the so-called blue copper proteins and to its variant, the purple copper proteins. The coordination properties of the metals in the various protein environments are summarized below, and are described in more detail in specific sections of this chapter (heme and iron–sulfur cofactors already have been described in Chapter IV).

The redox states of copper and iron most relevant for electron transfer are the copper(II)/copper(I) and the iron(III)/iron(II) pairs. Inorganic chemistry teaches us that these two metals have intrinsically different standard reduction potentials ($E^{\circ}_{Cu^{2+}/Cu^{+}} = 153$ mV; $E^{\circ}_{Fe^{3+}/Fe^{2+}} = 771$ mV in aqueous media). However, these potentials are for the aqua ions, and they tell us little of the redox properties of these metals when coordinated by protein ligands in a less hydrophilic environment. As an example, the reduction potential of iron from the ferric to the ferrous state in a heme protein may vary in the range of hundreds of millivolts, from positive to negative, due to a combination of factors (Fig. X.1.2). These factors include the number and type of axial ligands; accessibility to solvent; and integer and fractional charges of surrounding protein residues. The influence of each of these factors will be discussed here. In addition, more than one metal may be clustered together, and/or more than one redox center may be present in the same protein, introducing an additional way of tuning the reduction potential.

Figure X.1.2 summarizes the ranges of reduction potentials covered by the most common electron-transfer proteins. All together, the potentials span over 1 V. The ranges covered are modest for blue copper proteins, larger for cytochromes, and largest for iron–sulfur proteins, which is not surprising. The most important factor affecting the reduction potential of a given metal is its *coordination sphere,* and the blue copper proteins have the smallest variability. For cytochromes, the nature of the heme can change and, most importantly, the number and type of axial ligands is varied. Iron–sulfur proteins have the greatest variability because they can occur with different numbers of iron and sulfur ions, so that the overall charge of the cluster may change from one system to another. For Fe_4S_4 clusters, as described below, either $[Fe_4S_4]^{3+}$ or $[Fe_4S_4]^{2+}$ can act as electron acceptors, therefore providing different ranges of reduction potentials. In addition, the protein can tune the reduction potential in essentially three ways. First, the protein may create an electrostatic field around the redox center, which interacts through space with the reduction site. This electrostatic field is the combined *effect of all the fractional and unit charges of the protein atoms* around the center, and can be positive or negative at the redox site depending on whether there is an excess of positive or negative fractional charges in the neighborhood of the site. The second mode of tuning the potential involves the *overall charge of the coordination polyhedron.* A delivered electron can either decrease the positive charge of the redox center or increase its negative charge. In either case, addition of the electron can be electrostatically favored or disfavored depending on the sign of the protein electrostatic field at the reduction site. A third mode involves controlling *solvent accessibility,* so as to modulate the dielectric constant of the medium through which the electrostatic effects described above are exerted.

Quantitative integration of all of the above effects is difficult.[1–11] However, some hints are given.

X.1.2.1. The Coordination Sphere

The nature of the ligand donors and the geometry of the coordination polyhedron determine the energy of the *d* orbital that accepts or donates an electron. For the uptake of an electron, the energy of the orbital that hosts the incoming electron plays a fundamental role. This energy can well vary by thousands of reciprocal centimeters (cm^{-1}), that is, on the order of 1 eV (1 eV = 8065 cm^{-1}). The first determinant of the reduction potential is the choice of the metal ion, its ligands, and their geometry. A good example is provided by the rubredoxins (Rdx in Fig. X.1.2). Many representatives of this class of proteins are now known, all possessing the same protein fold, that is, all proteins in this class have a similar or almost superimposable backbone arrangement, although their sequence similarity may not be high (see Section X.1.3). In these proteins, a single iron ion is coordinated by four cysteine sulfur atoms, provided by two different loops, in a pseudo-tetrahedral arrangement. The reduction potentials of all members of the class span a relatively narrow range of $\sim$100 mV around zero. The value is typical of this type of protein construct, which enforces a well-defined coordination sphere. Bacteria use the rubredoxin construct in many instances where this particular reduction potential is needed. Such reasoning can be extended (with some caveats, Section X.1.2.6) to the blue copper proteins and to other systems.

X.1.2.2. Solvent Accessibility

Ignoring for the moment other energetic considerations, addition of one electron to a metal center can be favored or disfavored by electrostatics, depending on whether the metal complex that receives the electron has a positive or a negative charge. In general, reduction is favored when addition of the electron reduces or abolishes a positive charge, and disfavored when it creates or increases a negative charge. An excess of charge can be tolerated better when the dielectric of the medium is high. The dielectric constant of water is 80, whereas the dielectric constant of the inner parts of a protein is much smaller (reasonable estimates are from 2 to 8). (Strictly speaking, one can only use the concept of "dielectric" for a continuum, whereas a protein cannot be considered a continuum. Even the water molecules that hydrate a protein cannot be considered as part of the continuum constituted by the bulk water. Nevertheless, the concept of dielectric in a loose sense is useful, and helps us to understand the factors influencing reduction potentials.) Therefore, excess charge in a protein is not easily dissipated when the solvent is far, whereas it is more easily dissipated when the solvent is close, that is, when the metal site is solvent accessible. *An increase in hydrophilicity makes it easier to build up charge (either positive or negative) at the metal site.* As a result, all other things being equal, an increase in solvent exposure *increases* the reduction potential when the metal center becomes more negatively charged upon reduction, but *decreases* the reduction potential when the metal center becomes less positively charged upon reduction.

There are good examples of changes of reduction potential effected by changing the hydrophilicity of the metal environment. One is rubredoxin: In this protein, the total charge of the metal site is -1 in the oxidized state and becomes -2 in the reduced state (the four cysteine ligands are deprotonated when coordinated to the iron). It has been proposed that reduction is favored by a movement of a gate residue (Leu 41), which opens up upon reduction and increases the exposure of the metal to the solvent with respect to the oxidized form.[12] Other examples will be discussed below.

X.1.2.3. The Effect of Overall Charge of the Metal Coordination Polyhedron in the Presence of More Metals

The same local peptide fold around the metal site in rubredoxin, with the important substitution of two histidines for two cysteines, allows two Fe ions to be coordinated instead of one (see Section X.1.3). This coordination is achieved by involving two sulfide ions which, together with the two iron ions, form the so-called diamond structure (see also Chapter IV). The resulting proteins, called Rieske proteins, span a reduction potential range that is larger than that of rubredoxins and is on the more positive side (Fig. X.1.2). On the other hand, two-iron two-sulfur ferredoxins, where the diamond is coordinated to four cysteines arranged in a different protein fold (see Section X.1.3), also span a larger reduction potential range, but centered around a much more negative potential (Fig. X.1.2). In both cases, the physiologically relevant redox states of the two Fe ions in the diamond are the ferric–ferric and the ferric–ferrous states (i.e., the diamond cycles between the $[Fe_2S_2]^{2+}$ and the $[Fe_2S_2]^{+}$ states). Again, the difference arises from the different total charge of the metal center, which holds two cysteinates and two neutral histidines in the Rieske case and four cysteinates in the two-iron, two-sulfur ferredoxins. The overall charges of the chromophores pass, upon reduction, from 0 to -1 in the Rieske case and from -2 to -3 in the ferredoxin case. All other things being equal, one expects a much more negative reduction potential for the ferredoxins, as observed. For the rubredoxins discussed above, the charge passes from -1 to -2 upon reduction, and indeed the potential is intermediate. This reasoning can be extended to clusters where three or four Fe ions are involved. There are ferredoxins containing three or four metals, arranged in $[Fe_3S_4]^{0/+}$ or $[Fe_4S_4]^{+/2+}$ clusters, with three or four cysteine ligands, respectively (see Chapter IV). In both cases, the overall charges pass, upon reduction, from -2 to -3, and the ranges of potentials are on the negative side (Fig. X.1.2), although less negative for the three-metal cluster because of the different distribution of the charge between the metal atoms and the sulfur atoms.

In all the iron–sulfur proteins discussed above, the metal centers are coordinated by relatively external loops of the protein, and the solvent exposure of the metal center is, by and large, similar. It is interesting to note that, within a given cluster, different solvent exposure (see preceding paragraph) dictates which of the iron ions is more reducible (i.e., which side of the cluster can tolerate a buildup of negative charge). Thus, in Fe_2S_2 ferredoxins, the more reducible iron is the one that is more exposed to the solvent.[4,13] In clusters with >2 iron ions, partial valence delocalization over *pairs* of irons is observed (see Chapter IV), so we can only refer to more reducible iron pairs. Even there, solvent exposure seems to be relevant in determining the location of the more reducible pair.[4]

Among the parameters affecting the orbital energies in polymetallic centers, magnetic coupling may also be relevant. Although no studies have addressed this point, magnetic coupling affects the orbital energies to the extent of hundreds and sometimes thousands of reciprocal centimeters, and therefore can be another tool to tune the reduction potentials.

X.1.2.4. The Effect of Fractional Protein Charges

A positive charge nearby a metal center increases its reduction potential, and a negative charge decreases it (Fig. X.1.3*a*). Proteins contain charged groups, but these are almost always located at the surface of the molecule. Broadly speaking, proteins fold by maximizing intramolecular hydrophobic interactions and, at the same time, by maximizing hydrophilic solute–solvent interactions. As a rule, hydrophobic side chains are often found in the interior of the protein, and hydrophilic and charged side chains are at the surface. On the other hand, polar groups bearing fractional charges do appear in the interior of proteins. The most important are the CO–NH peptide groups, which can be viewed as electric dipoles with the positive end on the amide hydrogen and the negative end on the carbonyl oxygen (Fig. X.1.3*b*). Although their fractional charge is not very large, when they are in the interior of the protein these dipoles are immersed in a low dielectric constant medium, and exert a sizable electrostatic effect on the nearby metal center. The peptide bond dipoles have been shown to stabilize positive or negative charges, such as those contained in metal centers, according to whether they face the center with their negative or positive end.[1] In particular, the negative charge of the clusters in ferredoxins is stabilized by

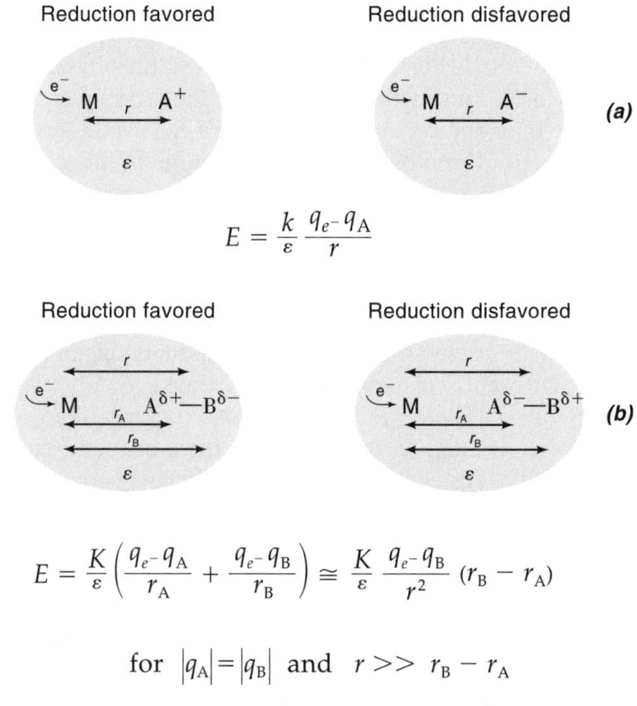

Fig. X.1.3.
Electrostatic effects of charges (*a*) and dipoles (*b*) interacting with a metal center.

a relatively large number of CO–NH dipoles oriented with their positive end toward the cluster.

The importance of the combined effect of solvent exposure and protein dipoles is illustrated by the different redox behavior of 4Fe4S ferredoxins and the so-called high-potential iron proteins (HiPIPs). In both cases, a cubane iron–sulfur cluster coordinated by four cysteines is present (see Chapter IV). However, while in ferredoxins the combined protein–solvent environment is capable of stabilizing the $[Fe_4S_4]^{2+/+}$ cluster charges (i.e., overall charges of $-2/-3$), in HiPIPs the stabilized cluster charges are the $[Fe_4S_4]^{3+/2+}$ (i.e., overall charges of $-1/-2$). This difference is due to a protein environment in HiPIPs that is (1) much more shielded from solvent and (2) much less rich in CO–NH dipoles oriented with their positive end toward the cluster.[1,14] As a result, HiPIPs span a range of potentials that is on the positive side and is, overall, the highest among all iron–sulfur proteins (Fig. X.1.2). It has been shown that "superreduction" of HiPIPs can be achieved *in vitro,* and the reduction potential is a few hundreds of millivolts more negative than that of 4Fe4S ferredoxins with which it is directly comparable.[15] It has been estimated that the two redox steps of 4Fe4S clusters are separated by $\sim 1.2\,\text{V}$.[2] A single CO–NH dipole facing a 4Fe4S cluster with its positive end, and being hydrogen bonded to a coordinated S atom, contributes $\sim 100\,\text{mV}$ to the reduction potential of the cluster. This effect has been illustrated by using a proline mutation, which removes the peptide NH proton (Ref. 16 and references cited therein). It is interesting to note that changes in the donor atoms are much less relevant for the overall reduction potential, unless they alter the total charge as happens when Rieske proteins are compared with 2Fe2S ferredoxins. For example, substitution of a cysteine by a serine in an iron–sulfur cluster, as long as the Ser deprotonates upon coordination as Cys does, changes the reduction potential only modestly (a few tens of millivolts).[17] The change is in the direction of making that particular iron less reducible, that is, it stabilizes the ferric state as would be qualitatively expected from hard–soft considerations.

X.1.2.5. The Effect of Unit Charges on the Protein Surface

As discussed above, unit charges are almost always found at the surface of proteins. Unit charges arise from ionized side chains and can be either positive or negative. Positive charges may arise from lysines ($<$pH 10), arginines ($<$pH 9), or histidines ($<$pH 6). Negative charges generally arise from aspartates or glutamates ($>$pH 5). Being at the surface, these charges are more or less immersed in the high-dielectric medium provided by water (Fig. X.1.3*a*), and their effect on reduction potentials of metal centers, for a given distance, is not as strong as it could be if they were inside the protein. Nevertheless, the effect is measurable.

A good example is given by the class of HiPIPs, which share a very conserved protein fold and therefore most individual HiPIP clusters are equally shielded from solvent and interact with the same number of CO–NH dipoles. On the other hand, the sequence similarity within the HiPIP class is rather low, and many surface residues change from one member of the class to another. Several charged residues on the surface vary considerably. The overall charges of the proteins span from -14 to $+3$. These changes result in a range of reduction potentials spanning $\sim 400\,\text{mV}$ (Fig. X.1.2). There is a reasonably good correlation between total charge and reduction potential, the less positive potentials belonging to the more negatively charged proteins and the most positive potentials to the positively charged proteins.[4] It is estimated that a charged surface residue contributes between 15 and 30 mV to the reduction potential, depending on how close it is to the metal center. For heme proteins, effects of surface charges in the range of 5–15 mV[18] and 0–10 mV[5] have been reported. So, the effect of surface charges can, in general, be considered to be well understood.

X.1.2.6. Further Considerations on Heme Iron and Blue Copper Proteins

Most of the considerations above have used iron–sulfur proteins as examples. However, similar considerations extend to heme proteins and blue copper proteins. Of course, the effects will be scaled by the intrinsic difference in reduction potential between iron-tetrathiolates, iron-hemes, and blue copper complexes. Such intrinsic differences are not easily assessed, because every model suffers from some kind of bias (imperfect mimic of the coordination site, different exposure to water, etc.). Nevertheless, it is safe to say, qualitatively, that the average reduction potentials is blue copper > iron-hemes > iron-tetrathiolates.

As far as hemes are concerned, the reduction potentials are tuned mainly by the number and nature of the axial ligands and, to a minor extent, by the nature of the heme itself. In electron-transfer proteins, the heme is of the *a, b,* or *c* type (see Chapter IV) and has either two axial ligands, which can be two histidines or a histidine and a methionine, or one axial ligand, which is a His. All other factors being equal, cytochromes that have one His and one Met as axial ligands are more easily reducible than cytochromes that have two histidines as axial ligands. This is expected on chemical grounds, as Met is a softer ligand than His and favors the softer ferrous iron versus the harder ferric iron.

Another factor that may influence the reduction potential typical of heme proteins is the exposure of the propionate side chains of the heme. Besides being possible gates for the electron-transfer process (see Section X.1.7), their exposure to the solvent changes the influence of their two negative charges (propionates are negatively charged >pH 5) on the overall charge of the heme-iron center. At variance with iron–sulfur proteins, where the metal-donor atom cluster bears a negative charge, the heme would be neutral in the ferrous and positively charged in the ferric state if it were not for the two propionates, which bring the total charge to -2 or -1, respectively. The effective charge of the metal center is thus dependent on how much the propionate charge is shielded by the solvent. As a matter of fact, the propionates are always significantly exposed, so much so that it is better to neglect their charge as part of the metal site and to include it as if it were part of the charged surface residues. This effect has been emphasized by using hemes where the propionates are substituted with their methylesters.[19] The reduction potential increases by ~60 mV, as expected from the deletion of two negative surface charges relatively close to the metal (see Section X.1.2.5).

According to the above considerations, the overall charge of heme moieties is $+1/0$, that is, about two units higher than for iron–sulfur clusters, so that the ranges of potentials are more to the positive than to the negative side (Fig. X.1.2). If this interpretation is correct, an increase in solvent exposure should cause a decrease in heme-iron reduction potential, which is indeed the case. A typical example comes from the comparison of reduction potentials of cytochrome b_5 with those of the multiheme cytochromes c_3 or c_7. All of these proteins have bis(His) axial ligation and the histidines all form hydrogen bonds to backbone carbonyls. However, the solvent exposure is much larger in multi-heme systems due to the higher heme/protein ratio. These latter systems have reduction potentials ~300 mV lower than that of cyt b_5.

Analogous to what has been seen for iron–sulfur proteins, hydrogen bonds to iron ligands also modulate the metal reduction potential. In iron–sulfur proteins, these hydrogen bonds are provided by peptide NH groups. In heme proteins, it is common that the axial His is hydrogen bonded to some amino acid side chain or backbone CO through its Nδ1 hydrogen. The presence and strength of this hydrogen bond may be an important factor, especially when the hydrogen-bond acceptor is an Asp or Glu side chain. As already mentioned, negatively charged side chains are rarely found embedded in the interior of proteins. When they are, it is because they are engaged in salt bridges or hydrogen bonded to metal ligands. In such cases, their negative charge is poorly shielded from the solvent, and their effect on the reduction potential can be sizable. This is the case for some heme-containing enzymes. Substitution of a hydrogen-bonded Asp with an Ala in a heme-containing peroxidase resulted in an increase of reduction potential as large as 200 mV.[20] Finally, the effect of surface charges has also been seen in heme proteins. For example, within the three heme-cytochrome c_7 and the four-heme cytochrome c_3, the heme iron with the higher reduction potential is heme IV, which is surrounded by several Lys residues, its solvent accessibility being essentially the same as heme I, which has the expected lower reduction potential.[21,22]

As far as blue copper proteins are concerned, their behavior is somewhat similar to rubredoxins, in the sense that the protein fold hosting the copper ion (the cupredoxin fold) is rather invariant, and therefore there are no dramatic changes in peptide dipoles or solvent exposure within the class. Despite the similarity in fold, changes in coordination geometry do take place. As will be seen in more detail in Section X.1.5, the characteristic feature of all blue copper centers is the presence of a copper ion trigonally coordinated by two histidines and a cysteinate. A weakly coordinated axial Met or Glu, and, in some cases, an even weaker peptide carbonyl oxygen, complete the coordination sphere. The total charge of the metal center is therefore $+1$ for the oxidized Cu(II)-containing form and 0 for the reduced Cu(I)-containing form. Considerations of the effect of total charge on the reduction potential of the metal center suggested that blue copper proteins should have a strongly positive reduction potential. However, Cu(II) has an intrinsically lower tendency than Fe(III) to be reduced, so the nature of the metal balances the difference in charge, and only modestly more positive reduction potentials are observed for blue copper proteins. Reversing the argument, it may be concluded that Nature has chosen a less negative environment for Cu than for Fe in electron-transfer proteins to compensate for the lower reduction potential of Cu(II) to Cu(I) [cf. Fe(III) to Fe(II)]. The differences in reduction potential among the different blue copper proteins (Fig. X.1.2) are not so much ascribed to methionine–glutamine substitution, but rather to modest changes in solvent accessibility. Indeed, rusticyanin, which has by far the highest reduction potential, is also the protein where copper is more shielded from the solvent.[23]

Finally, we consider the purple copper centers where, as will be described in more detail in Section X.1.5, a diamond of two Cu ions bridged by two Cys sulfurs is present. The purple copper site has been shown to originate from the same cupredoxin fold as blue copper proteins by the presence of an extra Cys. The trigonal coordination sphere of the two coppers is completed by one His for each of them. A weakly bound Met on one Cu site and a weakly bound peptide carbonyl on the other site provide a fourth axial ligand to each Cu. Purple copper sites cycle between an oxidized state consisting of a Cu(II) and a Cu(I) ion, which constitutes a mixed-valence pair, and a reduced state consisting of two Cu(I) ions. The total charge of the site therefore passes from $+1$ to 0 and, accordingly, the reduction potential is similar to that of blue copper proteins.

X.1.2.7. The Effect of the Presence of More Than One Redox Center

In the following sections, many examples of proteins containing more than one redox center will be encountered both in the iron–sulfur and in the heme–protein classes. The presence of more than one redox center provides Nature with a further tool to modulate the reduction potential, and, of course, allows the protein to exchange more than one electron if needed. Generally speaking, the effect of the presence of two centers of the same type on the reduction potential may be easily predicted. If the centers bear, for example, a negative charge, each center will generally have the effect of a buried charge in the protein, and will decrease the reduction potential of the other. Once one of the two centers is reduced, its negative charge increases by one unit, and the reduction potential of the second is further decreased. This effect is commonly observed in ferredoxins containing two 4Fe4S clusters (see Section X.1.3). The effect is well known, and should not be mistaken for an intrinsic strong difference in the (microscopic) reduction potentials of the two clusters. For several proteins of this class, it could be shown that the microscopic reduction potentials are actually equal within a few tens of millivolts, whereas the two reduction steps differ by a few hundred millivolts.[24,25] In other words, addition of the first electron produces a protein wherein one of the two clusters is reduced, and there is a ~50:50 mixture of molecules having one or the other cluster in the reduced state. There is a dynamic equilibrium, with the intramolecular electron transfer being at least of the order of microseconds.[26] The situation is different when the two clusters are different, as it is in the case of the ferredoxins containing one 3Fe4S and one 4Fe4S cluster. In this case, the three-iron cluster is intrinsically more reducible and is reduced first.[27] The reduction step corresponding to the 4Fe4S cluster is more negative than it is in the second step for the 8Fe8S ferredoxins. The intramolecular electrostatic interaction, of course, depends on distance, but, given the low dielectric of proteins it is always non-negligible, even in large proteins when the centers are relatively far apart.

Other good examples of proteins with multiple redox centers are certain cytochromes containing either three or four heme groups. However, the overall absolute charge of the heme redox center is smaller than that of iron–sulfur proteins (as, to a first approximation, the propionate side chains are not counted). Therefore, the electrostatic interactions among centers are more subtle to interpret. Complete studies on the microscopic and macroscopic reduction potentials of multi-heme cytochromes are available.[21,28,29]

A summary of the electrostatic effects discussed in this section is provided in Table X.1.1.

Table X.1.1.
Electrostatic Effects in Electron-Transfer Proteins

Coordination sphere	The largest determinant is the metal ion. Then the nature and geometry of ligands and the charge of the coordination polyhedron. In polymetallic centers, metal–metal interactions.
Solvent accessibility	Affects the dielectric constant, and therefore the energy of electron uptake, up to more than an order of magnitude.
Partial charges and CONH dipoles	Partial charges of atoms affect the overall electrostatic field around the metal center. Sometimes atoms with opposite partial charges are conveniently grouped into clusters of atoms that constitute dipoles (e.g., CONH groups).
Charges inside the protein	Charges inside the proteins have an important role as the dielectric constant inside the protein is small (2–8). Typical examples are given by a negative carboxylate hydrogen-bonded to the proximal His in heme proteins.
Surface charges	Solvation (in high dielectric media) reduces the effects of these charges and they are meaningful only when all other parameters are essentially constant.

X.1.3. Iron–Sulfur Proteins

An important class of electron-transfer proteins utilizes iron–sulfur clusters as redox centers.[10,30,31] These clusters may be viewed as special inorganic cofactors, and as such have been described in Chapter IV. Here, we present the proteins bearing them; the presentation is limited to those proteins having solely electron-transfer function, while other iron–sulfur proteins with different functions are discussed in other chapters in this book, as summarized in Table IV.3.

Electron-transfer iron–sulfur proteins have been classified according to several criteria, that is, the type and number of clusters, the range of reduction potentials covered, or the biochemical cycles in which they are involved. Here, we give priority to the structural features, and group the proteins into classes according to the secondary and tertiary structural features of their polypeptide chains, termed types of "folds".[32] This classification still maintains a reasonable organization of the subject in terms of types of clusters and of redox properties.

All known proteins can be grouped into 10 broad classes. Out of these, iron–sulfur proteins are represented in the 5 most important, called (1) small proteins, (2) all β proteins, (3) all α proteins, (4) $\alpha + \beta$ proteins, and (5) α/β proteins, according to the absence (small proteins) or presence of secondary structure elements of prevailing α-helix or β-strand nature. Electron-transfer iron–sulfur proteins only belong to classes: (1) (rubredoxins and HiPIPs), (2) (Rieske proteins), or (4) (all ferredoxins). A more refined grouping places the proteins according to the types of folds, and into superfamilies. Within small proteins, rubredoxins constitute a fold termed "rubredoxin-like" and represent a superfamily by themselves, which is further divided into the family of rubredoxins and desulforedoxins. High potential iron–sulfur proteins constitute another superfamily, with one typical "HiPIP fold" that contains only the HiPIP family. Within all β proteins, Rieske proteins also constitute a unique fold, a superfamily, and a family. On the other hand, within $\alpha + \beta$ proteins, there is the superfamily of 2Fe–2S ferredoxins, belonging to the Beta-Grasp or ubiquitin-like fold, that comprises the families of 2Fe–2S ferredoxins and of putidaredoxin, and the superfamily of 4Fe–4S ferredoxins, belonging to the ferredoxin-like $(\beta\alpha\beta)_2$ fold, that comprises the families of the short-chain ferredoxins, the 7Fe–8S ferredoxins, the archaeal ferredoxins, and the single 4Fe–4S cluster ferredoxins.

X.1.3.1. Rubredoxins

Rubredoxins are bacterial proteins of low molecular weight (~6000 Da) containing a single Fe ion coordinated by four cysteinate sulfurs arranged in a distorted tetrahedral environment. Although rubredoxins do not contain sulfide ions, and therefore should not be considered, strictly speaking, iron–sulfur proteins, they are treated here for a number of reasons that will become apparent in the remainder of this section. As of the beginning of this millennium, there are >20 rubredoxin structures deposited in the Protein Data Bank (PDB), and they all share very similar features. An illustrative example of a rubredoxin is shown in Fig. X.1.4.[33] The rubredoxin fold is characterized by a relatively short three-stranded antiparallel β sheet and two or three short helical turns. Two strands of the β sheet come from the N-terminus and the third strand is provided by the C-terminus. The Fe ion is located close to the β sheet on the surface of the protein. Two cysteines are provided by the N-terminus, while the other two are in a hydrogen-bonded turn in the C-terminus.

Fig. X.1.4.
The X-ray crystal structure of rubredoxin from *Pyrococcus furiosus* (PDB code: 1BRF). [See color insert.]

As with all iron–sulfur proteins, rubredoxins have their own consensus sequence that recognizes the metal ion(s) and folds the protein around it in a particular active site conformation. The consensus sequence for the rubredoxin superfamily is

$$\text{Cys Pro X Cys Gly X}_n \text{ Cys (X X) Cys}$$

where the two residues in parentheses are present in the rubredoxin but not in the desulforedoxin family.

It is worth examining the fold around the metal center in some detail (see Fig. X.1.4), as it begins to instruct us as to how Nature exploits the steric requirements of polypeptides in metalloproteins to produce a large variety of patterns. The four ligands originate from two different parts of the protein. In this way, the metal also acts as a structural element. A metal bridging two cysteines from distal regions of a sequence plays a role similar to a disulfide bridge. On the other hand, the presence of two metal ligands separated by only two other residues along the chain is a common motif in metalloprotein biochemistry. As far as cysteines are concerned, a survey over all proteins of known sequence shows that cysteine residues occur close in sequence (i.e., separated by from one to four X residues) much more often than expected on purely statistical grounds.[34]

The Cys X X Cys sequence is quite common in iron–sulfur proteins, although there are no particular stereochemical requirements for the number of X residues being two. Interestingly, this backbone conformation is very similar to that of the analogous residues in the zinc-binding domains of Zn-finger proteins (see Section XIV.1.2). Even more interestingly, the whole polypeptide fold around the iron in rubredoxin is similar to that around the Fe_2S_2 cluster in Rieske proteins (see Section X.1.3.3). For example, a single point mutation in *Clostridium pasteurianum* rubredoxin converts it into a Fe_2S_2 cluster-containing protein, although the structural details of the resulting artificial chromophore are not yet clear.[35,36] Finally, several proteins that are rich in cysteines, for example, some thioneins (see Chapter V and Section VIII.4), are used by living organisms to bind a number of divalent metals for storage or detoxification. In several cases, the metal-binding stoichiometry is variable depending on the types and relative concentrations of metal ions present in the medium. All of the above considerations point to an intrinsic flexibility of cysteine-containing proteins that is exploited in the class of iron–sulfur proteins to produce a variety of structural motifs from relatively few ingredients.

As seen in Section X.1.2.1, the reduction potentials of rubredoxins are ~ 0 V and are spread within a range of ~ 100 mV, slightly more on the negative side. Little is known about the physiological electron donors and acceptors of rubredoxins. In some cases, rubredoxins are found in electron-transfer chains that also involve other iron–sulfur proteins, as is the case for rubredoxin from *Chlorobium tepidum,* which acts as the electron acceptor for pyruvate–ferredoxin oxidoreductase. Other rubredoxins are found in the electron-transfer pathway from a rubredoxin reductase to an alkane monooxygenase. These proteins and enzymes belong to a degradation pathway that some bacteria have developed to metabolize alkanes as their sole source of carbon. Similar degradation pathways occur for arenes, where the corresponding enzymes are the arene dioxygenases and the electron transfer is performed by either 2Fe–2S ferredoxins or Rieske proteins (see Section X.1.3.3).[37] These rubredoxins have possibly evolved through gene duplication, which leads to the presence of two similar Fe centers in the molecule. The biological significance of the presence of two redox centers instead of one is still unclear. However, in other cases, dimeric rubredoxins in the family of desulforedoxins have also been found, suggesting that the presence of two redox centers may be significant to the biological function. In any case, these centers are relatively far from each other, so their reciprocal influence on the reduction potentials (see Section X.1.3.1) should be modest.

X.1.3.2. High-Potential Iron–Sulfur Proteins

High-potential iron–sulfur proteins are a well-defined superfamily encountered in photosynthetic anaerobic bacteria. These proteins, with molecular weights of

~6000–10,000 Da, are probably the most typical representatives of the class of small proteins (i.e., of those proteins that have little or no secondary structure). They coordinate a 4Fe–4S cluster with the typical cubane structure also found in 4Fe–4S ferredoxins (Chapter IV). However, HiPIPs contain one less electron in both oxidized and reduced states with respect to ferredoxins, that is, they formally contain three out of four irons in the ferric state when oxidized, and two ferric ions (as in the oxidized form of ferredoxins) when reduced (see Section X.1.2.4).[38,39] The characteristic fold of this class of proteins allows the cluster to be almost completely embedded within the protein in a rather hydrophobic environment. The latter environment, as seen in Section X.1.2, favors the presence of metal centers with lower total charge.

More than 20 different HiPIPs have been described, and for several of them the three-dimensional (3D) structure is known. Although the sequence identity within the HiPIP family is low, the tertiary structures are very similar. A representative structure is shown in Fig. X.1.5. The metal cluster is very much shielded from the solvent. Indeed, this feature is essential for the existence of the protein. If the cluster becomes more accessible to the solvent when in the oxidized state, then hydrolysis occurs. On the other hand, the protein containing a reduced cluster can even be partially unfolded, without release of the cluster, and reversibly folds to the native state.[40]

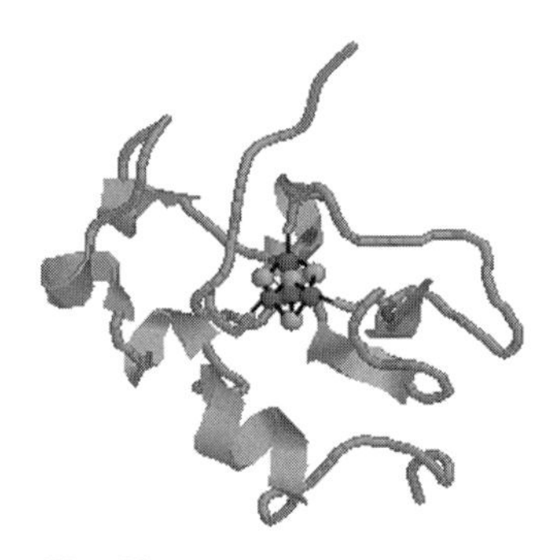

Fig. X.1.5.
The X-ray crystal structure of *Chromatium vinosum* HiPIP (PDB code: 1CKU). [See color insert.]

The structure of all HiPIPs is characterized by a series of turns and loops all around the cluster. Aromatic side chains from the C-terminus and a strictly conserved Tyr from the N-terminus form a hydrophobic pocket that further protects the cluster. The consensus sequence that binds the cluster in HiPIPs is

$$\text{Cys X X Cys } X_{8-16} \text{ Cys } X_{10-13} \text{ Gly Z Cys}$$

where Z is either a Trp or a Tyr. The entire consensus sequence is contained in the second half of the full sequence. It appears that only the first two cysteines maintain a well-defined sequential relationship with each other. Despite the variability in the loops connecting the other cysteines, the fold around the cluster is essentially identical. However, this is only one of many consensus sequences that can host a 4Fe–4S cluster. Another totally different example will be seen in ferredoxins (Section X.1.3.5), and several other examples are found in 4Fe–4S containing proteins and enzymes that do not have electron-transfer function (e.g., aconitase; see Section IX.4).

Due to the peculiar electrostatic properties of the protein environment (discussed in Section X.1.2) the reduction potentials of HiPIPs are all positive and range between ~100 and 500 mV. Most HiPIPs known to date are found in purple photosynthetic anaerobic bacteria, such as *Chromatium, Ectothiorhodospira,* and *Rhodocyclus.* A few aerobic bacteria have also been found to contain HiPIP. In photosynthetic bacteria, HiPIPs are periplasmic proteins that act as soluble electron shuttles in the cyclic electron flow between two transmembrane complexes: the photosynthetic reaction center (RC) and the cytochrome bc_1 complex. Reduced HiPIP is the immediate electron donor to a RC-bound *c*-type tetraheme cytochrome.[41] In aerobic organisms HiPIPs seem to play an important role in the respiratory chain.[42] Other functions for HiPIPs such as: electron acceptor for a thiosulfate-oxidizing enzyme; electron donor to a nitrate reductase; Fe(II) oxidase; or electron donor to a caa_3 terminal oxidase, have also been proposed.

Both rubredoxins and HiPIPs seem to have appeared rather early in evolution, although probably not as early as some ferredoxins found in archaea (see Section X.1.3.5).[31] However, at variance with ferredoxins, which have continued to evolve and are present in humans, both rubredoxins and HiPIPs seem to be restricted to the bacterial kingdom. Their function has been replaced in "higher" eukaryotes by either more evolved forms of iron–sulfur proteins, by heme proteins, or blue copper proteins. The narrow phylogenetic range of HiPIPs may result from the proteins lack of extended elements of secondary structure, which may be advantageous as "preformed" building blocks for evolution.

X.1.3.3. Rieske Proteins

In contrast to the rubredoxins and HiPIPs, Rieske proteins occur in both prokaryotes and eukaryotes. Another mark of distinction is that the characterized systems up to now all belong to larger molecular entities, of which the Rieske center constitutes a domain. This domain is usually water soluble, so it can be considered separately. However, strictly speaking, the Rieske proteins described below are electron-transfer domains of more complex proteins rather than electron-transfer proteins by themselves. True single electron-transfer Rieske proteins do exist, although they have not been structurally characterized yet. The cluster in Rieske proteins is a Fe_2S_2 diamond anchored to the protein by two cysteines and two histidine ligands (see Chapter IV). The presence of two histidines is peculiar to Rieske proteins, and provides different redox properties with respect to the all-cysteine ferredoxins, as discussed in Section X.1.2.3.

The typical consensus sequence of Rieske proteins is as follows:

$$\text{Cys X His X}_n \text{ Cys X X His}$$

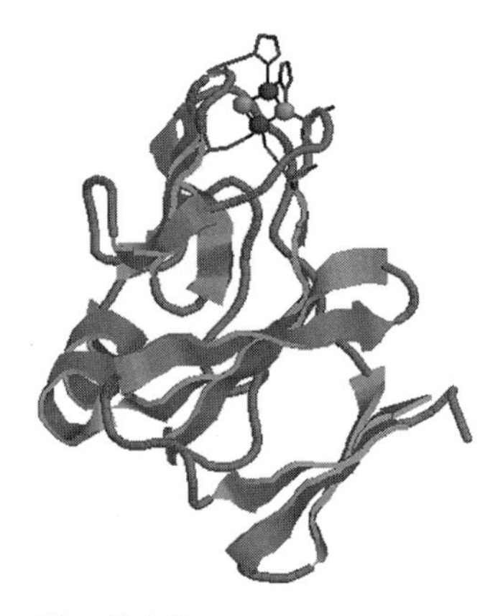

Fig. X.1.6.
The X-ray crystal structure of a water soluble fragment of the Rieske iron–sulfur protein of the bovine heart mithochondrial cytochrome bc_1 complex (PDB code: 1RIE). [See color insert.]

The consensus sequence is qualitatively different from that of the classical 2Fe–2S ferredoxins (see Section X.1.3.4), and more related to that of rubredoxins (see Section X.1.3.1). The topology can be viewed as forming a Z, where the first and third ligands (cysteines) coordinate one Fe and the second and fourth ligands (histidines) coordinate the other Fe. The structure of the Rieske subunit of the bovine heart mitochondrial bc_1 complex is shown in Fig. X.1.6. The Rieske fold consists of three antiparallel β sheets formed by a total of 10 β strands, thus justifying the classification of Rieske proteins as all β proteins. The last β sheet and its loops form the cluster-binding subdomain. The cluster itself lies near the surface on the "tip" of the protein, similarly to the iron center in rubredoxins. The iron coordinated to the two histidines is more exposed to solvent. This cluster geometry is strictly conserved in the three Rieske proteins structurally characterized to date.

The biological electron-transfer role of the structurally characterized Rieske proteins is related to their presence within more complex protein architectures.[43] Rieske domains are found in the widespread bacterial and mitochondrial bc_1 complexes, as well as in plastid b_6f complexes. The bc_1 and b_6f complexes are membrane proteins that contain four redox centers in three subunits. The subunits are a cytochrome containing two heme *b* centers (see Section X.1.4.2), a cytochrome *c* or *f*, and the Rieske domain. These complexes oxidize hydroquinones and transfer electrons to their respective acceptors, cytochrome *c* or plastocyanin. Rieske domains are also present in bacterial arene dioxygenases, and some true electron-transfer Rieske proteins may be involved in donating electrons to these enzymes.[37]

In terms of reduction potential, the three superfamilies of proteins discussed so far, that is, the rubredoxins, the HiPIPs, and the Rieske proteins range from slightly negative to strongly positive. Therefore, partial overlap exists in reduction potential with other electron-transfer proteins such as blue proteins or mono-heme cytochromes of the *c* type. It should not be surprising that: (1) redox partners exist with similar reduction potentials, such as an iron–sulfur protein and a heme protein, an iron–sulfur protein and a blue or purple copper protein, or a heme and a blue or purple copper protein (see Section X.1.5); and (2) sometimes cytochromes may replace HiPIPs, or HiPIPs may replace blue copper proteins, for the same function.

X.1.3.4. 2Fe–2S Ferredoxins

Ferredoxins that contain two Fe and two S atoms are widespread electron-transfer proteins, found in many metabolic reactions from bacteria to humans.[31] They belong to the $\alpha + \beta$ class and have molecular weights of ~8–10 kDa. More than 20 3D structures are available in the PDB, most of them of 2Fe–2S ferredoxins of plant-type. Figure X.1.7 illustrates the structural features of a representative member of the family, the ferredoxin from *Haloarcula marismortui.* The structural motif features two α helices and a four-stranded β sheet in which the first helix lies across the β sheet. The tertiary structure is stabilized by a conserved hydrophobic core. Similarly to rubredoxins and Rieske proteins, the Fe_2S_2 diamond is near the surface of the molecule in a loop between the first α-helix and the third β strand.[44] The consensus sequence is

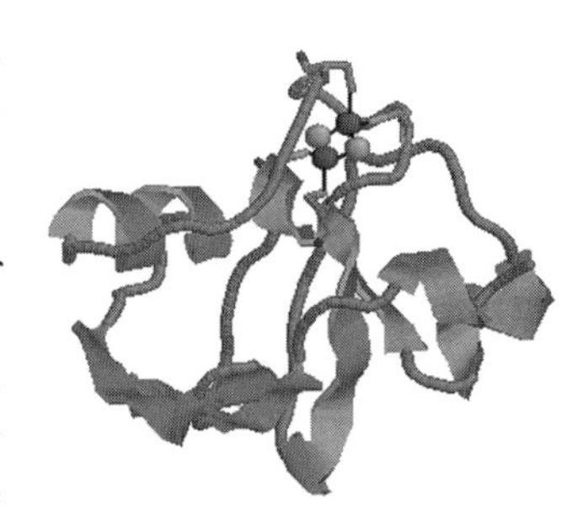

Fig. X.1.7.
The X-ray crystal structure of the oxidized 2Fe–2S ferredoxin from *Haloarcula marismortui* (PDB code: 1DOI). [See color insert.]

$$\text{Cys X}_4 \text{ Cys X X Cys X}_n \text{ Cys}$$

with the first two cysteines coordinating the first Fe ion and the last two cysteines coordinating the second Fe ion. This arrangement is thus of U-, as opposed to the Z-type arrangement found in Rieske clusters. The reducible iron, which is the one more exposed to solvent, is the one bound to the first two cysteines in the sequence.[13] Vertebrate ferredoxins, or adrenodoxins, have a similar overall fold, but are richer in secondary structure elements.

The 2Fe2S ferredoxins have large negative reduction potentials (see Section X.1.2) and are widely employed as terminal electron acceptors of photosystem I in many oxygenic photosynthetic organisms. The electron is provided by a Fe_4S_4 cluster of the PsaC subunit of photosystem I (see Section X.3.8) and transferred to ferredoxin-$NADP^+$ reductase. In addition, 2Fe2S ferredoxins act as electron transfer in many other metabolic pathways such as thioredoxin oxidoreduction; glutamate synthesis; sulfite reduction; lipid desaturation; nitrite reduction; and nitrogen fixation.[31] Finally, many redox enzymes contain domains or subunits similar to 2Fe2S ferredoxins. For example, vertebrate ferredoxins are found in mitochondrial monooxygenase systems where they transfer electrons from reduced nicotinamide adenine dinucleotide phosphate (NADPH)-ferredoxin reductase to membrane-bound cytochrome P450 enzymes and in xanthine oxidases, where they mediate electron transfer between the Moco and flavin centers (see Section XII.6).

X.1.3.5. The 4Fe–4S Ferredoxins

Ferredoxins containing 4Fe4S cubane clusters are mainly found in bacteria, from anaerobes to photosynthetic and hyperthermophilic organisms, and in archaea, even if they are also found in eukaryotes.[24,31] They are low molecular weight proteins of the $\alpha + \beta$ class. All of these proteins share a common structural motif that can accommodate three, four, seven, or eight iron ions in one or two clusters. The structural features can be discussed in relation to the 8Fe8S ferredoxins of the clostridial type, as illustrated in Fig. X.1.8. The molecule has an ellipsoidal shape characterized by a pseudo-twofold symmetric arrangement of the two cubanes and of the secondary structure elements that constitute a βαβ-βαβ motif. The first antiparallel β sheet is formed by the N- and C-termini, and the second consists of residues in between the two clusters. Short helical segments are formed by the sequential stretches between Cys III of one cluster and Cys IV of the other cluster.[45] The consensus sequences for the two clusters also reflect the pseudo-twofold symmetry:

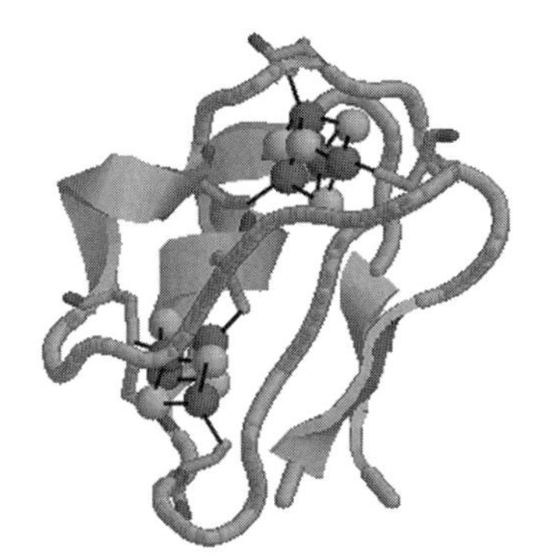

Fig. X.1.8.
The X-ray crystal structure of the 8Fe–8S *Clostridium acidiurici* ferredoxin (PDB code: 2FDN). [See color insert.]

$$\text{Cys X X Cys X X Cys X}_n \text{ Cys} \qquad \text{A}$$

$$\text{Cys X}_m \text{ Cys X X Cys X X Cys} \qquad \text{B}$$

The cluster(s) arrangements are shown schematically in Fig. X.1.9. Several variations on this general theme lead, for example, to: the deletion of cysteine II coordinating cluster A, and the consequent transformation of cluster A into a Fe_3S_4 cluster; or to the total loss of cluster B, with the formation of a mono-cubane protein; or to the loss of both cysteine II and cluster B, with the formation of a 3Fe4S protein. In some cases, additional secondary structure elements may replace the lost cluster and confer stability to the resulting protein.

Like 2Fe2S ferredoxins, cubane ferredoxins have rather negative reduction potentials and are involved in a large variety of metabolic reactions. Examples are the transfer of electrons during pyruvate oxidation in the nitrogen-fixation reaction system and to cytochrome P450. There seems to be no clear distinction between the role of two-cluster or single cluster ferredoxins, and there is no evidence of two-electron-

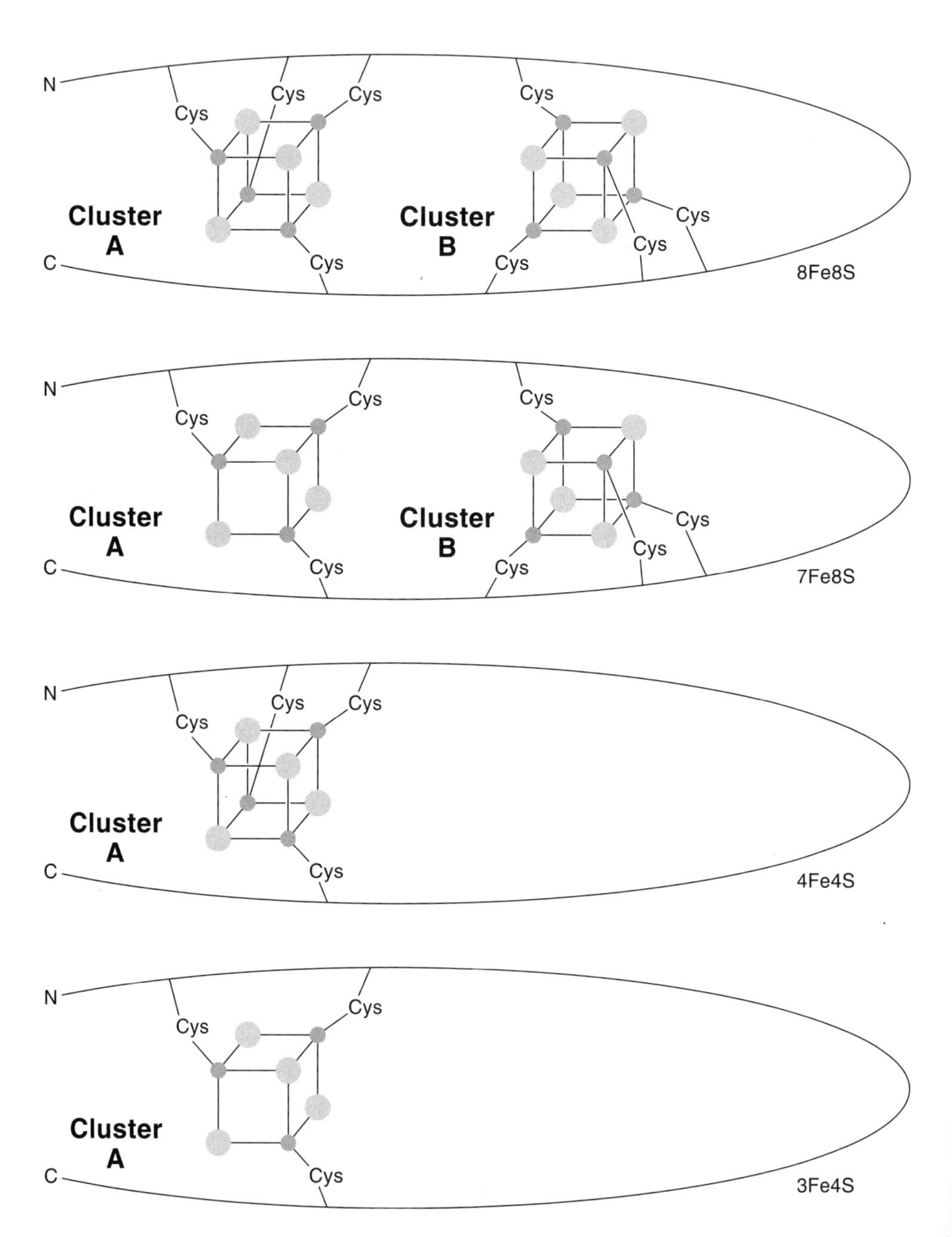

Fig. X.1.9.
Consensus sequences in ferredoxins.

transfer reactions being physiologically significant. The possibility exists that the presence of two clusters in a small molecule can help provide alternative electron-transfer pathways along the molecule. This versatility may be relevant for membrane ferredoxins such as the 8Fe8S ferredoxin PsaC, which is part of photosystem I and could accept the electron from one side of the molecule and release it from the other side taking advantage of the presence of two clusters in the electron-transfer protein, which rapidly equilibrate.

X.1.4. Cytochromes

Only hemes *a, b,* and *c* are relevant for the electron-transfer function of cytochromes.[46] Independently of the heme type, several axial ligations are possible for the heme iron in cytochromes, as reported in Table X.1.2. All the *c*-type cytochromes, however, have at least one His ligand and, with few exceptions, are characterized by the presence of the sequence motif Cys X Y Cys His, where the two Cys residues are those covalently bound to the heme and the His is the fifth axial ligand. A common feature of all electron-transfer cytochromes, except cytochrome c', is that their Fe ion is hexacoordinate. Cytochrome c' contains pentacoordinate iron and the sixth coordination position may be available for binding to small exogenous molecules like CO, NO, or CN^-. However, the affinity toward such molecules is much smaller than that of pentacoordinate heme *enzymes.* It has been proposed that cytochrome c' functions as an electron-transfer protein.

For *c*-type cytochromes, a classification into four broad classes is often used based on sequence similarities.[46] Class I includes the monoheme cytochromes with the heme-attachment site close to the N-terminus; class II constitutes the mono-heme cytochromes with heme attachment close to the C-terminus; class III comprises the multi-heme cytochromes with bis(His) axial ligation. Usually, class IV is taken as the class of multi-heme cytochromes with both bis(His) and His/Met axial ligation.

A large number of cytochromes contain a *b*-type heme. All of these contain a six-coordinate heme iron with bis(His) or His/Met axial ligation. No covalent bonds exist between the protein matrix and the heme, whose placement and conformation are only determined by hydrophobic interactions and hydrogen bonds with the surrounding residues. The lack of covalent bonds is reflected in the presence of two possible isomers of *b*-type cytochromes that differ by a 180° rotation of the heme about

Table X.1.2.
Cytochromes by Heme Iron Axial Coordination

Donor Atoms	Protein Class
Nε His/Nε His	Cytochrome *c* oxidase heme *a* (cyt *a*)
	Cytochrome b/b_6
	Cytochrome b_5
	Cytochromes *c* (class III and IV)
	Cytochrome c_{554} (hemes 3 and 4)
	Cytochrome cd_1 nitrite reductase (*c*-domain)
	Hydroxylamine oxidoreductase (heme *c*)
Nε His/Nδ His	Cytochrome c_{554} (heme 1)
Nε His/Sδ Met	Soluble cytochrome b_{562}
	Cytochromes *c* (cyt c_1, class I, class II besides c', and class IV)
Nε His/Amide N of Tyr1	Cytochrome *f*
Sδ Met/Sδ Met	Bacterioferritin (cyt b_1; cyt b_{577})

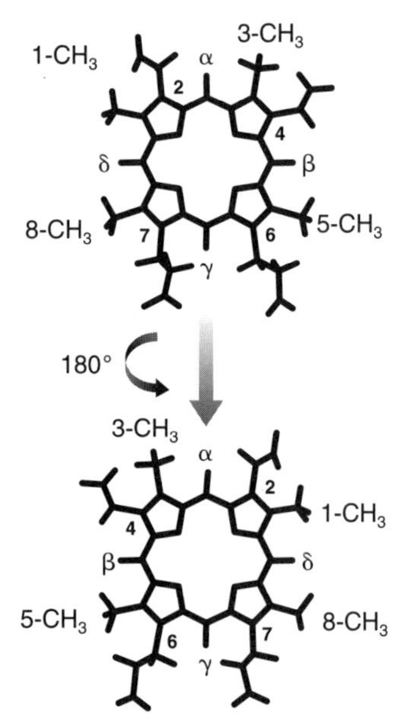

Scheme X.1.1.

an axis defined by the α-γ meso carbons (Scheme X.1.1). The relative ratio depends on the protein and, for the same proteins, on the isoenzyme.

Heme *a* is the electron-transfer heme of cytochrome *c* oxidase (Chapter XI.6); it receives an electron from the Cu_A center and transfers it to the oxygen-binding site. The heme *a* iron is six coordinate with two axial His ligands. The cytochrome *c* oxidase heme a_3 (cytochrome a_3), as well as cytochrome c_{554} (heme 2), and hydroxylamine oxidoreductase (heme P460) have the Nε of a His in the fifth coordination position, while the sixth position does not contain any protein ligand, but is accessible to exogenous molecules. Heme a_3 appears to be involved in the catalysis, in contrast to the other heme group in the same proteins, which participates in electron transfer.

Cytochromes can also be classified according to their protein fold.[47] They fall in the all α, α + β, and all β classes. The structural features of cytochromes are discussed in terms of the number of hemes and the protein fold. The "historical" classification of *c*-type cytochromes into four classes, which is based on sequence similarity, is still valid when we look at 3D structures and at folds and/or superfamilies of the fold classes.

X.1.4.1. Mono-Heme Cytochromes

Among the all α proteins, the so-called cytochrome *c* fold is characterized by a common core constituted by at least three helices, organized in a folded leaf structure where α helices are wrapped around a single hydrophobic core in such a way that the core becomes relatively solvent accessible.[48] An essential element of the hydrophobic core is the covalently bound heme. The presence of such a core and of the three helices define the cytochrome *c* superfamily. Members of this superfamily are the monodomain cytochromes *c*, which include the mitochondrial cyt *c*, cyt c_6, cyt c_2, cyt c_5, cyt c_{551}, cyt c_{555}, and cytochrome *c* domains in larger proteins. Examples of this folding type are shown in Fig. X.1.10. Soluble cytochromes *c* in this class have a heme-attachment site toward the N-terminus, and the sixth ligand is provided by a methionine residue ~40 residues further on toward the C-terminus.[49] The total number of amino acids ranges between 80 and 120. As seen in Fig. X.1.10, the "large" cytochromes in this class have an extra loop spatially located close to the heme side that bears the propionates. Due to the presence of this loop in large cytochromes, the two propionate groups are essentially buried in the protein matrix, while in small cytochromes the 6-propionate (Scheme X.1.1) is largely solvent exposed. In all cases, the two heme propionate groups occupy a local environment that is clearly hydrophilic in character and form a number of hydrogen bonds with nearby polar groups. Redox-dependent changes in the hydrogen-bond network involving propionates have been proposed for several of these proteins.

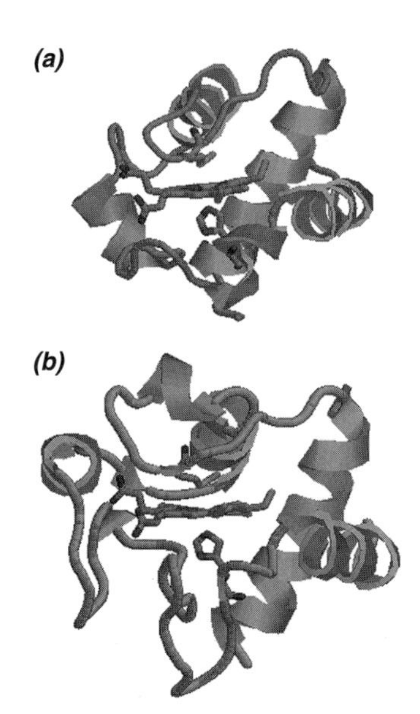

Fig. X.1.10.
The structure of cytochrome c_{553} from *Desulfovibrio vulgaris* (strain Miyazaki; constituted by 79 amino acids; PDB code: 1C53) (*a*) is compared with the structure of horse heart cytochrome *c* (containing 105 amino acids; PDB code: 1HRC) (*b*). The presence of an extra loop facing the heme propionates in the larger cytochrome is evident from this protein orientation. [See color insert.]

In all of these cytochromes, the heme group has the edge bearing the 4-vinyl substituent exposed to the solvent. The protein surface region that contains this heme edge has been found to be involved in the interaction with the redox partner cytochrome *c* peroxidase (see Section X.I.7). The heme deviates from planarity, with the two pyrrole rings involved in thioether linkages showing the greatest deviations; this suggests that covalent bond attachment of the heme to the protein chain is an important factor in establishing the heme conformation.

The presence of heme is an essential feature of cytochrome *c* folding, indicating the important structural role of the Cys X Y Cys His sequence motif. Indeed, apocytochrome *c* does not fold spontaneously. In contrast, experimental results indicate that the bond between the iron and the distal methionine is not an essential feature for the maintenance of the overall structure: methionine can be easily replaced by exogenous ligands [e.g., CN^- in the case of the Fe(III) form and CO for the Fe(II) form] without altering the protein fold; the structure of the mutant obtained by replacing methionine with alanine is essentially that of the native protein.[50]

Another type of all α protein is the four-helical up-and-down bundle fold that comprises a core of four helices oriented roughly along the same (bundle) axis leaving the hydrophobic core closed or partly solvent accessible. The helices are organized in an up-and-down structure in which consecutive helices are adjacent and antiparallel, and are arranged with respect to the bundle axis such that the structure is slightly twisted. Cytochrome b_{562}[51,52] and cytochrome c'[53–58] belong to this fold. In cytochrome b_{562} (which has 106 amino acid residues), a *b*-type heme group is inserted between the helices near one end of the molecule. The heme plane is roughly parallel to the bundle axis and one heme face is partially exposed to solvent (see Fig. X.1.11*a*). Histidine on helix 4 and methionine on helix 1 provide the fifth and sixth heme ligands. The heme prosthetic group can be readily removed and the resulting apoprotein has a topology similar to the holoprotein (Fig. X.1.11*b*).[59] However, while helices II and III are essentially preserved, helix I is less defined and only the first residues constituting helix IV in the holoprotein maintain a helical conformation in the apocytochrome. Cytochrome c' usually exists as a dimer, with ~110–130 amino acids per monomer. The four-helix bundle incorporates a covalently bound pentacoordinate heme group. The histidine axial ligand and the heme-bound cysteines are located close to the C-terminus. Other cytochromes sharing this fold are the so-called class II cytochrome *c*, which are six-coordinate cytochromes, for example, cyt *c*556 (120–130 amino acids), whose heme attachment site is close to the C-terminus and has His/Met axial ligation.

(a)

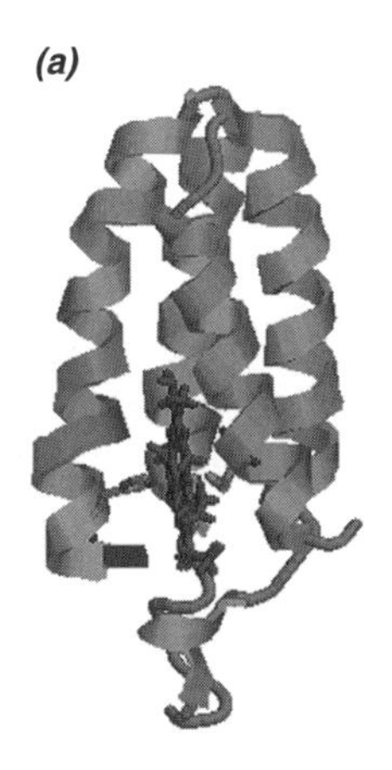

(b)

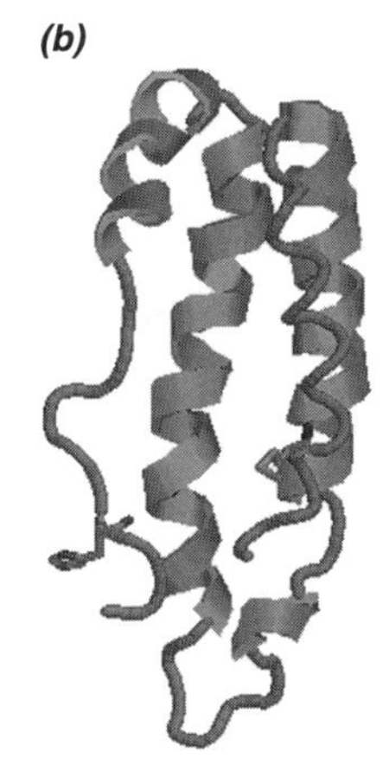

Fig. X.1.11.
The solution structures of the (*a*) holo- and (*b*) apo-form of cytochrome b_{562} from *E. coli* (PDB codes: 1QPU and 1APC, respectively). [See color insert.]

The α + β class contains mainly antiparallel β-sheets with segregated α and β regions. The cytochrome b_5 fold belongs to this class, which is formed by small proteins binding a heme group.[60,61] In cytochrome b_5, two hydrophobic cores are present on each side of a β-sheet. The larger hydrophobic core constitutes the heme-binding pocket, closed off on each side by a pair of helices connected by a turn. The smaller hydrophobic core may have only a structural role and is formed by spatially close N-terminal and C-terminal segments. Histidines provide the fifth and sixth heme ligands and the propionate edge of the heme group lies at the opening of the heme crevice (Fig. X.1.12). The microsomal and mitochondrial cytochromes b_5 are membrane bound, while those from erythrocytes and other animal tissues are soluble. In the membrane-bound cytochrome b_5, the 40 residues that bind the protein to the membrane may be cleaved by proteolysis from the C-terminus. The soluble fragment, containing ~95 amino acids, retains the heme and its biological function. The solution structure of the apo-form of the soluble fragment has also been determined. It contains most of the original secondary structure, and the smaller core is still well organized; the rest of the protein can be viewed as a partly folded, fluctuating ensemble of residues that make up the larger part of the heme pocket.[62]

From the above description it is clear that, in contrast with cytochrome *c*, the hydrophobic core of *b*-type cytochromes represents a very stable structural feature that does not require the presence of heme to give rise to the proper fold. The family of b_5-like cytochromes comprises, besides cytochrome b_5 itself, heme protein domains covalently associated with other redox centers as in flavocytochrome b_2, sulfite oxidase, and assimilatory nitrate reductase.

Cytochrome *f*, which contains a *c*-type heme, belongs to the all β class and includes different structural domains.[63] The largest domain lies on top of the membrane-bound domain and has a sandwich fold made of nine strands organized in two β-sheets with some interstrand connections between β-sheets, which is typical of an immunoglobin-like fold. This domain also contains a short heme-binding peptide. The small domain is inserted between two β strands of the large domain and contains a sandwich of half-barrel shaped β-sheets. The heme nestles between two

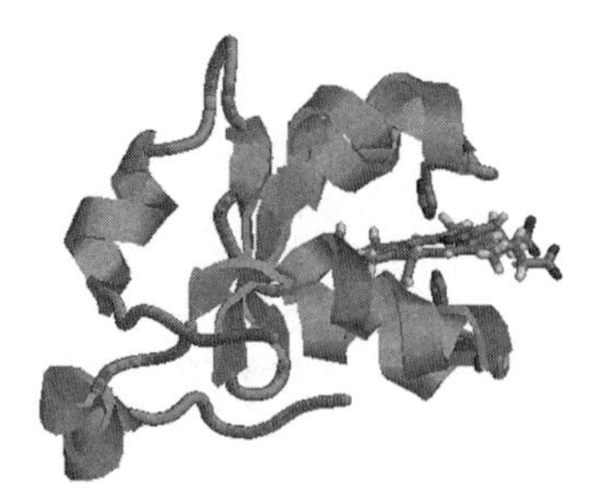

Fig. X.1.12.
The solution structure of the soluble fragment of oxidized rat microsomal cytochrome b_5 (PDB code: 1BFX). [See color insert.]

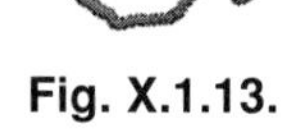

Fig. X.1.13.
The X-ray crystal structure of the lumen-side domain of reduced cytochrome *f* (PDB code: 1HCZ). [See color insert.]

short helices at the N-terminus of cytochrome *f* (Fig. X.1.13). The second helix contains the sequence motif for the *c*-type cytochromes, C-X-Y-C-H, which spans residue 21 to 25. The unusual heme coordination is noteworthy: while His25 acts as the fifth heme–iron ligand, the sixth heme–iron ligand is the amino group of Tyr41 in the first helix. Cytochrome *f* has an internal network of water molecules that appears to be a conserved structural feature and it has been hypothesized that this water chain functions as a proton wire.

X.1.4.2. Multi-Heme Cytochromes

Multi-heme cytochromes are all *c* type and contain multiple Cys X Y Cys His motifs. They belong to the class of all α proteins and are characterized by the presence of a variable number of helices and little β structure.

Those with bis(His) axial ligation are present in sulfate- and sulfur-reducing bacteria, such as *Desulfovibrio* and *Desulfuromonas,* and can be classified into two families: cytochrome c_3 (containing four heme groups)[64] and cytochrome $c_{551.5}$ or c_7 (containing three heme groups).[65] The elements of secondary structure are essentially the same for cytochrome c_3 and cytochrome c_7, if one takes into account the deletion of residues containing amino acids bound to the second heme (i.e., the one that is lacking in the three-heme protein) and contain one or two antiparallel β sheets at the N-terminus and three to four α-helices. The individual heme groups are structurally and functionally nonequivalent. In all the cytochromes c_3 and in cytochrome c_7 the spatial arrangement of the hemes is substantially conserved (Fig. X.1.14). The distances between the heme–iron atoms are in the range of 11.5–19 Å for cytochrome c_7 and 11–18 Å for cytochrome c_3. In cytochrome c_7, hemes I and IV are almost parallel, whereas heme III forms an angle of $\sim 50°$ with heme I. In cytochromes c_3 the values of the angles between hemes I, III, and IV are essentially the same as in c_7; hemes II and III are almost parallel (the angle between the two heme planes being $\sim 30°$). These cytochromes have only 30–40 residues per heme group, which gives rise to a relatively open and solvent accessible structure. Solvent exposure has been taken as the main contributor to the low reduction potentials found in these proteins (i.e., -90 to -400 mV). In both c_7 and c_3 proteins, electrostatic potential calculations reproduce the experimental finding that heme IV has the highest reduction potential.[21,28,29,64,66] The high reduction potential of heme IV is mainly due to the electrostatic term arising from its location in a Lys-rich protein region. The presence of many positively charged residues around the heme with the highest reduction potential is a common feature for cytochromes c_3 and has also been proposed to be important for the molecular recognition of the partner hydrogenase (heme IV being the site of interaction of the proteins).

The oxidation of molecular hydrogen catalyzed by hydrogenase constitutes a central step in the metabolism of *Desulfovibrio* organisms (see Sections XII.1 and XII.5),

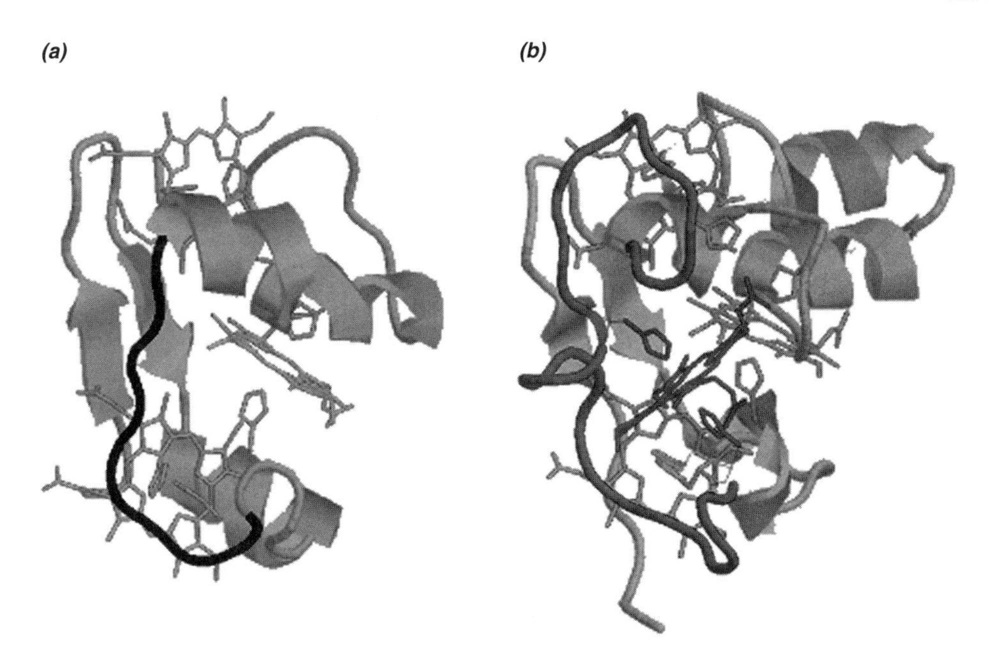

Fig. X.1.14.
Comparison of the structures of the (*a*) three-heme cytochrome c_7 from *Desulfuromonas acetoxidans* and of the (*b*) four-heme *Desulfovibrio desulfuricans* cytochrome c_3. Regions that differ due to the different length of the protein and the different number of hemes are shown in black (PDB codes: 1EHJ and 3CYR, respectively). [See color insert.]

Scheme X.1.2.

Transmembrane electron carrier
Acceptor
H_2A_1
H_2A_1
$2e^-$
H_2A_3
H_2A_3
$nH^+(pK = 7.6)$
Hydrogenase
B_1
B_1
$nH^+(pK = 6.4)$
$2e^-$
Transmembrane proton carrier
B_3
B_3
Donor

but this enzymatic activity is quite low at physiological pH. It has been proposed that hydrogenase can have high efficiency at physiological pH in the presence of cytochrome c_3 through the concerted capture of electrons and protons that can be achieved through an extensive cooperativity between redox and ionizable centers. Indeed, cytochrome c_3 constitutes a complex redox system where the reduction potential of each heme–iron is affected by the oxidation state of the others (redox cooperativity) and the heme–irons have pH dependent reduction potentials (redox-Bohr effect). A scheme has been developed that takes into account all these effects and gives a proposed mechanism of action of cyt c_3 (Scheme X.1.2).[28b] According to this scheme, the two electrons and the two protons produced in the oxidation of molecular hydrogen are received by cyt c_3. The two electrons are subsequently released to a transmembrane electron-transfer complex and utilized for the reduction of sulfate to sulfide. The two protons are released and used for the production of adenosine triphosphate (ATP). In this process, the protein switches between a state with three oxidized hemes (B_3 in Scheme X.1.2) and a state with a single oxidized heme (B_1). As the pK_a for the triply oxidized state is lower than that of the single-oxidized state, upon reduction there is proton uptake (conversion from B_1 to H_2A_1), while upon oxidation (from H_2A_1 to H_2A_3) there is proton release (from H_2A_3 to B_3). The identity of the two ionizable residues is still in doubt, although it has been proposed that they could be the two propionates of heme I.[67]

Tetraheme cytochromes containing *c*-type hemes with both bis(His) and His(Met) coordination also exist. Their 3D structure is exemplified by cytochrome *c* of purple bacteria of the photosynthetic reaction center (PRC), and they form a structurally homogeneous family.[68] The PRCs are membrane-spanning complexes of polypeptide chains and cofactors that catalyze the first steps in the conversion of light energy to chemical energy during photosynthesis (see Section X.3). The PRCs of photosynthetic purple bacteria consist of at least three protein subunits termed L (light), M (medium), and H (heavy). In *Rhodopseudomonas viridis* and *Thiocapsa pfennigii,* the four-heme cytochrome *c* is the fourth (and largest) protein subunit. *Rhodopseudomonas viridis* PRC cytochrome *c* is a lipoprotein where the lipid-binding group is the N-terminal Cys, which is linked to a diglyceride that helps in attaching the cytochrome *c* to the surface of the membrane. Most of cytochrome *c* is located in the aqueous periplasmic compartment. The PRC cytochrome *c* consists of an N-terminal segment and two pairs of heme-binding segments connected by a loop. Each heme-binding segment consists of an α helix with an average length of 17 residues followed by a turn and the Cys X Y Cys His motif. The hemes are connected to the Cys residues via thioether linkages in such a way that the heme planes are parallel to the helix axes. The sixth ligands to three of the four hemes are the S atoms of Met residues within the helices. Heme 2 is bis(His) coordinated. The two pairs of heme-binding segments, containing hemes 1, 2 and 3, 4, respectively, are related by a local twofold symmetry (Fig. X.1.15). The heme irons are 14–16 Å apart. The four reduction potentials are −60 (heme 4), 20 (heme 2), 310 (heme 3), and 380 (heme 1) mV. Electrostatic calculations on these systems have been able to reproduce the 440-mV spread in reduction potentials and have allowed assignment of reduction potentials to specific hemes as follows:[69] heme 1 is 380 mV, heme 3 is 310 mV, heme 2 is 20 mV, and heme 4 is −60 mV. Pairwise electrostatic interactions with charged amino acids and heme propionates, as well as the nature of the axial ligands, are the dominant contributions. The reduction potential of heme 2 is lowered, relative to the other hemes, by its bis(His) axial ligation. The reduction potential of heme 4 is lowered by the contribution of charged protein groups.

X.1.5. Copper Proteins

Copper is the only metallic element other than iron used by Nature for electron-transfer purposes. The appearance of copper as a bioelement occurred later in the history of life than that of iron, probably triggered by the increased concentration of O_2 in the atmosphere. Since then (see Chapter II), copper proteins have progressively replaced some (but not all) iron proteins in their redox tasks. In several

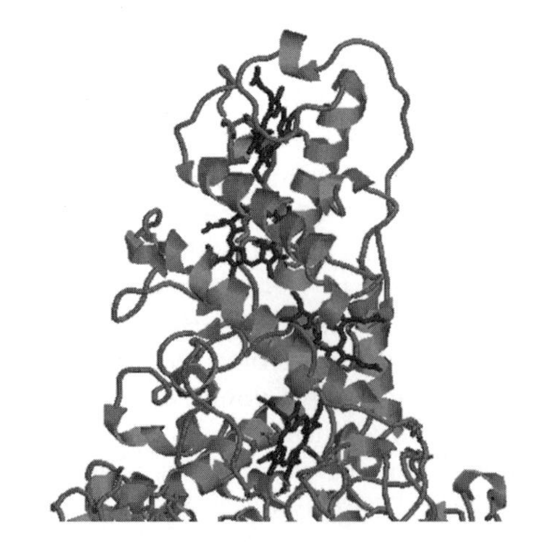

Fig. X.1.15.
Detail of the structure of the photosynthetic reaction center from *R. viridis* obtained by X-ray diffraction (PDB code: 1PRC). The heme with bis (His) axial coordination is the second from the bottom. [See color insert.]

Type 1 copper

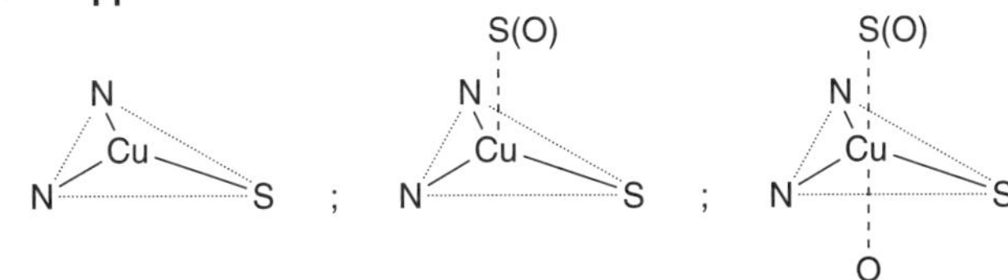

Type 2 copper

Type 3 copper

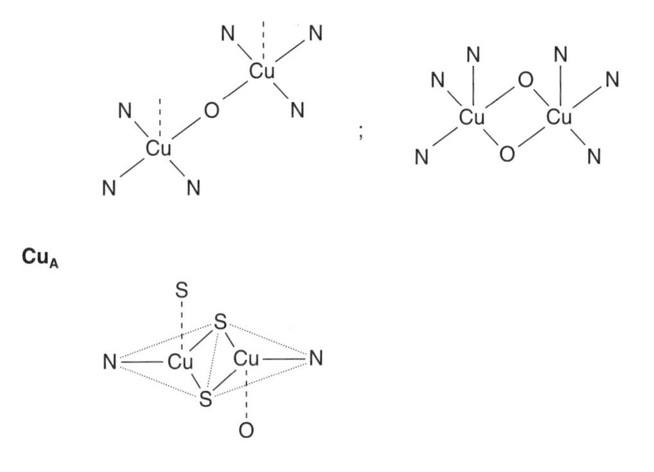

Fig. X.1.16.
Summary of relevant copper coordination properties in the four classes of copper proteins.

Table X.1.3.
A Summary of Relevant Features of the Four Classes of Copper in Proteins

	Nuclearity	Donor Atoms
Type 1	1	3 Donors (N, N, S) + 0, 1, or 2 Axial Donors (S, O)
Type 2	1	4 Donors (N_x, O_{4-x}) + 0 or 1 Axial Donors (N, O)
Type 3	2	3 Donors (N) + 1 or 2 Bridging Donors (O)
Cu_A	2	3 Donors (N, S, S) + 1 Axial Donor (S, O)

organisms, iron–sulfur proteins and copper proteins dedicated to the same electron-transfer purpose even coexist.

In contrast to iron, copper does not need special cofactors to be utilized as a redox center in biology, which may be due to the greater versatility of copper in adapting to the various and often irregular coordination geometries imposed by proteins. Copper sites in proteins are usually classified into four classes according to the spectroscopic properties of the metal center which, in turn, depend on the type, number, and geometry of the host protein ligands. The four classes are type 1 (or blue), type 2 (mononuclear catalytic), type 3 [antiferromagnetically coupled binuclear Cu(II)], and Cu_A [binuclear Cu(I) or mixed valence].[70] The relevant coordination properties of copper in the four classes are summarized in Fig. X.1.16 and Table X.1.3. Electron-transfer centers are only either of the type 1 or of the Cu_A class, and these will be discussed here. In more complex systems, copper ions belonging to different classes may also be simultaneously present. These systems will be discussed in Section XI.7, but the present considerations about their electron-transfer centers are also valid for them.

X.1.5.1. The Cupredoxin Fold

In striking contrast with the variety of folds encountered in iron–sulfur proteins and heme proteins, copper proteins devoted to electron transfer essentially adopt only one fold, termed the cupredoxin fold. The typical cupredoxin fold is constituted by a single domain polypeptide of ~90–150 residues, hosting one copper ion per molecule. Polypeptides of this type were first identified in the class of blue copper proteins.[70c] Similar cupredoxin folds are found as subdomains of more complex constructs and can host either one blue copper center or one two-copper center of the Cu_A type.

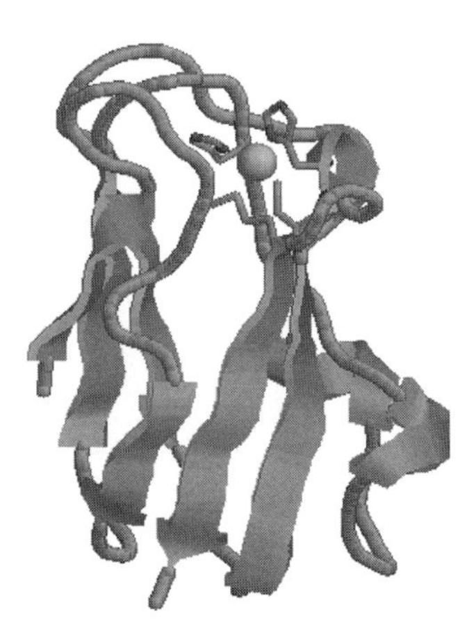

Fig. X.1.17.
The X-ray crystal structure of the plastocyanin from spinach (PDB code: 1AG6). [See color insert.]

The cupredoxin fold belongs to the all β class and consists of a β-barrel fold defined by two β sheets. These sheets may contain from 6 to 13 strands that follow a Greek key motif. A few α-helical segments may sometimes be present. Figure X.1.17 shows the structure of the blue copper protein plastocyanin (see below) as a typical example. The copper-binding site is located in the so-called "northern" region of the molecule, in a depression close to the protein surface, but is not solvent exposed. Three copper ligands, a cysteine and two histidines, are conserved in all cupredoxins (see also Table X.1.3). A fourth (axial) ligand, either a methionine or a glutamine, can also be present. Three of these four ligands (Cys, His, axial) are located in a loop linking two β strands. The other His ligand is located upstream in the sequence, in an adjacent β strand in the interior of the protein. The general consensus sequence for the three residues in the loop is

$$\text{Cys } X_m \text{ His } X_n \text{ Met (Gln)}$$

with the Cys and Met (Gln) residues located in the transition region from the loop and the preceding or following sandwiched β strands, respectively.

X.1.5.2. Blue Copper Proteins

Soluble copper proteins with the cupredoxin fold and one copper ion are found in bacteria and plants, and are grouped under the heading of "blue copper proteins".[71] They perform electron transfer by cycling copper between the Cu(I) and Cu(II) states. Blue copper proteins are involved in photosynthesis (e.g., carrying electrons from cytochrome *f* to photosystem I); in respiration (e.g., as electron donors for bacterial terminal oxidases); in oxidative deamination of primary amines; and in reduction of nitrite reductase. Their name reflects the most striking characteristic of the copper chromophore in its oxidized form, that is, its intense blue color. This characteristic (together with the peculiarly low value of the hyperfine splitting of the g_{zz} electron paramagnetic resonance (EPR) feature) has attracted the interest of inorganic chemists since the beginning of biological inorganic chemistry. The blue color is due to an intense electronic absorption at ~600 nm, assigned to a ligand-to-metal charge-transfer transition. The high intensity is attributed to a particularly strong π bond between the *p* orbital of the cysteine sulfur in the trigonal plane and the $d_{x^2-y^2}$ orbital of copper, which contains the unpaired electron and also lies in the trigonal plane (Fig. X.1.18*a*). Indeed, this transition is the salient characteristic of these systems.[72] It has been estimated that, due to this strong bond, the unpaired electron is actually almost equally shared between the copper $d_{x^2-y^2}$ orbital and the cysteine sulfur *p* orbital. In other words, the redox center of blue copper proteins in its oxidized form might be better described as containing $Cu^{1.5+}$ and $S^{1.5-}$ atoms.

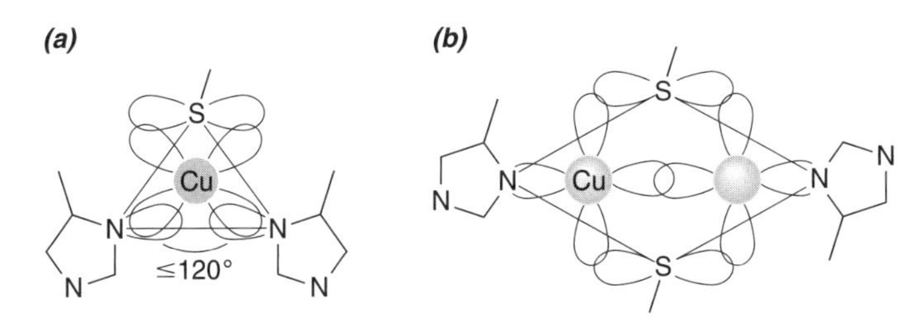

Fig. X.1.18.
Schematic representation of π bonds between the metal ion and its ligands in blue copper (*a*) and Cu_A (*b*).

Blue copper proteins may be grouped into five different families according to phylogenetic analyses. These are (1) the plastocyanin family (plastocyanin, amicyanin, pseudoazurin, and halocyanin); (2) auracyanin; (3) azurin; (4) rusticyanin; and (5) the phytocyanin family (plantacyanin, stellacyanin, and uclacyanin).[73] As far as the active Cu site is concerned, these families differ essentially by the axial ligand. Plastocyanins and rusticyanins have one methionine sulfur at 2.8–2.9 Å, azurin has a methionine sulfur and a peptide carbonyl oxygen, both at ~3.1 Å and, among phytocyanins, stellacyanin has a glutamine oxygen at ~2.2 Å from the copper ion. In the azurin case, the coordination geometry can be described as a strongly axially elongated trigonal bipyramid, and in stellacyanin as a trigonally flattened tetrahedron. No high-resolution structural characterization of auracyanin is available.

The cupredoxin fold is characterized by an extended secondary structure motif, so that the correct folding of the protein does not depend on the presence of the metal ion. Nevertheless, and similarly to the situation encountered in the rubredoxin case, the Cu atom bridges two different parts of the protein (e.g., the upstream His on one side and the three other ligands on the other, Fig. X.1.17), so that an additional stabilization, comparable to that exerted by a disulfide bridge, is provided by the metal.

The reduction potentials of blue copper proteins are positive,[74] in the range 180–370 mV, with the exception of rusticyanin, which has a much higher potential of 680 mV. As discussed in Section X.1.2, the potential is modulated by the axial ligand and by hydrogen bonding from a backbone NH to the coordinating cysteine. Rusticyanin is believed to have a higher potential because the copper lies in a more hydrophobic environment and is shielded from the solvent.

X.1.5.3. Cu_A

In contrast to blue copper proteins, the cupredoxin fold hosting the Cu_A dimetallic center is never found alone, but only as a domain in multidomain proteins with catalytic activity.[70a] Examples are cytochrome *c* oxidase, where the Cu_A domain constitutes the entry port for electrons donated by cytochrome *c* (see Sections X.1.4, X.3, and XI.3), and nitrous oxide reductase (Section XII.4). The features of the Cu_A domain are described here in comparison to the blue copper proteins. The structure of the Cu_A domain of bovine heart cytochrome *c* oxidase is shown in Fig. X.1.19. The similarity with the overall structure of blue copper proteins (Fig. X.1.17) is clear. The "northern" region contains a very similar copper-binding motif. There is an upstream histidine and the same downstream Cys-His-Met sequence. However, one more cysteine is inserted between Cys and His. This extra Cys is able to act as a metal ligand, and allows the site to host two Cu ions rather than one. The resulting consensus sequence is

Cys X_l **Cys** X_m His X_n Met

with the extra Cys shown in bold. The fact that the Cu_A site is so similar to the blue copper site explains why, for many years, it was believed that Cu_A was just an

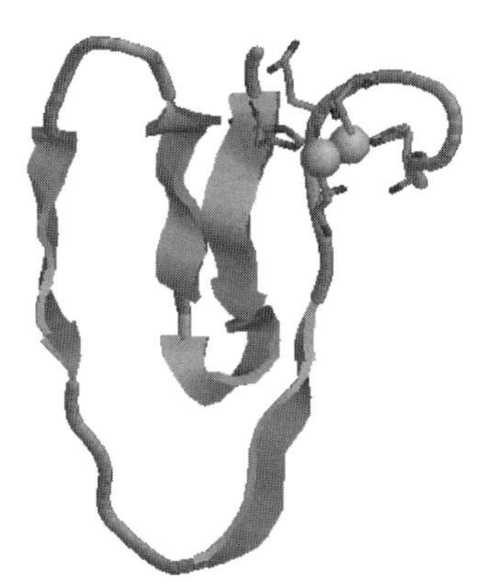

Fig. X.1.19.
Structure of the Cu_A site in bovine heart cytochrome *c* oxidase (PDB code: 1OCC). [See color insert.]

irregular type of blue copper. Its color in the oxidized state is actually purple, on account of a blue shift of the charge-transfer band with respect to blue copper proteins. Upon one-electron reduction, the protein becomes colorless, as expected for the Cu(I) state. The situation became clear in the 1990s, when it was first proposed, and then experimentally verified, that the oxidized form contained two Cu ions in a mixed-valence $Cu^{1.5+}$ state, bridged by two Cys ligands. A possible scheme for the overlap of the relevant atomic orbitals is shown in Fig. X.1.18*b*. The two Cu ions and the two Cys sulfurs form a characteristic "diamond" structure, similar to the ferredoxin diamond described in Section IV.2, but involving Cys sulfurs rather than sulfide ions. The unpaired electron in the oxidized form is shared between the two Cu centers due to the overlap of the two $d_{x^2-y^2}$ orbitals in the trigonal planes of each Cu, overlap that occurs both directly and through the sulfur ligands. The fully reduced form contains two colorless Cu(I) ions. The two Cu ions are similar, but not equivalent. Each Cu has a weak axial ligand, a Met sulfur for one and a peptide carbonyl for the other. These two ligands correspond to the two axial ligands of Cu in azurin. As in the blue copper proteins, there is a substantial amount of unpaired electron delocalization onto the two Cys ligands. A reasonable qualitative description of the mixed-valence system is that the unpaired electron is ~25% on each of the four atoms in the diamond structure. In this case, the "oxidation states" of the diamond atoms are $2xCu^{1.25+}$ and $2xS^{1.75-}$. Each atom is thus more reduced than in an oxidized blue copper protein, but the whole system is still able to take up one full electron upon reduction.

Perhaps, the most salient feature of these Cu-based electron-transfer systems is the high degree of electron delocalization onto the Cys sulfur ligands, which is not observed, for example, in iron–sulfur proteins. Qualitatively, this high degree of delocalization is explained by a higher covalency of the Cu–S bond with respect to the Fe–S bond, as expected from the softer character of Cu. In terms of electron-transfer ability, delocalizing the extra negative charge acquired upon reduction over more than one atom may help to reduce the reorganization energy. Iron–sulfur proteins achieve this by increasing the number of Fe atoms to three and four, while blue and purple copper proteins seem to exploit more efficiently the presence of the S atoms.

X.1.6. A Further Comment on the Size of the Cofactor

As described in Section X.2, electron transfer from an atom A to an atom B occurs at rates that differ dramatically depending on the nature of the intervening medium. If A and B are separated by a vacuum, the rate decays rapidly and exponentially to zero with increasing distance. On the other hand, if the medium is a conducting material, the electron can travel very fast over very large distances. In the reality of the biological world, the medium is constituted by relatively light protein atoms connected by covalent bonds, some hydrogen bonds, van der Waals contacts, and a few (small) vacuum gaps. In intermolecular electron transfer, a few water molecules may also be involved at the interface between the donor and the acceptor. Overall, this medium has properties intermediate between a vacuum and a conducting material.

Typically, an electron takes off from a metal center contained in the donor, travels a distance across the donor protein matrix from that center to the protein surface, jumps onto the acceptor protein through some intermolecular contact, travels across the acceptor protein matrix, and finally lands on the acceptor metal center. Coordination chemistry tells us that the metal centers involved in this process are constituted not only by the metal atom, but also by its ligands. Indeed, we know that the molecular orbital (MO) of the donor containing the leaving electron and, likewise, the MO of the acceptor that will host the incoming electron, consist of atomic orbitals of both the metal and its donor atoms, such that the electron is actually shared among the metal and a few O, N, or S atoms. Therefore, we can approximate the electron transfer from the metal to its donors as an instantaneous process; in other words, the coordination polyhedron can be considered as a conductor. This reasoning can be extended to a larger number of atoms in some cases. For example, a His ligand ring whose π system is involved in the MO containing the electron to be transferred is a conductor (this phenomenon is analogous to electrical conduction along the planes of graphite). Therefore, the electron transfer will slow down only when the electron moves away from that ligand. A heme has an extended π system that can also be considered a conductor. So, among the reasons why Nature has chosen heme as a cofactor for electron-transfer proteins, is that there could be the advantage of short-circuiting part of the electron-transfer pathway, provided that the direction of the transfer required is along the plane of the heme and not perpendicular to it. The diameter of the heme is ~ 12 Å, corresponding to a gain of ~ 6 Å from the metal to the periphery of the heme in any direction along the plane.

Analogous considerations may hold for iron–sulfur proteins. The "conducting" part in rubredoxin is only a couple of angstroms long (i.e., overall the length of the covalent bond between Fe and each of the Cys sulfur donors, and the size of the FeS_4 tetrahedron is <4 Å). On passing to 2Fe2S ferredoxins, the conducting part may extend >6 Å, if measured between the Cys sulfur coordinated to one Fe and one of those coordinated to the other Fe. This distance is further increased in Rieske centers, due to the presence of two His ligands on one Fe. The Fe_4S_4 clusters also extend >6 Å in all directions. As mentioned in Section X.1.3.5, the two Fe_4S_4 clusters in clostridial ferredoxin are relatively close. There is a small gap in between the β-CH_2 protons of two cysteines that are ligands of the two different clusters. Apart from this gap, which may somewhat slow down the rate, an electron can travel over the whole protein for >17 Å if both clusters are involved in the electron transfer by way of intramolecular redox.

Analogous gain is observed when one compares mononuclear blue copper proteins with dinuclear purple copper proteins (Section X.1.5). The longest conducting distance from the tip of one His to the end of the other is ~ 9 Å in the former and almost 12 Å in the latter.

In conclusion, it is conceivable that Nature has exploited the properties of the metal assemblies described in this chapter in order to shorten the electron-transfer pathways as much as possible. Depending on whether the protein delivers the electron through the same surface port through which it has entered (i.e., there is only one electron-exchange port on the surface) or through a different one, the particular geometry of the conducting part of the metal center may be exploited in different ways. For example, some two Fe_4S_4 ferredoxins are involved in electron transfer in the photosynthetic system. These cylindrical proteins are immobilized at the surface of the membrane, and a possible electron-transfer pathway from the donor to the acceptor could actually be along the cylinder axis (i.e., through the two clusters). Similarly, the Cu_A domain in cytochrome *c* oxidase might actually exploit the full length of the conducting cluster, as the electron entry port seems to be close to one of the two His ligands, while the other His points to the interior of the molecule in the direction of the heme *a* acceptor.

X.1.7. Donor–Acceptor Interactions

The pathways of electron transfer in a protein are discussed in Section X.2. A special redox metal complex attached to an electron-transfer protein provides a convenient way to measure the electron-transfer rate between the two metals. The case of self-exchange between the oxidized and reduced partners of the same protein also is quite interesting. However, the most interesting process is the electron transfer between physiological partners. The electron-transfer mechanism and the factors determining the interactions between two partners are still quite controversial.[74–79] For sure, the two proteins have to get in touch with one another. One simple mechanism may be based on stochastic collisions between partners. Electron transfer may occur upon collision. The observation that Nature may have optimized the electron-transfer rate, but not the formation of stable biomolecular systems, which would play an adverse role with respect to efficient electron transfer, may be in favor of this model. It is possible that not all collisions are efficient with respect to the electron-transfer process. Another model, however, favors "oriented" collisions that would optimize the electron-transfer pathway. The collision, therefore, would not be stochastic anymore but, for example, positively charged regions of one protein would interact with negatively charged regions of its protein partner. Much of the data on electron-transfer rates would favor the latter model, as ionic strength or site directed mutagenesis affect the rates in a predictable way using such a model. There is also a third model by which the partners collide, adjust their reciprocal orientation, and then electron transfer takes place. The long-range attractive forces seem to be electrostatic dopolar in nature, which depend on r^{-2}. Hydrophobic interactions are short-range interactions and depend on r^{-6}. Hydrogen bonds between noncharged residues also play a role.

Despite some studies on the dynamical properties of single proteins, no dynamical study has yet been reported for a protein–protein complex.

Mitochondrial cytochrome *c* is one of the most extensively studied electron-transfer proteins and can be taken as an example to explain protein–protein interactions.[80] It has an almost spherical shape and is characterized by a large α-helical content. Cytochrome *c* contains 17 lysines (*Saccharomyces cerevisiae* iso-1), which provide an overall positive charge to the protein (+8 in the oxidized state). The lysines are present over the entire molecule, but particularly Lys 13, 27, 72–73 and 86–87 have a key role, forming a ring around the solvent-exposed edge of the heme (pyrrole *c* ring).[81] This protein is involved in the mitochondrial respiratory chain, where it interacts with cytochrome *c* oxidase. In yeast, cytochrome *c* interacts with and is oxidized by cytochrome *c* peroxidase. Cytochrome *c* also interacts with the cytochrome bc_1 complex as well as being involved in many other electron-transfer processes. Cytochrome *c* can be considered a case study to further the discussion of interprotein electron transfer.

In its function, cyt *c* interacts with cyt *c* oxidase (CcP),[82] cyt bc_1 complex, and cyt *c* peroxidase.[83] The X-ray structures of the complexes between cyt *c* and CcP[74] and between cyt *c* and bc_1 complex[75] are known.

Cytochrome *c* peroxidase from yeast (see Section XI.3) is a heme-containing enzyme that catalyzes the oxidation of cytochrome *c* by using H_2O_2, through a two one-electron step mechanism. In the complexes with either cytochrome *c* from yeast or from horse, CcP does not experience any significant structural change while a few charged side chains in cytochrome *c* undergo some sizable changes, which further support the role of electrostatic interactions in complex formation. In the structure of the complex of CcP with the yeast isoenzyme of cytochrome *c*,[74] only a weak (3.3 Å) hydrogen bond is detected, even if other potential hydrogen bonds can be formed with minor side-chain rearrangements that might occur in solution. Some of these potential hydrogen bonds could easily be formed as the two donor–acceptor atoms are only 4 Å apart and belong to charged residues. At lower ionic strengths than in the crystallization medium, that is, at physiological ionic strengths, it is likely that the charged groups would get closer as a result of electrostatic attraction. The two proteins also experience hydrophobic contacts, which involve residues of cytochrome *c* just adjacent to the axial methionine iron ligand. The iron ions of the two hemes, that is, the two centers between which the electron-transfer reaction occurs are 26.5 Å apart. The propionate groups of both hemes are pointing in the opposite directions with respect to the other heme moiety. In CcP, the heme is essentially buried having only the δ meso edge exposed to the solvent. Also, in cytochrome *c* the heme is quite buried, with the exception of the pyrrole *c* ring that is bearing the 3-CH_3 group.

On the basis of the X-ray structure of the cytochrome *c*–CcP complex, a possible efficient electron-transfer pathway has been proposed that connects the 3-CH_3 group or the thioether group of cytochrome *c* (its exposed heme edge), through some residues of cytochrome *c* to Trp191 of CcP, which is at van der Waals contact with the heme of CcP, and which bears the cation radical of compound I (Fig. X.1.20).[75]

The electron-transfer rates have been found to be dependent on the ionic strength, which is consistent with the relevance of electrostatic interactions in complex formation. Interpretation of the kinetics studies has been and still is a matter of scientific debate. (See Sections X.2 and X.3.) Nevertheless, tight and specific interactions are mediated mainly by nonpolar forces. The importance of van der Waals interactions involving the cyt *c* residue Thr12, Arg13, Val28, and Ala81 is further confirmed by the fact that the very same residues are also involved in van der Waals interactions with the bc_1 complex,[75] suggesting a preferential electron-transfer pathway for cyt *c* that passes through the three-methyl heme edge.

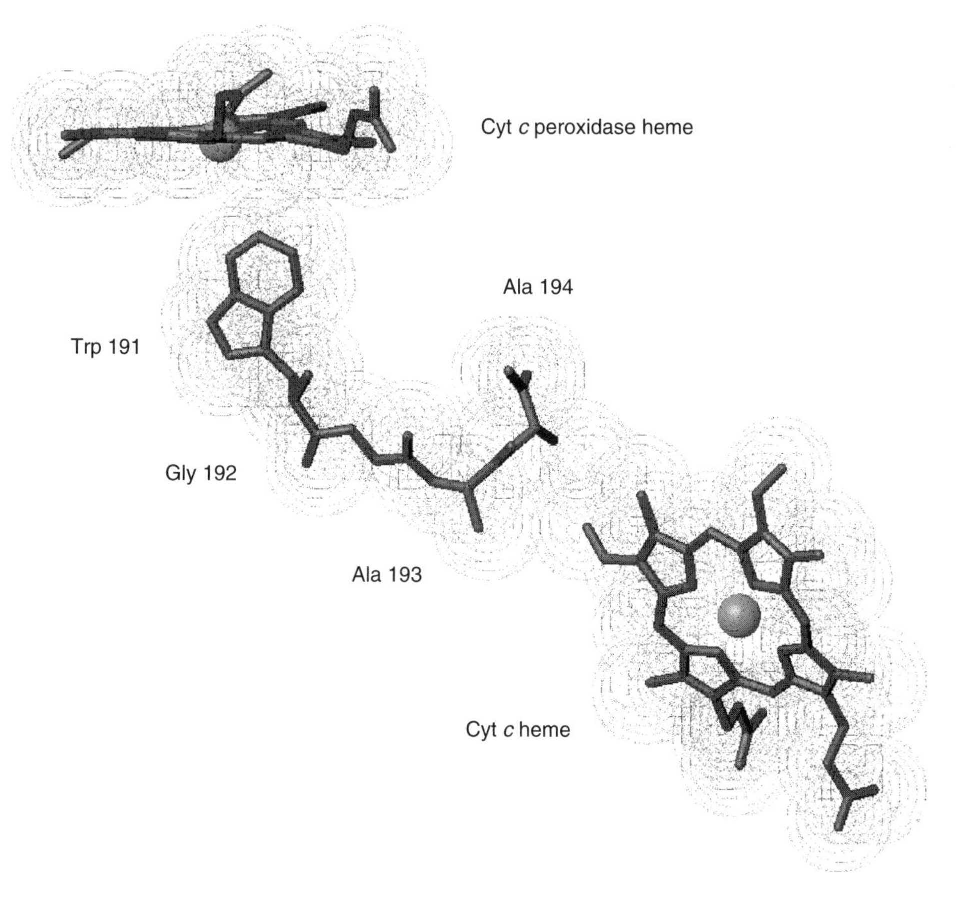

Fig. X.1.20.
The proposed electron-transfer pathway from the cytochrome *c* heme to the CcP heme, based on the X-ray structure of the adduct (PDB code: 2PCC).[75]

References

General References

1. Stephens, P. J., Jollie, D. R., and Warshel, A., "Protein Control of Redox Potentials of Iron–Sulfur Proteins", *Chem. Rev.*, **96,** 2491–2513 (1996).
2. Capozzi, F., Ciurli, S., and Luchinat, C., "Coordination sphere versus protein environment as determinants of electronic and functional properties of iron–sulfur proteins", *Struct. Bonding,* **90,** 127–160 (1998).
3. Zhou, H. X., "Control of reduction potential by protein matrix: lesson from a spherical protein model", *J. Biol. Inorg. Chem.*, **2,** 109–113 (1997).
4. Banci, L., Bertini, I., Gori Savellini, G., and Luchinat, C., "Individual Reduction Potentials of the Iron-Ions in Fe_2S_2 and high potential Fe_4S_4 ferredoxins", *Inorg. Chem.*, **35,** 4248–4253 (1996).
5. Mauk, A. G. and Moore, G. R., "Control of metalloprotein redox potentials: what does site-directed mutagenesis of hemoproteins tell us?", *J. Biol. Inorg. Chem.*, **2,** 119–115 (1997).
6. Gunner, M. R., Alexov, E., Torres, E., and Lipovaca, S., "The importance of the protein in controlling the electrochemistry of hememetalloproteins: methods of calculations and analysis", *J. Biol. Inorg. Chem.*, **2,** 126–134 (1997).
7. Naray-Szabo, G., "Electrostatic modulation of electron transfer in the active site of heme peroxidases", *J. Biol. Inorg. Chem.*, **2,** 135–138 (1997).
8. Armstrong, F. A., "Evaluations of reduction potential data in relation to coupling, kinetics and function", *J. Biol. Inorg. Chem.*, **2,** 139–142 (1997).
9. Warshel, A., Papazyan, A., and Muegge, I., "Microscopic and semimacroscopic redox calculations: what can and cannot be learned from continuum models", *J. Biol. Inorg. Chem.*, **2,** 143–152 (1997).
10. Luchinat, C., Capozzi, F., and Bentrop, D., "Iron–Sulfur Proteins", in *Handbook on Metalloproteins,* Bertini, I., Sigel, A., and Sigel, H., eds., Marcel Dekker, New York, 2001.
11. Sharp, K. and Honig, B., "Electrostatic Interactions in Macromolecules: Theory and Applications", *Annu. Rev. Biophys. Biophys. Chem.*, **19,** 301–332 (1990).

Specific References

12. Min, T., Ergenekan, C. E., Eidsness, M. K., Ichiye, T., and Kang, C., "Leucine 41 is a gate for water entry in the reduction of *Clostridium pasteurianum* rubredoxin", *Protein Sci.*, **10,** 613–621 (2001).
13. Dugad, L. B., La Mar, G. N., Banci, L., and Bertini, I., "Identification of localized redox states in plant-type two-iron ferredoxins using the nuclear overhauser effect", *Biochemistry,* **29,** 2263–2271 (1990).
14. Backes, G., Mino, Y., Loehr, T. M., Meyer, T. E., Cusanovich, M. A., Sweeney, W. V., Adman, E. T., and Sanders-Loehr, J., "The environment of Fe_4S_4 cluster in ferredoxins and high-potential iron proteins. New information from X-ray crystallography and resonance Raman spectroscopy", *J. Am. Chem. Soc.*, **113,** 2055–2064 (1991).
15. Heering, H. A., Bulsink, Y. B. M., Hagen, W. R., and Meyer, T. E., "Influence of Charge and Polarity on the Redox Potentials of High-Potential Iron-Sulfur Proteins: Evidence for the Existence of Two Groups", *Biochemistry,* **34,** 14675–14686 (1995).
16. Babini, E., Borsari, M., Capozzi, F., Eltis, L. D., and Luchinat, C., "Experimental evidence for the role of buried polar groups in determining the reduction potential of metalloproteins: the S79P variant of *Chromatium vinosum* HiPIP", *J. Biol. Inorg. Chem.*, **1,** 692–700 (1999).
17. Babini, E., Bertini, I., Borsari, M., Capozzi, F., Dikiy, A., Eltis, L. D., and Luchinat, C., "A serine → cysteine ligand mutation in the High Potential Iron–Sulfur Protein from *Chromatium vinosum* provides insight into the electronic structure of the [4Fe–4S] cluster", *J. Am. Chem. Soc.*, **118,** 75–80 (1996).
18. Schejter, A., Aviram, I., and Goldkorn, T., "*Electron transport and oxygen utilisation*", Ho, C., ed., North-Holland, Amsterdam, The Netherlands, 1982, pp. 95–99.
19. Hunter, C. L., Lloyd, E., Eltis, L. D., Rafferty, S. P., Lee, H., Smith, M., and Mauk, A. G., "Role of the Heme Propionates in the Interaction of Heme with Apomyoglobin and Apocytochrome b_5", *Biochemistry,* **36,** 1010–1017 (1997).
20. Goodin, D. B. and McRee, D. E., "The Asp-His-Fe triad of cytochrome *c* peroxidase controls the reduction potentials, electronic structure, and coupling of the tryptophan free radical to the heme", *Biochemistry,* **32,** 3313–3324 (1993).
21. Assfalg, M., Banci, L., Bertini, I., Bruschi, M., Giudici-Orticoni, M. T., and Turano, P., "A proton-NMR investigation of the fully reduced cytochrome c_7 from *Desulfuromonas acetoxidans:* comparison between the reduced and the oxidized forms", *Eur. J. Biochem.*, **266,** 634–643 (1999).
22. Soares, C. M., Martel, P. J., and Carrondo, M. A., "Theoretical studies on the redox-Bohr effect in cytochrome c_3 from *Desulfovibrio vulgaris* Hildenborough", *J. Biol. Inorg. Chem.*, **2,** 714–727 (1997).
23. Walter, R. L., Ealick, S. E., Friedman, A. M., Blake, R. C. I., Proctor, P., and Shoam, M., "Multiple Wavelength Anomalous Diffraction (MAD) Crystal Structure of Rusticyanin: a Highly Oxidizing Cupredoxin with Extreme Acid Stability", *J. Mol. Biol.*, **263,** 730–751 (1996).
24. Sweeney, W. V. and Rabinowitz, J. C., "Proteins containing 4Fe–4S clusters: an overview", *Annu. Rev. Biochem.*, **49,** 139–161 (1980).
25. Bertini, I., Briganti, F., Luchinat, C., and Scozzafava, A., "^{1}H NMR studies of the oxidized and partially reduced 2(4Fe–4S) ferredoxin from *Clostridium pasteurianum*", *Inorg. Chem.*, **29,** 1874–1880 (1990).
26. Aono, S., Bentrop, D., Bertini, I., Cosenza, G., and Luchinat, C., "Solution structure of an artificial Fe_8S_8

ferredoxin: the D13C variant of *Bacillus schlegelii* Fe_7S_8 ferredoxin", *Eur. J. Biochem.*, **258,** 502–514 (1998).
27. Armstrong, F. A., "Dynamic electrochemistry of iron–sulfur proteins", *Adv. Inorg. Chem.*, **38,** 117–163 (1992).
28. (a) Santos, H., Moura, J. J. G., Moura, I., LeGall, J., and Xavier, A. V., "NMR studies of electron transfer mechanisms in a protein with interacting redox centers: *Desulfovibrio gigas* cytochrome c_3", *Eur. J. Biochem.*, **141,** 283–296 (1984). (b) Louro, R. O., Catarino, T., Turner, D. L., Picarra-Pereira, M. A., Pacheco, I., LeGall, J., and Xavier, A. V., "Functional and mechanistic studies of cytochrome c_3 from *Desulfovibrio gigas:* thermodynamics of a "Proton Thruster", *Biochemistry,* **37,** 15,808–15,815 (1998). (c) Coutinho, I. B., Turner, D. L., LeGall, J., and Xavier, A. V., "NMR studies and redox titration of the tetraheme cytochrome c_3 from *Desulfomicrobium baculatum*. Identification of the low-potential heme", *Eur. J. Biochem.*, **230,** 1007–1013 (1995).
29. Assfalg, M. Bertini, I. Bruschi, M. Michel, C. and Turano, P., "The metal reductase activity of some multiheme cytochrome *c*: NMR structural characterization of the reduction of chromium(VI) to chromium(III) by cytochrome c_7, *Proc. Natl. Acad. Sci. U.S.A.*, **99,** 9750–9754 (2002).
30. Beinert, H. and Kiley, P. J., "Fe–S proteins in sensing and regulatory functions", *Curr. Opin. Chem. Biol.*, **3,** 152–157 (1999).
31. Tsukihara, T., Fukuyama, K., and Katsube, Y., "Structure–function relationship of [2Fe–2S] ferredoxins", in *Iron–sulfur protein research,* Matsubara, H., Katsube, Y., and Wada, K., eds., Springer-Verlag, Berlin, 1986, pp. 59–68.
32. Todd, A. E., Orengo, C. A., and Thornton, J. M., "Evolution of protein function, from a structural perspective", *Curr. Opin. Struct. Biol.*, 548–556 (1999).
33. Bau, R., Rees, D. C., Kuntz, D. M., Scott, R. A., Huang, H. S., Adams, M. W. W., and Eidsness, M. K., "Crystal structure of rubredoxin from *Pyrococcus furiousus* at 0.95 Å resolution, and the structures of N-terminal methionine and formylmethionine variants of Pf Rd. Contributions of N-terminal interactions to thermostability", *J. Biol. Inorg. Chem.*, **3,** 484 (1998).
34. Cootes, A. P., Curmi, P. M. G., Cunningham, R., Donnelly, C., and Torda, A. E., "The dependence of amino acid pair correlations on structural environment", *Proteins Struct. Funct. Genet.*, **32,** 175–189 (1998).
35. Liang, C. and Mislow, K., "Topological chirality of iron–sulfur proteins", *Biopolymers,* **42,** 411–414 (1997).
36. Meyer, J., Gagnon, J., Gaillard, J., Lutz, M., Achim, C., Munck, E., Petillot, Y., Colangelo, C. M., and Scott, R. A., "Assembly of a $[2Fe–2S]^{2+}$ cluster in a molecular variant of *Clostridium pasteurianum* rubredoxin", *Biochemistry,* **36,** 13,374–13,380 (1997).
37. Mason, J. R. and Cammack, R., "The electron-transport proteins of hydroxylating bacterial dioxygenases", *Annu. Rev. Microbiol.*, **46,** 277–305.
38. Carter, C. W. J., Kraut, J., Freer, S. T., Alden, R. A., Sieker, L. C., Adman, E. T., and Jensen, L. H., "A comparison of Fe_4S_4 Clusters in High Potential Iron Protein and in Ferredoxin", *Proc. Natl. Acad. Sci. U.S.A.*, **69,** 3526–3529 (1972).
39. Cowan, J. A. and Lui, S. M., "Structure–Function Correlations in High-Potential Iron Proteins", *Adv. Inorg. Chem.*, **45,** 313–350 (1998).
40. Bentrop, D., Bertini, I., Iacoviello, R., Luchinat, C., Niikura, Y., Piccioli, M., Presenti, C., and Rosato, A., "Structural and dynamical properties of a partially unfolded Fe_4S_4 protein: the role of the cofactor in protein folding", *Biochemistry,* **38,** 4669–4680 (1999).
41. Hochkoeppler, A., Ciurli, S., Venturoli, G., and Zannoni, D., "The high potential iron–sulfur protein (HiPIP) from *Rhodoferax fermentans* is competent in photosynthetic electron transfer", *FEBS Lett.*, **357,** 70–74 (1995).
42. Pereira, M. M., Carita, J. N., and Teixeira, M., "Membrane-bound electron transfer chain of the thermohalophilic bacterium *Rhodothermus marinus*: characterization of the iron-sulfur centers from the dehydrogenases and investigation of the high-potential iron-sulfur protein function by in vivo reconstitution of the respiratory chain", *Biochemistry,* **38,** 1276–1283 (1999).
43. Link, T. A., "The structures of Rieske and Rieske-type proteins", *Adv. Inorg. Chem.*, **47,** 83–157 (1999).
44. Ikemizu, S., Bando, M., Sato, T., Morimoto, Y., Tsukihara, T., and Fukuyama, K., "Structure of [2Fe–2S] ferredoxin I from *Equisetum arvense* at 1.8 A resolution", *Acta Crystallogr.*, **D50,** 167–174 (1994).
45. Moulis, J.-M., Sieker, L. C., Wilson, K. S., and Dauter, Z., "Crystal structure of the 2[4Fe–4S] ferredoxin from *Chromatium vinosum:* evolutionary and mechanistic inferences for [3/4Fe–4S] ferredoxins", *Protein Sci.*, **5,** 1765–1775 (1996).
46. Moore, G. R. and Pettigrew, G. W., *Cytochromes c; Evolutionary, Structural and Physicochemical Aspects,* Springer-Verlag, Berlin, 1990.
47. Turano, P. and Lu, Y., "Iron in heme and related proteins", in *Handbook on Metalloproteins,* Bertini, I., Sigel, A., and Sigel, H., eds., Marcel Dekker, New York, 2001, pp. 269–356.
48. Brayer, G. D. and Murphy, M. E. P., "Structural studies of eukariotic cytochromes *c*. In *Cytochrome c. A multidisciplinary approach*", Scott, R. A. and Mauk, A. G., eds., University Science Books, Sausalito, CA, 1996, pp. 103–166.
49. Banci, L. and Assfalg, M., "Mitochondrial cytochrome *c*", in *Handbook of Metalloproteins,* Wieghart, K., Huber, R., Poulos, T., and Messerschmidt, A., eds., John Wiley & Sons, Ltd, New York, pp. 32–43.
50. Banci, L., Bertini, I., Bren, K. L., Gray, H. B., Sompornpisut, P., and Turano, P., "Three dimensional solution structure of the cyanide adduct of *Saccharomyces cerevisiae* Met80Ala-*iso*-1-cytochrome *c*. Identifi-

cation of ligand-residue interactions in the distal heme cavity", *Biochemistry,* **34,** 11,385–11,398 (1995).
51. Arnesano, F., Banci, L., Bertini, I., Faraone-Mennella, J., Rosato, A., Barker, P. D., and Fersht, A. R., "The solution structure of oxidized *Escherichia coli* cytochrome b_{562}", *Biochemistry,* **38,** 8657–8670 (1999).
52. Lederer, F., Glatigny, A., Bethge, P. H., Bellamy, H. D., and Mathews, F. S., "Improvement of the 2.5 Angstroms Resolution Model of Cytochrome b_{562} by Redetermining the Primary Structure and Using Molecular Graphics", *J. Mol. Biol.,* **148,** 427 (1981).
53. Moore, G. R., "Bacterial 4-alpha-helical bundle cytochromes", *Biochim. Biophys. Acta,* **1058,** 38–41 (1991).
54. Archer, M., Banci, L., Dikaya, E., and Romao, M. J., "Crystal structure of cytochrome c' from *Rhodocyclus gelatinosus* and comparison with other cytochromes c'", *J. Biol. Inorg. Chem.,* **2,** 611–622 (1997).
55. Shibata, N., Iba, S., Misaki, S., Meyer, T. E., Bartsch, R. G., Cusanovich, M. A., Morimoto, Y., Higuchi, Y., and Yasuoka, N., "Basis for monomer stabilization in *Rhodopseudomonas palustris* cytochrome c' derived from the crystal structure", *J. Mol. Biol.,* **284,** 751 (1996).
56. Ren, Z., Meyer, T. E., and McRee, D. E., "Atomic structure of a cytochrome c' with an unusual ligand-controlled dimer dissociation at 1.8 A resolution", *J. Mol. Biol.,* **234,** 433–445 (1993).
57. Dobbs, A. J., Anderson, B. F., Faber, H. R., and Baker, E. N., "Three-dimensional structure of cytochrome c' from two *Alcaligenes* species and the implications for four-helix bundle structures", *Acta Crystallogr. D Biol. Crystallogr.,* **52,** 356 (1996).
58. Tahirov, T. H., Misaki, S., Meyer, T. E., Cusanovich, M. A., Higuchi, Y., and Yasuoka, N., "High-resolution crystal structures of two polymorphs of cytochrome c' from the purple phototrophic bacterium *Rhodobacter capsulatus*", *J. Mol. Biol.,* **259,** 467–479 (1996).
59. Feng, Y. Q., Sligar, S. G., and Wand, A. J., "Solution structure of apocytochrome b_{562}", *Nat. Struct. Biol.,* **1,** 30–35 (1994).
60. Lederer, F., "The cytochrome b_5 fold: an adaptable module", *Biochimie,* **76,** 674–692 (1994).
61. Mathews, F. S., "The structure, function and evolution of cytochromes", *Prog. Biophys. Mol. Biol.,* **45,** 1–56 (1985).
62. Falzone, C. J., Mayer, M. R., Whiteman, E. L., Moore, C. D., and Lecomte, J. T. J., "Design Challenges for Hemoproteins: The Solution Structure of Apocytochrome b_5", *Biochemistry,* **35,** 6519–6526 (1996).
63. Martinez, S. E., Huang, D., Ponomarev, M., Cramer, W. A., and Smith, J. L., "The heme redox center of chloroplast cytochrome *f* is linked to a buried five-water chain", *Protein Sci.,* **5,** 1081–1092 (1996).
64. Coutinho, I. B. and Xavier, A. V., "Tetraheme cytochromes", *Methods Enzymol.,* **243,** 119–140 (1994).
65. Banci, L. and Assfalg, M., "Cytochrome *c7*", in *Handbook of Metalloproteins,* Wieghardt, K., Huber, R., Poulos, T., and Messerschmidt, A., eds., John Wiley & Sons, Ltd., New York, pp. 109–118.
66. Banci, L., Bertini, I., Bruschi, M., Sompornpisut, P., and Turano, P., "A NMR characterization and solution structure determination of the oxidized cytochrome c_7 from *Desulfuromonas acetoxidans*", *Proc. Natl. Acad. Sci. U.S.A.,* **93,** 14,396–14,400 (1996).
67. Brennan, L., Turner, D. L., Messias, A. C., Teodoro, M. L., LeGall, J., Santos, H., and Xavier, A. V., "Structural basis for the network of functional cooperativities in cytochrome c_3 from *Desulfovibrio gigas*: solution structures of the oxidised and reduced states", *J. Mol. Biol.,* **298,** 61–82 (2000).
68. Deisenhofer, J. and Michel, H., "High-resolution crystal structures of bacterial photosynthetic reaction centers", in *Molecular Mechanisms in Bioenergetics*", Ernster, L., ed., Elsevier, Amsterdam, The Netherlands, pp. 103–120.
69. Gunner, M. R. and Honig, B., "Electrostatic control of midpoint potentials in the cytochrome subunit of the *Rhodopseudomonas viridis* reaction center", *Proc. Natl. Acad. Sci. U.S.A.,* **88,** 9151–9155 (1991).
70. (a) Halcrow, M. A., Knowles, P. F., and Phillips, E. V., "Copper proteins in the transport and activation of dioxygen, and the reduction of inorganic molecules", in *Handbook on Metalloproteins,* Bertini, I., Sigel, A., and Sigel, H., eds., Marcel Dekker, New York, 2001, pp. 709–762. (b) Lindley, P. F., "Multi-copper oxidases", in *Handbook on Metalloproteins,* Bertini, I., Sigel, A., and Sigel, H., eds., Marcel Dekker, New York, 2001, pp. 763–812. (c) Vila, A. J. and Fernandez, C. O., "Copper in electron-transfer proteins", in *Handbook on Metalloproteins,* Bertini, I., Sigel, A., and Sigel, H., eds., Marcel Dekker, New York, 2001, pp. 813–856.
71. Adman, E. T., "Copper protein structures", *Adv. Prot. Chem.,* **42,** 144–197 (1991).
72. Holm, R. H., Kennepohl, P., and Solomon, E. I., "Structural and Functional Aspects of Metal Sites in Biology", *Chem. Rev.,* **96,** 2239–2314 (1996).
73. Ryden, L., *Copper Proteins and Copper Enzymes, Vol. II,* Lontie, R., ed., CRC Press, Boca Raton, FL, pp. 207–240.
74. Pelletier, H., and Kraut, J., "Crystal structure of a complex between electron transfer partners, cytochrome *c* peroxidase and cytochrome *c*", *Science,* **258,** 1748–1755 (1992).
75. Lange, C. and Hunte, C., "Crystal structure of the yeast cytochrome bc_1 complex with its bound substrate cytochrome *c*", *Proc. Natl. Acad. Sci. U.S.A.,* **99,** 2800–2805 (2002).
76. Northrup, S. H., Boles, J. O., and Reynolds, J. C. L., "Brownian Dynamics of Cytochrome *c* and Cytochrome *c* Peroxidase Association", *Science,* **241,** 67–71 (1988).
77. Williams, P. A., Fülöp, V., Leung, Y.-C., Moir, J. W. B., Howlett, G., Ferguson, S. J., Radford, S. E., and Hajdu, J., "Pseudospecific docking surfaces on electron transfervproteins as illustrated by pseudoazurin, cytochrome c_{550}, and cytochrome cd_1 nitrite reductase", *Nat. Struct. Biol.,* **2,** 975–982 (1995).

78. Erman, J. E., Kreschek, G. C., Vitello, L. B., and Miller, M. A., "Cytochrome *c*/cytochrome *c* peroxidase complex: effect of binding-site mutations on the thermodynamics of complex formation", *Biochemistry,* **36,** 4054–4060 (1997).
79. Osyczka, A., Nagashima, K. V. P., Sogabe, S., Miki, K., Shimada, K., and Matsuura, K., "Compariosn of the binding sites for high-potential Iron–sulfur protein and cytochrome *c* on the tetraheme cytochrome subunit bound to the bacterial photosynthetic reaction center", *Biochemistry,* **38,** 15,779–15,790 (1999).
80. Nocek, J. M., Zhou, J. S., De Forest, S., Priyadarshy, S., Beratan, D. A., Onuchic, J. N., and Hoffman, B. M., "Theory and Practice of Electron Transfer within Protein–Protein Complexes:application to the multi-domain binding of cytochrome *c* by cytochrome *c* peroxidase", *Chem. Rev.,* **96,** 2459–2490 (1996).
81. Witt, H., Malatesta, F., Nicoletti, F., Brunori, M., and Ludwig, B., "Cytochrome *c*-binding site on cytochrome oxidase in *Paracoccus denitrificans*", *Eur. J. Biochem.,* **251,** 367–373 (1998).
82. Tsukihara, T., Aoyama, H., Yamashita, E., Tomizaki, T., Yamaguchi, H., Shinzawa-Itoh, K., Nakashima, R., Yaono, R., and Yoshikawa, S., "The whole structure of the 13-subunit oxidized cytochrome *c* oxidase at 2.8 Å", *Science,* **272,** 1136–1144 (1996).
83. Finzel, B. C., Poulos, T. L., and Kraut, J., "Crystal structure of yeast cytochrome *c* peroxidase at 1.7 Å resolution", *J. Biol. Chem.,* **259,** 13,027–13,036 (1984).

X.2. Electron Transfer through Proteins

Harry B. Gray and Jay R. Winkler
Beckman Institute
California Institute of Technology
Pasadena, California 91125

Contents

X.2.1. Introduction

Aerobic respiration and photosynthesis work in concert: the oxygen that is evolved by photosynthetic organisms is the oxidant that sustains life in aerobic microbes and animals; and, in turn, the end products of aerobic respiratory metabolism (CO_2 and H_2O) nourish photosynthetic organisms. Section X.1 dealt with the structures and dynamics of redox-active proteins and complexes that take part in these and other biological processes. In this section, we develop a framework for interpretation of electron flow between donors and acceptors that are separated by large molecular distances in folded polypeptides, as such distant electron transfers are fundamental steps in both respiration and photosynthesis (which are discussed in Sections X.3 and X.4).[1–11]

X.2.2. Basic Concepts

A bimolecular electron-transfer (ET) reaction involves three steps: formation of a precursor complex, Eq. 1; electron transfer from donor to acceptor, Eq. 2; and dissociation of the successor complex to products, Eq. 3. In Eqs. 1 and 2, K is the equilibrium constant for the formation of the precursor complex $[A_{ox}, B_{red}]$, and k_{ET} is the forward electron transfer rate constant for producing the successor complex $[A_{red}, B_{ox}]$.

$$\mathrm{A_{ox}} + \mathrm{B_{red}} \overset{K}{\rightleftharpoons} [\mathrm{A_{ox}}, \mathrm{B_{red}}] \quad (1)$$

$$[\mathrm{A_{ox}}, \mathrm{B_{red}}] \overset{K_{\mathrm{ET}}}{\rightleftharpoons} [\mathrm{A_{red}}, \mathrm{B_{ox}}] \quad (2)$$

$$[\mathrm{A_{red}}, \mathrm{B_{ox}}] \overset{\mathrm{fast}}{\rightleftharpoons} \mathrm{A_{red}} + \mathrm{B_{ox}} \quad (3)$$

For the electron-transfer step, Eq. 2, the transition state is at the intersection of reactant and product surfaces that are usually approximated by parabolas (Fig. X.2.1). Some degree of electronic interaction, or coupling, is required if the redox system is to pass from the precursor to the successor complex. If the degree of electronic interaction is sufficiently small, perturbation theory can be used to obtain the energies of the new surfaces, which do not cross. The splitting at the intersection is equal to $2H_{\mathrm{AB}}$, where H_{AB} is the electronic coupling matrix element. Two cases can be distinguished. First, H_{AB} is very small; for these so-called "nonadiabatic" reactions, once the transiton state nuclear configuration is formed, there is a low probability that the reactants will "jump" to the product potential energy surface. If the electronic interaction is sufficiently large, as it is for "adiabatic" reactions, the reactants will always cross to the product potential energy surface upon passage through the transition state region.

The term adiabatic (Greek: *a-dia-bainein,* not able to go through) is used in both thermodynamics and quantum mechanics, and the uses are analogous. In the former, it indicates that there is no heat flow in or out of the system. In the latter, it indicates that the system makes no transition to other states. Hence, for an adiabatic reaction, the system remains on the same (i.e., lower) first-order electronic surface for the entire reaction. The probability of electron transfer occurring when the reactants reach the transition state is unity. The degree of adiabaticity of the reaction is given by a transmission coefficient, κ, whose values range from 0 to 1. For systems whose H_{AB} is sufficiently large ($>k_{\mathrm{B}}T$, where k_{B} is the Boltzmann constant), $\kappa = 1$. Often H_{AB} is

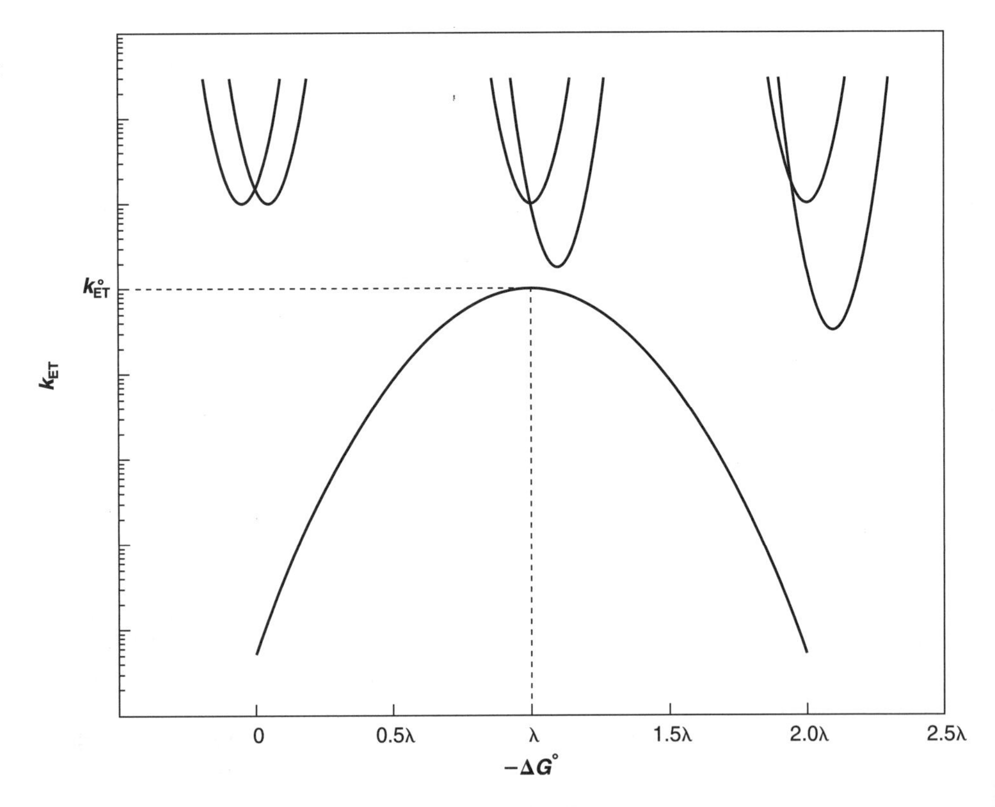

Fig. X.2.1.
Electron-transfer rate constant versus $-\Delta G^{\circ}$ plot and 3 ET reaction surfaces (see Eqs. 9 and 10).

large when the reacting centers are close, as in this case there can be direct overlap of donor and acceptor orbitals. The transmission coefficient is generally very small ($\kappa < 1$) for reactions of metalloproteins, owing to the long distances involved.

X.2.2.1. Marcus Theory

According to transition state theory, the rate constant for a bimolecular reaction in solution is given by Eq. 4:

$$k = \kappa \nu_N \exp\left(-\frac{\Delta G^*}{RT}\right) \tag{4}$$

where ν_N, the collisional, or diffusion-limited rate, is $\sim 10^{11}\ M^{-1}\,s^{-1}$ for small molecules, and ΔG^* is the Gibbs free energy difference between the activated complex and the precursor complex. In an adiabatic process, the transmission coefficient κ is unity. In this case, the problem of calculating the rate constant involves the calculation of ΔG^*, which R. A. Marcus partitioned into several parameters:[4]

$$\Delta G^* = w^r + \frac{(\lambda + \Delta G^{\circ\prime})^2}{4} \tag{5}$$

$$\Delta G^{\circ\prime} = \Delta G^\circ + w^p - w^r \tag{6}$$

Here w^r is the electrostatic work involved in bringing the reactants to the mean reactant separation distance in the activated complex, and w^p is the analogous work term for dissociation of the products. These terms vanish in situations where one of the reactants (or products) is uncharged. The parameter ΔG° is the Gibbs free energy change when the two reactants and products are an infinite distance apart and $\Delta G^{\circ\prime}$ is the free energy of the reaction when the reactants are at a distance r apart in the medium; ΔG° is the standard free energy of the reaction, obtainable from electrochemical measurements (the quantity $-\Delta G^\circ$ is called the *driving force* for the reaction).

The reorganization energy λ is a parameter that contains both inner-sphere (λ_i) and outer-sphere (λ_o) components; $\lambda = \lambda_i + \lambda_o$. The inner-sphere reorganization energy is the free energy change associated with changes in the bond lengths and angles of the reactants, often approximated by Eq. 7:

$$\lambda_i = \frac{1}{2}\sum_j k_j(\Delta x_j)^2 \tag{7}$$

where k_j values are normal-mode force constants, and the Δx_j values are differences in equilibrium bond lengths between the reduced and oxidized forms of a redox center.

Marcus used a dielectric continuum model to estimate the outer-sphere reorganization energy, according to Eq. 8:

$$\lambda_o = e^2\left[\frac{1}{2r_A} + \frac{1}{2r_B} - \frac{1}{d}\right]\left[\frac{1}{D_{op}} - \frac{1}{D_s}\right] \tag{8}$$

where d is the distance between centers in the activated complex, generally taken to be the sum of the reactant radii r_A and r_B; D_{op} is the optical dielectric constant of the medium (or, equivalently, the square of the refractive index); and D_s is the static dielectric constant.

Variations in λ can have enormous effects on electron-transfer rates. Some of the possible variations are apparent from inspection of Eq. 8. First, λ_o decreases with increasing reactant size. Second, the dependence of the reaction rate on separation distance attributable to λ_o occurs via the $1/d$ term. Third, λ_o decreases markedly as the solvent polarity decreases. For nonpolar solvents, $D_s \simeq D_{op} \simeq 1.5$–$4.0$. Protein interiors are nonpolar with $D_s \simeq 4$, whereas $D_s \simeq 78$ for water. A very important conclusion is that metalloproteins that contain buried redox cofactors will have very small outer-sphere reorganization energies.

A key result of Marcus theory is that the free energy of activation displays a quadratic dependence on ΔG° and λ (ignoring work terms). Hence, the reaction rate constant may be written as in Eq. 9:

$$k_{ET} = \nu_n \kappa \exp\left(-\frac{(\lambda + \Delta G^\circ)^2}{4\lambda RT} \right) \tag{9}$$

For intramolecular reactions, the nuclear frequency factor (ν_n) is $\sim 10^{13}\ s^{-1}$. One of the most striking predictions is the so-called "inverted effect": as the driving force of the reaction increases, the reaction rate increases, reaching a maximum at $-\Delta G^\circ = \lambda$; but when $-\Delta G^\circ$ is $> \lambda$, the rate decreases as the driving force increases (Fig. X.2.1). Two free energy regions, depending on the relative magnitudes of $-\Delta G^\circ$ and λ, are thus distinguished. The normal free energy region is defined by $-\Delta G^\circ < \lambda$. In this region, ΔG^* decreases if $-\Delta G^\circ$ increases or if λ decreases. If $-\Delta G^\circ = \lambda$, there is no free energy barrier to the reaction. In the inverted region, defined by $-\Delta G^\circ > \lambda$, ΔG^* increases if λ decreases or if $-\Delta G^\circ$ increases.

When a donor (**D**) and an acceptor (**A**) are separated by a long distance, as in protein electron transfer, the **D**–**A** electronic coupling is very small such that H_{AB} becomes a critical factor in the rate constant, k_{ET}. In this case, a semiclassical theory is employed, as discussed in Section X.2.3, which includes an analysis of distant electron-transfer reactions.

X.2.3. Semiclassical Theory of Electron Transfer

According to semiclassical theory (Eq. 10), the rate of ET from a donor (**D**) to an acceptor (**A**) held at fixed distance and orientation is a function of

$$k_{ET} = \sqrt{\frac{4\pi^3}{h^2 \lambda k_B T}} H_{AB}^2 \exp\left\{ -\frac{(\Delta G^\circ + \lambda)^2}{4\lambda k_B T} \right\} \tag{10}$$

temperature (T), reaction driving force ($-\Delta G^\circ$), a nuclear reorganization parameter (λ), and an electronic coupling matrix element (H_{AB}).[1,2,4] The reorganization parameter reflects the changes in structure and solvation that result when an electron moves from **D** to **A.** A balance between nuclear reorganization and reaction driving force determines both the transition state configuration and the height of the barrier associated with the ET process. At the optimum driving force ($-\Delta G^\circ = \lambda$), the reaction is activationless, and the rate ($k_{ET}{}^\circ$) is limited only by the strength of the **D**/**A** electronic coupling (Fig. X.2.1). When **D** and **A** are in van der Waals contact, the coupling strength is usually so large that the ET reaction is adiabatic, that is, it occurs every time the transition state configuration is formed.[2] In this adiabatic limit, the ET rate is independent of H_{AB} and the prefactor depends only on the frequency of motion along the reaction coordinate. An ET reaction is nonadiabatic (Eq. 10) when the **D**/**A** interaction is weak and the transition state must be reached many times before an electron is transferred. The electronic coupling element determines the frequency of crossing from reactants (**D** + **A**) to products ($\mathbf{D}^+ + \mathbf{A}^-$) in the region of the transition state.

The barriers to electron exchange between hydrated transition metal ions are readily interpreted in terms of semiclassical theory.[4] The 0.66-eV activation energy for electron exchange between aqua ferrous and ferric complexes, for example, implies a 2.7-eV reorganization energy. The major contribution (1.5 eV) is attributed to the 0.14-Å difference in Fe–O bond lengths in the ferric and ferrous ions. The remainder (1.2 eV) arises from repolarization of the aqueous solvent upon ET. When the ferrous and ferric ions are in contact, this large reorganization energy leads to a 100-ms time constant for electron exchange. Even at short range, then, the barriers to ET between hydrated metal ions are too great for the demands of biological electron flow. To reduce the reorganization energy, proteins must sequester the redox-active metals in hydrophobic cavities, away from the polar aqueous solvent. In this way, a threefold decrease in reorganization energy can be achieved, decreasing the time constant for electron exchange by nine orders of magnitude (Fig. X.2.2).

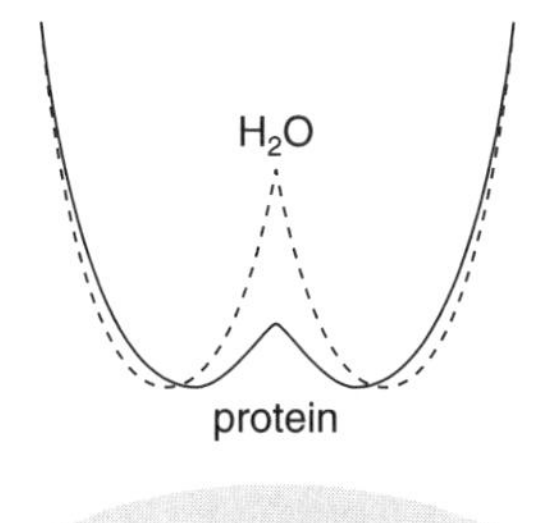

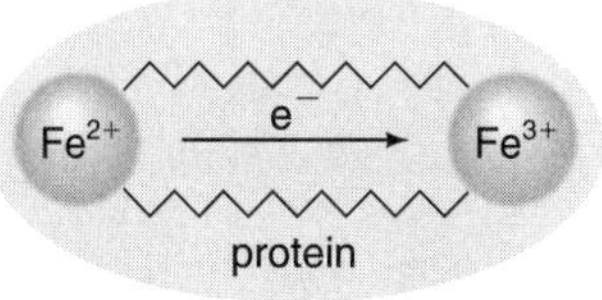

Fig. X.2.2.
Schematic representation of the free energy surfaces for electron exchange between hydrated ferrous and ferric ions in aqueous solution and between the ions complexed in a protein.

The singular feature of ET is that reactions can proceed at very high rates when **D** and **A** are separated by long distances. The electron tunnels through a potential barrier between **D** and **A**; for a square barrier, H_{AB} displays an exponential dependence on the distance (**R**) between the reactants.[12] The medium between redox centers is vitally important for long-range ET. Owing to a 3.5-Å^{-1} distance decay constant (β), the time required for electron exchange between hydrated ferrous and ferric ions is estimated to be 10^{17} years if the complexes are separated by 20 Å in a vacuum.[1] Superexchange coupling via hole and electron states of the intervening medium enhances the **D**/**A** electronic interaction and produces a more gradual decrease in rate with distance. Fill the void between hydrated ferrous and ferric ions with water ($\beta = 1.65\,\text{Å}^{-1}$)[13] and the time constant for 20-Å electron exchange decreases dramatically (5×10^4 years), but the reaction is still far too slow to support biological activity. If the distance decay factor for ET across a polypeptide is comparable to that found for electron tunneling across hydrocarbon bridges ($\beta = 0.8$–$1.0\,\text{Å}^{-1}$),[1] then the time for a 20-Å electron exchange between complexed ferrous and ferric ions in the hydrophobic interior of a protein could be in the millisecond to microsecond range. It is clear, then, that in addition to lowering reorganization barriers, the protein plays an important electronic coupling role.

Investigations of the dependences of ET rates on reaction driving force and temperature can be used to evaluate reorganization energies. Although conceptually straightforward, both methods have difficulties. In order to vary the reaction driving force, changes in one or both redox sites are required. These chemical modifications must be chosen with care to ensure that λ does not change along with $\Delta G°$. Studies of the temperature dependence of k_{ET} are easier in principle, but will not provide accurate λ values unless the temperature dependence of $\Delta G°$ is determined as well.[4]

X.2.3.1. Reorganization Energies of Ru-Modified Metalloproteins

Investigations of protein–protein ET reactions have led to a better understanding of biological electron flow.[5,6,14] Natural systems, however, often are not amenable to the systematic studies that are required to elucidate the factors that control biological ET reactions. A successful alternative approach involves measurements of ET in metalloproteins that have been labeled with redox-active molecules.[9] Ruthenium complexes have been employed as probes in a great many investigations, as Ru(II)-aquo reagents react readily with surface His residues to form stable protein derivatives.

Ru–cytochrome *c*. Electron transfer in a Ru-modified protein was first measured in $Ru(NH_3)_5(His\ 33)^{3+}$–ferricytochrome *c*.[15] The rate of Ru^{2+} to Fe^{3+} electron transfer over a distance of 18 Å at a driving force of 0.2 eV is $30\,s^{-1}$. Replacement of the native Fe center in the heme with Zn facilitated ET measurements at higher driving forces.[10] The driving-force dependence of ET rates in $Ru(NH_3)_4L(His\ 33)$-Zn-cyt *c* (L = NH_3, pyridine, isonicotinamide) yielded the parameters $\lambda = 1.15$ eV

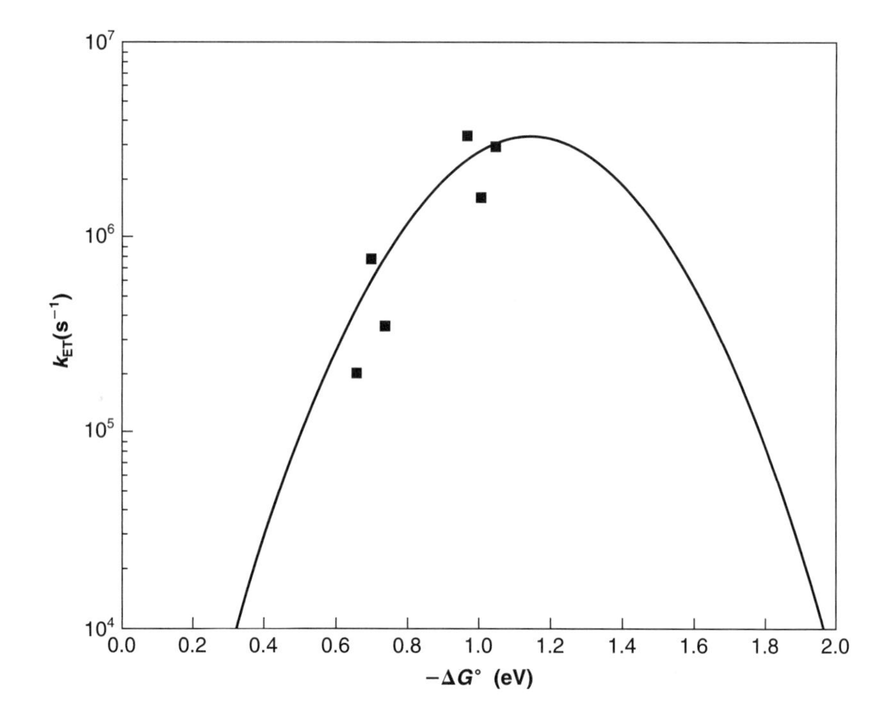

Fig. X.2.3.
Driving-force dependence of intramolecular ET rates in Ru-ammine-His-33 modified Zn-substituted cytochrome *c* (■). Solid lines were generated using Eq. 10 and the following parameters: $\lambda = 1.15\,\text{eV}$, $H_{AB} = 0.10\,\text{cm}^{-1}$.

and $H_{AB} = 0.1\,\text{cm}^{-1}$ (Fig. X.2.3).[10] According to the Marcus cross-relation $(\lambda_{12} = \frac{1}{2}[\lambda_{11} + \lambda_{22}])$, the reorganization energy for the cross-reaction between **D** and **A** (λ_{12}) is the mean of the reorganization energies for electron exchange between **D** and $\mathbf{D}^+$ (λ_{11}) and exchange between **A** and $\mathbf{A}^-$ (λ_{22}).[4] The Fe-cyt *c* self-exchange reorganization energy has been estimated to be 0.7 eV.[16] If the Zn-cyt *c* reorganization energy is comparable to that of the Fe protein, then it is apparent that the hydrophilic Ru–ammine complex is responsible for about two-thirds of the total reorganization energy in Ru–ammine-modified cyt *c* ET reactions $[\lambda(Ru^{3+/2+}) = 1.7\,\text{eV}]$. This result is in good agreement with estimates of the self-exchange reorganization energy in $Ru(NH_3)_6{}^{3+/2+}$.[4]

A great deal of work on Ru-modified proteins has employed the $Ru(bpy)_2(im)(HisX)^{2+}$ (bpy = 2,2′-bipyridine; Im = imidazole) label (Fig. X.2.4).[9] In addition to favorable ET properties, these Ru–bpy complexes have long-lived, luminescent metal-to-ligand charge-transfer excited states that can be prepared with short laser pulses. These excited states enable a wider range of ET measurements than is possible with nonluminescent complexes. Furthermore, the bpy ligands raise the $Ru^{3+/2+}$ reduction potential [>1 V versus normal hydrogen electrode (NHE)] so that observed ET rates are close to $k_{ET}°$, improving the reliability of H_{AB} and λ determinations.

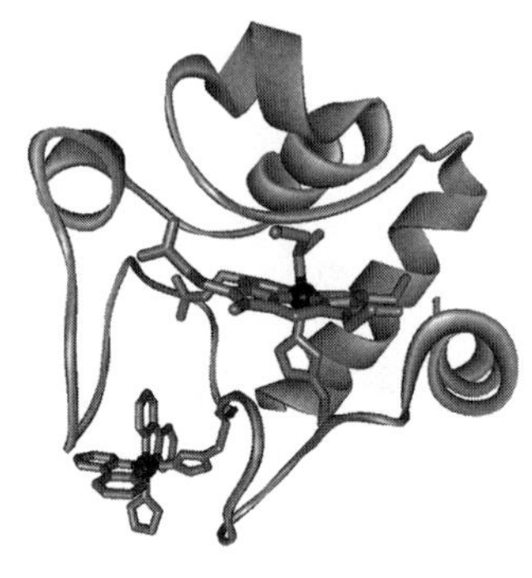

Fig. X.2.4.
Ribbon structure representation of the structure of $Ru(bpy)_2(Im)(His33)^{2+}$-labeled cytochrome *c*.

A study of the driving-force dependence of $Fe^{2+} \rightarrow Ru^{3+}$ ET rates in $Ru(LL)_2(Im)(His33)$-Fe-cyt *c* [LL = bpy, 4,4′-$(CH_3)_2$-bpy, 4,4′,5,5′-$(CH_3)_4$-bpy, 4,4′-$(CONH(C_2H_5))_2$-bpy] gave $\lambda = 0.74\,\text{eV}$ and $H_{AB} = 0.095\,\text{cm}^{-1}$ (Fig. X.2.5).[17,18] The 0.4 eV decrease in reorganization energy resulting from replacement of the Ru–ammine complex with a Ru–bpy label is in excellent agreement with estimates from cross-reactions of model complexes.[4]

The large difference in reorganization energy between Ru–ammine and Ru–bpy modified cytochromes highlights the important role of water in protein ET. The bulky bpy ligands shield the charged metal center from the polar aqueous solution, reducing the solvent reorganization energy. In the same manner, the medium surrounding a metalloprotein active site will affect the reorganization energy associated

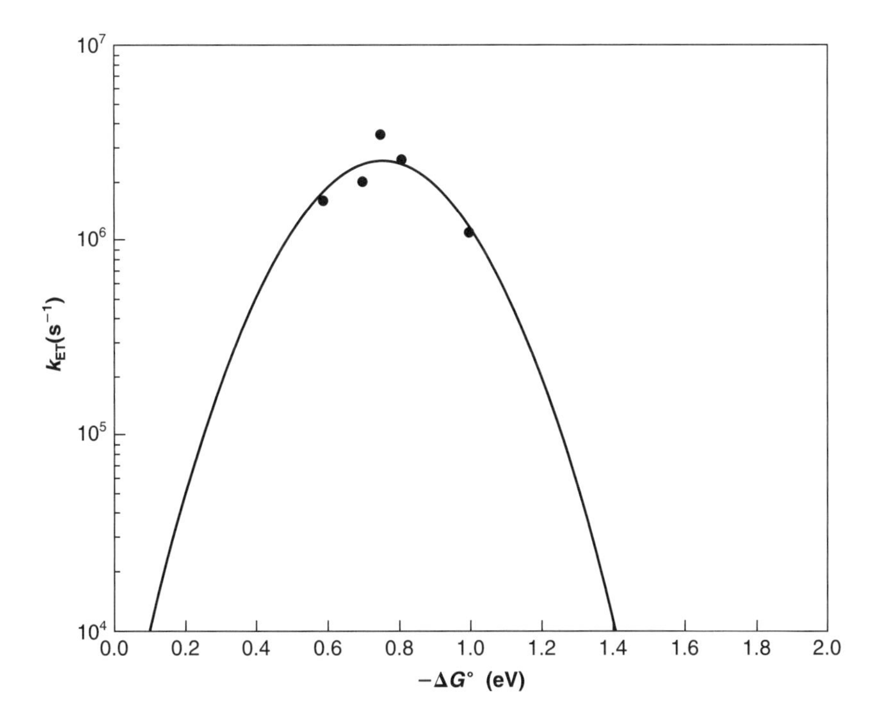

Fig. X.2.5. Driving-force dependence of intramolecular ET rates in Ru-bpy-His33 modified Fe–cytochrome *c* (●). Solid lines were generated using Eq. 10 and the following parameters: $\lambda = 0.74\,\text{eV}$, $H_{AB} = 0.095\,\text{cm}^{-1}$.

with its ET reactions. A hydrophilic active site will lead to larger reorganization energies than a hydrophobic site. Consequently, the kinetics of protein ET reactions will be very sensitive to the active-site environment.

The rates of heme reduction by $^*Ru^{2+}$ (electronically excited Ru^{2+}) and Ru^+ in Ru(His33)cyt *c* have been examined at very high driving forces ($1.3\,\text{eV} \leq -\Delta G° \leq 1.9\,\text{eV}$).[18] The semiclassical theory predicts significant inverted effects in this driving force regime, but at driving forces $>1.3\,\text{eV}$, rates leveled at a value eightfold below the maximum ET rate. Rate–energy leveling is a common phenomenon, particularly in excited-state ET reactions; formation of electronically excited products is a likely explanation for the absence of inverted effects.[18] In the case of cytochrome *c* heme reduction by Ru^+, reactions that form low-lying metal-to-ligand charge transfer excited states of the ferroheme [Fe($d\pi$) → Porphyrin (π^*); 1.05–1.3 eV] should be faster than reactions forming ground-state products. The low-lying excited states of ferro- and ferrihemes are likely to mask inverted driving-force effects in the ET reactions of heme proteins.

Ru–azurin. Electron tunneling in Ru-modified *Pseudomonas aeruginosa* azurin has been investigated both in solutions and in single crystals. The reorganization energy for electron exchange in this blue copper protein has been obtained from analysis of the temperature dependence of the rate of Cu^+ to Ru^{3+} ET in $Ru(bpy)_2(Im)$(His 83)azurin between 170 and 308 K (Fig. X.2.6). Data over this wide temperature range, along with determinations of the temperature dependence of $\Delta G°$, constrain λ to a value of $0.7 \pm 0.1\,\text{eV}$. The reorganization energy for $Cu^+ \rightarrow Ru(LL)_2(Im)(His\ 83)^{3+}$ ET in azurin is the same, within error, as that for $Fe^{2+} \rightarrow Ru(LL)_2(Im)(His\ 33)^{3+}$ ET in cytochrome *c*, indicating that the two redox proteins have comparable self-exchange reorganization energies ($0.7 \pm 0.1\,\text{eV}$). The fact that the reorganization energy for electron exchange in $Cu(phen)_2^{2+/+}$ (phen = 1,10-phenanthroline)[19] is almost 2 eV greater than that for azurin illustrates the important role played by the folded polypeptide in lowering barriers to ET reactions.[20]

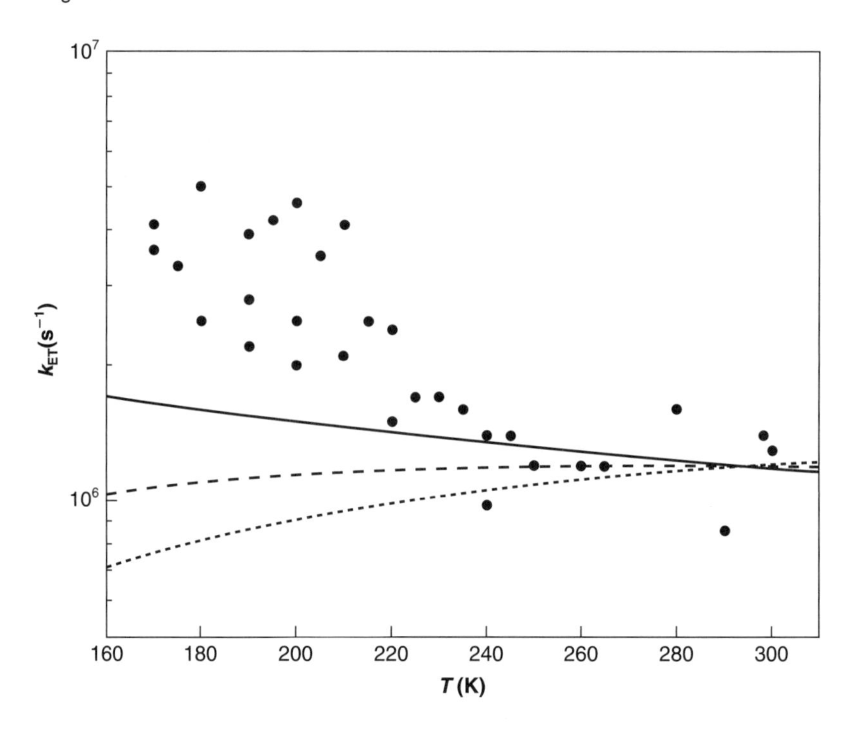

Fig. X.2.6.
Temperature dependence of Cu^+ to Ru^{3+} ET rates in $Ru(bpy)_2(Im)(His83)^{2+}$azurin. Lines were calculated using Eq. 10 with $H_{AB} = 0.07\,cm^{-1}$ and $\lambda = 0.7\,eV$ (solid line); 0.5 eV (dashed line); and 1.0 eV (dotted line).

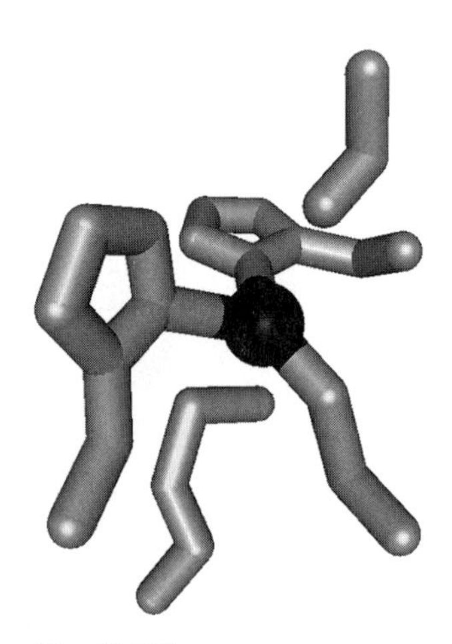

Fig. X.2.7.
Structure of the active-site region of *P. aeruginosa* azurin.

Hydrated Cu ions and synthetic Cu complexes tend to be rather sluggish redox reagents because of large structural changes that accompany ET.[21] The structure of the active site in blue copper proteins, by contrast, is the same in both oxidation states of the protein (Fig. X.2.7).[21] Whether this structural rigidity is an intrinsic property of the ligand set or is enforced by the protein fold has been a hotly debated topic.[21–23] Regardless of its origin, the key consequence of the oxidation-state invariant active-site structure is a low reorganization energy for ET.[20]

X.2.3.2. Electronic Coupling

Nonadiabatic ET reactions are characterized by weak electronic interaction between the reactants and products at the transition state nuclear configuration ($H_{AB} \ll k_B T$). This coupling is directly related to the strength of the electronic interaction between the donor and acceptor.[24] When donors and acceptors are separated by long distances (> 10 Å), direct overlap of their wave functions is vanishingly small; the material between the two redox sites must mediate the coupling.

For electron tunneling through a square potential barrier, the electronic coupling matrix element (H_{AB}) drops off exponentially with increasing **D**/**A** separation.[3,12] The height of the tunneling barrier relative to the energies of the **D**/**A** states determines the distance decay constant (β). Hopfield estimated $\beta \sim 1.4\,Å^{-1}$ on the basis of measurements of the temperature dependence of ET from a cytochrome to the oxidized special pair of chlorophylls in the photosynthetic reaction center of *Chromatium vinosum*.[12] His analysis predicted that the heme edge of the cytochrome would be 8 Å from the nearest edge of the special pair; later structural studies revealed that the actual distance was somewhat greater (12.3 Å).[25]

Superexchange Coupling. Coupling **D** to **A** through electronic interactions with the intervening bridge is called superexchange. If oxidized states of the bridge mediate the coupling, the process is referred to as "hole transfer"; mediation by reduced bridge states is known as "electron transfer". In 1961, McConnell developed a nearest-neighbor superexchange coupling model to describe charge-transfer interactions between donors and acceptors separated by spacers comprised of identical repeat units.[26] The total coupling (H_{AB}) is given as the product of the coupling decays for each bridge site (ε). For a bridge built from identical repeat units separated by m bonds, H_{AB} will be proportional to ε^m. In this model, the ET rate constant will exhibit an exponential dependence on the number of bonds separating **D** and **A**; experimental studies on synthetic **D**-*br*-**A** complexes support this prediction, br = bridge.[1]

Tunneling Pathways in Proteins. The McConnell superexchange model is too simplistic for a protein intervening medium because of the complex array of bonded and nonbonded contacts between **D** and **A.** An important advance was made by Beratan Onuchic, and co-workers[24] who developed a generalized tunneling pathway (TP) superexchange coupling model that reduces the diverse interactions between the atoms in a folded polypeptide to a set of three types of contacts: covalent bonds, hydrogen bonds, and through space contacts. Each type of contact is assigned a coupling decay value (ε_C, ε_H, and ε_S), which permits implementation of a search algorithm for finding optimal coupling pathways through proteins. The total coupling of a single pathway is given as a product of the couplings for the individual links (Eq. 11).[11]

$$H_{AB} \propto \prod \varepsilon_C \prod \varepsilon_H \prod \varepsilon_S \tag{11}$$

A tunneling pathway can be described in terms of an effective covalent tunneling path comprised of n (nonintegral) covalent bonds, with a total length equal to σ_l (Eq. 12). The relationship between σ_l and the direct **D**–**A** distance (**R**) reflects

$$H_{AB} \propto (\varepsilon_C)^n \tag{12a}$$

$$\sigma_l = n \times 1.4\,\text{Å/bond} \tag{12b}$$

the coupling efficiency of a pathway. The variation of ET rates with **R** is expected to depend on the coupling decay for a single covalent bond (ε_C); the magnitude of ε_C depends critically on the energy of the tunneling electron relative to the energies of the bridge hole and electron states.

Rate–Distance Dependence. The **D**–**A** distance decay of protein ET rate constants depends on the capacity of the polypeptide matrix to mediate distant electronic couplings. If dominant coupling pathways mediate long-range ET in proteins, then single-site mutations could have profound effects on enzyme function. In addition, if single pathways operate in biological ET reactions, then they have presumably been optimized through natural selection. These consequences of tunneling pathways impart a certain lack of robustness into the protein structure–function relationship. Concerns about this issue led Dutton and co-workers[25] to propose that a folded polypeptide matrix behaves like a glassy solvent, imposing a uniform barrier (UB) to electron tunneling. Analysis of a variety of ET rates, especially those from the photosynthetic reaction center, produced a universal distance decay constant for protein ET that was in remarkable agreement with Hopfield's estimate ($1.4\,\text{Å}^{-1}$).[25] Disagreements over the appropriate **D**–**A** distance measure (edge-to-edge vs. center-to-center) fueled disputes about whether the large body of protein ET data supports a homogeneous barrier model, or whether a structure-dependent model is necessary. Recently, the UB model has been amended to include the packing density of the protein matrix.[27] Although this model ignores bond connectivity, it does embody many of the same elements as the TP model by accounting for, in a rudimentary fashion, the protein structure dependence of long-range couplings.

The great strength of the TP model is that a straightforward analysis of a protein structure identifies residues that are important for mediating long-range coupling. Employing this model, Beratan, Betts, and Onuchic predicted in 1991 that proteins comprised largely of β-sheet structures would be more effective at mediating long-range couplings than those built from α helices.[28] This analysis can be taken a step further by comparing the coupling efficiencies of individual protein secondary structural elements (β sheets, α helices). The coupling efficiency can be determined from the variation of σ_l as a function of **R**. A linear $\sigma_l/\mathbf{R}$ relationship implies that k°_{ET} will be an exponential function of **R**; the distance decay constant is determined by the slope of the $\sigma_l/\mathbf{R}$ plot and the value of ε_C.[9]

A β sheet is comprised of extended polypeptide chains interconnected by hydrogen bonds; the individual strands of β sheets define nearly linear coupling pathways along the peptide backbone spanning 3.4 Å/residue. The tunneling length for a β strand exhibits an excellent linear correlation with β-carbon separation ($\mathbf{R}_\beta$); the best linear fit with zero intercept yields a slope of 1.37 $\sigma_l/\mathbf{R}_\beta$ (distance decay constant = 1.0 Å^{-1}) (Fig. X.2.8). Couplings across a β sheet depend on the ability of hydrogen bonds to mediate the **D**–**A** interaction. The standard parameterization of the TP model defines the coupling decay across a hydrogen bond in terms of the heteroatom (i.e., N, O) separation. If the two heteroatoms are separated by twice the 1.4-Å covalent-bond distance, then the hydrogen-bond decay is assigned a value equal to that of a covalent bond. Longer heteroatom separations lead to weaker predicted couplings but, as yet, there is no experimental confirmation of this relationship.

In the coiled α-helix structure, a linear distance of just 1.5 Å is spanned per residue along the helix axis. In the absence of mediation by hydrogen bonds, σ_l is a very steep function of $\mathbf{R}_\beta$, implying that an α helix is a poor conductor of electronic coupling (2.7 $\sigma_l/\mathbf{R}_\beta$, distance decay constant = 1.97 Å^{-1}). If the hydrogen-bond networks in α helices mediate coupling, then the Beratan–Onuchic parameterization of hydrogen-bond couplings suggests a $\sigma_l/\mathbf{R}_\beta$ ratio of 1.72 (distance decay constant = 1.26 Å^{-1}) (Fig. X.2.8). Treating hydrogen bonds as covalent bonds further reduces this ratio (1.29 $\sigma_l/\mathbf{R}_\beta$, distance decay constant = 0.94 Å^{-1}). Hydrogen-bond interactions, then, will determine whether α helices are vastly inferior to or are slightly

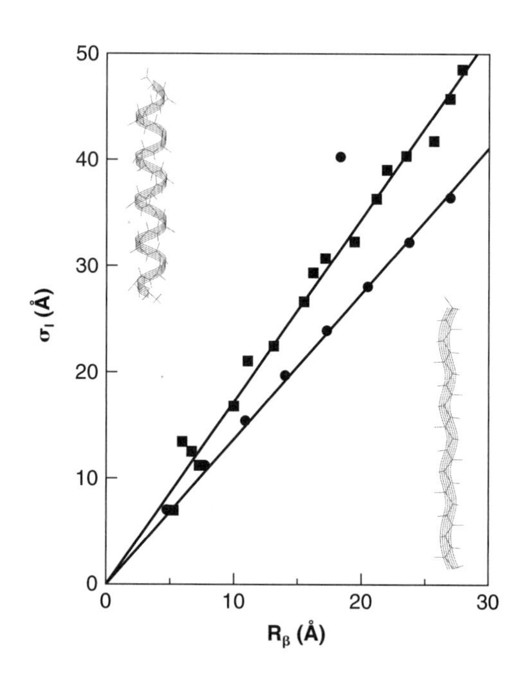

Fig. X.2.8. Plot of calculated tunneling path length (σ_l) versus β-carbon separation ($\mathbf{R}_\beta$) for ET along an idealized α helix (■, using the standard hydrogen-bond coupling parameters in the TP model) and a β strand (●).

better than β sheets in mediating long-range electronic couplings. It is important to note that the coiled helical structure leads to poorer $\sigma_l/\mathbf{R}_\beta$ correlations, especially for values of $\mathbf{R}_\beta$ under 10 Å. In this distance region, the TP model predicts little variation in coupling efficiencies for the different secondary structures.

X.2.3.3. Tunneling Timetables

Plots of coupling limited tunneling times ($1/k_{ET}°$) versus distance (**R**) are called tunneling timetables.[9] When comparing tunneling times from systems with different donors and/or acceptors, it can be difficult to identify a proper distance measure. So-called edge-to-edge distances are often employed, but there are many ambiguities, not the least of which is defining the set of atoms that constitute the edges of **D** and **A.** For planar aromatic molecules (e.g., chlorophylls, pheophytins, quinones), edge–edge separations are usually defined on the basis of the shortest distance between aromatic carbon atoms of **D** and **A.** In transition metal complexes (e.g., Fe–heme, Ru–ammine, Ru–bpy), however, atoms on the periphery are not always well coupled to the central metal, and empirical evidence suggests that metal–metal distances are more appropriate.

Blue Copper Proteins. *Pseudomonas aeruginosa* azurin is a prototypal blue protein: embedded in its β-barrel tertiary structure is a Cu atom that is coordinated to Cys 112 (S), His 117 (N), and His 46 (N) donor atoms in a trigonal-planar structure, with weakly interacting Met 121 (S) and Gly 45 (carbonyl O atom) ligands above and below the plane.[21] Individual β strands extend from these ligands forming a β sheet. Coupling-limited tunneling times ($1/k_{ET}°$) have been obtained for Ru-modified azurin mutants with His residues at different sites on the β strands extending from Met 121 (His 122, His 124, His 126) and Cys 112 (His 109, His 107).[29–31] The variation of tunneling time with the Cu–Ru separation (exponential decay constant of 1.1 Å^{-1}) is in remarkably good agreement with the predicted value of 1.0 Å^{-1} for a strand of an ideal β sheet (Fig. X.2.9). Detailed electronic structure calculations indicate that the S atom of Cys 112 has by far the strongest coupling to the Cu center;

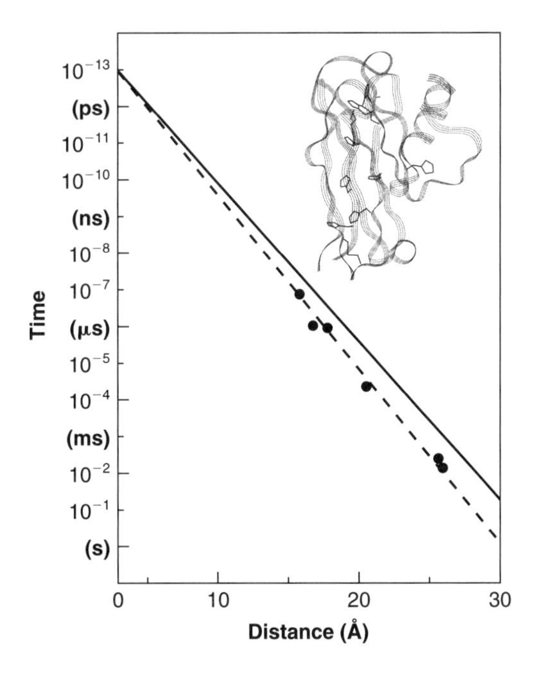

Fig. X.2.9.
Tunneling timetable for ET in Ru-modified azurin. The solid line is the TP prediction for coupling along a β strand ($\beta = 1.0$ Å^{-1}); the dashed line illustrates a 1.1 Å^{-1} distance decay.

the His (Im) couplings are only one-third that of the Cys ligand, and the Met 121 (S) and Gly 45 (O) couplings are just a tenth of the Cys coupling.[32] These highly anisotropic ligand interactions strongly favor pathways that couple to the Cu through Cys112. Couplings along different β strands would be expected to have the same distance-decay constants, but different intercepts at close contact. Relatively strong Cu/Ru couplings also have been found for ET reactions involving Ru-modified His 83.[33]

One explanation for uniform distance dependence of couplings along the Met 121 and Cys 112 strands is that strong interstrand hydrogen bonds serve to direct all of the distant couplings through the Cys 112 ligand.[34] A hydrogen bond between Met 121(O) and Cys 112(NH) could mediate coupling from the Ru complex bound to His 122. A second hydrogen bond [Gly 123(O)–Phe 110(NH)] would provide a coupling link for His 124 and His 126 ET reactions. The importance of the pathways that cross from the Met 121 strand to the Cys 112 strand depends on the coupling efficiencies of the hydrogen bonds. Model-complex studies have demonstrated efficient ET across hydrogen-bonded interfaces.[1] In the standard Beratan–Onuchic pathway model, hydrogen-bond couplings are distance scaled and generally afford weaker couplings than covalent bonds.[11] This procedure for calculating hydrogen-bond couplings cannot explain the similar distance dependences of ET along the Met 121 and Cys 112 strands in Ru-modified azurins. Treating the hydrogen bonds as covalent bonds in the TP model ($\varepsilon_H = \varepsilon_C^2$), however, does lead to better agreement with experiment.[33]

Long-range ET from the Cys 3–Cys 26 disulfide radical anion to the Cu in azurin has been studied extensively by Farver and Pecht.[35] Estimates based on experimental rate data indicate that the S_2/Cu coupling is unusually strong for a donor–acceptor pair separated by 26 Å. Interestingly, both the Cys 3–Cys 26 and His 83 tunneling times fit on the 1.1-$Å^{-1}$ distance decay defined by the couplings along the Met 121 and Cys 112 strands.

Rates of Cu^+ to Ru^{3+} electron transfer also have been measured in modified mutants of spinach plastocyanin, a blue copper protein from the photosynthetic ET chain.[36] The Ru–bpy complexes were introduced at surface sites with Cu–Ru distances ranging from 13 to 24 Å. The ET rate constants, measured using laser flash-quench techniques, vary from 10^4 to $10^7\ s^{-1}$. Electron transfer in Ru-modified plastocyanin is not activationless as it is in Ru-modified azurin, suggesting a slightly greater reorganization energy for the photosynthetic protein. The distance dependence of ET in Ru-modified plastocyanin is exponential with a distance decay factor identical with that reported for Ru-modified azurin (1.1 $Å^{-1}$).

The reduced forms of several blue copper proteins are known to undergo a pH dependent change in Cu ligation; the pK_a values for these transformations lie in the 5–7 range. In the low-pH forms of these proteins an imidazole ligand dissociates from the Cu center and is protonated.[37] The reduction potential of the Cu site in the low pH form of the protein, owing to the trigonal-coordination geometry composed of S(Cys), N(His), and S(Met) ligands, is substantially greater than that of the neutral pH form. The pH dependence of Cu^+ to Ru^{3+} ET has been examined in *Scenedesmus obliquus* plastocyanin labeled with a Ru complex at His 59 (Ru(tpy)(bpy)(His 59)$^{2+}$) (tpy = 2,2′:6,2″-terpyridine).[38] The pK_a for His 87 dissociation in the reduced form of this protein is 5.5.[39] At pH 7 the Cu^+ to Ru^{3+} ET rate is $2.9(2) \times 10^7\ s^{-1}$. Biphasic ET kinetics are observed in acidic solutions (pH 5.6–4.1). The rate for the faster phase agrees well with that measured at pH 7 and has been attributed to oxidation of a population of protein in the high pH form. The slower phase corresponds to oxidation of the low-pH, trigonal Cu sites. The strong temperature and pH dependences of the slower reaction suggest that Cu^+ rearranges to the high-pH tetrahedral geometry prior to electron transfer to Ru^{3+}.[37]

Heme Proteins. Electron-transfer rates have been measured in eight different $Ru(bpy)_2(Im)(HisX)^{2+}$ derivatives of wild-type and mutant cytochromes *c*. Maximum ET rates do not correlate well with a simple exponential distance dependence (Fig. X.2.10). Two modified proteins, for example, have comparable ET rates [Ru(His 72), $9.0 \times 10^5\ s^{-1}$; Ru(His 39), $3.2 \times 10^6\ s^{-1}$], yet the Ru–Fe distances differ by 6.5 Å (His 72, 13.8 Å; His 39, 20.3 Å). Moreover, the **D**–**A** distances in the Ru(His 39) and Ru(His 62) derivatives are nearly identical (20.3 and 20.2 Å, respectively), yet their maximum ET rates differ by a factor of 300 (3.2×10^6 and $1.0 \times 10^4\ s^{-1}$, respectively). The scatter in the data illustrates conclusively that the UB model does not adequately describe long-range couplings in proteins; a model that takes into account the structure of the bridging medium is required to explain the data.

Donor–acceptor pairs separated by α helices include the heme–Ru redox sites in two Ru-modified myoglobins (Mb), $Ru(bpy)_2(Im)$(His X)–Mb (X = 83, 95).[40] The tunneling pathway from His 95 to the Mb–heme is comprised of a short section of α helix terminating at His 93, the heme axial ligand. The coupling for the [$Fe^{2+} \rightarrow Ru^{3+}$(His 95)]-Mb ET reaction[40] is of the same magnitude as that found in Ru-modified azurins with comparable **D**–**A** spacings. This result is consistent with the TP model, which predicts very little difference in the coupling efficiencies of α helices and β sheets at small **D**–**A** separations. The [$Fe^{2+} \rightarrow Ru^{3+}$(His 83)]-Mb tunneling time,[40] however, is substantially longer than those found in β-sheet structures at similar separations, in accord with the predicted distance-decay constant for an α helix (Fig. X.2.11).

Electron-transfer rate data are available for nine Ru-modified derivatives of cytochrome b_{562}, a four-helix-bundle protein.[41] The tunneling times for Ru-modified b_{562} exhibit far more scatter than was found for Ru-modified azurin. Two derivatives exhibit ET rates close to those predicted for coupling along a simple α helix, and several others lie close to the β-strand decay (Fig. X.2.11). In these proteins, as in Ru(His 70)Mb, the intervening medium is not a simple section of α helix. Coupling across helices, perhaps on multiple interfering pathways, is likely to produce a complex distance dependence.

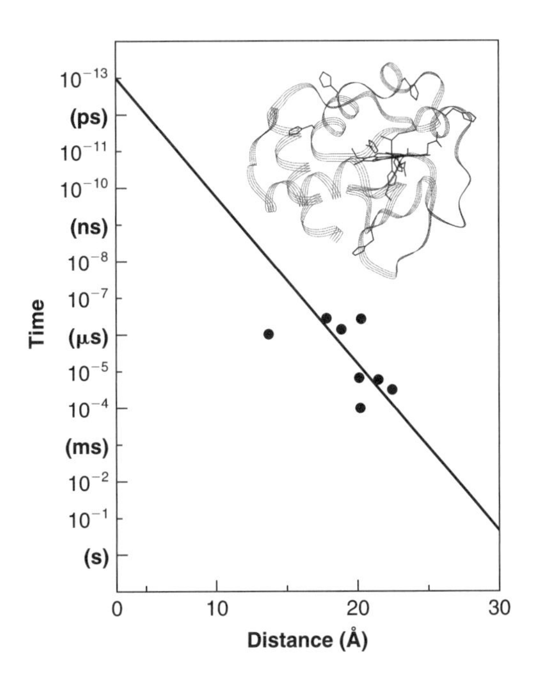

Fig. X.2.10.
Tunneling timetable for ET in Ru-modified cytochrome *c*. The solid line illustrates a 1.05-Å^{-1} distance decay.

Water. The relevant solvent for protein ET is water. Indeed, aqueous-solution redox processes pervade chemistry and biology and ET reactions in water have been among the most intensively studied.[1,2,4] In 1984, Larsson[42] suggested that long-range ET in water would be inefficient ($\beta = 2.4\,Å^{-1}$) because of the large energy gap between the hole states of water and those of **D** and **A.** More recent theoretical treatments, however, have produced β values in the 1.0–1.8-$Å^{-1}$ range.[43,44]

An experimental investigation of $[Ru(tpy)_2]^{2+}$ luminescence quenching by $[Fe(OH_2)_6]^{3+}$ in aqueous acidic glasses has produced a well-defined value of β.[13] In the absence of quenchers, the luminescence lifetime of $[Ru(tpy)_2]^{2+}$ is 8.0 μs in H_2SO_4/H_2O and HSO_3F/H_2O glasses (25% v/v) at 77 K, and 10.2 μs in D_2SO_4/D_2O (25% v/v). Addition of moderate concentrations of the powerfully oxidizing $[Fe(OH_2)_6]^{3+}$ ion (0.01–0.5 *M*, $E° = 0.77$ V vs. NHE) to the glasses leads to accelerated and highly nonexponential $[{}^*Ru(tpy)_2]^{2+}$ decay kinetics. Distance decay constants for $[Fe(OH_2)_6]^{3+}$ quenching of $[{}^*Ru(tpy)_2]^{2+}$ in aqueous glasses were extracted from measurements of luminescence decay kinetics as functions of quencher concentration. The luminescence lifetime of $[{}^*Ru(tpy)_2]^{2+}$ in aqueous glasses is long enough to allow a significant distance range (~25 Å) to be probed. Quantum yield data were used to scale the intensities of the decay kinetics so that just two parameters were required to fit the data. A distance decay constant of $1.65 \pm 0.05\,Å^{-1}$ adequately describes ET in the three different glasses, and the rate constants for ET at van der Waals contact are $> 10^{11}\,s^{-1}$. Although large concentrations of acid (25% v/v) are required for vitrification, water is still the dominant component in these matrices. On a molar basis, the acidic glasses are >90% H_2O/H_3O^+. The fact that the distance-decay parameter in the HSO_3F glass is virtually identical with that obtained in the H_2SO_4 glass provides additional evidence that the oxo anions are not playing an important coupling role.

The distance decay factor for tunneling through water provides an interesting comparison to results from **D**-br-**A** complexes. Electron transfer across saturated alkane spacers is best described by an exponential distance decay constant of $0.9\,Å^{-1}$.[45] The 1.6-$Å^{-1}$ β value reported for tunneling in 2-methyl tetrahydrofuran

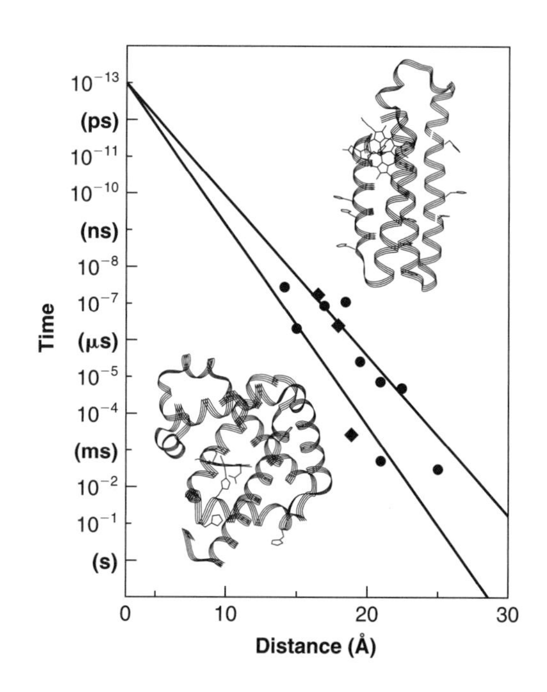

Fig. X.2.11.
Tunneling timetable for ET in Ru-modified myoglobin (♦) and cytochrome b_{562} (●). The solid lines illustrate the TP predictions for coupling along β strands ($\beta = 1.0\,Å^{-1}$) and α helices ($\beta = 1.3\,Å^{-1}$).

glasses[46] can be explained by weaker coupling between solvent molecules than between adjacent carbon atoms in the covalently bonded alkane chain. The average distance decay constant found for tunneling through proteins is 1.1 Å^{-1}, although there is significant deviation from a simple exponential distance dependence because proteins do not provide a homogeneous tunneling barrier. The region representing the distance decay for coupling through water ($\beta = 1.6$–1.7 Å^{-1}) demonstrates that, although better than a vacuum ($\beta = 3$–4 Å^{-1}), tunneling 20 Å through water is at least 100 times slower than tunneling through protein or hydrocarbon bridges.

Water and Proteins. The tunneling timetable for water and proteins is shown in Fig. X.2.12. Virtually all of the observed protein ET rates fall in a zone bound by the predicted distance decays for α helices (1.3 Å^{-1}) and β strands (1.0 Å^{-1}). The data provide compelling support for coupling mediated by the sigma-bonded framework of the protein. The relatively large β value for water indicates that, in addition to large reorganization barriers, this ubiquitous biological solvent also imposes a large tunneling barrier to long-range ET. The poor coupling efficiency of water suggests that pathways involving interstitial water molecules in proteins may not be as effective as all-peptide pathways.

Simple theoretical models (e.g., uniform barrier, tunneling pathway) do not capture all of the critical factors that control the rates of protein ET reactions. Refined pathway models are being developed that, in most cases, aim to identify the atoms most responsible for mediating donor–acceptor electronic couplings.[47] With increasing levels of sophistication in both theory and experiment, new questions are emerging from the study of protein ET. The TP model was based on the static structure of a protein, but it is clear that protein structures are dynamic. How protein dynamics affect long-range couplings is an issue of great current interest. Very long range reactions proceed too slowly to sustain many biological transformations. Multistep tunneling processes, even with endergonic intermediate steps, can compete effectively with single-step long-range reactions.[27,41] Long-range ET reactions via a series of

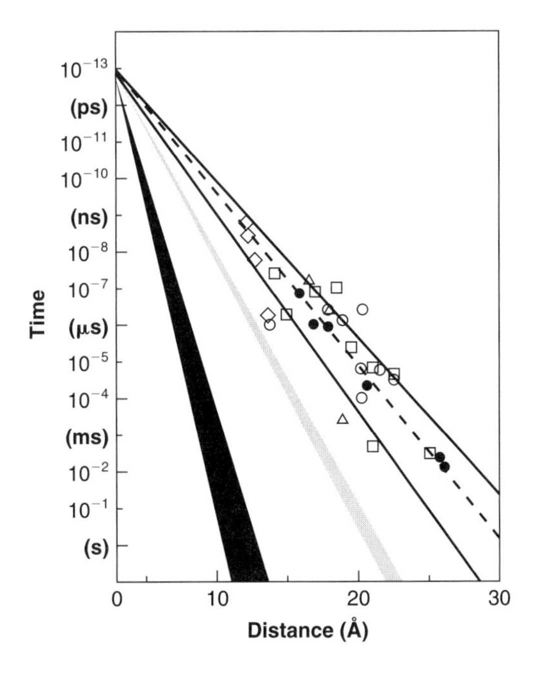

Fig. X.2.12. Tunneling timetable for ET in Ru-modified proteins: azurin (●); cytochrome *c* (○); myoglobin (△); cytochrome b_{562} (□); and HiPIP (HiPIP = high-potential iron–sulfur proteins) (♦). The solid lines illustrate the TP predictions for coupling along β strands ($\beta = 1.0$ Å^{-1}) and α helices ($\beta = 1.3$ Å^{-1}); the dashed line illustrates a 1.1-Å^{-1} distance decay. Distance decay for electron tunneling through water is shown as a gray wedge. Estimated distance dependence for tunneling through vacuum is shown as the black wedge.

real intermediates, rather than the virtual intermediates implicated in pathway models, may play important roles in many biochemical transformations.[48]

Photosynthesis is the process that leads to the storage of solar energy in molecules (ATP and NADPH) that are used in a series of enzymatic reactions to convert CO_2 into organic compounds. The extraction of energy during respiration involves the oxidation of these compounds to CO_2 and H_2O with the concomitant production of water-soluble reductants (NADH and succinate). Electron-transfer reactions lie at the heart of the photosynthesis and respiratory machinery. The structures of the proteins that play key roles in photosynthesis and respiration, as well as mechanisms of their redox reactions, are the subjects of the next section of this chapter (Section X.3).

References

General References

1. Balzani, V. and Balzani, V., Eds., *Electron Transfer in Chemistry,* Wiley-VCH, Weinheim, 2001; Vol. III, pp. 710.
2. Gray, H. B., Ellis, W. R., Jr., Bertini, I., Gray, H. B., Lippard, S. J., and Valentine, J. S., Eds., "Electron Transfer", in *Bioinorganic Chemistry,* University Science Books, Mill Valley, CA, 1994, pp. 315–363.
3. Wasielewski, M. R., Fox, M. A., and Chanon, M., Eds., *Distance Dependencies of Electron-Transfer Reactions,* Elsevier, Amsterdam, The Netherlands, 1988.
4. Marcus, R. A. and Sutin, N., "Electron Transfers in Chemistry and Biology", *Biochim. Biophys. Acta,* **811,** 265–322 (1985).
5. Nocek, J. M., Zhou, J. S., DeForest, S., Priyadarshy, S., Beratan, D. N., Onuchic, J. N., and Hoffman, B. M., "Theory and Practice of Electron Transfer within Protein-Protein Complexes: Application to the Multidomain Binding of Cytochrome *c* by Cytochrome *c* Peroxidase", *Chem. Rev.,* **96,** 2459–2489 (1996).
6. Davidson, V. L., "What Controls the Rates of Interprotein Electron-Transfer Reactions", *Acc. Chem. Res.,* **33,** 87–93 (2000).
7. Gray, H. B. and Winkler, J. R., "Electron Tunneling through Proteins", *Q. Rev. Biophys.,* **36,** 341–372 (2003).
8. Mei, H., Wang, K., Peffer, N., Weatherly, G., Cohen, D. S., Miller, M., Pielak, G., Durham, B., and Millett, F., "Role of Configurational Gating in Intracomplex Electron Transfer from Cytochrome *c* to the Radical Cation in Cytochrome *c* Peroxidase", *Biochemistry,* **38,** 6846–6854 (1999).
9. Gray, H. B. and Winkler, J. R., "Electron Transfer in Proteins", *Annu. Rev. Biochem.,* **65,** 537–561 (1996).
10. Winkler, J. R. and Gray, H. B., "Electron Transfer in Ruthenium-Modified Proteins", *Chem. Rev.,* **92,** 369–379 (1992).
11. Onuchic, J. N., Beratan, D. N., Winkler, J. R., and Gray, H. B., "Pathway Analysis of Protein Electron Transfer Reactions", *Annu. Rev. Biophys. Biomol. Struct.,* **21,** 349–377 (1992).

Specific References

12. Hopfield, J. J., "Electron Transfer between Biological Molecules by Thermally Activated Tunneling", *Proc. Natl. Acad. Sci. U.S.A.,* **71,** 3640–3644 (1974).
13. Ponce, A., Gray, H. B., and Winkler, J. R., "Electron Tunneling through Water: Oxidative Quenching of Electronically Excited $Ru(tpy)_2^{2+}$ (tpy = 2,2′:6,2″-terpyridine) by Ferric Ion in Aqueous Glasses at 77 K", *J. Am. Chem. Soc.,* **122,** 8187–8191 (2000).
14. Millett, F., Miller, M. A., Geren, L., and Durham, B., "Electron Transfer between Cytochrome *c* and Cytochrome *c* Peroxidase", *J. Bioenerg. Biomemb.,* **27,** 341–351 (1995).
15. Winkler, J. R., Nocera, D. G., Yocom, K. M., Bordignon, E., and Gray, H. B., "Electron-Transfer Kinetics of Pentaammineruthenium(III)(histidine-33)-Ferricytochrome *c*. Measurement of the Rate of Intramolecular Electron Transfer between Redox Centers Separated by 15 Å in a Protein", *J. Am. Chem. Soc.,* **104,** 5798–5800 (1982).
16. Andrew, S. M., Thomasson, K. A., and Northrup, S. H., "Simulation of Electron Transfer Self-Exchange in Cytochrome *c* and Cytochrome b_5", *J. Am. Chem. Soc.,* **115,** 5516–5521 (1993).
17. Bjerrum, M. J., Casimiro, D. R., Chang, I.-J., Di Bilio, A. J., Gray, H. B., Hill, M. G., Langen, R., Mines, G. A., Skov, L. K., Winkler, J. R., and Wuttke, D. S., "Electron Transfer in Ruthenium-Modified Proteins", *J. Bioenerg. Biomemb.,* **27,** 295–302 (1995).
18. Mines, G. A., Bjerrum, M. J., Hill, M. G., Casimiro, D. R., Chang, I.-J., Winkler, J. R., and Gray, H. B., "Rates of Heme Oxidation and Reduction in Ru(His 33)cytochrome *c* at Very High Driving Forces", *J. Am. Chem. Soc.,* **118,** 1961–1965 (1996).
19. Augustin, M. A. and Yandell, J. K., "Rates of Electron-Transfer Reactions of Some Copper(II)-Phenanthroline Complexes with Cytochrome *c*(II) and Tris(phenanthroline)cobalt(II) Ion", *Inorg. Chem.,* **18,** 577–583 (1979).
20. Winkler, J. R., Wittung-Stafshede, P., Leckner, J., Malmström, B. G., and Gray, H. B., "Effects of Folding on Metalloprotein Active Sites", *Proc. Natl. Acad. Sci. U.S.A.,* **94,** 4246–4249 (1997).

21. Gray, H. B., Malmström, B. G., and Williams, R. J. P., "Copper Coordination in Blue Proteins", *J. Biol. Inorg. Chem.*, **5,** 551–559 (2000).
22. Ryde, U., Olsson, M. H. M., Roos, B. O., De Kerpel, J. O. A., Pierloot, K., "On the Role of Strain in Blue Copper Proteins", *J. Biol. Inorg. Chem.*, **5,** 565–574 (2000).
23. Randall, D. W., Gamelin, D. R., LaCroix, L. B., and Solomon, E. I., "Electronic structure contributions to electron transfer in blue Cu and Cu_A", *J. Biol. Inorg. Chem.*, **5,** 16–29 (2000).
24. Newton, M. D., "Electronic Structure Analysis of Electron-Transfer Matrix Elements for Transition Metal Redox Pairs", *J. Phys. Chem.*, **92,** 3049–3056 (1988).
25. Moser, C. C., Keske, J. M., Warncke, K., Farid, R. S., and Dutton, P. L., "Nature of Biological Electron Transfer", *Nature (London)*, **355,** 796–802 (1992).
26. McConnell, H. M., "Intramolecular Charge Transfer in Aromatic Free Radicals", *J. Chem. Phys.*, **35,** 508–515 (1961).
27. Page, C. C., Moser, C. C., Chen, X., and Dutton, P. L., "Natural Engineering Principles of Electron Tunnelling in Biological Oxidation-Reduction", *Nature (London)*, **402,** 47–52 (1999).
28. Beratan, D. N., Betts, J. N., and Onuchic, J. N., "Protein Electron Transfer Rates Set by the Bridging Secondary and Tertiary Structure", *Science*, **252,** 1285–1288 (1991).
29. Di Bilio, A. J., Hill, M. G., Bonander, N., Karlsson, B. G., Villahermosa, R. M., Malmström, B. G., Winkler, J. R., and Gray, H. B., "Reorganization Energy of Blue Copper: Effects of Temperature and Driving Force on the Rates of Electron Transfer in Ruthenium- and Osmium-Modified Azurins", *J. Am. Chem. Soc.*, **119,** 9921–9922 (1997).
30. Skov, L. K., Pascher, T., Winkler, J. R., and Gray, H. B., "Rates of Intramolecular Electron Transfer in $Ru(bpy)_2(im)$(His 83)-Modified Azurin Increase below 220 K", *J. Am. Chem. Soc.*, **120,** 1102–1103 (1998).
31. Langen, R., Chang, I.-J., Germanas, J. P., Richards, J. H., Winkler, J. R., and Gray, H. B., "Electron Tunneling in Proteins: Coupling through a β-Strand", *Science*, **268,** 1733–1735 (1995).
32. Guckert, J. A., Lowery, M. D., and Solomon, E. I., "Electronic Structure of the Reduced Blue Copper Active Site—Contributions to the Reduction Potentials and Geometry", *J. Am. Chem. Soc.*, **117,** 2817–2844 (1995).
33. Regan, J. J., Di Bilio, A. J., Langen, R., Skov, L. K., Winkler, J. R., Gray, H. B., and Onuchic, J. N., "Electron Tunneling in Azurin: Coupling across a β Sheet", *Chem. Biol.*, **2,** 489–496 (1995).
34. Daizadeh, I., Gehlen, J. N., and Stuchebrukhov, A. A., "Calculation of Electronic Tunneling Matrix Element in Proteins: Comparison of Exact and Approximate One-Electron Methods for Ru-Modified Azurin", *J. Chem. Phys.*, **106,** 5658–5666 (1997).
35. Farver, O. and Pecht, I., "Copper Proteins as Model Systems for Investigating Intramolecular Electron-Transfer Processes", *Adv. Chem. Phys.*, **107,** 555–589 (1999).
36. Sigfridsson, K., Ejdeback, M., Sundahl, M., and Hansson, O., "Electron Transfer in Ruthenium-Modified Spinach Plastocyanin Mutants", *Arch. Biochem. Biophys.*, **351,** 197–206 (1998).
37. Guss, J. M., Harrowell, P. R., Murata, M., Norris, V. A., and Freeman, H. C., "Crystal Structure Analysis of Reduced (Cu(I)) Poplar Plastocyanin a 6 pH Values", *J. Mol. Biol.*, **192,** 361–387 (1986).
38. Di Bilio, A. J., Dennison, C., Gray, H. B., Ramirez, B. E., Sykes, A. G., and Winkler, J. R., "Electron Transfer in Ruthenium-Modified Plastocyanin", *J. Am. Chem. Soc.*, **120,** 7551–7556 (1998).
39. Sykes, A. G., "Structure and Electron Transfer Reactivity of the Blue Copper Protein Plastocyanin", *Chem. Soc. Rev.*, **14,** 283–315 (1985).
40. Langen, R., Colón, J. L., Casimiro, D. R., Karpishin, T. B., Winkler, J. R., and Gray, H. B., "Electron Tunneling in Proteins. Role of the Intervening Medium", *J. Biol. Inorg. Chem.*, **1,** 221–225 (1996).
41. Winkler, J. R., Di Bilio, A., Farrow, N. A., Richards, J. H., and Gray, H. B., "Electron Tunneling in Biological Molecules", *Pure Appl. Chem.*, **71,** 1753–1764 (1999).
42. Larsson, S., "Electron-Exchange Reactions in Aqueous Solution", *J. Phys. Chem.*, **88,** 1321–1323 (1984).
43. Miller, N. E., Wander, M. C., and Cave, R. J., "A Theoretical Study of the Electronic Coupling Element for Electron Transfer in Water", *J. Phys. Chem. A*, **103,** 1084–1093 (1999).
44. Benjamin, I., Evans, D., and Nitzan, A., "Electron Tunneling through Water Layers: Effect of Layer Structure and Thickness", *J. Chem. Phys.*, **106,** 6647–6654 (1997).
45. Smalley, J. F., Feldberg, S. W., Chidsey, C. E. D., Linford, M. R., Newton, M. D., and Liu, Y.-P., "The Kinetics of Electron Transfer through Ferrocene-Terminated Alkanethiol Monolayers on Gold", *J. Phys. Chem.*, **99,** 13141–13149 (1995).
46. Wenger, O. S., Leigh, B. S., Villahermosa, R. M., Gray, H. B., and Winkler, J. R., "Electron Tunneling through Organic Molecules in Frozen Glasses", *Science*, **307,** 99–102 (2005).
47. Daizadeh, I., Guo, J.-X., and Stuchebrukhov, A. A., "Vortex Structure of the Tunneling Flow in Long-Range Electron Transfer Reactions", *J. Chem. Phys.*, **110,** 8865–8868 (1999).
48. Jortner, J., Bixon, M., Langenbacher, T. L., and Michel-Beyerle, M. E., "Charge Transfer and Transport in DNA", *Proc. Natl. Acad. Sci. U.S.A.*, **95,** 12759–12765 (1998).

X.3. Photosynthesis and Respiration

Shelagh Ferguson-Miller
Biochemistry and Molecular Biology
Michigan State University
East Lansing, MI 48824

Gerald T. Babcock
Chemistry
Michigan State University
East Lansing, MI 48828

Charles Yocum
Chemistry and MCD Biology
University of Michigan
Ann Arbor, MI 48109

Contents

X.3.1. Introduction

In plants, animals, and microorganisms, the energy source for processes within the cell that maintains viability and produces new proteins, new genes, and new cells is adenosine triphosphate (ATP). The cleavage of one of the phosphate groups to produce adenosine diphosphate (ADP) and the cleaved inorganic phosphate group (P_i), according to Eq. 1, releases ~8 kcal of energy.

$$\text{ATP} \rightarrow \text{ADP} + \text{P}_i \tag{1}$$

This energy release is coupled to many synthetic reactions to drive them to product formation. For example, the chemistry involved in adding a new amino acid to a growing polypeptide that is on its way to being a new protein can be written as

$$\text{Polypeptide} + \text{amino acid} \rightarrow \text{polypeptide elongated by one amino acid} \tag{2}$$

By itself, reaction 2 is not spontaneous; rather, it is endergonic by ~32 kcal (the equivalent of 4 high-energy phosphate bonds) and will not proceed to the right to any significant extent. However, if we couple the endergonic reaction to the energy-releasing (exergonic) splitting of ATP into ADP and P_i, the overall reaction becomes spontaneous and significant product formation occurs.

$$\text{Polypeptide} + \text{amino acid} + 4\ \text{ATP} \rightarrow \text{new polypeptide} + 4\ \text{ADP} + 4\ \text{Pi} \tag{3}$$

From this brief example, we see that ATP is the "energy currency" of the cell. Its enzyme-catalyzed cleavage to ADP and P_i is used to drive the diverse chemistry that leads to cell maintenance, growth, and proliferation. But, where and how is the generation of ATP carried out in the cell? That is, how do plants, animals, and microorganisms derive energy from their environments? Two major processes—photosynthesis (in plants, algae, and photosynthetic bacteria) and respiration (in those organisms as well as animals and bacteria)—are used to drive ATP formation. There are fundamental principles that are common to both processes. We will consider these first, before turning to the unique aspects of photosynthesis and respiration. For general references on biological energy production see Refs. 1–13.

X.3.2. Qualitative Aspects of Mitchell's Chemiosmotic Hypothesis for Phosphorylation

The initial models for how phosphorylation occurs invoked an intermediate (**I**) that formed a high-energy bond with inorganic phosphate (I$\sim$P), as found in other enzyme-catalyzed energy-transfer reactions. In this model, the energy source for the formation of I$\sim$P is the electron- and proton-transfer chemistry that we discuss below. In the second step of the process, I$\sim$P is postulated to react with ADP to transfer the phosphate and form ATP. Despite an intense effort to identify I and I$\sim$P, this work was unsuccessful. At the same time, Peter Mitchell, who had a background in the membrane ion gradients that are important for neuronal function, proposed a revolutionary model for phosphorylation.[7,14] Mitchell noted that in eukaryotes, the subcellular organelles: mitochondria and chloroplasts that carry out respiration and photosynthesis are closed, heavily invaginated vesicles. In prokaryotes, which do not have subcellular organelles, the cell membrane forms a similar, sometimes invaginated vesicle. Mitchell incorporated these closed vesicles into his model for phosphorylation; that is, he proposed that a necessary condition for both respiratory and photosynthetic phosphorylation was sealed vesicles relatively impermeable to charged ions, with a clearly defined inside and outside space.

The second tenet of his model was that the proton and electron motions that provided the driving force for phosphorylation occur with a distinct asymmetry. For example, the electrons donated by cytochrome *c* to the terminal respiratory complex, cytochrome oxidase, come from a unique side of the mitochondrial (or cell) membrane in which this enzyme is embedded; conversely, the protons that are required for the conversion of O_2 to water in this reduction process come from the other side of the membrane. Thus, the first two principles of Mitchell's theory for phosphorylation in respiration and photosynthesis, the Chemiosmotic Hypothesis, invoked a requirement for sealed membrane vesicles and the asymmetric organization of the proteins involved so that the proton and electron motions they catalyzed were also asymmetric and vectorial (directional). Before we can explore Mitchell's theory further, however, we must first take a brief detour into the theory of reduction potentials so that we will be able to understand the quantitative aspect of chemiosmosis in photosynthesis and respiration.

X.3.3. An Interlude: Reduction Potentials

Oxygen participates in respiration and photosynthesis, as a substrate that is reduced to water in the former and a product that is made from water in the latter. Similarly, the soluble cofactor NADH (or NADPH) is used in both systems as a carrier of electrons and protons. If we ask whether it is easier to transfer electrons and protons to NAD^+ to make NADH, or to O_2 to make two molecules of H_2O, we are asking a quantitative reduction potential question. To answer this question, experience has taught us that the best way is to define a standard state to which all redox couples (e.g., NAD^+/NADH; O_2/2 H_2O) are referred. By convention, that standard state is the reduction of a proton to one-half of an H_2 molecule at pH 0:

$$H^+ + e^- \rightarrow \tfrac{1}{2}H_2 \qquad E^\circ = 0 \tag{4}$$

The reduction potential, E°, for this process is set, by definition, to zero. In biological processes, the pH 0 state is not very useful, as few relevant reactions are able to proceed under such acidic conditions. However, we can maintain this pH 0 condition as our standard state and slide our reference state to the appropriate pH, keeping in mind that 59 mV/pH unit must be subtracted or added to the standard state, depending on whether protons are the reactant or product of the reaction. We illustrate this procedure for adjusting our pH 0 standard state to a more relevant pH. In the standard state expression for reduction of a proton to hydrogen, a proton is the reactant, and, as we lower its concentration by increasing the pH of the solution, we expect that it will become progressively harder to reduce this proton to $\frac{1}{2}\,H_2$. This is the case, and the potential for H^+ at pH 7 is equal to $-0.413\,V$ as derived from $7 \times 0.059\,V$:

$$H^+ + e^- \rightarrow \tfrac{1}{2}\,H_2 \qquad E^{\circ\prime} = -0.413\,V \tag{5}$$

where the superscript prime indicates that this is the reduction potential for the hydrogen couple at pH 7 (an alternative notation is E'_{m},[7] which is often used in bioenergetic research), and the negative sign is due to the fact that we are subtracting the pH effect since H^+ is a reactant. Measured relative to our pH 0 standard state and then adjusted to the "biological standard state" of pH 7, the reduction potentials for NAD^+ and O_2 are as follows:

$$\tfrac{1}{2}\,NAD^+ + e^- + H^+ \rightarrow \tfrac{1}{2}\,NADH \qquad E^{\circ\prime} = -0.33\,V \tag{6}$$

$$\tfrac{1}{4}\,O_2 + e^- + H^+ \rightarrow \tfrac{1}{2}\,H_2O \qquad E^{\circ\prime} = +0.815\,V \tag{7}$$

(Note that we tabulate these reduction potentials on a *per electron* basis, and the more positive the E° the more readily the reactant accepts electrons and the more favorable is the reaction in the forward direction as written.) Thus, at pH 7, both NAD^+ and O_2 are easier to reduce than H^+, and O_2 is much easier to reduce than NAD^+.

In both biology and chemistry, free electrons do not occur. Thus, when one component is oxidized, another is reduced, and we need to consider the energy of the overall reaction. How do we use the half-cell reactions tabulated above to calculate this? Keeping to our NAD^+/O_2 example, either of two reactions is possible when we mix the NAD^+/NADH couple in air-saturated buffer at pH 7. In the first, H_2O is oxidized and NAD^+ is reduced,

$$\tfrac{1}{2}\,NAD^+ + \tfrac{1}{2}\,H_2O \rightarrow \tfrac{1}{2}\,NADH + \tfrac{1}{4}\,O_2 \tag{8}$$

and in the second, the opposite occurs:

$$\tfrac{1}{2}\,NADH + \tfrac{1}{4}\,O_2 \rightarrow \tfrac{1}{2}\,NAD^+ + \tfrac{1}{2}\,H_2O \tag{9}$$

How do we tell which of the two reactions will occur? First, let us answer this question intuitively: We know, by experience, that when NAD^+ is added to buffer at pH 7, bubbles of O_2 do not appear. Thus, our intuition tells us that the second of the two reactions, not the first, is the spontaneous one. This is, in fact, correct; and now let us see how to predict this by using half-cell potentials. For the first of the two reactions, the half-cell reactions are

$$\tfrac{1}{2}\,NAD^+ + H^+ + e^- \rightarrow \tfrac{1}{2}\,NADH + \tfrac{1}{2}\,H^+ \qquad E^{\circ\prime} = -0.33\,V \tag{10}$$

$$\tfrac{1}{2}\,H_2O \rightarrow \tfrac{1}{4}\,O_2 + H^+ + e^- \qquad -E^{\circ\prime} = -0.815\,V \tag{11}$$

$$\tfrac{1}{2}\,NAD^+ + \tfrac{1}{2}\,H_2O \rightarrow \tfrac{1}{2}\,NADH + \tfrac{1}{4}\,O_2 + \tfrac{1}{2}\,H^+ \qquad \Delta E^{\circ\prime} = -1.145\,V \tag{12}$$

where we have written the second of the two reactions as an oxidation and, accordingly, changed the sign of $E^{\circ\prime}$. To get the $E^{\circ\prime}$ of the overall reaction, we simply total the two half-cell potentials after we have rewritten the appropriate half-cell reaction as an oxidation and changed the sign of its $E^{\circ\prime}$. For the reaction above, we see that an overall $E^{\circ\prime}$ of $-1.145\,V$ is predicted. Because $E^{\circ\prime}$ must be positive for the reaction to be spontaneous, this $E^{\circ\prime}$ value clearly indicates that the reaction is not spontaneous. For the second possibility, we write

$$\tfrac{1}{2}\,\mathrm{NADH} \rightarrow \tfrac{1}{2}\,\mathrm{NAD^+} + \mathrm{H^+} + \mathrm{e^-} \qquad -E^{\circ\prime} = +0.33\,\mathrm{V} \qquad (13)$$

$$\tfrac{1}{4}\,\mathrm{O_2} + \mathrm{H^+} + \mathrm{e^-} \rightarrow \tfrac{1}{2}\,\mathrm{H_2O} \qquad E^{\circ\prime} = +0.815\,\mathrm{V} \qquad (14)$$

$$\tfrac{1}{2}\,\mathrm{NADH} + \tfrac{1}{4}\,\mathrm{O_2} \rightarrow \tfrac{1}{2}\,\mathrm{NAD^+} + \tfrac{1}{2}\,\mathrm{H_2O} \qquad \Delta E^{\circ\prime} = 1.145\,\mathrm{V} \qquad (15)$$

and here we see the prediction that the oxidation of NADH by O_2 is the spontaneous process. The NADH is a very good reductant; O_2 is a very good oxidant; when they react, 1.145 V of free energy is released. This is the free energy that is released in respiration and used to drive phosphorylation of ADP. To make ATP from ADP and $P_i \sim 8$ kcal mol^{-1} of energy is required, but how does that relate to the redox energy in volts? The energy required for chemical synthesis can be compared with the redox energy produced in photosynthesis and respiration using the conversion factor of 23 kcal V^{-1}. Thus, the 1.145 V that we calculate for the NADH/O_2 reaction releases 26.3 kcal per electron transferred through the respiratory chain for ATP synthesis; formally, ($\Delta G^\circ = -n\mathscr{F}\Delta E^{\circ\prime}$).

X.3.4. Maximizing Free Energy and ATP Production

With this brief introduction to reduction potentials, we are now in a position to explore the quantitative aspects of these processes and of Mitchell's chemiosmotic hypothesis.[6] The first lesson that emerges is that the reduction potential difference between the most reducing substrate and the most oxidizing determines the overall energy available for the formation of ATP. In respiration, this is straightforward, as biochemical oxidation of food stuffs provides a continuing supply of the strong reductant, NADH. As we have seen above, this is a molecule that readily donates its electrons to oxygen, and we can generate $\sim$1.145 V = 26.3 kcal of energy for ATP formation with each mole of electrons that we transfer from NADH to O_2. The mitochondrial energetics also highlight the importance of O_2. Because O_2 readily accepts electrons ($E^{\circ\prime} = 0.815$ at pH 7), its use as the terminal respiratory oxidant maximizes the redox potential drop, and, hence, the free energy made available for ATP formation. If we were to use Cl_2 ($E^{\circ\prime} = 1.35$ V) as the terminal acceptor in oxidizing NADH, even more free energy would be available for ATP production (1.35 V + 0.33 V = 1.68 V). Chlorine gas, however, is toxic, owing to the fact that it reacts rapidly and lethally with cell tissue. Moreover, Cl_2 is seldom if ever available in natural systems. Dioxygen, on the other hand, is kinetically inert, owing to its triplet ground state and, hence, is relatively benign in a biological milieu, and used extensively in biological processes (Fig. X.3.1).

To function effectively in respiration, however, O_2 must be regenerated by other biological processes. This takes us into the realm of photosynthesis, where oxygenic organisms have developed a light-generated oxidizing species, a form of chlorophyll that is called P680$^+$ because of its visible light absorption spectrum in the 680-nm region. It is a sufficiently strong oxidant to take electrons from H_2O and produce O_2 as a byproduct (Section X.4.6). The net reaction (Eq. 16)

$$\mathrm{P680^+} + \tfrac{1}{2}\,\mathrm{H_2O} \rightarrow \mathrm{P680} + \tfrac{1}{4}\,\mathrm{O_2} + \mathrm{H^+} \qquad (16)$$

is spontaneous. This chemistry is remarkable, as H_2O is one of the most stable molecules in Nature, hence its abundance. Below, we delve more deeply into the actual chemistry of O_2 production; but for now, we see that we have completed the water–oxygen cycle in biology by using light energy to generate oxygen and reducing equivalents to supply respiration:

$$2\,\mathrm{H_2O} \rightarrow \mathrm{O_2} + 4\,\mathrm{H^+} + 4\,\mathrm{e^-}$$

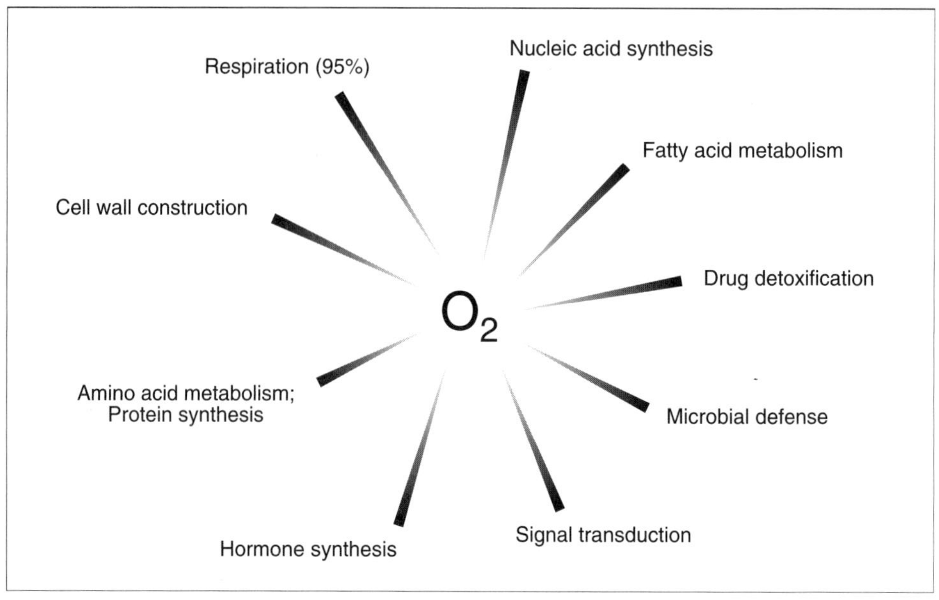

Fig. X.3.1. Some biological reactions in cells involving molecular oxygen.

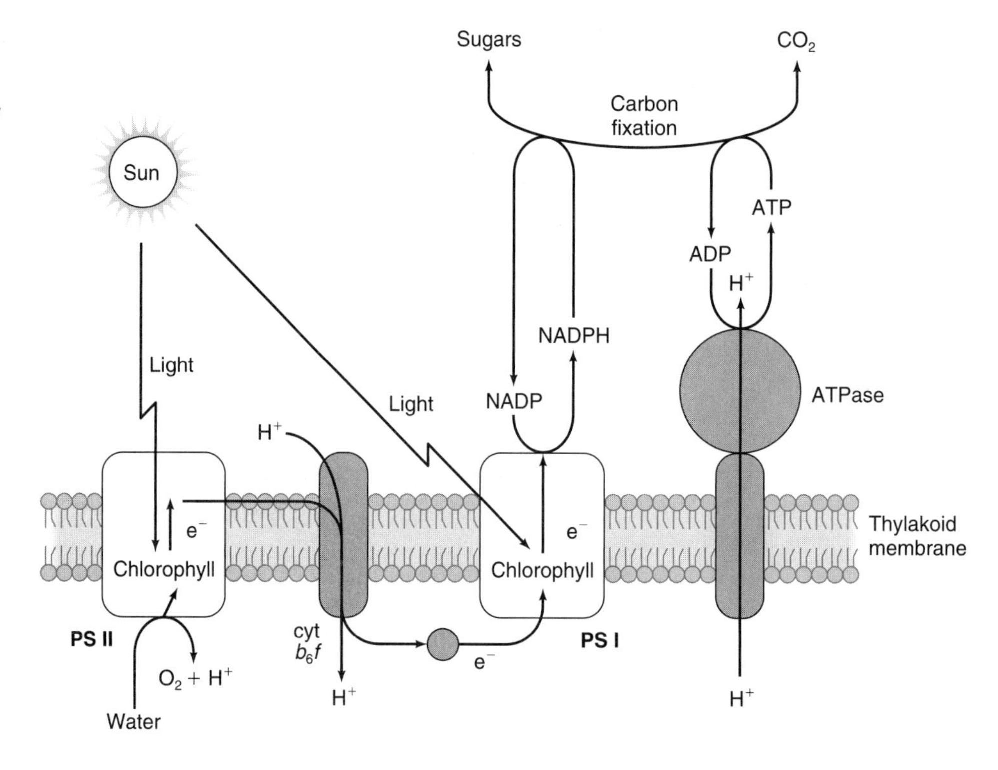

Fig. X.3.2. Overview of photosynthesis. [From Wim Vermaas, http://photoscience.la.asu.edu/photosyn/education/photointro.html]

Only photosynthetic organisms are able to generate O_2 and thus provide the O_2 that we breathe, of which 95% is consumed in the respiratory process to make energy. The remaining 5% is used in vital processes such as protein, lipid, and steroid synthesis, cell signaling, drug metabolism, and a variety of other reactions in which the highly oxidizing potential of O_2 is used to drive the overall process.

Photosynthesis generates O_2 and the ultimate reducing substrate, NADPH, by arranging two light reactions in series (Fig. X.3.2). In the first step, catalyzed by

Photosystem II (PSII), light is used to create chlorophyll $P680^+$, oxidize water, and reduce a quinone molecule, Q_A, not shown in Fig. X.3.2, but discussed below. This lipid-soluble quinone is the recipient of the light-activated electrons from chlorophyll and can transfer them to cytochrome b_6f. In the second light-driven step, catalyzed by Photosystem I (PSI), light oxidizes a second, specialized chlorophyll, P700, and reduces NADP to NADPH. The transfer of electrons from H_2O through PSII and cytochrome b_6f to PSI and $P700^+$ is spontaneous and this process is coupled to proton pumping and ATP formation. Thus, both photosynthesis and respiration are capable of generating useful free energy that is used in phosphorylation.

X.3.5. Quantitative Aspects of Mitchell's Chemiosmotic Hypothesis for Phosphorylation

Rather than take the free energy that results from photosynthetic and respiratory electron transfer in one fell swoop, both systems use a chain of redox cofactors that gradually drop the potential from the most reducing compound (NADH in respiration, H_2O in photosynthesis) to the most oxidizing terminal component (O_2 in respiration, $P700^+$ in oxygenic photosynthesis). In Chapter IV and Section X.1, we learned the identities of these redox-active cofactors; but now we need to consider the underlying rationale and consequences of graded electron–proton transfer for the mechanism by which ATP is generated.

We saw above that the chemiosmotic hypothesis postulates that phosphorylation requires closed membrane vesicles with an asymmetric placement of the membrane-bound cofactors such that the electron- and proton-transfer reactions take place on discrete sides of the membrane. Mitchell suggested that the electron-transport chains in respiration and photosynthesis alternated between electron-only and electron–proton carriers. If we consider three such redox centers, A, B, and C, we can illustrate the essential features of this mechanism. Centers A and C are electron-only components, whereas B requires that protons are directly involved in its redox chemistry. Center A is the most reducing species and C is the most oxidizing; A can only be oxidized on the inner side of the membrane, and C can only be reduced at the outer side. Center B is diffusible and can move from one side of the membrane to the other, as, in both its oxidized and reduced states, it is charge neutral and, therefore, is membrane permeable. The vectorial e^-/H^+ chemistry that takes place can be represented as:

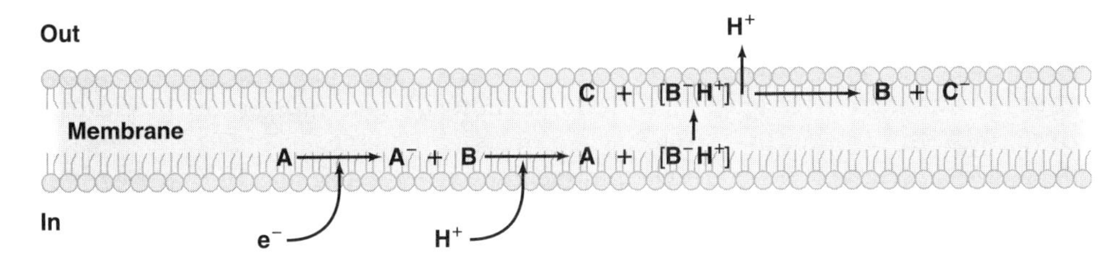

In this process, the net chemistry is (where the negative charge indicates a reduced form)

$$A^- + C + \{H^+\}_{in} \rightarrow A + C^- + \{H^+\}_{out} \tag{17}$$

and redox chemistry has been accompanied by the vectorial transport of protons from the inside to the outside of the membrane: mitochondrial, bacterial, or chloroplast. Energy has been stored in this process by two means: (1) the pH has risen on the inside, while it has dropped on the outside, and (2) the charge on the inside has decreased by +2, while that on the outside has risen by +2. Mitchell represented this energy storage as $\Delta\mu$, the chemiosmotic gradient. In millivolts, it is given by

$$\Delta\mu = \Delta\mu^{H^+} + \Delta\psi \tag{18}$$

where $\Delta\mu^{H^+}$ represents the pH difference and $\Delta\psi$ the charge imbalance across the membrane.

In biological systems that use this mechanism for creating a useful form of energy, the relative contributions of the pH and charge gradients may vary due to electrically neutral proton–ion exchange systems in the membrane, but the net storage of energy as $\Delta\mu$ is maintained. The membrane potential is used in all cases to drive a variety of energy-requiring transport processes and, most importantly, to synthesize ATP.

The system used to synthesize ATP has recently been elucidated by crystal structures and by biochemical and single-molecule measurements. The results confirm what had seemed originally a fairly radical proposal, that the proton–electrical gradient drives a rotating molecular motor. A large body of evidence supports the idea that this amazing machine, ATP synthase, uses the proton gradient to induce rotational motion in the membrane portion of the enzyme, resulting in conversion of bound ADP and P_i into ATP, and the release of ATP. The rotation causes sequential conformational changes at three nucleotide-binding sites, driving the synthetic chemistry and ATP release.[10]

Thus, energy conservation in living organisms uses light or reducing power from food to create a separation of charge and drive the mechanochemical synthesis of ATP. In the following sections, we will consider in more detail the structural and mechanistic basis of the charge-separation processes that provide the energy required, both at the level of the organelles and the metalloproteins involved.

X.3.6. Cellular Structures Involved in the Energy Transduction Process: Similarities among Bacteria, Mitochondria, and Chloroplasts

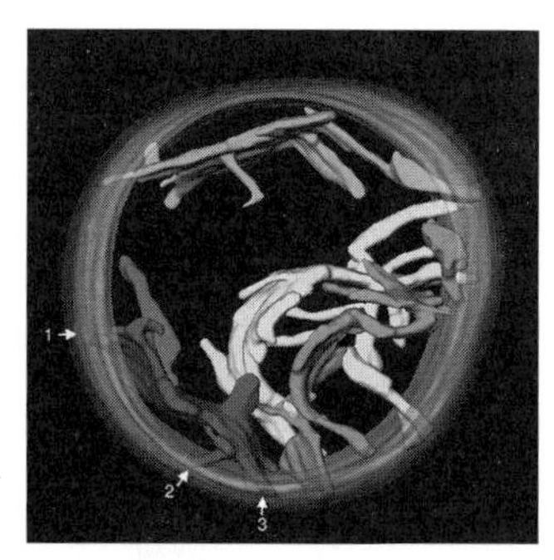

Fig. X.3.3.
Mitochondrial structure showing the cristae invaginations, like sacks with tubular attachments to intermembrane space. [From C. Manella, Wadsworth Center, Albany, NY 12201.] This membrane organization has some resemblance to the interior structure of the thylakoid membranes of chloroplasts.

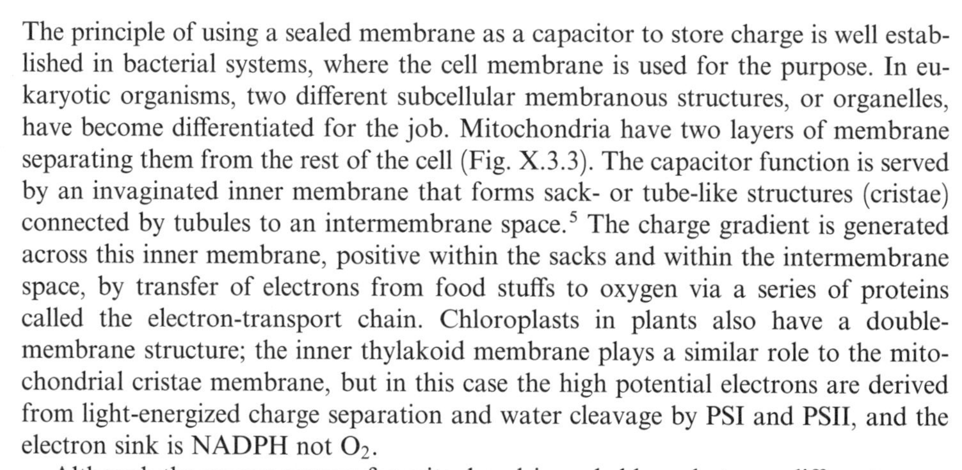

The principle of using a sealed membrane as a capacitor to store charge is well established in bacterial systems, where the cell membrane is used for the purpose. In eukaryotic organisms, two different subcellular membranous structures, or organelles, have become differentiated for the job. Mitochondria have two layers of membrane separating them from the rest of the cell (Fig. X.3.3). The capacitor function is served by an invaginated inner membrane that forms sack- or tube-like structures (cristae) connected by tubules to an intermembrane space.[5] The charge gradient is generated across this inner membrane, positive within the sacks and within the intermembrane space, by transfer of electrons from food stuffs to oxygen via a series of proteins called the electron-transport chain. Chloroplasts in plants also have a double-membrane structure; the inner thylakoid membrane plays a similar role to the mitochondrial cristae membrane, but in this case the high potential electrons are derived from light-energized charge separation and water cleavage by PSI and PSII, and the electron sink is NADPH not O_2.

Although the energy sources for mitochondria and chloroplasts are different, some of the components are similar, including the cytochrome bc_1 (mitochondria) and cytochrome b_6f (chloroplasts), as well as some smaller electron-carrier molecules, including quinols, cytochrome c, and plastocyanin. Other mechanistic and molecular similarities between photosynthesis and respiration will be discussed in Section X.3.8, but it is worth noting here the structural similarity between mitochondria and chloroplasts. The protons are pumped into the mitochondrial cristae and the chloroplast thylakoid space; the internal acidification is then used by the ATP synthase to

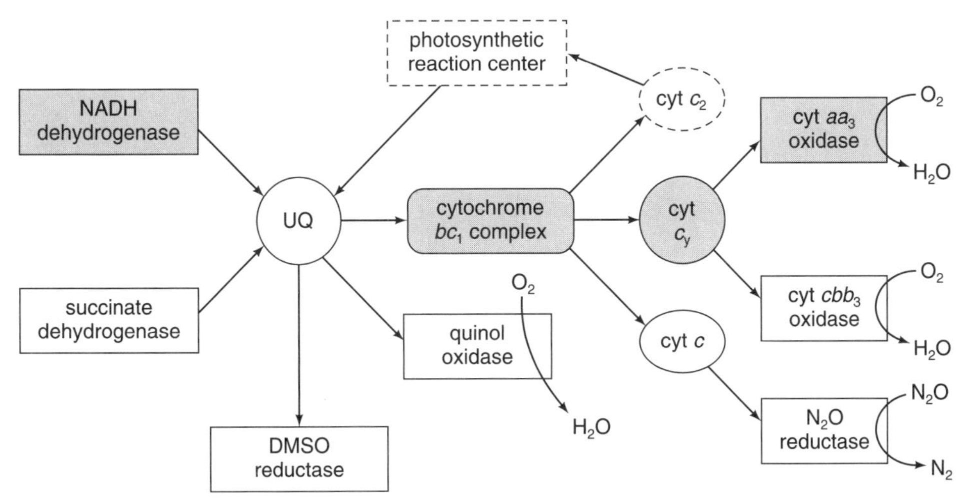

Fig. X.3.4.
An overview of the respiratory pathways in *R. sphaeroides.* A variety of oxidases and of cytochromes *c* (cyt *c*), as well as sources and sinks for electrons are present and induced under different growth conditions. UQ = ubiquinone; DMSO = dimethyl sulfoxide. The pathway with components most homologous to the mitochondrial respiratory chain is highlighted in gray. Connections with photosynthesis highlighted with dashed lines. [Adapted from Yuejun Zhen, Ph.D. thesis, Michigan State University, 1998.]

make ATP. The ATP is formed in the mitochondrial matrix and the chloroplast stroma; in both cases, a transport system is required to get the ATP, or the energy derived from it, out to the rest of the cell.

In some bacteria, such as *Rhodobacter sphaeroides* (Fig. X.3.4), the photosynthetic and respiratory functions are carried out in the same membrane, giving them the flexibility to use whatever source of energy is available, whether light or food stuffs in their environment, to generate the membrane potential. The photosynthetic and respiratory electron-transfer chains of prokaryotes can even share electrons through the small, mobile carriers such as quinols and cytochromes *c*. Many of the components of the bacterial energy transducing respiratory system, highlighted in red in Fig. X.3.4, have strong evolutionary similarity to their eukaryotic counterparts. Since mutational techniques can be applied more easily to the prokaryotic systems, bacteria have proved to be powerful tools for dissecting the mechanisms involved.

The electron-transfer processes used to conserve energy employ a large number and wide variety of metalloproteins, relying on the ability of metals to carry and transfer electrons and the ability of proteins to control and direct the process. Here, we will focus on the structural features of these metalloproteins that allow them to couple energy yielding electron transfer to pumping protons across a membrane, first considering the mitochondrial respiratory chain.

X.3.7. The Respiratory Chain

How do electron-coupled proton pumps work? The short answer is, in many different ways. One of the fascinating aspects of the mitochondrial respiratory chain is the diversity in the chemistry that has been recruited to the common task of creating a membrane potential. Three completely different metalloprotein complexes (and many more variants in bacteria) are used in series to pump protons across a membrane. Because they work in series, each functions at a different redox potential and hands over its "used" electrons to the next complex for further extraction of energy. It is this efficient in-series approach that dictates that different metal centers must be involved and thus different mechanistic strategies are used for pumping protons. Our understanding of the processes involved has increased dramatically with the advent of high-resolution structures of many of these complexes (Fig. X.3.5).

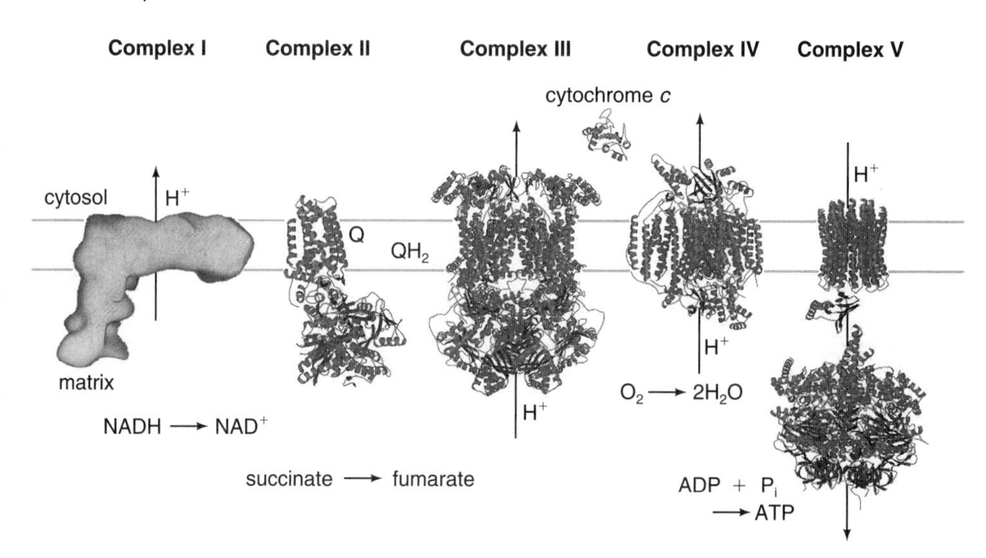

Fig. X.3.5.
Components of the mitochondrial electron-transfer chain. Four of the five complexes shown are represented by their actual crystal structures (see text) as they would sit in the inner-mitochondrial membrane. Complex **I**, or NADH-ubiquinone(Q) oxidoreductase, is represented by the current low-resolution picture of its overall shape (not to scale). Complex **II**, or succinate dehydrogenase, is not a part of the proton-pumping machinery, but is a supplier of electrons and protons in the form of ubiquinol (QH_2). [Adapted from Schultz, B. E. and Chan, S. I. *Annu. Rev. Biophys. Biomolec. Structure* 30, 23–65 (2001) with permission from Annual Reviews, Palo Alto, CA.] [See color insert.]

Complex I: NADH-Ubiquinone Oxidoreductase As discussed above, the highest potential (most negative reduction potential) electron source for the respiratory chain is NADH, $E^{\circ\prime} = -0.33$ V, which is produced by the oxidation of foodstuffs such as fat, protein, or sugar. The NADH-ubiquinone oxidoreductase, as indicated by its full name, ultimately delivers the electrons to the small, lipid-soluble acceptor, ubiquinone (CoQ). Complex **I** is the largest of the three proton-pumping complexes, the mammalian version has at least 43 subunits and a molecular weight >900,000 (Fig. X.3.5). The prosthetic groups that transfer electrons between NADH and CoQ include a flavin mononucleotide (FMN) and nine different iron–sulfur centers.[15] No crystal structure has been obtained of this multisubunit enzyme and thus its mechanism of harnessing the electrical potential energy to pump protons is the subject of much speculation and few firm conclusions. Although spectral and genetic analysis is giving interesting clues concerning possible pumping models, the daunting array of iron–sulfur centers and quinone interaction possibilities have made this the most difficult of the respiratory complexes to address experimentally.

Nevertheless, it is established that Complex **I** translocates 2 protons per electron and produces reduced ubiquinol. The ubiquinone–ubiquinol couple has a reduction potential $E^{\circ\prime} = 0.04$ V. As expected, there is a substantial loss of potential energy during transit through Complex **I**, ~0.3 V. From this value we can calculate (as described in Section X.3.3) a loss of 6.9 kcal of free energy for each electron transferred. This energy has not been wasted of course, but used to pump protons across the mitochondrial inner membrane, from the matrix side into the intermembrane–cristae space.

Complex II: Succinate-Ubiquinone Oxidoreductase This inner-membrane complex is not a proton pump, but is another source of ubiquinol for Complex **III**, deriving its reducing equivalents from oxidation of succinate to fumarate ($E^{\circ\prime} = 0.1$ V). Atomic resolution structural information is available for this metalloprotein and related bacterial enzymes (Fig. X.3.5),[16] but its characteristics will not be discussed here as the focus is on the energy conserving process.

Complex III: Ubiquinol-Cytochrome* c *Oxidoreductase The "used" electrons produced by Complex **I** (and **II**) in the form of ubiquinol still have plenty of potential energy as well as the ability to bind to the next energy transducing complex, Com-

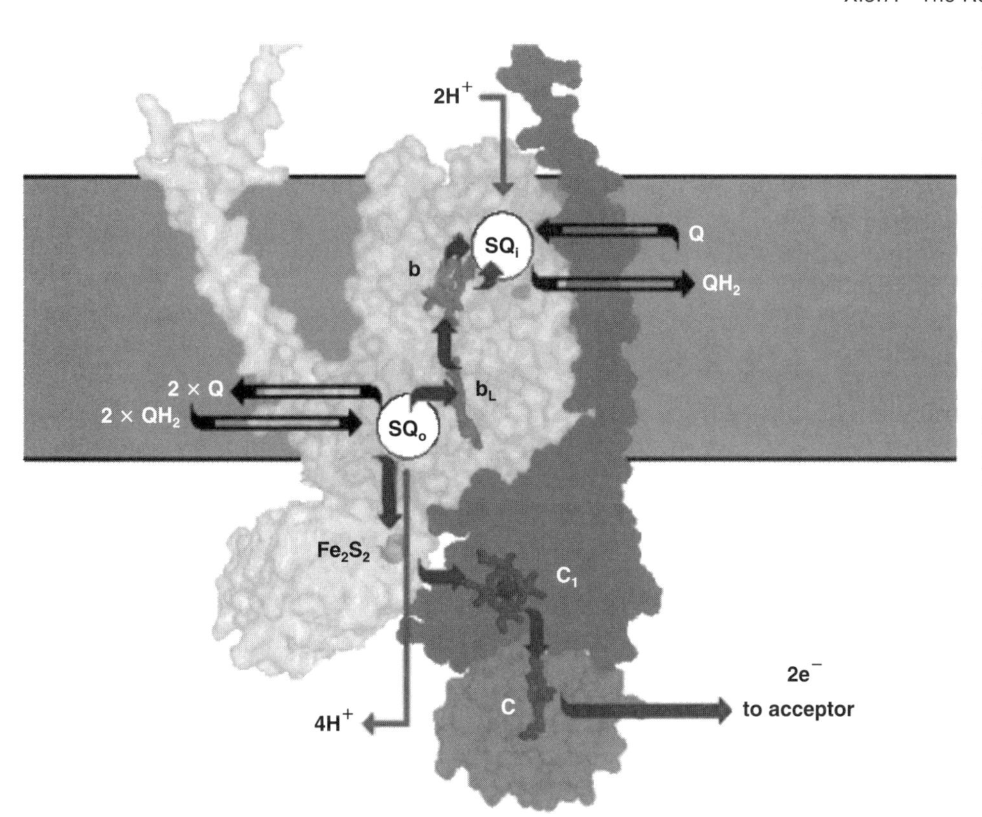

Fig. X.3.6.
Representation of the Q-cycle mechanism of the bc_1 complex. A monomer of the functional dimer is shown embedded in the membrane; the outside is toward the bottom, where protons are pumped. Q = quinone; QH_2 = quinol; SQ = semiquinone (1 electron reduced). FeS = Rieske iron–sulfur protein. For details see text. (From Tony Crofts, U.Illinois-Urbana, Website. http://www.life.uiuc.edu/crofts/bc-complex_site/index.html) [See color insert.]

plex **III,** or ubiquinol-cytochrome *c* oxidoreductase. Because of its readily visible heme prosthetic groups, this complex is often referred to in terms of its *b*- and *c*-type hemes, as cytochrome bc_1. High-resolution crystal structures of various forms of this enzyme have been obtained (Fig. X.3.6)[17–19] in spite of the daunting obstacles provided by its large size (240 kDa) and transmembrane character, as well as its 11 different subunits per monomer and further association into a dimer.

The prosthetic groups of this assembly are much more spectrally accessible than those of Complex **I,** contributing to greater success in understanding its mechanism. They include two *b*-type hemes, a single iron–sulfur center (the Rieske center, named after its discoverer, see Section X.1) and a *c*-type heme, cytochrome c_1. The complex has at least two binding sites for quinones, one close to the inside of the membrane (Q_i) and one close to the outside (Q_o). The mechanism by which Complex **III** uses electron potential energy to pump protons was proposed originally by Peter Mitchell and called the Q-cycle. The Q-cycle principles have been born out, clarified, and refined by mutational, spectral, and X-ray studies. The mechanism depends on the fact that the lipid-soluble ubiquinone is able to move across the membrane either in the oxidized (deprotonated) or reduced (protonated) states (Fig. X.3.5) and *must* take up protons whenever it accepts electrons and releases protons when it gives electrons up. By strategically locating the sites at which the quinone can gain or lose its electrons, it can be used to accomplish net movement of protons across the membrane, an example of compound "**B**" in the diagram given in Section X.3.5. Thus Complex **I** (or **II**) is designed to reduce ubiquinone on the inner side of the membrane, where it must pick up two protons from the matrix space in order to take on two electrons and become ubiquinol. In sequence, Complex **III** provides a binding and delivery site for the electrons and protons near the outer side of the membrane, so that reduction of Complex **III** by ubiquinol results in release of protons on the exterior, a net pumping of protons. In this process, the externally situated acceptor of the first electron from the ubiquinol at the Qo site is the Rieske iron–sulfur center ($E^{\circ\prime} = +0.2\,\text{V}$). It passes the electron on to heme c_1 ($E^{\circ\prime} = +0.23\,\text{V}$) and then to the water-soluble acceptor cytochrome c ($E^{\circ\prime} = +0.26\,\text{V}$), the product of the reaction. The second electron from the now one-electron reduced semiquinone goes to the low-potential heme b_L group ($E^{\circ\prime} = -0.03\,\text{V}$) in the complex. The electron on b_L is sent back toward the inside of the membrane via heme b_H ($E^{\circ\prime} = +0.05\,\text{V}$) and from there reduces an oxidized quinone bound at the Qi site close to the matrix side of the membrane, reducing it to a semiquinone. When a second pair of electrons is donated by ubiquinol at the Qo site, the first electron again goes to cytochrome c via the Rieske Fe–S–cytochrome c_1 path. The second follows the b_L–b_H–Qi path, completing the reduction of the semiquinone and the uptake of two protons at the inside of the membrane.

The overall reaction for Complex **III** thus results in the release of four protons on the outside, the uptake of two protons from the matrix (plus two more, counting the two used by Complex **I** or **II** to reduce the quinone to begin with), and the transfer of two electrons to cytochrome c (Fig. X.3.6).

One key element of this cyclic mechanism was not understood until several crystal structures were completed, revealing different positions of the Rieske Fe–S domain: In one structure, a position close to the ubiquinone-binding site is found;[20] in another, the Fe–S domain is closer to cytochrome c_1.[17] With the Rieske protein in the former location, it is the kinetically and thermodynamically favored recipient of the first electron from QH_2; in the latter more distant location, heme c_1 becomes the kinetically favored acceptor from the Rieske center. Note that the reduction potentials of the redox centers of the bc_1 complex are likely altered during the electron-transfer process due to conformational and electrostatic changes in their environments, and that kinetics *and* thermodynamics influence the pathways of the electrons.

The proton-pumping mechanism of Complex **III** depends on the ability of the protein to capture the mobile quinone on the right side of the membrane in the right oxidation state, so that protons are released on the outside and taken up on the inside. Equally important are the redox potentials and the positions, with respect to the two sides of the membrane and to each other, of its metal centers, providing both kinetic and thermodynamic guidance to the flow of electrons. The fact that a major conformational change supports this process was unforeseen, but may be an important principle to consider in other systems.

The electrons released from Complex **III** are in a much lower energy state, in the form of reduced cytochrome c, $E^{\circ\prime} = 0.26\,\text{V}$. The energy change, 0.23 V (= 5.3 kcal/e^-) has again accomplished significant work in transporting two protons per electron across the membrane. Even at this potential, the electron in cytochrome c has plenty of energy to support further work, especially since the final electron sink is provided by oxygen at +0.82 V. But a whole new set of protein-embedded metal centers is required to make use of it.

Complex IV: Cytochrome* c *Oxidase The employment of oxygen as the final electron sink for the respiratory chain in most aerobic organisms is an excellent "choice", since O_2 is both relatively unreactive, and yet has a high affinity for electrons, as indicated by its high positive redox potential. Many details and some speculations on the mechanism of this large multisubunit membrane protein are discussed in Section XI.6. Here, we will provide an overview of its unique oxygen reduction and proton-pumping functions, the former being much better defined than the latter.

The structure and metal centers of the three core subunits of cytochrome oxidase are shown in Fig. X.3.7, as well as the position of the protein in the membrane. A pathway for protons is indicated as a dotted line, since there are several and their exact routes are not established, especially on the outside of the protein. But for each four electrons donated from cytochrome *c,* four protons must be provided to the buried heme a_3–Cu_B center where the O_2 is reduced to H_2O. Concomitantly, another four protons are translocated across the membrane, as indicated in the overall reaction summarized in Eq. 19:

$$4 \text{ cytochrome } c \text{ (reduced, on the outside)} + O_2 + 8\ H^+\text{(from inside)}$$
$$\rightarrow 4 \text{ cytochrome } c \text{ (oxidized, on outside)} + 2\ H_2O + 4\ H^+\text{(outside)} \quad (19)$$

It is important to note that, unlike Complex **III,** the metal centers and redox cofactor-binding sites are not found on both sides of the membrane: the cytochrome *c*-binding site and its electron acceptor site, Cu_A, are on the outside, while heme *a* and the oxygen-binding site, heme a_3 and Cu_B, are in the middle (Fig. X.3.7). Furthermore, none of the centers directly binds protons and there is no lipid-soluble mobile cofactor like ubiquinone to move protons across the membrane. These facts led Peter Mitchell to propose that cytochrome *c* oxidase was not a proton pump.[21] Other studies showed this was not correct,[22] but even now with a detailed molecular structure available, the mechanism of operation of this pump is not understood. However, as in the case of Complex **III,** the many spectrally accessible metal centers have provided powerful approaches to address the problem.

The drop in potential energy for the electron traveling from cytochrome *c* (+0.23 V, when bound to the oxidase) to oxygen (+0.82 V) is Δ0.59 V. By the calculation method described previously (Section X.3.3), this will give us 13.6 kcal/e^-, or 54 kcal per reduction of one O_2 to 2 H_2O (4 electrons used). Plenty of free energy is thus released, but where does it go? How does cytochrome oxidase capture it and use it to push protons across the membrane against a concentration and electrical gradient?

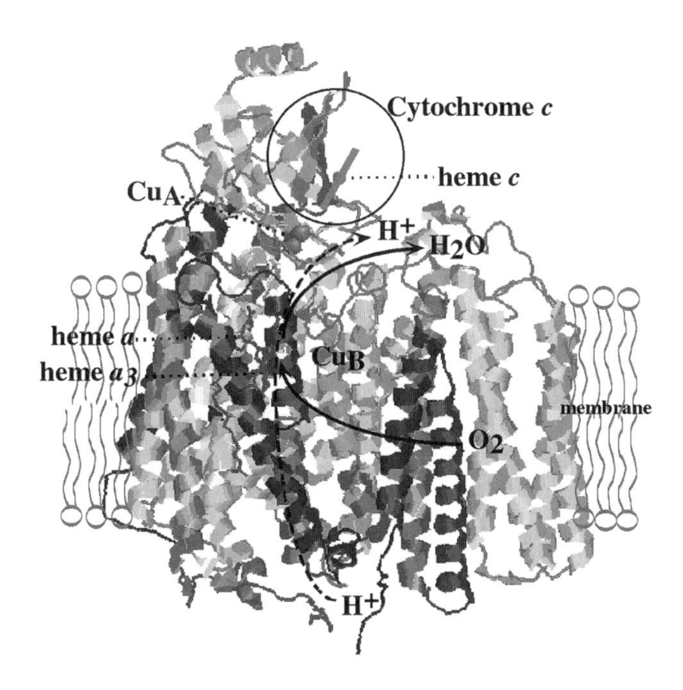

Fig. X.3.7.
Cytochrome *c* oxidase oriented in the membrane. The three largest subunits of the bovine oxidase are shown, with two hemes and three copper ions. Cytochrome *c* is represented at its interaction site on the outside surface, where electrons are donated and protons are released. (Drawing made in Rasmol and Canvas from PDB code: 1OCO.) [See color insert.]

Looking at how Nature accomplishes ion pumping in general, we see that chemical energy from ATP (e.g., the Ca^{2+} ATPase in muscle)[4] or light energy (e.g., bacteriorhodopsin)[3] is often used to induce a conformational change in a protein that allows alternate release of ions on one side of a membrane and binding and uptake on the other. When the ion is a proton, changes in binding can be accomplished by small conformational changes that differentially affect the pK values of protonatable groups on two sides of the membrane, resulting in proton transfer between them if the groups are connected by a path the protons can follow. If redox reactions are the source of energy, a similar "redox Bohr effect" could occur, where the change in redox state of one or more redox-active centers alters the protein conformation and pK values at a distance.[23] The analogy is with the Bohr effect in hemoglobin, where O_2 binding to one heme influences O_2 affinity and pK values at some distance (see Section X.4). As discussed in Section XI.6, a conformational change between oxidized and reduced states has been observed in the crystal structures of bovine cytochrome *c* oxidase,[24] and a mechanism whereby it could participate in proton-pumping activity is discussed. A similar conformational change has not yet been observed in the bacterial oxidases, and some of the key residues involved in the bovine enzyme are not conserved in the bacterial oxidases. Thus the validity and generality of the proposed mechanism remain to be established.

Another model of proton pumping that has been seriously considered in cytochrome *c* oxidase involves the possibility that direct ligands of the metal centers may undergo protonation–deprotonation during certain stages of the O_2 reduction chemistry. Protonation could result in release of the ligand from the metal in the protonated state and subsequent release of the proton at a new externally accessible location. In a different redox state of the metal center, rebinding of the deprotonated ligand could reset the proton pump. A particularly well-developed version of such a "ligand shuttle" mechanism called the Histidine Shuttle has been advanced by Wikstrom et al.[25]

An additional important general concept that has influenced the thinking in this field is the idea of proton movement being driven by the need to maintain electroneutrality.[26] An electron-transfer event that reduces a metal center in a low dielectric medium is energetically unfavorable unless there is movement of a positive charge for charge compensation. This fact suggests that every electron taken up by the buried metal centers in cytochrome oxidase, heme *a*, a_3, or Cu_B, could require a proton to be pulled into the low dielectric through a proton-conducting channel to a position close to the reduced center. If protons can only access the metal centers from the inside of the membrane when charge neutralization is required, this could form the basis for a pump. A highly conserved, buried glutamate (bovine oxidase E242, *R. sphaeroides* E286, *Pseudomonas denitrificans* E278) has been suggested as a location for such a proton to reside. The propionates of the hemes have also been invoked as additional proton-accepting sites.[27] But the charge compensation need not involve a protonation per se; other rearrangements of charged or polar groups are possible, such as a movement of a region of the polar backbone closer to the metal center. Conformational changes elsewhere might follow, as invoked by the "redox Bohr mechanism" discussed above.

To determine the validity of any of these mechanisms, efforts are being made to correlate the electron- and proton-transfer events in time and space. The timing and sequence of electron-transfer events, and the nature of the resulting oxygen intermediates, are reasonably well understood,[1,28] but their correlation with proton movements remains controversial. The overall process can be written in a way that indicates where protons are used in the oxygen reduction chemistry, and possible steps where they are pumped, as shown in Fig. X.3.8 from Ref. (1).

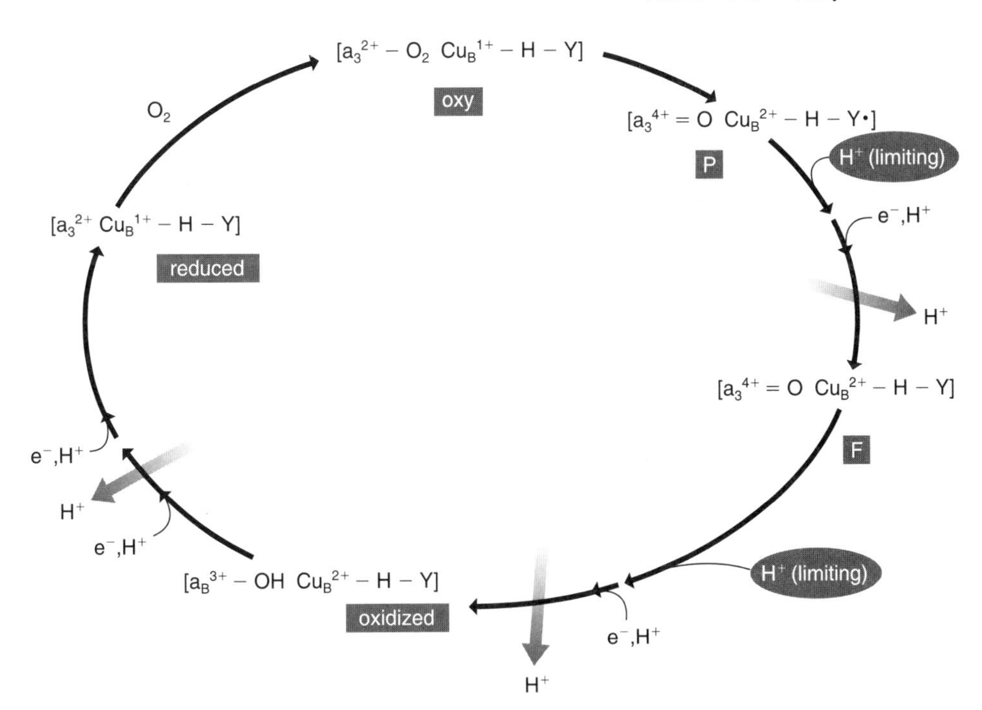

Fig. X.3.8.
Simplified scheme for the reaction between cytochrome oxidase and O_2. The active site, which contains heme a_3, Cu_B, and a cross-linked His-Tyr, H-Y, is shown. Reduction (e^-) and protonation (H^+) of the oxidized form of the center produces the reduced site. This binds oxygen (O_2) to form initially the oxy species, which reacts further to produce P (for peroxy) and F (for ferryl) intermediates, before regenerating the oxidized form of the enzyme. The reduction of P and F are limited by proton-transfer reactions (H^+-limiting), as indicated. The steps between P and the reduced form of the site have been implicated in proton-pumping processes (thick outward-pointing arrows).

In summary, the mechanism of the redox energy-driven proton pump in cytochrome *c* oxidase is still to be solved. Yet, the respiratory chain as a whole is gradually yielding to the combined tools of atomic-resolution structure, spectroscopy, and genetic analysis, giving us an ever clearer understanding of how energy can be extracted from the environment to support life. The implications of this understanding in the field of human health are great, since many states of disease and aging appear to involve failure of energy production.[8,9]

X.3.8. The Photosynthetic Electron-Transfer Chain

As we have seen, mitochondria obtain reducing power from oxidation of carbon substrates derived from protein, carbohydrate, or fat. The resulting NADH and $FADH_2$ are used in turn as substrates for the electron-transfer reactions of the respiratory chain that we have just described. The supply of reduced carbon in the biosphere is not infinite, and without a continuing renewal of this nutrient, much life on Earth would eventually come to an end. The supply of reduced carbon comes from the process of oxygenic photosynthesis in plants, algae, and cyanobacteria. Because energy is lost in converting reduced carbon into CO_2 and H_2O (respiration), an alternate source of energy is needed to drive the reduction of CO_2 into sugars and other metabolites; that energy source is light, and the molecule that absorbs and converts this energy into redox reactions is chlorophyll.

The redox activity associated with photosynthetic reactions is localized in the chloroplast. As we have seen already, this organelle has structural similarities to the mitochondrion. Electron-transfer reactions are localized in an internal membrane system that is comprised of sac-like disks called thylakoids; the thylakoids are surrounded by a solution of enzymes, which is called the stroma. Like the mitochondrial matrix, the chloroplast stroma is the site of carbon metabolism, but rather than breaking down sugar to CO_2 and reducing equivalents, in this case CO_2 is reduced to glucose and other sugars, using energy in the form of ATP and NADPH that are the products of the photosynthetic activity of the thylakoid membranes.[10]

Light absorption by chlorophyll triggers photosynthetic redox reactions. The structure of chlorophyll *a* is shown in Fig. X.3.9. The chlorin ring is the site of light absorption, and of oxidation–reduction activity of this pigment molecule. The aliphatic side chain (called the phytyl tail) of the molecule is used to assist in orienting the pigment molecule when it binds to protein. Before we examine the redox reactions of photosynthesis, it is useful to discuss briefly the mechanisms by which chlorophyll converts light energy into redox energy. This reaction is based on the energy gap between the ground and excited states of the molecule. In oxygenic photosynthesis, these energy gaps lie in the blue to far-red portion of the spectrum (400–700 nm, or a range of $\sim$300–170 kJ mol^{-1}, or 72–40 kcal mol^{-1}). Absorption of a photon by a chlorophyll molecule causes the promotion of an electron from the ground state to the excited singlet state. This reaction conserves the photon energy in the form of the difference in energies between the ground and excited states. Pigment excited states can have a number of fates: decay to the ground state with light emission as fluorescence; exciton transfers from one chlorophyll to another; or loss of an electron to initiate electron-transfer reactions. In this last case, the resulting oxidized chlorophyll cation radical is now highly reactive, and seeks to regain an electron in the ground state. The result is oxidation of a donor molecule and regeneration of the pigment's ground state so it can enter into another photocycle of redox activity. The first two reactions of chlorophyll (decay or exciton transfer) can be described by the following equations:

1. Excitation and decay to the ground state:

$$\mathrm{Chl} + h\nu_1 \rightarrow \mathrm{Chl}^* \rightarrow \mathrm{Chl} + h\nu_2$$

   (the emitted light, $h\nu_2$, has a longer wavelength than the absorbed light, $h\nu_1$ owing to energy lost by radiationless deexcitation within the excited state of the chlorophyll molecule)
2. Exciton transfer reactions:

$$\mathrm{Chl}_1 + h\nu \rightarrow \mathrm{Chl}_1{}^* + \mathrm{Chl}_2 \rightarrow \mathrm{Chl}_1 + \mathrm{Chl}_2{}^* + \mathrm{Chl}_3$$
$$\rightarrow \mathrm{Chl}_1 + \mathrm{Chl}_2 + \mathrm{Chl}_3{}^* \rightarrow \text{etc.}$$

   (ultimately, the exciton must either decay to the ground state, as in (1) or be absorbed by a reaction center chlorophyll to initiate electron transfer).

Most of the chlorophylls *a* and *b* in a photosynthetic membrane carry out exciton transfers, which enable them to function as antennae, transferring excitation energy throughout a bed of pigments. Eventually, much of this energy will arrive at a special chlorophyll *a* molecule or molecules that are capable of catalyzing a redox reaction; absorption of the excitation energy initiates redox chemistry that produces stable oxidized and reduced species that can participate in chemical redox reactions. In a typi-

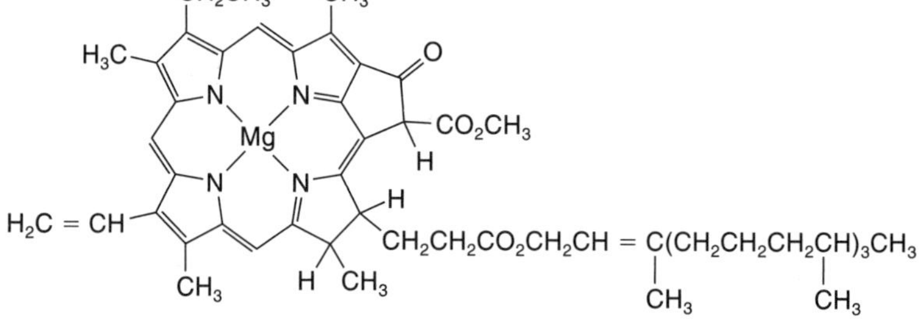

Fig. X.3.9.
The structure of chlorophyll *a*. See the text for details.

cal photosynthetic membrane, ~99% of the chlorophyll molecules will be engaged in excitation energy-transfer reactions, functioning as antennae molecules. One percent, or less, of the remaining chlorophyll molecules will function as redox-active catalysts that initiate electron-transfer reactions.

The antenna chlorophylls *a* and *b* are ligated to a family of proteins that are called light-harvesting chlorophyll proteins, or LHCPs.[11] The pigment–protein stoichiometry among various LHCPs is variable, the largest number being 13–14 chlorophylls *a* and *b* bound to an antenna protein found in higher plants (a three-dimensional structure of this protein is presented in Ref. 29). In contrast, the chlorophylls and other cofactors that are involved in light-catalyzed redox activity are associated with different types of complexes made up of several proteins. These protein complexes are called reaction centers, and like the electron-transfer complexes of the mitochondrial inner membrane, reaction centers are embedded in the lipid bilayer of the chloroplast thylakoid membranes. Reaction center protein complexes are more complicated than are the LHCPs. This complexity is due to a greater diversity of cofactors that must be ligated to the proteins of reaction centers, and the necessity for the redox activity of a reaction center to produce a directional, or vectorial electron-transfer reaction similar to that observed in the mitochondrial electron-transfer complexes. Electrons are transferred across the reaction center, in this case from the lumen side of the thylakoid membrane to the stroma side. There are two reaction centers in plants and green algae. They are called PSII and PSI. Electron transfer within these photosystems can be modeled using a set of hypothetical redox components:

D = Electron donor

P = Special chlorophyll molecules that generate redox reactions upon absorption of light

I = Intermediate electron carrier

A = First stable electron acceptor

Light absorption by the special redox active chlorophyll molecules of the reaction center catalyzes the following sequence of reactions:

$$\mathrm{DPIA} + h\nu \rightarrow \mathrm{DP^{*}IA} \rightarrow \mathrm{DP^{+}I^{-}A} \rightarrow \mathrm{DP^{+}IA^{-}} \rightarrow \mathrm{D^{+}PIA^{-}}$$

The final product contains the two reactive species, the oxidized donor molecule (D^+) and the reduced acceptor (A^-) [Fig. X.3.10 *b*] whose redox potentials are widely separated by ~0.8 V or more. These strong oxidants and reductants created by the activity of reaction centers are used to catalyze redox reactions that produce the stable forms of chemical energy (ATP, NADPH) that are used for CO_2 reduction.

The identities of the cofactors in reaction centers are listed in Table X.3.1, along with their estimated $E^{\circ\prime}$ values. The electron donor in PSII is a multicomponent complex that employs a Mn cluster and a redox active tyrosine to oxidize water to O_2 (see Section X.4.6). The electron donor in PSI is the soluble blue copper protein, plastocyanin. The first stable electron acceptor for PSII is a quinone, while for PSI this acceptor is a low-potential 4Fe–4S cluster. As we shall see, this low redox potential is needed to catalyze the subsequent reactions of PSI that result in formation of NADPH ($E^{\circ\prime} = -0.32\,\mathrm{V}$).

The operation of the photosynthetic electron-transfer chain has a strong functional similarity to the mitochondrial respiratory chain, and some marked differences as well. One of the similarities is the presence of a cytochrome bc_1 type complex, called the b_6f complex. Another is the presence of an ATP synthase that is very similar to the mitochondrial enzyme. The organization of electron carriers in photosynthetic membranes, like that in the mitochondrial inner membrane, leads to the

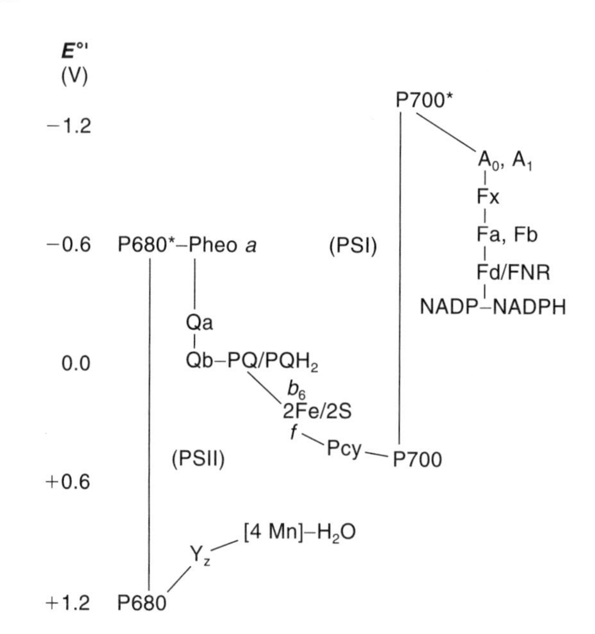

Fig. X.3.10.
Schematic diagram of the redox potentials of intermediates in photosynthetic electron transfer. Abbreviations used in the diagram are Y_Z = a redox active tyrosine; P680 = the redox-active chlorophyll of the PSII reaction center: pheo *a*, pheophytin *a;* Q_A and Q_B, bound plastoquinone molecules; PQ = plastoquinone; b_6 and *f*, cytochromes; Pcy = plastocyanin, a soluble blue copper protein; P700 = the redox-active chlorophyll *a* dimer of the PSII reaction center; Ao = a chlorophyll *a* molecule; Fx, Fa, Fb, 4Fe–4S are clusters; Fd = ferredoxin; FNR = ferredoxin:NADP reductase.

Table X.3.1.
Electron-Transfer Cofactors of PSII and PSI

	PSII			PSI
(D) Electron donor	[H_2O →	Mn_4 cluster →	Tyrosine]	Plastocyanin (Cu^+)
$E^{o\prime}$	+0.815 V	ca. +0.9 V	ca. +1.0 V	+0.375 V
(P) Reaction center chlorophyll	P680/$P680^+$			P700/$P700^+$
$E^{o\prime}$	ca. +1.2 V			+0.45 V
(I) Intermediate electron carrier	Pheophytin *a*			Chlorophyll *a*, vitamin K
$E^{o\prime}$	−0.66 V			(?) (ca. −1.0 V)
(A) Stable electon acceptor	Quinone A (Q_A)			Fx (4Fe–4S)
$E^{o\prime}$	−0.13 V			(−0.7 V)

formation of the proton gradient that drives ATP synthesis. The major differences between respiratory and photosynthetic redox activities become apparent when the unique function of the photosystems is integrated into the photosynthetic electron-transfer chain, but even these have analogies with the mitochondrial respiratory chain. Photosystem II replaces mitochondrial complex **I** (NADH-ubiquinone oxidoreductase) as the enzyme system that reduces quinone to quinol, the substrate for the cytochrome b_6f complex. Photosystem I replaces mitochondrial complex **IV** (cytochrome *c* oxidase) as the enzyme that functions as the oxidant for electrons that exit the b_6f complex by way of plastocyanin, which is also a functional analogue of cytochrome *c*. The plastoquinone that is used in thylakoid membranes is very similar chemically to the ubiquinone found in the mitochondrial inner membrane. The sequence of electron carriers in oxygenic photosynthesis can be diagrammed as follows:

$$H_2O \rightarrow \text{PSII} \rightarrow \text{plastoquinone} \rightarrow \text{cytochrome } b_6f$$
$$\rightarrow \text{plastocyanin} \rightarrow \text{PSI} \rightarrow \text{ferredoxin} \rightarrow \text{NADP} \quad (20)$$

A schematic diagram of the photosynthetic electron-transfer chain, which emphasizes the reduction potentials of the redox reactions of the chain, is shown in Fig. X.3.10.

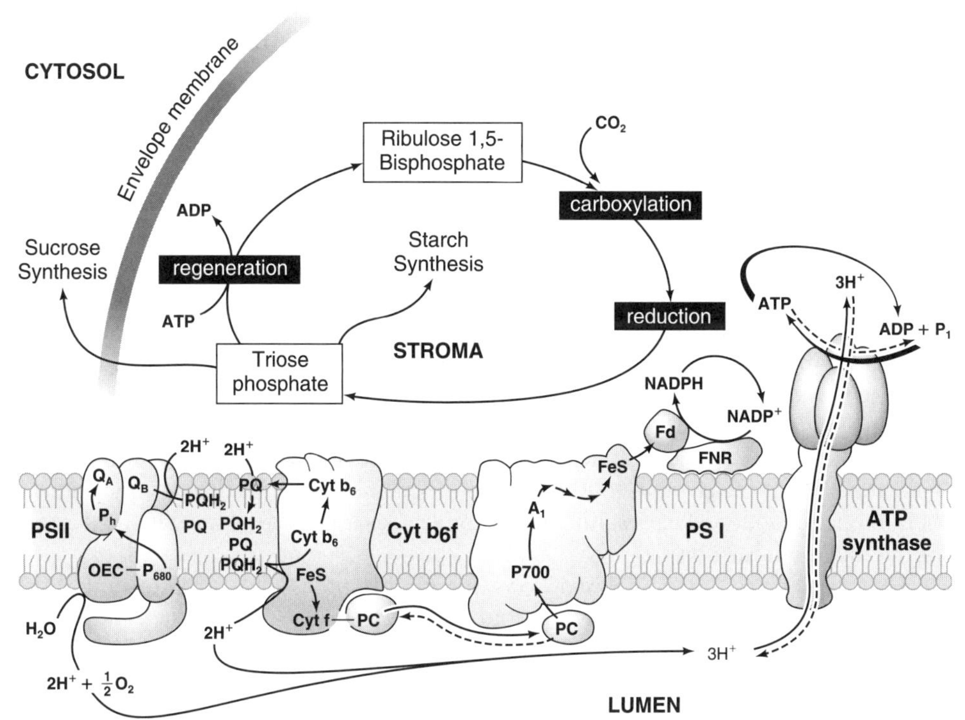

Fig. X.3.11.
Schematic diagram of the photosynthetic electron-transfer chain in the thylakoid membrane. The lumen is the interior side of the membrane, the stroma surrounds the thylakoids and contains the enzymes for carbon reduction, as shown in the figure. Electron transfer deposits protons in the lumen space, and the resulting ΔpH is used to drive ATP synthesis by the ATP synthase. [See color insert.]

If we look at the path of electron transfer, we can see that the function of PSII is to absorb photons and to initiate oxidation–reduction reactions that catalyze H_2O oxidation to O_2, and reduction–protonation of plastoquinone. The electrons from PSII, like those produced by NAD-ubiquinone oxidoreductase, possess a relatively negative $E^{o\prime}$ ($\sim 0\,V$ in the case of reduced plastoquinone). This potential is sufficient to reduce the b_6f complex; oxidation of plastoquinol releases protons into the thylakoid lumen, as does oxidation of water by PSII. Transfer of electrons from reduced plastoquinone through the b_6f complex that connects the two photosystems, releases energy ($\Delta E^{o\prime} = +0.38\,V$ or 36.7 kJ (8.7 kcal mol^{-1})) and causes reduction of plastocyanin. As Fig. X.3.10 shows, PSI transfers electrons to the stromal side of the membrane through several carriers to a series of 4Fe–4S clusters. These are likely to be reactions with some of the most negative $E^{o\prime}$ values in biology, and this is necessary because the next electron acceptor is ferredoxin, or Fd, a water soluble 2Fe–2S protein ($E^{o\prime} = -0.432\,V$). Electrons from Fd reduce a membrane-associated flavoprotein containing FAD that is called ferredoxin-NADP reductase, or FNR, and this reductase catalyzes the reduction of NADP to NADPH in the stroma. Figure X.3.11 translates this information on redox potentials and electron-transfer reactions into a schematic diagram of a segment of a chloroplast thylakoid membrane. Proton-coupled redox reactions produce an acidification of the thylakoid lumen, and the energy of this ion gradient is utilized by an ATP synthase enzyme, CF_0CF_1, in which the CF_1 caps the proton channel CF_0 on the stromal side of the membrane. Rotations of CF_0 cause conformational changes to occur in the catalytic subunits of CF_1, and this produces ATP.

A more detailed discussion of the photosynthetic electron-transfer chain begins with PSII. Structure and function of this enzyme are covered in Section X.4, so only a brief description of its activity is given here. Photosystem II is comprised of at least nine major intrinsic, membrane spanning polypeptides, and three extrinsic proteins that are attached to the lumenal side of the enzyme. The estimated molecular mass of PSII is $\sim 400\,kDa$, and a crystal structure at low resolution has recently been published (Ref. 30; see also Section X.4.6). Chlorophyll-catalyzed redox activity of the enzyme is coupled, by a redox active tyrosine on one of the intrinsic membrane proteins of the complex, to a cluster of four Mn atoms. The Mn cluster is oxidized in response to the light-induced oxidation of the reaction center chlorophyll P680 in one-electron steps. Calcium and Cl^- are essential cofactors for Mn oxidation. Extraction of four electrons from the cluster results in release of an O_2, and 4 H^+ are also released into the thylakoid lumen as a consequence of Mn oxidation. Water oxidation reduces the Mn atoms, and the cycle repeats. Electrons are transferred rapidly [on the ps (10^{-12} s) time scale] from the reaction center chlorophyll, P680, through a pheophytin a molecule to the first quinone acceptor, Q_A. In a slower step (~100 μs), Q_A^- reduces a second quinone acceptor, Q_B. Two cycles of reduction, coupled to binding of protons from the stroma, produce Q_BH_2, which is released into the thylakoid membrane bilayer, where it functions as the electron donor to the cytochrome b_6f complex. In PSII, the transfer of four electrons from H_2O releases four H^+ into the thylakoid lumen, and reduces two molecules of Q_B, which must bind four H^+ at the stromal (outer) side of the membrane as part of the mechanism of quinone reduction. The resulting Q_BH_2 molecules are released into the lipid phase of the thylakoid membrane.

The cytochrome b_6f complex of oxygenic photosynthesis has striking similarities to the corresponding bc_1 mitochondrial complex. A schematic diagram of the complex is shown in Fig. X.3.12. There are two *b*-type hemes [called b_6, $E^{o\prime} = -150$ (b_L) and -50 mV (b_H)], a c_1-type heme protein called cytochrome f ($E^{o\prime} = +340$ mV), and a Rieske 2Fe–2S iron–sulfur cluster ($E^{o\prime} \sim 300$ mV).[12] The b_6f complex contains five major protein subunits, whose masses range from 17 to 35 kDa. Like the mitochondrial system, the functional form of cytochrome b_6f is a dimer of complexes in the membrane. Oxidation of reduced plastoquinone by the b_6f complex results in the release of two protons into the lumen of the thylakoid membrane. Two crystal structures of b_6f complexes have been solved to ~3-Å resolution. The first of these[31] is from a thermophilic cyanobacterium, *Mastigocladus laminosus,* and has a PDB code: 1UM3. The second structure, from the green alga *Chlamydomonas reinhardtii,* has a PDB code of 1Q90.[32] A comparison of these structures with the cytochrome bc_1 complex reveals two interesting differences. First, the photosynthetic complexes contain a molecule of chlorophyll, which is unlikely to be

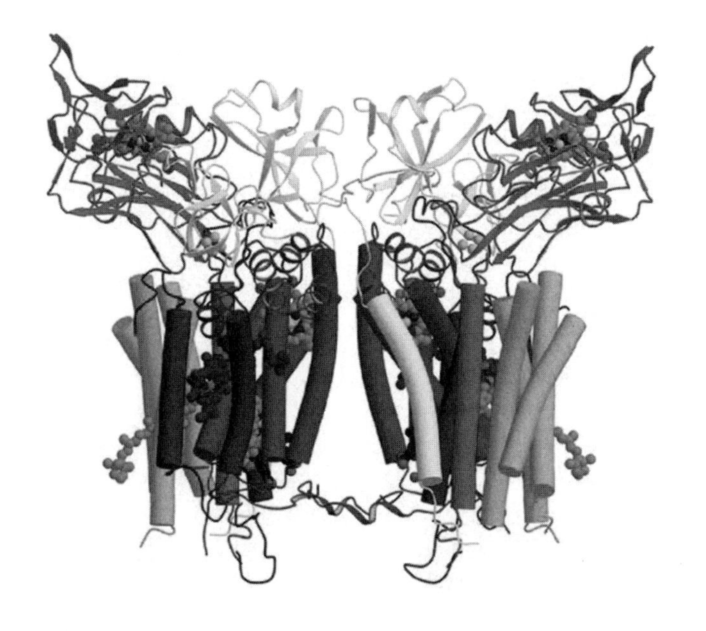

Fig. X.3.12.
Structure of the cytochrome b_6f complex. Shown is an illustration of the cytochrome b_6f protein complex, which is critical for photosynthesis. The eight shades represent the eight protein components of the cytochrome complex; the cylinders are the 26 segments of the complex that cross the photosynthetic membrane; the rings made of little balls that are embedded in protein are the groups that actually carry the electrons stimulated by light absorbed in photosynthesis. [Adapted from a model by H. Zhang/W. A. Cramer] (http://news.uns.purdue.edu/UNS/html4ever/031002.Cramer.photo.html) Crystal structures of two photosynthetic b_6f complexes are available (PDB codes: 1UM3 and 1Q90). [See color insert.]

involved in light absorption or electron transfer, and second, the surprising discovery has been made of the presence of an additional heme group in these complexes, which is likewise of unknown function at the present time.

At this point in the sequence of electron-transfer reactions, a substantial proton gradient has been formed. Four H^+ were released from 2 H_2O oxidized by PSII, and four H^+ from the stroma were bound upon formation of two Q_BH_2 in PSII. Thus, the $H^+/2e^-$ (or O) ratio from electron transfer in thylakoids, up to this point, is 4 (1 H^+/e^- each from H_2O oxidation, and from plastoquinol oxidation), which suggests that for this type of electron transfer, a Q-cycle may not be operating in the photosynthetic electron-transfer chain. Plastoquinol oxidation by the b_6f complex also differs from ubiquinol oxidation by the mitochondrial bc_1 complex, in that electron-transfer activity between the two photosystems, catalyzed by the b_6f complex, is insensitive to the quinone analogue inhibitor antimycin A, which is a potent inhibitor of cytochrome bc_1. Another notable difference between the chloroplast and mitochondrial complexes is the presence of a chlorophyll *a* molecule in b_6f.[12,31–33] Finally, the b_6f complex isolated from spinach contains bound ferredoxin-NADP reductase, the flavoprotein that is also involved in ferredoxin-catalyzed NADP reduction.[33] The oxidation reactions discussed here are catalyzed by the blue copper protein plastocyanin ($E^{\circ\prime} = +0.375\,V$). This water soluble, 10-kDa polypeptide contains a single Cu atom ligated by two His imidazole nitrogens, a Cys sulfur, and the sulfur atom of a Met residue. More information on plastocyanin and other blue copper proteins is given in Chapter IV and Section X.1. By analogy to the activity of mitochondrial cytochrome *c*, reduced plastocyanin is the electron donor to the photooxidized reaction center of PSI, P700+ (a dimer of chlorophyll a molecules), which is the first component of the terminal oxidase for electrons generated by PSII. It is believed that like cytochrome *c*, plastocyanin functions as a mobile electron carrier, shuttling between the b_6f complex, and PSI.

A crystal structure of PSI at a resolution of 2.5 Å has recently appeared, and is shown in Fig. X.3.13.[34] This structure is currently the most highly resolved of an electron-transfer complex from the oxygenic electron-transport chain. Three polypeptides, with masses of 83 (psaA), 82.4 (psaB), and 8.9 kDa (psaC) contain the

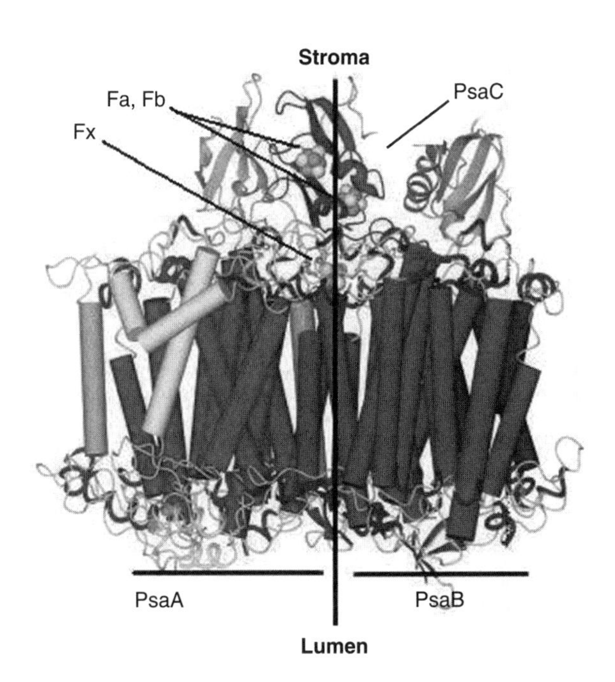

Fig. X.3.13.
Structure of the PSI reaction center. Only the major polypeptide subunits are shown; membrane spanning α helices are represented by cylinders. The approximate positions of the psaA and psaB subunits are indicated by underlining on the lumen side of the reaction center. These large subunits form the "core" of the reaction center, and ligate the chlorophylls, vitamins K (menaquinones), and the 4Fe–4S cluster called Fx. The approximate twofold axis of symmetry is indicated in this figure, as is the position of the psaC subunit on the stromal side of the reaction center, that binds two additional 4Fe–4S clusters that are participants in the electron-transfer reactions catalyzed by this reaction center. [Structure adapted from Ref. 34.] The complete structure is available (PDB code: 1BJ0). [See color insert.]

essential electron-transfer cofactors of PSI. Many additional polypeptides in the range from 4 to 18 kDa are also present (although they are not shown in Fig. X.3.13), and have diverse functions, such as plastocyanin and ferredoxin binding, and regulation of assembly and stabilization of the membrane protein complex. The molecular mass of the complete assembly of PSII polypeptides and cofactors is estimated to be 356 kDa. Spectroscopic characterizations of PSI electron transfer identified the pathway of electron transfer within the protein complex as: P700 $\rightarrow$ Chl *a* $\rightarrow$ Vitamin K $\rightarrow$ 4Fe–4S (see Table X.3.1).[35] The crystal structure has revealed that the electron-transfer chain after P700 exits as symmetric pairs of cofactors (chlorophyll *a* and vitamin K) on the two large ($\sim$80 kDa) protein subunits (psaA and psaB) of the complex, as shown in Fig. X.3.14. Thus, all photosynthetic reaction centers contain bifurcated electron-transfer pathways that diverge after the reaction center chlorophylls, but only in PSI does the sequence of carriers converge at the acceptor (there are two quinone acceptors, one on each "side", of the reaction centers of PSII and bacteria, but only one "leg" of the pathway is active). Research is underway to determine whether one, or both, "legs" of the PSI electron-transfer chain are active. Photosystem I is also unique in possessing an additional pair of 4Fe–4S clusters on the stromally located 8.9-kDa subunit psaC of the enzyme.[36] These centers, called F_A and F_B, function as a conduit for the transfer of electrons from the first iron–sulfur center acceptor, Fx, to water-soluble ferredoxin. Ferredoxin (11 kDa) is a key electron-transfer protein in chloroplasts. Although our primary interest is in the protein's role in NADP reduction, ferredoxin is also the electron donor to nitrite reductase, glutamate synthase, thioredoxin reductase, and betaine monooxygenase.[13,37] Ferredoxin reduces a flavoenzyme (FNR), and FNR in turn catalyzes the reduction of NADP to NADPH.

An additional function of reduced ferredoxin is to act as the catalyst of cyclic electron transfer in chloroplasts. In this reaction, ferredoxin transfers electrons from PSI to the b_6f complex, reducing it either directly, or perhaps through the recently discovered FNR that is bound to the stromal side of the b_6f complex.[33] This pathway of electron transfer results in additional proton pumping, very probably by a Q-cycle, which increases the H^+/e^- ratio, and therefore the efficiency of ATP synthesis by

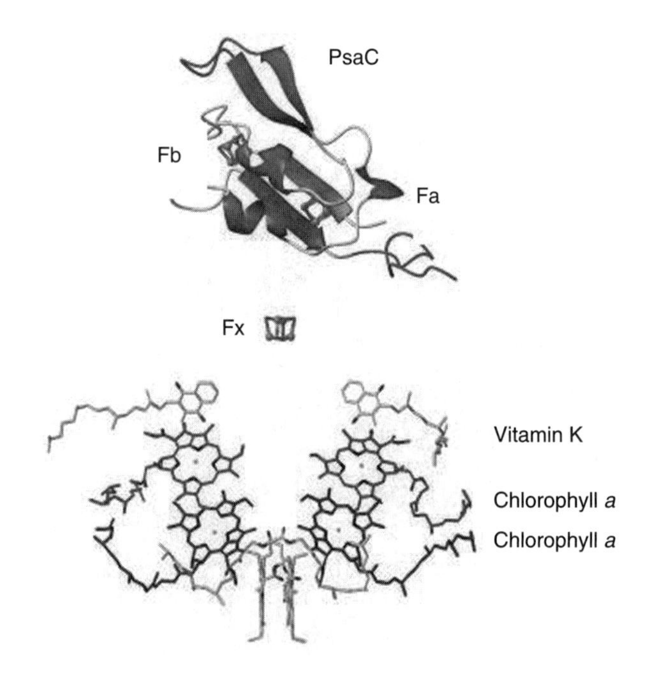

Fig. X.3.14.
Organization of the electron-transfer cofactors in the PSI reaction center. The two sides of the electron-transfer chain are shown here; the pigments are ligated by the psaA and psaB subunits (see Fig. X.3.13), as is the primary electron acceptor, Fx. The discovery of a second chlorophyll *a* in close proximity to the chlorophyll *a* special pair, P700, was an unanticipated finding. So far as is known, only one of these two chlorophylls is engaged in electron transfer from P700 to Fx. The psaC polypeptide subunit is shown with its pair of 4Fe–4S clusters. [See color insert.]

chloroplasts. It is likely that the NADP/NADPH ratio in chloroplasts, as well as the concentration of reduced ferredoxin, regulate the switching of electron transfer between the noncyclic and cyclic pathways, which in turn affect the yield of ATP. In its role in NADP reduction, ferredoxin forms an electrostatically stabilized 1:1 complex with FNR, the 35-kDa FAD-containing enzyme that is associated with the thylakoid membrane as well as with the b_6f complex. Reduction of the FAD by two cycles of electron transfer from reduced ferredoxin yields $FADH_2$, which catalyzes the reduction of NADP. It will be interesting to see whether a similar ferredoxin–FNR complex can be demonstrated for the fraction of reductase that is associated with the b_6f complex.

The products of photosynthetic electron transfer are O_2, ATP, and NADPH. Oxygen is a waste product of the redox reactions, but the ATP and NADPH are essential cofactors for the reduction of carbon by the Calvin, or reductive pentose phosphate, cycle. Both NADPH and ATP are consumed in the reduction of CO_2 to glyceraldehyde-3-phosphate, which is converted into sugars (fructose, glucose, sucrose) and into starch.[10] Thus, in the same way that carbon oxidation in the mitochondrion is coupled to the redox activity of the respiratory chain, carbon reduction in the chloroplast is coupled to the redox activity of the photosynthetic electron-transport chain.

X.3.9. A Common Underlying Theme in Biological O_2/H_2O Metabolism: Metalloradical Active Sites

At first sight, mechanisms of O_2 reduction by cytochrome oxidase and of H_2O oxidation by PSII would seem to have little in common. For example, it has been obvious for a number of years that the O_2-evolving reaction is not simply cytochrome oxidase "running backwards". As our knowledge of the components of these complex enzymes has increased, however, attention has been drawn to features of both enzymes that argue for certain common properties (Fig. X.3.15). First, it is very probable that both oxidase and PSII utilize a pair of metals (Fe/Cu or Mn/Mn) to carry out the actual reduction or oxidation of bound substrates. Second, in both of these enzymes, redox transformation of the bound substrate is coupled either directly (PSII) or indirectly (oxidase) to formation of the part of the proton gradient that drives ATP synthesis. A third observation, that redox active tyrosines are components of the active sites of both enzymes, has triggered a reconsideration of the possible mechanisms of these reactions.

In the case of PSII, the redox-active tyrosine called Y_Z is the intermediate electron carrier between the Mn cluster and the reaction center chlorophyll P680. Rereduction of $Y_Z\bullet$ by the Mn cluster assures an increase in the oxidation state of a Mn atom ($Mn^{3+} \rightarrow Mn^{4+}$). This creates a problem, in that sequential oxidations of high-valent Mn atoms would result in a large increase in the charge on the Mn cluster. One of G. T. Babcock's many contributions to the field of biological redox chemistry was his insistence that such charge accumulations had to be circumvented. One possible solution to the problem proposed by Babcock and co-workers[38] was that hydrogen atom abstraction from H_2O bound to the Mn cluster would constitute a "charge neutral" mechanism for removing electrons from water without raising the charge on the cluster: $Mn^{3+}{-}OH_2 + \bullet O{-}Tyr \rightarrow Mn^{4+}{-}OH + HO{-}Tyr$. This reaction at first looks to be thermodynamically uphill. However, calculations of O–H bond energies for the phenolic OH of Tyr, and for an OH group bound to a Mn^{3+}, showed that these energies were in fact close to one another. If the actual energies are close to the calculated bond energies, then thermodynamic barriers to a hydrogen atom transfer model would not be significant.

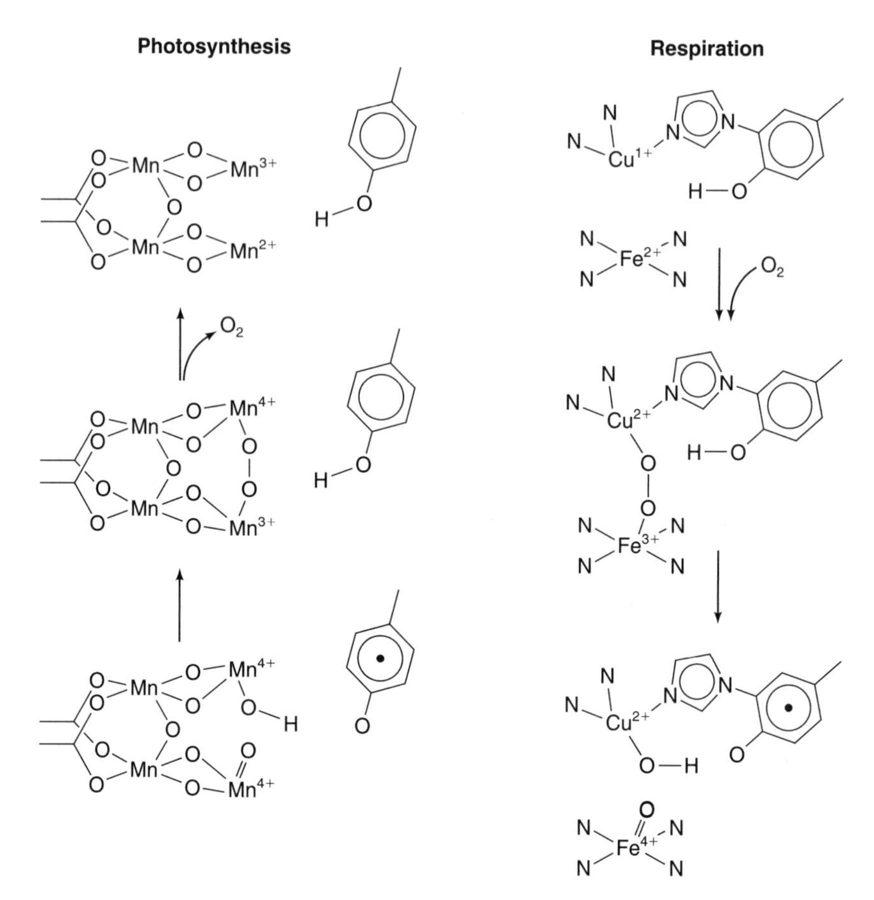

Fig. X.3.15.
Comparison of the mechanisms of oxygen reduction by respiration and water oxidation by photosynthesis. This model was proposed by Babcock and co-workers[38] prior to the determination of the crystal structure of PSII.

This interesting model for electron transfer, which could apply equally well to the oxidase mechanism, is probably not feasible because the distance between Mn atoms and the Y_z tyrosine in PSII is currently estimated to be 7 Å, a distance that is prohibitive for hydrogen atom transfer. However, an alternative mechanism is available that accomplishes the same end, the concerted transfer of an electron and a proton. Such mechanisms are called proton-coupled electron-transfer reactions, and like hydrogen atom transfers, they maintain charge neutrality on the redox center that is involved in electron transfer. In the case of PSII, such a reaction could be written as follows: $Mn^{3+}–OH_2 + \cdot O–Tyr \rightarrow Mn^{4+}–OH_2 + {}^{-}O–Tyr$ (electron transfer) $\rightarrow Mn^{4+}–OH +$ HO–Tyr (H^+ transfer). The important feature arising out of Babcock's models of cytochrome *c* oxidase and PSII is that electron-transfer reactions in these important enzymes are catalyzed by an active site made up of a redox-active amino acid in close proximity to a transition metal cluster that is capable of binding and reacting with either water or oxygen. A critical function of this metalloradical active site is that it preserves charge neutrality through a cycle of several redox intermediates, using the transfers of both electrons and protons to accomplish this end (see Chapter XIII for other examples of metalloradical enzymes).

References

General References

1. Babcock, G. T., "How Oxygen is Activated and Reduced in Respiration", *Proc. Natl. Acad. Sci. U.S.A.*, **96,** 12971–12973 (1999).
2. Ferguson-Miller, S. and Babcock, G. T., "Heme/Copper Terminal Oxidases", *Chem Rev.* **96,** 2889–2908 (1996).
3. Lanyi, J. K. and Luecke, H., "Bacteriorhodopsin", *Curr. Opin. Struct. Biol.*, **11,** 415–419.
4. MacLennan, D. H. and Green, N. M. 2000. Structural biology. Pumping ions. *Nature,* **405,** 633–634 (2001).
5. Mannella, C. A., Marko, M., and Buttle, K., "Reconsidering mitochondrial structure: new views of an old organelle", *Trends Biochem. Sci.*, **22,** 37–38 (1997).
6. Nicholls, D. G. and Ferguson, S. J., *Bioenergetics 2,* Academic Press, New York, 1992.
7. Racker, E., "From Pasteur to Mitchell: a hundred years of bioenergetics", *Fed. Proc.*, **39,** 210–215 (1980).
8. Wallace, D. C., "Aging and degenerative diseases: A mitochondrial paradigm", in *Frontiers of Cellular Bioenergetics, Pe al,* Ed., Kluwer Academic/Plenum, New York, 1999, p. 751.
9. Wallace, K. B. and Starkov, A. A., "Mitochondrial targets of drug toxicity", *Annu. Rev. Pharmacol. Toxicol.*, **40,** 353–388 (2000).
10. Voet, D. and Voet, J., *Biochemistry,* 3rd ed., John Wiley & Sons, New York, 2003, pp. 871–908.
11. Simpson, D. J. and Knoetzel, J., "Light-harvesting complexes of plants and algae: Introduction, survey and nomenclature", in *Oxygenic photosynthesis: The light reactions,* Ort, D. R. and Yocum, C. F., Eds., Kluwer Academic Publishers, Dordrecht, The Netherlands, 1996, pp. 493–506.
12. Hauska, G., Schütz, M., and Büttner, M., "The cytochrome b_6f complex-composition, structure and function", in *Oxygenic photosynthesis: The light reactions,* Ort, D. R. and Yocum, C. F., Eds., Kluwer Academic Publishers, Dordrecht, The Netherlands, 1996, pp. 377–398.
13. Knaff, D. B., "Ferredoxin and ferredoxin-dependent enzymes", in *Oxygenic photosynthesis: The light reactions,* D. R. Ort and C. F. Yocum, Eds., Kluwer Academic Publishers, Dordrecht, The Netherlands, 1996, pp. 333–361.

Specific References

14. Mitchell, P. and Moyle, J., "Stoichiometry of proton translocation through the respiratory chain and adenosine triphosphatase systems of rat liver mitochondria", *Nature* (*London*), **208,** 147–151 (1965).
15. Friedrich, T., "Complex I: a chimaera of a redox and conformation-driven proton pump?", *J. Bioenerg. Biomembr.*, **33,** 169–177 (2001).
16. Lancaster, C. R. and Kroger, A., "Succinate: quinone oxidoreductases: new insights from X-ray crystal structures", *Biochim. Biophys. Acta,* **1459,** 422–431 (2000).
17. Iwata, S., Lee, J. W., Okada, K., Lee, J. K., Iwata, M., et al., "Complete structure of the 11-subunit bovine mitochondrial cytochrome bc_1 complex", *Science,* **281,** 64–71 (1998).
18. Lange, C. and Hunte, C., "Crystal structure of the yeast cytochrome bc_1 complex with its bound substrate cytochrome *c*", *Proc. Natl. Acad. Sci. U.S.A.*, **99,** 2800–2805 (2002).
19. Xia, D., Yu, C. A., Kim, H., Xia, J. Z., Kachurin, A. M., Zhang, L., Yu, L., and Deisenhofer, J., "Crystal structure of the cytochrome bc_1 complex from bovine heart mitochondria", *Science,* **277,** 60–66 (1997).
20. Yu, C. A., Xia, D., Kim, H., Deisenhofer, J., Zhang, L., Kachurin, A. M., and Yu, L., "Structural basis of functions of the mitochondrial cytochrome bc_1 complex", *Biochim. Biophys. Acta,* **1365,** 151–158 (1998).
21. Moyle, J. and Mitchell, P., "Cytochrome *c* oxidase is not a proton pump", *FEBS Lett.*, **88,** 268–272 (1978).
22. Wikström, M., "Proton Pump Coupled to Cytochrome *c* Oxidase in Mitochondria", *Nature* (*London*), **266,** 271–273 (1977).
23. Mills, D. A. and Ferguson-Miller, S., "Understanding the mechanism of proton movement linked to oxygen reduction in cytochrome *c* oxidase: lessons from other proteins", *FEBS Lett.*, **545,** 47–51 (2003).
24. Yoshikawa, S., Shinzawa-Itoh, K., Nakashima, R., Yaono, R., Yamashita, E., Inoue, N., Yao, M., Fei, M. J., Libeu, C. P., Mizushima, T., Yamaguchi, H., Tomizaki, T., and Tsukihara, T., "Redox-coupled crystal structural changes in bovine heart cytochrome *c* oxidase", *Science,* **280,** 1723–1729 (1998).
25. Wikström, M., Bogachev, A., Finel, M., Morgan, J. E., Puustinen, A., Raitio, M., Verkhovskaya, M., and Verkhovsky, M. I., "Mechanism of Proton Translocation by the Respiratory Oxidases. The Histidine Cycle", *Biochim. Biophys. Acta,* **1187,** 106–111 (1994).
26. Rich, P., Meunier, B., Mitchell, R., and Moody, R., "Coupling of Charge and Proton Movement in Cytochrome *c* Oxidase", *Biochim. Biophys. Acta,* **1275,** 91–95 (1996).
27. Michel, H., "The Mechanism of Proton Pumping by Cytochrome *c* Oxidase", *Proc. Natl. Acad. Sci. U.S.A.*, **95,** 12819–12824 (1998).
28. Proshlyakov, D. A., Pressler, M. A., Babcock, G. T., "Dioxygen activation and bond cleavage by mixed-valence cytochrome *c* oxidase", *Proc. Natl. Acad. Sci. U.S.A.*, **95,** 8020–8025 (1998).
29. Kuhlbrandt, W. and Wang, D. N., "Atomic model of plant light-harvesting complex by electron crystallography", *Nature* (*London*), **350,** 326–331 (1994).
30. Zouni, A., Witt, H.-T., Kern, J., Fromme, P., Krauss, N., Saenger, W., and Orth, P., "Crystal structure of photosystem II from *Synechococcus elongatus* at 3.8 Å resolution", *Nature* (*London*), **409,** 739–743 (2001).

31. Kurisu, G., Ahang, H., Smith, J. L., and Cramer, W. A., "Structure of the cytochrome b_6f complex of oxygenic photosynthesis: tuning the cavity", *Science,* **302,** 1009–1014 (2003).
32. Stroebel, D., Choquet, Y., Popot, J.-L., and Picot, D., "An atypical haem in the cytochrome b_6f complex", *Nature* (*London*), **426,** 413–418 (2003).
33. Zhang, H., Whitelegge, J., and Cramer, W. A., "Ferredoxin:NADP$^+$ oxidoreductase is a subunit of the chloroplast cytochrome b_6f complex", *J. Biol. Chem.,* **276,** 38159–38165 (2001).
34. Jordan, P., Fromme, P., Witt, H.-T., Klukas, O., Saenger, W., and Krauss, N., "Three-dimensional structure of cyanobacterial photosystem I at 2.5 Å resolution", *Nature* (*London*), **411,** 909–917 (2001).
35. Brettel, K. and Leibl, W., "Electron transfer in photosystem I", *Biochim. Biophys. Acta,* **1507,** 100–114 (2001).
36. Vassiliev, I. R., Antonkine, M. L., and Golbeck, J. H., "Iron–sulfur clusters in type I reaction centers", *Biochim. Biophys. Acta,* **1507,** 139–160 (2001).
37. Broquisse, R., Weigel, P., Rhodes, D., Yocum, C. F., and Hanson, A. D., "Evidence for a ferredoxin-dependent choline monooxygenase from spinach chloroplast stroma", *Plant Physiol.,* **90,** 322–329 (1989).
38. Hoganson, C. W., Pressler, M. A., Proshlyakov, D. A., and Babcock, G. T., "From water to oxygen and back again: mechanistic similarities in the enzymatic redox conversions between water and dioxygen", *Biochim. Biophys. Acta,* **1365,** 170–174 (1998).

X.4. Dioxygen Production: Photosystem II

Charles Yocum
Chemistry and
MCD Biology
University of Michigan
Ann Arbor, MI 48109

Gerald T. Babcock
Department of Chemistry
Michigan State University
East Lansing, MI 48828

Contents

X.4.1. Introduction

Electron transfer in oxygenic photosynthesis, the process described in Section XI.3, is dependent on an abundant oxidizable substrate, H_2O, as the source for the electrons that are ultimately used to reduce CO_2 to sugars. The enzyme system that catalyzes H_2O oxidation is called photosystem II, or PSII (see Refs. 1–14). The redox activity of this protein complex can be summarized as follows:

$$2\,H_2O + 2\text{ Plastoquinone (PQ)} + 4h\nu \rightarrow O_2 + 2\,PQH_2 \tag{1}$$

Light absorption provides the energy to catalyze the four-electron oxidation of H_2O to O_2, a reaction that is carried out by an inorganic ion cluster comprised of four Mn atoms, and one atom each of Ca^{2+} and Cl^-. The reduced quinone is the electron donor to the interphotosystem electron-transport chain that is described in Section X.3. At 55.5 M concentration, H_2O is an ideal physiological source of electrons to reduce CO_2. However, the chemistry of oxidation of H_2O to O_2 presents thermodynamic challenges that will be discussed prior to a more detailed description of the redox components and protein structure of PSII, and of the mechanisms for the electron-transfer reactions that it catalyzes. In Section X.3, principles of oxidation–reduction, or redox chemistry, were introduced and this discussion needs to be taken up briefly here. Information about the free energy associated with redox reactions can be obtained from the following equation:

$$\Delta G^{o\prime} = -n\mathscr{F}\Delta E^{o\prime} \tag{2}$$

where n = number of electrons transferred in the redox reaction

$\mathscr{F}$ = the Faraday (96.50 kJ V^{-1} mol^{-1} or 23.06 kcal V^{-1} mol^{-1})

$\Delta E^{o\prime}$ = difference in redox potentials between the electron acceptor and the electron donor of a redox reaction

The redox potential for H_2O oxidation is

$$2\,H_2O \rightarrow O_2 + 4\,H^+ + 4\,e^- \qquad E^{o\prime} = +0.815\,V \tag{3}$$

This potential is, by convention, given *per electron* and four electrons are removed in the oxidation of 2 H_2O to O_2. The wavelengths of red light used to drive chlorophyll-catalyzed redox reactions that lead to H_2O oxidation are centered ~680 nm, so the available redox energy is in theory quite large, on the order of 1.8 V/photon. The electron that is expelled by light from a special redox-active chlorophyll *a* molecule, which is called P680, must reduce a pheophytin, a molecule whose potential is estimated to be about −0.6 V (pheophytin *a* is a chlorophyll *a* molecule lacking the Mg^{2+} ion ligated in the center of the chlorin ring). This means that the redox potential of the P680/P680$^+$ species is limited to ~+1.2 V, which is ~0.4 V more positive than the potential needed to oxidize H_2O.

A consideration of the possible mechanisms for oxidation of H_2O reveals some impressive stumbling blocks. A straightforward mechanism would involve sequential, one-electron oxidations of the substrate. The first product in such a mechanism would be the hydroxyl radical (OH•) shown in Eq. 4:

$$H_2O \rightarrow OH\bullet + H^+ + e^- \qquad E^{o\prime} = +1.8\,V \tag{4}$$

The barrier to this reaction is posed by the restriction that only one photon is required for each electron-transfer event in PSII (Eq. 2). From the information given in Eq. 4, OH• formation does not appear to be feasible because the required redox potential is ~0.6 V greater than the potential that is available in the P680/P680$^+$ couple (+1.2 V). A second possible mechanism for H_2O oxidation might consist of a concerted two-electron oxidation reaction that would bypass the one-electron thermodynamic barrier in Eq. 4 and produce H_2O_2 as an intermediate:

$$2\,H_2O \rightarrow H_2O_2 + 2\,H^+ + 2\,e^- \qquad E^{o\prime} = +1.35\,V \tag{5}$$

$$H_2O_2 \rightarrow O_2 + 2\,H^+ + 2\,e^- \qquad E^{o\prime} = +0.295\,V \tag{6}$$

Unfortunately, in this sequence the redox potential of reaction 5 exceeds the +1.2 V limit imposed by the redox potential of the P680/P680$^+$ couple. A last possibility is that H_2O oxidation actually occurs as written in Eq. 3, that is, as a concerted four-electron reaction that might bypass the thermodynamic barriers encountered in multistep reactions leading from H_2O to O_2. Thermodynamic barriers like those just discussed are indeed side-stepped by PSII. How this happens is a question of intense interest. Possible mechanisms for H_2O oxidation that are now under discussion will be considered after a review of the activity, structure, and components of PSII.

X.4.2. Photosystem II Activity: Light-Catalyzed Two- and Four-Electron Redox Chemistry

When PSII is exposed to a series of short (<10 μs) light pulses, a sensitive O_2 electrode detects the oscillatory pattern of oxygen release shown in Fig. X.4.1. This "fingerprint" of the enzyme's activity was discovered by Bessel Kok and by Pierre

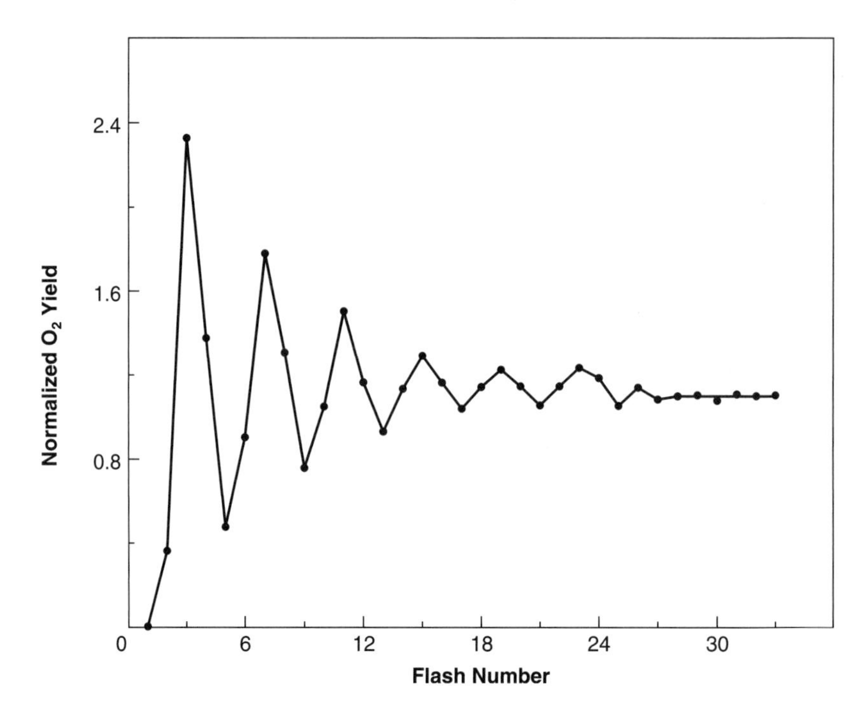

Fig. X.4.1.
Dioxygen release from PSII as a function of the number of single-turnover flashes of light. Peak release of O_2 occurs on flashes 3, 7, and 11. The damping of the O_2 oscillations with increasing numbers of flashes is caused by imperfections in the photochemical reactions of PSII "misses" (failure of a photon to excite a chlorophyll) and "double hits" (one chlorophyll absorbs a photon, reacts, and then absorbs a second photon from the same flash) are important causes of damping.

Joliot[3,15] >30 years ago; Kok formulated a mechanism to explain this behavior,[15] invoking a set of states called "S" with sub- and superscripts to identify the hypothetical differences in oxidation level among the states, as shown below:

$$2\,H_2O + S_0{}^0 \xrightarrow{h\nu} S_1{}^{+1} \xrightarrow{h\nu} S_2{}^{+2} \xrightarrow{h\nu} S_3{}^{+3} \xrightarrow{h\nu} S_4{}^{+4} \rightarrow S_0{}^0 + O_2 + 4\,H^+ \qquad (7)$$

The peculiar behavior of the O_2 oscillations (a spike of O_2 on the third flash and on every fourth flash thereafter) led Kok to propose that the S_1 state predominates in the dark-adapted enzyme. The short pulses of light that are absorbed by PSII result in single turnovers of the photochemical machinery that is composed of the redox carriers associated with the photosystem. These turnovers also catalyze the sequential two-electron reduction of protein-bound quinone molecules (see also Section XI.3). Two molecules of these quinones are present in PSII. One quinone, which is very tightly bound to the enzyme, is called Q_A, while a second plastoquinone, called Q_B, is much less tightly bound and can exchange in and out of its binding site in the enzyme to become a free quinone, in either the oxidized (PQ_{ox}) or reduced (PQH_2) form. This pair of quinones functions in PSII electron transfer as follows:[2]

$$h\nu + \text{PSII } Q_A/Q_B \rightarrow \text{PSII } Q_A{}^-/Q_B \rightarrow \text{PSII } Q_A/Q_B{}^- \qquad (8)$$

$$h\nu + \text{PSII } Q_A/Q_B{}^- \rightarrow \text{PSII } Q_A{}^-/Q_B{}^- \rightarrow \text{PSII } Q_A/Q_B{}^{2-} + 2\,H^+ \rightarrow \text{PSII } Q_A/Q_BH_2 \qquad (9)$$

$$\text{PSII } Q_A/Q_BH_2 + PQ_{ox} \rightarrow \text{PSII } Q_A/Q_B + PQH_2 \quad \text{(PQ is used here to represent the free, membrane-soluble form of the quinone)} \qquad (10)$$

Equations 7 and 8–10 reveal an interesting feature of the redox behavior of PSII; four- and two-electron redox reactions can occur simultaneously in the same protein complex, without recombination of the oxidants and reductants that are generated by these reactions. This is remarkable, given that the $\Delta E^{\circ\prime}$ for the $H_2O/\frac{1}{2}\,O_2 - PQ^-/PQ$ couple is about $+0.8\,V$, corresponding to $-77.2\,kJ\,mol^{-1}$ ($-18.5\,kcal\,mol^{-1}$).

X.4.3. Photosystem II Protein Structure and Redox Cofactors

Because PSII is a membrane protein complex, it is not surprising that the majority of its polypeptides are intrinsic, containing a large number of nonpolar amino acid residues.[1] The polypeptides of the complex are oriented in the membrane so that water oxidation occurs on the lumenal (interior) side of the membrane. This assures that the H^+ generated by redox chemistry can be trapped and used to generate a proton gradient to drive ATP synthesis. Quinone reduction occurs toward the membrane outer surface (the stromal face) to allow for binding of solvent H^+ to the semiquinone anion form of Q_B from this side of the membrane. More information on the organization of protein complexes in membranes, and on how redox-linked proton pumping occurs is given in Section X.3.

The polypeptide composition of PSII is subdivided into three domains. The first two domains are comprised of intrinsic polypeptides that span the membrane bilayer. Proteins called "D1" and "D2" containing the redox catalysts of the enzyme system form a core domain in the center of the complex.[4] The amino acid sequences of these polypeptides contain redox-active tyrosine residues called Y_Z and Y_D on the D1 and D2 subunits, respectively. One of these tyrosine residues (Y_Z) is directly in the pathway of electron transfer in PSII. The other exists as a dark-stable radical ($Y_D\cdot$) that does not undergo reduction or oxidation during light-driven, steady-state electron transfer.[5] The D1 and D2 subunits also bind the redox active chlorophyll *a*, P680, and several subsidiary chlorophyll *a* molecules that act as antennae to funnel excitation energy to P680. Two pheophytin *a* molecules are bound to the D1 and D2 proteins. The quinones Q_A and Q_B are also bound to D1 and D2, as is an atom of nonheme Fe^{2+} that is redox inactive and does not participate in PSII electron transfer reactions. The other intrinsic subunit of this core PSII domain is a cytochrome, b_{559}, whose Fe atom is axially ligated by histidine residues of the two polypeptides that comprise the cytochrome.[16] There are a number of hypotheses concerning the role of b_{559} in PSII,[6] but its function remains uncertain.

The second domain of PSII is also comprised of membrane-associated intrinsic proteins. The most important of these is a pair of polypeptides called CP47 and CP43, which ligate chlorophyll *a* molecules.[1] These additional chlorophylls function as accessory antenna pigments; they absorb light and transfer the excitation energy to the chlorophylls bound to the D1 and D2 subunits. As will be seen later, CP43 may also contribute a ligand to the Mn cluster that catalyzes H_2O oxidation. The third, and last, domain of PSII is comprised of extrinsic, water-soluble proteins that are bound to the intrinsic polypeptides of the photosystem on the lumenal side of the complex,[7] which is the site of H_2O oxidation. In eukaryotic organisms, these polypeptides have molecular masses of 26.5, 20, and 17 kDa, and contain no known prosthetic groups. In cyanobacteria, only the largest extrinsic subunit (26.5 kDa), called PsbO or manganese stabilizing protein, matches an extrinsic subunit found in eukaryotes. The 20- and 17-kDa subunits, found in eukaryotes, are replaced in prokaryotes by cytochrome c_{550} (also called PsbV), which has a negative (-0.2 V) redox potential, and by a 12-kDa extrinsic protein, called PsbU. The function, if any, of the heme cofactor in PsbV is uncertain. In both eukaryotes and prokaryotes, the inorganic ion cofactors (Mn, Ca^{2+}, and Cl^-) that are the components of the "*S*" state cycle are shielded from the external medium by the extrinsic polypeptides of PSII. These inorganic cofactors will be discussed in more detail later.

The structure of PSII and the relationships among the various polypeptides and the organic and inorganic cofactors is becoming clearer. There are now three crystal structures of PSII from thermophilic cyanobacteria.[17–19] Figure X.4.2 presents a diagram of PSII derived from the 3.5-Å crystal structure of the enzyme that was isolated from a thermophilic cyanobacterium, *Thermosynechococcus elongatus.*[19] The

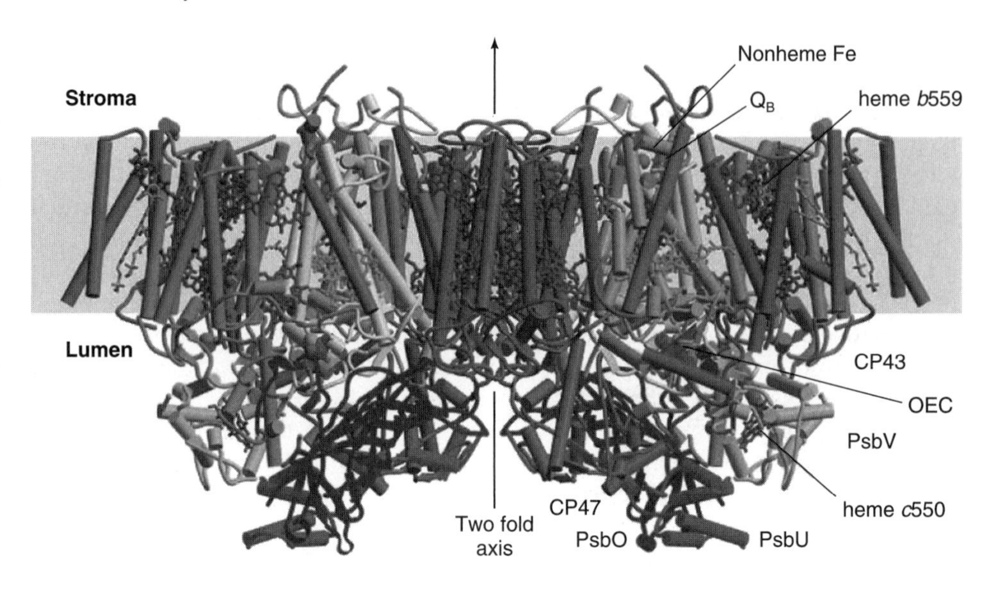

Fig. X.4.2.
The 3.5-Å resolution structure of PSII from the thermophilic cyanobacterium *T. elongatus.*[19] The structure consists of a dimer of reaction centers of the enzyme, split by the twofold axis shown in the figure. The major integral membrane polypeptides shown here are colored as follows: D1; D2; CP47; CP43; and cytochrome b_{559}. The extrinsic polypeptides are manganese stabilizing protein, or PsbO, PsbU, cytochrome c_{550}, or PsbV. The position of the Mn cluster is indicated by "OEC" (O_2 evolving complex). Descriptions of the functions of the polypeptides are given in the text (PDB code: 1S5L). The PDB codes for the other structures are 1FE1[17] and 1IZL.[18] [See color insert.]

X-ray crystal structure shown in Fig. X.4.2 provides an overall picture of the structure of PSII that coincides with much of the biochemical and biophysical data that has been generated on the enzyme, and presents a clearer picture of the organization of the protein complex. For example, the membrane spanning α helices form the "backbone" of the intrinsic structure of the enzyme, and the largest extrinsic protein (PsbO) stands out from the membrane spanning helices. A great deal of information is also emerging about the structures of the large extrinsic amino acid sequences on the lumenal side of PSII, that make up part of the accessory antenna proteins, CP47 and CP43. The structure of the Mn cluster is also visible in the crystal structure, and will be discussed in more detail in Section X.4.4, which deals with the roles of the inorganic ion cofactors that catalyze H_2O oxidation.

By analogy with the structure of the reaction centers of photosynthetic bacteria, and of PSI, the disposition of organic redox cofactors in PSII is believed to generate a pseudo-C_2 symmetry, as shown in Fig. X.4.3. A pair of electron-transfer chains diverges after P680 into the two pheophytin a molecules on D1 and D2; however, only a single pathway, through one of the pheophytins leading to Q_A, is functional in electron transfer.[2] Although they may be subject to later revisions, estimated distances between cofactors in the PSII structure are available, and some of these are shown schematically in Fig. X.4.3.

At this point, we can examine the kinetics of electron transfer between the various redox cofactors that are bound to PSII. The forward reaction is catalyzed by the following sequence of electron carriers:

$$S_N \rightarrow Y_Z \rightarrow \text{P680} \rightarrow \text{Pheo a} \rightarrow Q_A \rightarrow Q_B$$

$$(S_N \text{ refers to the } S \text{ states described in Eq. 7}) \qquad (11)$$

To see how this pathway functions, it must be broken down into two parts, the first of which is the photochemically driven charge-separation step, which initiates the forward electron-transfer reactions in PSII, as shown in Table X.4.1. Also shown in

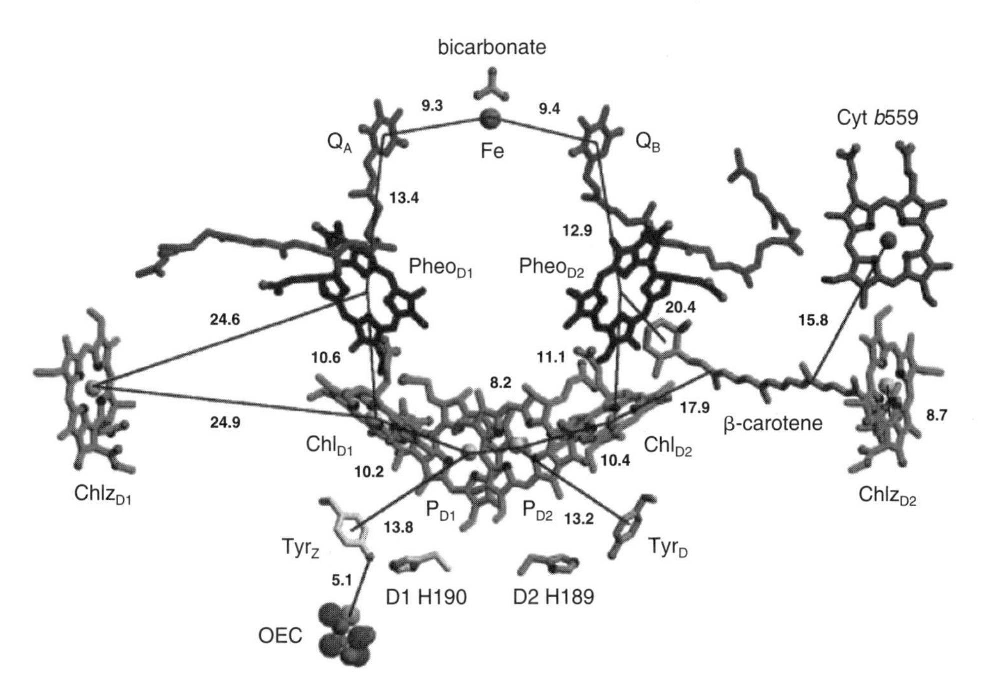

Fig. X.4.3. Distances between the principal cofactors of the PSII reaction center.[19] The numbers next to the lines connecting molecules are the distances in angstroms. Both Chl_{D1} and $Pheo_{D1}$ are used to identify the pigments ligated by the D1 polypeptide of the reaction center. A similar notation is used for the pigments associated with D2. Likewise, D1 H190 and D2 H189 identify two His residues that that are proposed to hydrogen bond to tyrosine residues Tyr_Z and Tyr_D. [See color insert.]

Table X.4.1.
Electron-Transfer Kinetics in Photosystem II

Forward Electron-Transfer Reaction	$t_{1/2}$	Recombination Reaction	$t_{1/2}$
P680/Pheo $a \rightarrow$ P680$^+$/Pheo a$^-$	<1 ps		
P680$^+$/Pheo a^-/Q_A $\rightarrow$ P680$^+$/Pheo a/Q_A^-	250 ps	Q_A^-/P680$^+$ $\rightarrow$ Q_A/P680	150 μs
Y_Z/P680$^+$/Q_A^- $\rightarrow$ $Y_Z\cdot$/P680/Q_A^-	50–250 ns	$Y_Z\cdot/Q_A^-$ $\rightarrow$ Y_Z/Q_A	200 ms
$Y_Z\cdot/Q_A^-/Q_B$ $\rightarrow$ $Y_Z\cdot/Q_A/Q_B^-$	100 μs	$Y_Z\cdot/Q_B^-$ $\rightarrow$ Y_Z/Q_B	400 ms
$S^N/Y_Z\cdot/Q_B^-$ $\rightarrow$ $S^{N+1}/Y_Z/Q_B^-$	30 μs–1 ms	S^{N+1}/Q_B^- $\rightarrow$ S^N/Q_B	30 s

the table are rates of the succeeding events, which consist of a series of reactions that result in the trapping of oxidizing and reducing equivalents in relatively stable, long-lived chemical intermediates (an advanced *S* state and a semiquinone anion). A list of the rates of recombination (reverse) reactions, also shown in Table X.4.1, completes a description of this pathway. A couple of points are revealed by the numbers. First, the primary photochemical reaction in PSII (P680 → Pheo), like those in other photosystems, occurs with a very short (ps) half-time. A second point is that without exception, the "forward" reactions are orders of magnitude more rapid than the rates of "reverse", or recombination reactions. This, in turn, produces a situation in which reactive intermediates generated in the course of forward electron-transfer reactions are quenched more rapidly by the succeeding forward reaction, which prevents recombination reactions from dominating electron transfer within PSII. For example, from the data in Table X.4.1 it can be seen that P680$^+$ is reduced by Y_Z with a $t_{1/2}$ in the range of 50–250 ns, whereas P680$^+$ is reduced by Q_A^- with an estimated half-time of 150 μs. The ratio of $t_{1/2}$ values is $\sim 10^3$ in favor of the forward reaction, which assures that under most conditions, Y_Z, rather than Q_A^-, will be oxidized by P680$^+$. The extremely fast forward reactions in PSII are facilitated by the relatively close distances (Fig. X.4.3) estimated for the participating redox cofactors in the

Table X.4.2.
Inorganic Ion Cofactors of the Dioxygen-Evolving Complex

Cofactor	Substitutes	Inhibitors
4 Mn^{n+} ($n > 2$)	None	Reductants (Mn^{2+})
1 Ca^{2+}	Sr^{2+}	Ln^{3+},[a] Cd^{2+}, $Na^+/K^+/Cs^+$
1 Cl^-	$Br^- > NO_3^- > I^- > NO_2^-$	Lewis bases (RNH_2, ^-OH, F^-)

[a] $Ln^{3+} = La^{3+}$, Pr^{3+}, Dy^{3+}, Lu^{3+}, Yb^{3+}.

T. elongatus crystal structure. (See Section X.2 for a discussion of distance dependence of ET rates.)

The reactions listed in Table X.4.1 are associated with the intrinsic protein domain of PSII, where the participating organic cofactors are ligated. The inorganic cofactors that form the site for catalysis of H_2O oxidation appear to be bound at the interface between the intrinsic and extrinsic proteins of the enzyme, as shown in Fig. X.4.2. This combination of inorganic ions (4 Mn, Ca^{2+}, and Cl^-) forms the site of H_2O oxidation. Taken together with the intrinsic and extrinsic polypeptide components, this part of PSII is often referred to as the O_2-evolving complex, or OEC. This component of PSII has been the subject of extensive characterization. Among the investigations that have been carried out are studies to determine the extent to which other inorganic ions can replace those that occur naturally in PSII. These are summarized in Table X.4.2.

The only redox-active ion in this list is Mn, which forms a tetranuclear cluster that functions as the site of H_2O oxidation (but, see below). A wide range of biochemical and spectroscopic results support this assertion. The requirements for Ca^{2+} and Cl^- are not easily rationalized by way of comparisons with other metalloenzyme systems. Both cofactors are essential activators of the H_2O oxidation reaction, and both are relatively easy to extract from the enzyme, provided that the smaller extrinsic subunits are first removed. In the absence of either Ca^{2+} or Cl^-, the $S_1 \rightarrow S_2$ transition is observed, but subsequent oxidation state advancements of the S-state system are blocked. Calcium can be replaced by Sr^{2+}, but the resulting steady-state electron-transfer activity is lower by 50% than that of the rate assayed in the Ca^{2+}-reconstituted enzyme. A few anions can replace Cl^- (Table XI.6.2); of these, only Br^- is capable of producing steady-state activity approaching that of the native system. Each inorganic cofactor of the OEC is discussed in more detail in Section X.4.4.

X.4.4. Inorganic Ions of PSII

X.4.4.1. Manganese

Many research groups have been actively engaged in studying the behavior of the Mn atoms in PSII (see Ref. 8 for a general review). Treatments with either small reducing agents, like NH_2OH, or exposure to high concentrations of chaotropic agents like tris (tris-hydroxymethylaminomethane) produce $[Mn(H_2O)_6]^{2+}$ in parallel with inactivation of the enzyme.[9] The reductant-catalyzed release of Mn^{2+} from PSII is likely due to the absence of ligand-field stabilization energy associated with the high-spin $3d^5$ Mn^{2+} ion. Restoration of O_2 evolution activity by illumination of metal-depleted PSII in the presence of Mn^{2+} suggests, but does not prove that the active site of the enzyme system contains higher oxidation states of Mn (Mn^{3+}, Mn^{4+}), most probably bound to PSII proteins by ligation that is predominantly ionic in nature.

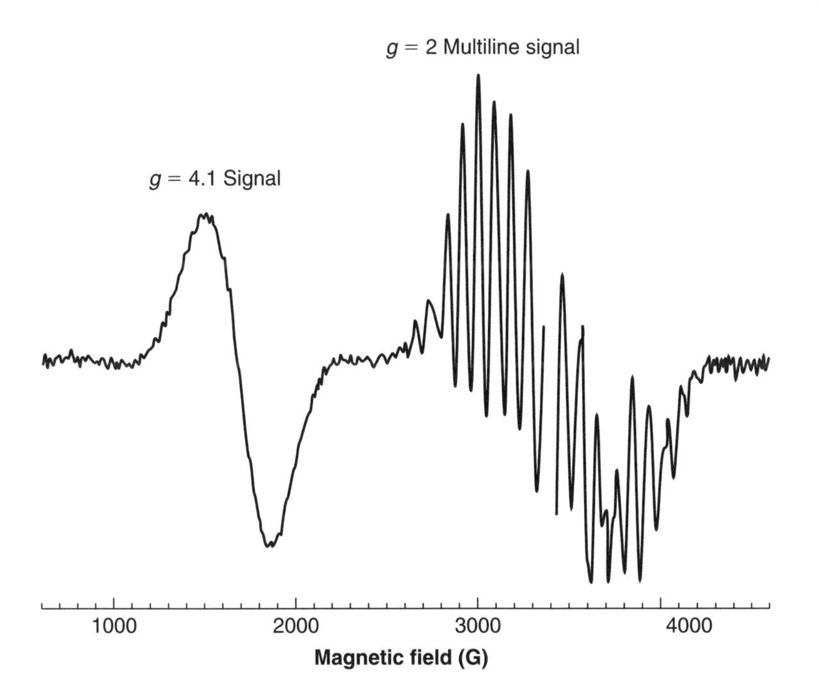

Fig. X.4.4.
The EPR signals associated with the S_2 state of PSII. The appearance of both the S_2 multiline signal and $g = 4.1$ signal are shown as detected at 10 K in a single sample. These signals are produced by continuous illumination of PSII at 200 K, which limits the system to a single electron-transfer reaction that produces Q_A^- and the S_2 state. See the text for further details. [Spectrum courtesy of R. D. Britt.]

An explosion of research beginning in 1980 and continuing to the present day has greatly expanded knowledge of Mn function in PSII. Quantification of the Mn contents of intact photosynthetic membranes, as well as those of isolated PSII, established the Mn stoichiometry at four atoms per active site.[9] Upon illumination of PSII at 200 K (which assures that a single turnover generates S_2/Q_A^-), a $g = 2$, $S = \frac{1}{2}$ EPR multiline signal, detectable only at cryogenic (<10 K) temperatures was discovered, as shown in Fig. X.4.4. The signal spans ~1600 G, exhibits ~18 lines, has been shown to resemble the EPR signals associated with synthetic mixed-valence dioxo-bridged Mn^{3+}–Mn^{4+} dimers, and oscillates with a periodicity of four.[8] The structure of the S_2 signal is rationalized as follows. The EPR signal from one Mn ion should give rise to six lines, due to the six possible m_I values from the spin $I = \frac{5}{2}$ ^{55}Mn nucleus. An antiferromagnetically coupled synthetic Mn^{3+}–Mn^{4+} dimer produces ~16 hyperfine lines arising from the EPR transitions from the 36 m_I combinations of the two ^{55}Mn nuclei in the synthetic complex. (The actual number of hyperfine lines is much larger, but their superposition reduces the actual number detected to ~16.) However, in the case of the signal from PSII, further investigations of its properties uncovered additional features that were not evident in the multiline signal from synthetic mixed-valence Mn dimers. Foremost among these was the discovery of another EPR signal, at lower field ($g = 4.1$), in samples also exhibiting the multiline signal. This signal is also shown in Fig. XI.6.4. The $g = 4.1$ species is proposed to arise from a $S = \frac{5}{2}$ excited state of the Mn cluster.[8]

At first, it was thought that the $g = 4.1$ signal might be associated with a separate Mn cluster from the one that gives rise to the $g = 2$ multiline signal. These speculations were removed when it was discovered that a special form of PSII sample in which NH_3 is bound to the Mn cluster yielded a $g = 4.1$ signal that had Mn hyperfine structure.[8] This finding is explained as follows: First, to generate a signal with the number of hyperfine lines associated with the $S = \frac{1}{2}$ multiline species, it is necessary to postulate the existence of a mixed-valence Mn dimer (either Mn^{3+}/Mn^{4+} or Mn^{2+}/Mn^{3+}), as discussed above. The case of the $g = 4.1$ signal is more complicated. Synthetic complexes containing monomeric Mn^{4+} are known to produce low-field signals that resemble the PSII $g = 4.1$ signal. However, the appearance of the

Mn hyperfine structure on the $g = 4.1$ signal that was detected in NH_3-derivatized PSII samples requires a multinuclear metal center containing at least three Mn atoms. The reasoning behind this assertion is as follows:

1. The hyperfine structure requires an interaction of the unpaired electron spin with more than one Mn nucleus;
2. Because a binuclear center would produce an $S = \frac{1}{2}$ multiline signal, it is necessary to postulate a minimum of a trinuclear Mn cluster to explain the origin of the modified $g = 4.1$ signal.
3. There are four Mn atoms in the PSII OEC, rather than the five atoms that would be required if the $g = 4.1$ and 2 signals were to arise from separate trimeric and dimeric clusters.

For these reasons, the current view is that both S_2 EPR signals ($g = 2$ and 4.1) originate from the same tetranuclear Mn cluster.

Additional information on the S_2 EPR signals supports the hypothesis that they have a common origin in the tetranuclear Mn cluster. While both signals can be detected after 200 K illumination of a PSII sample, a number of experimental manipulations can increase the yield of either signal, or convert one species into the other. The yield of the $g = 4.1$ signal is enhanced by the presence of sucrose in the sample, as well as by illumination at temperatures ~160 K. Depletion of Cl^-, addition of F^-, and treatment of PSII with NH_3 will produce enhanced yields of the $g = 4.1$ signal. The amount of $g = 2$ multiline signal can be greatly enhanced by the addition of methanol to PSII before illumination. Conversion of the $g = 4.1$ signal to the multiline signal can be achieved in several ways. Samples illuminated at 160 K can be warmed and refrozen to convert the signal to the multiline species, addition of Cl^- to Cl^- depleted PSII after 200 K illumination produces the multiline signal, and irradiation of PSII samples containing the multiline signal with near IR ($\lambda = 820$ nm) converts the $g = 2$ signal to the $g = 4.1$ species.[20] This result is proposed to arise from an intervalence electron-transfer reaction *within* the Mn cluster, as follows:

$$Mn_A{}^{3+}\text{–}Mn_B{}^{4+} + \text{IR } h\nu \rightarrow Mn_A{}^{4+}\text{–}Mn_B{}^{3+} \tag{12}$$

While the chemical bases for such observations are not well understood, they do provide evidence for multiple paramagnetic forms of the S_2 Mn cluster.

Investigations of the Mn cluster by X-ray absorption spectroscopy have revealed certain structural features; two types of experiments have been carried out on PSII. One of these, XANES (X-ray absorption near-edge structure) is a sensitive indicator of metal oxidation states. The second experiment (EXAFS, or extended X-Ray absorption fine structure) can provide precise information about metal–ligand and metal–metal distances. The technique is specific for Mn at the energies employed, but it is not selective; all Mn atoms present in the sample are detected indiscriminately. The X-ray absorption technique is likewise unable to distinguish between adjacent atoms in the same row of the periodic table, so, for example, O ligands cannot be distinguished from N ligands. In spite of these limitations, the XANES studies have provided useful data to indicate that the most likely oxidation states for PSII Mn in the S_1 state are $Mn^{4+}/Mn^{4+}/Mn^{3+}/Mn^{3+}$.[8] The EXAFS experiments have detected three types of scattering atoms at various distances from the Mn atoms in the S_1 state. Because this technique cannot distinguish among Mn atoms, the data represent an average over all such atoms in the cluster. The first group of scatterers is assigned to Mn–ligand (O or N) distances of ~1.8 Å. A second group of scattering atoms at 2.7 Å has been fit to a pair of Mn–Mn distances typical of those found in *oxobridged* binuclear Mn complexes. Lastly, a single scatterer has been detected at ~3.3 Å, which has been assigned to the presence of Mn, or to a combination of Mn and Ca.[21,22]

Positive absorbance changes in the region of 290–300 nm are detected in PSII preparations subjected to brief flashes of light. These absorption changes oscillate with a periodicity of 4, that is, they track the *S*-state transitions of PSII. The positive changes on the transitions from $S_0 \rightarrow S_3$ are interpreted as indicating $Mn^{2+} \rightarrow Mn^{3+}$ ($S_0 \rightarrow S_1$) or $Mn^{3+} \rightarrow Mn^{4+}$ ($S_1 \rightarrow S_2$) redox transitions.[10] The accumulated absorbance changes collapse on the $S_3 \rightarrow S_0$ transition, as would be predicted if Mn reduction were to accompany H_2O oxidation. The molecular origin of these absorbance changes is not entirely certain; the most likely explanation is that they arise from ligand to Mn charge-transfer bands.[10]

As mentioned earlier, the Mn cluster of PSII is susceptible to reduction. A classic example is the use of NH_2OH; at relatively high ($\geq 1\,mM$) concentrations, this compound produces nearly complete reduction of the Mn cluster to Mn^{2+}, which is released from its protein-binding sites as $[Mn(H_2O)_6]^{2+}$. Lower concentrations of NH_2OH ($\leq 100\,\mu M$) do not produce extraction of Mn^{2+} from the enzyme. Instead, the number of short flashes needed to produce the first "gush" of O_2 is increased. One stable intermediate is termed S_{-1} because five rather than three flashes are required to produce the first release of O_2 from PSII.[23] Further reduction is more difficult to monitor, because eventually the Mn^{2+} associated with these "*S*-minus" states dissociates from the protein, causing a loss of activity. Another state that can be formed by reduction is S_0; this state has an associated EPR multiline signal that is broader, with 26 more lines than the ~16–18 lines associated with the S_2 signal. The S_0 signal is attributed to the presence of a Mn^{3+}–Mn^{2+} interaction rather than to Mn^{4+}–Mn^{3+}.[24]

Taken together, the data now available on the PSII Mn cluster point to a sequence of PSII-catalyzed Mn oxidation reactions like that given below:

$$\begin{array}{ccccccccc} S_0{}^{0} & \rightarrow & S_1{}^{+1} & \rightarrow & S_2{}^{+2} & \rightarrow & S_3{}^{+3} & \rightarrow & S_4{}^{+4} \\ (2\,Mn^{4+}/Mn^{3+}/Mn^{2+}) & \rightarrow & (2\,Mn^{4+}/Mn^{3+}/Mn^{3+}) & \rightarrow & (3\,Mn^{4+}/Mn^{3+}) & \rightarrow & (Mn^{4+})_4 & \rightarrow & ((Mn^{4+})_3/Mn^{5+}(?))_4 \end{array} \quad (13)$$

There is a continuing discussion about the oxidation state assignments of Mn in S_1, mainly because it is difficult to simulate the S_2 multiline signal using the proposed oxidation states for Mn that are shown above for S_2. As indicated, in this scheme the oxidation states of Mn in S_4 are also uncertain. Another current topic of debate concerns whether Mn oxidation accompanies *S*-state transitions beyond S_2. Two lines of evidence, some XANES edge measurements on samples exposed to short flashes, and the UV–vis absorption changes described above, point convincingly to Mn oxidation on all states up to formation of S_3. However, other X-ray absorption experiments are interpreted as excluding Mn oxidation beyond S_2; the origin of this disagreement is unclear.

Higher oxidation states of Mn in model compounds are characterized by high charges, small ionic radii ($Mn^{4+} = 0.60$ Å; $Mn^{3+} = 0.66$ Å), and slow ligand exchange rates.[11] One would therefore predict that the coordination environment for Mn in PSII would be predominantly ionic in nature, that is, the majority of ligands would be O atoms. Candidates for Mn ligating amino acids are Glu and Asp residues that are exposed on the lumenal side of the complex; attention has focused on the D1 protein, which contains the redox-active tyrosine Y_Z, and as a result, several site-directed mutants of D1 have been prepared.[4] This approach has yielded encouraging results, in that some amino acid residues have been identified as candidates for Mn ligation sites. A pulsed EPR technique that is capable of resolving nearby ligands has also been used to probe the Mn cluster. By using ^{15}N and ^{14}N histidine labeled PSII preparations in the S_2 state, a His imidazole group has been identified as a ligating species.[12] New information from three-dimensional (3D) crystals of PSII, when combined with the existing spectroscopic and molecular data, should lead to a firmer identification of the amino acids involved in Mn ligation. This is discussed in Section X.4.5.

X.4.4.2. Calcium

Extraction of the extrinsic 23- and 17-kDa polypeptides of PSII with high ionic strength ($\geq 1\ M$ NaCl) produces a strong inhibition of H_2O oxidation activity. This defect is not repaired by rebinding of the extrinsic proteins, but activity is restored (without rebinding of the polypeptides) simply by addition of nonphysiologically high (mM) concentrations of Ca^{2+}. A series of experiments to probe the relationship between Ca^{2+} and the extrinsic polypeptides showed that these proteins act as a shield around the OEC, preventing entry of potentially damaging species like reductants (e.g., PQH_2), and slowing the exchange of Ca^{2+} and Cl^- between their binding sites and the external medium.[13] Calcium-depleted PSII is unable to undergo S-state advancement beyond S_2, and Sr^{2+} is the only ion known to replace Ca^{2+} and restore the H_2O oxidizing reaction (see Table X.6.2).

Experiments to probe the Ca^{2+} site in PSII have utilized several approaches. In the first of these, the ability of a number of ions, such as those listed in Table X.6.2, to compete with Ca^{2+} for activation of the OEC was examined. The results are similar to those obtained in investigations of Ca^{2+}-binding sites in other proteins, in that the alkali metals, lanthanides, and Cd^{2+} are all able to occupy the Ca^{2+}-binding site, albeit without reactivating the H_2O oxidation reaction. These results are in agreement with results on the metal-binding preferences of other Ca^{2+}-binding proteins. Measured affinities of the site (K_d values in the range of $60\ \mu M$ for the majority of PSII reaction centers) are on the same order as the Ca^{2+} affinities (K_d values of ~ 20–$250\ \mu M$) found in a series of Ca^{2+} cofactor proteins (proteases, nucleases, lipases, e.g.).

An alternate model for the structure of the PSII Ca^{2+}-binding site would be the metal-binding sites in EF hand, or Helix–Loop–Helix (HLH) proteins like parvalbumin (Ca^{2+} $K_d = 0.001$–$0.01\ \mu M$), troponin C ($K_d = 0.1$–$10\ \mu M$), or calmodulin ($K_d = 1$–$10\ \mu M$) (see Section IV.2). At the present time, this group of protein-binding sites does not seem to provide a likely model, owing to the lower affinity of the PSII site, and to the absence of amino acid sequence homologies to EF hand sites in any of the proteins making up PSII.[13] However, recent data have added a number of metals (Mg^{2+}, Cu^{2+}, Ni^{2+}, Co^{2+}) to the list of ions known to occupy the Ca^{2+} site of PSII, and these data are interpreted as suggesting that an EF-hand-type binding site may be a useful model for the PSII binding site.[25]

The requirement for Ca^{2+} as a participant in H_2O oxidation is unprecedented in comparison to many other redox metalloenzymes, and there have been a number of experiments and proposals concerning its role in the enzyme. One role of Ca^{2+} in proteins is purely structural, for example, in stabilizing tertiary structures (thermolysin and trypsin, e.g.). Some evidence exists to support such a role in PSII; if Ca^{2+} is retained in its site during reduction of the Mn cluster, stable, active, reduced intermediates can be isolated, including derivatives of the cluster in which up to two Mn^{2+} are retained by the enzyme and reinserted into the cluster upon illumination.[26] In the absence of Ca^{2+}, the Mn^{2+} is released into solution with loss of activity. An alternative role for Ca^{2+} would be as a binding site for substrate H_2O. This possibility will be discussed in Section X.6.6.

X.4.4.3. Chloride Ion

As in the case of the Ca^{2+} cofactor requirement, depletion of Cl^- from PSII blocks advancement of the S-state cycle at S_2. A select group of anions ($Br^- \gg I^-$, $NO_3^- > NO_2^-$) replace Cl^- and restore H_2O oxidation in the order shown; Br^-

can produce rates nearly equivalent to those obtained with Cl^-.[9] Chloride depletion results in formation of the $g = 4.1$ EPR signal, rather than the multiline signal. The Cl^- requirement for H_2O oxidation is as puzzling as the Ca^{2+} requirement. While it is attractive to suspect that this is a general anion phenomenon of some sort, the limited number of substituting species, and the stoichiometry (1 Cl^-/PSII)[27] suggest a more intimate role for the ion. A number of Lewis bases have been shown to displace Cl^- from its PSII binding site. These species include RNH_2, F^-, OH^-, and CH_3COO^-; inhibitory potency is proportional to the pK_a of the attacking base.[28] These results suggest that the activation–inhibition interactions arise from a Lewis acid–Lewis base interaction, where Mn functions as the Lewis acid, so it has been proposed that Cl^- is ligated to Mn. If Cl^- were a bridging ligand, it might facilitate inner-sphere electron transfer between two Mn atoms. As a terminal ligand, Cl^- might function to increase the redox potential of a Mn atom, to make it a more effective oxidant. Because there is no direct spectroscopic evidence for Cl^- ligation to Mn, all proposals for its function in H_2O oxidation remain speculative. However, it has been shown that Cl^- is required for the *S*-state transitions from S_2 through $S_4 \rightarrow S_0$,[29] so whatever its mechanism of action, the presence of Cl^- is obligatory for Mn redox chemistry in the steps that lead to H_2O oxidation and O_2 release from the *S*-state cycle.

Additional investigations of the ability of Cl^- to interfere with binding of other ligands to the OEC reveal that NH_3 can bind not only in competition with Cl^-, but also to a second site that is insensitive to the presence of Cl^-.[12] Larger amines are excluded from this second site, suggesting that it is sterically restricted to small ligands, including, possibly, H_2O.[28] Binding of NH_3 to this second site blocks electron transfer beyond S_2 and modifies the S_2 multiline signal (narrowed lines). Reductants show a similar steric specificity for the two OEC-binding sites. Small reductants such as NH_2OH are not blocked by Cl^-, but larger ones (*N*-methyl and *N,N*-dimethyl NH_2OH derivatives, hydroquinone, and phenylenediamines) appear to attack Mn preferentially at the Cl^- site, based on the ability of the anion to impede the inhibitory action of these reductants on the Mn cluster.

Chloride binding to PSII occurs with high affinity in the dark-adapted enzyme ($K_d = 20\,\mu M$).[27] Loss of Cl^- from the intact enzyme occurs slowly in the dark, with a $t_{1/2}$ of ~ 1 h, so extensive depletion requires several hours. In the absence of the smaller (20 and 17 kDa) extrinsic polypeptides, slow Cl^- exchange is abolished, as is high affinity binding of the anion. Estimates of Cl^- affinity as a function of the *S* state have revealed that the anion exchanges more rapidly in the higher *S* states, possibly because smaller, harder Mn^{4+} atoms would have a lower affinity for the larger, soft Cl^- ion.[30]

X.4.5. Modeling the Structure of the PSII Mn Cluster

A wide array of experimental results has been applied to producing models of the Mn cluster in PSII. The most recent of these is the structure derived from 3D crystals. An expanded view of the crystal-derived structure of the cluster[19] is shown in Fig. X.4.5. The arrangement of the Mn atoms shown here is similar to a monomer–trimer structure constructed from magnetic resonance experiments on the S_2 state Mn cluster in eukaryotic PSII, in which the monomeric Mn atom is termed a "dangler".[31] Structures have also been proposed on the basis of EXAFS experiments on oriented PSII samples of eukaryotic PSII. These models are somewhat different from the structure shown in Fig. X.4.5, in that the EXAFS-derived models predict a dimer-of-dimers structure arranged in a "c-clamp" shape, rather than a monomer trimer arrangement of the metals.[14] The EXAFS experiments have produced highly

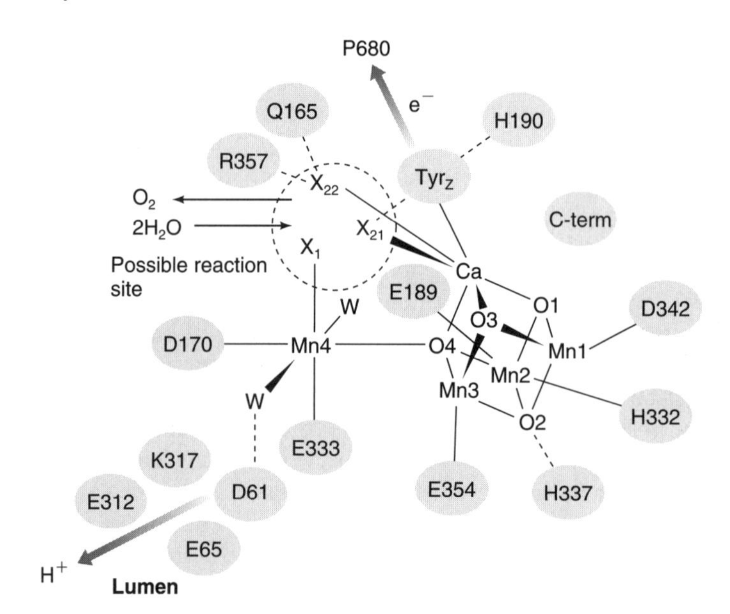

Fig. X.4.5.
A structural model for the Mn–Ca cluster of PSII with proposals for sites of water oxidation and H^+ release.[19] The cluster is comprised of a distorted cubane-like Mn_3Ca core and a Mn atom (labeled Mn4) that is close to, but not part of the inorganic core. Amino acid residues proposed to function as Mn ligands in this model are E354 and E357 (from the CP43 subunit) and D170, E189, H334, E333, and D342 (all from the D1 subunit). D61 and E65 from D1 and E312 and K317 from D2 are possible constitutents of a proton-transfer pathway.

accurate estimates of metal–metal distances, and these data give Mn–Mn distances of ~2.7 Å, and a longer Mn–Mn distance (3.3 Å) within the cluster. The longer distance has been proposed to contain contributions from both Mn and Ca^{2+}.[21] These distances are consistent with the model based on the crystal structure (Fig. X.4.5). An accompanying issue concerns the distribution of Mn oxidation states within the cluster. If one accepts the proposition that only two Mn atoms are involved in H_2O oxidation, and the oxidation states shown in Eq. (13) are correct, then it is likely that the pair of Mn(III) atoms in the S_1 oxidation state of the cluster would have to be in close proximity if they both participate directly in the mechanism of O–O bond formation.

X.4.6. Proposals for the Mechanism of Photosynthetic Water Oxidation

The *S*-state model (Eq. 7) predicts that photosynthetic O_2 evolution must occur by a linear four-step oxidation of H_2O, in which electrons are removed by PSII photochemistry one at a time from the site of H_2O oxidation. In addition, one must consider the implications of the observed pattern of H^+ release from PSII during the *S*-state cycle. These measurements are very difficult to execute (e.g., they can be influenced by the protein content of the sample), and there is currently some disagreement over the actual pattern. However, it appears likely that the release pattern is 1,1,1,1 for the *S*-state advancement, that is, one H^+ is released on every *S*-state transition. The origin of these protons is a subject of conjecture that is included in possible mechanisms of H_2O oxidation. At the present time, it is not clear whether protons are released directly into the solvent as a result of oxidation of bound substrate, indirectly, by way of a chain of amino acid residues (Glu, Asp, His) functioning as a proton-transfer pathway, or, perhaps, as Bohr protons expelled by structural changes in the enzyme that are caused by redox activity. Another constraint on mechanistic speculations is imposed by the effect of pH on the redox potential for H_2O oxidation. The lumen of the chloroplast thylakoid membrane is acidified by steady-state electron transfer, and the effect of this acidification would be to elevate the redox potential for H_2O oxidation by ~ 0.1 V, to $+0.915$ V/electron. If we combine this with the estimated redox potentials of other components of the OEC, we see the following:

$$\begin{array}{cccc} O_2/2\,H_2O \rightarrow [S] \rightarrow & Y_Z\cdot/Y_Z & \rightarrow P680^{+}/P680 \\ (+0.92\,V) & (ca.\ +1.0\,V) & (ca.\ +1.2\,V) \end{array}$$

The $\Delta E^{\circ\prime}$ values among the components of this chain are small, on the order of 0.1–0.2 V, or 9.6–19.2 kJ (2.3–4.6 kcal). This resurrects the problem illustrated earlier in Eqs. 4–6, which is further amplified if uncompensated charges are allowed to build up on the Mn cluster during its oxidation. In extreme cases, an increase in potential of as much as 0.3–0.4 V for a $Mn^{3+} \rightarrow Mn^{4+}$ transition might result.[32] A final, important experimental observation must be included in this background information. Measurements of H_2O exchange rates with PSII in the S states have shown that the first substrate H_2O is bound in S_0, but that the second does not bind until S_3.[33] Because H_2O can be shown to exchange with the S_3 state, it is likely that the reaction to form the first O–O bond does not occur until the S_4 state.

Contemporary speculations on the mechanism of H_2O oxidation focus on accommodating the experimentally observed facts about redox reactions in the OEC to the thermodynamic obstacles.[32,34–38] Common features of these models include binding of substrate H_2O to the Mn cluster prior to S_4 formation, as well as deprotonation of the substrate to generate the H^+ ions that are observed in the course of S-state cycling. Another general feature of these models is the prediction of Mn oxidation state changes on most ($S_0 \rightarrow S_3$) or on all of the transitions of the S-state cycle, and, in most cases, an attempt to avoid charge buildup during the redox cycle. The role of the redox-active tyrosine of the D1 protein, Y_Z, figures prominently in most models, since H_2O oxidation by either hydrogen-atom transfer, or proton-coupled electron transfer to $Y_Z^{\bullet}$ is a facile means to effect charge-accumulation reactions in the mechanism of H_2O oxidation. The viability of these mechanisms is supported by the finding that Y_Z may be hydrogen bonded to the imidazole N of His190 of D1 (Fig. X.4.3). Proton transfer from Y_Z to His190 during $Y_Z^{\bullet}$ formation clears the way for its re-reduction either by a hydrogen atom, or by coupling of transfer of an electron with transfer of a proton from the Mn cluster–substrate complex.[39] Although hydrogen-atom transfer reactions might appear to be thermodynamically unfavorable (see Eq. 4), substrate binding to a Mn(III) produces a leveling effect on O–H bond energies such that the phenolic O–H bond of tyrosine is of approximately the same energy as that of a Mn-bound O–H [$\sim 360\ kJ\,mol^{-1}$ ($86–87\ kcal\,mol^{-1}$)].[40] At the present time, both types of reactions may constitute plausible components of the mechanism for Mn-catalyzed H_2O oxidation, although the shortest Mn–Y_Z distance (~ 7 Å)[19] may be too long to favor hydrogen atom transfer reactions. Much of the current intense experimental scrutiny directed at PSII is aimed at testing alternative mechanisms for H_2O oxidation, and further refinements of the structure of the enzyme will play an important role in arriving at the mechanism of this reaction. The roles of Ca^{2+} and Cl^- in hypothetical mechanisms are not clearly defined. Chloride is portrayed as a Mn ligand, and Ca^{2+} as a site for ligation of either Cl^-, H_2O or OH^-.

One model for the mechanism of H_2O oxidation that draws on several features of current proposals for this reaction is shown in Fig. X.4.6. This depiction of the mechanism employs both proton-coupled electron transfer and hydrogen atom transfer steps as a means to assure that the overall cycle operates in a charge-neutral manner, that is, proton release or hydrogen atom transfers compensate for the gain of a positive charge on a Mn atom. The S_4 state could form as a result of either hydrogen atom transfer from the Mn-bound OH to $Y_Z\cdot$, as shown here, or by concerted H^+/e^- transfer. The chemistry and components of the S_4 transition state is the

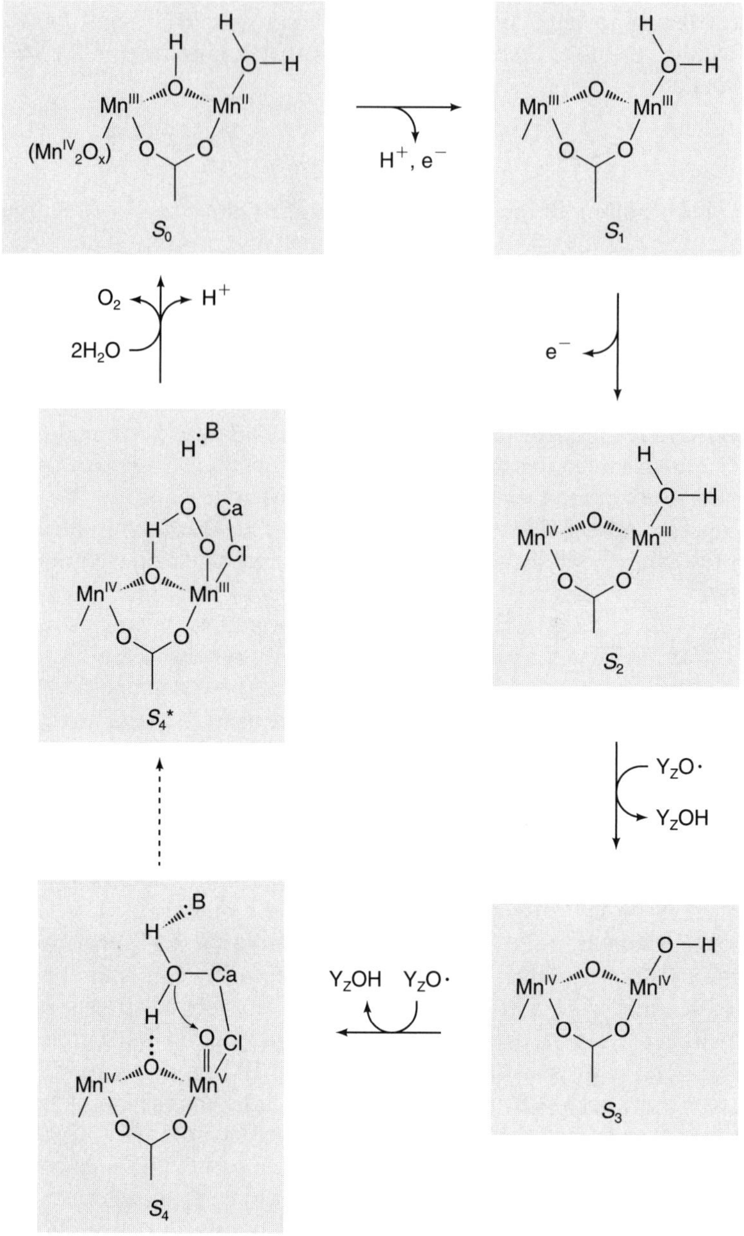

Fig. X.4.6.
A model for H_2O oxidation by the inorganic ion cluster of PSII.[41] See the text for details.

subject of a number of proposals. One area of great interest is the question of the redox state of S_4, which could contain a single $Mn^V{\equiv}O$ that undergoes nucleophilic attack by a Ca^{2+}-bound H_2O or $-OH$,[25] as shown in the model. Alternatively, S_4 could consist of the state S_3 $Y_Z\cdot$, which could catalyze a final proton-coupled electron-transfer reaction to oxidize Mn-bound substrate and generate the O–O bond that leads to O_2 formation. Many of the current proposals offer reasonable alternative mechanisms for the final step of the *S*-state cycle. These same proposals diverge on the issue of whether Mn oxidation occurs by concerted proton-/electron-transfer events, or by way of hydrogen atom abstraction reactions during the advancement of the lower *S* states. While enormous progress has been made in understanding the chemistry of PSII, it is equally clear that more information is needed to resolve these important remaining issues concerning the mechanism of H_2O oxidation.

References

General References

1. Bricker, T. M. and Ghanotakis, D. F., "Introduction to oxygen evolution", in *Oxygenic Photosynthesis: The light reactions,* Ort, D. R. and Yocum, C. F., Eds., Kluwer Academic Publishers, Dordrecht, The Netherlands, 1996, pp. 113–136.
2. Diner, B. A. and Babcock, G. T., "Structure, dynamics and energy conversion efficiency in photosystem II", in *Oxygenic Photosynthesis: The light reactions,* Ort, D. R. and Yocum, C. F., Eds., Kluwer Academic Publishers, Dordrecht, The Netherlands, 1996, pp. 213–247.
3. Joliot, P. and Kok, B., "Oxygen evolution in photosynthesis", in *Bioenergetics of Photosynthesis,* Govindjee, Ed., Academic Press, New York, 1975, pp. 387–411.
4. Debus, R. J., "The polypeptides of photosystem II and their influence on manganotyrosyl-based oxygen evolution", in *Metal Ions in Biological Systems: Manganese and its role in biological processes,* Sigel, A. and Sigel, H., Eds., Marcel Dekker, New York, 2000, pp. 657–711.
5. Barry, B. A., "The role of redox-active amino acids in the photosynthetic water oxidizing complex", *Photochem. Photobiol.,* **57,** 179–188 (1993).
6. Whitmarsh, J. and Pakrasi, H. B., "Form and function of cytochrome *b*-559", in: *Oxygenic Photosynthesis: The light reactions,* Ort, D. R. and Yocum, C. F., Eds., Kluwer Academic Publishers, Dordrecht, The Netherlands, 1996, pp. 249–264.
7. Seidler, A., "The extrinsic polypeptides of photosystem II", *Biochim. Biophys. Acta,* **1277,** 35–60 (1996).
8. Britt, R. D., "Oxygen Evolution", in *Oxygenic Photosynthesis: The light reactions,* Ort, D. R. and Yocum, C. F., Eds., Kluwer Academic Publishers, Dordrecht, The Netherlands, 1996, pp. 137–164.
9. Debus, R. J., "The manganese and calcium ions of photosynthetic oxygen evolution", *Biochim. Biophys. Acta,* **1102,** 269–352 (1992).
10. Dekker, J. P., "Optical studies on the oxygen-evolving complex of photosystem II", in *Manganese Enzymes,* Pecoraro, V. L., Ed., VCH Publishers, Inc., New York, 1992, pp. 85–104.
11. Pecoraro, V. L., "Structural proposals for the manganese centers of the oxygen evolving complex: An inorganic chemist's perspective", *Photochem. Photobiol.,* **48,** 249–264 (1988).
12. Britt, R. D., Peloquin, J. M., and Campbell, K. A., "Pulsed and parallel-polarization EPR characterization of the photosystem II oxygen-evolving complex", *Annu. Rev. Biophy. Biomol. Struct.,* **29,** 463–495 (2000).
13. Yocum, C. F., "Calcium activation of photosynthetic water oxidation", *Biochim. Biophys. Acta,* **1059,** 1–14 (1991).
14. Yachandra, V. K., Sauer, K., and Klein, M. P., "Manganese cluster in photosynthesis: Where plants oxidize water to dioxygen", *Chem. Rev.,* **96,** 2927–2950 (1996).

Specific References

15. Kok, B., Forbush, and McGloin, M., "Cooperation of charges in photosynthetic oxygen evolution. I. A linear four step mechanism", *Photochem. Photobiol.,* **11,** 457–475 (1970).
16. Babcock, G. T., Widger, W. R., Cramer, W. A., Oertling, W. A., and Metz, J. G., "Axial ligands of chloroplast cytochrome *b*-559: Identification and requirement for a heme-cross-linked polypeptide structure", *Biochemistry,* **24,** 3638–3545 (1985).
17. Zouni, A., Witt, H.-T., Kern, J., Fromme, P., Krauss, N., Saenger, W., and Orth, P., "Crystal structure of photosystem II from *Synechococcus elongatus* at 3.8 Å resolution" *Nature (London),* **409,** 739–743 (2001).
18. Kamiya, N. and Shen, J.-R., "Crystal structure of oxygen-evolving photosystem II from *Thermosynechococcus vulcanus* at 3.7 Å resolution", *Proc. Natl. Acad. Sci. U.S.A.,* **100,** 98–103 (2003).
19. Ferreira, K. N., Iverson, T. M., Maghlaoui, K., Barber, J., and Iwata, S., "Architecture of the photosynthetic oxygen-evolving center", *Science,* **303,** 1831–1838 (2004).
20. Boussac, A., Girerd, J.-J., and Rutherford, A. W., "Conversion of the spin state of the manganese complex in photosystem II induced by near-infrared light", *Biochemistry,* **35,** 6984–6989 (1996).
21. Cinco, R. M., Robblee, J. H., Rompel, A., Fernandez, C., Yachandra, V. K., Sauer, K., and Klein, M. P., "Strontium EXAFS reveals the proximity of calcium to the manganese cluster of oxygen-evolving photosystem II", *J. Phys. Chem. B,* **102,** 8248–8256 (1998).
22. Riggs-Gelasco, P., Mei, R., Ghanotakis, D. F., Yocum, C. F., and Penner-Hahn, J. E., "X-ray absorption spectroscopy of Ca^{2+} substituted derivatives of the oxygen evolving complex of photosystem II", *J. Am. Chem. Soc.,* **118,** 227–234 (1996).
23. Messinger, J., Pauly, S., and Witt, H.-T., "The flash pattern of photosynthetic oxygen evolution after treatment with low concentrations of hydroxylamine as a function of the previous S_1/S_0 ratio: Further evidence that NH_2OH reduces the water oxidizing complex in the dark", *Z. Naturforsch.,* **46c,** 1033–1038 (1991).
24. Messinger, J., Nugent, J. H. A., and Evans, M. C. W., "Detection of an EPR multiline signal for the $S_0{}^*$ state in photosynthesis", *Biochemistry,* **36,** 11055–11060 (1997).
25. Vrettos, J. S., Stone, D. A., and Brudvig, G. W., "Quantifying the ion selectivity of the Ca^{2+} site in photosystem II: Evidence for direct involvement of Ca^{2+} in O_2 formation", *Biochemistry,* **40,** 7117–7125 (2001).
26. Riggs, P., Mei, R., Yocum, C. F., and Penner-Hahn, J. E., "Reduced derivatives of the manganese cluster of the photosynthetic oxygen-evolving complex", *J. Am. Chem. Soc.,* **114,** 10650–10651 (1992).

27. Lindberg, K., Vänngård, T., and Andréasson, L. E., "Studies of the slow exchanging chloride in photosystem II of higher plants", *Photosyn. Res.*, **38,** 401–408 (1993).
28. Sandusky, P. O. and Yocum, C. F., "The chloride requirement for photosynthetic oxygen evolution. Analysis of the effect of chloride and other anions on the amine inhibition of the oxygen-evolving complex", *Biochim. Biophys. Acta*, **766,** 603–611 (1984).
29. Wincencjusz, H., van Gorkom, H. J., and Yocum, C. F., "The photosynthetic oxygen evolving complex requires chloride for its redox state $S_2 \rightarrow S_3$ and $S_3 \rightarrow S_0$ transitions but not for $S_0 \rightarrow S_1$ or $S_1 \rightarrow S_2$", *Biochemistry*, **36,** 3663–3670 (1997).
30. Wincencjusz, H., Yocum, C. F., and van Gorkom, H. J., "*S*-state dependence of chloride binding affinities and exchange dynamics in the intact and polypeptide-depleted O_2 evolving complex of photosystem II", *Biochemistry*, **37,** 8595–8604 (1998).
31. Peloquin, J. M., Campbell, K. A., Randall, D. W., Evanchik, M. A., Pecoraro, V. L., Armstrong, W. H., and Britt, R. D., "Mn-55 ENDOR of the S_2 state multiline EPR signal of photosystem II: Implications on the structure of the tetranuclear Mn cluster", *J. Am. Chem. Soc.*, **122,** 10926–10942 (2000).
32. Pecoraro, V. L., Baldwin, M. J., Caudle, T. M., Hsieh, W.-Y., and Law, N. A., "A proposal for water oxidation in photosystem II", *Pure Appl. Chem.*, **70,** 925–929 (1998).
33. Hillier, W. and Wydrzynski, T., "Oxygen ligand exchange at metal sites-implications for the O_2 evolving mechanism of photosystem II", *Biochim. Biophys. Acta*, **1503,** 197–209 (2001).
34. Hoganson, C., Lydakis-Simantiris, N., Tang, X. S., Tommos, C., Warncke, K., Babcock, G. T., Diner, B. A., McCracken, J., and Styring, S., "A hydrogen-atom abstraction model for the function of Y_Z in photosynthetic oxygen evolution", *Photosynth. Res.*, **46,** 177–184 (1995).
35. Hoganson, C. W. and Babcock, G. T., "A metalloradical mechanism for the generation of oxygen from water in photosynthesis", *Science*, **277,** 1953–1956 (1997).
36. Ahlbrink, R., Haumann, M., Cherepanov, D., Bogershausen, O., Mulkidjianian, A., and Junge, W., "Function of tyrosine Z in water oxidation by photosystem II: Electrostatical promotor instead of hydrogen abstractor", *Biochemistry*, **37,** 1131–1142 (1998).
37. Limburg, J., Szalai, V. A., and Brudvig, G. W., "A mechanistic and structural model for the formation and reactivity of a Mn^{V}=O species in photosynthetic water oxidation", *J. Chem. Soc., Dalton Trans.*, 1353–1361 (1999).
38. Nugent, J. H. A., Rich, A. M., and Evans, M. C. W., "Photosynthetic water oxidation: towards a mechanism", *Biochim. Biophys. Acta*, **1503,** 138–146 (2001).
39. Debus, R. J., "Amino acid residues that modulate the properties of tyrosine Y_Z and the manganese cluster in the water oxidizing complex of photosystem II", *Biochim. Biophys. Acta*, **1503,** 164–186 (2001).
40. Caudle, M. T. and Pecoraro, V. L., "Hydrogen atom abstraction of bound substrate is a viable mechanism of cluster oxidation in the oxygen evolving complex of photosystem II", *J. Am. Chem. Soc.*, **119,** 3415–3416 (1997).
41. Vrettos, J. S., Limburg, J., and Brudvig, G. W., "Mechanism of photosynthetic water oxidation: combining biophysical studies of photosystem II with inorganic model chemistry", *Biochim. Biophys. Acta*, **1503,** 229–245 (2001).

CHAPTER XI

Oxygen Metabolism

XI.1. Dioxygen Reactivity and Toxicity

Contents

Joan Selverstone Valentine
Department of Chemistry and Biochemistry
UCLA
Los Angeles, CA 90095

XI.1.1. Introduction

The energy that fuels much of nonphotosynthetic biology is obtained by enzymatically reducing the powerful oxidant molecule dioxygen (O_2) to water (see Sections X.3 and X.4); dioxygen must therefore be continuously supplied to respiring cells. Simple diffusion of O_2 may meet this need in single-celled and other small organisms, but it is not fast enough to supply each cell of multicellular organisms without help. Consequently, proteins such as hemoglobin or myoglobin that bind, transport, store, and release O_2 have evolved to aid in its rapid delivery (see Section XI.4). Dioxygen is also used as a source of oxygen atoms in a large variety of enzyme-catalyzed biosynthetic reactions of organic substrate molecules (see Section XI.5).[1–4]

The same oxidizing power of O_2 that is the basis of respiration, however, also makes O_2 simultaneously an agent of toxic oxidative stress. The interior of a living cell is a highly reducing environment, and the combination of O_2 and many of the vital components of the cell is thermodynamically unstable. Luckily, for reasons that are discussed below, most reactions of O_2 have high activation barriers in the absence of catalysts or free radical initiators and thus represent minor pathways of biological O_2 consumption. Otherwise, the cell would just "burn up", and aerobic life as we know it would be impossible. Nevertheless, there are small but significant amounts of products formed from non-enzymatic and enzymatic reactions of dioxygen that produce partially reduced forms of O_2, for example, superoxide (O_2^-) and hydrogen peroxide (H_2O_2) in aerobic cells. These forms of reduced O_2, and species derived from them by subsequent reactions, are loosely termed "reactive oxygen species" or "ROS", and they can cause oxidative damage *in vivo*.[1,4] Antioxidant enzymes and small molecule antioxidant molecules have been identified that protect against such hazards. These enzymes are, in the case of hydrogen peroxide, catalase and the peroxidase enzymes (see Section XI.3) and, in the case of superoxide, the superoxide dismutase enzymes (see Section XI.2).

The apparently simple diatomic molecule O_2 has a reactivity that is enormously complex, and this complexity is of fundamental importance in understanding how aerobic life works.[2] Luckily, a great deal is known about the factors that govern its reactivity and that of the ROS derived from it.[1–4] Much of the fascination of oxygen metabolism for those who study it stems from the fact that the mechanisms of the biological, enzyme-catalyzed reactions are very different from those of the uncatalyzed reactions of O_2 or even those of O_2 reactions catalyzed by a wide variety of nonbiological metal-containing catalysts.[3,4] To introduce this topic, therefore, we first consider the factors that determine the characteristics of uncatalyzed reactions of O_2 with electron-transfer agents as well as with typical organic and inorganic substrates.

XI.1.2. Chemistry of Dioxygen

XI.1.2.1. Thermodynamics

Dioxygen is a powerful oxidant and therefore has the thermodynamic ability to react with diverse substrates. For purposes of discussion, we divide the chemical reactions of O_2 into two categories, depending on the complexity of the products that are formed. The first category includes those electron-transfer reactions in which the products derived from O_2 contain only oxygen atoms or oxygen atoms plus hydrogen atoms (e.g., O_2^-, H_2O_2, and H_2O). The thermodynamics of such reactions can be estimated using tables of standard reduction potentials. The second category includes atom-transfer reactions in which at least one of the oxygen atoms ends up bonded to an element X other than oxygen or hydrogen. Estimating the thermodynamics of such reactions requires additional information, for example, the strengths of the O–X bonds that are formed. Both types of reactions are discussed below.

Electron-Transfer Reactions. The reduction potential for the four-electron reduction of O_2 (reaction 1) is a measure of the considerable oxidizing power of the O_2 molecule. However, the reaction involves the transfer of four electrons, a process that rarely, if ever, occurs in one concerted step. Since most reducing agents are capable of transferring at most one or two electrons at a time to an oxidizing agent, the thermodynamics of the one- and two-electron reductions of dioxygen must be considered in order to understand the overall mechanism.

$$\underset{\text{Dioxygen}}{O_2} \xrightarrow{e^-} \underset{\text{Superoxide}}{O_2^-} \xrightarrow{e^-,\,2\,H^+} \underset{\text{Hydrogen peroxide}}{H_2O_2} \xrightarrow{e^-,\,H^+} \underset{\text{Water}}{H_2O} + \underset{\text{Hydroxyl radical}}{OH\cdot} \xrightarrow{e^-,\,H^+} \underset{\text{Water}}{2\,H_2O} \qquad (1)$$

In aqueous solution, uncatalyzed electron-transfer reactions involving rapid and concerted transfer of more than one electron are relatively rare, and therefore the most common pathway for O_2 reduction by an electron-transfer agent, in the absence of any catalyst, is one-electron reduction to give superoxide. But, one-electron reduction is the thermodynamically least favorable of the reaction steps that comprise the full four-electron reduction of O_2, and it therefore requires a moderately strong reducing agent (see Table XI.1.1).

Thus, if only one-electron pathways are available for O_2 reduction, the low reduction potential for one-electron reduction of O_2 to O_2^- presents a barrier that protects vulnerable species from the full oxidizing power of O_2 because the reaction cannot get started. If O_2^- is actually formed, however, it disproportionates quite rapidly in aqueous solution (except at very high pH) to give H_2O_2 and O_2 so that the stoichiometry of the overall reaction is that of a net two-electron reduction. It is thus impossible under normal conditions to distinguish one- and two-electron reaction pathways for the reduction of O_2 in aqueous solution on the basis of stoichiometry alone.

One-electron reduction of dioxygen to give superoxide

$$O_2 + e^- \rightarrow O_2^- \tag{2}$$

Disproportionation of superoxide to give dioxygen and peroxide

$$2\,O_2^- + 2\,H^+ \rightarrow H_2O_2 + O_2 \tag{3}$$

Two-electron reduction of dioxygen to give peroxide

$$O_2 + 2\,e^- + 2\,H^+ \rightarrow H_2O_2 \tag{4}$$

Atom-Transfer Reactions. The atom-transfer reactions of O_2 that are of most importance in biology are those in which carbon–oxygen bonds are formed. To appreciate the thermodynamics of such reactions, consider the reactions of dioxygen with dihydrogen to give water ($2\,H_2 + O_2 \rightarrow 2\,H_2O$, $\Delta G^\circ = -58\,\text{kcal mol}^{-1}$), with methane to give methanol ($2\,CH_4 + O_2 \rightarrow 2\,CH_3OH$, $\Delta G^\circ = -30\,\text{kcal mol}^{-1}$), and of benzene to give phenol ($2\,C_6H_6 + O_2 \rightarrow 2\,C_6H_5OH$, $\Delta G^\circ = -43\,\text{kcal mol}^{-1}$). Despite the fact that all three are highly exergonic, none of them occurs spontaneously and rapidly at room temperature unless a catalyst or a free radical initiator is present. To understand the sluggishness of O_2 reactions with organic substrates, therefore, it is important to consider the kinetic barriers to these reactions in addition to the thermodynamics.

XI.1.2.2. Kinetics

The principal kinetic barrier to direct reaction of O_2 with an organic substrate arises from the fact that the ground state of the O_2 molecule contains two unpaired electrons and thus is a triplet. Typical organic molecules that are representative of biological substrates have ground states that contain no unpaired electrons, that is,

Table XI.1.1.
Standard Reduction Potential for Dioxygen Species in Water, pH 7, 25°C

Reaction	E° (V) vs. NHE[a]
$O_2 + e^- \rightarrow O_2^-$	-0.33[b]
$O_2^- + e^- + 2\,H^+ \rightarrow H_2O_2$	$+0.89$
$H_2O_2 + e^- + H^+ \rightarrow H_2O + OH$	$+0.38$
$OH + e^- + H^+ \rightarrow H_2O$	$+2.31$
$O_2 + 2\,e^- + 2\,H^+ \rightarrow H_2O_2$	$+0.281$[b]
$H_2O_2 + 2\,e^- + 2\,H^+ \rightarrow 2\,H_2O$	$+1.349$
$O_2 + 4\,H^+ + 4\,e^- \rightarrow 2\,H_2O$	$+0.815$[b]

[a] Normal hydrogen electrode = NHE.

[b] The standard state used here is unit pressure. If unit activity is used for the standard state of O_2, the redox potential for reactions of that species must be adjusted by $+0.17$ V.

they are singlets, and the products resulting from their oxygenation also are singlets. Triplet-to-singlet spin conversions are forbidden by quantum mechanics, and hence are slow. A collision between two molecules occurs much more rapidly than a spin flip, and so the reaction cannot be concerted. Instead, the number of unpaired electrons remains the same before and after each elementary step of a chemical reaction, and spin flips must be thought of as kinetically separate steps. For these reasons, we know that it is impossible for a spin-forbidden reaction to go in one concerted step.

$$^{3}O_2(\uparrow\uparrow) + {}^{1}X(\uparrow\downarrow) \not\rightarrow {}^{1}XO_2(\uparrow\downarrow) \quad (5)$$

[Arrows represent electron spins, i.e., $(\uparrow\downarrow)$ represents a singlet molecule with all electron spins paired, $(\uparrow\uparrow)$ represents a triplet molecule with two unpaired electrons; and $(\uparrow)$ represents a doublet molecule, also referred to as a free radical, with one unpaired electron.]

The pathways that do not violate the spin restriction are all costly with respect to energetics, resulting in high activation barriers. For example, the reaction of ground-state triplet dioxygen (i.e., $^{3}O_2$) with a singlet substrate to give the excited triplet state of the oxygenated product is spin allowed, and one could imagine a mechanism in which this process was followed by a spin conversion to a singlet product.

$$^{3}O_2(\uparrow\uparrow) + {}^{1}X(\uparrow\downarrow) \rightarrow {}^{3}XO_2(\uparrow\uparrow) \quad (6)$$

$$^{3}XO_2(\uparrow\uparrow) \rightarrow {}^{1}XO_2(\uparrow\downarrow) \quad (7)$$

But such a reaction pathway would give a high activation barrier because the excited triplet states of even unsaturated molecules are typically 40–70 kcal mol^{-1} less stable than the ground state and those of saturated hydrocarbons are much higher.

Likewise, a pathway in which O_2 is excited to a singlet state first and then reacts with the substrate might be possible. However, such a pathway for a reaction of O_2, which is initially in its ground triplet state, would also require a high activation energy since the lowest energy singlet excited state of O_2 is 22.5 kcal mol^{-1} less stable than ground-state triplet dioxygen.

$$^{3}O_2(\uparrow\uparrow) \rightarrow {}^{1}O_2(\uparrow\downarrow) \quad (8)$$

$$^{1}O_2(\uparrow\downarrow) + {}^{1}X(\uparrow\downarrow) \rightarrow {}^{1}XO_2(\uparrow\downarrow) \quad (9)$$

A similarly high activation barrier exists for a pathway in which the substrate X is first excited to the triplet state, followed by reaction with triplet dioxygen:

$$^{1}X(\uparrow\downarrow) \rightarrow {}^{3}X(\uparrow\uparrow) \quad (10)$$

$$^{3}O_2(\uparrow\uparrow) + {}^{3}X(\downarrow\downarrow) \rightarrow {}^{1}XO_2(\uparrow\downarrow) \quad (11)$$

A pathway for a direct reaction of triplet ground-state O_2 with a singlet ground-state organic substrate that can occur readily without a catalyst is one in which the first step is the one-electron oxidation of the substrate by dioxygen. The products of such a reaction would be two doublets, that is, superoxide and the one-electron oxidized substrate, each having one unpaired electron. These free radicals can diffuse apart and then recombine such that their spins pair.

$$^{3}O_2(\uparrow\uparrow) + {}^{1}X(\uparrow\downarrow) \rightarrow {}^{2}O_2^{-}(\uparrow) + {}^{2}X^{+}(\uparrow) \quad (12)$$

$$^{2}O_2^{-}(\uparrow) + {}^{2}X^{+}(\uparrow) \rightarrow {}^{2}O_2^{-}(\uparrow) + {}^{2}X^{+}(\downarrow) \quad (13)$$

$$^{2}O_2^{-}(\uparrow) + {}^{2}X^{+}(\downarrow) \rightarrow {}^{1}XO_2(\uparrow\downarrow) \quad (14)$$

Such a mechanism has been shown to occur for the reaction of dioxygen with reduced flavins.[5]

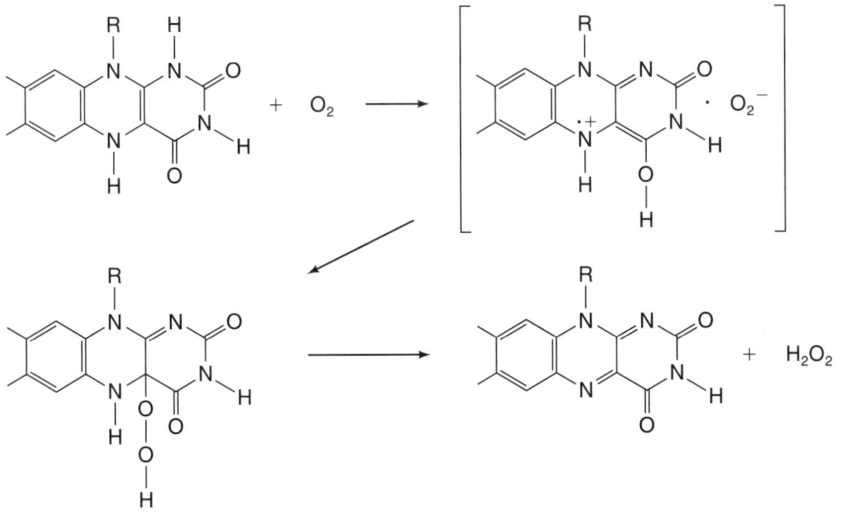

$$(15)$$

However, this type of pathway is relatively rare since it requires that the substrate be an unusually strong reducing agent (e.g., a reduced flavin) and thus be able to reduce dioxygen by one electron to superoxide. Typical organic substrates in enzymatic and non-enzymatic oxygenation reactions usually are not sufficiently strong one-electron reducing agents to reduce O_2 to O_2^-, and this pathway is therefore not commonly observed.

The result of these kinetic barriers to dioxygen reactions with most organic molecules is that uncatalyzed reactions of this type are usually quite slow. An exception to this rule is the oxidation pathway known as free radical autoxidation.

XI.1.2.3. Free Radical Autoxidation

The term, free radical autoxidation, describes a reaction pathway in which O_2 reacts with an organic substrate to give an oxygenated product in a free radical chain process that requires an initiator in order to get the chain reaction started. (A free radical initiator is a compound that yields free radicals readily upon thermal or photochemical decomposition.) The mechanism of free radical autoxidation is as follows:[6]

Initiation: $$X_2 \rightarrow 2\,\cdot X \tag{16}$$

$$X\cdot(\downarrow) + RH \rightarrow XH + R\cdot(\downarrow) \tag{17}$$

Propagation: $$R\cdot(\downarrow) + O_2(\uparrow\uparrow) \rightarrow ROO\cdot(\uparrow) \tag{18}$$

$$ROO\cdot(\uparrow) + RH \rightarrow ROOH + R\cdot(\uparrow) \tag{19}$$

Termination $$R\cdot + ROO\cdot \rightarrow ROOR \tag{20}$$

$$2\,ROO\cdot \rightarrow ROOOOR \rightarrow O_2 + ROOR \text{ [plus other oxidized products, e.g., ROOH, ROH, RC(O)R, RC(O)H]} \tag{21}$$

This reaction pathway results in oxygenation of a variety of organic substrates and is not impeded by the spin restriction because triplet ground-state O_2 can react with the free radical R· to give a free radical product ROO· in a spin-allowed process.

Since R· is regenerated in the propagation step, this chain reaction frequently occurs with long chain lengths prior to the termination steps, resulting in a very efficient pathway for oxygenation of some organic substrates. Inhibition of free radical autoxidation can be achieved by addition of radical scavengers that react rapidly with ROO·, thereby breaking the chain.

When free radical autoxidation is used for synthetic purposes, initiators are intentionally added. Common initiators are peroxides and other compounds capable of fragmenting readily into free radicals. Free radical autoxidation reactions are also frequently observed when no initiator has been intentionally added because organic substrates frequently contain peroxidic impurities, which may act as initiators. Investigators have sometimes been deceived in assuming that a metal-complex catalyzed reaction of O_2 with an organic substrate occurs by a nonradical mechanism. In such instances, upon further study, the reactions proved to be free radical autoxidations, the role of the metal complex having been to generate the initiating free radicals.

While often useful for synthesis of oxygenated derivatives of relatively simple hydrocarbons, free radical autoxidation suffers from a lack of selectivity and therefore, with more complex substrates, tends to give multiple products. In considering possible mechanisms for biological oxidation reactions used *in vivo* for biosynthesis or energy production, free radical autoxidation is not an attractive possibility because such a mechanism requires diffusion of highly reactive free radicals. Such radicals produced in the cell are expected to react indiscriminately with vulnerable sites on enzymes, substrates, and other cell components, causing serious damage. In fact, free radical autoxidation is believed to be the source of many of the deleterious reactions of O_2 that occur in biological systems, for example, in the peroxidation of lipids in membranes. It is also the process that causes fats and oils to become rancid.

Free radical autoxidation mechanisms per se do not appear to be used for activation of O_2 by oxygenase enzymes, but there are often similarities in the nature of the intermediates formed. The big difference between the enzymatic and non-enzymatic reactions is that reactive oxygen species or other reactive radical intermediates that are generated in enzymatic reactions are not allowed to diffuse far away from their site of generation and cause unselective oxidation. Instead, they are directed by steric means to react directly with nearby enzyme-bound substrates (see Section XI.5).

XI.1.2.4. How Do Enzymes Overcome These Kinetic Barriers?

Uncatalyzed reactions of O_2 are usually either slow or unselective. The functions of the metalloenzymes for which O_2 is a substrate are, therefore, to overcome the kinetic barriers imposed by spin restrictions or unfavorable one-electron reduction pathways and, in the case of the oxygenase enzymes, to direct the reactions and make them highly specific (Section XI.5). It is instructive to consider how such metalloenzymes function (1) to lower the kinetic barriers to dioxygen reactivity, and, in the case of the oxygenase enzymes, (2) to redirect the reactions along different pathways so that very different products are obtained.

Consider, for example, the monooxygenase enzyme cytochrome P450 (see Section XI.5). This enzyme catalyzes the reaction of O_2 with organic substrates. It binds O_2 at the paramagnetic metal ion at its active site, thus overcoming the spin restriction, and then carries out what can be formally described as a multielectron reduction of O_2 to give a highly reactive high-valent metal oxo species with reactivity similar to the hydroxyl radical. Unlike a free hydroxyl radical, however, which would be expected to be highly reactive but nonselective, the reaction that occurs at the active site of cytochrome P450 can be highly selective and stereospecific because the highly reactive metal oxo moiety is generated in close proximity to the substrate, which is bound to the enzyme in such a way as to direct the reactive oxygen atom to the correct position. Thus, metalloenzymes have evolved to bind O_2 and enhance its reactivity, but in a very controlled fashion. Such reactions are described in detail in Section XI.5.

XI.1.3. Dioxygen Toxicity

XI.1.3.1. Background

Most of the chemical reactions of O_2 that are known to occur *in vivo* are highly beneficial to the organisms in which they occur, but some of them are not because they lead to oxidative damage.[1,2] Organisms that can live in (or are exposed to) air have evolved a variety of strategies to cope with oxidative stress using both small molecule antioxidants and antioxidant enzymes (see Fig. XI.1.1). This section considers the chemistry of the oxidative damage caused by the agents of O_2 toxicity and of the small molecule antioxidant systems that defend against them. The major antioxidant enzymes, superoxide dismutases, superoxide reductases, catalase, and peroxidases, are discussed further in Sections XI.2 and XI.3.

Direct reactions with other molecules are generally slow in the absence of catalysts or radical initiators, as discussed above, and O_2 itself is therefore not the *primary* agent of oxidative stress. Instead, O_2^-, H_2O_2, hydroxyl radical (HO•), peroxynitrite ($ONOO^-$), and other energetic molecules derived from them are the ROS that are the agents of oxidative damage.[1,2]

XI.1.3.2. Production of Reactive Oxygen Species *In Vivo*

Most of the hydrogen peroxide and other ROS generated during the normal metabolism of a typical eukaryotic cell is derived from superoxide that is formed from reduction of O_2 by components of the mitochondrial electron-transport chain, primarily ubisemiquinone (QH•) in complex **III** and secondarily NADH dehydrogenase (complex **I**), in what are believed to be side reactions of electron transport (Section X.3). In addition, however, there also exist specialized systems whose primary purpose is to generate superoxide and ROS for use in defense systems that protect against pathogens. An example is the NADPH oxidase system in leukocytes, which catalyzes the one-electron reduction of O_2 by NADPH to form superoxide.[7]

Hydroxyl radical is one of the most reactive of the known ROS. It can be generated from reaction of H_2O_2 with reduced metal ions such as Fe^{2+} or Cu^+ in chemistry known as the Fenton reaction.[8] Hydroxyl radical formation involves high-valent metal oxo or hydroxo intermediates, for example, $(Fe^{IV}{=}O)^{2+}$ and $(Cu^{III}{-}OH)^{2+}$, which, along with free hydroxyl radical itself, are implicated as agents causing oxidative damage under conditions of oxidative stress. Hydroxyl radicals and high-valent metal oxo and hydroxo species can act as initiators of free radical autoxidation of lipids and can damage proteins, nucleic acids, carbohydrates, and other organic molecules when they are generated in close proximity to such molecules.

$$Fe^{2+} + H_2O_2 \longrightarrow [(Fe^{IV}{=}O)^{2+} + H_2O] \xrightarrow{+H^+} Fe^{3+} + H_2O + HO\cdot \qquad (22)$$

$$Cu^+ + H_2O_2 + H^+ \rightarrow [(Cu^{III}{-}OH)^{2+} + H_2O] \rightarrow Cu^{2+} + H_2O + HO\cdot \qquad (23)$$

Hydrogen peroxide itself is a strong oxidant thermodynamically, but its reactions tend to be quite slow in the absence of a catalyst. Very small traces of redox-active metal ions can catalyze such reactions, since hydrogen peroxide can act as a reductant as well as an oxidant, and hydroxyl radical can be generated as an intermediate.

$$Fe^{2+} + H_2O_2 + H^+ \rightarrow Fe^{3+} + H_2O + HO\cdot \qquad (24)$$

$$Fe^{3+} + \tfrac{1}{2}H_2O_2 \rightarrow Fe^{2+} + \tfrac{1}{2}O_2 + H^+ \qquad (25)$$

$$\tfrac{3}{2}H_2O_2 \rightarrow H_2O + \tfrac{1}{2}O_2 + HO\cdot \qquad (26)$$

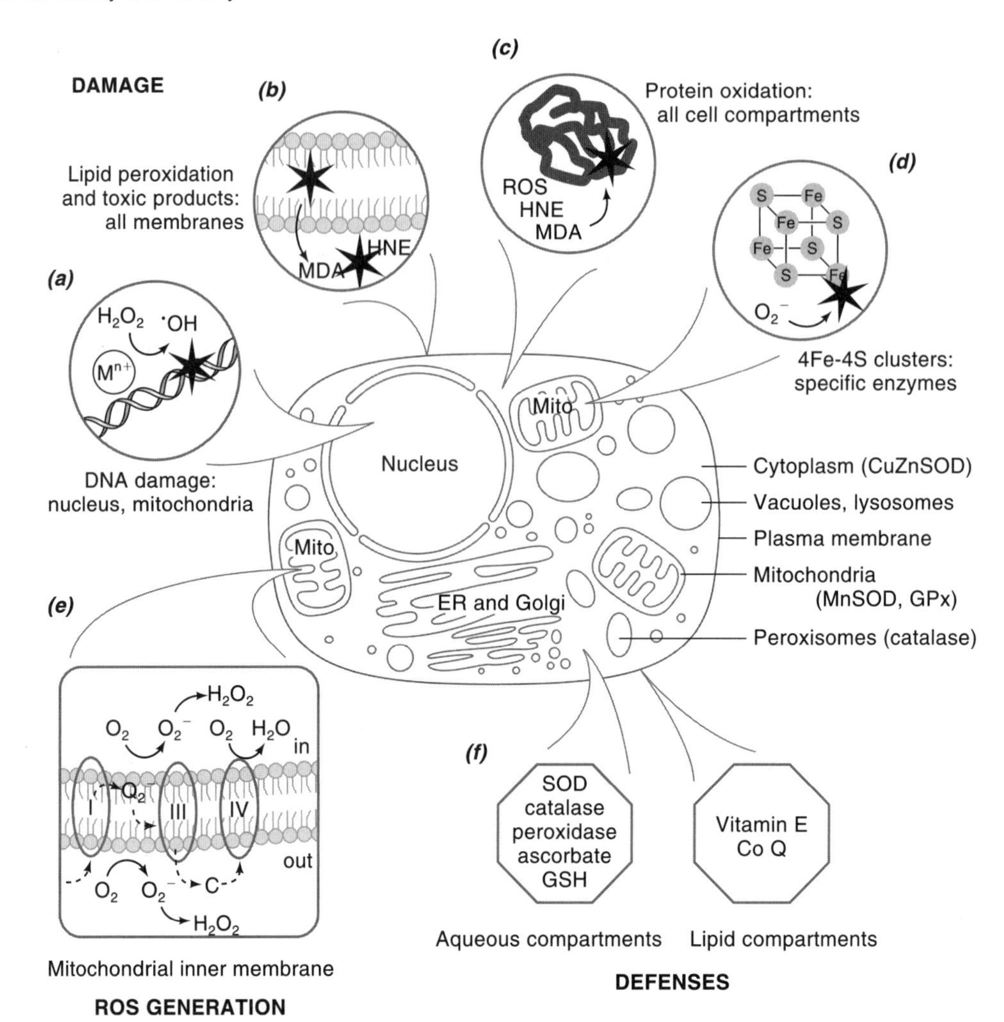

Fig. XI.1.1.
A schematic overview of some of the pathways leading to oxidative stress and of antioxidants that defend against them in a typical eukaryotic cell (center; mito, mitochondrion; ER, endoplasmic reticulum). There are four classes of oxidative damage: (*a*) Site-specific oxidative damage involving metal-catalyzed generation of hydroxyl radical from hydrogen peroxide that results in strand breaks and base damage in DNA. (Similar events could occur wherever metal ions bind adventitiously.) (*b*) Lipid peroxidation that damages membranes as well as producing toxic products such as MDA (malondialdehyde) and HNE (4-hydroxy-2-nonenal), which react with other cell components. (*c*) Damage to proteins resulting from direct oxidations by reactive oxygen species (ROS) or reactions with the products of lipid metabolism (e.g., HNE, MDA). (*d*) Direct reactions of superoxide itself with certain iron–sulfur cluster prosthetic groups in exposed positions, which result in full or partial disassembly of the cluster, inactivation of the enzyme, and release of iron. (Iron released in this manner may go on to catalyze more hydroxyl radical generation at specific locations.) (*e*) A schematic representing the major source of superoxide and hydrogen peroxide in the cell-leakage of electrons from the electron transport chain. Complexes I (NADH dehydrogenase), III (coenzyme Q: cytochrome *c* oxidoreductase), and IV (cytochrome oxidase) represent the electron transport chain. Q is coenzyme Q; C is cytochrome *c*. (*f*) Defensive molecules are listed according to whether they are present in aqueous or lipid compartments. Superoxide dismutase-SOD and reduced glutathione-GSH. (Reproduced with permission from Ref. 2.)

Superoxide itself is a much more sluggish oxidant than hydroxyl radical, and hence it is much more selective in the targets that it oxidizes. The best characterized of these targets is a class of iron–sulfur cluster-containing enzymes containing single labile iron atoms in their clusters. Such reactions are discussed in more detail below.

Peroxynitrite ($ONOO^-$) is a ROS formed from reaction of superoxide with nitric oxide ($O_2^- + NO \rightarrow ONOO^-$). It is widely believed to be an important agent in biological oxidative damage.[9] Recent evidence suggests that the reaction of peroxynitrite with CO_2 to give nitrosoperoxycarbonate ($ONOOCO_2^-$) may play a major role in the physiological chemistry of peroxynitrite *in vivo* and may be the more common ROS.[10]

XI.1.3.3. Low Molecular Weight Antioxidants

Low molecular weight antioxidant systems exist in both the nonaqueous and aqueous compartments of aerobic and aerotolerant cells and, for most multicellular organisms, in the extracellular fluids. They play an important role in allowing organisms to live in air.[1]

One class of molecules that is particularly susceptible to oxidative damage is the polyunsaturated lipids that are present in membranes of higher organisms. Unprotected, these molecules are highly susceptible to free radical autoxidation reactions (Fig. XI.1.2) that are a significant threat to membrane integrity and function. But the presence of abundant membrane-soluble free radical chain-breaking antioxidants such as α-tocopherol (vitamin E) and reduced ubiquinone (coenzyme Q), along

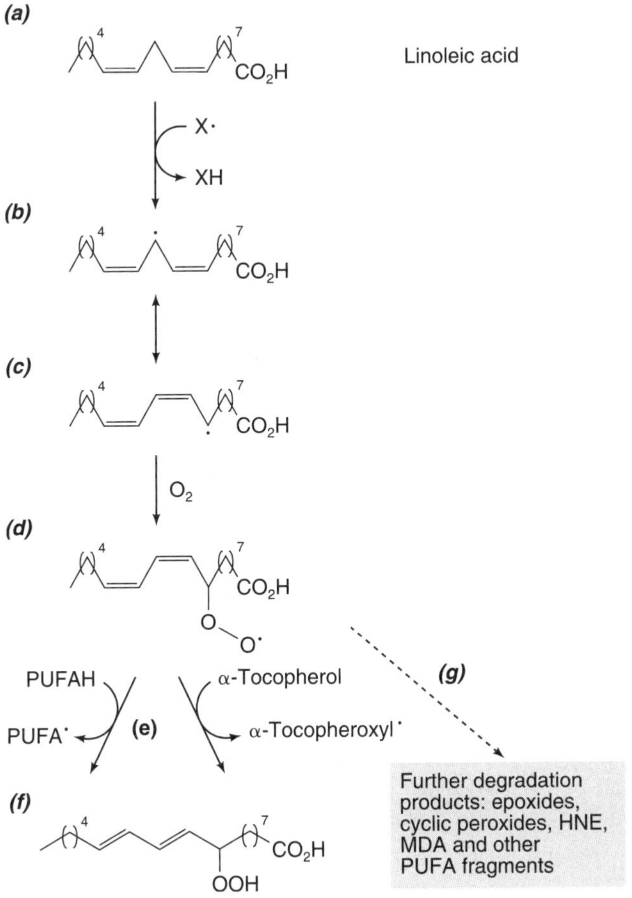

Fig. XI.1.2.
Free radical autoxidation reactions of polyunsaturated lipids leading to lipid peroxidation. (*a*) Various ROS, denoted X: are capable of hydrogen atom abstraction from polyunsaturated lipids such as linoleic acid. (*b*) The resulting doubly allylic radical rearranges to (*c*), the conjugated monoallylic system, which reacts readily with dioxygen to form (*d*), the peroxyl radical ROO•. (*e*) Reaction with other polyunsaturated fatty acids (PUFAH) or with vitamin E (α-tocopherol) transforms the peroxyl radical into (*f*), the lipid hydroperoxide ROOH. In some cases (*g*) the peroxyl radical reacts with its own conjugated diene system to produce epoxides and cyclic peroxides. Internal diene-peroxyl radical reactivity often leads to fragmentation of the polyunsaturated lipid to give HNE, MDA, and other carbonyl-containing compounds. (Reproduced with permission from Halliwell, B. and Gutteridge, J. M. C., *Free radicals in biology,* 2nd ed., Clarendon Press, Oxford, 1989.)

with coupled enzymatic systems that use reduced nicotinamide adenine dinucleotide phosphate (NADPH) to keep them reduced, provide excellent protection against such damage, which only occurs when these defenses are depleted or overwhelmed.[1]

The low molecular weight antioxidants in aqueous compartments, such as the cytosol and the extracellular fluids consist of water-soluble low molecular weight antioxidants, such as glutathione, ascorbate (vitamin C), and urate. These molecules are abundant and highly reactive with certain ROS and other strong oxidants, and therefore can protect other potentially vulnerable targets of oxidative damage. A detailed description of their antioxidant chemistry can be found in Refs. 1 and 3.

XI.1.3.4. Oxidative Damage to Biological Molecules

Oxidative stress is known to cause oxidative damage of proteins, lipids and lipoproteins, nucleic acids, carbohydrates, and other cellular components in living organisms, but the exact chemical identity of the particular damaging agent and the mechanism of its action in each case is often unknown.[2] Moreover, the agent causing the initial damage and the site of oxidative attack may be difficult to ascertain because some of the products of oxidative damage to lipids and sugars react readily with proteins and nucleic acids and thus, when formed, can propagate the damage from the initial site of oxidative attack.

Proteins. Chemical studies of radical-mediated protein oxidation have demonstrated oxidative modification of protein side chains, backbone cleavage, and protein–protein dimerization. Sulfur-containing side chains are particularly vulnerable to oxidation at sulfur, but most of the other oxidative pathways lead to carbonyl-containing products such as aldehydes and ketones, which are commonly measured using the 2,4-dinitrophenylhydrazine assay for carbonyls (see the Lipids and Carbohydrates section below for more discussion of the use of this reagent).

One of the most important discoveries in recent years concerning mechanisms of oxidative damage in biological systems is the facile reaction of superoxide with solvent exposed iron–sulfur clusters in enzymes, such as aconitase and other hydrolyase enzymes containing 4Fe–4S clusters (Fig. XI.1.3).[11] The reaction of superoxide with these centers has been demonstrated to inactivate the enzymes both *in vitro* and *in vivo* and to increase levels of free intracellular iron, which can itself promote additional oxidative damage to proteins and other substrates. This mechanism is the first well-documented example of a direct reaction of superoxide, rather than of a reactive oxygen species derived from it, leading to damage of a cellular component *in vivo*.

Lipids and Carbohydrates. Lipid peroxidation not only threatens the integrity and function of membranes and membranous proteins, but also produces a variety of toxic aldehydes and ketones, one of the worst of which, HNE, is produced in high yield.[1] The compounds HNE and MDA, and other common toxic products formed upon peroxidation of lipids, are known to react via a Michael addition with nucleophilic side chains of proteins and can result in protein cross-linking (Fig. XI.1.4). Oxidatively damaged carbohydrates can also produce products that are reactive with proteins and can result in further damage.[1]

Identification of the primary sites of oxidative damage in living organisms is a major challenge since antioxidant protection differs considerably, depending on the nature of the ROS and the site of attack. It is therefore important to note that products of lipid and/or carbohydrate oxidation can often react with proteins and that the resulting adducts contain carbonyl groups, and therefore are reactive with 2,4-dinitrophenylhydrazine in the assay for protein carbonyls. Thus detection of high levels of protein carbonyls does not necessarily indicate that the proteins themselves are being directly oxidized by ROS; the carbonyls may instead result from reactions of undamaged proteins with toxic products of lipid or carbohydrate oxidation.[1]

Nucleic Acids. Elevated levels of oxidative stress have long been known to result in oxidative damage to DNA.[1] The guanine bases are particularly vulnerable to such damage, which converts them to 8-oxo-7,8-dihydroguanine and subsequent oxidation products.[12] There is considerable evidence implicating "free" intracellular iron in this oxidation. Its source is believed to be iron liberated from iron–sulfur clusters as a consequence of elevated superoxide levels under conditions of oxidative stress. A widely accepted theory is that this "free" iron may bind loosely to various sites in the DNA, where it can act as a catalyst for the generation, from hydrogen peroxide, of very reactive species such as hydroxyl radical or high-valent iron oxo or hydroxo species that react with DNA in the immediate vicinity (see discussion of the Fenton reaction in Section X.1.3.2). Cellular reductants such as superoxide, ascorbate, and NADH may reduce the bound Fe^{3+} to the ferrous state, where Fenton chemistry may then occur in close proximity to DNA bases and sugars. This type of chemistry is also possible for Cu ions adventitiously bound to DNA; it is expected to result in modified bases and/or single strand breaks in the local DNA.[11]

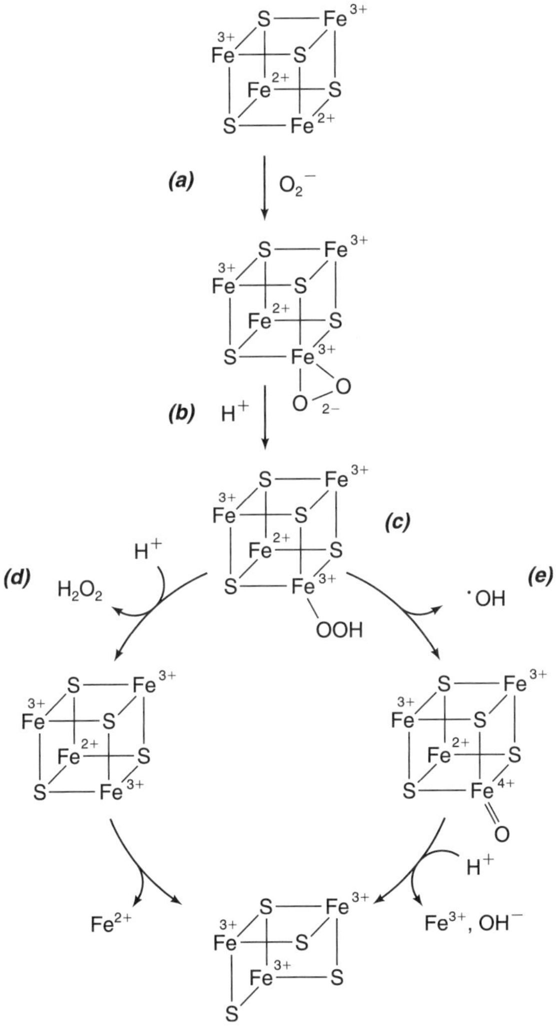

Fig. XI.1.3.
Hypothetical mechanism for the reaction of $[Fe_4S_4]^{2+}$ clusters with superoxide. Individual charges have been assigned to iron atoms in the figure for convenience in keeping track of redox changes; but, it should be emphasized that electron density in Fe–S clusters is known to be highly delocalized (see Chapter IV). (*a*) Reaction of O_2^- with the solvent-exposed iron center at one corner of the cube produces a ferric peroxo intermediate, $[Fe_4S_4(O_2)]^+$, and (*b*) protonation of the ferric peroxo yields (*c*) a ferric hydroperoxide, $[Fe_4S_4(OOH)]^{2+}$. Decomposition of the cluster might occur by one of two indicated pathways: (*d*) protonation and loss of hydrogen peroxide, forming an $[Fe_4S_4]^{3+}$ cluster that loses Fe^{2+} to give the $[Fe_3S_4]^+$ cluster, or (*e*) homolytic cleavage of the hydroperoxo ligand to give hydroxyl radical and a ferryl-containing cluster, $[Fe_4S_4(O)]^{2+}$, which could also give the $[Fe_3S_4]^+$ cluster upon protonation and loss of Fe^{3+} and hydroxide. [Adapted from Flint, D. H. and Allen, R. M., *Chem. Rev.*, **96,** 2315–2334 (1996). Reproduced with permission from Ref. 2.]

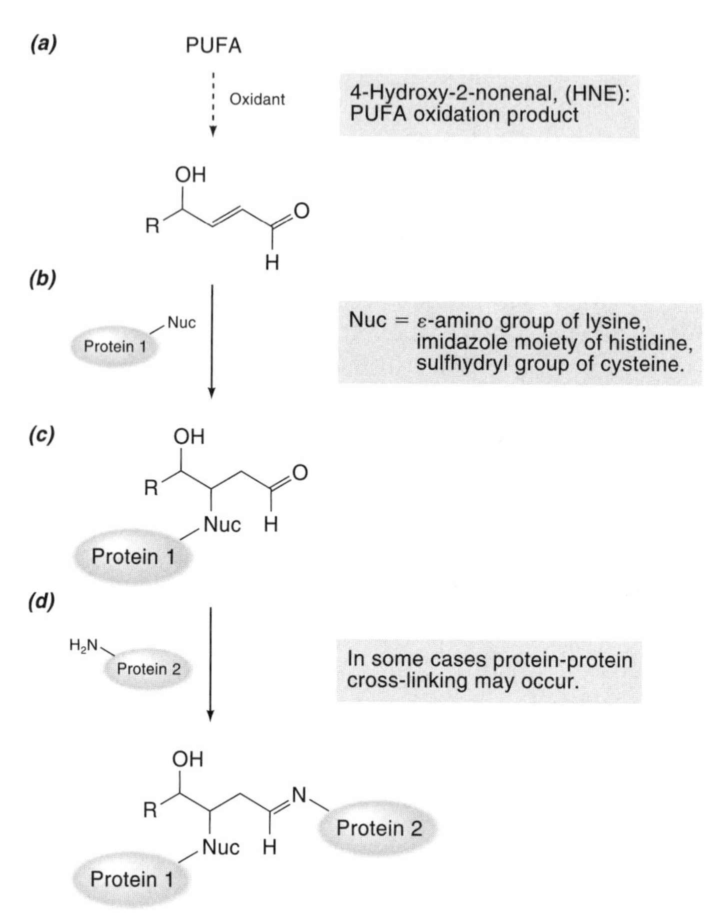

Fig. XI.1.4.
Reactions of nucleophilic protein side chains with HNE. (*a*) The degradation of PUFA often produces toxic aldehydes and ketones such as HNE. (*b*) This compound is detrimental to proteins as a result of its ability to function as a Michael accepter for various nucleophilic protein side chains. The resulting HNE–protein adduct (*c*) may have an altered hydrophobicity as well as an increased carbonyl content, which will cause it to be reactive with 2,4-dinitrophenylhydrazine in assays for protein carbonyls. (*d*) The HNE–protein adduct is capable of reacting further with the amino groups of other proteins to produce cross-linked protein–protein adducts. [Adapted from Stadtman ER, Berlett BS: *Chem. Res. Toxicol.*, **10,** 485–494 (1997). Reproduced with permission from Ref. 2.]

XI.1.3.5. Links between Nitric Oxide and Dioxygen Biochemistry

The physiological chemistries of nitric oxide (NO) and O_2 appear to be intimately related. For example, peroxynitrite, formed from reaction of nitric oxide and superoxide, is an ROS that is believed to be generated *in vivo* under some conditions and to react with tyrosine residues on proteins to give nitrotyrosine. Another example is *S*-nitrosylation of thiols *in vivo,* the mechanism of which is currently unknown, but in which O_2 or ROS probably serves as the oxidant(s) in the generation of nitrosating species.[13]

$$RSH + NO + Oxidant \rightarrow RSNO \tag{27}$$

Nitric oxide has also been reported to react with proteins containing iron–sulfur clusters.[14]

References

General References

1. Halliwell, B. and Gutteridge, J. M. C., "Free radicals in biology and medicine", Oxford University Press, New York, 1989.
2. Ho, R. Y. N., Liebman, J. F., and Valentine, J. S., "Overview of Energetics and Reactivity of Oxygen", in Foote, C. S., Valentine, J. S., Greenberg, A., and Liebman, J., Eds., *Active Oxygen in Chemistry*, Blackie Academic & Professional, Glasgow, 1995.
3. Ho, R. Y. N., Liebman, J. F., and Valentine, J. S., "Biological reactions of dioxygen: an introduction", in Valentine, J. S., Foote, C. S., Greenberg, A., and Liebman, J., Eds., *Active Oxygen in Biochemistry*, Blackie Academic & Professional, Glasgow, 1995.
4. Valentine, J. S., Wertz, D. L., Lyons, T. J., Liou, L. L., Goto, J. J., and Gralla, E. B., "The dark side of dioxygen biochemistry", *Curr. Opin. Chem. Biol.*, **2,** 253–262 (1998).

Specific References

5. Massey, V., "Activation of molecular oxygen by flavins and flavoproteins", *J. Biol. Chem.*, **269,** 22459–22462 (1994).
6. Porter, N. A., Caldwell, S. E., and Mills, K. A., "Mechanisms of free radical oxidation of unsaturated lipids", *Lipids*, **30,** 277–290 (1995).
7. Babior, B. M., Lambeth, J. D., and Nauseef, W., "The neutrophil NADPH oxidase", *Arch. Biochem. Biophys.*, **397,** 342–344 (2002).
8. Buettner, G. R., "The pecking order of free radicals and antioxidants: lipid peroxidation, alpha-tocopherol, and ascorbate", *Arch. Biochem. Biophys.*, **300,** 535–543 (1993).
9. Groves, J. T., "Peroxynitrite: reactive, invasive and enigmatic", *Curr. Opin. Chem. Biol.*, **3,** 226–235 (1999).
10. Lymar, S. V. and Hurst, J. K., "Carbon dioxide: physiological catalyst for peroxynitrite-mediated cellular damage or cellular protectant?" *Chem. Res. Toxicol.*, **9,** 845–850 (1996).
11. Gardner, P. R. and Fridovich, I., "Superoxide sensitivity of the *Escherichia coli* aconitase", *J. Biol. Chem.*, **266,** 19328–19333 (1991).
12. Burrows, C. J. and Muller, J. G., "Oxidative nucleobase modifications leading to strand scission", *Chem. Rev.*, **98,** 1109–1152 (1998).
13. Stamler, J. S., Simon, D. I., Osborne, J. A., Mullins, M. E., Jaraki, O., Michel, T., Singel, D. J., and Loscalzo, J., "S-nitrosylation of proteins with nitric oxide: synthesis and characterization of biologically active compounds", *Proc. Natl. Acad. Sci. U.S.A.*, **89,** 444–448 (1992).
14. Gardner, P. R., Costantino, G., Szabo, C., and Salzman, A. L., "Nitric oxide sensitivity of the aconitases", *J. Biol. Chem.*, **272,** 25071–27076 (1997).

XI.2. Superoxide Dismutases and Reductases

Joan Selverstone Valentine
Department of Chemistry
and Biochemistry
UCLA
Los Angeles, CA 90095

Contents

XI.2.1. Introduction

Superoxide anion (O_2^-), the one-electron reduction product of dioxygen (O_2), is a toxic reactive oxygen species (see Section IX.1).[1] In eukaryotes, its principal source is the mitochondrial inner membrane, both sides of which release superoxide as a byproduct of aerobic metabolism (see Section IX.1). Two classes of superoxide detoxification enzymes are known: superoxide dismutase (SOD) enzymes[1–5] and superoxide reductase (SOR) enzymes.[6]

Superoxide dismutases catalyze the disproportionation of superoxide to give O_2 and H_2O_2.[1,2,4,7] Thus superoxide is acting both as an oxidant (accepting electrons) and as a reductant (giving up electrons) in this reaction.

$$2\,O_2^- + 2\,H^+ \rightarrow O_2 + H_2O_2 \quad (1)$$

By contrast, the SOR enzymes catalyze the one-electron reduction of superoxide to give H_2O_2, using NADH or NADPH as the ultimate source of electrons. Thus in this case, superoxide is acting only as an oxidant.

$$O_2^- + 2\,H^+ + e^- \rightarrow H_2O_2 \quad (2)$$

Three classes of SOD enzymes have been characterized: copper–zinc superoxide dismutase (CuZnSOD), the structurally related iron and manganese superoxide dismutases (FeSOD and MnSOD), and nickel superoxide dismutase (NiSOD). Only one class of SOR enzymes is known to date; it has a mononuclear nonheme iron center at its active site.

XI.2.2. Superoxide Chemistry[2]

The equilibrium between HO_2 and its conjugate base, O_2^-, occurs with a pK_a of 4.8 in aqueous solution. Thus the predominant species present in solution at physiological pH is the unprotonated superoxide anion itself.

$$HO_2 \rightleftharpoons O_2^- + H^+ \qquad K_a = 1.6 \times 10^{-5}\,M \quad (3)$$

Superoxide disproportionates spontaneously to yield H_2O_2 and O_2 via a pH-dependent mechanism involving reactions 4 and 5. Reaction 6 does not occur at all in the pH range of 0.2–13.

$$HO_2 + HO_2 \rightarrow H_2O_2 + O_2 \qquad k = 8.3 \times 10^5\,M^{-1}\,s^{-1} \quad (4)$$

$$HO_2 + O_2^- \xrightarrow{H^+} H_2O_2 + O_2 \qquad k = 9.7 \times 10^7\,M^{-1}\,s^{-1} \quad (5)$$

$$O_2^- + O_2^- \rightarrow \text{No reaction} \quad (6)$$

The pH-rate profile for the spontaneous disproportionation of superoxide is shown in Fig. XI.2.1, where it is compared with those of the SOD-catalyzed reactions.

The relative rates of reactions 4–6 illustrate important features of superoxide reactivity. With respect to the thermodynamics of their reactions, HO_2 and O_2^- are fairly similar in their abilities to act either as reducing or as oxidizing agents (see Table XI.1.1), but the kinetics of their reactions are quite different. Both HO_2 and O_2^- are kinetically competent one-electron reducing agents but, for most substrates, only HO_2 is a kinetically competent one-electron oxidant, due to the instability of the peroxide dianion, O_2^{2-}. Those rare cases in which O_2^- is observed to oxidize substrates at high rates occur only when a proton transfer is simultaneous with electron transfer, resulting in formation of HO_2^- (reaction 7) rather than O_2^{2-}.

$$X\cdots O_2^-\cdots H\text{–}Y \rightarrow X^+ + HO_2^- + Y^- \quad (7)$$

An example of a fast oxidation by superoxide, in which such proton-coupled electron transfer to superoxide is likely occurring, is the rapid oxidation of hydroquinones by superoxide (reaction 8).

$$HO\text{–}C_6H_4\text{–}OH + O_2^- \longrightarrow HO\text{–}C_6H_4\text{–}O\cdot + HO_2^-$$

$$k = 1.7 \times 10^7\,M^{-1}\,s^{-1} \quad (8)$$

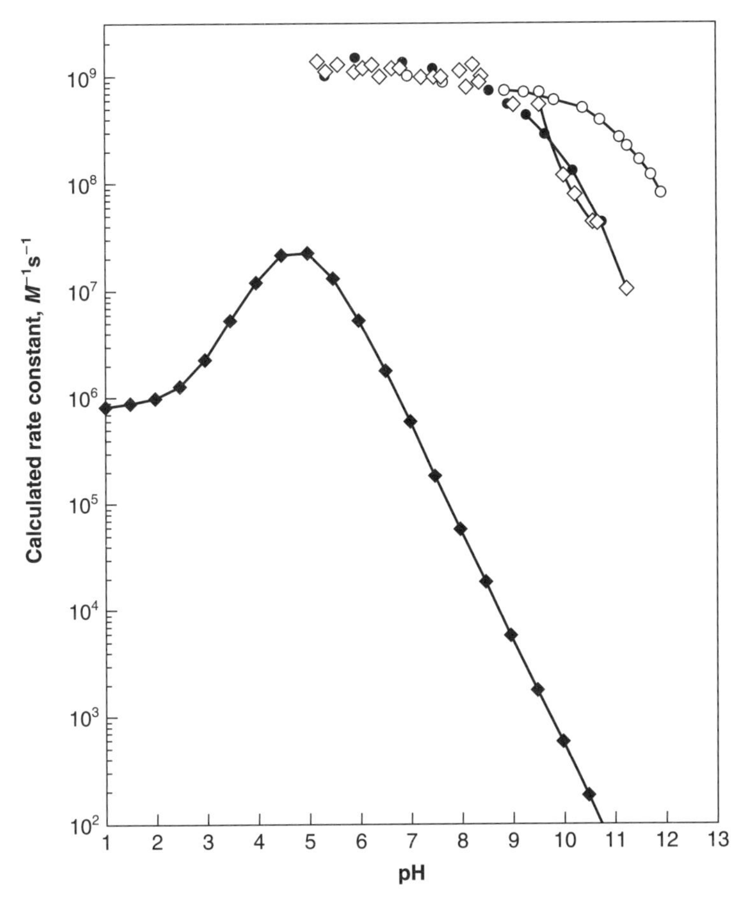

Fig. XI.2.1.
Rate constants for superoxide dismutation; filled diamonds, no catalyst; open circles, CuZnSOD, open diamonds, *Escherichia coli* MnSOD; filled circles, NiSOD. Rate constants for the SODs are per concentration of Cu, Mn, and Ni, respectively. (Figure courtesy of Dr. Diane E. Cabelli.)

Alternatively, a metal ion may be oxidized by superoxide in an oxidative addition reaction to give a metal–peroxo complex, where the peroxide is stabilized by coordination to the metal ion rather than by protonation. Subsequently, the peroxide dissociates, resulting in overall oxidation of the metal ion. In this case, the electron transfer to form a metal-bound peroxide can precede the protonation step because the metal ion stabilizes the $O_2{}^{2-}$ ligand as it is formed (reaction 9).

$$M^{n+} + O_2{}^- \longrightarrow M^{(n+1)+}(OO^{2-}) \xrightarrow{H^+} M^{(n+1)+} + H_2O_2 \tag{9}$$

XI.2.3. Superoxide Dismutase and Superoxide Reductase Mechanistic Principles[2]

Superoxide dismutase mechanisms may be conveniently divided into two separate steps, the first involving oxidation of $O_2{}^-$ by the oxidized enzyme, producing O_2 (reaction 10), and the second involving reduction of $O_2{}^-$ by the reduced enzyme, yielding H_2O_2 (reaction 11).

$$O_2{}^- + \text{Enzyme}_{ox} \rightarrow O_2 + \text{enzyme}_{red} \tag{10}$$

$$O_2{}^- + 2\,H^+ + \text{Enzyme}_{red} \rightarrow H_2O_2 + \text{enzyme}_{ox} \tag{11}$$

The SOR mechanism uses only reaction 11; the oxidized enzyme is then returned to the reduced state by reducing agents other than superoxide. In the following discussion, reactions 10 and 11 are discussed separately.

XI.2.3.1. Oxidation of Superoxide To Give Dioxygen

Studies of reactions of superoxide with transition metal complexes have led to the conclusion that superoxide can probably reduce metal ions or complexes by either inner- or outer-sphere pathways, but also that it is often difficult to distinguish these two pathways. At issue is whether or not the superoxide anion enters the first coordination sphere of the metal center prior to electron transfer (reaction 12, top). Alternatively, superoxide may be held close to the metal center by hydrogen bonding to a bound ligand, such, as water and transfer its electron without actually displacing that ligand (reaction 12, bottom).

$$M^{(n+1)+}(H_2O) \xrightarrow{-H_2O} M^{(n+1)+} \xrightarrow{O_2^-} M^{(n+1)+}(O_2^-) \xrightarrow{H_2O} M^{n+}(H_2O) + O_2$$
$$M^{(n+1)+}(H_2O) \xrightarrow{O_2^-} M^{(n+1)+}(H_2O) \cdots\cdots O_2^- \longrightarrow M^{n+}(H_2O) + O_2 \quad (12)$$

XI.2.3.2. Reduction of Superoxide To Give Hydrogen Peroxide

As discussed Section IX.2, because of the instability of a naked peroxide O_2^{2-} anion, superoxide can be reduced rapidly to the peroxide level only when proton-coupled electron transfer can occur, that is, when a proton transfers to the O_2^- moiety simultaneously with transfer of an electron. For the case where the reducing agent is a redox metal center, proton-coupled electron transfer to superoxide could conceivably occur with superoxide in the first coordination sphere of the metal ion (reaction 13). Such an inner-sphere pathway requires a vacant coordination site on the metal, and may be limited by the rate of ligand exchange on the metal ion.

$$M(H_2O)^{n+} \xrightarrow{-H_2O} M^{n+} \xrightarrow{O_2^-} M^{n+}(O_2^-) \xrightarrow[-X^-]{X-H} M^{(n+1)+}(OOH^-) \xrightarrow{H^+} M^{(n+1)+} + H_2O_2 \quad (13)$$

X—H = H_2O or amino acid side chain

Outer-sphere pathways for reduction of superoxide by transition metal complexes have also been reported. A particularly interesting example is in the case of some low molecular weight coordination complexes of manganese that are currently in testing for use as SOD mimetic drugs.[8] Outer-sphere mechanistic pathways may have a considerable kinetic advantage in some cases, since they bypass formation of stable metal hydroperoxide complexes, $M^{(n+1)+}(OOH^-)$, whose rates of dissociation can slow down SOD reactions.

An example of an outer-sphere mechanism for reduction of superoxide is shown in reaction 14. It involves a coordinated water molecule, which is the hydrogen-atom donor for a proton-coupled electron transfer (reaction 14).

$$M^{n+}\text{–}O(H)\text{–}H\cdots\cdots O^-\text{–}O\cdots\cdots H\text{–}X \longrightarrow M^{(n+1)+}\text{–}O^-(H) \quad H\text{–}O\text{–}O\text{–}H \quad X^- \quad (14)$$

Such a mechanism is reminiscent of the proton-coupled electron-transfer pathway postulated for the self-exchange reaction for $Fe(H_2O)_6{}^{2+/3+}$, that is, hydrogen-atom transfer from $[Fe^{II}(H_2O)_6]^{2+}$ to $[Fe^{III}(H_2O)_5(OH)]^{2+}$ (reaction 15).[9]

$$Fe^{2+}\text{—}\bar{O}(H)\cdots H\text{—}O(H)\text{—}Fe^{3+} \longrightarrow Fe^{3+}\text{—}O(H)\text{—}H\cdots\bar{O}(H)\text{—}Fe^{2+} \qquad (15)$$

Another outer-sphere mechanism that is theoretically possible requires that superoxide be hydrogen bonded to two proton donors and not be bonded to the metal ion prior to electron transfer (reaction 16). Such a transition state would require a high degree of spatial organization prior to electron transfer. This seems feasible in the active site of an enzyme that evolved for the function of catalyzing conversion of superoxide to hydrogen peroxide, but much less likely in the case of reactions of superoxide with model compounds.

$$M^{n+} + Y\text{—}H\cdots O_2^-\cdots H\text{—}Y \longrightarrow M^{(n+1)+} + H_2O_2 + 2Y^- \qquad (16)$$

Reactions 14 and 16 absolutely require that a proton donor or donors be present and in the correct orientation to allow simultaneous proton and electron transfer. If this is not the case, fast reduction of superoxide might still occur via a pH-dependent reaction that is fast at low pH because the oxidant is actually HO_2 rather than O_2^- (reaction 17).

$$M^{n+} + HO_2\cdots H\text{—}Y \longrightarrow M^{(n+1)+} + H_2O_2 + Y^- \qquad (17)$$

XI.2.4. Superoxide Dismutase and Superoxide Reductase Enzymes

XI.2.4.1. Copper–Zinc Superoxide Dismutase

Copper–zinc superoxide dismutase[1,2,4,5] is found in almost all eukaryotic cells and in many prokaryotes. The metal-binding site of CuZnSOD consists of six histidines and an aspartate (Fig. XI.2.2).[5]

From the point of view of protein structure, the overall structures of the reduced and oxidized enzymes are extremely similar; the few big changes are localized to a very small region close to the Cu ion, which goes from three coordinate in Cu^IZnSOD to five coordinate in $Cu^{II}ZnSOD$. In Cu^IZnSOD, Cu(I) is bound by three histidyl imidazoles in a nearly trigonal-planar arrangement while Zn(II) is

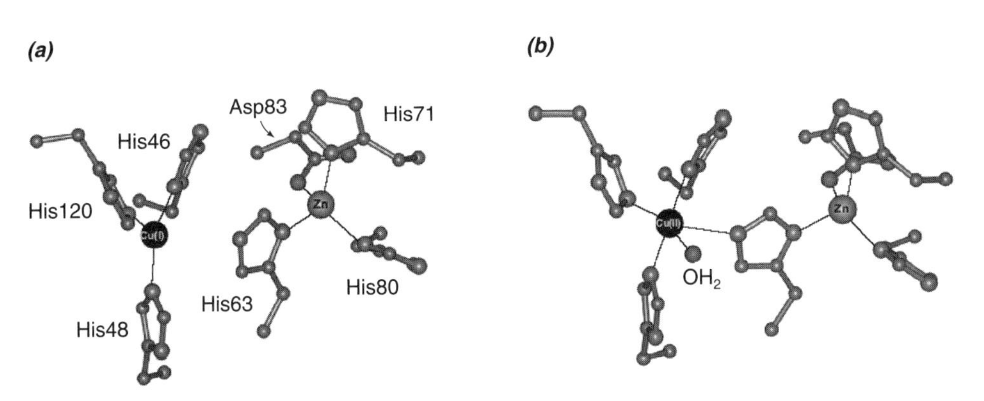

Fig. XI.2.2. Metal-binding site regions of reduced yeast CuZnSOD (*a*, PDB code: 2JCW) and oxidized human CuZnSOD (*b*, PDB code: 1HL5).

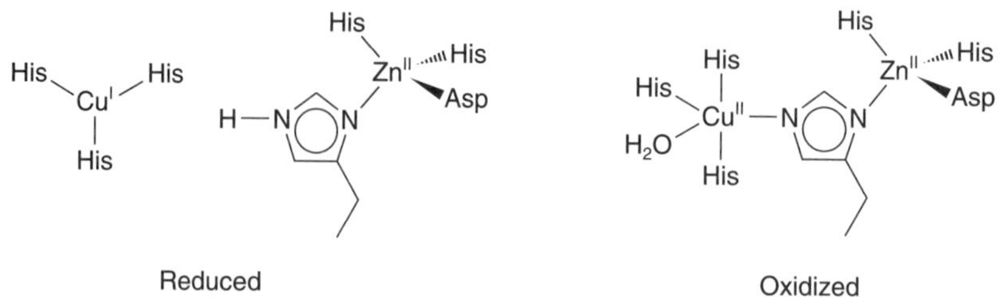

Fig. XI.2.3.
Schematic representation of the metal-binding sites in CuZnSOD.

bound by three histidyl imidazoles and an Asp carboxylate. In Cu^{II}ZnSOD, the structure of most of the protein, as well as the zinc site and the positions of the three His ligands to Cu, are changed very little from those in the reduced enzyme. The large structural changes near the copper are (1) The Cu(II) in Cu^{II}ZnSOD is displaced ~ 1 Å toward the Zn site, relative to its position in Cu^{I}ZnSOD; (2) His63 (human numbering), one of the Zn ligands tilts, deprotonates, and binds to the Cu(II) as a bridging ligand; and (3) a water ligand is added to the first coordination sphere of Cu(II) (see Fig. XI.2.3).[10,11]

One particularly important structural feature that is virtually identical in both the oxidized and reduced enzymes is the funnel-shaped channel that provides access to the Cu ion. This channel is 24 Å wide at the surface of the protein, but becomes progressively narrower until it is only 4 Å across directly above the Cu ion. This narrowing of the bottom of the channel has the effect of limiting ligand binding to Cu^{2+} in the oxidized form of the enzyme to small anions similar in size and charge to superoxide, for example, cyanide, azide, and fluoride, and thus makes the enzyme highly specific for reaction with superoxide anion.

The active site channel also contains a conserved arginine ~ 6 Å above the Cu ion, which plays an important role in generating the electrostatic field that draws superoxide to the mouth of the active site channel. In addition, the positively charged guanidinium group of that residue is believed to be well positioned to assist in guiding the superoxide anion down the channel and hydrogen bonding with it as it approaches the Cu center in either the oxidized or reduced enzyme. The active-site channel also contains a conserved chain of hydrogen-bonding waters that probably plays a role in the mechanism by interacting with superoxide as it traverses the channel and acting as a proton donor when superoxide is reduced to peroxide.

The rate of reaction of superoxide with CuZnSOD is strongly dependent on ionic strength, consistent with the dominance of the electrostatic field that attracts superoxide to the metalloenzyme, but is independent of the oxidation state of the Cu ion, as evidenced by the identical rate constants for reaction of superoxide with the oxidized and the reduced forms of CuZnSOD (i.e., $2 \times 10^9\ M^{-1}\ s^{-1}$). Thus, the rate constants for superoxide entry into the substrate channel in the reduced enzyme and oxidized enzymes must be the same, and any differences in the rates of reactions of superoxide with the Cu(I) or the Cu(II) forms of the enzyme after the superoxide enters the channel do not influence the overall rates.

A mechanistic model for the catalytic reaction of CuZnSOD with superoxide, based on numerous crystal structures of CuZnSOD and on kinetic and spectroscopic data,[5,7,10] is summarized in Fig. XI.2.4.

Oxidation of Superoxide by Cu^{II}ZnSOD. Anions, such as azide (N_3^-), cyanide (CN^-), and fluoride (F^-), are known to bind directly to Cu(II) in Cu^{II}ZnSOD, and the latter two ligands have been shown to exchange rapidly. Therefore, it is likely that superoxide also enters into the first coordination sphere of the Cu(II) ion, forming a $Cu^{II}(O_2^-)$ complex, prior to the electron transfer that forms the prod-

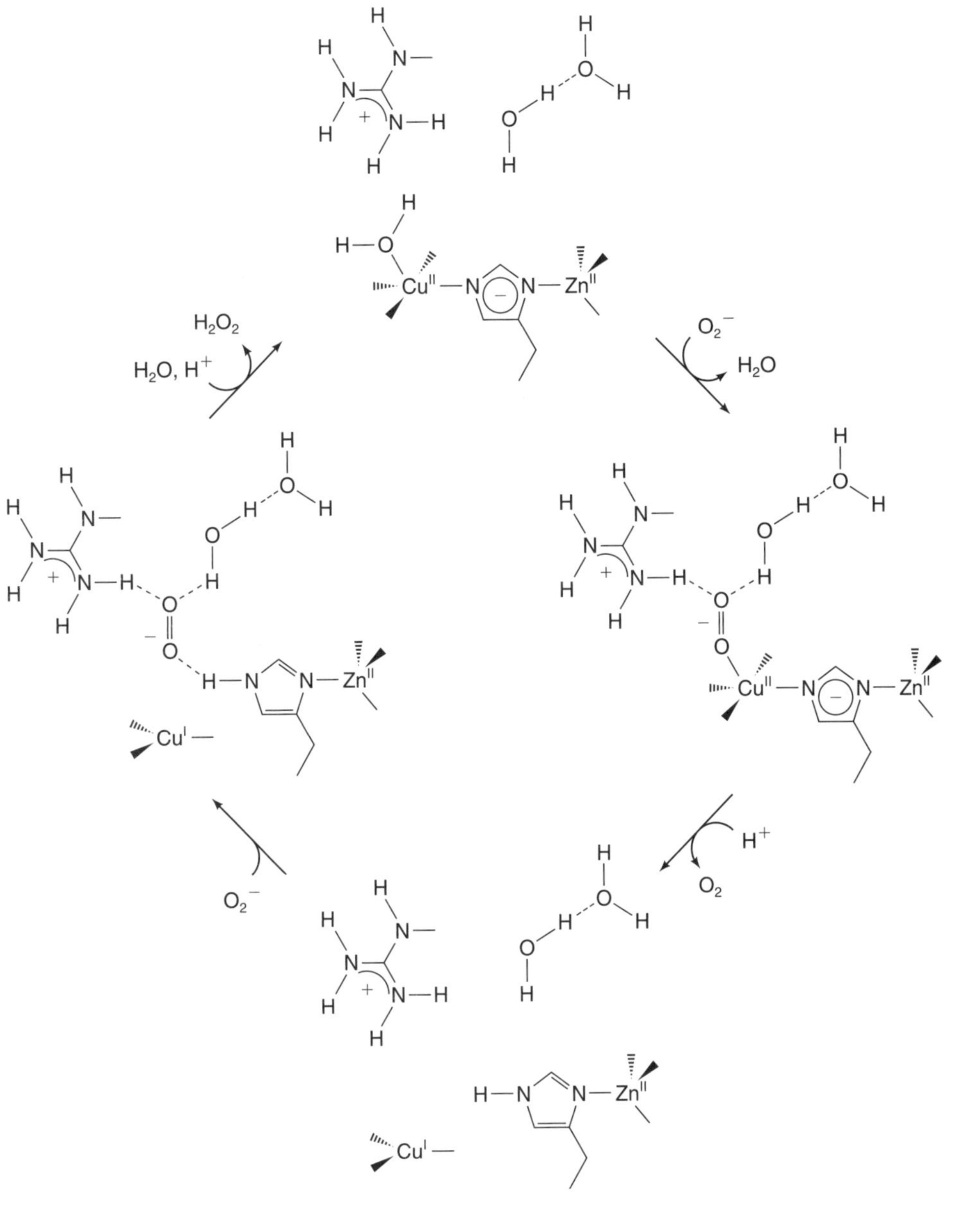

Fig. XI.2.4.
Schematic diagram of the catalytic cycle for CuZnSOD. The right-hand side describes the oxidation of superoxide by Cu^{II}ZnSOD to give O_2, and the left-hand side describes the reduction of superoxide by Cu^{I}ZnSOD to give H_2O_2.

ucts, Cu^{I}ZnSOD and O_2. It seems likely that O_2^- enters the active-site channel of Cu^{II}ZnSOD, interacts with the conserved chain of water molecules and the active-site Arg on its way, ultimately displacing the water ligand on the Cu(II) center. The electron transfer then occurs, O_2 is released and diffuses out of the channel, the imidazolate bridge is broken, and His63 is protonated (Fig. XI.2.4, right).

Reduction of Superoxide by Cu^{I}ZnSOD. The reduction of superoxide by Cu^{I}ZnSOD appears to proceed by an outer-sphere mechanism. The Cu^{I}ZnSOD active site binds small anions, but apparently not to the Cu(I) center itself. This result is not surprising because the trigonal-planar $Cu^{I}(His)_3$ center is expected to have little affinity for binding a fourth ligand, based on comparisons with model cuprous complexes. A study of azide binding to both oxidized and reduced CuZnSOD using Fourier transform infrared spectroscopy (FTIR)[12] led to the conclusion that azide binds to Cu(II) in the oxidized enzyme but that it binds elsewhere in the reduced as well as in the copper-free enzyme, probably at the guanidinium group of the conserved Arg residue in the active-site channel and possibly also interacting with the conserved chain of waters in the channel and the N–H of His63 (Fig. XI.2.4, left).

For superoxide to be reduced rapidly by $Cu^{I}ZnSOD$, there must be a proton intimately associated with the O_2^- as it accepts an electron to become peroxide. Presumably, the proton on His63 plays the role of proton donor and the product is the hydroperoxide anion, HOO^-. This oxy anion is likely to complex to Cu^{2+}, if it is formed at this point (reaction 18).

$$Cu^{I} — H{-}N\langle\text{His 63}\rangle N{-}H \xrightarrow{O_2^-} H{-}O{-}O{-}Cu^{II}{-}N\langle\text{His 63}\rangle N{-}H \qquad (18)$$

Protonation and loss of H_2O_2 would then follow.

Alternatively, HOO^- might accept a proton from the chain of conserved waters to form H_2O_2 directly, thus bypassing formation of a Cu(II)–hydroperoxide complex. It is this latter possibility that is described in Fig. XI.2.4.

The imidazole ring of His63 binds solely to Zn(II) in $Cu^{I}ZnSOD$, whereas it is deprotonated to form imidazolate, which bridges Cu(II) and Zn(II) in $Cu^{II}ZnSOD$. The imidazolate–Zn(II) moiety clearly plays an important role in the CuZnSOD mechanism, since removal of Zn from the enzyme makes the SOD activity pH dependent. Studies of the SOD activity of the zinc-deficient enzyme by pulse radiolysis established that the reduction of the Cu(II) form of the zinc-deficient enzyme by superoxide remained fast and pH independent. Therefore, it could be concluded that it is the oxidation of the Cu(I) form of the zinc-deficient enzyme by superoxide that had become rate determining and pH dependent. These observations suggest that the imidazolate-Zn^{2+} moiety plays an important role in determining the rate of reduction of superoxide by $Cu^{I}ZnSOD$ and its lack of dependence on pH. It has been postulated that the coordination of the imidazolate-Zn^{2+} moiety to Cu(II), creating the imidazolate bridge, prevents strong binding of the hydroperoxy anion to Cu by directing it to an axial coordination position on the Cu(II), thereby facilitating rapid loss of the peroxide product.[13]

XI.2.4.2. Manganese Superoxide Dismutase and Iron Superoxide Dismutase

Manganese superoxide dismutase[1,2,4,7] occurs with great frequency in both prokaryotic and eukaryotic cells,[4,14] while FeSOD occurs mainly in prokaryotes.[15] X-ray crystal structural studies have shown that the two enzymes have very similar structures and metal-binding sites (see Figs. XI.2.5 and XI.2.6). Despite their strong similarities, most Mn- and FeSODs show high metal ion specificity, that is, they have high SOD activities with only one of the two metal ions. The exceptions to this rule are the so-called "cambialistic" SOD proteins that are active with either Fe or Mn.[14]

Three histidines, one aspartate, and a water molecule bind to the manganese or iron ion of Fe- or MnSOD, forming a nearly trigonal-bipyramidal geometry. In the case of the oxidized enzymes, $Fe^{III}SOD$ and $Mn^{III}SOD$, the water ligand is believed to be the hydroxo ligand OH^- rather than the aqua ligand H_2O. The metal-binding site is further stabilized by a highly conserved extended hydrogen-bonding network involving Tyr and Glu, which are hydrogen bonded to the water and aspartate ligands. This proton linkage network is involved in binding the ligands to the metal center as well as in proton transfer, both of which are critical to the efficient turnover of the catalytic reaction.

The reduction of $M^{III}SOD$ to $M^{II}SOD$ ($M = Fe$ or Mn) by superoxide and the release of O_2 is fast for both enzymes ($\sim 10^9\ M^{-1}\ s^{-1}$). Studies of the anion-binding properties of $M^{III}SOD$ suggest that superoxide approaches the metal center and binds to the M^{3+} ion prior to electron transfer, that is, that the reduction step follows an inner-sphere mechanism (reaction 19).

$$O_2^{-} + M^{III}SOD \rightarrow (O_2^{-})M^{III}SOD \rightarrow M^{II}SOD + O_2 \quad (19)$$

In contrast, the principal pathway for oxidation of $M^{II}SOD$ by superoxide is believed to occur by an outer-sphere mechanism: First by docking of superoxide by hydrogen bonding to the water ligand bound to the M^{2+} ion via hydrogen bonding; followed by proton-coupled electron transfer to form the free hydroperoxide anion, thus avoiding the formation of a M(III)–hydroperoxide complex (reaction 20).[2,15]

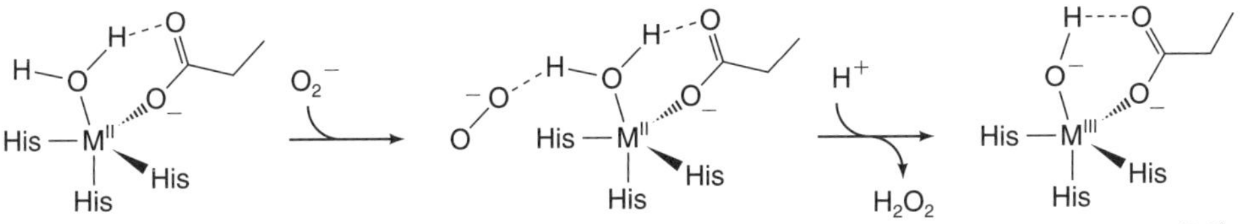

(20)

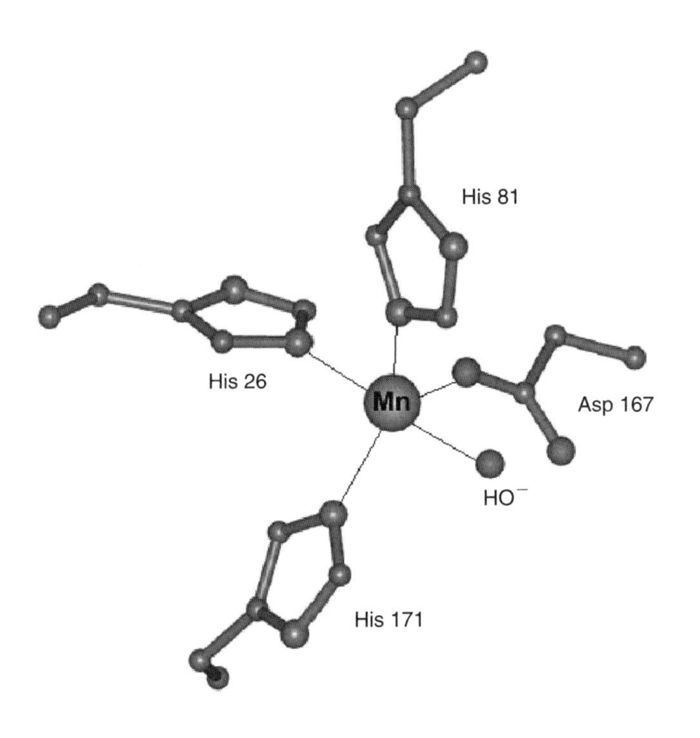

Fig. XI.2.5. Metal site in *Escherichia coli* MnSOD (PDB code: 1VEW). Three equatorial ligands (His81, Asp167, and His171) and two axial ligands (His26 and a solvent molecule, modeled as hydroxide ion) form a trigonal-bipyramidal arrangement.

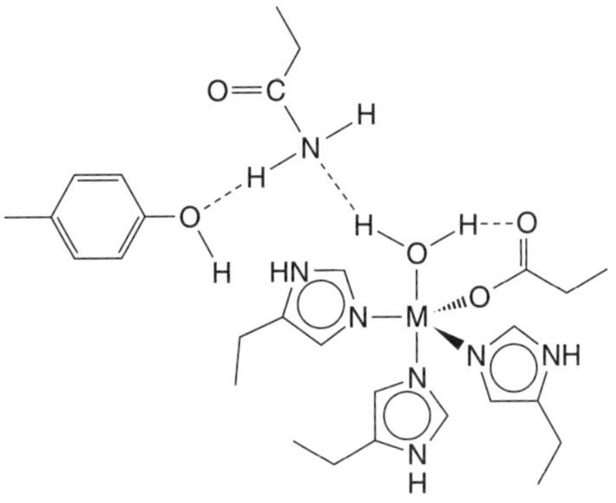

Fig. XI.2.6. Active-site configuration of $Mn^{II}SOD$ and $Fe^{II}SOD$.

The oxidation of $Fe^{II}SOD$ by superoxide appears to occur exclusively by the outer-sphere pathway described above, but the same reaction for Mn(II) appears to be more complex, involving a slower inner-sphere pathway (reaction 21) in addition to the faster outer-sphere pathway described above. This difference is not so surprising when we consider it in light of the observation that Mn(II)SOD, unlike Fe(II)SOD, has a significant affinity for the direct binding of small anionic ligands to the metal center.

$$(\text{His})_3\text{M}^{II}(\text{OH}_2)(\text{O}_2\text{CR}) \xrightarrow{\text{O}_2^-} (\text{His})_3\text{M}^{III}(\text{OO H})(\text{O}^-)(\text{O}_2\text{CR}) \xrightarrow[\text{H}_2\text{O}_2]{\text{H}^+} (\text{His})_3\text{M}^{III}(\text{O}^-)(\text{O}_2\text{CR}) \quad (21)$$

XI.2.4.3. Nickel Superoxide Dismutase

Nickel superoxide dismutase (NiSOD)[3,16] has been found in some *Streptomyces* species and in cyanobacteria. As seen in Fig. XI.2.7, X-ray crystal structural studies[17] reveal a five-coordinate square-pyramidal environment for the oxidized Ni(III) form, with two cysteine thiolates, the N-terminal amino group, and a backbone amide nitrogen in a near planar arrangement around the Ni center and an axial His ligand. Upon reduction to the Ni(II) form, the axial His ligand detaches from the Ni center, leaving the remaining four ligands in a square-planar array.

From the structure, one might predict that superoxide oxidizes the Ni(II) enzyme by an inner-sphere mechanism (reaction 12, top), but calculations of the surface accessibility[17] of the Ni(II) center suggest that the metal center is quite buried and inaccessible to solvent. Thus, it seems likely that oxidation occurs via an outer-sphere mechanism (reaction 12, bottom), with superoxide docking on a nearby proton donor prior to accepting an electron.

The buried nature of the active site of NiSOD may explain an apparent contradiction concerning this enzyme. It is clear that the thiolate ligation to nickel is required to move the Ni(II)/Ni(III) redox potential into a range suitable for an SOD, but it is

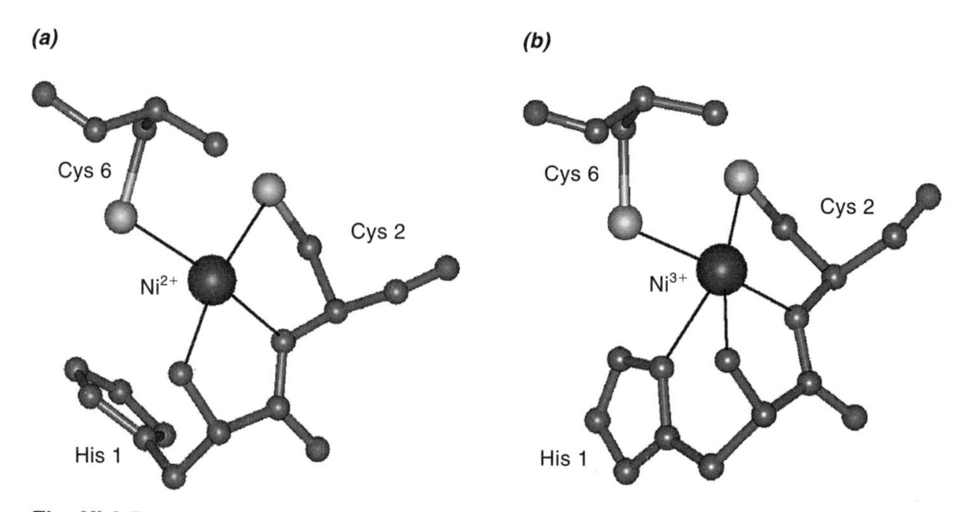

Fig. XI.2.7.
Nickel binding site of *Streptomyces seoulensis* SOD in oxidized, (*a*) Ni^{3+} (PDB code: 1Q0D) and (*b*) thiosulfate reduced, Ni^{2+} (PDB code: 1Q0K) forms.

not clear how the oxidatively sensitive metal thiolate moieties are protected from oxidative damage in the presence of both superoxide and hydrogen peroxide. Possibly, the buried nature of the center protects the thiolate ligands from damage, particularly if the reactions with superoxide are outer sphere in nature.

XI.2.4.4. Superoxide Reductase

Superoxide reductases[6,18–20] have been found in some anaerobic and microaerophilic bacteria. They are nonheme iron-containing enzymes that catalyze the one-electron reduction of superoxide to give H_2O_2. The ferrous state has a [Fe(His)$_4$(Cys)] metal-binding site, while the ferric state has an additional glutamate ligand, that is, [Fe(His)$_4$(Cys)(Glu)] (see Fig. XI.2.8). The mechanism of oxidation of the ferrous enzyme by superoxide is believed to be inner sphere in character (reaction 22).

$$\text{Cys–S}^-\text{–Fe}^{II}(\text{His})_4 \xrightarrow{O_2^-,\,H^+} \text{Cys–S}^-\text{–Fe}^{III}(\text{His})_4\text{–O–OH} \xrightarrow[-H_2O_2]{\text{Glu–CO}_2^-,\,H^+} \text{Cys–S}^-\text{–Fe}^{III}(\text{His})_4\text{–O–C(=O)–Glu} \xrightarrow[-\text{Glu–CO}_2^-]{e^-} \text{Cys–S}^-\text{–Fe}^{II}(\text{His})_4 \quad (22)$$

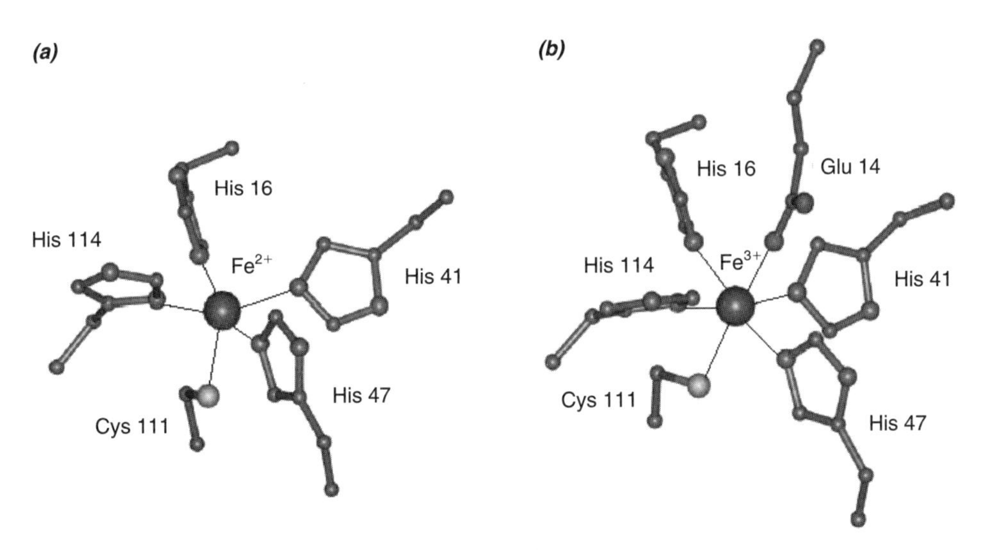

Fig. XI.2.8.
Iron center of *Pyrococcus furiosus* SOR in ferric (*a*, PDB code: 1DQI) and ferrous (*b*, PDB code: 1DQK) states.

One particularly interesting mechanistic question about the FeSORs is why they are not SODs, in other words, why is the ferric form reduced so slowly by superoxide? Based on the crystal structure, it has been hypothesized that the Glu ligand present in the ferric form may block the access to the ferric ion. Another interesting functional question is Why some organisms have SORs and others have SODs?[20] That question remains to be answered.

References

General References

1. Fridovich, I., "Superoxide anion radical (O_2^-), superoxide dismutases, and related matters", *J. Biol. Chem.*, **272,** 18515–18517 (1997).
2. Cabelli, D. E., Riley, D., Rodriguez, J. A., Valentine, J. S., and Zhu, H., "Models of superoxide dismutases", in *Biomimetic Oxidations Catalyzed by Transition Metal Complexes,* Meunier, B., Ed., World Scientific Publishing Company, 1999, pp. 461–508.
3. Bryngelson, P. A., Arobo, S. E., Pinkham, J. L., Cabelli, D. E., and Maroney, M. J., "Expression, reconstitution, and mutation of recombinant *Streptomycescoelicolor* NiSOD", *J. Am. Chem. Soc.*, **126,** 460–461 (2004).
4. Whittaker, J. W., "Manganese superoxide dismutase", *Met. Ions Biol. Syst.*, **37,** 587–611 (2000).
5. Bertini, I., Mangani, S., and Viezzoli, M. S., "Structure and properties of copper-zinc superoxide dismutases", *Adv. Inorg. Chem.*, **45,** 127–250 (1998).
6. Niviere, V. and Fontecave, M., "Discovery of superoxide reductase: an historical perspective", *J. Biol. Inorg. Chem.*, **9,** 119–123 (2004).

Specific References

7. Noodleman, L., Lovell, T., Han, W. G., Li, J., and Himo, F., "Quantum chemical studies of intermediates and reaction pathways in selected enzymes and catalytic synthetic systems", *Chem. Rev.*, **104,** 459–508 (2004).
8. Muscoli, C., Cuzzocrea, S., Riley, D. P., Zweier, J. L., Thiemermann, C., Wang, Z.-Q., and Salvemini, D., "On the selectivity of superoxide dismutase mimetics and its importance in pharmacological studies", *Br. J. Pharmacol.*, **140,** 445–460 (2003).
9. Hudis, J. and Dodson, R. W., "Rate of ferrous-ferric exchange in D_2O", *J. Am. Chem. Soc.*, **78,** 911 (1956).
10. Hart, P. J., Balbirnie, M. M., Ogihara, N. L., Nersissian, A. M., Weiss, M. S., Valentine, J. S., and Eisenberg, D., "A structure-based mechanism for copper-zinc superoxide dismutase", *Biochemistry,* **38,** 2167–2178 (1999).
11. Ogihara, N. L., Parge, H. E., Hart, P. J., Weiss, M. S., Goto, J. J., Crane, B. R., Tsang, J., Slater, K., Roe, J. A., Valentine, J. S., Eisenberg, D., and Tainer, J. A., "Unusual trigonal-planar copper configuration revealed in the atomic structure of yeast copper-zinc superoxide dismutase", *Biochemistry,* **35,** 2316–2321 (1996).
12. Leone, M., Cupane, A., Militello, V., Stroppolo, M. E., and Desideri, A., "Fourier Transform Infrared Analysis of the Interaction of Azide with the Active Site of Oxidized and Reduced Bovine Cu, Zn Superoxide Dismutase", *Biochemistry,* **37,** 4459–4464 (1998).
13. Ellerby, L. M., Cabelli, D. E., Graden, J. A., and Valentine, J. S., "Copper-Zinc Superoxide Dismutase—Why Not pH-Dependent", *J. Am. Chem. Soc.*, **118,** 6556–6561 (1996).
14. Whittaker, J. W., "The irony of manganese superoxide dismutase", *Biochem. Soc. Trans.*, **31,** 1318–1321 (2003).
15. Lah, M. S., Dixon, M. M., Pattridge, K. A., Stallings, W. C., Fee, J. A., and Ludwig, M. L., "Structure–function in *Escherichia coli* iron superoxide dismutase: comparisons with the manganese enzyme from *Thermus thermophilus*", *Biochemistry,* **34,** 1646–1660 (1995).
16. Szilagyi, R. K., Bryngelson, P. A., Maroney, M. J., Hedman, B., Hodgson, K. O., and Solomon, E. I., "S K-edge X-ray absorption spectroscopic investigation of the Ni-containing superoxide dismutase active site: new structural insight into the mechanism", *J. Am. Chem. Soc.*, **126,** 3018–3019 (2004).
17. Wuerges, J., Lee, J. W., Yim, Y. I., Yim, H. S., Kang, S. O., and Carugo, K. D., "Crystal structure of nickel-containing superoxide dismutase reveals another type of active site", *Proc. Natl. Acad. Sci. U. S. A.*, **101,** 8569–8574 (2004).
18. Adams, M. W., Jenney, F. E., Jr., Clay, M. D., and Johnson, M. K., "Superoxide reductase: fact or fiction?", *J. Biol. Inorg. Chem.*, **7,** 647–652 (2002).
19. Clay, M. D., Jenney, F. E., Jr., Hagedoorn, P. L., George, G. N., Adams, M. W., and Johnson, M. K., "Spectroscopic studies of *Pyrococcus furiosus* superoxide", *J. Am. Chem. Soc.*, **124,** 788–805 (2002).
20. Imlay, J. A., "What biological purpose is served by superoxide reductase", *J. Biol. Inorg. Chem.*, **7,** 659–663 (2002).

XI.3. Peroxidase and Catalases

Contents

Thomas L. Poulos
Departments of Molecular Biology and Biochemistry, Chemistry, and Physiology and Biophysics
University of California, Irvine
Irvine, CA 92617

XI.3.1. Introduction

Hydrogen peroxide (H_2O_2) is a useful oxidizing agent in biology and is employed by peroxidases to oxidize a variety of biologically important molecules (Fig. XI.3.1).[1–4] The two-electron reduction of H_2O_2 to water has a redox potential of 1.349 V versus normal hydrogen electrode (NHE) (pH 7, 25°C), which is equivalent to ~31 kcal mol^{-1}. Therefore, H_2O_2 is a good choice as an oxidizing agent and our everyday experience suggests that this is the case. If you were to accidentally get a few drops of a 30% peroxide solution on your hand, it will burn. Hydrogen peroxide is, however, a long-lived molecule primarily because the peroxide O–O bond energy in H_2O_2 is quite high (51 kcal mol^{-1}). To unleash the oxidizing power of peroxide requires a mechanism for overcoming this barrier and cleaving the peroxide O–O bond. Breaking this bond can occur in two ways: homolytically or heterolytically (Fig. XI.3.1). In the homolytic reaction, each O atom retains an equal number of valence electrons to give two highly reactive hydroxyl radicals. In the heterolytic reaction, one oxygen departs at the oxidation state of water with its full complement of eight valence electrons. The remaining oxygen atom, often called the "oxene"

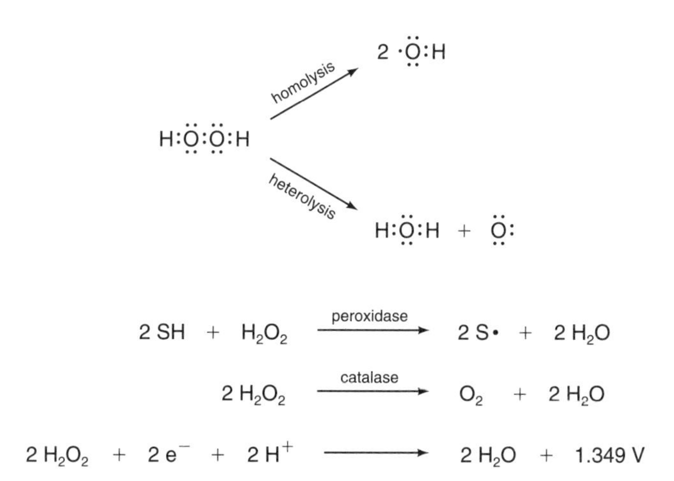

Fig. XI.3.1.
Hydrogen peroxide can cleave by two mechanisms: homolytic or hetereolytic. Peroxidases and catalases are designed such that only heterolytic fission of the peroxide O–O bond occurs. The reduction of peroxide to water is a thermodynamically favorable process but does not proceed readily since the peroxide O–O bond is relatively strong. Therefore, peroxidases and catalases have evolved a precise catalytic mechanism to overcome this kinetic barrier and ensure that the peroxide O–O bond is cleaved heterolytically.

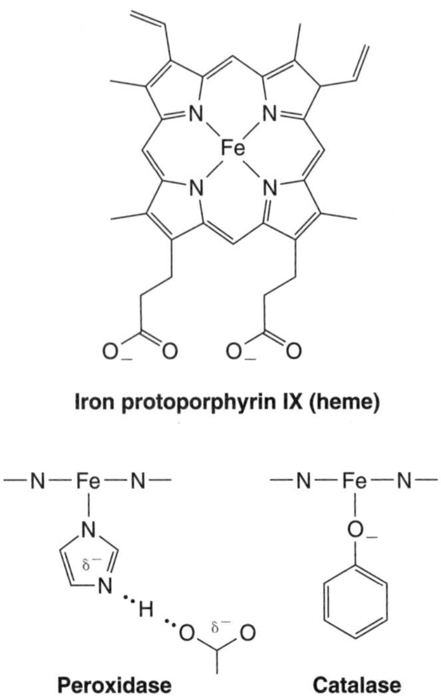

Fig. XI.3.2.
Peroxidases and catalases use iron protoporphyrin IX as the prosthetic group, which is the most common porphyrin and is used by a wide range of heme proteins including cytochrome P450, globins, and many cytochromes. The axial protein ligand in peroxidases is a His residue, while in catalases, Tyr serves this purpose. In peroxidases, the His ligand is hydrogen bonded to an Asp residue, which helps to impart considerable anionic character to the ligand.

Iron protoporphyrin IX (heme)

Peroxidase **Catalase**

O atom,[2] contains only six valence electrons, and is a highly unstable reactive oxidizing agent.

Transition metals like Fe^{2+} react with H_2O_2 in a reaction known as Fenton chemistry to give hydroxyl radicals (see Section XI.1). The reactivity of such radicals is impossible to control and their generation in biological systems can cause severe damage to macromolecules. As a result, peroxidases operate by a heterolytic mechanism, thus avoiding the formation of toxic hydroxyl radicals. In order to control cleavage of the peroxide O–O bond and effectively harness the oxidizing power of peroxides, peroxidases have evolved a precise catalytic mechanism that ensures heterolytic cleavage of the O–O bond with the highly reactive oxene O atom retained at the peroxidase active site. Stabilization is achieved by the oxene O atom binding to peroxidase heme iron. The heme group (Fig. XI.3.2) utilized by peroxidases for this purpose is exactly the same as that used by the globins, yet clearly these two classes of proteins exhibit very different reactivities. On the one hand, the globins are excellent at reversibly binding O_2 (see Section XI.4), but are very poor peroxidases, while just the opposite is true for peroxidases. The protein environment surrounding the heme is what controls this difference. A fairly recent book provides an in-depth overview of peroxidases.[1]

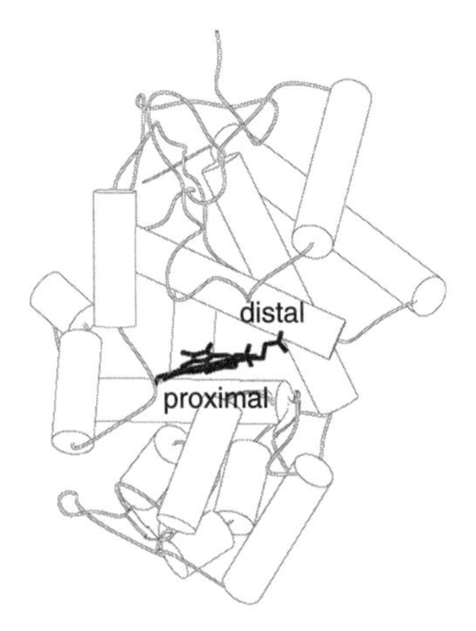

Fig. XI.3.3.
The crystal structure of the most well known peroxidase, horseradish peroxidase or HRP. All non-mammalian heme peroxidases look very similar and consist of a common 10 helical core. The proximal and distal helices provide key residues important for catalysis.

XI.3.2. Overall Structure

The crystal structures of several heme peroxidases are known.[5–14] These peroxidases are either extra- or intercellular and derive from various sources ranging from fungi to plants. In some cases, the sequences are very similar and in others quite distant. Nevertheless, the overall fold of all peroxidases is remarkably conserved and is built around a common 10 helical bundle motif (Fig. XI.3.3).

This structural similarity of heme peroxidases is not reflected in the primary sequences. For example, cytochrome *c* peroxidase (CcP) and peanut peroxidase (PNP) share only 17% sequence identity. It might be expected that for proteins with this low level of sequence identity would exhibit far less structural similarity than is observed in the peroxidases. Peroxidases, however, reflect a common pattern in structural biology, where tertiary structure is more conserved than primary structure. One way of representing this similarity is to carry out a least-squares superimposition of α-carbon atoms and calculate the root-mean-square (rms) distance between common α-carbon atoms. Peanut and horseradish peroxidase[6] are very similar and exhibit an rms deviation of α-carbons in structurally conserved regions of 0.81 Å, while lignin peroxidase and PNP give a 2.31 Å rms deviation. Ascorbate peroxidase appears to be the most "average" peroxidase structure, since it exhibits an rms deviation between these extremes with values ranging from 1.68 to 1.93 Å.

While the helical core is highly conserved, the various peroxidases differ primarily in elements of structure on the surface not involved in forming the helical core. For example, yeast mitochondrial cytochrome *c* peroxidase has a three stranded β-sheet, where the extracellular plant peroxidases have two additional helices, F′ and F″. The extracellular fungal peroxidases are larger than other peroxidases, primarily due to an extended C-terminal tail.

XI.3.3. Active-Site Structure

All peroxidases have a distal and proximal helix close to the heme. The distal helix contains residues that form the peroxide binding site, while the proximal helix contains the His heme ligand (Figs. XI.3.3 and XI.3.4). Although the catalase structure[15] is predominantly helical as well, the overall connectivity and fold are different from the peroxidases (Fig. XI.3.5). The active-site structure of catalase exhibits both similarities and differences to peroxidases (Fig. XI.3.6). Like peroxidases, there is a His residue (His74 in Fig. XI.3.6) in the distal pocket, but here the His is parallel to the heme plane rather than perpendicular. Catalase also has Asn147, which is roughly equivalent to Arg38 in HRP.

In all peroxidases, the His ligand forms a hydrogen bond with the conserved Asp residue, which is thought to impart considerable anionic character to the His.[5,16] The negative charge on the His ligand stabilizes the extra charge on Fe^{3+} as opposed to

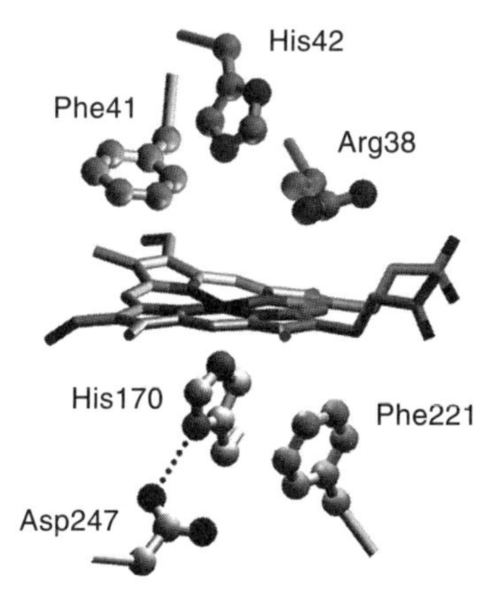

Fig. XI.3.4.
The active-site structure of HRP. This architecture is conserved in all non-mammalian heme peroxidases except that in some peroxidases the two Phe residues are replaced by Trp and in others Phe221 is replaced by an aliphatic side chain. However, His42 and Arg48, which form the peroxide binding pocket, are strictly conserved.

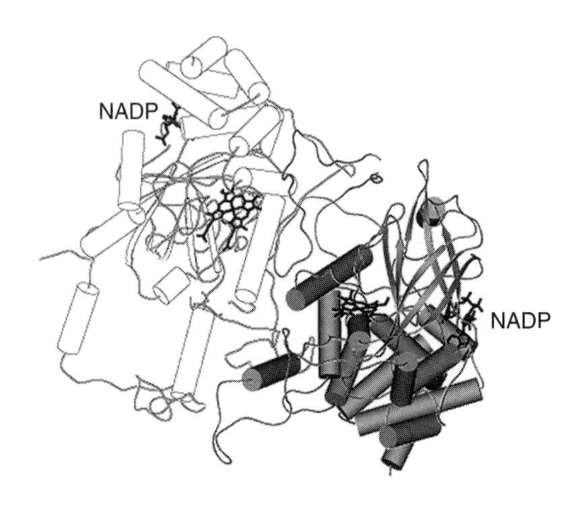

Fig. XI.3.5.
The crystal structure of catalase. Catalase is a tetramer and shown here are two of the four subunits. The overall fold is quite different than in peroxidases. An unusual feature of the catalase structure is a bound nicotinamide adenine diphosphate (NADP).

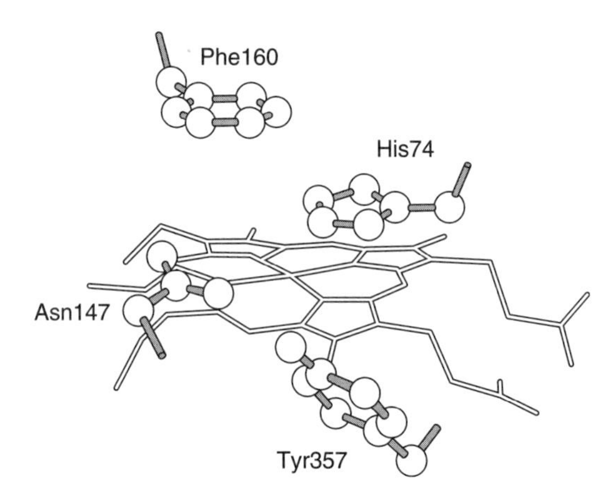

Fig. XI.3.6.
The active-site structure of catalase. Notice that the catalytic His74 is positioned parallel to the heme, while in peroxidases, the catalytic distal His is perpendicular to the heme. Catalase also does not have an Arg in the active site as do peroxidases, but instead uses Asn147.

Fe^{2+}. In sharp contrast, the His ligand in the globins hydrogen bonds with the oxygen atom of a peptide carbonyl group; therefore, in the globins the His ligand does not carry a partial or full negative charge as in peroxidases, which is one important reason why the Fe^{3+}/Fe^{2+} redox potential is lower in peroxidases compared to the globins.

The ability to selectively change amino acid side chains using site directed mutagenesis has provided a powerful tool to probe the question of redox potential control. For example, the conserved Asp residue hydrogen bonded to the His ligand (Fig. XI.3.2) has been altered to other side chains[17] and the His ligand itself also has been replaced with other residues that cannot carry a negative charge.[18] The change in redox potential in these mutants is what one would predict (see Section X.1) if a full or partial negative charge on the His ligand is responsible for the low redox potential in peroxidases. Catalases provide the first example of heme proteins where the phenolate oxygen atom from a Tyr residue coordinates to the heme iron (Fig. XI.3.2). Here, too, the negatively charged ligand helps to lower the redox potential of the Fe^{3+}/Fe^{2+} couple.

XI.3.4. Mechanism

The mechanism of the peroxide reaction with peroxidases has important historical significance. Spectrally distinct and relatively stable intermediates form and as a result, many of the modern methods of enzymology were developed using peroxidases as model enzyme systems. For example, rapid reaction stopped flow methods were developed by Britton Chance back in the 1940s[19] by examining the rate at which peroxidases react with peroxides. The well-defined change in color upon oxidation of the enzyme with peroxides provided a clearly defined spectral signal. Since then, nearly every sophisticated biophysical probe available has been applied to peroxidases to work out the structure of the various intermediates. These include EPR,[20] Mossbauer,[20] extended X-ray absorption fine structure (EXAFS),[21] and crystal structures.[22–24] From this wealth of data, the following three step mechanism has been developed (where R is the porphyrin ring or amino acid side chain).

Step 1. $[Fe^{3+}\ R] + H_2O_2 \rightarrow \underset{\textbf{Compound I}}{[Fe^{4+}{=}O\ R\cdot]} + H_2O$

Step 2. $\underset{\textbf{Compound I}}{[Fe^{4+}{=}O\ R\cdot]} + \text{Substrate} \rightarrow \underset{\textbf{Compound II}}{[Fe^{4+}{=}O\ R]} + \text{substrate}\cdot$

Step 3. $\underset{\textbf{Compound II}}{[Fe^{4+}{=}O\ R]} + \text{Substrate} \rightarrow [Fe^{3+}\ R] + \text{substrate}\cdot + H_2O$

In step 1, the peroxide O–O bond is cleaved, leaving behind the single O atom with only six valence electrons. The O atom removes one electron from the iron and a second electron from the porphyrin ring or amino acid side chain (R in the above scheme). In most peroxidases, R· is a porphyrin cation radical, although in cytochrome *c* peroxidase, the radical is centered on a Trp residue.[25] Compound **I** exhibits distinct spectral characteristics (Fig. XI.3.7) and is easily distinguished by its green color compared to the brownish-red color of the resting enzyme. Depending on the

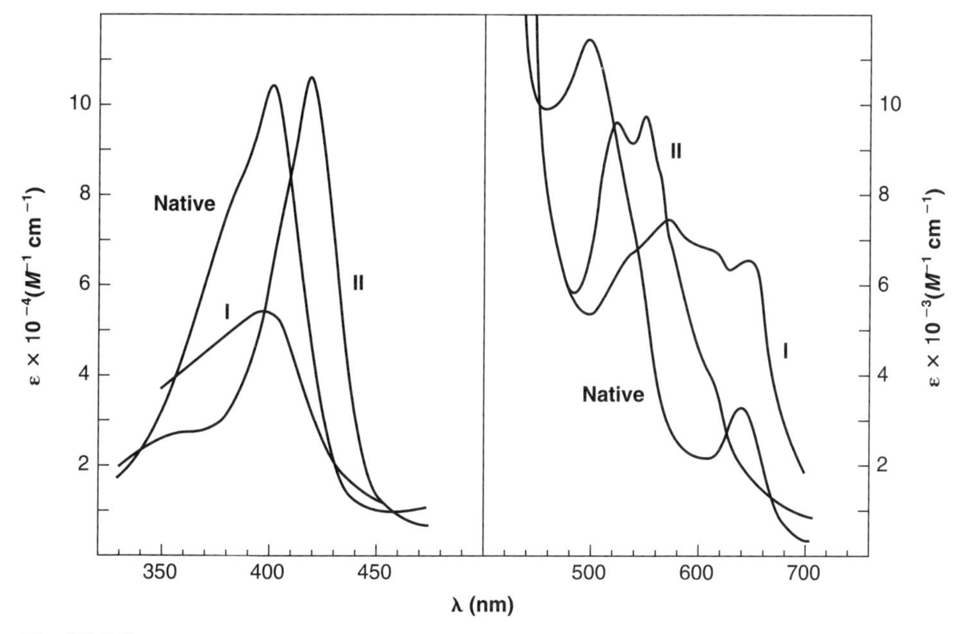

Fig. XI.3.7.
The various spectral intermediates of HRP. The native enzyme in the Fe^{3+} oxidation state gives a typical high-spin spectrum. Upon oxidation with peroxide to give Compound **I,** the main Soret absorbance band decreases to give green Compound **I.** The one-electron reduction of the porphyrin radical in Compound **I** gives red Compound **II** with a Soret band shifted to longer wavelengths relative to the resting native enzyme. These easily distinguished and stable intermediates have enabled a wealth of biophysical tools to be utilized in an attempt to work out the electronic structures of the various intermediates.

peroxidase, Compound **I** can be quite stable and has been amenable to detailed studies, including crystal structure determinations.[22–24] In the second step, a reducing substrate delivers one electron to R•, giving the red Compound **II** and a substrate radical. In the last step, the Fe^{4+}=O center of Compound **II** is reduced by a second substrate molecule. Under steady-state conditions, the reduction of Compound **II** to resting-state enzyme normally is the slow and, therefore, rate-limiting step in the reaction. Both the formation and reduction of Compound **I** are normally fast, second-order reactions.

The stereochemical mechanism of Compound **I** formation based on the crystal structures and a wealth of biochemical data is outlined in Fig. XI.3.8. The key fea-

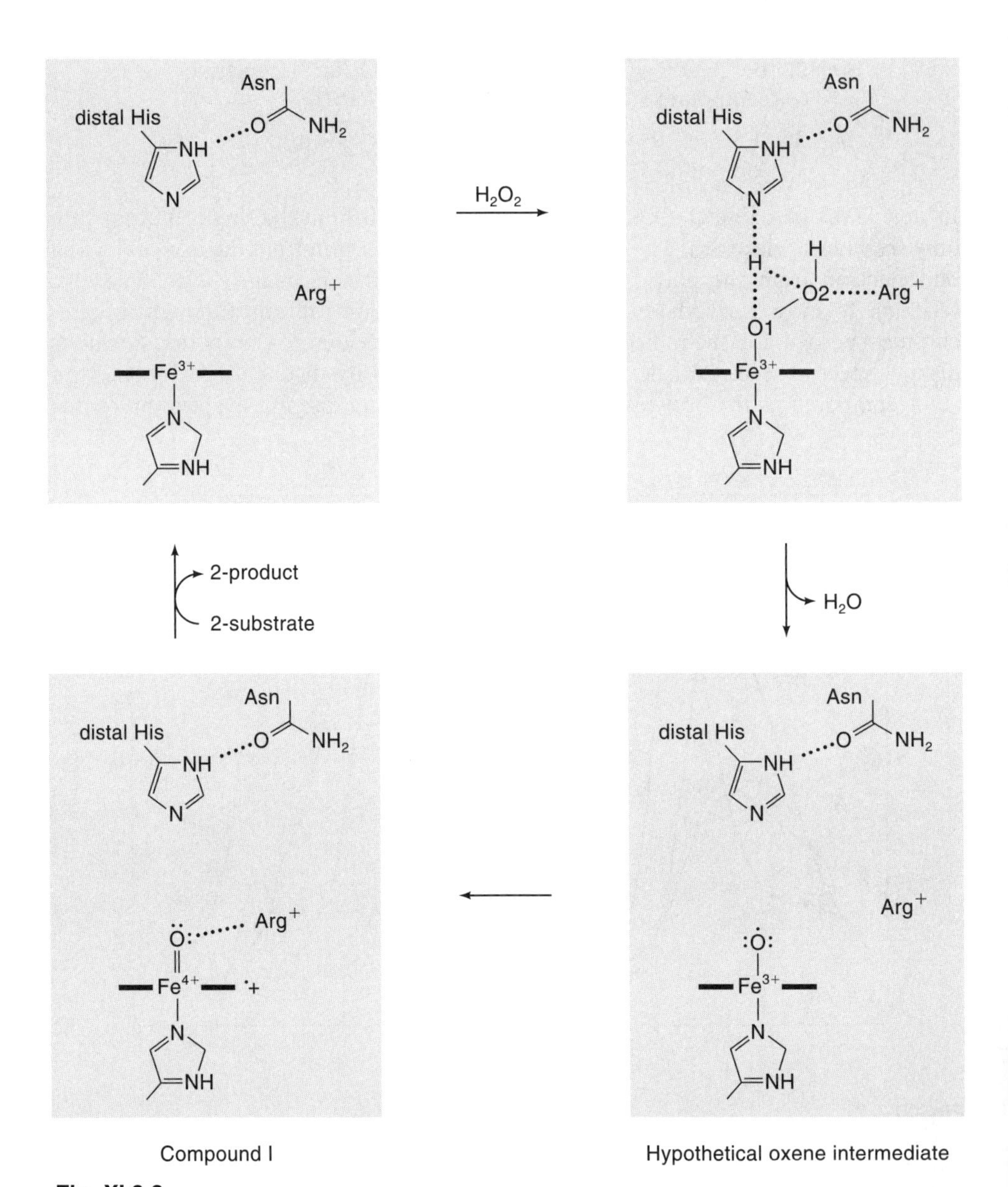

Fig. XI.3.8.
The stereochemical mechanism of Compound **I** formation. The distal His operates as an acid–base catalyst by removing a proton from the peroxide O atom directly bonded to the Fe and delivering a proton to the leaving O atom. This leaves behind an O atom that rapidly removes one electron from the iron and one from the porphyrin to give Compound **I.**

ture is the distal His, which serves as an acid–base catalytic group that ensures that the peroxide O–O bond is cleaved heterolytically. The distal side Arg is now known to move in and form a hydrogen bond with the Fe^{4+}=O group that possibly adds additional stability to Compounds **I** and **II.** Catalase is similar to peroxidases in forming Compound **I,** but differs substantially in how Compound **I** is reduced. In catalase, a second peroxide molecule serves as the reductant of Compound **I** in a concerted two-electron reduction process, so no Compound **II** builds-up under steady-state conditions. The generally accepted catalase mechanism[26] is outlined in Fig. XI.3.9.

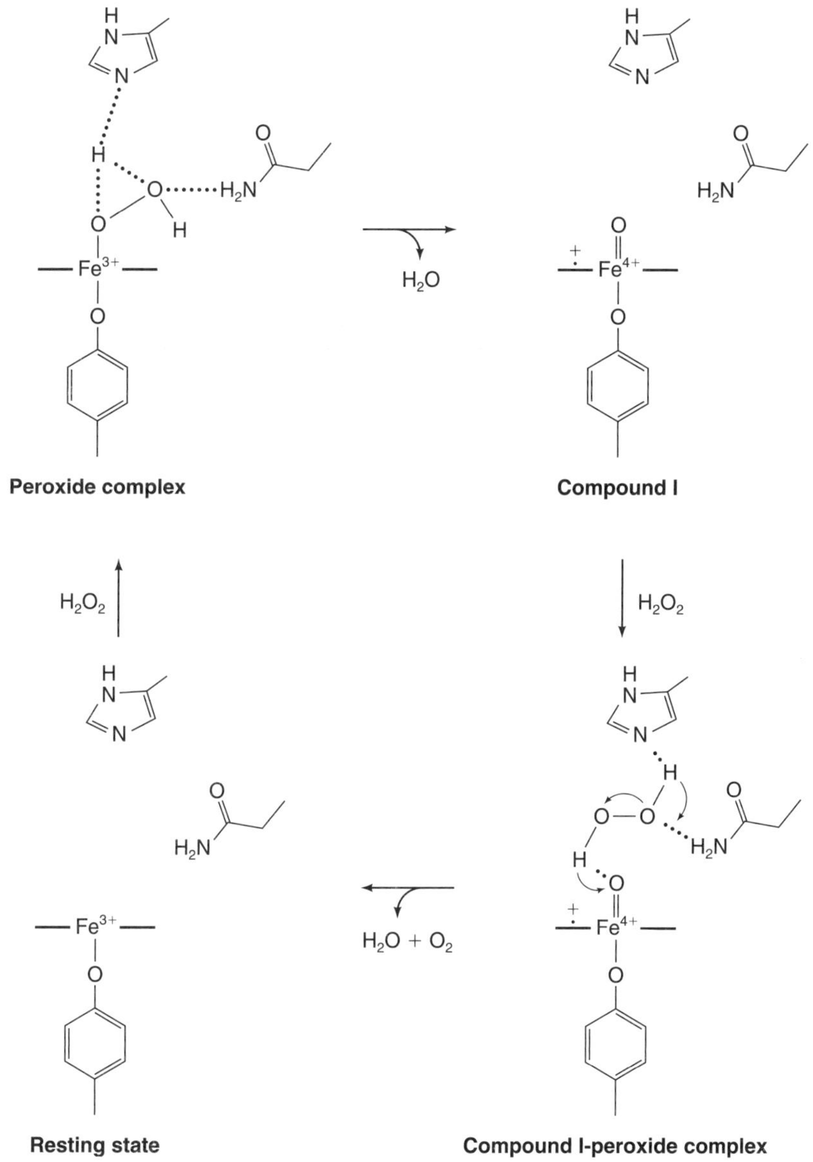

Fig. XI.3.9.
The proposed catalytic mechanism for catalase.[24] The formation of Compound **I** is similar to peroxidases except the active-site Asn in catalase serves the function of the Arg in peroxidases. What separates catalases from peroxidases is that H_2O_2 is the reductant of Compound **I**, giving H_2O and O_2. Since reduction of Compound **I** by H_2O_2 is a two-electron process, catalases do not form Compound **II** as part of the normal catalytic cycle.

Table XI.3.1.
The Effect on Activity of Key Proximal and Distal Side Residues in Various Peroxidases[a]

Enzyme	Mutant	Rate of Compound **I** formation ($M^{-1}s^{-1}$)	% Max. Steady-State Activity
CcP	WT	3.2×10^7	
CcP	Distal His-to-Leu	2.9×10^2	
CcP	Distal Arg-to-Leu	1.5×10^5	
CcP	Distal Asp-to-Asn	6.0×10^6	
CcP	Distal Trp-to-Phe	5.7×10^7	<1
APX	Distal Trp-to-Phe		100
CcP	Proximal His-to-Gln	1.2×10^7	100
CcP	Proximal His-to-Glu		764
HRP	WT	1.5×10^7	
HRP	Distal Asn-to-Val	0.12×10^7	6.4
HRP	Distal Asn-to-Asp	0.15×10^7	10.3

[a] Wild-type enzyme = WT; APX = ascorbate peroxidase; CcP = cytochrome *c* peroxidase. The distal His and Arg have been proposed to be involved in the acid–base catalytic mechanism required for cleavage of the peroxide O–O bond. The distal Asn directly hydrogen bonds to the distal His and may play a role in correctly orienting the distal His. The proximal His is the ligand directly coordinated to the heme iron while the proximal Trp contacts and is parallel to the His ligand. The distal Asp hydrogen bonds with the proximal His ligand. Refer to Fig. XI.3.4 for details. Data were taken from Refs. 27–29.

Site directed mutagenesis has proven to be an invaluable tool in directly testing the proposed roles of the various active-site side chains in the formation of Compound **I** (Table XI.3.1). Of the various residues altered, the most important is the distal His (His42 in Fig. XI.3.4). Changing this residue to Leu lowers the rate of Compound **I** formation by 10^5, which clearly illustrates the important role that His plays as an acid–base catalyst in Compound **I** formation. Somewhat surprisingly, changing the distal Arg (Arg38 in Fig. XI.3.4) to other residues lowers the rate of Compound **I** formation by only 10^2. This Arg may play a more important role in stabilizing Compound **I** once formed by making a hydrogen bond with the $Fe^{4+}{=}O$ center O atom.

XI.3.5. Reduction of Compounds I and II

The typical peroxidase reducing substrate is an aromatic phenol. Each substrate molecule donates one electron to either Compound **I** or **II**, leaving behind two substrate radicals for every cycle of enzyme catalysis. The location of the substrate-binding site has been surprisingly difficult to determine. Even when crystal structures became available, there was no active site typical of many enzymes that was an obvious choice for the binding of aromatic substrates. Investigations using NMR implicated key groups lining the exposed heme edge as the most likely site of interaction.[30] Another approach utilized suicide substrate inhibitors,[31] leading to similar conclusions. In this method (Fig. XI.3.10), the inhibitor reacts with Compound **I** to form a reactive radical that attaches to the closest available heme atom. Normally, this is the δ-meso heme carbon that is oriented toward the molecular surface in all known nonmammalian heme peroxidase structures. More recently, crystal structures[7,8,32] of various complexes have been solved that show the binding of substrates at the exposed heme edge, as expected (Fig. XI.3.11).

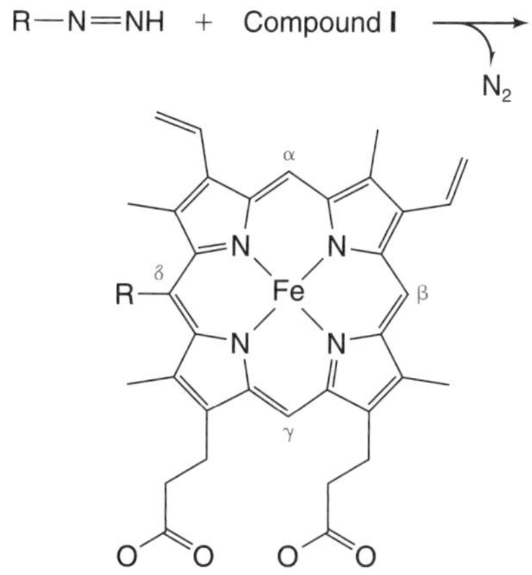

$$R{-}N{=}NH + \text{Compound } \mathbf{I} \xrightarrow{-N_2} R\cdot$$

Fig. XI.3.10.
The mechanism of peroxide suicide inhibition by alkyl and aryl diazines. The diazine reacts with Compound **I** to generate an organic radical. The radical rapidly reacts with the closest heme atom, which in most cases is the δ-meso carbon. This finding is taken as evidence that substrates must bind near the exposed heme edge near the δ-meso carbon.

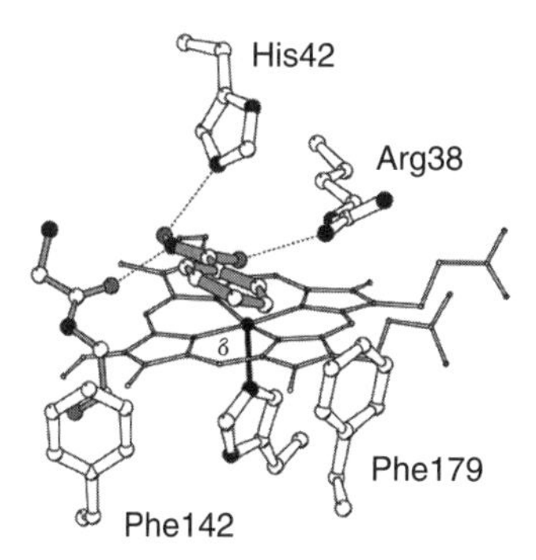

Fig. XI.3.11.
The crystal structure of HRP complexed with the well-known active-site probe and substrate, benzhydroxamic acid. The aromatic group sits over the δ-meso edge of the heme, which was the anticipated binding site for aromatic substrates.

There is, however, growing evidence that some peroxidases may have alternate sites for physiologically important substrates. For example, lignin peroxidase is able to oxidize 3,4-dimethoxybenzyl (veratryl) alcohol as well as other more traditional phenolic peroxidase substrates. The Trp171 residue in lignin peroxidase is now known to be essential for veratryl alcohol oxidation, but not for phenolic substrates.[33] This Trp171 is part of the same helix that contains the proximal heme ligand and is ~12 Å from the heme, which is well removed from the exposed heme edge. In addition, mutagenesis and chemical modification studies[34,35] predicted that the physiologically important substrate, ascorbate, binds closer to the heme propionates than the exposed heme edge, which was later confirmed by the crystal structure of the substrate complex.[36]

While most peroxidases accept electrons from small aromatic molecules, CcP is unusual in that its reducing substrate is cytochrome *c*. Cytochrome *c* peroxidase is a yeast mitochondrial enzyme and one possible function is to serve as an alternate sink for cytochrome *c* electrons in addition to cytochrome *c* oxidase, as well as serving as a peroxide scavenger. The CcP enzyme has proven to be an especially important model system for studying interprotein electron-transfer reactions. An intermolecular complex of CcP and cytochrome *c* forms prior to the transfer of an electron. The crystal structure of the noncovalent complex has been solved[37] and is shown in Fig. XI.3.12. The most interesting feature of this complex is the location of Trp191 in CcP relative to the cytochrome *c* heme. Unlike other heme peroxidases, the free radical in CcP Compound **I** is not centered on the prophyrin, but instead is centered on Trp191.[25,38] The cytochrome *c* heme directly contacts CcP at residue 194, so one could envision the transfer of the cytochrome *c* electron directly to the section of polypeptide containing the Trp191 radical, which leads to the mechanism outlined in Fig. XI.3.13.[39,40] In this mechanism, the cytochrome *c* electron is delivered directly to the Trp191 radical. This step is followed by an internal electron-transfer reaction, where the Fe^{4+}–O center oxidizes Trp191 giving, once again, a Trp191 radical that can receive a second cytochrome *c* electron. This mechanism requires that the same binding site on CcP be used for both electron-transfer steps, which is not universally accepted, since CcP is known to contain at least two sites for cytochrome *c* binding.[41,42] What remains to be established is whether or not both sites are competent in electron transfer.

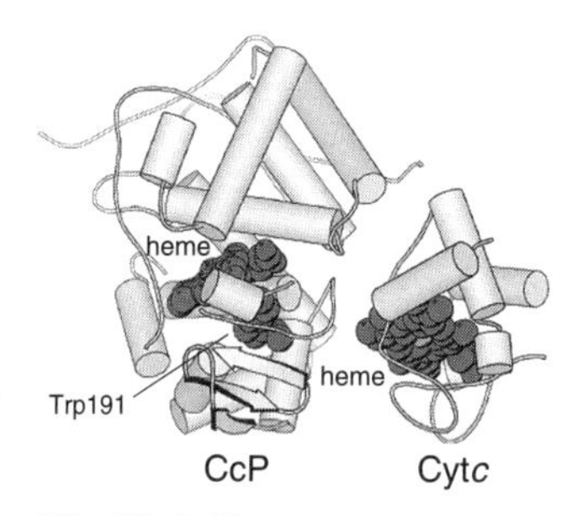

Fig. XI.3.12.
The crystal structure of the CcP–cytochrome *c* complex. Notice that the cytochrome *c* heme directly contacts CcP near Trp191, which provides a direct electron-transfer route from the cytochrome *c* heme to the Trp191 radical.

Fig. XI.3.13.
Proposed mechanism for reduction of CcP Compound **I.** In this mechanism, cytochrome *c* delivers the first electron directly to the Trp191 radical. This reaction is followed by an internal electron transfer where the Fe^{4+}–O center is reduced by Trp191 to give, once again, the Trp191 radical. This reaction enables a second cytochrome *c* molecule to deliver a second electron to the Trp191 radical, thereby completing the catalytic cycle.

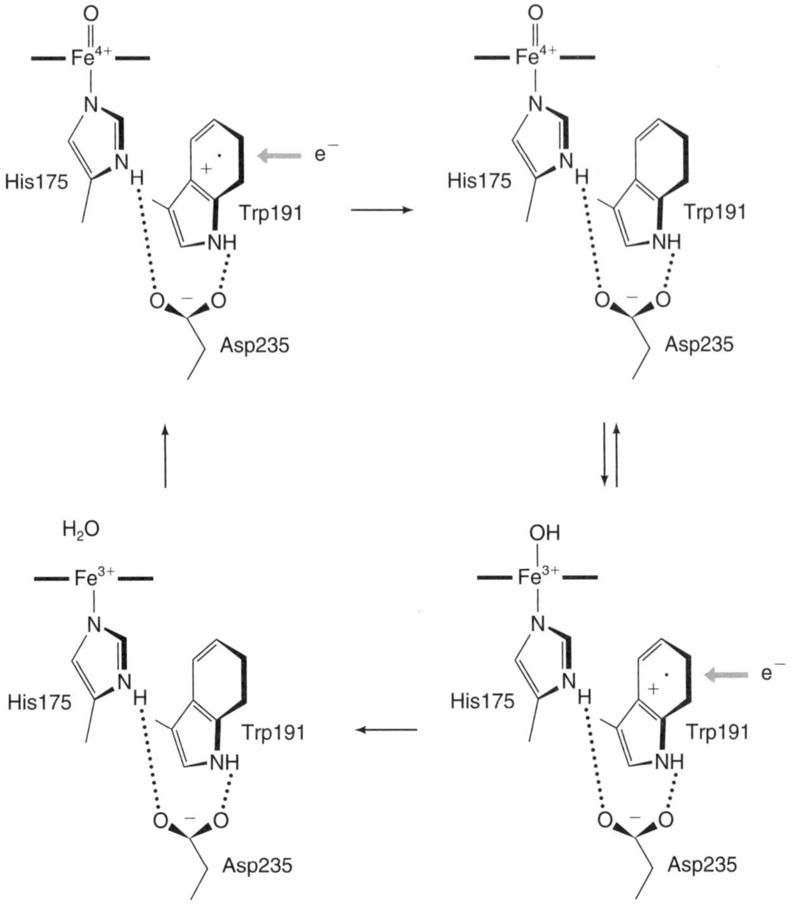

References

General References

1. Dunford, H. B., in *Peroxidases in Chemistry and Biology,* Everse, J. E., Everse, K. E., and Grisham, M. B., Eds., CRC Press, Boca Raton, FL, 1991, pp. 1–24.
2. Hamilton, G. A., "Chemical models and mechanisms for oxygenases", in *Molecular Mechanisms of Oxygen Activation,* Hayaishi, O. Ed., Academic Press, New York, 1974, pp. 405–451.
3. Poulos, T. L. and Fenna, R., "Peroxidases: Structure, function, and engineering", *Metals Biol.,* **30,** 25–75.
4. Poulos, T.L. and Finzel, B. C., "Heme enzyme structure and function", in *Peptide and Protein Reviews,* Mearn, M. T. Ed., Marcel Dekker, New York, 1984, pp. 115–171.

Specific References

5. Finzel, B. C., Poulos. T. L., and Kraut, J., "Crystal structure of yeast cytochrome c peroxidase refined at 1.7-A resolution", *J. Biol. Chem.,* **259,** 13027–13036 (1984).
6. Gajhede, M., et al., "Crystal structure of horseradish peroxidase C at 2.15 Å resolution", *Nat. Struct. Biol.,* **4,** 1032–1038 (1997).
7. Henriksen, A., et al., "Structural interactions between horseradish peroxidase C and the substrate benzhydroxamic acid determined by X-ray crystallography", *Biochemistry,* **37,** 8054–8060 (1998).
8. Henriksen, A., Welinder, K. G., and Gajhede, M., "Structure of barley grain peroxidase refined at 1.9-Å resolution. A plant peroxidase reversibly inactivated at neutral pH", *J. Biol. Chem.,* **273,** 2241–2248 (1998).
9. Kunishima, N., et al., "Crystal structure of the fungal peroxidase from *Arthromyces ramosus* at 1.9 Å resolution. Structural comparisons with the lignin and cytochrome c peroxidases", *J. Mol. Biol.,* **235,** 331–344 (1994).
10. Patterson, W. R. and Poulos, T. L., "Crystal structure of recombinant pea cytosolic ascorbate peroxidase", *Biochemistry,* **34,** 4331–4341 (1995).
11. Petersen, J. F., Kadziola, A., and Larsen, S., "Three-dimensional structure of a recombinant peroxidase from *Coprinus cinereus* at 2.6 Å resolution", *FEBS Lett.,* **339,** 291–296 (1994).
12. Piontek, K., Glumoff, T., and Winterhalter, K., "Low pH crystal structure of glycosylated lignin peroxidase from *Phanerochaete chrysosporium* at 2.5 Å resolution", *FEBS Lett.,* **315,** 119–124 (1993).

13. Poulos, T. L., Edwards, S. L., Wariishi, H., and Gold, M. H., "Crystallographic refinement of lignin peroxidase at 2 Å", *J. Biol. Chem.*, **268,** 4429–4440 (1993).
14. Sundaramoorthy, M., Kishi, K., Gold, M. H., and Poulos, T. L., "The crystal structure of manganese peroxidase from *Phanerochaete chrysosporium* at 2.06-Å resolution", *J. Biol. Chem.*, **269,** 32759–32767 (1994).
15. Murthy, M. R., et al., "Structure of beef liver catalase", *J. Mol. Biol.*, **152,** 465–499 (1981).
16. Valentine, J. S., Sheridan, R. P., Allen, L. C., and Kahn, P. C., "Coupling between oxidation state and hydrogen bond conformation in heme proteins", *Proc. Natl. Acad. Sci. U. S. A.*, **76,** 1009–1013 (1979).
17. Goodin, D. B. and McRee, D. E., "The Asp-His-Fe triad of cytochrome c peroxidase controls the reduction potential, electronic structure, and coupling of the tryptophan free radical to the heme", *Biochemistry,* **32,** 3313–3324 (1993).
18. Choudhury, K., et al., "Role of the proximal ligand in peroxidase catalysis. Crystallographic, kinetic, and spectral studies of cytochrome c peroxidase proximal ligand mutants", *J. Biol. Chem.*, **269,** 20239–20249 (1994).
19. Chance, B., "The kinetics of the enzyme-substrate compound of peroxidase", *J. Biol. Chem.*, **151,** 553–577 (1943).
20. Rutter, R. and Hager, L. P., "The detection of two electron paramagnetic resonance radical signals associated with chloroperoxidase compound I", *J. Biol. Chem.*, **257,** 7958–7961 (1982).
21. Chance, B., et al., "X-ray absorption studies of intermediates in peroxidase activity", *Arch. Biochem. Biophys.*, **235,** 596–611 (1984).
22. Bonagura, C. A., et al., "High-resolution crystal structures and spectroscopy of native and compound I cytochrome c peroxidase", *Biochemistry,* **42,** 5600–5608 (2003).
23. Edwards, S. L., Nguyen, H. X., Hamlin, R. C., and Kraut, J., "Crystal structure of cytochrome c peroxidase compound I", *Biochemistry,* **26,** 1503–1511 (1987).
24. Fulop, V., et al., "Laue diffraction study on the structure of cytochrome c peroxidase compound I", *Structure,* **2,** 201–208 (1994).
25. Sivaraja, M., Goodin, D. B., Smith, M., and Hoffman, B. M., "Identification by ENDOR of Trp191 as the free-radical site in cytochrome-*c* peroxidase compound ES", *Science,* **245,** 738–740 (1989).
26. Fita, I. and Rossmann, M. G., "The active center of catalase", *J. Mol. Biol.*, **185,** 21–37 (1985).
27. Erman, J. E., et al., "Histidine 52 is a critical residue for rapid formation of cytochrome c peroxidase compound I", *Biochemistry,* **32,** 9798–9806 (1993).
28. Rodriguez-Lopez, J. N., Smith, A. T., and Thorneley, R. N., "Recombinant horseradish peroxidase isoenzyme C: the effect of distal haem cavity mutations (His42 → Leu and Arg38 → Leu) on compound I formation and substrate binding", *J. Biol. Inorg. Chem.*, **1,** 136–142 (1996).
29. Nagano, S., et al., "Catalytic roles of the distal site asparagine-histidine couple in peroxidases", *Biochemistry,* **35,** 14251–14258 (1996).
30. Veitch, N. C., Gao, Y., Smith, A. T., and White, C. G., "Identification of a critical phenylalanine residue in horseradish peroxidase, Phe179, by site-directed mutagenesis and 1H-NMR: implications for complex formation with aromatic donor molecules", *Biochemistry,* **36,** 14751–14761 (1997).
31. Ortiz de Montellano, P. R., "Catalytic sites of hemoprotein peroxidases", *Annu. Rev. Pharmacol. Toxicol.*, **32,** 89–107 (1992).
32. Tsukamoto, K., et al., "Binding of salicylhydroxamic acid and several aromatic donor molecules to *Arthromyces ramosus* peroxidase, investigated by X-ray crystallography, optical difference spectroscopy, NMR relaxation, molecular dynamics, and kinetics", *Biochemistry,* **38,** 12558–12568 (1999).
33. Doyle, W. A., et al., "Two substrate interaction sites in lignin peroxidase revealed by site-directed mutagenesis", *Biochemistry,* **37,** 15097–15105 (1998).
34. Bursey, E. H. and Poulos, T. L., "Two substrate binding sites in ascorbate peroxidase: The role of arginine 172", *Biochemistry,* **39,** 7374–7379 (2000).
35. Mandelman, D., Jamal, J., and Poulos, T. L., "Identification of two electron-transfer sites in ascorbate peroxidase using chemical modification, enzyme kinetics, and crystallography", *Biochemistry,* **37,** 17610–17617 (1998).
36. Sharp, K. H., Mewies, M., Moody, P. C. E., and Raven, E. L., "Crystal structure of the ascorbate peroxidase-ascorbate complex", *Nat. Struct. Biol.*, **10,** 303–307 (2003).
37. Pelletier, H. and Kraut, J., "Crystal structure of a complex between electron transfer partners, cytochrome c peroxidase and cytochrome *c*", *Science,* **258,** 1748–1755 (1992).
38. Houseman, A. L., Doan, P. E., Goodin, D. B., and Hoffman, B. M., "Comprehensive explanation of the anomalous EPR spectra of wild-type and mutant cytochrome *c* peroxidase compound ES", *Biochemistry,* **32,** 4430–4443 (1993).
39. Hahm, S., Geren, L., Durham, B., and Millett, F., "Reaction of cytochrome-*c* with the radical in cytochrome-*c* peroxidase compound-I", *J. Am. Chem. Soc.*, **115,** 3372–3373 (1993).
40. Hahm, S., et al., "Reaction of horse cytochrome-c with the radical and the oxyferryl heme in cytochrome-c peroxidase compound-I", *Biochemistry,* **33,** 1473–1480 (1994).
41. Mauk, M. R., Ferrer, J. C., and Mauk, A. G., "Proton linkage in formation of the cytochrome *c*–cytochrome *c* peroxidase complex: electrostatic properties of the high- and low-affinity cytochrome binding sites on the peroxidase", *Biochemistry,* **33,** 12609–12614 (1994).
42. Zhou, J. S. and Hoffman, B. M., "Stern–Volmer in reverse: 2:1 stoichiometry of the cytochrome *c*–cytochrome *c* peroxidase electron-transfer complex", *Science,* **265,** 1693–1696 (1994).

XI.4. Dioxygen Carriers

Geoffrey B. Jameson
Centre for Structural Biology
Institute of Fundamental Sciences, Chemistry
Massey University
Palmerston North
New Zealand

James A. Ibers
Department of Chemistry
Northwestern University
Evanston, IL 60208

Contents

XI.4.1. Introduction: Biological Dioxygen Transport Systems

Most organisms require molecular oxygen (O_2) in order to survive. To increase the dioxygen-carrying capacity of aqueous media, in which O_2 is sparingly soluble, highly soluble proteins have evolved to transport O_2 from the external environment of relatively high O_2 abundance to regions of relatively low O_2 abundance, such as actively respiring cells. In addition, many multicellular aerobes also include a pump (heart), associated plumbing (arteries, veins, and capillaries), and a high surface-area phase-transfer device (lungs for gas–water transfer; gills for water–water transfer) to improve delivery of O_2 and to facilitate removal of carbon dioxide (CO_2).

Three chemically distinct dioxygen-carrier proteins are found today, the hemoglobins (Hb),[1–6] hemocyanins (Hc),[7,8] and hemerythrins (Hr)[9,10]—in this context the prefix *hem* denotes blood. The active site in these proteins, the site where O_2 binds, is a complex of either Cu(I) or Fe(II). The Hc and Hr carriers feature pairs of Cu and Fe atoms at their respective active sites. All Hb, from bacteria to plants to humans, share the same basic structure illustrated in Fig. XI.4.1 in which an Fe(II) porphyrin (heme) is embedded in the protein. When O_2 is not bound, the prefix *deoxy* is used, as in deoxyhemoglobin. When O_2 is bound, the prefix *oxy* is often added, as in oxyhemocyanin. The prefix *met* denotes the inactive oxidized form of the O_2 carrier, as in methemoglobin. All Hc and many Hb and Hr actually comprise two or more subunits for which the binding of O_2 to one subunit affects the binding of O_2 to remaining subunits—a phenomenon of great physiological importance known as *cooperativity*. Some basic properties of these O_2 carriers are summarized in Table XI.4.1. Many organisms contain several O_2 carriers in order to accommodate transient or periodic, internal or external changes in O_2 availability, temperature, hydrostatic pressure, pH, and salinity. In mammalian cells subject to sudden and high O_2 demand, such as those of skeletal and heart muscles, a monomeric Hb, myoglobin (Mb), facilitates O_2 diffusion to the mitochondria, where cytochrome *c* oxidase uses O_2 as the terminal electron acceptor (Sections X.3 and XI.6). Although the prefix *myo* comes from the Greek root *mys* for muscle, the designation Mb is used loosely for monomeric single subunit Hb. In this section, structures and properties of biological O_2 carriers are described with particular attention to the Hb family. This family has been studied in more detail than any other group of proteins and as a result a deeper more quantitative understanding of the relationships among structure, properties, and biological functions (i.e., physiology) exists. Some of this understanding has arisen through parallel studies on a host of synthetic systems, which have

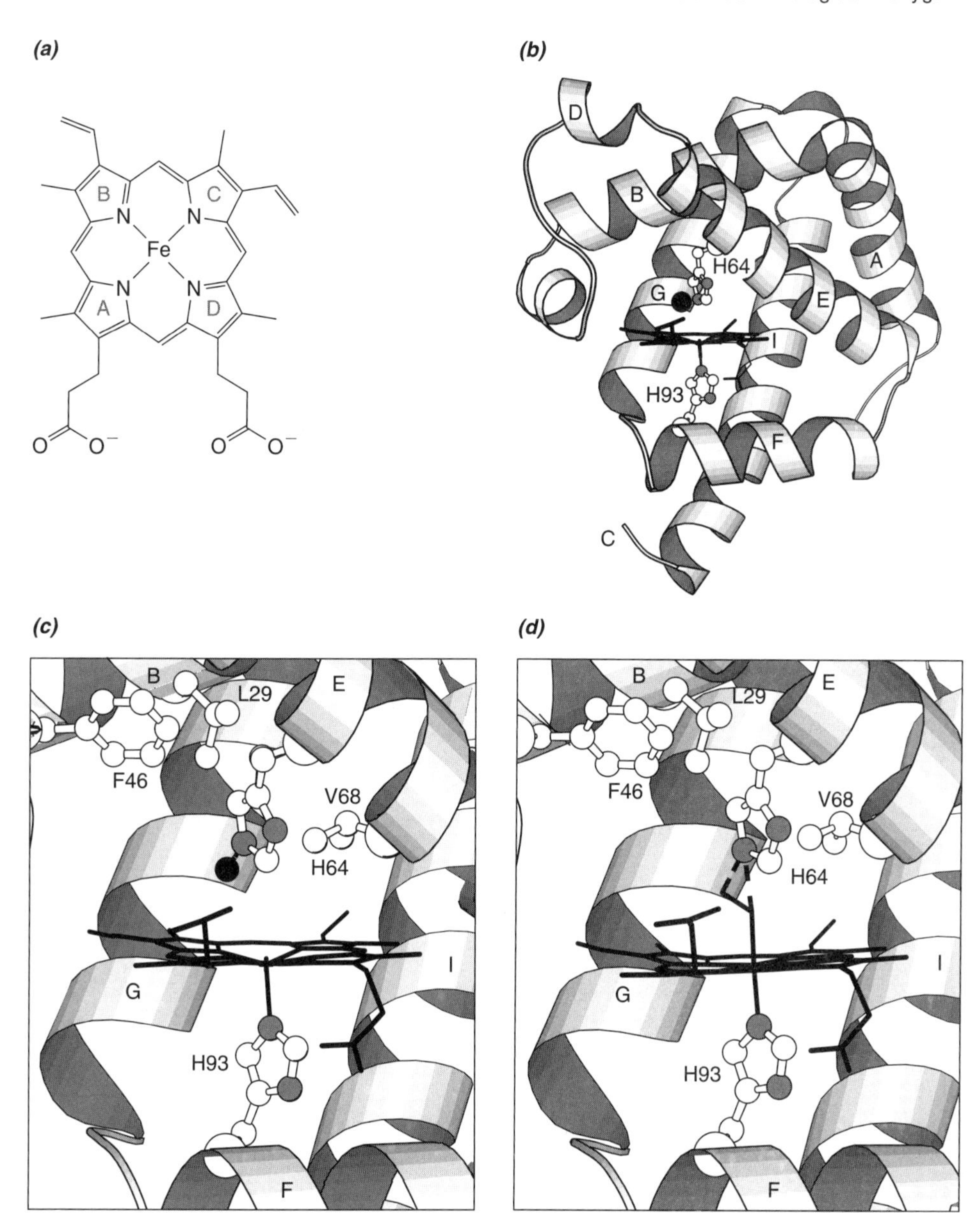

Fig. XI.4.1.
The heme group in sperm whale myoglobin. (*a*) The heme *b* (iron protoporphyrin IX) group is the active site of myoglobin. (*b*) The heme *b* group sits in a cleft formed by helices E and F. The side chains of the proximal His (H93) and the distal His (H64) are shown. For deoxyMb a water molecule is hydrogen bonded to the distal His and sits inside the ligand-binding pocket. The view is directly into the ligand-binding pocket. (*c*) Closeup of the active site of deoxyMb (PDB code: 1A6N). In addition to the distal His, a Leu (L29), Val (V68), and a Phe (F46) line the ligand-binding pocket. (*d*) Closeup of the active site of oxyMb (PDB code: 1A6M). The coordinated O_2 hydrogen bonds to the distal His. Note how the Fe atom moves into the plane of the porphyrin on binding of O_2. The distal and proximal histidines are often designated by their position on the helix, a manner that is independent of the species of Mb or Hb: The distal His is the seventh residue along helix E and is denoted E7His; analogously, the proximal His is denoted F8His.

Table XI.4.1.
General Features of Dioxygen-Carrier Proteins[a]

Metalloprotein	Active Site of Deoxy	Color Change Deoxy → Oxy	Active Site of Oxy	M_r (Da)	No. Subunits	Average Subunit M_r (Da)
HEMOGLOBINS						
Vertebrate myoglobin and hemoglobin						
Human A	Heme, Fe(II)	Purple → red	heme, $Fe^{III}–O_2^{-I}$	64,000	4	16,000
Sperm whale myoglobin	Heme, Fe(II)	Purple → red	heme, $Fe^{III}–O_2^{-I}$	17,800	1	17,800
Invertebrate erythrocruorin and chlorocruorin						
Earthworm (*Lumbricus terrestris*)	Heme, Fe(II)	Purple → red	heme, $Fe^{III}–O_2^{-I}$	$\sim 3.3 \times 10^6$	192	17,000
Eudistylia vancouveri	Chloroheme, Fe(II)	Purple → green	heme, $Fe^{III}–O_2^{-I}$	3.1×10^6	192	15,000
Truncated hemoglobins (plants, bacteria, protozoa)						
Mycobacterium tuberculosis	Heme, Fe(II)	Purple → red	heme, $Fe^{III}–O_2^{-I}$	28,800	2	14,400
Minihemoglobins (nemertean)						
Nemertean (*Cerebratualus lacteus*)	Heme, Fe(II)	Purple → red	heme, $Fe^{III}–O_2^{-I}$	45,600	4(deoxy)	11,400
HEMOCYANINS						
Arthropod (e.g., lobsters, crabs, centipedes; spiders, scorpions, horseshoe crab, but not insects)						
Crab (*Cancer magister*)	$Cu^{I} \cdots Cu^{I}$	Colorless → blue	$Cu^{I}–O_2^{-II}–Cu^{II}$	$\sim 9 \times 10^5$	12	76,600
Mollusc (e.g., octopus, bivalves, chitons, some snails)						
Edible snail (*Helix pomatia-α*)	$Cu^{I} \cdots Cu^{I}$	Colorless → blue	$Cu^{II}–O_2^{-II}–Cu^{II}$	$\sim 8.7 \times 10^6$	20	$8 \times 52,700$
HEMERYTHRINS (e.g., sipunculids, brachiopods, priapulids, some annelids)						
[*Phascolopsis* (*Golfingia*) *gouldii*]	$Fe^{II} \cdots Fe^{II}$	Colorless → burgundy	$Fe^{III} \cdots Fe^{III}–O_2(H)^{-II}$	108,000	8	13,500

[a] See Refs. 1–3, 7, and 9.

provided detailed information on the chemistry, and structural, spectroscopic, and magnetic properties of Fe porphyrin species in the absence of protein.[2,4,5,11] Indeed, by time-resolved X-ray diffraction techniques, ligand dissociation and reassociation processes have been mapped for the interaction of sperm whale Mb and CO at 1.5-Å spatial resolution and nanosecond (ns) temporal resolution.[12,13] There is accumulating evidence that many members of the Hb family have biological functions additional to, perhaps even more important than, their dioxygen-carrier function, especially with regard to metabolism of nitric oxide (NO).[6,14–16]

XI.4.2. Thermodynamic and Kinetic Aspects of Dioxygen Transport

In order for O_2 transport to be more efficient than simple diffusion through cell membranes and fluids, it is not sufficient that a metalloprotein merely binds O_2. There exists not only an optimal affinity of the carrier for O_2, but also optimal rates of O_2 binding and release. These thermodynamic and kinetic aspects are illustrated in Fig. XI.4.2, a general diagram of energy versus reaction coordinate for the (oversimplified) process, where M, an O_2 carrier (e.g., deoxyHb) binds O_2.

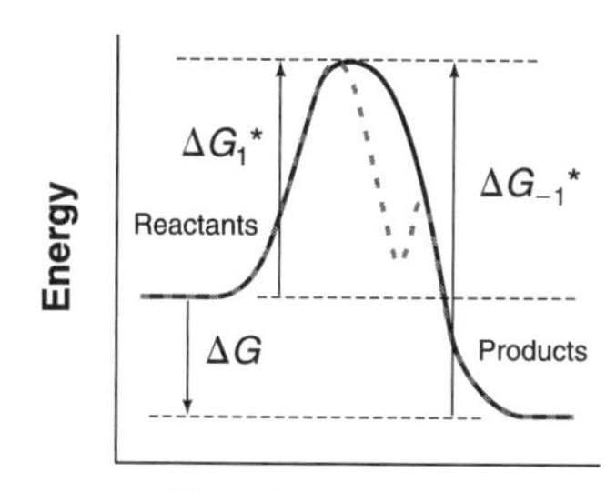

Fig. XI.4.2. Schematic diagram of free energy changes in O_2 binding. If the ligand-binding process follows the dotted line, the assumption that $K = k_1/k_{-1}$ need not hold.

$$M + O_2 \underset{k_{-1}}{\overset{k_1}{\rightleftharpoons}} MO_2 \tag{1}$$

XI.4.2.1. Thermodynamic Aspects of Dioxygen Binding

Thermodynamic or equilibrium aspects are summarized by ΔG in Fig. XI.4.2. As illustrated in Fig. XI.4.2, ΔG is negative and thus the forward reaction, O_2 binding, is spontaneous. The affinity is composed of enthalpic (ΔH) and entropic (ΔS) components with

$$\Delta G = \Delta H - T\Delta S \tag{2}$$

Within a family of O_2 carriers the values of $\Delta S^{\circ\prime}$ and $\Delta H^{\circ\prime}$ are usually similar: the superscripts denoting biological standard conditions of 298.15 K and activity of unity for all species except H^+ for which activity is 10^{-7} (i.e., pH 7). Large deviations (e.g., a change of sign) are therefore indicative of a change in the nature of the dioxygen-binding process. The equilibrium constant (K) is related to the change in free energy at biological standard conditions by

$$\Delta G^{\circ\prime} = -RT\ln(K/Q^{\circ\prime}) \tag{3}$$

where $Q^{\circ\prime}$ is unity if no protons are involved. Two reference points for $\Delta G^{\circ\prime}$ and $Q^{\circ\prime}$ exist: one where $a(O_2) = 1$ molal (crudely $[O_2] = 1\,M$), leading to K_C; the other where $p(O_2) = 1$ bar, leading to K_p. Respectively,

$$K_C = \frac{a(MO_2)}{a(M)a(O_2)} \approx \frac{[MO_2]}{[M][O_2]} \quad \text{and} \quad K_p \approx \frac{[MO_2]}{[M]p(O_2)}. \tag{4}$$

When the concentrations of MO_2 and M are equal, one-half of the dioxygen-binding sites is occupied; the partial pressure of O_2 is then denoted $p_{1/2}(O_2)$ or $p_{50}(O_2)$ and

$$p_{1/2}(O_2) = 1/K_p \tag{5}$$

Partial pressures of O_2 are often expressed in the non-SI unit of millimeters of mercury (mmHg or Torr), where 1 atm of pressure (1.013 bar) corresponds to 760 Torr. Thus, the partial pressure of O_2 in air at 1 atm is ~160 Torr. The advantage of $p_{1/2}$

or K_p compared to K_C is that $p_{1/2}$ is solvent *in*dependent, as well as being intuitively more accessible. The equilibrium constants K_C and K_p are linked by Henry's law, which relates solubility of a gas to its partial pressure:

$$[O_2] = K_H(O_2)p(O_2) \tag{6}$$

leading to

$$K_C K_H = K_p \tag{7}$$

XI.4.2.2. Kinetic Aspects of Dioxygen Binding

With regard to Fig. XI.4.2, the rate constant of the forward reaction (k_1) is related to the free energy of activation, $\Delta G_1{}^*$, and the rate constant of the reverse reaction (k_{-1}) is related to $\Delta G_{-1}{}^*$, leading to

$$K_C = k_1/k_{-1} \tag{8}$$

Measurements of the kinetics of ligand binding in solution involve the *concentrations* of components, rather than their partial pressures. Thus, to compare rate constants for ligand binding (k_1 or k_{on}) made in one solvent (e.g., water) with those made in another solvent (e.g., toluene where the solubilities of O_2 and CO are 1000 times greater) requires consideration of the different solubilities of O_2.

$$K_p = K_C K_H = (k_1/k_{-1})K_{H,H_2O} = (k_1{}'/k_{-1}{}')K_{H,\text{toluene}}{}' \tag{9}$$

leading to

$$k_1 K_H \approx k_1{}' K_H{}' \tag{10}$$

assuming reasonably that the first-order dissociation process, k_{-1} or k_{off}, is solvent independent. The reader is cautioned that different ligand solubilities have been overlooked in some attempts to establish linear free energy relationships between biological systems studied in aqueous environments and synthetic systems studied in nonaqueous solvents.

Whereas measurements of equilibrium give little or no molecular information, rather more molecular information may be inferred from kinetic data. A change in equilibrium constant may result from a change in either k_1, k_{-1}, or both. The processes of binding and release can be examined by a variety of techniques with time scales down to the nanosecond range.[12,13] The temperature behavior of the rate constants gives information on the heights of energy barriers that are encountered as O_2 molecules arrive at or depart from the binding site. The quantitative interpretation of kinetic data generally requires a molecular model of some sort.

XI.4.3. Cooperativity and Dioxygen Transport

Many O_2 carriers exist not as monomers, but as oligomers made up of two or more similar or identical subunits. The subunits may be held together by van der Waals' forces or by stronger interactions, such as hydrogen bonding or salt bridges, or even by covalent bonds, as in molluscan Hc and the Hb from the perienteric fluid of the parasite *Ascaris suum*. In a process known as cooperativity, the binding of O_2 to one subunit affects the binding to remaining subunits. As will be seen, cooperative ligand binding confers considerable physiological advantage to an organism. But first we need to analyze noncooperative binding of O_2.

XI.4.3.1. Noncooperative Binding of Dioxygen

If the dioxygen-binding sites M are mutually independent and noninteracting, then a plot of the fractional saturation of dioxygen-binding sites, θ, where

$$\theta = \frac{[MO_2]}{[M] + [MO_2]} = \frac{K_p p(O_2)}{1 + K_p p(O_2)} \tag{11}$$

versus $p(O_2)$ gives the hyperbolic curve labeled "noncooperative" in Fig. XI.4.3*a*. Alternatively, a so-called Hill plot,[17]

$$\log(\theta/(1 - \theta)) = n \log p(O_2) - \log p_{1/2}(O_2) \tag{12}$$

gives a line with a slope of unity ($n = 1$) and intercept of $-\log p_{1/2}(O_2)$ (Fig. XI.4.3*b*) for noncooperative binding of O_2.

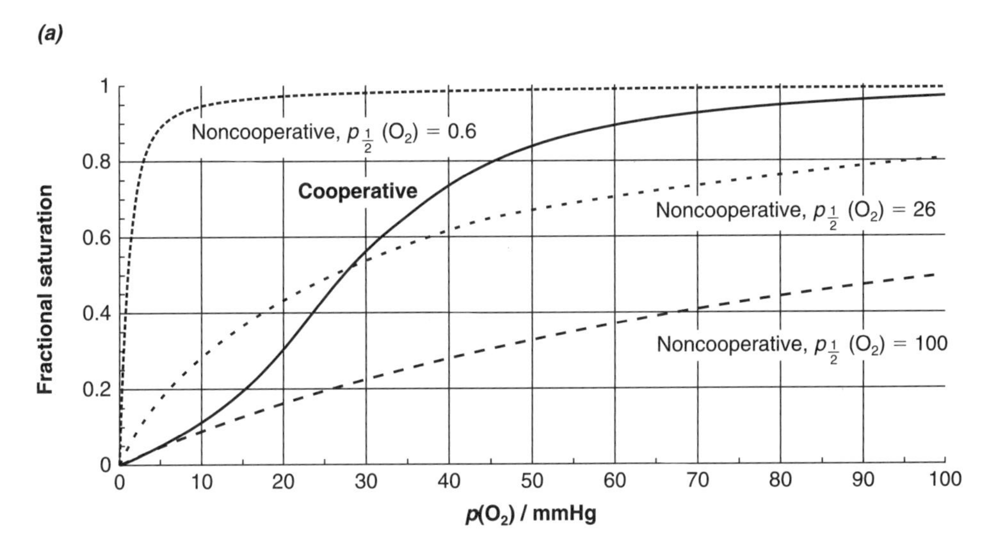

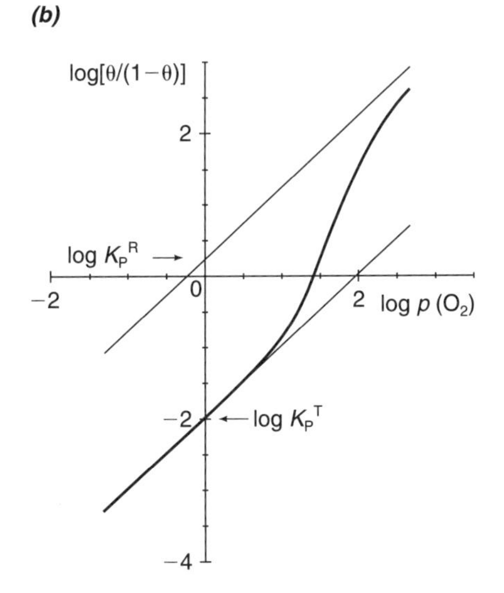

Fig. XI.4.3. Noncooperative and cooperative binding of O_2, illustrating the advantages of cooperative binding. The curve for cooperative binding is calculated using $K^T = 1/100\,\text{mmHg}^{-1}$, $K^R = 1/0.6\,\text{mmHg}^{-1}$, and $L_0 = 2 \times 10^{-9}$ (see Fig. XI.4.4 and text for definition of these parameters). These values of the allosteric constants give a curve very similar to that observed for O_2 binding to HbA. (*a*) Fractional saturation of the binding sites versus partial pressure of O_2. (*b*) Hill plot of ligand binding. Noncooperative binding leads to straight lines with a slope of unity. The vertical separation between the two extrapolated lines of the limiting behavior of O_2 binding is related to the free energy of cooperativity, $\Delta(\Delta G)/2.303RT = \log(K_p{}^R) - \log(K_p{}^T)$.

XI.4.3.2. Cooperative Binding of Dioxygen

Most mammalian Hb is tetrameric, comprising two pairs $[\alpha\beta]_2$ of Mb-like subunits denoted as α and β. The binding or release of O_2 at one site *may* affect the affinity and kinetics of O_2 binding and release at a neighboring site. As a result, the binding curve may become sigmoidal in shape, as illustrated by the curve labeled "cooperative" in Fig. XI.4.3*a*. When cooperativity is positive, the affinity of a vacant site is apparently increased by occupancy of an adjacent one.[18] This behavior, where the binding of one molecule influences the binding of successive molecules *of the same kind,* is referred to as a *homotropic allosteric* interaction. A *heterotropic allosteric* interaction occurs when the interaction with the protein of a second *unlike* molecule, for example, an organic polyphosphate for human Hb (HbA), influences the binding of the first molecule (e.g., O_2). Such molecules are termed *allosteric effectors.* Another commonly observed heterotropic allosteric interaction is the Bohr effect.[19] This effect relates the change in partial pressure of O_2 to a change in pH at constant saturation of binding sites (θ). Generally, O_2 binding is accompanied by the release of protons. A very large Bohr effect, where O_2 affinity decreases sharply with decreasing pH, is often called the Root effect.[20] It is physiologically important for fish such as trout in maintaining buoyancy.

The degree of cooperativity can be characterized in a number of ways. By means of a Hill plot (Eq. 12), the limiting slopes (which should be unity) at high O_2 pressure and low O_2 pressure are extrapolated, as shown in Fig. XI.4.3*b* to $\log[\theta/(1-\theta)] = 0$, where $\theta = 0.5$. Two limiting values for $p(O_2)$ are obtained: one at high $p(O_2)$, where the O_2 affinity is high (for the case illustrated of positive cooperativity); the other at low $p(O_2)$, where affinity is relatively low. This difference in affinities can be converted into a difference between the free energy change upon O_2 binding to the low-affinity state (K_p^T) and to the high-affinity state (K_p^R): The designations T and R will be described shortly.

$$\Delta\Delta G^\circ = -RT \ln(K_p^T/K_p^R) \tag{13}$$

A second way to characterize cooperativity involves fitting the dioxygen-binding data at intermediate saturations ($0.2 < \theta < 0.8$) to Eq. (12). A value for the Hill coefficient (n) greater than unity indicates positive cooperativity. If O_2 binding is an all or nothing affair, where dioxygen-binding sites are either all occupied or all vacant, the value of n equals the number of subunits in the molecule. For tetrameric Hb, a maximum Hill coefficient of ~ 3.0 is found, whereas for Hc n may be as high as 9 for 20-mer assemblies. These values, like $p_{1/2}(O_2)$ values, are sensitive to the nature and concentration of allosteric effectors. Weak or zero cooperativity in an oligomeric O_2 carrier may simply result from the absence of the appropriate allosteric effector.

XI.4.3.3. Physiological Benefits of Cooperative Binding of Dioxygen

Oligomerization and cooperative binding confer enormous physiological benefits to an organism. Dioxygen carriers either form small oligomers that are encapsulated into cells or erythrocytes (such Hb is referred to as intracellular) or associate into large oligomers of 100 or more subunits (such Hb are referred to as extracellular). This encapsulation and association reduce by orders of magnitude the number of independent particles in the blood, with consequent reductions in the osmotic pressure of the solution and in strain on vascular membranes.

The second benefit derives from cooperative binding of ligands and the abilities of heterotropic allosteric effectors to optimize exquisitely the dioxygen-binding behavior in response to the external and internal environment. The situation is illustrated in general terms in Fig. XI.4.3. Noncooperative O_2 binding to Hb with an affinity corresponding to the low-affinity state [${p_{1/2}}^T(O_2)$ or K_p^T] leads to a low fraction of subunits carrying O_2 even when O_2 availability is high, such as in the lungs, where at sea level the effective $p(O_2)$ is ~100 Torr. Noncooperative O_2 binding to Hb with an affinity corresponding to the high-affinity state [$p_{1/2}{}^R(O_2)$ or K_p^R] leads to limited release of O_2 in tissues where O_2 demand is high and where the concentration of O_2 corresponds to $p(O_2)$ of ~35–40 Torr. On the other hand, cooperative binding of O_2 to Hb, where $p_{1/2}{}^R(O_2) = 0.5$ Torr and $p_{1/2}{}^T(O_2) = 100$ Torr [giving overall $p_{1/2}(O_2) \approx 26$ Torr] leads to higher saturation of the dioxygen-binding sites at high $p(O_2)$ in the lungs and greater release of O_2 at low $p(O_2)$ in tissues. The net result is that whole blood, which contains ~15 g HbA/100 mL, can carry the equivalent of 20 mL O_2 (at 760 Torr)/100 mL, whereas blood plasma (no Hb) has a carrying capacity of only 0.3 mL O_2/100 mL.

XI.4.3.4. The Monod–Wyman–Changeux Model for Cooperativity

A simple model for analyzing cooperative ligand binding was proposed by Monod, Wyman, and Changeux in 1965 and is usually referred to as the MWC two-state concerted model.[21] Molecules are assumed to be in equilibrium between two conformations or quaternary structures, one that has a low ligand affinity and a second that has a high ligand affinity. The low-affinity conformation is often designated the T or tense state and the high-affinity conformation the R or relaxed state. The equilibrium between the two conformations is characterized by the allosteric constant

$$L_0 = [R_0]/[T_0] \tag{14}$$

where the subscript zero denotes the unliganded R and T states. The free energy change upon binding a ligand to the R state, irrespective of saturation, is assumed to be a constant and the associated equilibrium constant is designated K^R; a third constant K^T characterizes binding to the T state. Fig. XI.4.4 illustrates this model, and introduces the terminology conventionally used. To a reasonable approximation, the cooperative binding of O_2 can be summarized by these three parameters, L_0, K^R, and K^T. For HbA, the switching between R and T quaternary conformations is moderately rapid (at $4 \times 10^3\ s^{-1}$), and hence physiologically beneficial.

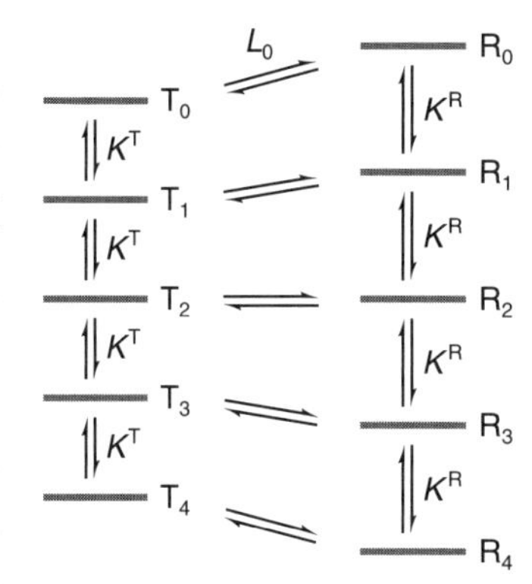

Fig. XI.4.4.
The MWC two-state model for cooperative ligand binding, showing the free energy relationships between the R and T states. The subscript denotes the number of O_2 molecules bound.

Human Hb are a heterogeneous group and many naturally occurring mutants are known, as well as a host of site-directed mutants. Perutz,[22] through the determination of the crystal structures of a variety of Hb derivatives, has given a sound structural basis to the MWC model of two basic quaternary states. Whereas there is no *a priori* reason for O_2 binding to Hb to follow the MWC model, this model has withstood many challenges.[23–28] As more structural data have accumulated, considerable conformational plasticity within the two basic quaternary structures is evident. Although a given crystal structure usually belongs to the ensemble of conformations of lowest energy in the solution, the pattern of hydrogen bonds and frozen conformations gives an enthalpic bias to relating structural data to kinetic and thermodynamic data. Entropic contributions to thermodynamic ΔG, latent in isolated X-ray structures, but explicit in NMR data, are often overlooked. Time-resolved X-ray crystallographic techniques give access to entropic contributions to the activation free energy ΔG^*.[12,13]

XI.4.4. Biological Dioxygen Carriers

XI.4.4.1. The Hemoglobin Family

Hemoglobins are the most evolutionarily diverse family of O_2 carriers.[1–3,29,30] Hemoglobins are found in bacteria (often as one domain of a multidomain multifunctional protein), plants (e.g., symbiont Hb found in the dinitrogen-fixing nodules of legumes, as well as non-symbiont Hb), invertebrates (including some insect larvae), crustaceans, molluscs (especially bivalves and snails), almost all annelid worms, and *all* vertebrates (with the exception of some fish species that are restricted to Antarctic waters and whose Hb genes no longer code for functional protein; the sluggish lifestyle of these species is satisfied by the enhanced solubility of O_2 at subzero temperatures).

Monomeric and oligomeric Hb all share a basically similar building block in which a single heme group is embedded in an almost entirely α-helical polypeptide with a molecular weight of ~16,000 Da, corresponding to ~150 amino acids (see Fig. XI.4.1). Both longer (190 residue) and shorter (109–130 residue) variations are known, the latter being designated *truncated hemoglobins*[30] in which two of the helices are largely replaced by extended strands to give a two-over-two helical sandwich in place of the canonical three-over-three sandwich illustrated in Fig. XI.2.1*b*. The heme group is anchored to the protein by coordination of the Fe center to an imidazole ligand from an absolutely conserved His residue—the *proximal histidine,* the fourth residue on helix F (denoted F4). The heme group is oriented by a host of van der Waals' interactions with aromatic and nonpolar residues. On the side of the heme opposite the proximal His, *distal residues* control access of molecules, such as O_2, CO, and NO, into the ligand-binding pocket. Sperm whale Mb is often taken as the archetypal Mb (see Fig. XI.4.1 and Table XI.4.1). The archetypal oligomeric Hb that shows cooperative binding of O_2 is the tetrameric human adult Hb, HbA. In HbA, the four heme groups are mutually separated by >25 Å. In the Hb of some high-flying birds, deoxyHb is associated as octamers that separate into tetramers on ligand binding, giving enhanced cooperativity and O_2 delivery. In some invertebrate Hb,[3] especially those of annelids, aggregates may contain as many as 192 binding sites to give a molecular weight of $\sim 3 \times 10^6$ Da. These and other high molecular weight Hb of arthropods are referred to as erythrocruorins (Er). A few annelid worms use chlorocruorins, which turn green upon oxygenation (*chloros,* Greek for green). Here, the otherwise ubiquitous heme *b* is replaced by chlorocruorin, in which one of the vinyl groups ($-CH{=}CH_2$) of heme *b* is replaced by a formyl group (–CHO). Some organisms, for example, the clam *Scapharca inaequivalvis,* feature a cooperative homodimeric Hb, where in contrast to vertebrate Hb the hemes are in close contact.

Kinetic and thermodynamic data for O_2 binding and release from a variety of Hb are summarized in Table XI.4.2. Notice that for HbA, which is composed of two pairs of slightly dissimilar subunits, the α and β chains bind O_2 with slightly different affinities and rate constants. Isolated chains of HbA behave similarly to monomeric vertebrate Mb, such as sperm whale Mb, and to R-state HbA. The chlorocruorins have a low affinity compared to other erythrocruorins. For proteins that bind O_2 cooperatively, ligand affinities and rate constants are sensitive to pH, ionic strength, water activity, specific anions and cations (allosteric effectors), and laboratory.

Where measured, the enthalpy change (ΔH) for the binding of O_2 is usually in the range -50 to $-75\,kJ\,mol^{-1}$ O_2 and roughly follows affinity. The entropy change (ΔS, reference state $[O_2] = 1\,M$) is about -80 to $-145\,J\,K^{-1}\,mol^{-1}$, consistent with loss of translational and rotational degrees of freedom of O_2 upon binding. For chlorocruorins, the enthalpy change is substantially lower at about $-19\,kJ\,mol^{-1}$ O_2, requiring that the entropy change be slightly positive. The intermediate steps of O_2 binding to HbA also appear to involve a positive entropy change, perhaps resulting from the release of protein-bound protons and water to the bulk solvent.[26–28]

Some nematode and trematode Hb have extraordinarily high affinities for O_2 compared to those typically found for vertebrate Hr and Mb; they also display NO–dioxygenase activity. The very high affinity for O_2 of the perienteric Hb from the nematode *A. suum* derives in large part from an extremely slow rate of O_2

Table XI.4.2.
Kinetics and Thermodynamics of Dioxygen Binding to Hemoglobins and Their Models[a]

(A) THERMODYNAMICS AND KINETICS OF O_2 AND CO BINDING TO FIVE-COORDINATE MONOMERIC HEMOGLOBINS[b,c]

Carrier (PDB code: ID)	Notable Feature	Dioxygen Binding				Carbon Monoxide Binding				Discrimination
		$p_{1/2}(O_2)$ (Torr)	k_{on} ($\mu M^{-1}s^{-1}$)	k_{off} (s^{-1})	K (μM^{-1})	$p_{1/2}(CO)$ (Torr)	k_{on} ($\mu M^{-1}s^{-1}$)	k_{off} (s^{-1})	K (μM^{-1})	$p_{1/2}(O_2)/p_{1/2}(CO)$
Sperm whale (SW) Mb (1A6N)	Standard Mb for comparison	0.51	14.(3)	12.(2)	1.2(3)	0.018	0.51(6)	0.019(5)	27.(8)	28.
SW Mb$^{E7His\rightarrow Gly}$ (2MGA)	Distal His → Gly; H_2O in pocket	(6.1)	140.	1600.	0.088	(0.0048)	5.8	0.038	150.	(1300.)
SW Mb$^{E7His\rightarrow Ala}$ (102M)	Distal His → Ala; no H_2O in pocket	(23.)	53.	2300.	0.023	(0.011)	4.2	0.061	69.	(2100.)
SW Mb$^{E7His\rightarrow Leu}$ (2MGC)	Distal His → Leu; no H_2O in pocket	(22.)	98.	**4100.**	0.023	(0.00068)	26.	0.024	1100.	**(32,000.)**
SW Mb$^{B10Leu\rightarrow Tyr;\ E7His\rightarrow Gln;\ E10Thr\rightarrow Arg}$ (1F65)	Same distal residues as nematodes	(0.28)	1.9	1.0	1.9	(0.22)	0.047	0.014	3.4	(1.3)
Hb *Ascaris* (perienteric fluid) (1ASH)	Very small k_{off}; high O_2 affinity	**0.0047**	**1.5**	**0.0041**	370.	0.063	0.17	0.018	9.4	0.07
Hb *Ascaris* (D1B10Tyr → Leu)	Loss of distal $O_2\cdots$Tyr H bond	0.82	9.0	5.0	1.8	0.75	~7.			
Hb *Paramphistomum epiclitum* (1H97)	Highest recorded O_2 affinity	(0.00016)	108.	0.033	**3300.**	28.	~0.001			
Leg Hb (symbiont Hb of soybean) (1GDJ)	Very large k_{on} leads to high O_2 affinity	0.047	150.	11.	14.	**0.00074**	14.	0.012	1100.	64.
Leg Hb$^{E7His\rightarrow Leu}$	Very high CO affinity	(0.032)	**400.**	24.	17.	(0.000010)	**170.**	0.0024	**72,000**	(3200.)
Hb *Vitreoscilla* (1VHB)	Bacterial Hb	17.	78.	**5600.**	**0.014**	0.17	**0.0007**	**0.16**	0.0044	100.
Chironomus Mb (1ECD)	Insect Mb (erythrocruorin)	0.40	300.	218.	1.4	0.0019	0.27	0.095	2.8	210.
Glycera dibranchiata Hb II (1HBG)	Marine blood worm; no distal His	5.2	190.	1800.	0.103	0.00089	27.	0.042	640.	5800.
Human neuroglobin (hNgb) (1OJ6)	Distal His coord.; sq. brackets denote coordination of O_2 or CO to six-coordinated species	1.	250.	0.8	310. [0.70]	(0.00016)	65.	0.014	4600. [10.]	(6300.)
hNgb$^{E7His\rightarrow Leu}$	Loss of His increases k_{off}	(68.)	700.	200.	3.5		200.			
HbA (isolated α chain)	Similar to Mb	0.74	50.(8)	28.(3.)	1.8	0.0025	4.0(6)	0.013(2)	310.	(300.)
HbA (isolated β chain)		0.42	60.(10)	18.(2.)	3.3	0.0016	4.5(7)	0.008(1)	560.	(260.)

Table XI.4.2.
(Continued)

(B) THERMODYNAMICS AND KINETICS OF O_2 AND CO BINDING TO MULTIMERIC HEMOGLOBINS[d,e]

Carrier (PDB code: ID)	Notable Feature	Dioxygen Binding				Carbon Monoxide Binding				Discrimination
		$p_{1/2}(O_2)$ (Torr)	k_{on} ($\mu M^{-1}s^{-1}$)	k_{off} (s^{-1})	K (μM^{-1})	$p_{1/2}(CO)$ (Torr)	k_{on} ($\mu M^{-1}s^{-1}$)	k_{off} (s^{-1})	K (μM^{-1})	$p_{1/2}(O_2)/p_{1/2}(CO)$
HbA R (α chain)	High-affinity form of Hb	0.15–1.5[f]	28.(9)	12.(3)	2.3(9)	0.001	2.9(5)	0.0046(15)	630.(240)	(190.)
HbA R (β chain)			100.(13)	22.(8)	4.5(17)	0.004	7.1(24)	0.0072(28)	990.(550)	(170.)
HbA T (α chain)	Low-affinity form of Hb	9–160[f]	2.9	183.	0.016	0.10–2.8[f]	0.099[f]	0.09[f]		
HbA T (β chain)			11.8	2500.	0.0047					

(C) THERMODYNAMICS AND KINETICS OF O_2 AND CO BINDING TO SELECTED SYNTHETIC SYSTEMS[g,h]

Carrier (See Fig. XI.4.7 for ligand sketches)	Notable Feature	Dioxygen Binding				Carbon Monoxide Binding				Discrimination
		$p_{1/2}(O_2)$ (Torr)	k_{on} ($\mu M^{-1}s^{-1}$)	k_{off} (s^{-1})	K (μM^{-1})	$p_{1/2}(CO)$ (Torr)	k_{on} ($\mu M^{-1}s^{-1}$)	k_{off} (s^{-1})	K (μM^{-1})	$p_{1/2}(O_2)/p_{1/2}(CO)$
PICKET FENCE PORPHYRIN DERIVATIVES										
Fe(PF-Im)	Similar affinity to Mb; larger k values	0.58	430.	2900.	0.15	(0.000022)	36.	0.0078	4600.	26,000.
Fe(PF)(1,2-Me_2Im)	Bulky base reduces ligand affinity	38.	106.	46,000.	0.0023	0.0089	1.4	0.14	10.	4300.
Fe(C_6Strap-PF)(1-MeIm)	Two pickets buttress strap	0.70	2.2	2.0	1.2	0.0097	0.080	0.082	9.8	7.
Fe(Poc-PF)(1-MeIm)	Discrim. to CO; small k values	0.36	2.2	9.	0.24	0.0015	0.58	0.0086	67.	240.
Fe(Poc-PF)(1,2-Me_2Im)	Discrim. to CO; small k values	13.	1.9	280.	0.0068	0.026	0.098	0.055	1.8	220.
Fe(TPh_3PP)(1,2-Me_2Im)	Apolar binding pocket	508.				0.0091				56,000.
Fe(Bis-neoPePF)(1,2-Me_2Im)	Ester links; no NH···O_2 interaction	230.	45.	79,000.	0.00057	0.017	0.90	0.15	6.0	14,000.
Fe(C_2Cap)(1-MeIm)	Unfavorable O_2···ester contacts	23.				0.0054	0.95	0.05	19.	4300.
Fe(Cyclam-Cap)(1,5DCIm)	Very strong discrim. against CO	~3.				>3500.				<0.003
Fe(PhU-PF)(1,2-Me_2Im)	Phenylurea picket; O_2···HN??	(2.6)	38.	1000.	0.035		132.			

(C) THERMODYNAMICS AND KINETICS OF O_2 AND CO BINDING TO SELECTED SYNTHETIC SYSTEMS[g,h]

Carrier (See Fig. XI.4.7 for ligand sketches)	Notable Feature	Dioxygen Binding				Carbon Monoxide Binding				Discrimination
		$p_{1/2}(O_2)$ (Torr)	k_{on} ($\mu M^{-1}s^{-1}$)	k_{off} (s^{-1})	K (μM^{-1})	$p_{1/2}(CO)$ (Torr)	k_{on} ($\mu M^{-1}s^{-1}$)	k_{off} (s^{-1})	K (μM^{-1})	$p_{1/2}(O_2)/p_{1/2}(CO)$
BIS-STRAPPED PORPHYRINS										
Fe(Amide-Strap-Py)	$O_2\cdots$H-N interactions	2.0	360.	5000.	0.072	0.00009	35.	0.03	1200.	22,000.
Fe(Ether-Strap-Py)	$O_2\cdots$ether affects $k_{off}(O_2)$ only	18.	300.	40,000	0.0075	0.0001	68	0.069	1000.	180,000.
DENDRITIC PORPHYRINS										
Fe(G1)(1,2-Me_2Im)	Polyether pickets (1° dendrite)	0.035				0.35				0.10
Fe(G2)(1,2-Me_2Im)	Polyether pickets (2° dendrite)	0.016				0.19				0.08
FLAT OPEN PORPHYRINS										
Fe(PPIX-Im)	No polar moiety near O_2	5.6	62.	4200.	0.016	0.00025	11.	0.025	440.	22,000.
FLAT OPEN PORPHYRINS; H_2O, ALKYLAMMONIUM MICELLES, pH 7.3										
Fe(PPIX-Im)	Polar H_2O moieties near O_2	1.0	26.	4.7	5.5	0.002	3.6	0.009	400.	500.

[a] Values in parentheses are derived by means of the relationships: $p(X)\ K_H(X) = 1/K(X)$, where X = CO, O_2; $M = p_{1/2}(O_2)/p_{1/2}(CO) = K(CO)K_H(CO)/K(O_2)\ K_H(O_2)$; $K = k_{on}/k_{off}$. Solubility of O_2 in water: $K_H(O_2, H_2O) = 1.86 \times 10^{-6}\ M\,Torr^{-1}$; solubility of O_2 in toluene: $K_H(O_2, \text{toluene}) = 1.02 \times 10^{-5}\ M\,Torr^{-1}$; solubility in benzene very similar. Solubility of CO in water: $K_H(CO, H_2O) = 1.36 \times 10^{-6}\ M\,Torr^{-1}$; solubility of CO in toluene: $K_H(CO, \text{toluene}) = 1.05 \times 10^{-5}\ M\,Torr^{-1}$.

[b] Taken from Refs. 1–3, 31–33. Values in boldface are unusually high and low values with respect to the reference of wild-type sperm whale myoglobin.

[c] At 20–25°C and buffered at pH 6.5–8.5.

[d] Taken from Ref. 2.

[e] At 20–25°C, buffered at pH 6.5–8.5, and without allosteric effectors. Values for $p_{1/2}(O_2)$ show the effects of allosteric effectors in lowering affinity.

[f] Average value of α and β chains.

[g] Taken from Refs. 2, 4, 5, and 11.

[h] At 20–25°C, benzene or toluene, except where stated. The porphyrins and their abbreviations are illustrated in Fig. XI.4.7.

release. The Hb from the trematode *P. epiclitum* has an affinity for O_2 that is yet another order of magnitude higher. Another carrier, LegHb, has a high affinity for O_2, largely because of a very high rate of O_2 binding that is close to the diffusion limit. LegHb is found in the root nodules of legumes, which have nitrogen-fixing bacteria (Section XII.3). Dioxygen is a poison for the nitrogenase enzyme, but the nodules also require O_2. The high affinity and rapid ligand binding of LegHb facilitate diffusion of O_2, lowering to nanomolar levels the concentration of free O_2 in the vicinity of dinitrogen-fixing sites.

Carbon monoxide (CO) is present in cells, where it has important neuronal and cell-signaling functions, at a concentration of $\sim 1\ \mu M$ as a result of heme catabolism. In general, the CO affinity of a dioxygen-binding protein is much greater than the corresponding O_2 affinity, leading to physiological stress in environments where moderate concentrations of CO are present. Given the physiological importance of CO, selected data for CO binding are also presented in Table XI.4.2. Other σ-donor–π-acceptor molecules, such as NO, alkyl isocyanides (R–NC), and alkyl nitroso species (R–NO) also bind strongly, in the case of NO even more strongly than O_2 and CO.

Nitric oxide is also present in cells, where, in vertebrates, it plays important roles in controlling vasculature and, through macrophage-generated reactive-nitrogen species, in biological defenses, against which microorganisms in turn have developed counterattacks. Whereas the reactive cysteine of HbA forms thionitroso species, the metal center mediates NO–dioxygenase activity:

$$Fe^{II}\text{–}O_2 + NO \rightarrow Fe^{III} + NO_3^- \quad \text{or} \quad Fe^{II}\text{–}NO + O_2 \rightarrow Fe^{III} + NO_3^- \qquad (15)$$

In addition to a wealth of kinetic and thermodynamic data on ligand binding, the Hb family is very well characterized structurally. High-resolution structures are now available for Hb from many species and for a host of recombinantly produced wild-type and mutant Mb of, in particular, sperm whale, pig, and human, in deoxy, oxy, carbonmonoxy, and met forms. Recombinant DNA techniques have obviated any ethical problems associated with the requirement for muscle tissue, making, for example, the Mb from *Homo sapiens* more readily available. In addition, hybrid Hb have been prepared, in which Co^{II}, Mn^{II}, Mn^{III}, Ni^{II}, Zn^{II}, and Mg^{II} hemes are substituted for the native (Fe^{II}) heme of the α or β subunits of HbA.

Dioxygen binds to Fe of the heme in an end-on manner with an Fe–O–O angle of ~125° (Fig. XI.4.1*d*).[34,35] There is considerable transfer of electron density from the Fe to the O_2 moiety, justifying a formal description of the Fe–O_2 moiety as a ferric superoxo species, $Fe^{III}\text{–}O_2^{I-}$. The Co(II) hybrid Hb also binds O_2, but less well than Fe(II) and with greater transfer of charge than for the comparable Fe species. The $Co^{III}\text{–}O_2^{I-}$ species is low-spin $(S = \frac{1}{2})$ and amenable to electron paramagnetic resonance (EPR) spectroscopy. The strong hydrogen bond between the coordinated O_2 and the H–NE2 atom of the distal histidine is physiologically important and is discussed in depth in Section XI.4.6, along with mechanisms that lead to O_2 affinities for physiologically competent O_2 carriers spanning more than six orders of magnitude. Ligand binding and structural data on the biological systems are complemented by corresponding data from a host of model systems, a selection of which is included in Table XI.4.2.

XI.4.4.2. The Hemocyanin Family

Hemocyanins, the Cu-containing O_2 carriers, are distributed erratically in two large phyla, *Mollusca* (e.g., octopus and snails) and *Arthropoda* (e.g., lobsters and scorpions).[7] The functional form of Hc consists of large assemblies of subunits, summarized in Table XI.4.1. In the mollusc family, the subunit comprises eight (occasionally seven) covalently linked domains (often called functional subunits). Each domain has a molecular weight of ~52,000 Da and contains a pair of Cu atoms. Electron microscopy reveals that molluscan Hc molecules are cylindrical assemblies ~190 or 380 Å long and 350 Å in diameter, made up of 10 or 20 subunits, respectively, leading to molecular weights as high as $\sim 9 \times 10^6$ Da. In the arthropod family, individual subunits have a molecular weight of ~70,000 Da and contain two Cu atoms. These subunits form into hexamers, which then form higher order oligomers containing 6, 12, 24, 36, or 48 subunits. Not all subunits are identical. Upon oxygenation, the colorless protein becomes blue (hence cyanin from *cyanos,* Greek for blue). Spectral changes upon oxygenation, oxygen affinities, and kinetics of oxygen binding (Table XI.4.3), anion binding, other chemical reactions, and the tertiary structures of a domain or subunit show that the active sites of Hc from the phyla *Arthropoda* and *Mollusca* are not identical.

No monomeric Hc, analogous to Mb and myoHr (Section XI.4.4.3), are known. For some Hc, the binding of O_2 is highly cooperative, if Ca^{2+} or Mg^{2+} ions are present, with Hill coefficients as high as $n \sim 9$. However, the free energy of interaction per subunit can be small in comparison with that for tetrameric Hb (−16–46 kJ $(mol{\cdot}O_2)^{-1}$) . Allosteric effects, at least for a 24-subunit tarantula Hc, can be separated into those within a dodecamer (12 subunits)—the major contributor to overall allostery—and those between dodecamers. This phenomenon has been termed *nested allostery*. In contrast to the Hb family, isolated chains of arthropod Hc have affinities typical of the T-state conformation for Hc. The binding of CO, which binds to only one Cu, is at best weakly cooperative. The overall enthalpy and entropy changes for O_2 binding to Hc are variable, in part owing to the wide range of O_2 availability in the environments in which marine animals that use Hc live. Overall enthalpy changes for O_2 binding range from about +13 to −67 kJ $(mol{\cdot}O_2)^{-1}$; concomitantly, entropy changes range from substantially positive (+125 J K $(mol{\cdot}O_2)^{-1}$) to negative (−125 J K $(mol{\cdot}O_2)^{-1}$). For Hc, preassembly of the dioxygen-binding site offsets the very high entropic cost associated with a highly ordered mode of O_2 binding between the two Cu atoms. With so little information available on the three-dimensional structures of Hc at near-atomic resolution, there is little understanding

Table XI.4.3.
Thermodynamics and Kinetics of O_2 Binding to Hemocyanins and Hemerythrins[a,b]

Carrier	Notable Feature	$p_{1/2}(O_2)$ (Torr)	k_{on} ($\mu M^{-1}s^{-1}$)	k_{off} (s^{-1})	K (μM^{-1})
MOLLUSCAN Hc					
Helix pomatia R	Common snail; complex Hc	2.7	3.8	10.	0.38
Helix pomatia T		55.	1.3	300.	0.0043
ARTHROPOD Hc					
Panulirus interruptus R	Spiny lobster, hexameric Hc	1.0	31.	60.	0.52
P. interruptus monomer		9.3	57.	100.	0.57
HEMERYTHRINS					
Phascolopsis gouldii	Octamer	2.0	7.4	56.	0.13
Themiste zostericola	Octamer; noncooperative	6.0	7.5	82.	0.092
Themiste zostericola	Monomer	2.2	78.	320.	0.25

[a] At 20–25°C and buffered at pH 6.5–8.5.

[b] Taken from Refs. 2 and 7.

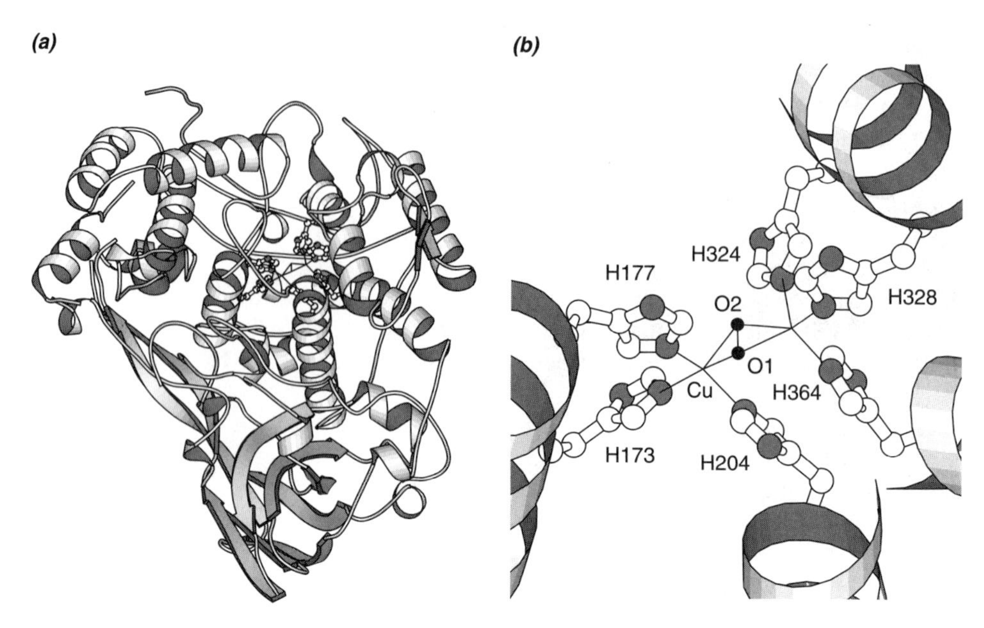

Fig. XI.4.5. Oxyhemocyanin of the arthropod horse crab, *L. polyphemus* [PDB code: 1NOL (oxy), 1LLA (deoxy)]. (*a*) One subunit with the side chains of the coordinated His residues shown. (*b*) Closeup of the active site showing the manner by which O_2 is bound by the pair of Cu atoms.

yet of the mechanisms by which these giant molecular assemblies tune their affinity for O_2.

The structure of the active site has been extensively probed by extended X-ray absorption fine structure (EXAFS) and X-ray absorption near-edge spectroscopy (XANES) methods and the X-ray crystal structures of representative Hc from molluscs and arthropods have been reported.[8,36] Despite different secondary and tertiary structures, each Cu atom is coordinated to three imidazole (Im) functions from His residues in a trigonal-pyramidal manner. The domains containing the active sites are shown in Fig. XI.4.5 for Hc from the arthropod *Limulus polyphemus*. The Cu(I) state of deoxyHc has electronic configuration d^{10}, and this diamagnetic Cu(I) species is colorless. On binding O_2, the dicopper(I) moiety is effectively oxidized to dicopper(II), concomitant with the development of an intense blue color. Although a blue color for a Cu(II) complex is not unexpected, its intensity is unusual and derives from a peroxo-to-Cu(II) charge-transfer transition. Laser excitation into this transition affords a resonance Raman spectrum that shows the (O–O) stretching mode at $750\,cm^{-1}$, which is an unusually low energy when compared to the $880\,cm^{-1}$ observed for H_2O_2. (For further discussion of this low O–O stretching frequency, see Section XI.5.1.) The two Cu(II) ions in this adduct are so strongly coupled ($-J > 600\,cm^{-1}$) that at room temperature and below the system is effectively diamagnetic and oxyHc is EPR silent. The properties of oxyHc are associated with a distinctive side-on bridging (μ-η^2:η^2-peroxo) mode of O_2 binding to the dicopper active site of Hc. This unique structure was finally established in 1992 by means of a model system that very closely reproduced the highly distinctive spectroscopic features of oxyHc.[37] The crystal structure of oxyHc itself was solved shortly thereafter.

XI.4.4.3. The Hemerythrin Family

The biological occurrence of Hr, the third class of O_2 carriers, is relatively rare, being restricted to the sipunculid family (nonsegmented worms) and selected members of the annelid (segmented worm), brachiopod (shrimps), and priapulid families.[9] The dioxygen-binding site contains, like Hc, a pair of metal atoms, in this case Fe. Upon oxygenation, the colorless protein becomes purple-red. Monomeric (myoHr), trimeric, and octameric forms of Hr are known; all appear to be based upon a similar subunit of ~13,500 Da. When Hr is extracted from the organism, its O_2 binding is often only weakly cooperative, with Hill coefficients in the range 1.1–2.1. In coelomic

(a)

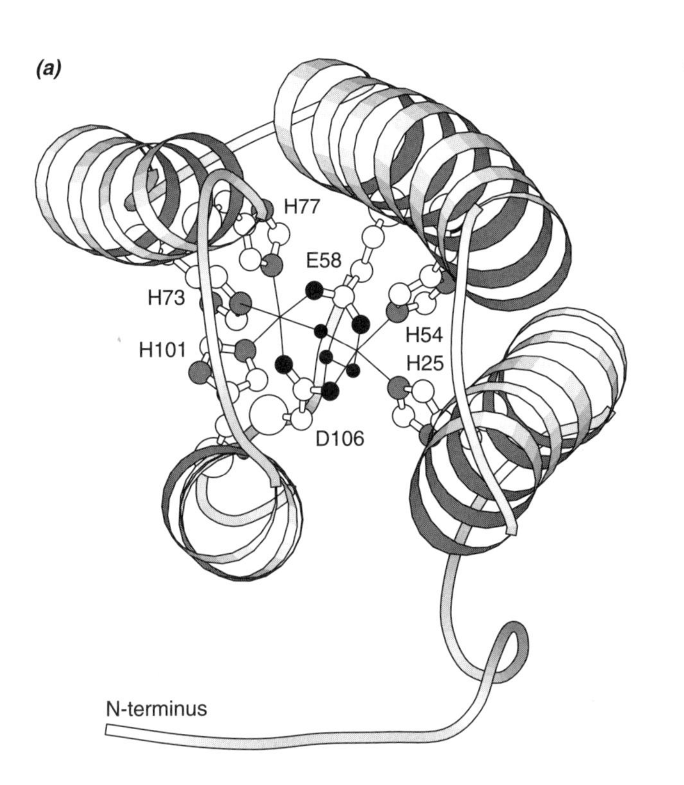

(b)

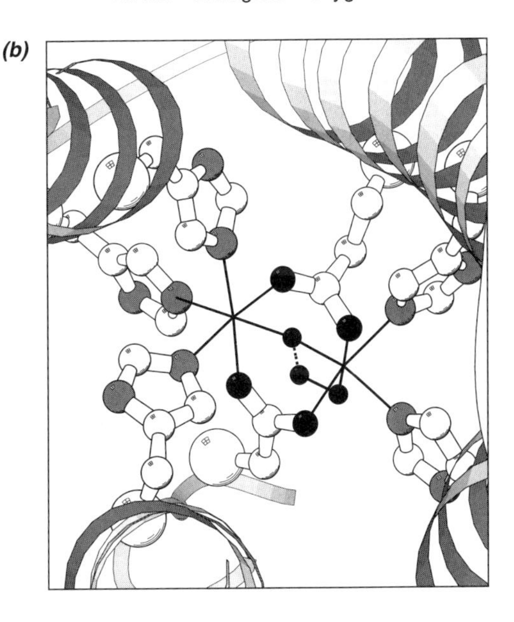

(c)

H
His101
His54
Fe^{II}
Fe^{II}
His77
His25
His73
Asp106
Glu58

Deoxyhemerythrin

O_2

His101
His54
Fe^{III}
Fe^{III}
His77
His25
His73
Asp106
Glu58

Oxyhemerythrin

Fig. XI.4.6.
Oxyhemerythrin of the sipunculid worm, *Themiste dyscrita* [PDB code: 1HMO (oxy), 1HMD (deoxy)]. (*a*) One subunit of the octameric protein. (*b*) Closeup of the active site showing the manner by which O_2 is bound to a single Fe atom and is hydrogen bonded to the bridging oxo moiety. (*c*) Scheme depicting the reversible binding of O_2 to the di-iron center of hemerythrin.

cells (the tissue between the inner membrane lining the digestive tract and the outer membrane of the worm—analogous to flesh in vertebrates), O_2 apparently binds with higher cooperativity ($n \sim 2.5$). Perchlorate ions have been observed to induce cooperativity: since ClO_4^- has no biological role, it appears that in protein purification the biological allosteric effector is lost. No Bohr effect is observed. The binding of O_2 occurs with a very large negative enthalpy change (ΔH ca. $-170\,\text{kJ}\,(\text{mol}\cdot O_2)^{-1}$).

As a result, the O_2 affinity of Hr is very sensitive to temperature, a physiological disadvantage. A selection of dioxygen-binding data is given in Table XI.4.3.

The structure of Hr in a variety of derivatives (oxy, azido, met, and deoxy) is now well characterized (Fig. XI.4.6*a* and *b*). The two Fe centers are triply bridged by a water-derived species (OH^- or O^{II-}) and two μ-1,3-carboxylato bridges from Asp and Glu residues. In deoxyHr, five of the six remaining coordination sites are occupied by Im moieties from His residues, the remaining site being available for exogenous ligand binding. This site is also vacant in metHr but can be occupied by anions such as azide as in metHrN_3. This site is indeed occupied by O_2 in the structure of oxyHr. The coordinated O_2 moiety bends back toward the di-iron core and the terminal oxygen atom appears to be within hydrogen-bonding distance of the solvent-derived bridge between the two Fe ions.[38,39]

The crystallographic information can be interpreted within the context of spectroscopic and magnetic measurements of the various Hr complexes (Fig. XI.4.6*c*). In deoxyHr, the metal ions are in the high-spin Fe^{II} state; they exhibit weak antiferromagnetic coupling ($-J \sim 10\,cm^{-1}$) that is mediated by the bridging ligands. On the other hand, in oxyHr and metHr derivatives, the metal ions are in the high-spin Fe(III) state and are strongly coupled ($-J \sim 100\,cm^{-1}$), owing to the presence of an oxo bridge.[10] Thus O_2 binding converts the di-iron(II) center of deoxyHr into a di-iron(III) center in oxyHr; furthermore, two electrons are transferred from the di-iron unit to the bound O_2, thereby reducing it to the peroxide oxidation state and generating an intense purple chromophore ($\lambda_{max} \sim 480\,nm$, $\varepsilon_{max} \sim 4000\,M^{-1}\,cm^{-1}$). The peroxide assignment is supported by the observation of an O–O stretch at $844\,cm^{-1}$ in the resonance Raman spectrum of oxyHr.[40] Moreover, the slight sensitivity of this mode to solvent change from H_2O to D_2O indicates protonation of the terminal oxygen atom; this proton is very likely transferred from the hydroxo bridge to form the hydroperoxide ligand and generate the (μ-oxo)di-iron(III) unit of oxyHr.

At first glance, the triply bridged di-iron core of Hr may appear to be very difficult to assemble in the absence of the protein, but closer scrutiny reveals that it is actually a wedge of the familiar basic metal acetate $[M_3(\mu_3\text{-}O)(\mu\text{-}O_2CR)_6]$ core.[10] Indeed, the use of a variety of nitrogen ligands capable of facile tridentate coordination allows the (μ-oxo)bis(μ-carboxylato)di-iron(III) core of oxyHr and metHr to be assembled rather easily. Some examples of complexes with the (μ-hydroxo)bis(μ-carboxylato)di-iron(II) core of deoxyHr are also known. Thus, this core appears to be a thermodynamically very stable structural motif and can be synthesized by "self-assembly" methods. Initially, model systems were symmetrical, coordinatively saturated species with six donor atoms to each iron, but more recent models reproduce the asymmetric six- and five-coordinate centers of Hr. However, structural characterization of a model complex with an O_2-binding mode mimicking that found for oxyHr (i.e., $Fe^{III}–O^{-II}–Fe^{III}–OOH$) remains a challenge for bioinorganic chemists.

XI.4.5. Protein Control of the Chemistry of Dioxygen, Iron, Copper, and Cobalt

Dioxygen is a powerful oxidant and the utilization of this property is a major subject of this chapter. Control of this property and prevention of unwanted side reactions by O_2 carriers is the subject of this subsection. In addition to the provision of ligand(s) to the metal center in an appropriate stereochemistry, the protein has two other roles: protection of the metal–O_2 moiety from oxidation and competitive ligands, and modulation of O_2 affinity through nonbonded interactions of the M–O_2 species with its surroundings.

XI.4.5.1. Role of the Protein in Protection of the M–O_2 Moiety

In the presence of O_2, Cu(I) and Fe(II) species are readily oxidized to, respectively, Cu(II) and Fe(III) species. In the presence of water, Fe(III) species frequently associate into Fe^{III}–O^{-II}–Fe^{III} species, a motif that turns up in oxy- and metHr. Several routes for the autoxidation of Fe(II) porphyrin species (PFe^{II}) exist.[41,42]

Bimolecular contact of two Fe porphyrin moieties

$$PFeO_2 + PFe^{II} \rightarrow\rightarrow\rightarrow PFe^{III}\text{–O–}Fe^{III}P \qquad (16)$$

Nucleophilic displacement (both by S_N1 and S_N2 mechanisms)

$$PFeO_2 + X^- \rightarrow PFe^{III}X + O_2^{-\cdot} \qquad (17)$$

$$\text{Electron transfer} \qquad PFe^{II} \rightarrow PFe^{III} + e^- \qquad (18)$$

Oxidation by the first route takes place in milliseconds at room temperature for simple isolated Fe(II) porphyrins. In biological systems, protection of the Fe–O_2 moiety from bimolecular attack is provided by the protein. The formation of metHb occurs *in vivo*, probably by a nucleophilic displacement mechanism, at the rate of ~3% of total Hb per day. In synthetic systems, protection may be provided by low temperature (less than −40°C, stabilizing both Fe–O_2 and Fe^{III}–O_2^{-II}–Fe^{III} species) or by porphyrin substituents that provide protection analogous to that provided by the protein. A selection of these porphyrins is shown in Fig. XI.4.7.[4,5,11]

The active-site motifs of Hb, Hc, and Hr also occur in slightly modified forms in oxygenases, such as cytochrome P450 (heme *b* group), tyrosinase [dicopper(I) motif similar to Hc], and methane monooxygenase [di-iron(II) motif similar to Hr] (see Section XI.5). Thus the protein matrix of Hb (usually), Hc, and Hr prevents oxygenase activity by the M–O_2 moiety. In addition, the protein also prevents the self-oxygenase activity that has been observed in a number of synthetic systems, especially for Cu.

Finally, the protein controls the environment in Hc, Hr, and many Hb so as to leave the metal coordinatively unsaturated, facilitating rapid binding of O_2. For isolated open Fe(II) porphyrins, a second axial ligand, such as pyridine or 1-methylimidazole (1-MeIm), binds 10–30 times more avidly than the first to give the thermodynamically stable and kinetically inert (d^6, $S = 0$) six-coordinate hemochrome species. For axial ligands bearing a bulky substituent at the 2-position, such as 2-methylimidazole (2-MeIm) and 1,2-dimethylimidazole (1,2-Me_2Im), the five-coordinate species predominates at room temperature even with a mild excess of ligand. In many deoxyHb, the protein chain restrains the distal His from coordination. However, coordination of this His to give low-spin hexacoordinated species is observed for both Fe(II) and Fe(III) states of nonsymbiont plant Hb, of Hb from unicellular eukaryotes, and of two recently characterized vertebrate proteins, neuroglobin and cytoglobin.[14,31] Neuroglobin [PDB code: 1OJ6] is expressed at low levels in neuronal cells, and cytoglobin (also called histoglobin) [PDB code: 1UMO] is ubiquitously located in the cell nucleus. Both are weakly related to the canonical vertebrate Mb and Hb. Ligand affinities and kinetics of these hexacoordinated deoxyHb lie within the ranges observed for the better known pentacoordinated deoxyHb.

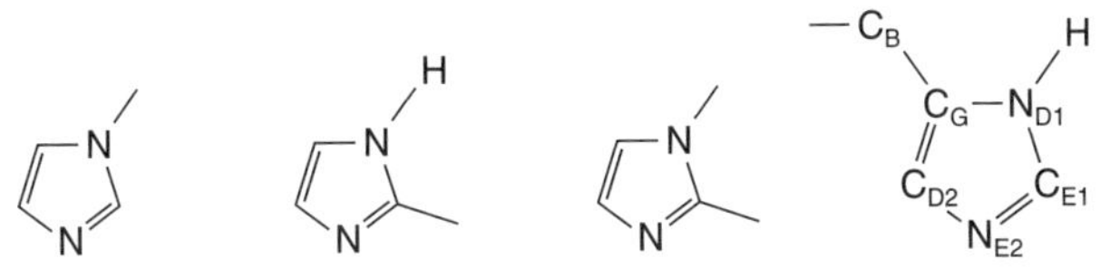

(19)

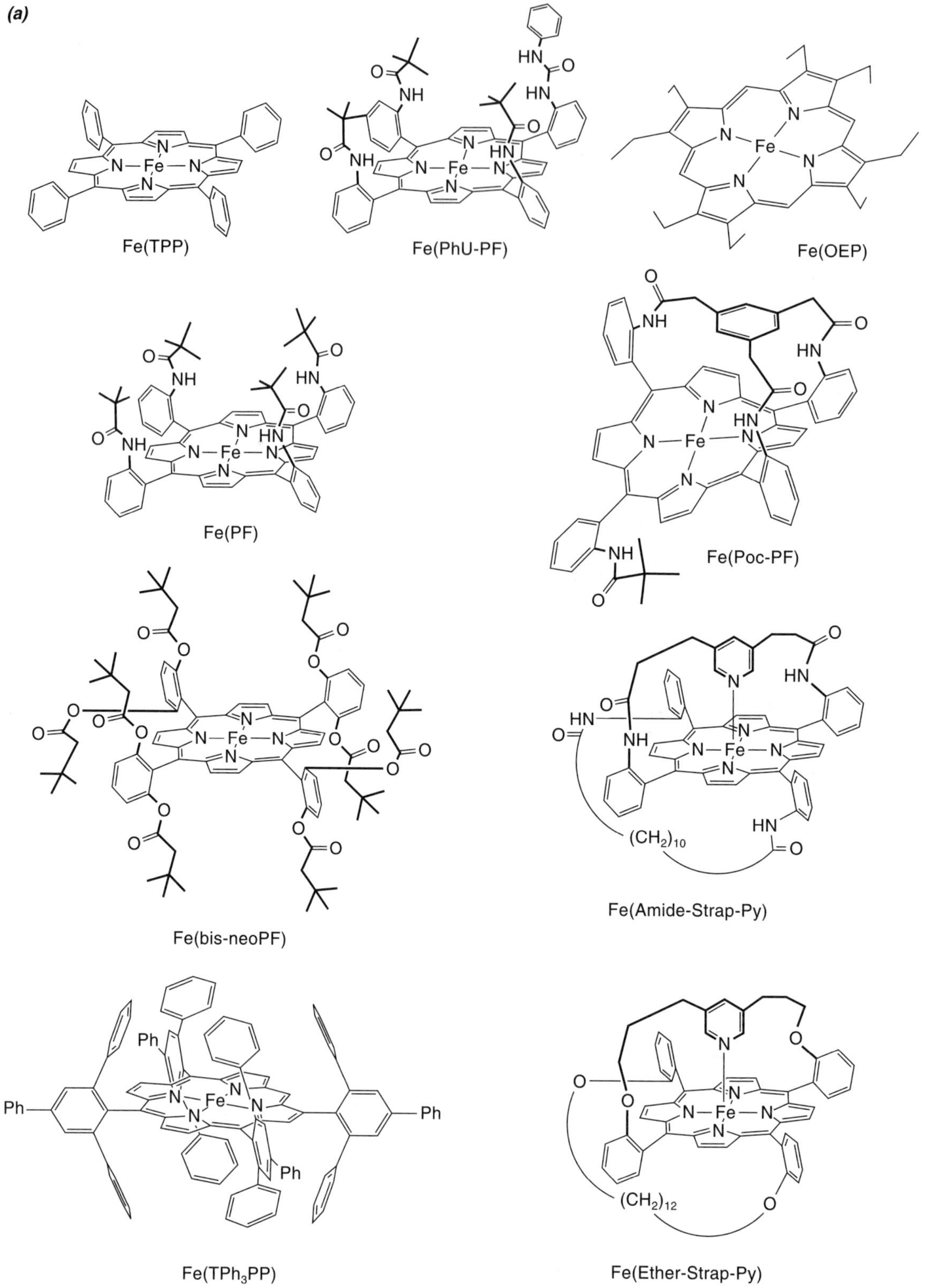
(a)
Fe(TPP)
Fe(PhU-PF)
Fe(OEP)
Fe(PF)
Fe(Poc-PF)
Fe(bis-neoPF)
Fe(Amide-Strap-Py)
(CH$_2$)$_{10}$
Fe(TPh$_3$PP)
Fe(Ether-Strap-Py)
(CH$_2$)$_{12}$

Fig. XI.4.7.
Selection of synthetic porphyrins used to model dioxygen- and carbon monoxide-binding processes. The superstructure that provides protection to a coordinated O_2 is drawn in bold.

(b)

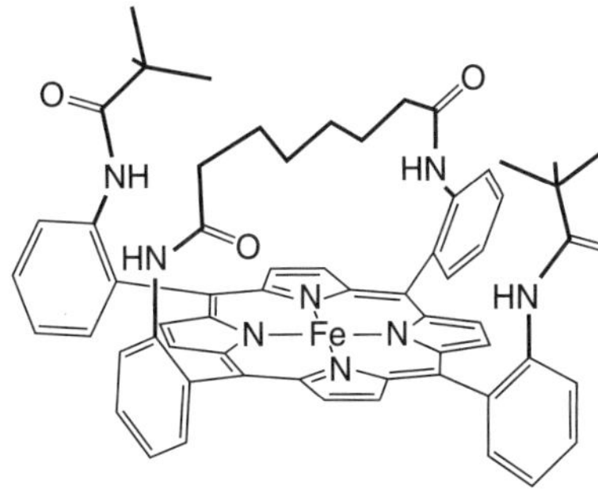

Fe(C_6Strap-PF)

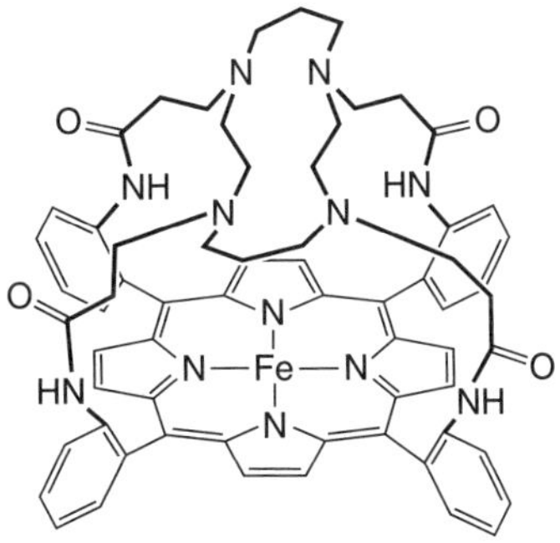

Fe(Cyclam-Cap)

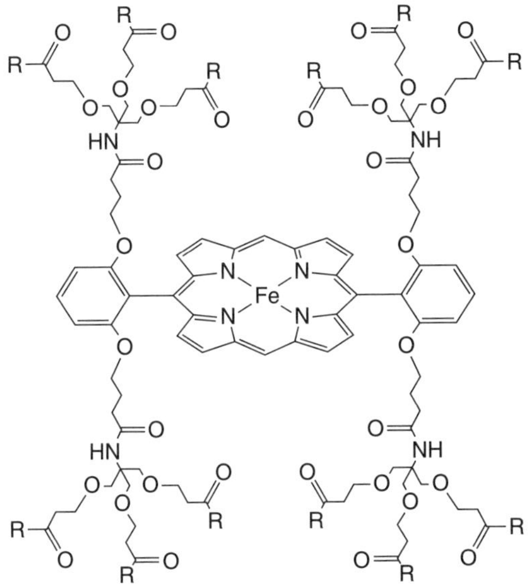

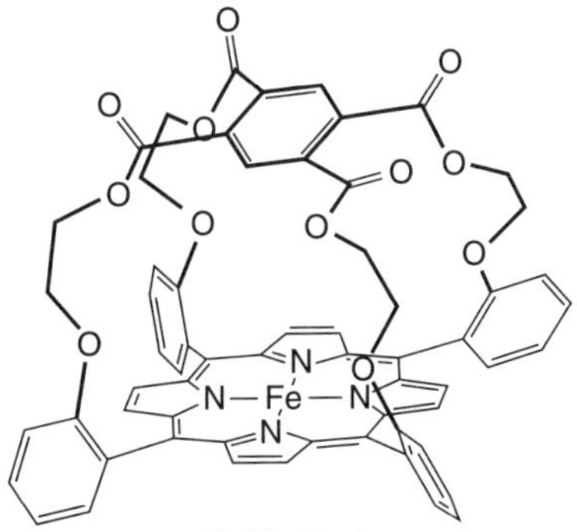

Fe(C_2-Cap)

Fe(G1), R = R_1 =

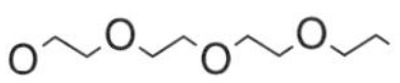

Fe(G2), R = R_2 =

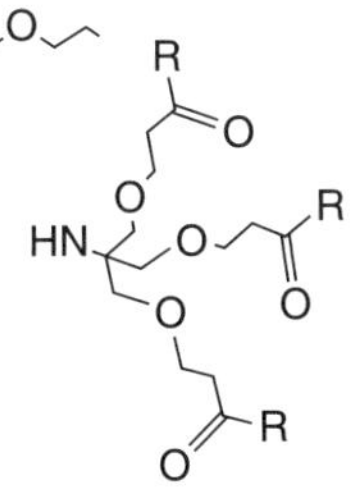

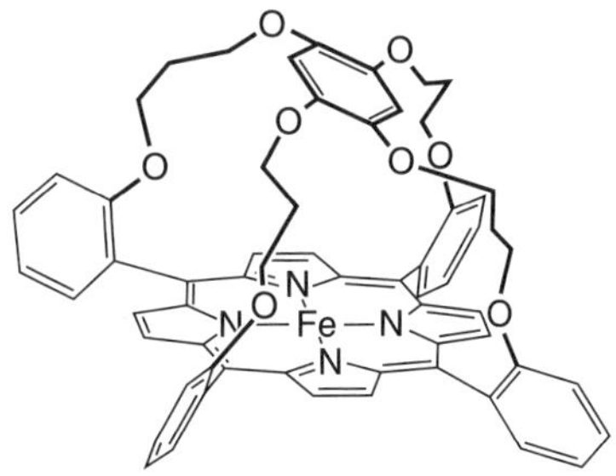

Fe(OC_3O-Cap)

Fig. XI.4.7.
(Continued)

XI.4.5.2. Modulation of Ligand-Binding Properties by the Protein

Few molecules have been as exhaustively characterized for their conformational properties as metalloporphyrins. The model systems provide yardsticks against which the usually less precisely determined protein structures may be compared. The ligand affinity of an Fe porphyrin may be perturbed either by modulating the structure of the deoxy species or by modulating the structure and surroundings of the liganded species, or both. It is convenient to divide these into two groups, referred to as distal and proximal effects. *Proximal* effects are associated with the stereochemistry of the metal porphyrin moiety and the coordination of the axial base, and thus their influence on O_2 and CO affinity is indirect. *Distal* effects pertain to *van der Waals' interactions* of the metal porphyrin skeleton and the sixth ligand (O_2, CO, etc.) with neighboring solvent molecules, with substituents, such as pickets or caps, on the porphyrin, and with the surrounding protein chain. The distal groups that hover over, or even bind to, the dioxygen-binding site engender the most important distal effects. A selection of parameters describing these effects is illustrated in Fig. XI.4.8, quantified in representative structures of Table XI.4.4, and described below.[2]

Proximal Effects: Porphyrin Conformation and M–N$_P$ Separations. The cyclic aromatic 24-atom porphyrin skeleton offers a tightly constrained metal-binding site. The conformation of least strain is planar and the radius of the hole of the dianion is close to 2.00 Å. Porphyrin substituents, crystal packing, and the protein matrix may cause substantial deviations from planarity. The 2.00-Å radius hole neatly accommodates low-spin Fe(II) ($S = 0$), Co(II), and Co(III) ions. Only rarely do M–N$_p$ bonds show small but significant scatter about their mean value.

Proximal Effects: Fe···Porphyrin Displacement. The displacement of the Fe atom from the plane of the four porphyrin nitrogen atoms, Fe···P$_N$, is sensitive to the number and type of axial ligands. A positive number denotes displacement toward the proximal His or axial base. Generally, displacement of the metal from the plane of the porphyrin nitrogen atoms, plane P$_N$ in Table XI.4.4, is within 0.04 Å of

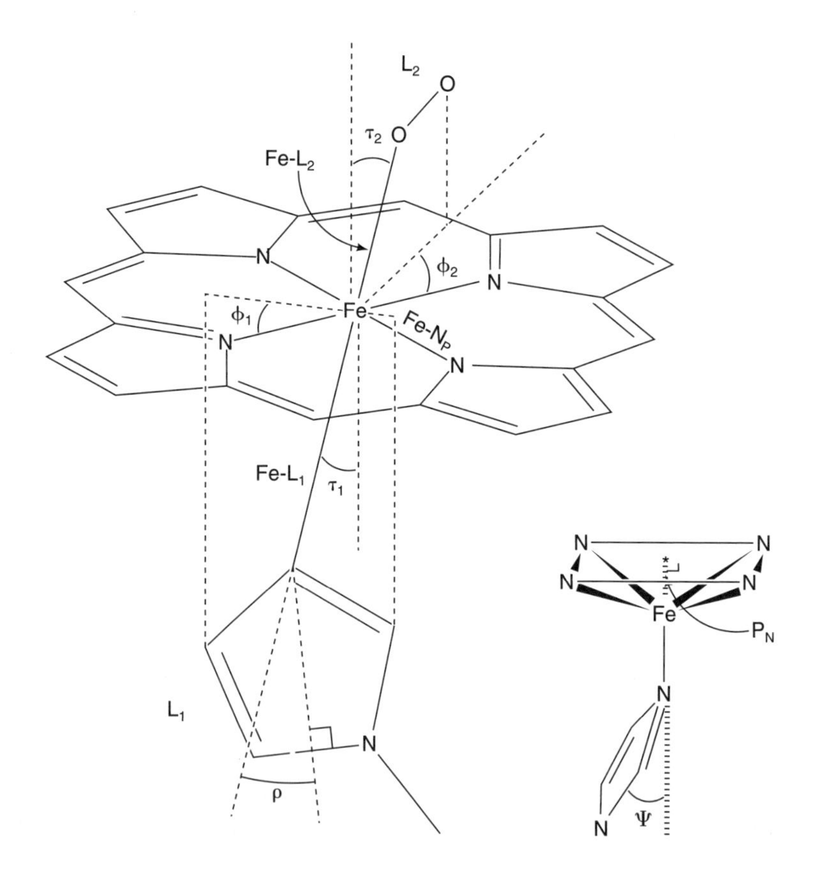

Fig. XI.4.8. Schematic diagram showing the orientation parameters of the porphyrin moiety and its ligands that determine ligand affinity.

Compound	PDB code		Resol. (Å)	Fe–Np (Å)				Fe···P_N[c] (Å)	Fe···P[d] (Å)	Fe–L_1 (Å)	Fe–L_2 (Å)	Fe–XY (deg)	φ_1[e] (deg)	φ_2[f] (deg)	τ_1[g] (deg)	τ_2[h] (deg)	ρ_1[i] (deg)	ψ[j] (deg)	
Fe(C_6Strap-PF)(1-MeIm)				2.05	2.10	2.08	2.07	0.31	0.34	2.13			30		5		6	98	
Fe(C_6Strap-PF)(1-MeIm)(CO)				1.98	1.97	1.99	1.99	−0.01	0.08	2.04	1.73	178	32	N/D	2	1	3	90	
Fe(PF)(1-MeIm)(O_2)				1.97	1.99			−0.03	0.02	2.07	1.75	131	20	45	0[k]	0[k]	0[k]	90[k]	
Fe(TPP)(1-MeIm)(CO)				2.01	2.00	2.01	2.01	−0.03	−0.00	2.07	1.79	179	29	N/D	2	0	3	85	
Fe(OEP)(1-MeIm)(CO)				2.00	2.00	2.00	2.00	−0.03	−0.00	2.07	1.74	175	1	N/D	0	5	−1	92	
Fe(OC_3O-Cap)(1-MeIm)(CO)				2.00	1.99	1.98	2.01	0.01	0.06	2.03	1.75	174	15	N/D	1	6	1	92	
Fe(C_2Cap)(1-MeIm)(CO)				2.00	1.98	1.99	2.00	−0.01	0.02	2.04	1.74	173	28	N/D	2	5	−1	95	
				1.97	2.00	1.99	1.99	−0.02	−0.00	2.04	1.75	176	28	N/D	2	2	−1	95	
Fe(PF)(2-MeIm)				2.07	2.08			0.40	0.43	2.09			23		10		6	90	
Fe(PF)(2-MeIm)(O_2)				2.00	2.00			0.09	0.02	2.11	1.90	129	22	43	7	0[k]	12	90	
Fe(Poc-PF)(1,2-Me_2Im)(CO)				1.98	1.97	1.97	1.98	−0.02	−0.01	2.08	1.77	172	35	N/D	4	4	13	73	Very ruffled porphyrin
DeoxyMb···H_2O	1A6N		1.15	2.11	2.07	2.05	2.06	0.36	0.39	2.14	(3.53)		3		5		−5	93	30% aquometMb
MbO_2	1A6M		1.00	2.00	2.00	2.00	2.02	0.02	0.09	2.06	1.81	123	2	−159	2	2	0	92	Eclipsed proximal His
MbCO	1A6G		1.15	2.00	2.02	1.94	1.98	−0.00	0.05	2.06	1.82	171	0	N/D	1	7	−2	91	Eclipsed proximal His
Hb(O_2) *Ascaris*	1ASH		2.15	1.94	2.01	1.98	1.96	−0.02	0.00	2.20	1.90	130	−62	−161	12	5	−14	79	Partly staggered prox His
Deoxy LegHb	1GDJ		1.7	1.83	2.11	2.10	2.08	0.30	0.32	2.25			−113		5		−8	111	Disordered prox His
$LegHbO_2$	2GDM		1.7	1.86	2.02	2.11	2.09	0.02	0.01	2.09	1.77	152	−30	−176	5	3	−6	96	Prox His rotates ~90°
LegHbCO	1GDI		1.8	1.81	2.09	2.11	1.91	−0.06	−0.00	2.09	2.17	149	−26	N/D	8	6	−8	93	C≡O 1.38 Å
DeoxyHbA	1A3N	α_1···H_2O	1.74	2.03	2.16	2.15	2.06	0.45	0.56	2.15	(3.39)		−17		3		−19	89	H_2O···His(distal)
(T state)		α_2···H_2O		2.08	2.09	2.03	2.03	0.34	0.50	2.17	(3.45)		−16		8		−1	87	H_2O···His(distal)
		β_1		2.05	2.00	1.99	2.06	0.33	0.43	2.15			−21		4		−21	88	
		β_2		2.13	2.03	2.09	2.06	0.38	0.46	2.03			−25		9		−1	88	
DeoxyHbHorse	1IBE	α···H_2O	1.80	2.03	2.07	2.01	2.04	0.40	0.41	2.19	(3.20)		−13		5		+2	92	H_2O···His(distal)
(R state)		β		2.02	2.04	2.06	2.06	0.34	0.39	2.28			−18		9		+21	87	
$HbAO_2$	1HGC	α_1-O_2	1.50	2.04	2.06	2.11	2.08	0.28	0.43	2.28	1.98	141	−19	172	5	3	−15	96	
(T state)		α_2-O_2		2.02	2.07	2.08	2.06	0.32	0.47	2.26	2.05	144	−14	164	2	2	−17	86	40% O_2; H_2O
		β_1		2.08	1.98	2.08	1.97	0.30	0.42	2.25			−25		3		−15	87	
		β_2		2.06	2.02	2.07	1.97	0.29	0.40	2.18			−22		4		+3	92	
$HbAO_2$	1HHO	α	2.10	2.02	2.04	1.95	1.93	0.12	0.15	1.94	1.66	153	−12	−165	3	2	−9	90	Good O_2···His(distal)
(R state)		β		1.90	1.94	2.06	1.95	−0.11	−0.06	2.07	1.87	159	−27	130	5	9	+9	87	Long O_2···His(distal)

[a] Taken from Ref. 2 and for proteins, calculated directly from the coordinates deposited in the PDB (http://www.rcsb.org).

[b] See Fig. XI.4.7 for diagrams and abbreviations of synthetic porphyrins and Fig. XI.4.8 for definitions of symbols.

[c] Fe···P_N: displacement of Fe from plane of the four porphyrin nitrogen atoms.

[d] Fe···P: displacement of Fe from plane of the 24 porphyrin atoms.

[e] φ_1: torsion angle NA-Fe-NE2-CE1 (standard PDB labeling scheme); for synthetic porphyrins, the unsigned average minimum torsion angle given.

[f] φ_2: torsion angle NA-Fe-O1-O2 (standard PDB labeling scheme); for synthetic porphyrins unsigned torsion angle given. Not determinable = ND, as Fe–CO bond is linear.

[g] τ_1: tilt of Fe–NE2 vector to the normal to plane of the four porphyrin nitrogen atoms.

[h] τ_2: tilt of Fe–O1 (or Fe–C) vector to the normal to plane of the four porphyrin nitrogen atoms.

[i] ρ_1: difference between Fe–NE2–CE1 and Fe–NE2–CD2 angles.

[j] ψ: dihedral angle between the plane of the Im ligand and P_N.

[k] These values are constrained by symmetry.

the displacement from the 24-atom mean plane of the entire porphyrin skeleton, plane P. On occasions, this second displacement is much larger, for example, in Fe(TPP)(2-MeIm), where it is 0.15 Å larger (TPP is 5,10,15,20-tetraphenylporphyrin). This effect is called *doming* and it is usually attributed to intermolecular interactions, such as crystal packing forces or heme···protein interactions. For five-coordinate, high-spin ($S = 2$) Fe(II) porphyrin species, such as deoxyMb, Fe(C_6Strap-PF)(1-MeIm), and Fe(PF)(2-MeIm), the large ionic radius, the presence of an electron in the $3d_{x^2-y^2}$ orbital (directed along M–N_p bonds), and minimization of steric clash between porphyrin and axial base lead to values for Fe···P_N of ~0.4 Å. On binding O_2 or CO, the Fe atom moves into the plane of the porphyrin, concomitant with a change in spin state from high spin ($S = 2$) to diamagnetic ($S = 0$). This movement of the Fe atom and its attached proximal His is the key to the stereochemical mechanism of cooperativity in the proteins.[22,23]

Proximal Effects: M–L Separations and Ligand Orientation Angle ϕ. The metal–axial ligand separations, M–L, are dependent on the nature of the ligand L. (Here L_1 denotes the heterocyclic axial base in the event that there are two axial ligands.) The angle between the plane of the axial base (e.g., Im) and the plane defined by atoms N_p, M, and L_1 is denoted ϕ_1 (Fig. XI.4.8). When $\phi_1 = 0$, the axial base eclipses a pair of M–N_p bonds; contacts with the porphyrin are maximized. When $\phi_1 = 45°$, contacts are minimized. The orientation angle ϕ_2 of coordinated O_2 is determined by distal effects, in particular the distal histidine, if present. In vertebrate Hb and Mb, a Ser or Thr hydroxy group hydrogen bonds to atom H–ND1 of the proximal His, enforcing an eclipsed conformation not seen in the higher affinity Hb *Ascaris* and legHb, where the side chain is nonpolar (see Table XI.4.4).

Proximal Effects: Imidazole Tilt Angle τ and Rotation Angle ρ. For 2-MeIm, contacts between the 2-methyl substituent and the porphyrin are relieved by tilting the M–N_{L1} vector so that it is no longer perpendicular to the porphyrin plane (tilt angle τ), and by rotating the Im group so that the M–N_{L1} vector no longer approximately bisects the C–N–C bond angle (rotation angle ρ), as illustrated in Fig. XI.4.8. The conformation of the protein chain is such that the proximal His in deoxyHbA and T state oxyHbA coordinates in a slightly tilted manner. Clearly, coordination of the His to the heme in a symmetrical manner, as would be expected in the absence of the protein constraints, does not produce the conformation of lowest free energy for the whole molecule.

Proximal Effects: Imidazole Tilt Angle ψ. Crystal-packing effects in model systems and nonbonded intramolecular interactions with the protein can tilt the plane of the Im ligand from perpendicularity to the plane of the four porphyrin nitrogen atoms, P_N.

Distal Effects. Distal effects arise from noncovalent interactions of the coordinated O_2, CO, or other ligand with its surroundings. Interaction of the coordinated O_2 or CO molecule with solvent molecules or with the protein has a profound influence on the kinetics and thermodynamics of these systems (see Table XI.4.2). As discussed earlier, there is accumulation of negative charge on the O_2 ligand. The possibility then arises for stabilization of coordinated O_2 through hydrogen bonding or dipolar interactions with solute molecules, porphyrin substituents (e.g., amide groups in the picket-fence porphyrins), or with protein residues (e.g., His). Destabilization of coordinated ligands and lowered affinity may result if the coordinated ligand is unable, through steric clash, to achieve its optimum stereochemistry or close neighboring groups are electronegative, as are the ether and ester linkages on capped porphyrins. The presence of hydrophobic cavities in the vicinity of the heme in Hb and, in particular, Mb, modulates and complicates the kinetics of ligand association and dissociation.[43,44] The most extreme distal effect is coordination of the distal His, which renders the first-order dissociation of the His the rate-determining step for exogenous ligand association. In the next subsections, we describe in detail the fascinating variety of means by which ligand binding is modulated by distal amino acid residues.

There is a convenient index to summarize the extent to which CO (or O_2) binding is discriminated against for a given Fe porphyrin system. Parameter M is defined as the ratio of O_2 affinity to CO affinity, both as $p_{1/2}$, for a particular system and experimental conditions:

$$M = p_{1/2}(O_2)/p_{1/2}(CO) \qquad (20)$$

With reference to Table XI.4.2, the M values calculated may be somewhat arbitrarily divided into three classes:

$M > 1 \times 10^4$ (no discrimination against CO or relatively good CO binder);

$1 \times 10^2 < M < 1 \times 10^4$;

$M < 1 \times 10^2$ (strong discrimination against CO or relatively good O_2 binder).

XI.4.6. Structural Basis of Ligand Affinities of Dioxygen Carriers

The affinities for O_2 of Hb (both native and their recombinantly produced wild-type forms) span nearly six orders of magnitude, which at room temperature (293 K) corresponds to a difference in the free energy of O_2 binding of $\sim 25\ kJ\ mol^{-1}$:

$$\Delta\Delta G = -RT\ \ln(K_{max}/K_{min}) = -RT\ \ln(p_{1/2\,min}/p_{1/2\,max}) \qquad (21)$$

This range can be substantially expanded at the low-affinity end by constructing appropriate site-directed mutants that further discourage ligand binding. For many synthetic systems, the CO affinities are orders of magnitude greater than those for biological systems that have an O_2 affinity similar to that of the synthetic system (e.g., compare the entries for the picket-fence porphyrin with those for native sperm whale Mb). The factors by which ligand affinities are modulated, generally to the benefit of the organism, are subtle and varied, and their elucidation requires the *precise* structural information that is currently available only from X-ray diffraction experiments on proteins and their models.

We now zoom in closer to the active site to answer three questions concerning protein control of ligand binding and ligand affinities, questions that have attracted considerable attention over many years. First, how do Hb discriminate against the binding of CO relative to that of O_2? Second, how are very high O_2 affinities achieved? Third, what are the restraints that cause the T state of multimeric Hb to have lowered affinity for CO and O_2 relative to the R state? In recent years, integrated studies on ligand-binding kinetics and thermodynamics of a host of Mb mutants have been complemented by high-resolution structural studies. There are now in excess of 300 structures from the Hb family, including >12 structures of various oxyMbs, >24 of carbonmonoxyMb, and >12 of deoxyMb. The principles enunciated here are applicable generally to Hr and Hc; however, we currently lack the thermodynamic, kinetic and, especially, structural data for these systems, compared to the Hb family.

XI.4.6.1. Discrimination of Dioxygen and Carbon Monoxide Binding to Myoglobins

The manner by which O_2 and CO bind to Hb has been a subject of considerable controversy, involving seemingly incompatible results from a variety of techniques applied to both biological and synthetic systems.[45–52]

Stereochemistry of the Fe–CO Moiety and O_2/CO Discrimination. Studies on synthetic systems have established unequivocally that in the absence of steric restraints and in an approximately fourfold symmetric electrostatic field, the Fe–C≡O moiety is linear and perpendicular to the porphyrin plane. Under such conditions CO binds extremely avidly, $p_{1/2}$(CO) < 0.0001 Torr, leading to discrimination factors >10,000 [e.g., $M = 28{,}000$ for the Fe(PF)(1-MeIm) system]. Were CO to bind this well in the biological systems, 20% of our Hb and Mb would be in the carbonmonoxy form, causing serious physiological stress. Fortunately, almost all Hb bind CO with an affinity orders of magnitude poorer than those observed in sterically unrestrained model systems [e.g., $p_{1/2}$(CO) = 0.018 Torr for sperm whale Mb, leading to $M \sim 25$]. The distal His, which hovers over the ligand-binding site in almost all native Mb and Hb, was proposed to play a key role in the discrimination against CO binding.[4,5,35] The position and orientation of the distal His allows O_2, which binds in an angular manner, to bind in a sterically unrestricted manner and to hydrogen bond to the N–H moiety of the Im ring of the distal His. On the other hand, for CO binding, the distal His offers steric restraints to the preferred Fe–CO geometry, that is an Fe–C≡O angle of 180° and a tilt angle of 0°, thereby leading to lowered affinity. This steric hindrance hypothesis was reinforced by the presence in the literature of many reports of substantially bent Fe–CO moieties, with angles as acute as 120°: a fictitious geometry that has made its way into widely used biochemistry textbooks. The structure of MbCO is now available at a resolution of 1.15 Å: the Fe–C≡O angle is 171° and the tilt of the Fe–C vector to the heme normal is 7° (PDB code: 1ABG). The structure of MbO_2 is also available at ultrahigh resolution (1.00 Å), and the Fe–O–O angle is 123° (PDB code: 1A6M). As resolution of Hb and Mb structures has improved, deviation of the Fe–C≡O moiety from linearity has decreased, and regularity in the Fe–N_P bond lengths has improved to approach results for model systems.

There has been considerable testing of the hypothesis of steric hindrance, both through model systems[11,47] and through studies of many native Mbs and their mutants, especially those of sperm whale Mb.[32,33,51] Model systems have been synthesized that offer steric restraint sufficient to all but eliminate CO binding, but permit O_2 binding with reasonable affinity. For example, Fe(Cyclam-Cap)(1,5-DCIm) (1,5-DCIm = 1,5-dicyclohexylimidazole) has a discrimination factor of < 0.003. When structurally characterized, the Fe–C≡O moiety in the severely constrained environments of several synthetic systems shows only modest departures from a linear and perpendicular geometry ($\beta + \tau < 10°$). For example, in Fe(Poc-PF)(1,2-Me_2Im)(CO), O_2 affinity (but not the kinetics of binding and release) is little changed from that for Fe(PF)(1,2-Me_2Im). However, the phenyl lid of Poc-PF leads to discrimination against CO. Here, the Fe–C≡O angle is 172.5(6)°; modest tilting of the Fe–CO group and substantial buckling of the porphyrin ring are apparent. The Fe(C_2Cap)(1-MeIm) system has low affinity for both O_2 and CO. The modestly bent and tilted Fe–C≡O moiety that is observed in the structure of Fe(C_2Cap)(1-MeIm)(CO) is accompanied by considerable distortion of the porphyrin.

Theoretical calculations have had a checkered history in providing insight into the nature of Fe–O_2 and Fe–C≡O moieties. However, density functional calculations perhaps provide a resolution of the conundrum of essentially linear Fe–C≡O moieties in the presence of severe steric stress, and significantly tilted and bent Fe–C≡O moieties in the absence of steric stress.[45,46,48,52] The tilting and bending modes of the Fe–C≡O moiety may combine synergistically, possibly in conjunction with porphyrin distortions and tilting of the axial base, to render the Fe–C≡O moiety susceptible to distortion. A combined tilting of 10° and bending of 6°, which leads to the O of the CO molecule being displaced 0.6 Å from the heme normal, has a modest energy cost of $\sim 4\,kJ\,mol^{-1}$—a displacement slightly larger than that seen in several model systems. A 0.9-Å displacement costs $< 10\,kJ\,mol^{-1}$ in energy.

A remaining conundrum was to reconcile the X-ray and theoretical results on Fe–CO deformation with an apparent tilt of the CO vector of $< 7°$ from the heme normal as derived from single-crystal polarized infrared (IR) spectroscopy of MbCO. These experiments in fact measured not the orientation of the CO moiety, but the orientation of the CO transition dipole vector for the C≡O vibration. Because of coupling of the CO moiety with electron density of the porphyrin, the transition dipole vector essentially bisects the angle that the CO moiety makes with the heme normal. Thus, a transition dipole tilted 7° to the heme normal corresponds to a combined tilting and bending of 16° or a 0.6-Å displacement of the O atom from the heme normal.[46,48,50,52]

Whereas steric effects are important in the discrimination against CO binding offered in several specifically designed synthetic systems studied under anhydrous conditions, other factors appear to be determining the discrimination against CO binding in biological systems and in model systems where the presence of water has not been rigorously excluded.

Electrostatic Interactions and O_2/CO Discrimination. As more structural data have accumulated, especially on the biological systems, it has become clear that the ligand-binding process in synthetic systems differs from that in biological systems in several key respects. As will be described in more detail below, water plays a crucial and structurally well-characterized role in O_2/CO discrimination in biological systems. For many synthetic systems, water is inimical to the stability of the Fe–O_2 moiety, and thus ligand binding is typically studied in rigorously nonaqueous aprotic solvents (e.g., toluene). The ligand-binding process then involves the simple entry of the ligand into the binding pocket, followed by binding to the Fe center. The coordinated ligand may be under steric strain or may also make favorable (e.g., $Fe{-}O_2^{\delta-}\cdots{}^{\delta+}H{-}N$) or unfavorable ($Fe{-}O_2^{\delta-}\cdots{}^{\delta-}O{-}C$) electrostatic interactions. When the metal-O_2 moiety is unprotected, formation of the highly polarized Co–O_2 moiety is strongly disfavored in apolar solvents. Formation of the less polarized Fe–O_2 moiety is still favored in the somewhat aqueous environment when heme is encapsulated in micelles, compared to anhydrous benzene. A new class of synthetic porphyrins, Fe(G1) and Fe(G2), features dendritic superstructure. Even with the sterically bulky 1,2-Me_2Im as the axial base, O_2 affinity greatly exceeds that of picket-fence porphyrin derivatives with a less sterically bulky 1-MeIm axial base. In contrast, affinity for CO is more than an order of magnitude lower than that for Fe(PF)(1,2-Me_2Im), leading to discrimination factors of 0.10 or less.[4,11] Although nominally prepared under anhydrous conditions, cryptated water molecules that hydrogen bond to the coordinated O_2 ligand may be responsible for the high affinity for O_2 and low affinity for CO, analogous to that seen in the structurally characterizable biological system. Unfortunately, definitive characterization of the ligand-binding site of these very large dendritic porphyrins by means of single-crystal X-ray diffraction methods poses a severe challenge.

In contrast, water plays a key and very well-characterized role in the discrimination between O_2 and CO binding to mammalian Mb.[32,33,51] To illustrate the interplay between a buried water and the polar distal residue, the ligand-binding process can be broken down into the series of steps illustrated in Fig. XI.4.9. The first step, loss of water from the ligand-binding pocket, occurs only for wild-type or native protein and those mutants with polar distal groups. The second step involves entry of ligand into the binding pocket to give the geminate complex.[43,44] For CO, this complex is a metastable intermediate long known to spectroscopists, as the Fe–CO moiety is readily photolyzed. The photolyzed CO molecule has been revealed by X-ray diffraction methods to lurk in a hydrophobic pocket near but not bound to the Fe atom. Binding studies using xenon have revealed four hydrophobic pockets in the vicinity of the heme. In an experimental *tour de force,* taking advantage of the ultraintense polychromatic pulsed X-ray sources of third-generation synchrotrons, the reverse of the third step—the breakage of the Fe–CO bond, movement of the CO molecule into its docking site, and relaxation of the heme and surroundings—has been observed crystallographically over time scales ranging from nano- to milli-seconds, revealing correlated protein motions that sweep the dissociated CO ligand from the iron site.[12,13] The fourth step in the ligand-binding process is the stabiliza-

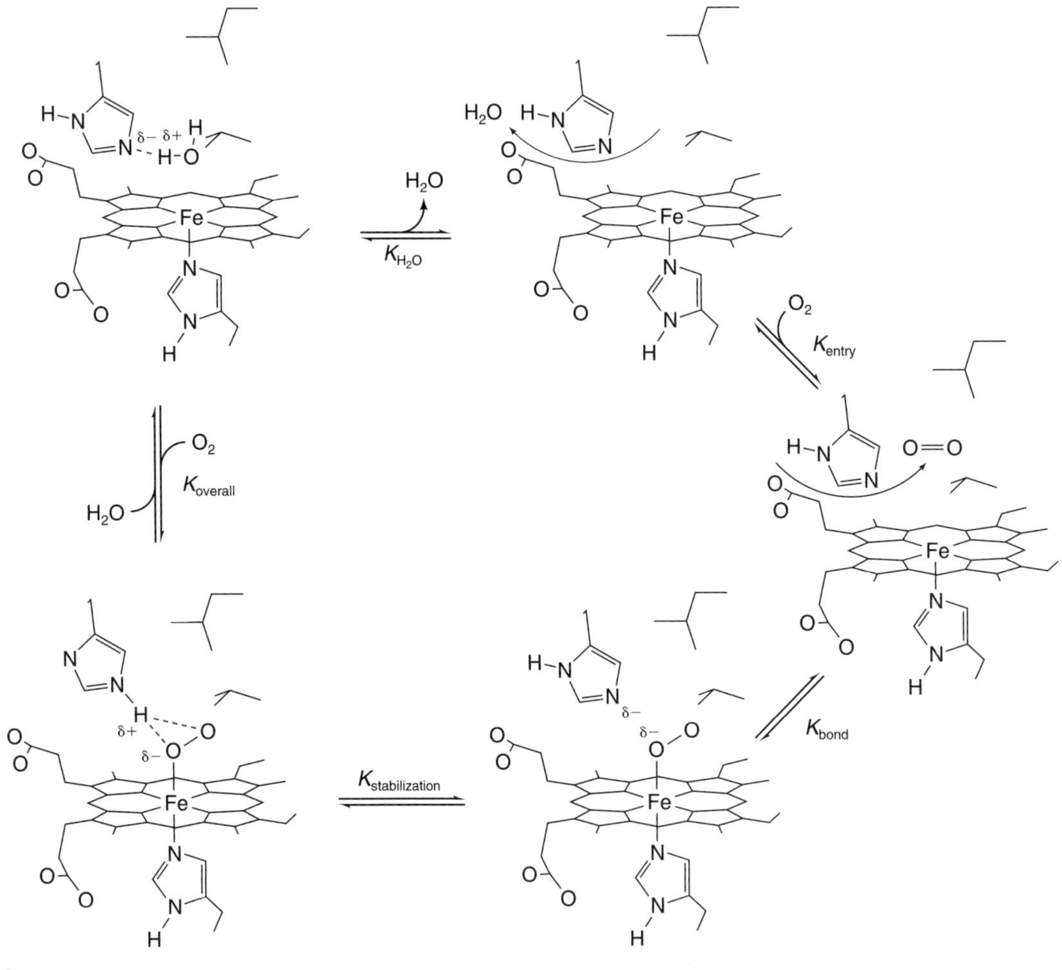

Fig. XI.4.9.
Schematic diagram of the ligand-binding process to myoglobin. [Adapted from Ref. 51.] A key feature of the discrimination shown by Mb against CO comes from the very weak dipolar interaction between the coordinated CO and the distal His, compared to the strong hydrogen bond that exists between the water and distal His for deoxyMb and for oxyMb between coordinated O_2 and the distal His. The entrance to the ligand-binding pocket is from the left-hand side of each panel. The view is orthogonal to that of Fig. XI.4.1*b–d*. Double bonds and nitrogen atoms on the porphyrin core are omitted for clarity.

tion of the bound ligand by interaction with the amino moiety (H)NE2 of the polar distal His. For ligand binding, loss of H_2O has a free energy cost of $\sim +6\,kJ\,mol^{-1}$, whereas for the His$\cdots O_2$ hydrogen bond the gain is $\sim -17\,kJ\,mol^{-1}$ for a net gain of $\sim -11\,kJ\,mol^{-1}$. This hydrogen bond is observed crystallographically both in wild-type and Co-substituted forms: the $O_2\cdots$(H)–NE2 separations are $\sim$2.8 Å.[49] Studies of $CoMbO_2$ by EPR provide further confirmation of a strong M–$O_2\cdots$His interaction.

The relatively nonpolar Fe–CO moiety enjoys an electrostatic gain of only $\sim -3\,kJ\,mol^{-1}$ for a net cost from loss of water and binding of CO of $+3\,kJ\,mol^{-1}$. This corresponds to a change in the discrimination factor of $\sim$100 from electrostatic factors alone. Earlier neutron diffraction studies of MbCO and MbO_2 showed that for MbCO the proton (deuteron) sits on imidazole atom ND1 and is therefore remote to the CO moiety, whereas for MbO_2 the alternative tautomer was observed.[53,54] More recent evidence, however, indicates that for both MbCO and MbO_2 the proton sits on atom NE2 and interacts with the coordinated ligand.[55–57]

The importance of electrostatic interactions on the Fe–CO moiety, and of protein dynamics in general, is revealed by the extreme sensitivity of the CO stretching vibration to environment.[57] Three modes are typically seen; these are designated A_0 (the most prominent band at pH < 5; 1966 cm^{-1}), A_1 (the most prominent band at pH $\sim$7; 1946 cm^{-1}), and A_3 (1933 cm^{-1}). Mode A_0 corresponds to the distal His swung away from the CO, leaving the CO moiety in a less polar environment. Mode A_0 becomes the only mode in E7His $\rightarrow$ X mutants, where X is a hydrophobic residue, such as Val or Leu. Moreover, mutants in which nonpolar Leu replaces the distal His have discrimination factors in excess of 15,000 for sperm whale, human, and pig Mb, mostly as a result of a much larger increase in CO affinity than a decrease in O_2 affinity compared to other nonpolar mutants. The rate constant of CO binding is greater, whereas that for CO release is smaller than corresponding values for other hydrophobic mutants. This Leu mutant illustrates the problem that not all changes in affinity can be traced neatly and directly to the active site. Empty hydrophobically lined internal spaces in protein molecules carry an entropic cost. When CO binds, much of this space becomes filled. For the Leu mutant, a better filling of this space would lead to the observed differences from other hydrophobic mutants, such as Ala. The origin of the anomalously small decrease in O_2 affinity and increase in CO affinity for the E7His $\rightarrow$ Gly mutant was revealed crystallographically to result from a water molecule that remained in the binding pocket of the deoxyMb and MbO_2 species.

XI.4.6.2. Structural Basis of Very High Dioxygen Affinity

If one hydrogen bond between coordinated O_2 and a polar distal residue stabilizes HbO_2 moieties relative to HbCO moieties, what then are the effects of additional hydrogen bonds? The Hb from the perienteric fluid of the parasitic nematode *A. suum* has an affinity for O_2 orders of magnitude higher than that of any vertebrate Hb, together with an unusually low affinity for CO. This very high O_2 affinity is largely a consequence of a very small rate constant for the dissociation of O_2. The structure of subunit D1 of the two-domain protein reveals that O_2 is hydrogen bonded not only to a distal Gln from a site analogous to that of the distal histidine of most vertebrate Hb, but also to a Tyr, which is located on helix B at a site corresponding to a Leu residue in vertebrate Hb, as illustrated in Fig. XI.4.10.[58]

One can estimate that if the $O_2\cdots$Gln hydrogen bond is comparable to the $O_2\cdots$His hydrogen bond of Mb ($\sim 11\,kJ\,mol^{-1}$), a second hydrogen bond of comparable strength will lead to a 100-fold increase in affinity, assuming no change in protein entropy. The CO affinity of Hb *Ascaris* is slightly larger than that of sperm

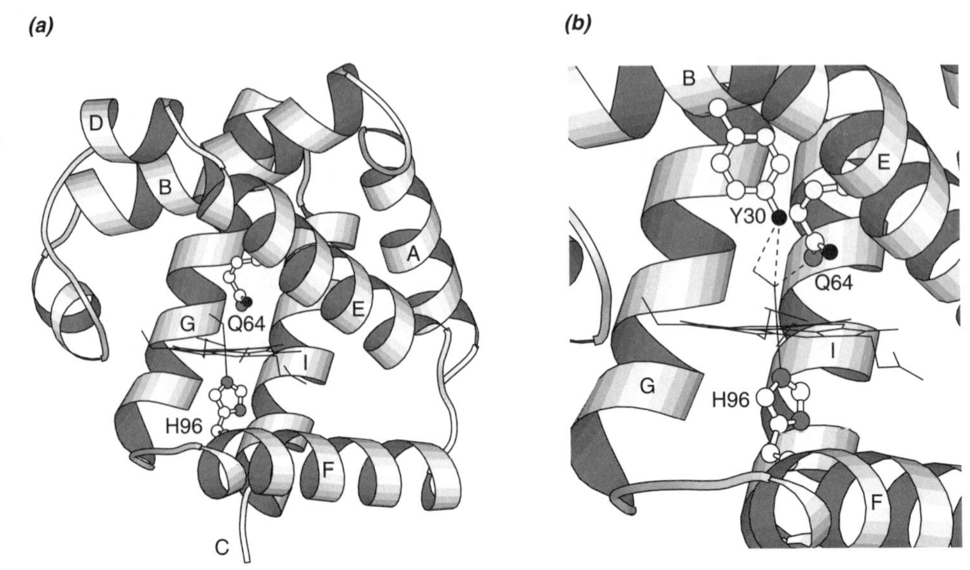

Fig. XI.4.10.
Closeup of the active site of the O_2-avid Hb from *A. suum* (PDB code: 1ASH). The distal His is replaced by a Gln (Q64), which hydrogen bonds to the coordinated O_2 through its $-NH_2$ moiety. The Leu29 at helix position B10 of most vertebrate Hb is replaced by a Tyr (Y30), which strongly hydrogen bonds to the coordinated O_2.

whale Mb, consistent with the small stabilizing role identified in Mb for the CO···(H)–N electrostatic interaction. In addition, the E7Tyr → Leu mutant of Hb *Ascaris* has a dramatically lower affinity for O_2, mostly owing to a large increase in k_{off}. The k_{on} for CO is largely unchanged: To a crude approximation O_2 affinity is largely determined by k_{off}, whereas CO affinity is largely determined by k_{on}. Note that sperm whale Mb mutated to contain the same distal residues as Hb *Ascaris* shows an affinity for O_2 only a little greater than that of wild-type sperm whale Mb, but a CO affinity an order of magnitude smaller. Although all Hb share a common tertiary structure, significant differences occur, for example, in the angle between helices E and F that form the cleft in which the heme group sits, and in the orientation of the proximal His, which in sperm whale Mb eclipses an Fe–N_p bond, disfavoring movement of the Fe (and its attached His···Ser) into the heme plane.

LegHb is also structurally well characterized in its deoxy, oxy, and carbonmonoxy forms. The origin of its high ligand association rate constant for both CO and O_2 appears to lie in a more accessible ligand-binding pocket than that for vertebrate Hb. Replacement of the distal His by Leu leads to a higher value of the discrimination constant similar to that for mammalian Mb, and also to a very high CO affinity similar to that for CO binding to sterically uncongested pockets of low polarity in model systems.

XI.4.6.3. Structural Basis of Cooperative Ligand Binding in Mammalian Hemoglobins

Relative to R-state HbA, ligand affinities are reduced nearly 1000-fold for T-state HbA in the presence of heterotropic allosteric effectors (e.g., 2,3-DPG). For O_2 binding, k_{on} is reduced and k_{off} is increased. Relative to mammalian Mb, R-state Hb and isolated chains show higher CO affinities, the result of increased rate constants for CO association, leading to a 10-fold increase in the CO/O_2 discrimination parameter M.

Cooperative ligand binding to Hb requires that the structural changes occurring on ligand binding to the heme be communicated to the interfaces between subunits. All mammalian Hb that have been structurally characterized can be categorized as having either the T- or R-state quaternary conformation, although considerable conformational plasticity within each quaternary conformation is observed. However,

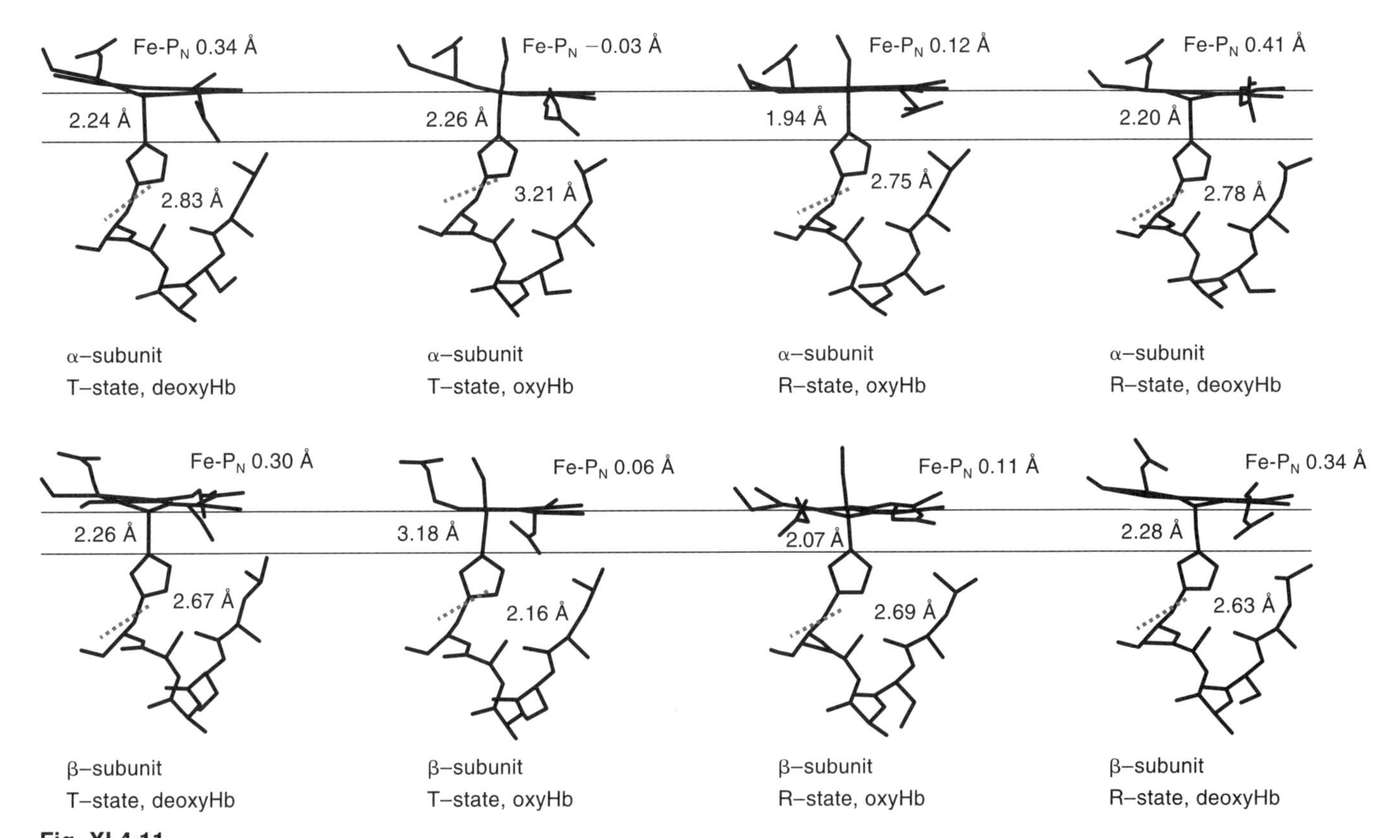

Fig. XI.4.11.
Structural changes occurring upon ligand binding to the four subunits of mammalian Hb. (Adapted from Ref. 23.) In the α subunits of T-state deoxyHbA (PDB code: 1A3N) and T-state oxyHbA (PDB code: 1GZX), the heme group is substantially domed. On change to the R state of oxyHbA (PDB code: 1HHO) and deoxyHb(horse) (PDB code: 1IBE), the heme group flattens. On the other hand, in the β subunits, the whole porphyrin moiety moves in the change from the R to T state. For T-state oxyHbA [and also the partly oxygenated structure (PDB code: 1HGC), Table XI.4.4], the hydrogen bond between the carbonyl oxygen atom of F4(Leu) and the H-ND1 atom of F8(His) is substantially lengthened for both α and β subunits compared to R-state structures. The very high resolution structure (1.25 Å) of R-state carbonmonoxyHbA (PDB code: 1IRD) shows very similar structural features to R-state oxyHbA.

the switch is not and cannot be a simple flip–flop from one state to the other, as the ligand-binding process is tied to a loss of protons in a sequential manner, the Bohr effect, and also for vertebrate Hb to a gain of some 65–70 water molecules.[28] The structures of T-state *liganded* HbA and R-state *unliganded* HbA have revealed much about tertiary structural changes that ligand binding or ligand loss cause within a given quaternary structure. The tertiary structural changes observed are part of the way along the path to R-state oxyHbA, and this leads to some changes and strain on salt bridges in the critical $\alpha_1 \cdots \beta_2$ interface.[23] Thus, oxyHbA trapped in the T state by crystal packing shows significant differences compared to deoxyHbA in the T state. When trapped in silica gel, T-state HbA binds ligands noncooperatively and with low affinity. Conversely, the structure of deoxyHb (horse) trapped in the R state shows tertiary structural changes partway along the path to the T-state deoxyHb. Again, when trapped in silica gel, R-state deoxyHbA binds ligands noncooperatively and with high affinity. The large structural differences that exist between deoxy (T) and oxy (R) HbA and the much smaller differences, which do not propagate beyond the immediate environment of the heme, between deoxy (T) and oxy (T) Hb and between deoxy (R) and oxy (R) are shown in Fig. XI.4.11 for the α and β subunits.

The role of the Fe-proximal His bond in cooperative ligand binding (termed proximal coupling) has been demonstrated by means of several mutants in which the proximal His in either the α-, the β-, or both chains was replaced by Gly, and the Im axial base then supplied exogenously. Dioxygen and *n*-butylisocyanide binding for each of the three mutants is weakly cooperative—isocyanide adducts are resistant to the oxidation that renders O_2 binding partly irreversible. The ligand affinity is much higher than the average affinity for wild-type Hb, especially for the two mutants lacking the proximal His in the α subunit. Response to the strong heterotropic allosteric effector, inositol hexaphosphate (IHP), is greatly reduced, indicating that the IHP-binding site at the interface between β-subunits, as well as the heme sites, is proximally coupled. The $[\alpha^{His87\rightarrow Gly(Im)}\beta]_2$ mutant remains trapped in the T state, leading to independent ligand binding at a high affinity site $[\alpha^{His87\rightarrow Gly(Im)}]$ and a low affinity site (β). Ligand binding to the $[\alpha\beta^{His92\rightarrow Gly(Im)}]_2$ mutant, however, is weakly cooperative (Hill $n = 1.5$) and the ligated molecule shows no sign of the T-state conformation. The role of the proximal histidine in controlling quaternary structure is confirmed in the double mutant, $[\alpha^{His87\rightarrow Gly(Im)}\text{-}\beta^{His92\rightarrow Gly(Im)}]_2$, which also shows weak cooperativity. The vestiges of the T state seen in the fully ligated double mutant indicate that quaternary switching is largely, but not entirely, controlled by the Fe-proximal His moiety of the α subunit.[59,60]

Are there structural features in the active site that can be associated with the low ligand affinity of the T state? Model systems have identified a simple mechanism for reducing O_2 and CO affinity by more than an order of magnitude. A methyl group at the 2-position on an Im ring prevents the Fe atom and its attached Im group from moving into the plane of the porphyrin on the binding of O_2 (and presumably CO). In the structures of five-coordinate Fe(C_6Strap-PF)(1-MeIm) and Fe(PF)(2-MeIm), a greater displacement of the Fe atom from the plane of the four porphyrin nitrogen atoms, P_N, for Fe(PF)(2-MeIm) (0.40 compared to 0.31 Å) is partly offset by a shorter Fe–N_{Im} bond (2.09 compared to 2.13 Å). On binding O_2 to Fe(PF)(1-MeIm), the Fe atom moves completely into the P_N plane and the Fe–N_{Im} bond shortens slightly; for the 2-MeIm analogue, the Fe atom remains substantially out of the plane and the Fe–N_{Im} lengthens slightly. For both Fe(PF)(2-MeIm) and Fe(PF)(2-MeIm)(O_2), the 2-MeIm group is tilted ($\tau_1 = 10°$ and 7°, respectively) and twisted ($\rho_1 = 6°$ and 12°), in order to reduce steric clash of the 2-methyl substituent with the porphyrin. In metrical terms, the lowered affinity of the 2-MeIm species is reflected in an increase in the sum of the axial bond lengths (Fe–O + Fe–N_{Im}) from $1.75 + 2.07 = 3.82$ Å for the 1-MeIm species to $1.90 + 2.11 = 4.01$ Å for the 2-MeIm species.[2,11]

In the binding of O_2 and CO to mammalian Mb, structural changes for the heme–His moiety are similar to those for model systems. The structures of oxyHbA and carbonmonoxyHbA in the R state also show that the Fe atom has moved into the plane of the heme, that the heme has flattened, and that the His is coordinated symmetrically (no tilt or twist), at least for the α subunit. On the other hand, the structures of T-state oxyHbA show that the restraints of the T-state quaternary structure restrict movement of the Fe–His moiety, especially for the α subunits. For the β subunits, both the structural changes observed and the ligand-binding behavior of the $[\alpha\beta^{His87\rightarrow Gly(Im)}]_2$ mutant indicate that the distal side plays an important role in controlling affinity. Although access to the ligand-binding site for the β chains is apparently blocked by groups at the entrance to the cavity above the Fe center, this does not prevent facile access to the binding site—the rate of O_2 binding is in fact faster by a factor of ~4 relative to the α subunit for both R and T state HbA. In unligated Hb, the distal His, one edge of which is exposed to the solvent, is much more mobile than other distal residues buried inside the ligand-binding pocket. Moreover, studies with hybrid Hb, in which a functional heme group is replaced by a nonfunctional

Co-, Mn-, or Ni-substituted equivalent, identify the β subunit as that with the higher rate constant for O_2 binding.[23,59–67]

XI.4.7. Final Remarks

XI.4.7.1. The Future of Model Systems

Since ~1994, there have been major advances in and rethinking of the mechanism of O_2 and CO binding to the Hbs. Insight into factors important to O_2 binding increasingly comes from differences rather than similarities between synthetic and biological systems. Notwithstanding the wealth of structural data on biological systems, there remains still a critical need for very precise parameters for the Fe–CO and, especially, the Fe–O_2 moiety as benchmarks for the still less precise information obtained from determinations of protein structures. For example, can electrostatic factors important in the biological system be recreated in a synthetic system, which can be crystallized, to allow determination of the Fe–CO angle to a precision an order of magnitude better than that achieved in the best protein structures? Can model systems that bind O_2 with affinities equal to, or greater than that displayed by some trematode Hbs be designed, synthesized and, most importantly, structurally characterized?

XI.4.7.2. Is There Anything Left to Know about Biological Dioxygen Transport and Storage?

The study of ligand binding to the Hb stands as a paradigm of scientific inquiry. Understanding has progressed from a simple qualitative description of biological function to a semiquantitative understanding of the ways in which ligand-binding properties of a given structural motif, in this case a heme and coordinated His, may be optimized for a particular environment. Along the way models have been developed to rationalize observations and to predict possible outcomes of experiments. Perutz's stereochemical model of cooperativity in Hb, linking ideas of Monod, Wyman, and Changeux with structural results, has formed the basis of many investigations; investigations that have led, in many instances, to refinements and clarifications of the model, but that have also spawned still hotly debated alternative models.

The design of blood substitutes based upon Hb relies upon accurate and quantitative knowledge of the physical chemistry and physiology of Hb. For example, ligated Hb, free of red blood cells, dissociates into dimers to a significant extent. Whereas Hb tetramers are retained in passage through the kidneys, dimers are not and, moreover, cause considerable damage. How then does one enforce a tetrameric association of Hb? Hemoglobin subunits may be covalently cross-linked, but how does this affect cooperativity and affinity? Moreover, with regard to multiple functionality of proteins, how can hypertensive shock associated with depletion of the vasodilatory nitric oxide be mitigated on administration of extracellular synthetic O_2 transporters? Such questions require quantitative answers if effective nontoxic blood substitutes are to be developed.

It is unlikely that the depth of knowledge on ligand binding to hemoglobins will be duplicated for the hemocyanins and hemerythrins any time soon. The nature of such ligand binding will certainly be different from that in the hemoglobins because the active sites are so different. Details on ligand binding to hemocyanins and hemerythrins are important to obtain, not only to serve in contrast to the Hb but also to increase our understanding of biological O_2 transport and storage and of protein–substrate interactions in general.

References

General References

1. Antonini, E. and Brunori, M., "Hemoglobin and Myoglobin in Their Reactions with Ligands", North Holland, 1971. The classic text on hemoglobins and myoglobins.
2. Jameson, G. B. and Ibers, J. A., "Biological and synthetic dioxygen carriers", in *Bioinorganic Chemistry*. Bertini, I., Gray, H. B., Lippard, S. J., and Valentine, J. S., Eds., University Science Books, Mill Valley, CA, pp. 167–252, 1994. This chapter has a comprehensive bibliography.
3. Weber, R. E. and Vinogradov, S. N., "Non vertebrate hemoglobins. Functions and molecular adaptations", *Physiol. Rev.*, **81,** 569–628 (2001). Comprehensive and insightful review, with comparisons to vertebrate hemoglobins.
4. Collman, J. P., Boulatov, R., Sunderland, C. J., and Fu, L., "Functional analogs of cytochrome c oxidase, myoglobin and hemoglobin", *Chem. Rev.*, **104,** 561–588 (2004).
5. Momenteau, M. and Reed, C. A., "Synthetic heme dioxygen complexes", *Chem. Rev.*, **94,** 659–698 (1994).
6. Moller, J. K. S. and Skibsted, L. H., "Nitric oxide and myoglobins", *Chem. Rev.*, **102,** 1167–1178 (2002).
7. Van Holde, K. E. and Miller, K. I., "Hemocyanins", *Adv. Protein Chem.*, **47,** 1–81 (1995).
8. Magnus, K. A., Ton-That, H., and Carpenter, J. E., "Structural work on the oxygen-transport protein hemocyanin", *Chem. Rev.*, **94,** 727–735 (1994).
9. Stenkamp, R. E., "Dioxygen and hemerythrin", *Chem. Rev.*, **94,** 715–726 (1994).
10. Kurtz, Jr, D. M., "Oxo- and hydroxo-bridged diiron complexes: a chemical perspective on a biological unit", *Chem. Rev.*, **90,** 585–606 (1990).

Specific References

11. Collman, J. P. and Fu, L., "Synthetic models for hemoglobin and myoglobin", *Acc. Chem. Res.*, **32,** 455–463 (1999).
12. Bourgeois, D., Vallone, B., Schotte, F., Arcovito, A., Miele, A. E., Sciara, G., Wulff, M., Anfinrud, P., and Brunori, M., "Complex landscape of protein structural dynamics unveiled by nanosecond Laue crystallography", *Proc. Natl. Acad. Sci. U.S.A.*, **100,** 8704–8709 (2003).
13. Schotte, F., Lim, M., Jackson, T. A., Smirnov, A. V., Soman, J., Olson, J. S., Phillips, G. N., Jr., Wulff, M., and Anfinrud, P. A., "Watching a protein as it functions with 150-ps time-resolved X-ray crystallography", *Science,* **300,** 1944–1947 (2003).
14. Frauenfelder, H., McMahon, B. H., and Fenimore, P. W., "Myoglobin: the hydrogen atom of biology and a paradigm of complexity", *Proc. Natl. Acad. Sci. U.S.A.*, **100,** 8615–8617 (2003).
15. Wittenberg, J. B. and Wittenberg, B. A., "Myoglobin function revisited", *J. Exp. Biol.*, **206,** 2011–2020 (2003).
16. Flögel, U., Merx, M. W., Gödecke, A., Decking, U. K. M., and Schrader, J., "Myoglobin: scavenger of NO", *Proc. Natl. Acad. Sci. U.S.A.*, **98,** 735–740 (2001).
17. Hill, A. V., "The possible effects of the aggregation of the molecules of haemoglobin on its dissociation curves", *J. Physiol.*, **40,** iv–vii (1910).
18. Adair, G. S., "The hemoglobin system. VI: The oxygen dissociation curve of hemoglobin", *J. Biol. Chem.*, **63,** 529–545 (1925).
19. Bohr, C., Hasselbach, K., and Krogh, A., "Ueber einen in biologischer Beziehung wichtigen Einfluss, den die Kohlensäurespannung des Blütes auf dessen Säurestoffbindung übt", *Skandinavisches Arc. Physiol.*, **16,** 402–412 (1904).
20. Root, R. W., "The respiratory function of the blood of marine fishes", *Biol. Bull.*, **61,** 427–456 (1931).
21. Monod, J., Wyman, J., and Changeux, J.-P., "On the nature of allosteric transitions: a plausible model", *J. Mol. Biol.*, **12,** 8–118 (1965).
22. Perutz, M. F., "Stereochemistry of cooperative effects in haemoglobin", *Nature (London)*, **228,** 726–739 (1970).
23. Perutz, M. F., Wilkinson, A. J., Paoli, M., and Dodson, G. G., "The stereochemical mechanism of the cooperative effects in hemoglobin revisited", *Ann. Rev. Biophys. Biomol. Structure,* **27,** 1–34 (1998).
24. Eaton, W. A., Henry, E. R., Hofrichter, J., and Mozzarelli, A., "Is cooperative binding by hemoglobin really understood?", *Nat. Struct. Biol.*, **6,** 351–357 (1999).
25. Ackers, G. K., "Deciphering the molecular code of hemoglobin allostery", *Adv. Protein Chem.*, **51,** 185–253 (1998). (A controversial alternative to the Perutz model of cooperativity.)
26. Bucci, E., Gryczynski, Z., Razynska, A., and Kwansa, H., "Entropy-driven intermediate steps of oxygenation may regulate the allosteric behavior of hemoglobin", *Biophys. J.*, **74,** 2638–2648 (1998).
27. Xu, C., Tobi, D., and Bahar, I., "Allosteric changes in protein structure computed by a simple mechanical model: hemoglobin T $\rightarrow$ R2 transition", *J. Mol. Biol.*, **333,** 153–168 (2003).
28. Colombo, M. F. and Bonilla-Rodriguez, G. O., "The water effect on allosteric regulation of hemoglobin probed in water/glucose and water/glycine solutions", *J. Biol. Chem.*, **271,** 4895–4899 (1996). This topic has generated controversy: see also Colombo et al. *Science,* **256,** 655–659 (1992); and correspondence, Bulone et al. *Science,* **259,** 1335–1336 (1993) and Colombo et al. *Science,* **259,** 1336 (1993).
29. Kundu, S., Trent, III, J. T., and Hagrove, M. S., "Plants, humans and hemoglobin", *Trends Plant Sci.*, **8,** 387–393 (2003).
30. Wittenberg, J. B., Bolognesi, M., Wittenberg, B. A., and Guertin, M., "Truncated hemoglobins: a new family of hemoglobins widely distributed in bacteria, unicellular eukaryotes, and plants", *J. Biol. Chem.*, **277,** 871–874 (2002).

31. Pesce, A., Bolognesi, M., Brocedi, A., Ascenzi, P., Dewilde, S., Moens, L., Hankeln, T., and Burmester, T., "Neuroglobin and cytoglobin. Fresh blood for the vertebrate globin family", *EMBO Rep.*, **3,** 1146–1151 (2002). See also from this group: Pesce et al. "Human neuroglobin structure reveals a distinct mode of controlling oxygen affinity", *Structure,* **11,** 1087–1095 (2003).
32. Springer, B. A., Sligar, S. G., Olson, J. S., and Phillips, G. N., "Mechanisms of ligand recognition in myoglobin", *Chem. Rev.*, **94,** 699–714 (1994).
33. Draghi, F., Miele, A. E., Travglini-Allocatelli, C., Vallone, B., Brunori, M., Gibson, Q. H., and Olson, J. S., "Controlling ligand affinity in myoglobin by mutagenesis", *J. Biol. Chem.*, **277,** 7509–7519 (2002).
34. Pauling, L., "Nature of the iron-oxygen bond in oxyhaemoglobin", *Nature (London),* **203,** 182–183 (1964).
35. Collman, J. P., "Synthetic models for oxygen binding hemoproteins", *Acc. Chem. Res.*, **10,** 265–272 (1977). [Summary of early work on picket-fence porphyrin models for O_2-binding hemoproteins.]
36. Cuff, M. E., Miller, K. I., van Holde, K. E., and Hendrickson, W. A., "Crystal structure of a functional unit from *Octopus* hemocyanin", *J. Mol. Biol.*, **278,** 855–870 (1998).
37. Kitajima, N., Fujisawa, K., Fujimoto, C., Yoshihiko, M.-o., Hashimoto, S., Kitagawa, T., Toriumi, K., Tatsumi, K., and Nakamura, A., "A new model for dioxygen binding in hemocyanin. Synthesis, characterization, and molecular structure of the μ-η^2:η^2 peroxo dinuclear copper(II) complexes, $[Cu(HB(3,5\text{-}R_2pz)_3]_2(O_2)$ (R = *i*-Pr and Ph)", *J. Am. Chem. Soc.*, **114,** 1277–1291 (1992).
38. Holmes, M. A., Le Trong, I., Turley, S., Sieker, L. C., and Stenkamp, R. E., "Structures of deoxy and oxy hemerythrin at 2.0 Å resolution", *J. Mol. Biol.*, **218,** 583–593 (1991).
39. Mizoguchi, T. J., Kuzelka, J., Spingler, B., DuBois, J. L., Davydov, R. M. K., Hedman, B., Hodgson, K. O., and Lippard, S. J., "Synthesis and spectroscopic studies of non-heme diiron(III) species with a terminal hydroperoxide ligand: models for hemerythrin", *Inorg. Chem.*, **40,** 4662–4673 (2001).
40. Shiemke, A. K., Loehr, T., and Sanders-Loehr, J., "Resonance Raman study of oxyhemerythrin and hydroxomethemerythrin. Evidence for hydrogen bonding of ligands to the Fe-O-Fe center", *J. Am. Chem. Soc.*, **108,** 2437–2443 (1986).
41. Brantley, R. E., Jr., Smerdon, S. J., Wilkinson, A. J., Singleton, E. W., and Olson, J. S., "The mechanism of autooxidation of myoglobin", *J. Biol. Chem.*, **268,** 6995–7010 (1993).
42. Shikama, K., "The molecular mechanism of autoxidation for myoglobin and hemoglobin: A venerable puzzle", *Chem. Rev.*, **98,** 1357–1373 (1998).
43. Scott, E. E., Gibson, Q. H., and Olson, J. S., "Mapping pathways for O_2 entry and exit from myoglobin", *J. Biol. Chem.*, **276,** 5177–5178 (2001).
44. de Sanctis, D., Dewilde, S., Pesce, A., Moens, L., Ascenzi, P., Hankeln, T., Burmester, T., and Bolognesi, M., "Mapping protein matrix cavities in human cytoglobin through Xe atom binding", *Biochem. Biophys. Res. Commun.*, **316,** 1217–1221 (2004).
45. Sigfridsson, E. and Ryde, U., "Theoretical study of the discrimination between O_2 and CO by myoglobin", *J. Inorg. Biochem.*, **91,** 101–115 (2002).
46. Spiro, T. G. and Kozlowski, P. M., "Will the real FeCO please stand up?", *J. Biol. Inorg. Chem.*, **2,** 516–520 (1997).
47. Slebodnick, C. and Ibers, J. A., "Myoglobin models and the steric origins of the discrimination between O_2 and CO", *J. Biol. Inorg. Chem.*, **2,** 521–525 (1997).
48. Vangberg, T., Bocian, D. F., and Ghosh, A., "Deformability of Fe(II)CO and Fe(III)CN groups in heme protein models: nonlocal density functional theory calculations", *J. Biol. Inorg. Chem.*, **2,** 526–530 (1997).
49. Lim, M., Jackson, T. A., and Afinrud, P. A., "Modulating carbon monoxide binding affinity and kinetics in myoglobin: roles of the distal histidine and the heme pocket docking site", *J. Biol. Inorg. Chem.*, **2,** 531–536 (1997).
50. Sage, J. T., "Myoglobin and CO: structure, energetics, and disorder", *J. Biol. Inorg. Chem.*, **2,** 537–543 (1997).
51. Olson, J. S. and Phillips, G. N., "Myoglobin discriminates between O_2, CO, and NO by electrostatic interactions with the bound ligand", *J. Biol. Inorg. Chem.*, **2,** 544–552 (1997).
52. Spiro, T. G. and Kozwolski, P. M., "Discordant results on FeCO deformability in heme proteins reconciled by density functional theory", *J. Am. Chem. Soc.*, **120,** 4524–4525 (1998).
53. Cheng, X. and Schoenborn, B. P., "Neutron diffraction study of carbonmonoxymyoglobin", *J. Mol. Biol.*, **220,** 381–399 (1991).
54. Phillips, S. E. V. and Schoenborn, B. P., "Neutron diffraction reveals oxygen-histidine hydrogen bond in oxymyoglobin", *Nature (London),* **292,** 81–82 (1981).
55. Unno, M., Christian, J. F., Olson, J. S., Sage, J. T., and Champion, P. M., "Evidence for hydrogen bonding effects in the iron ligand vibrations of carbonmonoxy myoglobin", *J. Am. Chem. Soc.*, **120,** 2670–2671 (1998). This work, and other work cited therein, contradict the neutron diffraction analysis (53) that there is no proton (deuteron) between the proximal histidine and the coordinated CO ligand.
56. Ray, G. B., Li, X. Y., Ibers, J. A., Sessler, J. L., and Spiro, T. G., "How far can proteins bend the FeCO unit? Distal polar and steric effects in heme proteins and models", *J. Am. Chem. Soc.*, **116,** 162–176 (1994).
57. Franzen, S., "An electrostatic model for the frequency shifts in the carbonmonoxy stretching band of myoglobin: correlation of hydrogen bonding and Stark tuning rate", *J. Am. Chem. Soc.*, **124,** 13271–13281 (2002).
58. Yang, J., Kloek, A. P., Goldberg, D. E., and Mathews, F. S., "The structure of *Ascaris* hemoglobin domain-I at

2.2 Å resolution: molecular features of oxygen avidity", *Proc. Natl. Acad. Sci. U.S.A.*, **92,** 4224–4228 (1995).
59. Barrick, D., "Replacement of the proximal ligand of sperm whale myoglobin with free imidazole in the mutant His-93 → Gly", *Biochemistry,* **33,** 6546–6554 (1994).
60. Barrick, D., Ho, N. T., Simplaceanu, V., Dahlquist, F. W., and Ho, C., "A test of the role of the proximal histidines in the Perutz model for cooperativity in hemoglobin", *Nat. Struct. Biol.*, **4,** 78–83 (1997).
61. Samuni, U., Juszczak, L., Dantsker, D., Khan, I., Friedman, A. J., Perez-Gonzalez-De-Apodaca, J., Bruno, S., Hui, H. L., Colby, J. E., Karasik, E., Kwiatkowski, L. D., Mozzarelli, A., Noble, R., and Friedman, J. M., "Functional and spectroscopic characterization of half-liganded iron-zinc hybrid hemoglobin: Evidence for conformational plasticity within the T state", *Biochemistry,* **42,** 8272–8288 (2003).
62. Blough, N. V. and Hoffman, B. M., "Carbon monoxide binding to the ferrous chains of manganese, iron [Mn,Fe(II)] hybrid hemoglobins: pH dependence of the chain affinity constants associated with specific hemoglobin ligation pathways", *Biochemistry,* **23,** 2875–2882 (1984).
63. Bowen, J. H., Shokhirev, N. V., Raitsimring, A. M., Buttlaire, D. H., and Walker, F. A., "EPR studies of the dynamics of rotation of dioxygen in model cobalt(II) hemes and cobalt-containing hybrid hemoglobins", *J. Phys. Chem. B,* **101,** 8683–8691 (1997).
64. Venkatesh, B., Manoharan, P. T., and Rifkind, J. M., "Metal ion reconstituted hybrid hemoglobins", *Prog. Inorg. Chem.*, **47,** 563–684 (1998).
65. Klinger, A. L. and Ackers, G. K., "Analysis of spectra from multiwavelength oxygen-binding studies of mixed metal hybrid hemoglobins", *Methods Enzymol.*, **295,** 190–207 (1998).
66. Miyazaki, G., Morimoto, H., Yun, K.-M., Park, S. Y., Nakagawa, A., Minagawa, H., and Shibayama, N., "Magnesium(II) and zinc(II)-protoporphyrin IX's stabilize the lowest oxygen affinity state of human hemoglobin even more strongly than deoxyheme", *J. Mol. Biol.*, **292,** 1121–1136 (1999).
67. Unzai, S., Eich, R., Shibayama, N., Olson, J. S., and Morimoto, H., "Rate constants for O_2 and CO binding to the α and β subunits within the R and T states of human hemoglobin", *J. Biol. Chem.*, **273,** 23150–23159 (1998).

XI.5. Dioxygen Activating Enzymes

Lawrence Que, Jr.
Department of Chemistry and Center for Metals in Biocatalysis
University of Minnesota
Minneapolis, MN 55455

Contents

XI.5.1. Introduction: Converting Carriers into Activators

The activation of dioxygen (O_2) is an extremely important reaction in biology. As a consequence, Nature has evolved a diverse array of metalloenzymes to activate O_2 and utilize its oxidative power to drive transformations critical to the metabolisms of aerobic organisms.[1–8] Such enzymes are involved in the conversion of methane to methanol in methanotrophic bacteria, in amino acid modification and peptide processing, in the biosynthesis of neurotransmitters, hormones, and antibiotics, and in the detoxification of xenobiotics. These enzymes catalyze the functionalization of aliphatic C–H bonds into alcohols, alkenes, or heterocycles, the epoxidation of olefins, and the hydroxylation or cis-dihydroxylation of arenes. The efficiency and specificity of the enzymes that activate O_2 have raised important mechanistic questions as to the nature of the chemical species that carry out such reactions.

The three known types of O_2 carriers, namely, myoglobin (Mb)–hemoglobin (Hb), hemocyanin, and hemerythrin (Hr) (Fig. XI.5.1*a*; see also Section X.4) serve as an appropriate starting point for a discussion of oxygen-activating metalloenzymes, since oxygen binding to a metal center is the first step in oxygen activation. In the discussion below, these oxygen carriers will be compared to corresponding oxygen-activating enzymes, cytochrome P450,[1] tyrosinase,[4] and methane monooxygenase (MMO),[2,3] respectively (Fig. XI.5.1*b*), to understand how Nature has converted an oxygen-binding center into one that promotes O–O cleavage. Indeed, the challenge of oxygen activation is to harness the tremendous oxidizing power associated with the scission of the O–O bond. As readily evident from an examination of the redox properties of O_2 (Section XI.1), the reduction of O_2 is thermodynamically unfavorable unless facilitated by the participation of protons to neutralize the negative charge of the incipient anions. For example, the conversion of superoxide anion to peroxide dianion has a very negative potential in aprotic media ($E_{1/2} < -2.00$ V), but becomes quite favorable at pH 7 ($E_{1/2} = +0.89$ V). Figure XI.5.2 shows a common mechanistic framework that links the three enzymes in question. There are two key intermediates: a peroxo species and a high-valent metal–oxo species. In this section we will discuss the principles by which an O_2 adduct is converted to an oxidizing species.

XI.5.1.1. Cytochrome P450: The Heme Paradigm

Due to its many critical functions in mammalian metabolism, including the biosynthesis of steroids, the detoxification of xenobiotics, drug metabolism, and carcinogenesis, cytochrome P450 is the most studied and best understood oxygen-activating metalloenzyme.[1] The active site of cytochrome P450 differs from that of hemoglobin in that the former has a proximal cysteinate ligand, instead of a His. Due to this ligand, the Soret band of the iron^{II}–CO adduct is found near 450 nm (instead of 420 nm in Mb–CO), which gives rise to its name.

Substantial mechanistic insight from a variety of approaches has been obtained for the role of the heme cofactor. Fig. XI.5.3 summarizes the principal features of the cytochrome P450 mechanism, deduced mainly from spectroscopic and crystallographic studies of P450$_{cam}$, the enzyme from *Pseudomonas putida*, which hydroxylates

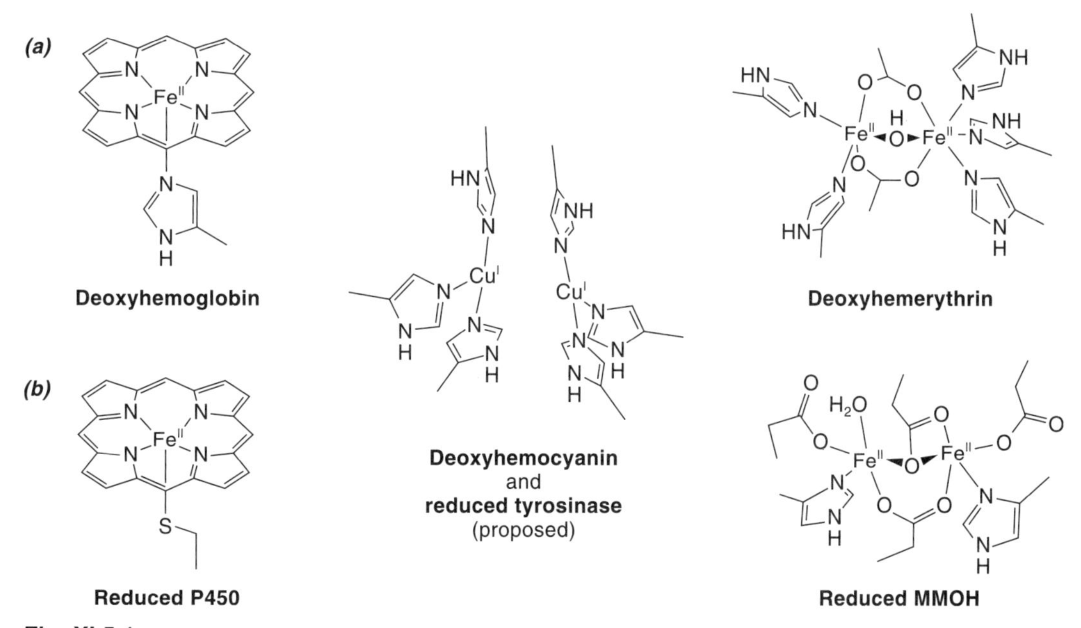

Fig. XI.5.1.
Active sites of the three known O_2 carriers (*a*) and their counterparts in oxygen activation (*b*). (MMOH = methane monooxygenase hydroxylase)

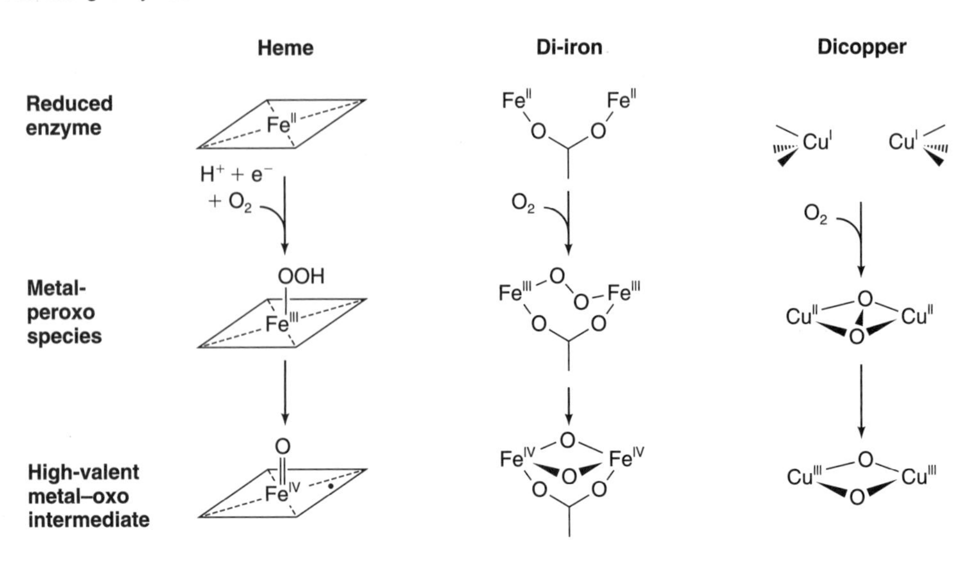

Fig. XI.5.2. A common framework for the proposed mechanisms of cytochrome P450, MMOH, and tyrosinase.

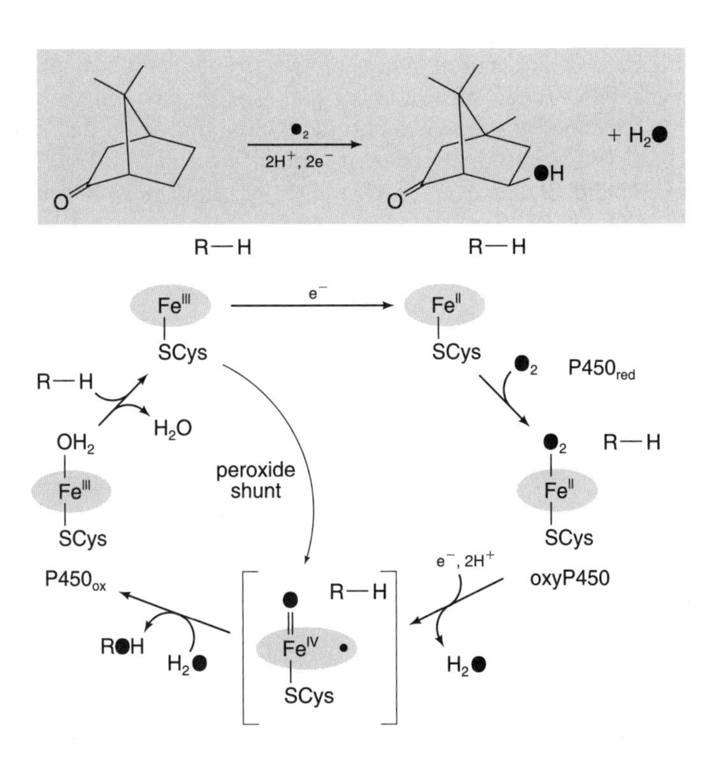

Fig. XI.5.3. Catalytic mechanism for cytochrome P450.

camphor at the C5 carbon atom in the exo position. The starting point is the as isolated low-spin Fe(III) form, $P450_{ox}$. Substrate binding converts it to a high-spin form, which allows it to be reduced by putidaredoxin, its electron-donor partner, to its high-spin Fe(II) form, $P450_{red}$. Dioxygen then binds to form oxyP450 in a mode analogous to that found in oxyhemoglobin. The O_2 ligand is found to be hydrogen bonded to the conserved Thr252 and to a water molecule that becomes well ordered in this structure (Fig. XI.5.4*a*).[9] The complex is now poised for the key step in the catalytic mechanism.

This next step is the reduction of oxyP450 by a second electron from putidaredoxin, which initiates O–O bond cleavage and substrate oxidation. The transient nature of the species involved in this step has made its characterization challenging. To

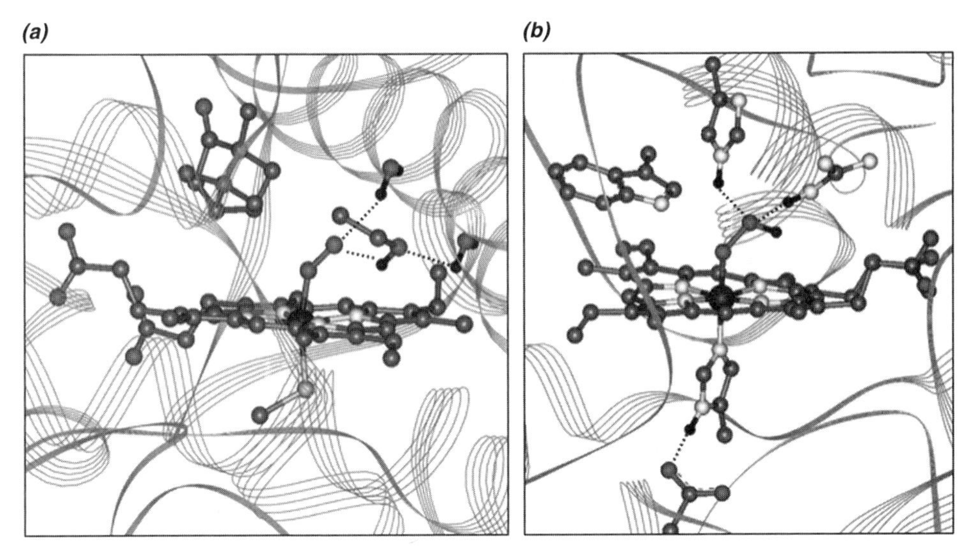

Fig. XI.5.4.
(*a*) Active site of oxyP450.
(*b*) Postulated H_2O_2 adduct to cytochrome *c* peroxidase (CcP) based on the crystal structure of the native enzyme (PDB code: 2CYP). Dashed lines show key hydrogen-bonding interactions important for the heterolysis of the O–O bond.

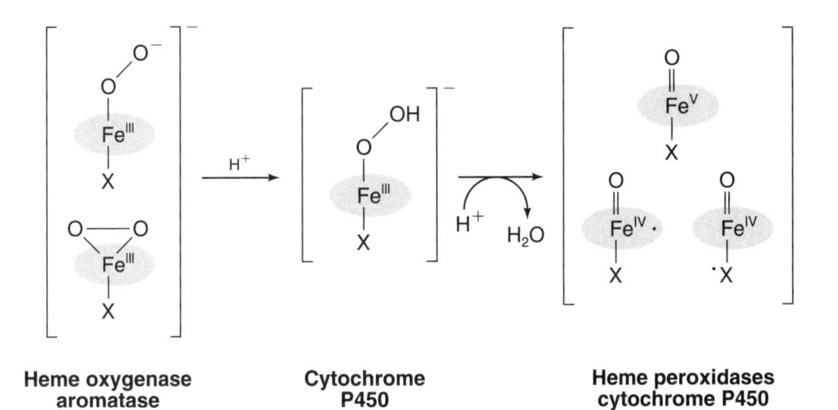

Fig. XI.5.5.
Proposed active species that may form upon one-electron reduction of oxyheme enzymes.

serve as a starting point for a discussion of this step, Fig. XI.5.5. illustrates some of the possible structures that have been considered for the intermediates. The addition of the second electron to oxyP450 can in principle generate in sequence an iron(III)–peroxo species, an iron(III)–hydroperoxo species by protonation, and a formally iron(V)–oxo species derived from heterolysis of the O–O bond. These three species are all plausible candidates and may all participate in the catalytic mechanism. Insights into these three postulated structures, discussed below, derive from model compounds, computational studies, cryo-reduction experiments, and analogies to the better characterized peroxide activation mechanism of heme peroxidases (see Section XI.3).

Fe(III)–Peroxo. The injection of an electron into oxyP450 should afford, at least initially, an $[Fe^{III}–O_2]^-$ moiety. Such a species may adopt the end-on (η^1) bent dioxy structure found in oxyMb and oxyHb or it could adopt a side-on (η^2) mode. Synthetic high-spin $[(P)Fe^{III}–O_2]^-$ complexes have been obtained in aprotic solvents.[10] It is proposed that the peroxide anion is coordinated in an η^2 mode, pulling the Fe center out of the porphyrin plane, by analogy to the crystallographically characterized Mn analogue.[11] Reactivity studies suggest that such η^2-peroxo species cannot act as electrophilic oxidants. Instead they serve as sources of a nucleophilic peroxide that can attack electron-poor centers. Such peroxo species are thus unlikely to be the key oxidizing species for alkane hydroxylation or olefin epoxidation in the P450 mechanism.

Fe(III)–Hydroperoxo. Protonation of the distal peroxo oxygen very likely generates a six-coordinate low-spin Fe^{III}–OOH intermediate. Such a species has recently been observed by cryo-reduction of oxyP450.[12] By using γ-rays to generate solvated electrons, an oxyheme center can be converted to its one-electron reduced form at cryogenic temperatures. Thus irradiation of oxymyogolobin at 77 K affords an EPR signal typical of a low-spin Fe(III) heme center that is associated with the one-electron reduced Fe^{III}–(η^1-O_2) species. Subsequent annealing at 200 K converts this species to a species with a different low-spin iron(III) EPR signal that is attributed to the Fe^{III}–(η^1-OOH) intermediate. When oxyP450 is cryoreduced, the species observed at 77 K has an EPR spectrum corresponding to that of cryoreduced oxyMb annealed at 200 K. Thus a proton is delivered immediately to the O_2 ligand upon reduction of oxyP450, a conclusion confirmed by electron–nuclear double resonance (ENDOR) measurements. Gas chromatography (GC) analysis of this sample upon warming to room temperature shows the expected hydroxylated camphor product, demonstrating that cryoreduction generates a catalytically competent species.

Computational studies of the low-spin Fe^{III}–OOH intermediate suggest that this species has electronic properties that move the peroxo species along a trajectory leading to O–O bond cleavage.[13] It is found that the Fe–O bond is strengthened, while the O–O bond is weakened in such a species, that is,

$$Fe^{III}\text{–O–O}^- \rightarrow Fe^{III}\underset{\cdots}{–}\text{O}\cdots\text{OH} \rightarrow Fe^{V}\text{=O} + OH^-$$

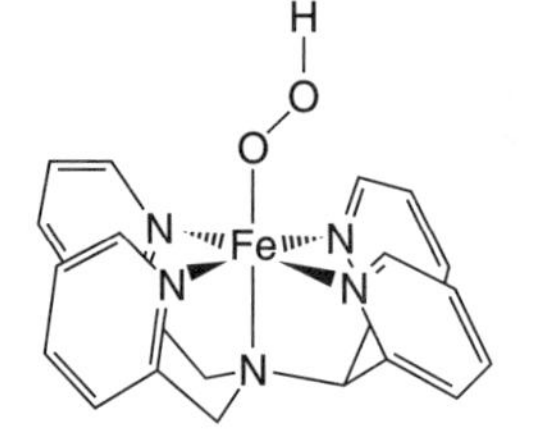

Fig. XI.5.6.
The complex $[Fe^{III}(N4Py)(\eta^1\text{-}OOH)]^{2+}$ is a spectroscopically characterized low-spin Fe^{III}–OOH species (Py = pyridine).

While it has not yet been possible to test whether this description applies to heme complexes, resonance Raman studies on related nonheme low-spin iron(III) complexes, such as $[Fe(N4Py)(\eta^1\text{-}OOH)]^{2+}$ (Fig. XI.5.6), support this notion.[14] This complex exhibits a ν(Fe–O) of 630 cm^{-1} and a ν(O–O) of 790 cm^{-1}, compared to values of 503 and 844 cm^{-1}, respectively, for the high-spin Fe^{III}–OOH species in oxyhemerythrin.[15] So the low-spin Fe(III) center does appear to make the Fe–O bond stronger and the O–O bond weaker, thus promoting O–O bond cleavage.

High-Valent Iron–Oxo. A high-valent intermediate has not yet been unequivocally observed in the catalytic cycle of cytochrome P450, but there is strong indirect evidence for its involvement. First, Fe(III)–cytochrome P450 can react with peroxides and oxygen atom donors such as C_6H_5IO and ClO^- to generate a species capable of performing the same reactions as the native enzyme.[16] This reaction bypasses the Fe(II) oxidation state in the P450 cycle and is known as the peroxide shunt (Fig. XI.5.3). Second, related heme peroxidases react with H_2O_2 to afford a transient high-valent intermediate called Compound **I**. Compound **I** is formally Fe^{V}=O, but has been spectroscopically characterized for horseradish peroxidase (HRP) as an oxoiron(IV) porphyrin radical complex.[1]

The crystal structures of several heme peroxidases show a number of conserved residues in the heme pocket, which are postulated to be involved in the catalytic mechanism (Fig. XI.5.4*b*) (see Section X.3).[1] On the distal side is a conserved His residue that is proposed to accept a proton from the proximal oxygen of H_2O_2 as it coordinates to the Fe(III) center. The resultant imidazolium ion, together with the nearby conserved Arg residue, then protonates the distal oxygen of the bound hydroperoxide to promote the expulsion of H_2O. On the proximal side, the His ligand is hydrogen bonded to a conserved Asp residue, which serves to make the His a better electron donor to stabilize the incipient high-valent metal center. Thus the heterolysis of the O–O bond to afford compound **I** is promoted by a "push–pull" mechanism, whereby the low-spin Fe(III) center, together with the axial His ligand, strengthens the Fe–O bond to provide the "push" and the conserved His-Arg motif "pulls" off a water molecule by protonating the distal hydroperoxo oxygen.

A similar mechanism may be envisioned for cytochrome P450; but what takes the place of the peroxidase His-Arg motif to deliver the protons needed for O–O bond cleavage? As pointed out earlier, a water molecule becomes well ordered in the active site upon O_2 binding (Fig. XI.5.4*a*). It is postulated that this water molecule is polarized by the highly conserved Asp251 and Thr252 residues that are hydrogen bonded to it and thus serves as the needed acid. It is proposed that the weaker "pull" due to the lower acidity of the water molecule relative to the imidazolium ion in peroxidases may be compensated for by a stronger "push" from the more basic Cys in place of His as the proximal ligand.

Crystallographic evidence has now been obtained for the O–O bond cleavage step of cytochrome P450. Schlichting et al.[9] accomplished this feat by cryoreduction of oxyP450 crystals with long wavelength X-rays from a synchrotron. In a comparison of oxyP450 and cryoreduced oxyP450 crystals, they have observed the loss of electron density from the distal oxygen of the O_2 ligand (Fig. XI.5.7). This exciting result must be interpreted cautiously, since the experiments are challenging and conversion from oxyP450 to its one-electron reduced form is very likely incomplete. Nevertheless, the authors favor an interpretation wherein an iron(IV)–oxo species ($r_{Fe-O} \sim 1.65$ vs. 1.80 Å for oxyP450) has been formed. To support this interpretation, the cryoreduced oxyP450 crystals, when warmed up to allow the intermediate to decay, are converted to crystals of the product complex. Thus, cryoreduction of oxyP450 generates an intermediate catalytically competent to hydroxylate camphor.

The last point to be addressed in the cytochrome P450 mechanism is the hydroxylation of substrate. Such a reaction can be thought of as a one-step direct oxygen-atom insertion into the target C–H bond or a two-step process known as oxygen rebound (Fig. XI.5.8).[1] The latter mechanism involves hydrogen-atom abstraction by the formally iron(V)–oxo species to form an iron(IV)–hydroxo species and a short-lived alkyl radical, rapidly followed by C–O bond formation. The second step must occur faster than the epimerization of the alkyl radical to account for the

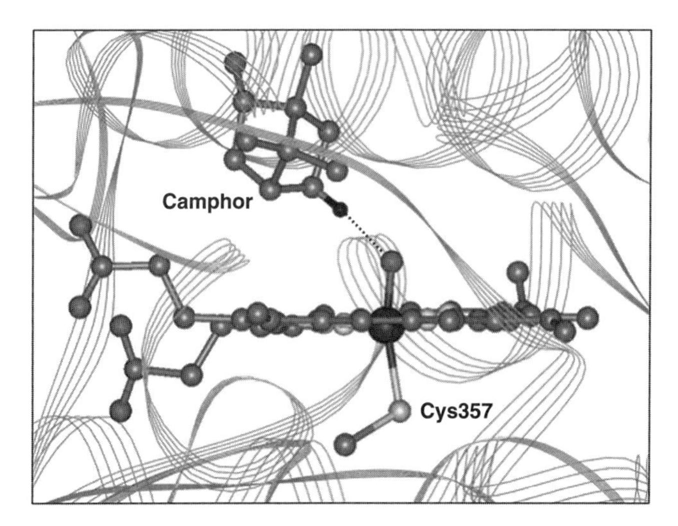

Fig. XI.5.7.
Structure of putative ferryl species in cytochrome P450 obtained by cryoreduction of oxyP450. Note that the ferryl oxygen is poised only 2.05 Å away from the 5-exo-hydrogen of camphor.

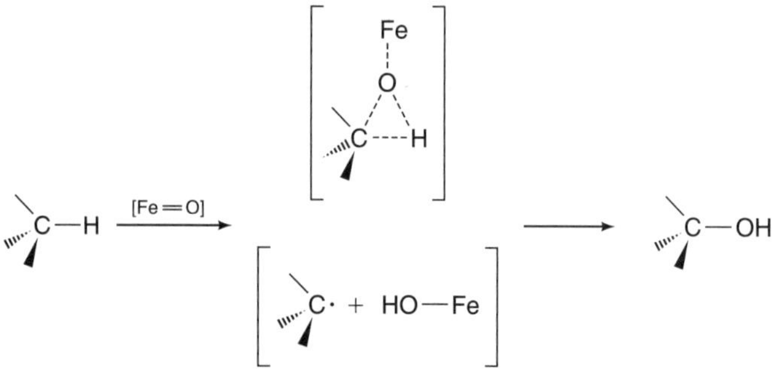

Fig. XI.5.8.
Postulated C–H hydroxylation mechanisms.

stereo- and enantioselectivity of hydroxylation. The oxygen rebound mechanism has been generally accepted on the basis of the accumulated evidence, which includes: (1) large isotope effects for substrate C–H bond cleavage ($k_H/k_D > 11$), (2) stereochemical scrambling in the hydroxylation of deuterated camphor and norbornane, and (3) observation of rearranged products in the hydroxylation of certain radical clock substrates.

However, more recent studies using ultrafast radical clock substrates with rearrangement times in the femtosecond time scale have reopened discussion on this question.[17] Because the short rearrangement times of the ultrafast probes are on the time scale of a vibration, the question has been raised to whether a two-step mechanism is viable on so short a time scale. Furthermore, there is increasing evidence to suggest that there is more than one viable oxidant possible in heme enzyme mechanisms. Besides the high-valent iron–oxo species, the iron(III)–hydroperoxo intermediate has been proposed to have the electrophilicity required to carry out the substrate hydroxylation.[17] A similar electrophilic iron(III)–peroxo species is proposed to play a key role in the mechanism of heme oxygenase (involved in mammalian heme degradation),[18] while a nucleophilic iron(III)–peroxo species is suggested to carry out the final step of estrone biosynthesis by the aromatase enzyme.[1] Clearly, the heme enzyme active site is a versatile cavity within which a number of reactive iron–dioxygen species can be generated and utilized to carry out a remarkable range of oxidative transformations.

XI.5.1.2. Monooxygenases with Dinuclear Active Sites

The other examples discussed in this section, tyrosinase and MMOH, are enzymes that have dinuclear active sites. The introduction of a second metal ion to an active site for oxygen activation has important consequences. First, the second metal ion serves as an additional source of electrons to reduce O_2 to the peroxide level without a requirement for an external electron donor. [Recall in cytochrome P450 that an electron from putidaredoxin is needed to reduce oxyP450 and generate the iron(III)–peroxo intermediate.] Second, the second metal ion can serve as an additional Lewis acid in place of a proton to help cleave the O–O bond and stabilize the incipient oxide ions that form upon O–O bond cleavage.

A Monooxygenase with a Dicopper Active Site. Tyrosinase is the only monooxygenase thus far known to have an active site closely related to that of the oxygen carrier hemocyanin. This enzyme catalyzes the hydroxylation of tyrosine to dopa [3-(3,4,-dihydroxyphenyl)alanine] and the oxidation of dopa to dopaquinone to initiate the browning reaction observed in freshly cut fruits and vegetables as well as the biosynthesis of melanin for vertebrate pigmentation.[4]

COO$^-$ NH$_3^+$ OH $\xrightarrow[2\,H^+,\,2\,e^-]{O_2}$ COO$^-$ NH$_3^+$ OH OH $\longrightarrow$ COO$^-$ NH$_3^+$ O O

$+ H_2O$ $+ 2\,H^+ + 2\,e^-$

Although a crystal structure is not yet available for tyrosinase, spectroscopic comparisons show that it has an active site similar to that of hemocyanin.[19] Oxytyrosinase exhibits spectroscopic features that closely match those of oxyhemocyanin.[4] First, its UV–vis spectrum exhibits an intense absorption at 345 nm and a much weaker band at 590 nm. Second, resonance Raman studies show a ν(O–O) at 755 cm^{-1} that is associated with a μ-η^2:η^2-bound peroxide, indicating the presence of a weakened O–O bond. Finally, EXAFS studies of the oxytyrosinase from *Neurospora crassa* show a Cu–Cu distance of 3.6 Å. Thus, oxytyrosinase and oxyhemocyanin are likely to have nearly identical Cu_2O_2 core structures. The two sites may differ in the ability of tyrosinase to bind substrate and allow it to come into close contact with the active oxygen species. Such a substrate-binding pocket is found in the crystal structure of catechol oxidase, another dicopper enzyme with an active site very similar to that of hemocyanin that only catalyzes the oxidation of catechols to quinones.[20] It has been proposed that the phenolic substrate can bind to one of the Cu ions in the catechol oxidase active site. Further support for this notion comes from the observation of tyrosinase activity for the hemocyanin from tarantula after limited proteolysis to provide access to the dicopper active site.[21]

Many of the mechanistic insights into O_2 activation at a dicopper center have come from biomimetic efforts, due to the large number of crystal structures now available of oxygen intermediates derived from the reactions of a Cu(I) complex with O_2.[22–26] Three core structures have been observed, all with the basic Cu_2O_2 formula (Fig. XI.5.9). The Cu_2(*trans*-μ-1,2-O_2) core is found in the complex with the tetradentate ligand TPA.[22] Here, the Cu ions are five-coordinate and the peroxide acts as an η^1-ligand. The use of the sterically bulky, but tridentate ligand $Tp^{i\text{-}Pr_2}$ affords an O_2 adduct with a $Cu_2(\mu$-η^2:η^2-$O_2)$ core.[23] With the less sterically bulky tridentate ligand Bn_3TACN[24] or the bidentate diamine ligand Me_4CHD,[25] a $Cu_2^{III}(\mu$-$O)_2$ core is obtained. These three basic structural motifs show the range of copper–dioxygen interactions possible with a dinuclear center. The Cu–Cu distance goes from 4.4 Å in the first motif to 2.8 Å in the third, and the O–O bond becomes weakened and then broken along this trajectory. In the μ-1,2-peroxo complex, the η^1-ligand simply acts as a σ donor. The side-on binding mode in the μ-η^2:η^2-peroxo complexes allows the peroxide to act both as a σ donor and a π acceptor. The back-bonding implied in the latter interaction transfers some electron density from the Cu $d_{x^2-y^2}$ orbitals to the peroxo σ^* orbital, resulting in the weaker O–O bond observed in the Raman spectra of these complexes (Fig. XI.5.10).[27] In the complexes with a $Cu_2^{III}(\mu$-$O)_2$ core, the metal-to-ligand charge transfer (MLCT) is complete, so the Cu ions are in the Cu(III) state and the O–O bond is cleaved.[28]

A perusal of the properties of the various Cu_2O_2 complexes suggests that the availability of coordination sites and steric bulk control which core structure is adopted. The bulky isopropyl groups on the Tp ligand likely protect the bound O_2 and stabilize the $Cu_2^{II}(\mu$-η^2:η^2-$O_2)$ core to allow its crystallographic characterization. Moreover, the bulky substituents further prevent the Cu ions from getting even closer and thus hinder O–O bond scission. The TACN and TPA complexes illustrate additional principles. For example, a μ-η^2:η^2-peroxo species can be obtained with

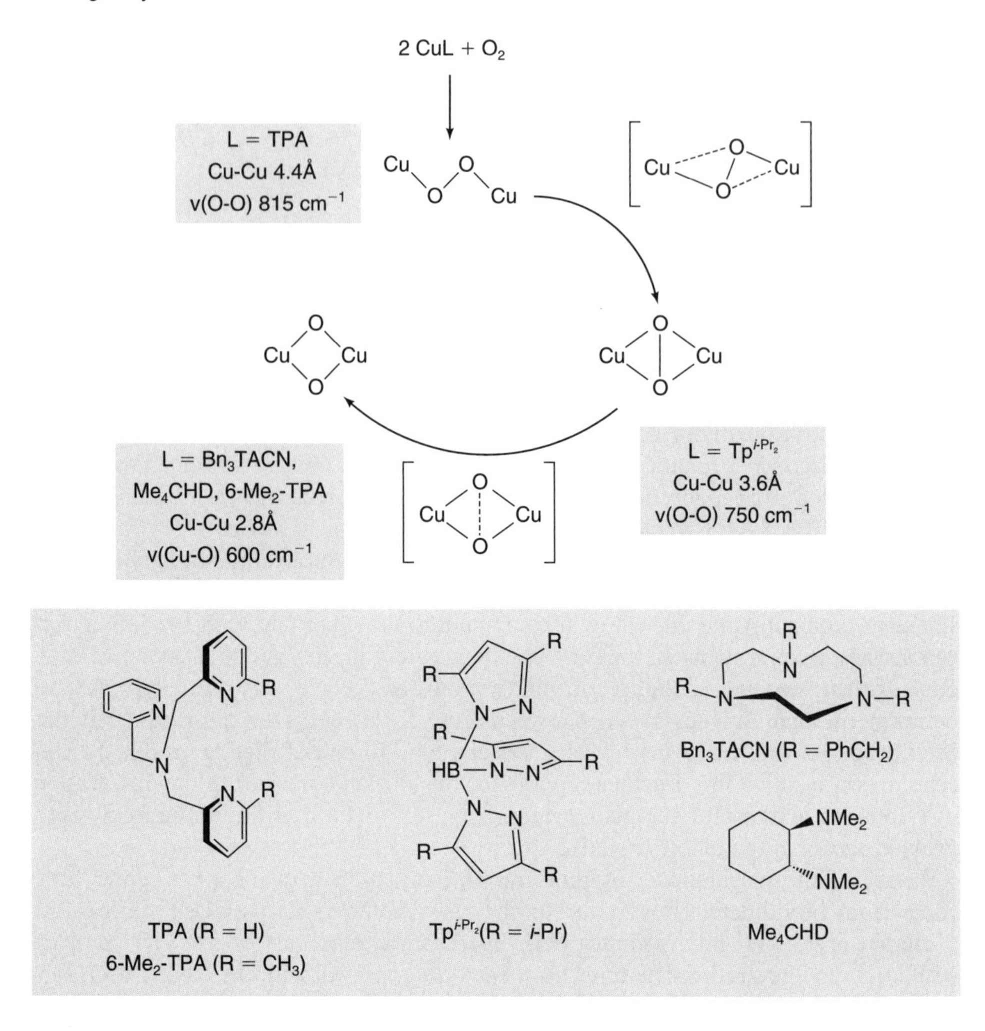

Fig. XI.5.9.
The reaction of a Cu(I) precursor with O_2. (TPA = tris (2-pyridylmethyl) amine; Tp = hydridotris (pyrazolyl) borate monoanion; TACN = 1,4,7-triazacyclononane; Bn = benzyl; Me = CH_3)

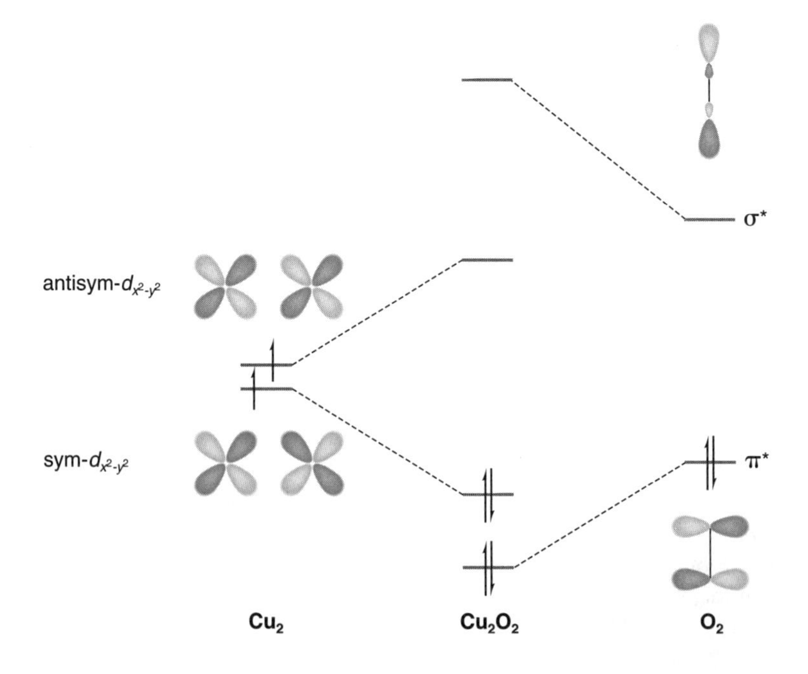

Fig. XI.5.10.
Simple molecular orbital picture for the interaction of a Cu unit with dioxygen.

i-Pr_3TACN, but, when the isopropyl groups are replaced by less bulky benzyl groups, a complex with a $Cu_2^{III}(\mu\text{-}O)_2$ core is obtained.[24] Even more interesting is the finding that the *i*-Pr_3TACN complex can adopt either core structure depending on the solvent and the counterion and that the two cores are in equilibrium with each other.[24] In the cases of TPA and 6-Me_2-TPA, the former favors the $Cu_2(\mu\text{-}1,2\text{-}O_2)$ core,[22] while the latter adopts the $Cu_2(\mu\text{-}O)_2$ core.[26] The crystal structure of the latter shows that the sterically bulkier 6-Me-pyridine pendant arms occupy the weaker axial coordination sites and have much longer Cu–N bond lengths to accommodate the formation of the $Cu_2(\mu\text{-}O)_2$ core. This complex can be converted back to its Cu(I) precursor by treatment with $P(C_6H_5)_3$, showing that the O–O bond broken in the formation of the $Cu_2(\mu\text{-}O)_2$ core can be re-formed, as found for the *i*-Pr_3TACN complex.[24] These observations illustrate how shallow the potential energy surface for the various Cu_2O_2 cores can be. Assuming that the reduced form of tyrosinase has an active-site structure akin to that of deoxyhemocyanin with a Cu–Cu distance of 4.6 Å, it may not be too farfetched to imagine that the dicopper active site traverses through all three core structures in the course of oxygen activation.

As with the heme enzymes, the question of which is the active oxygen species in the tyrosinase mechanism has not yet been resolved.[29] While a $Cu_2(\mu\text{-}1,2\text{-}O_2)$ moiety can easily be eliminated due to the observation of a $Cu_2(\mu\text{-}\eta^2{:}\eta^2\text{-}O_2)$ core for oxytyrosinase, tyrosine hydroxylation may occur by direct attack of the $Cu_2(\mu\text{-}\eta^2{:}\eta^2\text{-}O_2)$ core on substrate or via the isomeric $Cu_2^{III}(\mu\text{-}O)_2$ core. Model studies indicate that either mechanism is plausible. There are examples of intramolecular arene hydroxylation in which one or the other intermediate has been observed prior to the attack of the arene (Fig. XI.5.11).[29]

Finally, it is interesting to note that all the biomimetic reactions discussed above occur in aprotic solvents. Thus, unlike heme systems, these dicopper complexes do not appear to require the participation of protons for O–O cleavage. The metal ions themselves are sufficient to promote O–O bond lysis. Because the proposed copper–dioxygen species all have rather symmetric structures, it appears likely that O–O bond cleavage proceeds by homolysis.

A Monooxygenase with a Nonheme Di-iron Active Site. Methane monooxygenase is a multicomponent enzyme from methanotrophs that catalyzes the hydroxylation of methane to methanol.[2] This enzyme consists of a hydroxylase, a reductase, and a third component that regulates the activities of the first two. Dioxygen activation

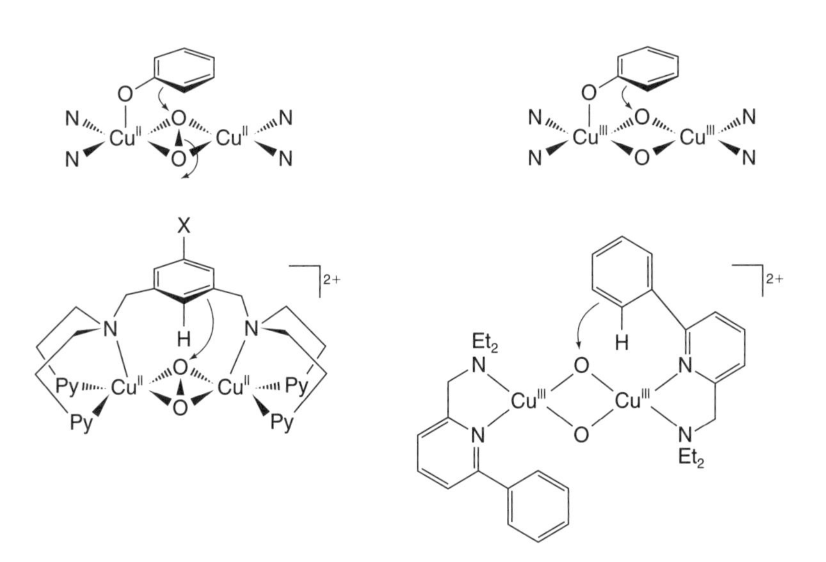

Fig. XI.5.11.
Intermediates proposed for arene hydroxylation in tyrosinase and biomimetic Cu complexes (Et = ethyl).

occurs at the hydroxylase component (MMOH), using electrons from NADH that are funneled in through the reductase component.

The MMOH component has a nonheme di-iron center whose core is similar to that of the oxygen carrier hemerythrin, but the two ligand environments are quite different (Fig. XI.5.1). This oxygen carrier–oxygenase pair is thus unlike the hemocyanin–tyrosinase pair, where the first coordination spheres are probably identical, and the hemoglobin–cytochrome P450 pair, where the two first coordination spheres differ only in the nature of the axial ligand. While deoxyhemerythrin has an $Fe_2(\mu\text{-}OH)(\mu\text{-}RCO_2)_2$ core with five terminal His ligands,[30] reduced MMOH has an $Fe_2(\mu\text{-}RCO_2)_2$ core with two terminal His and two terminal carboxylate ligands.[31] Furthermore, deoxyhemerythrin has only one site for exogenous ligand binding, while reduced MMOH has an available site on each iron. Note that closely related nonheme di-iron active sites are found in the R2 protein of the ribonucleotide reductase from *Escherichia coli* and soluble fatty acid desaturases.[31] In all three enzymes, there is a pair of (D/E)XXH sequence motifs that characterize this subclass of oxygen-activating nonheme di-iron active sites. The di-iron sites of ribonucleotide reductase and fatty acid desaturase activate O_2 to carry out the oxidation of a tyrosine residue to its catalytically essential radical form and the dehydrogenation of fatty acids, respectively. The carboxylate-rich ligand environment found in these enzymes may be important for the activation of O_2 at nonheme di-iron sites.

Rapid kinetics studies of the reaction of reduced MMOH with O_2 show the participation of two key intermediates that have been spectroscopically characterized. The first intermediate, called **P** or $\mathbf{H_{peroxo}}$, is associated with a broad absorption centered near 700 nm, suspected to be a peroxo-to-iron(III) charge-transfer band. Similar intermediates have been observed in the oxygen activation cycles of a mutant ribonucleotide reductase and a fatty acid desaturase. No reliable resonance Raman data are yet available for MMOH-**P**, but the other two intermediates have been found to exhibit ν(Fe–O) and ν(O–O) features associated with (μ-1,2-peroxo)diiron(III) species.[32,33] All three intermediates exhibit Mössbauer isomer shifts of $\sim 0.66\ \text{mm s}^{-1}$, a value that appears to be characteristic of such species, but one that is higher than normally observed for high-spin iron(III) centers.[34–36]

Interestingly, this unusual Mössbauer isomer shift has been observed for a synthetic complex, $[FeTp^{i\text{-}Pr_2}(O_2CCH_2Ph)]_2O_2$, (Fig. XI.5.12*a*), which is obtained from the oxygenation of the precursor iron(II) complex.[37] This O_2 adduct has been crystallized and shown to have a (μ-1,2-peroxo)bis(μ-carboxylato)di-iron(III) core. It is

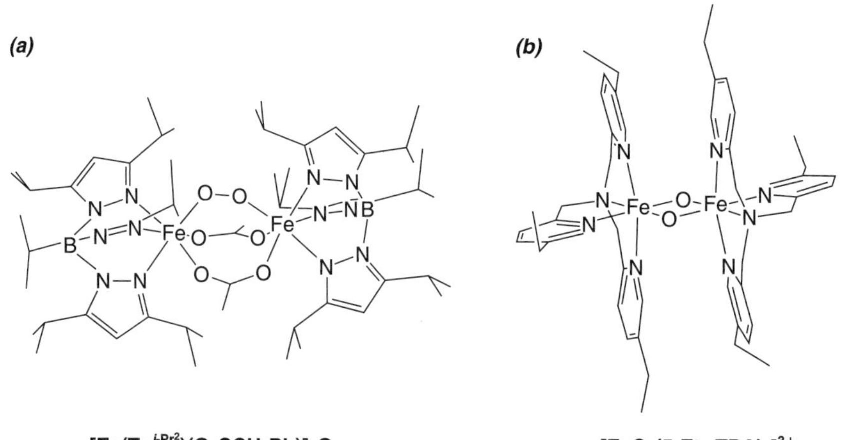

(a) $[Fe(Tp^{i\text{-}Pr^2})(O_2CCH_2Ph)]_2O_2$

(b) $[FeO_2(5\text{-}Et_3\text{-}TPA)_2]^{3+}$

Fig. XI.5.12. Synthetic precedents for the core structures proposed for MMOH-**P** (*a*) and MMOH-**Q** (*b*).

plausible that the three enzyme intermediates have a similar bridged peroxo structure, which differs from the Fe(III)-η^1-OOH structure found for oxyhemerythrin.[30] This structural difference may be due to the availability of coordination sites on both metal ions in the active sites of the oxygen-activating enzymes.

The second intermediate observed in the MMOH cycle is intermediate **Q,** which has been shown to be kinetically competent to oxidize methane to methanol.[2] This intermediate has a Mössbauer spectrum indicative of an antiferromagnetically coupled di-iron(IV) center, the Fe(IV) oxidation state suggested by the relatively small isomer shift observed [$\delta = 0.17(4)\,\mathrm{mm\,s^{-1}}$].[34,38] The change in iron oxidation state implies that the O–O bond present in intermediate **P** has been cleaved upon conversion to this intermediate. The EXAFS analysis of this intermediate shows an Fe–Fe distance of 2.5 Å,[39] which has been used to implicate the presence of an $Fe_2(\mu\text{-}O)_2$ core. Such a core structure is precedented in a synthetic $Fe^{III}Fe^{IV}$ complex (Fig. XI.5.12*b*)[40] and the viability of an $Fe_2^{IV}(\mu\text{-}O)_2$ core is supported by computational studies.[41,42]

An important unanswered question is the mechanism for the conversion of **P** to **Q.** Two possible pathways are shown in Fig. XI.5.13. Given the structural similarities between the proposed high-valent metal–oxo species for di-iron and dicopper centers, it is tempting to consider a (μ-η^2:η^2-peroxo)di-iron transition state (Fig. XI.5.13(top)). However, the conversion of **P** to **Q** appears to be facilitated by a proton.[43] Thus an alternative pathway involves the isomerization of the (μ-1,2-peroxo)di-iron(III) intermediate to a (μ-1,1-hydroperoxo)di-iron(III) species (Fig. XI.5.13(bottom)), which subsequently undergoes heterolytic cleavage to afford MMOH-**Q.** Since the two mechanisms predict different extents of incorporation from $^{18}O_2$ (Fig. XI.5.13), resonance Raman spectroscopy of intermediate **Q** would be helpful in distinguishing between the two mechanisms, but such data are unfortunately not yet available.

Lastly, as with cytochrome P450, the reaction of MMOH-**Q** with substrate is the subject of intense discussion.[2,3] When chiral ethane is used as substrate, there is partial retention of configuration at the hydroxylated chiral center, implicating an intermediate that can lose some of its stereochemistry. Similar considerations as in the P450 mechanism apply in this case. The oxygen rebound mechanism involving a transient alkyl radical (Fig. XI.5.8) could at first glance be used to rationalize the results. But, because the lifetime of an ethyl radical is thought to be too short for a two-step mechanism, a concerted insertion mechanism has been suggested (Fig. XI.5.8), with C–O and O–H bond formation steps being nonsynchronous to allow for partial epimerization of the chiral center. It is clear that further work is needed to resolve these mechanistic questions.

In summary, the mechanisms for oxygen activation at three distinct active sites have been compared. Although the three metal centers appear quite distinct, there

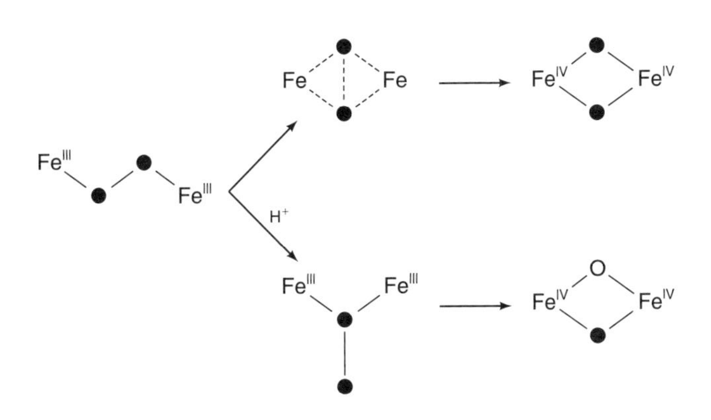

Fig. XI.5.13. Alternative modes for the conversion of MMOH-**P** to MMOH-**Q.**

are common features among the proposed mechanisms, namely, the participation of metal–peroxo and high-valent metal–oxo intermediates. This unified scheme may also apply to other oxygen activating metalloenzymes, such as those with mononuclear nonheme iron and copper centers covered in Section XI.5.2. This array of active sites demonstrates the versatility of Nature in using available metal ions to carry out metabolically important transformations.

XI.5.2. Mononuclear Nonheme Metal Centers That Activate Dioxygen

Mononuclear nonheme metal centers in biology that carry out oxygen activation may be expected to follow the mechanistic paradigm defined in the previous section. The key features of this scheme are the involvement of a metal–peroxo species and a high-valent metal–oxo species subsequent to O_2 binding to the metal center. Principal variations among the different systems discussed below are (1) the source of electrons needed for oxygen activation; (2) the metal oxidation states involved, particularly for the putative high-valent metal–oxo species; and (3) whether the metal–peroxo species is reactive enough to carry out the oxidation reaction directly without the intermediacy of a metal–oxo species.

XI.5.2.1. The Antitumor Drug Bleomycin

Bleomycins (BLMs) are a family of antitumor glycopeptide drugs that target double helical DNA and effect cleavage of the polynucleotide chain in concert with Fe(II) and O_2.[44–46] Though it is technically not an enzyme, the chemistry of Fe^{II}BLM with O_2 is typically discussed within the context of oxygen activating metalloenzymes. From a number of spectroscopic investigations, the generally accepted structure for metallobleomycins consists of a square-pyramidal metal center coordinated to five nitrogen atoms of the drug, namely, a pyrimidine nitrogen, an imidazole nitrogen, an amidate, and two amines, as shown in Fig. XI.5.14, with the possibility of an easily displaceable ligand at the sixth site. The metal center in Fe^{II}BLM thus resembles that found in deoxymyoglobin and deoxyhemoglobin, and its oxygen-activation chemistry follows the mechanistic paradigm for heme centers discussed in the previous section.

As shown in Fig. XI.5.14, the reaction cycle of Fe^{II}BLM begins with O_2 binding to the available sixth coordination site on the high-spin Fe(II) center, affording a diamagnetic Fe(BLM)–O_2 adduct, like oxyhemoglobin and oxyP450. The anionic amidate ligand very likely plays a key role in increasing electron density at the iron(II) center to allow it to bind O_2, while strong backbonding from the pyrimidine ligand is suggested to stabilize the O_2 complex with respect to dissociation into Fe^{III}BLM and superoxide.[8]

The next step in the cycle is the addition of an electron to Fe^{II}BLM–O_2, converting it to "activated BLM". Since it is the last intermediate detectable prior to DNA cleavage, activated BLM has been the subject of extensive spectroscopic investigation. These studies conclude that activated BLM is best described as a low-spin Fe^{III}–OOH species. The low-spin Fe(III) assignment for the metal center derives from the observation of an $S = \frac{1}{2}$ EPR spectrum ($g = 2.26, 2.17$, and 1.94) and Mössbauer features typical of a low-spin Fe(III) center. While it has not been possible to obtain Raman evidence for the hydroperoxide moiety, electrospray ionization mass spectrometry provides clear evidence that the Fe^{III}–OOH unit remains intact in activated BLM.[47]

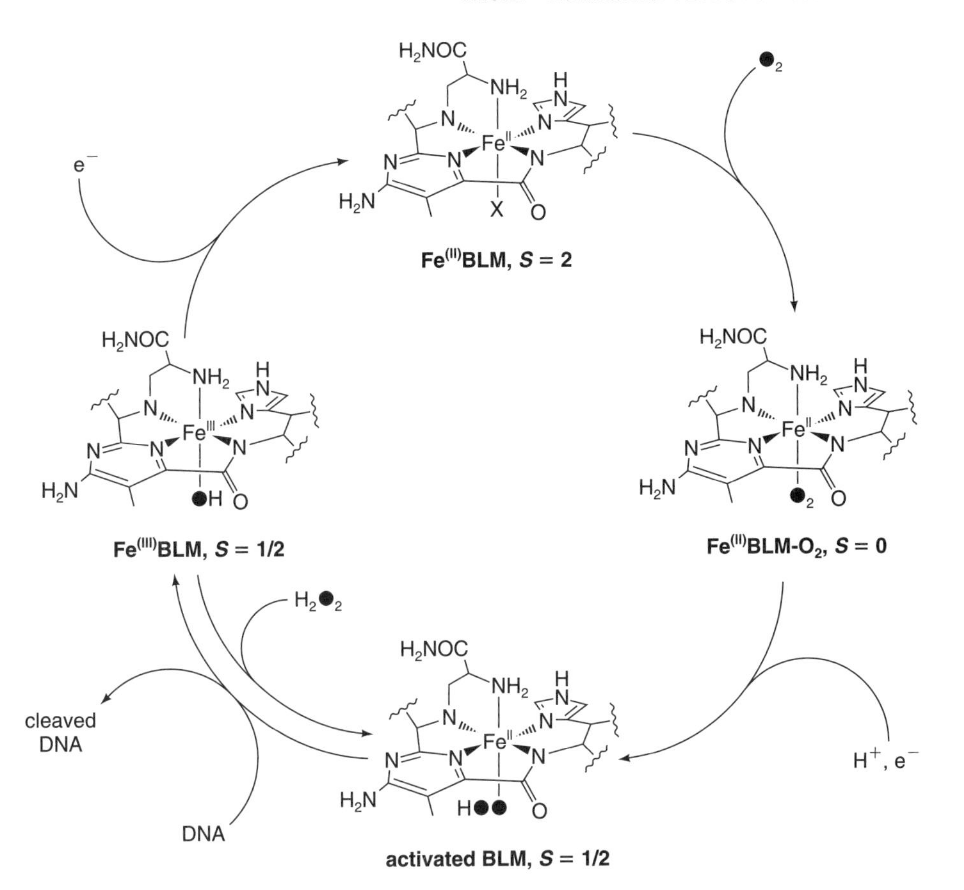

Fig. XI.5.14.
Oxygen activation cycle of the antitumor drug bleomycin.

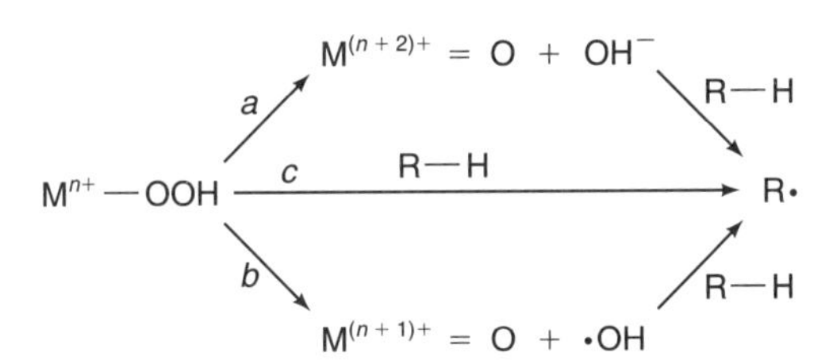

Fig. XI.5.15.
Decomposition pathways for the metal–peroxo intermediate.

The Fe^{III}–OOH formulation for activated BLM implies that the oxidizing equivalents needed for the oxidation chemistry are localized on the dioxygen moiety at this stage. It is clear from the magnitude of the $^{18}O_2$ isotope effect on the decay of activated BLM that O–O bond cleavage is involved in the rate-determining step,[47] so activated BLM may react with DNA via one of three pathways: (*a*) O–O heterolysis to generate an Fe^{V}=O oxidant, (*b*) O–O bond homolysis to afford Fe^{IV}=O and HO· oxidants, or (*c*) direct attack (Fig. XI.5.15).

Density functional theory (DFT) calculations suggest that O–O bond heterolysis to generate a formally Fe^{V}=O species following the heme paradigm (Fig. XI.5.2) is energetically unfavorable.[46] The O–O bond homolysis prior to C–H bond cleavage is also disfavored, because the large isotope effects observed for hydrogen abstraction from C4′-tritiated DNA exclude a highly reactive hydrogen-abstraction agent (e.g., HO·).[45] Thus the only mechanistic option that remains is for the Fe^{III}–OOH intermediate itself to be the oxidant, and this notion appears to be gaining acceptance.[46]

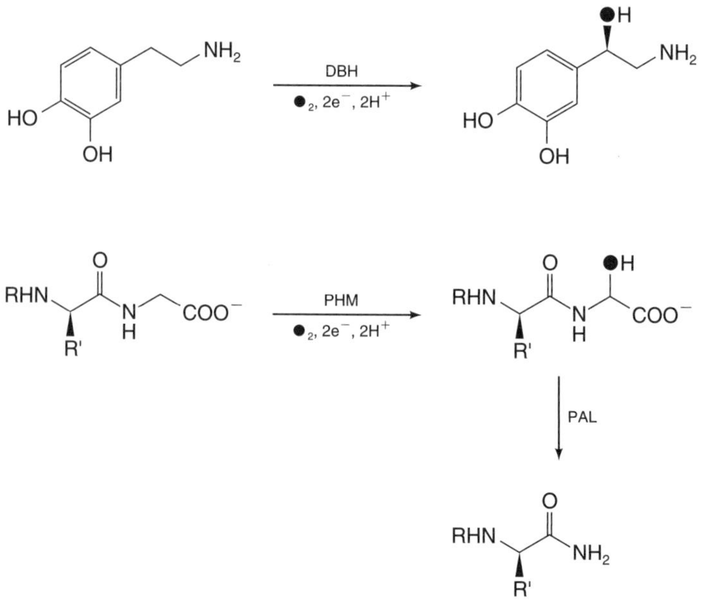

Fig. XI.5.16.
Oxygenation reactions catalyzed by DBH and PHM.

XI.5.2.2. Copper Hydroxylases: DBH and PHM

Besides tyrosinase and catechol oxidase discussed in an earlier section, there are two other known dicopper enzymes that activate O_2 and carry out substrate hydroxylation, namely, dopamine β-hydroxylase (DBH) and peptidylglycine α-hydroxylating monooxygenase (PHM).[5] Both enzymes have strong physiological significance. β-Hydroxylase, found within the neurosecretory vesicles of the adrenal gland and the sympathetic nervous system, catalyzes the hydroxylation of dopamine to norepinephrine and controls the relative levels of these two neurotransmitters. The PHM enzyme is part of a two-enzyme unit that converts peptide hormones into their bioactive C-terminal amidated forms; PHM hydroxylates the α-CH of a terminal glycyl residue of the pro-hormone, while its partner (peptidyl-α-hydroxyglycine-α-amidating lyase = PAL) facilitates the hydrolysis of the nascent geminal aminoalcohol to form the peptide hormone and glyoxalic acid (Fig. XI.5.16). Both DBH and PHM are involved in stereospecific hydroxylations of activated C–H bonds, abstracting the *pro-S* hydrogen of the target CH_2 group and replacing it with an oxygen atom from dioxygen.

Unlike tyrosinase and catechol oxidase, the two Cu centers in DBH and PHM are not close enough in proximity to allow an O_2 molecule to bridge between them. Indeed, the oxidized forms of these enzymes exhibit EPR signals characteristic of magnetically isolated Type 2 Cu(II) centers. Crystallographic studies of PHM[48] reveal the Cu centers to be separated by 11 Å (Fig. XI.5.17). One Cu center, designated Cu_A, has three His ligands, while the other, designated Cu_B, is bound to two histidines, a methionine, and a solvent molecule. The structure of the oxidized enzyme–substrate complex shows that the substrate *N*-acetyl-3′,5′-diidotyrosylglycine binds close to Cu_B in a cleft between the two metal-binding domains. The Cu_B site appears to be the hydroxylating center, as it is only 4.3 Å away from the glycine C_α that is hydroxylated and the *pro-S* hydrogen that is abstracted points toward the bound solvent atom on this metal center. Sequence comparisons and EXAFS studies indicate that DBH very likely has analogous Cu centers. Mechanistic studies have established the following reaction sequence: (1) the oxidized dicopper(II) form of the enzyme is first reduced by two electrons to its dicopper(I) state by ascorbate; (2) substrate bind-

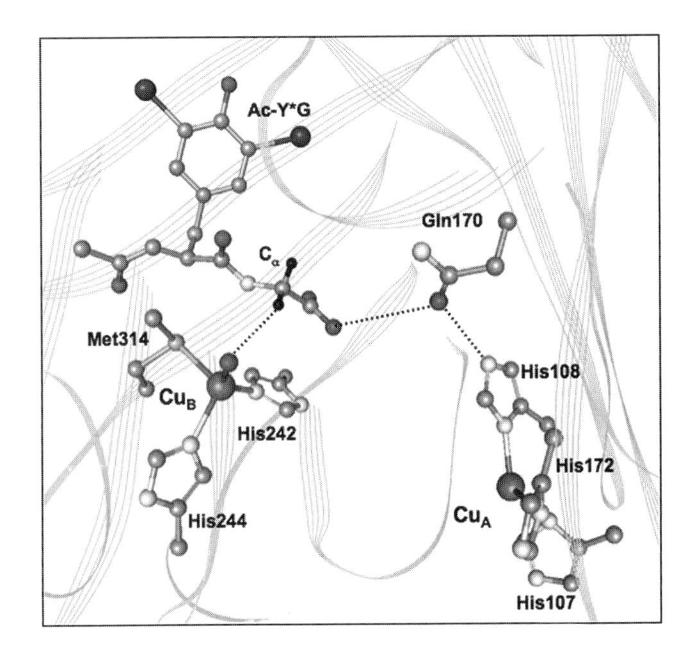

Fig. XI.5.17.
Dicopper active site of oxidized PHM with the substrate *N*-acetyl-3,5-diiodotyrosylglycine (Ac-Y*G) (PDB code: 1OPM).

ing then triggers dioxygen binding and activation; (3) the substrate is then hydroxylated and the dicopper(II) form is regenerated.

A key puzzle posed by the reaction sequence of these enzymes is how the needed electrons get together for oxygen activation. The mechanism must be distinct from that of tyrosinase, because of the difference in metal–metal separations, and is likely to diverge from the mechanistic paradigm for oxygen activation by metalloenzymes discussed in previous sections. It is relatively easy to envision the first step of the mechanism as the binding of O_2 to one Cu(I) center. Indeed crystallographic evidence for an end-on bound O_2 at the Cu_B site has been obtained by diffusion of O_2 into crystals of PHM soaked with another substrate analogue IYT.[49] What happens next remains in question.

Following the heme paradigm (Fig. XI.5.2), the next step should be the formation of a Cu(II)–peroxo intermediate to form the oxidant responsible for C–H bond cleavage. However, such a step is problematic, as the 11-Å distance separating the two Cu centers poses a substantial barrier for electron transfer from the other Cu center to the Cu–O_2 adduct. Furthermore, this electron-transfer step needs to be rapid for efficient catalytic turnover. Thus, the possibility of the Cu(II)–superoxo moiety acting as the hydrogen atom abstraction agent prior to electron transfer from Cu_A has also been considered.[50] Indeed DFT calculations suggest that the Cu(II)–superoxo species may be a better hydrogen atom abstracting agent than the copper–hydroperoxo species. Fig. XI.5.18 illustrates these two possibilities. A similar mechanism may be envisioned for PHM, but much work remains to establish the credibility of this hypothesis.

XI.5.2.3. Iron(II) Enzymes with a 2-His-1-carboxylate Facial Triad Motif

Mononuclear nonheme iron enzymes that activate O_2 are a large and diverse group that catalyze a broad range of key metabolic oxidations.[6] The many crystal structures that have recently become available for this class of enzymes reveal the 2-His-1-carboxylate facial triad active site[6,51] as a common feature of these enzymes (Fig. XI.5.19). This motif, consisting of two histidines and one carboxylate arrayed on one face of an octahedron, serves as the recurring platform for binding the divalent metal

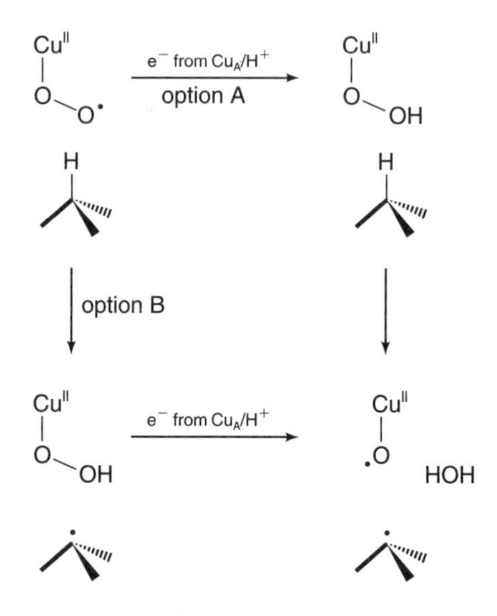

Fig. XI.5.18.
Proposed mechanisms for substrate reaction in DBH and PHM.

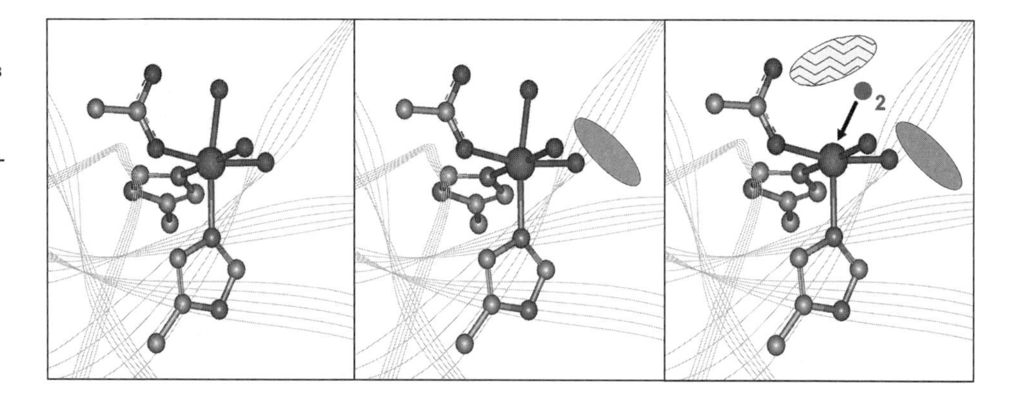

Fig. XI.5.19.
General mechanistic scheme proposed by Solomon et al.[7,8] for a mononuclear Fe(II) enzyme with a 2-His-1-carboxylate facial triad active-site motif. The gray ellipse at the right represents cofactor, while the wavy-lined ellipse represents substrate.

center. The three amino acids of the triad can be found in a number of sequence motifs that are conserved within each subclass but differ from subclass to subclass. Convergent evolution appears to have generated these metal-binding sites.

This large group of enzymes catalyzes two- and four-electron oxidations (Fig. XI.5.20). Examples of the latter are the reactions carried out by the extradiol cleaving catechol dioxygenases and isopenicillin N-synthase (IPNS). The extradiol cleaving catechol dioxygenases are bacterial enzymes involved in the biodegradation of aromatic molecules.[52] Cleavage of a carbon–carbon bond of the catechol substrate adjacent to the enediol moiety results in the incorporation of both atoms of O_2 into the product, thus the designation of these enzymes as dioxygenases. On the other hand, IPNS, isolated from microorganisms such as *Cephalosporium, Penicilium,* and *Streptomyces,* reduces O_2 to water in the course of converting δ-(L-α-aminoadipoyl)-L-cysteinyl-D-valine (ACV) to isopenicillin N.[53] Two C–H bonds are cleaved at the expense of forming C–N and C–S bonds to assemble the bicyclic ring structure. For enzymes involved in the two-electron substrate oxidations, a coreactant contributes the two other electrons needed for the reduction and activation of O_2. These two electrons can come from (1) tetrahydropterin, for the mammalian aromatic amino acid hydroxylases that convert Phe to Tyr, Tyr to dopa, and Trp to 5-hydroxytryptophan in a number of neurologically important transformations;[54] (2) 2-oxoglutarate

Extradiol cleaving catechol dioxygenases

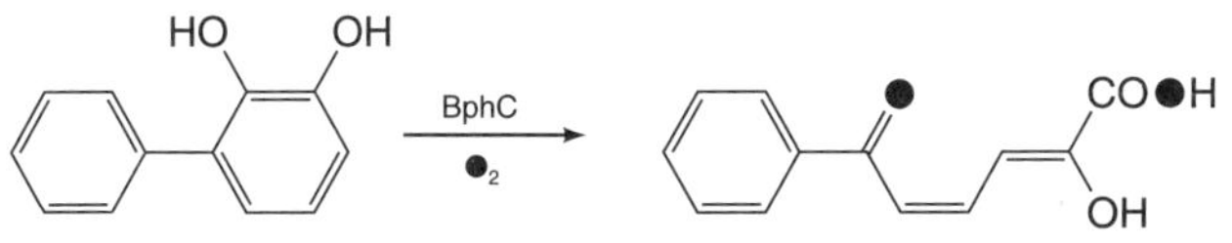

BphC

Isopenicillin N synthase

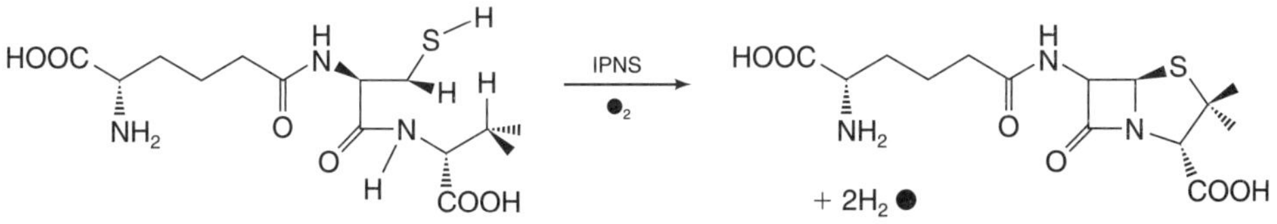

IPNS

$+ 2H_2O$

Pterin-dependent hydroxylases

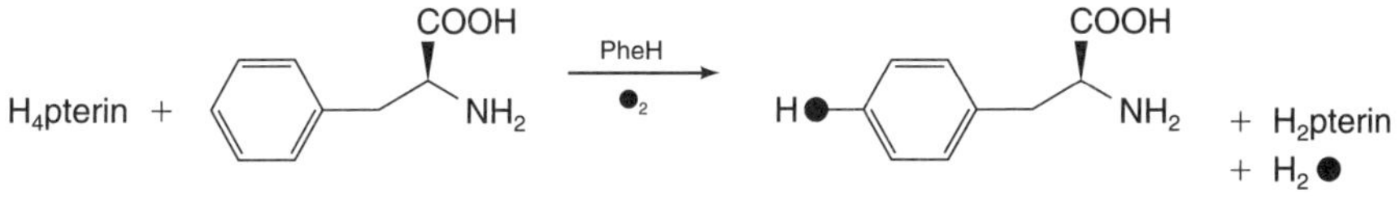

H_4pterin + ... PheH ... + H_2pterin + H_2O

2-Oxoglutarate-dependent oxidases and hydroxylases

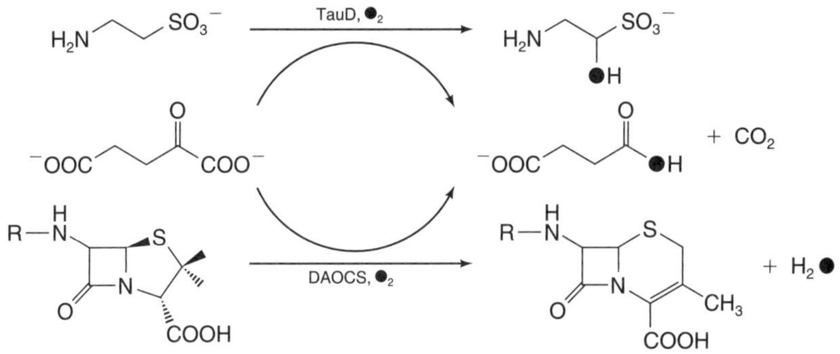

TauD, O_2

$+ CO_2$

DAOCS, O_2

$+ H_2O$

Rieske dioxygenases

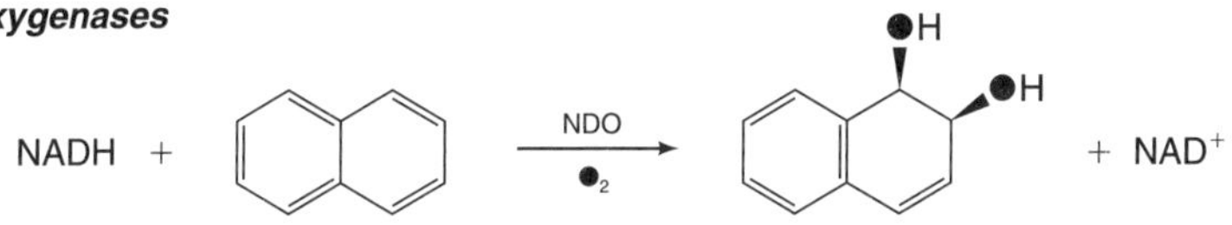

NADH + ... NDO ... + NAD^+

1-Aminocyclopropane-1-carboxylate oxidase

Ascorbate + ... ACCO ... $H_2C{=}CH_2$ + $HC{\equiv}N$ + CO_2 + Dehydroascorbate + 2 H_2O

Fig. XI.5.20.
Reactions catalyzed by mononuclear nonheme iron enzymes with the 2-His-1-carboxylate facial triad motif. BphC = biphenyl-2,3-diol catechol dioxygenase; PheH = phenylalanine hydroxylase; NDO = naphthalene-1,2-dioxygenase; ACCO = 1-aminocyclopropane-1-carboxylate oxidase; TauD = taurine: alpha-ketoglutarate dioxygenase; DAOCS = deacetoxycephalosporin *C* synthase.

(2-OG), for 2-OG-dependent enzymes that mainly functionalize aliphatic C–H bonds in key metabolic reactions;[55] (3) NADH, for the Rieske dioxygenases that carry out the cis-dihydroxylation of arenes in the first step of the biodegradation of aromatics in the soil;[56] and (4) ascorbate, for 1-aminocyclopropane-1-carboxylate oxidase (ACCO) in the formation of ethylene.[57]

The 2-His-1-carboxylate triad only occupies one face of the octahedral metal coordination environment; consequently, the remaining three coordination sites are available for binding exogenous ligands. These sites are usually occupied by as many as three solvent molecules in the resting enzyme, which can be replaced by O_2 and substrate or cofactor in the course of the catalytic cycle. This arrangement thereby provides the active site the flexibility to catalyze the diverse reactions observed, as the chemistry of the metal center can be modulated by the nature of the exogenous ligands. However, despite the wide range of reactions catalyzed, a basic mechanistic paradigm has emerged that serves to unify the oxygen activation mechanisms of these enzymes (Fig. XI.5.19). From detailed spectroscopic studies of a number of these enzymes, Solomon et al.[7,46] have noted that the metal center in resting states of the enzymes is generally six coordinate and relatively unreactive toward O_2 binding. The binding of substrate and/or cofactor to the active site then triggers the conversion of the six-coordinate metal center to a five-coordinate site by loss of a solvent ligand, thereby facilitating the binding of O_2. In support, Lipscomb has found a dramatic increase in the affinity of NO (an O_2 surrogate) for the iron(II) center in IPNS and the extradiol dioxygenases in the presence of the substrate.[58] This elegant triggering strategy ensures formation of the reactive oxygen species only when the substrate is bound.

Figure XI.5.21 compares the structures of O_2 adducts proposed for a number of these enzymes. In four of the six types shown, there is crystallographic or spectroscopic evidence for coordination of substrate or cosubstrate to the metal center.[6] The introduction of such a monoanionic ligand to the Fe(II) coordination sphere is expected to lower the $Fe^{III/II}$ potential and promote the binding of O_2 at the open coordination site. Interestingly, the O_2-binding site can be trans to the carboxylate residue, as found for IPNS and the extradiol dioxygenase BphC, or to a His residue, as observed for 2-OG dependent enzymes like DAOCS, CAS, and TauD. This variability provides Nature with yet another mode for tuning the reactivity of the metal center in this diverse group of enzymes, but there is little insight currently available as to what effect the trans ligand exerts. The formation of the O_2 adduct juxtaposes the reactants on the metal center, thereby facilitating the flow of electrons from substrate (or cosubstrate) to O_2 to form the appropriate peroxo intermediate.

The situation is somewhat different for the pterin-dependent hydroxylases and Rieske dioxygenases, since neither substrate nor cosubstrate binds to the metal center. As a consequence, the carboxylate residue of the facial triad becomes bidentate to occupy a fourth coordination site on the metal center. The active sites must thus be properly configured to allow facile delivery of the electrons required for the reductive activation of O_2. For the pterin-dependent hydroxylases, the pterin cofactor is positioned just 4.5 Å away from the iron center, about the right distance for a proposed Fe^{II}-η^1-O-O-pterin intermediate to form. For the Rieske dioxygenases, the cosubstrate NADH transfers its electrons one at a time via a Rieske Fe_2S_2 cluster near the mononuclear iron center. The O_2 adduct of naphthalene dioxygenase has in fact been crystallographically characterized, the first example for a nonheme iron enzyme.[59] It shows a novel side-on bound O_2 moiety, which is best interpreted as the key iron(III)–hydroperoxo intermediate that transfers both oxygen atoms to the arene substrate in the cis-dihydroxylation mechanism. The different ways that O_2 can interact with the Fe(II) active site shown in Fig. XI.5.21 vividly emphasize the mechanistic flexibility afforded by the 2-His-1-carboxylate motif.

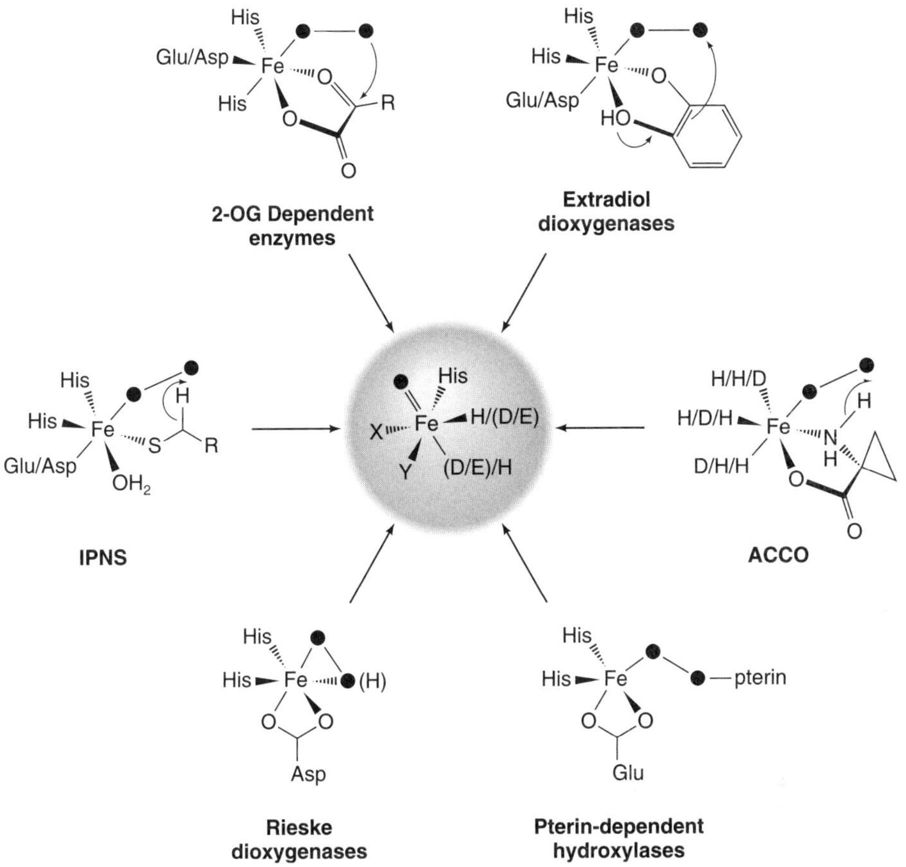

Fig. XI.5.21.
Putative O_2 adducts of various Fe(II) enzymes with a 2-His-1-carboxylate triad.

Following the heme paradigm, the peroxo intermediates derived from the O_2 adducts in Fig. XI.5.21 are proposed to convert to high-valent iron–oxo species upon O–O bond cleavage. Allowing for variations in the X and Y ligands, it could be argued that this intermediate is the common oxidant for this superfamily of enzymes. While the Rieske dioxygenases are proposed to use an $Fe^{III}/Fe^{V}{=}O$ couple like cytochrome P450, most of the enzymes represented in Fig. XI.5.21 must utilize an $Fe^{II}/Fe^{IV}{=}O$ couple to carry out the more difficult oxidative step of the transformation. Indirect evidence for the participation of an iron–oxo moiety usually consists of either the demonstration that a particular active site can carry out a range of oxidations typically associated with a metal–oxo oxidant and/or the observed incorporation of ^{18}O from $H_2{}^{18}O$ into the products, a result that cannot be accounted for by direct attack of a peroxo intermediate on a substrate (Fig. XI.5.15*c*). Direct evidence for such a high-valent species is scarce, but the first example of a kinetically competent Fe(IV)–oxo intermediate has recently been trapped for the 2-OG dependent TauD.[60]

Biomimetic complexes that model reactivity properties of the metalloenzyme active sites have provided some important mechanistic insights. One example is $[Fe(Tp^{Ph_2})(O_2CC(O)Ph)]$ (Fig. XI.5.22), which serves as a model for 2-OG dependent enzymes. The Tp^{R2} ligand is a monoanionic facially coordinating tridentate ligand that serves as a convenient C_3-symmetric analogue for the 2-His-1-carboxylate facial triad. This five-coordinate complex readily reacts with O_2 to afford the oxidative decarboxylation of the 2-oxo acid ligand as well as the hydroxylation of a phenyl ring on the ligand. The $^{18}O_2$ labeling studies show ^{18}O incorporation into both the hydroxylated arene and benzoate products, thereby demonstrating the dioxygenase

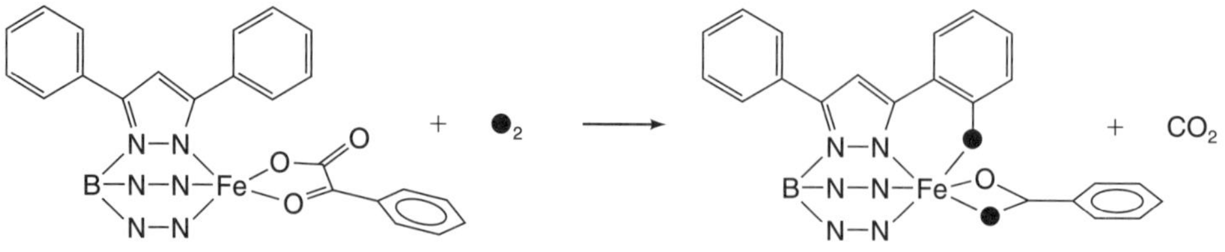

Fig. XI.5.22.
Model for 2-OG dependent enzymes that mimics the dioxygenase nature of the oxygenation reaction.

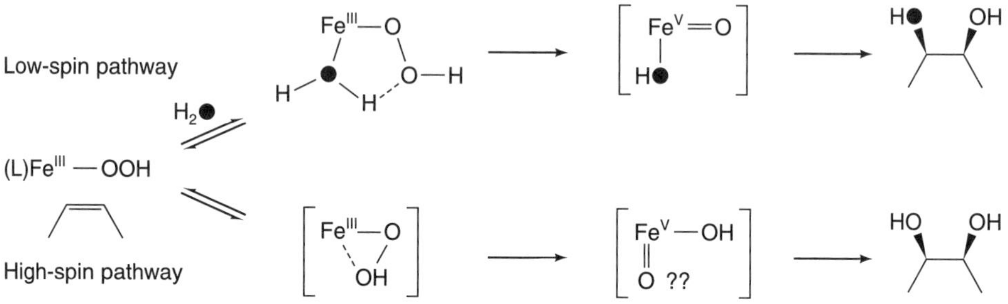

Fig. XI.5.23.
Biomimetic olefin cis-dihydroxylation mechanisms.

nature of the model reaction.[61] Comparison of the reactivities of a series of complexes with substituted aryl α-keto acid ligands shows that the rate of reaction increases as the substituent becomes more electron withdrawing, supporting the notion that an important step of the mechanism is the nucleophilic attack of the bound O_2 moiety on the electrophilic keto carbon of the coordinated α-keto acid, as shown in Fig. XI.5.21.

A second example consists of the first synthetic iron complexes to catalyze the cis-dihydroxylation of olefins (Fig. XI.5.23),[62] mimicking the arene cis-dihydroxylation reactions of Rieske dioxygenases. The biomimetic complexes use H_2O_2 as the oxidant to convert olefins to cis-diols with yields as high as 95%. The effectiveness of tetradentate tripodal N4 ligands suggests that a key requirement is a metal complex with two available cis coordination sites for the activation of a bound hydroperoxo moiety. Interestingly, ligands that give rise to high-spin iron centers yield diol products with both oxygen atoms from H_2O_2, via a proposed Fe^{III}-η^2-OOH intermediate (Fig. XI.5.23) analogous to that observed for naphthalene dioxygenase (Fig. XI.5.21). However, related ligands that give rise to low-spin centers afford diol products with one oxygen atom from H_2O_2 and the other from H_2O; this outcome strongly implicates an HO–Fe^{V}=O oxidant (Fig. XI.5.23). These intriguing results educate us on the possible modes of action in which metal–peroxo species can participate (Fig. XI.5.15) and illustrate the role of spin state in modulating the outcome.

Fig. XI.5.24.
Crystal structure of $[Fe^{IV}(O)(TMC)(NCMe)]^{2+}$ (TMC = 1,4,8,11-tetramethyl-1,4,8,11-tetraazacyclotetradecane).

An additional key biomimetic example is the high-yield generation and isolation of synthetic oxoiron(IV) complexes.[63–65] As illustrated in Fig. XI.5.21, a high-valent iron–oxo species is invoked in almost every catalytic mechanism of a mononuclear iron enzyme involved in O_2 activation. The identification of reaction conditions that form such complexes in high yield has led to their detailed spectroscopic characterization and, in one case, a crystallographic determination (Fig. XI.4.24). The existence of such complexes demonstrates that nonheme ligand environments can indeed access the Fe(IV) oxidation state and will allow the reactivity properties of such complexes to be investigated in detail.

XI.5.2.4. Iron(III) Dioxygenases: An Exception to the Mechanistic Paradigm

The intradiol cleaving catechol dioxygenases differ from the nonheme iron enzymes discussed in Section XI.5.2.3 in their requirement for an Fe(III) center. These bacterial enzymes act in the last step of transforming aromatic molecules into aliphatic products and catalyze the cleavage of the enediol carbon–carbon bond to form *cis,cis*-muconic acid with the elements of O_2 incorporated into the product. Along with the Fe(III) requirement comes a different coordination chemistry and a distinct mechanism for oxygen activation. These enzymes are among the most thoroughly investigated of the nonheme iron oxygenases.[6,58]

Crystal structures of catechol 1,2-dioxygenase (CTD) and protocatechuate 3,4-dioxygenase (PCD) show an iron(III) center in a trigonal-bipyramidal coordination environment consisting of two histidines, two tyrosinates, and a solvent molecule (Fig. XI.5.25). One Tyr and one His serve as axial ligands, while the other Tyr, the other His, and solvent define the trigonal plane. Studies using EXAFS show that the solvent must be a hydroxide, thus affording a charge-neutral Fe(III) active site. The tyrosine ligands give rise to ligand/Fe(III) charge-transfer transitions, which account for the rich burgundy color associated with these enzymes and serve as a convenient spectroscopic probe. Substrate-binding results in the bidentate coordination of the catecholate and the displacement of the solvent molecule and the axial tyrosinate (Fig. XI.5.25), affording a five-coordinate metal center that is now poised to react with dioxygen.

The key mechanistic question is what role the high-spin Fe(III) center plays in the reaction with O_2. Given the phenolate-rich nature of the coordination environment in the enzyme–substrate complex, the Fe(III) oxidation state would be strongly favored, so the heme paradigm does not apply. Consistently, the O_2 surrogate NO does not react with the enzyme substrate (ES) complex, as observed for Fe(II) centers. An alternative mechanism involving substrate activation has thus been proposed in which the coordination of catechol to the Fe(III) center activates the catechol for direct attack by O_2 (Fig. XI.5.26).

A critical feature in oxygenase mechanisms is to provide a means for triplet O_2 to react with the singlet substrate. The high barrier to this reaction is usually overcome by the initial coordination of O_2 to the metal center and subsequent activation. In the substrate activation mechanism proposed for the catechol dioxygenases, it is postulated that unpaired spin density from the high-spin Fe(III) center is delocalized into the frontier molecular orbitals of the bound catecholate via ligand-to-metal

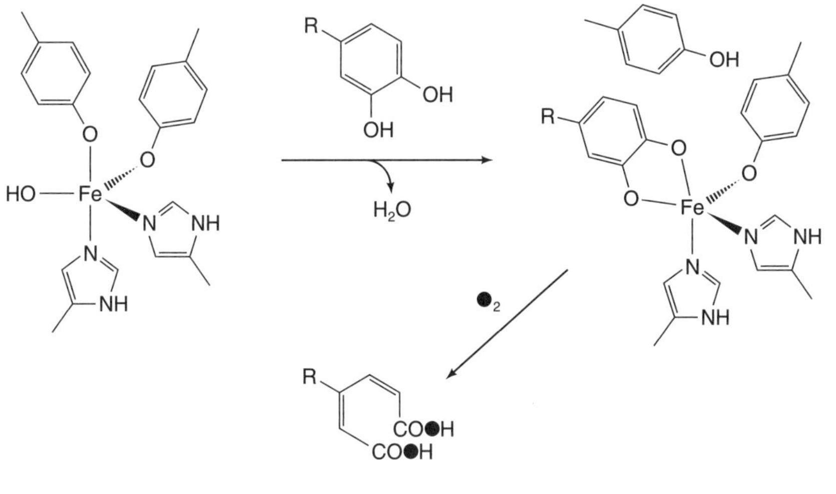

Fig. XI.5.25.
Active site of PCD before and after substrate binding.

Fig. XI.5.26.
Proposed mechanism for intradiol cleaving catechol dioxygenases.

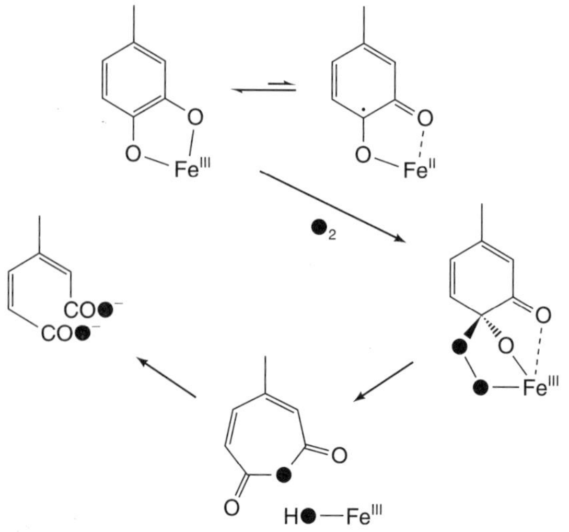

charge transfer. This provides an alternative means for relaxing the high activation barrier for O_2 reaction by introducing radical character into the substrate, thereby allowing O_2 to attack the substrate directly. An Fe(III)–peroxo adduct is then formed, which decays via the muconic anhydride to afford the dioxygenated product. The putative tridentate peroxy intermediate is precedented in the crystal structure of the O_2 adduct of an $[Ir^{III}(triphos)(catecholate)]^+$ complex[66] and can be readily accommodated into the structure of the PCD ES complex by molecular modeling methods.[67]

Support for the novel substrate activation mechanism has been obtained from studies of synthetic $[Fe^{III}(L)catecholate]$ complexes (L = tetradentate ligand).[68] These complexes react with O_2 to afford intradiol cleavage of the bound catecholates, some in nearly quantitative yield, thus serving as excellent functional models of the enzymes. In a systematic investigation of a series of complexes,[69,70] the rate of intradiol cleavage was found to correlate with the Lewis acidity of the metal center. The more Lewis acidic the metal center, the lower the energy of the catecholate/Fe(III) charge-transfer band, the greater the amount of unpaired spin density that is delocalized onto the catecholate, the faster the reaction with O_2. Thus the induction of radical character onto the substrate promotes the direct attack of O_2 on the substrate. The proposed substrate activation mechanism works for this particular system because, unlike most of the substrates for the Fe(II) enzymes discussed in Section XI.5.2.3, catecholates are easily oxidizable in the first place, albeit usually to quinones. The role of the Fe(III) active site is thus to channel this oxidizability toward oxidative cleavage and thus contribute to the recycling of carbon in the environment.

References

General References

1. Sono, M., Roach, M. P., Coulter, E. D., and Dawson, J. H., "Heme-Containing Oxygenases", *Chem. Rev.*, **96,** 2841–2887 (1996).
2. Wallar, B. J. and Lipscomb, J. D., "Dioxygen Activation by Enzymes Containing Binuclear Non-Heme Iron Clusters", *Chem. Rev.*, **96,** 2625–2658 (1996).
3. Merkx, M., Kopp, D. A., Sazinsky, M. H., Blazyk, J. L., Müller, J., and Lippard, S. J., "Dioxygen Activation and Methane Hydroxylation by Soluble Methane Monooxygenase: A Tale of Two Irons and Three Proteins", *Angew. Chem. Int. Ed.*, **40,** 2782–2807 (2001).
4. Solomon, E. I., Sundaram, U. M., and Machonkin, T. E., "Multicopper Oxidases and Oxygenases", *Chem. Rev.*, **96,** 2563–2605 (1996).

5. Klinman, J. P., "Mechanisms Whereby Mononuclear Copper Proteins Functionalize Organic Substrates", *Chem. Rev.*, **96,** 2541–2561 (1996).
6. Costas, M., Mehn, M. P., Jensen, M. P., and Que, L., Jr., "Oxygen Activation at Mononuclear Nonheme Iron: Enzymes, Intermediates, and Models", *Chem. Rev.*, **104,** 939–986 (2004).
7. Solomon, E. I., Brunold, T. C., Davis, M. I., Kemsley, J. N., Lee, S.-K., Lehnert, N., Neese, F., Skulan, A. J., Yang, Y.-S., and Zhou, J., "Geometric and Electronic Structure/Function Correlations in Non-Heme Iron Enzymes", *Chem. Rev.*, **100,** 235–349 (2000).
8. Solomon, E. I., Decker, A., and Lehnert, N., "Non-heme iron enzymes: Contrasts to heme catalysis", *Proc. Natl. Acad. Sci. U.S.A.*, **100,** 3589–3594 (2003).

Specific References

9. Schlichting, I., Berendzen, J., Chu, K., Stock, A. M., Maves, S. A., Benson, D. E., Sweet, R. M., Ringe, D., Petsko, G. A., and Sligar, S. G., "The Catalytic Pathway of Cytochrome $P450_{cam}$ at Atomic Resolution", *Science*, **287,** 1615–1622 (2000).
10. Wertz, D. L. and Valentine, J. S., "Nucleophilicity of Iron-Peroxo Porphyrin Complexes", *Struct. Bonding*, **97,** 37–60 (2000).
11. VanAtta, R. B., Strouse, C. E., Hanson, L. K., and Valentine, J. S., "[Peroxotetraphenylporphinato]-manganese(III) and [Chlorotetraphenylporphinato]-manganese(II) Anions. Syntheses, Crystal Structures, and Electronic Structures", *J. Am. Chem. Soc.*, **109,** 1425–1434 (1987).
12. Davydov, R., Macdonald, I. D. G., Makris, T. M., Sligar, S. G., and Hoffman, B. M., "EPR and ENDOR of Catalytic Intermediates in Cryoreduced Native and Mutant Oxy-Cytochromes P450cam: Mutation-Induced Changes in the Proton Delivery System", *J. Am. Chem. Soc.*, **121,** 10654–10655 (1999).
13. Loew, G. H. and Harris, D. L., "Role of the Heme Active Site and Protein Environment in Structure, Spectra, and Function of the Cytochrome P450s", *Chem. Rev.*, **100,** 407–419 (2000).
14. Roelfes, G., Vrajmisu, V., Chen, K., Ho, R. Y. N., Rohde, J.-U., Zondervan, C., la Crois, R. M., Schudde, E. P., Lutz, M., Spek, A. L., Hage, R., Feringa, B. L., Münck, E., and Que, L., Jr., "End-on and Side-on Peroxo Derivatives of Non-Heme Iron Complexes with Pentadentate Ligands: Models for Putative Intermediates in Biological Iron/Dioxygen Chemistry", *Inorg. Chem.*, **42,** 2639–2653 (2003).
15. Shiemke, A. K., Loehr, T. M., and Sanders-Loehr, J., "Resonance Raman Study of Oxyhemerythrin and Hydroxomethemerythrin. Evidence for Hydrogen Bonding of Ligands to the Fe–O–Fe Center", *J. Am. Chem. Soc.*, **108,** 2437–2443 (1986).
16. Groves, J. T. and Han, Y.-Z., "Models and Mechanisms of Cytochrome P450 Action", in *Cytochrome P450: Structure, Mechanism, and Biochemistry,* 2nd ed., Ortiz de Montellano, P. R., Ed., Plenum Press, New York, 1995, pp. 3–48.
17. Newcomb, M. and Toy, P. H., "Hypersensitive Radical Probes and the Mechanisms of Cytochrome P450-Catalyzed Hydroxylation Reactions", *Acc. Chem. Res.*, **33,** 449–455 (2000).
18. Ortiz de Montellano, P. R., "Heme Oxygenase Mechanism: Evidence for an Electrophilic, Ferric Peroxide Species", *Acc. Chem. Res.*, **31,** 543–549 (1998).
19. Magnus, K. A., Ton-That, H., and Carpenter, J. E., "Recent Structural Work on the Oxygen Transport Protein Hemocyanin", *Chem. Rev.*, **94,** 727–735 (1994).
20. Klabunde, T., Eicken, C., Sacchettini, J. C., and Krebs, B., "Crystal Structure of a Plant Catechol Oxidase Containing a Dicopper Center", *Nat. Struct. Biol.*, **5,** 1084–1090 (1998).
21. Decker, H. and Rimke, T., "Tarantula Hemocyanin Shows Phenoloxidase Activity", *J. Biol. Chem.*, **273,** 25889–25892 (1998).
22. Tyeklár, Z. and Karlin, K. D., "Copper-Dioxygen Chemistry: A Bioinorganic Challenge", *Acc. Chem. Res.*, **22,** 241–248 (1989).
23. Kitajima, N., Fujisawa, L., Fujimoto, C., Moro-oka, Y., Hashimoto, S., Kitagawa, T., Toriumi, K., Tatsumi, K., and Nakamura, A., "A New Model for Dioxygen Binding in Hemocyanin. Synthesis, Characterization, and Molecular Structure of the μ-η^2:η^2 Peroxo Dinuclear Copper(II) Complexes, $[Cu(HB(3,5\text{-}R_2pz)_3)]_2(O_2)$ ($R = i$-Pr and Ph)", *J. Am. Chem. Soc.*, **114,** 1277–1291 (1992).
24. Tolman, W. B., "Making and Breaking the Dioxygen O–O Bond: New Insights from Studies of Synthetic Copper Complexes", *Acc. Chem. Res.*, **30,** 227–237 (1997).
25. Mirica, L. M., Ottenwaelder, X., and Stack, T. D. P., "Structure and Spectroscopy of Copper-Dioxygen Complexes", *Chem. Rev.*, **104,** 1013–1045 (2004).
26. Hayashi, H., Fujinami, S., Nagatomo, S., Ogo, S., Suzuki, M., Uehara, A., Watanabe, Y., and Kitagawa, T., "A Bis(μ-oxo)dicopper(III) Complex with Aromatic Nitrogen Donors: Structural Characterization and Reversible Conversion between Copper(I) and Bis(μ-oxo)dicopper(III) Species", *J. Am. Chem. Soc.*, **122,** 2124–2125 (2000).
27. Baldwin, M. J., Root, D. E., Pate, J. E., Fujisawa, K., Kitajima, N., and Solomon, E. I. "Spectroscopic Studies of Side-On Peroxide-Bridged Binuclear Copper(II) Model Complexes of Relevance to Oxyhemocyanin and Oxytyrosinase", *J. Am. Chem. Soc.*, **114,** 10421–10431 (1992).
28. Henson, M. J., Mukherjee, P., Root, D. E., Stack, T. D. P., and Solomon, E. I., "Spectroscopic and Electronic Structural Studies of the $Cu(III)_2$ Bis-μ-oxo Core and Its Relation to the Side-On Peroxo-Bridged Dimer", *J. Am. Chem. Soc.*, **121,** 10332–10345 (1999).
29. Decker, H., Dillinger, R., and Tuczek, F., "How Does Tyrosinase Work? Recent Insights from Model Chemis-

try and Structural Biology", *Angew. Chem. Int. Ed.*, **39,** 1591–1595 (2000).
30. Stenkamp, R. E., "Dioxygen and Hemerythrin", *Chem. Rev.*, **94,** 715–726 (1994).
31. Kurtz, D. M., Jr., "Structural Similarity and Functional Diversity in Diiron-Oxo Proteins", *J. Biol. Inorg. Chem.*, **2,** 159–167 (1997).
32. Moënne-Loccoz, P., Baldwin, J., Ley, B. A., Loehr, T. M., and Bollinger, J. M., Jr., "O_2 Activation by Non-Heme Diiron Proteins: Identification of a Symmetric μ-1,2-Peroxide in a Mutant of Ribonucleotide Reductase", *Biochemistry,* **37,** 14659–14663 (1998).
33. Broadwater, J. A., Ai, J., Loehr, T. M., Sanders-Loehr, J., and Fox, B. G., "Peroxodiferric Intermediate of Stearoyl-Acyl Carrier Protein Δ^9 Desaturase: Oxidase Reactivity during Single Turnover and Implications for the Mechanism of Desaturation", *Biochemistry,* **37,** 14664–14671 (1998).
34. Liu, K. E., Valentine, A. M., Wang, D., Huynh, B. H., Edmondson, D. E., Salifoglou, A., and Lippard, S. J., "Kinetic and Spectroscopic Characterization of Intermediates and Component Interactions in Reactions of Methane Monooxygenase from *Methylococcus capsulatus* (Bath)", *J. Am. Chem. Soc.*, **117,** 10174–10185 (1995).
35. Bollinger, J. M., Jr., Krebs, C., Vicol, A., Chen, S., Ley, B. A., Edmondson, D. E., and Huynh, B. H., "Engineering the Diiron Site of *Escherichia coli* Ribonucleotide Reductase Protein R2 to Accumulate an Intermediate Similar to H_{peroxo}, the Putative Peroxodiiron(III) Complex from the Methane Monooxygenase Catalytic Cycle", *J. Am. Chem. Soc.*, **120,** 1094–1095 (1998).
36. Broadwater, J. A., Achim, C., Münck, E., and Fox, B. G., "Mössbauer Studies of the Formation and Reactivity of a Quasi-Stable Peroxo Intermediate of Stearoyl-Acyl Carrier Protein Δ^9-Desaturase", *Biochemistry,* 12197–12204 (1999).
37. Kim, K. and Lippard, S. J., "Structure and Mössbauer Spectrum of a (μ-1,2-Peroxo)bis(μ-carboxylato)diiron(III) Model for the Peroxo Intermediate in the Methane Monooxygenase Hydroxylase Reaction Cycle", *J. Am. Chem. Soc.*, **118,** 4914–4915 (1996).
38. Lee, S.-K., Fox, B. G., Froland, W. A., Lipscomb, J. D., and Münck, E., "A Transient Intermediate of the Methane Monooxygenase Catalytic Cycle Containing an $Fe^{IV}Fe^{IV}$ Cluster", *J. Am. Chem. Soc.*, **115,** 6450–6451 (1993).
39. Shu, L., Nesheim, J. C., Kauffmann, K., Münck, E., Lipscomb, J. D., and Que, L., Jr., "An $Fe_2{}^{IV}O_2$ Diamond Core Structure for the Key Intermediate **Q** of Methane Monooxygenase", *Science,* **275,** 515–518 (1997).
40. Hsu, H.-F., Dong, Y., Shu, L., Young, V. G., Jr., and Que, L., Jr., "Crystal Structure of a Synthetic High-Valent Complex with an $Fe_2(\mu\text{-}O)_2$ Diamond Core. Implications for the Core Structures of Methane Monooxygenase Intermediate Q and Ribonucleotide Reductase Intermediate X", *J. Am. Chem. Soc.*, **121,** 5230–5237 (1999).
41. Siegbahn, P. E. M., "Theoretical Model Studies of the Iron Dimer Complex of MMO and RNR", *Inorg. Chem.*, **38,** 2880–2889 (1999).
42. Dunietz, B. D., Beachy, M. D., Cao, Y., Whittington, D. A., Lippard, S. J., and Friesner, R. A., "Large Scale *ab initio* Quantum Chemical Calculation of the Intermediates in the Soluble Methane Monooxygenase Catalytic Cycle", *J. Am. Chem. Soc.*, **122,** 2828–2839 (2000).
43. Lee, S.-K. and Lipscomb, J. D., "Oxygen Activation Catalyzed by Methane Monooxygenase Hydroxylase Component: Proton Delivery during the O–O Bond Cleavage Steps", *Biochemistry,* **38,** 4423–4432 (1999).
44. Stubbe, J. and Kozarich, J. W., "Mechanisms of Bleomycin-Induced DNA Degradation", *Chem. Rev.*, **87,** 1107–1136 (1987).
45. Burger, R. M., "Cleavage of Nucleic Acids by Bleomycin", *Chem. Rev.*, **98,** 1153–1169 (1998).
46. Sam, J. W., Tang, X.-J., and Peisach, J., "Electrospray Mass Spectrometry of Iron Bleomycin: Demonstration that Activated Bleomycin is a Ferric Peroxide Complex", *J. Am. Chem. Soc.*, **116,** 5250–5256 (1994).
47. Burger, R. M., "Nature of Activated Bleomycin", *Struct. Bonding,* **97,** 287–303 (2000).
48. Prigge, S. T., Kolhekar, A. S., Eipper, B. A., Mains, R. E., and Amzel, L. M., "Substrate-Mediated Electron Transfer in Peptidylglycine α-Hydroxylating Monooxygenase", *Nat. Struct. Biol.*, **6,** 976–983 (1999).
49. Prigge, S. T., Eipper, B. A., Mains, R. E., and Amzel, L. M., "Dioxygen Binds End-On to Mononuclear Copper in a Precatalytic Enzyme Complex", *Science,* **304,** 864–867 (2004).
50. Chen, P. and Solomon, E. I., "Oxygen Activation by the Noncoupled Binuclear Copper Site in Peptidylglycine-α-Hydroxylating Monooxygenase. Reaction Mechanism and Role of the Noncoupled Nature of the Active Site", *J. Am. Chem. Soc.*, **126,** 4991–5000 (2004).
51. Hegg, E. L. and Que, L., Jr., "The 2-His-1-Carboxylate Facial Triad: An Emerging Structural Motif in Mononuclear Non-Heme Iron(II) Enzymes", *Eur. J. Biochem.*, **250,** 625–629 (1997).
52. Bugg, T. D. H. and Lin, G., "Solving the Riddle of the Intradiol and Extradiol Catechol Dioxygenases: How Do Enzymes Control Hydroperoxide Rearrangements", *Chem. Commun.*, **11,** 941–952 (2001).
53. Schenk, W. A., "Isopenicillin N Synthase: An Enzyme at Work", *Angew. Chem. Int. Ed.*, **39,** 3409–3411 (2000).
54. Flatmark, T. and Stevens, R. C., "Structural Insight into the Aromatic Amino Acid Hydroxylases and Their Disease-Related Mutant Forms", *Chem. Rev.*, **99,** 2137–2160 (1999).
55. Prescott, A. G. and Lloyd, M. D., "The Iron(II) and 2-Oxoacid-Dependent Dioxygenases and Their Role in Metabolism", *Nat. Prod. Rep.*, **17,** 367–383 (2000).

56. Gibson, D. T. and Parales, R. E., "Aromatic Hydrocarbon Dioxygenases in Environmental Biotechnology", *Curr. Opin. Biotechnol.*, **11,** 236–243 (2000).
57. John, P., *Physiologia Plantarum,* **100,** 583–592 (1997).
58. Lipscomb, J. D. and Orville, A. M., "Mechanistic Aspects of Dihydroxybenzoate Dioxygenases", *Metal Ions Biol. Syst.*, **28,** 243–298 (1992).
59. Karlsson, A., Parales, J. V., Parales, R. E., Gibson, D. T., Eklund, H., and Ramaswamy, S., "Crystal Structure of Naphthalene Dioxygenase: Side-on Binding of Dioxygen to Iron", *Science,* **299,** 1039–1042 (2003).
60. (a) Price, J. C., Barr, E. W., Tirupati, B., Bollinger, J. M., Jr., and Krebs, C., "The First Direct Characterization of a High-Valent Iron Intermediate in the Reaction of an α-Ketoglutarate-Dependent Dioxygenase: A High-Spin Fe(IV) Complex in Taurine/α-Ketoglutarate Dioxygenase (TauD) from *Escherichia coli*", *Biochemistry,* **42,** 7497–7508 (2003). (b) Price, J. C., Barr, E. W., Glass, T. E., Krebs, C., and Bollinger, J. M., Jr., "Evidence for Hydrogen Abstraction from C1 of Taurine by the High-Spin Fe(IV) Intermediate Detected during Oxygen Activation by Taurine:α-Ketoglutarate Dioxygenase (TauD)", *J. Am. Chem. Soc.*, **125,** 13008–13009 (2003). (c) Proshlyakov, D. A., Henshaw, T. F., Monterosso, G. R., Ryle, M. J., and Hausinger, R. P., "Direct Detection of Oxygen Intermediates in the Non-Heme Fe Enzyme Taurine/α-Ketoglutarate Dioxygenase", *J. Am. Chem. Soc.*, **126,** 1022–1023 (2004). (d) Riggs-Gelasco, P. J., Price, J. C., Guyer, R. B., Brehm, J. H., Barr, E. W., Bollinger, J. M., Jr., and Krebs, C., "EXAFS Spectroscopic Evidence for an Fe=O Unit in the Fe(IV) Intermediate Observed during Oxygen Activation by Taurine:α-Ketoglutarate Dioxygenase", *J. Am. Chem. Soc.*, **126,** 8108–8109 (2004).
61. Mehn, M. P., Fujisawa, K., Hegg, E. L., and Que, L., Jr., "Oxygen Activation by Nonheme Iron(II) Complexes: α-Keto Carboxylate versus Carboxylate", *J. Am. Chem. Soc.*, **125,** 7828–7842 (2003).
62. Chen, K., Costas, M., and Que, L., Jr., "Spin State Tuning of Non-Heme Iron-Catalyzed Hydrocarbon Oxidations: Participation of Fe^{III}–OOH and Fe^{V}=O Intermediates", *J. Chem. Soc., Dalton Trans.*, 672–679 (2002).
63. Rohde, J.-U., In, J. H., Lim, M. H., Brennessel, W. W., Bukowski, M. R., Stubna, A., Münck, E., Nam, W., and Que, L., Jr., "Crystallographic and Spectroscopic Evidence for a Nonheme Fe^{IV}=O Complex", *Science,* **299,** 1037–1039 (2003).
64. Lim, M. H., Rohde, J.-U., Stubna, A., Bukowski, M. R., Costas, M., Ho, R. Y. N., Münck, E., Nam, W., and Que, L., Jr., "An Fe^{IV}=O Complex of a Tetradentate Tripodal Nonheme Ligand", *Proc. Natl. Acad. Sci. U.S.A.*, **100,** 3665–3670 (2003).
65. Kaizer, J., Klinker, E. J., Oh, N. Y., Rohde, J.-U., Song, W. J., Stubna, A., Kim, J., Münck, E., Nam, W., and Que, L., Jr., "Nonheme Fe^{IV}O Complexes That Can Oxidize the C–H Bonds of Cyclohexane at Room Temperature", *J. Am. Chem. Soc.*, **126,** 472–473 (2004).
66. Barbaro, P., Bianchini, C., Linn, K., Mealli, C., Meli, A., Vizza, F., Laschi, F., and Zanello, P., "Dioxygen Uptake and Transfer by Co(III), Rh(III) and Ir(III) Catecholate Complexes", *Inorg. Chim. Acta,* **198–200,** 31–56 (1992).
67. Orville, A. M., Lipscomb, J. D., and Ohlendorf, D. H., "Crystal Structures of Substrate and Substrate Analog Complexes of Protocatechuate 3,4-Dioxygenase: Endogenous Fe^{3+} Ligand Displacement in Response to Substrate Binding", *Biochemistry,* **36,** 10052–10066 (1997).
68. Krüger, H.-J., "Iron-Containing Models of Catechol Dioxygenases", in *Biomimetic Oxidations Catalyzed by Metal Complexes,* Meunier, B., Ed., Imperial College Press, London, 2000, pp. 363–413.
69. Cox, D. D. and Que, L., Jr., "Functional Models for Catechol 1,2-Dioxygenase. The Role of the Iron(III) Center", *J. Am. Chem. Soc.*, **110,** 8085–8092 (1988).
70. Jang, H. G., Cox, D. D., and Que, L., Jr., "A Highly Reactive Functional Model for the Catechol Dioxygenases. Structure and Properties of [Fe(TPA)DBC]BPh_4", *J. Am. Chem. Soc.*, **113,** 9200–9204 (1991).

XI.6. Reducing Dioxygen to Water: Cytochrome *c* Oxidase

Contents

Shinya Yoshikawa
Department of Life Science
University of Hyogo
Kamigohri Akoh
Hyogo 678-1297, Japan

XI.6.1. Introduction

Cytochrome *c* oxidase is a four-electron oxidase that reduces dioxygen (O_2) to water with electrons transferred from ferrocytochrome *c* (reaction 1).

$$O_2 + 4 \text{ Ferrocytochrome } c \ (Fe^{2+}) + 4 \ H^+ \rightarrow 2 \ H_2O + 4 \text{ ferricytochrome } c \ (Fe^{3+}) \quad (1)$$

The enzyme is a large membrane protein found in the mitochondrial inner membrane of eukaryotic cells and in the plasma membrane of prokaryotic cells. Biological membranes are composed of a phospholipid bilayer with a thickness of ~50 Å, which includes various membrane proteins to endow the membrane with specific physiological functions (see Tutorial I and Section X.3). Typical membrane proteins such as cytochrome *c* oxidase have hydrophilic moieties protruding out of both sides of the membrane as well as a transmembrane (hydrophobic) moiety. The space enclosed by the mitochondrial membrane is called the matrix space. The mitochondrial outer membrane surrounds the inner membrane, partitioning a space between the two membranes, which is called the intermembrane space. The periplasmic and cytoplasmic spaces in prokaryotic cells, respectively, correspond to the intermembrane and matrix spaces in mitochondria.

Bovine heart cytochrome *c* oxidase contains two hemes and two copper centers at its catalytic sites. Dioxygen reduction by the enzyme is coupled with proton pumping across the mitochondrial inner membrane.[1–4] Electrons destined for reduction of O_2 are transferred to the O_2 reduction site from ferrocytochrome *c* molecules located in the intermembrane space, while the protons come from the matrix space. That is, reduction of each O_2 molecule by this enzyme is coupled to the translocation of 4 equiv of net positive charge from the matrix space to the intermembrane space. The direction of this positive charge translocation due to O_2 reduction is the same as that of proton pumping driven by O_2 reduction. Thus, both O_2 reduction and proton pumping contribute to the formation of an electrochemical potential across the mitochondrial inner membrane that is used for ATP formation by ATP synthase.

Cytochrome *c* oxidase is one of the most important and one of the most extensively investigated enzymes. Its physiological importance is due to the crucial role it plays as the terminal oxidant in aerobic respiration, which is ~20 times more effective than glycolysis in adenosine triphosphate (ATP) formation. The enzyme is also extremely interesting chemically in its ability to catalyze the four-electron reduction of O_2 to H_2O and to couple this reduction to proton pumping.

Three-dimensional (3D) structural information on this complex enzyme (13 subunits in the bovine enzyme) is indispensable for the elucidation of its mechanism. A major breakthrough in cytochrome *c* oxidase research was the determination of the three-dimensional (3D) structures of bovine and bacterial enzymes by X-ray crystallography in 1995.[5,6] The most challenging step in obtaining an X-ray crystal structure had been the crystallization of the enzyme. Membrane proteins are notoriously difficult to crystallize, since they have both hydrophilic and hydrophobic surfaces and are thus not stable in either aqueous solutions or in organic solvents. Only a small number of membrane proteins have been crystallized, compared to the large number of water-soluble proteins with solved X-ray structures.

To date, the X-ray structure of bovine heart cytochrome *c* oxidase has been determined at 2.3-Å resolution in the fully oxidized state (PDB code: 2OCC).[7] In this X-ray structure, the O–O bond distance of peroxide bound at the O_2-binding site can be determined at an accuracy of 0.2 Å. However, much higher accuracy is required for effective evaluation of the chemical reactivity of the peroxide. Furthermore, the accuracy of electron density determination for proteins by X-ray crystallography is usually not high enough for the determination of the oxidation states of the transition metal ions in proteins. Spectroscopic methods have therefore been invaluable in providing information about the details of active-site structures. When this information is combined with that obtained by X-ray crystallography, it becomes possible to deduce the details of structural changes of the enzyme during catalytic turnover.

XI.6.2. Lessons from the X-Ray Structures of Bovine Heart Cytochrome *c* Oxidase

XI.6.2.1. Structure of the Protein Moiety

Figure XI.6.1 shows that this enzyme is found in a dimeric state in crystals of the fully oxidized enzyme at 2.8-Å resolution.[8] All 13 different subunits, which had been suggested by chemical analysis,[2,9] are clearly seen in each monomer. In the 2.8-Å structure, 1780 out of 1803 amino acid residues are identified in each monomer. The assembly of 28 α-helices clearly indicates a transmembrane moiety in the middle section of the enzyme. There is an extramembrane moiety protruding into the intermembrane space that contains one of the copper sites, Cu_A, which is the direct electron acceptor from cytochrome *c* found in the intermembrane space.

As shown in Fig. XI.6.2, the three biggest subunits encoded by mitochondrial genes form a core including all the redox-active metal sites. The core is surrounded by the other 10 smaller subunits encoded by nuclear genes.[8] The 3D structure of the core portion is essentially identical to that of the bacterial enzyme.[6] The biggest subunit contains two hemes and the Cu_B site, and the third biggest subunit contains the Cu_A site. It has been shown that a bacterial enzyme preparation containing these two subunits has the basic function of this enzyme, that is, O_2 reduction to water coupled with proton pumping.[10] Physiological roles for the other 11 subunits have not been established at this point in time.

Locations of the metal sites in the enzyme molecule are shown in Fig. XI.6.3.[5] The Cu_A and the magnesium sites are in the extramembrane region protruding into the intermembrane space, while a Zn ion is ligated by four Cys residues in tetrahedral coordination in a nuclear-coded subunit placed on the matrix side. As described below, the Cu_A site has two Cu atoms in close proximity. The other metal centers, hemes *a* and a_3, and Cu_B, are at the same level in the transmembrane region.

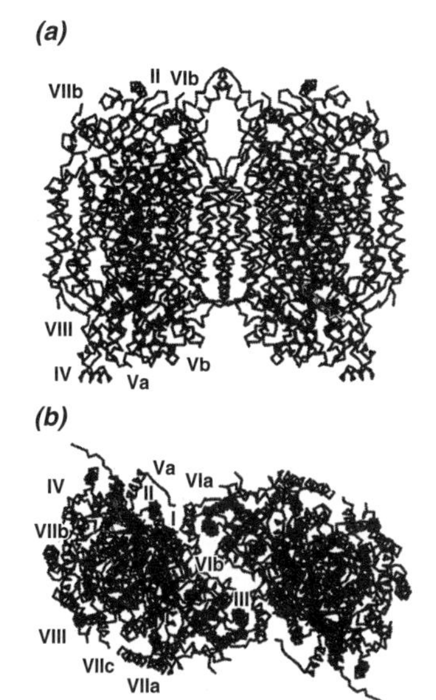

Fig. XI.6.1. The Cα backbone trace of bovine heart cytochrome *c* oxidase in the fully oxidized form at 2.8-Å resolution. (*a*) A view to the transmembrane surface and (*b*) a view from the intermembrane space. Roman numerals and alphabetic subscripts denote names of subunits.

XI.6.2.2. Structures of the Metal Centers in the Largest Subunit

Cytochrome *c* oxidase has two heme A cofactors, iron porphyrins that are characterized by a hydroxyfarnesylethyl group at position 2 of the heme periphery and a formyl group at position 8 (Fig. XI.6.4). The electron-withdrawing character of the formyl group provides the characteristic bright blue-green color for the reduced form of the heme. One of the hemes, designated heme *a*, is coordinated by two His imidazoles and the other, designated heme a_3, is ligated by only a single His imidazole. The latter heme is the binding site for O_2 and for respiratory inhibitors such as CO, CN^-, and N_3^-. Distal to the Fe center of heme a_3 is the mononuclear Cu_B site, which is coordinated by three histidyl imidazole groups.

Bovine heart cytochrome *c* oxidase as prepared under aerobic conditions is in the fully oxidized state. Reduction of heme a_3 in this fully oxidized enzyme, when ferrocytochrome *c* is used as the reductant under anaerobic conditions, is much slower than the enzymatic turnover rate of the enzyme.[11,12] Furthermore, the rate of CN^- binding to the fully oxidized enzyme as prepared is also much slower than the rate of CN^- binding to the enzyme under turnover conditions.[13] These results suggest that the enzyme, as prepared, is not directly involved in the catalytic turnover. The physiological significance of the oxidized enzyme as prepared is still not known.

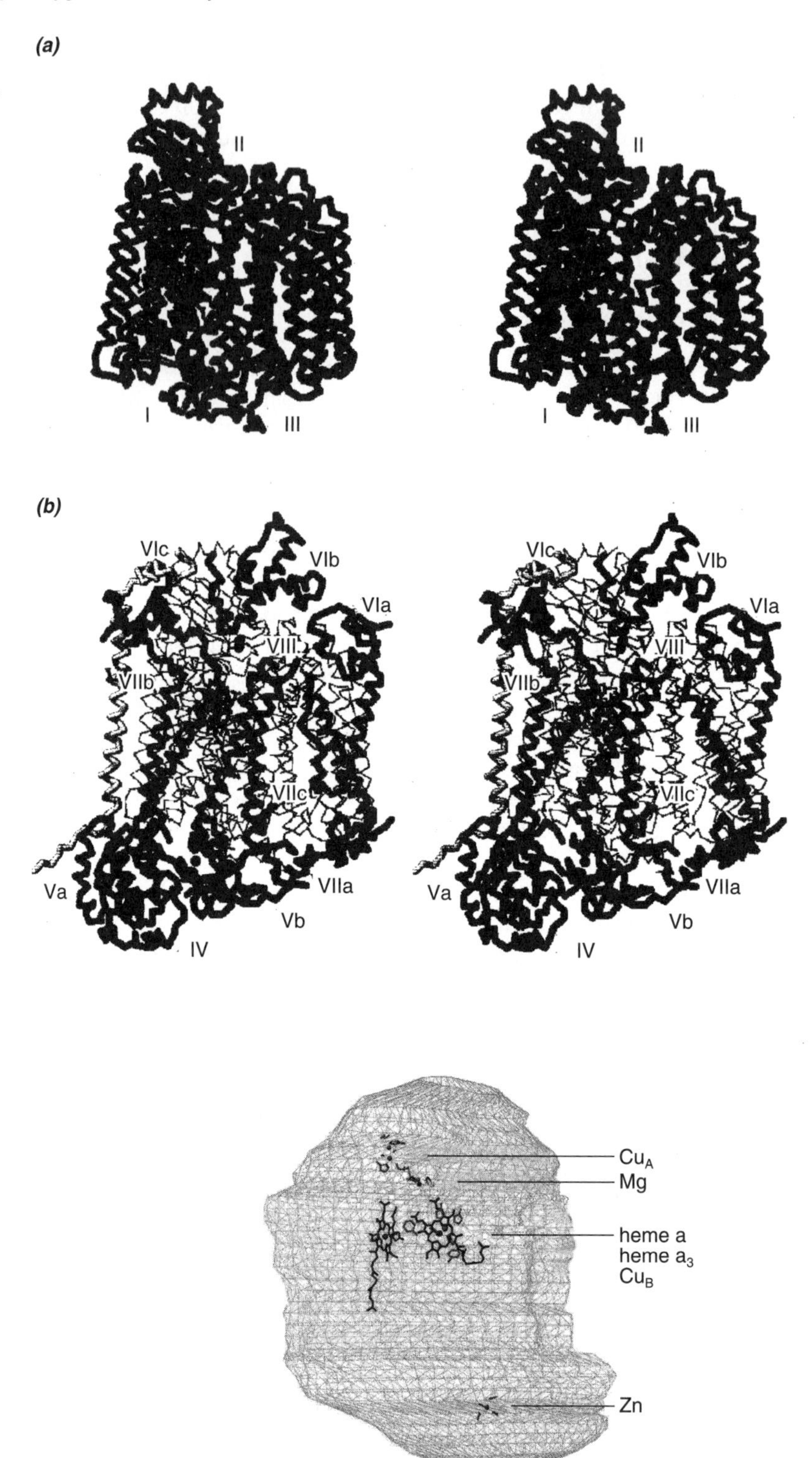

Fig. XI.6.2.
Stereoscopic drawings of Cα backbone trace for a monomer in the dimeric structure shown in Fig. XI.5A.1*a*. The biggest three subunits encoded by mitochondrial gene and (*b*) all 13 subunits in which the biggest three subunits given in (*a*) are shown by thin stick models. Roman numerals and alphabetic subscripts denote names of subunits.

Fig. XI.6.3.
Location of the metal sites in bovine heart cytochrome *c* oxidase. Molecular surface shown by the cage in a view to the transmembrane surface is defined with the electron density map at 5-Å resolution. Top and bottom sides show the intermembrane and matrix side of the mitochondrial inner membrane, respectively. The Cu_A, Mg, and Zn sites are outside the transmembrane region. Two heme irons and Cu_B are at the same level in the transmembrane region.

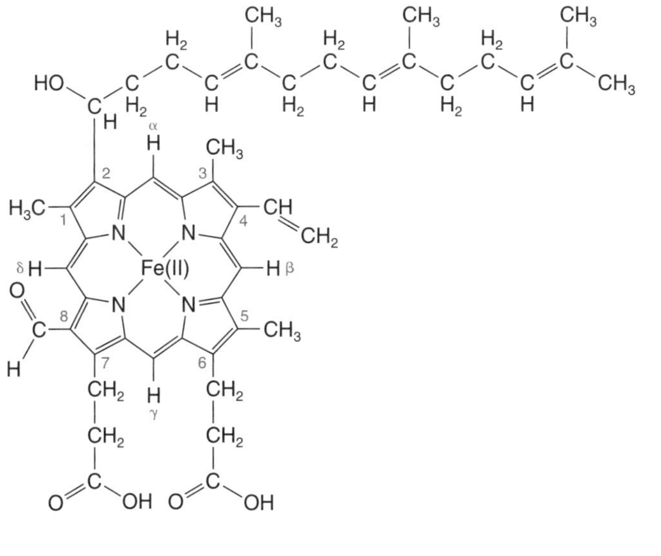

Heme A

Fig. XI.6.4.
Structure of heme A.

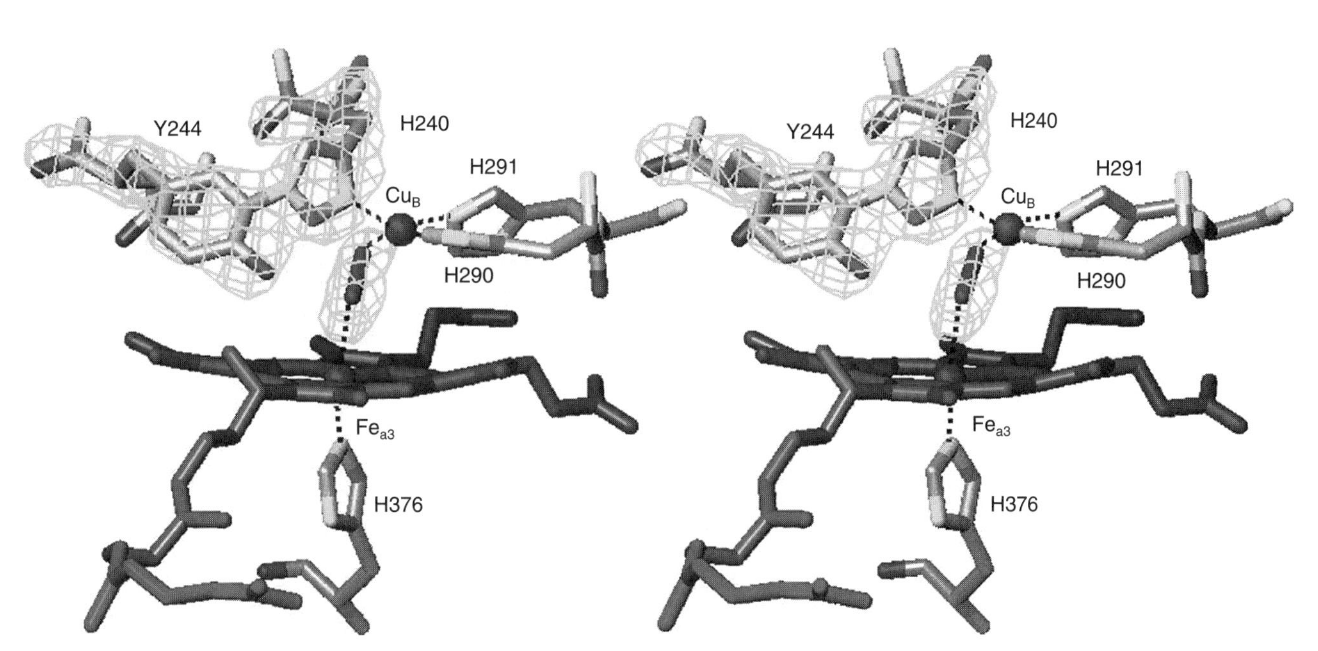

Fig. XI.6.5.
X-Ray structure of the O_2 reduction site in the fully oxidized (resting) form at 2.3-Å resolution in stereoscopic drawing. The difference Fourier map is drawn for the bridging peroxide at the 7σ level ($1\sigma = 0.0456\,e^-/Å$).

Electron spin resonance and magnetic susceptibility measurements on oxidized cytochrome *c* oxidase as prepared have shown that there is antiferromagnetic coupling between the high-spin Fe^{3+} and the $Cu_B{}^{2+}$ ions.[14,15] Several candidates for the bridging ligand that mediates the coupling between the two metal centers have been proposed,[3,16,17] but none of these suggestions anticipated the bridging peroxide found in the 2.3-Å structure of the fully oxidized form (Fig. XI.6.5).[7] The peroxide bridges between heme a_3 and Cu_B with a Cu–O distance of 1.9 Å and an Fe–O distance of 2.4 Å. The latter distance is, surprisingly, much longer than the typical Fe–O coordination bond. The peroxide in the X-ray structure is consistent with the fact that two extra electron equivalents besides the four electron equivalents for reduction of the four redox active metal sites are required for complete reduction of the oxidized oxidase as prepared.[18]

The Cu_B-binding site consists of three imidazole nitrogen atoms in a triangular array (Fig. XI.6.5). This coordination environment is observed also in the structures of the fully reduced (PDB code: 1OCR), the CO-bound fully reduced (PDB code: 1OCO) and the azide-bound fully oxidized (PDB code: 1OCZ) states of the enzyme.[7] In the fully reduced state, there is no bridging ligand between heme a_3 Fe and Cu_B, while, in the CO-bound structure, the CO carbon is coordinated to heme a_3 and the CO oxygen is 2.47 Å away from the Cu_B center. These observations suggest that the Cu_B site is relatively inert relative to the heme a_3 site.

The 2.3-Å fully oxidized structure also shows a direct covalent linkage between one of the imidazole nitrogen atoms of His240 ligated to Cu_B and the 3′-carbon atom of Tyr244 (Fig. XI.6.5). The angle between the phenol and imidazole planes is ~60°. This covalently linked structure is also consistent with the electron density maps in other states of the bovine enzyme determined thus far,[7] as well as in the bacterial enzyme.[19] Although the physiological significance of this covalent bond is still under discussion, one possible role is to fix the OH group of Tyr244 in a position close enough to the O_2 reduction site to hydrogen bond to the O_2 bound at heme a_3Fe^{2+}. This hydrogen atom could then be donated to the bound O_2 in the course of dioxygen reduction.

XI.6.2.3. Structure of the Cu_A Site

The second copper site, Cu_A, is buried inside a β-barrel structure in the extramembrane moiety of the third biggest subunit. This is a binuclear Cu site with a core structure similar to that of the 2Fe2S type iron–sulfur cluster in ferredoxins, with the two Cu ions bridged by the two Cys thiolates forming an M_2S_2 rhomb. Each Cu ion has two other amino acid ligands to complete a tetrahedral coordination environment, as shown in Fig. XI.6.6.[5] The fully oxidized Cu_A site has an EPR spectrum that indicates a valence-delocalized Cu^ICu^{II} center, which becomes EPR silent upon reduction to the fully reduced Cu^ICu^I form.[20] Thus, this center acts only to transfer one electron. No conformational change is detectable on reduction of the Cu_A site. Why this site is dinuclear rather than mononuclear is an intriguing question yet to be answered.

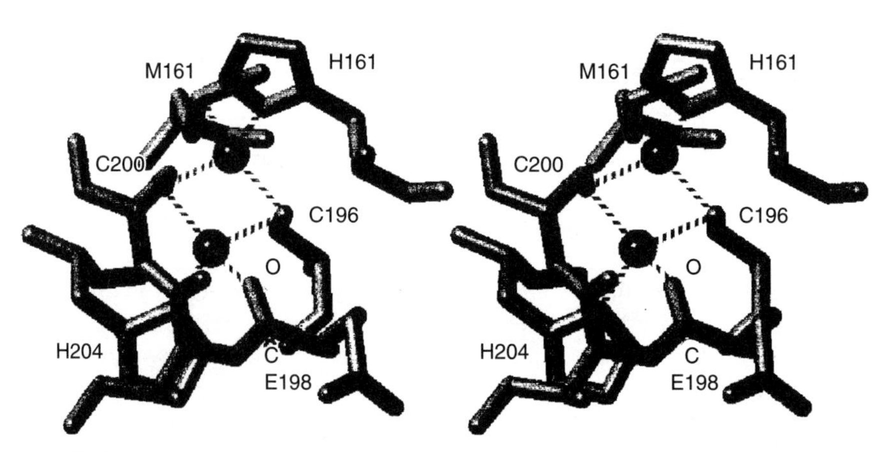

Fig. XI.6.6.
Structure of the Cu_A site in the fully oxidized state at 2.8-Å resolution in stereoscopic drawing. Two balls show positions of copper atoms. Broken lines denote coordination bonds.

XI.6.3. Reaction Mechanism

XI.6.3.1. Electron Transfer within the Enzyme

Figure XI.6.7 shows the four redox-active metal sites and the amino acids connecting these sites. The Cu_A center, which is the primary electron-acceptor site, is found in the extramembrane moiety, providing ready access for ferrocytochrome *c*, the water-soluble electron donor. One of the ligands of the Cu site, His204, is connected to a propionate group of heme *a* via a hydrogen-bond network including a peptide bond between the two Arg residues. The Cu site is also connected to heme a_3 by another hydrogen bond and coordination bond network including Glu198, a Mg^{2+} ion, and His368. The direct distances between Cu_A and heme *a*Fe and between Cu_A and heme a_3Fe, measured from the central point of the two copper atoms of Cu_A, are 20.6 and 23.2 Å, respectively.

The above structure seems to allow direct electron transfer from Cu_A to both hemes *a* and a_3. However, theoretical calculation of electron-transfer rates for the two pathways based on the X-ray structure[21] shows that electron transfer to heme *a* should be much faster than to heme a_3, consistent with kinetic results.[22] As seen in Fig. XI.6.7, heme *a* is quite close to heme a_3, with the shortest distance between the heme peripheries being ~4.0 Å. Furthermore, His376, the fifth ligand of heme a_3, is separated by only one residue from His378, one of the histidine ligands of heme *a*. These structures strongly suggest that heme *a* could serve as an effective electron donor to heme a_3. Thus, electron transfer from Cu_A to heme a_3 is likely to occur through heme *a*, consistent with the results of extensive kinetic investigations[22] that date back to the pioneering work of Gibson and Greenwood in 1963.[23] However, this X-ray structure raises new questions. If the direct electron transfer from Cu_A to heme a_3 is too slow to be physiologically relevant, what is the role of the network connecting the two metal sites directly?

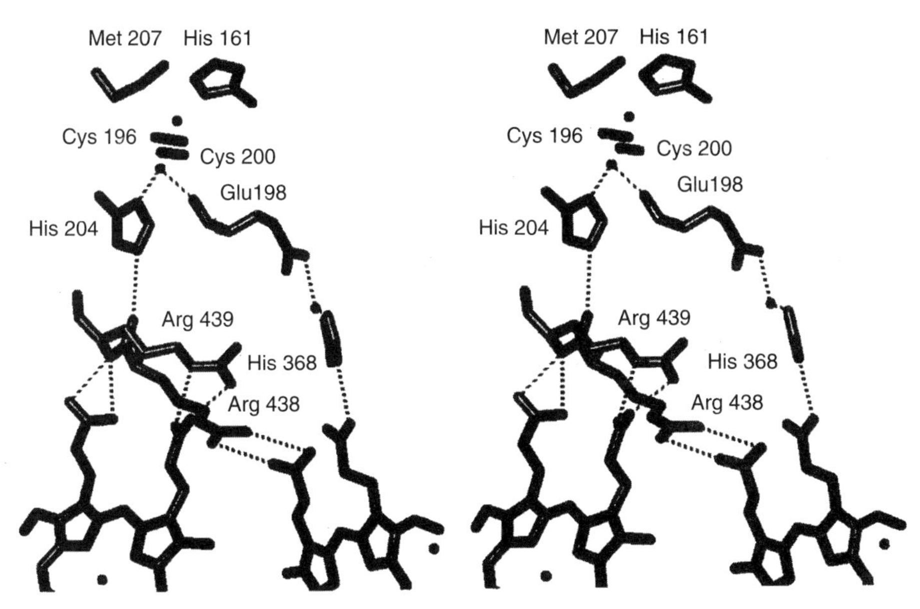

Fig. XI.6.7.
Possible electron-transfer paths from Cu_A to heme *a* and to heme a_3 in stereoscopic drawing. Two balls near Cys196 and Cys200 denote the positions of the two copper atoms in the Cu_A site. Hemes on the left and right sides denote hemes *a* and a_3, respectively. A small ball between Glu198 and His368 indicates the position of Mg ion. Dotted lines are hydrogen bonds or coordination bonds.

XI.6.3.2. Dioxygen Reduction

The first step in O_2 reduction by cytochrome oxidase is the binding of O_2 to heme a_3. This step has been readily observed in time-resolved resonance Raman measurements of the reaction of the fully reduced enzyme with O_2.[24–26] The appearance of an ($Fe–O_2$) feature at 571 cm^{-1}, essentially identical to those found in oxyhemoglobins and oxymyoglobins,[27,28] shows that O_2 is formally reduced by one electron in this step.

The next step should be the transfer of a second electron to the bound O_2. The close proximity of Cu_B to heme a_3 makes it a logical donor, forming the heme $a_3Fe^{3+}–O–O–Cu_B^{2+}$ μ-peroxo intermediate, as observed in the structure of the resting enzyme. Though expected, such a species is in fact not observed in the time-resolved Raman experiments,[4] thereby questioning the kinetic competence of the peroxo species observed in the crystal structure. Indeed, the fact that such a species is observed in crystals of the oxidized enzyme as prepared obtained under ambient conditions should already raise doubts as to its relevance to the O_2 reduction mechanism.

An alternative reduction mechanism, suggested by the X-ray structure of the fully reduced enzyme, involves Tyr244, whose phenolic OH is found to be fixed near the expected position of the distal atom of the bound dioxygen. This residue could serve as a hydrogen-atom donor to afford a hydroperoxo intermediate, $Fe^{3+}–O–O–H$. The Cu_B^+ ion could then readily reduce the resulting tyrosyl radical via the cross-link between Tyr244 and His240. In this mechanism, Cu_B in fact serves indirectly as the donor of the second electron required for O_2 reduction.

However, no $Fe^{3+}–O–O–H$ intermediate has been detected to date in time-resolved experiments. The intermediate designated as P observed after the decay of the heme a_3-O_2 adduct exhibits a resonance Raman band at 804 cm^{-1}, which has been assigned to an $Fe^{4+}=O$ species.[4] The putative $Fe^{3+}–O–O–H$ intermediate, if formed at all, must form at a slower rate than the subsequent O–O bond cleavage to afford $Fe^{4+}=O$. The electron inventory for the cleavage of the $Fe^{3+}–O–O–H$ species to $Fe^{4+}=O$ requires the introduction of a proton and an electron. The His coupled Tyr244 residue has again been proposed to act as a hydrogen-atom donor for this step,[4] but convincing evidence for its involvement is not yet available.

To complete the O_2 reduction process, two electrons are required to bring the O_2 reduction site to its oxidized state (a_3Fe^{3+}, Cu_B^{2+}); these two electrons are provided by heme *a* and Cu_A. The fact that CO reacts with partially reduced cytochrome oxidase has suggested that O_2 reacts with the enzyme only when both heme a_3Fe and Cu_B are in the reduced state. Thus, the O–O bond-cleavage step, which requires two electron equivalents, is not necessarily coupled to electron transfer from heme *a* and Cu_A, since these two electron equivalents are available from Fe^{3+} and Tyr244. Two protons for formation of the first water molecule should be taken up at or before the O–O bond cleavage step. The subsequent electron-transfer step to form the oxidized state involves uptake of another two protons for the formation of water from the oxide in $Fe^{4+}=O$. It has been proposed that only the latter process is coupled to proton pumping,[29] but this notion has been challenged.[30,31] The coupling mechanism between O_2 reduction and proton pumping is still under debate.[32,33]

XI.6.3.3. Proton Transfers through Cytochrome *c* Oxidase

Cytochrome *c* oxidase must transfer protons for the reduction of O_2 to water and for proton pumping. In general, the hydrophobic environment inside membrane proteins strongly inhibits proton transfer (or any other ion transfer). However, a hydrogen-bond network within the membrane protein could provide an effective proton-transfer path. For example, in a hydrogen bond between two OH groups, ei-

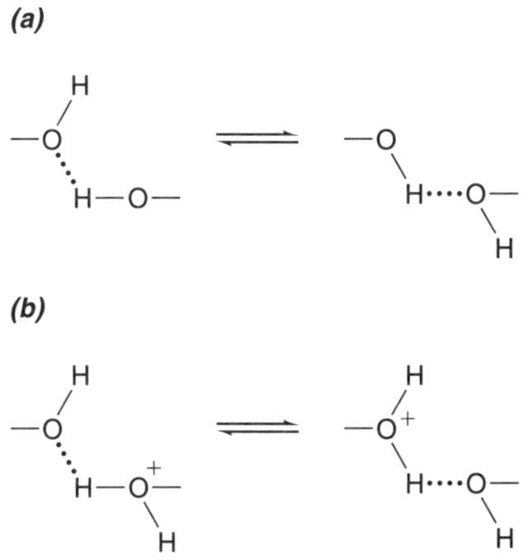

Scheme XI.6.1.

(a)

(b)

ther group can act as the hydrogen-bond donor or acceptor, as shown in Scheme XI.6.1*a*. When one of the OH groups is protonated, the positive charge can readily be transferred to the other OH group, via the hydrogen bond (Scheme XI.6.1*b*).

The X-ray structure of bovine heart cytochrome *c* oxidase has many cavities, namely, spaces without any associated electron density, but large enough to accommodate solvent water molecules. The absence of electron density indicates that the waters are quite mobile inside the cavity. The water molecules can transfer protons inside the cavity, as quickly as in the water phase outside the protein. In the X-ray structure of bovine heart cytochrome *c* oxidase, there are pairs of amino acid residues that could be involved in hydrogen-bonding interactions, but are not located close enough to each other to form a hydrogen bond between them. However, a hydrogen bond could be formed within each pair if small conformational changes in the side chains of these residues without movement in the peptide backbone are induced, possibly by changes in oxidation or ligand-binding states of the redox-active centers.

Figure XI.6.8 shows a putative hydrogen-bonding network composed of these three types of possible proton-transfer paths. The network would provide a proton-transfer path from the matrix space to Tyr244, which is the possible proton donor to the bound O_2 as described above. The two hydrogen bonds to Thr490 are drawn from the two fixed waters in Fig. XI.6.8, which indicates that the OH group of Thr490 forms a hydrogen bond with one of the two waters and the OH group forms another hydrogen bond with another fixed water when the methyl group of Thr490 is replaced with the OH group by a rotation around the C_α–C_β bond axis of Thr490. Similarly, Asn491 has two hydrogen bonds from two fixed waters. However, since the carbonyl group is a hydrogen-bonding acceptor, a rotation in the acid-amide group of Asn491 is required for proton transfer through the residue. Furthermore, an exchange in the protonation site in the imidazole group of His256 or a rotation of the imidazole plane is also required for proton transfer through the imidazole group. The conformational changes in Lys319 and Thr316 are indispensable for proton transfer between the two residues, as indicated by a double headed arrow in Fig. XI.6.8.

All of the above conformational changes are required for proton transfer. This requirement indicates that proton transfer through the channel for water formation is not a simple passive transport, but is likely to be tightly controlled by the O_2 reduction process at the redox-active sites. The conformational changes facilitate unidirectional proton transfer. Conformational change in the network that depends on the oxidation and ligand-binding states has not been detected. This network may be only transiently active during O_2 reduction, thereby preventing spontaneous penetra-

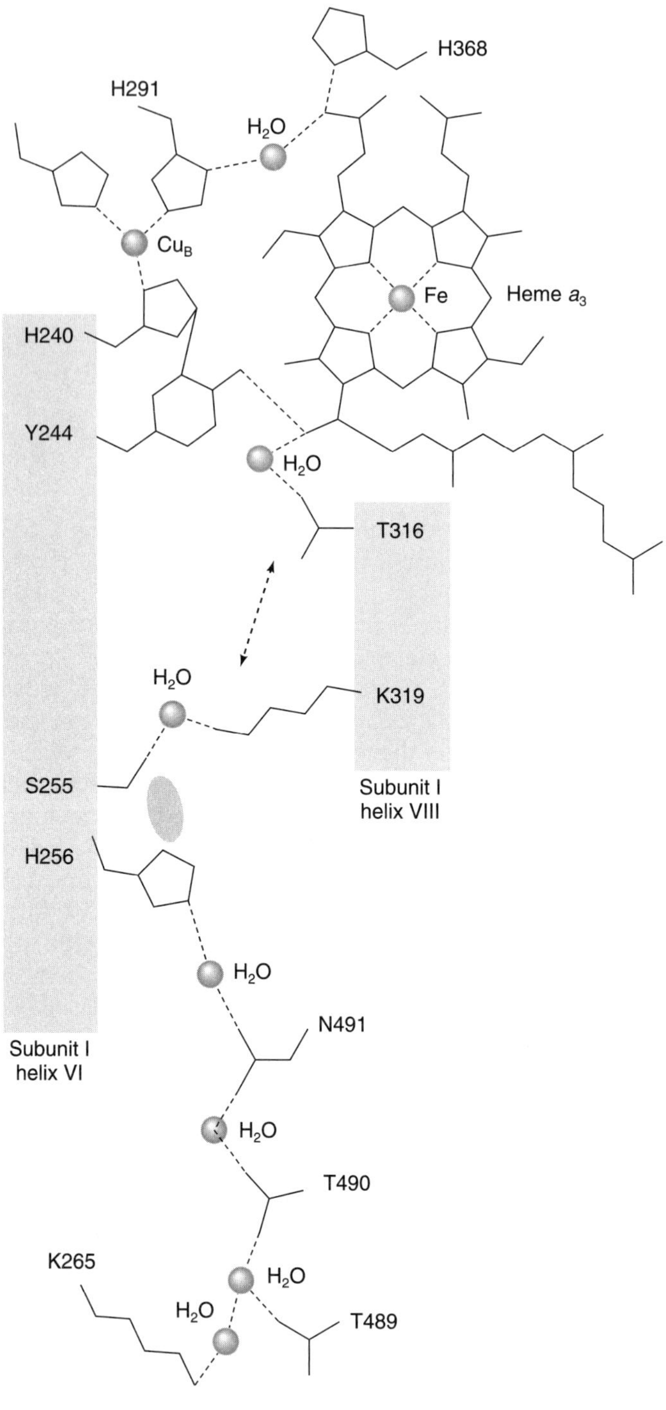

Fig. XI.6.8.
A schematic representation of a hydrogen-bond network between the O_2 reduction site and the molecular surface exposed to the matrix space. Dotted lines are hydrogen or coordination bonds. A dotted line with arrowheads on both ends denotes the space for conformational change for the proton transfer between the two residues. An oval shows a cavity including mobile waters.

tion of protons, which could collapse the electrochemical potential across the mitochondrial inner membrane.

XI.6.3.4. Identification of Proton-Transfer Paths by Site-Directed Mutagenesis

The X-ray structures of bovine and bacterial cytochrome *c* oxidase have stimulated site-directed mutagenesis efforts to attempt to identify possible proton-transfer paths. These studies, using cytochrome *c* oxidases from *Paracoccus denitrificans* and *Rhodobactor spheroides*, as well as quinol oxidase (cytochrome b_{o_3}) from *Escherichia coli*, have revealed two possible proton-transfer paths, designated the K and D pathways, which include well-conserved Lys and Asp residues, respectively.[34] The mutant enzymes in which the lysine corresponding to Lys319 in the bovine enzyme is replaced by Met have essentially no enzymatic activity under turnover conditions.[35]

The extremely low enzymatic activity of the mutated protein is due to a decrease of several orders of magnitude in the rate of reduction of the O_2 binding site (heme a_3Fe and Cu_B). However, once the O_2 bound to the mutant enzyme is reduced completely under anaerobic conditions, the reduced enzyme reacts with O_2 as readily as the wild-type enzyme, both in O_2 reduction and in proton uptake.[35] Moreover, it has been shown that reduction of the O_2 binding site is coupled with proton uptake.[36] These results show that proton transfer through the K pathway is coupled to the reduction of the O_2 binding site and that the K pathway is not involved in proton uptake during O_2 reduction. Thus, protons taken up during reduction of the O_2 binding site are likely to be used for producing the initial water molecule. However, this channel does not provide the other two protons required for producing the second water molecule.

The mutation of glutamate corresponding to Glu242 in bovine heart enzyme to Gln impairs enzymatic activity under turnover conditions almost completely.[37] In contrast to the effects of K-pathway mutants, this D-pathway mutant enzyme has a normal reduction rate of the O_2 binding site, and no significant effect is detectable in the initial two-electron reduction of O_2.[37] However, the second two-electron reduction process seems to be inhibited.[37] This result strongly suggests that the two protons required for producing the second water are transferred via the D pathway and that proton transfer is tightly coupled to electron transfer for the second process in the O_2 reduction mechanism, that is, reduction of the O_2 reduction site in the high oxidation state. Mutation of the residue corresponding to Asp91 in the bovine heart enzyme also has similar effects on electron and proton transfers.[38] These mutagenesis results clearly show that this enzyme has two proton-transfer paths for water production and that proton transfer through the D channel is coupled with the second two-electron reduction.

XI.6.3.5. Proton Pumping

In order to utilize the free energy produced by O_2 reduction to pump protons across the mitochondrial inner membrane, the proton-pumping site must change its affinity for protons (pK_a) and its accessibility to water phases on both sides of the membrane when the redox state of the metal sites changes. Thus, a conformational change is a prerequisite for the pump function. As shown in Fig. XI.6.9, an extramembrane loop region connecting the two transmembrane α-helices shows a fairly large conformational change upon complete reduction of the fully oxidized enzyme.[7] The Asp51 residue, which is completely buried inside the protein in the oxidized state, becomes exposed to the intermembrane aqueous phase upon reduction of the enzyme. In the fully oxidized state, Asp51 is connected to the bulk water phase on the matrix side of the mitochondrion by a network including hydrogen bonds, a large

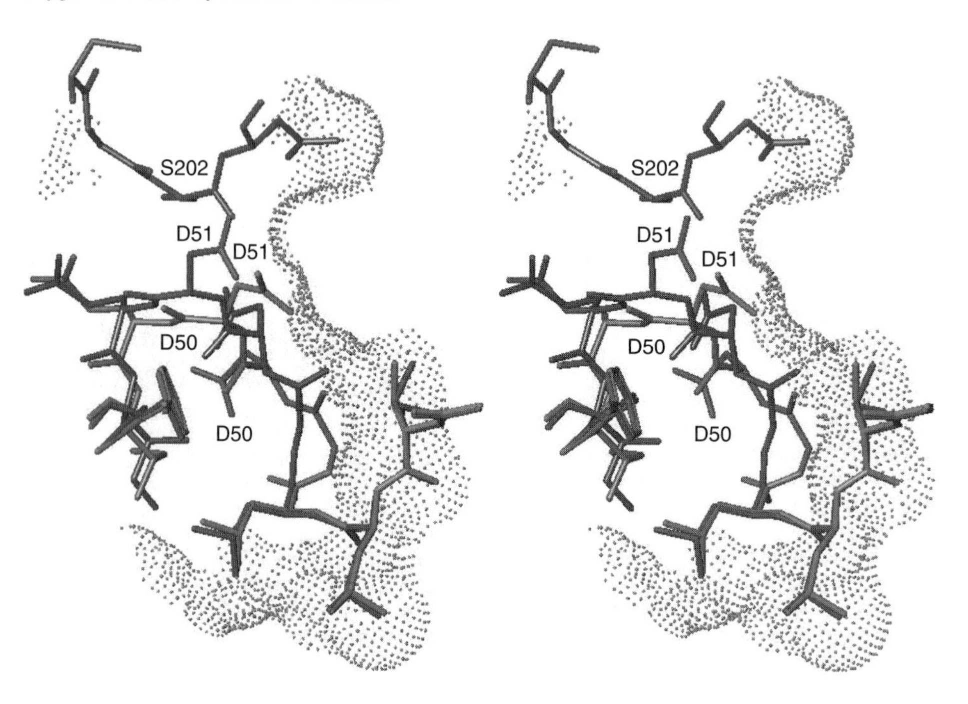

Fig. XI.6.9.
Redox-coupled conformational change in the segment between Gly49 and Asn55 in a stereoscopic drawing. Dark gray and light gray structures are in the fully oxidized and reduced states, respectively. The molecular surface in the fully oxidized state is defined by the accessibility of water molecules in the intermembrane space is shown by dots.

cavity, and a water path. Water molecules in the matrix space of mitochondria have access to Arg38, which is connected to Asp51 by a hydrogen-bond network. Thus, Asp51 can take up protons from the matrix space through the network. In this sense, Asp51 has accessibility to the matrix space in the fully oxidized state. However, upon reduction, Asp51 is released from the network, thus losing its accessibility to the matrix space, but it gains accessibility to the intermembrane space as described above. Furthermore, the pK_a of Asp51 must decrease significantly on reduction of the enzyme, owing to an increase in the dielectric constant of the microenvironment of its COOH group by the redox-coupled conformational change. This redox-coupled conformational change strongly suggests that Asp51 is critical for proton-pumping in this enzyme.[7]

The hydrogen-bond network between Asp51 and Arg38 includes a peptide bond between Tyr440 and Ser441. This peptide bond is likely to promote a unidirectional proton transfer to Asp51. It has been well established that exchange of NH protons with bulk water protons is triggered by protonation on the peptide carbonyl to form an imidic acid intermediate, $-C(OH)=N^{+}H-$.[39] Thus, the COO^{-} group of Asp51, which is close to the peptide NH, forming a hydrogen bond, could extract a proton from the nitrogen atom, when the imidic acid is formed even transiently. Once the COOH group is formed, the resulting peptide in the enol form will tautomerize back to the keto form. In this tautomerization, the proton on the C–OH will be transferred to the peptide nitrogen. The migration of the proton from C–OH to –N= requires a small conformational change in the protein surrounding the peptide bond. The difference in the stability between the keto and enol forms of the peptide seems to be sufficient to produce the free energy needed to induce the small conformational change near the peptide. Thus, this property of the peptide provides the unidirectional proton transfer to Asp51.

The evolution of biological species is reflected in the history of changes in amino acid sequence. The mutation of an amino acid residue indispensable for its physiological function would be lethal. Thus, a comparison of amino acid sequences in homologous proteins often provides important information for identification of the amino acid residues directly involved in the physiological function. For example, Asp51 is conserved only in the animal kingdom; bacterial and plant enzymes do not have this residue. On the other hand, amino acid residues involved in O_2 reduction and the heme *a* sites are strictly conserved. The greater extent of amino acid residue conservation in the O_2 reduction site compared to that of the proton-pumping site is likely related to the fact that O_2 reduction is a much more complex chemical process than proton pumping. Thus various amino acids could form the proton-pumping site, while only histidines can provide the appropriate environment for these metal centers to reduce O_2 to water.

However, various proton-pumping mechanisms at the O_2 reduction site have been proposed, assuming that both the proton-pumping site and the O_2 reduction site in cytochrome *c* oxidase are conserved in all biological species. For example: a redox-coupled conformational change in one of the histidyl imidazoles liganded to Cu_B;[40] proton pumping driven by electrostatic repulsion between pumping protons and chemical protons for producing water;[31] and conformational change of a glutamate residue in the D pathway, which determines the proton-transfer direction toward the O_2 reduction site or toward the proton-pumping site.[41] None of these proposals provides a mechanism for disallowing access of pumping protons to the O_2 reduction site. This avoidance is indispensable for preventing pumping protons from being consumed in the production of water. In the indirect coupling mechanism, including the movement of Asp51 described above, the proton-pumping pathway is completely separated from the proton pathways for water formation.

References

General References

1. Ferguson-Miller, S. and Babcock, G. T., "Heme/copper terminal oxidases", *Chem. Rev.*, **96,** 2889–2907 (1996).
2. Buse, G., Steffens, G. J., Steffens, G. C. M., Meinecke, L., and Hensel, S., "Sequence analysis of complex membrane proteins (cytochrome *c* oxidase)", in "Advanced Methods in Protein Microsequence Analysis", Wittman-Liebold et al., Eds. Springer-Verlag, Berlin, 1986, pp. 340–351.
3. Scott. R. A., "X-ray absorption spectroscopic investigations of cytochrome *c* oxidase structure and function", *Annu. Rev. Biophys. Biophys. Chem.*, **18,** 137–158 (1989).
4. Kitagawa, T. and Ogura, T., "Oxygen activation mechanism at the binuclear site of heme-copper oxidase superfamily as revealed by time-resolved resonance raman spectroscopy", *Progr. Inorg. Chem.*, **45,** 431–479 (1997).

Specific References

5. Tsukihara, T., Aoyama, H., Yamashita, E., Tomizaki, T., Yamaguchi, H., Shinzawa-Itoh, K., Nakashima, R., Yaono, R., and Yoshikawa, S., "Structures of metal sites of oxidized bovine heart cytochrome *c* oxidase at 2.8 Å", *Science*, **269,** 1069–1074 (1995).
6. Iwata, S., Ostermeier, C., Ludwig, B., and Michel, H., "Structure at 2.8 Å resolution of cytochrome *c* oxidase from *Paracoccus denitrificans*", *Nature* (*London*), **376,** 660–669 (1995).
7. Yoshikawa, S., Shinzawa-Itoh, K., Nakashima, R., Yaono, R., Yamashita, E., Inoue, N., Yao, M., Fei, M. J., Peters Libeu, C., Mizushima, T., Yamaguchi, H., Tomizaki, T., and Tsukihara, T., "Redox-coupled crystal structural changes in bovine heart cytochrome *c* oxidase", *Science*, **280,** 1723–1729 (1998).
8. Tsukihara, T., Aoyama, H., Yamashita, E., Tomizaki, T., Yamaguchi, H., Shinzawa-Itoh, K., Nakashima, R., Yaono, R., and Yoshikawa, S., "The whole structure of the 13-subunit oxidized cytochrome *c* oxidase at 2.8 Å", *Science*, **272,** 1136–1144 (1996).
9. Kadenbach, B., Ungibauer, M., Jarausch, J., Büge, U., and Kuhn-Nentwig, L., "The complexity of respiratory complexes", *Trends Biochem. Sci.*, **8,** 398–400 (1983).
10. Ludwig, B. and Schatz, G., "A two-subunit cytochrome *c* oxidase (cytochrome aa_3) from *Paracoccus denitrificans*", *Proc. Natl. Acad. Sci., U.S.A.*, **77,** 196–200 (1980).
11. Gibson, Q. H., Greenwood, C., Wharton, D. C., and Palmer, G., "The reaction of cytochrome oxidase with cytochrome *c*", *J. Biol. Chem.*, **240,** 888–894 (1965).
12. Antonini E., Brunori, M., Colosimo, A. Greenwood, C., and Wilson, M. T., "Oxygen "pulsed" cytochrome *c* oxidase: Functional properties and catalytic relevance", *Proc. Natl. Acad. Sci. U.S.A.*, **74,** 3128–3132 (1977).
13. Baker, G. M., Noguchi, M., and Palmer, G., "The reaction of cytochrome oxidase with cyanide. Preparation of the partially rapidly reacting form and its conversion to the slowly reacting form", *J. Biol. Chem.*, **262,** 595–604 (1987).

14. Van Gelder, B. F. and Beinert, H., "Studies of the heme components of cytochrome *c* oxidase by EPR spectroscopy", *Biochim. Biophys. Acta,* **189,** 1–24 (1969).
15. Tweedle, M. F., Wilson, L. J., Garcia-Iniguez, L., Babcoch, G. T., and Palmer, G., "Electronic state of heme in cytochrome oxidase. III. The magnetic susceptibility of beef heart cytochrome oxidase and some of its derivatives from 7-200 K. Direct evidence for an antiferromagnetically coupled Fe(III)/Cu(II) pair", *J. Biol. Chem.,* **253,** 8065–8071 (1978).
16. Reed, C. A. and Landrum, J. T., "Structural models for the heme a_3-copper active site of cytochrome *c* oxidase", *FEBS Lett.,* **106,** 265–267 (1979).
17. Giuffre, A., Stubauer, G., Brunori, M., Sarti, P., Torres, J., and Wilson, M. T., "Chloride Bound to Oxidized Cytochrome *c* Oxidase Controls the Reaction with Nitric Oxide", *J. Biol. Chem.,* **273,** 32475–32478 (1998).
18. Mochizuki, M, Aoyama, H., Shinzawa-Itoh, K., Usui, T., Tsukihara, T., and Yoshikawa, S., "Quantitative reevaluation of the redox active sites of crystalline bovine heart cytochrome *c* oxidase", *J. Biol. Chem.,* **274,** 33403–33411 (1999).
19. Ostermeier, C., Harrenga, A., Eromler, U., and Michel, H., "Structure at 2.7 Å resolution of the *Paracoccus denitrificans* two-subunit cytochrome *c* oxidase complexed with an antibody Fv fragment", *Proc. Natl. Acad. Sci. U.S.A.,* **94,** 10547–10553 (1997).
20. Kroneck, P. M. H., Antholine, W. A., Riester, J., and Zumft, W. G., "The cupric site in nitrous oxide reductase contains a mixed-valence [Cu(II), Cu(I)] binuclear center: a multifrequency electron paramagnetic resonance investigation", *FEBS Lett.,* **242,** 70–74 (1988).
21. Regan, J. J., Ramirez, B. E., Winkler, J. R., Gray, H. B., and Malmström, B. G., "Pathways for electron tunneling in cytochrome *c* oxidase", *J. Bioenerg. Biomembr,* **30,** 35–39 (1998).
22. Hill, B. C., "Modeling the Sequence of Electron Transfer Reactions in the Single Turnover of Reduced, Mammalian Cytochrome *c* Oxidase", *J. Biol. Chem.,* **269,** 2419–2425 (1994).
23. Gibson, Q. H. and Greenwood, C., "Reactions of cytochrome oxidase with oxygen and carbon monoxide", *Biochem. J.,* **86,** 541–555 (1963).
24. Varotsis, C., Woodruff, W. H., and Babcock, G. T., "Time-resolved Raman detection of ν(Fe–O) in an early intermediate in the reduction of O_2 by cytochrome oxidase", *J. Am. Chem. Soc.,* **112,** 1297 (1990).
25. Ogura, T., Takahashi, S., Shinzawa-Itoh, K., Yoshikawa, S., and Kitagawa, T., "Observation of the $Fe^{II}–O_2$ stretching Raman band for cytochrome oxidase compound A at ambient temperature", *J. Am. Chem. Soc.,* **112,** 5630–5631 (1990).
26. Han, S., Ching, Y.-C., and Rousseau, D. L., "Primary intermediate in the reaction of oxygen with fully reduced cytochrome *c* oxidase", *Proc. Natl. Acad. Sci. U.S.A.,* **87,** 2491–2495 (1990).
27. Nagai, K., Kitagawa, T., and Morimoto, H., "Quaternary structures and low frequency molecular vibrations of haems of deoxy and oxyhaemoglobin studied by resonance raman scattering", *J. Mol. Biol.,* **136,** 271–289 (1980).
28. Van Çart, H. E. and Zimmer, J., "Resonance Raman evidence for the activation of dioxygen in horseradish peroxidase", *J. Biol. Chem.,* **260,** 8372–8377 (1985).
29. Wikström, M., "Identification of the electron transfers in cytochrome oxidase that are coupled to proton-pumping", *Nature (London),* **338,** 776–778 (1989).
30. Michel, H., "The mechanism of proton pumping by cytochrome *c* oxidase", *Proc. Natl. Acad. Sci. U.S.A.,* **95,** 12819–12824 (1998).
31. Michel, H., "Cytochrome *c* Oxidase: Catalytic Cycle and Mechanisms of Proton Pumping-A Discussion", *Biochemistry,* **38,** 15129–15140 (1999).
32. Verkhovsky, M. I., Jasaitis, A., Verkhovskaya, M. L., Morgan, J. E., and Wikström, M., "Proton translocation by cytochrome *c* oxidase", *Nature (London),* **400,** 480–483 (1999).
33. Ruitenberg, M., Kannt, A., Bamberg, E., Fendler, K., and Michel, H., "Reduction of cytochrome *c* oxidase by a second electron leads to proton translocation", *Nature (London),* **417,** 99–102 (2002).
34. Gennis, R. B., "Multiple proton-conducting pathways in cytochrome oxidase and a proposed role for the active-site tyrosine", *Biochim. Biophys. Acta,* **1365,** 241–248 (1998).
35. Gennis, R. B. and Zaslavsky, D., "Substitution of Lysine-362 in a Putative Proton-Conducting Channel in the Cytochrome *c* Oxidase from *Rhodobacter sphaeroides* Blocks Turnover with O_2 but Not with H_2O_2", *Biochemistry,* **37,** 3062–3067 (1998).
36. Mitchell, R. and Rich, P. R., "Proton uptake by cytochrome *c* oxidase on reduction and on ligand binding", *Biochim. Biophys. Acta,* **1186,** 19–26 (1994).
37. Ädelroth, P., Ek, M. S., Mitchell, D. M., Gennis, R. B., and Brzezinski, P., "Glutamate 286 in Cytochrome aa_3 from *Rhodobacter sphaeroides* Is Involved in Proton Uptake during the Reaction of the Fully-Reduced Enzyme with Dioxygen", *Biochemistry,* **36,** 13824–13829 (1997).
38. Smirnova, I. A., Ädelroth, P., Gennis, R. B., and Brzezinski, P., "Aspartate-132 in Cytochrome *c* Oxidase from *Rhodobacter sphaeroides* Is Involved in a Two-Step Proton Transfer during Oxo-Ferryl Formation", *Biochemistry,* **38,** 6826–6833 (1999).
39. Perrin, C. L., "Proton exchange in amides: surprises from simple system", *Acc. Chem. Res.,* **22,** 268–275 (1989).
40. Wikström, M., Bogachev, A., Finel, M., Morgan, J. E., Puustinen, A., Raitio, M., Verkhovskaya, M., and Verkhovsky, M. I., "Mechanism of proton translocation by the respiratory oxidases. The histidine cycle", *Biochim. Biophys. Acta,* **1187,** 106–111 (1994).
41. Wikstrom, M., Verkhovsky, M. I., and Hummer, G., "Water-gated mechanism of proton translocation by cytochrome *c* oxidase", *Biochim. Biophys. Acta,* **1604,** 61–65 (2003).

XI.7. Reducing Dioxygen to Water: Multi-Copper Oxidases

Peter F. Lindley
Instituto de Tecnologia
Química e Biológica
Universidade Nova de
Lisboa
2781-901 Oeiras, Portugal

Contents

XI.7.1. Introduction

The multi-copper oxidases comprise a family of enzymes whose prime members are the laccases (benzenediol oxygen oxidoreductase), ascorbate oxidase (L-ascorbate oxygen oxidoreductase) and ceruloplasmin [Fe(II) oxygen oxidoreductase].[1–3] These enzymes normally contain a characteristic configuration of four Cu ions; a mononuclear Cu ion some 12–13 Å distant from a trinuclear Cu center. The one-electron oxidation of substrates at the mononuclear Cu center occurs concomitantly with a four-electron reduction of molecular O_2 to two molecules of water at the trinuclear center. Nitrite reductase (Cu containing nitrite reductase) is also a member of the family although it lacks the trinuclear Cu center. Most interestingly, the blood coagulation factors V and VIII may also be more distant members of the family, although their Cu content may be at most one Cu ion.

XI.7.2. Occurrence and General Properties

XI.7.2.1. Ascorbate Oxidase

Ascorbate oxidase (AO) is found in higher plants, but cucumber and green zucchini squash are the most common sources.[4] The monomer molecule comprises some 552 amino acid residues and four Cu ions. At the cellular level, the enzyme is most abundant in the cell wall and the cytoplasm.

XI.7.2.2. Laccases

Most of the known laccases (LAC) have fungal (e.g., white-rot fungi) or plant origins, although a few laccases have recently been identified and isolated from bacteria.[5–8] Although the LACs are roughly the same size as AO (a single polypeptide chain of some 500 residues), their molecular weights are often inflated by an extensive carbohydrate coating; the molecular weight of the Japanese lacquer tree enzyme *Rhus vernicifera* approaches 110 kDa.

XI.7.2.3. Ceruloplasmin

Ceruloplasmin (CP) is a plasma protein that binds most of the 5–6 mg of Cu circulating at any instant in the blood plasma. The name is derived from its "sky-blue" color when in the pure state. Ceruloplasmin, first isolated by Holmberg and Laurell in the 1940s,[9,10] is a monomeric glycoprotein with an overall molecular weight of ~132 kDa, comprising a single polypeptide chain of 1046 amino acid residues and a carbohydrate content of 7–8% by weight. Ceruloplasmin is synthesized in the liver hepatocyte cells and in normal circumstances six Cu ions are incorporated into the protein before it is secreted into the plasma.

XI.7.2.4. Nitrite Reductase

Nitrite reductase (NR) participates in the denitrification pathway in which nitrate is converted sequentially to nitrite, nitric oxide, nitrous oxide, and, finally, nitrogen.[11] Denitrification results in the loss of plant nutrients from soil. An understanding of this process is therefore as important as understanding other key components of the nitrogen cycle including nitrogen fixation and photosynthesis (see Chapter II; Sections X.3 and XII.3). There are two classes of NR depending on whether a heme group or Cu is used at the active site and Cu-containing enzymes have been isolated from a number of organisms including *Achromobacter cycloclastes.*

XI.7.3. Functions

XI.7.3.1. Ascorbate Oxidase

The *in vivo* role of AO is still a matter of debate. The enzyme reduces ascorbic acid (vitamin C) to dehydro-ascorbic acid, but it also reduces other compounds that contain a lactone ring with an enediol group adjacent to a carbonyl group. Since *in vitro* catechols and phenols are also substrates, AO may be involved in the ripening of fruit, but a role in plant respiration has also been suggested.[4]

XI.7.3.2. Laccases

The LACs have been implicated in many diverse physiological functions such as morphogenesis, pathogenesis, lignin synthesis, and lignolysis.[6,12] Chemically, all these functions are related to the oxidation of a range of aromatic substrates, such as polyphenols, methoxy-substituted phenols, diamines, and even some inorganic compounds. The initial substrate reaction products are oxygen-centered radicals or cation radicals that usually react further through non-enzymatic routes for the oxidative coupling of monomers or the degradation of polymers. Owing to their high relatively nonspecific oxidation capacity, laccases have been found to be useful biocatalysts for diverse biotechnological applications.[13] Their biotechnological importance has shown a marked increase after the discovery that the oxidizing reaction substrate range could be further extended in the presence of the so-called mediators, small readily oxidizable molecules,[14] by a mechanism that remains, as yet, quite unclear. The most extensively investigated laccase mediator is 2,2′-azinobis-(3-ethylbenzothiazoline-6-sulfonic acid) (ABTS), which allows the oxidation of nonphenolic lignin model compounds and pulp delignification by laccase.[14,15] Several other effective redox mediators have also been extensively reported and the applications of laccase–mediator systems have been described that target lignocellulosics and other insoluble materials.[13]

XI.7.3.3. Ceruloplasmin

Despite extensive studies, the precise function(s) of CP remain unclear and it has been termed the "enigmatic" copper protein.[16] It is likely to be multifunctional and most of the functions ascribed to CP focus on the presence of the Cu centers. These include

1. Ferroxidase activity either to eliminate free iron from the plasma, thus protecting blood and membrane lipids from peroxidative damage, and/or to mobilize iron from cells for transport via the plasma protein, transferrin.
2. Antioxidant activity to remove oxygen based and other free radicals from the plasma.
3. Amine oxidase activity to control levels of biogenic amines in the plasma, cerebral spinal, and interstitial fluids.
4. Copper transport to deliver the metal to extrahepatic tissues.

Defects in hepatic biosynthesis have been associated with diseases, such as Wilson's disease (hepatolenticular degradation), which results in Cu deposition in body tissues, particularly liver, brain, and kidney. Incomplete expression of CP has been associated with hemosiderosis (iron overload),[17] thus emphasizing the important existence of the synergy between Cu and Fe metabolism in the body. The CP protein is an acute-phase reactant and in the case of tissue damage or infection can exhibit a two- to threefold increase over the normal concentration of $300\,\mu g\,mL^{-1}$ in adult plasma. Useful reviews include Refs. 16 and 18–20.

XI.7.3.4. Nitrite Reductase

The NR from *A. cycloclastes* participates in the reduction of nitrite to NO with the consumption of two H^+ and one electron,

$$NO_2^- + 2\,H^+ \text{ and } 1\,e^- \rightarrow NO + H_2O$$

The electron may initially be provided by ascorbate, but indirectly through a pseudoazurin molecule.

XI.7.4. X-Ray Structures

XI.7.4.1. Overall Molecular Organization

The multi-copper oxidases are multidomain proteins and Fig. XI.7.1*a–c* shows schematically the arrangements of the domains in the AO (and LAC), CP, and NR structures, respectively, and the relationships among the structures; Cu ions are designated by black circles. Both AO and LAC have three domains with a mononuclear copper ions in domain 3 and a trinuclear center between domains 1 and 3. Ascorbate oxidase is in reality a dimer comprising two complete molecules. The dimer contains nine Cu ions in total; four in each of the two molecules in the dimer and an additional Cu ions located between the two molecules. Ceruloplasmin has six domains with mononuclear copper ions in domains 2, 4 and 6 and a trinuclear center between domains 1 and 6; domains 1, 2 and 6 are equivalent to domains 1, 2 and 3 in AO. Nitrite reductase is a trimer of two domains with a mononuclear Cu ions in domains 1, 1′, and 1″ with additional mononuclear centers between domain 1 of one dimer and domain 2 of a second.

X-ray crystallography shows that all the domains are based on the cupredoxin fold.[21] This folding pattern is typified by the structures of the small blue Cu proteins, azurin and plastocyanin, and comprises two β-sheets arranged in a sandwich configuration. Figure XI.7.2 shows the structures of domains 1 and 6 in ceruloplasmin; in both domains, the two β-sheets are built from strands 2a, 1, 3, and 6, and 4, 7, 8, and 2b, respectively, but in domain 6 strand 5 can also be considered as part of the second sheet. Figure XI.7.2*b* also shows the mononuclear Cu in domain 6 bound by two histidines, a cysteine, and a methionine residue.

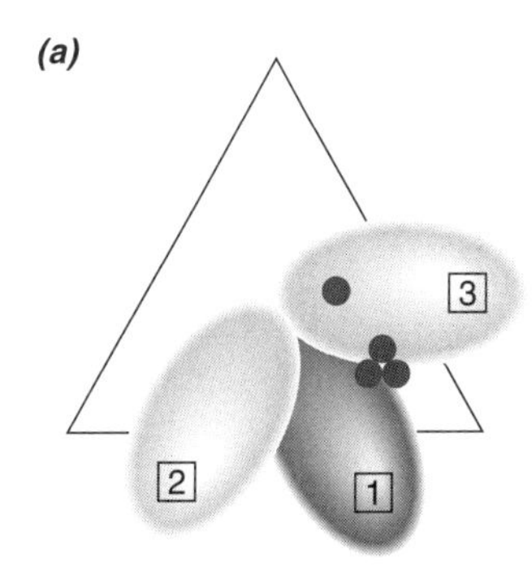

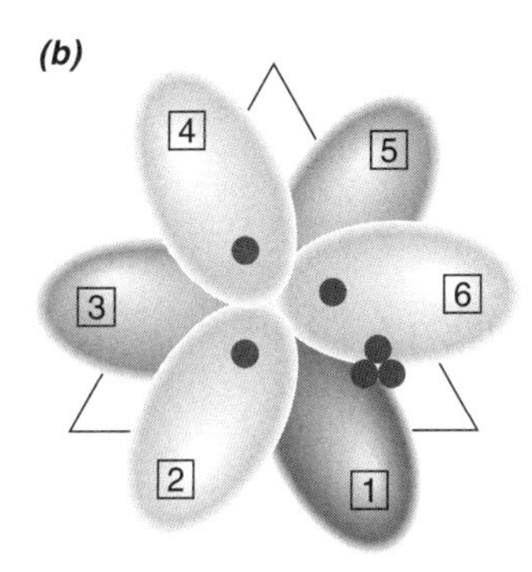

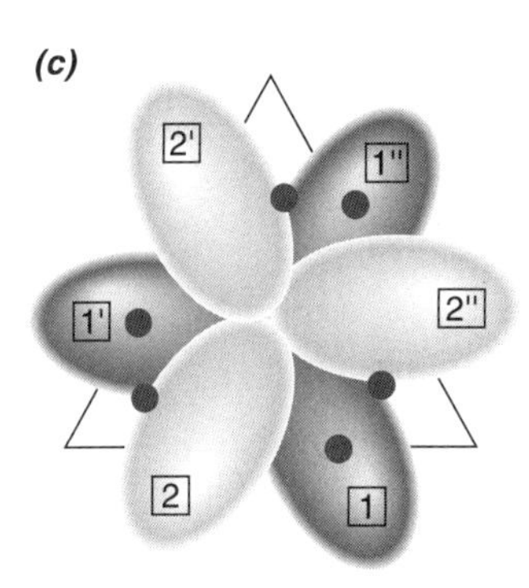

Fig. XI.7.1a–c. Schematic representations of the organization of cupredoxin domains in AO, CP, and NR, respectively, showing relationships among the structures. The copper centers are shown by full black circles. In AO (and LAC), domain 3 (equivalent to domain 6 in CP) contains a mononuclear Cu center and a trinuclear center is located between domains 1 and 3. In CP, each of the even domains contains a mononuclear Cu center and the trinuclear center is sited between domains 1 and 6. In the trimeric NR structure, the odd domains have mononuclear Cu centers and there is a second mononuclear center between each pair of dimers in the trimer.

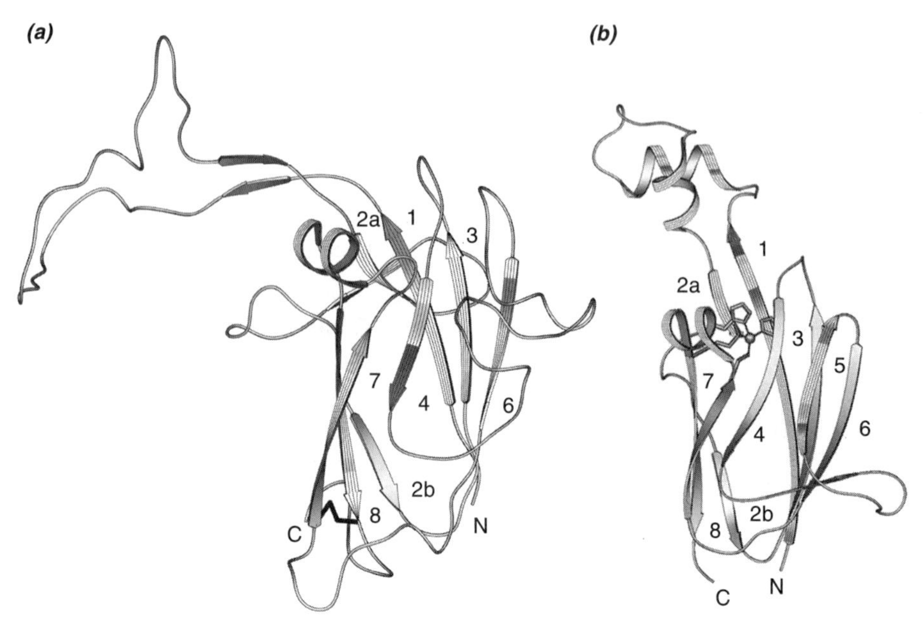

Fig. XI.7.2.
Typical cupredoxin domains in CP represented by ribbon diagrams.
(*a*) *Domain 1.* The β-sheets comprise strands 2a, 1, 3, and 6, and 4, 7, 8, and 2b. All the domains in CP, with the exception of C-terminal domain 6, contain a disulfide bridge, which appears to hold the final strand to the domain. In the odd domains, this bridge occurs between strands 7 and 8, as shown in the figure; for the even domains 2 and 4, it occurs between strands 2b and 8.
(*b*) *Domain 6.* Strand 5 can be considered as part of the second sheet. The mononuclear copper in this domain is shown bound to its four ligands, two histidines, a cysteine, and a methionine residue.

X-ray structures have been determined as follows;

AO from green zucchini squash[22]	(PDB code[23]: 1AOZ)
Fungal LAC, from *Coprinus cinereus*[24]	(PDB codes: 1A65, 1HFU)
Trametes versicolor[25,26]	(PDB codes: 1KYA, 1GYC)
Melanocarpus albomyces[27]	(PDB code: 1GE0)
Bacterial LAC, CotA[a] from *Bacillus subtilis*[28]	(PDB code: 1GSK)
CP from human plasma[29]	(PDB code: 1KCW)
NR[b] from *A. cycloclastes*[30]	(PDB code: 1AS6)

[a] A 65-kDa protein that is an abundant component of the outer-coat layer of the bacterial endospore from *B. subtilis.*

[b] Other NRs have also been studied, but the structure from *A. cycloclastes* was the first to be reported.

Only the overall structure of CP will be described in this section. The X-ray structure of the human enzyme at a resolution of 3.0 Å[29] shows that the molecule features a triangular array of six cupredoxin-type domains. The odd and even numbered domains comprise ~200 and 150 residues, respectively, and there is considerable structural similarity within them; the main differences between the odd and even domains lie in the nature of the loop regions (Fig. XI.7.2*a* and *b*) between strands 1

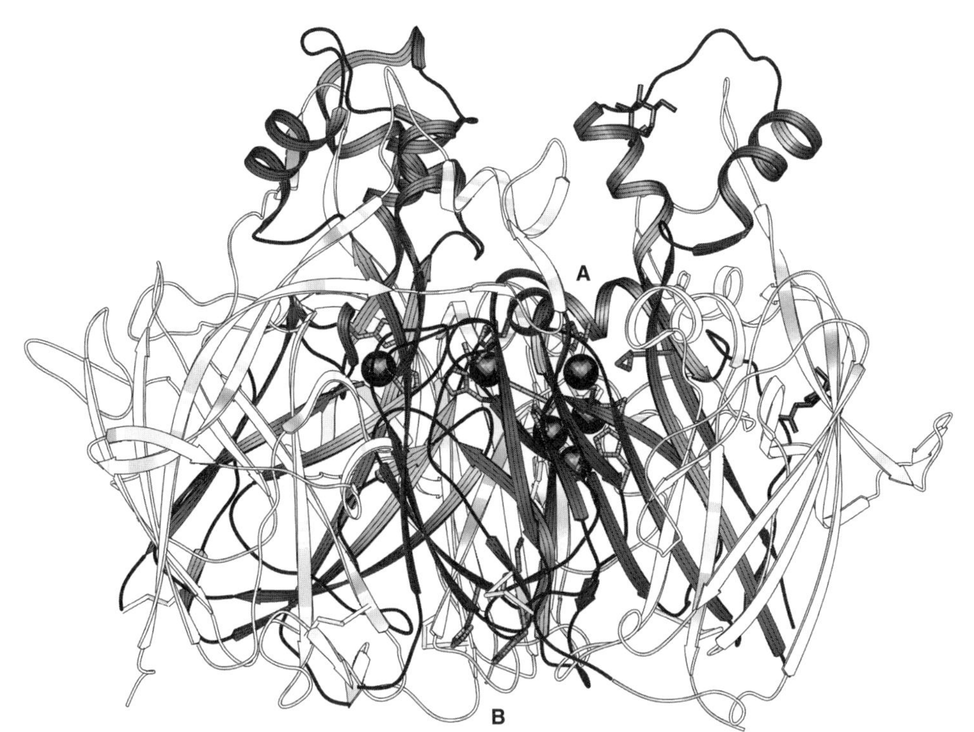

Fig. XI.7.3.
A view of the CP molecule showing how the large loops between strands 1 and 2a in each domain cover the top surface of the molecule and restrict access to the mononuclear copper ions (shown as black circles). The bottom surface of the CP molecule is relatively flat. The binding sites for metal cations and biogenic amines, and aromatic diamines, are designated by A and B, respectively (see Section XI.7.5).

and 2a. All six domains are oriented in the same sense, but the even domains point inward toward the pseudo-threefold axis of the molecule, whereas the odd domains point outward. Such an arrangement has a pronounced effect on the separation of the mononuclear Cu-binding sites in the even domains, giving separations of ~ 18 Å that are well within the range of electron transfer between the Cu centers. The large loop regions occur at the top of the molecule as shown in Fig. XI.7.3 and appear to restrict access to the mononuclear copper sites. On the other hand, the bottom of the molecule is comparatively flat.

XI.7.4.2. Copper Sites

The multi-copper oxidases contain at least three different types of Cu^{2+} ions from the visible, ultraviolet (UV), and EPR spectroscopic viewpoint,[31] as listed below.

T1 Cu^{2+} — T1 copper ions are normally coordinated by three strong ligands, a Cys and two His residues, and may also have one or two weaker ligands (e.g., Met sulfur or oxygen). They show a high absorption in the visible region, $\varepsilon > 3000\ M^{-1}\,cm^{-1}$ at 600 nm, giving rise to the blue color of the enzyme, and the EPR spectrum exhibits a hyperfine splitting of $A_{||} < 95 \times 10^{-4}\,cm^{-1}$.

T2 Cu^{2+} — T2 copper is normally three or four coordinate with histidines and water (hydroxyl) ligands. It exhibits no detectable absorption in the UV/vis. The EPR line shape is typical of normal low molecular weight Cu complexes with $A_{||} > 140 \times 10^{-4}\,cm^{-1}$.

T3 Cu^{2+} — Three histidines and a bridging ligand such as oxygen or hydroxyl usually coordinate T3 coppers. They show strong absorption in the near UV with $\lambda_{max} = 330$ nm, but no EPR signal since they occur in pairs of copper ions, which are antiferromagnetically coupled.

Reduction of Cu^{2+} to Cu^{+} by suitable electron donors causes loss of the optical and EPR signals.

The Copper Centers in AO, LAC, and CP. In normal circumstances, AO, LAC, and CP contain a trinuclear Cu center located some 12–13 Å from a T1 copper center, as shown for CP in Fig. XI.7.4. However, Cu depletion of the T2 and T3 centers can readily occur, as indicated in the section on copper centers for the LAC from *C. cinereus.* The trinuclear center contains a pair of T3 copper ions and one T2. The T3 copper ions are normally bridged by a hydroxyl group and a second hydroxyl group links the T2 copper ion to the protein either through a solvent atom (in AO) or a tyrosine residue in CP. This trinuclear center is linked to the T1 center through a Cys and two His residues (Cys1021, His1020, and His1022 for CP). Substrate molecules bind near the T1 center and each substrate in turn donates an electron (thereby becoming oxidized) to the copper ion, which is then reduced. The electron is then transferred via the Cys and His ligands to the trinuclear center. A molecule of O_2 binds at the trinuclear center and after the transfer of four electrons is reduced to two molecules of water. A plausible mechanism has been described in detail[32,33] and involves loss of the bridging hydroxyl moiety and an increase in the separation of the T3 copper atoms from 3.7 to 5.1 Å as reduction takes place. Four pairs of histidine residues, two pairs each from domains 1 and 6 in CP, bind the trinuclear center to the protein with Cu–N distances of ~2.1 Å.

Ceruloplasmin contains three mononuclear copper centers, one in each of the even domains. Those in domains 4 and 6 are typical T1 blue centers, coordinated by a Cys and two His residues; a Met residue is less strongly coordinated with a Cu–S distance of some 3.0 Å. However, in the domain 2 copper center the Met residue is replaced by Leu. There is no evidence that this Cu ion participates in substrate binding and, indeed, the center may be in a permanently reduced state.[34]

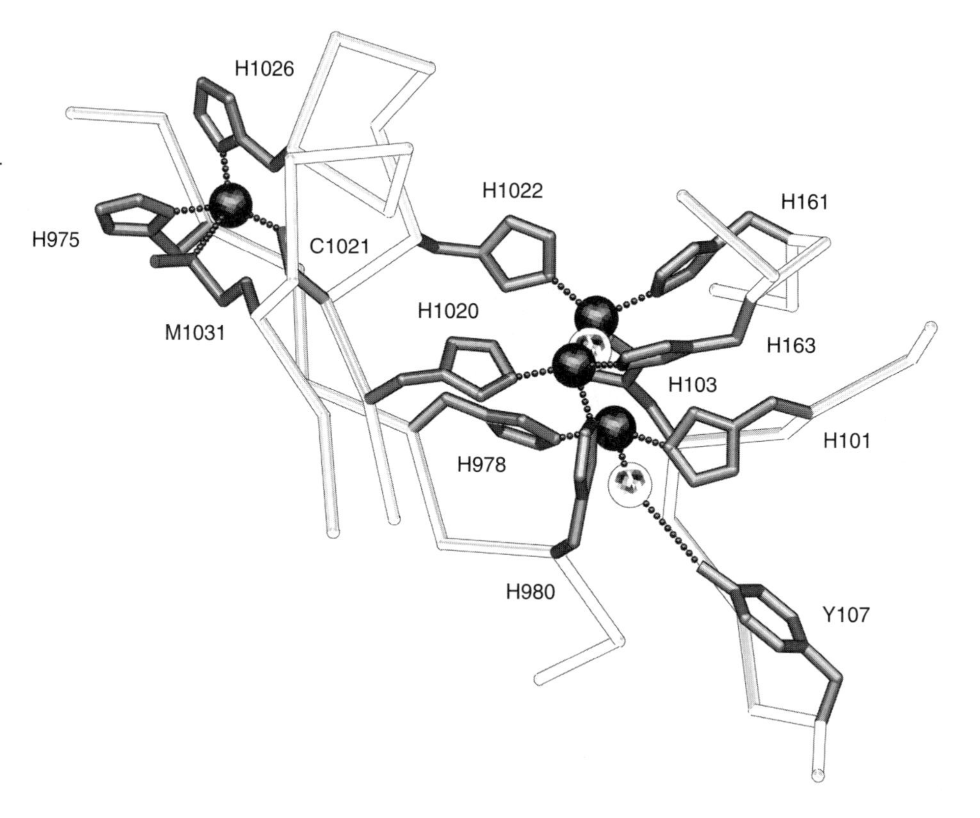

Fig. XI.7.4.
The relationship of the T1 copper center in domain 6 of CP and the trinuclear cluster situated at the interface of domains 1 and 6. This arrangement is almost identical with that found in ascorbate oxidase. The closest distance between the domain 6 copper and the trinuclear center is ~12.5 Å. Within the trinuclear cluster the Cu–Cu distances range between 3.4 and 4.0 Å. The 2 T3 copper ions are bridged by an oxygen atom (OH^-) and a second oxygen (OH^-) is liganded to the T2 copper and forms a hydrogen bond with a nearby tyrosine residue (bottom right in the figure). Divalent metal cations and biogenic amines bind close to the domain 6 copper (and also the domain 4 copper for cations) and release electrons, thereby becoming oxidized. (See Tutorial I for one-letter residue codes.)

The domain arrangement in CP gives rise to distances between the mononuclear centers of some 18 Å, well within the electron-transfer range, and clear electron-transfer pathways exist between them.[35] For the domain 4 and 6 copper ions, this pathway is defined by

Cu4–His685. . .H-bond. . .Glu971–Ile972–Asp973–Leu974–His975–Cu6

Thus, the Cu ion in domain 4 is also a potential acceptor of electrons, which can be transferred to the domain 6 copper prior to reducing O_2 at the trinuclear Cu center.

The Copper Centers in the Laccase from *C. cinereus*. The LACs are often difficult to crystallize in a form suitable for an X-ray structure analysis, probably because the carbohydrate moieties cover the surface of the molecules in a partially disordered manner. The first structure to be reported was that from *C. cinereus*[24] using crystals grown from the enzyme in a deglycosylated form and in the presence of 20 m*M* ethylenediaminetetracetic acid (EDTA). The resulting structure, although very similar to AO overall, has two important differences. First, the weaker Met ligand is absent at the T1 mononuclear Cu center (cf. the domain 2 center in CP). Second, the trinuclear center is found to be T2 copper depleted, presumably as a result of the use of EDTA in the preparation stage. This loss of Cu has both structural and spectroscopic implications. From the structural viewpoint, the two T3 copper ions are separated by some 4.9 Å (cf. 3.7 and 5.1 Å in fully oxidized and reduced AO, respectively) and the bridging oxygen moiety (OH^-) is asymmetrically placed between them, as shown in Fig. XI.7.5, forming a hydrogen bond with a nearby solvent molecule.

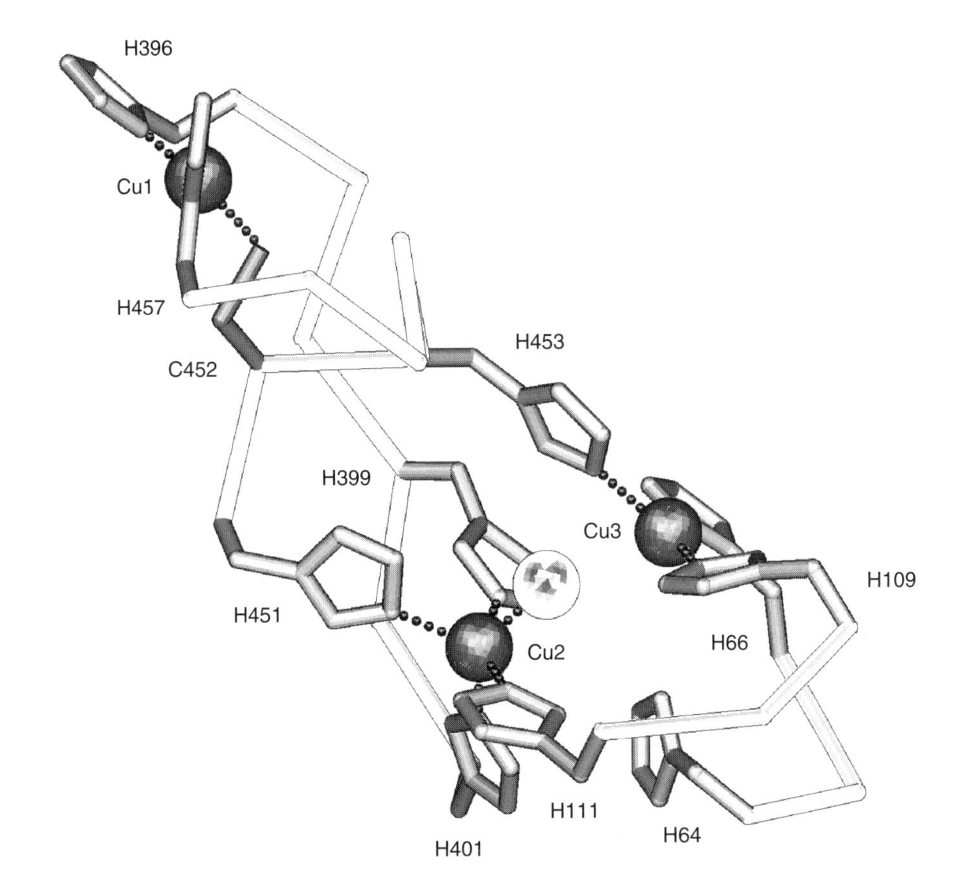

Fig. XI.7.5.
A view of the Cu centers in the LAC structure of the enzyme as isolated from *C. cinereus.* Both Cu2 and Cu3 are T3 copper atoms, whereas Cu1 is a T1 copper, but lacking the methionine ligand. The structure is T2 copper depleted, which has both structural and spectroscopic implications. The T3 copper ions are further apart than in the normal trinuclear center found in AO and CP, and the bridging hydroxyl moiety is asymmetrically placed with respect to them. At the site normally occupied by the T2 center, two histidine residues are in van der Waals' contact. In addition, the T1 copper center lacks the methionine ligand, having the same configuration as the domain 2 copper center in CP. (See Tutorial I for one-letter residue codes.)

From the spectroscopic viewpoint, spin-pairing between the two T3 coppers is unlikely with this arrangement and the lack of a corresponding EPR signal suggests that T2 copper depletion is accompanied by reduction of both T3 copper centers. One of the His residues, His399, which would normally bind to the T2 copper, rotates across to form a fourth His ligand to a T3 copper ion. The second T3 copper is three coordinate with three histidines bound in a trigonal planar environment, typical of Cu^{+}. The T2 copper ion is not simply replaced by a solvent molecule, as might be expected. The residues His64 and His399, which would normally constitute T2 copper ligands, are in van der Waals contact with a closest approach of 3.3 Å, leaving no space. It is therefore likely that both main and side-chain structural rearrangements would be required to accommodate a T2 copper ion. Recent structure determinations of fungal laccases[25–27] and a bacterial CotA laccase[28] show that all four Cu sites are occupied.

The Copper Centers in Nitrite Reductase from *A. cycloclastes*. Nitrite reductase has a T1 copper center in each of the domains 1 in the trimer, in marked contrast to CP. In NR these centers are close to the external surface of the trimer and far more accessible than in CP. This arrangement is consistent with the function of NR, whereby a large molecule such as pseudoazurin has to transfer an electron so that nitrite can be converted into NO (see the section on Nitrite Reductase).

In NR the T2 centers, located between domain 1 of one dimer and domain 2 of a second, have an almost regular tetrahedral geometry (Fig. XI.7.6). Two His residues, His100 and His135, from domain 1 of one subunit, and a third, His306, from domain 2 of a different monomer in the trimer, constitute three of the ligands. The fourth

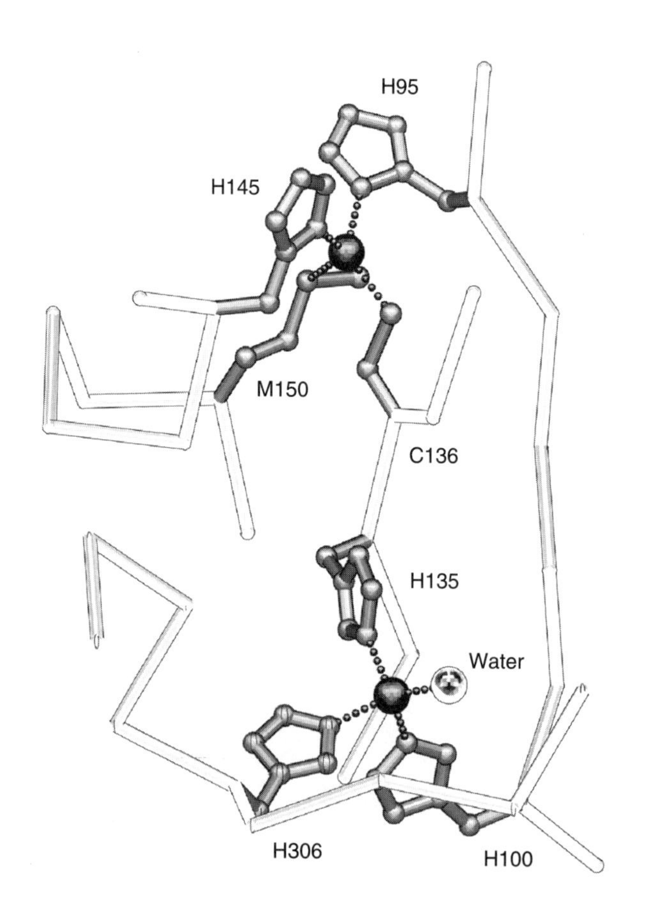

Fig. XI.7.6.
The Cu centers in NR. At the T2 center, two histidine residues, His100 and His135, arise from domain 1 of one monomer, whereas His306 is located in domain 2 of a different monomer in the trimer; a water molecule is the fourth ligand in an almost regular tetrahedral geometry. The distance between the copper centers is 12.7 Å. (See Tutorial I for one-letter residue codes.)

ligand is a water molecule oriented toward a solvent access channel. It appears to be replaced by nitrite under suitable soaking conditions in the crystal, indicating that the T2 copper site is the active center for reduction.[36] The separation between the T1 and T2 centers is 12.7 Å.

XI.7.5. Structure–Function Relationships

XI.7.5.1. Ascorbate Oxidase and the Laccases

Dioxygen Binding. In all the multi-copper oxidases so far studied, there are water channels that link the trinuclear Cu center with the external environment. Oxygen approaches the trinuclear Cu center through these channels, but the precise nature of its subsequent interaction with the center is the subject of debate. The presence of an O_2 moiety amid the three copper atoms of the trinuclear cluster is reported in the crystal structure of the laccase from *Melanocarpus albomyces.*[27] Thus, the distance between the O1 atom of the O_2 moiety and the T2 copper atom is 2.5 Å, and all the distances between the O1 and O2 atoms and the T3 copper atoms lie in the range 2.4–2.6 Å. There is no bridging hydroxyl group as observed in the fully oxidized AO structure[22] and the two T3 copper atoms are separated by 4.8 Å. This value is far closer to the value of 5.1 Å observed in the dithionite reduced structure of AO[32] than that of 3.7 Å in the corresponding structure of the oxidized enzyme.

In the crystal structure of the *M. albomyces* enzyme, the C-terminus residues block access to the channel through which it is believed the O_2 molecule approaches the trinuclear site. Although this arrangement may account for the stability of the O_2 adduct, it also raises the question of whether this structure represents the active state of the enzyme. A conformational change has been suggested to allow entry of the oxygen molecule. The absorption spectra of the enzyme in the range 300–700 nm are similar to those obtained from the laccase from *R. vernicifera,*[37–39] and may support the structure of an O_2 moiety bound to all three Cu ions in the trinuclear center. However, these studies contrast strongly with those undertaken on AO,[32] whereby peroxide (an oxygen substitute) and azide (an inhibitor) were found to bind only to one of the T3 copper ions with a concomitant loss of the bridging hydroxide moiety and an increase in the separation of the two T3 copper ions from 3.7 to 4.8 Å and 5.1 Å, for peroxide and azide, respectively. In the case of azide, two azide moieties were found to bind to the same T3 copper ion in the trinuclear cluster in ascorbate oxidase. The addition of azide to CP produced only one binding position under the soaking conditions used, but again this involved only one T3 copper ion, equivalent to that used by ascorbate oxidase.[40] Thus the precise nature of oxygen binding to the trinuclear center in the multi-copper oxidases appears to remain undefined, and indeed, may be different for different members of this enzyme family. One possibility is that O_2 binding occurs in two stages, the first involving only one of the T3 copper atoms, and the second involving the trinuclear cluster as a whole.

Substrate Binding. The structure of AO clearly reveals a binding pocket (Fig. XI.7.7) for the organic substrate in the vicinity of the T1 copper. Within this pocket two tryptophan residues, Trp163 and Trp362, and a histidine, His512, seem to be important. The histidine residue lies at the bottom of the pocket with its Nδ1 atom bound to the Cu; the Nε2 is therefore available for substrate binding. The aromatic ring of Trp362 acts as a wall to the pocket, being approximately parallel to the imidazole ring of His512 and with its Nε1 atom accessible for hydrogen bonding. The aromatic ring of Trp163 can be considered as forming another wall of the pocket. Modeling studies have shown that an L-ascorbate molecule readily fits into the pocket with its

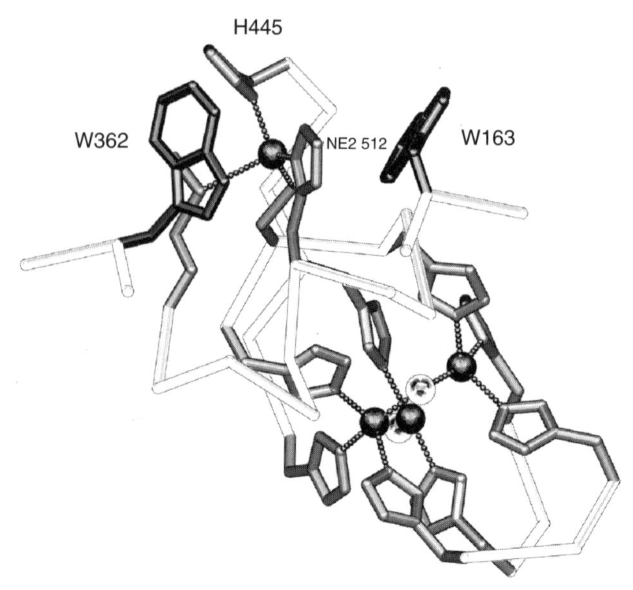

Fig. XI.7.7.
The putative organic substrate binding site in AO. The Nε1 and Nε2 atoms of Trp362 and His512, respectively, are available for hydrogen bonding to the hydroxyl groups of the substrate prior to oxidation. The Trp163 residue forms part of the wall of the cavity.

lactone ring oriented parallel to the aromatic ring of Trp163 and the side chain pointing toward the outside of the AO molecule. In this model, the hydroxyl oxygen atoms of the L-ascorbate could form hydrogen bonds with the Nε2 of His512 and the Nε1 of Trp362. The transfer of an electron, leading to oxidation of the substrate, to the T1 copper and afterward to the trinuclear center, can be readily envisaged.

With respect to the LAC, only the binding of 2,5-xylidine, a LAC inducer, has been observed in the LAC from the white rot fungus *T. versicolor.*[25] In the most abundant isozyme present in the crystal structure of this LAC, the aromatic ring system of the 2,5-xylidine is almost perpendicular to one of the histidine residues liganding the T1 copper site. The amino group of the 2,5-xylidine is 2.61 Å from the Nε2 atom of His458 and 3.20 Å from the OD2 atom of Asp206. Note that a phenyl residue replaces the weak ligand methionine at the T1 center in the LAC from *T. versicolor*.

XI.7.5.2. Ceruloplasmin

The X-ray structure provides no clear evidence that CP is a specific transporter of copper. The six Cu ions appear to be an integral part of the structure, although catabolism of the protein would result in the release of Cu into the environment. However, there exist two additional metal-binding sites,[35] one each in the vicinity of the domain 4 and 6 copper ions, as shown for domain 6 in Fig. XI.7.8. In crystals of CP, these "labile" sites are partially occupied by copper, but whether this represents a mechanism for copper transport is not clear. The labile binding sites can serve, at least *in vitro,* as sites of oxidation for Fe^{2+} cations. After oxidation, the Fe^{3+} cations appear to migrate to a "holding" site, a rather diffuse entity that could hold more than one cation, and from which the cations could be available for transfer to, for example, apo-transferrin. It also appears that biogenic amines such as serotonin, epinephrine, and 3-(3,4-dihydroxyphenyl) alanine (Dopa) bind in the vicinity of the labile sites (marked A in Fig. XI.7.3), but a different binding site has been identified for aromatic diamines. This site is situated at the base of domain 4 and in-

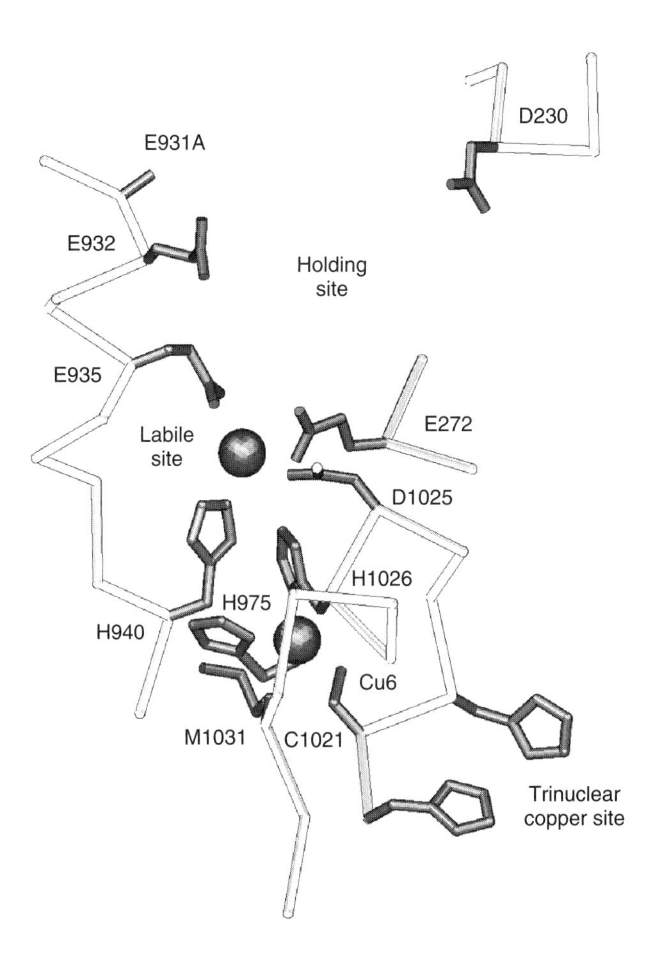

Fig. XI.7.8.
The domain 6 "labile" binding site in human CP. This site is surrounded by three negatively charged residues, Glu272 from domain 2, Glu935 and D1025 from domain 6, and a His, His940, also from domain 6. The His, H940, is some 3–5 Å away from one of the His residues, His1026, that is bound to the mononuclear integral domain 6 copper, thus facilitating electron transfer. In the native enzyme, as crystallized for the X-ray studies, the labile sites are partially occupied by copper.

This figure also shows the domain 6 "holding" site in human CP. The "holding" site is rather diffuse and could probably accommodate more than one Fe atom. It is possible that iron cations could be oxidized at the labile site and then transferred to the holding site for transfer, for example, to apo-transferrin. The X-ray studies indicate that Glu935 rotates away from the labile site toward the holding site and may play a central role in the movement of the metal cations. The electron density for Glu931 is poorly defined and this residue has been modeled as an alanine. (See Tutorial I for one-letter residue codes.)

volves an exposed tryptophan residue, Trp669;[40] site B in Fig. XI.7.3. The current X-ray studies clearly indicate that CP is likely to be multifunctional.

XI.7.6. Perspectives

X-ray structural studies have enabled a significant number of questions to be answered regarding the multi-copper oxidase family of enzymes and have provided a structural basis for understanding structure–function relationships. In particular, the number, location, and functions of the various Cu sites have been clarified.

The structural studies have also enabled a study of evolutionary relationships.[3] It is likely that a small blue ancestral cupredoxin protein first underwent domain duplication to give a two-domain structure, corresponding to domains 1 and 2 in both AO and CP in Fig. XI.7.1. A further development then took place to give an NR-like trimer. The larger members of the family then evolved from this six-domain trimer. The formation of the oxidase function in AO and CP would arise from four mutations to give a total of eight His residues at the NR trimer interfaces required to bind the trinuclear Cu centers. After CP evolved, two of the three oxidase sites were lost and vestigial histidine residues, His280 at the interface between domains 2 and 3, and His816 and His818 at the interface of domains 4 and 5, are remnants of these oxidase sites. The hydrogen-bond interactions between CP domains are more conserved between domains 1 and 2, 3 and 4, and 5 and 6 than between 2 and 3, and so on. This result is also consistent with a two-domain precursor consisting of domains 1 and 2 in Fig. XI.7.1. In the case of the three-domain AO structure, this evolution involved the replacement of three domains by a short linker peptide. A contrasting scheme[41] had proposed a single domain added to the ancestral two-domain structure. The fact that both the second and third domains in AO belong to the same structural similarity class is consistent with the more recent hypothesis.[3]

However, despite the wealth of structural information that has been obtained over the last decade and the numerous biochemical and spectroscopic studies, many key questions remain to be answered concerning structure–function relationships in the multi-copper oxidase family. These include: the mechanisms of electron transfer; the precise mechanism of oxygen reduction; the role of the T2 copper and its equivalence with respect to the T3 coppers during oxygen reduction; the details of the substrate and oxygen-binding sites; and the interpretation of spectroscopic and kinetics measurements. For a update on O_2 reduction see Ref. 42. In particular, CP remains enigmatic with many, as yet, unresolved questions, ranging from the reasons for the existence of three mononuclear Cu centers to the physiological roles of the protein, especially the role as a Cu transporter. Ceruloplasmin is often difficult to study, since it is particularly sensitive to proteolytic cleavage, change of valency of the Cu atoms, loss of Cu, and carbohydrate heterogeneity. From the structural viewpoint, an increase in resolution in the X-ray studies would be highly desirable and this may be achievable using cryogenic methods, finding a new crystal form, using protein from a different species (e.g., dolphin), or a combination of these factors.

Clearly, the multi-copper oxidases, and particularly CP, still present a real challenge for future research studies from molecular biological, biochemical, spectroscopic, and structural perspectives.[43,45]

References

General References

1. Messerschmidt, A. Ed., *Multi-copper Oxidases,* World Scientific Publishing Co. Pte. Ltd., Singapore, 1997. This book was edited by A. Messerschmidt and contains reviews on ascorbate oxidase, laccases, and ceruloplasmin by leading workers in the field.
2. Lindley, P. F., "Multi-copper Oxidases", in *Handbook on Metalloproteins,* Bertini, I., Sigel, A., and Sigel, H., Eds., Marcel Dekker Inc., New York, Basel, 2001, pp. 763–811.
3. Murphy, M. E., Lindley, P. F., and Adman, E. T., "Structural comparison of cupredoxin domains: domain recycling to construct proteins with novel functions", *Protein Sci.,* **6,** 761–770 (1997).

Specific References

4. Avigliano, L. and Finazzi-Agrò, A., "Biological Function and Enzyme Kinetics of Ascorbate Oxidase", in *Multi-Copper Oxidases,* Messerschmidt, A., Ed., World Scientific Publishing Co. Pte. Ltd., Singapore, 1997, Chapt. 9, pp. 251–284.
5. Reinhammar, B., "Kinetic Studies on Polyporus and Tree Laccases", in *Multi-Copper Oxidases,* Messerschmidt, A., Ed., World Scientific Publishing Co. Pte. Ltd., Singapore, 1997, Chapt. 6, pp. 167–200.
6. Gianfreda, L., Xu, F., and Bollag, J. M., "Laccases: A Useful Group of Oxidoreductive Enzymes", *Bioremediation J.,* **3,** 1–25 (1999).
7. Martins, L. O., Soares, C. M., Pereira, M. M., Teixeira, M., Costa, T., Jones, G. H., and Henriques, A. O., "Molecular and biochemical characterization of a

highly stable bacterial laccase that occurs as a structural component of the *Bacillus subtilis* endospore coat", *J. Biol. Chem.*, **277,** 18849–18859 (2002).

8. Claus, H., "Laccases and their occurrence in prokaryotes", *Arch. Microbiol.*, **179,** 145–150 (2003).
9. Holmberg, G. C. and Laurell, C.-B., "Investigations in serum copper. I. Nature of serum copper and its relation to the iron-binding protein in human serum", *Acta Physiol. Scand.*, **1,** 944–950 (1947).
10. Holmberg, G. C. and Laurell, C.-B., "Investigations in serum copper. II. Isolation of the copper containing protein and a description of some of its properties", *Acta Physiol. Scand.*, **2,** 550–556 (1948).
11. Ferguson, S. J., "Nitrogen cycle enzymology", *Curr. Opin. Chem. Biol.*, **2,** 182–193 (1998).
12. Mayer, A. M. and Staples, R. C., "Laccase: new functions for an old enzyme", *Phytochemistry,* **60,** 551–565 (2002).
13. Xu, F., "Recent Progress in Laccase Study: Properties, Enzymology, Production and Applications", in *The Encyclopedia of Bioprocess Technology: Fermentation, Biocatalysis and Bioseparation,* Flinkinger, M. C. and Drew, S. W., Eds., John Wiley & Sons Inc., New York, 1999, pp. 1545–1554.
14. Bourbonnais, R. and Paice, M. G., "Oxidation of non-phenolic substrates. An expanded role for laccase in lignin biodegradation", *FEBS Lett.*, **267,** 99–102 (1990).
15. Bourbonnais, R., Paice, M. G., Reid, I. D., Lanthier, P., and Yaguchi, M., "Lignin oxidation by laccase isozymes from *Trametes versicolor* and role of the mediator 2,2′-azinobis(3-ethylbenzthiazoline-6-sulfonate) in kraft lignin depolymerization", *Appl. Environ. Microbiol.*, **61,** 1876–1880 (1995).
16. Laurie, S. N. and Mohammed, E. S., "Caeruloplasmin: the enigmatic copper protein", *Coord. Chem. Rev.*, **33,** 279–312 (1980).
17. Yoshida, K., Furihata, K., Takeda, S., Nakamura, A., Yamamoto, K., Morita, H., Hiyamuta, S., Ikeda, S., Shimizu, N., and Yanagisawa, N., "A mutation in the ceruloplasmin gene is associated with systemic hemosiderosis in humans", *Nat. Genet.*, **9,** 267–272 (1995).
18. Frieden, E. and Hsieh, H. S., "Ceruloplasmin: the copper transport protein with essential oxidase activity", in *Advances in Enzymology and Related Areas in Molecular Biology,* vol. 44, John Wiley, & Sons, Inc., New York, 1976, pp. 187–236.
19. Saenko, E. L., Yaropolov, A. I., and Harris, E. D., "Biological Functions of Ceruloplasmin Expressed Through Copper-Binding Sites and a Cellular Receptor", *J. Trace Elements Expt. Med.*, **7,** 69–88 (1994).
20. Lindley, P., Card, G., Zaitseva, I., and Zaitsev, V., "Ceruloplasmin: the beginning of the end of an enigma", in *Perspectives on Bioinorganic Chemistry,* vol. 4, JAI Press Inc., Greenwich, CN, pp. 51–89 (1999).
21. Adman, E. T., "Copper protein structures", *Adv. Prot. Chem.*, **42,** 145–197 (1991).
22. A. Messerschmidt, R. Ladenstein, R. Huber, M. Bolognesi, L. Avigliano, R. Petruzzelli, A. Rossi, and A. Finazzi-Agró, "Refined crystal structure of ascorbate oxidase at 1.9 Å resolution", *J. Mol. Biol.*, **224,** 179–205 (1992).
23. Berman, H. M., Westbrook, J., Feng, Z., Gilliland, G., Bhat, T. N., Weissig, H., Shindyalov, I. N., and Bourne, P. E., "The Protein Data Bank", *Nucleic Acids Res.*, **28,** 235–242 (2000).
24. Ducros, V., Brzozowski, A. M., Wilson, K. S., Ostergaard, P., Schneider, P., Svendson, A., and Davies, G. J., "Structure of the laccase from *Coprinus cinereus* at 1.68 Å resolution: evidence for different 'T2 Cu-depleted' isoforms", *Acta Crystallogr. D Biol. Crystallogr.* **57,** 333–336 (2001).
25. Bertrand, T., Jolivalt, C., Briozzo, P., Caminade, E., Joly, N., Madzak, C., and Mougin, C., "Crystal structure of a four-copper laccase complexed with an arylamine: insights into substrate recognition and correlation with kinetics", *Biochemistry,* **41,** 7325–7333 (2002).
26. Piontek, K., Antorini, M., and Choinowski, T., "Crystal structure of a laccase from the fungus *Trametes versicolor* at 1.90 Å resolution containing a full complement of coppers", *J. Biol. Chem.*, **277,** 37663–37669 (2002).
27. Hakulinen, N., Kiiskinen, L. L., Kruus, K., Saloheimo, M., Paananen, A., Koivula, A., and Rouvinen, J., "Crystal structure of a laccase from *Melanocarpus albomyces* with an intact trinuclear copper site", *Nat. Struct. Biol.*, **9,** 601–605 (2002).
28. Enguita, F. J., Martins, L. O., Henriques, A. O., and Carrondo, M. A., "Crystal structure of a bacterial endospore coat component. A laccase with enhanced thermostability properties", *J. Biol. Chem.*, **278,** 19416–19425 (2002).
29. Zaitseva, I., Zaitsev, V., Card, G., Moshkov, K., Bax, B., Ralph, A., and Lindley, P. F., "The X-ray structure of human serum ceruloplasmin at 3.1 Å: nature of the copper centres", *J. Biol. Inorg. Chem.*, **1,** 15–23 (1996).
30. J. W. Godden, S. Turley, D. C. Teller, E. T. Adman, M.-Y. Liu, W. J. Payne, and J. LeGall, "The 2.3 angstrom X-ray structure of nitrite reductase from *Achromobacter cycloclastes*", *Science,* **253,** 438–442 (1991).
31. Malmström, B. G., "Enzymology of oxygen", *Annu. Rev. Biochem.*, **51,** 21–59 (1982).
32. Messerschmidt, A., Luecke, H., and Huber, R., "X-ray structures and mechanistic implications of three functional derivatives of ascorbate oxidase from zucchini. Reduced, peroxide and azide forms", *J. Mol. Biol.*, **230,** 997–1014 (1993).
33. Messerschmidt, A., "Spatial structures of ascorbate oxidase, laccase and related proteins: implications for the catalytic mechanism", in *Multi-Copper Oxidases,* Messerschmidt, A., Ed., World Scientific Publishing Co. Pte. Ltd., Chapt. 2, pp. 23–79 (1997).
34. Machonkin, T. E., Zhang, H. H., Hedman, B., Hodgson, K. O., and Solomon, E. I., "Spectroscopic and magnetic studies of human ceruloplasmin: identification of a redox-inactive reduced Type 1 copper site", *Biochemistry,* **37,** 9570–9578 (1998).

35. Lindley, P. F., Card, G., Zaitseva, I., Zaitsev, V., Reinhammar, B., Selin-Lindgren, E., and Yoshida, K., "An X-ray structural study of human ceruloplasmin in relation to ferroxidase activity", *J. Biol. Inorg. Chem.*, **2,** 454–463 (1997).
36. Adman, E. T., Godden, J. W., and Turley, S., "The structure of copper-nitrite reductase from *Achromobacter cycloclastes* at five pH values, with NO_2-bound and with type II copper depleted", *J. Biol. Chem.*, **270,** 27458–27474 (1995).
37. Huang, H., Zoppellaro, G., and Sakurai, T., "Spectroscopic and kinetic studies on the oxygen-centred radical formed during the four-electron reduction process of dioxygen by *Rhus vernicifera* laccase", *J. Biol. Chem.*, **274,** 32718–32724 (1999).
38. Sundaram, U. M., Zhang, H. H., Hedman, B., Hodgson, K. O., and Solomon, E. I., "Spectroscopic investigation of peroxide binding to the trinuclear copper cluster site in laccase: correlation with the peroxy-level intermediate and relevance to catalysis", *J. Am. Chem. Soc.*, **119,** 12525–12540 (1997).
39. Lee, S.-K., George, S. D., Antholine, W. E., Hedman, B., Hodgson, K. O., and Solomon, E. I., "Nature of the intermediate formed in the reduction of O_2 to H_2O at the trinuclear copper cluster active site in native laccase", *J. Am. Chem. Soc.*, **124,** 6180–6193 (2002).
40. Zaitsev, V. N., Zaitseva, I., Papiz, M., and Lindley, P. F., "An X-ray crystallographic study of the binding sites of the azide inhibitor and of organic substrates to ceruloplasmin, a multi-copper oxidase in the plasma", *J. Biol. Inorg. Chem.*, **4,** 579–587 (1999).
41. Rydén, L. G. and Hunt, L. T., "Evolution of protein complexity: the blue copper-containing oxidases and related proteins", *J. Mol. Evol.*, **36,** 41–66 (1993).
42. Bento, I., Martins, L. O., Lopes, G. G., Carrondo, M. A., and Lindley, P. F. "Dioxygen reduction by multi-copper oxidases; a structural perspective", *Dalton Trans.*, **21,** 3507–3513 (2005).
43. Many of the figures have been drawn with the SETOR suite of programs[44] using atomic coordinates extracted from the Protein Data Bank.[23]
44. S. V. Evans, "SETOR: hardware-lighted three-dimensional solid model representations of macromolecules", *J. Mol. Graphics,* **11,** 134–138 (1993).
45. There are many colleagues who deserve acknowledgment for this section, but I would particularly like to thank Albrecht Messerschmidt, Elinor Adman, Slava and Irina Zaitsev, Graeme Card, and my colleagues at the European Synchrotron Radiation Facility, Grenoble, France and the Instituto de Tecnologia Química e Biológica in Oeiras, Portugal.

XI.8. Reducing Dioxygen to Water: Mechanistic Considerations

Lawrence Que, Jr.
Department of Chemistry and Center for Metals in Biocatalysis
University of Minnesota
Minneapolis, MN 55455

The mechanisms for dioxygen (O_2) reduction by cytochrome *c* oxidase (after referred to as cytochrome oxidase or sometimes simply as 'oxidase') and the blue copper oxidases follow the paradigm discussed in Section XI.4, unifying the mechanisms of O_2 activation at heme, dicopper, and nonheme di-iron centers (Fig. XI.4.1). Cytochrome oxidase has four redox centers to mediate the reduction of O_2 to H_2O. Dioxygen binding occurs in the pocket harboring the heme a_3 and Cu_B centers, with Cu_A and heme *a* serving as electron-transfer sites. From time-resolved resonance Raman studies,[1] it is clear that O_2 initially binds to heme a_3, forming an O_2 adduct as in hemoglobin. The next step should be the formation of a peroxo intermediate, either an Fe^{III}–OOH intermediate by analogy to cytochrome P450 and oxyhemerythrin, or a bridged peroxo species by analogy to oxyhemocyanin and the peroxo intermediates of nonheme di-iron enzymes.

As discussed in Section XI.5.1.1, the crystal structure of resting cytochrome *c* oxidase in fact shows the presence of a peroxide bridging heme a_3 and Cu_B. But the fact that this is an inactive form of the enzyme suggests that this peroxo-bridged species may not be catalytically relevant. Indeed, a peroxo intermediate has not been observed in rapid kinetics studies of the reaction of reduced cytochrome oxidase with O_2. Instead, intermediate P, so-called because it had been anticipated to be a peroxo species, is actually a species in which the O–O bond has already been cleaved, that is, a high-valent iron–oxo intermediate analogous to Compound **I** of HRP or Compound ES of cytochrome *c* peroxidase. A Raman feature associated with the ν(Fe=O) of an oxo–iron(IV) porphyrin has been observed, while the other oxidizing equivalent may be stored on a nearby oxidized amino acid residue. This residue has been suggested to be Tyr244, the residue that is covalently coupled to a His ligand on Cu_B, but experimental evidence for this notion is not yet firm. Subsequent transfer of two electrons to intermediate P affords in sequence intermediate F, wherein the proposed amino acid is reduced, and then the Fe(III)–heme $a_3/Cu_B(II)$ state.

Due to its relative simplicity, laccase, among the "blue copper" oxidases, has been the subject of most studies focused on the catalytic cycle. With its four copper centers (one Type 1 center, one Type 2 center, and two Type 3 centers), it is easy to imagine the four-electron reduction of O_2 to H_2O, with electrons funneling through the Type 1 center to the trinuclear Type 2/Type 3 copper cluster that serves as the site of O_2 binding.[2] Although a peroxo intermediate has not been observed for the reaction of fully reduced native laccase with O_2, such an intermediate can be trapped when the redox-active copper ion in the Type 1 copper site is replaced by redox-inactive Hg(II). From a number of spectroscopic experiments, it has been established that the Type 3 copper centers are both in the Cu(II) state, while the Type 2 copper center remains in the Cu(I) state. This intermediate exhibits a UV absorption band at 340 nm and, by EXAFS, a Cu scatterer at 3.4 Å, which are properties not found for reduced or oxidized forms of laccase. Since the laccase intermediate is also distinct from oxyhemocyanin and oxytyrosinase, the peroxo moiety is proposed to bridge between the Type 2 and Type 3 centers, in contrast to the μ-η^2:η^2 configuration associated with the latter two proteins. The decomposition of the laccase intermediate exhibits an

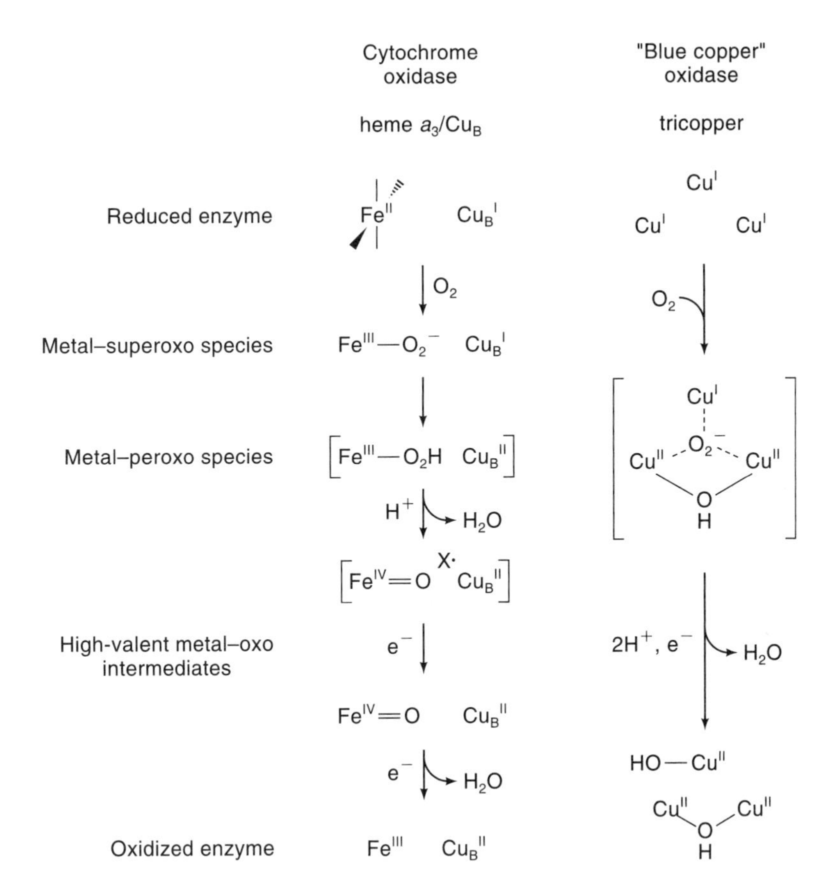

Fig. XI.8.1.
The mechanistic paradigm for dioxygen activation (see Fig. XI.4.2) as applied to cytochrome oxidase and "blue copper" oxidases.

$^{16}O_2/^{18}O_2$ isotope effect of 1.1, suggesting that the rate-determining step involves O–O bond cleavage.[3] The long lifetime of this intermediate in the Hg-substituted enzyme can be rationalized by the existence of a large Franck–Condon barrier to the one-electron reduction of this intermediate. However, its two-electron reduction in native laccase is facile, so the intermediate does not accumulate in native laccase.

Both cytochrome oxidase and laccases are thus well set up for the four-electron reduction of O_2 to H_2O, using O_2-binding centers related to hemoglobin and hemocyanin, respectively. Cleavage of the O–O bond in the course of reduction in all likelihood produces reactive oxidizing species analogous to the high-valent metal–oxo species proposed in the mechanisms in corresponding oxygenases (Fig. XI.4.1). However, the efficient transfer of electrons from electron storage sites to the O_2^- binding sites assures that the lifetimes of these reactive species are short and prevents production of potentially harmful reactive oxygen species (see Section XI.1).

References

1. Kitagawa, T., "Structures of reaction intermediates of bovine cytochrome *c* oxidase probed by time-resolved vibrational spectroscopy", *J. Inorg. Biochem.*, **82,** 9–18 (2000).
2. Solomon, E. I., Sundaram, U. M., and Machonkin, T. E., "Multicopper Oxidases and Oxygenases", *Chem. Rev.*, **96,** 2563–2605 (1996).
3. Palmer, A. E., Lee, S. K., and Solomon, E. I., "Decay of the Peroxide Intermediate in Laccase: Reductive Cleavage of the O–O Bond", *J. Am. Chem. Soc.*, **123,** 6591–6599 (2001).

CHAPTER XII

Hydrogen, Carbon, and Sulfur Metabolism

XII.1. Hydrogen Metabolism and Hydrogenase

Michael J. Maroney
Department of Chemistry
University of Massachusetts, Amherst
Amherst, MA 01003

Contents

XII.1.1. Introduction: Microbiology and Biochemistry of Hydrogen

The capability to produce or utilize hydrogen gas (H_2) is involved in a variety of physiological roles.[1] Production of H_2 is a feature of fermentative anaerobes that utilize the proton as a terminal electron acceptor. Other organisms use H_2 as a reductant by catalyzing its oxidation. The electrons produced from H_2 oxidation can be used to generate energy, and are ultimately used to catalyze the reduction of various inorganic species including CO_2 (e.g., acetogenic, methanogenic, and photosynthetic bacteria), Fe^{3+} (iron-reducing bacteria), N_2 (nitrogen-fixing bacteria), NO_3^- (denitrifying bacteria), and SO_4^{2-} (sulfate-reducing bacteria).

Although most of the organisms involved in hydrogen metabolism are anaerobic prokaryotes, examples are known from eukaryotes and from aerobes (e.g., *Ralstonia eutropha*) that can derive energy from the reaction of H_2 with O_2 to form water.

The redox chemistry involved in the production and utilization of H_2 (Eq. 1) is catalyzed by the enzyme called hydrogenase (H_2ase).[2] The H_2ases are likely to be some of the most primitive enzymes in biology. The vectorial use of this enzymology (e.g., across a membrane) may represent a primitive mechanism for generating and maintaining proton gradients. Homology between some H_2ases and the subunits of mitochondrial complex **I** (Section X.3) suggest that H_2ase and complex **I** are evolutionarily related. In fact, hydrogen metabolism may have been the link that lead to eukaryotes from prokaryotes.[3]

$$H_2 \rightleftharpoons 2\,H^+ + 2\,e^- \qquad (1)$$

All known H_2ases are metalloenzymes that contain Fe–S clusters, with the exception of a H_2ase found in methanogens that utilizes an organic cofactor as a hydride carrier (H_2-forming N^5,N^{10}-methenyl-5,6,7,8-tetrahydromethanopterin dehydrogenase, Hmd) and contains a single Fe center.[4,5] The Fe–S-containing H_2ases are further distinguished from Hmd by their ability to utilize artificial redox donors and acceptors (e.g., viologens), to catalyze H/D exchange, to catalyze the interconversion of ortho and para H_2, and to be inhibited by CO and other small metal-binding molecules. The Fe–S-containing H_2ases may be further classified on the basis of the active site metal content into Ni-Fe and Fe-Fe H_2ases; structural classes that are both biochemically and immunologically distinct.[6] The Fe-Fe H_2ases are a small group of enzymes frequently found in obligate anaerobes (e.g., *Clostridium pasteurianum*) and are generally associated with physiological H_2 production.[7] They have high rates of both H_2 evolution and oxidation *in vitro* and are typically rapidly and irreversibly oxidized by air. Nucleotide sequences are now known for >100 Ni-Fe H_2ases, making this class by far the most common and most studied type of H_2ase. The Ni-Fe H_2ases are generally associated with the physiological oxidation of H_2 (uptake H_2ases) and are more O_2 tolerant in that they are reversibly oxidized by exposure to air.[8] The activity of the Ni-Fe enzymes is $<10\%$ of the Fe-Fe class, but the Ni-Fe H_2ases have higher substrate affinity and thus can catalyze H_2 oxidation at lower H_2 partial pressures.

Some Ni-Fe hydrogenases are distinguished by incorporating a selenocysteine (SeCys) ligand in the terminal position analogous to Cys530 in the *Desulfovibrio gigas* sequence (see below). The incorporation of SeCys is specifically encoded in the DNA and involves a complex system specifically designed to incorporate SeCys in Ni-Fe-Se H_2ases.[9,10] The resulting Se-containing enzymes often have distinct redox and catalytic properties but generally resemble Ni-Fe H_2ases.

XII.1.2. Hydrogenase Structures

XII.1.2.1. Fe-Fe Hydrogenases

One of the triumphs of protein crystallography in the 1990s was the elucidation of the structures of representative examples of the Fe-Fe and Ni-Fe classes of H_2ases (Fig. XII.1.1). The structure of the cytoplasmic monomeric Fe-Fe H_2ase from the obligate anaerobe *C. pasteurianum*[11] reveals that the polypeptide chain may be divided into four domains that together contain 20 Fe atoms in various Fe–S clusters (three Fe_4S_4 clusters and one Fe_2S_2 cluster in addition to the active site). The distal Fe_4S_4 cluster is distinguished from the other ferredoxin-like clusters by virtue of having one Fe bound to the protein by His (Cys_3His cluster ligation). The putative active site, a six-Fe cluster called the H-cluster, is composed of a dinuclear Fe subcluster bridged to an Fe_4S_4 subcluster via a cysteinate ligand bound to one of its Fe atoms (designated Fe_A). The remaining three Fe atoms in the Fe_4S_4 subcluster are bound to the protein by Cys residues. The Fe atoms in the dinuclear subcluster are six coordinate, bridged by two S donor ligands and separated by a distance of 2.6 Å. The exact nature of these bridging S donor ligands is uncertain, but the only attachment to the protein is the bridging cysteinate ligand to Fe_A. The structure of the *C. pasteurianum* enzyme models the electron density between the Fe atoms as two S^{2-} ligands that are involved in hydrogen bonding to a water molecule. The remaining ligands bound to the dinuclear subcluster are five diatomic ligands (likely CO and CN^-), two to each Fe atom, and one forming a third bridge between them and a water molecule bound to Fe_B. The involvement of CO/CN^- ligation in the active site of H_2ases was first detected by Fourier transform infrared spectroscopy (FTIR) in a NiFe H_2ase.[12] The FTIR studies subsequently showed that these ligands are a feature common to all classes of Fe-S-containing-H_2ases,[13] and that both CO and CN^- ligands are involved in Ni-Fe H_2ase.[14]

A generally similar structure with respect to the inorganic cofactors was obtained for the heterodimeric periplasmic Fe-Fe H_2ase from *Desulfovibrio desulfuricans.*[15] This enzyme contains 14 Fe atoms distributed among two Fe_4S_4 clusters and the H-cluster. Again, the distal Fe_4S_4 cluster features Cys_3His ligation. The H-cluster in the *D. desulfuricans* enzyme is also composed of a dinuclear Fe subcluster bound to a Fe_4S_4 cluster via a bridging Cys ligand. The structure of the dinuclear subcluster differs from that described for *C. pasteurianum* in that S donors bridging the two Fe atoms are modeled as a 1,3-propanedithiolate ligand. The distribution of the remaining ligands is also different in that the bridging CO/CN^- ligand is replaced by an oxo/hydroxo bridge, and the site analogous to the one occupied by water on Fe_B is vacant.

The structural similarity of the dinuclear Fe subcluster to known organometallic small molecules has inspired a great deal of synthetic chemistry designed to model the putative active site.[16–19] Since it is unlikely that C, N, and O could be distinguished, it has been proposed that the ligand modeled in the crystal structure as 1,3-propanedithiolate is actually a di(thiomethyl)amine moiety. This would position a base for protonation in the active site, a proposal that has strong support from theory.[20]

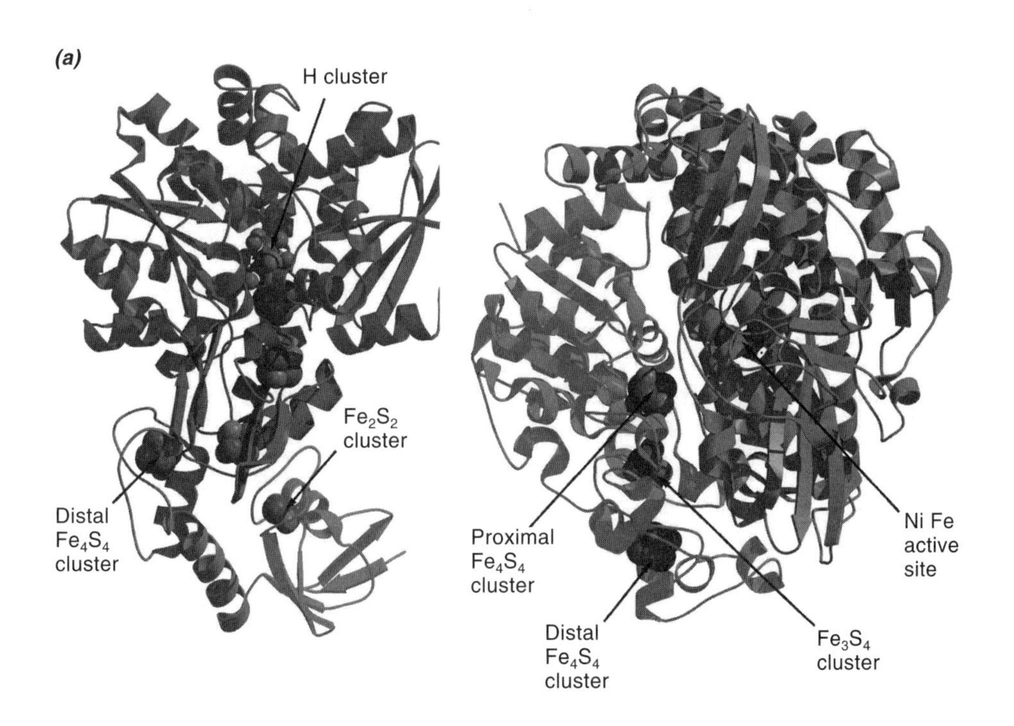

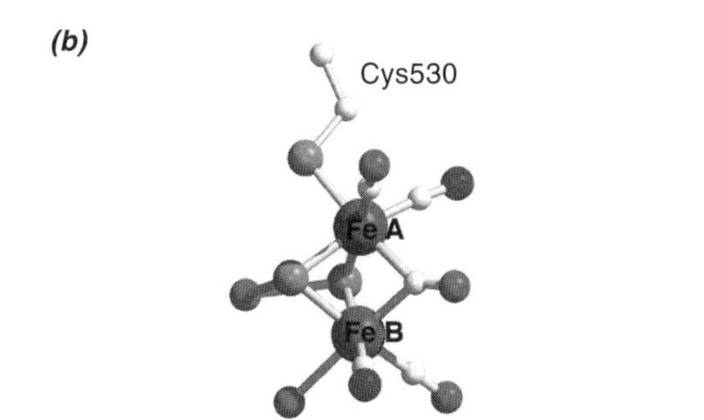

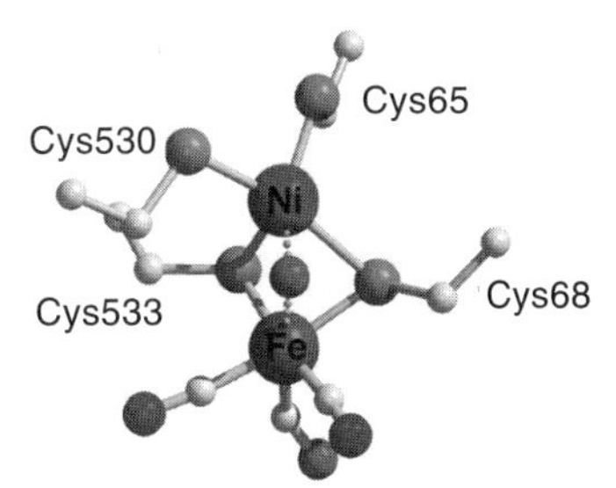

Fig. XII.1.1.
Structures of *C. pasteurianum* Fe-Fe H_2ase-I (PDB code: 1FEH) and *D. gigas* Ni-Fe H_2ase (PDB code: 2FRV). (*a*) Ribbon diagrams summarizing the protein structures and indicating the relative locations of the inorganic cofactors. (*b*) The structures of the dinuclear H_2-oxidizing and producing sites. [See color insert.]

XII.1.2.2. Ni-Fe Hydrogenases

Ni-Fe hydrogenases are typically heterodimers composed of large (~46–72 kDa) and small (~23–38 kDa) subunits. The structure of *D. gigas* Ni-Fe H_2ase is prototypical of this class.[21,22] The crystals used in this study contain enzymes in several oxidation states, but correspond to mostly oxidized enzyme. The structure reveals that the enzyme contains 12 Fe atoms and 1 Ni atom. The Fe atoms are distributed between three Fe–S clusters, two Fe_4S_4 clusters, and one Fe_3S_4 cluster, all found in the small subunit, and the Ni-Fe heterodinuclear active site located in the large subunit. Like the Fe-Fe enzymes, the distal Fe_4S_4 cluster has one His ligand. The six-coordinate active site Fe atom provided the first structurally characterized example of natural CO and CN^- ligation in biology, and the ligands are of biogenic origin.[23] The active site Fe atom is bridged to the Ni center at a distance of 2.9 Å via two Cys ligands and a low-Z atom (oxo/hydroxo/aqua) bridge. The remaining ligands on the Ni center are provided by two Cys sulfurs. The terminal and bridging Cys ligands are derived from two rigorously conserved Cys-X-X-Cys sequences, one near the N-terminus and one near the C-terminus of the large subunit. Each of these motifs provide one terminal (Cys65 and Cys530) and one bridging (Cys68 and Cys533) cysteine ligand. The proximal Fe_4S_4 cluster is a ferredoxin-like cluster with cysteinate ligation that makes its closest approach to the active site via the S atom of a terminal cysteine ligand (Cys65).

The structure of one member of the Ni-Fe-Se subclass is also known.[24] Crystals of *Desulfomicrobium baculatum* periplasmic H_2ase in a reduced state reveal a structure quite similar to that of the Ni-Fe enzyme described above. The enzyme is a heterodimer containing a total of 14 Fe atoms. The main distinguishing features are the absence of the low-Z bridging atom; the replacement of the Fe_3S_4 cluster by an Fe_4S_4 cluster; and the presence of an additional Fe atom in a site occupied by Mg^{2+} in the Ni-Fe H_2ase. The absence of the oxo/hydroxo/aqua bridge is associated with a shorter Ni–Fe distance in this enzyme (2.5 Å) and may be functionally important. The loss of this bridge and the shortening of the Ni–Fe distance have been shown to be features of the reductive activation of the Ni-Fe H_2ase from *Chromatium vinosum*,[25] and may be involved in the creation of a possible binding site for H_2 or H^-.

Although the proteins involved in the Fe-Fe and Ni-Fe classes are genetically and structurally unrelated, the inorganic cofactors suggest that a common strategy guided the convergent evolution of the active sites of these two classes of H_2ase. First, the active sites contain dinuclear complexes where the metals are bridged by two S-donor ligands. Second, the active site Fe atoms feature CO and CN^- ligation. The likely function of this unusual ligand environment is to lock down the oxidation state as ferrous and the spin state of the Fe as low spin. This notion is supported by ^{57}Fe electron–nuclear double resonance (ENDOR) studies, which show a low-spin ferrous center in three distinct states of a Ni-Fe enzyme.[26] Third, the dinuclear centers are in close proximity to a Fe_4S_4 cluster, and the interaction between them is mediated by an active-site Cys residue. Since Fe_A is the center in the dinuclear Fe subcluster of Fe-Fe H_2ases with the Cys ligand and is linked to the Fe_4S_4 subcluster, it is structurally analogous to the Ni center in the Ni-Fe enzymes. Hydrophobic channels connecting Ni or Fe_A to the protein surface that may provide H_2 access to the active site have been identified,[15,27] and they further suggest that these two metal centers may also be functionally analogous. Fourth, all of the enzymes contain at least three Fe-S clusters with the distal cluster having unusual Cys_3His coordination. These clusters are spaced roughly 10 Å apart and constitute an efficient electron-transfer chain in and out of the dinuclear active site cluster, which is the likely site of H_2 oxidation or production.

XII.1.3. Biosynthesis

The assembly of the active site of Ni-Fe hydrogenases in the large subunit is a tightly choreographed process that in *Escherichia coli* requires seven auxiliary proteins, adenosine triphosphate (ATP), guanosine triphosphate (GTP) and carbamoyl phosphate (the precursor to the CN^-, and possibly also the CO, ligands).[28] The proteins HypA and HypB (or homologues) are involved in Ni insertion, the latter having GTPase activity. The proteins HypC and HypD form a complex that is involved in Fe insertion; HypD is an Fe–S protein. It is likely that the $Fe(CN)_2CO$ center is assembled in the HypC–D complex and inserted into the hydrogenase large subunit. The proteins HypE and HypF form a complex that is involved in CN^- biosynthesis.[23] The HypF protein is a carbamoyl phosphatase that catalyzes the formation of a carbamoyl moiety. The carbamoyl group is subsequently transferred to the S atom of the C-terminal Cys residue of HypE, which then catalyzes an ATP-dependent dehydration to form a thiocyanate group and subsequently transfers CN^- to Fe. In addition, a protease is required to process the C-terminus of the enzyme. The sequence of events involved in the assembly of the active site is Fe insertion, followed by Ni insertion, followed by C-terminal processing. The small subunit associates with the mature large subunit to form active H_2ase. Nickel cannot be inserted prior to Fe insertion, and Ni insertion requires the presence of the C-terminal extension, which is cleaved by the protease only after Ni binding. Binding of other metals to the Ni site prevents processing of the C-terminus; this processing is required before association with the small subunit can occur.

XII.1.4. Hydrogenase Reaction Mechanism

Despite the excellent structural information now available for H_2ases, the intimate mechanism for how the enzyme catalyzes H_2 redox chemistry is not yet fully defined. With the advent of the structures, significant mechanistic insight has been obtained from theory.[29,30] The mechanism has been perhaps best explored by theory and experiment in the case of Ni-Fe enzymes. One hypothetical scheme is shown in Fig. XII.1.2. Given the structural analogy between the Fe-Fe and Ni-Fe classes, a similar mechanism can be written for the Fe-Fe H_2ase. The scheme features reaction of H_2 at the Fe center (analogous to Fe_B in Fe-Fe H_2ases), followed by heterolytic cleavage of H_2 to form a bridging hydride and thiol ligands. The hydride is then oxidized by two electrons to produce the second proton, and the active site is reoxidized by electron transfer to the Fe–S clusters.

Much of the difficulty in understanding the intimate mechanism of Ni-Fe H_2ase lies in correlating the one-electron redox chemistry exhibited by the enzyme (Fig. XII.1.3) with the oxidation of H_2, which is a two-electron process, and in attempting to assign spectroscopically characterized states to intermediates in the reaction. The Ni-Fe enzymes are typically isolated in air as a mixture of two oxidized and catalytically inactive forms (forms A and B). These oxidized forms can be distinguished by the $S = \frac{1}{2}$ EPR spectra that they exhibit, and by the kinetics of their reductive activation by H_2. Form A exhibits a lag phase, and is thus unready (u), while form B is in a ready (r) configuration and rapidly reduces. Both forms A and B are in equilibrium with EPR-silent states (SI_u and SI_r) that equilibrate at room temperature. The SI_r state is in equilibrium with an active configuration (SI_a). Each of the three EPR silent states can be distinguished on the basis of the FTIR spectra arising from the CO and CN^- ligands.[31,32] The SI_a state can be reduced further to another EPR active level, form C, and ultimately to the fully reduced enzyme, R, which is EPR silent. Studies

Fig. XII.1.2.
A mechanistic scheme for the oxidation–production of H_2 at the active site of Ni-Fe H_2ase.

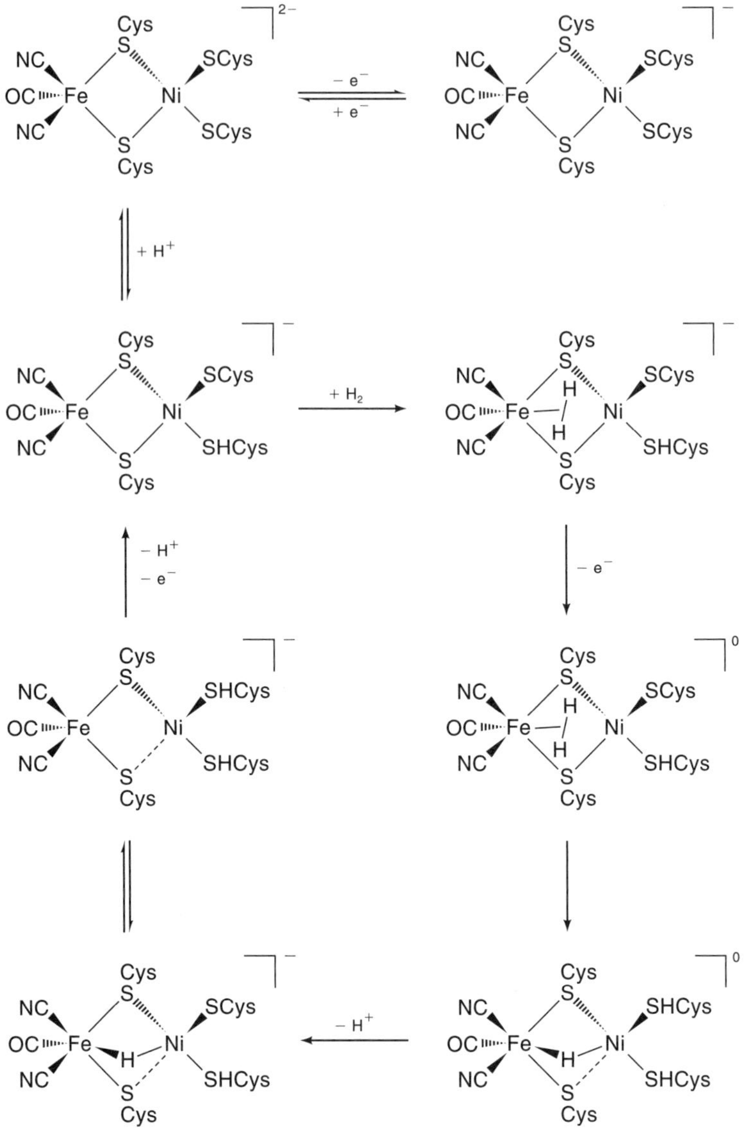

of the kinetics of the interconversion of forms C and R using rapid freeze–quench techniques show that this conversion occurs in <8 ms, and is thus kinetically competent to support H_2 oxidation or evolution.[33]

The structures of the spectroscopically accessible states of several H_2ases have been extensively investigated by Ni K-edge X-ray absorption spectroscopy (XAS).[34,35] Extended X-ray absorption fine structure (EXAFS) and X-ray absorption near-edge spectroscopy (XANES) analysis show that usually very little structural change occurs at the Ni site in response to changes in the redox state of the enzyme. Edge energy shifts are small, indicating that the electron density changes at the Ni site that are associated with the redox processes can accommodate a one-electron process at best. When correlated with FTIR data,[36] the spectroscopically accessible states fit best with a two-state redox model of the active site, in which the cluster exists in an oxidized $S = \frac{1}{2}$ state in all EPR active forms and in a one-electron reduced state in all of the EPR silent forms. The small changes in electron density revealed by Ni K-edge energy shifts and by shifts in the Fe–CO ν_{CO} frequencies indicate that the changes in

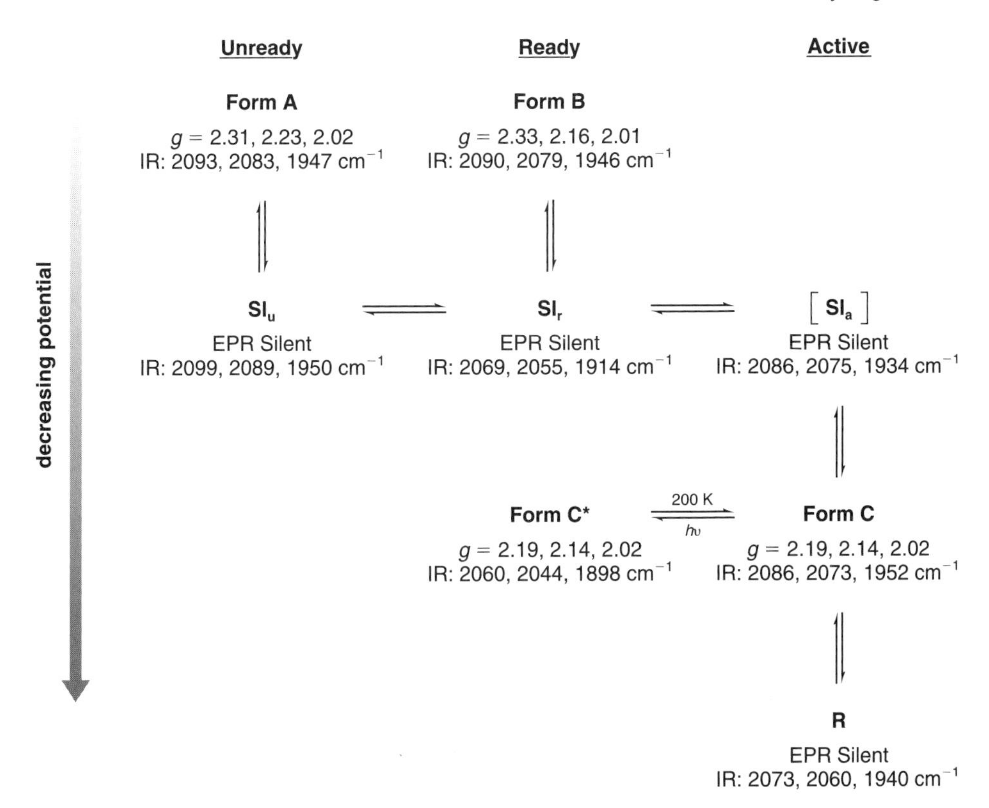

Fig. XII.1.3. Summary of the redox chemistry of Ni-Fe H_2ase and the electron paramagnetic resonance (EPR) and IR spectroscopic signatures of each state.

electron density are distributed over the cluster and are moderated by the S donor ligands.

The metal site involved in binding H_2 is not clear. There is no data indicating that a dihydrogen ligand exists in any species that has been spectroscopically probed. However, binding of H_2 to the Fe center is consistent with expectations from small molecule chemistry, where low-spin d^6 centers dominate the known examples of H_2 complexes.[37,38] Theoretical justification also exists for binding H_2 to Fe.[39] However, the channel by which H_2 likely accesses the active site is more closely associated with the Ni site,[27] and exogenous CO binds to Ni.[33] Moreover, a bridging H_2 interaction is also possible.

Based on the enzyme's ability to interconvert ortho and para H_2, the mechanism of H_2 oxidation is believed to proceed via heterolytic cleavage of H_2.[40] Attention has focused on form C as a possible hydride intermediate; a resonance assigned to a solvent-exchangeable proton that exists in form C, but not in either form A or B, has been identified by ENDOR spectroscopy.[41,42] Form C is light sensitive, and the EPR signal associated with form C is altered in the photoproduct. The ENDOR signal assigned to the solvent-exchangeable proton is also not observed in the photoproduct. The photoprocess is reversible, and the original spectrum is obtained upon annealing the sample at ~200 K. One possible interpretation of this data is that form C contains a Ni–H bond that is reversibly lost upon photolysis (e.g., Eq. 2).

$$Ni^{III}H^- \underset{200\,K}{\overset{h\nu}{\rightleftharpoons}} Ni^{I} + H^+ \tag{2}$$

However, the proton involved is not strongly coupled to the spin in form C (20 MHz). Furthermore, Ni K-edge studies of the photoprocess fail to detect any large structural or charge density perturbations at the Ni site.[42] Given the weak coupling, it may be that this resonance corresponds to a thiol proton in the active site. Workers using ENDOR and HYSCORE have identified an exchangeable proton with a small isotropic hyperfine coupling constant (−3.5 MHz) as a bound hydride.[43] This proton is shown to lie in a direction that places it in a bridging position in the active site, and its small hyperfine coupling constant is proposed to derive from the fact that it lies perpendicular to the orbital containing the unpaired spin (d_{z^2}).

XII.1.5. Regulation by Hydrogen

The hydrogen metabolism of microorganisms may be regulated by H_2. A H_2-sensing protein that regulates the transcription of H_2ase genes has been demonstrated for *Ralstonia eutropha*[44] and *Rhodobacter capsulatus.*[45,46] The nature of the H_2-sensing protein and the mechanism by which it regulates transcription has been examined in detail in *Alcaligenes eutrophus.*[47,48] The regulation involves four proteins, where HoxA is a transcriptional regulator of H_2ase genes. The HoxJ protein is a His kinase that phosphorylates Asp55 on HoxA to repress expression of H_2ase in the absence of H_2. Both HoxB and HoxC are subunits of the H_2 sensor, which inhibits the HoxJ-mediated phosphorylation of HoxA in the presence of H_2 by an unknown mechanism. The H_2 sensor binds H_2 and can catalyze H/D exchange, but it is not a good catalyst for H_2 redox chemistry and does not complement the deletion of H_2ase structural genes. The nucleotide sequences reveal that the H_2 sensor is structurally analogous to heterodimeric Ni-Fe H_2ases. Spectroscopic studies further support the similarity, but point to one important difference. The sensor exhibits an EPR signal analogous to form C and has one EPR silent state analogous to the SI level. Infrared spectroscopy reveals the absorptions associated with the Fe CO/CN^- ligands in the active site. Thus, the H_2 sensor appears to have an active site that is closely related to those of Ni-Fe H_2ases. The difference is that the H_2 sensor cannot be reduced to the R level, consistent with the role of this state in redox catalysis.

References

General References

1. Adams, M. W., Mortenson, L. E., and Chen, J. S., "Hydrogenase", *Biochim. Biophys. Acta,* **594,** 105–176 (1980).
2. Cammack, R., Frey, M., and Robson, R., Eds., *Hydrogen as a Fuel: Learning from Nature,* Taylor and Francis, Inc., London, 2001.
3. Martin, W. and Müller, M., "The hydrogen hypothesis for the first eukaryote", *Nature (London),* **392,** 37–41 (1998).

Specific References

4. Lyon, E. J., Shima, S., Buurman, G., Chowdhuri, S., Batschauer, A., Steinbach, K., and Thauer, R. K., "UV-A/blue-light inactivation of the 'metal-free' hydrogenase (Hmd) from methanogenic archaea", *Eur. J. Biochem.,* **271,** 195–204 (2004).
5. Thauer, R. K., Klein, A. R., and Hartmann, G. C., "Reactions with molecular hydrogen in microorganisms: evidence for a purely organic hydrogenation catalyst", *Chem. Rev.,* **96,** 3031–3042 (1996).
6. Fauque, G., Peck, H. D., Jr., Moura, J. J. G., Huynh, B. H., Berlier, Y., DerVartanian, D. V., Teixeira, M., Przybyla, A. E., Lespinat, P. A., Moura, I., and LeGall, J., "The three classes of hydrogenases from sulfate-reducing bacteria of the genus *Desulfovibrio*", *FEMS Microbiol. Rev.,* **54,** 299–344 (1988).
7. Adams, M. W. W., "The structure and mechanism of iron-hydrogenases", *Biochim. Biophys. Acta,* **1020,** 115–145 (1990).
8. Albracht, S. P. J., "Nickel hydrogenases: in search of the active site", *Biochim. Biophys. Acta,* **1188,** 167–204 (1994).
9. Voordouw, G., Menon, N. K., LeGall, J., Choi, E. S., Peck, H. J., and Przybyla, A. E., "Analysis and comparison of nucleotide sequences encoding the genes for [NiFe] and [NiFeSe] hydrogenases from *Desulfovibrio gigas* and *Desulfovibrio baculatus*", *J. Bacteriol.,* **171,** 2894–2899 (1989).
10. Stadtman, T. C., "Selenocysteine", *Annu. Rev. Biochem.,* **65,** 83–100 (1996).
11. Peters, J. W., Lanzilotta, W. N., Lemon, B. J., and Seefeldt, L. C., "X-ray crystal structure of the Fe-only hydrogenase (CpI) from *Clostridium pasteurianum* to 1.8 angstrom resolution", *Science,* **282,** 1853–1858 (1998).
12. Bagley, K. A., Van Garderen, C. J., Chen, M., Woodruff, W. H., Duin, E. C., and Albracht, S. P. J., "Infrared Studies on the Interaction of Carbon Monoxide with Divalent Nickel in Hydrogenase from *Chromatium vinosum*", *Biochemistry,* **33,** 9229–9236 (1994).

13. Van der Spek, T. M., Arendsen, A. F., Happe, R. P., Yun, S., Bagley, K. A., Stufkens, D. J., Hagen, W. R., and Albracht, S. P. J., "Similarities in the architecture of the active sites of Ni-hydrogenases and Fe-hydrogenases detected by means of infrared spectroscopy", *Eur. J. Biochem.*, **237,** 629–634 (1996).
14. Pierik, A. J., Roseboom, W., Happe, R. P., Bagley, K. A., and Albracht, S. P. J., "Carbon monoxide and cyanide as intrinsic ligands to iron in the active site of [NiFe]-hydrogenases. $NiFe(CN)_2CO$, biology's way to activate H_2", *J. Biol. Chem.*, **274,** 3331–3337 (1999).
15. Nicolet, Y., Piras, C., Legrand, P., Hatchikian, C. E., and Fontecilla-Camps, J. C., "*Desulfovibrio desulfuricans* iron hydrogenase: the structure shows unusual coordination to an active site Fe binuclear center", *Structure,* **7,** 13–23 (1999).
16. Darensbourg, M. Y., Lyon, E. J., and Smee, J. J., "The bio-organometallic chemistry of active site iron in hydrogenases", *Coord. Chem. Rev.*, **206–207,** 533–561 (2000).
17. Georgakaki, I. P., Thomson, L. M., Lyon, E. J., Hall, M. B., and Darensbourg, M. Y., "Fundamental properties of small molecule models of Fe-only hydrogenase: computations relative to the definition of an entatic state in the active site", *Coord. Chem. Rev.*, **238–239,** 255–266 (2003).
18. Lawrence, J. D., Li, H. X., Rauchfuss, T. B., Benard, M., and Rohmer, M. M., "Diiron azadithiolates as models for the iron-only hydrogenase active site: Synthesis, structure, and stereoelectronics", *Angew. Chem.-Int. Ed.*, **40,** 1768–1771 (2001).
19. Gloaguen, F., Lawrence, J. D., Schmidt, M., Wilson, S. R., and Rauchfuss, T. B., "Synthetic and Structural Studies on $[Fe_2(SR)_2(CN)_x(CO)_{6-x}]_x^-$ as Active Site Models for Fe-Only Hydrogenases", *J. Am. Chem. Soc.*, **123,** 12518–12527 (2001).
20. Fan, H. J. and Hall, M. B., "A capable bridging ligand for Fe-only hydrogenase: Density functional calculations of a low-energy route for heterolytic cleavage and formation of dihydrogen", *J. Am. Chem. Soc.*, **123,** 3828–3829 (2001).
21. Volbeda, A., Charon, M. H., Piras, C., Hatchikian, E. C., Frey, M., and Fontecilla-Camps, J. C., "Crystal structure of the nickel-iron hydrogenase from *Desulfovibrio gigas*", *Nature (London),* **373,** 580–587 (1995).
22. Volbeda, A., Garcin, E., Piras, C., de Lacey, A. L., Fernandez, V. M., Hatchikian, E. C., Frey, M., and Fontecilla-Camps, J. C., "Structure of the [NiFe] hydrogenase active site: Evidence for biologically uncommon Fe ligands", *J. Am. Chem. Soc.*, **118,** 12989–12996 (1996).
23. Reissmann, S., Hochleitner, E., Wang, H., Paschos, A., Lottspeich, F., Glass, R. S., and Boeck, A., "Taming of a Poison: Biosynthesis of the NiFe-Hydrogenase Cyanide Ligands", *Science,* **299,** 1067–1070 (2003).
24. Garcin, E., Vernede, X., Hatchikian, E. C., Volbeda, A., Frey, M., and Fontecilla-Camps, J. C., "The crystal structure of a reduced [NiFeSe] hydrogenase provides an image of the activated catalytic center", *Structure,* **7,** 557–566 (1999).
25. Davidson, G., Choudhury, S. B., Gu, Z. J., Bose, K., Roseboom, W., Albracht, S. P. J., and Maroney, M. J., "Structural examination of the nickel site in *Chromatium vinosum* hydrogenase: Redox state oscillations and structural changes accompanying reductive activation and CO binding", *Biochemistry,* **39,** 7468–7479 (2000).
26. Huyett, J. E., Carepo, M., Pamplona, A., Franco, R., Moura, I., Moura, J. J. G., and Hoffman, B. M., "57Fe Q-band pulsed ENDOR of the hetero-dinuclear site of nickel hydrogenase: Comparison of the NiA, NiB, and NiC states", *J. Am. Chem. Soc.*, **119,** 9291–9292 (1997).
27. Montet, Y., Amara, P., Volbeda, A., Vernede, X., Hatchikian, E. C., Field, M. J., Frey, M., and Fontecilla-Camps, J. C., "Gas access to the active site of Ni-Fe hydrogenases probed by X-ray crystallography and molecular dynamics", *Nat. Struct. Biol.*, **4,** 523–526 (1997).
28. Blokesch, M., Paschos, A., Theodoratou, E., Bauer, A., Hube, M., Huth, S., and Boeck, A., "Metal insertion into NiFe-hydrogenases", *Biochem. Soc. Trans.*, **30,** 674–680 (2002).
29. Liu, Z.-P. and Hu, P., "A density functional theory study on the active center of Fe-only hydrogenase: Characterization and electronic structure of the redox states", *J. Am. Chem. Soc.*, **124,** 5175–5182 (2002).
30. Cao, Z. X. and Hall, M. B., "Modeling the active sites in metalloenzymes. 3. Density functional calculations on models for [Fe]-hydrogenase: Structures and vibrational frequencies of the observed redox forms and the reaction mechanism at the diiron active center", *J. Am. Chem. Soc.*, **123,** 3734–3742 (2001).
31. Bagley, K. A., Duin, E. C., Roseboom, W., Albracht, S. P. J., and Woodruff, W. H., "Infrared-Detectable Group Senses Changes in Charge Density on the Nickel Center in Hydrogenase from *Chromatium vinosum*", *Biochemistry,* **34,** 5527–5535 (1995).
32. de Lacey, A. L., Hatchikian, E. C., Volbeda, A., Frey, M., Fontecilla-Camps, J. C., and Fernandez, V. M., "Infrared-spectroelectrochemical characterization of the [NiFe] hydrogenase of *Desulfovibrio gigas*", *J. Am. Chem. Soc.*, **119,** 7181–7189 (1997).
33. Happe, R. P., Roseboom, W., and Albracht, S. P. J., "Pre-steady-state kinetics of the reactions of [NiFe]-hydrogenase from *Chromatium vinosum* with H_2 and CO", *Eur. J. Biochem.*, **259,** 602–608 (1999).
34. Bagyinka, C., Whitehead, J. P., and Maroney, M. J., "An x-ray absorption spectroscopic study of nickel redox chemistry in hydrogenase", *J. Am. Chem. Soc.*, **115,** 3576–3585 (1993).
35. Gu, Z., Dong, J., Allan, C. B., Choudhury, S. B., Franco, R., Moura, J. J. G., Moura, I., LeGall, J., Przybyla, A. E., Roseboom, W., Albracht, S. P. J., Axley, M. J., Scott, R. A., and Maroney, M. J., "Structure of the Ni Sites in Hydrogenases by X-ray Absorption Spectroscopy. Species Variation and the Effects of Redox Poise", *J. Am. Chem. Soc.*, **118,** 11155–11165 (1996).

36. Maroney, M. J., Davidson, G., Allan, C. B., and Figlar, J., "The structure and function of nickel sites in metalloproteins", *Struct. Bonding,* **92,** 1–65 (1998).
37. Kubas, G. J., "Molecular Hydrogen Complexes: Coordination of a [σ] Bond to Transition Metals", *Acc. Chem. Res.,* **21,** 120–128 (1988).
38. Heinekey, D. M. and Oldham, W. J., "Coordination chemistry of dihydrogen", *Chem. Rev.,* **93,** 913–926 (1993).
39. Niu, S., Thomson, L. M., and Hall, M. B., "Theoretical Characterization of the Reaction Intermediates in a Model of the Nickel-Iron Hydrogenase of *Desulfovibrio gigas*", *J. Am. Chem. Soc.,* **121,** 4000–4007 (1999).
40. Krasna, A. I., *Enzyme Microb. Tech.,* **1,** 165–172 (1979).
41. Fan, C., Teixeira, M., Moura, J., Moura, I., Huynh, B. H., Le Gall, J., Peck, H. D., Jr., and Hoffman, B. M., "Detection and characterization of exchangeable protons bound to the hydrogen-activation nickel site of *Desulfovibrio gigas* hydrogenase: a proton and deuterium Q-band ENDOR study", *J. Am. Chem. Soc.,* **113,** 20–24 (1991).
42. Whitehead, J. P., Gurbiel, R. J., Bagyinka, C., Hoffman, B. M., and Maroney, M. J., "The hydrogen binding site in hydrogenase: 35-GHz ENDOR and XAS studies of the nickel-C (reduced and active form) and the Ni-L photoproduct", *J. Am. Chem. Soc.,* **115,** 5629–5635 (1993).
43. Foerster, S., van Gastel, M., Brecht, M., and Lubitz, W., "An orientation-selected ENDOR and HYSCORE study of the Ni-C active state of *Desulfovibrio vulgaris* Miyazaki F hydrogenase", *J. Biol. Inorg. Chem.,* **10** (1), 51–62 (2005).
44. Lenz, O., Strack, A., Tran-Betcke, A., and Friedrich, B., "A hydrogen-sensing system in transcriptional regulation of hydrogenase gene expression in *Alcaligenes* species", *J. Bacteriol.,* **179,** 1655–1663 (1997).
45. Elsen, S., Colbeau, A., and Vignais, P. M., "Purification and in vitro phosphorylation of HupT, a regulatory protein controlling hydrogenase gene expression in *Rhodobacter capsulatus*", *J. Bacteriol.,* **179,** 968–971 (1997).
46. Vignais, P. M., Dimon, B., Zorin, N. A., Colbeau, A., and Elsen, S., "HupUV proteins of *Rhodobacter capsulatus* can bind H_2: evidence from the H-D exchange reaction", *J. Bacteriol.,* **179,** 290–292 (1997).
47. Pierik, A. J., Schmelz, M., Lenz, O., Friedrich, B., and Albracht, S. P. J., "Characterization of the active site of a hydrogen sensor from *Alcaligenes eutrophus*", *FEBS Lett.,* **438,** 231–235 (1998).
48. Lenz, O. and Friedrich, B., "A novel multicomponent regulatory system mediates H_2 sensing in *Alcaligenes eutrophus*", *Proc. Natl. Acad. Sci. U. S. A.,* **95,** 12474–12479 (1998).

XII.2. Metalloenzymes in the Reduction of One-Carbon Compounds

Stephen W. Ragsdale

Department of Biochemistry
University of Nebraska
Lincoln, NE 68588

The author is grateful to Rudolf Thauer for his significant contributions in coauthoring the original draft of this section.

Contents

XII.2.1. Introduction: Metalloenzymes in the Reduction of One-Carbon Compounds to Methane and Acetic Acid

Several general reviews on the topics covered in this section have been published.[1–14]

One-carbon reductions are important in the global carbon cycle. Carbon dioxide, the most oxidized one-carbon compound, is produced by the oxidation of organic carbon by heterotrophic organisms, including humans. The return of CO_2 to the carbon cycle is accomplished by one of the four known reductive CO_2 fixation pathways: the Calvin–Benson–Basham cycle, the reductive tricarboxylic acid cycle (TCA) cycle, the Wood–Ljungdahl (acetyl-CoA) pathway (Eq. 1), or the 3-hydroxypropionate[15] cycle. The Calvin–Benson–Bashem cycle requires energy, and is driven by photosynthesis or chemoautotrophy. On the other hand, the Wood–Ljungdahl pathway, in coupling H_2 oxidation to CO_2 reduction, is exergonic. It has been suggested that this reaction sequence is the extant survivor of the earliest metabolic pathway.[16] The Wood–Ljungdahl pathway and the methanogenesis pathway (Eq. 2), which are the focus of this section, conserve energy for the organism by electron-transfer-linked phosphorylation. Figure XII.2.1 shows the redox states of the various one-carbon compounds that undergo interconversion in the Wood–Ljungdahl and methanogenesis pathways.

$$4\,H_2 + 2\,CO_2 \rightarrow CH_3COO^- + H^+ + 2\,H_2O \qquad \Delta G^{o\prime} = -95\,\text{kJ mol}^{-1} \tag{1}$$

$$4\,H_2 + CO_2 \rightarrow CH_4 + 2\,H_2O \qquad \Delta G_0 = -131\,\text{kJ mol}^{-1} \tag{2}$$

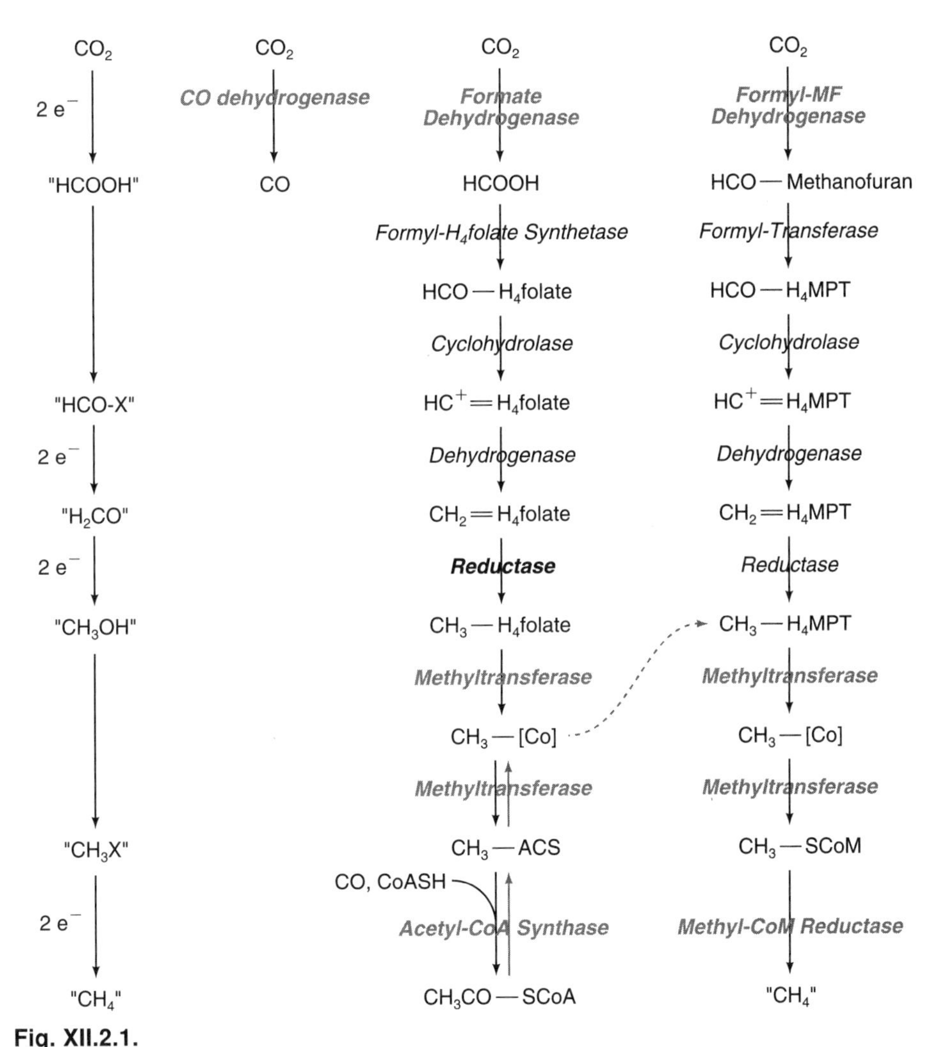

Fig. XII.2.1.
Pathways of one-carbon reduction discussed in this article: CO_2 to acetate and CO_2 to methane. The metalloenzymes covered here are underlined and outlined. The first methyltransferase in the methanogenic and acetogenic pathway are not metalloenzymes, but use B_{12} as a substrate. The methylenetetrahydrofolate reductase in some acetogens contains a [4Fe–4S] cluster.

Enzymes involved in one-carbon metabolism present a trove of interesting metal cofactors (Fig. XII.2.2). Molybdenum or tungsten are redox catalysts at the active site of several CO_2 reducing enzymes (see Sections XII.6 and XII.7).[23] Cobalt in vitamin B_{12} is a nucleophilic catalyst that forms an organometallic alkylcobalt species during the catalytic cycle of enzymes that catalyze methyl group transfers.[2–4] In this same enzyme class, Zn aids in stabilizing a nucleophilic thiolate anion at the active site.[2,3] Iron is present in heme and nonheme iron proteins and, in several cases, iron is bridged to other transition metals. In 1985, based on spectroscopic studies, a metal cluster containing Ni and Fe was discovered in the bifunctional enzyme CODH/ACS;[24] subsequently, the crystal structure of hydrogenase[25] revealed the first atomic level description of a Ni–Fe cluster (see Section XII.1). In methyl-CoM reductase, Ni is part of a tetrapyrrole structure called Factor F_{430}.[1]

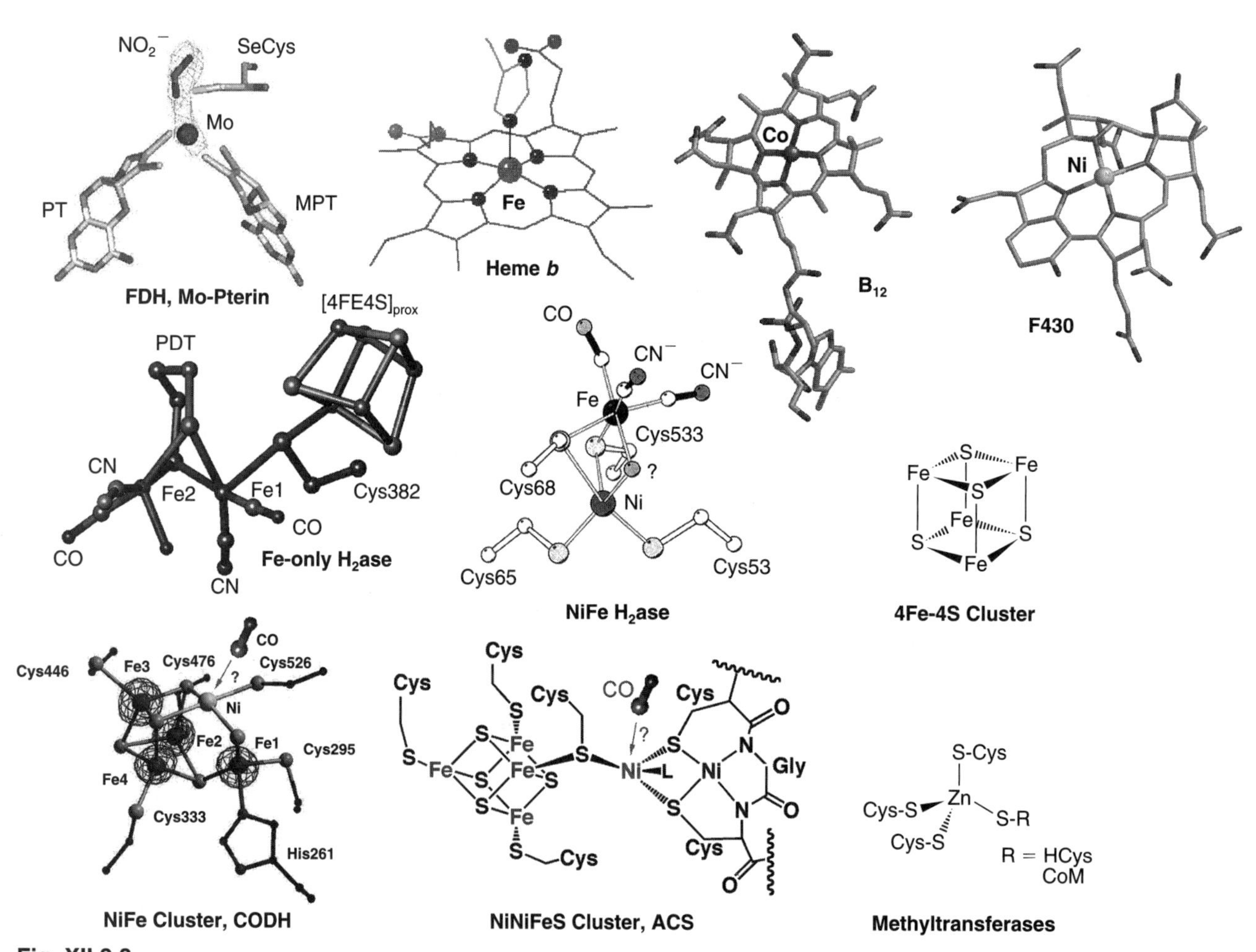

Fig. XII.2.2.
Metal sites in the pathways of one-carbon reduction. The structure of molybdopterin guanine dinucleotide (MPTppG) in FDH is from Ref. 17. The structure of B_{12} is drawn from the coordinates of methionine synthase;[18] those of the Fe-only and NiFe hydrogenases are from Ref. 19. The structure of structure C (NiFe cluster) from carbon monoxide dehydrogenase (CODH) is based on the recent 1.63-Å structure[20] and that of the A cluster (NiNi FeS cluster of ACS is from a recent 2.2-Å structure with Ni replacing Cu.[21] The structure of factor F430 is from the coordinates of MCR.[22] (Acetyl-CoA synthase = ACS; coenzyme M = CoM; formate dehydrogenase = FDH.) [See color insert.]

XII.2.2. Electron Donors and Acceptors for One-Carbon Redox Reactions

XII.2.2.1. Hydrogen as Electron Donor for One-Carbon Redox Reactions

Dihydrogen is the most common electron donor for one-carbon reduction reactions (Eqs. 1 and 2). Hydrogenotrophic organisms are extremely important in the global carbon cycle because elevated H_2 levels inhibit the anaerobic biodegradation of biomass. The ability to oxidize H_2 is conferred by hydrogenase (see Section XII.1) (Eq. 3). Besides serving as a reductant for energy generation, the hydrogenase reaction allows organisms to siphon off excess reducing equivalents by transferring electrons to protons. The structures of the catalytic centers of the Fe-only and Ni–Fe hydrogenases are shown in Fig. XII.2.2. Both active centers are binuclear metal clusters with CO and cyanide ligands. Aspects of hydrogen metabolism and enzymology are covered in detail in Chapter II and Section XII.1. The third class of hydrogenase, the so-called "metal-free hydrogenase" from methanogens, was recently shown to contain an Fe-cofactor.[26]

$$2\,H^+ + 2\,e^- \leftrightarrows H_2 \qquad \Delta E^{\circ\prime} = -414\,\text{mV} \tag{3}$$

XII.2.2.2. Electron Acceptors

The pathways of CO_2 reduction depend on efficient and rapid electron transfer (ET) from the ultimate electron donor to the redox catalyst. The ET is accomplished by a redox mediator(s), which can be an organic cofactor, such as nicotinamide adenine dinucleotide phosphate [NAD(P)], flavin mononucleotide (FMN), flavin adenine dinucleotide (FAD), coenzyme F_{420} (an 8-hydroxy-5-deazariboflavin derivative),[27] methanophenazine[28] or a metal center, (see Chapter IV). A common electron carrier is ferredoxin, which is usually found as a small (~6 kDa) protein containing 2, 3, or 4 irons in a cluster in which the Fe atoms are bridged by inorganic sulfide and bound to the protein by thiolate groups of Cys[29] (see Chapter IV). Methanogens, however, contain polyferredoxins with up to 14 [4 Fe–4 S] clusters.[30,31] In addition, some methanogens and some acetogens, but not all, contain cytochromes.[5,6,32]

XII.2.3. Conversion to the "Formate" Oxidation Level by Two-Electron Reduction of Carbon Dioxide

XII.2.3.1. Carbon Monoxide Dehydrogenase

The ability of some microbes to use CO as a sole source of carbon and energy is conferred by the catalytic activity of CODH (for reviews, see Refs. 7–10). Every year, $\sim 10^8$ tons of CO are removed from the lower atmosphere of Earth by bacterial oxidation.[33] There are three classes of CODH: the Mo–CODH, the Ni–CODH, and the bifunctional Ni–CODH/ACS. Each of these enzymes can reversibly oxidize CO to CO_2 (Eq. 4). The nickel enzyme, working together with hydrogenase, catalyzes the formation of CO_2 and H_2 from CO and H_2O, which is a well-known metal-catalyzed industrial reaction called the water–gas shift (Eq. 5). The CODH/ACS protein is bifunctional and contains a separate activity on each subunit (Fig. XII.2.3). In microbes that use the Wood–Ljungdahl pathway, CO is produced by CO_2 reduction at the CODH subunit; then CO, a CH_3 group, and CoA condense to form acetyl-CoA on the ACS subunit. The ACS reaction is covered later.

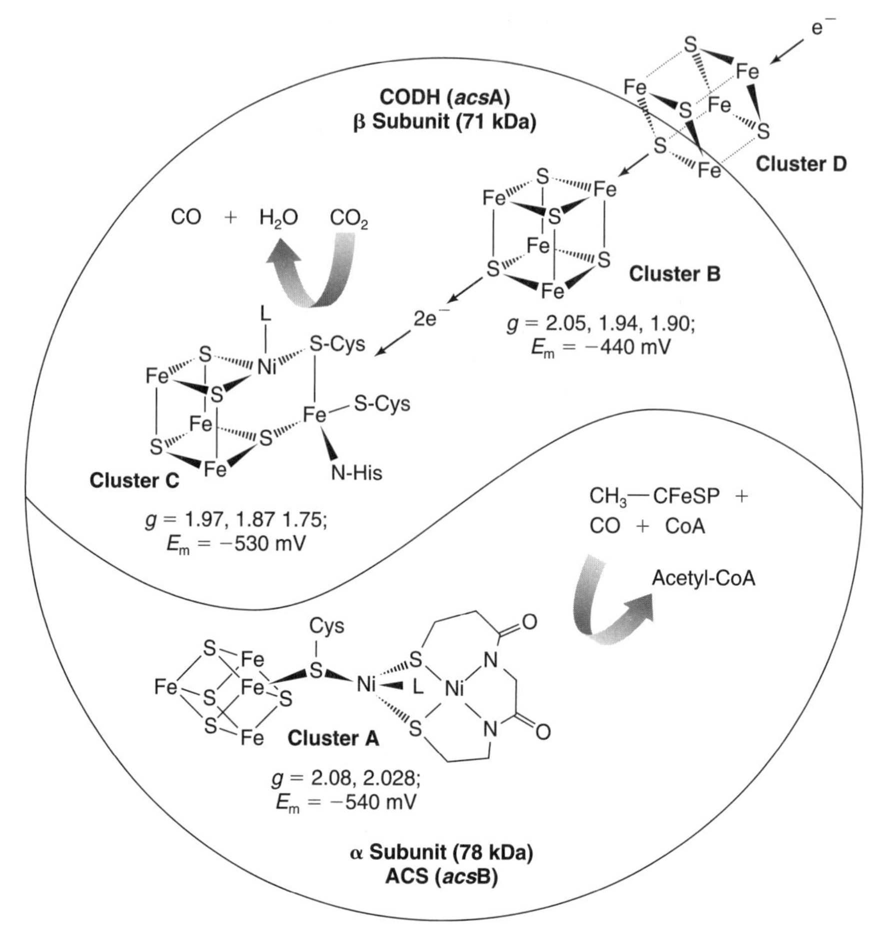

Fig. XII.2.3.
The bifunctional CODH/ACS. The C cluster of CODH subunit generates CO, which is delivered through a channel to the A cluster of the ACS subunit, which catalyses the condensation of CO, CoA, and a CH_3 group derived from the methyl-Co state of a corrinoid iron–sulfur protein.

$$CO_2 + 2\,H^+ + 2\,e^- \rightarrow CO + H_2O \qquad \Delta E^{o\prime} = -520\,\text{mV} \tag{4}$$

$$CO_2 + H_2 \rightarrow CO + H_2O \qquad \Delta E^{o\prime} = -106\,\text{mV} \tag{5}$$

The Mo-CODH is found in aerobic bacteria that oxidize CO with O_2 (Eq. 5).[34] The electrons derived from CO oxidation pass through an electron-transport chain within the enzyme and finally reduce dioxygen. The CO_2 generated is fixed into cell carbon by the Calvin–Benson–Basham cycle. The Mo–CODH, which is related to the Mo hydroxylases, contains FAD, [2Fe–2S] centers, Cu, and 2 Mo atoms, bound by molybdopterin cytosine dinucleotide (MPTppC), per mol of enzyme. The Mo–CODH structure,[35,36] is discussed in Section XII.6.

The most extensively studied nickel CODHs are from *Moorella thermoacetica* (formerly, *Clostridium thermoaceticum*) and *Rhodospirillum rubrum*. These enzymes can catalyze the oxidation of 3200 molecules of CO every second; in the opposite direction, they can catalyze the reduction of 10 molecules of CO_2 to CO every second.[37] The X-ray crystal structure, which is known for the CODHs from *R. rubrum*,[38] *Carboxydothermus hydrogenoformans*,[20] and *M. thermoacetica*[21,39] reveals a mushroom-shaped homodimeric enzyme containing five metal clusters. The two

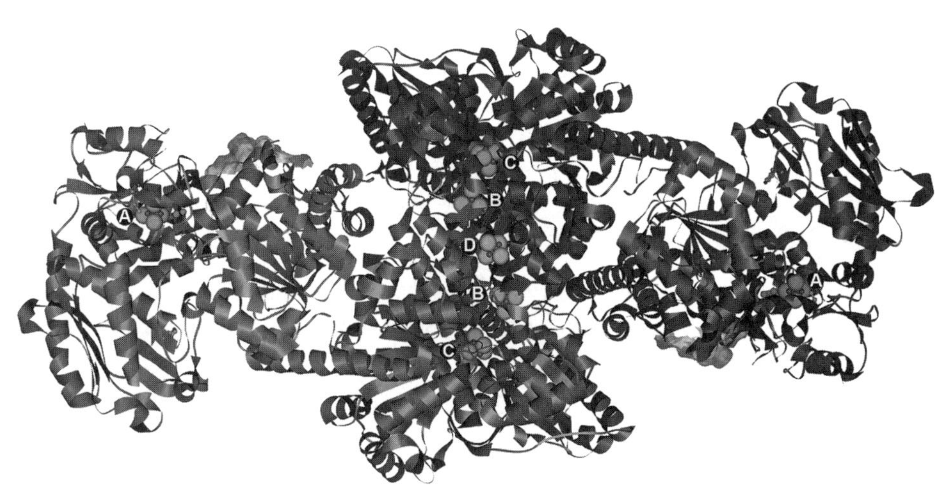

Fig. XII.2.4.
Structure of CODH/ACS. The CODH subunit is composed of the two core subunits, while the ACS subunits are attached to each central CODH subunit. (Adapted from Ref. 21.) [See color insert.]

CODH subunits are the central blue and purple molecules shown at the central core of the CODH/ACS structure in Fig. XII.2.4. Tethered to the dimeric CODH core are the ACS subunits, which will be discussed below. Each CODH subunit contains a NiFeS cluster, called cluster C, which is buried 18 Å below the surface and is the catalytic site for CO oxidation, and a typical [4Fe–4S] ferredoxin-like cluster named Cluster B. In addition, a $[4Fe–4S]^{2+/1+}$ cluster covalently links the two identical subunits, similar to the FeS cluster in the iron protein of nitrogenase (see Section XII.3). Cluster C is a Ni–4Fe–5S cluster that can be viewed as a [3Fe–4S] cluster bridged to a heterobinuclear Ni–Fe cluster (Fig. XII.2.2).

It came as a shock to learn that Ni is <3 Å from two of the Fe atoms. On the basis of spectroscopic studies, it had been concluded that Cluster C must be a [4Fe–4S] cube with one of the Fe atoms bridged to a distant (>3.3 Å) Ni atom. Converging just above a slightly distorted square-planar Ni center are a hydrophobic channel, which is proposed to facilitate CO delivery to the free apical Ni coordination site, and a solvent channel, which contains >40 water molecules. These tunnels could direct the two substrates (CO and H_2O) to the Ni–Fe active center and facilitate product (protons and CO_2) release. Located in the hydrophilic tunnel is a series of His residues that are proposed to deliver protons produced during CO oxidation to the solvent.[38] Electrons generated during CO oxidation at this catalytic center are transferred to a wire consisting of clusters B and D. Interestingly, it is cluster B of the adjacent subunit that is in a suitable position to mediate the electron transfer between clusters C and D. Since cluster D is nearest the molecular surface and solvent exposed, it is proposed to mediate electron transfer between CODH and the terminal electron acceptor (ferredoxin, flavodoxin, etc.). The reduced electron acceptors then couple to other energy requiring cellular processes.

XII.2.3.2. Formate Dehydrogenase

The enzyme that is called formate dehydrogenase (FDH) usually catalyzes the reverse reaction, reduction of CO_2 to formate, and thus could be called CO_2 reductase (Eq. 6). We use the FDH designation only because it is so engrained in the literature that to do otherwise might confuse even the experts. Enzymes in this class exhibit a diverse metal composition and some (like the plant enzyme) lack metals. The crystal structure of the Mo, Se, and Fe–S containing FDH from *E. coli* was recently published.[3] Formate dehydrogenases contain Mo or W coordinated to a pterin dithiolene prosthetic group (MPTppG, see Fig. XII.2.2) and [2Fe–2S] and [4Fe–4S] clusters. The *C. thermoaceticum* FDH was the first to be described as a tungsten enzyme.[40] The Mo and W enzymes are discussed in Sections XII.6 and XII.7, respectively. Some, but not all, FDHs contain Se as selenocysteine (SeCys). In methanogens, coenzyme F_{420} is the electron acceptor for formate oxidation[41] and, in acetogens, reduced NADP (NADPH) or NADH is the electron donor for CO_2 reduction. These enzymes additionally contain FAD to provide two- or one-electron redox capability, which effectively couples redox flow from the Fe–S centers to nicotinamide dinucleotides.

$$CO_2 + 2\,H^+ + 2\,e^- \rightarrow HCOOH \qquad \Delta E^{\circ\prime} = -420\,\text{mV} \tag{6}$$

XII.2.3.3. Formyl-Methanofuran Dehydrogenase

Like FDH, formyl-methanofuran dehydrogenase (FMD) is a Mo or W iron–sulfur enzyme and a CO_2 reductase. The enzyme catalyzes the reversible reduction of CO_2 plus methanofuran to formylmethanofuran (Eq. 7).[1] FMD is involved in methane formation from CO_2 in methanogenic archaea and in methyl group oxidation to CO_2 in methanogenic and sulfate-reducing archaea. Recent evidence indicates that this enzyme is also present in methylotrophic bacteria.[42]

$$\text{R-furan-CH}_2\text{-NH}_3^+ \;(\textbf{Methanofuran}) + CO_2 + 2[H] \rightleftharpoons \text{R-furan-CH}_2\text{-NH-C(=O)-H} \;(\textbf{Formylmethanofuran}) + H_2O + H^+ \qquad \Delta E^{\circ\prime} = -530\text{ mV} \tag{7}$$

Formylmethanofuran dehydrogenase is a cytoplasmic enzyme composed of at least three different subunits. The γ-subunit harbors the active site containing Mo or W coordinated to MPTppG and to a Cys or SeCys residue of the peptide chain. This subunit exhibits sequence similarity to the dimethyl sulfoxide reductase family of molybdoenzymes (see Section XII.6).[23,43] The physiological electron donor–acceptor of the enzyme appears to be a polyferredoxin.[30] Formylmethanofuran dehydrogenases from mesophilic organisms are generally Mo enzymes, while those from hyperthermophilic organisms are W enzymes.[23] In thermophilic methanogens, both a Mo and a W enzyme are found, with the W enzyme expressed constitutively and the Mo enzyme induced by molybdate.[44]

XII.2.4. Conversion from the "Formate" through the "Formaldehyde" to the "Methanol" Oxidation Level

As shown in Fig. XII.2.1, one-carbon metabolic enzymes catalyzing the conversion of formate oxidation level intermediates to the oxidation state of methanol require tetrahydrofolate in bacteria and tetrahydromethanopterin in archaea (Fig. XII.2.5). In methanogens, these enzymes include formylmethanofuran/tetrahydromethanopterin formyltransferase, methenyltetrahydromethanopterin cyclohydrolase, methylenetetrahydromethanopterin dehydrogenase, and methylenetetrahydromethanopterin reductase. The corresponding enzymes in acetogens are 10-formyltetrahydrofolate synthetase, methenyltetrahydrofolate cyclohydrolase, methylenetetrahydrofolate dehydrogenase, and methylenetetrahydrofolate reductase. The reactions

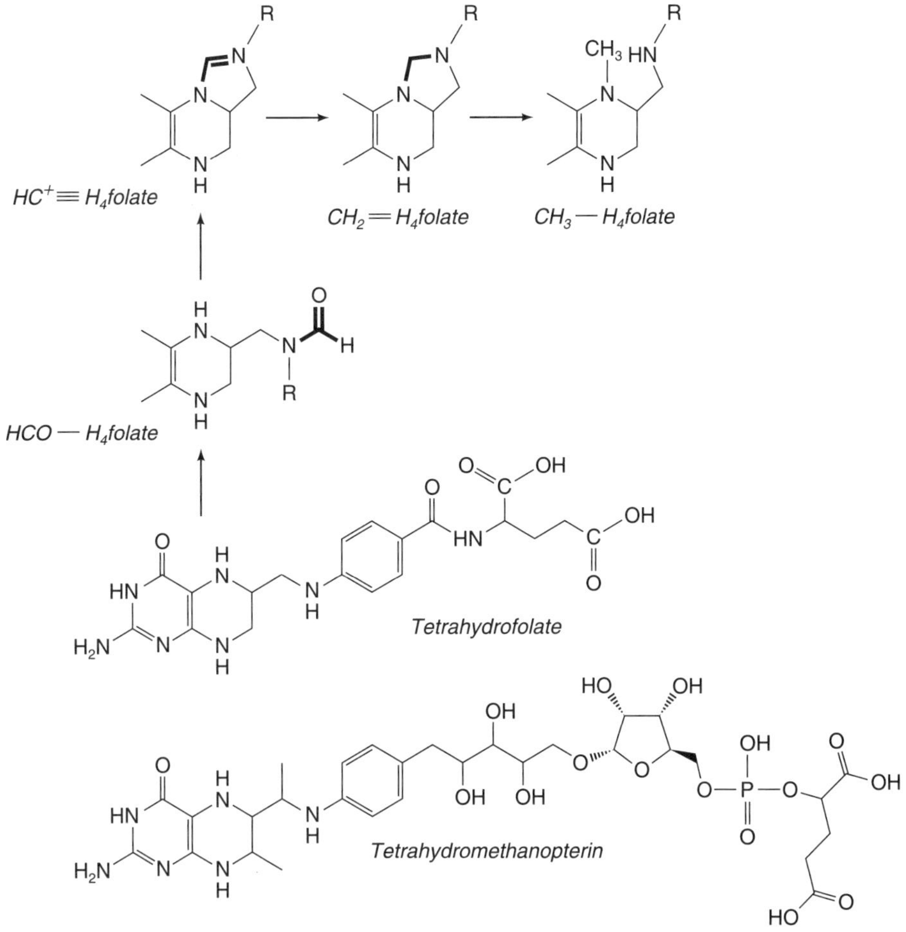

Fig. XII.2.5.
Conversion of "formate" to "methanol" by tetrahydrofolate-dependent enzymes. Reviews of this reaction sequence are available in the Wood–Ljungdahl[45] and methanogenic[1] pathways.

for the acetogenic enzymes are diagrammed in Fig. XII.2.4. Of these, only the acetogenic methylenetetrahydrofolate reductase is a metalloenzyme, containing a [4Fe–4S] cluster. The products of this reaction sequence are CH_3–H_4MPT in methanogens and CH_3–H_4folate in acetogens.

Of special interest is the hydrogen-forming methylenetetrahydromethanopterin dehydrogenase (see Fig. XII.2.1). This enzyme, which catalyzes the reduction of methenyltetrahydromethanopterin by H_2, was thought for many years to lack a redox-active transition metal;[46] and various mechanisms were proposed to account for this activity, including formation of a methenyl group centered carbocation that then probably spontaneously reacts with H_2.[47,48] However, recent studies demonstrate that this protein harbors an Fe-containing cofactor that is likely involved in catalysis.[26]

XII.2.5. Interconversions at the Methyl Level: Methyltransferases

Vitamin B_{12}-dependent methyltransferases play an important role in one-carbon metabolism. Their mechanisms and roles in methanogenesis,[2] acetogenesis,[3] and

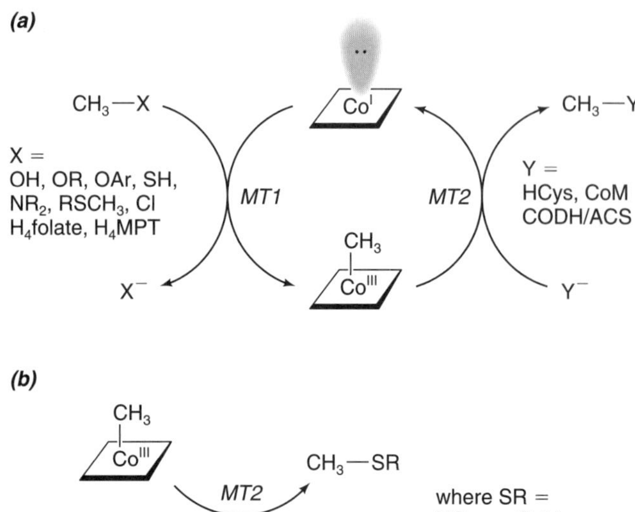

Fig. XII.2.6.
An overview of methyl-transfer biochemistry. The corrinoid iron–sulfur protein: CH_3-H_4folate methyltransferase from acetogenic bacteria is AcsE. The "R" at the Zn site of the MT2 proteins can be homocysteine or CoM.

methionine synthesis[11,12] have recently been reviewed. A review of B_{12} chemistry including methylcobalamin-dependent reactions, including methyltransferases, was recently published (see Section XII.2).[13] The substrates for the methyltransferases include methanol, methylated amines, methylated thiols, acetyl-CoA, CH_3–H_4folate, CH_3–H_4MPT, methoxylated aromatics, and methylated heavy metals (Fig. XII.2.6).

Three components are required for the methyl-transfer (MT) reaction; usually each component is found on a different polypeptide. The first component (MT1) transfers the methyl group from a methylated substrate (CH_3OH, CH_3–H_4folate, etc.) to the second component, which contains bound cobamide, to form CH_3–B_{12} (Eq. 8). The third component (MT2) catalyzes transfer of the Co-bound methyl group to an acceptor (HS–CoM, acetyl–CoA synthase, H_4MPT, etc.; Eq. 9). In these equations, X and Y are the leaving group and the nucleophile, respectively. The methylamine methyltransferases contain a novel amino acid, pyrrolysine, which is encoded by a UAG (stop) codon within the coding regions of the *mtm*B, *mtb*B, and *mtt*B genes.[49,50] Thus, pyrrolysine may be considered the twenty-second amino acid (see Tutorial I).

$$CH_3\text{–}X + Co^I \rightarrow CH_3\text{–}Co^{III} \quad (8)$$

$$CH_3\text{–}Co^{III} + Y^- \rightarrow CH_3\text{–}Y + Co^I \quad (9)$$

In the acetogenic methyltransferase system, the methyl acceptor is a metal site on the enzyme acetyl–CoA synthase (below). When the acceptor is a thiol-containing substrate, a Zn site binds and activates the thiol group (e.g., CoMSH) facilitating transfer of the methyl group from Co to form the methyl thioether product (e.g., CH_3–SCoM; see Fig. XII.2.6).[2] In *M. thermoautotrophicum* and *M. mazei* strain Gö1, this reaction is catalyzed by a membrane-bound multisubunit complex[51] and involves sodium ion translocation, which is linked with ATP synthesis.[6] Within this class of proteins, the X-ray structures of the acetogenic CH_3–H_4folate-dependent methyltransferase[52] and the B_{12}[18] and S-adenosylmethionine[53] binding domains of methionine synthase have been determined.

$$Co^{I} \leftrightarrow Co^{II} + 1\,e^{-} \tag{10}$$

A benzimidazole group, appended to the corrin ring of B_{12}, is coordinated to Co in solution. However, as was first shown with the acetogenic corrinoid Fe–S protein,[54] when cobalamin binds to many proteins, it undergoes a conformational change, which releases the benzimidazole ligand. In some enzymes, a His residue contributed by the protein replaces benzimidazole; in others, the lower axial position remains vacant. Maintaining Co in the 1+ oxidation state is essential; oxidation to the Co(II) state (Eq. 10) requires reductive activation by electron transfer through an Fe–S cluster.[55] In some cases, coupling ATP hydrolysis[56] or S-adenosylmethionine[11] cleavage to the unfavorable reduction of Co(II) to Co(I) facilitates the activation reaction. Although Co(I) is a strong nucleophile, the methyl group is not very electrophilic. Methyl group activation is achieved by reaction of the adjacent heteroatom with a proton[57] or a Lewis acid like Zn^{2+} for the methanol methyltransferase system.[2]

XII.2.6. Methyl Group Reduction or Carbonylation

Once the one-carbon substrates have been reduced to the redox level of methanol, the pathways of methanogenesis and acetogenesis radically diverge. Methanogens reduce CH_3CoM to methane (Eq. 11). On the other hand, acetogens perform a carbonylation reaction to convert enzyme-bound CH_3-Co(III) to acetyl-CoA (Eq. 12).

$$CH_3\text{–SCoM} + \text{HS–CoB} \rightarrow \text{CoM–S–S–CoB} + CH_4 \qquad \Delta G^{\circ\prime} = -45\,\text{kJ}\,\text{mol}^{-1} \tag{11}$$

$$CH_3\text{–Co}^{III} + CO + \text{HSCoA} \rightarrow Co^{I} + \text{acetyl–SCoA} + H^{+} \tag{12}$$

XII.2.6.1. Acetyl-CoA Synthase

In the final steps of the autotrophic Wood–Ljungdahl pathway, a CH_3 group, CO, and CoA are condensed to form acetyl-CoA (see Refs. 19, 57 for a review). Recently, a series of reviews that focus on the remaining controversial mechanistic issues related to ACS has been published.[58–61] In a mechanistically related industrial process (the "Monsanto" or "Reppe" process), a Rh complex catalyzes the formation of acetate from CH_3OH and CO in the presence of HI through CH_3–Rh, Rh–CO, and acetyl–Rh organometallic intermediates. In 1985, the acetogenic CO dehydrogenase was discovered to be the catalyst for the biological synthesis of acetyl-CoA and was dubbed "acetyl-CoA synthase".[62]

It is now known that the ACS and the CODH activities occur at separate NiFeS clusters on different subunits, so this macromolecular machine is called CODH/ACS. These subunits must coordinate their individual catalytic events since CO generated at cluster C by CODH is the substrate for cluster A (see Fig. XII.2.3), which catalyzes acetyl-CoA synthesis on the ACS subunit. The crystal structure of CODH/ACS (Fig. XII.2.4) reveals a 70-Å channel between the C and A cluster,[34,35] which facilitates delivery of CO to the ACS active site, prevents escape of this energy-rich substrate, and, since these microbes inhabit the gastrointestinal (GI) tract of animals, protects the host from the toxic effects of CO. This channel ends just above the proximal metal (M_p in Fig. XII.2.3 where M_p is the prominent metal A cluster, which is shown in gray and bridges the FeS cluster and the distal Ni), which forms a metal–carbonyl adduct, as revealed by FTIR[63] and EPR[24] spectroscopy. The IR studies are most consistent with formation of a Ni^{I}–CO. This metal–CO complex is only one of several organometallic intermediates formed during the final steps in acetyl-CoA synthesis by this enzyme. In this respect, the biological process resembles the Monsanto process for industrial acetate synthesis since metal–carbonyl, methyl–metal, and acetyl–metal intermediates are involved. Methylation of ACS involves the reaction of one organometallic species (CH_3–Co^{III}) to form another (CH_3–Ni^{III}). There is evidence that the CH_3 group binds to Ni, not Fe.[64] Although evidence supports an S_N2 biochemical reaction,[55] inorganic model studies implicate a CH_3 radical transfer from CH_3–Co^{III} to Ni(I).[65]

In methanogenic archaea that convert acetate to methane as a source of energy, the Wood–Ljungdahl pathway runs in reverse. After conversion of acetate to acetyl-CoA, ACS disassembles acetyl-CoA to produce CoA and CO (which is oxidized to CO_2), and catalyzes methylation of a corrinoid iron–sulfur protein. The methyl group is transferred to tetrahydromethanopterin and then to CoM. Subsequently, CH_3–CoM is reduced to CH_4 (Eq. 11) as described in Section XII.2.6.2.

XII.2.6.2. Methyl-CoM Reductase

This nickel enzyme, which was recently reviewed,[14] is a cytoplasmic 300-kDa protein composed of three different subunits in a $\alpha_2\beta_2\gamma_2$ arrangement.[1] Figure XII.2.7 shows the crystal structure with each subunit separately colored. It contains 2 mol Ni/mol enzyme (circled in yellow) in the form of a tightly, but not covalently, bound nickel porphinoid Co F_{430} (see Fig. XII.2.2). Various states of the enzyme have been described: ox1, ox2, red1, red2, and Ni(II). The enzyme exhibits activity only when its prosthetic group is in the red1 Ni(I) oxidation state.[66,67] Recent spectroscopic results indicate that the red1 and red2 state contains Ni(I), while the ox1 and ox2 states are more oxidized (Fig. XII.2.7).[68–70]

Recently, the crystal structures of the enzyme–substrate complex and of the enzyme–product complex in the inactive Ni(II) oxidation state have been resolved.[22] Interestingly, the electron-density map revealed five modified amino acids located in subunits α and α' at or very near the active site region, 1-*N*-CH_3His (α 257), 5-(*S*)-CH_3Arg (α 271), 2-CH_3Gln (α 400), *S*-CH_3Cys (α 452), and, most unusually, thioglycine (α 445), where the carbonyl oxygen is substituted by sulfur. These modifications have been confirmed by chemical analysis. Evidence has been presented that the CH_3 groups are derived from the methyl group of Met and are introduced by *S*-adenosylmethionine dependent post-translational or cotranslational modifications.[71]

The catalytic mechanism of CH_3–CoM reductase has not yet been resolved. As shown schematically in Fig. XII.2.8, the prosthetic group of CH_3CoM reductase is accessible only through a narrow channel that is completely locked once CoB binds. Thus, first CH_3CoM and then CoB enter the 30-Å long channel. The sulfonate group of CH_3CoM is anchored to the protein matrix with its sulfonate group such that both the methyl group and the thioether sulfur could directly interact with the Ni(I). Model building studies indicate that the two sulfur atoms of CoM and CoB come into van der Waals contact when the methyl group of CH_3CoM is placed at van der Waals distance to the potentially attacking Ni(I). Thus, a Ni–CH_3 intermediate, proposed based on free CoF_{430} studies,[72,73] appears to be compatible with the steric requirements of the active site. Methane would then be formed by protonolysis of the CH_3–Ni species. This mechanism is consistent with the requirement that reduction of the CH_3 group of CH_3CoM to CH_4 proceeds with inversion of stereoconfiguration.[74] The inversion is expected to occur when the CH_3–Ni intermediate is formed; whereas, the subsequent protonolysis should proceed with retention of configuration, resulting in net inversion. Other intermediates not shown in the scheme have been proposed and could include a thiyl radical and either a disulfide

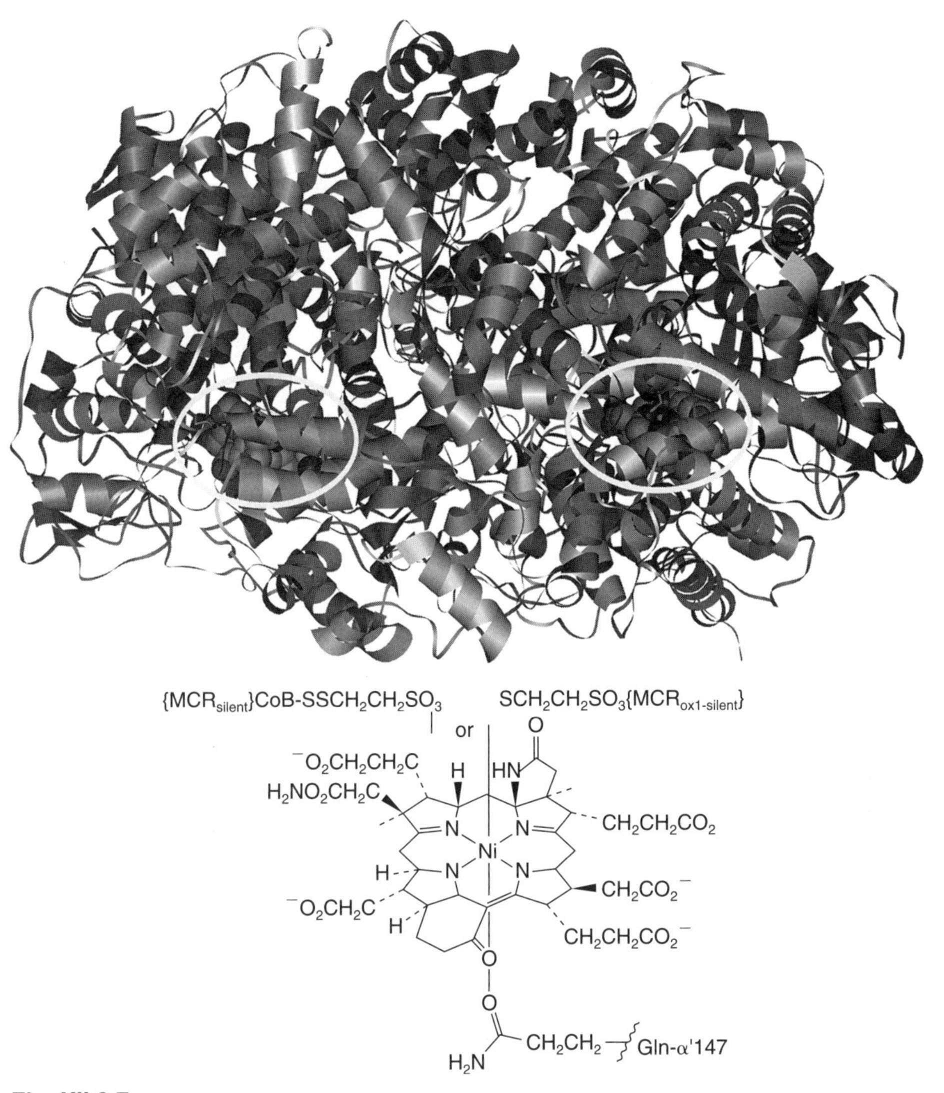

Fig. XII.2.7.
The structure of CH_3-CoM reductase. Each of the subunits are differentially shaded and CoF_{430} is circled as a space-filling model and shown diagrammatically below the structure. The structure of MCR has PDB code: 1HBM. [See color insert.]

anion radical, RS–S•–R, or methylsulfuranyl radical, RS–S•–R(CH_3), intermediate.[15] Model studies have established an important chemical precedent for a mechanism in which the Ni(I) state of F_{430} and a thiyl radical can react with a methyl thioether to yield methane and the corresponding disulfide.[75] Substrate-induced conformational changes in MCR may also play an important role in catalysis.[76] Researchers are also considering the possibility that the early steps in the MCR mechanism involve a CH_3 radical intermediate that abstracts a hydrogen atom from CoB to form CH_4.[77]

XII.2.6.3. Heterodisulfide Reductase

This membrane-associated Fe–S enzyme catalyzes the reduction of the heterodisulfide of CoM and CoB (Eq. 13).[1] Heterodisulfide reductases consist of at least two

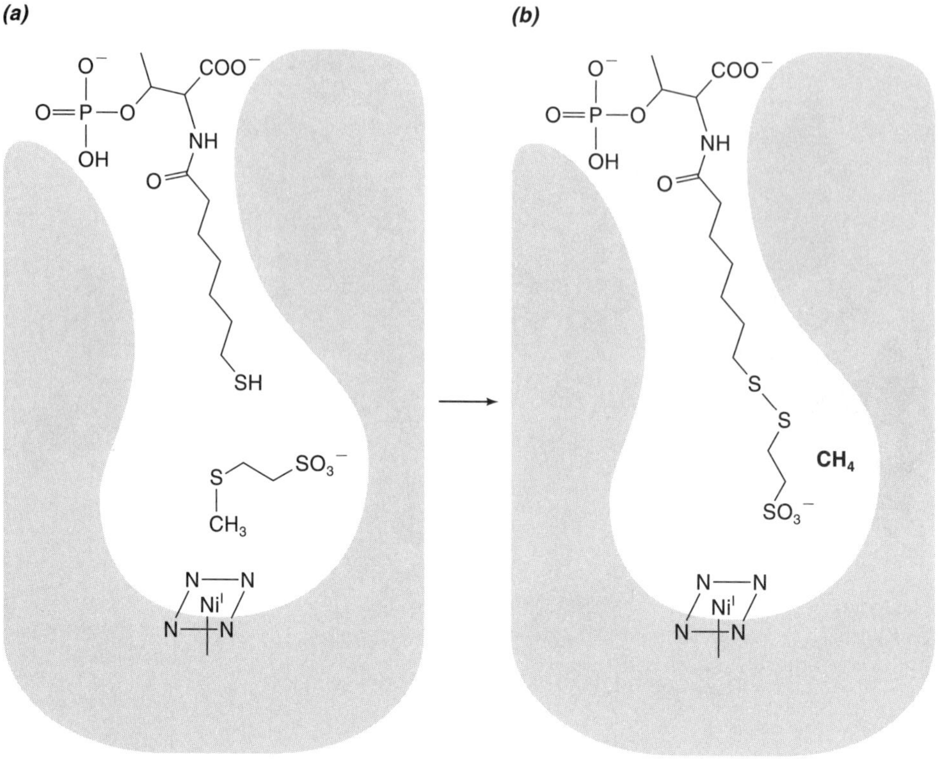

Fig. XII.2.8.
Diagram of the enzyme–substrate complex (*a*) and of the enzyme–product complex (*b*) of active CH_3CoM reductase containing F_{430} in the Ni(I) oxidation state.[1]

subunits; one contains Fe–S clusters and another contains either cytochrome *b* or flavin and an iron–sulfur cluster, which suggests that heterodisulfide reduction occurs at the common iron–sulfur subunit.[32] If so, heterodisulfide reduction would likely proceed via two consecutive one-electron steps, and therefore involve a thiyl radical intermediate(s).

$$\text{CoM–S–S–CoM} + 2\,e^- + 2\,H^+ \leftrightarrow \text{CoMSH} + \text{CoBSH} \qquad \Delta E^{o\prime} = -200\,\text{mV} \quad (13)$$

The iron–sulfur cluster subunit of heterodisulfide reductase does not show sequence similarity to other disulfide reductases such as the pyridine nucleotide-dependent disulfide reductases or the ferredoxin-dependent disulfide reductases.[32] However, it exhibits sequence similarity to the soluble fumarate reductase from methanogens,[78] which uses CoM and CoB as electron donors for fumarate reduction.

XII.2.7. Summary

Redox-active metal centers are at the catalytic center and the electron-transfer sites of one-carbon reduction reactions. We have recently witnessed the unfolding of many surprising atomic level descriptions of these catalytic and redox centers. Now, mechanistic hypotheses with a structural underpinning are being tested to better understand how these microbes catalyze one-carbon reductions. Studies of the enzymes that catalyze these reactions are revealing novel ways that metals function in biology. Such investigations are also important in understanding the biogeochemistry of the Earth and the evolution of life. Perhaps, further research on these processes will return more practical benefits, such as development of a strategy to reduce the levels of two greenhouse gases—CO_2 and CH_4.

References

General References

1. Thauer, R. K., "Biochemistry of methanogenesis: a tribute to Marjory Stephenson", *Microbiology UK*, **144**, 2377–2406 (1998).
2. Sauer, K. and Thauer, R. K., *Chemistry and biochemistry of* B_{12}, Banerjee, R., Ed., Vol. 1, 1 vol., John Wiley & Sons, Inc., New York, 1999, pp. 655–680.
3. Ragsdale, S. W., *Chemistry and biochemistry of* B_{12}, Banerjee, R., Ed., Vol. 1, 1 vol., John Wiley & Sons, Inc., New York, 1999, pp. 633–654.
4. Ludwig, M. L. and Matthews, R. G., "Structure-based perspectives on B12-dependent enzymes", *Annu. Rev. Biochem.*, **66**, 269–313 (1997).
5. Hugenholtz, J. and Ljungdahl, L. G., "Metabolism and energy generation in *homoacetogenic clostridia*", *FEMS Microbiol. Rev.*, **7**, 383–389 (1990).
6. Deppenmeier, U., Muller, V., and Gottschalk, G., "Pathways of energy conservation in methanogenic archaea", *Arch. Microbiol.*, **165**, 149–163 (1996).
7. Ragsdale, S. W. and Kumar, M., "Ni containing carbon monoxide dehydrogenase/acetyl-CoA synthase", *Chem. Rev.*, **96**, 2515–2539 (1996).
8. Ragsdale, S. W., *Enzyme-catalyzed electron and radical transfer*, Holzenburg, A. and Scrutton, N., Eds., Vol. 35, Plenum Press, New York, 2000, pp. 487–518.
9. Ragsdale, S. W., "Life with carbon monoxide", *CRC Crit. Rev. Biochem. Mol. Biol.*, **39**, 165–195 (2004).
10. Lindahl, P. A., "The Ni-containing carbon monoxide dehydrogenase family: light at the end of the tunnel?", *Biochemistry*, **41**, 2097–2105 (2002).
11. Matthews, R. G., *Vitamin* B_{12}, Banerjee, R., Ed., Vol. 1, 2 vols., John Wiley & Sons, Inc., New York, 1999, pp. 681–706.
12. Ludwig, M. L. and Matthews, R. G., "Structure-based perspectives on B-12-dependent enzymes", *Annu. Rev. Biochem.*, **66**, 269–313 (1997).
13. Banerjee, R. and Ragsdale, S. W., "The Many Faces of Vitamin B_{12}: Catalysis by Cobalamin-dependent Enzymes", *Ann. Rev. Biochem.*, **72**, 209–247 (2003).
14. Ragsdale, S. W., *The Porphyrin Handbook*, Kadish, K. M., Smith, K. M., and Guilard, R., Eds., Vol. 11, 20 vols., Academic Press, New York, 2003, pp. 205–228.

Specific References

15. Menendez, C., Bauer, Z., Huber, H., Gad'on, N., Stetter, K. O., and Fuchs, G., "Presence of acetyl coenzyme A (CoA) carboxylase and propionyl-CoA carboxylase in autotrophic Crenarchaeota and indication for operation of a 3-hydroxypropionate cycle in autotrophic carbon fixation", *J. Bacteriol.*, **181**, 1088–1098 (1999).
16. Wächtershäuser, G. and Huber, C., "Activated acetic acid by carbon fixation on (Fe,Ni)S under primordial conditions", *Science*, **276**, 245–247 (1997).
17. Boyington, J. C., Gladyshev, V. N., Khangulov, S. V., Stadtman, T. C., and Sun, P. D., "Crystal structure of formate dehydrogenase H: catalysis involving Mo, molybdopterin, selenocysteine, and an Fe4S4 cluster", *Science*, **275**, 1305–1308 (1997).
18. Drennan, C. L., Huang, S., Drummond, J. T., Matthews, R. G., and Ludwig, M. L., "How a protein binds B_{12}: A 3.0 Å X-ray structure of B_{12}-binding domains of methionine synthase", *Science*, **266**, 1669–1674 (1994).
19. Fontecilla-Camps, J.-C. and Ragsdale, S. W., *Advances in Inorganic Chemistry*, Sykes, A. G. and Cammack, R., Eds., Vol. 47, Academic Press, Inc., San Diego, CA, 1999, pp. 283–333.
20. Dobbek, H., Svetlitchnyi, V., Gremer, L., Huber, R., and Meyer, O., "Crystal structure of a carbon monoxide dehydrogenase reveals a [Ni-4Fe-5S] cluster", *Science*, **293**, 1281–1285 (2001).
21. Doukov, T. I., Iverson, T., Seravalli, J., Ragsdale, S. W., and Drennan, C. L., "A Ni-Fe-Cu center in a bifunctional carbon monoxide dehydrogenase/acetyl-CoA synthase", *Science*, **298**, 567–572 (2002).
22. Ermler, U., Grabarse, W., Shima, S., Goubeaud, M., and Thauer, R. K., "Crystal structure of methyl-Coenzyme M reductase: the key enzyme of biological methane formation", *Science*, **278**, 1457–1462 (1997).
23. Vorholt, J. A., Vaupel, M., and Thauer, R. K., "A selenium-dependent and a selenium-independent formylmethanofuran dehydrogenase and their transcriptional regulation in the hyperthermophilic *Methanopyrus kandleri*", *Mol. Microbiol.*, **23**, 1033–1042 (1997).
24. Ragsdale, S. W., Wood, H. G., and Antholine, W. E., "Evidence that an iron–nickel–carbon complex is formed by reaction of CO with the CO dehydrogenase from *Clostridium thermoaceticum*", *Proc. Natl. Acad. Sci. U.S.A.*, **82**, 6811–6814 (1985).
25. Volbeda, A., Charon, M. H., Piras, C., Hatchikian, E. C., Frey, M., and Fontecilla-Camps, J. C., "Crystal structure of the nickel-iron hydrogenase from *Desulfovibrio gigas*", *Nature (London)*, **373**, 580–587 (1995).
26. Lyon, E. J., Shima, S., Buurman, G., Chowdhuri, S., Batschauer, A., Steinbach, K., and Thauer, R. K., "UV-A/blue-light inactivation of the 'metal-free' hydrogenase (Hmd) from methanogenic archaea", *Eur. J. Biochem.*, **271**, 195–204 (2004).
27. Westenberg, D. J., Braune, A., Ruppert, C., Muller, V., Herzberg, C., Gottschalk, G., and Blaut, M., "The F420H2-dehydrogenase from *Methanolobus tindarius*: cloning of the ffd operon and expression of the genes in *Escherichia coli*", *FEMS Microbiol. Lett.*, **170**, 389–398 (1999).
28. Abken, H. J., Tietze, M., Brodersen, J., Bäumer, S., Beifuss, U., and Deppenmeier, U., "Isolation and characterization of methanophenazine and the function of phenazines in membrane-bound electron transport of *Methanosarcina mazei* Go1", *J. Bacteriol.*, **180**, 2027–2032 (1998).

29. Stiefel, E. I. and George, G. N., *Bioinorganic Chemistry*, Bertini, I., Gray, H. B., Lippard, S. J., and Valentine, J. S., Eds., University Science Books, Mill Valley, CA, 1994, pp. 365–463.
30. Vorholt, J. A., Vaupel, M., and Thauer, R. K., "A polyferredoxin with eight [4Fe–4S] clusters as a subunit of molybdenum formylmethanofuran dehydrogenase from *Methanosarcina barkeri*", *Eur. J. Biochem.*, **236,** 309–317 (1996).
31. Tersteegen, A. and Hedderich, R., "*Methanobacterium thermoautotrophicum* encodes two multisubunit membrane-bound [NiFe] hydrogenases. Transcription of the operons and sequence analysis of the deduced proteins", *Eur. J. Biochem.*, **264,** 930–943 (1999).
32. Hedderich, R., Klimmek, O., Kröger, A., Dirmeier, R., Keller, M., and Stetter, K. O., "Anaerobic respiration with elemental sulfur and with disulfides", *FEMS Microbiol. Rev.*, **22,** 353–381 (1998).
33. Bartholomew, G. W. and Alexander, M., "Microbial metabolism of carbon monoxide in culture and in soil", *Appl. Environ. Microbiol.*, **37,** 932–937 (1979).
34. Meyer, O., Frunzke, K., and Mörsdorf, G., *Microbial growth on C_1 compounds,* Murrell, J. C. and Kelly, D. P., Eds., Intercept, Ltd., Andover, MA, pp. 433–459.
35. Dobbek, H., Gremer, L., Meyer, O., and Huber, R., "Crystal structure and mechanism of CO dehydrogenase, a molybdo iron–sulfur flavoprotein containing S-selenylcysteine", *Proc. Natl. Acad. Sci. U. S. A.*, **96,** 8884–8889 (1999).
36. Gnida, M., Ferner, R., Gremer, L., Meyer, O., and Meyer-Klaucke, W., "A novel binuclear [CuSMo] cluster at the active site of carbon monoxide dehydrogenase: characterization by X-ray absorption spectroscopy", *Biochemistry,* **42,** 222–230 (2003).
37. Seravalli, J., Kumar, M., Lu, W.-P., and Ragsdale, S. W., "Mechanism of carbon monoxide oxidation by the carbon monoxide dehydrogenase/acetyl-CoA synthase from *Clostridium thermoaceticum:* Kinetic characterization of the intermediates", *Biochemistry,* **36,** 11241–11251 (1997).
38. Drennan, C. L., Heo, J., Sintchak, M. D., Schreiter, E., and Ludden, P. W., "Life on carbon monoxide: X-ray structure of *Rhodospirillum rubrum* Ni-Fe-S carbon monoxide dehydrogenase", *Proc. Natl. Acad. Sci. U. S. A.*, **98,** 11973–11978 (2001).
39. Darnault, C., Volbeda, A., Kim, E. J., Legrand, P., Vernede, X., Lindahl, P. A., and Fontecilla-Camps, J. C., "Ni-Zn-[Fe(4)-S(4)] and Ni-Ni-[Fe(4)-S(4)] clusters in closed and open alpha subunits of acetyl-CoA synthase/carbon monoxide dehydrogenase", *Nat. Struct. Biol.*, **10,** 271–279 (2003).
40. Ljungdahl, L. G. and Andreesen, J. R., "Tungsten, a component of active formate dehydrogenase from *Clostridium thermoaceticum*", *FEBS Lett.*, **54,** 279–282 (1975).
41. Schauer, N. L. and Ferry, J. G., "Composition of the coenzyme F420-dependent formate dehydrogenase from *Methanobacterium formicicum*", *J. Bacteriol.*, **165,** 405–411 (1986).
42. Chistoserdova, L., Vorholt, J. A., Thauer, R. K., and Lidstrom, M. E., "C-1 transfer enzymes and coenzymes linking methylotrophic bacteria and methanogenic Archaea", *Science,* **281,** 99–102 (1998).
43. Hille, R., "The mononuclear molybdenum enzymes", *Chem. Rev.*, **96,** 2757–2816 (1996).
44. Hochheimer, A., Hedderich, R., and Thauer, R. K., "The formylmethanofuran dehydrogenase isoenzymes in *Methanobacterium wolfei* and *Methanobacterium thermoautotrophicum:* induction of the molybdenum isoenzyme by molybdate and constitutive synthesis of the tungsten isoenzyme", *Arch. Microbiol.*, **170,** 389–393 (1998).
45. Ragsdale, S. W., "The Eastern and Western branches of the Wood/Ljungdahl pathway: how the East and West were won", *BioFactors,* **9,** 1–9 (1997).
46. Zirngibl, C., Hedderich, R., and Thauer, R. K., "Metal-free hydrogenase", *FEBS Lett.*, **261,** 112–116 (1990).
47. Thauer, R. K., Klein, A. R., and Hartmann, G. C., "Reactions with molecular hydrogen in microorganisms. Evidence for a purely organic hydrogenation catalyst", *Chem. Rev.*, **96,** 3031–3042 (1996).
48. Geierstanger, B. H., Prasch, T., Griesinger, C., Hartmann, G., Buurman, G., and Thauer, R. K., "Catalytic mechanism of the metal-free hydrogenase from methanogenic archaea: Reversed stereospecificity of the catalytic and noncatalytic reaction", *Angew. Chem. Int. Ed. Engl.*, **37,** 3300–3303 (1998).
49. Paul, L., Ferguson, D. J., and Krzycki, J. A., "The trimethylamine methyltransferase gene and multiple dimethylamine methyltransferase genes of *Methanosarcina barkeri* contain in-frame and read-through amber codons", *J. Bacteriol.*, **182,** 2520–2529 (2000).
50. Srinivasan, G., James, C. M., and Krzycki, J. A., "Pyrrolysine encoded by UAG in Archaea: charging of a UAG-decoding specialized tRNA", *Science,* **296,** 1459–1462 (2002).
51. Hippler, B. and Thauer, R. K., "The energy conserving methyltetrahydromethanopterin:coenzyme M methyltransferase complex from methanogenic archaea: function of the subunit MtrH", *FEBS Lett.*, **449,** 165–168 (1999).
52. Doukov, T., Seravalli, J., Stezowski, J., and Ragsdale, S. W., "Crystal structure of a methyltetrahydrofolate and corrinoid dependent methyltransferase", *Structure,* **8,** 817–830 (2000).
53. Dixon, M. M., Huang, S., Matthews, R. G., and Ludwig, M., "The structure of the C-terminal domain of methionine synthase: Presenting S-adenosylmethionine for reductive methylation of B-12", *Structure,* **4,** 1263–1275 (1996).
54. Ragsdale, S. W., Lindahl, P. A., and Münck, E., "Mössbauer, EPR, and optical studies of the corrinoid/Fe-S protein involved in the synthesis of acetyl-CoA by *Clostridium thermoaceticum*", *J. Biol. Chem.*, **262,** 14289–14297 (1987).

55. Menon, S. and Ragsdale, S. W., "The role of an iron–sulfur cluster in an enzymatic methylation reaction: methylation of CO dehydrogenase/acetyl-CoA synthase by the methylated corrinoid iron–sulfur protein", *J. Biol. Chem.*, **274,** 11513–11518 (1999).
56. Wassenaar, R. W., Keltjens, J. T., and van der Drift, C., "Activation and reaction kinetics of the dimethylamine/coenzyme M methyltransfer in *Methanosarcina barkeri* strain Fusaro", *Eur. J. Biochem.*, **258,** 597–602 (1998).
57. Seravalli, J., Zhao, S., and Ragsdale, S. W., "Mechanism of transfer of the methyl group from (6*S*)-methyltetrahydrofolate to the corrinoid/iron–sulfur protein catalyzed by the methyltransferase from *Clostridium thermoaceticum*: a key step in the Wood–Ljungdahl pathway of acetyl-CoA synthesis", *Biochemistry,* **38,** 5728–5735 (1999).
58. Drennan, C. L., Doukov, T. I., and Ragsdale, S. W., "The Metalloclusters of Carbon Monoxide Dehydrogenase/Acetyl-CoA Synthase: A Story in Pictures", *J. Biol. Inorg. Chem.*, **9,** 511–515 (2004).
59. Brunold, T. C., "Spectroscopic and Computational Insights into the Geometric and Electronic Properties of the A Cluster of Acetyl-Coenzyme A Synthase", *J. Biol. Inorg. Chem.,* **9,** 533–541 (2004).
60. Riordan, C., "Synthetic Chemistry and Chemical Precedents for Understanding the Structure and Function of Acetyl Coenzyme A Synthase", *J. Biol. Inorg. Chem.,* **9,** 542–549 (2004).
61. Lindahl, P. A., "Acetyl-Coenzyme A Synthase: The Case for a $Ni_p{}^0$-Based Mechanism of Catalysis", *J. Biol. Inorg. Chem.,* **9,** 516–524 (2004).
62. Ragsdale, S. W. and Wood, H. G., "Acetate biosynthesis by acetogenic bacteria: evidence that carbon monoxide dehydrogenase is the condensing enzyme that catalyzes the final steps of the synthesis", *J. Biol. Chem.*, **260,** 3970–3977 (1985).
63. Chen, J., Huang, S., Seravalli, J., Jr., H. G., Swartz, D. J., Ragsdale, S. W., and Bagley, K. A., "Infrared Studies of Carbon Monoxide Binding to Carbon Monoxide Dehydrogenase/Acetyl-CoA Synthase from *Moorella thermoacetica*", *Biochemistry,* **42,** 14822–14830 (2003).
64. Shin, W., Anderson, M. E., and Lindahl, P. A., "Heterogeneous nickel environments in carbon monoxide dehydrogenase from *Clostridium thermoaceticum*", *J. Am. Chem. Soc.,* **115,** 5522–5526 (1993).
65. Ram, M. S., Riordan, C. G., Yap, G. P. A., Liable-Sands, L., Rheingold, A. L., Marchaj, A., and Norton, J. R., "Kinetics and mechanism of alkyl transfer from organocobalt(III) to nickel(I): Implications for the synthesis of acetyl coenzyme A by CO dehydrogenase", *J. Am. Chem. Soc.,* **119,** 1648–1655 (1997).
66. Goubeaud, M., Schreiner, G., and Thauer, R. K., "Purified methyl-coenzyme-M reductase is activated when the enzyme-bound coenzyme F_{430} is reduced to the nickel(I) oxidation state by titanium(III) citrate", *Eur. J. Biochem.*, **243,** 110–114 (1997).
67. Becker, D. F. and Ragsdale, S. W., "Activation of methyl-SCoM reductase to high specific activity after treatment of whole cells with sodium sulfide", *Biochemistry,* **37,** 2639–2647 (1998).
68. Telser, J., Davydov, R., Horng, Y. C., Ragsdale, S. W., and Hoffman, B. M., "Cryoreduction of methyl-coenzyme M reductase: EPR characterization of forms, MCR(ox1) and MCR(red1)", *J. Am. Chem. Soc.,* **123,** 5853–5860 (2001).
69. Telser, J., Horng, Y.-C., Becker, D., Hoffman, B., and Ragsdale, S. W., "On the assignment of nickel oxidation states of the Ox1 and Ox2 Forms of Methyl-Coenzyme M Reductase", *J. Am. Chem. Soc.,* **122,** 182–183 (2000).
70. Craft, J. L., Horng, Y.-C., Ragsdale, S. W., and Brunold, T. C., "Nickel Oxidation States of F_{430} Cofactor in Methyl-Coenzyme M Reductase", *J. Am. Chem. Soc.,* **126,** 4068–4069 (2004).
71. Selmer, T., Kahnt, J., Goubeaud, M., Shima, S., Grabarse, W., Ermler, U., and Thauer, R. K., "On the biosynthesis of methylated amino acids in the active site region of methyl-coenzyme M reductase", *J. Biol. Chem.,* **275,** 3755–3760 (2000).
72. Lin, S.-K. and Jaun, B., "Coenzyme F430 from methanogenic bacteria: mechanistic studies on the reductive cleavage of sulfonium ions catalyzed by F430 pentamethyl ester", *Helv. Chim. Acta,* **75** (1992).
73. Jaun, B., *Metal Ions in Biological Systems,* Sigel, H. and Sigel, A., Eds., Vol. 298, Marcel Dekker, New York, 1993, pp. 287–337.
74. Ahn, Y., Krzycki, J. A., and Floss, H. G., "Stereochemistry of methane formation", *J. Am. Chem. Soc.,* **113,** 4700–4701 (1991).
75. Signor, L., Knuppe, C., Hug, R., Schweizer, B., Pfaltz, A., and Jaun, B., "Methane formation by reaction of a methyl thioether with a photo-excited nickel thiolate-a process mimicking methanogenesis in archaea," *Chem. Eur. J.*, **6,** 3508–3516 (2000).
76. Grabarse, W. G., Mahlert, F., Duin, E. C., Goubeaud, M., Shima, S., Thauer, R. K., Lamzin, V., and Ermler, U., "On the mechanism of biological methane formation: Structural evidence for conformational changes in methyl-coenzyme M reductase upon substrate binding", *J. Mol. Biol.,* **309,** 315–330 (2001).
77. Pelmenschikov, V., Blomberg, M. R. A., Siegbahn, P. E. M., and Crabtree, R. H., "A Mechanism from Quantum Chemical Studies for Methane Formation in Methanogenesis", *J. Am. Chem. Soc.,* **124,** 4039–4049 (2002).
78. Heim, S., Kunkel, A., Thauer, R. K., and Hedderich, R., "Thiol:Fumarate reductase (Tfr) from *Methanobacterium thermoautotrophicum*—Identification of the catalytic sites for fumarate reduction and thiol oxidation", *Eur. J. Biochem.,* **253,** 292–299 (1998).

XII.3. Biological Nitrogen Fixation and Nitrification

William E. Newton
Department of Biochemistry
The Virginia Polytechnic Institute and State University
Blacksburg, VA 24061

Contents

XII.3.1. Introduction

All living things depend on usable, "fixed" nitrogen (fixed-N) for incorporation into the molecules of life, namely, DNA, ribonucleic acid (RNA), and proteins. Perhaps life's greatest paradox is that only some of the smallest members of Earth's community (bacteria and archaea) can produce fixed-N from the otherwise inert, molecular nitrogen (N_2) that surrounds and saturates us. At any one time, only ~0.0007% of the nitrogen available on Earth and in its atmosphere is in "fixed" forms, such as ammonia and nitrate. The biogeochemical cycling of nitrogen (see Chapter II) allows interconversion of the inert atmospheric pool and the usable fixed-N terrestrial pool. Within this cycle, biological nitrogen fixation drives the conversion of atmospheric N_2 to ammonia,[1–17] whereas nitrification,[18–21] together with denitrification (see Section XII.4), returns "fixed" forms of nitrogen to the atmosphere as N_2.

In the modern world, N_2 is "fixed" by several processes. Abiological natural processes, which include lightning, fires, and volcanic eruptions, account for <10% of the total annual amount of fixed-N made available to the biosphere, whereas man-made processes, mainly industrial ammonia production through the Haber–Bosch process, contribute ~30% of the annual total. Biological nitrogen fixation at ~60% of total annual fixation is, therefore, the main provider of fixed-N.[1,2]

Nitrification is the first step in reversing nitrogen fixation. It converts reduced-N compounds, such as NH_3, in the environment into oxidized-N forms, such as NO_2^- and NO_3^-. The production of these negatively charged ions allows rapid movement of fixed-N through the negatively charged soil. Although such movement facilitates the uptake of oxidized-N compounds by plant roots, it also results in a loss of fixed-N through the leaching of these materials into ground waters and their increased denitrification to gaseous N-oxides, like NO, N_2O, and N_2 (see Chapter II; Section XII.4). Nitrification is also a vital component of modern wastewater treatment, where the substantial amounts of reduced-N compounds in urban waste water are degraded to the less toxic oxidized-N forms before release into rivers and lakes.

To sustain life on Earth, the biological world must prevent an overall nitrogen deficiency, which would occur if the nitrification–denitrification rate exceeded the fixation rate. It apparently does so, but by only a slim margin.

XII.3.2. Biological Nitrogen Fixation: When and How Did Biological Nitrogen Fixation Evolve?

Nitrogen fixation is unlikely to have arisen before the geochemical reserves of fixed-N in the biosphere were depleted or its availability became limiting. Only then would selective pressure appear for the fixation of atmospheric N_2, because organisms prefer to use available sources of fixed-N before fixing N_2 themselves. Unfortunately, there is no agreement as to when this depletion of the fixed-N reserves occurred in geological time, leading to considerable uncertainty about when microbes acquired the ability to fix N_2. Even so, we can be sure that once free O_2 appeared, any geochemical ammonia would react to form N_2 and nitrogen oxides and, eventually, the availability of the nitrogen oxides would limit the growth of organisms. This situation would then lead to selective pressure and the appearance of biological nitrogen fixation (also known as diazotrophy).

XII.3.2.1. Biological Nitrogen Fixation and Photosynthesis

Does the above scenario indicate that O_2-producing photosynthesis predated diazotrophy? Not necessarily! Free O_2 (albeit at a much lower level than present-day levels produced photosynthetically) could have been produced abiologically by photolysis of water much earlier in geological time. However, the appearance of non-O_2-producing phototrophy, as practiced by some bacteria, would have contributed to the pressure for diazotrophy because the photosynthetically driven carbon–sulfur/carbon–oxygen cycles would have out-paced the inputs from the sources of fixed-N. This suggestion does not mean that phototrophic organisms were necessarily the first N_2 fixers because early organisms, including phototrophs, might simply assimilate the products of photosynthesis from other organisms. No matter what the situation was vis-à-vis photosynthesis, the high sensitivity to free O_2 of the biological catalyst for nitrogen fixation, called nitrogenase, suggests that it evolved on Earth under O_2-free conditions where such sensitivity would not be a liability.[3,4]

XII.3.2.2. Types of Nitrogenase

We now know of four genetically distinct nitrogenases, three of which are closely related. These three form the "classical" nitrogenase group, which is made up of a molybdenum-based enzyme system (Mo-nitrogenase), a vanadium-based enzyme system (V-nitrogenase), and an enzyme system apparently based on iron-only (Fe-only nitrogenase). Even though each enzyme has a different "heterometal" (Mo, V, or Fe), they are otherwise so similar that they must have arisen from a common ancestor. In contrast, the fourth nitrogenase is so different (see Section XII.3.4.3) that it may well be an evolutionary "independent invention" and be completely unrelated to the "classical" enzymes.

Some controversy exists as to which of the "classical" nitrogenases is the oldest. One evolutionary viewpoint suggests the existence of an aboriginal "nitrogenase", in the form of a pyrite-forming Fe–S cluster, which produced C-bound nitrogen rather than ammonia.[5] This view of the origin of life has "nitrogenase" fixing N_2 prior to the existence of enzymes and templates and suggests that nitrogen fixation is very ancient indeed. A logical extension of the "pyrite theory" is that the Fe-only nitrogenase is the forerunner of the three "classical" nitrogenases recognized today. Support for this view comes from a consideration of the chemistry of Mo. If indeed the early Earth were hot and anoxic, Mo would have been deposited as molybdenite (MoS_2), which is insoluble in water and so unavailable. If nitrogenase were appearing under these same conditions, Mo would not be available and an Fe-only nitrogenase might have been the compromise choice.[6]

Another possibility is that nitrogenase originally arose as an assimilatory cyanide reductase to detoxify local ancient environments.[7] If so, Mo-nitrogenase would be the most likely forerunner because it is superior to the other two enzymes for reducing cyanide. A third suggestion cites one primal nitrogenase from which the three "classical" nitrogenases developed in more recent time. This suggestion is supported by the observation that, although separately encoded genetically, the three sets of structural genes, which encode the polypeptide subunits of the proteins, were likely formed by gene iteration. Moreover, the products of certain nitrogen-fixation-specific genes support all three nitrogenases, the heterometallic cofactors found in all three nitrogenases are transferable among the proteins, and all three nitrogenases have similar catalytic properties.[6,7]

Very little is known about the origin and evolution of the nitrogen-fixation genes and the mechanisms that were involved in shaping the process itself.[22] Similarly, ideas are only now solidifying on how diazotrophy became distributed among the relatively few genera of bacteria and archaea that fix N_2.[8] From a classical evolutionary point of view, the haphazard distribution of diazotrophy among microbes may be thought of as a common ancestral property that was lost randomly during divergent evolution. Alternatively, the nitrogen-fixation genes could be of more recent origin, having spread laterally, like antibiotic resistance, among diverse prokaryotic genera.[23] If the former explanation were true, phylogenies (relatedness among microbes) based on the sequences of both the 16S rRNA (ribosomal) genes and the genes encoding the subunits of nitrogenase should match. In fact, sometimes the phylogenies match and sometimes they do not. Early on, some researchers saw in the gene (DNA) sequences support for an ancient origin, whereas others saw evidence for more recent lateral gene transfer. As more and more genomes are sequenced and the consequences of the duplication of genes during evolution are considered, it appears that there may have been either multiple losses or multiple transfers of the nitrogen-fixation genes or a combination of both. Whatever the case, the genetic relatedness of the "classical" nitrogenases, including even the order in which the genes are found in the genome, supports a common ancient ancestry.

XII.3.3. Nitrogen-Fixing Organisms and Crop Plants

After diazotrophy appeared, fixed-N was no longer the limiting nutrient in balanced ecosystems. However, when such ecosystems are perturbed, the nutrients recycle biogeochemically and fixed-N usually becomes limiting again. Agriculture is a major and persistent perturbation of natural balanced ecosystems. So, it is no surprise to find that, with agriculture being a common practice in human societies, the global availability of fixed-N is again a major limitation of agricultural production.

Although farmers recognized the benefits of crop rotations centuries ago, the source of much of that benefit was unknown to them. The first report of nitrogen fixation by Boussingault in 1838[9,10] was based on the comparative growth and nitrogen content of cereals with leguminous plants (mainly clover in rotations with wheat and tuber crops) in both greenhouse and field experiments. The result "... that azote (nitrogen) may enter the living frame of the plants directly ..." was received with skepticism. It was not until almost 50 years later that Boussingault's work was convincingly confirmed by Hellriegel and Wilfarth, who also solved the perplexing question of the source of the fixed-N by localizing the activity to the bacteria-filled nodules on the roots of their pea plants.[9,10] Only prokaryotes, that is, those living things without an organized nucleus (Eubacteria and Archaea, see Tutorial 1), can fix N_2 biologically. The ability to fix N_2 is widely spread among microbial genera and, despite a number of claims, no eukaryote has been clearly established to fix N_2.

The most agriculturally important symbiotic relationship involves leguminous plants (peas and beans) and certain bacteria, collectively called rhizobia, but many other nitrogen-fixing symbiotic relationships are known. Although this tight, nodule-based, symbiotic arrangement is very successful, it does not extend to most of the important food crops, like the cereal grains (rice, wheat, and corn), and root and tuber crops. So, for the productivity of these crops to reach commercially acceptable levels, extensive augmentation by commercially fixed nitrogen is usually necessary.

Most nitrogen-fixing microbes are free living and fix nitrogen for their own benefit. The most extensively studied free-living bacterial species are *Azotobacter vinelandii* (an obligate aerobe), *Clostridium pasteurianum* (an obligate anaerobe), *Klebsiella pneumoniae* (a facultative anaerobe), *Rhodobacter capsulatus* (a photosynthetic bacterium), and *Anabaena sp.* 7120 (a heterocyst-forming cyanobacterium).

XII.3.4. Relationships among Nitrogenases

With a single exception (see Section XII.3.4.3), all nitrogen-fixing organisms have the Mo-nitrogenase. As mentioned above, there are two other types of "classical" nitrogenase, which are structurally and functionally related to the Mo-nitrogenase, but the distribution of the V-nitrogenase and the Fe-only nitrogenase is completely haphazard. Some organisms (e.g., *K. pnemoniae*) have only Mo-nitrogenase, whereas others (e.g., *A. vinelandii*) have all three. Other combinations are also found; for example, *Azotobacter chroococcum* has both a Mo- and a V-nitrogenase, whereas *R. capsulatus* has a Mo- and an Fe-only nitrogenase.[11]

Which of these nitrogenases is expressed at any time depends on the availability of the metal ions (either Mo or V) in the growth medium.[12] Whenever Mo is available, expression of the Mo-dependent nitrogenase is stimulated and the expression of the other two nitrogenases is repressed. Similarly, when V is available and Mo is absent, expression of only the V-nitrogenase occurs. If both metals are absent, then just the Fe-only nitrogenase is expressed. Such control by metal availability is physiologically reasonable because Mo-nitrogenase is the most efficient catalyst for N_2 reduction, followed by V-nitrogenase, with the Fe-nitrogenase being the least efficient because it evolves large amounts of H_2 gas and produces much less ammonia.

All three "classical" nitrogenases consist of two metalloproteins,[6,13,14] which can be separated and purified individually, but which have no activity alone (see Sections XII.3.4.1 and XII.3.4.2). When either component protein from Mo-nitrogenase is mixed with the complementary protein component from V-nitrogenase, an active hybrid nitrogenase is produced. But, when either component protein from the Fe-only nitrogenase is crossed with the complementary protein from either the Mo- or V-nitrogenase, the resulting hybrid nitrogenases are completely inactive.

XII.3.4.1. Mo-Nitrogenase

The individual component proteins of Mo-nitrogenase are called the Fe protein (or component 2 or, sometimes, dinitrogenase reductase) and the MoFe protein (or component 1 or, sometimes, dinitrogenase). The trivial names for these proteins are derived from their metal compositions. The Fe protein is a homodimer with a molecular weight of ~64,000 Da. It contains two MgATP (adenosine triphosphate = ATP) binding sites and a single [4Fe–4S] cluster bridging the two subunits. The MoFe protein is an $\alpha_2\beta_2$ heterotetramer with a molecular weight of ~230,000 Da. It contains two pairs of two different metalloclusters, called the P-cluster and the iron–molybdenum-cofactor (or FeMo-cofactor or FeMoco or the M-center). The Fe protein serves as a specific reductant for the MoFe protein, which contains the site(s) of substrate binding and reduction (Fig. XII.3.1). The three-dimensional (3D) structures of the Fe protein, the MoFe protein, the complex of the two proteins, and variant MoFe proteins from mutant bacterial strains are all known (see Section XII.3.5).

Molybdenum-nitrogenase catalyzes the biological nitrogen-fixation reaction, which is usually described as in Eq. 1, where ADP = adenosine diphosphate.

$$N_2 + 8\,H^+ + 8\,e^- + 16\,MgATP \rightarrow 2\,NH_3 + H_2 + 16\,MgADP + 16\,P_i \qquad (1)$$

In addition to N_2 and H^+ as substrates, nitrogenase catalyzes the reduction of many other small-molecule "alternative" substrates, all of which have the same requirements as for N_2 reduction, namely, a supply of MgATP, a low-potential reductant, and an anaerobic environment.[11,15] *In vivo,* nitrogenase uses either a ferredoxin or a flavodoxin as the reductant of the Fe protein, whereas *in vitro,* the artificial reductant sodium dithionite is most often used. With sodium dithionite as reductant, nitrogenase hydrolyzes about four molecules of MgATP for each pair of electrons that is transferred to substrate and this ratio is independent of the substrate being reduced. The most often used of these "alternative substrates" is acetylene, which is reduced by only two electrons to ethylene. Other "alternative substrates" are shown in Fig. XII.3.2 together with carbon monoxide (CO), which is not a substrate of Mo-nitrogenase, but is a potent inhibitor of all nitrogenase-catalyzed substrate reductions except for the reduction of H^+ to H_2.[14]

Dihydrogen has a unique involvement with Mo-nitrogenase and with N_2 reduction, in particular. Via the action of hydrogenase, H_2 can be the source of reducing equivalents for nitrogenase. Dihydrogen is the sole product in the absence of any other reducible substrate. Moreover, H_2 is a specific inhibitor of N_2 reduction, affecting neither the reduction of any other substrate nor its own evolution. Under a mixed atmosphere of N_2 and D_2, HD is formed in a reaction that has all the requirements of a nitrogenase-catalyzed reaction and is inhibited by CO.[14]

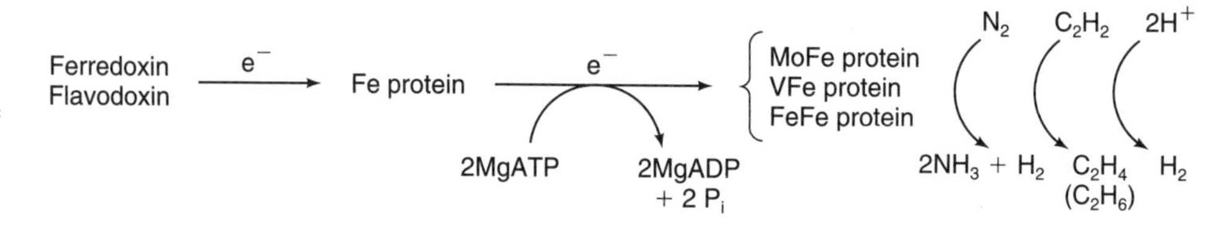

Fig. XII.3.1.
A general electron-transfer pathway through the nitrogenases. (Inorganic phosphate = P_i.)

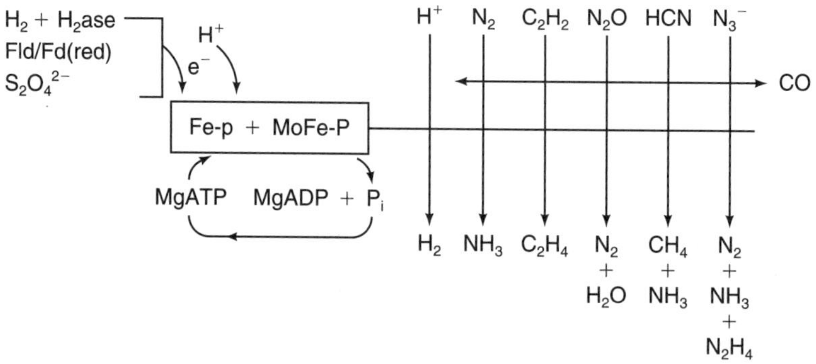

Fig. XII.3.2.
Electron donors, substrates, and their products for wild-type Mo-nitrogenase catalysis. Carbon monoxide is a potent reversible inhibitor of all nitrogenase-catalyzed substrate reductions except for that of protons to H_2. The Fe protein is represented by Fe-p, MoFe-p is the MoFe protein, H_2ase is hydrogenase, Fld is flavodoxin, and Fd is ferredoxin.

The Mo-nitrogenases from a variety of bacterial genera exhibit a high level of primary (amino acid) sequence identity. The sequence conservation is particularly high in the regions of the MgATP- and metallocluster-binding sites. With the notable exception of *C. pasteurianum*, the component proteins from all Mo-based nitrogenases form catalytically active enzymes when crossed with one another.[11,15]

Each of the component proteins of Mo-nitrogenase exhibits an EPR spectrum when reduced with sodium dithionite. Dithionite reduction of the [4Fe–4S] cluster of the Fe protein produces a mixture of $S = \frac{1}{2}$ and $S = \frac{3}{2}$ spin states. The EPR signal $\sim g = 2$, which arises from the $S = \frac{1}{2}$ spin state, is much sharper than the EPR signal at $\sim g = 4$ from the $S = \frac{3}{2}$ spin state. This difference in line shape caused the $S = \frac{3}{2}$ component to go unrecognized until relatively recently, even though the measured intensity of the $S = \frac{1}{2}$ component never fully accounted for the spin of the unpaired electron. The dithionite-reduced MoFe protein exhibits a complex EPR signal with two g values ~ 4 ($g = 4.3$ and 3.7) and one $\sim g = 2$ ($g = 2.01$). This EPR signal arises from the FeMo-cofactor in a $S = \frac{3}{2}$ spin state. These EPR signals have proved to be vital in determining the direction of electron flow between the component proteins and in monitoring many of their reactions (Fig. XII.3.3).

XII.3.4.2. V-Nitrogenase and Fe-Only Nitrogenase

These so-called "alternative nitrogenases"[6,11] also consist of two protein components. Each has its own specific Fe-protein component. The larger of the two component proteins of the V-nitrogenase contains a VFe-cofactor, with a V atom in place of the Mo atom found in the Mo-nitrogenase. This replacement, together with its distinct polypeptide, produces a VFe protein. Similarly, the Fe-only nitrogenase appears to contain a cofactor in which the Mo atom is substituted by Fe to produce a FeFe-cofactor in a FeFe protein. The high level of primary sequence identity recognized among the Mo-nitrogenases also extends to the V- and Fe-only nitrogenases. This identity strongly suggests that all nitrogenases share common structural features and have mechanistic similarities. A major difference, however, is that the VFe protein and FeFe protein have six subunits, with an $\alpha_2\beta_2\gamma_2$ composition, rather than four ($\alpha_2\beta_2$) subunits as found for the MoFe protein. The small "extra" γ subunits apparently bind to the apoprotein during biosynthesis of the VFe and FeFe proteins and remain bound thereafter. In contrast, the equivalent of the γ subunit for the MoFe-protein is lost in the late stages of its maturation.

The Fe-protein component of both alternative nitrogenases, when reduced with sodium dithionite, exhibits an EPR spectrum essentially identical to that of the Fe protein from the Mo-nitrogenase. Similarly, the dithionite-reduced VFe protein exhibits an EPR spectrum that is consistent with the presence of a FeV-cofactor. The $S = \frac{3}{2}$ EPR signal observed from purified FeFe protein again supports the presence of a FeFe-cofactor.

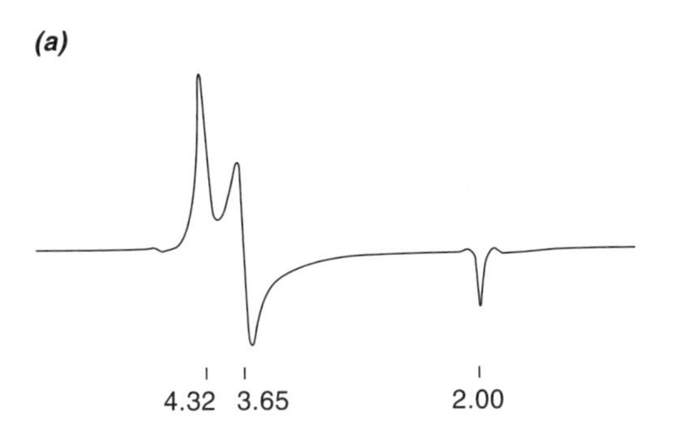

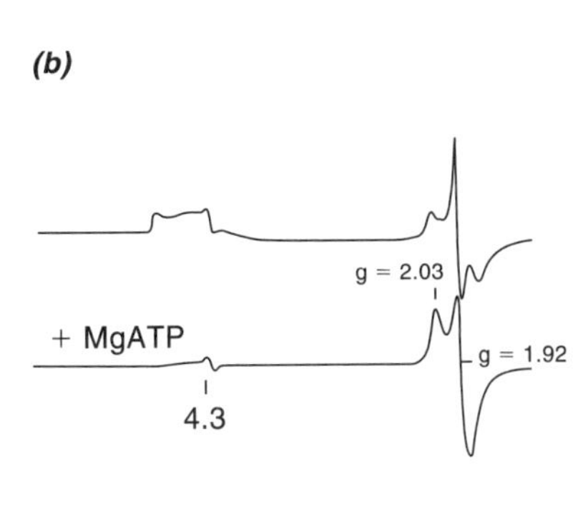

Fig. XII.3.3.
The EPR signals of the as-isolated dithionite-reduced state of both (*a*) the MoFe protein and (*b*) the Fe protein in the absence or presence of MgATP in frozen solution at ~ 10 K.[15,17]

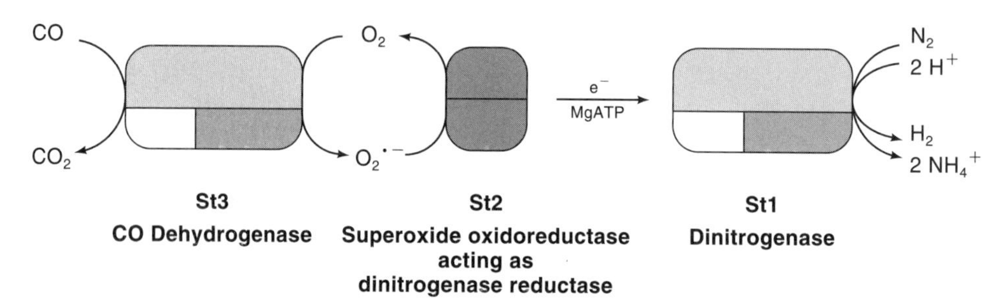

Fig. XII.3.4.
The N_2-fixation system of *S. thermoautotrophicus*.[24] Dioxygen is reduced to superoxide ($O_2^{\cdot -}$) by electrons released from the oxidation of CO as catalyzed by the MoFeS-containing CO dehydrogenase (St3). Superoxide is subsequently reoxidized to O_2 by a Mn-containing superoxide oxidoreductase (St2) that then delivers the electrons to the MoFeS- and pterin-dithiolene-containing dinitrogenase (St1), where N_2 (or H^+, but not C_2H_2) is reduced.

XII.3.4.3. *Streptomyces thermoautotrophicus* Nitrogenase

The fourth type of nitrogenase is the recently discovered,[24] unique nitrogenase from a thermophile, *Streptomyces thermoautotrophicus*. Although it also consists of two component proteins, the larger of which contains Mo, Fe, and sulfide, the similarity to the "classical" Mo-nitrogenase ends there. There is no Fe-protein component as found in the classical nitrogenase. Rather, a manganese-containing superoxide oxidoreductase oxidizes $O_2^{\cdot -}$ to O_2 and transfers the electron to the MoFeS-containing protein. This protein furnishes the site at which N_2 is reduced to two molecules of NH_3 accompanied by the evolution of one molecule of H_2 (Fig. XII.3.4). This eight-electron reaction appears to require the hydrolysis of less MgATP per N_2 reduced than does the classical Mo-nitrogenase. Also, in contrast to the classical Mo-nitrogenase, the electron donor is a Mo-containing carbon monoxide dehydrogenase, which couples the oxidation of CO to the reduction of O_2 to produce $O_2^{\cdot -}$.

The N_2-reducing MoFeS-containing protein is a $\alpha\beta\gamma$ heterotrimer quite different from the $\alpha_2\beta_2$ composition of the classical Mo-nitrogenase. The *S. thermoautotrophicus* nitrogenase also has several functional features that are unique. It is completely insensitive to the presence of O_2, CO, and H_2, all of which are potent inhibitors of nitrogen fixation in the "classical" system, and it does not catalyze the reduction of acetylene to ethylene. Clearly, further research is required to establish the structural and mechanistic properties of this unusual enzyme.

XII.3.5. Structures of the Mo-Nitrogenase Component Proteins and Their Complex

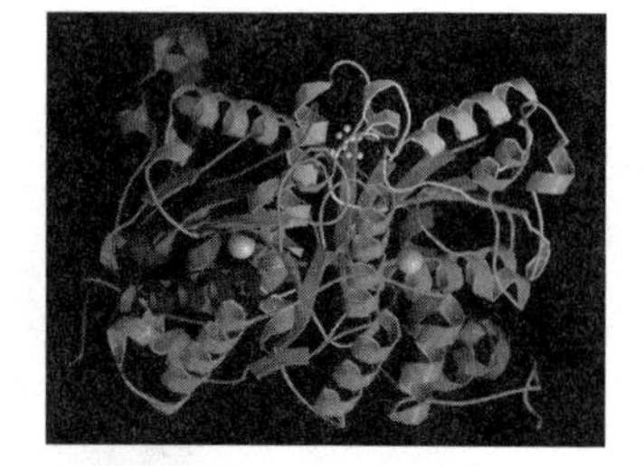

Fig. XII.3.5.
The overall polypeptide fold of the Fe-protein dimer complexed with MgADP, showing the [4Fe–4S] cluster bound between the two subunits and with the MgADP and the switch regions individually indicated (Ref. 27; PDB code: 1FP6). [See color insert.]

XII.3.5.1. The Fe Protein

The Fe protein from *A. vinelandii* Mo-nitrogenase is a homodimer of subunits encoded by the *nifH* gene. Its X-ray derived structure reveals that each identical subunit is made up of a single domain, which involves an eight-stranded β-sheet that is flanked by nine α helices (Fig. XII.3.5). Its single [4Fe–4S] cluster is bridged symmetrically between the two subunits. Each subunit provides two cysteinyl (numbered 97 and 132 in the *A. vinelandii* primary sequence) ligands to the [4Fe–4S] cluster, which occupies a solvent-exposed position at one end of the dimer interface.

Each subunit has a nucleotide-binding site, which is located in the channel between the subunits. These sites were first recognized by the presence of two consensus amino acid sequences that are common to other nucleotide-binding proteins. The first sequence, GXXGXGKS (where X represents any amino acid residue), for residues 9–16 is known as the Walker A motif (or P-loop), and forms a β-strand–loop–α-helix structure. This region provides all of the direct interactions of the Fe protein with the nucleotide phosphates. The second sequence, DXXG, for residues 125–128 is the Walker B motif, with the Asp125 of the Fe protein interacting with the nucleotide-bound Mg^{2+} ion.

The original Fe-protein structure showed only partial occupancy, equivalent to about one-half of an MgADP molecule per Fe protein. The presence of nucleotide was unexpected because none had been added during crystallization, and so it must have co-purified with the Fe protein. A second surprise was that the nucleotide is bound *across* the subunit–subunit interface, which includes the twofold symmetry axis, with the adenosine at one end of the ADP molecule interacting with one subunit and the terminal PO_4^{3-} with the other subunit.[25]

Two later structures of the Fe protein show a different orientation of the bound nucleotide. The first of these structures is from a 2:1 complex of the Fe protein with the MoFe protein.[26] This complex had been stabilized as a "transition state" analogue by adding MgADP and tetrafluoroaluminate (AlF_4^-), so it had a full complement of nucleotide present during crystallization. This structure shows two MgADP•AlF_4^- entities bound to each Fe protein with one nucleotide associated primarily with each Fe-protein subunit. Both MgADP•AlF_4^- entities were oriented approximately *parallel* to, and not across, the subunit–subunit interface. The second of these later structures also involved crystallization in the presence of excess MgADP but, this time, of the Fe protein alone.[27] This structure shows the same full occupancy and parallel orientation of nucleotide as observed with the complex. Apparently, the orientation assumed by the nucleotide in the structure depends on when it was added and whether sufficient is present to saturate the binding sites with two nucleotides per Fe-protein molecule.

Nucleotide binding to the Fe protein is cooperative and it induces several changes in the properties of the [4Fe–4S] cluster.[14,16] These include a change in the EPR spectral line shape and a lowering of the redox potential by ca. −100 to ca. −400 mV. Even with the several recent structures, it is still unclear exactly how these changes are achieved. The crystal structures show that the nucleotide does not contact the [4Fe–4S] cluster. However, they do show that a region of the Fe-protein backbone undergoes a significant structural change when nucleotide binds. This region, called switch II by analogy to the nomenclature used with G proteins, involves the sequence from the Asp125 residue, which interacts with the nucleotide-bound Mg^{2+}, to the Cys132 residue, which is a ligand to the [4Fe–4S] cluster. The effects of binding nucleotide are propagated through this region of the polypeptide backbone and result in a change in conformation at the cluster and a change in its electronic and redox properties.[16]

A similar mechanism, through a loop called switch I, might also allow communication between the nucleotide-binding site and that part of the Fe-protein surface that interacts with the MoFe protein during complex formation. This switch involves the region from residue Asp39 to a loop region of the Fe protein composed of residues 59–68, which also undergoes a nucleotide-dependent structural change. This loop may communicate to the nucleotide-binding site that contact has occurred with the MoFe-protein surface, thereby initiating MgATP hydrolysis, which induces electron transfer to the MoFe protein, followed by dissociation of the Fe protein from the MoFe protein.

XII.3.5.2. The MoFe Protein

This $\alpha_2\beta_2$ heterotetramer is encoded by the *nifDK* genes. Early spectroscopic and X-ray anomalous scattering studies of the MoFe protein, together with the ability to

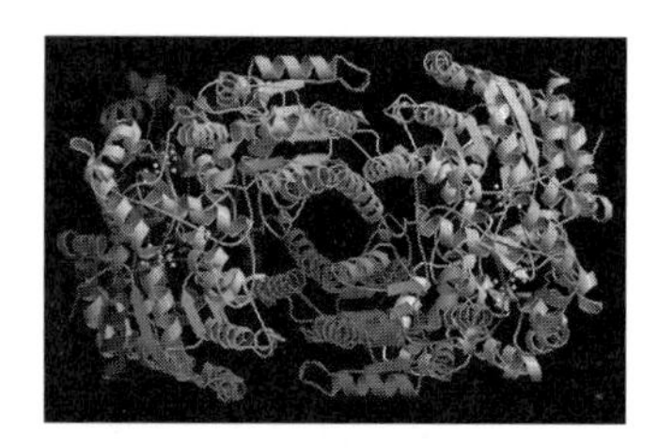

Fig. XII.3.6.
The overall polypeptide fold of the *A. vinelandii* nitrogenase MoFe protein tetramer as viewed along its twofold axis, showing the lack of interactions between the α-subunits. The α- and β-subunits and prosthetic groups are individually colored. Each FeMo-cofactor is situated completely within an α-subunit, whereas each P cluster sits at an interface of the α- and β-subunits (Ref. 33; PDB code: 3MIN). [See color insert.]

extrude metal clusters specifically, showed that the MoFe protein contained two types of prosthetic group, neither of which had been previously recognized. Each of these unique cluster types contains ~50% of both the Fe and S^{2-} content of the MoFe protein. They are called the P cluster and the FeMo-cofactor. The exact composition and distribution of these clusters within the protein was firmly established only after X-ray techniques had revealed the structures of both cluster types within the MoFe protein from both *A. vinelandii*[13,28] and *K. pneumoniae* (Ref. 29; PDB codes: 1QGU, 1QH1, and 1QH8). The clusters are distributed in pairs comprised of one FeMo-cofactor, which is a $[Mo-Fe_7-S_9]$ cluster with an attached (*R*)-homocitrate molecule, and one P cluster, which has a $[Fe_8-S_7]$ composition. One pair of prosthetic groups resides within each αβ-subunit pair and is separated by ~70 Å from the other pair (Fig. XII.3.6).

The MoFe protein is often treated as a dimer of dimers with each αβ-dimer functioning independently of the other, even though active αβ dimers have never been isolated and there is evidence of long-range interactions between its two Fe-protein binding sites. It is generally agreed that the Fe protein, as the specific electron donor to the MoFe protein, first associates with the MoFe protein, which contains the substrate-reduction site. Initially, an electron is transferred from the [4Fe–4S] cluster of the Fe protein to the P cluster. The electron then traverses the 15-Å distance to the FeMo-cofactor, where substrate is bound and reduced. It is still unclear how, when, and where the eight electrons necessary for the reduction of each N_2 are stored within the MoFe protein and how the required protons are delivered to this site.

The two αβ dimers interface primarily through interactions of helices in the two β-subunits. An ~8-Å wide channel, which contains another twofold rotation axis, passes through the center of the tetramer. The α- and β-subunits have similar polypeptide folds. Both subunits consist of three domains composed of α-helices and parallel β-sheets. In the α-subunit, these three domains meet to form a shallow cleft within which the FeMo-cofactor resides ~10 Å below the protein's surface. The FeMo-cofactor is buried within the α-subunit. It is covalently bound to only two amino acid ligands (Cys275 and His442, using the *A. vinelandii* numbering scheme) from the α-subunit and it has no close involvement with the β-subunit. In contrast, the P cluster is located at the interface of the α- and β-subunits with each subunit providing an equal number of ligating Cys residues. Each P cluster is bisected by a pseudo-twofold axis that relates the α- and β-subunits.

XII.3.5.3. The FeMo-cofactor Prosthetic Group

The FeMo-cofactor can be extruded intact from the MoFe protein. After isolation, it still shows the $S = \frac{3}{2}$ EPR signal, although significantly broadened, but it is ineffective in catalyzing the reduction of N_2. It has been recalcitrant to crystallization. A definitive description of its composition and structure had to await a high-resolution structure of the MoFe protein. Now, it is clear that the FeMo-cofactor consists of two subclusters, one $[Mo-Fe_3-S_3]$ and one $[Fe_4-S_3]$. Each subcluster may be visualized as missing one sulfide from either a $[Mo-Fe_3-S_4]$ or a $[Fe_4-S_4]$ thiocubane cluster, respectively. The original descriptions of the structure indicated that the two subclusters were bridged to one another by three non-protein-based sulfides.[30,31] As a result, only one of the Fe atoms (the terminal one) had tetrahedral geometry and the other six (central) Fe atoms had apparent trigonal geometry. This very unusual Fe geometry caused considerable consternation, particularly in the bioinorganic-synthesis community. More recently, a very high-resolution structure[32] has provided evidence for a single light atom, generally thought to be a nitrogen atom, within the central cavity of the FeMo-cofactor and equidistant to all six of the central Fe atoms

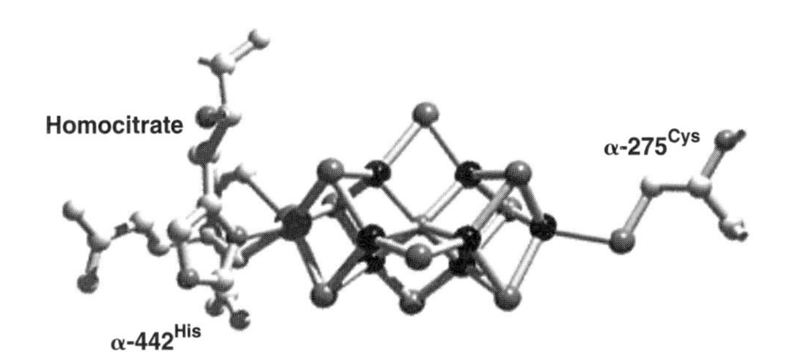

Fig. XII.3.7. The structure of the FeMo-cofactor of *A. vinelandii* nitrogenase MoFe protein with its α-subunit-based ligating amino acid residues (α-Cys275 and α-His442) and homocitrate. The identity of the central atom has been tentatively assigned as N (Ref. 32; PDB code: 1M1N).

(Fig. XII.3.7). It is uncertain whether this light atom has a mechanistic or a structural role, or both.

As mentioned above, the FeMo-cofactor has only two covalent bonds to the protein. These are an Fe–S bond, which is provided by the γ-S of α-cysteinyl275 to the terminal Fe atom, and a Mo–N, which is provided by the imidazole δ-N of α-histidinyl442. The octahedral coordination of the Mo atom is completed by the three cubane μ_3-sulfides and by ligation from the 2-hydroxyl and 2-carboxyl groups of (*R*)-homocitrate. The structure shows that the α-cysteinyl275-Fe apex of the FeMo-cofactor has no associated water molecules, whereas the homocitrate-Mo apex is surrounded by water molecules. Moreover, the homocitrate is positioned between the FeMo-cofactor and the P cluster and may be involved in the transfer of electrons and/or protons to bound substrate.

In addition to the covalent bonding provided by α-cysteinyl275 and α-histidinyl442, there are many hydrogen-bonding interactions between the FeMo-cofactor and the surrounding amino acid residues. For example, the side chains of both α-glutaminyl191 and α-glutaminyl440 hydrogen bond to the homocitrate (one at each of homocitrate's terminal carboxyl groups), whereas the α-histidinyl195, α-argininyl96, and α-argininyl359 hydrogen bond to various sulfides of the cluster.

XII.3.5.4. The P Cluster Prosthetic Group

The P cluster also has a biologically unique structure. Both its location at the α/β-subunit interface and its formulation as a Fe–S cluster containing eight Fe atoms were confirmed by the original X-ray crystal structure of the MoFe protein. Each P cluster is ligated by six Cys residues, three from each subunit. The P cluster consists of a [4Fe–4S] subcluster that shares one of its sulfides with a [4Fe–3S] partial cube. This shared sulfide is hexacoordinated by the six central Fe atoms, a very unusual situation for sulfide. The [4Fe–4S] subcluster is terminally ligated by the γ-S of both α-cysteinyl62 and α-cysteinyl154, whereas the [4Fe–3S] partial cube has terminal ligation from the equivalent residues of the β-subunit, namely, the γ-S of both β-cysteinyl70 and β-cysteinyl153 as terminal ligands. In addition, two other cysteinyl residues, α-cysteinyl88 and β-cysteinyl95, form μ_2-sulfide bridges between the subclusters (Fig. XII.3.8).[33]

On oxidation by redox-active dyes, the P cluster structurally rearranges to a more open structure. Two of the four Fe atoms located in the [4Fe–3S] partial cube lose contact with the central hexacoordinated sulfide and undergo a change in ligation. One Fe atom becomes ligated by the γ-O of β-serinyl188 and the other Fe atom bonds to the deprotonated backbone amide-N of the already bound and bridging α-cysteinyl88. Because both of these latter ligands are protonated in the unbound state and deprotonated in the bound state, these redox-induced ligand changes raise the possibility that a two-electron oxidation of the P cluster during catalysis will also release two protons from the P cluster ligands.[33]

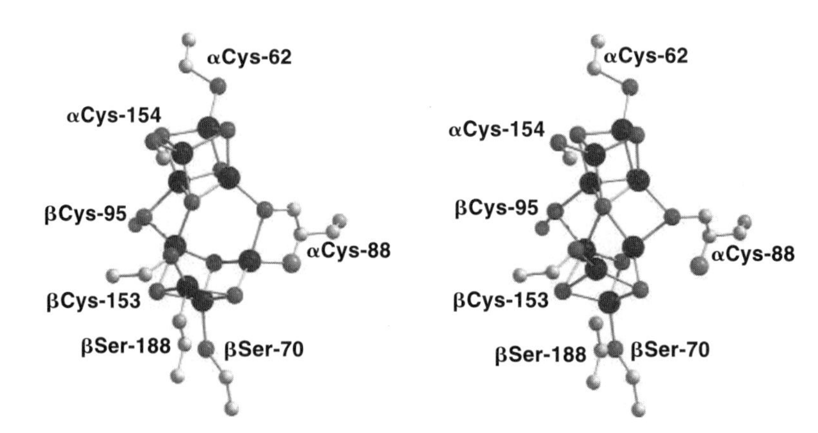

Fig. XII.3.8. The structure of the P cluster of *A. vinelandii* nitrogenase MoFe protein in its dithionite-reduced (P^N; at right) and oxidized (P^{ox}; at left) states with its ligating amino acid residues (α-Cys62, α-Cys88, α-Cys154, β-Cys70, β-Cys95, β-Cys153, and β-Ser188), which are provided by both the α- and β-subunits. (Ref. 33; PDB code: 3MIN and 2MIN).

The VFe protein has P clusters with similar properties to those found in the MoFe protein.[6] The presence of P clusters in the FeFe protein has not been demonstrated but, because of both the relatively high sequence identity and the similar genetic basis of its biosynthesis, their occurrence seems highly likely. The catalytic role assigned to the P cluster involves accepting electrons from the Fe protein for storage and future delivery to the substrate via the FeMo-cofactor centers, but this role has yet to be proved.

The crystal structures of three variant MoFe proteins have been solved. The first was of the MoFe protein that has its α-His195 residue replaced by glutamine (the αH195Q MoFe protein; PDB code: 1FP4). This variant was chosen because it has been used widely in reactivity studies. Its structure is virtually identical with that of the native MoFe protein with the single exception that the >NH → S hydrogen bond between a central μ_2-sulfide of the FeMo-cofactor and the ε-N of the imidazole side chain of α-His195 is replaced by a similar bond with the amide-N of glutamine.[34] The second structure solved was of the variant MoFe protein from a *nifV*$^-$ mutant that had, therefore, lost the ability to biosynthesize homocitrate (Ref. 35; PDB code: 1H1L). Again, the structure was essentially identical to that of the native protein, except that now citrate replaces homocitrate as a ligand to the Mo atom of the FeMo-cofactor. This structure proved that citrate is not simply present but actually replaces homocitrate in the FeMo-cofactor. Both structures confirm that the substitutions introduced, at least, in these variant MoFe proteins have only a local and not a global effect.

The third structure solved was of a "so-called" apo-MoFe protein, which is actually a FeMo-cofactor-deficient form of the MoFe protein (Ref. 36; PDB code: 1L5H). Not surprisingly, the three domains of the β-subunits are essentially unchanged, but so are two of the three domains of the α-subunit. However, the third α-subunit domain shows significant structural changes, which create a funnel that leads to the FeMo-cofactor-binding site within the α-subunit. This funnel is lined with positively charged residues from its entrance all the way down to α-His442, which likely serves both as the initial docking point for the negatively charged FeMo-cofactor and to trigger the conformational changes that close the funnel and bury the FeMo-cofactor within the α-subunit.

XII.3.5.5. The MoFe-Protein–Fe-Protein Complex

Because the primary structure of the Fe protein shows considerable similarities to those of "nucleotide switch" proteins, like ATPases and GTPases, similar trapping techniques, using AlF_4^- together with ADP, have been applied to nitrogenase. The result is a stable nitrogenase complex comprised of two Fe-protein molecules with one MoFe-protein molecule. The 3-Å-resolution crystal structure of this complex (Fig. XII.3.9), which may approximate the transition state for intercomponent electron transfer, confirmed predictions of earlier modeling studies. The Fe protein had indeed undergone a significant conformational change, whereas the MoFe protein was little changed from its uncomplexed structure.[26]

The conformational change of the Fe protein results from a 13° rotation of both subunits, resulting in a more compact structure. Interestingly, small-angle X-ray scattering data indicate that the complex maintains this structure in solution. Docking of the two proteins occurs along the Fe protein's twofold symmetry axis, which bisects its single [4Fe–4S] cluster, and the pseudosymmetric αβ-interface of the MoFe protein. The more compact conformation of the Fe protein allows its [4Fe–4S] cluster to become buried in the protein–protein interface and to approach to within ~14 Å of the P cluster of the MoFe protein. This approach is ~4 Å closer than predicted from computer modeling studies of the complex and results in the P cluster being situated equidistant between the Fe-protein's [4Fe–4S] cluster and the FeMo-cofactor. This arrangement suggests that electrons are transferred from the [4Fe–4S] cluster through the P cluster to the FeMo-cofactor, where substrate reduction occurs.

About 20 intimate contacts occur between the [4Fe–4S] cluster and adjacent amino acid residues of the Fe protein and stretches of both the α- and β-main chains of the MoFe protein. A significant interaction involves the Arg100 residues of the Fe protein. In some bacteria, the Arg100 residue is modified in a facile, reversible manner to regulate nitrogenase activity. The structural data shows that each Arg100 is in the first turn of a pair of helices that extend symmetrically on both sides of the [4Fe–4S] cluster. The side chain of each Arg100 residue protrudes into a small depression in the surface of the MoFe protein. Here, the side chain interacts through multiple hydrogen bonds, van der Waals contacts, and ionic bonds with the side chains of Glu residues from both subunits of the MoFe protein (α-Glu120, α-Glu184, and β-Glu120). Obviously, modification of these Arg100 residues would introduce steric hindrance to the protein–protein interface, prevent complex formation, and so suppress activity and regulate the enzyme.

Each of the four Fe-protein subunits in the complex has an associated Mg^{2+}–ADP–AlF_4^- moiety, which is bound parallel to the interface between the subunits. Although different from the orientation of the bound MgADP in the original structure of the isolated Fe protein, this orientation is compatible with that in the more recent crystal structures of the Fe protein (see Section XII.3.5.1).

Two other structures of the 2:1 Fe-protein–MoFe-protein complex have been solved. The first structure involved a variant (L127Δ) Fe protein, which has Leu127 deleted from the switch-II region (see Section XII.3.5.1). The effect of this deletion is to mimic the binding of nucleotide and it causes the Fe protein to be permanently in a state equivalent to an MgATP-bound state. As such, the L127Δ Fe protein forms a

Fig. XII.3.9.
The structure of the 2:1 Fe protein–MoFe protein complex of the *A. vinelandii* nitrogenase stabilized by MgADP plus AlF_4^-. The Fe-protein molecules and the α- and β-subunits of the MoFe protein are individually shaded. Each Fe-protein component docks over a MoFe-protein α/β-subunit interface and is juxtaposed with a P cluster (Ref. 26; PDB code: 1N2C). [See color insert.]

tightly bound inactive 2:1 complex with the MoFe protein (Ref. 37; PDB code: 1G20 and PDB code: 1G21 for the structures with or without MgATP present). This structure closely resembles that of the AlF_4^--stabilized complex as far as the MoFe-protein component and the protein–protein interfaces are concerned. The one significant difference is in the more open conformation adopted by the complexed L127Δ Fe protein, which more closely resembles the conformation of the uncomplexed nucleotide-free, rather than the complexed nucleotide-bound, native Fe protein. The native Fe protein in the second 2:1 complex structure, which resulted from chemical cross-linking the two native component proteins through Glu112 and β-Lys400, adopted a structure that is even more open than when it is uncomplexed (Ref. 38; PDB code: 1M1Y). In contrast, the structure of the MoFe-protein component is again effectively unchanged. What is changed is the relative orientation of the two components and the presence of a completely different interface area. It is possible that the cross-linked complex represents an "initial encounter" state that then proceeds through a series of conformational changes on the Fe protein to reorient the Fe protein in the "electron-transfer competent" state, which is likely represented by the AlF_4^--stabilized complex.

XII.3.6. Mechanism of Nitrogenase Action

During catalysis *in vitro* using sodium dithionite as the reductant, the Fe protein delivers electrons, one-at-a-time, to the MoFe protein in a process that couples MgATP binding and hydrolysis to the association and dissociation of the two component proteins and concomitant electron transfer. Although the Fe protein alone is capable of binding MgATP, both component proteins are required for MgATP hydrolysis and neither component protein alone, with or without MgATP and/or reductant, will reduce substrate.

XII.3.6.1. The Lowe–Thorneley Model

A computational model has been developed to describe the process by which electrons are sequentially delivered to the MoFe protein and then to substrate.[39] This model treats the MoFe protein as a dimer of dimers with each αβ-dimer operating independently and being serviced by Fe protein. It involves two interconnecting processes, which are called the Fe-protein cycle (Fig. XII.3.10*a*) and the MoFe-protein cycle (Fig. XII.3.10*b*). The Fe-protein cycle describes the series of reactions that allows the Fe protein's [4Fe–4S] cluster to cycle between its reduced 1+ and its oxidized 2+ redox states. The reduced Fe protein, with MgATP bound, associates with the MoFe protein and delivers one electron coupled with MgATP hydrolysis. The complex then dissociates. The liberated oxidized Fe protein exchanges its bound MgADP for MgATP and is re-reduced by dithionite.

The MoFe-protein cycle is necessarily more complex because it involves the progressive reduction of the MoFe protein by up to eight electrons for N_2 binding and reduction, which therefore requires eight turns of the Fe-protein cycle. This model indicates that N_2 is bound to the active site only after three electrons have been accumulated within the MoFe protein. How and where these electrons are stored prior to the binding and reduction of substrate is unknown.

XII.3.6.2. The Role(s) of MgATP in Catalysis

The overall reduction of N_2 to yield two molecules of NH_3 is thermodynamically favorable. So, if MgATP binding and hydrolysis during nitrogenase catalysis is not a thermodynamic requirement, it must be used for kinetic purposes. Most likely,

(a)

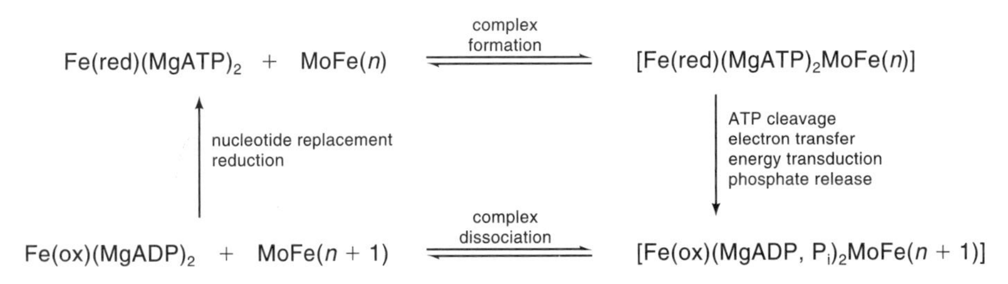

(b)

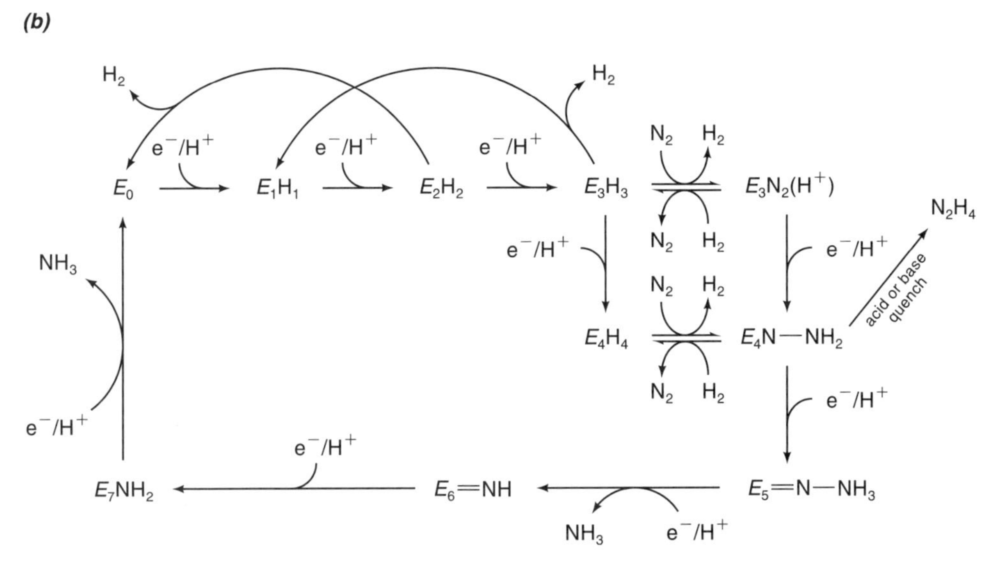

Fig. XII.3.10.
A modified Lowe–Thorneley kinetic scheme for nitrogenase catalysis. (*a*) The Fe-protein cycle involves both the nucleotide-bound and MoFe-protein complexed states of the Fe protein. This cycle describes the kinetics of one-electron oxidation–reduction of the [4Fe–4S] cluster, nucleotide exchange of the spent MgADP and phosphate (P_i) for MgATP, and complex formation with, and electron transfer to, the MoFe protein. Fe represents the Fe protein in its oxidized (ox) or one-electron reduced (red) states; MoFe represents the MoFe protein with the number of electrons accepted shown as (*n*).[39] (*b*) The MoFe-protein cycle for N_2 reduction with the dithionite-reduced resting state of the MoFe protein designated as E_0. Because each turn of the Fe-protein cycle transfers only one electron, the MoFe protein is treated as going through a succession of increasingly reduced states ($E_1, E_2, \ldots, E_7$) as electrons are transferred from the Fe protein until sufficient electrons (and accompanying protons) are accumulated to effect substrate reduction. An important concept of this scheme is that different substrates are regarded as reversibly binding to different MoFe-protein redox states, for example, N_2 binds at either E_3 or E_4, whereas C_2H_2 binds at either E_1 or E_2. Dihydrogen may be evolved from several redox states. The reduced-nitrogen intermediates shown are postulated and not proved. No intermediates are released during N_2 reduction, although hydrazine (N_2H_4) is released if the turning-over enzyme is quenched with either acid or base.[39]

MgATP hydrolysis serves to drive electron transfer toward substrate reduction and to ensure the irreversibility of the reaction by preventing the back flow of electrons to the Fe protein. This situation may be envisioned as a "gating" mechanism.[13,16]

When MgATP binds to the Fe protein, the resulting conformational change allows the Fe protein to complex with the MoFe protein. A further substantial decrease in the reduction potential of the Fe protein's [4Fe–4S] cluster occurs on complex formation. The additional lowering is ca. −200 mV to a final value of ca. −600 mV and might be due to limiting solvent (water) accessibility to the cluster. Similarly, on complex formation, the redox potential of the P cluster is also decreased, but by only ca. −100 to ca. −400 mV. So, complexation of the MgATP-bound Fe protein with the MoFe protein increases the driving force (by increasing the difference in redox potential of the donor and acceptor) for electron transfer from the Fe protein's [4Fe–4S] cluster to the P cluster of the MoFe protein. Complex formation, however, has no effect on the redox potential of the FeMo cofactor.[40]

In addition to promoting electron transfer, complex formation triggers MgATP hydrolysis, although it is not clear whether hydrolysis occurs shortly before, concomitantly with, or shortly after electron transfer or whether the timing varies depending on other factors. It is clear, however, that phosphate release (which is usually when energy transduction occurs in other systems) from the complex follows electron transfer and that phosphate release does *not* drive the dissociation of this complex into its component proteins.[41] Rather, the resulting formation of the MgADP-bound state of the Fe protein causes the conformational change to relax and the complex dissociates, thus preventing any flow of electrons back onto the Fe protein. This last process (complex dissociation) is the rate-limiting step in nitrogenase catalysis.[39] In this way, multiple electrons may be accumulated within the MoFe protein prior to their delivery to substrate. In support of this concept, primary amino acid sequence and structural comparisons show that the Fe protein is a member of a large class of signal-transduction proteins that undergo conformational changes upon MgATP binding and hydrolysis.

XII.3.6.3. Where and How do Substrates and Inhibitors Bind?

Because the substrates and inhibitors of nitrogenase are invariably small multiple-bonded molecules, like N_2 and CO, it has always been assumed that binding and any subsequent reduction occur at one or more metal atoms. However, direct evidence to support this assumption has been hard to come by. The major problem is that neither component protein alone binds either substrates or inhibitors. Both of the component proteins, plus reductant and MgATP, must be present for substrate binding, but then enzymatic turnover occurs immediately! If no added substrate is present, for example, even under an Ar atmosphere, the enzyme turns over by reducing protons to H_2 gas.

Circumstantial evidence for a role for the FeMo-cofactor in substrate binding comes from constructing mutant organisms and studying the effects of the mutation on the resulting variant nitrogenases. First, these studies show that the FeMo-cofactor is biosynthesized separately from the rest of the MoFe protein and is then inserted as a late step in the maturation of the MoFe protein. Moreover, mutant strains, which are unable to biosynthesize FeMo-cofactor, are also unable to catalyze nitrogen fixation. However, when isolated FeMo-cofactor is added to crude extracts prepared from such mutant strains, the ability to fix N_2 is restored. Second, when citrate replaces homocitrate as the organic constituent of an altered FeMo-cofactor in a variant MoFe protein that is produced by a different mutant strain, it changes the reactivity of that MoFe protein toward various substrates. Third, amino acid substitutions within the FeMo-cofactor's polypeptide environment produce variant MoFe proteins that have altered FeMo-cofactor-based spectroscopic properties and different catalytic activities.[17] Although definitive details of the interaction of substrates with the FeMo-cofactor are currently lacking, both the Mo atom and the centrally located Fe atoms are potential sites for substrate binding.

Theoretical calculations, which are compromised both by the physical size and chemical nature of the FeMo-cofactor and by the limited knowledge of its total spin state and the oxidation states of its constituent metal atoms during catalysis, favored

N_2 binding by bridging either two or four of the central Fe atoms early on.[42,43] Other later calculations have shown that, after dissociation of the carboxylate arm of homocitrate, the Mo atom could be the preferred N_2-binding site.[44] All of the proposed metal–N_2 interactions result in a small N_2-binding energy and, therefore, a weak bond. Further, direct N_2 cleavage to metal–nitrides is unfavored, suggesting that protonation of bound N_2 must occur. But all of these calculations were performed before the discovery of the central light atom within the FeMo-cofactor. Very recently,[45,46] this central light atom (assumed to be nitrogen) has been included in the calculations. The major consequence is that all of the various combinations of either two or four of the six central Fe atoms are less able to accommodate bridging Fe–N_2–Fe interactions, so much so that N_2 binding at a Fe_4 face was ruled out in favor of either an Fe_2 edge or a single Fe atom.

Direct spectroscopic evidence indicates that substrates and inhibitors only bind to nitrogenase under turnover conditions. Carbon monoxide is a potent inhibitor of the reduction of all substrates, except the proton, and so it likely remains bound to its site while nitrogenase is producing H_2 from protons. When wild-type nitrogenase begins to turn over under CO, the $S = \frac{3}{2}$ EPR signal arising from the FeMo-cofactor within the MoFe protein disappears and either of two new $S = \frac{1}{2}$ EPR signals appear. One signal appears when lower (<0.1 atm) CO concentrations are used and the second at higher (>0.5 atm) CO concentrations. The technique of ^{13}C and ^{57}Fe electron nuclear double resonance (ENDOR) was then applied to these signals, using ^{13}CO and ^{57}FeMo-cofactor. The results indicate that, at the lower CO concentration, only one CO is bound and likely forms a bridge between two Fe atoms of the FeMo-cofactor. At higher CO concentrations, two CO molecules are present, each terminally bound most likely to a different Fe atom of FeMo-cofactor. The two $S = \frac{1}{2}$ EPR signals interconvert when CO is added or removed, suggesting that the single bridging CO found under lower CO concentrations converts to a terminal CO under higher concentrations.[47] These results are entirely consistent with a stopped-flow FTIR study of wild-type nitrogenase turning over under CO.[48] Here, two CO vibrations, which were assigned to terminally bound CO molecules, were found under high CO concentration, whereas only a single lower frequency band was observed in the CO vibrational region under low CO conditions. Similar EPR–ENDOR studies, under turnover conditions, have shown that CS_2 interacts with wild-type nitrogenase, probably at the FeMo-cofactor, and that C_2HR (where R = H or CH_2OH) and CN^- can each interact with the FeMo-cofactor of variant MoFe proteins.[17]

XII.3.6.4. How Are Electrons and Protons Delivered?

The current dogma is that electrons are delivered by the Fe protein to the P cluster, then through the protein matrix to the FeMo-cofactor, and finally to bound substrate. Again, there is little direct evidence to support this suggested electron-transfer pathway. Two early spectroscopic observations clouded this proposed role for the P clusters. The first observation was that electrons transferred to the MoFe protein from the Fe protein are quickly relocated to the FeMo-cofactor as shown by the rapid loss of intensity of the $S = \frac{3}{2}$ EPR signal. The second was that no change occurred in the ^{57}Fe Mössbauer spectrum of the P clusters during nitrogenase turnover.[11,15] Neither of these observations demonstrates an electron-transfer role for the P clusters.

In contrast, the X-ray crystal structure of the nitrogenase complex is certainly indicative of such a role. It shows the P cluster of the MoFe protein located midway between the [4Fe–4S] cluster of the Fe protein, the proximal electron donor, and the FeMo-cofactor, the intended destination of the electrons. Although this arrangement could be coincidental, the redox-driven structural rearrangement of the P cluster supports an electron-transfer role (see Section XII.3.5.4). When the two Fe atoms undergo the ligation change on oxidation, they cause two unbound protonated ligands to become bound and deprotonated.[33] These ligand changes raise the possibility that a two-electron oxidation of the P cluster will also release two protons and may be the mechanism by which electron and proton transfer are coupled for delivery to the FeMo-cofactor for substrate reduction.

Other support comes from substituting the β-Cys153 ligand of the P cluster with serine (see Fig. XII.3.8). The resulting β-Ser153 MoFe protein has a normal FeMo-cofactor, gives normal substrate-reduction products, and interacts normally with the Fe protein. It cannot, however, match the wild-type's maximum rate of substrate reduction and so it is likely that intra-MoFe-protein electron transfer has been compromised. In addition, substitution of either α-Cys88 by Gly or α-Ser188 by Cys results in an additional $S = \frac{1}{2}$ signal in the EPR spectrum of the isolated variant MoFe proteins. This signal disappears when Fe protein, MgATP, and reductant are added, indicating that the P cluster oxidation state is changing during turnover.

So, if the P cluster is involved in intra-MoFe-protein electron transfer, how do the electrons traverse the protein matrix between the P cluster and the FeMo-cofactor? It is unclear whether there is only one or several electron-transfer pathways and also whether all substrates are serviced by the same pathway. A potential pathway involves [P cluster Fe]–α-Cys62–α-Gly61···α-Gln191···homocitrate–[Mo of FeMo-cofactor] (where···represents a hydrogen bond) and a considerable body of evidence indicates that disruption of the α-Gln191···homocitrate system decreases substrate-reduction activity. Other possibilities are the four helices (from α63-α74, α88-α92, α191-α209, and β93-β106), which are oriented parallel to each other between the P cluster and the FeMo-cofactor. The result of mutagenesis of the β-Tyr98 residue suggests that this last helix may mediate electron transfer through the protein matrix between the P cluster and the FeMo-cofactor.[11,14,15]

Finally, we have the question of how the protons are delivered to complete the conversion of substrate to product. In the absence of definitive evidence, qualitative molecular modeling has identified three likely proton-transfer routes.[49] An interstitial channel, filled with water molecules, runs between the α- and β-subunits from the surface of the MoFe protein to the pool of water molecules around the homocitrate of the FeMo-cofactor. These water molecules could act as a "proton wire" and deliver protons rapidly to bound substrate. This channel might also provide a pathway for N_2 to diffuse into and/or for NH_4^+ to diffuse away from the reduction site.

In addition, two possible proton-relay systems have been identified. The first involves the hydrogen bond by which the ε-N of the imidazole side chain of α-His195 interacts with a central μ_2-sulfide of the FeMo-cofactor. This strictly conserved residue is known to play an essential role in N_2 reduction. The δ-N of this same imidazole ring forms a hydrogen bond through an intervening water molecule to the OH group on α-Tyr281, which is close to the protein's surface and flanked by two potential proton-capturing histidines, α-His196 and α-His383. This system could provide protons as demanded by the redox state of the protein–substrate complex. The second potential proton relay is more complicated and involves a hydrogen-bonded series of three water molecules and three His residues, one of which (α-His362) is on the surface and may capture protons for this relay. This path terminates in a different central bridging sulfide of FeMo-cofactor. These proton relay systems may play a role in substrate reduction, but they probably cannot deliver the multiple protons required to complete substrate reduction. Proton delivery would require the relay to switch to an alternative hydrogen-bonding network as each proton is delivered and then realign itself to deliver the next proton. This operation would be much more difficult for the His residues than for the water molecules in the interstitial channel, suggesting that the interstitial channel may be the primary route of protons to be delivered to substrates.

XII.3.7. Future Perspectives for Nitrogen Fixation

Although tremendous progress has been made, many mysteries remain concerning biological nitrogen fixation, especially regarding the intimate details of substrate binding and reduction. The requirements for N_2 binding and reduction are likely to be so stringent that they are satisfied by only a single site on the MoFe protein. However, it is becoming clear that many of the alternative more easily reduced substrates may have more than one binding and reduction site. Sorting out these different sites and their possible relevance to the overall chemistry conducted by the enzyme should prove useful. Altered nitrogenases are now known that will bind but not reduce N_2. In these variants, N_2 now acts as a reversible inhibitor of electron flow through the enzyme resulting in a decreased H_2-evolution rate. What if we constructed an altered nitrogenase where N_2 binds as an irreversible inhibitor? Maybe then, we would be able to determine exactly where and how N_2 was bound to the FeMo-cofactor!

The second major challenge is to find out why the P clusters are present. Are they really involved in electron transfer or do they store electrons for later delivery to the FeMo-cofactor and substrate? And finally, will it be possible to apply what we learn about nitrogenase to enhance the nitrogen-fixation capabilities of microorganisms or to endow new organisms with this ability or to develop new commercial nitrogen-fertilizer production systems? These questions await answers that can only come from insightful research progress.

XII.3.8. Biological Nitrification: What Is Nitrification?

Nitrification is a series of oxidation reactions that microorganisms use to convert ammonia (NH_3) into nitrite (NO_2^-) or nitrate (NO_3^-). In effect, it reverses all the hard work done to reduce N_2 to NH_3 but, to do so, it uses a completely different pathway. The nitrogen-fixation reaction converts N in its 0 oxidation state (as N_2) to N in its -3 oxidation state (as NH_3). In so doing, the nitrogenase enzyme has to deal with an accumulation of as many as eight electrons to reduce each N_2 molecule. The processes that constitute nitrification cover an even wider range of oxidation states by converting NH_3 into NO_3^-, which entails another eight-electron change from the -3 to the $+5$ oxidation state at a *single* N atom. Again, the involved enzymes must deal with multiple electrons. In both cases, multiple prosthetic groups are present to do the job, but in contrast to nitrogen fixation, which uses a single enzyme system, nitrification uses three distinct enzymes.

Nitrification occurs both in autotrophic soil bacteria, which are self-sufficient and grow on CO_2 as their sole carbon source, and in heterotrophic soil microorganisms (bacteria, actinomycetes, and fungi), which require complex carbon-containing compounds for growth.[18,19,50] However, these two types of microorganisms use quite different enzymes to carry out nitrification. Further, with the autotrophic nitrifiers, nitrification produces all the energy they require for growth, most of which is used to fix CO_2 by the Calvin cycle. In contrast, the heterotrophic nitrifiers, which use complex sources of fixed nitrogen (usually amino-N), accrue no apparent energy gain from nitrification but do use the carbon to support growth.

XII.3.9. Enzymes Involved in Nitrification by Autotrophic Organisms

Nitrification takes place in three major steps. However, no bacterium is known that can catalyze all three steps. The NH_3-oxidizing bacteria, all of which have a genus

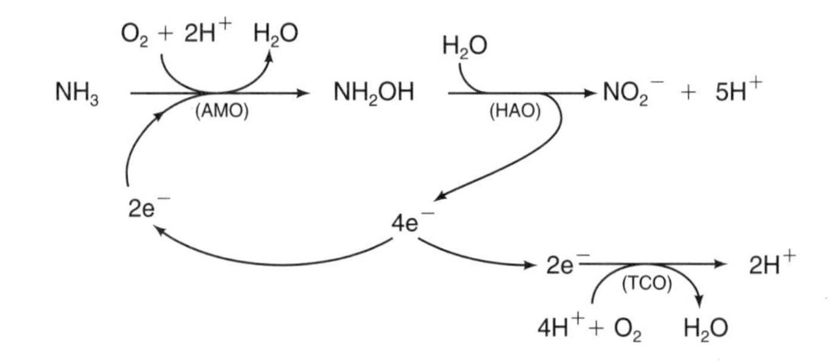

Fig. XII.3.11.
A general scheme for the oxidation of ammonia to nitrite by the enzymes, AMO and HAO, as it occurs in autotrophic organisms, showing the partitioning of the resulting electrons between the terminal cytochrome oxidase (TCO; for energy generation) and further ammonia oxidation.

name starting with *Nitroso* as in *Nitrosomonas,* have enzymes, namely, ammonia monoxygenase (AMO) and hydroxylamine oxidoreductase (HAO), which catalyze steps one and two (Eqs. 2 and 3), respectively, of the process. These steps provide four electrons for each hydroxylamine oxidized and they are the only source of reducing power for energy generation and biosynthesis available to these organisms.

$$NH_3 + O_2 + 2\,H^+ + 2\,e^- \rightarrow NH_2OH + H_2O \quad (2)$$

$$NH_2OH + H_2O \rightarrow NO_2^- + 5\,H^+ + 4\,e^- \quad (3)$$

Because AMO is a monooxygenase and so inserts only one atom of O_2 into the substrate (ammonia, but not ammonium), it requires a source of reducing power to reduce the other oxygen atom to water. The only source available to these organisms is from the subsequent oxidation of the product of the AMO-catalyzed reaction, hydroxylamine. In a remarkable series of reactions, two of the four electrons from hydroxylamine oxidation by HAO are "recycled" to AMO in order for the next NH_3 substrate molecule to be reduced. How just two of the four electrons are delivered for this function is still unclear (but see below). The other two electrons are transported to the terminal (Cu_2 cytochrome-aa_3) oxidase, where O_2 is reduced to H_2O and protons are pumped out of the cytoplasm of the cell. The resulting proton gradient serves to generate ATP through ATP synthase.[51] The interrelationship of these two processes is outlined in Fig. XII.3.11.

The third step (Eq. 4) is performed by NO_2^--oxidizing bacteria, all of which have genus names starting with *Nitro* as in *Nitrobacter*. These organisms have a nitrite oxidoreductase (NIO) that catalyzes reaction 4.

$$NO_2^- + H_2O \rightarrow NO_3^- + 2\,H^+ + 2\,e^- \quad (4)$$

XII.3.9.1. Ammonia Monooxygenase

The AMO enzyme is membrane-bound and catalyzes the production of hydroxylamine. It has not yet been purified in an active form from any autotrophic organism, so most of our functional information comes from studies with intact cells. In addition to ammonia, AMO catalyzes the oxidation of a broad range of nonpolar substrates comparable to those of cytochrome P450 and soluble methane monooxygenase (sMMO). However, none of these other substrates provide energy for cell growth, although the oxidation of methane (and maybe CH_3OH and CO) may provide a source of either carbon for incorporation into the cellular material or CO_2 for carbon fixation. The nonpolar nature of its substrates suggests that the active site of AMO is hydrophobic in character.[20]

The AMO protein is composed of at least two subunits, AmoA and AmoB. The genes (*amoA* and *amoB*) encoding these subunits have been sequenced[50] and the inferred amino acid sequence shows that both subunits have membrane-spanning segments and large loops that could extend into the periplasm of the cells. The

AMO protein resembles a particulate methane monooxygenase (pMMO, which is different from sMMO) compositionally (see Section XI.5), functionally, and evolutionarily. Both AMO and pMMO may contain as many as three Cu ions plus one (or more) Fe ions. Together, they constitute a new class of Cu-containing monooxygenases.[18,50] By analogy with other monooxygenases, AMO catalysis may involve initial O_2 activation to form an M=O species of similar structure and reactivity to the ferryl species of cytochrome P450. Then, abstraction of a hydrogen atom from substrate to form hydroxide, followed by recombination, would produce NH_2OH.[20]

XII.3.9.2. Hydroxylamine Oxidoreductase

In contrast to AMO, HAO is not membrane bound, but is water soluble and found in the periplasmic space between the inner membrane (within which AMO is bound) and the outer membrane of cells. Hydroxylamine is released from AMO into the periplasm, where it is reduced by HAO in two two-electron steps as shown below with the likely bound intermediate(s) in parentheses (Eqs. 5 and 6). If the O_2 concentration falls and slows the rate of electron transfer away from the active site of HAO, significant amounts of N_2O are produced by HAO action and a lower yield of electrons results (Eq. 7). Hydrazine is also a substrate of HAO with N_2 being the likely product (Eq. 8).

$$NH_2OH \rightarrow NOH\ (\text{or } NO^-) + (2 \text{ or } 3)\ H^+ + 2\ e^- \tag{5}$$

$$NOH\ (\text{or } NO^-) + H_2O \rightarrow NO_2^- + (3 \text{ or } 2)\ H^+ + 2\ e^- \tag{6}$$

$$2\ NH_2OH \rightarrow N_2O + H_2O + 4\ H^+ + 4\ e^- \tag{7}$$

$$N_2H_4 \rightarrow N_2 + 4\ H^+ + 4\ e^- \tag{8}$$

The *hao* gene has been sequenced[52] and functional insights into how HAO accomplishes these reactions have come from the crystal structure (Fig. XII.3.12) solved by X-ray techniques.[53] Purified HAO is a trimeric molecule with each of its subunits having eight covalently attached *c*-type hemes, including one called heme P460 that is uniquely five coordinated to allow for hydroxylamine binding. A very unusual feature of the structure is the cross-linkages between the three subunits. Each P460 heme is bound to the carbon of a Tyr residue from a second subunit. Whether these intersubunit interactions have a mechanistic function in addition to the likely stabilization of the trimer is unknown.

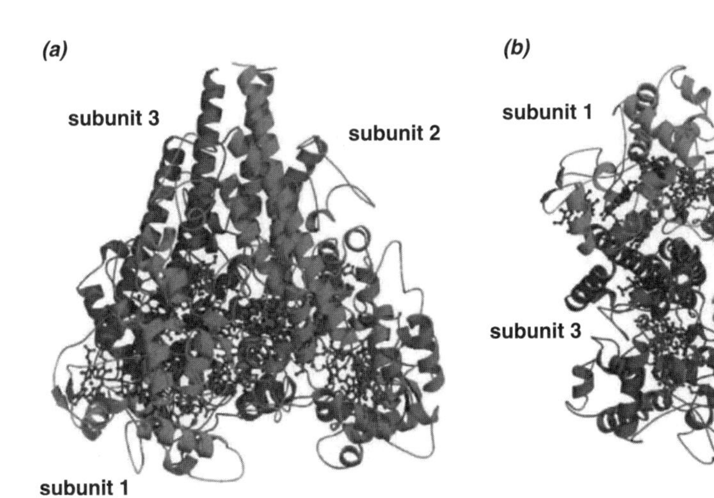

Fig. XII.3.12. The structure of HAO from *Nitrosomonas europaea.* (*a*) The complete trimeric molecule showing the eight heme groups per subunit with each subunit individually shaded. (*b*) A view at approximately right angles to the view in (*a*) along the molecular threefold axis (Ref. 53; PDB code: 1FGJ). [See color insert.]

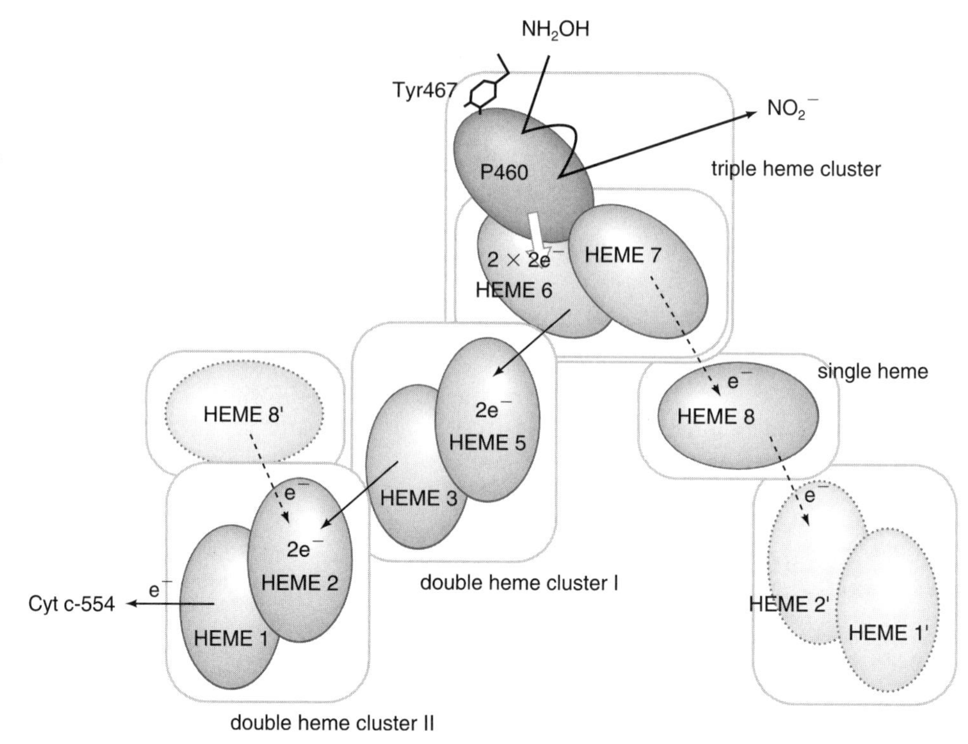

Fig. XII.3.13.
A model of the possible electron-transfer pathways within HAO capable of moving two electrons at a time. The dark gray ellipsoid is the single P460 heme; the light gray ellipsoids are the *c*-type hemes in the same subunit; the dotted ellipsoids are *c*-type hemes in an adjacent subunit. The likely major flow of electron pairs is through the two double-heme clusters to be accepted by cytochrome *c*-554 (Cyt *c*-554),[53] but see text and Fig. XII.3.14.

The eight hemes in each subunit are distributed in a way that suggests possible electron-transfer routes out of HAO. Three of the hemes, the P460 hydroxylamine-binding heme plus hemes No. 6 and No. 7, are clustered together. Hemes P460 and No. 6, with each accepting one electron, could accommodate the first two electrons released from bound substrate. Furthermore, these two hemes are aligned with two double-heme clusters (hemes No. 3 and No. 5 make up one pair and hemes No. 2 and No. 1 the second pair) as shown in Fig. XII.3.13. Because each heme usually accommodates only one electron at a time, this arrangement suggests a likely pathway to move both electrons quickly out of the active site of the molecule. The heme No. 1 of the second pair appears positioned within the structure to reduce the soluble electron acceptor, Cyt *c*-554. A similar sequence could also remove the second pair of electrons. Cytochrome *c*-554 would then be involved in electron donation to AMO.

If, however, HAO has the additional responsibility of partitioning the electrons between reducing ammonia at AMO and generating energy at the terminal cytochrome oxidase, another electron-transfer pathway might be available through which electrons are donated to the terminal cytochrome oxidase. Two other potential electron-transfer pathways exist; one possibility is through the bound Tyr residue and another is through heme No. 7.[18,21,53] If hemes No. 6 and No. 7 are the initial electron acceptors from heme P460, then only one electron from each two-electron step would be transported through the double-heme systems as described above. The other electron could travel from heme No. 7 through heme No. 8 to reduce an alternative electron acceptor, possibly the abundant Cyt *c*-552. These two pathways would allow HAO to control and maintain the required electron partitioning.[21,53] However, because heme No. 8 lies close to the heme No. 1-heme No. 2 pair in an adjacent subunit, another pathway could involve intersubunit electron transfer, but then it would terminate at the same acceptor as for electrons moving through heme No. 6.

If all four electrons travel through the double-heme system and are accepted by Cyt *c*-554 molecules, another possible partitioning point is at the inner membrane

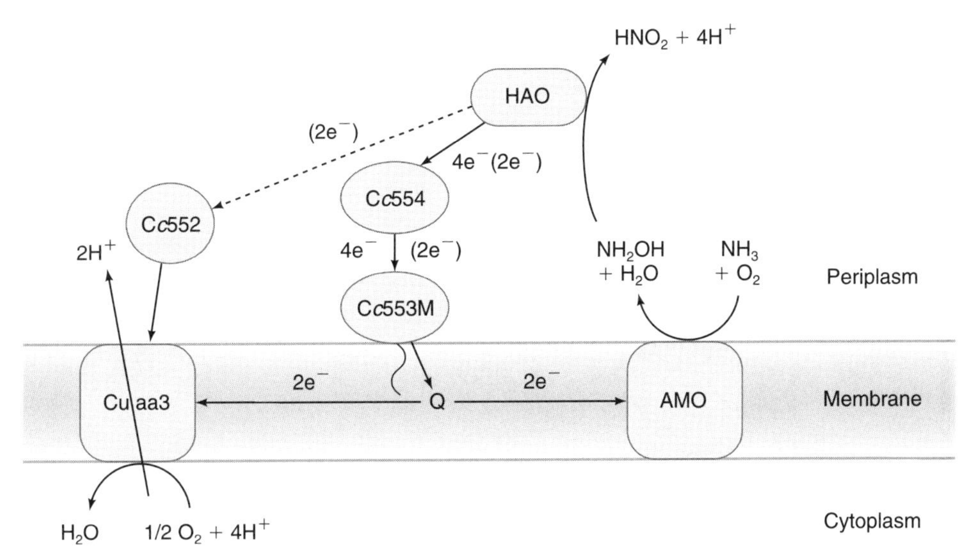

Fig. XII.3.14. Enzymes and possible electron-transfer pathways in the first two steps in autotrophic nitrification. The solid arrows represent the transfer of all four electrons (shown without parentheses) from hydroxylamine oxidation to ubiquinol, which could then partition them among energy generation and further ammonia oxidation. The dotted arrows and electron pairs (in parentheses) represent an alternative electron-transfer pathway if HAO itself is responsible for electron partitioning to these two processes. Terminal cytochrome oxidase, Cu aa3; ubiquinol, Q; various *c*-type cytochrome electron carriers, CcXXX; membrane-bound cytochrome c_M-553, Cc553M.

(Fig. XII.3.14). Here, Cyt *c*-554 is thought to discharge its electron to the membrane-bound cytochrome c_M-553.[20,51] This suggestion is supported by the genetic organization of *Nitrosomonas*. The *cycB* gene, which encodes cytochrome c_M-553, is located immediately downstream from (and is cotranscribed with) *cycA,* which encodes Cyt *c*-554. Both genes are components of a cluster that also includes *hao,* which encodes the HAO enzyme. Moreover, the inferred amino acid sequence of Cyt c_M-553 indicates that it is a member of a family of membrane-anchored cytochromes that normally function as electron-transfer agents between membrane quinols and periplasmic electron acceptors. Therefore, Cyt c_M-553 may function similarly, but in the reverse direction, and transfer electrons from the soluble reduced Cyt *c*-554 to ubiquinol in the membrane. Ubiquinol may then partition the electrons by some unknown mechanism between AMO and the respiratory electron-transfer chain.

XII.3.9.3. Nitrite Oxidoreductase

The final step in nitrification, the oxidation of nitrite to nitrate (Eq. 4), is the major energy-generating process for *Nitrobacter* and related NO_2^--oxidizing bacteria. This step is catalyzed by NIO in a reaction that incorporates an oxygen atom from H_2O into product. The released electron pair is then shuttled to a Cu-containing Cyt *c* oxidase, where O_2 reacts as the terminal electron acceptor to produce H_2O. The NIO is membrane bound and has not yet been crystallized. Its inferred amino acid sequence[54] suggests, however, that its composition and structure resemble that of the membrane-bound type of nitrate reductase, which catalyzes the exact reverse reaction and which is known as Nar to differentiate it from the soluble periplasmic type called Nap. Like Nar, NIO contains a Mo-containing subunit, a [Fe–S] cluster-containing subunit, and associated heme *a* and heme *c*.[55]

The NIO probably functions like Nar, but in reverse, that is, instead of reducing NO_3^- to NO_2^-, it oxidizes NO_2^- to NO_3^- (Fig. XII.3.15). If so, then NIO will be located on the cytoplasmic side of the membrane. Moreover, NO_2^- will bind and be oxidized at the Mo center, which is likely ligated by a pterin–dithiolene system (see Section XII.6). The electrons released will then pass through the [Fe–S] cluster in the second subunit to a third membrane-spanning subunit, which contains the hemes *a* and *c*. A periplasmic Cyt *c*-550 then shuttles the electrons to the terminal Cu-heme a_1 oxidase for energy generation.[55]

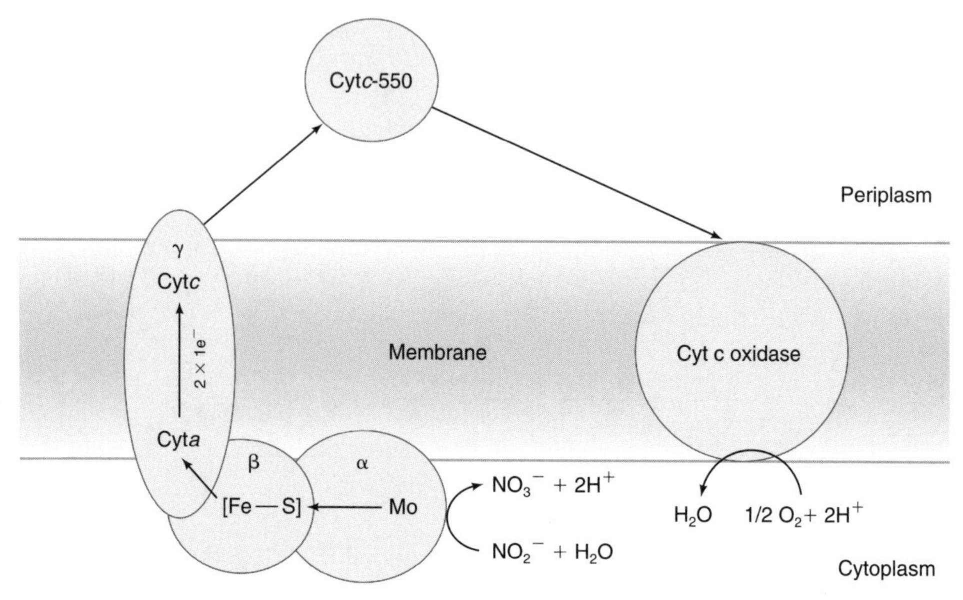

Fig. XII.3.15.
A possible electron-transfer pathway for the oxidation of nitrite by a NIO bound on the cytoplasmic side of the cell inner membrane of autotrophic organisms. The α, β, and γ proteins are the likely components of NIO; Cyt*c*550 is the soluble Cyt *c*-550 known to be reduced by NIO; Cyt *c* oxidase is the Cu- and heme *a* containing terminal oxidase.

XII.3.10. Nitrification by Heterotrophic Organisms

Other organisms, including fungi and actinomycetes as well as bacteria—none of which are dependent on nitrification for growth and energy, also have enzymes capable of catalyzing the individual reactions that comprise nitrification. These organisms more often use organic N sources, like amine-N, rather than inorganic NH_3.

In most cases, the nitrifying enzymes from heterotrophs, where they have been studied, are quite different from those from autotrophs. However, in one case, that of AMO isolated and purified from membranes in *Thiosphaera pantotropha* (also known as *Paracoccus denitrificans*), the enzyme appears quite similar to the AMO from the autotrophic *N. europaea*. Both have subunits of ~46- and 38-kDa molecular mass (although *T. pantotropha* may have a third subunit), both are dependent on Cu ions for activity, and both oxidize quinols to drive the conversion of NH_3 to NH_2OH.[19] Unfortunately, no crystal structure is available from this source either.

In contrast, the HAO from *T. pantotropha* (and several other heterotrophs) has no similarity to HAO from *N. europaea*. Not only is it a small (~18 kDa) protein, it contains no hemes (although nonheme iron is present) and only abstracts two electrons from each hydroxylamine.[19] To complete the formation of NO_2^-, the bound nitroxyl (NOH) intermediate reacts with O_2 (reactions 9 and 10) rather than with H_2O as occurs with HAO from autotrophs (reactions 5 and 6). The HAO from heterotrophs also oxidizes hydrazine to N_2, but it requires O_2 to do so (see Eq. 11 and compare it with Eq. 8). Under anaerobic conditions, this HAO can also oxidize hydroxylamine to produce nitrous oxide, not nitrite, as the product (see Eq. 7).

$$NH_2OH \rightarrow NOH + 2\,H^+ + 2\,e^- \tag{9}$$

$$NOH + \tfrac{1}{2}\,O_2 \rightarrow NO_2^- \tag{10}$$

$$N_2H_4 + O_2 \rightarrow N_2 + 2\,H_2O \tag{11}$$

In a typical heterotroph, *T. pantotropha*, hydroxylamine oxidation is not coupled to ATP synthesis and growth of the organism. Therefore, the two abstracted electrons are sufficient to support further AMO activity. Alternatively and as required, the two electrons could be used to reduce, via nitrite reductase, the nitrite formed and so couple nitrification to denitrification. Then, the organism would have the ability to catalyze the complete dissimilation of NH_3 to N_2, which could be important in disposing of excess reducing capacity and so alleviating redox stress.

XII.3.11. Anaerobic Ammonia Oxidation (Anammox)

This interesting and far from understood phenomenon was first encountered in studies of waste water treatment, where it was found that, *under anaerobic conditions*, the removal of ammonia from effluent streams correlated with both consumption of nitrate and gas production.[56] Apparently, ammonia is being oxidized by a mixed unidentified microbial culture that uses nitrate (or nitrite) as the electron acceptor (instead of O_2) to produce N_2. This so-called Anammox process is autotrophic in nature with the released electrons being used to fix CO_2 for incorporation into the cells. The mechanism of this process is unclear, but a similar phenomenon may be catalyzed by *Nitrosomonas eutropha*.[20] A possible pathway (Eqs. 12 and 13), which employs known enzymes, might include: (1) NO production from nitrate (or nitrite) by the appropriate reductases; (2) reaction of NO with ammonia to produce hydrazine catalyzed by an AMO-type enzyme; and (3) oxidation of hydrazine to N_2 by HAO.

$$NO + NH_3 + 3\,H^+ + 3\,e^- \rightarrow N_2H_4 + H_2O \tag{12}$$

$$N_2H_4 \rightarrow N_2 + 4\,H^+ + 4\,e^- \tag{13}$$

XII.3.12. Future Perspectives for Nitrification

As we found for nitrogenase research, progress in understanding nitrification continues to be made, but many mysteries remain and will not be solved until more crystal structures are available, especially for the membrane-bound AMO and NIO proteins. Even then, full mechanistic insight will be lacking. A crystal structure is only a snapshot of the protein at one particular time in its functional cycle. Witness the structure of HAO—with its 24 bound hemes—and we still do not know how the electrons are partitioned between AMO activity and energy generation! In fact, the electron-transfer pathways in most of the steps in nitrification are poorly understood and barely characterized. The interactions of the components in these pathways constitute the major challenge in this research area.

Finally, how can we apply what we learn about nitrification to everyday practical problems in agriculture? A major concern in nitrogen fertilization of crop plants is that most (as much as 60–70%) of the applied nitrogen never reaches the crop! Why? As we've seen, nitrification converts the positively charged ammonium ion into the negatively charged nitrite and nitrate ions. Then, because these anions have no affinity for the negatively charged soil, they are leached out of the soil and into ground water. In addition, nitrate is lost by denitrification (see Section XII.4), which produces volatile N_2 and nitrogen oxides. To combat this loss, a common agricultural practice is to add nitrification inhibitors to maximize use by the crop. Can research give the insights required either to produce new, more effective nitrification inhibitors or to develop new methods of minimizing fixed nitrogen loss? Only time and creative research will tell!

References

General References

1. Fisher, K. and Newton, W. E., "Nitrogen Fixation—A General Overview", in *Nitrogen Fixation at the Millennium,* Leigh, G. J., Ed., Elsevier Science B. V., Amsterdam, The Netherlands, 2002, pp. 1–34.
2. Smil, V., *Enriching the Earth: Fritz Haber, Carl Bosch, and the Transformation of World Food Production,* MIT Press, Cambridge, MA, 2001, 338p.
3. Sprent, J. I. and Sprent, P., *Nitrogen Fixing Organisms,* Chapman & Hall, New York, 1990, 256p.
4. Newton, W. E., "Nitrogen Fixation in Perspective", in *Nitrogen Fixation: From Molecules to Crop Productivity,* Pedrosa, F., Hungria, M., Yates, M. G., and Newton, W. E., Eds., Kluwer Academic, Dordrecht, The Netherlands, 2000, pp. 3–8.
5. Wächsterhäuser, G., "Before Enzymes and Templates: Theory of Surface Metabolism", *Microbiol. Rev.,* **52,** 452–488 (1988).
6. Eady, R. R., "Structure–Function Relationships of Alternative Nitrogenases", *Chem. Rev.,* **96,** 3013–3030 (1996).
7. Postgate, J. R. and Eady, R. R., "The Evolution of Biological Nitrogen Fixation", in *Nitrogen Fixation: Hundred Years After,* Bothe, H., DeBruijn, F. J., and Newton, W. E., Eds., Gustav Fischer, Stuttgart, 1988, pp. 31–40.
8. Young, J. P. W., "Phylogenetic Classification of Nitrogen-fixing Organisms", in *Biological Nitrogen Fixation,* Stacey, G., Burris, R. H., and Evans, H. J., Eds., Chapman & Hall, New York, 1992, pp. 43–86.
9. See Wilson, P. W., "The Biochemistry of Symbiotic Nitrogen Fixation", University of Wisconsin Press, Madison, WI, 1940, 302p.
10. Nutman, P. S., "Centenary Lecture on Nitrogen Fixation", *Philos. Trans. R. Soc. London,* **B317,** 69–106 (1987).
11. Newton, W. E., "Nitrogenases: Distribution, Composition, Structure and Function", in *New Horizons in Nitrogen Fixation,* Palacios, R., Mora, J., and Newton, W. E., Eds., Kluwer Academic, Dordrecht, The Netherlands, 1993, pp. 5–18.
12. Pau, R. N., *Metals and Nitrogenase* in "Advances in Inorganic Biochemistry", Eichhorn, G. L. and Marzilli, L. G., Eds., PTR Prentice Hall, New Jersey, Vol. 10, 1994, pp. 49–70.
13. Howard, J. B. and Rees, D. C., "Structural Basis of Biological Nitrogen Fixation", *Chem. Rev.,* **96,** 2965–2982 (1996).
14. Burgess, B. K. and Lowe, D. J., "Mechanism of Molybdenum Nitrogenase", *Chem. Rev.,* **96,** 2983–3011 (1996).
15. Smith, B. E., "Structure, Function, and Biosynthesis of the Metallosulfur Clusters in Nitrogenases", in *Advances in Inorganic Chemistry,* Sykes, A. G., Ed., Academic Press, New York, Vol. 47, 1999, pp. 159–218.
16. Seefeldt, L. C. and Dean, D. R., "Role of Nucleotides in Nitrogenase Catalysis", *Acc. Chem. Res.,* **30,** 260–266 (1997).
17. Igarashi, R. Y. and Seefeldt, L. C., "Nitrogen Fixation: The Mechanism of the Mo-Dependent Nitrogenase", *Crit. Rev. Biochem. Mol. Biol.,* **38,** 351–384 (2003).
18. Ferguson, S. J., "Nitrogen Cycle Enzymology", *Curr. Opinion Struct. Biol.,* **2,** 182–193 (1998).
19. Richardson, D. J., Wehrfritz, J.-M., Keech, A., Crossman, L. C., Roldan, M. D., Sears, H. J., Butler, C. S., Reilly, A., Moir, J. W. B., Berks, B. C., Ferguson, S. J., Thomson, A. J., and Spiro, S., "The Diversity of Redox Proteins Involved in Bacterial Heterotrophic Nitrification and Aerobic Denitrification", *Biochem. Soc. Trans.,* **26,** 401–408 (1998).
20. Hooper, A. B., Vanelli, T., Bergmann, D. J., and Arciero, D. M., "Enzymology of the Oxidation of Ammonia to Nitrite by Bacteria", *Antonie van Leeuwenhoek,* **71,** 59–67 (1997).
21. Prince, R. C. and George, G. N., "The Remarkable Complexity of Hydroxylamine Oxidoreductase", *Nat. Struct. Biol.,* **4,** 247–250 (1997).

Specific References

22. Fani, R., Gallo, R., and Lio, P., "Molecular Evolution of Nitrogen Fixation: The Evolutionary History of the *nifD, nifK, nifE,* and *nifN* Genes", *J. Mol. Evol.,* **51,** 1–11 (2000).
23. Postgate, J. R., "Evolution within Nitrogen-fixing Systems", *Symp. Soc. Gen. Microbiol.,* **24,** 263–292 (1974).
24. Ribbe, M., Gadkari, D., and Meyer, O., "N_2 Fixation by *Streptomyces thermoautotrophicus* involves a Molybdenum-Dinitrogenase and a Manganese-Superoxide Oxidoreductase that couple N_2 Reduction to the Oxidation of Superoxide produced from O_2 by a Molybdenum-CO Dehydrogenase", *J. Biol. Chem.,* **272,** 26627–26633 (1997).
25. Georgiadis, M. M., Komiya, H., Chakrabarti, P., Woo, D., Kornuc, J. J., and Rees, D. C., "Crystallographic Structure of the Nitrogenase Iron Protein from *Azotobacter vinelandii*", *Science,* **257,** 1653–1659 (1992).
26. Schindelin, H., Kisker, C., Schlessman, J. L., Howard, J. B., and Rees, D. C., "Structure of ADP•AlF_4^--stabilized Nitrogenase Complex and its Implications for Signal Transduction", *Nature (London),* **387,** 370–376 (1997).
27. Jang, S. B., Seefeldt, L. C., and Peters, J. W., "Insights into Nucleotide Signal Transduction in Nitrogenase: Structure of an Iron Protein with MgADP Bound", *Biochemistry,* **39,** 14745–14752 (2000).
28. Kim, C. and Rees, D. C., "Crystallographic Structure and Functional Implications of the Nitrogenase Molybdenum-Iron Protein from *Azotobacter vinelandii*", *Nature (London),* **360,** 553–560 (1992).
29. Mayer, S. M., Lawson, D. M., Gormal, C. A., Roe, S. M., and Smith, B. E., "New Insights into Structure–Function Relationships in Nitrogenase: A 1.6 Å Resolution X-ray Crystallographic Study of *Klebsiella pneumoniae* MoFe Protein", *J. Mol. Biol.,* **292,** 871–891 (1999).

30. Kim, C. and Rees, D. C., "Structural Models for the Metal Centers in the Nitrogenase Molybdenum-Iron Protein", *Science,* **257,** 1677–1682 (1992).
31. Chan, M. K., Kim, J., and Rees, D. C., "The Nitrogenase FeMo-cofactor and P-Cluster Pair: 2.2 Å Resolution Structures", *Science,* **260,** 792–794 (1993).
32. Einsle, O., Tezcan, A., Andrade, S. L. A., Schmid, B., Yoshida, M., Howard, J. B., and Rees, D. C., "Nitrogenase MoFe-Protein at 1.16 Å Resolution: A Central Ligand in the FeMo-cofactor", *Science,* **297,** 1696–1700 (2002).
33. Peters, J. W., Stowell, M. H. B., Soltis, S. M., Finnegan, M. G., Johnson, M. K., and Rees, D. C., "Redox-Dependent Structural Changes in the Nitrogenase P-Cluster", *Biochemistry,* **36,** 1181–1187 (1997).
34. Sørlie, M., Christiansen, J., Lemon, B. J., Peters, J. W., Dean, D. R., and Hales, B. J., "Mechanistic Features and Structure of the Nitrogenase α-Gln^{195} MoFe Protein", *Biochemistry,* **40,** 1540–1549 (2001).
35. Mayer, S. M., Gormal, C. A., Smith, B. E., and Lawson, D. M., "Crystallographic Analysis of the MoFe Protein of Nitrogenase from a *nifV* Mutant of *Klebsiella pneumoniae* Identifies Citrate as a Ligand to the Molybdenum of Iron Molybdenum Cofactor (FeMoco)", *J. Biol. Chem.,* **277,** 35263–35266 (2002).
36. Schmid, B., Ribbe, M. W., Einsle, O., Yoshida, M., Thomas, L. M., Dean, D. R., Rees, D. C., and Burgess, B. K., "Structure of a Cofactor-Deficient Nitrogenase MoFe Protein", *Science,* **296,** 352–356 (2002).
37. Chiu, H.-J., Peters, J. W., Lanzilotta, W. N., Ryle, M. J., Seefeldt, L. C., Howard, J. B., and Rees, D. C., "MgATP-Bound and Nucleotide-Free Structures of a Nitrogenase Protein Complex between Leu127Δ-Fe Protein and the MoFe Protein", *Biochemistry,* **40,** 641–650 (2001).
38. Schmid, B., Einsle, O., Chiu, H.-J., Willing, A., Yoshida, M., Howard, J. B., and Rees, D. C., "Biochemical and Structural Characterization of the Cross-Linked Complex of Nitrogenase: Comparison to the ADP-AlF_4^--Stabilized Structure", *Biochemistry,* **41,** 15557–15565 (2002).
39. Lowe, D. J. and Thorneley, R. N. F., "The Mechanism of *Klebsiella pneumoniae* Nitrogenase Action", *Biochem. J.,* **224,** 877–909 (1984).
40. Lanzilotta, W. N. and Seefeldt, L. C., "Changes in Midpoint Potentials of the Nitrogenase Metal Centers as a Result of Iron Protein–MoFe protein Complex Formation", *Biochemistry,* **36,** 12976–12983 (1997).
41. Lowe, D. J., Ashby, G. A., Brune, M., Knights, H., Webb, M. R., and Thorneley, R. N. F., *ATP Hydrolysis and Energy Transduction by Nitrogenase.* In: "Nitrogen Fixation: Fundamentals and Applications", Tikhonoich, I. A., Provorov, N. A., Romanov, V. I., and Newton, W. E., Eds., Kluwer Academic, Dordrecht, The Netherlands, 1995, pp. 103–108.
42. Deng, H. and Hoffmann, R., "How N_2 might be Activated by the FeMo-cofactor in Nitrogenase", *Angew. Chem., Intl. Edit. Engl.,* **32,** 1062–1065 (1993).
43. Dance, I., "The Binding and Reduction of Dinitrogen at an Fe_4 face of the FeMo cluster of Nitrogenase", *Aust. J. Chem.,* **47,** 979–990 (1994).
44. Durrant, M., "An Atomic Level Mechanism for Molybdenum Nitrogenase. Part 1. Reduction of Dinitrogen", *Biochemistry,* **41,** 13934–13945 (2002).
45. Dance, I., "The Consequences of an Interstitial N atom in the FeMo cofactor of Nitrogenase", *Chem. Commun.,* 324–325 (2003).
46. Lovell, T., Liu, T., Case, D. A., and Noodleman, L., "Structural, Spectroscopic, and Redox Consequences of a Central Ligand in the FeMoco of Nitrogenase: A Density Functional Theoretical Study", *J. Am. Chem. Soc.,* **125,** 8377–8383 (2003).
47. Lee, H.-I., Cameron, L. M., Hales, B. J., and Hoffman, B. M., "CO Binding to the FeMo cofactor of CO-Inhibited Nitrogenase: ^{13}CO and ^{1}H Q-Band ENDOR Investigation", *J. Am. Chem. Soc.,* **119,** 10121–10126 (1997).
48. George, S. J., Ashby, G. A., Wharton, C. W., and Thorneley, R. N. F., "Time-Resolved Binding of Carbon Monoxide to Nitrogenase Monitored by Stopped-Flow Infrared Spectroscopy", *J. Am. Chem. Soc.,* **119,** 6450–6451 (1997).
49. Durrant, M. C., "Controlled Protonation of Iron-Molybdenum Cofactor by Nitrogenase: A Structural and Theoretical Analysis", *Biochem. J.,* **355,** 569–576 (2001).
50. Norton, J. M., "Nitrification", in *Handbook of Soil Science,* Sumner, M. E., Ed., CRC Press, Boca Raton, FL, 2000, pp. C160–C181.
51. Whittaker, M., Bergmann, D., Arciero, D., and Hooper, A. B., "Electron Transfer during the Oxidation of Ammonia by the Chemolithotrophic Bacterium *Nitrosomonas europaea*", *Biochim. Biophys. Acta,* **1459,** 346–355 (2000).
52. Sayavedra-Soto, L. A., Hommes, N. G., and Arp, D. J., "Characterization of the Gene Encoding Hydroxylamine Oxidoreductase in *Nitrosomonas europaea*", *J. Bacteriol.,* **176,** 504–510 (1994).
53. Igarashi, N., Moriyama, H., Fujiwara, T., Fukumora, Y., and Tanaka, N., "The 2.8 Å Structure of Hydroxylamine Oxidoreductase from the Nitrifying Bacterium, *Nitrosomonas europaea*", *Nat. Struct. Biol.,* **4,** 276–284 (1997).
54. Berks, B. C., Ferguson, S. J., Moir, J. W. B., and Richardson, D. J., "Enzymes and Associated Electron Transport Systems that Catalyse the Respiratory Reduction of Nitrogen Oxides and Oxyanions", *Biochim. Biophys. Acta,* **1231,** 97–173 (1995).
55. Yamanaka, T., "Mechanisms of Oxidation of Inorganic Electron Donors in Autotrophic Bacteria", *Plant Cell Physiol.,* **37,** 569–574 (1996).
56. Jetten, M. S. M., Logemann, S., Muyzer, G., Robertson, L. A., de Vries, S., van Loosdrecht, M. C. M., and Kuenen, J. G., "Novel Principles in the Microbial Conversion of Nitrogen Compounds", *Antonie van Leeuwenhoek,* **71,** 75–93 (1997).

XII.4. Nitrogen Metabolism: Denitrification

Bruce A. Averill
Department of Chemistry
University of Toledo
Toledo, OH 43606

Contents

XII.4.1. Introduction

As discussed in Chapter II, the inorganic nitrogen cycle consists of several linked biological processes and two nonbiological processes. The latter involve the reaction of N_2 and O_2 in lightning discharges and internal combustion engines to produce NO_x, which is ultimately oxidized to nitrate, and the industrial reduction of N_2 to ammonia via the Haber process. Denitrification is the anaerobic use by certain bacteria of nitrogen oxide species (NO_3^-, NO_2^-, NO, and N_2O) as terminal electron acceptors instead of O_2;[1–6] the ultimate product of the reductive process is usually N_2. Because the reduced nitrogen-containing product is released into the atmosphere rather than being assimilated (and used as a source of nitrogen for growth), the process is referred to as *dissimilatory* reduction of nitrate. Although most denitrifying organisms prefer to use oxygen as an electron acceptor, denitrification provides them with an alternative that allows them to survive in the absence of oxygen. Under anaerobic conditions, they are able to use the energy obtained from reduction of N–O species to drive electron-transport coupled phosphorylation and store metabolic energy in the form of ATP. Because of to their metabolic flexibility, denitrifying bacteria are able to thrive in a wide range of natural habitats, including soil, water, foods, and the digestive tract.[7] From a global perspective, denitrification is important because it is the only process that returns large amounts of fixed nitrogen to the atmosphere, thereby completing the terrestrial nitrogen cycle. Unfortunately, denitrification is a "leaky" process that releases large amounts of N_2O into the atmosphere; N_2O is a greenhouse gas and has been implicated in depletion of the ozone layer.[7] Denitrification is also important commercially for two reasons: (1) in conjunction with nitrification, denitrification can result in the loss of a large fraction of nitrogen fertilizer that is applied to fields (up to 30%);[8] and (2) denitrifiers have significant potential for use in bioremediation and waste water remediation. Finally, the well-studied enzymes of denitrification provide structural and spectroscopic models for the mammalian enzymes that produce and utilize NO in a variety of signal transduction pathways (Section XIV.3).

XII.4.2. The Enzymes of Denitrification

The overall pathway of denitrification is outlined in Fig. XII.4.1. Four enzymatic steps are involved in most organisms. A few organisms are known that appear to lack the last enzyme; consequently, they can only produce N_2O rather than N_2 as the major product. The NO_2^- and N_2O reductases are generally soluble, reasonably

Fig. XII.4.1.
The pathway of denitrification, with NO as an obligatory free intermediate.

$$NO_3^- \longrightarrow NO_2^- \longrightarrow NO \longrightarrow N_2O \longrightarrow N_2$$

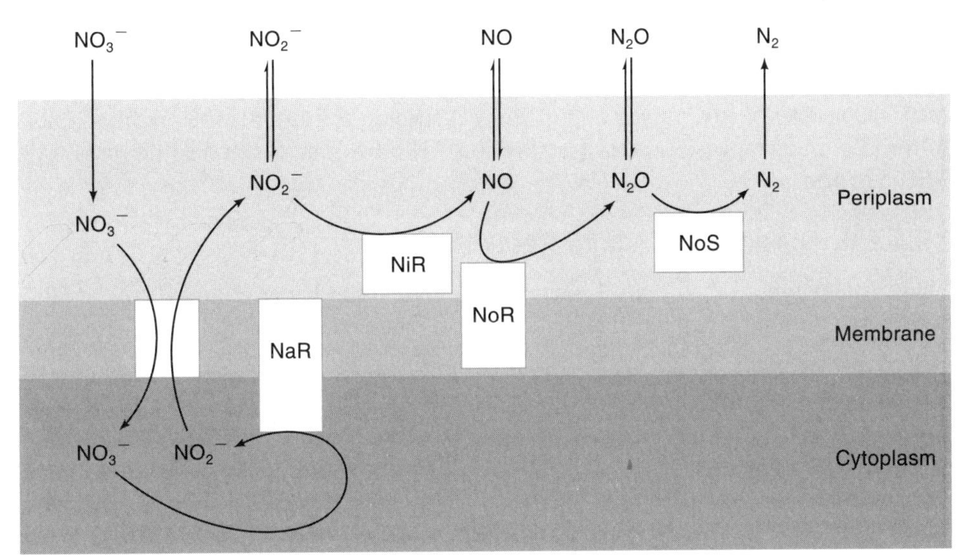

Fig. XII.4.2.
The arrangement of the enzymes of denitrification in Gram-negative bacteria: NaR-nitrate reductase; NiR-nitrite reductase; NoR-nitric oxide reductase; NoS-nitrous oxide reductase.

easy to obtain in pure form, and exhibit only moderate complexity in terms of their metal cofactor contents. Consequently, their structures and mechanisms are beginning to be rather well understood. In contrast, the NO_3^- and NO reductases are membrane-bound enzymes, and only recently have pure preparations of the latter been obtained for detailed study. Hence, much of what is known about these two enzymes is based upon experiments carried out with partially purified samples and upon parallels to apparently similar and better characterized systems.

The organization of the enzymes of denitrification in Gram-negative bacteria, as determined by cell fractionation, antibody labeling, and electron microscopy studies,[1–3] is shown schematically in Fig. XII.4.2. The first enzyme in the pathway, NaR, resides in the cytoplasmic membrane with its active site accessible from the cytoplasmic side. This arrangement requires that nitrate be transported across both the periplasmic and cytoplasmic membranes. The product of this enzyme is nitrite, which is transported back into the periplasm, where it is reduced by NiR, a soluble enzyme. The next enzyme, NoR, is located in the cytoplasmic membrane and releases its product, N_2O, back into the cytoplasmic space, where it is converted to N_2 by NoS, a soluble enzyme.

XII.4.2.1. Dissimilatory Nitrate Reductases

The first step in denitrification is carried out by the dissimilatory nitrate reductases. Although their membrane-bound nature has made it difficult to obtain homogeneous preparations for detailed physical and mechanistic studies, they appear to be fairly typical Mo cofactor enzymes (Section XII.6). All preparations appear to contain at least two types of subunit (α, 104–150 kDa; β, 52–63 kDa), together with Mo, both heme and nonheme iron, and acid-labile sulfur.[9] These components are consistent with the presence of an internal electron-transfer chain containing both cytochromes and iron–sulfur clusters.

The balanced enzymatic half-reaction is

$$NO_3^- + 2\,H^+ + 2\,e^- \rightarrow NO_2^- + H_2O \qquad (1)$$

At least on the surface, this appears to be a typical molybdenum-catalyzed oxygen atom transfer reaction (cf. Section XII.6). Although direct evidence is lacking, the reaction indicated in Eq. 2 is a plausible representation of the reduction of nitrate by the enzyme.

$$NO_3^- + Mo^{IV}{=}O \rightarrow NO_2^- + Mo^{VI}({=}O)_2 \tag{2}$$

The key step is abstraction of an oxygen atom from nitrate by an enzyme-bound $Mo^{IV}{=}O$ species to give nitrite and a Mo(VI) species; the latter is then reduced back to Mo(IV) in a stepwise process (see Section XII.6 for a more detailed discussion of Mo enzymes).

XII.4.2.2. Dissimilatory Nitrite Reductases

Two distinct types of NiR are known: those containing hemes *c* and d_1 (cd_1-NiRs) and those containing copper (Cu-NiRs).[4–6,9,10] Organisms containing the former appear to be more abundant in Nature, but organisms containing the latter occupy a wider range of ecological niches and exhibit more physiological diversity.[11] There is no obvious correlation between the type of enzyme present and the genus/species of the organism, and no organism is known to contain both types of NiR.

The overall reaction catalyzed by both types of enzyme is the same:

$$NO_2^- + e^- + 2\,H^+ \rightarrow NO + H_2O \tag{3}$$

Insights into the mechanism of reduction of nitrite by these enzymes have been limited by difficulties in performing classical enzyme kinetics measurements. The major problem is the fact that the product NO is a potent inhibitor of nitrite reduction.[12] As a result, plots of NO production versus time are nonlinear, and reliable initial velocity studies have not yet been reported. Consequently, not even the order of binding of substrates is known (i.e., does nitrite bind to the oxidized enzyme followed by electron transfer from the physiological electron donor, or does productive binding of nitrite require prior reduction of the enzyme?). Nonetheless, crystal structures of examples of both types of nitrite reductase are available, as is a wealth of results from spectroscopic measurements and isotope labeling studies. Thus, despite the lack of good kinetics data, the nitrite reductases are the best understood of the enzymes of denitrification, and a reasonable picture of their strucutre and function has been achieved.

Heme cd_1-Containing Nitrite Reductases. Heme cd_1-NiR enzymes have been isolated from a large number of bacteria; the enzymes typically consist of two identical subunits with a molecular mass of 60 kDa. Each subunit contains one heme *c* covalently linked to the polypeptide chain and one noncovalently bound heme d_1. The amino acids that bind heme *c* are located near the N-terminus of the protein.[13] A high degree of amino acid sequence homology is observed for these enzymes, and polyclonal antibodies raised against the cd_1-NiR from one representative denitrifier cross-react strongly with the cd_1-NiR enzymes from other denitrifiers.[11] These results strongly suggest the existence of a common structure and, presumably, a common mechanism for all cd_1-NiR enzymes.

The presence of the unusual green heme d_1 is unique to denitrifiers containing cd_1-NiR enzymes. The porphyrindione (dioxoisobacteriochlorin) structure shown in Fig. XII.4.3 was proposed for heme d_1 based solely on spectroscopic studies,[14] and confirmed by reconstitution of apoenzyme lacking heme d_1 with synthetic heme d_1 with full restoration of enzymatic activity.[15] It is reasonable to assume that the chemical nature of this unusual heme plays an important role in the ability of these enzymes to convert nitrite to NO, but the precise reason for the use of heme d_1 rather than, for example, protoheme (see Chapter IV), remains unclear.

A wide variety of spectroscopic and ligand-binding studies have been carried out on various examples of the cd_1 NiR enzymes. Many of the studies date from the days when the enzyme was referred to as "bacterial cytochrome oxidase" because of its ability to reduce O_2 to water (albeit much more slowly than does mammalian cyto-

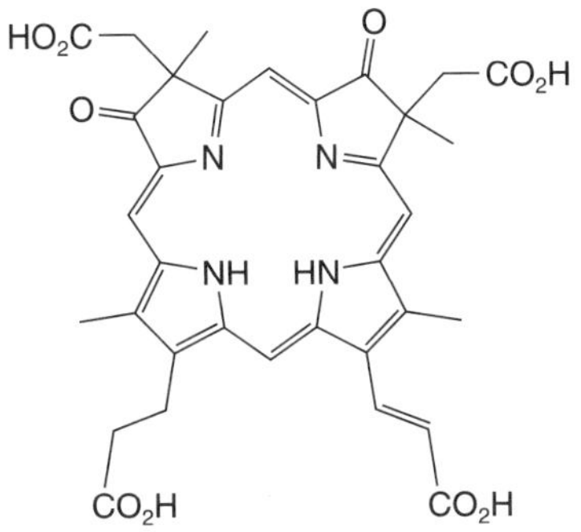

Fig. XII.4.3.
The structure of heme d_1.

chrome *c* oxidase) and the presence of four electron acceptor groups per dimer. The mechanistic and structural similarities to mammalian cytochrome *c* oxidase are now recognized to be weak, and the physiological role of the enzyme in nitrite reduction is clear. Nonetheless, these spectroscopic studies revealed some important points.[16] First, the resting oxidized enzyme contains both a low-spin ferric heme (assigned to the *c* heme) and a mixture of low- and high-spin ferric heme, both of which are assigned to the heme d_1. Second, the high-spin heme is converted to a low-spin form by added ligands such as cyanide. Third, in the reduced enzyme, the heme d_1 readily binds strong acceptor ligands such as CO and NO. The ferrous heme-NO complexes are analogous electronically to such well-studied species as nitrosylhemoglobin; they have an $S = \frac{1}{2}$ ground state and are readily detected by EPR. Thus, the early spectroscopic work resulted in a model in which the enzyme contained one heme (d_1) that could be either high spin (five coordinate?) or low spin (six coordinate) and was capable of binding exogenous ligands in addition to one six-coordinate, low-spin heme (*c*). By analogy to other heme proteins, these hemes presumably provided the site at which nitrite binds and is reduced (heme d_1) and played a role in electron transfer (heme *c*) to the active site, respectively. As we shall see, these conclusions were right on target.

The report of a 1.55-Å resolution crystal structure of a heme cd_1 NiR[17] reveals that the enzyme is a single polypeptide of at least 559 amino acids arranged in two domains: the smaller domain (residues 1–134) contains the covalently bound heme *c*, and the larger domain (residues 135–567) contains the heme d_1 (Fig. XII.4.4). This arrangement supports the idea of separate electron-transfer (heme *c*) and substrate-binding (heme d_1) sites. Unexpectedly, however, the coordination to the iron of the heme *c* is atypical, and the presumed substrate-binding site on the heme d_1 is occupied by a coordinated amino acid side chain.

The heme *c* domain is largely α-helical and bears a general structural resemblance to other structurally characterized cytochromes *c*. In contrast to the expected His–Met ligation, however, the protein ligands to the heme *c* are two His imidazoles. The larger domain consists of an eight-bladed β-propeller structure that surrounds the heme d_1. This heme also exhibits unusual ligation, in that one axial ligand is a conventional histidine imidazole, but the second axial ligand is a phenoxide provided by a Tyr residue in the heme *c* domain. Tyrosine ligation to hemes is relatively unusual, and its presence normally results in major effects on the reduction potential of the heme[18] (e.g., catalases[19] and pathological hemoglobin mutants such as HbM[20]). The tyrosinate ligand to the heme d_1 should significantly influence the chemistry of the enzyme, and, indeed, the structural report postulates a role for Tyr25 in eliminating NO from the enzyme–product complex.

A consideration of amino acid sequences of cd_1-NiR enzymes, however, shows that the Tyr residue in this position is not strictly conserved, and that at least two

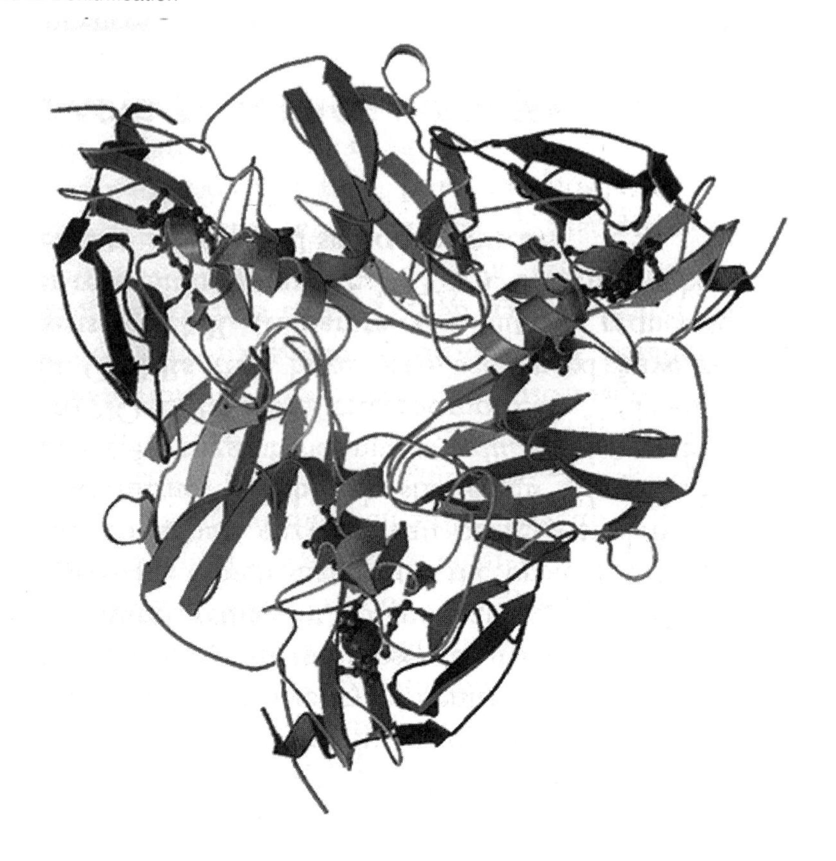

Fig. XII.4.4.
The structure of the dimeric heme cd_1-nitrite reductase from *Pseudomonas pantotrophus.* (PDB code: 1QKS). [See color insert.]

functional cd_1-NiR enzymes contain no Tyr residues in the entire heme *c* domain. These points raise the possibility that the Tyr ligand may be present only in the crystal, and is absent or replaced by a weak ligand, such as water, in solution. The two hemes within each monomer of the cd_1-NiR are relatively close to one another. Although the Fe···Fe distance is 20.6 Å, the closest contact between the edges of the two hemes is ~11 Å, which is consistent with the proposed role of heme *c* in electron transfer to heme d_1. The distances between hemes in adjacent subunits are much longer (>40 Å), consistent with the view that each subunit functions independently in nitrite reduction. Finally, the reaction catalyzed by the enzyme (Eq. 3) requires that two protons be added to one of the oxygen atoms of nitrite in order for it to be released as water. The crystal structure shows the presence of two His imidazoles near the Tyr coordinated to heme d_1. If the position of these two His residues does not change significantly in the active enzyme, they are ideally located to provide the required protons.

A plausible mechanism of action of the heme cd_1 NiR enzymes is shown in Fig. XII.4.5. A variety of studies have established the existence of an electrophilic nitrosyl intermediate derived from nitrite via a simple protonation–dehydration reaction.[1–3,7] This intermediate can be trapped by nucleophiles, such as azide and hydroxylamine. The reaction of this electrophilic intermediate with water to give bound nitrite is simply the reverse of the initial dehydration reaction, and results in exchange of ^{18}O from $H_2{}^{18}O$ into the product NO. Use of $^{15}NO_2{}^-$ has allowed the nitrosyl intermediate to be trapped with $^{14}N_3{}^-$ or $C_6H_5{}^{14}NH_2$ to give $^{14,15}N_2O$, which can be detected by mass spectrometry. This electrophilic nitrosyl species is the key intermediate in the mechanism shown in Fig. XII.4.4. It can be formulated as either a ferrous heme d_1–NO^+ complex or as a ferric heme d_1–NO complex. In either case,

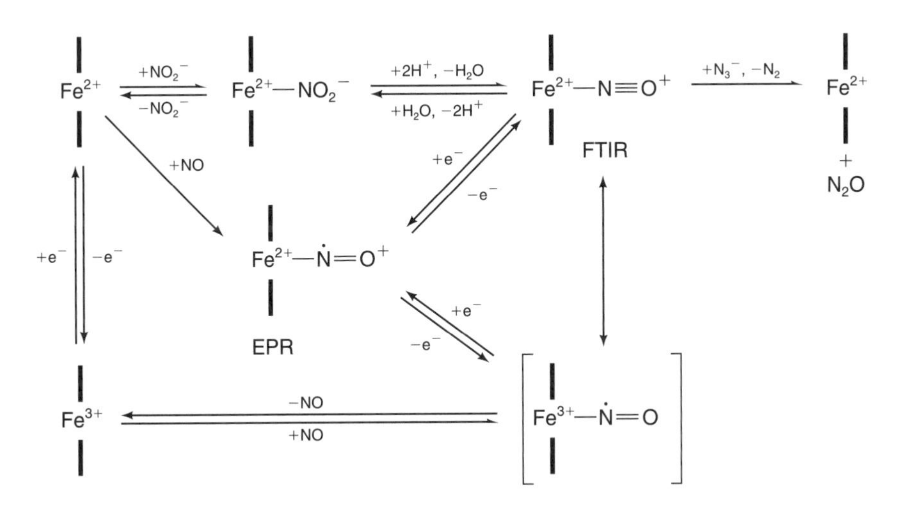

Fig. XII.4.5. Proposed mechanism for reduction of nitrite by the heme cd_1-containing nitrite reductases.

it has a total of eight electrons derived from Fe and NO, making it one electron more oxidized than the paramagnetic ferrous heme–NO complex found in nitrosylmyoglobin or nitrosylhemoglobin.[21] Studies on synthetic model compounds suggest that the ferrous heme d_1–NO^+ species should decompose rapidly via an internal electron-transfer reaction to produce NO and the ferric heme d_1, as shown at the bottom of Fig. XII.4.4. Electron transfer from the ferrous heme *c* then reduces the heme d_1 back to the ferrous state.

Although the ferrous heme d_1–NO^+ intermediate shown in Fig. XII.4.4 is diamagnetic, and hence not detectable by EPR, it has recently been generated by the back reaction of NO with the oxidized enzyme, and its N–O stretching band was observed directly by FTIR.[22] The position of this band, at $\sim 1910\,cm^{-1}$, is consistent with the presence of a nitrogen–oxygen triple bond and an Fe^{2+}–$N{\equiv}O^+$ formulation. For example, it is similar in position to the N–O stretch observed for the NO complex of metHb ($1925\,cm^{-1}$), but very different from the $1615\,cm^{-1}$ observed for the NO complex of deoxyHb.[23] Known heme Fe^{2+}–NO^+ species are highly reactive,[24] and decompose rapidly to NO and the Fe^{3+} heme if the NO^+ species is not trapped by a nucleophile or reduced by one electron to the very stable heme Fe^{2+}–NO species. The latter species has been observed by EPR many times upon treatment of cd_1–NiR enzymes with NO_2^- and has been suggested to be an intermediate in the reduction of nitrite to NO.[9] This suggested intermediate is unlikely for a very simple reason: The half-life for dissociation of NO from ferrous heme proteins is on the order of days, making the reaction much too slow to be part of the catalytic cycle of the nitrite reductase.[25] Instead, the ferrous heme d_1–NO complex is apparently an inactive enzyme–NO complex that is formed if NO is allowed to accumulate. Under physiological conditions, however, NO is rapidly consumed by the next enzyme in the pathway, NO reductase.

Copper-Containing Nitrite Reductases. Denitrifiers with copper NiR enzymes account for one-third of the denitrifiers isolated from soil,[8] and include species from a variety of genera. Most Cu NiR enzymes cross-react with polyclonal antibodies raised against well-studied enzymes such as that from *Achromobacter cycloclastes*,[11] suggesting substantial structural similarity. Originally, many of these enzymes were reported to be dimers or tetramers of identical subunits with subunit molecular weights in the 30–40-kDa range. The publication of a crystal structure of the *A. cycloclastes* NiR in 1991, however, showing that the enzyme was a *trimer* with Cu sites

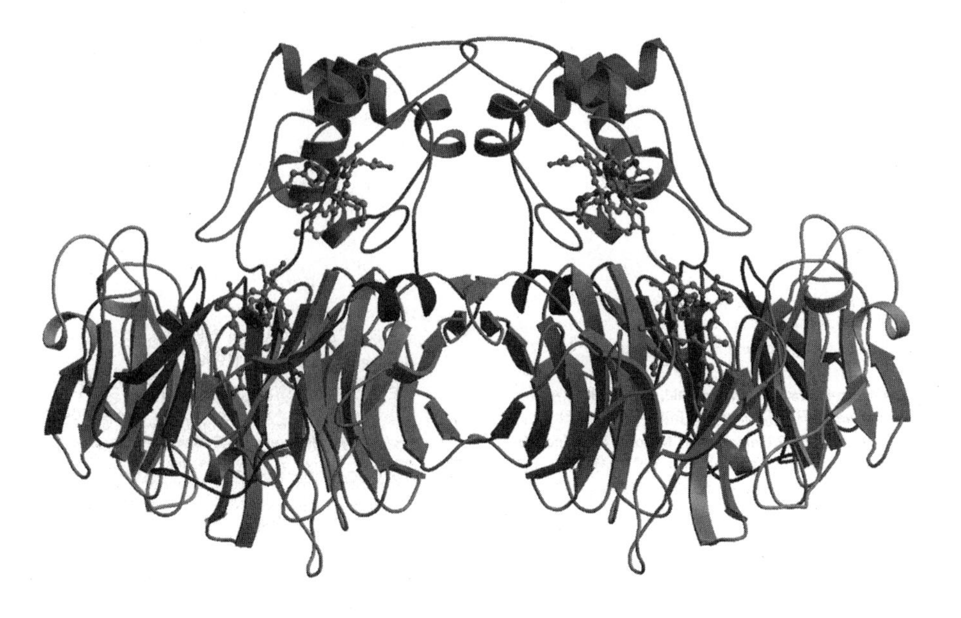

Fig. XII.4.6.
The structure of the trimeric Cu NiR from *Alcaligenes faecalis* S-6 (PDB code: 1AS6).

located between the subunits, forced a reassessment.[26] In view of the strong sequence homologies, it seems likely that most if not all Cu NiR enzymes are actually trimers.

The copper-containing nitrite reductases are similar to the heme cd_1 enzymes in that they contain two types of metal center. Crystallographic studies have shown that the enzymes are trimers of 37-kDa subunits, and that each subunit contains two Cu atoms in distinct sites (Fig. XII.4.6). One is an unusual green variant of the "blue" or Type 1 Cu center familiar from plastocyanin or azurin (Section XI.1), with one methionine thioether, one cysteine thiolate, and two histidine imidazole ligands. The structure of the Type 1 center is, even at high resolution, essentially indistinguishable from the Cu sites in plastocyanin and pseudoazurin, and reveals no obvious structural basis for the strong band at 458 nm that is responsible for the intense green color. Spectroscopic studies indicate, however, a substantially different description of the bonding in the green Type 1 centers than in the well-studied blue Type 1 Cu centers,[27,28] with a stronger Cu–S(Met) interaction. The other Cu atom, a "non-blue" or Type 2 Cu, is coordinated in a roughly tetrahedral fashion by three His imidazoles (two from the same subunit that ligates the Type 1 Cu and one from an adjacent subunit) and one molecule of water. The two Cu atoms are linked by a His and a Cys that are adjacent in the polypeptide chain, resulting in a Cu···Cu distance of 12.5 Å. A schematic drawing illustrating these essential features of the structure is shown in Fig. XII.4.7. The overall structure of the protein and the arrangement of the Cu atoms in the structure is very similar to that observed in the structure of ascorbate oxidase, where the Type 2 Cu is part of a trinuclear center lying at the subunit interface (Section XI.7).[29]

Both crystallographic[30] and spectroscopic[31] studies of nitrite binding to oxidized Cu NiR enzymes and studies of Type 2 Cu-depleted enzyme,[32] where a linear correlation between Type 2 Cu content and specific activity was observed, have shown that the Type 2 Cu center is the site at which NO_2^- binds and is reduced. It is thus analogous to the heme d_1 center in the cd_1–NiR enzymes. The role of the Type 1 Cu is therefore presumably that of an electron-transfer center, analogous to the role of the heme c in the cd_1–NiR enzymes.

A mechanism similar to that discussed above for the cd_1–NiR enzymes has been proposed for the Cu NiR enzymes[10,33] (Fig. XII.4.8, lower half). In this mechanism, nitrite displaces the water bound to the Type 2 Cu site to generate a cuprous–nitrite

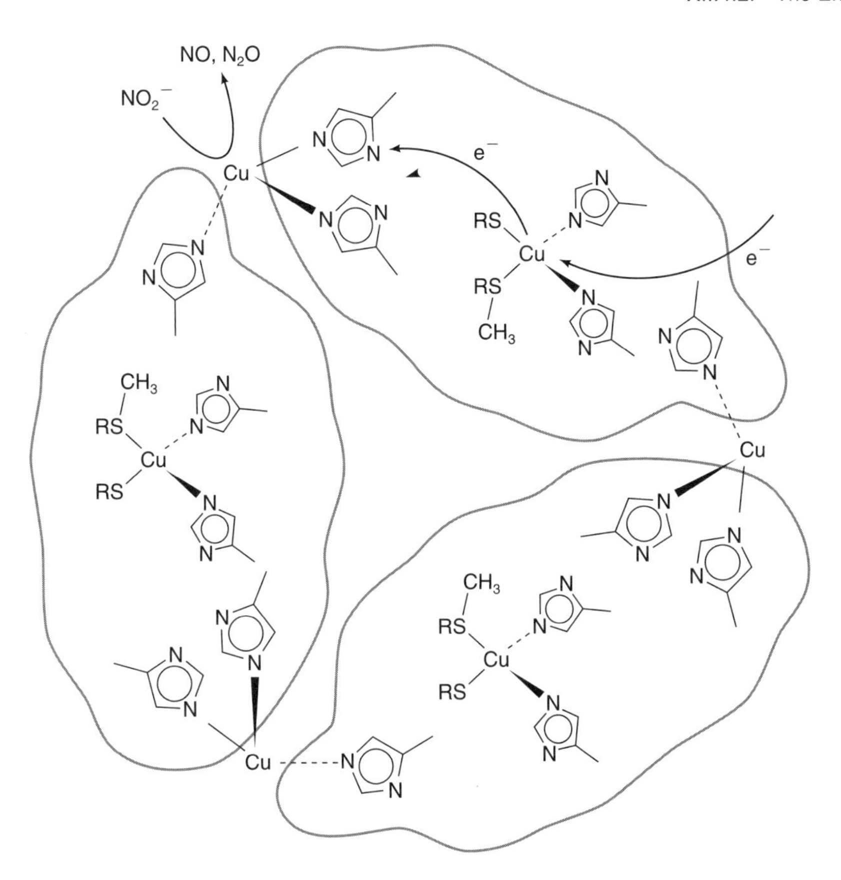

Fig. XII.4.7.
A schematic arrangement of the copper centers in the trimeric Cu-containing nitrite reductase from *A. cycloclastes.*

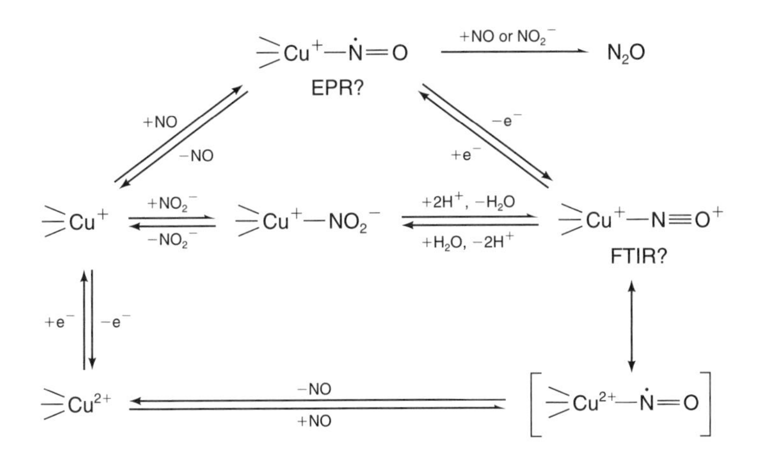

Fig. XII.4.8.
Proposed mechanism for reduction of nitrite to nitric oxide (bottom) or nitrous oxide (top) by the Cu-containing nitrite reductases.

complex. The coordinated nitrite ion is drawn in Fig. XII.4.8 as N bound, even though the crystal structure of the oxidized enzyme–nitrite complex shows the nitrite to be coordinated asymmetrically through the two oxygen atoms. The available structural data on synthetic copper–nitrite complexes[34] suggest that O coordination is favored for Cu(II),[35] while the only known Cu(I)–nitrite complex is N bound.[36] Since it is not yet known whether nitrite binding to Cu occurs prior to or subsequent to reduction of the Type 2 Cu center, both possibilities must be kept in mind, and in fact an equilibrium between both N- and O-bound forms is plausible. The N-bound form would allow an N-bound NO derivative to be generated by protonation of one of the uncoordinated oxygen atoms, followed by dehydration to give a cuprous–NO^+ species analogous to the well-characterized ferrous heme–NO^+ species of the cd_1–NiR enzymes. The imidazole side chain of a nearby His residue and an ordered water molecule bridging the His and an Asp carboxylate are potential sources of the protons required in the reaction (Eq. 4). The Cu^+–NO^+ complex would then decompose to NO and Cu(II) via an internal electron-transfer reaction analogous to that proposed for the cd_1–NiR enzymes. This sequence of reactions is supported by the fact that protonation of synthetic Cu^I–NO_2^- compounds results in smooth conversion of nitrite to NO.[36]

The results of site-directed mutagenesis studies on the Cu–NiR from *A. faecalis* are consistent with the distinct roles of the two types of Cu center. Thus, replacement of one of the His imidazole ligands to the Type 2 Cu by a Lys amino group results in an altered Type 2 site and an enzymatically inactive protein, even though the Type 1 Cu site was unperturbed.[37] Conversely, replacement of the Met thioether ligand to the Type 1 Cu by a Glu carboxylate resulted in a protein that bound Zn^{2+} in place of Cu in the Type 1 site and contained an unperturbed Type 2 center.[38] This mutant protein was catalytically inactive using reduced pseudo-azurin, the physiological electron donor, as reductant, but a nonphysiological electron donor, the methyl viologen cation, was apparently able to "short-circuit" the normal electron-transfer pathway and reduce the Type 2 Cu directly, resulting in significant nitrite reductase activity.

A variety of evidence, however, suggests that there are significant mechanistic differences between the heme and copper NiR enzymes. First, no direct evidence for a Cu^+–NO^+ species has been obtained via $H_2{}^{18}O$ exchange or trapping experiments with the purified enzyme. Second, unlike the cd_1–NiR enzymes, purified Cu–NiR enzymes produce a significant amount of N_2O from NO_2^-, but only after substantial levels of NO have built up.[39] Third, reaction of $^{15}NO_2^-$ with the enzyme in the presence of ^{14}NO produces large amounts of $^{14,15}N_2O$.[39] These results suggest that a labile Cu–NO species is formed, which can react with NO_2^- (or with NO derived from NO_2^-) to produce N_2O (Fig. XII.4.6, upper half). The net charge on such a species could range from $+1$ to $+3$, depending on the oxidation states of both Cu and NO, but a Cu^+–NO^+ formulation seems most likely.

XII.4.2.3. Nitric Oxide Reductases

Bacterial NoR is an integral membrane protein that defied purification for many years. Homogeneous preparations have now been obtained from several sources, using a variety of detergents to solubilize the enzyme. All consist of a heterodimer with subunits of ~53 and 17 kDa, containing a heme *b* and a heme *c*, respectively, which give rise to EPR spectra typical of a ~1:1 mixture of low- and high-spin hemes in the oxidized enzyme. In addition, however, significant quantities of nonheme iron (≥ 3 Fe mol^{-1}) are present.[40,41]

The enzyme catalyzes the reductive dimerization of NO to N_2O via the reaction:

$$2\,NO + 2\,e^- + 2\,H^+ \rightarrow N_2O + H_2O \qquad (4)$$

A feature that has complicated interpretation of mechanistic studies is the existence of a strong coupling between the nitrite reductase and nitric oxide reductase activities at all levels: transcriptional, translational, and even mechanistic.[3–6] These results suggest that at least in some organisms the two enzymes may form a multienzyme complex, in which the NO produced by NiR is directly channeled to the active site of NoR. In addition, at high concentrations NO is a potent inhibitor of NO reductase activity,[3–6,42] suggesting the presence of multiple NO binding sites in the enzyme. Although no firm structural data are yet available for a NoR, mechanistic studies are relatively well advanced.

Reduction of NO by NoR poses an intriguing mechanistic problem:[7] How are two NO molecules brought together to form the N=N bond of N_2O? If the enzyme utilizes a high-spin, five-coordinate heme group as the site of NO binding and reduction, how can a heme, normally capable of binding only a single small molecule such as NO, catalyze the dimerization reaction? A number of possibilities have been suggested to overcome this problem, including: reduction of NO to NO^- (HNO), which then dimerizes rapidly and spontaneously in a nonenzymatic reaction to produce N_2O and H_2O;[43] reductive protonation of the heme Fe–NO complex, followed by elimination of water and formation of a high-valent iron-nitride intermediate that can react with a second NO;[44] reductive coupling of two NO molecules at a non-heme iron atom via a variant of the mechanisms well established for reduction of NO by synthetic metal complexes;[45,46] and reduction of NO at a binuclear heme–non-heme Fe center in a manner analogous to reduction of O_2 at the binuclear heme a_3–Cu center of cytochrome *c* oxidase. All of these mechanistic possibilities are potentially consistent with both the data indicating ^{18}O exchange during reduction of NO and the observed inhibition by NO of its own reduction.

The last of these possibilities is perhaps the most intriguing, and is supported by analysis of the deduced amino acid sequences of the NoR subunits. There is clear evidence from the sequences for an evolutionary relationship between the subunits of NoR and subunits I and II of the heme–copper oxidase superfamily, which includes mammalian cytochrome *c* oxidase.[47,48] Both NoR C and subunit II of cytochrome oxidase are predicted to be membrane-anchored electron-transfer proteins with similar topologies, with in one case a heme *c* and in the other a binuclear Cu_A center as the redox active chromophores.[47,48] More significantly, there exists significant sequence similarity between the NoR B protein and subunit I of cytochrome oxidase. Both are postulated to contain a total of 12 membrane-spanning helices, and, in particular, the six His residues that are known to provide ligands to hemes *a* and a_3 and Cu_{a3} in cytochrome *c* oxidase are conserved in NoR B. The analogy to a more primitive cytochrome oxidase, cytochrome cbb_3, is even stronger, in that the latter possesses a subunit that contains a cytochrome *c* rather than a binuclear Cu_A center. Based on these similarities, it has been suggested that NoR contains a binuclear heme–nonheme Fe center similar to the heme a_3–Cu_{a3} unit of cytochrome *c* oxidase (see Section XII.6), and that NO is bound and reduced at this binuclear center in a reaction analogous to the reduction of O_2 by cytochrome *c* oxidase.[47,48]

XII.4.2.4. Nitrous Oxide Reductase

The final step in the denitrification process is the reduction of nitrous oxide to dinitrogen:

$$N_2O + 2\,e^- + 2\,H^+ \rightarrow N_2 + H_2O \qquad (5)$$

This reaction is catalyzed by the soluble enzyme NoS, which is unusual in a number of ways. Nitrous oxide reductase is usually obtained as a homodimer of ~74-kDa subunits with ≥4 Cu/subunit. As isolated, the enzyme is bright purple or pink, depending on the isolation procedure; the typical blue color expected for Cu proteins is observed only after (irreversible) reduction with dithionite, which inactivates the enzyme.[49] A variety of spectroscopic studies strongly suggest that the enzyme contains a thiolate-bridged Cu^I–Cu^{II} unit that is very similar to the dinuclear Cu_A center found in cytochrome *c* oxidase (cf. Section XII.6).[50,51] By analogy to cytochrome *c* oxidase, the Cu_A center was proposed to be involved in electron transfer to the Cu_Z site at which N_2O is bound and reduced.

This proposal was confirmed by the solution of the structures of NoSs from two organisms, *Pseudomonas nautica*[52] and *Paracoccus denitrificans.*[53] As shown in

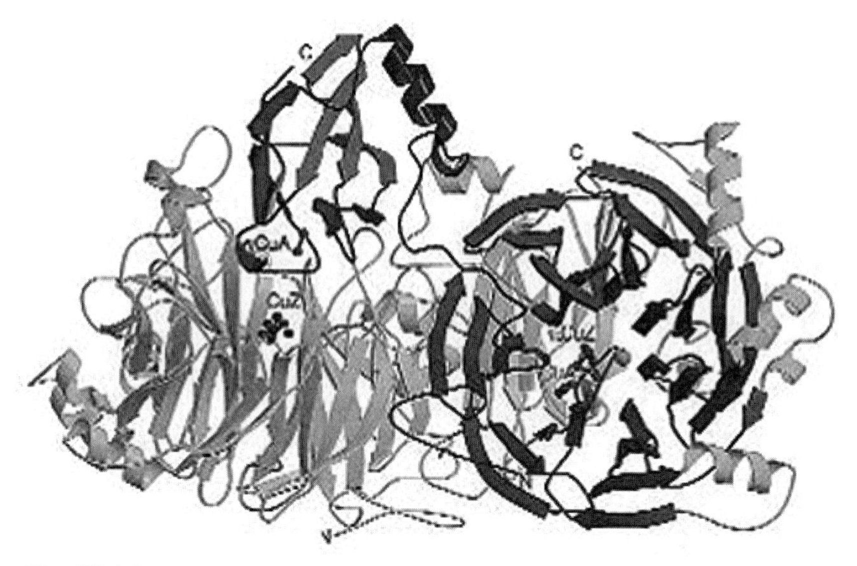

Fig. XII.4.9.
The dimeric structure of the nitrous oxide reductase from *P. nautica* (PDB code: 1QNI). The cupredoxin domain containing the dinuclear Cu_A center of one subunit (top center) is near the Cu_Z center of the other monomer (shown in gray); the seven-bladed propeller domain that contains the catalytic Cu_z site of the same subunit is shown at right.

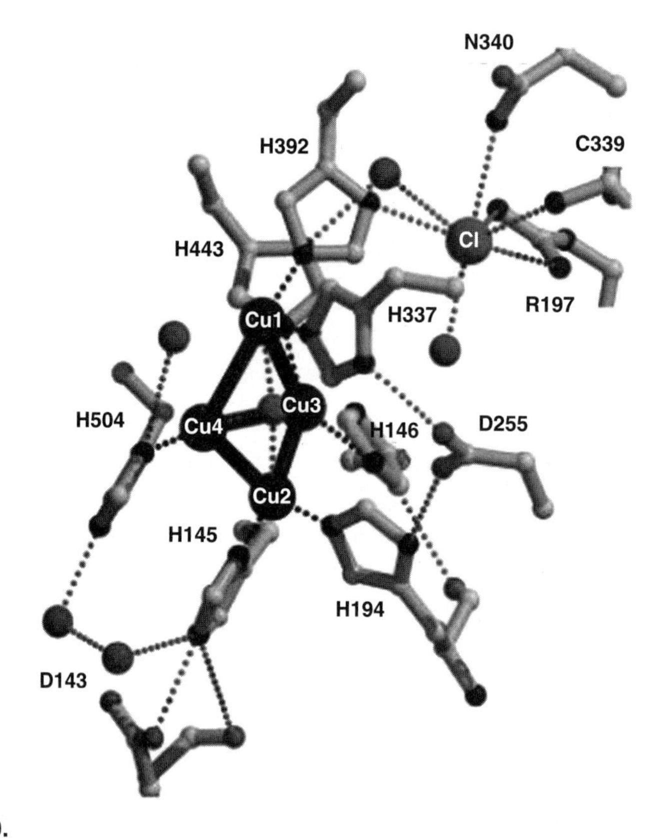

Fig. XII.4.10.
The structure of the tetranuclear Cu_Z center in the nitrous oxide reductase from *P. denitrificans* (PDB code: 1FWX). Note the presence of a sulfide ion bridging all four Cu atoms, a hydroxide–oxide ion bridging Cu1 and Cu4, and a chloride binding site adjacent to Cu1.

Fig. XII.4.9, the enzymes are dimers of identical subunits that consist of two distinct domains. One is a typical cupredoxin domain that contains the dinuclear Cu_A center, while the other is a seven-bladed β-propeller that contains the tetranuclear Cu_Z center. The Cu_A center of one subunit is only ~10 Å from the Cu_Z center of the subunit, while the intrasubunit $Cu_A \cdots Cu_Z$ distance is ~40 Å. The structure thus suggests that electrons enter the enzyme via the Cu_A centers and are then transferred to the Cu_Z centers of the other subunit.

The structure of the tetranuclear Cu_Z site was both unexpected and unprecedented. The structure was initially formulated as containing an oxide–hydroxide bridging all four Cu atoms,[52] but resonance Raman spectra[54] and more detailed analysis of the X-ray data[53,55] have shown convincingly that the quadruply bridging ligand is in fact a sulfide ion (Fig. XII.4.10). In addition, two of the copper atoms (Cu1 and Cu4) are bridged by a water-derived ligand that may well occupy the site at which N_2O binds and is reduced, while a well-defined Cl^- binding site is adjacent to Cu1.[53,55]

Although the reaction catalyzed by NoS is deceptively simple, no definite insights into the mechanism have yet been obtained.

XII.4.3. Summary

The enzymes of denitrification provide a lovely illustration of a relatively general phenomenon in biochemistry: The occurrence of multiple, chemically equivalent but evolutionarily unrelated solutions to important biological problems. Denitrification is one solution to the problem of how organisms can survive and grow in the absence of dioxygen. Denitrifying bacteria have taken advantage of the fact that many anaerobic environments contain relatively high concentrations of oxidized nitrogen–oxygen species, and these organisms have adapted their metabolism to use reduction of nitrogen oxides and oxyanions rather than O_2 to drive formation of a transmembrane proton gradient. This proton gradient is then used to drive formation of ATP by the ATPase, a highly efficient machine that synthesizes ATP as the protons are transported back across the membrane. In the case of NO and N_2O reduction, the enzymes appear to utilize variants of the machinery adapted for reduction of O_2. In addition, denitrifying bacteria demonstrate the existence of two independent solutions to the problem of nitrite reduction, one based on Cu and one based on heme Fe. Other examples of multiple solutions to a biological problem include: the classic example of three unrelated types of reversible oxygen carrier (hemoglobin–myoglobin, hemerythrin, and hemocyanin; Section XII.4); at least three distinct types of both ribonucleotide reductase (containing cobalamin, dinuclear nonheme Fe, and an Fe–S cluster; Section XIII.3); superoxide dismutase (containing Fe, Mn, Ni or Cu and Zn Section XII.2); and two types of catalase (heme Fe and Mn; Section XII.3).

References

General References

1. Revsbech, N. P. and Sørensen, J., Eds., *Denitrification in Soil and Sediment*, FEMS Symposium Series 56, Plenum Press, New York, 1990.
2. Eady, R. R. and Hasnain, S. S., "Denitrification", in *Comprehensive Coordination Chemistry II*, McCleverty, J. A. and Meyer, T. J., Eds., Vol. 8, *Bio-coordination Chemistry*, Que, L. Jr. and Tolman, W. B., Eds., Elsevier Pergamon, Amsterdam, The Netherlands, 2004, Chapt. 8.28, pp. 759–786.
3. For concise summaries of the properties and structures of many of the individual enzymes mentioned, see *Handbook of Metalloproteins*, Vols. 1 and 2, Messerschmidt, A., Huber, R., Poulos, T., and Wieghardt, K., Eds., John Wiley & Sons, Chichester, 2001.

Specific References

4. Ye, R. W., Averill, B. A., and Tiedje, J. M., "Denitrification: production and consumption of nitric oxide", *Appl. Environ. Microbiol.*, **60,** 1053–1058 (1994).
5. Berks, B. C., Ferguson, S. J., Moir, J. W. B., and Richardson, D. J., "Enzymes and associated electron transport systems that catalyze the respiratory reduction of nitrogen oxides and oxyanions", *Biochim. Biophys. Acta,* **1232,** 97–173 (1995).
6. Zumft, W. G., "Cell biology and molecular basis of denitrification", *Microbiol. Mol. Biol. Rev.*, **61,** 533–616 (1997).
7. Tiedje, J. M., "Ecology of denitrification and dissimilatory nitrate reduction to ammonium", in *Biology of anaerobic microorganisms,* Zehnder, A. J. B., Ed., John Wiley & Sons, Inc. New York, 1988, pp. 179–243.
8. Siegel, R. S., Hauck, R. D., and Kurtz, L. T., "Determination of dinitrogen-30 and application to measurement of dinitrogen evolution during denitrification", *Soil Sci. Soc. Am. J.*, **46,** 68–74 (1982).
9. Hochstein, L. I. and Tomlinson, G. A., "The enzymes associated with denitrification", *Ann. Rev. Microbiol.*, **42,** 231–261 (1988).
10. Averill, B. A., "Dissimilatory nitrite and nitric oxide reductases", *Chem. Rev.*, **96,** 2951–2964 (1996).
11. Coyne, M. S., Arunakumari, A., Averill, B. A., and Tiedje, J. M., "Immunological identification and distribution of dissimilatory heme cd_1 and nonheme copper nitrite reductase in denitrifying bacteria", *Appl. Environ. Microbiol.*, **55,** 2924–2931 (1989).
12. Dhesi, R. and Timkovich, R., "Patterns of product inhibition for bacterial nitrite reductase", *Biochem. Biophys. Res. Commun.*, **123,** 966–972 (1984).
13. Smith, G. B. and Tiedje, J. M., "Isolation and characterization of a nitrite reductase gene and its use as a probe for denitrifying bacteria", *Appl. Environ. Microbiol.*, **58,** 376–384 (1992).
14. Chang, C. K., "On the structure of heme d_1", *J. Biol. Chem.*, **260,** 9520–9522 (1985).
15. Weeg-Aerssens, E., Wu, W., Ye, R. W., Tiedje, J. M., and Chang, C. K., "Purification of cytochrome cd_1 nitrite reductase from *Pseudomonas stutzeri* JM300 and reconstitution with native and synthetic heme d_1", *J. Biol. Chem.*, **266,** 7496–7502 (1991).
16. Sutherland, J., Greenwood, C., Peterson, J., and Thomson, A. J., "An investigation of the ligand-binding properties of *Pseudomonas aeruginosa* nitrite reductase", *Biochem. J.*, **233,** 893–898 (1986).
17. Fülöp, V., Moir, J. W. B., Ferguson, S. J., and Hajdu, J., "The anatomy of a bifunctional enzyme: structural basis for reduction of oxygen to water and synthesis of nitric oxide by cytochrome cd_1", *Cell,* **81,** 360–377 (1995).
18. Cowan, J. A., *Inorganic Biochemistry. An Introduction,* VCH Publisher, Inc., New York, 1993, p. 265.
19. Fito, I. and Rossman, M. G., "The active center of catalase", *J. Mol. Biol.*, **87,** 21–37 (1985).
20. Weatherall, D. J., Clegg, J. B., Higgs, D. R., and Wood, W. G., "The hemoglobinapathies", in *The Metabolic Basis of Inherited Disease,* 6 ed., Scriver, C. R., Beaudet, A. L., Sly, W. S., and Valle, D., Eds., McGraw-Hill, 1989, pp. 2281–2339.
21. Antonini, E. and Brunori, M., *Hemoglobin and myoglobin in their reactions with ligands,* North-Holland Publishing Co., Amsterdam, The Netherlands, 1971.
22. Wang, Y. and Averill, B. A., "Direct observation by FTIR spectroscopy of the ferrous heme NO^+ intermediate in reduction of nitrite by a dissimilatory heme cd_1 nitrite reductase", *J. Am. Chem. Soc.*, **118,** 3972–3973 (1996).
23. Sampath, V., Zhao, X.-j., and Caughey, W. S., "Characterization of interactions of nitric oxide with human hemoglobin A by infrared spectroscopy", *Biochem. Biophys. Res. Commun.*, **198,** 281–287 (1994).
24. Olson, L. W., Schaeper, D., Lacon, D., and Kadish, K. M., "Characterization of several novel iron nitrosyl porphyrins", *J. Am. Chem. Soc.*, **104,** 2042–2044 (1982).
25. Traylor, T. G. and Sharma, V. S., "Why NO?", *Biochemistry,* **31,** 2847–2849 (1992).
26. Godden, J. W., Turley, S., Teller, D. C., Adman, E. T., Liu, M. Y., Payne, W. J., and LeGall, J., "The 2.3 Å X-ray structure of nitrite reductase from *Achromobacter cycloclastes*", *Science,* **253,** 438–442 (1991).
27. Han, J., Loehr, T. M., Lu, Y., Selverstone-Valentine, J., Averill, B. A., and Sander-Loehr, J., "Resonance Raman excitation profiles indicate multiple Cys → Cu charge transfer transitions in type I copper proteins", *J. Am. Chem. Soc.*, **115,** 4256–4263 (1993).
28. LaCroix, L. B., Shadle, S. E., Wang, Y., Averill, B. A., Hedman, B., Hodgson, K. O., and Solomon, E. I., "Electronic Structure of the Perturbed Blue Copper Site in Nitrite Reductase: Spectroscopic Properties, Bonding, and Implications for the Entatic/Rack State", *J. Am. Chem. Soc.*, **118,** 7755–7768 (1996).
29. Messerschmidt, A., Rossi, A., Ladenstein, R., Huber, R., Bolognesi, M., Gatti, G., Marchesini, A., Petruzzelli, R., and Finazzi-Agró, A., "X-ray crystal structure of the blue oxidase ascorbate oxidase from *Zucchini.* Analysis of the peptide fold and a model of the copper sites and ligands", *J. Mol. Biol.*, **206,** 513–529 (1989).
30. Adman, E. T., Godden, J. W., and Turley, S., "The structure of copper-nitrite reductase from *Achromobacter cycloclastes* at five pH values, with NO_2^- bound and with type II copper depleted", *J. Biol. Chem.*, **270,** 27458–27474 (1995).
31. Howes, B. D., Abraham, Z. H. L., Lowe, D. J., Bruser, T., Eady, R. R., and Smith, B. E., "EPR and Electron Nuclear Double Resonance (ENDOR) Studies show nitrite binding to Type 2 copper centres of the dissimilatory nitrite reductase of *Alcaligenes xylosoxidans* (NCIMB 11015)", *Biochemistry,* **33,** 3171–3177 (1994).
32. Libby, E. and Averill, B. A., "Evidence that the type 2 copper centers are the site of nitrite reduction by

Achromobacter cycloclastes nitrite reductase", *Biochem. Biophys. Res. Commun.*, **187,** 1529–1535 (1992).

33. Hulse, C. L. and Averill, B. A., "Evidence for a copper-nitrosyl intermediate in denitrification by the copper-containing nitrite reductase of *Achromobacter cycloclastes*", *J. Am. Chem. Soc.*, **111,** 2322–2323 (1989).
34. Averill, B. A., "Novel copper nitrosyl complexes: contributions to the understanding of dissimilatory copper-containing nitrite reductases", *Angew. Chem. Int. Ed. Engl.*, **33,** 2003–2004 (1994).
35. Tolman, W. B., "A model for the substrate adduct of copper nitrite reductase and its conversion to a novel tetrahedral copper (II) triflate complex", *Inorg. Chem.*, **30,** 4877–4880 (1991).
36. Halfen, J. A., Mahapatra, S., Wilkinson, E. C., Gengenbach, A. J., Young, V. G., Jr., Que, L., Jr., and Tolman, W. B., "Synthetic Modelling of Nitrite Binding and Activation by Reduced Copper Proteins. Characterization of Copper(I)–Nitrite Complexes That Evolve Nitric Oxide", *J. Am. Chem. Soc.*, **118,** 763–776 (1996).
37. Kukimoto, M., Nishiyama, M., Murphy, M. E. P., Turley, S., Adman, E. T., Horinouchi, S., and Beppu, T., "X-ray structure and site-directed mutagenesis of a nitrite reductase from *Alcaligenes faecalis* S-6: Roles of two copper atoms in nitrite reduction", *Biochemistry,* **33,** 5246–5252 (1994).
38. Murphy, M. E. P., Turley, S., Kukimoto, M., Nishiyama, M., Horinouchi, S., Sasaki, H., Tanokura, M., and Adman, E. T., "Structure of *Alcaligenes faecalis* nitrite reductase and a copper site mutant, M150E, that contains zinc", *Biochemistry,* **34,** 12107–12117 (1995).
39. Jackson, M. A., Tiedje, J. M., and Averill, B. A., "Evidence for an NO-rebound mechanism for the production of N_2O from nitrite by the copper-containing nitrite reductase from *Achromobacter cycloclastes*", *FEBS Lett.*, **291,** 41–44 (1991).
40. Kastrau, D. H. W., Heiss, B., Kroneck, P. M. H., and Zumft, W. G., "Nitric oxide reductase from *Pseudomonas stutzeri,* a novel cytochrome *bc* complex. Phospholipid requirement, electron paramagnetic resonance and redox properties", *Eur. J. Biochem.*, **222,** 293–303 (1994).
41. Girsch, P. and de Vries, S., "Purification and initial kinetic and spectroscopic characterization of NO reductase from *Paracoccus denitrificans*", *Biochim. Biophys. Acta,* **1318,** 202–216 (1997).
42. Dermastia, M., Turk, T., and Hollocher, T. C., "Nitric oxide reductase. Purification from *Paracoccus denitrificans* with use of a single column and some characteristics", *J. Biol. Chem.*, **266,** 10899–10905 (1991).
43. Turk, T. and Hollocher, T. C., "Oxidation of dithiothreitol during turnover of nitric oxide reductase: evidence for generation of nitroxyl with the enzyme from *Paracoccus denitrificans*", Biochem. Biophys. Res. Commun., **183,** 983–988 (1992).
44. Zhao, X.-J., Sampath, V., and Caughey, W. S., "Cytochrome *c* oxidase catalysis of the reduction of nitric oxide to nitrous oxide", *Biochem. Biophys. Res. Commun.*, **212,** 1054–1060 (1995).
45. Paul, P. P. and Karlin, K. D., "Functional modeling of copper nitrite reductases: reactions of NO_2^- or NO with copper (I) complexes", *J. Am. Chem. Soc.*, **113,** 6331–6332 (1991).
46. Bottomley, F., "Reactions of nitrosyls", in *Reactions of Coordinated Ligands*, Braterman, B. S., Ed., Plenum Publishing Corporation, Vol. 2, New York, 1989, pp. 115–222.
47. Oost van der, J., Boer de, A. P. N., Gier de, J.-W., L., Zumft, W. G., Stouthamer, A. H., and Spanning van, R. J. M., "The heme-copper oxidase family consists of three distinct types of terminal oxidases and is related to nitric oxide reductase", *FEMS Microbiol. Lett.*, **121,** 1–10 (1994).
48. Saraste, M. and Castresana, J., "Cytochrome oxidase evolved by tinkering with denitrification enzymes", *FEBS Lett.*, **341,** 1–4 (1994).
49. Antholine, W. E., Kastrau, D. H. W., Steffens, G. C. M., Buse, G., Zumft, W. G., and Kroneck, P. M. H., "A comparative EPR investigation of the multicopper proteins nitrous-oxide reductase and cytochrome *c* oxidase", *Eur. J. Biochem.*, **209,** 875–881 (1992).
50. Kroneck, P. M. H., Antholine, W. A., Riester, J., and Zumft, W. G., "The nature of the cupric site in nitrous oxide reductase and of Cu_a in cytochrome *c* oxidase", *FEBS Lett.*, **248,** 212–213 (1989).
51. Andrew, C. R., Han, J., de Vries, S., van der Oost, J., Averill, B. A., Loehr, T. M., and Sanders-Loehr, J., "Cu_A of cytochrome *c* oxidase and the A site of N_2O reductase are tetrahedrally distorted type 1 Cu cysteinates", *J. Am. Chem. Soc.*, **116,** 10805–10806 (1994).
52. Brown, K., Tegoni, M., Prudêncio, M., Pereira, A. S., Besson, S., Moura, J. J., Moura, I., and Cambillau, C., "A novel type of catalytic copper cluster in nitrous oxide reductase", *Nat. Struct. Biol.*, **7,** 191–195 (2000).
53. Brown, K., Djinovic-Carugo, K., Haltia, T., Cabrito, I., Saraste, M., Moura, J. J., Moura, I., Tegoni, M., and Cambillau, C., "Revisiting the catalytic Cu_Z cluster of nitrous oxide (N_2O) reductase. Evidence of a bridging inorganic sulfur", *J. Biol. Chem.*, **275,** 41133–41136 (2000).
54. Alvarez, M. L., Ai, J., Zumft, W., Sanders-Loehr, J., and Dooley, D. M., "Characterization of the Copper–Sulfur Chromophores in Nitrous Oxide Reductase by Resonance Raman Spectroscopy: Evidence for Sulfur Coordination in the Catalytic Cluster", *J. Am. Chem. Soc.*, **123,** 576–587 (2001).
55. Haltia, T., Brown, K., Tegoni, M., Cambillau, C., Saraste, M., and Djinovic-Carugo, K., "Crystal structure of nitrous oxide reductase from *Paracoccus denitrificans* at 1.6 Å resolution", *Biochem. J.*, **369,** 77–88 (2003).

XII.5. Sulfur Metabolism

António V. Xavier and Jean LeGall
Instituto de Tecnologia Química e Biológica
Universidade Nova de Lisboa
2780-156 Oeiras, Portugal

Contents

XII.5.1. Introduction

Sulfur is capable of existing in a wide array of chemical forms and has always been an element available to biological systems.[1–5] In the reducing atmospheres prevalent in prebiotic and early periods of biological evolution, sulfur was available either as H_2S ($pK_{a1} = 6.9$, favoring HS^- at physiological pH) or even as FeS [$K_{SP} = 6 \times 10^{-18}$, a solubility product only approximately two orders of magnitude smaller than that of $Fe(OH)_2$, thus not causing very serious precipitation problem at $pK < 7$]. The increase in oxygen concentration in the atmosphere changed this situation and at present sulfate is the main source of sulfur to the biosphere. However, sulfur is also available in colloidal elemental form (S_8) obtained by spontaneous oxidation of some reduced organic sulfur compounds produced by microbial metabolism and H_2S liberated by volcanic activity.

Both the availability and the physicochemical properties of sulfur make it one of the elements universally utilized in biological systems, although the least abundant of the six primordial elements. In fact, within the restrictions imposed by aqueous solutions there are very few elements that can engage in redox reactions with such a wide range of stable oxidation states, −II to +VI in two-electron steps (Fig. XII.5.1). Moreover, these redox reactions are coupled with acid–base reactions. Thus, the dependence of the various equilibria upon protons makes the reduction potentials pH

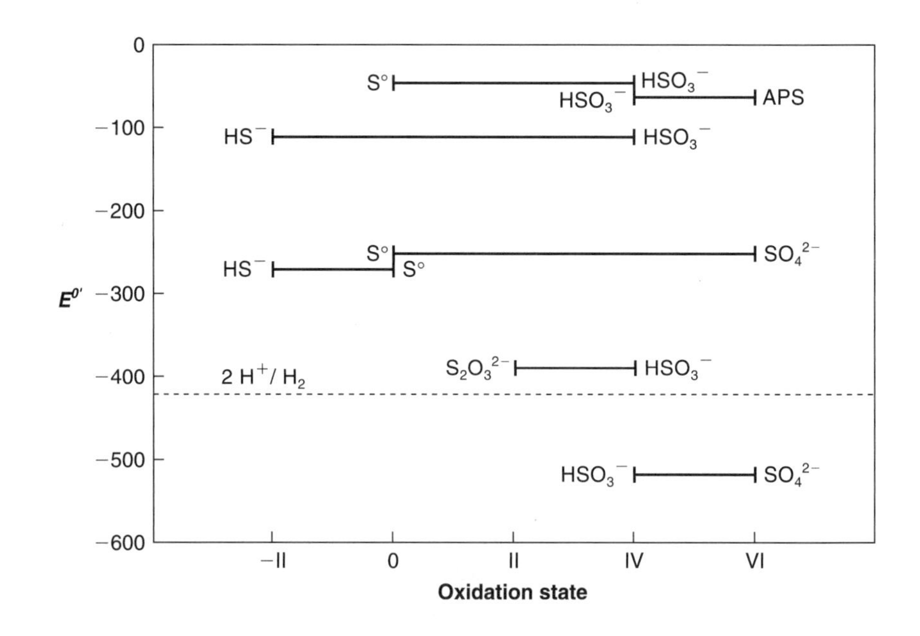

Fig. XII.5.1. Reduction potentials ($E^{o\prime}$) of the most relevant redox reactions used in the biological sulfur cycle. The dashed line indicates the value of $E^{o\prime}$ for the 2 H^+/H_2 redox couple ($E^{o\prime} = -420$ mV). Adenylphosphosulfate = APS.

dependent for all reactions depicted in this figure. In many cases, the dependence extends throughout the relevant physiological pH ranges.

In contrast to the stability of the C–H and N–H bonds in organic compounds, which strongly limits carbon and nitrogen reactivity, this stability is not the case for S–H. Thus, sulfur can form labile covalent bonds with organic molecules, and thiolate (RS^-) is a good attacking and leaving group that readily forms intermediates of low energy. Furthermore, as thiolate can perform both redox and acid–base chemistry, it is often the group of choice to catalyze condensation reactions. Moreover, owing to sulfur's high polarizability, which decreases the ionic character of its bonds, sulfur ligands will not easily displace oxygen ligands from the more electropositive metal ions (e.g., Mg^{2+}, Ca^{2+}, Mn^{2+}, and even Fe^{3+}). Nevertheless, sulfur has considerable affinity for metal ions and it is widely accepted that metal sulfides (clusters) have played a central role in biological chemistry since the early periods of evolution, when competition from oxygen was very low.

With all the essential functional physico-chemical properties described above, it is not surprising that sulfur is a very prolific element in the protoplasm.

XII.5.2. Biological Role of Sulfur Compounds

Clearly, sulfur compounds pervade life chemistry, where they play many irreplaceable roles. In particular, its compounds are important in the bioenergetic metabolism of specialized microorganisms and are used either as electron donors, in anaerobic phototrophy and aerobic chemotrophy, or as terminal electron acceptors, in anaerobic sulfate respiration (see Section XII.5.3.1).

Reacting as a base, thiolate readily binds H^+ and can bind all the transition metal ions used by biological systems, with the exception of manganese. Together with inorganic sulfide (S^{2-}), the thiolate group of cysteinyl residues is responsible for holding together some of the most important transition metal clusters found in electron-transfer metalloproteins (e.g., ferredoxins, Chapter IV) and metalloenzymes (e.g., nitrogenase, Section XII.3 and hydrogenase, Section XII.1). The attacking and leaving characteristics of thiolate (see above) gives it an important role in acid–base catalysis, performed by enzymes that use a cysteinyl residue in their active site (e.g., proteases). On a less positive note, these active sites are the target of toxic heavy metal ions, which bind to thiolate inhibiting these enzymes.

Thiols are readily oxidized to disulfides, forming a relatively weak bond that is easily reduced. In particular, the equilibrium between the RS–SR/2RSH redox couple of glutathione (the tripeptide δ-glutamylcysteinylglycine) is used as a redox potential buffer in the cytoplasm of eukaryotes, keeping the redox potential at ca. $-100\,mV$. This equilibrium is maintained by glutathione reductase; electrons are shuttled from NADPH to redox active thiol groups in the active site of the enzyme. Since glutathione reductase is an NADPH-dependent enzyme, it can be used to link the glutathione redox equilibrium with the reductive anabolic metabolism. Also, reduction of disulfides to sulfhydryl groups is responsible for the activation of several enzymes (e.g., some enzymes of the photosynthetic carbon reduction cycle are activated this way). The mechanism is mediated by thioredoxin, a small protein that contains two thiol groups reversibly oxidizable to disulfide, after reduction by ferredoxin. The weakness of the RS–H bond also makes thiols good hydrogen donors to aggressive organic radicals (see Chapter XIII), and gives glutathione a protective role against these radicals. The mechanism used by ribonucleoside reductase illustrates another important role of the thiol group of Cys (via a thiyl radical, –S•) in the active site of the enzyme (Section XIII.3).

An important structural function of this redox reaction is the formation of disulfide bridges in proteins, resulting in cross-linking of polypeptide chain(s). Further structural aspects involving sulfur chemistry are the covalent binding of type *c* hemes by thioether bonds to cysteinyl residues of the apoprotein, and the use of the thioether of Met or the thiol of Cys residues as axial ligands to heme proteins (Chapter IV). Also, the thiolates of cysteinyl residues bind Zn to impose specific structural motifs (e.g., in zinc fingers, which bind DNA and mRNAs and have regulatory functions, see Chapter III and Sections XIV.1 and XIV.2).

The importance of sulfur for group transfer in biological reactions includes the methylation processes performed by S-adenosylmethionine (SAM, also known as Adomet) bound to appropriate enzymes. The highly reactive methyl group, originating from the thioether of the Met moiety of SAM, is transferred by nucleophilic displacement from the trialkylsulfonium ion obtained when Met reacts with ATP, generating a compound with high group-transfer energy. S-Adenosylmethionine-dependent transmethylations are involved in numerous biological processes (see Sections XIII.2 and XIII.4), including synthesis (e.g., hemes, epinephrine, and phosphatidylcholine) and regulation (e.g., in methylation of protein amino acid residues of signal transducers for chemotaxis and specific guanines of nucleic acids in protective mechanisms, and control of gene expression).

A more specific role of sulfur compounds is found in one of the metabolic steps in methanogenesis (see Section XII.2). This step uses the thiolate group of coenzyme M (a sulfonated thiol, $H{-}SCH_2CH_2SO_3^-$, only found in methanogenic archaea) that accepts a methyl group from N5-methyltetrahydromethanopterin, another C_1 carrier, to produce the thioester $H_3C{-}SCH_2CH_2SO_3^-$. This reaction is so far unique in that it is used for energy conservation, coupling the methyl transferase activity with the generation of a sodium gradient.

XII.5.3. Biological Sulfur Cycle

The transformation of sulfur compounds constitutes one of the major elemental biological cycles (Chapter II). The anaerobic sulfur cycle (Fig. XII.5.2) is particularly important in the food degradation chain performed by sulfate-reducing bacteria. These bacteria decompose small organic acids (e.g., lactate) and alcohols (e.g., ethanol) excreted by fermentative bacteria that use more complex nutrients. Sulfate-reducing bacteria are particularly important in marine habitats, where they direct the food chain toward the production of CO_2, rather than CH_4, as observed in environments where methanogens dominate. Sulfur is involved in the anaerobic production of biomass, where some of its compounds are used as electron donors in the light-dependent carbon fixation associated with the anoxygenic photosynthesis performed by anaerobic purple and green sulfur bacteria. Thus, it is likely that the redox equilibria of sulfur compounds were crucial during the earlier steps of evolution, similar to the way that the redox pair O_2/H_2O is now critical for aerobic biomass cycling.

The cycling of sulfur is also extremely important in keeping the global balance of environmental sulfur by connecting the anoxic and oxic zones. Indeed, while the anaerobic sulfate respiration of sulfate-reducing bacteria generates hydrogen sulfide, the accumulation of this highly toxic compound is prevented by the aerobic sulfide oxidation performed by colorless sulfur bacteria. Even local alterations of this redox balance can have dramatic ecological (e.g., massive kills of organisms), environmental (e.g., acid rain and acidification of soil), and economic (e.g., metal corrosion) implications. Unfortunately, most of these alterations are a direct consequence of the increase of highly deleterious (nonbiological) human activities.

Curiously, although the importance of sulfur has been recognized since the early days of biochemical research, characterization of the sulfur cycle remains rather incomplete. Nevertheless, the major steps of this biological cycle can be outlined (Fig. XII.5.2). The metabolic pathways of the sulfur cycle include both assimilatory reactions (through which inorganic compounds are reduced to be incorporated as organic sulfur, i.e., sulfur fixation) and dissimilatory ones (where inorganic sulfur compounds are used either as terminal electron acceptors or electron donors in energy conservation processes and then excreted).

The eight-electron redox interconversion between HS^- and SO_4^{2-} (outer cycle of Fig. XII.5.2) and the two-electron interconversion between HS^- and S^0 (inner cycle) are achieved by specialized microorganisms growing in a wide variety of habitats. These include most of the extreme conditions where life has been shown to be possible. Of particular importance are the habitats where syntrophic associations perform a full dissimilatory cycle, known as sulfureta. These associations can be achieved either by sulfate-reducing bacteria associated with sulfide-oxidizing microorganisms (outer cycle of Fig. XII.5.2), or by sulfur reducers (e.g., *Desulfuromonas*) associated with phototrophic green sulfur bacteria (inner cycle). Sulfureta can be found in marine sediments, hot sulfur springs, volcanic environments, and salt marshes. In some of these extreme habitats, the biological sulfur cycle is in permanent interaction with the interconversions of especially active geochemical sulfur cycles.

The interactions between the bio- and the very active geocycle may be responsible for some of the difficulties in distinguishing whether intermediate sulfur compounds are true metabolites, or a contribution from *in vitro* nonenzymatic conversions. Thus, extrapolations to *in vivo* enzymatic activities may result in proposals for pathways whose importance remains the subject of much debate. Also, the widely different environmental habitat conditions of the microorganisms directly involved in the sulfur cycle has resulted in a great metabolic diversity. It is expected that a substantial ecological change was an important evolutionary pressure that drove the evolution of

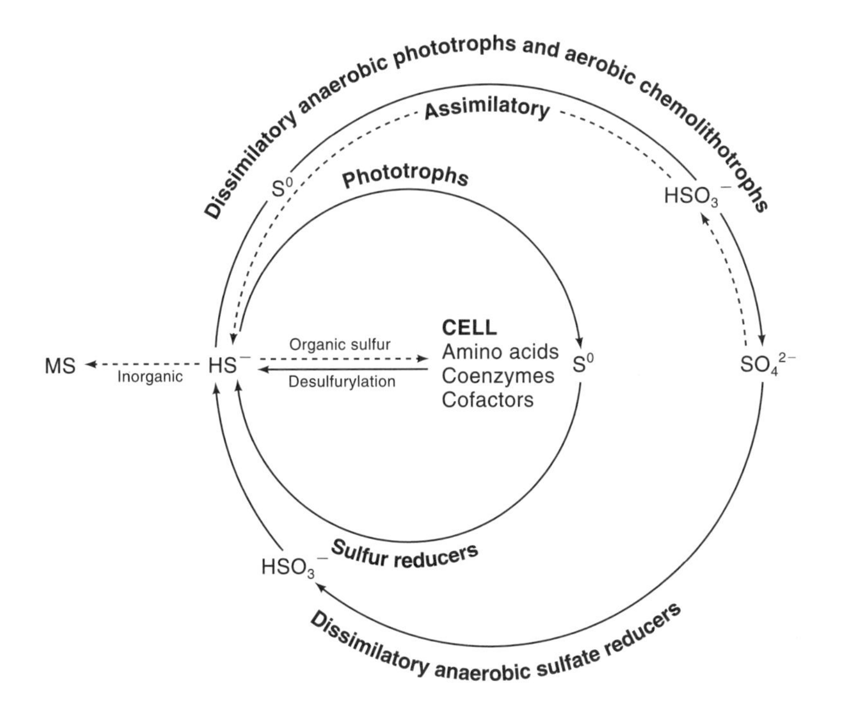

Fig. XII.5.2.
Biological cycle of sulfur.

microorganisms adapted to the peculiarities of their habitats, both in terms of operative pathways and enzymatic systems utilized. However, the recently obtained genome sequences of sulfate-reducing bacteria are a hopeful step toward the clarification of some of the more relevant pathways (e.g., the Archaean *Archaeoglobus fulgidus* and *Desulfovibrio vulgaris* Hildenborough obtained at the Institute of Genomic Research,[6] as well as that of *Desulfovibrio desulfuricans* G20 which is obtained at the Department of Energy (DOE) Joint Genome Institute).[7] Nevertheless, the structural similarity between the oxyanions of sulfur and nitrogen, including the similarities of some redox steps of their biological cycles, may still add to possible confusion. Indeed it has been shown that some enzymes can use either nitrogen or sulfur oxyanions[8,9] and it may be difficult to distinguish *in vivo* whether these enzymes are truly bifunctional.

Still, the utilization of elemental sulfur in both directions, that is, oxidation and reduction (inner "small" cycle of Fig. XII.5.2) is well established. In addition, the bidirectional use of the two-electron conversion of sulfate to sulfite and the six-electron conversion of sulfite to sulfide (outer "large" cycle) can occur. Nevertheless, evidence is accumulating for the involvement of thiosulfate ($S_2O_3^{2-}$), trithionate ($S_3O_6^{2-}$), and even tetrathionate ($S_4O_6^{2-}$) as possible intermediates in this cycle. Also, disproportionation reactions may be involved, using large (membrane linked multi-) enzyme complexes, or even nonenzymatic reactions.

XII.5.3.1. Dissimilatory Pathways

With all the above precautionary remarks in mind, and leaving details to more specialized literature, it is possible to reduce the description of the major steps of the anaerobic chemotrophic dissimilatory pathway of sulfur to three enzymes: (1) ATP sulfurylase containing both Zn and Co in a unique structural arrangement;[10] (2) APS reductase (APS = Ad–P_i–O–SO_3^-, where Ad stands for adenosine), an iron–sulfur flavoenzyme; and (3) sulfite reductase (SIR), an iron–sulfur and siroheme containing enzyme (see Section XII.5.3.3). While this reductionistic approach is still a matter of active debate, the importance of these enzymes in life and evolution is undisputed. Actually, phylogenetic studies suggest that both APS and sulfite reductases may have evolved from a common ancestor of bacteria and archaea, possibly from a protogenotic period that predates the evolution of the photosynthetic apparatus.[11]

The requirement for the first two enzymes results from the chemical stability of sulfate and the thermodynamic constraint for its direct reduction to sulfite ($E^{\circ\prime} = -516\,\text{mV}$). This reduction is facilitated after sulfate energization by ATP, forming the phosphoester APS,

$$SO_4^{2-} + ATP + 2\,H^+ \rightarrow APS + PP_i \qquad \Delta G^{\circ\prime} = +46\,\text{kJ mol}^{-1}$$

Although this reaction is highly endergonic, both products are immediately removed by two highly exergonic reactions: Pyrophosphate is hydrolyzed ($\Delta G^{\circ\prime} = -22\,\text{kJ mol}^{-1}$), and the sulfate moiety of APS is reduced to bisulfite ($E^{\circ\prime} = -60\,\text{mV}$) and released together with adenosine monophosphate (AMP),

$$APS + H^+ + 2\,e^- \rightarrow HSO_3^- + AMP$$

$$\Delta G^{\circ\prime} = -69\,\text{kJ mol}^{-1} \text{ (assuming } H_2 \text{ as the reductant)}$$

Thus, at the cellular steady-state equilibrium, generated by this ingenious mechanism, the net process becomes thermodynamically favorable. Bisulfite can then be reduced in a six-electron step to give bisulfide, which is excreted,

$$HSO_3^- + 6\,e^- + 6\,H^+ \rightarrow HS^- + 3\,H_2O$$

$$\Delta G^{\circ\prime} = -174\,\text{kJ mol}^{-1} \text{ (assuming } H_2 \text{ as the reductant)}$$

This dissimilatory reduction of sulfate is an anaerobic inorganic respiration, where oxidized sulfur compounds are used as electron acceptors and through which the bacteria obtain ATP by oxidative phosphorylation. Here, the energy is supplied by the oxidation of dihydrogen ($H_2 \rightarrow 2\,H^+ + 2\,e^-$) via hydrogenase (see the section Bioenergetic Characteristics of Sulfate Respiration; Section XII.1).

A different anaerobic dissimilatory pathway involves the oxidation of sulfur compounds by anoxygenic photosynthesis, which has been studied since the nineteenth century, culminating with the classic experiments of van Niel in the 1930s. In this process, performed by the purple and green sulfur bacteria, reduced sulfur compounds (such as sulfide, elemental sulfur, sulfite, thiosulfite, polythionate, and dimethyl sulfide) are used as electron donors to photosystem I, generating ATP by cyclic photophosphorylation (without the need for water splitting and the subsequent generation of oxygen by photosystem II). Thus, reduced sulfur compounds can be used as a source of the reducing power necessary for a light-dependent CO_2 fixation. The oxidized sulfur compounds resulting from this anoxygenic phototrophic process can be further oxidized up to the level of sulfate, to generate additional ATP by oxidative phosphorylation, as well as additional reducing power (NADPH). Some cyanobacteria (e.g., *Oscillatoria limnetica*) can also combine the anoxygenic photoassimilation of CO_2, coupled to the oxidation of sulfide to S_8 (during the daylight period), with the respiration of this sulfur (during the night) to obtain polyglucose, which is stored.

The chemolithotrophic colorless sulfur bacteria also oxidize reduced sulfur compounds as an energy source, but in aerobic conditions. The stepwise oxidation is linked to oxidative phosphorylation by a complex system of electron-transfer metalloproteins. Here, some microorganisms use an alternative enzyme for the oxidation of sulfite. Instead of using APS reductase in reverse, they use sulfite oxidase, a molybdoprotein (see Section XII.5.3.3; Section XII.6). Sulfite oxidase is also used by animals for the final step of the oxidative degradation of sulfur amino acids.

An interesting combination of aerobic sulfur oxidation and anaerobic sulfate respiration capabilities is found in some extremophile archaea (e.g., *Desulfurolobus ambivalens*).

In some marine environments dimethyl sulfide, produced by degradation of dimethylsulfonium propionate, can be oxidized to dimethyl sulfoxide (DMSO) by phototrophic bacteria to fix CO_2 (see above). The DMSO can then be reduced back ($E^{\circ\prime} = 160\,\text{mV}$) by anaerobic respiration using the DMSO reductase (Section XII.6). Dimethyl sulfide is the major sulfur compound in Nature (see Chapter II; Section XII.6).

Bioenergetic Characteristics of Sulfate Respiration. Owing to the quite negative redox potentials of the sulfur compounds used as final acceptors during sulfate respiration (see Fig. XII.5.1 and compare with the redox potential of the O_2/H_2O pair, $E^{\circ\prime} = +816\,\text{mV}$), there is an important energetic constraint for the energy conservation process of sulfate-reducing bacteria, since the free-energy change available for this process is quite small, even when using H_2 as an energy source (see below and Fig. XII.5.1). Furthermore, both sulfate activation (to APS), as well as its transport to the cytoplasm, are processes that consume significant amounts of ATP. Nevertheless, sulfate-reducing bacteria were the first nonphotosynthetic organisms for which energy conservation by oxidative phosphorylation was demonstrated, by H. D. Peck, Jr., and collaborators in the 1960s. In *Desulfovibrio* sp., oxidation of organic nutrients can be used to generate H_2 (a reaction catalyzed by a cytoplasmic hydrogenase). As also proposed by Peck, a process known as the hydrogen cycle mechanism can then be used to energize ATP synthase: The H_2 produced in the cytoplasm diffuses to the periplasm, where it can be reoxidized via an uptake hydrogenase to generate the protons necessary to induce ATP synthesis. However, this hydrogenase activity is considerably hindered as the proton concentration increases, thus introducing yet another thermodynamic constraint. This obstacle can be surpassed by an electroprotonic energy transduction step, performed by Type I tetraheme cytochrome c_3, 4HCc_3TpI (PDB code: 2CTH), another metalloprotein that stimulates the uptake activity of hydrogenase at lower pH values.

This tetraheme cytochrome (Fig. XII.5.3) is a small (~110 amino acid residues and 14 kDa), soluble, and very stable protein that has been studied in great detail, being one of the best characterized electron-transfer proteins.[12,13] All its hemes have bis-His axial ligation and very negative, pH-dependent reduction potentials (−200 to −400 mV).

Its energy transduction role is governed by an essential thermodynamic link (heterotropic cooperativity) between the electron affinity of its redox centers (i.e., redox potential) and the proton affinity of a protic center (i.e., pK_a), such that its reduction is coupled to the uptake of two protons. Furthermore, its redox chemistry is dominated by a positive homotropic cooperativity between the redox potentials of two of its hemes, an anti-Coulomb process that must involve a redox-linked structural change and results in a concerted 2 e^- step.[16] Thus, the network of pairwise homo- and heterocooperativities of the four heme Cytc_3 Type I (4HCc_3TpI) is designed to

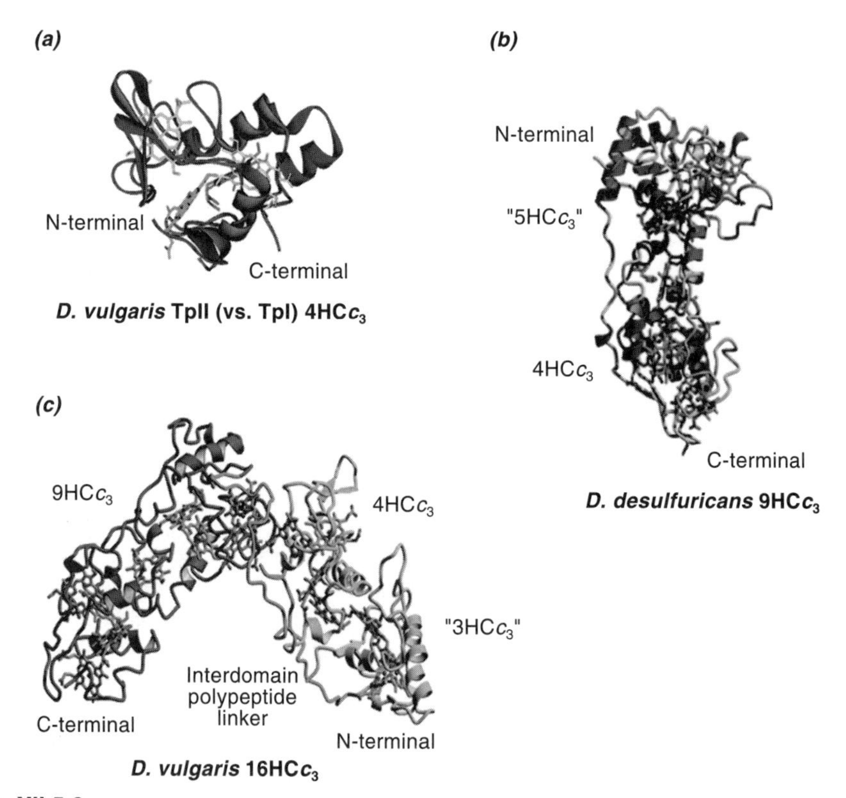

Fig. XII.5.3.
Multiheme cytochromes c_3. (*a*) Comparison of *D. vulgaris* 4HCc_3TpII and 4HCc_3TpI; By superimposing the hemes, although only one set of hemes is shown: Heme 1, heme 2, heme 3, heme 4. (PDB code: 2CTH) (*b*) *Desulfovibrio desulfuricans* 9HCc_3; the 5HCc_3 region refers to a 4HCc_3-like domain plus an extra heme in an extra loop.[14] (PDB code: 19HC) (*c*) *Desulfovibrio vulgaris* 16HCc_3; the 3HCc_3 region refers to a 4HCc_3-like domain missing one heme.[15] (PDB code: 1H29) [See color insert.]

receive the electrons and protons generated by the oxidation of H_2 and to perform an electroprotonic energy transduction process, consisting of the following steps: (1) accept 2 e^- of high reducing energy (i.e., very low redox potential), with the consequent increase of the pK_a of its protic center, which is then eager to receive basic protons; (2) accept these protons, which stabilize the two reduced hemes (i.e., increase their redox potentials); (3) deliver the electrons to an acceptor with centers of higher redox potentials, and concomitantly acidify the protic centers (redox-linked proton activation, also known as "proton thrusting"); which (4) can be used to trigger the ATP synthase activity. The deenergized electrons can be transported to the sulfate respiratory chain via a transmembrane electron-transfer complex.

In particular, some multiheme cytochromes *c* (Fig. XII.5.3) can be used as electron entry gates to such complexes[14,15,17] since 4HC*c*3TpI stimulates the transfer of electrons from hydrogenase to these membrane-linked cytochromes. As they contain one or more 4HC*c*3TpI like-domains, these cytochromes belong to the cytochrome c_3 family, which include the Type II tetraheme cytochrome c_3, 4HC*c*3TpII,[17] the 9 heme cytochrome c_3, 9HC*c*3 (PDB code: 19HC),[14] and the 16, 16HC*c*3 (PDB entry 1H29).[15] Their genes are included in a polycystronic operon that codes transmembrane electron-transfer membrane proteins.

XII.5.3.2. Assimilatory Pathways

Sulfate is the major nutrient source for the assimilative pathway leading to the synthesis of organic sulfur compounds. Many organisms use a similar strategy to that of the dissimilatory reduction of sulfate: Sulfate is activated to form the phosphoester APS (which in some organisms is further phosphorylated to phospho-APS, PAPS), and subsequently reduced to sulfite (2 e^-) and sulfide (6 e^-). Once reduced, sulfur can be incorporated into biological molecules (see above). Alternatively, some purple nonsulfur bacteria that cannot assimilate sulfate use reduced sulfur compounds directly.

XII.5.3.3. Metalloenzymes from Sulfate-Reducing Bacteria

The microorganisms utilizing sulfur compounds for respiration contain a wide variety of metal ion centers and homo- and heterometal clusters of which many examples can be found in specific sections of this book. In particular, the purification and characterization of several metalloproteins from sulfate-reducing bacteria have greatly contributed to the elucidation of the metabolic steps used by these microorganisms, thus explaining the growing interest in microbial sulfur metabolism. Also, several novel metal ion centers, clusters, and structural motifs were first found in proteins isolated from sulfate-reducing bacteria (although not necessarily used in processes directly involving reactions with sulfur compounds) and have served as models to help in elucidating the mechanisms used by proteins of more complex organisms. For example: (1) aldehyde dehydrogenase, a molybdopterin-containing enzyme, whose structure helped to explain the mechanism of xanthine oxidase (Section XII.6); (2) the first demonstration of the regulation of a redox chain by a calcium-binding protein was found in sulfate-reducing bacteria, *Desulfovibrio gigas*;[18] only several years later, a similar regulation was found for the mammalian endothelial nitric oxide synthase;[19] (3) the structural, thermodynamic, and kinetic characterization of tetraheme cytochromes c_3 provided models to explain the redox-linked mechanism for proton pumping by cytochrome *c* oxidase;[16] (4) the first example for the structural arrangement of [3Fe–4S] centers and the significance of its interconversion to [4Fe–4S] provided by the characterization of *D. gigas* ferredoxin[20] assisted in the characterization of a similar center of aconitase (Section IX.4); (5) The first crystal structure of a native di-iron site in an iron-storage protein was obtained for the *Desulfovibrio desulfuricans* bacterioferritin;[21] (6) Another diiron site was characterized in a novel bifunctional enzyme, rubredoxin–oxygen oxidoreductase, which catalyzes both the reduction of O_2 to H_2O[22] as well as the reduction of NO.[23]

Several other *Desulfovibrio* metalloproteins, containing unusual or novel metal cofactors and prosthetic groups, include the following examples: sulfite reductase, with a unique Fe–siroheme supercluster at the active site (see above); hydrogenase, which provided the first example for a Ni–Fe–S containing enzyme;[24] neelaredoxin, which contains a mononuclear iron center with a structural arrangement and activity reminiscent of superoxide dismutase;[25] rubrerythrin, a protein containing both a rubredoxin- and a hemerythrin-like center.[26]

Three-dimensional crystal structures are available for most of the above mentioned proteins. In contrast, from all the enzymes directly involved in redox processes of the sulfur cycle mentioned above, only a few X-ray structures are available: sulfite oxidase and sulfite reductase. Furthermore, none of these structures are of protein isolated from sulfate-reducing bacteria. In particular, it is rather unfortunate that no structure is yet available for the so-called dissimilatory bisulfite reductases, although these enzymes are responsible for the production of large amounts of hydrogen sulfide. In addition, some sulfite reductases, such as "desulfoviridin", contain 80% demetalized siroporphyrin instead of heme and the *in vitro* product of the isolated proteins is mainly trithionate instead of sulfide. No clear mechanism is available for these important enzymes. Other important metalloenzymes for reactions involving sulfur compounds (e.g., DMSO reductase) are covered in detail in elsewhere in this book (Section XII.6).

Sulfite Oxidase

Chicken liver sulfite oxidase is a 110-kDa homodimer for which a crystallographic structure was obtained at 1.9-Å resolution. It contains two domains: the smaller N-terminal domain has a cytochrome b_5 working as an electron acceptor, and the C-terminal domain binds a deeply buried active site molybdenum cofactor (Moco). Its pyranopterin dithiolate ligand is surrounded by an array of hydrogen bonds to the protein main chain and residue side chains in the binding pocket (Section XII.6).

Sulfite Reductase

Assimilatory sulfite reductases are rather diverse oligomeric enzymes (complexes), which have been characterized from a variety of organisms. They have variable monomeric compositions, either from membrane or soluble fractions, and show different responses to regulatory effectors. Despite this variety, all sulfite reductases have an active site that contains a siroheme (Fe-tetrahydroporphyrin, $E^{\circ\prime} = -340\,\text{mV}$) bridged to a [4Fe–4S] cluster ($E^{\circ\prime} = -405\,\text{mV}$) through the thiolate group of a common bridging cysteinyl residue, a unique active-site arrangement.

The *Escherichia coli* assimilatory SIR is an oligomer of eight 66-kDa flavoprotein and four 64-kDa hemoprotein subunits (PDB code: 1AOP). The octamer binds four flavin adenine dinucleotide (FAD) and four flavin mononucleotide (FMN) electron supplier cofactors and the tetramer binds four siroheme/[4Fe–4S] superclusters.

The supercluster works as a coupled electronic unit whose electronic properties are modulated by sulfite binding. Based on its 1.6-Å resolution X-ray structure, it is proposed that sulfite binds to the iron of siroheme, where it is stabilized by a cluster of positive charges supplied by two Arg and two Lys residues.[9] No release of intermediates has been detected during the three two-electron steps of the six-electron reduction of sulfite to sulfide. Thus, the intermediates produced by sequential reduction and dehydration of sulfite remain bound to the siroheme.

References

General References

1. Williams, R. J. P. and Fraústo da Silva, J. J. R., *The Natural Selection of the Chemical Elements—The Environment and Life's Chemistry,* Clarendon Press Oxford, 1996.
2. Peck H. D., Jr. and Le Gall, J., Eds., "Inorganic Microbial Sulfur Metabolism", *Methods in Enzymology,* Vol. 243 Academic Press, New York, 1994.
3. Barton, L. L., Ed., *"Sulfate-Reducing Bacteria",* Plenum Press, New York, 1995.
4. Shively, J. M. and Barton, L. L., Eds., *"Variations in Autotrophic Life",* Academic Press, New York, 1991.
5. Friedrich, C. G., *Advances in Microbial Physiology,* Vol. 39, Poole, R. K., Ed., Academic Press, New York, 1998, pp. 238–287.

Specific References

6. http://www.tigr.org
7. http://www.jgi.doe.gov/JGI_microbial/html/index.html
8. Crane, B. R., Siegel, L. M., and Getzoff, E. D., "Sulfite reductase structure at 1.6 Å: evolution and catalysis for reduction of inorganic anions", *Science,* **270,** 59–67 (1995).
9. Messerschmidt, A., Huber, R., Poulos, T., and Wieghardt, K., Eds., *Handbook of Metalloproteins,* John Wiley, & Sons, Inc., Chichester, 2001.
10. Gavel, O. Y., Bursakov, S. A., Calvete, J. J., George, G. N., Moura, J. J. G., and Moura, I., "ATP sulfurylases from sulfate-reducing bacteria of the genus *Desulfovibrio.* A novel metalloprotein containing cobalt and zinc", *Biochemistry,* **37,** 16225–16232 (1998).
11. Hipp, W. M., Pott, A. S., Thum-Schimtz, N., Faath, I., Dahl, C., and Trüper, H. G., "Towards the phylogeny of APS reductases and sirohaem sulfide reductases in sulfate-reducing and sulfur-oxidizing prokaryotes", *Microbiology,* **143,** 2891–2902 (1997).
12 Matias, P. M., Frazão, C., Morais, J., Coll, M., and Carrondo, M. A., "Structure analysis of cytochrome c_3 from *Desulfovibrio vulgaris* Hildenborough at 1.9 Å resolution", *J. Mol. Biol.,* **234,** 680–699 (1993).
13. Louro, R. O., Bento, I., Matias, P. M., Catarino, T., Baptista, A. M., Soares, C. M., Carrondo, M. A., Turner, D. L., and Xavier, A. V., "Conformational component in the coupled transfer of multiple electrons and protons in a monomeric tetrahaem cytochrome", *J. Biol. Chem.,* **276,** 44044–44051 (2001).
14. Matias, P. M., Saraiva, L. M., Soares, C. M., Coelho, A. V., LeGall, J., and Carrondo, M. A., "Nine-haem cytochrome *c* from *Desulfovibrio desulfuricans* ATCC 27774: primary sequence determination, crystallographic refinement at 1.8 Å and modeling studies of its interaction with the tetrahaem cytochrome c_3", *J. Biol. Inorg. Chem.,* **4,** 478–494 (1999).
15. Matias, P. M., Coelho, A. V., Valente, F. M. A., D., P., LeGall, J., Xavier, A. V., Pereira, I. A. C., and Carrondo, M. A., "Sulfate respiration in *Desulfovibrio vulgaris* Hildenborough: Structure of the 16-heme cytochrome *c* HmcA at 2.5 Å resolution and a view of its role in transmembrane electron transfer", *J. Biol. Chem.,* **277,** 47907–47916 (2002).
16. Xavier, A. V., "A mechano-chemical model for energy transduction in cytochrome *c* oxidase: The work of a Maxwell's god", *FEBS Lett.,* **532,** 261–266 (2002).
17. Valente, F. M., Saraiva, L. M., LeGall, J., Xavier, A. V., Teixeira, M., and Pereira, I. A., "A membrane-bound cytochrome c_3: a type II cytochrome c_3 from *Desulfovibrio vulgaris* Hildenborough", *ChemBioChem,* **2,** 895–905 (2001).
18. Chen, L., Liu, M.-Y., and LeGall, J., "Calcium is required for the reduction of sulfide from hydrogen in a reconstituted electron transfer chain from the sulfate-reducing bacterium *Desulfovibrio gigas*", *Biophys. Biochem. Res. Commun.,* **180,** 238–242 (1991).
19. Michel, J., B., Feron, O., Sacks, D., and Michel, T., "Reciprocal regulation of endothelial nitric-oxide synthase by Ca^{2+}-calmodulin and cavelin", *J. Biol. Chem.,* **272,** 15583–15586 (1997).
20. Moura, J. J. G., Xavier, A. V., Hatchikian, E. C., and LeGall, J., "Structural control of the redox potentials and of the physiological activity by oligomerization of ferredoxin", *FEBS Lett.,* **89,** 177–179 (1978).
21. Macedo, S., Romão, C. V., Mitchell, E., Matias, P. M., Liu, M. Y., Xavier, A. V., LeGall, J., Teixeira, M., Lindley, P., and Carrondo, M. A., "The nature of the diiron site in the bacterioferritin from *Desulfovibrio desulfuricans*", *Nat. Struct. Biol.,* **10,** 285–290 (2003).
22. Frazão, C., Silva, G., Gomes, C. M., Matias, P., Coelho, R., Sieker, L., Macedo, S., Liu, M. Y., Oliveira, S., Teixeira, M., Xavier, A. V., Rodrigues-Pousada, C., Carrondo, M. A., and LeGall, J., "Structure of a dioxygen reduction enzyme from *Desulfovibrio gigas*", *Nat. Struct. Biol.,* **7,** 1041–1045 (2000).
23. Gomes, C. M., Giuffre, A., Forte, E., Vicente, J. B., Saraiva, L. M., Brunori, M., and Teixeira, M., "A novel type of nitric-oxide reductase", *J. Biol. Chem.,* **277,** 25273–25276 (2002).
24. LeGall, J., Ljungdahl, P. O., Moura, I., H. D. Peck, J., Xavier, A. V., Moura, J. J. G., Teixeira, M., Huynh, B. H., and DerVartanian, D. V., "The presence of redox-sensitive nickel in the periplasmic hydrogenase from *Desulfovibrio gigas*", *Biochem. Biophys. Res. Commun.,* **106,** 610–616 (1982).
25. Chen, L., Sharma, P., LeGall, J., Mariano, A. M., Teixeira, M., and Xavier, A. V., "A blue non-heme iron protein from *Desulfovibrio gigas*", *Eur. J. Biochem.,* **226** (2), 613–618 (1994).
26. Sieker, L. C., Holmes, M., Le Trong, I., Turley, S., Liu, M.-Y., LeGall, J., and Stenkemp, R. E., "The 1.9 Å crystal structure of the "as isolated" rubrerythrin from *Desulfovibrio vulgaris:* some surprising results", *J. Biol. Inorg. Chem.,* **5,** 505–513 (2000).

XII.6. Molybdenum Enzymes

Jonathan McMaster and C. David Garner
The School of Chemistry
The University of Nottingham
Nottingham NG7 2RD
UK

Edward I. Stiefel
Department of Chemistry
Princeton University
Princeton, New Jersey
08544

Contents

XII.6.1. Introduction

Molybdenum is required as a trace element by virtually all forms of life and is unique in being the only 4*d* transition metal utilized by biological systems.[1–7] The high concentration of Mo in seawater (10^{-5} g dm^{-3}) makes this element readily available for incorporation into biological systems, despite its relatively low terrestrial abundance (see Chapter II). However, Mo may not have been as available in the early stages of the development of life on Earth since it occurs in the Earth's crust primarily as molybdenite (MoS_2), which is insoluble in water. However, in the presence of O_2 and H_2O, MoS_2 is converted (Eq. 1) into molybdate, $[MoO_4]^{2-}$, which is readily soluble in water. Therefore, following the development of oxygenic photosynthesis and the production of an oxidizing atmosphere on Earth (Chapter II), Mo would have become much more available for incorporation into biological systems.[8]

$$2\ MoS_2 + 7\ O_2 + 2\ H_2O \rightarrow 2[MoO_4]^{2-} + 4\ SO_2 + 4\ H^+ \tag{1}$$

The first evidence that molybdenum is required for the metabolism of living systems was obtained in the 1930s, through studies of plants, which established that this element is essential for nitrogen fixation. This special process is accomplished by the nitrogenase enzymes of bacteria and archaea (Chapter II; Section XII.3). Subsequently, in the 1950s, it was shown that Mo is required for the normal growth and health of animals, plants, and microorganisms. We now know that Mo is a key component of the active site in an extensive range of enzymes that catalyze important reactions in the carbon, nitrogen, sulfur, selenium, and arsenic cycles of the biosphere. Human metabolism involves several Mo enzymes, making Mo an essential trace element in our diet.

Table XII.6.1 summarizes the nature of the more prominent Mo enzymes and the principal reaction that each catalyzes at the Mo site.[9–24] The reactions are formulated as involving water, protons, and electrons. However, inspection of these reactions reveals that most Mo enzymes catalyze a conversion, the net effect of which is to add an oxygen atom to, or remove an oxygen atom from, the substrate (Eq. 2).

$$X + H_2O \rightarrow XO + 2\ H^+ + 2\ e^- \tag{2}$$

These enzymes have a considerable strategic significance in the biosphere, for example:

- Nitrate reductases occur in plants, fungi, and bacteria and catalyze the reduction of NO_3^- to NO_2^-, as the first step in either nitrogen assimilation or anaerobic respiration using NO_3^- as the terminal electron acceptor.[1,10,16] Therefore, Mo is essential for *both* routes to fixed nitrogen: from N_2 via the nitrogenases (Chapter XII.3) and from NO_3^- via the nitrate reductases (Chapter XII.4).

Table XII.6.1.
Representatives of the Molybdenum Enzymes

Enzyme and Substrate Half-Reaction	Typical Source
THE DMSO REDUCTASE FAMILY	
Arsenite oxidase[9]	*Alcaligenes faecalis*
$AsO_3^{3-} + H_2O \rightarrow AsO_4^{3-} + 2\,H^+ + 2\,e^-$	
Biotin-*S*-oxide reductase[18]	*Escherichia coli*
(see **I** below)	*Rhodobacter sphaeroides*
DMSO reductase (DMSO → DMS)[11,12]	*E. coli*
$CH_3S(O)CH_3 + 2\,H^+ + 2\,e^- \rightarrow CH_3SCH_3 + H_2O$	*Rhodobacter capsulatus*
	R. sphaeroides
Formate dehydrogenase[13,14]	*E. coli*
$HCOO^- \rightarrow CO_2 + H^+ + 2\,e^-$	*Desulfovibrio desulfuricans*
Formylmethanofurandehydrogenase[15]	*Methanobacterium wolfei*
(see **II** below)	*Methanobacterium thermoautotrophicum*
Nitrate reductase (dissimilatory)[16]	*E. coli*
$NO_3^- + 2\,H^+ + 2\,e^- \rightarrow NO_2^- + H_2O$	*D. desulfuricans*
Trimethylamine-*N*-oxide reductase[17]	*Shewanella massilia*
$(CH_3)_3NO + 2\,H^+ + 2\,e^- \rightarrow (CH_3)_3N + H_2O$	
THE SULFITE OXIDASE FAMILY	
Nitrate reductase (assimilatory)[10]	*Arabidopsis thaliana*
$NO_3^- + 2\,H^+ + 2\,e^- \rightarrow NO_2^- + H_2O$	Spinach
Sulfite oxidase[19,40,41]	Human liver
$SO_3^{2-} + H_2O \rightarrow SO_4^{2-} + 2\,H^+ + 2\,e^-$	Chicken liver
	Thiobacillus novellas
	A. thaliana
THE XANTHINE OXIDASE FAMILY	
Aldehyde oxidase[20]	Cow's liver
$RCHO + H_2O \rightarrow RCOOH + 2\,H^+ + 2\,e^-$	Human liver
Aldehyde oxido-reductase (dehydrogenase)[21]	*Desulfovibrio gigas*
$RCHO + H_2O \rightarrow RCOOH + 2\,H^+ + 2\,e^-$	
Xanthine dehydrogenase[22]	Chicken liver
(see **III** below)	Human liver
Xanthine oxidase[22]	Cow's milk
(see **III** below)	

Table XII.6.1.
(Continued)

Enzyme and Substrate Half-Reaction	Typical Source
THE CO DEHYDROGENASE FAMILY	
Carbon monoxide dehydrogenase (oxidoreductase)[23,24]	*Oligotropha carboxidovorans*
$CO + H_2O \rightarrow CO_2 + 2\,H^+ + 2\,e^-$	

I

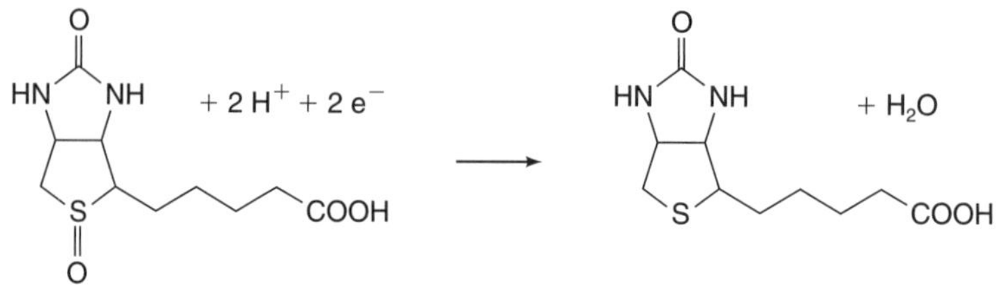

Biotin sulfoxide $+ 2\,H^+ + 2\,e^- \longrightarrow$ Biotin $+ H_2O$

Biotin sulfoxide **Biotin**

II

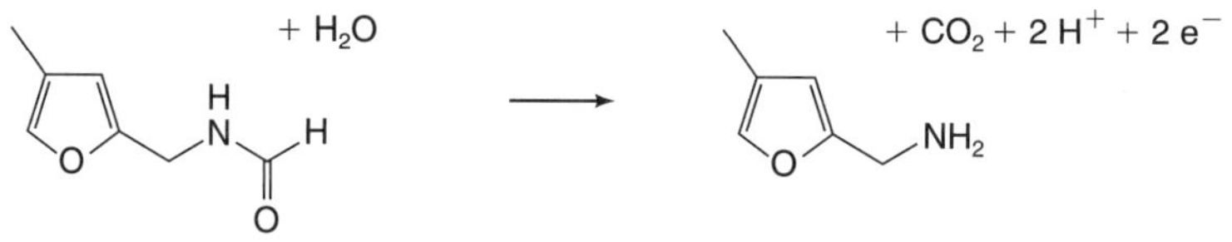

Formylmethanofuran $+ H_2O \longrightarrow$ Methanofuran $+ CO_2 + 2\,H^+ + 2\,e^-$

Formylmethanofuran **Methanofuran**

III

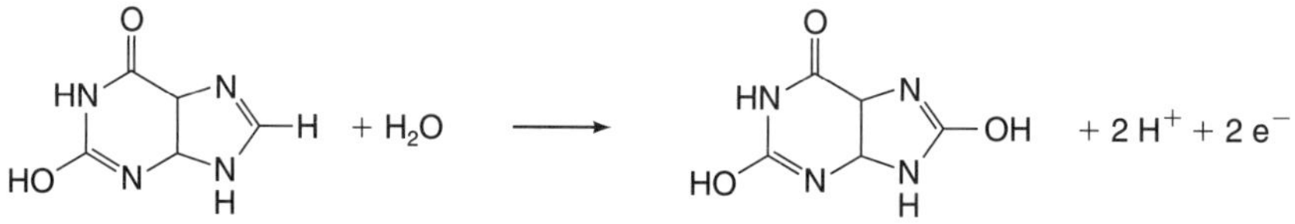

Xanthine $+ H_2O \longrightarrow$ Uric acid $+ 2\,H^+ + 2\,e^-$

Xanthine **Uric acid**

- Sulfite oxidases catalyze the oxidation of SO_3^{2-} to SO_4^{2-}, the final reaction in the oxidative degradation of the sulfur-containing amino acids Cys and Met and other sulfur compounds.[1,19] Sulfite oxidases are located in the liver of animals and humans; typically, an adult human excretes ~1 g of SO_4^{2-} each day. A rare genetic deficiency leads to a lack of an active form of sulfite oxidase, resulting in death in infancy.[19] Humans require the removal of toxic SO_3^{2-} and some of the SO_4^{2-} produced by sulfite oxidase is used in the synthesis of a wide variety of important molecules.
- Xanthine dehydrogenases–oxidases occur in the liver and kidneys of humans and animals and catalyze the final step in purine metabolism, the oxidation of xanthine to uric acid (see Section XII.6.3.3).[1,22]
- Aldehyde oxidases[1,20] are located in the liver of mammals and catalyze the oxidation of aldehydes (RCHO) to the corresponding carboxylic acid (RCO_2H). For $R = CH_3$, this represents the second step in the conversion of ethanol to acetic acid; the first step is achieved by the Zn enzyme, alcohol dehydrogenase.
- Dimethyl sulfoxide reductases occur in facultative anaerobic bacteria that are present in the oceans and salt marshes. These enzymes are located in the periplasm and function in a respiratory chain that uses DMSO as the terminal electron ac-

ceptor. The product of the reduction, dimethyl sulfide (DMS), is volatile and possesses a characteristic smell; photooxidation of DMS in air produces methylsulfonic acid, the salts of which acts as nucleation centers for cloud formation (see Chapter II).[1,11,12,25]

Tungsten enzymes (discussed in Section XII.7) are close relatives of the Mo enzymes. Each involves a single W atom bound to the same type of ligand as Mo, MPT (see below); W enzymes also catalyze conversions of Eq. 2.[1b,26]

Recent crystallographic and spectroscopic investigations of the molybdenum and tungsten enzymes, combined with studies of relevant chemical systems, have allowed us to begin to understand the nature and activity of their catalytic centers.

XII.6.2. The Active Sites of the Molybdenum Enzymes

Figure XII.6.1 shows the general nature of the catalytically active site of the Mo (and W) enzymes.[1] The material in darker print is present in each of the enzymes; the presence or absence of material in lighter print is characteristic of a particular enzyme. Each Mo enzyme possesses a catalytic center involving a single Mo atom bound to either one or two molecules of a special pyranopterin cofactor. This cofactor was originally called "molybdopterin",[6,27,28] but as it also binds W, it is better named as the **m**etal-binding **p**yranopterin ene-1,2-di**t**hiolate (MPT), which conveniently can use the same abbreviation as "**m**olybdo**pt**erin". The MPT is constituted as a pterin—comprising a pyrimidine and a pyrazine ring—linked to a pyran ring that carries an ene-1,2-dithiolate (dithiolene) and one $CH_2OPO_3^{2-}$ group. In each Mo (and W) enzyme, the structure of MPT is essentially the same with the dithiolene group coordinated as a bidentate ligand to the metal atom. The pyrazine and pyran rings are distinctly nonplanar and three chiral carbon atoms of the pyran ring are each in the (*R*)-configuration. However, relative to the pyrimidine ring of the pterin, there is conformational flexibility in the orientation of the $CH_2OPO_3^{2-}$ group and in the tilt of the pyran ring. In enzymes isolated from prokaryotes (bacteria and archaea), the phosphate group may be part of a diphosphate linkage forming a dinucleotide involving MPT and one of adenosine, guanosine, cytosine, or inosine as the second nucleotide.

XII.6.2.1. Families of Molybdenum Enzymes

Individual Mo enzymes are named on the basis of their principal metabolic role (see Table XII.6.1). Structural and spectroscopic studies of the oxidized Mo enzymes have identified four varieties of Mo center, as shown in Fig. XII.6.2, and these

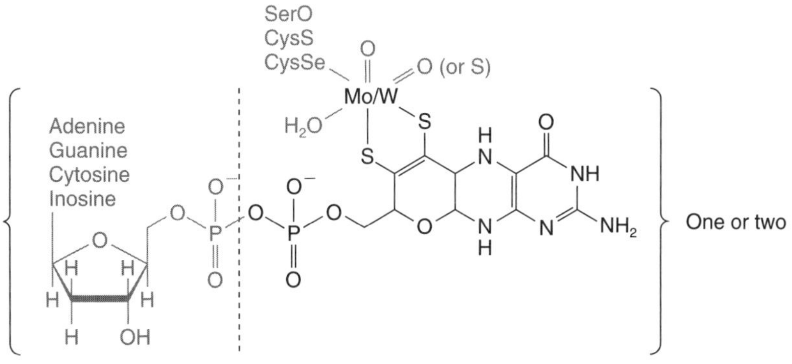

Fig. XII.6.1.
The nature of the oxidized form of the catalytic site of Mo (and W) enzymes. Groups (dark) are present in all Mo (and W) enzymes. Particular combinations of ligands (gray) are present or absent in individual enzymes and oxidation states.

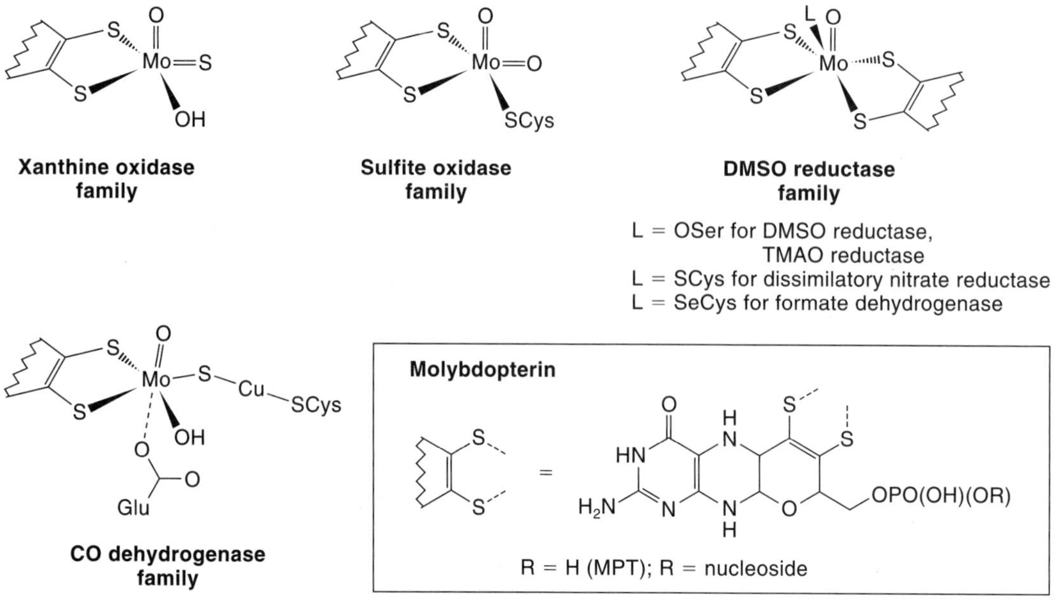

Xanthine oxidase family

Sulfite oxidase family

DMSO reductase family

L = OSer for DMSO reductase, TMAO reductase
L = SCys for dissimilatory nitrate reductase
L = SeCys for formate dehydrogenase

CO dehydrogenase family

Molybdopterin

R = H (MPT); R = nucleoside

Fig. XII.6.2.
A structural classification of the molybdenum enzymes based upon the nature of the oxidized form of their catalytic centers; the nature of the ligands represents a consensus of the results obtained from structural and spectroscopic investigations.

provide a good basis for the classification of Mo enzymes into families.[1] For the oxidized, Mo(VI), state of the enzyme:

- Members of the xanthine dehydrogenase–oxidase family, including the aldehyde oxidases, have a *cis*-oxo, sulfido (MoOS) center, a hydroxo group, and one MPT ligand.
- Members of the sulfite oxidase family, including the assimilatory nitrate reductases, possess a *cis*-dioxo (MoO_2) center that is bound to one hydroxo group, a cysteinyl residue, and one MPT ligand.
- Members of the DMSO reductase family usually have the Mo ligated by one oxo group, the donor atom from the side chain of an amino acid residue—Oγ of a serinyl residue [DMSO reductase or trimethylamine-oxide (TMAO) reductase], Sγ of a cysteinyl residue (dissimilatory nitrate reductase), or Seγ of selenocysteinyl residue (formate dehydrogenase), and two MPT ligands. In addition, members of this family may have one oxo or hydroxo group or no other ligand, depending on the enzyme.
- The Mo center of CO dehydrogenase is bound to an oxo, a sulfido, and a hydroxo group and one MPT ligand. The Mo site is weakly associated with the oxygen of a Glu residue. A novel aspect of this center is that the sulfido group bridges to a Cu^{I}(S cysteinyl) center.

Thus, unlike the molybdenum nitrogenases, each of which possesses the iron–molybdenum cofactor (FeMoco, Section XII.3), there is no unique "molybdenum cofactor" (or Moco) for the molybdenum enzymes. Rather there is a family of catalytic centers with the common features of a single Mo atom ligated by one or two MPTs, plus additional ligands, which can be an amino acid side chain, and/or one, two, or three small inorganic ligands, such as oxo, sulfido, or hydroxo groups.

Molybdenum enzymes may have evolved from a common ancestral gene and members of the same family generally possess significant sequence similarity. Some important chemical aspects, common to the active sites of all Mo enzymes, are considered in Section XII.6.2.2.

XII.6.2.2. Oxomolybdenum Centers and Oxygen Atom Transfer

Molybdenum enzymes (Table XII.6.1) possess a special ability to catalyze reactions of the type (Eq. 2) for a wide range of substrates, that is, to achieve a two-electron oxidation, or reduction, of the substrate, that is usually coupled to the net loss or gain of an oxygen atom.[2] The catalytic conversion of the substrate is executed at the Mo center and involves $Mo^{VI} \rightarrow Mo^{IV}$, for oxidases, and $Mo^{IV} \rightarrow Mo^{VI}$, for reductases. The catalytically active state is regenerated by oxidation of $Mo^{IV} \rightarrow Mo^{VI}$, for oxidases, and reduction of $Mo^{VI} \rightarrow Mo^{IV}$, for reductases; these regeneration processes proceed in two, one-electron steps that are coupled to proton loss or proton uptake, respectively (see the section The Electronic Structure of Dioxomolybdenum $[MoO_2]^{2+}$ Centers). In each case, a Mo(V) intermediate is generated.

The Electronic Structure of Oxomolybdenum $[MoO]^{2+}$ Centers. The oxidized state of the active site of most molybdenum enzymes (Fig. XII.6.2) involves at least one oxo group, often depicted as Mo=O. Members of the sulfite oxidase family involve a *cis*-dioxo $[MoO_2]^{2+}$ center and the xanthine dehydrogenase–oxidase family a *cis*-$[MoOS]^{2+}$ center in the Mo(VI) state. The O^{2-} ion has a closed-shell electronic configuration and behaves as a strong σ and π donor to the Mo center, which is in a high oxidation state (IV, V, or VI) and possesses few or no *d* electrons. The Mo(VI) ($4d^0$), Mo(V) ($4d^1$), and Mo(IV) ($4d^2$) compounds, like those formed by other early *d*-transition metals in their higher oxidation states, usually contain at least one oxo group. Reduction of Mo(VI) to Mo(V) or Mo(IV) generally involves the loss or protonation of an oxo group, and the reverse is true for an oxidation. The concomitant change in the oxidation state of the Mo and the number of oxo groups represents a fundamental aspect of Mo chemistry that may be employed in the catalysis effected by the Mo enzymes. For example, it has long been known that simple chemical systems such as the dithiocarbamate complexes shown in (Eq. 3) can engage in oxygen atom transfer.[2,29]

$$[Mo^{VI}O_2(S_2CNEt_2)_2] + Ph_3P \rightarrow [Mo^{IV}O(S_2CNEt_2)_2] + Ph_3PO \qquad (3)$$

A simple valence bond description restricts an oxo (and a sulfido) group to the formation of a double bond to a transition metal center. In contrast, a molecular orbital (MO) description of the bonding, with each O^{2-} ion possessing a filled $2p$ orbital of σ symmetry and two filled $2p$ orbitals of π-symmetry, allows an oxo group to have a bond order of ≤ 3. The MO treatment emphasizes the flexible nature of this bond since the π component can vary, depending on how effectively the p_π orbitals of the O^{2-} ion compete with other π-donor ligands for vacant d_π orbitals of the molybdenum. A partial MO diagram for an $[Mo{=}O]^{4+}$ center is shown in Fig. XII.6.3, illustrating how the π bonds formed between the p_x and p_y orbitals of the O^{2-} ion and the d_{xz} and d_{yz} orbitals of a Mo(VI) center produce two π-bonding and two π-antibonding (π^*) orbitals. The Mo(VI) center is in an approximate octahedral environment bound to six ligands, each by a σ bond, resulting in the splitting of the $4d$ orbitals. The two highest energy σ^* orbitals are d_{z^2} and $d_{x^2-y^2}$; since the Mo=O group is located on the *z* axis and, since this is generally the shortest metal–ligand bond, σ^*–d_{z^2} is at a higher energy than σ^*–$d_{x^2-y^2}$. This MO description shows how an "Mo=O" bond can attain a bond order of three; that is, be a triple bond with one σ and two π components. The separation between the d_{xy} and π^* (d_{xz}, d_{yz}) orbitals is generally large because of the strong π donation from the oxo group. Since the d_{xy} level is considerably lower in energy, $[Mo^{V}O]^{3+}$ centers have a $(d_{xy})^1$ configuration and are paramagnetic; $[Mo^{IV}O]^{2+}$ centers have $(d_{xy})^2$ configuration and are diamagnetic.[5]

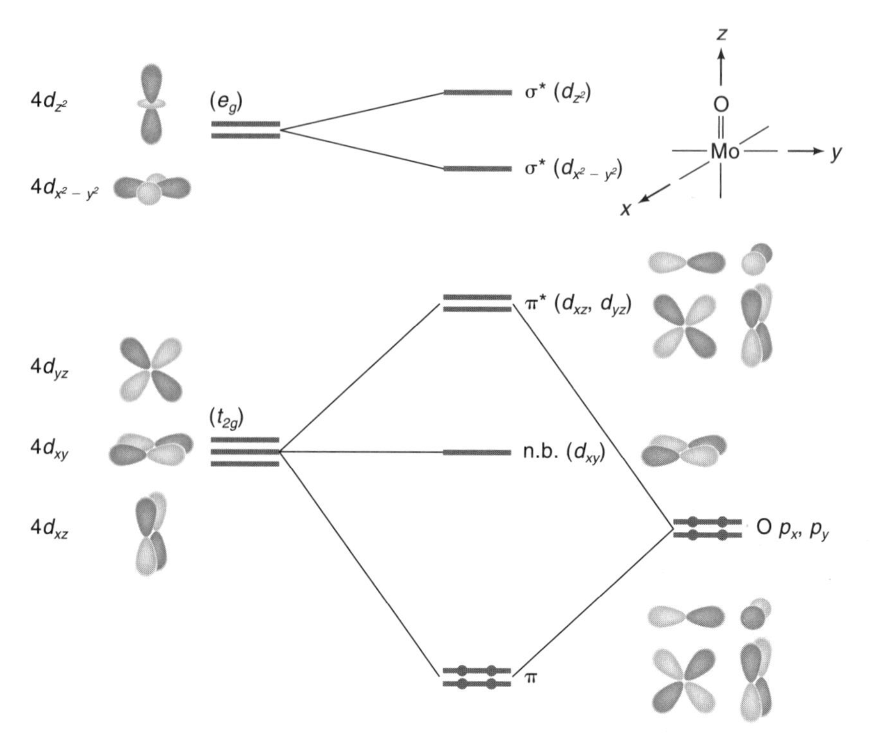

Fig. XII.6.3.
A partial MO diagram showing the π bonding of a $[Mo{=}O]^{4+}$ center.

The Electronic Structure of Dioxomolybdenum $[MoO_2]^{2+}$ Centers. The Mo(VI) complexes often involve a $[MoO_2]^{2+}$ center with two oxo groups that are mutually cis (i.e., the O=Mo=O angle is $\sim 105°$).[5] This orientation allows the most effective π overlap between the oxygen $2p_\pi$ orbitals and the d_π orbitals on the Mo. As shown by Fig. XII.6.4, a *cis*-$[MoO_2]^{2+}$ geometry allows two $O(p_\pi)$–$Mo(d_\pi)$ bonds to be formed involving all three metal d_π orbitals, resulting in each oxo group having a bond order $(\sigma + \pi)$ of 2.5. In contrast, a trans, O=Mo=O, arrangement would utilize only the d_{xz} and d_{yz} orbitals and lead to a maximum bond order of 2 for each oxo group.

Coupled Electron and Proton Transfer. Reduction of an $[MoO_2]^{2+}$ center, by the addition of one or two electrons, is often accompanied by coupled electron–proton transfer (CEPT) to form a $[Mo^{V}O(OH)]^{2+}$ or a $[Mo^{IV}O(H_2O)]^{2+}$ center.[1b] The CEPT process is common in Mo enzymes and is driven by the effect of oxidation state on the pK_a of coordinated ligands. Specifically, higher oxidation states tend to have more acidic (deprotonated) ligands, whereas lower oxidation states tend to have more basic (protonated) ligands. The effect is large.[1b]

Similar considerations apply to $[Mo^{VI}OS]^{2+}$ centers, as present in the oxidized state of xanthine dehydrogenases–oxidases and aldehyde oxidases. The lowest unoccupied molecular orbital (LUMO) of these centers has considerable Mo=S π^* character and coupled e^-/H^+ addition leads to protonation of the sulfur to form an $[Mo^{V}O(SH)]^{2+}$ center.

Oxygen Atom Transfer. Table XII.6.2 lists the redox potentials for the Mo(VI)/Mo(V) and Mo(V)/Mo(IV) couples of three molybdenum enzymes, to show that:

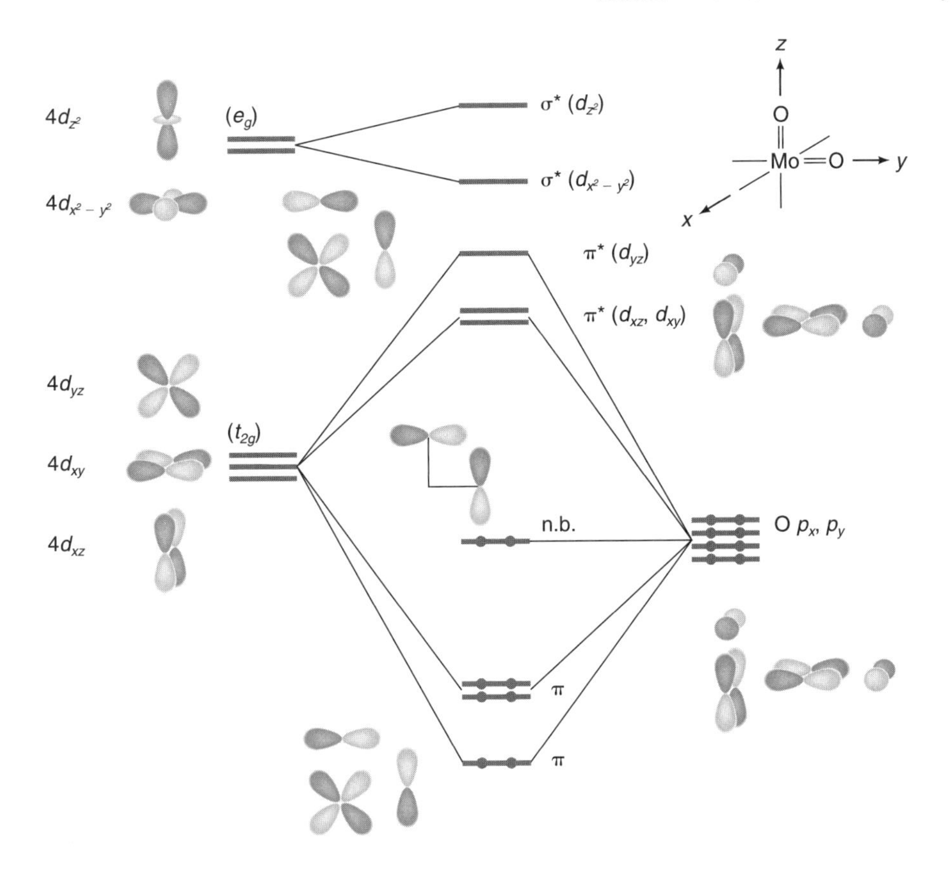

Fig. XII.6.4.
A partial MO diagram showing the π bonding of a *cis*-$[MoO_2]^{2+}$ center.

Table XII.6.2.
Redox Potentials of the Active Sites of Three Molybdenum Enzymes[a]

Protein	$E°$ vs. SHE[b] (mV)
Xanthine oxidase from milk, pH 7.7	
Mo(VI)/Mo(V)	−345
Mo(V)/Mo(IV)	−315
Sulfite oxidase from chicken liver, pH 7.0	
Mo(VI)/Mo(V)	+70
Mo(V)/Mo(IV)	−90
DMSO reductase from *R. sphaeroides*, pH 7.0	
Mo(VI)/Mo(V)	+144
Mo(V)/Mo(IV)	+160

[a] See Ref. 3.
[b] Standard hydrogen electrode = SHE.

- The redox potentials of the Mo center vary with the nature of the coordination sphere (see Fig. XII.6.2), but are within the biological potential window of −400 to +800 mV versus SHE at pH 7.[3]
- The Mo centers are poised for a *two*-electron process, that is, there is relatively little difference between the potentials of the Mo(VI)/Mo(V) and Mo(V)/Mo(IV) couples.

The oxidations catalyzed by the Mo enzymes are generally considered easier to accomplish than those that directly use O_2 as the oxidant. Such direct oxidation using O_2 (usually oxygenations) often involve more resistant substrates, such as alkyl chains or aromatic compounds, and are catalyzed by Cu proteins, cytochrome P450, or the di-iron monooxygenases—the active sites of which involve, for example, a chemically more aggressive $Fe^{IV}{=}O$ (cytochrome P450) or $Fe_2{}^{IV}{=}O$ (methane monooxygenase) group that attacks the substrate (Section XI.5). The catalytic centers of these enzymes possess considerable radical character and are produced by O_2 activation. In contrast, the oxygen atom transferred by the Mo enzymes derives from, or is converted to, a water molecule. However, the recent characterization of ethylbenzene dehydrogenase shows that Mo enzymes can catalyze reactions involving CH bond activation.

Molybdenum complexes can catalyze reactions that involve oxygen atom transfer. Fig. XII.6.5 shows the catalytic cycle for the oxidation of PPh_3 to Ph_3PO by H_2O and an external oxidant with a molybdenum hydrotrispyrazolylborate complex acting as the catalyst.[7] This catalysis involves the transfer of an oxygen atom from a $[Mo^{VI}O_2]^{2+}$ center to the phosphine, to produce a $[Mo^{IV}O]^{2+}$ center and the phosphine oxide. The nature of this first step of the catalytic cycle is readily described within the context of MO theory. The highest occupied molecular orbital (HOMO) of the nucleophilic phosphine "substrate" interacts with the LUMO of the electrophilic $[Mo^{VI}O_2]^{2+}$ center (Fig. XII.6.4), introducing electron density into the π^* orbital, weakening a Mo=O bond and facilitating oxygen atom transfer. The other "spectator" oxo group helps to drive the reaction by forming a stronger $(\sigma + \pi)$ bond to the resulting Mo(IV) center (see Fig. XII.6.3). The phosphine oxide then is replaced in the coordination sphere of the metal by a water molecule. The original $[(L\text{-}N_3)MoO_2(SPh)]$ $[(L\text{-}N_3) =$ hydrotris(3,5-dimethyl-1-pyrazolyl)borate$]$ complex can be regenerated by two CEPT processes that convert the $[Mo^{IV}O(OH_2)]^{2+}$ center, via a $[Mo^{V}O(OH)]^{2+}$ center, to a $[Mo^{VI}O_2]^{2+}$ center. The $[Mo^{V}O(OH)]^{2+}$ intermediate has been detected by EPR spectroscopy.[7,30]

Cycles similar to that shown in Fig. XII.6.5 have been postulated for the oxidations catalyzed by the Mo enzymes. For reductions, such cycles operate in reverse: that is, commencing from a $[Mo^{IV}O]^{2+}$ center, oxygen atom transfer produces a

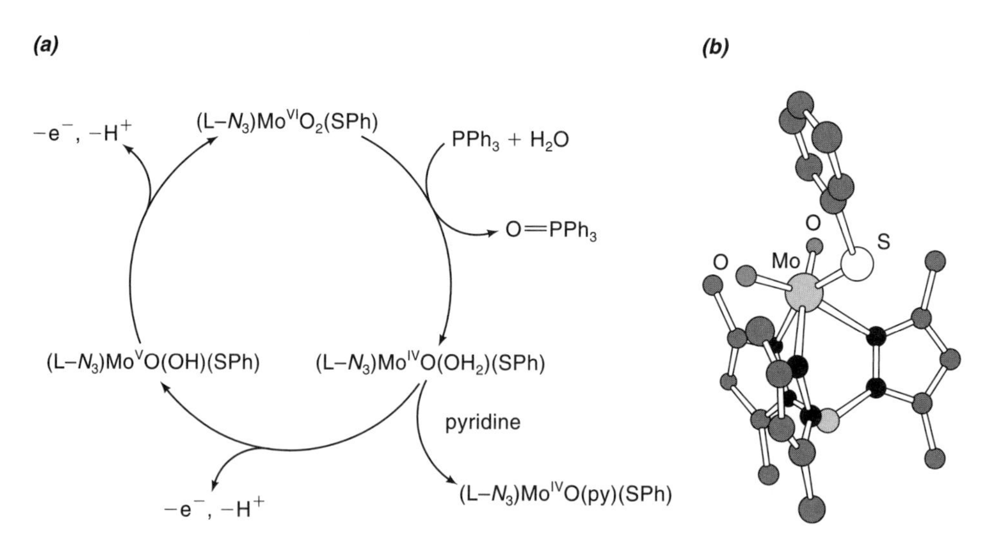

Fig. XII.6.5.
A catalytic cycle for the oxidation of PPh_3 to Ph_3PO catalyzed by $[(L\text{–}N_3)Mo^{VI}O_2(SPh)]$, where $L\text{–}N_3 =$ hydrotris(3,5-dimethyl-1-pyrazolyl)borate.

$[Mo^{VI}O_2]^{2+}$ center, CEPT reduction of which proceeds via a $[Mo^{V}O(OH)]^{2+}$ intermediate to a $[Mo^{IV}O]^{2+}$ center (with release of a H_2O molecule).

Given the variation in the nature of the Mo centers in enzymes (Fig. XII.6.2) and the range of substrates converted (Table XII.6.1), although the chemical principles described above are generally applicable, the mechanistic details of the catalysis may differ from enzyme to enzyme.

XII.6.2.3. MPT and Dithiolene Ligation of Molybdenum Centers

The active site of all molybdenum enzymes (Figs. XII.6.1 and XII.6.2) involves the metal coordinated to one or two MPTs. There are several ways in which MPT could participate in the catalysis, using its dithiolene, pyran, pterin, and other functional groups. However, it is probable that the whole entity (i.e., the unique combination of the particular functional groups) makes MPT the ideal ligand for supporting Mo as the catalytic center of an extensive family of enzymes.

Dithiolene Ligation. A key feature of MPT is that it provides an ene-1,2-dithiolate, or dithiolene, ligand[31] that binds molybdenum. Dithiolene ligands were initially used as analytical reagents since they form stable and intensely colored complexes with a range of metals. These ligands have a special electronic structure because of the C=C backbone that leads to a planar M–S–C–C–S metallocycle with a delocalized π system.[1d] A comparison of the electronic structures of ethane-1,2-dithiolate and ethene-1,2-dithiolate, a prototypical dithiolene ligand, is provided by Fig. XII.6.6. The former can act as a four electron σ donor to a metal with a lone pair of electrons donated by each sulfur atom; depending on the conformation of the ligand, this σ bonding can be augmented by π donation from lone pair(s) of electrons on one (or both) sulfur atom(s). For a dithiolene, the sulfur lone-pair π orbitals

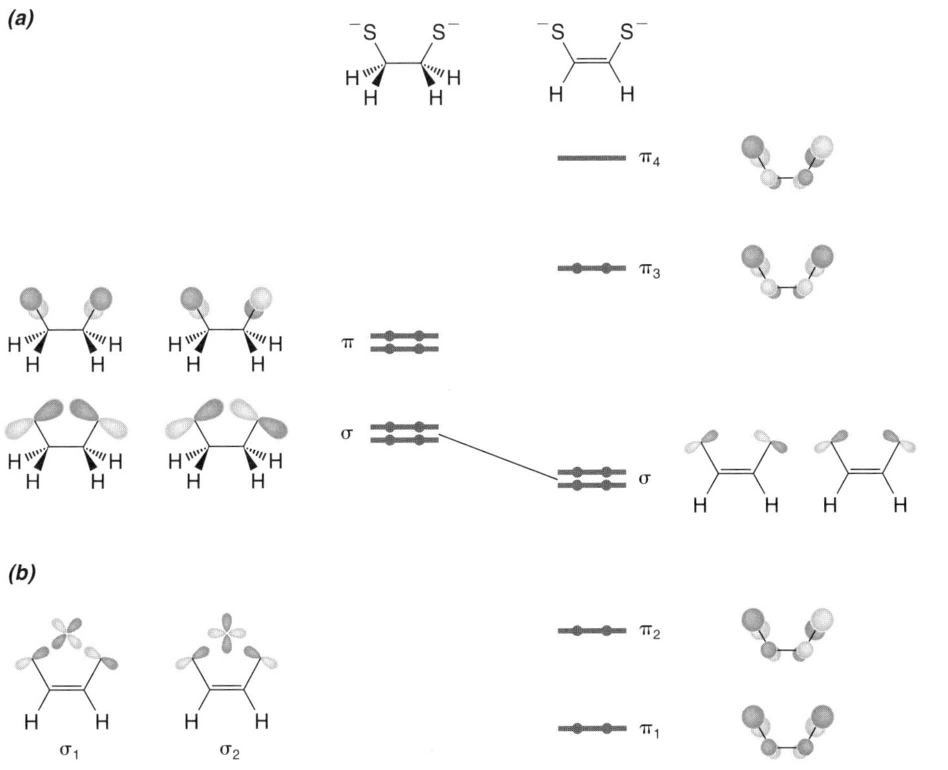

Fig. XII.6.6.
(*a*) A comparison of the donor orbitals of ethane-1,2-dithiolate and ethene-1,2-dithiolate and (*b*) bonding combinations between the ethane-1,2-dithiolate π_1 and metal $4d_{x^2-y^2}$ and the ethene-1,2-dthiolate π_2 and the metal $4d_{xy}$ orbitals.

perpendicular to the plane of the dithiolene framework are oriented for a more effective overlap with a metal's d_{π} orbitals than those of a saturated dithiolate ligand. The π orbitals of a dithiolene extend over the ligand framework, with the π_3 orbital being C–C bonding and C–S antibonding and the π_4 orbital both C–C and C–S antibonding. The π delocalization in dithiolene complexes:

- Reduces the negative charge on each sulfur atom, relative to that of an alkane-1,2-dithiolate complex; as a result, the σ-donor orbitals of the dithiolene are lowered in energy and it is a poorer σ-donor ligand and a poorer nucleophile than its saturated analogue. The σ-donor orbitals (σ_1 and σ_2) of a dithiolene can interact with metal d orbitals as shown in Fig. XII.6.6.
- Leads to a dithiolene being a stronger π donor than an alkane-1,2-dithiolate and, since the HOMO of the former (π_3) is C–S antibonding, there is a clear advantage in donating this electron density to a metal center.
- Introduces some double-bond character into the C–S bonds.

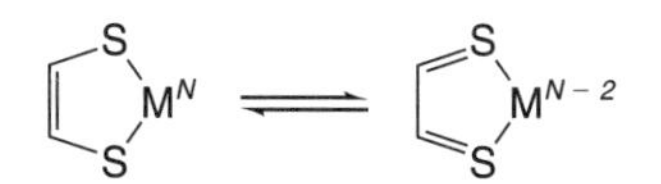

Fig. XII.6.7.
The extremes of the redox states available to a single metal–dithiolene moiety (N = oxidation number).

Thus, a very important aspect of the bonding of a dithiolene to a metal center is the very effective π donation from the ligand to the metal.[1d] The dithiolene group(s) at the Mo centers of enzymes will facilitate reactions, such as oxygen atom transfer (with the loss of an Mo=O group) and CEPT (converting a $[Mo^{VI}O_2]^{2+}$ center to a $[Mo^{V}O(OH)]^{2+}$ or $[Mo^{IV}O(H_2O)]^{2+}$ center) by competing effectively with an oxo group for the metal's empty π orbitals, thereby weakening an Mo=O bond. The π donation from a dithiolene to a metal means that it is not always reasonable to consider that the ligand possesses a 2^- charge. If π donation is extensive, the metal is formally reduced and the dithiolene will develop some dithioketone character, as illustrated in Fig. XII.6.7, and this can be manifest by a lengthening of the C–C and a shortening of the C–S bonds.

The extensive covalency ($\sigma + \pi$ donation) of a metal–dithiolene complex means that the charges associated with the metal and the ligand cannot be specified separately; that is, strictly speaking, it is only possible to consider the redox state of the complex as a whole. Ligands, such as the dithiolenes, which have their own redox capacity, are termed "noninnocent" and their complexes often undergo facile redox reactions in which the ligand may be intimately involved.

Extensive investigations of dithiolene complexes formed by d transition metals have been carried out since the 1960s.[1d,31] This research was stimulated by the behavior of complexes such as $[Mo(dithiolene)_3]^{n-}$ ($n = 0, 1,$ or 2). These complexes adopt a novel, trigonal-prismatic geometry and possess two reversible, one-electron redox couples. Square pyramidal $[MoO(dithiolene)_2]^{n-}$ ($n = 1$ or 2) complexes manifest a reversible, one-electron redox couple; also, the reduced state is capable of the reversible uptake of an oxygen atom (see Eq. 4) a reaction directly related to those catalyzed by Mo centers in enzymes.

$$[MoO(dithiolene)_2]^{2-} \underset{(C_6H_5)_3P}{\overset{(CH_3)_3NO}{\rightleftharpoons}} [MoO_2(dithiolene)_2]^{2-} \qquad (4)$$

The protein crystallographic characterization of Mo and W enzymes (see Figs. XII.6.1 and XII.6.2) has stimulated the development of chemical analogues of several of these sites. These include $[MoO_2(SR)(dithiolene)]^-$ complexes,[32] as chemical analogues of the oxidized form of the active site of sulfite oxidase and assimilatory nitrate reductase. Also, $[Mo(QR)(dithiolene)_2]^-$ complexes[33] have been prepared for Q = (1) O; (2) S, or (3) Se, as chemical analogues of the reduced form of the active site of (1) DMSO or TMAO reductase; (2) dissimilatory nitrate reductase; or (3) formate dehydrogenase. These $[Mo(QR)(dithiolene)_2]^-$ complexes accept an oxygen atom from molecules such as DMSO or $(CH_3)_3NO$, but the putative $[MoO(QR)(dithiolene)_2]^-$ product is rather unstable.

Clearly, the active centers of the Mo enzymes (Figs. XII.6.1 and XII.6.2) use the special properties of dithiolene ligands to good effect. In particular, these ligands facilitate the redox changes of the catalytic centers that are necessary for enzyme activity. Also, the dithiolene group(s) appear to provide an effective route for the intraprotein electron transfer that is necessary for regeneration of the catalytically active state of the center.[34]

Pyranopterin. Pterins were first identified as pigments in butterfly wings ("pteron" is Greek for wing). Not only are pterins widely distributed in Nature as pigments, but also they are involved in human metabolism, as the vitamin folic acid and as tetrahydrobiopterin, a vital cofactor for the synthesis of neurotransmitters.

Plants, animals, and prokaryotes use the same route to synthesize MPT, which appears to be an ancient biosynthetic pathway.[4,6] The biosynthesis of MPT commences with the nucleoside guanosine and the conversion involves a series of enzymes, including the enzyme MPT synthase that catalyzes the addition of sulfur to the organic framework. The final step is the incorporation of the metal. Molybdenum generally enters an organism as the $[MoO_4]^{2-}$ ion bound in a transport protein that locates the anion in a cavity of the appropriate size using a series of hydrogen bonds, formed between the Mo=O groups and HO or H_2N groups of the polypeptide chain. However, the mechanism for binding Mo to MPT is unknown, and at present, it is not clear how biosynthesis and protein assembly produce the various coordination environments shown in Fig. XII.6.2.

A complete understanding of the biosynthetic pathway that produces MPT is important for at least two reasons. First, defects in the pathway in humans lead to metabolic problems that produce severe neurological abnormalities, mental retardation, and death in infancy.[35] These conditions arise in infants born without an effective MPT synthase who, in consequence, are deficient in sulfite oxidase, xanthine dehydrogenase–oxidase, and aldehyde oxidase. A successful treatment of this profound metabolic problem requires a process that restores the MPT synthase activity. Second, a full knowledge of this pathway should allow modifications to be engineered, for example, by site-directed mutagenesis, to produce variants of MPT. These variants could provide valuable information about the nature of the catalytic process and/or alter the substrate specificity of a Mo enzyme.

The pyranopterin nucleus of MPT has several properties that are important for the structure and function of Mo enzymes. The crystal structures of the Mo (and W) enzymes show that the polar (O, N, NH, and NH_2) groups of MPT participate extensively in the formation of hydrogen bonds to complementary groups of the polypeptide chain. The extent of this hydrogen bonding is significantly increased when the phosphate is part of a dinucleotide (see Fig. XII.6.1). The resultant network of hydrogen bonds serves to precisely locate and orient the catalytically active center within the polypeptide chain with respect to:

- The "funnel" in the protein "sheath", through which the substrate enters and product leaves.
- The "pocket" that binds the substrate in an orientation suitable for catalytic conversion to the particular product.
- The electron-transfer relay, involving other redox-active prosthetic groups, that leads to the site on the exterior of the protein where the redox partner binds.

Redox reactions are an important aspect of the biochemistry of pterins. The pyrazine ring of a pterin can adopt one of three different oxidation levels: tetrahydro, dihydro, or fully oxidized, corresponding to four, two, or no additional hydrogen atoms being bound (see Fig. XII.6.8). The results of protein crystallographic characterizations of Mo (and W) enzymes imply that the pyrazine ring of MPT is in its

Fig. XII.6.8.
Oxidation levels of pterins.

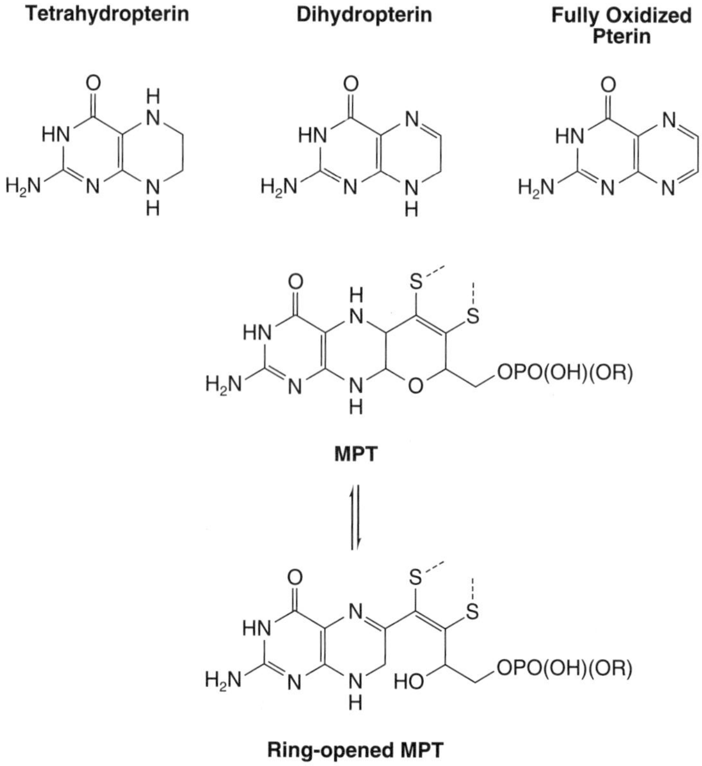

Tetrahydropterin Dihydropterin Fully Oxidized Pterin

MPT

Ring-opened MPT

fully reduced, or tetrahydropterin, level. However, there is a special aspect of the pyranopterin nucleus that should be considered in this respect; thus, opening of the pyran ring leads to the removal of two hydrogens from the pyrazine ring (see Fig. XII.6.8). Therefore, it may be more instructive to regard MPT (as depicted in Fig. XII.6.2) as a dihydropterin. Such an opening of the pyrazine ring is facile, at least in related chemical systems, and has the attraction of conjugating the metal to the pyrazine ring via the dithiolene group. A recent protein crystallographic determination of dissimilatory nitrate reductase has provided the first evidence for the ring-opened form of MPT.[36]

In some cases, the structure of a Mo (or a W) enzyme involves MPT hydrogen bonded to a redox-active prosthetic group, such as an Fe–S cluster. Therefore, MPT seems to be able to participate in the network used to relay electrons to (or from) the Mo center from (or to) the external reductant (or oxidant). Also, CEPT (see the section Coupled Electron Proton Transfer) is important for the operation of the active sites of the Mo enzymes, especially in the regeneration of the catalytically active form. The basic sites of the pterin framework may act as a local buffer, mediating the proton flux to (or from) the Mo center during CEPT. If the participation of MPT in electron transfer and the buffering proton flux *are combined,* this could lead to a redox change of the pterin during the catalytic action of the molybdenum–dithiolene center and be related to the opening and closing of the pyran ring.

XII.6.3. Molybdenum Enzymes

Although the nature of the catalytic center and the molecular architecture of the various families of the Mo enzymes differ significantly, some general considerations apply:

- Each catalytic center is located in the interior of the molecule and the substrate approaches the center through a "funnel" through which the reaction product is removed.
- Prior to reaction, the substrate binds in a "pocket" immediately adjacent to the catalytic center in an orientation that facilitates the subsequent catalysis.
- In addition to the catalytic center, almost all of the Mo enzymes involve one (and sometimes several) other prosthetic group(s). These enzymes are organized in a manner analogous to an electrochemical cell: the site where oxidation occurs acts like an anode and the site where reduction occurs acts like a cathode.[1b] The redox active prosthetic groups are arranged to conduct electrons from the "anode" to the "cathode" and the external medium that supports the enzyme is analogous to a salt bridge.[1b]
- Although information is available to support the mechanism proposed for the catalysis, in most cases further investigations are required to establish the precise details of the process.

XII.6.3.1. The DMSO Reductase Family

The DMSO reductases are the simplest Mo enzymes; they are isolated as a single subunit of a relatively low molecular weight (<90 kDa) that contains the catalytic center as the sole prosthetic group. The simplicity of these enzymes contrasts with other, larger Mo enzymes that contain Fe–S, heme, and/or FAD groups, which participate in the conduction of electrons to (or from) the Mo center. Members of the DMSO reductase family have two MPT ligands and an amino acid residue bound to the Mo and, in their oxidized form, an oxo group (see Fig. XII.6.2).[1] Many spectroscopic, kinetic, and crystallographic investigations have been carried out on the DMSO reductases from the purple phototrophic bacteria *R. sphaeroides* and *R. capsulatus*.

Occurrence and Biological Function. The DMSO reductases may play a significant role in the global sulfur cycle (see Chapter II and Section XII.5). These enzymes occur only in eubacteria and catalyze the anaerobic reduction of DMSO to DMS (Eq. 5).

$$\text{DMSO} + 2\,e^- + 2\,H^+ \rightarrow \text{DMS} + H_2O \qquad (5)$$

The action of these enzymes can be linked to bioenergetic processes, for which they serve as a terminal electron acceptor. However, DMSO respiration provides far less energy than O_2 respiration.

Structure. The DMSO reductases of *R. sphaeroides*[11] and *R. capsulatus*[37] each consist of a single polypeptide chain comprised of 780 residues. There is a 77% identity between the sequences of these polypeptides; therefore, it is not surprising that the two enzymes possess the same overall molecular architecture. The polypeptide chain is folded into four principal domains that are linked to form an ellipsoidal-shaped molecule; and a "funnel", which leads from the exterior of the protein to the active site, is formed at the junction of three of the domains (see Fig. XII.6.9). The size of this funnel and the lining of the interior surface by aromatic residues facilitate the passage of DMSO to, and DMS from, the active site.

In the oxidized form of DMSO reductase (see Fig. XII.6.10), the Mo is bound to an oxo group, the oxygen of Ser147, and the dithiolene groups of two MPTs, each of which is covalently linked to a guanine dinucleotide to produce an MGD group (MGD = MPT guanine dinucleotide, see Section XII.6.2). The two oxygen atoms and the four sulfur atoms bound to the Mo are arranged in an approximately trigonal-prismatic geometry.

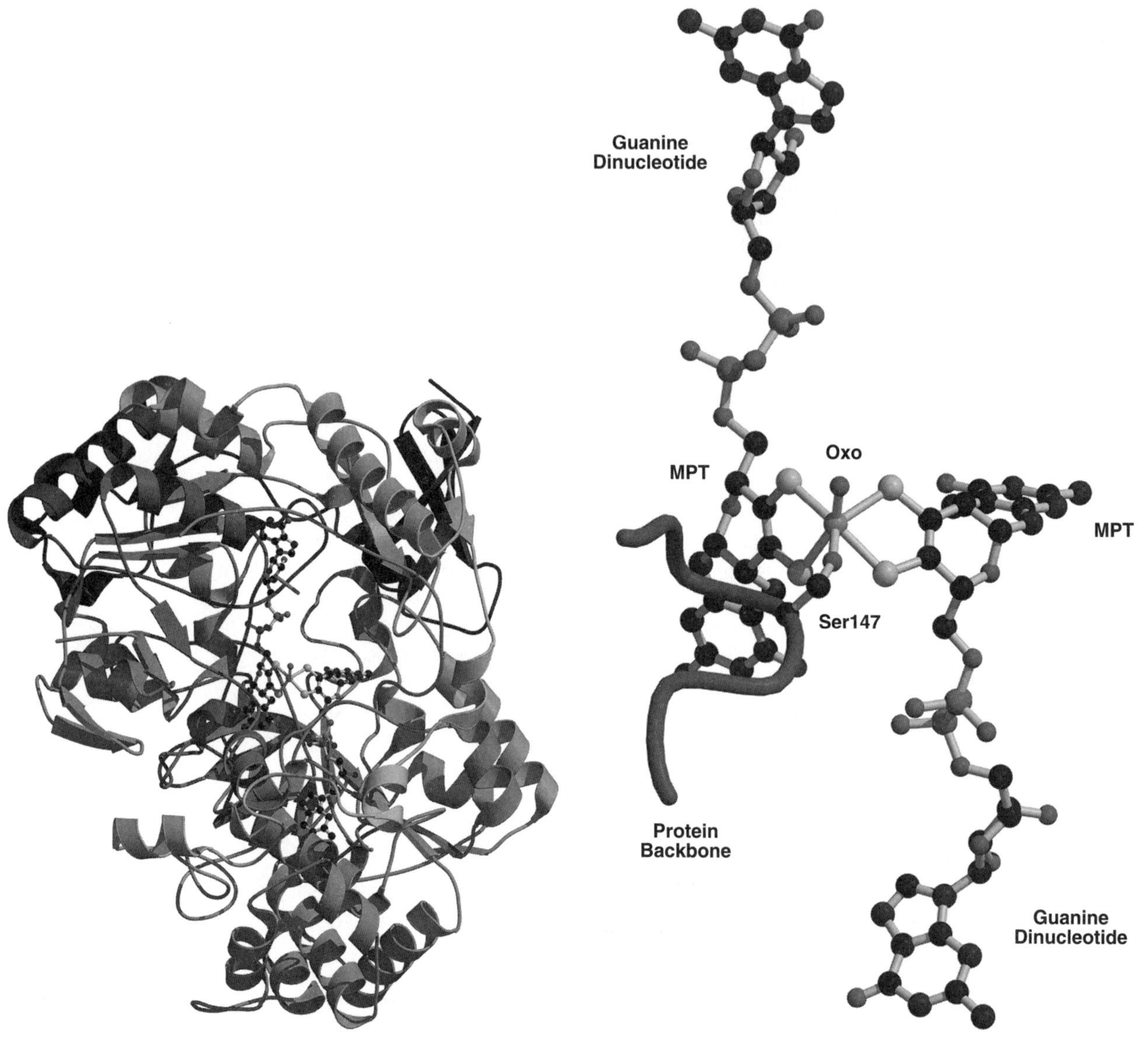

Fig. XII.6.9.
The structure of the DMSO reductase from *R. sphaeroides*, viewed from above the funnel through which the substrate passes to reach the active site (PDB code: 1EU1).

Fig. XII.6.10.
The structure of the oxidized form of the active site of the DMSO reductase from *R. sphaeroides* (PDB code: 1EU1).

Mechanism. Reaction 5 can be reversed; thus, the addition of DMS to the oxidized enzyme generates a center with DMSO bound to a reduced molybdenum center (Fig. XII.6.11). Protein crystallographic characterization of dithionite-reduced forms of DMSO reductases have revealed the loss of the oxo group and formation of a five-coordinated Mo center.[38] This structural information, together with complementary spectroscopic investigations of the native and reduced enzymes and the results of studies (especially resonance Raman[39]) using $DMS^{18}O$, are all consistent with a mechanism involving *direct* oxygen atom transfer, from the substrate to the molybdenum. Therefore, it is proposed that DMSO binds to the $[Mo^{IV}(OSer147)(MGD)_2]$ center and an oxygen atom is transferred to produce a $[Mo^{VI}O(OSer147)(MGD)_2]$ center (Fig. XII.6.10) and DMS. Reduction of the Mo(VI) center, in two CEPT

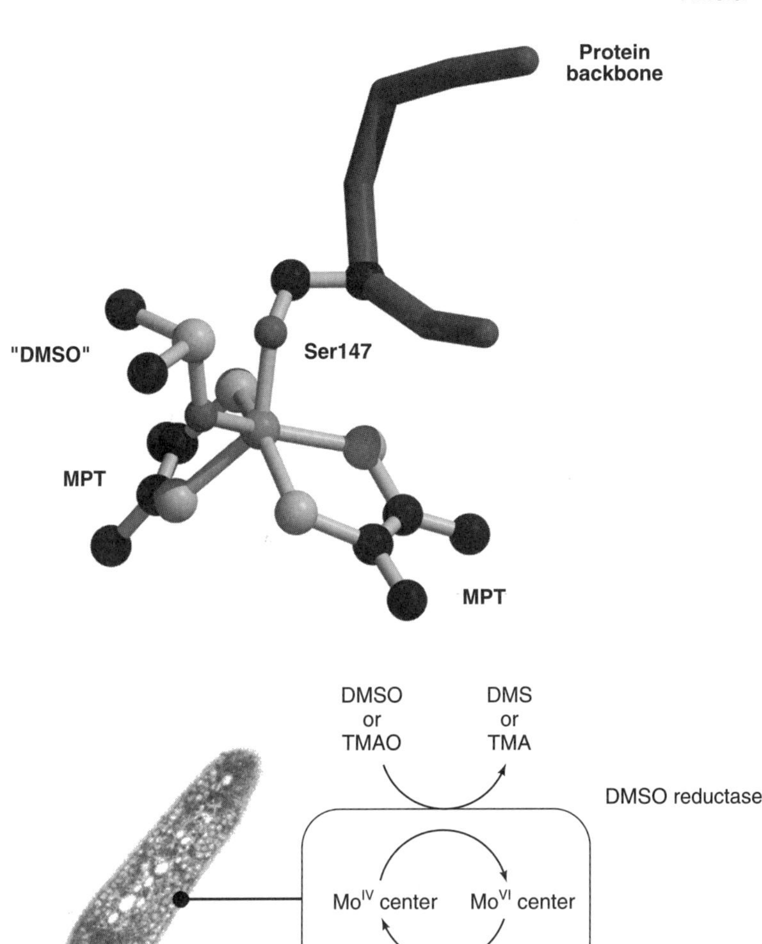

Fig. XII.6.11.
A representation of the active site of DMSO reductase from *R. capsulatus* with the substrate bound (PDB code: 4DMR).

Fig. XII.6.12.
A representation of the catalytic cycle of DMSO reductase, using either DMSO or TMAO as the electron acceptor with the pentaheme DorC as the reductant.

steps, converts the oxo group to an H_2O molecule (that is subsequently released from the coordination sphere of the metal) and regenerates the catalytically active Mo(IV) center. The electrons for this reduction are provided by an external donor. In the case of *R. capsulatus* DMSO reductase, this is the pentaheme cytochrome, DorC (see Fig. XII.6.12).

Alternate Substrates and Other Members of the Family. A membrane-bound DMSO reductase has been identified in *E. coli.*[1] This enzyme is composed of three subunits: one contains the catalytic center that is essentially the same as those of the DMSO reductases of *R. capsulatus* and *R. sphaeroides*; a second contains four [4Fe–4S] clusters, to conduct electrons to the catalytic center; and a third anchors the enzyme in the membrane and contains the binding site for the external reductant, menaquinol.

The DMSO reductases are capable of catalyzing the reduction of a wide range of substrates, including a range of sulfoxides. An interesting aspect of these reductions is the enantioselective reduction of racemic mixtures of chiral sulfoxides; this is doubtless a consequence of the chiral nature of the substrate-binding pocket. Also, these enzymes catalyze the loss of an oxygen atom from a variety of *N*-oxides, including TMAO (see Fig. XII.6.12), pyridine-*N*-oxides, and chlorate, $[ClO_3]^-$. A TMAO reductase has been isolated from the bacterium *S. massilia* and shown to possess a structure that is very similar to those of the DMSO reductases of *R. sphaeroides* and *R. capsulatus,* including an essentially identical catalytic center.[17] However, whereas the DMSO reductases catalyze the reduction of both sulfoxides and amine oxides, the TMAO reductases will not catalyze the reduction of sulfoxides. This difference in behavior appears to derive from some positively charged amino acid residues present on the surface of the funnel that leads to the active site in the TMAO reductase, which presumably allow amine oxides to pass, but not sulfoxides.

Other proteins of the DMSO reductase family that have been structurally characterized include the dissimilatory nitrate reductases from *D. desulfuricans* and *E. coli.*[16] Nitrate reductases occur in plants, fungi, and bacteria and catalyze the reduction of NO_3^- to NO_2^- (Eq. 6) either as the first step in nitrogen assimilation (*Note*: the assimilatory nitrate reductases belong to the sulfite oxidase family, see Section XII.6.3.2) or as nitrogen dissimilation, that is, anaerobic respiration using nitrate as the terminal electron acceptor to generate a transmembrane potential gradient. Only a few sulfate-reducing bacteria can grow with nitrate as the terminal electron acceptor of the respiratory chain; One example is *D. desulfuricans.* This nitrate reductase has a Mo center that, in its oxidized state, is bound to an oxo group plus five sulfur atoms, four from two MGD groups and one from Cys140. These six donor atoms are arranged in a trigonal prism centered on the molybdenum. Reduction of nitrate is suggested to occur by a mechanism that corresponds to that described for the DMSO reductases (See the Mechanism section from Section XII.6.3.1). Thus, NO_3^- binds to the $[Mo^{IV}(SCys140)(MGD)_2]$ center, and an oxygen atom is transferred to form NO_2^- and the $[Mo^{VI}O(SCys140)(MGD)_2]$ center. The latter is reduced to the catalytically active Mo(IV) center by two CEPT steps, with the electrons transferred to the Mo via a [4Fe–4S] cluster that is an integral component of the enzyme.

$$NO_3^- + 2\,e^- + 2\,H^+ \rightarrow NO_2^- + H_2O \qquad (6)$$

Formate dehydrogenases also belong to the DMSO reductase family. These enzymes catalyze the oxidation of formate to carbon dioxide (Eq. 7) and in some organisms carry out the reverse, the fixation of carbon dioxide. The formate dehydrogenase of *E. coli* has a structure very similar to that of the nitrate reductase of *D. sulfuricans,* but with a selenocysteine 140 residue bound to the molybdenum. The oxidized form involves an $[Mo^{VI}O(SeCys140)(MGD)_2]$ center and a [4Fe–4S] cluster.[14] Formate binds to the oxidized enzyme, and two electrons and a proton are transferred to the $[Mo^{VI}O]^{4+}$ center to form a $[Mo^{IV}(OH)]^{3+}$ center and CO_2. Reoxidation of the Mo(IV) center involves the removal of the two electrons via the [4Fe–4S] cluster, coupled to the loss of H^+. Note that the formate dehydrogenase does not (and cannot) involve oxygen atom transfer.

$$HCO_2^- \rightarrow CO_2 + H^+ + 2\,e^- \qquad (7)$$

XII.6.3.2. The Sulfite Oxidase Family

Occurrence and Biological Function. Sulfite oxidases occur in the liver, kidney, and hearts of animals where they catalyze the vital physiological oxidation of sulfite to sulfate (Eq. 8). This represents the terminal reaction in the degradation of the sulfur-containing amino acids Cys and Met and the means of removing excess sulfite from the body. The reducing equivalents associated with the oxidation of sulfite to sulfate are passed to cytochrome *c* and are then donated into the respiratory chain. A few individuals are born with a defective sulfite oxidase, because of either an impediment in the synthesis of MPT or a mutation in the gene that codes for the protein sequence of this enzyme. The lack of an effective sulfite oxidase leads to severe neurological abnormalities, dislocated ocular lenses, mental retardation, and invariably death in infancy.[19,35]

$$SO_3^{2-} + H_2O + 2\,Fe^{III}Cyt\ c \rightarrow SO_4^{2-} + 2\,H^+ + 2\,Fe^{II}Cyt\ c \qquad (8)$$

Sulfite oxidase has also been isolated from the prokaryote *T. novellus*[40] and the plant *A. thaliana.*[41]

Structure. Sulfite oxidase from chicken liver is a dimer of molecular weight 106 kDa.[19] Each monomer contains a Mo center and a cytochrome b_5 type heme group (Fig. XII.6.13). The Mo center and the heme reside in different domains of the polypeptide assembly that are joined by a loop; this section of the protein is weakly defined in the crystal structure, appears to be very flexible, and is easily cleaved enzymatically.

The active site of sulfite oxidase is shown in Fig. XII.6.14. The Mo has an approximately square-pyramidal coordination geometry: an oxo group is in the axial position; three sulfur atoms (two from an MPT and one from Cys185) and a second oxygen atom occupy the equatorial positions. The second oxygen is considered to be a water molecule because of its relatively long distance from the metal (Mo–O = 2.3 Å). This would imply that the active site structure represents the Mo(IV) form of the enzyme, even though the enzyme was purified in the fully oxidized, Mo(VI)/Fe(III), form; it is likely that the Mo center was photoreduced during the exposure to X-rays necessary for the crystallographic investigation. Resonance Raman spectroscopy suggests the presence of a $[Mo^{VI}O_2]^{2+}$ center in the oxidized enzyme. A combination of the information derived from the structural and spectroscopic studies was used to produce the active site depicted in Fig. XII.6.14.

Mechanism. There is an anion-binding site adjacent to the Mo center of chicken liver sulfite oxidase. In the crystal structure, SO_4^{2-} occupies this site (Fig. XII.6.15) and is bound by a network of hydrogen bonds to polar amino acid residues. It is proposed that, during turnover, SO_3^{2-} is bound in this site in an orientation that facilitates

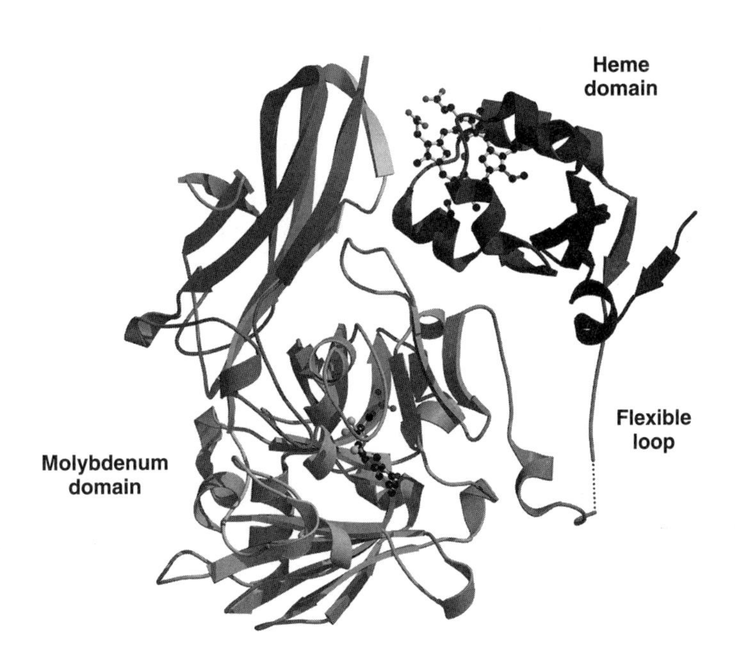

Fig. XII.6.13. Schematic representation of the structure of monomer of sulfite oxidase from chicken liver, showing the smaller heme domain and the larger Mo domain (PDB code: 1SOX). [See color insert.]

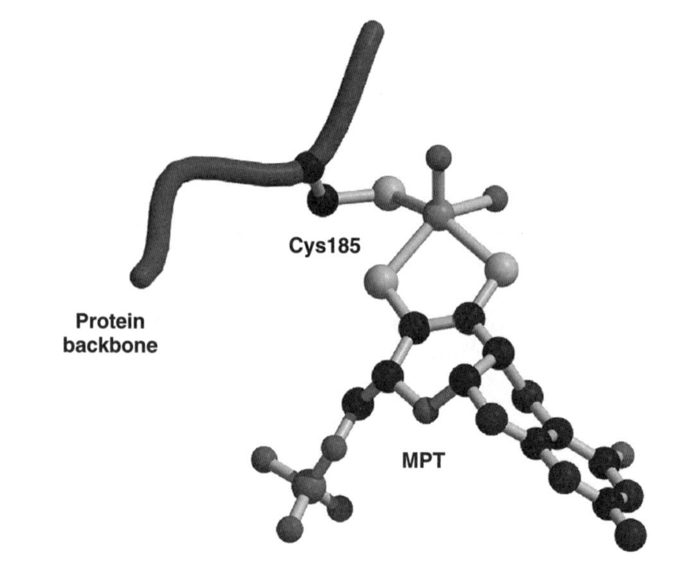

Fig. XII.6.14.
The structure of the oxidized form of the active site of sulfite oxidase from chicken liver (PDB code: 1SOX).

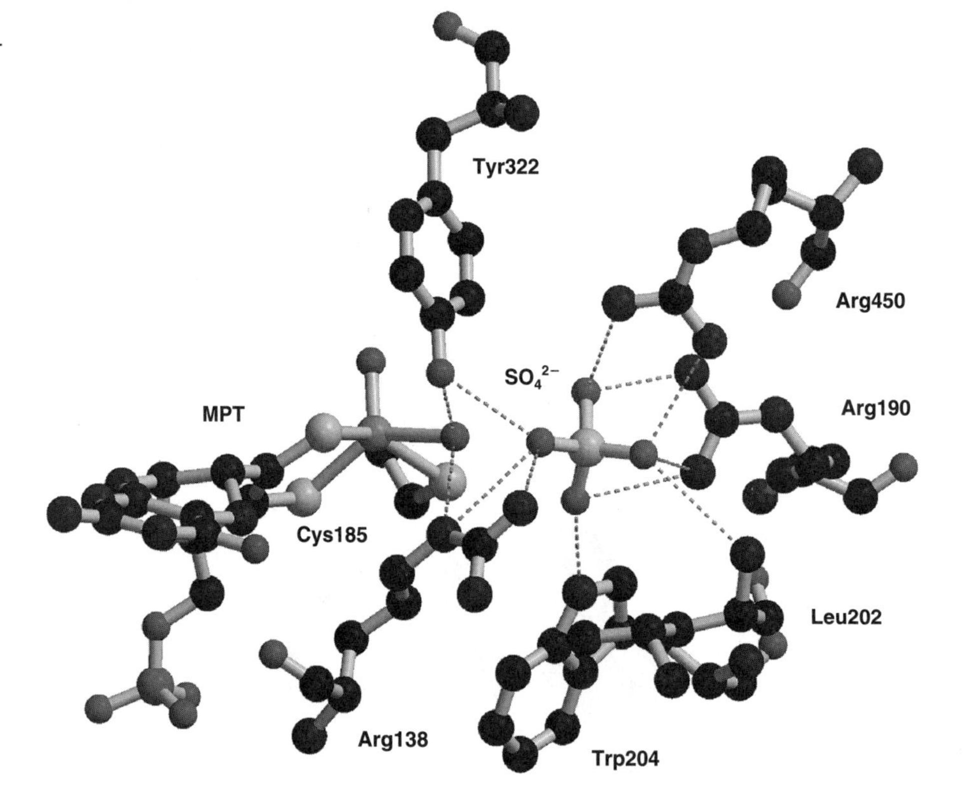

Fig. XII.6.15.
The structure of the substrate-binding site of sulfite oxidase from chicken liver with the substrate bound (PDB code: 1SOX).

catalysis with the lone pair of the S(IV) acting as a nucleophile for the electrophilic $[Mo^{VI}O_2]^{2+}$ center (cf. Section XII.6.2.2). This leads to a weakening of the equatorial Mo^{VI}=O group and the transfer of the oxygen atom to the sulfur, forming SO_4^{2-} and a $[Mo^{IV}O(SCys185)(MPT)]$ center. As noted in Section XII.6.2.2, the axial "spectator" oxo group may assist this oxygen atom transfer by forming a stronger Mo=O bond. Regeneration of the catalytically active Mo(VI) state proceeds by intramolecular electron transfer, from the Mo(IV) center to Fe^{III}Cyt *c*, to form a

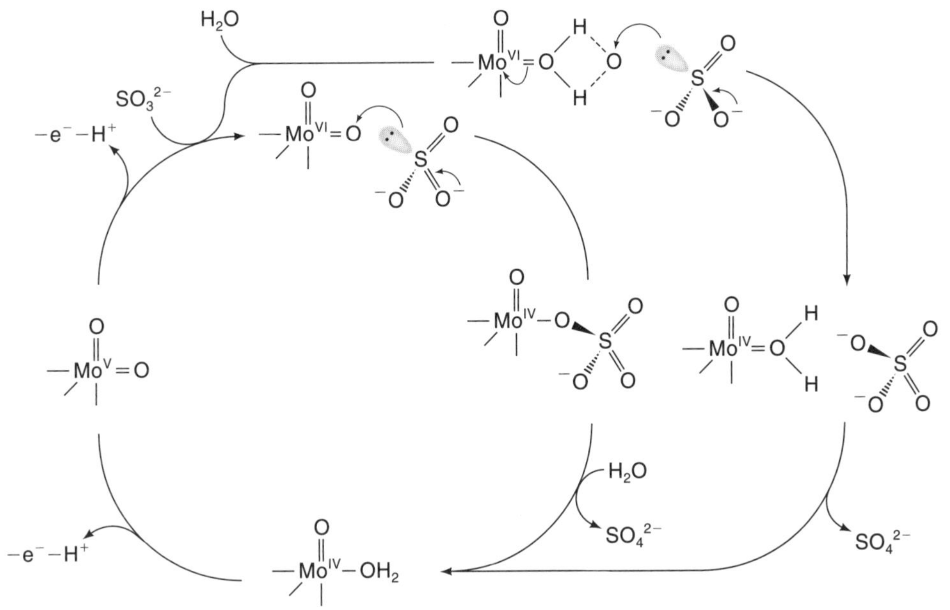

Fig. XII.6.16.
A proposed catalytic cycle for the oxidation of sulfite to sulfate by sulfite oxidase.

Mo(V)/Fe(II) state, spectroscopic studies of which indicate the presence of an $[Mo^VO(OH)]^{2+}$ center. The Fe^{II}Cyt *c* is oxidized by an external oxidant and the Mo(VI) center is then produced by CEPT, the electron being transferred to the Fe^{III}Cyt *c* with the loss of the hydroxide proton to produce an $[Mo^{VI}O_2]^{2+}$ center. Alternatively, there appears to be sufficient room in the active site for a water molecule to mediate the oxidation of SO_3^{2-}. Thus, CEPT to a $[Mo^{VI}O_2]^{2+}$ center through O–H bridges could lead to the formation of SO_4^{2-} and a $[Mo^{IV}O]^{2+}$ center. The regeneration of the oxidized state would be the same as in the first proposal and the two possible reaction sequences are summarized in Fig. XII.6.16. Suitable isotopic labeling and spectroscopic studies are required to distinguish between these two possible mechanisms.

The rate of the intramolecular electron transfer of sulfite oxidase ($\sim 10^3\ s^{-1}$) is not consistent with the long distance (32 Å) observed between the molybdenum and heme centers in the crystal structure (Fig. XII.6.13). An electron-transfer rate of $< 100\ s^{-1}$ would be expected for this distance (see Section X.2). Also, in the crystal structure, the MPT is oriented away from the heme, in contrast to structures of other Mo enzymes, in which MPT is oriented toward the accompanying redox center and connected to it by a series of hydrogen bonds. It is possible that the structure shown in Fig. XII.6.13 represents just one conformation for this enzyme and the flexible polypeptide loop may produce a different conformation, in which the molybdenum and heme centers are much closer together and the MPT is favorably oriented for electron transfer to the heme.

Other Members of the Family. The sulfite oxidase family includes the assimilatory nitrate reductases, which possess a high-sequence homology to the sulfite oxidases and are considered to involve the same type of catalytic center. The assimilatory nitrate reductases[10] accomplish the reduction of NO_3^- to NO_2^- (Eq. 7). The reduction of nitrate with the loss of one, and only one, oxygen atom is important since it prevents nitrogen being lost by the system as a gaseous product, such as N_2O, NO, or N_2. The second stage in nitrogen assimilation is catalyzed by nitrite reductases, enzymes that convert NO_2^- to NH_3 with the nitrogen bound to the metal center throughout the process (see Chapter XII.4).

XII.6.3.3. The Xanthine Dehydrogenase–Oxidase Family

Members of the xanthine dehydrogenase–oxidase family catalyze hydroxylation reactions of the type (Eq. 9) for purine, pyrimidine, pterin, and aldehyde substrates. All of these enzymes have a molecular weight of ~290 kDa and involve similar redox centers. Cow's milk is a good source of XO and this enzyme has been extensively investigated since its initial identification as an aldehyde oxidase in 1902. Xanthine oxidase is an important representative of a group of enzymes that involve a metal working in concert with a flavin (Chapter IV).

$$RH + H_2O \rightarrow ROH + 2\,H^+ + 2\,e^- \qquad (9)$$

The biochemical distinction between a xanthine dehydrogenase (XDH) and a xanthine oxidase (XO) is that XDH is oxidized by NAD^+ whereas the XO fails to react with NAD^+ and exclusively uses O_2 as its substrate, leading to the formation of the superoxide anion ($O_2^{\bullet -}$) and H_2O_2. In mammals, the XDH form of the enzyme is synthesized initially and is readily converted into the XO form by the oxidation of two sulfhydryl residues or by proteolysis.

The crystal structures of both the XDH and XO forms of this enzyme from cow's milk have been determined and a comparison of the details of the two structures has allowed identification of some aspects of the conformational change that converts one type of behavior to the other. However, the nature of the catalytic Mo centers and their location in the protein is identical for XDH and XO.

Occurrence and Biological Function. Both XO and XDH occur in the liver of animals and catalyze the last two steps of the formation of uric acid in mammals, the oxidation of hypoxanthine to xanthine and xanthine to uric acid (see Fig. XII.6.17). If produced in excess, uric acid or its monosodium salt is deposited in joints and tissues where it causes the inflammation known as gout; this condition can be relieved by the administration of allopurinol, an inhibitor of XO and XDH.

Structure. The active form of XDH is a homodimer of molecular weight ~290,000 Da involving ~1330 amino acids. Each monomer is comprised of three domains (see Fig. XII.6.18). The small N-terminal domain contains two [2Fe–2S] centers and is connected to a second, FAD-binding, domain by a long segment of the polypeptide chain; and the FAD domain is connected to a third domain by another segment of polypeptide. This large third domain contains the molybdenum center, which is bound near the interfaces of both the Fe–S and the FAD domains. Figure XII.6.19 shows the structure of the oxidized state of the catalytic center of

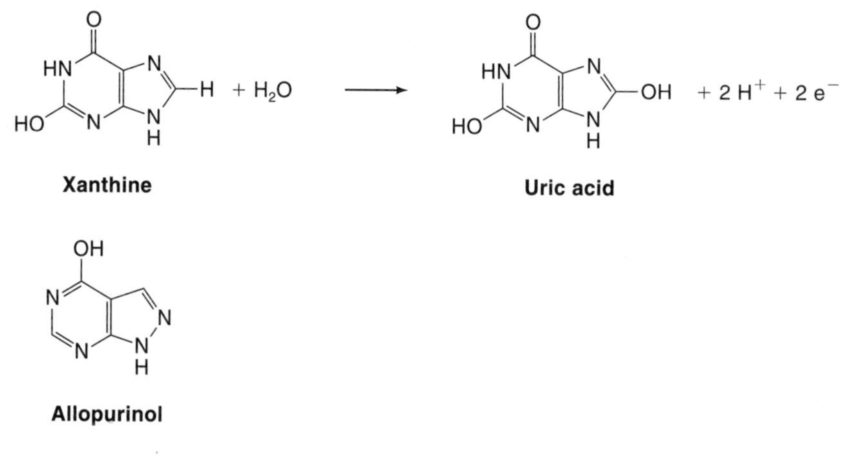

Fig. XII.6.17.
The oxidation of xanthine to uric acid and the structure of allopurinol.

Xanthine + H_2O → Uric acid + $2\,H^+ + 2\,e^-$

Xanthine

Uric acid

Allopurinol

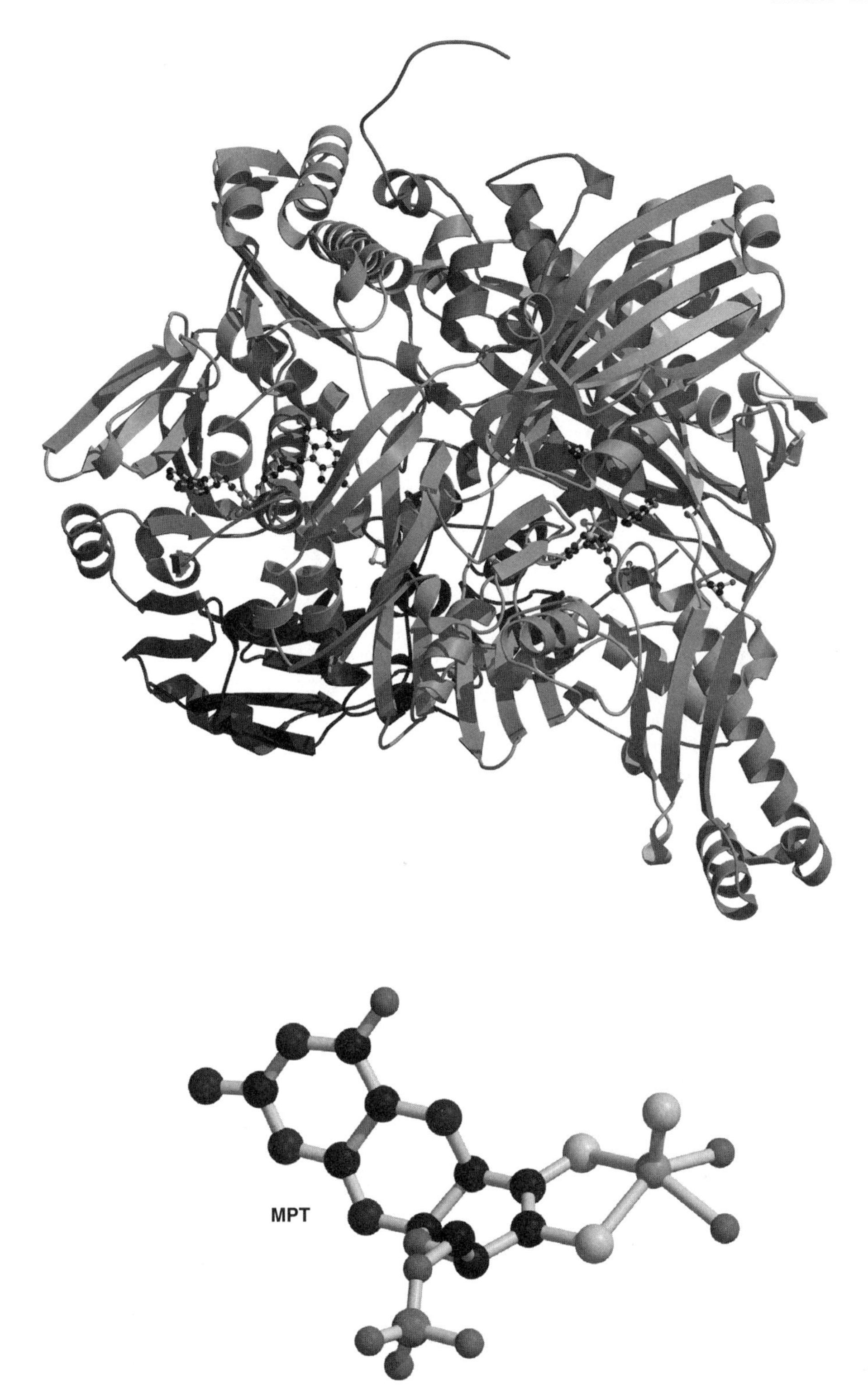

Fig. XII.6.18.
The structure of the monomer of bovine milk XDH (PDB code: 1FO4). [See color insert.]

Fig. XII.6.19.
The structure of the oxidized state of the active site of bovine milk XDH (PDB code: 1FO4). (*Note*: This study placed the Mo=S group in the axial position of the square-pyramidal coordination sphere with two oxygen atoms on equatorial sites, but the resolution of the structure does not allow the position of the sulfur to be located unequivocally. Recently, a crystallographic investigation of an intermediate in a hydroxylation reaction catalyzed by XO clearly identified an Mo–SH group in an equatorial site, thus implying such a location for the Mo=S group of oxidized XO, XDH, and related enzymes.)

XDH and related enzymes. The Mo is in a square-pyramidal coordination geometry, bound to two sulfur atoms of MPT, one sulfido (Mo=S), one oxo, and one hydroxo group. The presence of an Mo=S group is well established from spectroscopic and chemical studies; for example, treatment of the active enzyme with CN^- results in the formation of desulfo XO and NCS^-. The crystallographic characterization of XDH from cow's milk placed the Mo=S group in the axial position with two oxygen atoms on equatorial sites, but the resolution does not allow the position of the sulfur to be located unequivocally. Recently, a crystallographic investigation of an intermediate formed in a hydroxylation reaction catalyzed by XO clearly identified an Mo–SH group in an equatorial site, thus implicating such a location for the Mo=S group of oxidized XO, XDH, and related enzymes.

Mechanism. The mechanism of the oxidations catalyzed by XO, XDH, and related enzymes is more readily understood with an Mo=O group and the Mo=S group, respectively, being located in the axial and an equatorial position of the square-pyramidal coordination sphere, rather than the reverse locations. The mechanism is considered to involve deprotonation of the $[Mo^{VI}O(S)(OH)]^+$ center, and nucleophilic attack of the resultant Mo–O group on the carbon of the substrate, R(X=)C–H, to form an $[Mo^{IV}O(SH)(OC(=X)R)]$ intermediate. This latter step can be considered as involving the donation of the two electrons of the C–H bond into the Mo=S π^* orbital, linked to a lone pair of electrons on the O^- attacking the carbon atom to form a C–O bond. Hydrolysis releases the product, R(X=)COH, and subsequent CEPT with the loss of 2 e^-/2 H^+ re-forms the $[Mo^{VI}O(S)(OH)]^+$ active site. The electrons are conducted away from the Mo center via the relay of redox centers illustrated in Fig. XII.6.20, that is, the [2Fe–2S] centers and the FAD, to an external oxidant. This oxidation can proceed in two distinct, one-electron, steps leading to the formation of an Mo(V) intermediate, which has been investigated extensively by EPR spectroscopy.

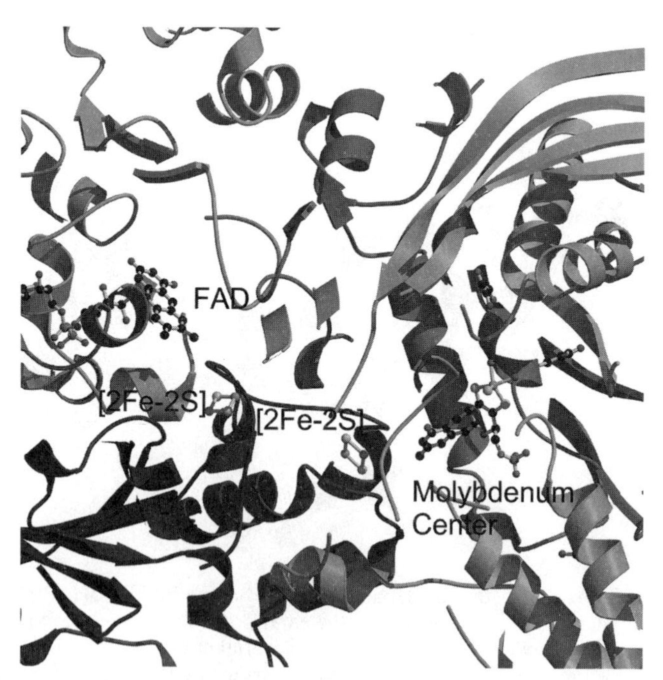

Fig. XII.6.20.
The organization of the redox centers in bovine milk XDH (PDB code: 1FO4).

Other Members of the Family. The aldehyde oxidase from *D. gigas* has Mo and two [2Fe–2S] domains that correspond to those of XDH and XO, but lacks the FAD domain. The mechanism of catalysis, for this and other aldehyde oxidases, is considered to be analogous to that described for XO and XDH, with the Mo=S group being crucial for the initial activation of the C–H bond, prior to conversion of RCHO to RCO_2H.

XII.6.3.4. Carbon Monoxide Dehydrogenases

The Mo bacterial CO dehydrogenases exhibit significant amino acid sequence homologies with the eukaryotic XO–XDH enzymes. However, a recent X-ray crystallographic study has shown that the active site of the CO dehydrogenase from *Oligotropha carboxidovorans* contains a unique MoSCu center (Fig. XII.6.21). Furthermore, X-ray fluorescence measurements on crystals of the enzyme confirm the presence of Cu; and anomalous difference Fourier maps, calculated from diffraction data collected below and above the Cu–K absorption edge, clearly indicate the presence of a Cu atom ~3.8 Å from the Mo.[23,24]

Occurrence and Biological Function. The molybdenum-dependent CO dehydrogenases are found in aerobic CO-oxidizing bacteria (e.g., *O. carboxidovorans*), that use CO as their sole source of carbon. These bacteria live in the top few centimeters of the soil, metabolize ~20% of the total CO in the atmosphere and contribute significantly to the regulation of its CO content. The energy released by the oxidation of CO to CO_2 (Eq. 10) is used to drive the synthesis of ATP.

$$CO + H_2O \rightarrow CO_2 + 2\,H^+ + 2\,e^- \qquad (10)$$

Structure. The CO dehydrogenase of *O. carboxidovarans* is a dimer with each monomer containing three subunits: one that binds the Mo center; one that binds two [2Fe–2S] clusters; and one that binds an FAD cofactor Fig. XII.6.22. Thus, the organization of the CO dehydrogenase is very similar to that of XO–XDH, except that instead of the domains, there are distinct subunits binding the prosthetic groups. The chain of redox active prosthetic groups facilitates the oxidation of the Mo center

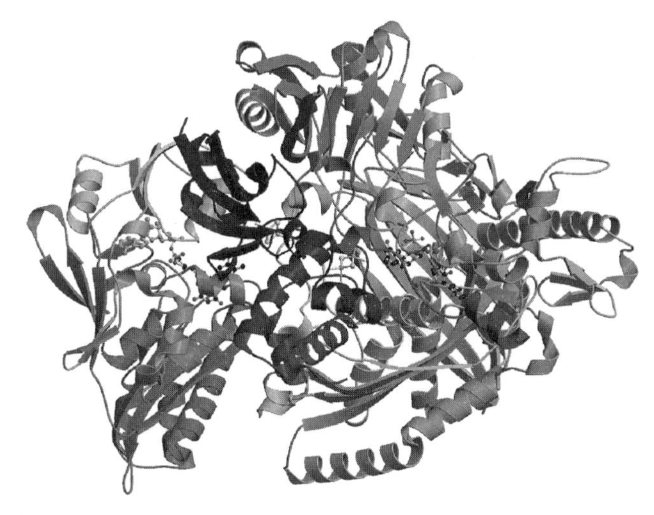

Fig. XII.6.21.
The structure of the CO dehydrogenase from *O. carboxidovarans* (PDB code: 1N5W). [See color insert.]

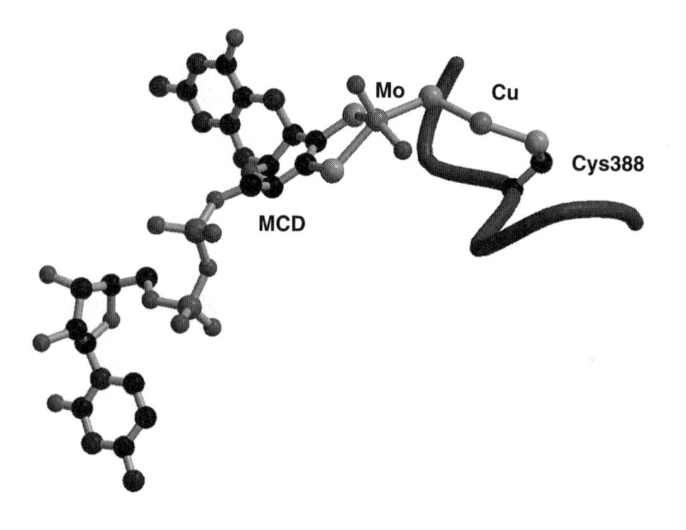

Fig. XII.6.22.
The active site of the oxidized form of the CO dehydrogenase from *O. carboxidovarans* (PDB code: 1N5W).

during enzyme turnover by providing a pathway for electron transfer, via the [2Fe–2S] clusters and then FAD, to the external redox partner, cytochrome b_{561}. The MPT ring is located between one of the [2Fe–2S] clusters and the Mo, lending support to the role of the former in electron transfer.

The active site of CO dehydrogenase (Fig. XII.6.21) is a unique, heterometallic, Mo–Cu center.[23,24] The Mo has a square-pyramidal coordination geometry, with two oxo groups, one in the axial and one in an equatorial position. The other equatorial sites are occupied by three sulfur atoms: A sulfido group that bridges to the Cu center and a dithiolene group of MPT that is bound to a cytosine dinucleotide (to form an MCD group). The coordination about the Cu is completed by Cys388, to give an approximately linear S–Cu–S geometry, typical of a Cu(I) (d^{10}) center.

Mechanism. The active site of CO dehydrogenase lies at the bottom of a hydrophobic channel ~17 Å deep and 6 Å wide that extends from the surface of the protein. This channel represents a pathway for CO to reach, and CO_2 to leave, the active site. The proposed reaction mechanism for CO dehydrogenase is uncertain, and definitive kinetic and spectroscopic studies need to be accomplished to define the nature of the process. Given the behavior of the catalytic centers of the other Mo enzymes, it is likely that redox reactions—involving Mo(VI/V/IV) states, oxygen atom transfer, and CEPT—participate in the catalysis effected by this enzyme. In addition, the Cu(I) site may be important for the binding of CO. The EPR spectroscopic studies on dithionite-reduced CO dehydrogenase have revealed characteristic resonances for the reduced [2Fe–2S] clusters and for a Mo(V) center that are similar to those observed for XO–XDH. Air-oxidized CO dehydrogenase is EPR silent, suggesting that the Cu center remains as Cu(I) throughout the reaction.

XII.6.4. Conclusions

This chapter has reviewed the nature of the catalytic sites in enzymes that involve Mo, the only 4*d*-transition element that is involved in biological systems. Molybdenum enzymes are found in all forms of life, from bacteria, through higher plants and animals, to humans. Most of the reactions catalyzed by the Mo enzymes involve the net transfer of an oxygen atom, either to or from the substrate. Important examples of these enzymes include the DMSO reductases, the nitrate reductases, the sulfite oxidases, the xanthine oxidases, the aldehyde oxidases, and the CO dehydrogenases. There are four types of catalytic center (see Figs. XII.6.1 and XII.6.2) with the special cofactor, "molybdopterin", being involved in each case.

The chemical nature of these catalytic centers and the redox changes involved in the catalytic cycle are now well defined. However, significant challenges remain with respect to developing an understanding of the role of "molybdopterin" in the catalytic processes, the changes in the electronic structure at the catalytic center during conversion, and why a particular type of Mo center is required for a particular catalytic reaction.

References

General References

1. (a) Hille, R., "The Mononuclear Molybdenum Enzymes", *Chem. Rev.*, **96,** 2757–2816 (1996). (b) Pilato, R. S. and Stiefel, E. I., "Molybdenum and Tungsten Enzymes", in *Bioinorganic Catalysis,* 2nd ed., by Reedijk, J. and Bouwman, E., Eds., Dekker, New York, 1999, pp. 81–152. (c) Sigel, A. and Sigel, H., Eds., "Molybdenum and Tungsten: Their Roles in Biological Systems", *Metal Ions in Biological Systems,* Vol. 39, Dekker, New York, 2002. (d) Stiefel, E. I., Ed., "Dithiolene Chemistry: Synthesis, Properties, and Applications", *Progr. Inorg. Chem.*, **52,** 1–72 (2004).
2. Holm, R. H., "The Biologically Relevant Oxygen Atom Transfer Chemistry of Molybdenum: From Synthetic Analog Systems to Enzymes", *Coord. Chem. Rev.*, **100,** 183–221 (1990).
3. McMaster, J. and Enemark, J. H., "The Active Sites of Molybdenum- and Tungsten-Containing Enzymes", *Curr. Opin. Chem. Biol.*, **2,** 201–207 (1998).
4. Mendel, R. R. and Schwarz, G., "Biosynthesis and Molecular Biology of the Molybdenum Cofactor (Moco)", *Metal Ions in Biological Systems,* Sigel, A. and Sigel, H., Eds., **39,** 2002, pp. 317–368.
5. Nugent, W. A. and Mayer, J. M., *Metal–Ligand Multiple Bonds: The Chemistry of Transition Metal Complexes Containing Oxo, Nitrido, Imido, Alkylidene, or Alkylidyne Ligands,* John Wiley & Sons, Inc., New York, 1988.
6. Rajagopalan, K. V. and Johnson, J. L., "The Pterin Molybdenum Cofactors", *J. Biol. Chem.*, **267,** 10199–10202 (1992).
7. Enemark, J. H. and Young, C. G., "Bioinorganic Chemistry of Pterin-Containing Molybdenum and Tungsten Enzymes", *Adv. Inorg. Chem.*, **40,** 1–88 (1994).

Specific References

8. Anbar, A. D. and Knoll, A. H., "Proterozoic Ocean Chemistry and Evolution: A Bioinorganic Bridge?", *Science,* **297,** 1137–1142 (2002).
9. Conrads, T., Hemann, C., George, G. N., Pickering, I. J., Prince, R. C., and Hille, R., "The Active Site of Arsenite Oxidase from *Alcaligenes faecalis*", *J. Am. Chem. Soc.*, **124,** 11276–11277 (2002).
10. Pollock, V. V., Conover, R. C., Johnson, M. K., and Barber, M. J., "Bacterial Expression of the Molybdenum Domain of Assimilatory Nitrate Reductase: Production of Both the Functional Molybdenum-Containing Domain and the Nonfunctional Tungsten Analog", *Arch. Biochem. Biophys.*, **403,** 237–248 (2002).
11. Li, H. K., Temple, C., Rajagopalan, K. V., and Schindelin, H., "The 1.3 Å Crystal Structure of *Rhodobacter sphaeroides* Dimethyl Sulfoxide Reductase Reveals Two Distinct Molybdenum Coordination Environments", *J. Am. Chem. Soc.*, **122,** 7673–7680 (2000).
12. Bray, R. C., Adams, B., Smith, A. T., Richards, R. L., Lowe, D. J., and Bailey, S., "Reactions of Dimethylsulfoxide Reductase in the Presence of Dimethyl Sulfide and the Structure of the Dimethyl Sulfide-Modified Enzyme", *Biochemistry,* **40,** 9810–9820 (2001).
13. Jormakka, M., Tornroth, S., Byrne, B., and Iwata, S., "Molecular Basis of Proton Motive Force Generation: Structure of Formate Dehydrogenase-N", *Science,* **295,** 1862–1868 (2002).
14. Boyington, J. C., Gladyshev, V. N., Khangulov, S. V., Stadtman, T. C., and Sun, P. D., "Crystal Structure of Formate Dehydrogenase H: Catalysis Involving Mo, Molybdopterin, Selenocysteine, and an Fe_4S_4 Cluster", *Science,* **275,** 1305–1308 (1997).
15. Hochheimer, A., Hedderich, R., and Thauer, R. K., "The Formylmethanofuran Dehydrogenase Isoenzymes in *Methanobacterium wolfei* and *Methanobacterium thermoautotrophicum:* Induction of the Molybdenum Isoenzyme by Molybdate and Constitutive Synthesis of the Tungsten Isoenzyme", *Arch. Microbiol.*, **170,** 389–393 (1998).
16. Dias, J., Than, M., Humm, A., Huber, R., Bourenkov, G. P., Bartunik, H. D., Bursakov, S., Calvete, J., Caldeira, J., Carneiro, C., Moura, J. J. G., Moura, I., and Romão, M. J., "Crystal Structure of the First Dissimilatory Nitrate Reductase at 1.9 Å Solved by MAD Methods", *Structure (London),* **7,** 65–79 (1999).
17. Czjzek, M., Dos Santos, J. P., Pommier, J., Giordano, G., Mejean, V., and Haser, R., "Crystal Structure of Oxidized Trimethylamine N-Oxide Reductase from *Shewanella massilia* at 2.5 Å Resolution", *J. Mol. Biol.*, **284,** 435–447 (1998).

18. Pollock, V. V. and Barber, M. J., "Kinetic and Mechanistic Properties of Biotin Sulfoxide Reductase", *Biochemistry,* **40,** 1430–1440 (2001).
19. Kisker, C., Schindelin, H., Pacheco, A., Wehbi, W. A., Garrett, R. M., Rajagopalan, K. V., Enemark, J. H., and Rees, D. C., "Molecular Basis of Sulfite Oxidase Deficiency from the Structure of Sulfite Oxidase", *Cell,* **91,** 973–983 (1997).
20. Terao, M., Kurosaki, M., Demontis, S., Zanotta, S., and Garattini, E., "Isolation and Characterization of the Human Aldehyde Oxidase Gene: Conservation of Intron/Exon Boundaries with the Xanthine Oxidoreductase Gene Indicates a Common Origin", *Biochem. J.,* **332,** 383–393 (1998).
21. Romão, M. J., Rosch, N., and Huber, R., "The Molybdenum Site in the Xanthine Oxidase-Related Aldehyde Oxidoreductase from *Desulfovibrio gigas* and a Catalytic Mechanism for This Class of Enzymes", *JBIC,* **2,** 782–785 (1997).
22. Enroth, C., Eger, B. T., Okamoto, K., Nishino, T., and Pai, E. F., "Crystal Structures of Bovine Milk Xanthine Dehydrogenase and Xanthine Oxidase: Structure-Based Mechanism of Conversion", *Proc. Natl. Acad. Sci. U. S. A.,* **97,** 10723–10728 (2000).
23. Dobbek, H., Gremer, L., Meyer, O., and Huber, R., "Crystal Structure and Mechanism of CO Dehydrogenase, a Molybdo Iron–Sulfur Flavoprotein Containing S-Selenylcysteine", *Proc. Natl. Acad. Sci. U. S. A.,* **96,** 8884–8889 (1999).
24. Dobbek, H., Gremer, L., Kiefersauer, R., Huber, R., and Meyer, O., "Catalysis at a Dinuclear [CuSMo(=O)OH] Cluster in a CO Dehydrogenase Resolved at 1.1 Å Resolution", *Proc. Natl. Acad. Sci. U. S. A.,* **99,** 15971–15976 (2002).
25. Stewart, L. J., Bailey, S., Bennett, B., Charnock, J. M., Garner, C. D., and McAlpine, A. S., "Dimethylsulfoxide Reductase: An Enzyme Capable of Catalysis with Either Molybdenum or Tungsten at the Active Site", *J. Mol. Biol.,* **299,** 593–600 (2000).
26. Johnson, M. K., Rees, D. C., and Adams, M. W. W., "Tungstoenzymes", *Chem. Rev.,* **96,** 2817–2839 (1996).
27. Rajagopalan, K. V., "The Molybdenum Cofactors—Perspective from Crystal Structure", *JBIC,* **2,** 786–789 (1997).
28. Enemark, J. H. and Garner, C. D., "The Coordination Chemistry and Function of the Molybdenum Centres of the Oxomolybdoenzymes", *JBIC,* **2,** 817–822 (1997).
29. Holm, R. H. and Berg, J. M., "Mononuclear Active-Sites of Molybdoenzymes: Chemical Approaches to Structure and Reactivity", *Pure Appl. Chem.,* **56,** 1645–1657 (1984).
30. Young, C. G. and Wedd, A. G., "Metal Chemistry Relevant to the Mononuclear Molybdenum and Tungsten Pterin Enzymes", *Chem. Commun.,* 1251–1257 (1997).
31. McMaster, J., Tunney, J. M., and Garner, C. D., "Chemical Analogues of the Catalytic Centers of Molybdenum and Tungsten Enzymes", *Prog. Inorg. Chem.,* **52,** 539–583 (2004).
32. Lim, B. S., Willer, M. W., Miao, M. M., and Holm, R. H., "Monodithiolene Molybdenum(V, VI) Complexes: A Structural Analogue of the Oxidized Active Site of the Sulfite Oxidase Enzyme Family", *J. Am. Chem. Soc.,* **123,** 8343–8349 (2001).
33. Lim, B. S., Sung, K. M., and Holm, R. H., "Structural and Functional Bis(Dithiolene)-Molybdenum/Tungsten Active Site Analogues of the Dimethylsulfoxide Reductase Enzyme Family", *J. Am. Chem. Soc.,* **122,** 7410–7411 (2000).
34. Inscore, F. E., McNaughton, R., Westcott, B. L., Helton, M. E., Jones, R., Dhawan, I. K., Enemark, J. H., and Kirk, M. L., "Spectroscopic Evidence for a Unique Bonding Interaction in Oxo-Molybdenum Dithiolate Complexes: Implications for σ Electron Transfer Pathways in the Pyranopterin Dithiolate Centers of Enzymes", *Inorg. Chem.,* **38,** 1401–1410 (1999).
35. Johnson, J. L. and Wadman, S. K., "Molybdenum Cofactor Deficiency and Isolated Sulfite Oxidase Deficiency", *The Metabolic Basis of Inherited Disease,* Scriver, C. R., Beaudet, A. L., Sly, W. S., and Valle, D. L., Eds., McGraw-Hill, New York, 1989, pp. 1463–1475.
36. Bertero, M. G., Rothery, R. A., Palak, M., Hou, C., Lim, D., Blasco, F., Weiner, J. H., and Strynadka, N. C. J., "Insights into the Respiratory Electron Transfer Pathway from the Structure of Nitrate Reductase A", *Nat. Struct. Biol.,* **10,** 681–687 (2003).
37. McAlpine, A. S., McEwan, A. G., Shaw, A. L., and Bailey, S., "Molybdenum Active Center of DMSO Reductase from *Rhodobacter capsulatus:* Crystal Structure of the Oxidised Enzyme at 1.82 Å Resolution and the Dithionite-Reduced Enzyme at 2.8 Å Resolution", *JBIC,* **2,** 690–701 (1997).
38. McAlpine, A. S., McEwan, A. G., and Bailey, S., "The High Resolution Crystal Structure of DMSO Reductase in Complex with DMSO", *J. Mol. Biol.,* **275,** 613–623 (1998).
39. Garton, S. D., Hilton, J., Oku, H., Crouse, B. R., Rajagopalan, K. V., and Johnson, M. K., "Active Site Structures and Catalytic Mechanism of *Rhodobacter sphaeroides* Dimethyl Sulfoxide Reductase as Revealed by Resonance Raman Spectroscopy", *J. Am. Chem. Soc.,* **119,** 12906–12916 (1997).
40. Kappler, U., Bennett, B., Rethmeier, J., Schwarz, G., Deutzmann, R., McEwan, A. G., and Dahl, C., "Sulfite: Cytochrome *C* Oxidoreductase from *Thiobacillus novellus.* Purification, Characterization, and Molecular Biology of a Heterodimeric Member of the Sulfite Oxidase Family", *J. Biol. Chem.,* **275,** 13202–13212 (2000).
41. Schrader, N., Fischer, K., Theis, K., Mendel, R. R., Schwarz, G., and Kisker, C., "The Crystal Structure of Plant Sulfite Oxidase Provides Insights into Sulfite Oxidation in Plants and Animals", *Structure,* **11,** 1251–1263 (2003).

XII.7. Tungsten Enzymes

Contents

Roopali Roy and Michael W. W. Adams
Department of Biochemistry and Molecular Biology, Center for Metalloenzyme Studies
University of Georgia
Athens, GA 30602

XII.7.1. Introduction

Tungsten (W, atomic number 74) has long been valued for its high-melting point, great strength, and good conductivity. It is widely used in the manufacture of industrial products such as lubricating agents and catalysts in the petroleum industry, but it is perhaps best known for its use in light bulb filaments for incandescent light bulbs and as a component of steel. The element is relatively scarce in natural environments and is usually isolated from oxygen-rich mineral ores such as wolframite ($[Fe/Mn]WO_4$), scheelite ($CaWO_4$) or, less frequently, from the more reduced form tungstenite (WS_2). The concentration of W in soil and freshwater environments is very low, usually $< 0.5\,\text{n}M$ in most lakes and rivers.[1] However, significantly higher concentrations ($> 50\,\text{n}M$) are found in alkaline brine lakes, hot spring waters and, most spectacularly, in marine hydrothermal vents.[2]

Deep sea hydrothermal vents are formed when sea water circulates through newly formed oceanic crust during volcanic events at the sea floor and jets of water at temperatures approaching 400°C are discharged at these sites. The superheated water is rich in suspended metal sulfide precipitates giving rise to the term "black smokers". Black smoker "chimneys", formed from the transition metal precipitates, are porous structures with multiple channels through which sea water freely percolates. The vents, despite such extreme conditions, support complex ecosystems. Tungsten is present in samples of vent fluids at about 1000 times the level found in seawater.[2]

The elements molybdenum (Mo, atomic number 42) and tungsten have very similar chemical properties. The atomic and ionic radii of W and Mo, as well as their electron affinities, are virtually identical. In aqueous solution, both elements are stabilized in higher oxidation states (IV–VI). The reduction potential for their (V/VI) and (IV/V) redox couples differ by small amounts, being more negative for W. Tungsten has exploitable isotopes, with ^{185}W useful as a radioisotope tracer and the stable spin $\frac{1}{2}$ ^{183}W used for spectroscopic studies.

The central role that Mo[3,4] plays in biological systems has been known for many decades and Mo-containing or molybdoenzymes have been extensively studied (see Section XII.6). Tungsten, on the other hand, with its high atomic number (74), is an unlikely choice for a metal with a biological function.[2–5] In general, biological systems do not use elements with atomic numbers >35, with the exception of molybdenum, 42, and iodine, 53. The earliest biological work on W found it to be an antagonist of Mo. Because the chemistry of W and Mo is so similar, organisms grown in the presence of W typically try to incorporate W into their molybdoenzymes. The result is either molybdoenzymes that lack any metal (and therefore are not catalytically active), or W-substituted molybdoenzymes, which are typically not catalytically active. Thus, while a requirement for Mo in many biological systems has been known since the 1930s, only in the last decade or so has it been realized that W has a positive role in several distinct biological systems. Moreover, in at least one group of microorganisms, Mo cannot fulfill that role.

So far, a biological function for W has not been found in any eukaryote. In the early 1970s, W was reported to stimulate the growth of certain microorganisms that grow by producing acetate (acetogens) or gaseous methane (methanogens) (see Section XII.2). Subsequently, a W-containing enzyme was purified from an acetogen[6] and at present, roughly a dozen tungstoenzymes are known.[2,5] In addition to the acetogens and methanogens, the metabolism of W has been intensively studied in hyperthermophilic microorganisms that grow at temperatures near and even $>100°C$. Organisms with this property belong to the domain known as the archaea, and are distinct from conventional bacteria. Methanogens are also archaea. Tungstoenzymes have also been isolated from microorganisms that grow by reducing sulfate to hydrogen sulfide, or use the gas acetylene for growth.

All tungstoenzymes can be classified into one of three functional families, abbreviated AOR, F(M)DH, and AH.[2,5] The three families appear to be unrelated from an evolutionary perspective. The prototypical member of the AOR family is represented by aldehyde ferredoxin oxidoreductase (AOR). This class also includes formaldehyde ferredoxin oxidoreductase and glyceraldehyde-3-phosphate ferredoxin oxidoreductase (GAPOR) from the hyperthermophilic archaea, carboxylic acid reductase (CAR) from acetogens, and aldehyde dehydrogenase (ADH) from sulfate-reducing bacteria. The F(M)DH class has two members; formate dehydrogenase (FDH) from methanogens and acetogens and *N*-formylmethanofuran dehydrogenase (FMDH) from methanogens. The third family has only one member so far; acetylene hydratase (AH) from an acetylene-utilizing bacterium. All known tungstoenzymes have been isolated from strict anaerobes and accordingly are very sensitive to inactivation by oxygen. All are located in the cytoplasm of cells and none are membrane associated. The general properties of tungstoenzymes therefore differ from those of molybdoenzymes. The latter are ubiquitous in the aerobic world and, consequently, many of them are insensitive to oxygen. In addition, many molybdoenzymes are membrane bound.

XII.7.2. Biochemical Properties of Tungstoenzymes

XII.7.2.1. Aldehyde Ferredoxin Oxidoreductase Family

Most known tungstoenzymes belong to the AOR family, with AOR from the hyperthermophilic archaeon *Pyrococcus furiosus* being the most studied.[2,5,7] Its properties are summarized in Table XII.7.1, together with those of FOR and GAPOR, two other tungstoenzymes found in hyperthermophiles such as *P. furiosus.*[10] All three enzymes consist of a single type of subunit, ~70 kDa, but they differ in their quaternary structures. The GAPOR is monomeric, AOR is dimeric, and FOR is a tetramer. Each of their subunits contains a mononuclear W site and one iron–sulfur cluster. Comparison of complete amino acid sequences reveals that these three enzymes are also closed related in evolutionary terms. The AOR and FOR families are the most closely related (40% sequence identity) while GAPOR is more distant (23% identity with AOR or FOR).[10]

The three enzymes from *P. furiosus* catalyze the conversion of various types of aldehyde to the corresponding acid (Table XII.7.2). This two-electron oxidation occurs inside the cell with the redox protein ferredoxin (Fd) serving as the electron acceptor.

Table XII.7.1.
Molecular Properties of Tungstoenzymes

Organism/ Enzyme[a]	Holoenzyme M_r (kDa)	Subunits	Subunit M_r (kDa)	W Content[b]	Pterin Cofactor[c]	FeS or Cluster Content[d]
I. AOR-TYPE						
Pf AOR	136	α_2	67	2	Nonnuc	$2[Fe_4S_4] + 1$ Fe
Pf FOR	280	α_4	69	4	Nonnuc	$4[Fe_4S_4]$
Pf GAPOR	73	α	73	1	Nonnuc	$1[Fe_4S_4]$
Pf WOR-4	138	α_2	69	2	NR	~6 Fe
Ct CAR (form I)	86	$\alpha\beta$	64, 14	1	Nonnuc	~29 Fe, ~25 S
Ct CAR (form II)	300	$\alpha_3\beta_3\gamma$	64, 14, 43	3	Nonnuc	~82 Fe, ~54 S
Cf CAR	134	α_2	67	2	Nonnuc	~11 Fe, ~16 S
Dg ADH	132	α_2	62	2	NR	$2[Fe_4S_4]$
II. F(M)DH-TYPE						
Ct FDH	340	$\alpha_2\beta_2$	96, 76	2	NR	20–40 Fe, 2 Se
Mw FMDH	130	$\alpha\beta\gamma$	64, 51, 35	1	Nuc[e]	2–5 Fe
III. AH-TYPE						
Pa AH	73	α	73	1	Nuc[e]	$1[Fe_4S_4]$

[a] The abbreviations are Pf, *P. furiosus;* Ct, *C. thermoaceticum;* Cf, *C. formicoaceticum;* Dg *D. gigas*; Mw, *M. wolfei;* Pa, *Pe. acetylenicus.* The data are taken from Refs. 2, 8, and 9.

[b] Expressed as integer value per mole of holoenzyme.

[c] Indicates whether the pterin is with (Nuc) or without (Nonnuc) an appended nucleotide. NR, not reported.

[d] Cluster content expressed per mole of holoenzyme.

[e] Appended nucleotide is guanosine 5′-monophosphate (GMP).

The ferredoxin of *P. furiosus* contains a single [4Fe–4S] cluster that undergoes a one-electron redox reaction. Therefore, assuming each W-containing subunit functions independently, one catalytic turnover per subunit requires the reduction of two molecules of ferredoxin. It is likely that two ferredoxin molecules are reduced sequentially (rather than two at a time) although this has not been proven.

The three hyperthermophilic enzymes differ in the types of aldehyde that they are able to oxidize (Table XII.7.2). The AOR family is thought to play a role in the degradation of amino acids when *P. furiosus* grows on peptides. The enzyme shows the highest catalytic efficiencies with acetaldehyde (from alanine), isovaleraldehyde (from valine), indoleacetaldehyde (from tryptophan), and phenylacetaldehyde (from phenylalanine). These aldehydes are derived by transamination (conversion of an amino acid to a keto acid) followed by decarboxylation (conversion of the keto acid to an aldehyde) of the common amino acids, alanine, valine, tryptophan, and phenylalanine, respectively. The AOR family has low K_m values for these aldehydes ($< 100\ \mu M$), consistent with its proposed physiological role.

FOR from *P. furiosus* exhibits high activity (and low K_m values) with C4–C6 di- and semialdehydes (OHC–R–CHO and HOOC–R–CHO, where R is C2–C4). Such compounds are intermediates in the metabolic pathways of amino acids such as arginine, lysine, and proline, so FOR may also be involved in the breakdown of

Table XII.7.2.
Catalytic Properties of Tungstoenzymes

Enzyme/Source[a]	Reaction Catalyzed	K_m (mM)	V_{max}[b]	Optimum Temperature (°C)
AOR (*P. furiosus*)	$RCHO^c + H_2O + 2\ Fd^d(ox) \rightarrow RCOOH + 2\ H^+ + 2\ Fd(red)$	0.04[e]	67	>80
FOR (*P. furiosus*)	$RCHO^f + H_2O + 2\ Fd(ox) \rightarrow RCOOH + 2\ H^+ + 2\ Fd(red)$	25.00[g]	62	>80
GAPOR (*P. furiosus*)	$R'CHO^h + H_2O + 2\ Fd(ox) \rightarrow R'COOH + 2\ H^+ + 2\ Fd(red)$	0.03	350	70
CAR (*C. formicoaceticum*)	$RCHO + MV^{++i} + OH^- \leftrightarrow RCOO^- + 2\ H^+ + MV^+$	0.14[j]	500	40
ADH (*D. gigas*)	$RCHO + BV^{++k} + H_2O \rightarrow RCOOH + 2\ H^+ + BV^+$	0.01[l]	38	30
FDH (*C. thermoaceticum*)	$CO_2 + NADPH + H^+ \rightarrow HCOOH + NADP^+$	0.11	880	55
FMDH (*M. wolfei*)	$CO_2 + MFR^m + 2\ H^+ \rightarrow CHO\text{–}MFR + H_2O$	0.01	11	65
AH (*Pe. acetylenicus*)	$C_2H_2 + H_2O \rightarrow CH_3CHO$	0.01	69	30

[a] The sources of data are given in Table XII.7.1.

[b] Micromoles (μmol) of substrate utilized min^{-1} mg^{-1} of protein at the indicated temperature.

[c] R = H, C1–C4, Ar.

[d] Ferredoxin = Fd.

[e] The K_m value was obtained using crotonaldehyde as the substrate for AOR.

[f] R = H, C1–C3, C4–C6 di-semialdehydes.

[g] The K_m value was obtained using formaldehyde as the substrate for FOR.

[h] $R' = CHOHCH_2OPO_3^{2-}$.

[i] Methyl viologen.

[j] The K_m value was obtained using butyraldehyde as the substrate for CAR.

[k] Benzyl viologen.

[l] The K_m value was obtained using acetaldehyde as the substrate for ADH.

[m] *N*-formylmethanofuran.

amino acids.[10] However, the actual reaction that FOR catalyzes inside the cell is not known. The enzyme was discovered by its ability to oxidize formaldehyde (HCHO), but it has a very high K_m for this substrate (25 mM) and this reaction is not thought to be of physiological significance.

The third tungstoenzyme in hyperthermophiles, GAPOR, is absolutely specific for glyceraldehyde-3-phosphate ($HOC{\cdot}CHOH{\cdot}CH_2OPO_3^{2-}$), which is oxidized to 3-phosphoglycerate ($HOOC{\cdot}CHOH{\cdot}CH_2OPO_3^{2-}$). The GAPOR enzyme plays a central role in the unusual glycolytic pathway that *P. furiosus* uses to convert glucose to pyruvate ($CH_3{\cdot}CO{\cdot}COOH$). None of GAPOR, FOR, or AOR can utilize the nicotinamide nucleotides, NAD and NADP, as electron carriers and all show a high affinity for *P. furiosus* ferredoxin.

The advent of complete genome sequences of microorganisms permits insight into the total number of enzymes of a given type that might be present. Consequently, a search of the *P. furiosus* genome for genes homologous to those encoding AOR, GAPOR, and FOR revealed two additional members of this tungstoenzyme family.[8] The product of one of these genes, termed WOR-4, was subsequently purified on the basis of its W content, see Table XII.7.1.[8] The catalytic properties and function of WOR-4, and of the putative fifth member (WOR-5), are unknown.

CAR was discovered in acetate-producing bacteria by its ability to catalyze the reductive activation of carboxylic acids, although it can also catalyze the reverse reaction, aldehyde oxidation (Table XII.7.2). The acid–aldehyde couple has one of the lowest reduction potentials known in biology. For example, the $E_0^{\circ\prime}$ value for acetaldehyde–acetate is −580 mV (SHE), and, consequently, under most conditions, aldehyde oxidation is much more thermodynamically favorable than acid reduction. The molecular properties of CAR from *Clostridium formicoaceticum* are very similar to those of *P. furiosus* AOR, whereas CAR from *C. thermoaceticum* is purified in a more complex form (see Table XII.7.1). The N-terminal amino acid sequences of CAR and AOR are similar, but the complete amino acid sequence of CAR is not known. The molecular properties of the other member of the AOR family, ADH from sulfate-reducing bacteria, are also virtually identical to those of AOR (Table XII.7.1); and CAR, ADH, and AOR use a similar range of aldehyde substrates (Table XII.7.2). However, the physiological electron carriers for ADH and CAR are not known.

XII.7.2.2. Formate Dehydrogenase Family

Formate dehydrogenase, one of the two types of enzyme in the F(M)DH family, catalyzes the reversible reduction of CO_2 to formate (Table XII.7.2), which involves the transfer of two electrons from the appropriate electron carrier. This reaction has a very negative reduction potential (−420 mV, SHE). The FDHs from anaerobic microorganisms are typically molybdoenzymes, but growth studies and preliminary enzyme analyses indicate that several types of bacteria and also methanogens may harbor W-containing FDHs. The only purified FDH is that from an acetogenic bacterium, *C. thermoaceticum.*[6] As shown in Table XII.7.2, this very complex enzyme contains the unusual amino acid selenocysteine, in addition to W and iron–sulfur centers. Acetogens can grow on H_2 and CO_2 and the key first step in CO_2 activation is catalyzed by FDH, which reduces CO_2 to formate using NADPH as the physiological electron donor.

Like FDH, the other member of this class of tungstoenzyme, FMDH, is typically a molybdoenzyme. The FMDH enzyme is found only in the methanogens where it catalyzes the reduction and addition of CO_2 to the organic, methanogenic cofactor, methanofuran (Table XII.7.2, see Section XII.2). The physiological electron donor for this reaction is not known. Methanogens grow on H_2 and CO_2 and generate CH_4, in contrast to the acetogens where the product is acetate. The FMDH enzyme catalyzes the first step in the conversion of CO_2 to CH_4 and is usually a molybdoenzyme, but in several thermophilic and hyperthemophilic methanogens the enzyme contains W. These FMDH enzymes are complex multisubunit enzymes (Table XII.7.1) and are extremely oxygen sensitive.[12]

Although FDH and FMDH are typically Mo-containing enzymes, organisms that exhibit W-dependence do not contain W-substituted molybdoenzymes. Rather, the W-containing FDHs and FMDHs are encoded by specific genes that are distinct from the genes that encode the Mo-containing enzymes. However, these tungstoenzymes do show amino acid sequence similarity to their Mo-containing counterparts and it is clear that they are related in evolutionary terms. In contrast, the AOR of the family of tungstoenzymes discussed above, while showing a high degree of molecular and sequence similarity to each other, show no sequence similarity to the F(M)DH family nor to any class of molybdoenzyme. Therefore, the two main groups of tungstoenzyme diverged very early on the evolutionary time scale, if they are related at all, and only one of them shows any relationship to molybdoenzymes.

XII.7.2.3. Acetylene Hydratase Family

Acetylene hydratase from the acetylene-utilizing anaerobic bacterium, *Pelobacter acetylenicus,* consists of a single type of subunit that contains one W atom and a single iron–sulfur cluster (Table XII.7.1). Unfortunately, the complete amino acid sequence of AH is not yet available, so its relation if any to the AOR or F(M)DH families is not known. However, the reaction catalyzed by AH, in contrast to those catalyzed by the other two families, does not involve an overall oxidation or reduction. Acetylene hydratase catalyzes the hydration of acetylene to acetaldehyde (Table XII.7.2). Since AH is active only in the presence of a strong reducing agent, acetylene hydration might involve sequential reduction, hydration, and oxidation steps, wherein the W site and the FeS center are required to carry out the redox chemistry. A fully functional Mo-containing AH has been purified from *P. acetylenicus* when it was grown in the presence of Mo, so W is not essential for the catalysis of acetylene hydration.[11]

XII.7.3. Structural Properties of Tungstoenzymes

The W atom in tungstoenzymes is coordinated by the same nonprotein organic ligand (MPT, see Section XII.6.6) that coordinates the Mo in molybdoenzymes. The three-ring pterin cofactor in Mo and W enzymes was first revealed by the crystallographic analysis of the tungstoenzyme, *P. furiosus* AOR.[7] As shown in Fig. XII.7.1, the third ring contains a dithiolene group and its two sulfur atoms coordinate the W atom. As in Mo enzymes the ligand is not covalently attached to the protein. Analyses of the cofactor after extraction from the protein have shown that in molybdoenzymes from bacteria the pterin is usually modified with a mononucleotide [AMP, GMP, cytidine 5′-monophosphate (CMP), or inosine 5′-monophosphate (IMP)] attached to the terminal phosphate, which is typically lacking in molybdoenzymes from higher organisms. An appended nucleotide is also absent with members of the AOR family of tungstoenzymes (Table XII.7.1), but the dinucleotide form of the cofactor is present in FMDH and AH.

The crystal structure of *P. furiosus* AOR had two surprising features (PDB code: 1AOR).[7] First, it showed that the two subunits are bridged by a single iron atom that had escaped spectroscopic detection. This site is >20 Å from the W and is presumed to have a structural role. Second, it was found that the W atom is coordinated not by two S atoms from one pterin ligand, but by four S atoms from two dithiolene side

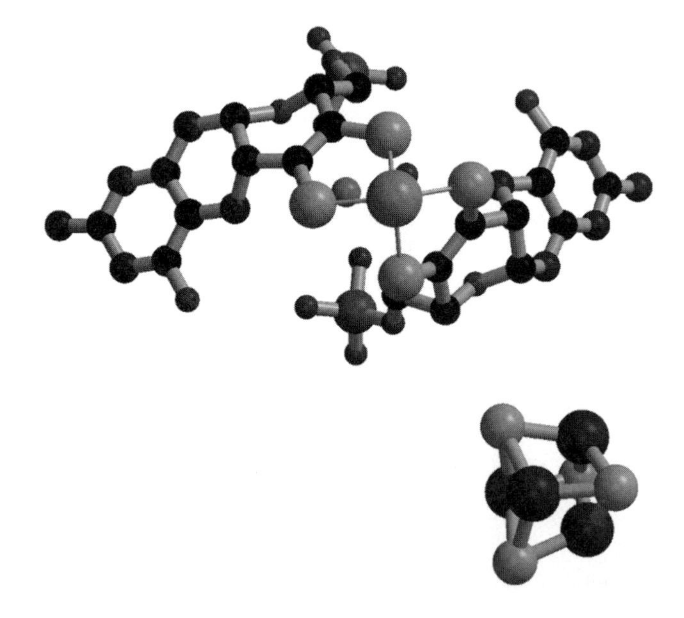

Fig. XII.7.1.
Three-dimensional structure of the bis(pterin)–tungsten site in *P. furiosus* AOR. The ball-and-stick figure depicts tungsten coordinated by bis(dithiolene) sulfurs from two pterins. The terminal phosphates and the Mg ion that link the two pterins via the phosphates are also shown, as is the [4Fe–4S] situated close to one of the pterin cofactors. (Adapted from Ref. 7.)

chains from two pterins (Fig. XII.7.1). The geometry of the W site in the enzyme displays a distorted square-pyramidal arrangement, with an angle between the planes of the pterin ligands of 97°. The two pterins are also linked to each other through their terminal phosphate groups, which coordinate the same Mg^{2+} ion. The so-called bis(pterin)-type site has also been found in molybdoenzymes of the DMSO reductase family.[3] In AOR, this W site is buried deep inside the protein ~10 Å away from a single [4Fe–4S] cluster. The iron cluster is closer to the protein surface, consistent with its proposed role as an intermediary in electron flow. The [4Fe–4S] cluster is coordinated to the protein via four cysteine ligands. One of these Cys residues is positioned to accept a hydrogen bond from a pterin ring nitrogen, providing evidence of the pterin's role in facilitating internal electron transfer in the enzyme. A hydrophobic channel of ~15 Å in length leads from the tungsten site to the surface of the protein. This channel is lined with apolar residues and is large enough to allow access of aromatic substrates, consistent with the ability of AOR to oxidize a broad range of substrates.

The crystal structure for a second tungstoenzyme in the AOR family, FOR from *P. furiosus*, is also available (FOR PDB code: 1B25 and FOR–glutarate complex PDB code: 1B4N).[11] The overall folding of the FOR subunit is virtually superimposable on that of AOR, consistent with their high-sequence similarity. However, FOR is a homotetramer with the four subunits arranged around a central cavity ~27 Å in diameter. Like AOR, each FOR subunit has one W atom coordinated to a bis(pterin) cofactor and one [4Fe–4S] cluster ~10 Å away. The distance between the two W atoms is ~50 Å and the two subunits are thought to be catalytically independent.

FOR lacks the subunit-bridging mononuclear Fe site found in AOR, but FOR does have an additional metal site. A Ca^{2+} ion is liganded to one pterin, but it appears to play a purely structural role. Cocrystallization of FOR and its physiological electron carrier, *P. furiosus* ferredoxin, shows that the binding site on FOR minimizes the distance between the [4Fe–4S] clusters of FOR and ferredoxin. This positioning suggests an electron-transfer pathway that begins at the W center, leads to the [4Fe–4S] cluster of FOR via one of the two coordinated pterins, and ends at the [4Fe–4S] cluster of Fd (see Fig. XII.7.2). The channel connecting the W site to the protein surface in AOR is replaced by a cavity in FOR. This cavity has two distinct parts: a large chamber at one end with a much narrower channel leading toward the protein surface. The bottom chamber, which contains the tungstopterin site, is lined with the bulky side chains of amino acids such as tyrosine, arginine, leucine, and valine. Consequently, the active-site cavity in FOR is much smaller than that in AOR, which presumably contributes to the difference in substrate specificity between the two enzymes.

A key question is what coordinates the W atom of these enzymes in addition to the four dithiolene sulfur atoms? Crystallographic studies of AOR and FOR suggest that in both enzymes there is one oxo ligand (W=O) to the W atom. However, it is difficult to ascertain with any degree of certainty the exact nature of ligands near the heavy W atom by crystallography because of the "ripple" around the W atom from

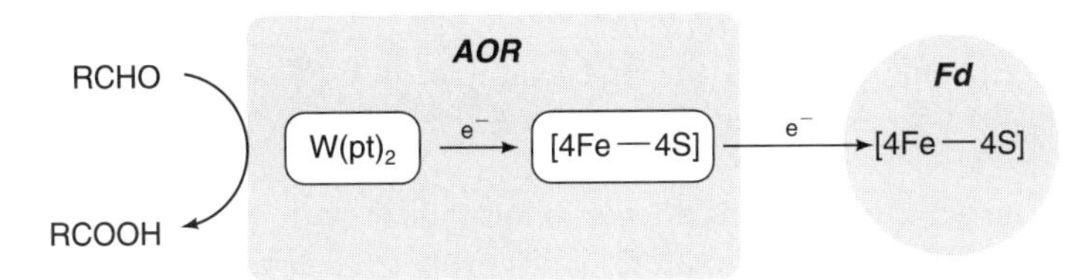

Fig. XII.7.2. Schematic representation of the pathway of electron flow in the subunit of the AOR family of tungstoenzyme. Ferredoxin is Fd, and $W(pt)_2$ represents the tungstobispterin site.

the series termination effects.[7,11] X-ray absorption spectroscopy (XAS) studies on the reduced form of AOR also reveal a single W=O, as well as four or five W–S and possibly a W–O/N group. An oxidized, catalytically inactive form of the enzyme, on the other hand, displays a coordination sphere with two W=O, three W–S, and possibly a W–O/N. The number and type of ligands around the W are an important indicator of the oxidation state of the metal center. For example, dioxo W coordination suggests that W is present in the W(VI) oxidation state in oxidized AOR, while monooxo is consistent with W(IV) or W(V) in the reduced enzyme. These assignments have been confirmed by other spectroscopic techniques.

Although the structure of the third and fourth tungstoenzymes from *P. furiosus*, GAPOR, and WOR-4 are not known, their amino acid sequence, size, and W and Fe content suggest that the two enzymes are closely related in structure to AOR and FOR. The same is probably true for ADH, whose quaternary structure, subunit size and W and Fe contents also match those of *P. furiosus* AOR (Table XII.7.1). This similarity extends to the smallest of the CAR enzymes, although its larger counterpart contains additional subunits, FAD, and much more Fe (Table XII.7.1). The relationship of the larger CAR enzyme to AOR is not clear. Structures have yet to be reported for the FDH and FMDH class of tungstoenzyme. However, the high similarity in their amino acid sequences and those of Mo-containing FDHs suggests that they have similar structures.[13,14]

It therefore appears that there is a fundamental difference in the structures of the active sites of the two major classes of tungstoenzyme. The W site of the AOR family is coordinated by the four S atoms of a bis(pterin) non-nucleotide cofactor and there is no direct coordination from the protein. On the other hand, the F(M)DH class contains the nucleotide form of a bis(pterin) cofactor and is coordinated by its four S atoms and by a (Se)S atom of a (seleno)cysteine residue of the protein, at least according to sequence comparisons with the crystal structure of the analogous Mo-containing enzyme. Interestingly, although the size and cofactor content of the Mo-containing FDH of *Escherichia coli* is very similar to that of AOR, the two enzymes show no similarities in their overall structures, providing further evidence that either the AOR family of tungstoenzyme and Mo-containing enzymes diverged early in evolutionary times, or they arose completely independently.

XII.7.4. Spectroscopic Properties of Tungstoenzymes

During catalysis the W center is thought to undergo a two-electron cycle, as is the case for many Mo enzymes which exhibit the (IV)/(V)/(VI) oxidation states.[3] Evidence for the W(V) state comes from EPR spectroscopy. The W(V) state has d^1 configuration, is paramagnetic, and exhibits EPR resonances with g values <2. These resonances are typically slow relaxing and observable at temperatures >100 K. The W(IV) and W(VI) states are diamagnetic and do not display EPR signals. Enzymes from all three classes of tungstoenzyme, AOR, FM(D)H, and AH, show resonances attributed to W(V) species.[5] The spectra are assigned to W(V) based on their relaxation properties and g values. Unambiguous assignment of the resonance comes from hyperfine interactions arising from ^{183}W (13% natural abundance), which has a nuclear spin $I = \frac{1}{2}$. Enzymes enriched in ^{183}W can be obtained from cells grown in the presence of ^{183}W-tungstate.[16]

The FDH enzyme provided the first example of W(V) EPR resonances from a biological system.[15] At lower temperatures, resonances are seen from reduced iron–sulfur centers (both [2Fe–2S] and [4Fe–4S]), whereas at higher temperatures the reduced enzyme exhibits a slow-relaxing signal with unusually high g values ($g = 2.1, 1.98, 1.95$). These anomalous EPR properties have been attributed to the

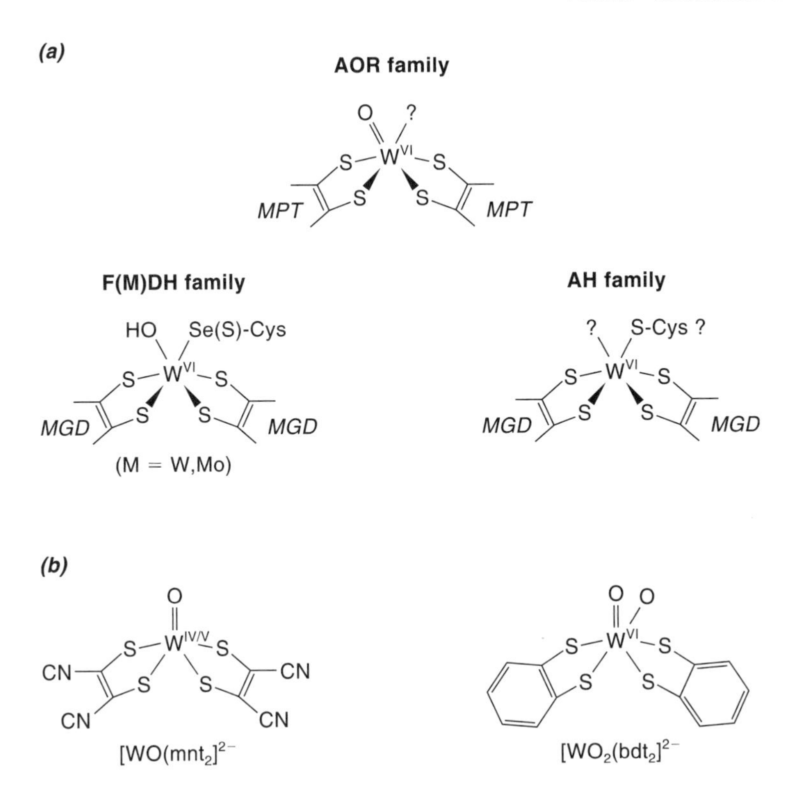

Fig. XII.7.3.
(*a*) Schematic representation of active sites of tungsten enzymes. Tungsten is shown coordinated to two pterin dithiolene ligands with or without appended nucleotide. The MPT (metal-binding pyronopterin dithiolate) is the non-nucleotide form, MGD (MPT guanine dinucleotide) has GMP attached. (*b*) Structures for W(IV), W(V), and W(VI) bis(dithiolene) model complexes with two distinct ligands that coordinate tungsten: 1,2-dicyanoethylene-1,2-dithiolate (mnt) and benzene-1,2-dithiolate (bdt). (Taken from Ref. 18 and references cited therein.)

S-rich environment of the W comprising four S from two dithiolene side chains, a coordinated selenocysteine and the lack of an oxo ligand in the W(V) state.[5] The EPR properties of FMDH have not been reported, but the sole member of the third class of tungstoenzyme, AH, also gives rise to a slow-relaxing, EPR signal with relatively high g values ($g = 2.04, 2.02, 2.00$) that is assigned to W(V).[9] These properties suggest that the enzyme contains a W site coordinated by the S atoms of two pterins and possibly a cysteine ligand.

Based on their EPR properties (Fig. XII.7.3*a*), tungstoenzymes can therefore be classified into two distinct groups. In the AOR class, the active site W is coordinated by the dithiolene side chains from two non-nucleotide pterin cofactors and an oxo (or possibly an hydroxyl) ligand in the W(V) state, but there is no direct interaction with any protein side chain. The AOR, FOR, GAPOR, ADH and CAR families all appear to fit in this group, despite considerable heterogeneity of their active sites by EPR analyses. In contrast, it appears that the FM(D)H class contains a W center coordinated by the nucleotide form of the pterin cofactor with an additional cysteine or selenocysteine ligand from the protein. The latter property leads to an increase in the g_{av} value of the W(V) EPR signal. This second category may also include AH.

XII.7.5. Mechanism of Action of Tungstoenzymes

The structural features of AOR and FOR give an overall idea of how these enzymes catalyze the oxidation of aldehydes. Substrates are thought to migrate via the hydrophobic channel or cavity to the W site, which is in the oxidized (VI) form. The two-electron oxidation of the aldehyde leads to the reduction of the W site to the (IV) form. As shown in Fig. XII.7.2, the W site is then oxidized one electron at a time (likely coupled with deprotonation) by sequential reduction of the adjacent 4Fe center, which in turn is oxidized by the external electron carrier, ferredoxin.

What do we know about the intimate mechanism of aldehyde oxidation? Detailed kinetic studies of pterin-based mononuclear molybdoenzymes[15] have shown that catalysis takes place by one of two mechanisms, coupled proton–electron transfer or oxygen atom transfer. Although similar kinetic data are not yet available for tungstoenzymes, comparison of the crystal structure of FOR with that of a Mo-containing, aldehyde-oxidizing enzyme known as aldehyde oxidoreductase from *Desulfovibrio gigas* (abbreviated here as DgAOX) gives some clues as to the possible mechanism of action of the AOR family.[8]

Like FOR and AOR, DgAOX is a pterin-containing enzyme that catalyzes the oxidation of aldehydes. However, its amino acid sequence and 3D structure is unrelated to that of the tungstoenzymes. Structural and mechanistic studies of DgAOX have identified a specific glutamic acid residue (Glu869) that plays an essential catalytic role.[16] Although the overall 3D structures of FOR and DgAOX are completely unrelated, when the structures of their active sites are superimposed (aligning their metal atoms and the single pterin of DgAOX with one of the two pterins of FOR), the side-chain Glu308 of FOR corresponds to Glu869 in DgAOX.[8] In addition, the O atom of the W=O group and one of the S atoms of the WS_4 site in FOR correspond to the O of a water molecule and the S of a Mo–S group, respectively, in DgAOX. Also, there is an overlap between the two enzymes of several amino acid side chains (of tyrosine and threonine) that might have catalytic roles. This may represent a remarkable case of convergent evolution or the enzymes may have diverged so long ago that the structural homology cannot be recognized from sequence comparisons. In any event, based on these unexpected structural similarities, it was proposed that FOR uses the coupled electron–proton-transfer mechanism of aldehyde oxidation of DgAOX.[8] Unfortunately, there is as yet no experimental evidence to support such a mechanism in the tungstoenzymes.

XII.7.6. Tungsten Model Complexes

Synthetic analogues that mimic the active sites of both tungsto- and molybdoenzymes are important in understanding the chemistry of the metal center and in interpreting spectroscopic results. In recent years, many synthetic Mo complexes have been synthesized and structurally and spectroscopically characterized,[3] but analogous W complexes, specifically with sulfur ligands, are rarer. One reason for this is the difficulty in reducing mononuclear W(VI) species to the corresponding W(IV) state.[17] Indeed, W(VI)–oxo complexes are relatively unstable and attempts to carry out oxo-transfer reactions lead, as is often the case in Mo chemistry, to the formation of W(V) dimers. It seems that bis(dithiolene) coordination of the W center via the two pterin cofactors is essential in tungstoenzymes in order to maintain the redox potentials [W(VI)/W(IV)] in the range required for catalysis.[5]

Monooxo and des-oxo W(IV) and W(V) complexes with bis(dithiolene) coordination have been synthesized.[17] The former adopts a distorted octahedral geometry whereas the latter has square-pyramidal stereochemistry. Stable monooxo W(IV)/W(V) and dioxo W(VI) complexes have also been reported with bdt, and mnt, ligation.[18] As shown in Fig. XII.7.2*b*, the monooxo W(IV) and W(V) complexes display square-pyramidal geometry, whereas the dioxo W(VI) compounds are in a distorted octahedral arrangement. The ligation modes seen in these model complexes appear to mimic the W sites of the tungstoenzymes. Moreover, a W(VI) complex with terminal oxo and bidentate sulfur ligands was shown to catalyze aldehyde oxidation in a reaction analogous to that of AOR. Similarly, $[W^{IV}O(mnt)_2]^{2-}$ catalyzes the reduction of CO_2/HCO_3^- to formate in aqueous media in the presence of reductant, analogous to W-containing FDH.[18] In a reaction analogous to AH, the model complex

$[W^{IV}O(mnt)_2]^{2-}$ has been reported to catalyze acetylene hydration to acetaldehyde, whereas the corresponding compound $[W^{VI}O_2(mnt)_2]^{2-}$ could not.[18] Interestingly, AH as purified is inactive unless reduced with a strong reductant, suggesting a possible close relationship between the synthetic complexes and the active sites of the enzymes.

XII.7.7. Tungsten versus Molybdenum

An important role for W in biological catalysis has slowly emerged over the last decade. However, the question arises as to why some microorganisms prefer W, rather than the chemically analogous and usually more abundant metal Mo. If one compares the properties of the mononuclear W and Mo synthetic complexes,[5] it is evident that the relevant W complexes are much more oxygen sensitive (oxophilic) than the equivalent Mo compounds, indicating that such a W site within an enzyme would be stable only under anaerobic conditions. Additionally, W complexes display enhanced thermal stability compared to Mo complexes, suggesting that the former rather than the latter would be preferred at temperatures near the normal boiling point of water. This enhanced bond strength of W complexes may also account for the observation that such complexes are generally kinetically slower than the equivalent Mo complexes. In addition, the much lower redox potentials (typically by >300 mV) for the W complexes compared to the Mo ones may make W better suited to catalyze lower potential redox processes.[5] Lastly, availability of these metals may play a major role in determining which of them is utilized in various ecosystems. For example, in freshwater and marine environments, the concentration of W is typically several orders of magnitude lower than that of Mo, although in deep sea hydrothermal vent systems W is much more prevalent than Mo.[2]

Hence, in pure chemical terms, it might be expected that W would only be utilized in biological systems to catalyze low-potential reactions under anaerobic conditions, and that significant catalytic rates would be observed only at high temperatures. Conversely, Mo complexes would be unstable at high temperatures, but at lower temperatures they would be catalytically competent over the whole biological range of potentials under both aerobic and anaerobic conditions.[5] To a large extent this appears to be the case. The tungstoenzymes known so far all catalyze reactions involving extremely low potential chemistry that require anoxic conditions. However, for organisms that grow at low or moderate temperatures, tungstoenzymes can be replaced by molybdoenzymes.[2] For example, functional Mo-containing isoenzymes exist for W-containing FMDH, CAR, FDH, and AH in respective mesophilic or thermophilic organisms.[19] In fact, the notable exception is the AOR family of tungstoenzymes found in hyperthermophiles such as *P. furiosus.*[2,5] So far they are the only known organisms that obligately require W for growth. They do not contain Mo-analogues of the AOR family, and they appear not to utilize Mo. Such organisms carry out chemical reactions near the limits of biology, at extremely low potentials and at high temperatures. It is possible that catalysis under such extreme conditions is a feat that can be accomplished by W, but not by Mo. In light of these considerations, one can also speculate on the evolution of pterin-containing enzymes.

Such considerations also have implications on the evolution of the pterin cofactor. If life first evolved on this planet under high-temperature anoxic conditions, as has been proposed by some researchers,[20] then the first pterin-type cofactor would likely have contained W coordinated by four sulfur atoms from two pterin molecules, analogous to the AOR catalytic site. This entity would be thermostable, could participate in two-electron redox chemistry in the absence of oxygen and, if incorporated as a protein cofactor, could have given rise to the precursors of the AOR family of tungstoenzymes.[20] However, as the Earth cooled and life adjusted to lower and lower temperatures, the same cofactor must have also been incorporated (or the AOR family diverged) into a second protein lineage that ultimately gave rise to present day molybdoenzymes, with Mo almost completely replacing W.[20] This metal substitution appears to have enabled enzymes to catalyze two-electron-transfer reactions more efficiently at low temperatures, and to do so under aerobic conditions, even when the metal (Mo) was coordinated by two sulfurs from a single pterin. At present, however, there is no direct evidence for this evolutionary scenario.[21]

References

General References

1. Krauspof, K. B., "Tungsten", in *Handbook of Geochemistry,* Wedepohl, K. H., Ed., Vol. II, Chap. 42, Springer Verlag, New York, 1972.
2. Kletzin, A. and Adams, M. W. W., "Tungsten in biology", *FEMS Microbiol. Rev.,* **18,** 5 (1996).
3. Chapter XII.6 in this volume.
4. Hille, R., "The mononuclear molybdenum enzymes", *Chem. Rev.,* **96,** 2757 (1996).
5. Johnson, M. K., Rees, D. C., and Adams, M. W. W., "Tungstoenzymes", *Chem. Rev.,* **96,** 2817 (1996).

Specific References

6. Yamamoto, I., Saiki, T., Liu, S.-M., and Ljungdahl, L. G., "Purification and properties of NADP-dependent formate dehydrogenase from *Clostridium thermoaceticum,* a tungsten-selenium-iron protein", *J. Biol. Chem.,* **258,** 1826 (1983).
7. Chan, M. K., Mukund, S., Kletzin, A., Adams, M. W. W., and Rees, D. C., "Structure of the hyperthermophilic tungstoprotein enzyme aldehyde ferredoxin oxidoreductase", *Science,* **267,** 1463 (1995).
8. Roy, R. and Adams, M. W. W., "Characterization of a fourth tungsten-containing enzyme from the hyperthermophilic archaeon *Pyrococcus furiosus*", *J. Bacteriol.,* **184,** 6952–6956 (2002).
9. Meckenstock, R. U., Kreiger, R., Ensign, S., Kroneck, P. M. H., and Schink, B., "Acetylene hydratase of *Pelobacter acetylenicus.* Molecular and spectroscopic properties of the tungsten iron–sulfur enzyme", *Eur. J. Biochem.,* **264,** 176 (1999).
10. Roy, R., Mukund, S., Schut, G. J., Dunn, D. M., Weiss, R., and Adams, M. W. W., "Purification and molecular characterization of the tungsten-containing formaldehyde ferredoxin oxidoreductase from the hyperthermophilic archaeon *Pyrococcus furiosus*: the third of a putative five member tungstoenzyme family", *J. Bacteriol.,* **181,** 1171 (1999).
11. Hu, Y., Faham, S., Roy, R., Adams, M. W. W., and Rees, D. C., "Formaldehyde ferredoxin oxidoreductase from *Pyrococcus furiosus:* the 1.85 Å resolution crystal structure and its mechanistic implications", *J. Mol. Biol.,* **286,** 899 (1999).
12. Vorholt, J. A., Vaupel, M., and Thauer, R. K., "A selenium-dependent and a selenium-independent formylmethanofuran dehydrogenase and their transcriptional regulation in the hyperthermophilic *Methanopyrus kandleri*", *Mol. Microbiol.,* **23,** 1033 (1997).
13. Boyington, J. C., Gladyshev, V. N., Khangulov, S. V., Stadtman, T. C., and Sun, P. D., "Crystal structure of formate dehydrogenase H: catalysis involving Mo, molybdopterin, selenocysteine, and an Fe_4S_4 cluster", *Science,* **275,** 1305 (1997).
14. Deaton, J. C., Solomon, E. I., Watt, G. D., Weatherbee, P. J., and Durfor, C. N., "Electron paramagnetic resonance studies of the tungsten-containing formate dehydrogenase from *Clostridium thermoaceticum*", *Biochem. Biophys. Res. Commun.,* **149,** 424 (1987).
15. Hille, R., Retey, J., Bartlewski-Hof, U., Reichenbecher, W., and Schink, B., "Mechanistic aspects of molybdenum-containing enzymes", *FEMS Microbiol, Rev.,* **22,** 489 (1998).
16. Romao, M. J. and Huber, R., "Crystal structure and mechanism of action of the xanthine oxidase-related aldehyde oxidoreductase from *Desulfovibrio gigas*", *Biochem. Soc. Trans.,* **25,** 755 (1997).
17. Sung, K. M. and Holm, R. H., "Substitution and oxidation reactions of bis(dithiolene)tungsten complexes of potential relevance to enzyme sites", *Inorg. Chem.,* **40,** 4518 (2001).
18. Yadav, J., Das, S. K., and Sarkar, S., "A functional mimic of the new class of tungstoenzyme, acetylene hydratase", *JACS,* **119,** 4315 (1997).
19. Kisker, C., Schindelin, H., Baas, D., Retey, J., Meckenstock, R. U., and Kroneck, P. M. H., "A structural comparison of molybdenum cofactor-containing enzymes", *FEMS Microbiol. Rev.,* **22,** 503 (1999).
20. Wiegel, J. and Adams, M. W. W., Eds., *Thermophiles: The Keys to Molecular Evolution and the Origin of Life?,* Taylor and Francis, Washington, DC, 1998, 352 pp.
21. Research on tungstoenzymes carried out in the authors' laboratory was supported by grants from the U.S. Department of Energy.

CHAPTER XIII

Metalloenzymes with Radical Intermediates

XIII.1. Introduction to Free Radicals

Contents

James W. Whittaker
Environmental and
Biomolecular Systems
Oregon Health and Science
University
Beaverton, OR 97006

XIII.1.1. Introduction

Free radicals[1,2] are a special type of molecular species that are involved in many essential biological processes,[3–5] serving as the organic chemical equivalent of a transition metal ion. The term "radical" originally referred to the *root* character of the functional group fragments of molecular structure, that is, a methyl functional group was defined as a methyl "radical", $CH_3•$, in early chemistry texts. In current usage, radicals (or free radicals) are molecules containing unpaired electrons in their valence shell. These open-shell molecules are ordinarily highly reactive, inducing a spectrum of molecular transformations not readily accessible via polar (heterolytic) mechanisms. As a result of their involvement in violent and unstable processes like explosive detonations and chain reactions, radicals have over time acquired an "out-of-control" connotation that is not always deserved. Their open-shell character makes them in some ways the organic counterpart of transition metal ions, allowing them to break in aggressively on the bonding of neighboring molecules, resulting in atom abstraction, rearrangements, and oxidation–reduction reactions. Each of these characteristic radical reactions has been harnessed in enzyme catalysis, as described in subsequent sections of this chapter.

Radicals fundamentally differ from other molecules in the unpairing of electrons. While closed-shell molecules always contain an even number of electrons, which may be conventionally described as Pauli paired within two-electron bonds, radicals generally contain an odd number of electrons, leaving one unpaired. However, there are exceptions, including dioxygen (O_2, an inorganic radical), which contains an even number of electrons yet is a radical because two valence electrons are unpaired in the ground state. The spin *multiplicity* $(= 2S + 1)$ of a radical is determined by the number of unpaired electrons, each contributing $S = \pm\frac{1}{2}$, according to whether the spins are parallel $\left(+\frac{1}{2}\right)$ or antiparallel $\left(-\frac{1}{2}\right)$. One unpaired electron defines a doublet radical $(2S + 1 = 2)$, two unpaired electrons defines a triplet radical, and so on. Dioxygen, having two unpaired electrons with parallel spin, is a triplet diradical.

Table XIII.1.1.
Examples of Free Radicals Found in Enzymes

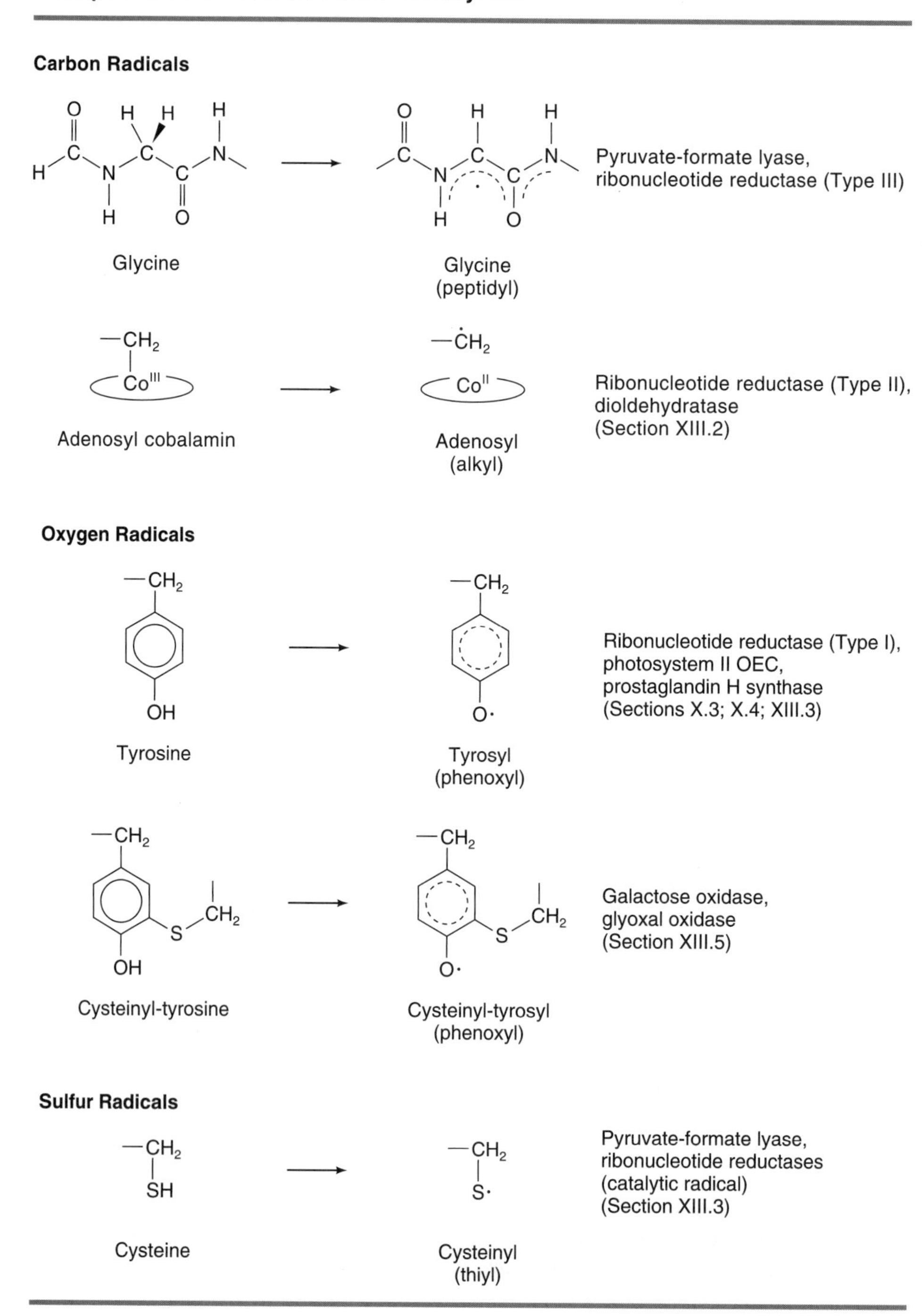

The distinctive character of radicals is difficult to represent in conventional diagrams of molecular geometry and connectivity, and the open-shell character of a radical is generally indicated by a dot representing an unpaired valence electron next to the structure as shown in Table XIII.1.1. The position of the dot can be misleading, however, especially when the unpaired electron is delocalized, and a broken conjugation path is sometimes drawn to show the extent of delocalization. Radicals are named by adding the suffix "-yl" to the root of the name of the parent compound. Thus, the radical formed from the amino acid glycine is the glycyl radical, and the radical formed from a thiol is a thiyl radical, while that formed from a phenol (or phenoxide) is a phenoxyl radical (Table XIII.1.1). In this scheme, the hydrogen *atom* may be regarded as the simplest radical.

XIII.1.2. Free Radical Stability and Reactivity

The stability of a radical depends on details of electronic structure that are not obvious from inspection of a molecular formula. In simple terms, radicals will form when the balance between electron–electron repulsion in the valence shell and nuclear binding makes loss of an electron energetically favorable. Clearly, the presence of high-lying filled orbitals containing electrons that repel each other strongly will favor the formation of open-shell structures. This principle is reflected in the predominance of open-shell transition metal complexes, whose valence shells contain a weakly bound and highly localized set of *d* electrons that exhibit large interelectron repulsions. Among organic compounds, these bonding features are often associated with conjugated π-systems, or with heteroatoms (e.g., sulfur) containing weakly bound valence electrons.

The chemistry of radicals is governed by a fundamental rule of spin algebra, based on the conservation of electronic angular momentum. A singlet molecule (with all electrons spin paired to give a molecular spin, $S = 0$) must form an even number of doublet $\left(S = \frac{1}{2}\right)$ radical products, and reaction of a radical $\left(S = \frac{1}{2}\right)$ with a singlet molecule ($S = 0$) must give rise to at least one radical product. Two radicals may react to form a singlet product, as occurs in the termination step of radical chain processes.

Radicals may be formed from a closed-shell parent molecule either by removing one electron from the highest occupied molecular orbital (HOMO) through oxidation, or by adding one electron to the lowest unoccupied molecular orbital (LUMO) through reduction (Fig. XIII.1.1). The MO that contains the unpaired electron in a radical is defined as the spin occupied molecular orbital (SOMO). Since the redox

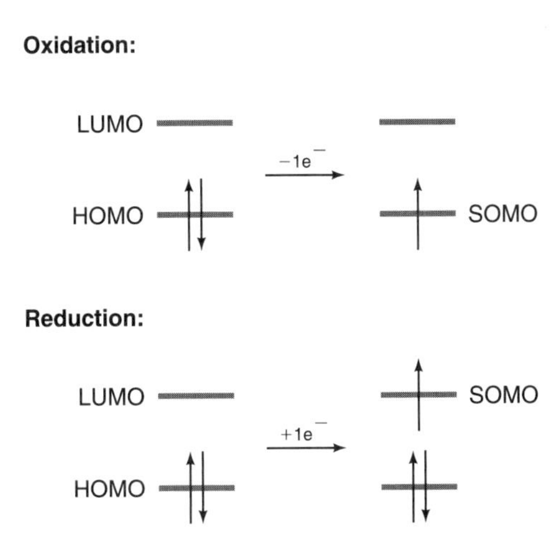

Fig. XIII.1.1. Molecular orbital representations of two radical-forming reactions.

properties of the radical are largely determined by the character of the SOMO wave function, it may be described as the redox orbital of the molecule. The bonding contribution of the SOMO is only one-half that of a conventional two-electron bond (or antibond), and this one-electron bonding in radicals contributes to their distinctive chemistry.

Radicals formed by oxidation (or reduction) are the result of loss (or gain) of one electron, changing the overall charge on the molecule by one unit (± 1) unless the change in electronic charge is compensated by a change in the net nuclear charge by ionization or protonation. This type of charge compensation is in fact commonly observed in free radical chemistry, where a shift in pK_a of Brønsted (protonic) acids drives a change in protonation state. Changes in protonation associated with oxidation and reduction are an important example of coupling between proton and electron transfer that is fundamental to many biological processes, including respiration and photosynthesis. This coupling tends to be very strong for organic radicals, allowing them to serve as key coupling elements in the respiratory machinery (see Section X.3). Strong coupling between electron and proton transfer also makes radicals very effective hydrogen atom abstraction agents in catalysis.

XIII.1.3. Electron Paramagnetic Resonance Spectroscopy

The presence of unpaired electrons in organic radicals results in molecular paramagnetism, allowing electron paramagnetic resonance (EPR) spectroscopy to be used to probe their electronic structure directly.[6] Electron paramagnetic resonance spectroscopy is particularly useful for detection of organic free radicals, which give rise to characteristically sharp resonance signals near the free electron g-value ($g = 2.00$). The integrated intensity of the EPR signal gives a quantitative measure of the amount of radical species present in a sample, and the fine structure in the spectrum contains information permitting identification of the radical. The intrinsic electronic orbital g-shifts or deviations from $g = 2.00$ often allows assignment of carbon, oxygen, or sulfur radicals, and the magnitude and multiplicity of nuclear hyperfine splittings reflect the number of protons or other magnetic nuclei interacting with the unpaired electron, giving more detailed information about the structure of a radical.

XIII.1.4. Biological Radical Complexes

Biological radicals[3–5] are often associated with specialized redox sites in enzymes, which may be amino acid side chains. These include (Table XIII.1): the redox-active Tyr residues in type I ribonucleotide reductase (Section XIII.3);[7] prostaglandin H synthase;[8] the oxygen-evolving complex of photosystem II (Sections X.3 and X.4);[9] the tryptophan residue in cytochrome *c* peroxidase (Section XI.3);[10] or the catalytic thiyl radical intermediates proposed for ribonucleotide reductase (Section XIII.3[11]); the peptide backbone itself (as in type III ribonucleotide reductase[12] or pyruvate-formate lyase[13]); or post-translationally modified amino acids, such as the cysteinyl-tyrosine found in the radical copper oxidases (Section XIII.5).[14] These sites are generally protected from reaction with potential reductants in solution by deep burial in the protein core. This restricted access controls reactivity and effectively harnesses the radical for catalysis.[15]

Radicals may be formed by a variety of paths, both thermal and photochemical, and, once formed, may give rise to secondary radicals.[1,2] Homolysis of the carbon–cobalt bond in adenosyl–cobalamin can occur either thermally or photochemically to generate a carbon radical[16] (see Table XIII.1.1 and Section XIII.2). Generation of a pair of radicals by photoionization of chlorophyll represents the primary photochemical event in photosynthesis.[17] This high-potential primary radical is capable of oxidizing protein side chains, and ultimately converts a specific Tyr to a phenoxyl radical that may be involved in hydrogen abstraction from water in the oxygen evolving complex (OEC)[18] (see Sections X.3 and X.4).

Metal cofactors are intimately involved in the formation and stabilization of protein radicals, being ideally suited for one-electron redox processes needed for radical generation. All known forms of ribonucleotide reductase (RNR, the key enzyme in DNA biosynthesis) make use of radicals,[19–21] which arise from reactions of a metallocofactor, whether a binuclear oxygen-activating complex (as in the Type I RNR), an organometallic complex (as in the vitamin B_{12}-dependent Type II RNR), or an FeS redox cluster (as in the Type III RNR; see Section XIII.4). Modified phenoxyl radicals in a number of oxidases, including the radical copper oxidases (galactose oxidase, and glyoxal oxidase, Section XIII.5), amine oxidase, Section XIII.6, and perhaps cytochrome *c* oxidase, are closely associated with mononuclear copper active sites that participate in their formation and control their reactivity. These metalloradical active sites are discussed in more detail in Sections XIII.5 and XIII.6.

References

General References

1. Parsons, A. F., *An Introduction to Free Radical Chemistry,* Blackwell Science, Oxford, UK, 2000.
2. Perkins, M. J., *Radical Chemistry–An Introduction,* Oxford University Press, Oxford, UK, 2000.
3. Stubbe, J. and van der Donk, W. A., "Protein radicals in enzyme catalysis", *Chem. Rev.,* **98,** 705–762 (1998).
4. Frey, P. A., "Radicals in enzymatic reactions", *Curr. Opin. Chem. Biol.,* **1,** 347–356 (1997).
5. Pedersen, J. Z. and Finazzi-Agro, A., "Protein-radical enzymes", *FEBS Lett.,* **325,** 53–58 (1993).

Specific References

6. Weil, J. A., Bolton, J. R., and Wertz, J. E., *Electron Paramagnetic Resonance*—Elementary Theory and Practical Applications, Wiley-Interscience, New York, 1994.
7. Sjoberg, B.-M., "Ribonucleotide reductase—A Group of Enzymes with Different Metallosites and a Similar Reaction Mechanism", *Struct. Bonding (Berlin),* **88,** 139–173 (1997).
8. Dorlet, P., Seibold, S. A., Babcock, G. T., Gerfen, G. J., Smith, W. L., Tsai, A. L., and Un, S., "High field EPR study of tyrosyl radicals in prostaglandin H(2) synthase-1", *Biochemistry,* **41,** 6107–6114 (2002).
9. Pujols-Ayala, I. and Barry, B. A., "Tyrosyl radicals in photosystem II", *Biochim. Biophys. Acta,* **1655,** 205–216 (2004).
10. Huyett, J. E., Dean, P. E., Gurbiel, R., Houseman, A. L. P., Sivaraja, M., Goodin, D. B., and Hoffman, B. M., "Compound ES of Cytochrome *c* Peroxidase Contains a Trp π-Cation Radical. Characterization by CW and Pulsed Q-Band ENDOR", *J. Am. Chem. Soc.,* **117,** 9033–9041 (1995).
11. Licht, S., Gerfen, G. J., and Stubbe, J., "Thiyl radicals in ribonucleotide reductase", *Science,* **271,** 477–481 (1996).
12. Sun, X., Ollagnier, S., Schmidt, P. P., Atta, M., Mulliez, E., Lepape, L., Eliason, R., Graslund, A., Fontecave, M., Reichard, P., and Sjöberg, B.-M., "The free radical of the anaerobic ribonucleotide reductase of *Escherichia coli* is at glycine 681", *J. Biol. Chem.,* **271,** 6827–6831 (1996).
13. Knappe, J. and Wagner, A. F., "Stable glycyl radical from pyrunvate formate-lyase and ribonucleotide reductase (III)", *Adv. Prot. Chem.,* **58,** 277–315 (2001).
14. Whittaker, J. W. and Whittaker, M. M., "Radical copper oxidases, one electron at a time", *Pure Appl. Chem.,* **70,** 903–910 (1998).
15. Banerjee, R., "Introduction: Radical Enzymology", *Chem. Rev.,* **103,** 2081–2083 (2003).
16. Banerjee, R., "Radical Carbon Skeleton Rearrangements: Catalysis by Coenzyme B_{12}-Dependent Mutases", *Chem. Rev.,* **103,** 2083–2094 (2003).
17. Diner, B. A. and Rappaport, F., "Structure, Dynamics and Energetics of the Primary Photochemistry of Oxygenic Photosynthesis", *Annu. Rev. Plant. Biol.,* **53,** 551–580 (2002).
18. Tommos, C. and Babcock, G. T., "Oxygen Production in Nature: A Light-Driven Metalloradical Complex", *Acc. Chem. Res.,* **31,** 18–25 (1998).
19. Stubbe, J., "Di-iron Tyrosyl Radical Ribonucleotide Reductases", *Curr. Opin. Chem. Biol.,* **7,** 183–188 (2003).
20. Eklund, H., Uhlin, U., Farnegardh, M., Logan, D. T., and Nordlund, P., "Structure and Function of the Radical Enzyme Ribonucleotide Reductase", *Prog. Biophys. Mol. Biol.,* **77,** 177–268 (2001).
21. Poole, A. M., Logan, D. T., and Sjöberg, B.-M., "The evolution of the ribonucleotide reductases: much ado about oxygen", *J. Mol. Evol.,* **55,** 180–196 (2002).

XIII.2. Cobalamins

JoAnne Stubbe
Departments of Chemistry and Biology
MIT
Cambridge, MA 02139

Contents

XIII.2.1. Introduction

The cobalamins (Cbls) are one of Nature's most complex and beautiful cofactors and have fascinated biochemists and chemists for decades.[1–7] There are two different forms of these cofactors: methylcobalamin (MeCbl) in which the axial ligand (R in Fig. XIII.2.1) is a methyl group and adenosylcobalamin (AdoCbl) in which the axial ligand is 5′-deoxyadenosine (5′-dA).

XIII.2.2. Nomenclature and Chemistry

Cyanocobalamin is Vitamin B_{12} (Fig. XIII.2.1, R = CN). The axial cyano group is an artifact of the isolation process, which starts initially from hydroxycobalamin (R = OH).[8] The cobalamins are corrins composed of A, B, C, and D rings that encapsulate the cobalt. In humans, the β face axial ligands are either CH_3 or 5′-dA- (Fig. XIII.2.1) and the lower, α face axial ligand, is the unusual α nucleotide dimethylbenzimidazole (DMB). The imidazole moiety of DMB has a pK_a of 5.6 in AdoCbl.[9] Thus under appropriate conditions the imidazole may be coordinated to the cobalt (base-on) or not coordinated to the cobalt (base-off). Cobamide is a ge-

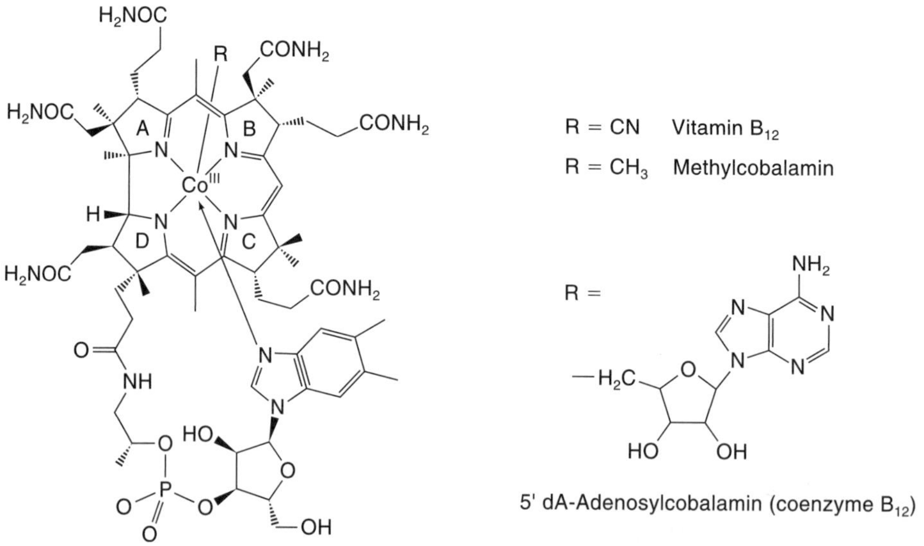

Fig. XIII.2.1.
The structures of MeCbl, coenzyme B_{12}, and Vitamin B_{12}. R = Me(CH_3), methylcobalamin; R = 5′-deoxyadenosyl, coenzyme B_{12}; R = CN, vitamin B_{12}.

R = CN Vitamin B_{12}
R = CH_3 Methylcobalamin

5' dA-Adenosylcobalamin (coenzyme B_{12})

neric term for cofactors with different nucleotide bases. Alternative nucleotides are found in a variety of methyltransferases.

The biosynthesis of B_{12} has long fascinated chemists and biochemists and with the recent revolution in molecular biology, all of the genes required for this process have been identified. There are two pathways for B_{12} biosynthesis, one observed in aerobic and one in anaerobic organisms.[10,11] Novel chemical transformations have been associated with many of the steps, including the formation of the D ring and its linkage to the A ring. The mechanism and timing of insertion of the cobalt into the corrin depend on the pathway.

The discovery that AdoCbl contained a carbon–cobalt bond resulted in the Nobel Prize in Chemistry for Dorothy Crowfoot Hodgkin in 1964 and led to the characterization of the spectroscopy and chemical reactivity of this complex molecule.[12] The cobalt in the corrin can exist in three oxidation states under physiological conditions: Co^{+}, Co^{2+}, and Co^{3+} (Fig. XIII.2.2). The resting oxidation state of the cofactors is 3+, and, in general, the cofactor coordinates to six ligands forming an octahedral complex. The cobalt in the 3+ oxidation state is low spin ($3d^6$) and is diamagnetic.

The carbon–cobalt bond is weak and in the case of AdoCbl, light sensitive. The bond-dissociation enthalpies for carbon–cobalt bond homolysis have been established to be 31 $kcal\,mol^{-1}$ for AdoCbl and 37 $kcal\,mol^{-1}$ for MeCbl.[13–16] As with the thermodynamics, kinetic differences between the cofactors govern their different chemical reactivities. In the case of AdoCbl, subsequent to homolysis within the solvent cage, the carbon–cobalt bond is re-formed with a rate constant of $1 \times 10^9\ s^{-1}$.[17] The pyramidyl geometry of 5′-deoxyadenosyl radical (5′-dA•), proposed to be controlled by the stereoelectronic effects from the ribose of adenosine, potentiates reformation. On the other hand, the methyl radical generated from homolysis of the CH_3–C bond of MeCbl rapidly becomes planar and escapes from the solvent cage more rapidly than its carbon–cobalt bond reforms.[17] These observations provide an explanation for the enzymatic mechanisms discussed subsequently. In general, AdoCbl is the ultimate radical chain initiator using homolytic cleavage of the carbon–cobalt bond to initiate the chemistry. On the other hand, MeCbl utilizes heterolytic chemistry to initiate its chemical transformations.

As noted above, Co^{+} and Co^{2+} in the corrin are both biochemically achievable oxidation states. Heterolysis of the MeCbl yields the Co^{+} state that contains two electrons in its d_{z^2} orbital perpendicular to the corrin ring and has very strong nucleophilic character (Fig. XIII.2.2). On the other hand, AdoCbl undergoes homolytic carbon–cobalt bond cleavage to generate cob(II)alamin and a 5′-dA•; it is thought to be the radical chain initiator in all AdoCbl-requiring reactions in Table XIII.2.1. The Co^{2+} state has one unpaired electron in the d_{z^2} orbital $\left(s = \frac{1}{2}\right)$ and is thus paramagnetic (Fig. XIII.2.2).

The differences in the visible spectra and the paramagnetic nature of Co^{2+} relative to Co^{3+} have played a major role in the elucidation of the chemistry of the enzymatic systems. All three cobalt oxidation states have been well characterized and exhibit distinctive ultraviolet (UV)–visible spectra (Fig. XIII.2.3). Since cobalt has a nuclear spin (I) of $\frac{7}{2}$, the EPR spectrum of the cob(II)alamin form of the cofactor has also been very useful (Fig. XIII.2.4). The spectrum has a g_z component centered at $g = 2.0$ and a g_{xy} component centered at $g = 2.3$.[18] On the high-field (right) side of

Fig. XIII.2.2.
The three biological accessible oxidation states of corrins. The box represents the corrin ring shown in Fig. XIII.1.1.

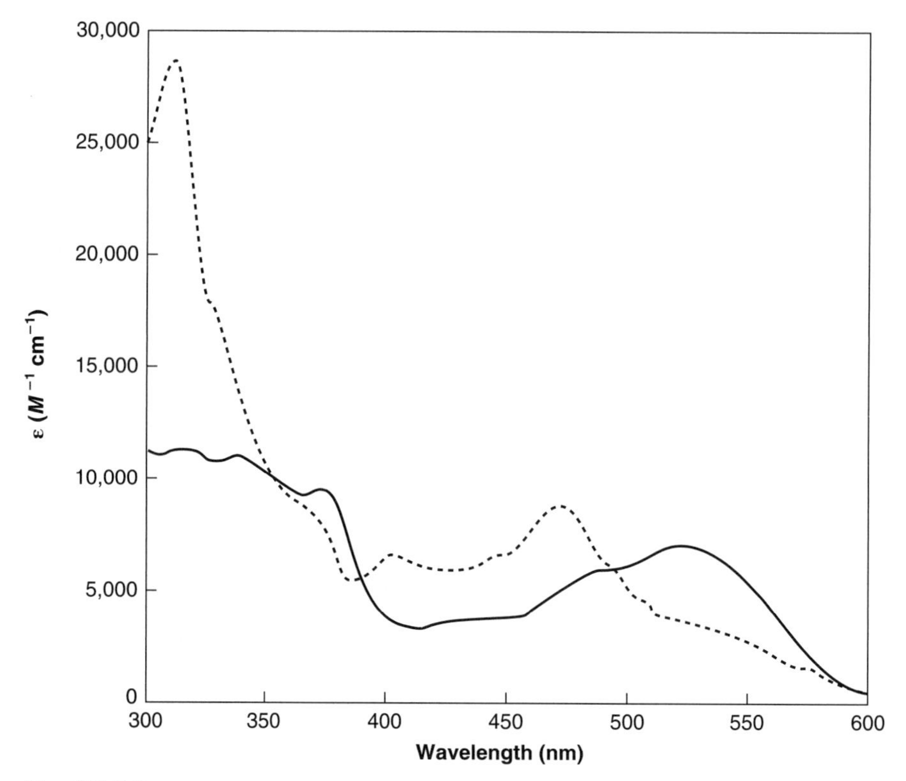

Fig. XIII.2.3.
The visible spectra of AdoCbl (—) and Cob(II)alamin (—).

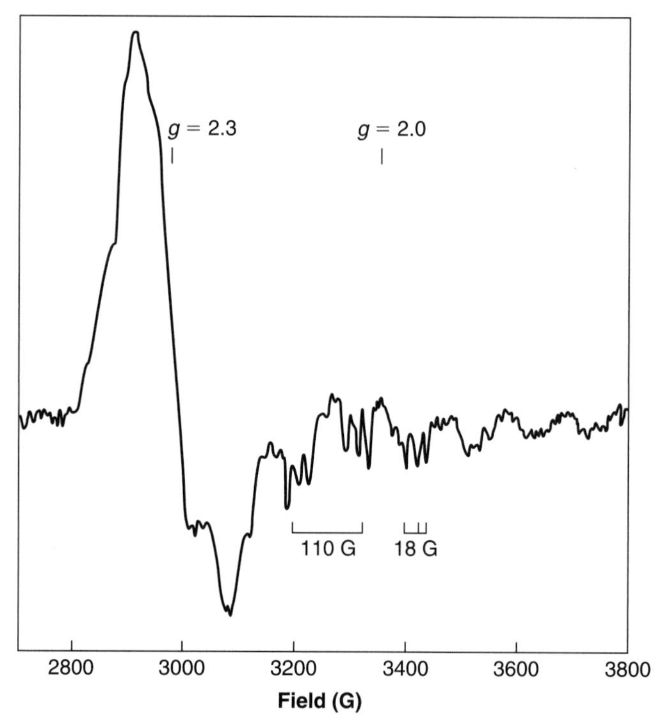

Fig. XIII.2.4.
The EPR (9 GHz) spectrum of cob(II)alamin in solution.

the EPR spectrum, associated with the g_z component, is the hyperfine structure associated with the Co nucleus as well as the hyperfine structure associated with the nitrogen nucleus ($I = 1$) of the DMB axial ligand.[19] The hyperfine splittings are 110 and 18 G, respectively.

XIII.2.3. Enzyme Systems Using AdoCbl

XIII.2.3.1. Generalizations about the Mechanism of AdoCbl Requiring Enzymes

Several excellent reviews have recently been published on the structures and function of enzymes containing cobalamin cofactors.[2,3,20–22] The diverse chemistry associated with these enzymes is summarized in Table XIII.2.1.[2,4,5,7]

The generic mechanism for all B_{12}-dependent rearrangements involves enzyme mediated homolytic cleavage of the carbon–cobalt bond of AdoCbl to form 5′-dA• and cob(II)alamin (Fig. XIII.2.5).

In all reactions, the rate acceleration of this process over the non-enzymatic process is a factor of 10^{10}–10^{11}. The 5′-dA• radical mediates hydrogen atom abstraction from the substrate (S) to form the symmetric, methyl containing, 5′-deoxyadenosine (5′-dA) intermediate and a substrate radical (S•). The substrate radical then undergoes a rearrangement reaction to generate the product radical (P•). The product radical is then reduced by hydrogen atom transfer from 5′-dA to generate product (P) and 5′-dA•, which can then reform the carbon–cobalt bond to generate AdoCbl.

Table XIII.2.1.
Enzymes that Use AdoCbl and MeCbl

Enzyme	Substrate	Product	Migrating Group	Reversible–Irreversible
ADOCBL COFACTOR				
Glutamate mutase	L-Glutamate	L-*threo*-3-Methylaspartate	2-Glycinyl	Reversible
2-Methyleneglutarate mutase	2-Methyleneglutarate	(*R*)-3-Methylitaconate	2-Acrylate	Reversible
Methylmalonyl-CoA mutase	Methylmalonyl-CoA	Succinyl-CoA	Formyl-CoA	Reversible
Isobutyryl-CoA mutase	Isobutyryl-CoA	*n*-Butyryl-CoA	Formyl-CoA	Reversible
β-Lysine-5,6-amino mutase	5-Aminolysine	6-Aminolysine	NH_3^+	Reversible
D-Ornithine-4,5-amino mutase	4-Aminoornithine	5-Aminoornithine	NH_3^+	Reversible
Diol dehydrase	1,2-Propanediol	Propionaldehyde	OH	Irreversible[a]
Glycerol dehydrase	Glycerol	3-Hydroxypropionaldehyde	OH	Irreversible[a]
Ethanolamine ammonia lyase	Ethanolamine	Acetaldehyde	NH_3^+	Irreversible[a]
Ribonucleotide reductase (RNR)	Ribonucleotide	2′-Deoxyribonucleotide	OH	Irreversible[b]
Toluate *cis*-dihydrodiol dehydrogenase	Toluene *cis*-2,3-dihydrodiol	Catechol	CH_3	Irreversible[c]
MECBL COFACTOR				
Methionine synthase	Homocysteine	Methionine	CH_3	Irreversible

[a] After the migration of group X, elimination of HX occurs to give the observed product.

[b] The reaction involves elimination of H_2O followed by reduction.

[c] The reaction appears to involve intermolecular migration of a CH_3 group followed by reduction.

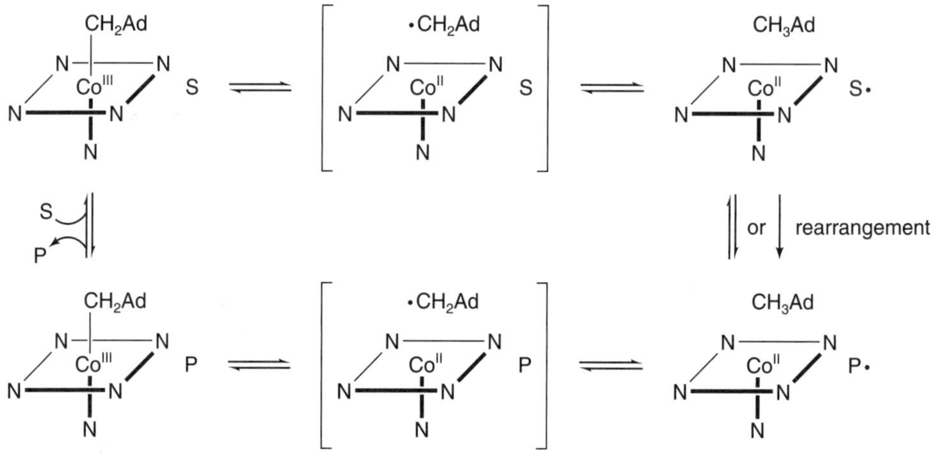

Fig. XIII.2.5.
The generic mechanism for AdoCbl-dependent rearrangement reactions. The "square" with 4N atoms is the corrin ring. The N below the corrin plane is one of the axial ligands, either the imidazole of DMB or a His residue from the protein. The substrate is represented by S, and P is the product.

XIII.2.3.2. Diol Dehydrase and Methylmalonyl-CoA Mutase as Paradigms

Diol dehydrase (DD), which catalyzes the conversion of propanediol to propanal, is one of the best characterized AdoCbl requiring enzymes (Fig. XIII.2.6*a*). Seminal experiments by Abeles and co-workers[23] on diol dehydrase in the 1960s and 1970s showed that deuterium, derived from stereospecifically deuteriated substrate propanediol, could be transferred to the cofactor derived intermediate, [5′-^{2}H] 5′-dA, and that this deuterium is eventually transferred to P• to give propanal. These transfers were demonstrated to occur in both a chemically and a kinetically competent fashion. While this mechanism is now well accepted, it was chemically unprecedented at the time it was discovered; these studies represent a landmark in mechanistic enzymology.

Model studies and recent structures of DD establish that the cob(II)alamin plays a spectator role in the rearrangement process,[3,22] that is, no C–Co bond between the substrate or product and cob(II)alamin is generated during the rearrangement. The details of this rearrangement are not understood. The working hypothesis, based on recent structural data and computational studies, is shown in Fig. XIII.2.6*a*.

MMCoA mutase catalyzes the reversible conversion of MMCoA to succinylCoA (Fig. XIII.2.6*b* and *c*). The mechanism of the rearrangement, as in the case of DD, involves the interconversion of a substrate radical (S•) to a product radical (P•). The detailed mechanism of the rearrangement, which has been proposed to proceed through a cyclopropyl oxy radical intermediate (Fig. XIII.2.6*b*) or an acrylate •COCoA intermediate via a fragmentation–recombination mechanism (Fig. XIII.2.6*c*), is still unresolved.[2]

The stereochemistry of each rearrangement reaction (Table XIII.2.1) has been studied in detail. There is no stereochemical imperative for these reactions: some take place with inversion,[24–26] some with retention,[27,28] and in one case, the reaction even goes with racemization.[29]

Mechanisms of these reactions have generally been studied by stopped flow (SF) visible spectroscopy as well as rapid freeze quench (RFQ) and steady-state EPR spectroscopy.[30–38] Cob(II)alamin formation can be readily monitored by both methods. Its formation occurs in a kinetically competent fashion. More amazing is that the amount of cob(II)alamin in the steady state varies from 0.1 to 0.7 equiv/equiv of enzyme, depending on the system. Given the activation barrier of $\sim 15\,\mathrm{kcal\,mol^{-1}}$ for the homolysis process, the enzyme cannot catalyze this reaction by uniquely sta-

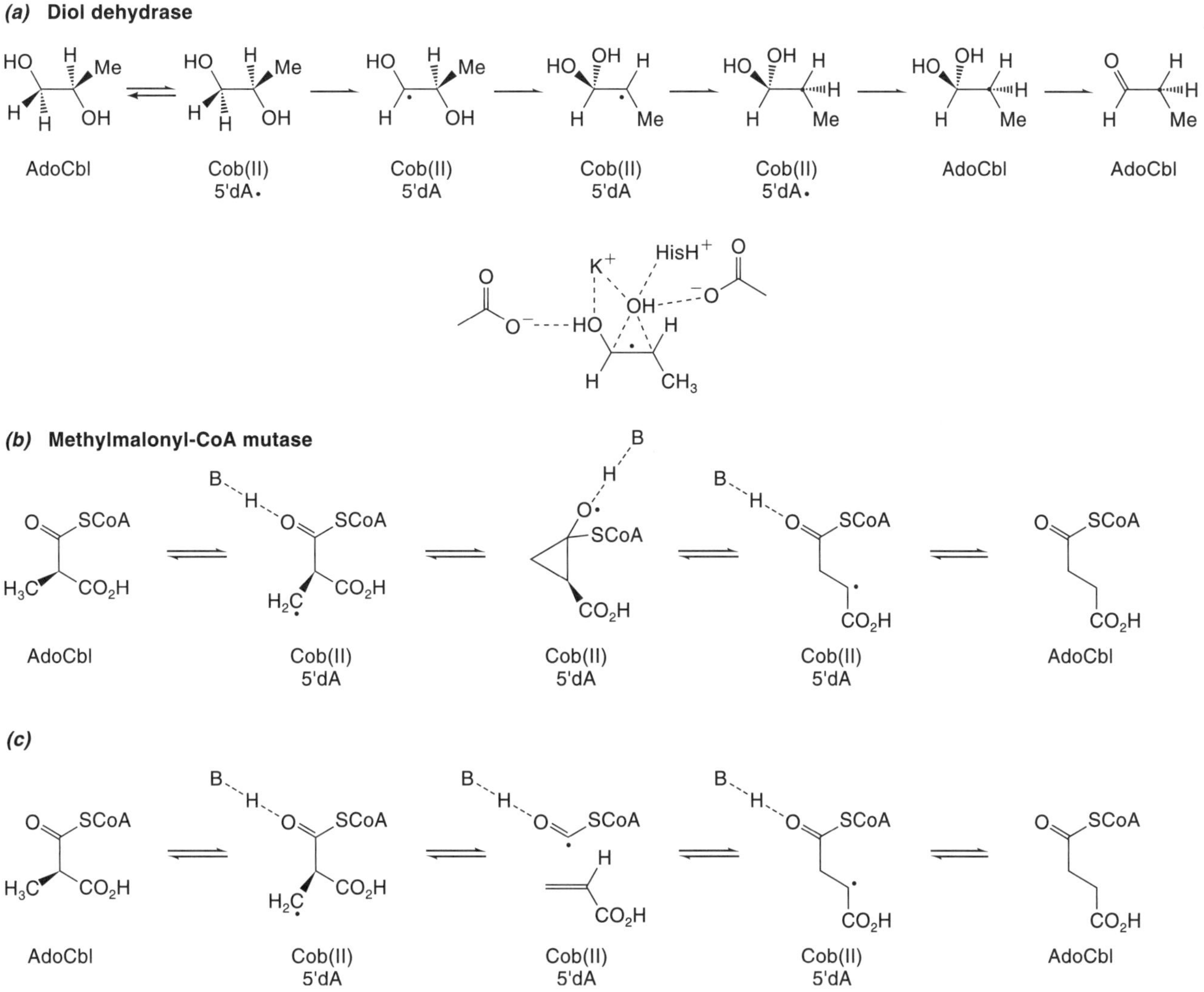

(a) **Diol dehydrase**

AdoCbl — Cob(II) 5'dA• — Cob(II) 5'dA — Cob(II) 5'dA — Cob(II) 5'dA• — AdoCbl — AdoCbl

(b) **Methylmalonyl-CoA mutase**

AdoCbl — Cob(II) 5'dA — Cob(II) 5'dA — Cob(II) 5'dA — AdoCbl

(c)

AdoCbl — Cob(II) 5'dA — Cob(II) 5'dA — Cob(II) 5'dA — AdoCbl

Fig. XIII.2.6.
Proposed mechanisms for the substrate rearrangements in DD and methylmalonylCoA (MMCoA) mutase: (*a*) DD. [Adapted from Ref. 3.] The barrier for stereospecific OH group migration is proposed to be lowered by partial protonation. The His or K^+ (proton surrogate) found in the active site could facilitate this process. Whether the migration is stepwise or concerted remains to be established. The transition state of the concerted OH migration is shown. (*b* and *c*) Methyl malonylCoA mutase mechanism. [Adapted from Ref. 2.] Two mechanisms have been considered: (*b*) involves a cyclopropyl-type intermediate while (*c*) involves a fragmentation/addition. The BH in the mechanism is also proposed to be a His based on the structures.

bilizing the transition state. The enzyme must also stabilize the cob(II)alamin and/or destabilize the ground state of AdoCbl.

By using steady-state and RFQ–EPR methods, substrate-derived radical intermediates have been identified in diol dehydrase, ethanolamine ammonia lyase, MMCoA mutase, glutamate mutase, and others.[33,39–44] In these enzymes, the radicals are generated in a kinetically competent fashion and isotopic labeling studies have been used to establish that they are substrate derived. Furthermore, the dipolar interaction, and in some cases the exchange coupled interactions between the substrate derived radical and cob(II)alamin, have been useful in establishing the distance between the two paramagnetic species, which have been shown to vary from 13 to 6 Å, depending on the system. Simulation of the EPR spectra gives structural insight about the orientation of these intermediates that can then be mapped into the active site using recent crystallographic information.[44]

XIII.2.3.3. Ethanolamine Ammonia Lyase: Paramagnetic Methods to Define Structures of Intermediates

Ethanolamine ammonia lyase (EAL) catalyzes the conversion of ethanolamine to acetaldehyde, a reaction very similar to DD (Fig. XIII.2.6*a*). Experiments using EPR and electron–nuclear double resonance (ENDOR) with [U-^{13}C] labeled AdoCbl indicate that the substrate radical is 11 Å from the cob(II)alamin (Fig. XIII.2.7*a*).[44] From the structure of the homologous DD, a model is proposed for EAL in which subsequent to carbon–cobalt bond homolysis to generate the 5′-dA•, rotation around the glycosidic bond of the adenosyl moiety (A) moves the 5′-carbon of the 5′-dA• from a position adjacent to the cob(II)alamin to a position near the substrate (Fig. XIII.2.7*a*), spanning the required 11 Å.

The structures of AdoCbl requiring MMCoA mutase[3,45] and glutamate mutase[20] have been reported and these structures, in conjunction with EPR data, present a different picture of the function of the cofactor in these carbon-skeleton rearrangement reactions.[44] In the case of both MMCoA mutase and glutamate mutase, EPR experiments of the enzymatic reaction in the steady state reveal that the substrate radical is 6–7 Å from cob(II)alamin (Fig. XIII.2.7*b*). The glutamate mutase structure suggests that a ribose pseudo-rotation provides the means to shuttle the radical between the 5′-dA• and the substrate, spanning the required 6–7 Å. In the C2′-endo configuration the C5′ of dA• is adjacent to the cob(II)alamin and in the C3′-endo configuration the C5′ of dA• is adjacent to the substrate (Fig. XIII.2.7*b*).

Structures of the Cbl-binding regions of DD, glutamate mutase, and MMCoA mutase reveal that the latter two are very similar and quite distinct from the former.[46] In the case of MMCoA mutase and glutamate mutase, the corrin ring of AdoCbl sits at the carboxy-terminal end of the α/β Rossmann fold-like structure, and a His residue (in a conserved DXHG motif) replaces the axial DMB ligand. This domain is very similar to the methylcobalamin-binding site in methionine synthase discussed below. In contrast, the AdoCbl binds to DD on the side of a β sheet in an $\alpha + \beta$-fold with DMB as the axial ligand. For all three enzymes, the cofactor binds between two subunits or between two domains.

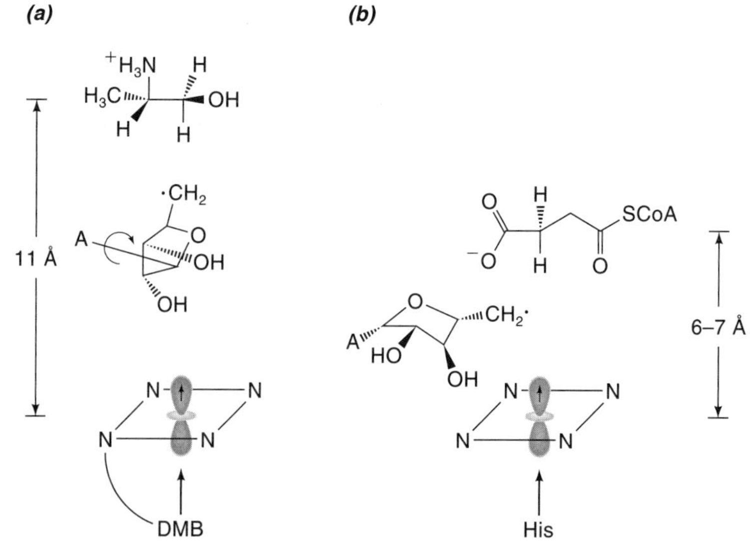

Fig. XIII.2.7.
Structure and EPR analysis of cob(II)alamin-S• intermediates give insight about the mechanism of S• formation from 5′-dA•. (*a*) A model for EAL using propanolamine, a poor substrate analogue. Rotation around the adenosyl glycosidic bond can juxtapose the 5′-dA• near cob(II)alamin or substrate. The distance is based on the EPR analysis.[44] (*b*) A model for MMCoA mutase using a normal substrate and EPR spectrum in the steady state. The model for the role of the cofactor is based on structural analysis of glutamate mutase. Change in the sugar pucker is thought to adjust the 5′-dA• adjacent to cob(II)alamin or substrate.

XIII.2.4. Unresolved Issues in AdoCbl Requiring Enzymes

XIII.2.4.1. The Mechanism of Carbon–Cobalt Homolysis Remains Unresolved

Despite the availability of many structures of AdoCbl-requiring enzymes with various bound ligands, the mechanism(s) by which the enzymes accelerate carbon–cobalt bond homolysis is (are) still unresolved. Model studies, from the past 20 years, had suggested the importance of DMB as the axial ligand. It was proposed that DMB, because of its steric bulk, and the corrin ring's flexibility, could stretch the carbon–cobalt bond by pushing the corrin ring upward.[47–54] Furthermore, the lower pK_a of the imidazole of DMB relative to the imidazole of His in the protein was also proposed to play a role in stabilizing cob(II)alamin in the former case. However, the enzymes involving carbon skeletal rearrangements, as noted above, have replaced the DMB ligand with a protein His residue. Thus, DMB serves to anchor the cofactor to the proteins and does not appear to play a direct role in catalysis. The recent structures suggest one mechanism for cob(II)alamin stabilization that could be important for catalysis. In MMCoA mutase, glutamate mutase, and DD, the cobalt–nitrogen bonds of the axial ligand appear to be long, 2.5 Å,[22,55] in comparison with the Co–N bond of AdoCbl and cob(II)alamin in solution, which vary from 1.9 to 2.1 Å. At present, however, many of the structures have cofactors with mixed oxidation states, unnatural axial ligands, or axial ligands missing or in multiple conformations. Thus these structural data should not be overinterpreted.

XIII.2.4.2. The Thermodynamics of Carbon–Cobalt Bond Homolysis Remain Unresolved

Thermodynamic measurements of the carbon–cobalt bond homolysis process catalyzed by these enzymes have been complicated by the requirement that substrate be present to effect carbon–cobalt bond cleavage. However, the unusual AdoCbl requiring enzyme ribonucleotide reductase RNR (Table XIII.2.1) does not require the presence of substrate to effect this homolysis, and yet the enzyme can still accelerate this process by a factor of 10^{10}-fold.[32,56] The first thermodynamic studies on the homolysis of AdoCbl in RNR suggest that the homolysis process in the enzyme is entropy driven ($\Delta H^{\ddagger} = 46\,\text{kcal mol}^{-1}$ and $\Delta S^{\ddagger} = 96\,\text{cal mol}^{-1}\,\text{K}^{-1}$), in sharp contrast to the non-enzymatic system ($\Delta H^{\ddagger} = 35\,\text{kcal mol}^{-1}$ and $\Delta S^{\ddagger} = 14\,\text{cal mol}^{-1}\,\text{K}^{-1}$).[30] This huge entropy effect implicates tremendous solvent reorganization during the homolysis and suggests that model systems may not be able to provide the key insights into the tremendous rate accelerations observed with the enzymes.

However, RNR is unique among AdoCbl utilizing enzymes for two reasons. First, RNR is the only AdoCbl requiring enzyme that does not catalyze a rearrangement during conversion of substrate to product. Second, the function of the protein is to generate a protein thiyl radical that is the radical initiatior rather than 5′-dA•.[57] The recent structures of MMCoA mutase in open and closed forms suggest that extensive structural reorganization occurs to allow substate binding and protection of radical intermediates. The role of the solvent in this reorganization has not yet been investigated. Unfortunately, the requirement for substrate to observe homolysis has precluded thermodynamic measurements on this process.

XIII.2.4.3. 5′-dA: An Intermediate or a Transition State?

A second unresolved mechanistic issue is the intermediacy of the 5′-dA• radical. In none of the enzymatic systems thus far examined, using steady-state or RFQ EPR methods, has this intermediate been detected. This lack of detection raises the possibility that the reaction is concerted, that is, 5′-dA forms concomitant with the substrate radical, or that if 5′-dA• exists, it is a high-energy intermediate (Fig. XIII.2.5). The latter mechanistic option is generally favored.

XIII.2.4.4. The Mechanism(s) of the Rearrangement Reactions Remain Unresolved

Finally, the actual mechanisms of the rearrangement reactions and the function of the enzymes in these processes have been investigated for years with a variety of outcomes. The recent structures have resolved some of the contentious issues, but many remain unresolved. The availability of these structures and sophisticated calculation methods[58] have provided new insight into possible rearrangement mechanisms. The structures reveal that the rearrangement mechanisms do not involve protein-based radicals in catalysis[22,55,58] such as that established for RNRs. They also reveal, as noted above, that the cob(II)alamin plays a spectator role and does not form a covalent bond with any of the intermediates. The active site cavity appears to be involved in "negative catalysis", preventing unwanted side reactions of highly reactive intermediates. Despite many model studies, enzymatic studies, and recent computational studies, many mechanistic issues remain unresolved.

XIII.2.5. MeCbl Using Methionine Synthase as a Case Study

The combination of mechanistic enzymology coupled to recent structural data and mutagenesis studies have given us new insight into the mechanism of one of the *E. coli* methionine synthases.[6] Methionine synthase (MS) plays a central role in one-carbon metabolism and is a wonderful example of a modular enzyme, which must undergo amazing reorganization during its catalytic cycle. It catalyzes the conversion of homocysteine to methionine and utilizes all three organic cofactors involved in methyl transfer reactions *in vivo:* MeCbl, N^5-methyl tetrahydrofolate (N^5-MeTHF), and S-adenosylmethionine (AdoMet) (Fig. XIII.2.8). The MeCbl transfers its methyl group to the thiolate of homocysteine (activated by Zn^{2+}), converting cob(III)alamin to cob(I)alamin and generating methionine. The Co^+ is an excellent nucleophile and reacts with N^5-MeTHF by a mechanism that remains to be determined, to regenerate MeCbl and THF. The mechanism of this methylation is energetically challenging and a number of mechanistic options have recently been described in detail.[6]

The requirement for the third methlyating agent, AdoMet (also known as SAM), results from the observation that in one out of every 2000 turnovers, MS inactivates itself and has consequently evolved a mechanism of self-repair. *In vitro* and *in vivo* the Co^+ form of the enzyme is oxidized by oxygen to generate superoxide and cob(II)alamin. The latter is an inactive form of the cofactor and must be reduced and methylated to regenerate MeCbl. Thus, MS has a domain that can interact with a reducing system and the methylating agent AdoMet. Reduction of Co^{2+} to Co^+ is thermodynamically difficult ($\Delta E° = -0.64\,V$), even in the base-off form ($\Delta E° = -0.45\,V$). In *E. coli,* flavodoxin and flavodoxin reductase reduce the Co^{2+} to Co^+, which rapidly reacts with the hottest of the methylating agents, AdoMet. The domain of MS that catalyzes this reaction acts as a protein chaperone that reactivates the inactive form of the cofactor.

It is interesting to note that protein chaperones have also recently been identified for DD.[3] AdoCbl in DD is frequently oxidized from cob(II)alamin to hydroxycobalamin (cobalt 3+), which remains tightly bound to the enzyme and is inactive. The chaperone protein allows exchange of the inactive form of the cofactor for the active form[59] and requires ATP.

Limited proteolysis studies of *E. coli* MS have led to division of this 134-kDa protein into four modules connected by flexible linkers. The first domain isolated was 27 kDa in mass and was crystallized in the presence of MeCbl. The structure of this fragment provided the first glimpse of a cobalamin-binding domain.[60] The most unexpected observation from the structure was the coordination of a His from the protein in place of the DMB. As discussed above, a similar binding motif has now been observed in MMCoA mutase as well, despite the fact that the cofactor and the chemistry are different.[55] The structure of the 27-kDa module of MS was also confusing in that access to the methyl group of the cobalamin in the structure was completely blocked by a helix across its top face, leaving no room for small molecule access, required for reaction.[60] This structure provided the first hint that this enzyme must undergo carefully orchestrated conformational changes to allow homocysteine and THF and their binding domains to access the corrin cofactor, or to allow AdoMet and its binding domain, which have been recently structurally characterized,[61] access to the cobalamin in order to regenerate the active form of the cofactor.[62]

The stereochemistry of the MS reaction has been investigated starting with a chiral methyl [^{1}H,^{2}H,^{3}H] attached to N5 of THF, followed by analysis of the chirality of the methyl group attached to Met.[63] This conversion was shown to occur with retention of configuration, consistent with the mechanism (Fig. XIII.2.8). The Co^{+} attacks the methyl group of N^{5}-MeTHF with inversion of configuration, generating MeCbl, and then the thiolate of homocysteine attacks the methyl group of MeCbl, perhaps by an S_N2 reaction, also with inversion of configuration. However, retention of configuration is also consistent with other mechanisms of methyl transfer recently considered.[6]

A detailed steady-state kinetic analysis defined the order of addition of substrates to MS.[64] Pre-steady-state stopped-flow studies revealed that MeCbl is converted to Co^{+} in a kinetically competent fashion in the presence of homocysteine. Structure

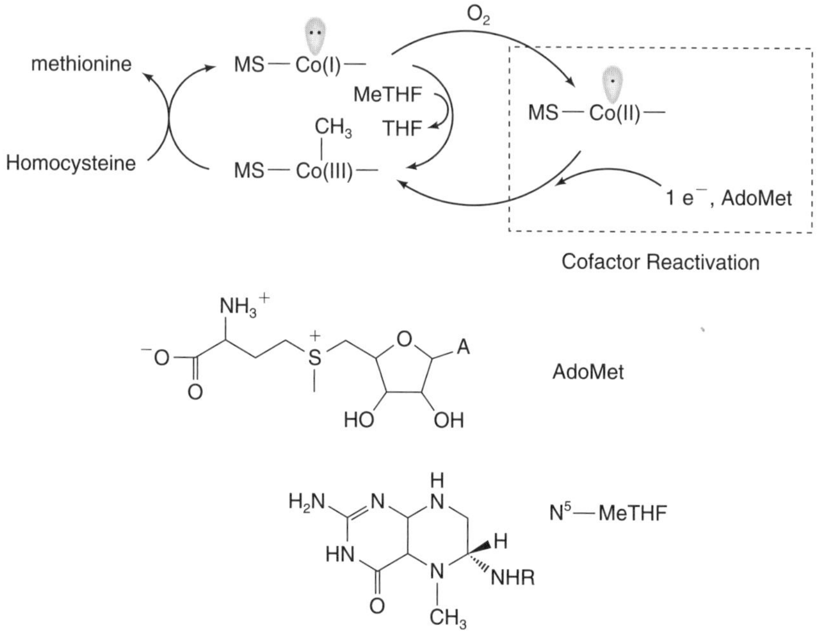

Fig. XIII.2.8.
The role of the three oxidation states of the MeCbl in MS. The reaction in the box indicates that on occasion, *in vivo*, the Co^{+} is oxidized to Co^{2+} and must be converted back into MeCbl by one-electron reduction and methylation by S-adenosylmethionine (AdoMet).

and mutagenesis studies indicate that this conversion involves transformation of the hexacoordinate MeCbl to the four-coordinate Co^{+}. The X-ray structure revealed an unusual conserved catalytic triad on the bottom face of the corrin cofactor composed of H759, D757, and S810.[60] The S810 is located on the surface of the protein and has access to the solvent. This triad is proposed to facilitate changes in the coordination chemistry of the cofactor during transformation of Co^{3+} to Co^{+}. The His is proposed to be ligated in the imidazolate form in the resting state and to become protonated to the imidazole form during the reduction process.

XIII.2.6. Unresolved Issues in Methyl Transfer Reactions with MeCbl

The mechanisms of methyl transfers to and from MeCbl still are not understood. Single electron-transfer reactions, oxidative additions, and S_N2 reactions are still viable mechanistic options.[6] The gross details of protein module movement to orchestrate these methyl transfers to the different acceptors have been elucidated by available structures of each module. The factors that govern the conformational equilibria between these various structures and the kinetics of their interconversion remain, however, to be elucidated. This protein will serve to set a paradigm, not only for methylation by MeCbl dependent enzymes, but also for controlling the chemistry of macromolecular machines.

References

General References

1. Krautler, B., Arigoni, D., and Golding, B. T., "*Vitamin B_{12} and B_{12}-Proteins*", Wiley-VCH, Weinheim, 1998.
2. Banerjee, R., "Radical carbon skeleton rearrangements: catalysis by coenzyme B12-dependent mutases", *Chem. Rev.*, **103,** 2083–2094 (2003).
3. Toraya, T., "Radical catalysis in coenzyme B12-dependent isomerization (eliminating) reactions", *Chem. Rev.*, **103,** 2095–2127 (2003).
4. Buckel, W. and Golding, B. T., "Glutamate and 2-methyleneglutarate mutase: From microbial curiosities to paradigms for coenzyme B-12-dependent enzymes", *Chem. Soc. Rev.*, **25,** 329 (1996).
5. Lawrence, C. C. and Stubbe, J., "The function of adenosylcobalamin in the mechanism of ribonucleoside triphosphate reductase from *Lactobacillus leichmannii*", *Curr. Opin. Chem. Biol.*, **2,** 650–655 (1998).
6. Matthews, R. G., "Cobalamin-dependent methyltransferases", *Acc. Chem. Res.*, **34,** 681–689 (2001).
7. Thauer, R. K., "Biochemistry of methanogenesis: a tribute to Marjory Stephenson. 1998 Marjory Stephenson Prize Lecture", *Microbiology*, **144,** 2377–2406 (1998).

Specific References

8. Rickes, E. L., Brink, N. G., Koniuszy, F. R., Wood, T. R., and Folkers, K., "Crystalline vitamin B12", *Science*, **107,** 396–397 (1948).
9. Brown, K. L. and Hakimi, J. M., "Heteronuclear Nmr-Studies of Cobalamins .4. Alpha-Ribazole-3′-Phosphate and the Nucleotide Loop of Base-on Cobalamins", *J. Am. Chem. Soc.*, **108,** 496–503 (1986).
10. Battersby, A. R., "B_{12}-Biosynthesis in an Aerobic organism: How the Pathway was Elucidated. In *Vitamin B_{12} and B_{12}-Proteins*", Krautler, B., Arigoni, D., Golding, B. T., Eds., Wiley-VCH, Weinheim, 1998, pp. 47–62.
11. Scott, A. I., "How Nature Synthesizes B_{12} Without Oxygen: Discoveries Along the Ancient, Anaerobic Pathway", in *Vitamin B_{12} and B_{12}-Proteins,* Krautler, B., Arigoni, D., and Golding, B. T., Wiley-VCH, Weinheim, 1998, pp. 81–100.
12. Lenhert, P. G. and Hodgkin, D. C., "Structure of the 5,6-dimethyl-benzimidazolylcobamide coenzyme", *Nature (London)*, **192,** 937–938 (1961).
13. Finke, R. G. and Hay, B. P., "Thermolysis of Adenosylcobalamin—a Product, Kinetic, and Co–C5′ Bond-Dissociation Energy Study", *Inorg. Chem.*, **23,** 3041–3043 (1984).
14. Halpern, J., Kim, S. H., and Leung, T. W., "Cobalt Carbon Bond-Dissociation Energy of Coenzyme-B12", *J. Am. Chem. Soc.*, **106,** 8317–8319 (1984).
15. Martin, B. D. and Finke, R. G., "Co–C Homolysis and Bond-Dissociation Energy Studies of Biological Alkylcobalamins—Methylcobalamin, Including a $\geq 10(15)$ Co–CH_3 Homolysis Rate Enhancement at 25°C Following One-Electron Reduction", *J. Am. Chem. Soc.*, **112,** 2419–2420 (1990).
16. Hung, R. R. and Grabowski, J. J., "Listening to reactive intermediates: Application of photoacoustic calorimetry to vitamin B-12 compounds", *J. Am. Chem. Soc.*, **121,** 1359–1364 (1999).

17. Natarajan, E. and Grissom, C. B., "Insight Into the Mechanism of B_{12} Dependent Enzymes: Magnetic Field Effects as a Probe of Reaction Mechanism and the Role of the Ribofuranose Ring Oxygen", in *Vitamin B_{12} and B_{12}-Proteins,* Krautler, B., Arigoni, D., and Golding, B. T., Eds., Wiley-VCH, Weinheim, 1998, pp. 403–416.
18. Pilbrow, J. R., "EPR of B12-Dependent Enzyme Reactions and Related Systems", in *B12,* Dolphin, D., Ed., Wiley-Interscience, New York, 1982, pp. 431–462.
19. Schrauzer, G. N. and Lian-Pin, L., "The molecular and electronic structure of vitamin B12r, cobaloximes(II), and related compounds", *J. Am. Chem. Soc.,* **90,** 6541–6543 (1968).
20. Reitzer, R., Gruber, K., Jogl, G., Wagner, U. G., Bothe, H., Buckel, W., and Kratky, C., "Glutamate mutase from *Clostridium cochlearium*: the structure of a coenzyme B12-dependent enzyme provides new mechanistic insights", *Structure,* **7,** 891–902 (1999).
21. Tollinger, M., Konrat, R., Hilbert, B. H., Marsh, E. N. G., and Krautler, B., "How a protein prepares for B12 binding: structure and dynamics of the B12-binding subunit of glutamate mutase from *Clostridium tetanomorphum*", *Structure,* **6,** 1021–1033 (1998).
22. Shibata, N., Masuda, J., Tobimatsu, T., Toraya, T., Suto, K., Morimoto, Y., and Yasuoka, N., "A new mode of B12 binding and the direct participation of a potassium ion in enzyme catalysis: X-ray structure of diol dehydratase", *Structure Fold Des.,* **7,** 997–1008 (1999).
23. Essenberg, M. K., Frey, P. A., and Abeles, R. H., "Studies on the mechanism of hydrogen transfer in the coenzyme B12 dependent dioldehydrase reaction II", *J. Am. Chem. Soc.,* **93,** 1242–1251 (1971).
24. Zagalak, B., Frey, P. A., Karabatsos, G. L., and Abeles, R. H., "The stereochemistry of the conversion of D and L 1,2-propanediols to propionaldehyde", *J. Biol. Chem.,* **241,** 3028–3035 (1966).
25. Retey, J. and Robinson, J. A., *Stereospecificity in Organic Chemistry and Enzymology,* Verlag Chemie, Weinheim, 1982.
26. Hartrampf, G. and Buckel, W., "On the Steric Course of the Adenosylcobalamin-Dependent 2-Methyleneglutarate Mutase Reaction in *Clostridium-Barkeri*", *Eur. J. Biochem.,* **156,** 301–304 (1986).
27. Reynolds, K. A., Ohagan, D., Gani, D., and Robinson, J. A., "Butyrate Metabolism in Streptomycetes—Characterization of an Intramolecular Vicinal Interchange Rearrangement Linking Isobutyrate and Butyrate in *Streptomyces Cinnamonensis*", *J. Chem. Soc. Perkin Trans. 1,* 3195–3207 (1988).
28. Moore, B. S., Eisenberg, R., Weber, C., Bridges, A., Nang, D., and Robinson, J. A., "On the Stereospecificity of the Coenzyme B-12-Dependent Isobutyryl-CoA Mutase Reaction", *J. Am. Chem. Soc.,* **117,** 11285–11291 (1995).
29. Hartrampf, G. and Buckel, W., "The stereochemistry of the formation of the methyl group in the glutamate mutase-catalysed reaction in *Clostridium tetanomorphum*", *FEBS Lett.,* **171,** 73–78 (1984).
30. Licht, S. S., Booker, S., and Stubbe, J. A., "Studies on the catalysis of carbon–cobalt bond homolysis by ribonucleoside triphosphate reductase: Evidence for concerted carbon–cobalt bond homolysis and thiyl radical formation", *Biochemistry,* **38,** 1221–1233 (1999).
31. Licht, S. S., Lawrence, C. C., and Stubbe, J. A., "Thermodynamic and kinetic studies on carbon–cobalt bond homolysis by ribonucleoside triphosphate reductase: The importance of entropy in catalysis", *Biochemistry,* **38,** 1234–1242 (1999).
32. Licht, S., Gerfen, G. J., and Stubbe, J. A., "Thiyl radicals in ribonucleotide reductases", *Science,* **271,** 477–481 (1996).
33. Valinsky, J. E., Abeles, R. H., and Fee, J. A., "Electron spin resonance studies on diol dehydrase. 3. Rapid kinetic studies on the rate of formation of radicals in the reaction with propanediol", *J. Am. Chem. Soc.,* **96,** 4709–4710 (1974).
34. Harkins, T. T. and Grissom, C. B., "The Magnetic-Field Dependent Step in Bit Ethanolamine Ammonia-Lyase Is Radical-Pair Recombination", *J. Am. Chem. Soc.,* **117,** 566–567 (1995).
35. Hollaway, M. R., White, H. A., Joblin, K. N., Johnson, A. W., Lappert, M. F., and Wally, O. C., "A spectrophotometric rapid kinetic study of reactions catalysed by coenzyme-B12-dependent ethanolamine ammonia-lyase", *Eur. J. Biochem.,* **82,** 143–154 (1978).
36. Marsh, E. N. and Ballou, D. P., "Coupling of cobalt–carbon bond homolysis and hydrogen atom abstraction in adenosylcobalamin-dependent glutamate mutase", *Biochemistry,* **37,** 11864–11872 (1998).
37. Meier, T. W., Thoma, N. H., and Leadlay, P. F., "Tritium isotope effects in adenosylcobalamin-dependent methylmalonyl-CoA mutase", *Biochemistry,* **35,** 11791–11796 (1996).
38. Padmakumar, R. and Banerjee, R., "Evidence that cobalt–carbon bond homolysis is coupled to hydrogen atom abstraction from substrate in methylmalonyl-CoA mutase", *Biochemistry,* **36,** 3713–3718 (1997).
39. Padmakumar, R. and Banerjee, R., "Evidence from electron paramagnetic resonance spectroscopy of the participation of radical intermediates in the reaction catalyzed by methylmalonyl-coenzyme A mutase", *J. Biol. Chem.,* **270,** 9295–9300 (1995).
40. Bothe, H., Darley, D. J., Albracht, S. P., Gerfen, G. J., Golding, B. T., and Buckel, W., "Identification of the 4-glutamyl radical as an intermediate in the carbon skeleton rearrangement catalyzed by coenzyme B12-dependent glutamate mutase from *Clostridium cochlearium*", *Biochemistry,* **37,** 4105–4113 (1998).
41. Babior, B. M., Moss, T. H., Orme-Johnson, W. H., and Beinert, H., "The mechanism of action of ethanolamine ammonia-lyase, a B-12-dependent enzyme. The participation of paramagnetic species in the catalytic deamination of 2-aminopropanol", *J. Biol. Chem.,* **249,** 4537–4544 (1974).
42. Tan, S. L., Kopczynski, M. G., Bachovchin, W. W., Orme-Johnson, W. H., and Bobior, B. M., "Electron

Spin–Echo Studies of the Composition of the Paramagnetic Intermediate Formed During the Deamination of Propanolamine by Ethanolamine Ammonia-Lyase, and Adocbl-Dependent Enzyme", *J. Biol. Chem.*, **261,** 3483–3485 (1986).
43. Silva, D. J., Stubbe, J., Samano, V., and Robins, M. J., "Gemcitabine 5′-triphosphate is a stoichiometric mechanism-based inhibitor of *Lactobacillus leichmannii* ribonucleoside triphosphate reductase: evidence for thiyl radical-mediated nucleotide radical formation", *Biochemistry,* **37,** 5528–5535 (1998).
44. Reed, G. H. and Mansoorabadi, S. O., "The positions of radical intermediates in the active sites of adenosylcobalamin-dependent enzymes", *Curr. Opin. Struct. Biol.,* **13,** 716–721 (2003).
45. Mancia, F. and Evans, P. R., "Conformational changes on substrate binding to methylmalonyl CoA mutase and new insights into the free radical mechanism", *Structure,* **6,** 711–720 (1998).
46. Marsh, E. N. and Drennan, C. L., "Adenosylcobalamin-dependent isomerases: new insights into structure and mechanism", *Curr. Opin. Chem. Biol.,* **5,** 499–505 (2001).
47. Grate, J. H. and Schrauzer, G. N., "Chemistry of Cobalamins and Related Compounds .48. Sterically Induced, Spontaneous Dealkylation of Secondary Alkylcobalamins Due to Axial Base Coordination and Conformational-Changes of the Corrin Ligand", *J. Am. Chem. Soc.,* **101,** 4601–4611 (1979).
48. Chemaly, S. M. and Pratt, J. M., "The Chemistry of Vitamin-B12 .19. Labilization of the Cobalt–Carbon Bond in Organocobalamins by Steric Distortions—Neopentylcobalamin as a Model for Labilization of the Vitamin-B12 Coenzymes", *J. Chem. Soc. Dalton Trans.,* 2274–2281 (1980).
49. Randaccio, L., Brescianipahor, N., Toscano, P. J., and Marzilli, L. G., "Angular Distortions at the Carbon Bound to Cobalt in Co-Enzyme B-12 Models—Implications with Regard to Co-C Bond-Cleavage in Co-Enzyme B-12 and Other Alkylcobalamins", *J. Am. Chem. Soc.,* **103,** 6347–6351 (1981).
50. Geno, M. K. and Halpern, J., "Why Does Nature Not Use the Porphyrin Ligand in Vitamin-B12", *J. Am. Chem. Soc.,* **109,** 1238–1240 (1987).
51. Krautler, B., Konrat, R., Stupperich, E., Gerald, F., Gruber, K., and Kratky, C., "Direct Evidence for the Conformational Deformation of the Corrin Ring by the Nucleotide Base in Vitamin-B12-Synthesis and Solution Spectroscopic and Crystal-Structure Analysis of Co-β-Cyanoimidazolylcobamide", *Inorg. Chem.,* **33,** 4128–4139 (1994).
52. Ng, F. T. T., Rempel, G. L., and Halpern, J., "Ligand Effects on Transition-Metal Alkyl Bond-Dissociation Energies", *J. Am. Chem. Soc.,* **104,** 621–623 (1982).
53. Marzilli, L. G., Summers, M. F., Bresciani-Pahor, N., Zangrando, E., Charland, J.-P., and Randaccio, L., "Rare Examples of Structurally Characterized 5-Coordinate Organocobalt Complexes—Novel Dynamic Nmr Evidence for Synergistic Enhancement of Cis and Trans Effects in B-12 Models", *J. Am. Chem. Soc.,* **107,** 6880–6888 (1985).
54. DeRidder, D. J. A., Zangrando, E., and Burgi, H. B., "Structural behaviour of cobaloximes: Planarity, an anomalous trans-influence and possible implications on Co–C bond cleavage in coenzyme-B-12-dependent enzymes", *J. Mol. Struct.,* **374,** 63–83 (1996).
55. Mancia, F., Keep, N. H., Nakagawa, A., Leadlay, P. F., McSweeney, S., Rasmussen, B., Bosecke, P., Diat, O., and Evans, P. R., "How coenzyme B12 radicals are generated: the crystal structure of methylmalonyl-coenzyme A mutase at 2 A resolution", *Structure,* **4,** 339–350 (1996).
56. Tamao, Y. and Blakley, R. L., "Direct spectrophotometric observation of an intermediate formed from deoxyadenosylcobalamin in ribonucleotide reduction", *Biochemistry,* **12,** 24–34 (1973).
57. Stubbe, J. and van Der Donk, W. A., "Protein Radicals in Enzyme Catalysis", *Chem. Rev.* **98,** 705–762 (1998).
58. Smith, D. M., Golding, B. T., and Radom, L., "Toward a consistent mechanism for diol dehydratase catalyzed reactions: An application of the partial-proton-transfer concept", *J. Am. Chem. Soc.,* **121,** 5700–5704 (1999).
59. Toraya, T. and Mori, K., "A reactivating factor for coenzyme B12-dependent diol dehydratase", *J. Biol. Chem.,* **274,** 3372–3377 (1999).
60. Drennan, C. L., Huang, S., Drummond, J. T., Matthews, R. G., and Ludwig, M. L., "How a Protein Binds B-12-a 3.0-Angstrom X-Ray Structure of B-12-Binding Domains of Methionine Synthase", *Science,* **266,** 1669–1674 (1994).
61. Dixon, M. M., Huang, S., Matthews, R. G., and Ludwig, M., "The structure of the C-terminal domain of methionine synthase: presenting S-adenosylmethionine for reductive methylation of B12", *Structure,* **4,** 1263–1275 (1996).
62. Bandarian, V., Ludwig, M. L., and Matthews, R. G., "Factors modulating conformational equilibria in large modular proteins: a case study with cobalamin-dependent methionine synthase", *Proc. Natl. Acad. Sci. U. S. A.,* **100,** 8156–8163 (2003).
63. Zydowsky, T. M., Courtney, L. F., Frasca, V., Kobayashi, K., Shimizu, H., Yuen, L.-D., Matthews, R. G., Benkovei, S. J., and Floss, H. G., "Stereochemical Analysis of the Methyl Transfer Catalyzed by Cobalamin-Dependent Methionine Synthase from *Escherichia-Coli*-B", *J. Am. Chem. Soc.,* **108,** 3152–3153 (1986).
64. Banerjee, R. V., Frasca, V., Ballou, D. P., and Matthews, R. G., "Participation of cob(I) alamin in the reaction catalyzed by methionine synthase from *Escherichia coli:* a steady-state and rapid reaction kinetic analysis", *Biochemistry,* **29,** 11101–11109 (1990).

XIII.3. Ribonucleotide Reductases

Contents

Marc Fontecave
Université Joseph Fourier
CNRS–CEA
CEA–Grenoble
38054 Grenoble
France

XIII.3.1. Introduction: Three Classes of Ribonucleotide Reductases

XIII.3.1.1. Different Metal Cofactors and Free Radicals

Deoxyribonucleic acid synthesis depends on a balanced supply of the four deoxyribonucleotides.[1] In all living organisms, with no exception to date, this is achieved by reduction of the corresponding ribonucleotides (Fig. XIII.3.1). This reaction is catalyzed by a fascinating family of allosterically regulated radical metalloenzymes, named ribonucleotide reductases (RNRs), extensively described in a number of excellent recent review articles.[2–7]

Class I RNRs are strictly aerobic α2β2 enzymes. They are found in mammals, plants, several viruses, and aerobic prokaryotes. The prototype for this class of RNR is the enzyme from *Escherichia coli.* Both proteins R1 (α2) and R2 (β2) from *E. coli* have been crystallized and their three-dimensional (3D) structures determined at high resolution. Protein R1 contains the binding sites for both substrates and allosteric effectors. The substrate site contains three conserved redox active cysteines (Cys439, Cys225, Cys462), which participate in ribonucleotide reduction. Each polypeptide chain of protein R2 contains a nonheme di-iron center, in which the ferric ions are linked by an oxo and a bidentate glutamate bridge, and a stable tyrosyl radical, on Tyr122 in *E. coli* RNR, (see Fig. XIII.3.2). This radical is essential for catalysis since it is required to generate on protein R1 a transient cysteinyl radical, Cys439, X•, in Fig. XIII.3.1, involved in the radical activation of the substrate. The metal center, in combination with molecular oxygen (O_2), serves to generate the tyrosyl radical.

Class II RNRs are found in bacteria and archaea. They are active both aerobically and anaerobically. What characterizes a class II RNR is the requirement for adenosylcobalamin (AdoCbl), supposed to function as a radical chain initiator (see

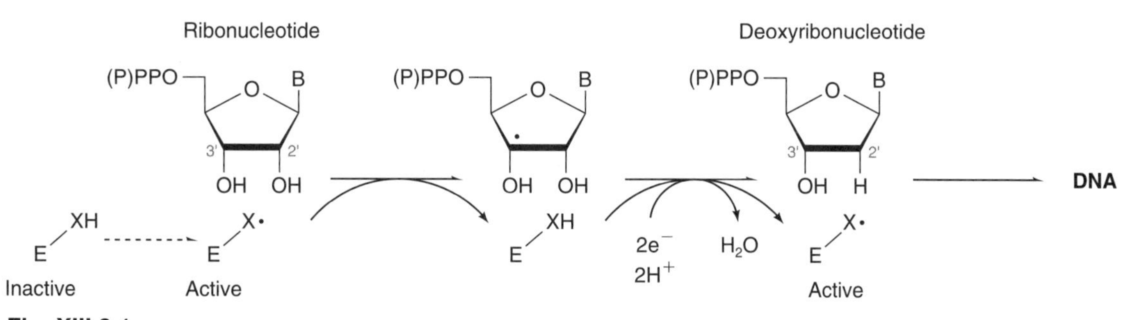

Fig. XIII.3.1.
The mechanism of the reaction catalyzed by RNR.
(P)PPO = Tri- or diphosphate
XH = cysteine 439
X• = Cys439, a transient cysteinyl (thiyl) radical.
B = Purine or pyrimidine base

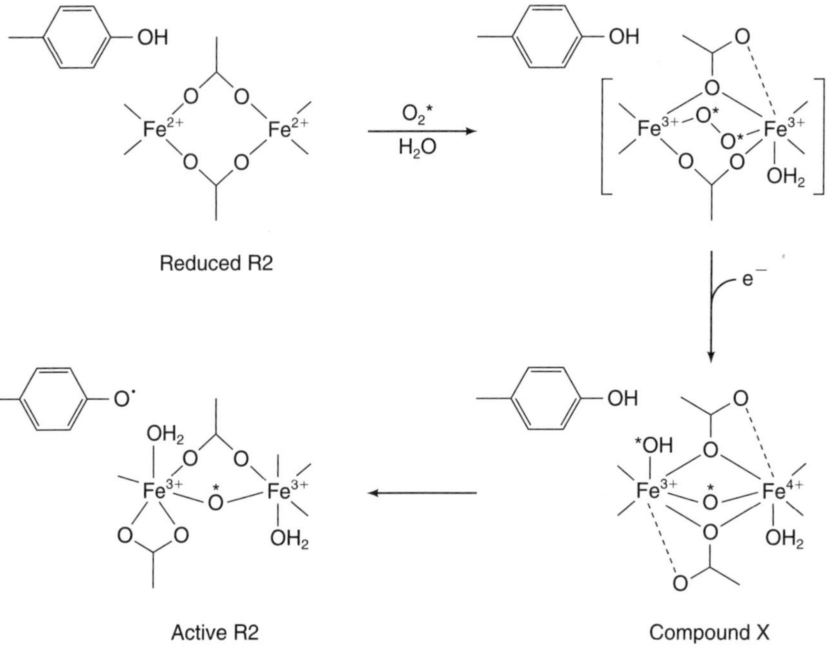

Fig. XIII.3.2.
Mechanism for the generation of the tyrosyl radical in class I RNR.

Section XIII.2). A 3D structure has been determined, confirming that the active site also contains three essential redox-active cysteines, in a spatial arrangement similar to that in class I RNRs.[8,9] It has been shown that the reaction proceeds much as in class I RNRs. In particular, in class II one of the cysteines is transiently converted into a cysteinyl radical (X• in Fig. XIII.3.1) and used to initiate nucleotide reduction. Only in the case of class II RNR can the intermediate thiyl radical be observed by EPR spectroscopy.[2,10]

Class III RNRs are oxygen-sensitive enzymes found in some facultative anaerobes and in methanogens.[7] The prototype is the $\alpha2\beta2$ enzyme from anaerobically growing *E. coli* cells. The large component $\alpha2$ contains the substrate and the allosteric effector-binding sites, and, in its active form, a glycyl radical (Gly681 in *E. coli*) absolutely required for catalysis. The 3D structure of a class III RNR (from bacteriophage T4) suggests that, during catalysis, a Cys in the active site is transiently transformed into a thiyl radical (X• in Fig. XIII.3.1) by reaction with the adjacent glycyl radical.[11] As for other RNRs, the thiyl radical is proposed to initiate the reduction of ribonucleotides. The small component $\beta2$ contains an iron–sulfur center, namely, a $(4Fe–4S)^{2+/+}$ cluster, which in combination with S-adenosylmethionine (also known as AdoMet) serves to generate the glycyl radical on protein $\alpha2$ (see Section XIII.4).[7] The β subunit is sometimes called the activase or the activating component of the anaerobic RNR.

XIII.3.1.2. The Chemical Basis for the Requirement of Radical Chemistry

All ribonucleotide reductases are radical enzymes: During catalysis they generate a protein radical, namely, a cysteinyl radical, which is absolutely required for initiating the enzyme reaction (Fig. XIII.3.1). This raises the obvious question: Why does ribonucleotide reduction require a radical mechanism?

The chemical reaction is a 2′-deoxygenation of nucleosides. This, at first sight, seems to be a rather trivial transformation. Numerous ionic reactions have been developed for the deoxygenation of simple alcohols. However, they have great limitations in the case of polyfunctional compounds with sterically hindered OH groups, such as glycols and ribonucleosides. The S_N2 displacement at C2′ of nucleosides is inhibited by steric and electronic factors. Cation formation (S_N1) at C2′ is precluded by bonding to the adjacent electron-deficient anomeric carbon. Generation of anionic character at C2′ results in the elimination of the base at C1′.[12] Thus radical reactions are the only feasible approach. For example, in model reactions, efficient deoxygenation of nucleosides was achieved by homolytic cleavage of an activated form of the 2′ C–O bond followed by reaction of the resulting alkyl radical with a hydrogen atom donor.[13] Thus it is quite obvious that Nature's choice of a radical mechanism for the biological reduction of ribonucleotides is dictated by purely chemical constraints.

The mechanism of ribonucleotide reduction will not be described here in detail. Excellent articles discuss it in depth.[2,14] Briefly, in all classes of RNR, the function of the transient cysteinyl radical, present in the substrate site, is to abstract the hydrogen atom at the 3′-position of the substrate ribose (Fig. XIII.3.1). Then a complex sequence of reactions converts the intermediate 3′-deoxynucleotide radical into the deoxyribonucleotide product with regeneration of the initiating protein-based radical. In both class I and II enzymes the two electrons are provided by a pair of cysteines in the active site. In class III, the two electrons are provided by formate, which is converted to CO_2 during reaction.[6,7]

XIII.3.2. Mechanisms of Radical Formation

The cysteine radicals of RNRs are only transiently generated for catalysis (Fig. XIII.3.1). In all cases, the enzymes carry a stable radical precursor (a tyrosyl radical in class I, AdoCbl in class II, and a glycyl radical in class III). A complex chemistry is used for the post-translational incorporation of these radicals into the polypeptide chains. *In all cases, a metal center is required.*

XIII.3.2.1. Class I: Activation of Molecular Oxygen and Formation of the Tyrosyl Radical

Class I reductase activity depends on the presence of the TyrO• radical on the small protein R2. Conversion of a tyrosine into a tyrosyl radical is a one-electron oxidation, requiring an oxidizing power of ~0.8–0.9 V. Dioxygen is used as the oxidant during the reaction. However, the kinetic barrier is large and a metal center (a diferrous center, here) combined with a source of electrons is required for activating O_2. The stoichiometry of the reaction is as follows:

$$\text{Tyr122–R2} + 2\,Fe^{2+} + O_2 + H^+ + e^- \text{ (from exogenous reductant)}$$
$$\rightarrow \text{Tyr122•–R2} + Fe^{3+}\text{–O–}Fe^{3+} + H_2O \qquad (1)$$

A four-electron reduction of O_2 generates one molecule of water and one oxo ion bridging the two iron ions in the active center. The two ferrous ions and the tyrosine each supply one electron. The fourth electron comes from an exogenous reductant.

The reaction starts with the binding of two ferrous ions on each polypeptide chain of protein R2.[2,15] The first intermediate, named reduced R2, has been characterized by X-ray crystallography and its structure is shown in Fig. XIII.3.2. The two irons are 3.8-Å apart and are bridged by two bidentate glutamates (Glu238 and Glu115 in *E. coli* R2). The Tyr122 residue is found 5 Å away from the closest iron atom. Reduced R2 then rapidly reacts with O_2 to produce a short-lived intermediate that, from spectroscopic studies, can be defined as a μ-1,2-peroxo diferric complex. The reaction is thus an oxidative addition during which two electrons from the diferrous center are transferred to O_2, generating an iron-bound peroxide. One-electron reduction then further weakens the O–O bond and cleavage of that bond generates a second intermediate, named compound X.

Compound X is paramagnetic and can be studied by stopped-flow UV–visible spectroscopy, rapid freeze quench (RFQ) EPR, Mössbauer, ENDOR, and extended X-ray absorption fine structure (EXAFS) experiments using wild type, mutant, and isotopically labeled proteins.[15,16] All of the data suggest that compound X is a spin-coupled Fe(III)/Fe(IV) center, in which the two irons, separated by a very short distance (2.49 Å), are bridged by two monodentate carboxylates and one oxo ion. One oxygen from O_2 supplies the oxo bridge and the second, in the form of water or hydroxide ligand, binds to Fe^{3+} (Fig. XIII.3.2). Compound X has the potential to convert the adjacent Tyr to a tyrosyl radical by H abstraction or coupled electron–proton transfer. In the final state, the two irons are bridged by one oxo ion and one bidentate glutamate. Clearly, important ligand reorganization occurs throughout the process.

XIII.3.2.2. Class II: Activation of Adenosylcobalamin and Formation of the Cysteinyl Radical

Adenosylcobalamin is required for the activity of class II RNRs. The function of AdoCbl in all AdoCbl-requiring enzymes has been presumed to involve generation of the carbon-centered 5′-deoxyadenosyl radical, through the homolytic cleavage of the Co–C bond, in order to initiate a radical-dependent reaction (see Section XIII.3). However, such a radical has never been observed during enzyme catalysis. Furthermore, the rate constant for homolysis of the cofactor, with a bond dissociation energy of 31 $kcal\,mol^{-1}$,[17] in solution is $10^{-9}\,s^{-1}$ while the thermodynamic constant for the homolysis reaction is in the range of 10^{-9}. These values cannot account for the fast enzyme reactions that are dependent on AdoCbl.

In the case of RNR, the enzyme accelerates the homolytic cleavage of the Co–C bond at a rate 10^{11}-fold faster than the uncatalyzed reaction, which corresponds to a transition state stabilization energy of 15 $kcal\,mol^{-1}$. Furthermore, the thermodynamic constant is now on the order of 1 ($\Delta G = 0\,kcal\,mol^{-1}$). However, the product of the reaction is not the 5′-deoxyadenosyl radical, but a cysteinyl radical on the polypeptide chain (see below).[10] Even though it has not been directly detected so far, there is strong evidence for the intermediate formation of the 5′-deoxyadenosyl radical (Fig. XIII.3.3). In particular, epimerization of (5′*R*)-[5′-^{2}H]adenosylcobalamin could be observed using a class II RNR in which the critical cysteine was substituted by a serine or an alanine.[18] Cob(II)alamin is also formed during the process, at the same rate as 5′-deoxyadenosine. The stoichiometry of the radical generation reaction is the following:

$$\text{AdoCbl} + \text{RSH} \rightarrow \text{AdCH}_3 + \text{Co}^{\text{II}} + \text{RS}\bullet \ (\text{AdCH}_3 = 5'\text{-deoxyadenosine}) \quad (2)$$

The dramatic shift in equilibrium and rate constants in AdoCbl homolysis is still not understood. The RNR protein does not seem to exploit binding energy to force the coenzyme into an unfavorable conformation, leading to a weakening of the Co–C bond through steric or electronic effects. The AdoCbl compound is not tightly bound and Co maintains its axial benzimidazole, in contrast to methionine synthase and methylmalonylCoA mutase in which a protein-bound His instead is in the axial position (see Section XIII.2).[9,19] Several factors, such as the tight binding of the products, cob(II)alamin and 5′-deoxyadenosine, or entropic factors, play an important role in the acceleration. Furthermore, endergonic coupling of the Co–C bond cleavage to C–H bond formation (to generate 5′-deoxyadenosine) and S–H bond cleavage (to generate the cysteinyl radical) is expected to reduce the enthalpic cost by 10 $kcal\,mol^{-1}$, since the homolytic bond dissociation energy of S–H is 88–91 $kcal\,mol^{-1}$, while that of C–H in R–CH_3 is 100 $kcal\,mol^{-1}$.

The reaction of AdoCbl with class II RNR was investigated by stopped flow and rapid freeze quench EPR techniques. A paramagnetic species could be trapped

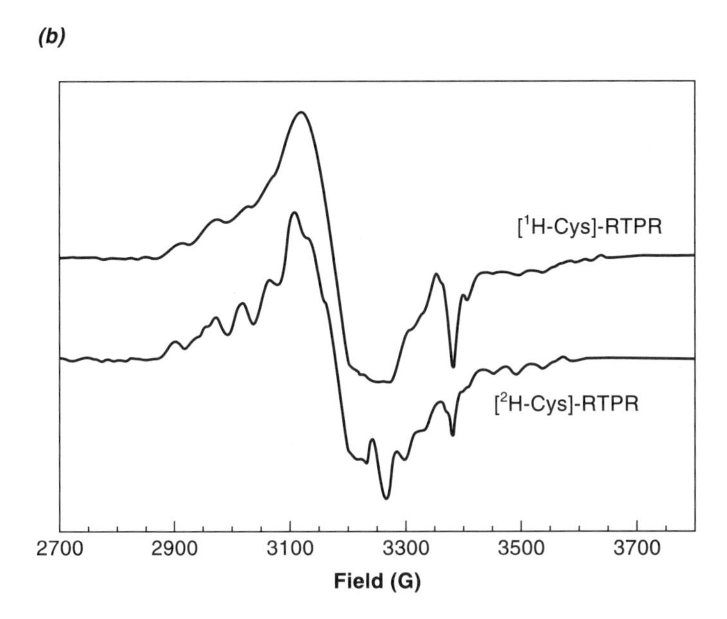

Fig. XIII.3.3.
(*a*) Mechanism for the generation of the cysteinyl radical in class II ribonucleoside triphosphate reductase (RTPR) and (*b*) the EPR spectrum of the coupled cob(II)alamin-radical center.[10]

during the first milliseconds and was identified, from its EPR spectrum, as a cob(II)alamin interacting with a cysteinyl radical by a combination of spin–spin exchange and spin–spin dipole mechanisms. When enzyme labeled with β-(^{2}H)cysteines was used, the EPR spectrum was significantly altered with a sharpening of the features associated with the Co hyperfine structure (Fig. XIII.3.3).[2,10] The distance between cob(II)alamin and the thiyl radical is suggested to be in the 5–8-Å range. This radical species is presumed to abstract the hydrogen atom at the 3′-position of the ribose moiety of the substrate during catalysis.

XIII.3.2.3. Class III: Activation of S-Adenosylmethionine and Formation of the Glycyl Radical

In class III RNR, the active enzyme contains a glycyl radical, highly sensitive to oxygen, in the large protein. Conversion of a Gly to its corresponding radical is a one-electron oxidation. The reaction, which requires a source of electrons (reduced flavodoxin, a FMN (flavin mononucleotide)-containing electron-transfer flavoprotein) and AdoMet as the oxidant,[6,7] can be written as

$$e^- + \text{GlyH} + \text{AdoMet} \rightarrow \text{AdCH}_3 + \text{Gly}\bullet + \text{methionine} \tag{3}$$

$$(\text{AdCH}_3 = 5'\text{-deoxyadenosine})$$

As a sulfonium ion, AdoMet ($RR'AdCH_2S^+$ in Eq. 4) is not a strong oxidant and needs to be activated, in order to be used for the one-electron oxidation of a Gly residue, which requires a redox potential of ~1.0 V. The working hypothesis is that this reactivity is achieved by injection of a single electron into the sulfonium ion, which

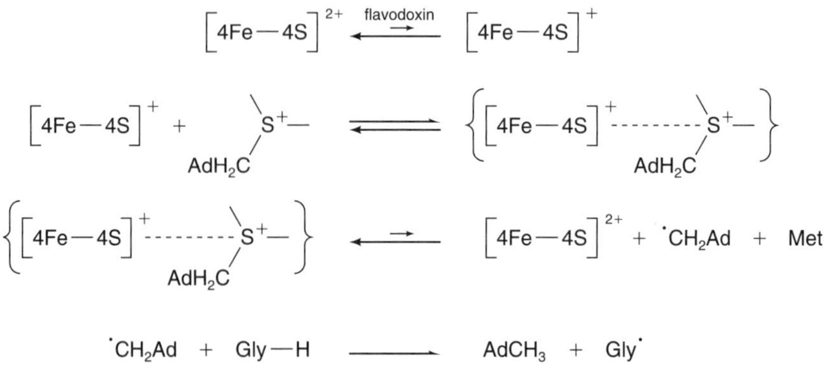

Fig. XIII.3.4.
Mechanism for the generation of the glycyl radical in class III RNR.

$$[4Fe{-}4S]^{2+} \xrightarrow{\text{flavodoxin}} [4Fe{-}4S]^{+}$$

$$[4Fe{-}4S]^{+} + AdH_2C{-}S^{+}{<} \rightleftharpoons \{[4Fe{-}4S]^{+} \cdots S^{+}(CH_2Ad){<}\}$$

$$\{[4Fe{-}4S]^{+} \cdots S^{+}(CH_2Ad){<}\} \rightleftharpoons [4Fe{-}4S]^{2+} + {}^{\bullet}CH_2Ad + Met$$

$${}^{\bullet}CH_2Ad + Gly{-}H \longrightarrow AdCH_3 + Gly^{\bullet}$$

would result in a short-lived sulfuranyl radical that decomposes homolytically into Met and the $AdCH_2\bullet$ carbon-centered radical. The latter has the potential to abstract the hydrogen atom of a Gly residue, as the C–H bond dissociation energy for $AdCH_3$ is larger than that for Gly ($80\,kcal\,mol^{-1}$).

$$RR'AdCH_2S^{+} + e^{-} \rightarrow RR'AdCH_2S\bullet \rightarrow RR'S + AdCH_2\bullet \quad (4)$$

$$AdCH_2\bullet + GlyH \rightarrow Gly\bullet + AdCH_3 \quad (5)$$

Such a mechanism is supported by the following observations: (1) during the formation of the glycyl radical, AdoMet is indeed converted to 5′-deoxyadenosine and Met; (2) deuterium is recovered into 5′-deoxyadenosine, in experiments using enzyme labeled with deuterated Gly.[7] However, no direct evidence for intermediate free radicals, such as the sulfuranyl radical or the 5′-deoxyadenosyl radical, is yet available. Nevertheless, it is well established that alkyl radicals can be produced by single-electron reduction of sulfonium salts with Mg metal, SmI_2, or electrochemically.[20] Furthermore, the chemical feasibility of the formation of a glycyl radical by H abstraction by a primary alkyl radical has been demonstrated.[21]

The problem is that sulfonium ions such as AdoMet have a very low redox potential and reduction by reduced flavodoxin is thermodynamically unfavorable. However, the reduction is coupled to subsequent irreversible steps (cleavage of the sulfuranyl radical and formation of the glycyl radical), which drive the reaction. Furthermore, the enzyme system contains a (4Fe–4S) center, located on the small protein, that mediates the electron transfer to AdoMet (Eq. 4). It has been established from single turnover experiments that the EPR active $S = \frac{1}{2}(4Fe{-}4S)^{+}$ form of the Fe center is competent for reduction and cleavage of AdoMet, release of Met, and generation of the glycyl radical, suggesting that the electron in Eq. 4 is provided by the reduced cluster (Fig. XIII.3.4).[7,22] It is likely that this electron transfer takes place within a [4Fe–4S]–AdoMet complex, as in pyruvate formate lyase, a related iron–sulfur system requiring a glycyl radical and AdoMet for activation.[23]

XIII.3.3. Conclusions

Free radicals are still believed to be highly unstable, reactive, and toxic species. Until recently, it was hardly conceivable that they might serve as intermediates during selective and regulated biological reactions. However, as revealed by the mechanism of RNRs, free radicals can in fact be tightly controlled. It is quite remarkable that DNA synthesis, a crucial process for living organisms, depends on radical chemistry. This dependence probably reflects the fact that a radical is the most attractive chemical solution for storing an oxidizing equivalent and using it for hydrogen-atom abstraction reactions, as required for initiating ribonucleotide reduction. In all RNRs, the key radical is a transient cysteinyl radical adjacent to the substrate, even though such a radical has been directly observed only in the case of class II. The difference between the three classes resides in the mechanism by which this radical is formed. In all cases, a metal center is absolutely required.

References

General References

1. Reichard, P., "Interactions between Deoxyribonucleotide and DNA Synthesis", *Annu. Rev. Biochem.*, **57,** 349 (1988).
2. Stubbe, J. and van der Donk, W., "Protein Radicals in Enzyme Catalysis", *Chem. Rev.*, **98,** 705 (1998).
3. Sjöberg, B.-M., "Ribonucleotide Reductases—A Group of Enzymes with Different Metallosites and Similar Reaction Mechanisms", *Struct. Bonding,* **88,** 139 (1997).
4. Reichard, P. and Jordan, A., "Ribonucleotide Reductases", *Annu. Rev. Biochem.*, **67,** 71 (1998).
5. Mulliez, E. and Fontecave, M., "Ribonucleotide Reductases: Metal and Free Radical Interplay", *Coord. Chem. Rev.*, **775,** 185 (1999).
6. Fontecave, M. and Mulliez, E., "Ribonucleotide Reductases", in *Chemistry and Biochemistry of B12,* Banerjee, R. Ed., John Wiley & Sons, Inc., New York, 1999, p. 31.
7. Fontecave, M., Mulliez, E., and Logan, D. T., "Deoxyribonucleotide Synthesis in Anaerobic Microorganisms: The Class III Ribonucleotide Reductase", *Prog. Nucleic Acid Res. Mol. Biol.*, **72,** 95 (2002).

Specific References

8. Booker, S., Licht, S., Broderick, J., and Stubbe, J., "Coenzyme B12-Dependent Ribonucleotide Reductase: Evidence for the Participation of Five Cysteine Residues in Ribonucleotide Reduction", *Biochemistry,* **33,** 12676 (1994).
9. Sintchak, M. D., Arjara, G., Kellogg, B. A., Stubbe, J., and Drennan, C. L., "The Crystal Structure of Class II Ribonucleotide Reductase Reveals How an Allosterically Regulated Monomer Mimics a Dimer", *Nat. Struct. Biol.*, **9,** 293 (2002).
10. Licht, S., Gerfen, G. J., and Stubbe, J., "Thiyl Radicals in Ribonucleotide Reductases", *Science,* **271,** 477 (1996).
11. Logan, D. T., Andersson, J., Sjöberg, B. M., and Nordlund, P., "A Glycyl Radical Site in the Crystal Structure of a Class III Ribonucleotide Reductase", *Science,* **283,** 1499 (1999).
12. Robins, M. J. and Wilson, J. S., "Smooth and Efficient Deoxygenation of Secondary Alcohols. A General Procedure for the Conversion of Ribonucleosides to 2′-Deoxynucleosides", *J. Am. Chem. Soc.*, **10,** 932 (1981).
13. Hartwig, W., "Modern Methods for the Radical Deoxygenation of Alcohols", *Tetrahedron,* **39,** 2609 (1983).
14. Stubbe, J. and Van der Donk, W., "Ribonucleotide Reductases: Radical Enzymes with Suicidal Tendencies", *Chem. Biol.*, **2,** 793 (1995).
15. Stubbe, J., "Di-Iron-Tyrosyl Radical Ribonucleotide Reductases", *Curr. Opin. Chem. Biol.*, **7,** 183 (2003).
16. Riggs-Gelasco, P. J., Shu, L. J., Chen, S. X., Burdi, D., Huynh, B. H., Que, L., and Stubbe, J., "EXAFS Characterization of the Intermediate X Generated During the Assembly of the *Escherichia coli* Ribonucleotide Reductase R2 Diferric Tyrosyl Radical Cofactor", *J. Am. Chem. Soc.*, **120,** 849 (1998).
17. Hay, B. P. and Finke, R. G., "Thermolysis of the Cobalt–Carbon Bond in Adenosylcorrins. 3. Quantification of the Axial Base Effect in Adenosylcobalamin by the Synthesis and Thermolysis of Axial Base-Free Adenosylcobinamide. Insights into the Energetics of Enzyme-Assisted Cobalt–Carbon Bond Homolysis", *J. Am. Chem. Soc.*, **109,** 8012 (1987).
18. Chen, D., Abend, A., Stubbe, J., and Frey, P. A., "Epimerization at Carbon-5′ of (5′*R*)-[5′-^{2}H]Adenosylcobalamin by Ribonucleoside Triphosphate Reductase: Cysteine 408-Independent Cleavage of the Co–C5′ Bond", *Biochemistry,* **42,** 4578 (2003).
19. Ludwig, M. L. and Matthews, R. G., "Structure-Based Perspectives on B12-Dependent Enzymes", *Annu. Rev. Biochem.*, **66,** 269 (1997).
20. Grimshaw, J., "Electrochemistry of the Sulphonium Group", in *The Chemistry of the Sulphonium Group,* Stirling, C. J. M. and Patai, S., Eds., J. Wiley & Sons, Inc., 1981, p. 141.
21. Baldwin, J. E., Brown, D., Scudder, P. H., and Wood, M. E., "A Chemical Model for the Activation of Pyruvate-Formate Lyase", *Tetrahedron Lett.*, **36,** 2105 (1995).
22. Ollagnier, S., Mulliez, E., Schmidt, P. P., Eliasson, R., Gaillard, J., Deronzier, C., Bergman, T., Gräslund, A., Reichard, P., and Fontecave, M., "Activation of the Anaerobic Ribonucleotide Reductase from *Escherichia coli.* The Essential Role of the Iron–Sulfur Center for S-Adenosylmethionine Reduction", *J. Biol. Chem.*, **272,** 24216 (1997).
23. Walsby, C. J., Hong, W., Broderick, W. E., Cheek, J., Ortillo, D., Broderick, J. B., and Hoffman, B. M., "Electron-Nuclear Resonance Spectroscopic Evidence that S-Adenosylmethionine Binds in Contact with the Catalytically Active (4Fe–4S) Cluster of Pyruvate Formate-Lyase Activating Enzyme", *J. Am. Chem. Soc.*, **124,** 3143 (2002).

XIII.4. Fe–S Clusters in Radical Generation

Joan B. Broderick
Department of Chemistry and Biochemistry
Montana State University
Bozeman, MT 59717

Contents

XIII.4.1. Introduction

Iron–sulfur clusters are ubiquitous in biological systems and serve a broad variety of functions. The first discovered function, and still the one most associated with iron–sulfur clusters, is electron transfer (see Section X.1). Many other roles for these fascinating clusters have now been identified, including enzymatic catalysis, structural stabilization, and regulation of gene expression. A particularly interesting and unexpected role for iron–sulfur clusters in biological systems is in the generation of radicals, a function that has now been confirmed or proposed in a significant number of systems.[1]

The iron–sulfur enzymes involved in radical generation all utilize S-adenosylmethionine (SAM) as a cofactor or cosubstrate and comprise the novel Radical-SAM protein superfamily.[2] This superfamily has hundreds of putative members and includes enzymes catalyzing glycyl radical formation, rearrangement reactions, cofactor biosynthesis, repair of UV-induced DNA damage, and many other functions, many of which are currently unknown. Some of the reactions catalyzed by members of the Radical-SAM superfamily are shown in Table XIII.4.1. Despite the diversity of reactions catalyzed, the Radical-SAM enzymes are thought to have in common key mechanistic steps that involve the reductive cleavage of SAM to generate a catalytically essential 5′-deoxyadenosyl radical intermediate. This adenosyl radical intermediate initiates each reaction by abstracting the hydrogen atom(s) indicated in Table XIII.4.1. The resulting radical is either the reaction product (as in the activating enzymes) or undergoes further reaction to generate the product, as will be discussed in more detail later in this section.

XIII.4.1.1. Utilization of SAM

The known Radical-SAM enzymes can be classified into two broad categories: those that use SAM as a substrate (e.g., the activating enzymes and biotin synthase) and those that use SAM as a catalytic cofactor (e.g., lysine 2,3-aminomutase and spore photoproduct lyase); these differences in utilization are indicated in Table XIII.4.1. The former enzymes convert SAM stoichiometrically to Met and 5′-deoxyadenosine, with incorporation of the hydrogen atom abstracted from substrate into the 5′-deoxyadenosine, while in the latter enzymes the 5′-deoxyadenosyl radical is regenerated after each turnover due to hydrogen atom abstraction from 5′-deoxyadenosine by the product radical. In other words, the enzymes that utilize SAM as a substrate catalyze a net redox reaction, in which SAM is reduced and the substrate is oxidized, while enzymes that use SAM as a cofactor, catalyze reactions that do not involve a

Table XIII.4.1.
Representative members of the Radical-SAM superfamily. The hydrogen atom(s) abstracted are shown in red for each reaction.

Enzyme	Reaction Catalyzed	Role of SAM
Activating enzymes	 E—Gly → E—Gly•	Substrate
Lysine aminomutase	 H_3^+N–(CH$_2$)$_3$–CH(H)–CH(NH_3^+)–COO^- ⇌ H_3^+N–(CH$_2$)$_3$–CH(NH_3^+)–CH(H)–COO^-	Cofactor
Biotin synthase	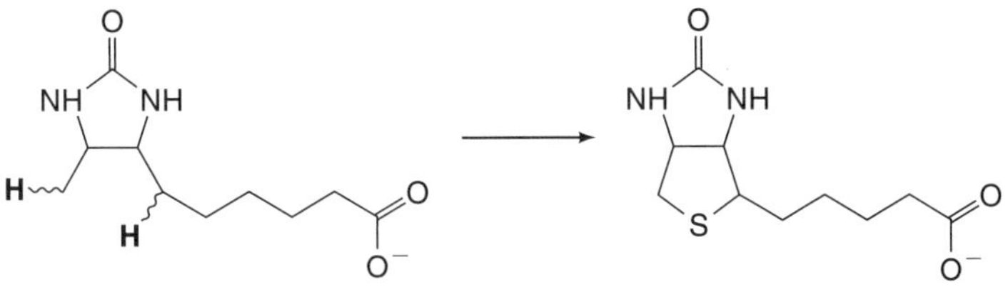 dethiobiotin → biotin	Substrate
Lipoate synthase	 octanoyl-NH → lipoyl-NH (S–S)	Substrate
Spore photoproduct lyase	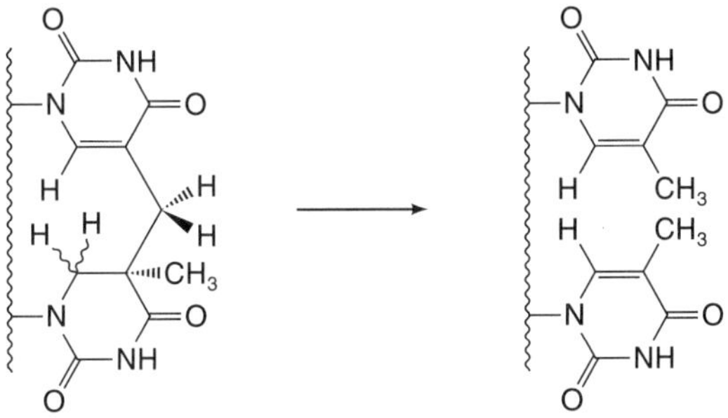 spore photoproduct → two thymines (CH_3)	Cofactor
Coproporphyrinogen III oxidase	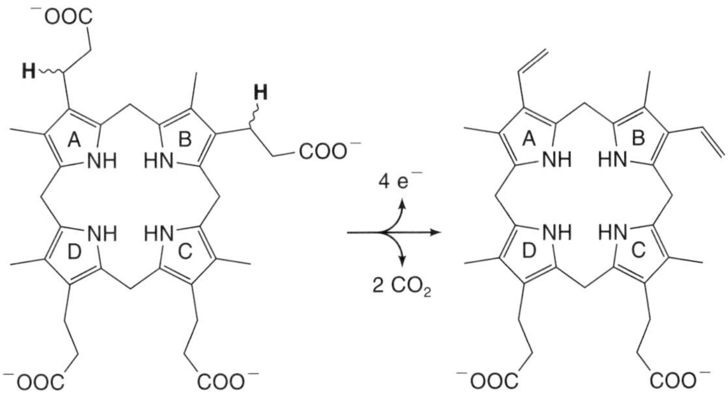 coproporphyrinogen III → protoporphyrinogen IX (4 e$^-$, 2 CO_2)	Substrate

net oxidation–reduction. Each of these general mechanisms will be discussed further below.

XIII.4.1.2. Iron–Sulfur Clusters

The iron–sulfur clusters in the Radical-SAM enzymes are generally both oxygen sensitive and labile, which has resulted in difficulty in cleanly identifying the cluster forms present in a given preparation and in determining which cluster form(s) are catalytically relevant. The characterized members of the Radical-SAM superfamily have been isolated with various combinations of $[4Fe–4S]^{2+/+}$, cuboidal and linear $[3Fe–4S]^{+}$, and $[2Fe–2S]^{2+}$ clusters, depending on the specific enzyme and the expression and purification conditions used. In some cases the cluster is not stable to the protein purification methods employed, and must be reconstituted in the purified protein. The various cluster forms have been identified by using a combination of UV–visible, EPR, resonance Raman, and Mössbauer spectroscopies.[3] Mössbauer spectroscopy of ^{57}Fe-enriched protein has proven to be particularly valuable in cluster characterization in these enzymes, particularly in those situations where multiple cluster forms exist, as it is able to detect both diamagnetic and paramagnetic iron species, even those present in relatively low quantities. Although the precise cluster forms found in purified and/or reconstituted Radical-SAM enzymes can vary considerably, all these clusters appear to convert to $[4Fe–4S]^{2+/+}$ under appropriate reducing conditions. Experiments on several of the Radical-SAM enzymes, including the pyruvate formate-lyase activating enzyme, the anaerobic ribonucleotide reductase activating enzyme, and lysine 2,3-aminomutase, have clearly demonstrated that the reduced $[4Fe–4S]^{+}$ state of the cluster is catalytically essential, and that this reduced cluster is the source of the electron necessary for reductive cleavage of SAM to generate the 5′-deoxyadenosyl radical intermediate.

The iron–sulfur clusters in the Radical-SAM enzymes are coordinated by a conserved cluster motif (CXXXCXXC) containing only three cysteines. The fact that there are only three cysteines in this conserved motif suggests that the catalytically relevant [4Fe–4S] cluster in these enzymes is a site-differentiated cluster, with one Fe coordinated by a noncysteine ligand. The apparent conservation throughout the superfamily suggests that this site-differentiated cluster is functionally significant.

XIII.4.1.3. Catalytic Mechanisms

Of considerable interest in the field is the mechanism by which an iron–sulfur cluster and SAM conspire to generate an adenosyl radical intermediate. The putative involvement of a 5′-deoxyadenosyl radical intermediate in the Radical-SAM reactions is reminiscent of AdoCbl enzymes (see Section XIII.2), the only other enzymes known to utilize adenosyl radical intermediates. The types of reactions catalyzed by the Radical-SAM enzymes are also similar to AdoCbl reactions; for example, both families of enzymes contain a ribonucleotide reductase (see Section XIII.3) and a lysine aminomutase. The similarities between the Radical-SAM and AdoCbl enzymatic reactions led many to speculate about possible organometallic intermediates in the Radical-SAM reactions (Fig. XIII.4.1). However, as will be discussed below, the Radical-SAM enzymes appear to catalyze adenosyl-radical mediated reactions by a mechanism quite distinct from that of AdoCbl enzymes.

The mechanisms of the Radical-SAM enzymes involve an electron transfer from a reduced $[4Fe–4S]^{+}$ cluster to SAM as an initial step. Given the estimated redox potentials of the clusters in this superfamily (-400 to -600 mV) and the typical redox potentials of sulfonium ions (-1 to -2 V), it seemed unlikely that this initial electron-transfer step would occur via simple long-range electron transfer. Rather, it was expected that the iron–sulfur cluster would interact directly with SAM to mediate

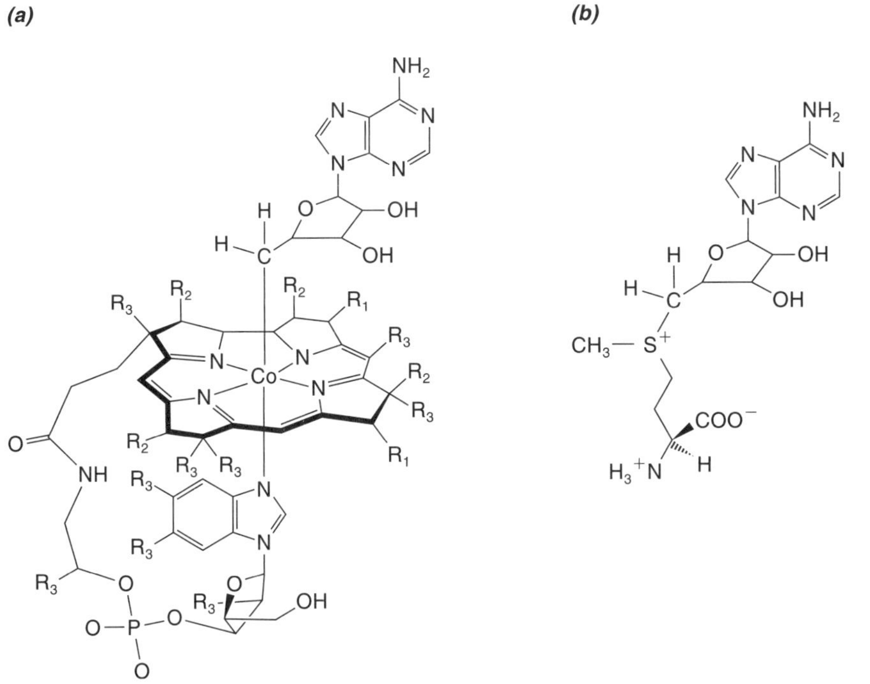

Fig. XIII.4.1.
The adenosyl radical generating cofactors, AdoCbl (*a*) and SAM (*b*).

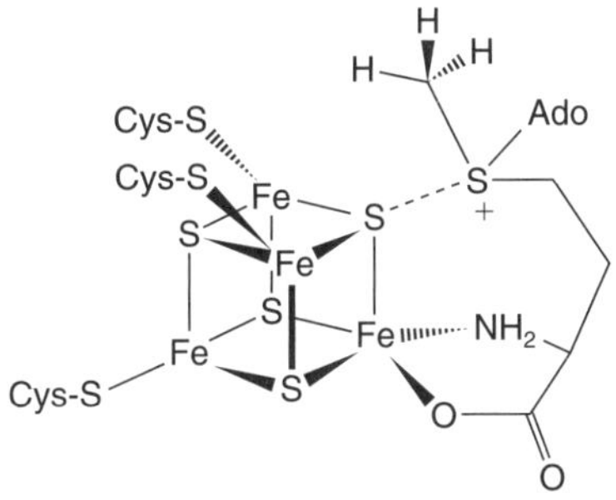

Fig. XIII.4.2.
Interaction of SAM with the [4Fe–4S] cluster of pyruvate formate-lyase activating enzyme as deduced from spectroscopic studies.

reductive cleavage. Such a direct interaction between SAM and a $[4Fe\text{–}4S]^+$ cluster was first shown for the pyruvate formate-lyase activating enzyme by using electron–nuclear double resonance (ENDOR) to look at coupling between NMR active nuclei on SAM and the EPR active iron–sulfur cluster.[3] The SAM–cluster interaction identified by these studies is shown in Fig. XIII.4.2. S-adenosylmethionine chelates the unique iron of the [4Fe–4S] cluster through the amino and carboxylate moieties of the Met portion of SAM, which in essence serves as an anchor to position the SAM for reaction. This direct chelation of the unique Fe by the substrate–cofactor SAM is reminiscent of the interaction of substrate with aconitase (see Section IX.4), another enzyme containing a site-differentiated [4Fe–4S] cluster. In aconitase, however, the chelated Fe does not serve merely as a structural anchor, but plays a key functional role as a Lewis acid in catalysis. The sulfonium sulfur of SAM has orbital overlap with the cluster, as demonstrated by observation of anisotropic coupling between the cluster and the deuterons of $^2H_3$-methyl-SAM.[3] This orbital overlap has been proposed to occur via a bridging sulfide of the cluster, and provides a likely pathway for inner-sphere electron transfer from the cluster to SAM to initiate the reductive cleavage. Similar ENDOR studies on lysine 2,3-aminomutase yield nearly identical results, suggesting a common mode of SAM interaction with the [4Fe–4S] clusters

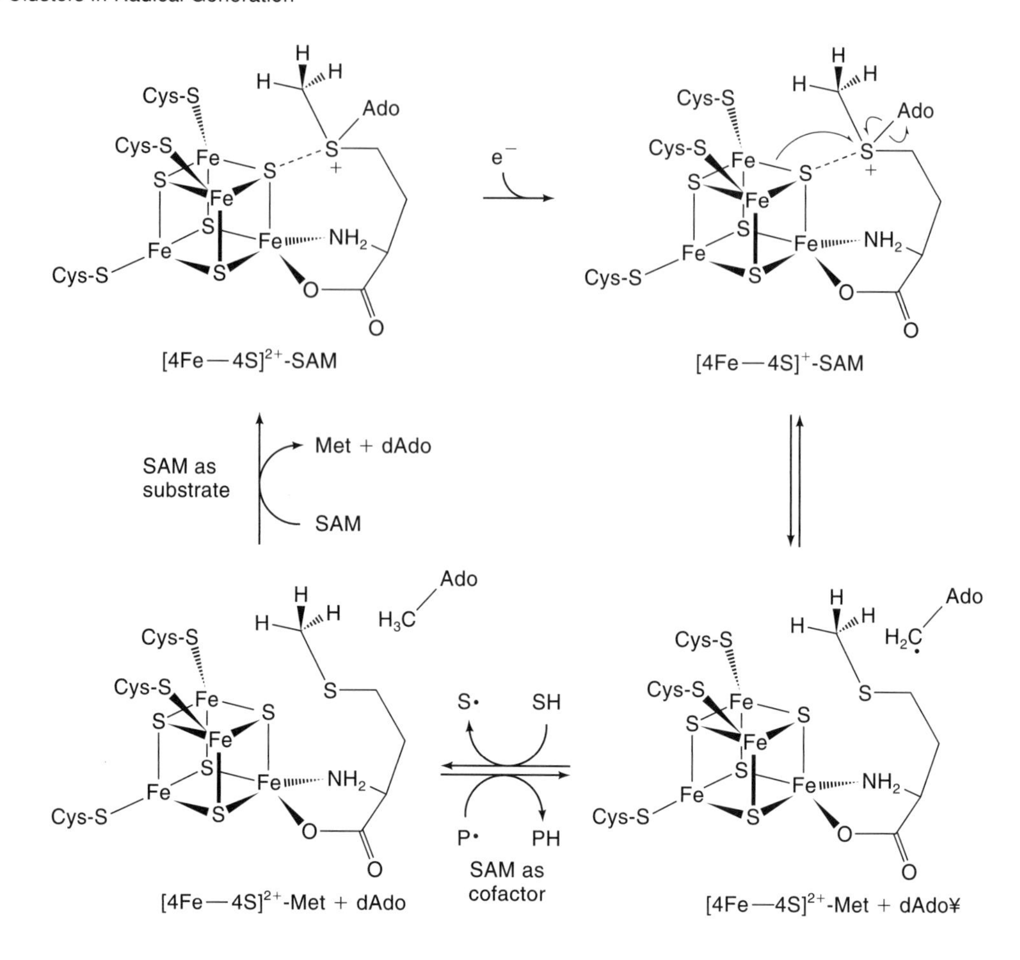

Fig. XIII.4.3.
A general mechanism for reactions catalyzed by the Radical-SAM enzymes.

of the Radical-SAM enzymes. Crystal structures of biotin synthase[4] and HemN[5] revealed similar cluster–SAM interactions in these two enzymes (see below).

Based on the information provided above, a general mechanism for the Radical-SAM enzymes can be proposed, as shown in Fig. XIII.4.3. In this mechanism, SAM binds to the $[4Fe–4S]^{2+}$ cluster, and this cluster is then reduced (e.g., by flavodoxin) to the $[4Fe–4S]^{+}$ state. The $[4Fe–4S]^{+}$ cluster transfers an electron to SAM, resulting in reductive cleavage of the S–C(5′) bond to generate Met and a 5′-deoxyadenosyl radical intermediate. The 5′-deoxyadenosyl radical intermediate abstracts a hydrogen atom from substrate to produce 5′-deoxyadenosine and a substrate radical. S-adenosylmethionine is either stoichiometrically cleaved, forming Met and 5′-deoxyadenosine as products, or it is regenerated by hydrogen atom abstraction by the product radical to regenerate the 5′-deoxyadenosyl radical, followed by loss of an electron to the iron–sulfur cluster and re-formation of the S–C(5′) bond.

XIII.4.2. Glycyl Radical Generation

Radical SAM enzymes involved in glycyl radical formation include the activating component of the anaerobic ribonucleotide reductase from *Escherichia coli* (see Section XIII.3), the pyruvate formate lyase activating enzyme, and the benzylsuccinate synthase activating enzyme. Each of these enzymes functions anaerobically to generate a glycyl radical on its substrate enzyme, and, in two cases, the glycyl radical has been shown to be located on structurally similar finger loops (Fig. XIII.4.4). The

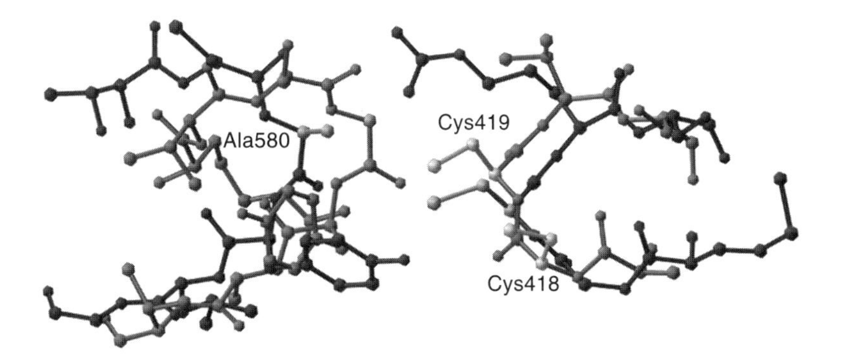

Fig. XIII.4.4.
Finger loops in the active sites of pyruvate formate-lyase (gray) and the anaerobic ribonucleotide reductase G580A (black). On the left loop is the glycyl residue (G734 or G580) that is converted to the radical form upon activation. On the right loop are catalytically essential Cys residues. (PDB codes: 2PFL and 1B8B.)

glycyl radicals are stable under anaerobic conditions, but react rapidly with dioxygen, consistent with the observation that all three of these glycyl radical enzymes function anaerobically, either in strict anaerobes or when facultative anaerobes experience anaerobic conditions. Despite the strong similarities among these activating enzymes, very little sequence homology is observed other than the conserved three-Cys motif (CXXXCXXC) involved in Fe–S cluster binding.

XIII.4.2.1. Pyruvate Formate-Lyase Activating Enzyme

The catalytically essential glycyl radical of pyruvate formate-lyase was first identified by Knappe et al.[6] using a combination of EPR spectroscopy, isotopic labeling, and analysis of the products of oxygenolytic cleavage. Their surprising results provided the first direct evidence for a stable glycyl radical in a protein. The glycyl radical of active pyruvate formate-lyase was generated stereospecifically by a specific activating enzyme in a reaction dependent on SAM and reducing equivalents (Fig. XIII.4.5).[7] Early studies on the pyruvate formate-lyase activating enzyme showed a strict dependence of enzymatic activity on added Fe in the assay, indicating that the glycyl radical generation reaction had its basis in bioinorganic chemistry.

Pyruvate formate-lyase activating enzyme was ultimately found to contain an iron–sulfur cluster.[1,3] The iron–sulfur cluster is oxygen sensitive and labile, and all cluster forms previously mentioned for the Radical-SAM enzymes have been found in this enzyme, including $[2Fe–2S]^{2+/+}$, cuboidal and linear $[3Fe–4S]^{+}$, and $[4Fe–4S]^{2+/+}$, with the relative proportions of these depending on the precise purification conditions. The [2Fe–2S] and [3Fe–4S] clusters can be converted to $[4Fe–4S]^{2+/+}$ clusters under anaerobic reducing conditions, even in the absence of added iron. The unusual plasticity in the cluster coordination environment in pyruvate formate-lyase activating enzyme, as well as the observation of facile cluster interconversion, are reminiscent of the iron–sulfur cluster of aconitase (see Section IX.4). It remains to be established, however, whether the cluster lability and cluster interconversions observed in pyruvate formate-lyase activating enzyme *in vitro* play any role *in vivo*. It is well established, however, that the $[4Fe–4S]^{+}$ state of the enzyme is the catalytically relevant form, as it generates exactly 1 equiv of glycyl radical per $[4Fe–4S]^{+}$ cluster in the absence of exogenous reductants.

The current proposed mechanism for the pyruvate formate-lyase activating enzyme is essentially that shown in Fig. XIII.4.3. Two points are of particular interest. First, spectroscopic studies suggest that the iron–sulfur cluster–SAM interaction shown in Fig. XII.4.3 is valid for both oxidized and reduced states of the [4Fe–4S] cluster. Second, the SAM complex with the active $[4Fe–4S]^{+}$ state of pyruvate formate-lyase activating enzyme is stable in the absence of pyruvate formate-lyase;

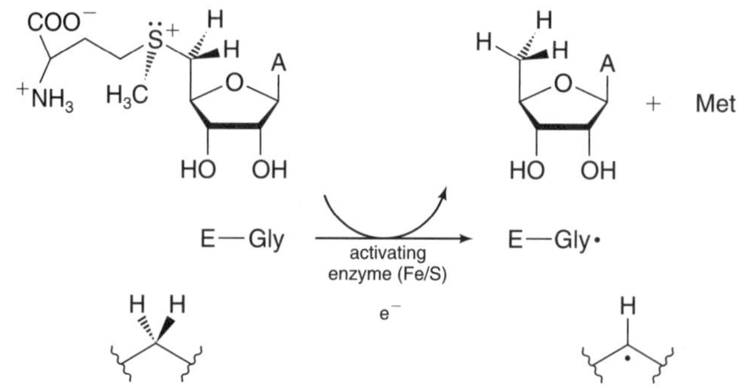

Fig. XIII.4.5.
Reaction scheme for the Radical-SAM activating enzymes.

in other words, reductive cleavage to generate the intermediate adenosyl radical occurs only in the presence of both substrates, suggesting that the presence of PFL produces a structural and/or electronic perturbation that promotes the inner-sphere electron-transfer reaction between the activating enzyme and SAM. How this electron-transfer reaction is triggered is not yet understood.

XIII.4.2.2. Anaerobic Ribonucleotide Reductase Activating Enzyme

This enzyme performs a function identical to that of pyruvate formate-lyase activating enzyme: Activation of an enzyme (anaerobic ribonucleotide reductase) by generation of a stable glycyl radical (see Section XIII.3). Because the anaerobic ribonucleotide reductase activating enzyme associates tightly with the anaerobic ribonucleotide reductase, it was thought for some time that the former was a subunit of the latter, and in most of the literature it is referred to as the β_2 subunit of the $\alpha_2\beta_2$ *E. coli* anaerobic ribonucleotide reductase. It has now been shown, however, that the β_2 protein can catalyze the introduction of the glycyl radical in several α_2 polypeptides, thereby characterizing β_2 as an enzyme. Like pyruvate formate-lyase activating enzyme, anaerobic ribonucleotide reductase activating enzyme has been shown to contain an iron–sulfur cluster that is unusually oxygen sensitive. In fact, most of the detailed characterization of the iron–sulfur cluster of this enzyme has been done on reconstituted enzyme, due to a loss of the native cluster during isolation.

Anaerobic ribonucleotide reductase activating enzyme, like pyruvate formate-lyase activating enzyme, can also accommodate a variety of interconvertible cluster forms. The $[4Fe–4S]^+$ cluster has been shown to be competent for reductive cleavage of SAM and for generation of the glycyl radical, and it is expected that the mechanism of this enzyme will be very similar to that shown in Fig. XIII.4.3. Although flavodoxin is the electron donor responsible for reducing the iron–sulfur cluster, which subsequently catalyzes the reductive cleavage of SAM, it has been found that the $[4Fe–4S]^{2+/+}$ couple of anaerobic ribonucleotide reductase activating enzyme is $\sim$300 mV more negative than that of the flavodoxin–semiquinone couple.[8] This observation suggests that the thermodynamically unfavorable cluster reduction is driven by coupling to two thermodynamically favorable reactions: the reductive cleavage of SAM and the generation of the glycyl radical.

XIII.4.2.3. Benzylsuccinate Synthase Activating Enzyme

Little is known about this putative member of the Radical-SAM activases. Benzylsuccinate synthase exhibits significant sequence similarity with pyruvate formate-lyase in the glycyl radical region, and an open reading frame adjacent to that for benzylsuccinate synthase codes for a protein with homology to the conserved cysteine region of the Radical-SAM enzymes. Preliminary studies indicate that the putative benzylsuccinate synthase activating enzyme will be similar in its properties and function to the activases described above.[1]

XIII.4.3. Isomerization Reactions

Lysine 2,3-aminomutase catalyzes the isomerization of L-lysine into L-β-lysine (Table XIII.4.1), a reversible rearrangement reaction directly analogous to those catalyzed by adenosylcobalamin-dependent enzymes. Like the activases, lysine 2,3-aminomutase requires SAM to catalyze this reaction, although in the case of lysine 2,3-aminomutase, SAM functions as a true cofactor in that it is not consumed stoichiometrically during turnover. Lysine 2,3-aminomutase has been shown to contain one $[4Fe–4S]^{2+/+}$ cluster and one divalent metal-binding site per subunit occupied by either Co^{2+} or Zn^{2+}.[1] In addition to an iron–sulfur cluster and SAM, lysine 2,3-aminomutase requires pyridoxal 5′-phosphate (PLP) for catalytic activity. Interestingly, a requirement for PLP is also found in the adenosylcobalamin-dependent aminomutases.

The 5′-deoxyadenosyl (5′-dAdo) moiety of SAM is responsible for mediating the transfer of the hydrogen atom of the substrate,[1d] a role directly analogous to that of the 5′-dAdo moiety of adenosylcobalamin in AdoCbl-dependent rearrangements (Fig. XIII.4.6). Evidence has been obtained for the presence of both L-lysine and L-β-lysine centered radical intermediates in the rearrangement reaction catalyzed by lysine 2,3-aminomutase, and the L-β-lysine radical has been shown to be kinetically competent. Use of the SAM analogue 3′,4′-anhydroadenosylmethionine has

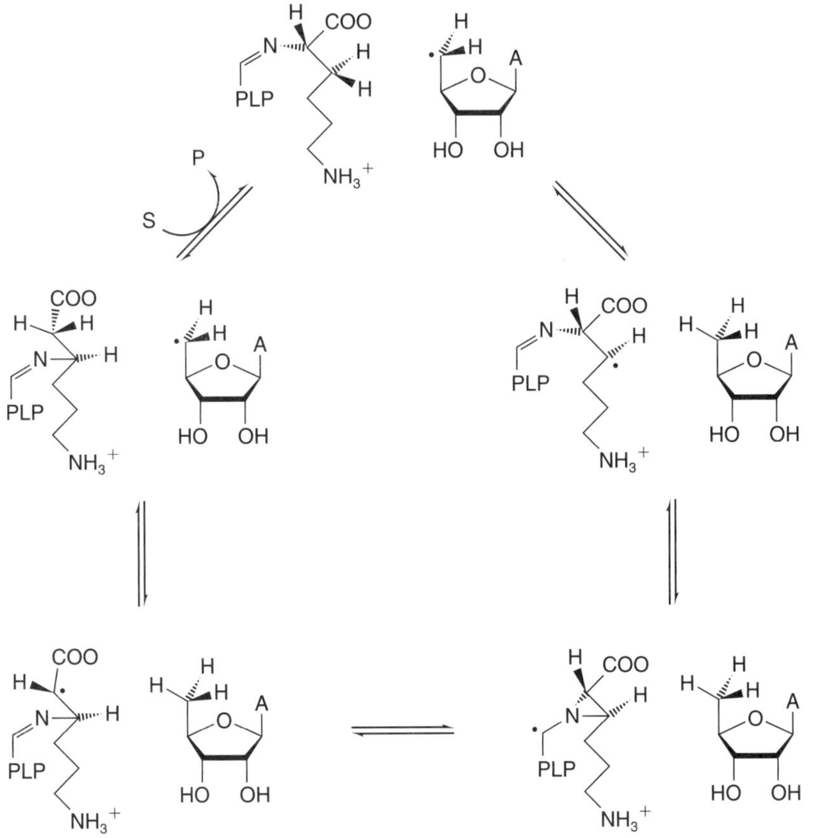

Fig. XIII.4.6.
Proposed reaction scheme for lysine 2,3-aminomutase involving adenosyl, substrate, and product-centered radicals.

allowed observation of the stabilized 3′,4′-anhydroadenosyl radical intermediate by EPR, thereby providing the most convincing evidence to date for an intermediate adenosyl radical in the Radical-SAM enzymes.[9] The $[4Fe–4S]^{+}$ form of lysine 2,3-aminomutase has been shown to be the catalytically active form, and SAM has been shown to interact with this cluster in much the same way as depicted for pyruvate formate lyase activating enzyme (Fig. VIII.4.2).[10] Therefore, although SAM is utilized as a substrate in the activating enzymes and a cofactor in lysine aminomutase, and although these enzymes catalyze very different reactions, they seem to do so by using a common mechanism, in which a reduced [4Fe–4S] cluster anchors SAM using the unique Fe site of the cluster, and then transfers an electron to initiate C–S bond cleavage.

XIII.4.4. Cofactor Biosynthesis

XIII.4.4.1. Biotin Synthase

Biotin synthase (BioB) catalyzes the insertion of an atom of sulfur into dethiobiotin to generate biotin, as the final step in the biotin biosynthetic pathway (Table XIII.4.1). Initial mechanistic studies are suggestive of a radical mechanism initiated by two sequential hydrogen atom abstractions, much like the reactions discussed previously in this section.[1,11] Catalytic activity has been shown minimally to require SAM, reduced nicotinamide adenine dinucleotide phosphate (NADPH), and an electron-transport system, much like the requirements for the activases. Full *in vitro* activity, however, has not yet been obtained for purified biotin synthase, with only a small number of turnovers being observed, and this has limited the mechanistic information available on this enzyme.

As purified under aerobic conditions, biotin synthase contains one $[2Fe–2S]^{2+}$ cluster per monomer, and two of these are converted stoichiometrically to $[4Fe–4S]^{2+/+}$ under anaerobic reducing conditions. Recently, it has been demonstrated that under appropriate conditions, biotin synthase can be reconstituted with one [4Fe–4S] and one [2Fe–2S] cluster per monomer. Not only does this form of the enzyme exhibit higher activity than other forms, but it has also been shown that the [2Fe–2S] cluster is destroyed concomitant with sulfur insertion.[12] These results led to the proposal that biotin synthase utilizes the [4Fe–4S] cluster to reductively cleave SAM to generate the 5′-deoxyadenosyl radical, and utilizes the [2Fe–2S] cluster as a sulfur donor. The [2Fe–2S] cluster would thus act not as a catalytic cluster, but rather as a substrate, and this cluster would need to be regenerated prior to the next turnover, which might explain the low activity observed for biotin synthase. However, other studies have suggested that biotin synthase does not require a [2Fe–2S] cluster for sulfur insertion, but rather that it binds pyridoxal phosphate and exhibits cysteine desulfurase activity, which provides the sulfide required for sulfur insertion into dethiobiotin.[13]

Biotin synthase and HemN (see below) are the first two members of the Radical-SAM family to be structurally characterized. The biotin synthase structure, solved to 3.4-Å resolution on protein reconstituted to contain one [4Fe–4S] and one [2Fe–2S] cluster per monomer, exhibits a triosephosphate isomerase (TIM) type $(\alpha/\beta)_8$ barrel-fold for each monomer of the dimeric structure (Fig. XIII.4.7).[4] A [4Fe–4S] cluster is bound at the C-terminal end of the TIM barrel, and a [2Fe–2S] cluster is buried in the barrel. The protein was crystallized in the presence of both SAM and dethiobiotin, and both are found bound in the active site between the two iron–sulfur clusters. S-Adenosylmethionine is coordinated via the amino and carboxylato groups to an iron of the [4Fe–4S] cluster, similar to the structure determined for the SAM–cluster

(a)

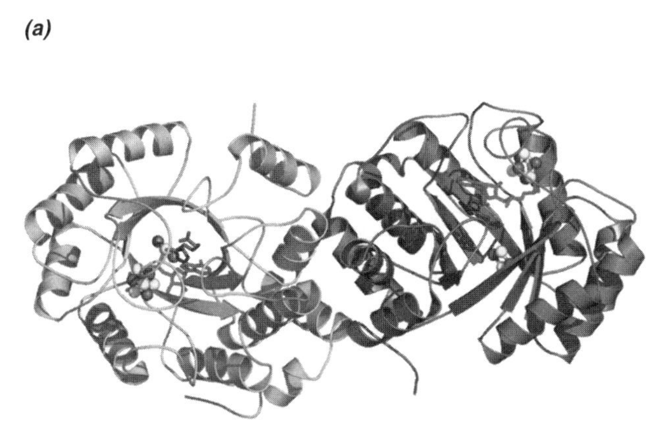

(b)

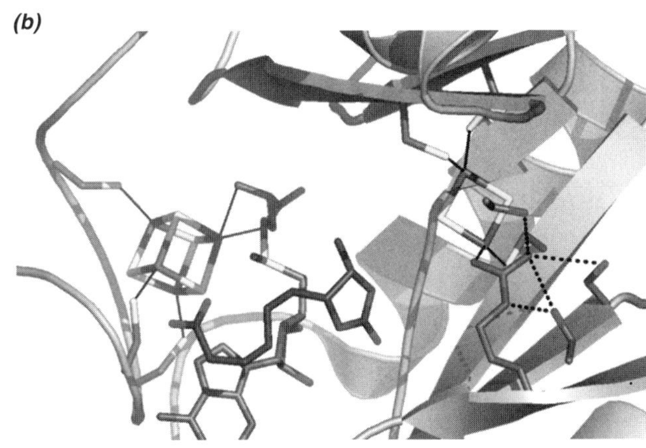

Fig. XIII.4.7.
(*a*) Overall X-ray crystal structure of the biotin synthase dimer, showing the [4Fe–4S] and [2Fe–2S] clusters, SAM, and dethiobiotin. (*b*) The active site of biotin synthase, showing SAM coordinated to an iron of the [4Fe–4S] cluster, dethiobiotin bound between SAM and the [2Fe–2S] cluster, and the unusual arginine ligand to the [2Fe–2S] cluster (PDB code: 1R30). [See color insert.] (Figures courtesy of F. Berkovitch and C. Drennan.)

interaction in the pyruvate formate-lyase activating enzyme. The sulfonium sulfur is ~4 Å away from the unique iron of the [4Fe–4S] cluster, and SAM is bound in an extended conformation, buried from the solvent and apparently well situated for reductive cleavage by the [4Fe–4S] cluster to generate the intermediate adenosyl radical. Dethiobiotin sits between the bound SAM and the [2Fe–2S] cluster, poised for hydrogen atom abstraction by the adenosyl radical intermediate and sulfur insertion from the [2Fe–2S] cluster. The relative positions of the two clusters, SAM, and dethiobiotin strongly support the proposed involvement of a [2Fe–2S] cluster as the sulfur donor in biotin biosynthesis. Another interesting aspect of the structure is the observation that the [2Fe–2S] cluster is coordinated by three Cys and one Arg residue; this structure represents the first biological example of Arg serving as a metal ligand.

XIII.4.4.1. Lipoic Acid Synthase

The biosynthesis of lipoic acid from octanoic acid, like the synthesis of biotin from dethiobiotin, involves the insertion of sulfur atoms into unactivated C–H bonds (Table XIII.4.1). The sulfur insertion reaction of lipoic acid biosynthesis utilizes octanoyl-acyl carrier protein as a substrate and is catalyzed by LipA, a dimeric enzyme with some amino acid sequence homology to BioB.[1] LipA contains a labile iron–sulfur cluster and utilizes SAM to catalyze two sequential hydrogen atom abstraction reactions, from C8 and C6 of octanoic acid, respectively. The source of the sulfur for lipoic acid biosynthesis, as well as more details about the structure and mechanism of LipA, have yet to be firmly established.

XIII.4.4.2. Coproporphyrinogen III Oxidase

A key step in heme and chlorophyll biosynthesis involves the oxidative decarboxylation of the propionate side chains of rings A and B of coproporphyrinogen III to produce protoporphyrinogen IX (Table XIII.4.1). This reaction is catalyzed by two unrelated enzymes, the oxygen-dependent (HemF) and the oxygen-independent (HemN) coproporphyrinogen III oxidases. Although few biophysical studies have

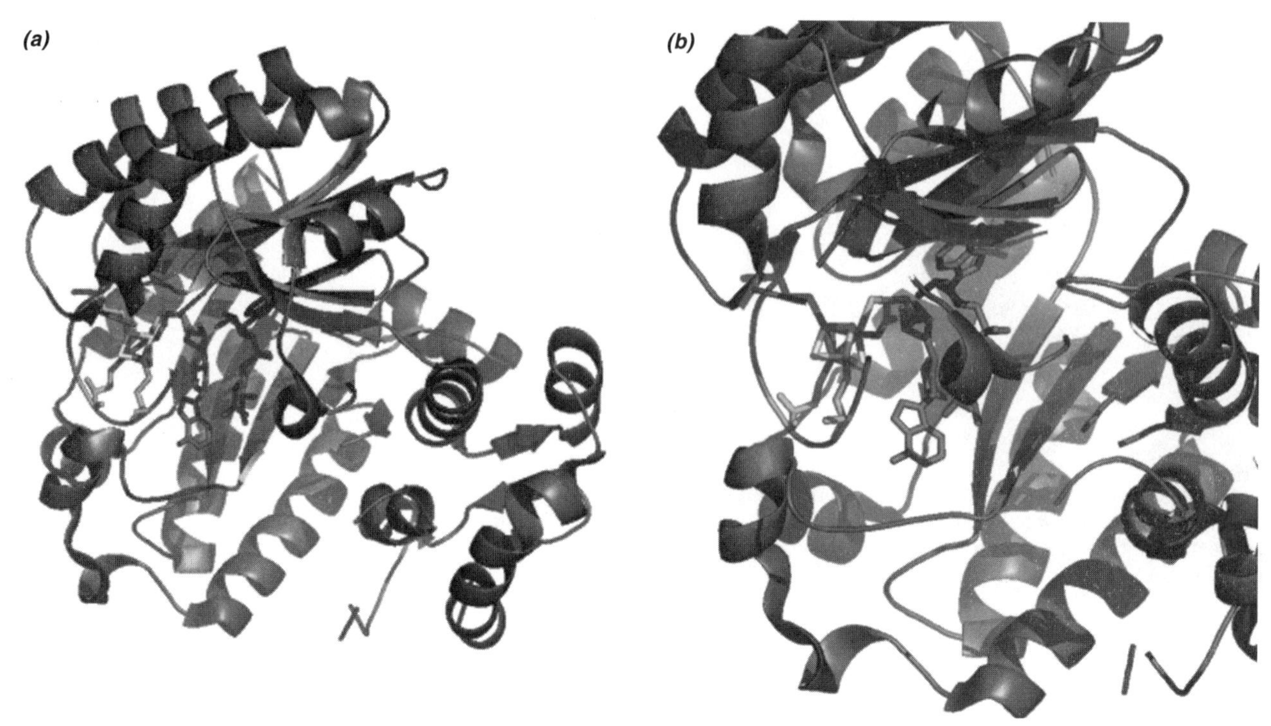

Fig. XIII.4.8.
The overall structure of HemN (*a*) and a close-up view of the SAM-[4Fe–4S] interaction (*b*). (PDB code: 1OLT). [See color insert.]

been reported for HemN, it was identified as a member of the Radical-SAM superfamily based on its amino acid sequence, including the characteristic CXXXCXXC motif. The decarboxylation reaction catalyzed by HemN is thought to proceed via hydrogen atom abstraction from the proprionate groups, as shown in Table XIII.4.1.

A crystal structure of HemN at 2.07-Å resolution reveals a $(\beta/\alpha)_6$ repeat that somewhat resembles the $(\beta/\alpha)_8$ TIM barrel domain seen in biotin synthase (Fig. XIII.4.8).[5] A [4Fe–4S] cluster is bound deep within the barrel, and is coordinated by the three cysteines of the CXXXCXXC Radical-SAM motif. Two SAMs are bound, one chelating the unique iron site of the [4Fe–4S] cluster through the carboxylate and amino groups, and another ~5 Å further away. Intricate interactions between the second SAM and several conserved residues suggest that the second SAM is functionally relevant rather than adventitious. However, further mechanistic studies are required to establish a possible catalytic role for the second SAM.

XIII.4.5. DNA Repair

Bacterial spores, when subjected to UV irradiation, accumulate a unique type of DNA damage known as spore photoproduct (SP). This SP, or 5-thyminyl-5,6-dihydrothymine, is repaired by a specific enzyme, SP lyase, during spore germination (Table XIII.4.1). The reaction catalyzed by SP lyase is reminiscent of the monomerization of thymine dimers catalyzed by DNA photolyase. Unlike DNA photolyase, however, SP lyase requires neither flavin cofactors nor visible light to repair DNA. The SP lyase was predicted to be a member of the Radical-SAM superfamily, and this has been supported by the identification of an iron–sulfur cluster in the purified

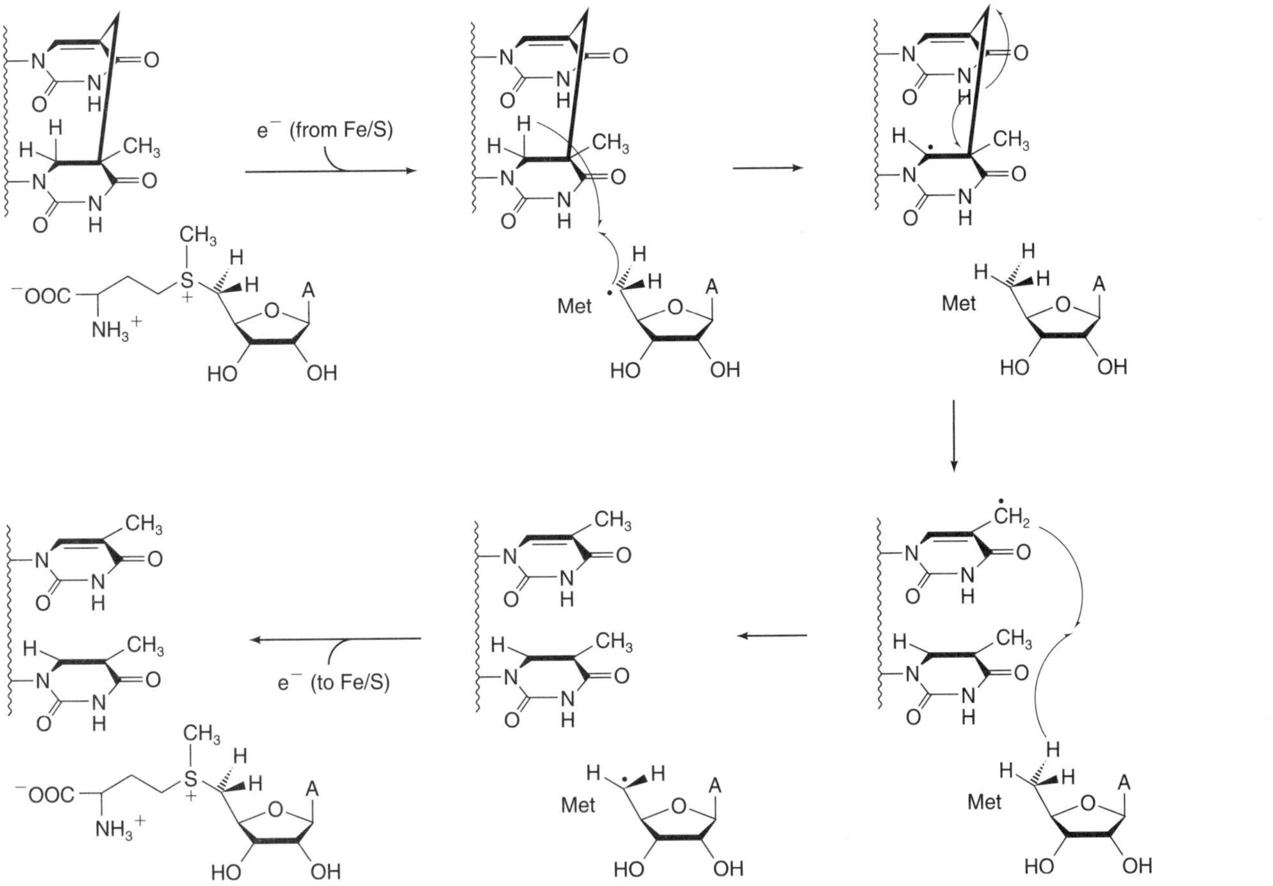

Fig. XIII.4.9.
Proposed mechanism for SP lyase.

enzyme and the dependence of enzyme activity on SAM.[1] Like lysine aminomutase, SP lyase uses SAM as a catalytic cofactor to reversibly generate an intermediate adenosyl radical. Label transfer studies have demonstrated that the repair reaction is initiated by C6 hydrogen atom abstraction from the damaged DNA, as shown in Fig. XIII.4.9.

XIII.4.6. Radical-SAM Enzymes: Unifying Themes

The role of iron–sulfur clusters in biological radical generation, initially a surprising discovery, is now well documented in a wide variety of enzyme systems, and bioinformatic studies have led to the identification of >500 putative members of the Radical-SAM superfamily, most of which have not yet been characterized.[2] Although the functions of the Radical-SAM enzymes are diverse, a few unifying themes are apparent that should guide future mechanistic studies.

First, as indicated in the previous sections, all of these enzymes require SAM, either as a cosubstrate or cofactor in the reaction. In reactions that have been studied in some detail, such as the pyruvate formate-lyase activating enzyme (PFL–AE) and lysine 2,3-aminomutase (LAM), evidence suggests that the SAM serves to generate an intermediate 5′-deoxyadenosyl radical that directly abstracts a hydrogen atom from substrate to initiate the reaction. Therefore SAM serves essentially the same role as AdoCbl in enzyme-catalyzed radical reactions: Both are innocuous precursors to the extremely reactive 5′-deoxyadenosyl radical.

Second, the Radical-SAM enzymes require a reduced $[4Fe–4S]^+$ cluster to generate the 5′-deoxyadenosyl radical intermediate by reductive cleavage of SAM. The [4Fe–4S] clusters are coordinated by a conserved 3-Cys motif, and therefore a site-differentiated cluster appears to be an obligate feature of the superfamily. The unique Fe site of the cluster plays a structural role by coordinating the amino and carboxylate groups of SAM in the enzyme–SAM complexes, thereby anchoring SAM in position to react with the reduced cluster to generate the 5′-deoxyadenosyl radical intermediate, believed to be common to all the Radical-SAM reactions. The 5′-deoxyadenosyl radical goes on to abstract a hydrogen atom from substrate, at which point the mechanisms diverge as a wide variety of radical transformations are initiated.

Third, several of these enzymes exhibit flexibility to coordinate a variety of iron–sulfur clusters, including [2Fe–2S], [3Fe–4S], and [4Fe–4S] clusters. In cases where [2Fe–2S] or [3Fe–4S] clusters are found, reduction results in facile conversion to the catalytically relevant [4Fe–4S] state. In addition, the [4Fe–4S] clusters of most of these enzymes appear to be quite oxygen sensitive, degrading to [3Fe–4S] and [2Fe–2S] clusters, even upon brief exposures to oxygen. Some of the Radical-SAM enzymes, such as the activating enzymes, function only under anaerobic conditions *in vivo,* and it is conceivable that the cluster sensitivity to oxygen might serve as a switch to turn the enzyme activity on or off under appropriate conditions, although this has yet to be demonstrated. However, most of the Radical-SAM enzymes studied to date, including biotin synthase, lipoate synthase, and the lysine aminomutase from *B. subtilis,* are not strictly anaerobic enzymes. Of these three, only the lysine 2,3-aminomutase from *B. subtilis* is stable to oxygen *in vitro.* How the other aerobic Radical-SAM enzymes stabilize their clusters *in vivo,* when they seem to be so sensitive *in vitro,* is an intriguing question.

Despite the differences in the details of chemical reactions catalyzed, forms of the iron–sulfur cluster observed, and mode of utilization of SAM, the accumulating evidence suggests that all of the Radical-SAM enzymes will exhibit a common mechanism of radical generation involving a SAM-derived adenosyl radical intermediate. These iron–sulfur/SAM enzymes therefore represent a new general paradigm for radical generation in biological systems, joining the better understood AdoCbl and binuclear Fe center/O_2/Tyr systems.

References

General References

1. For helpful reviews of the Radical-SAM superfamily, see the following: (a) Cheek, J. and Broderick, J. B., "Adenosylmethionine-Dependent Iron–Sulfur Enzymes: Versatile Clusters in a Radical New Role", *J. Biol. Inorg. Chem.*, **6,** 209–226 (2001). (b) Fontecave, M., Mulliez, E., and Ollagnier-de-Choudens, S., "Adenosylmethionine as a Source of 5′-Deoxyadenosyl Radicals", *Curr. Opin. Chem. Biol.*, **5,** 506–512 (2001). (c) Jarrett, J. T., "The generation of 5′-deoxyadenosyl radicals by adenosylmethionine-dependent radical enzymes", *Curr. Opin. Chem. Biol.*, **7,** 174–182 (2003). (d) Frey, P. A. and Magnusson, O. T., "S-Adenosylmethionine: A wolf in sheep's clothing, or a rich man's adenosylmethionine?", *Chem. Rev.*, **103,** 2129–2148 (2003). (e) Broderick, J. B., "Iron–Sulfur Clusters in Enzyme Catalysis", in *Comprehensive Coordination Chemistry II: From Biology to Nanotechnology,* Vol. 8, McCleverty, J. and Meyer, T. J., Eds., (Que, L. and Tolman, W. Eds.), Elsevier, London, 2003, pp. 739–757.
2. Sofia, H. J., Chen, G., Hetzler, B. G., Reyes-Spindola, J. F., and Miller, N. E., "Radical SAM, a novel protein superfamily linking unresolved steps in familiar biosynthetic pathways with radical mechanisms: functional characterization using new analysis and information visualization methods", *Nucleic Acids Res.*, **29,** 1097–1106 (2001).
3. For reviews on spectroscopic studies of Radical-SAM enzymes, see (a) Broderick, J. B., Walsby, C., Broderick, W. E., Krebs, C., Hong, W., Ortillo, D., Cheek, J., Huynh, B. H., and Hoffman, B. M., "Paramagnetic Resonance in Mechanistic Studies of Fe-S/Radical Enzymes", *ACS Symp. Ser.*, **858,** 113–127 (2003). (b) Hoffman, B. M., "ENDOR of Metalloenzymes", *Acc. Chem. Res.*, **36,** 522–529 (2003).

Specific References

4. Berkovitch, F., Nicolet, Y., Wan, J. T., Jarrett, J. T., and Drennan, C. L., "Crystal structure of biotin synthase, an S-adenosylmethionine-dependent radical enzyme", *Science*, **303,** 76–79 (2004).
5. Layer, G., Moser, J., Heinz, D. W., Jahn, D., and Schubert, W.-D., "Crystal structure of coproporphyrinogen III oxidase reveals cofactor geometry of Radical SAM enzymes", *EMBO J.*, **22,** 6214–6224 (2003).
6. Knappe, J., Neugebauer, F. A., Blaschkowski, H. P., and Gänzler, M., "Post-translational activation introduces a free radical into pyruvate formate-lyase", *Proc. Natl. Acad. Sci. U.S.A.*, **81,** 1332–1335 (1984).
7. Frey, M., Rothe, M., Wagner, A. F. V., and Knappe, J., "Adenosylmethionine-dependent synthesis of the glycyl radical in pyruvate formate-lyase by abstraction of the glycine C-2 pro-S hydrogen atom", *J. Biol. Chem.*, **269,** 12432–12437 (1994).
8. Mulliez, E., Padovani, D., Atta, M., Alcouffe, C., and Fontecave, M., "Activation of class III ribonucleotide reductase by flavodoxin: A protein radical-driven electron transfer to the iron–sulfur center", *Biochemistry*, **40,** 3730–3736 (2001).
9. Magnusson, O. T., Reed, G. H., and Frey, P. A., "Spectroscopic evidence for the participation of an allylic analogue of the 5′-deoxyadenosyl radical in the reaction of lysine 2,3-aminomutase", *J. Am. Chem. Soc.*, **121,** 9764–9765 (1999).
10. Chen, D., Walsby, C., Hoffman, B. M., and Frey, P. A., "Coordination and mechanism of reversible cleavage of S-adenosylmethionine by the [4Fe–4S] center in lysine 2,3-aminomutase", *J. Am. Chem. Soc.*, **125,** 11788–11789 (2003).
11. For reviews related to biotin synthase and lipoate synthase, see (a) Marquet, A., "Enzymology of carbon-sulfur bond formation", *Curr. Opin. Chem. Biol.*, **5,** 541–549 (2001). (b) Fontecave, M., Ollagnier-De-Choudens, S., and Mulliez, E., "Biological radical sulfur insertion reactions", *Chem. Rev.*, **103,** 2149–2166 (2003).
12. (a) Ugulava, N. B., Sacanell, C. J., and Jarrett, J. T., "Evidence from Mössbauer spectroscopy for distinct $[2Fe–2S]^{2+}$ and $[4Fe–4S]^{2+}$ cluster binding sites in biotin synthase from *Escherichia coli*", *Biochemistry*, **40,** 8352–8358 (2002). (b) Bui, B. T. S., Benda, R., Schünemann, V., Florentin, D., Trautwein, A. X., and Marquet, A., "Fate of the $(2Fe–2S)^{2+}$ cluster of *Escherichia coli* biotin synthase during reaction: A Mössbauer characterization", **42,** 8791–8798 (2003).
13. Ollagnier-de-Choudens, S., Mulliez, E., Hewitson, K. S., and Fontecave, M., "Biotin synthase is a pyridoxal phosphate-dependent cysteine desulfurase", *Biochemistry*, **41,** 9145–9152 (2002).

XIII.5. Galactose Oxidase

James A. Whittaker

Environmental and Biomolecular Systems
Oregon Health and Science University
Beaverton, OR 97006

Contents

XIII.5.1. Introduction

Galactose oxidase contains one of the most remarkable radicals in biology, a protein side chain radical that is also a metal ligand, forming a free radical coupled Cu complex at the catalytic active site.[1,2] Coupling between the free radical and the metal ion creates a two-electron redox site, capable of efficiently oxidizing organic substrates (primary alcohols) and reducing dioxygen (O_2). The strong coupling of electronic structures within the radical complex is achieved by the sharing of electrons between the free radical ligand and the metal ion, a feature that has recently been successfully applied in synthetic modeling and catalyst design.[3–5] Galactose oxidase provides an excellent example of how novel structures and catalytic principles uncovered in bioinorganic studies are inspiring new directions in chemistry.

Galactose oxidase catalyzes the oxidation of primary alcohols to the corresponding aldehydes:

$$RCH_2OH + O_2 \rightarrow RCHO + H_2O_2 \quad (1)$$

effectively transferring the elements of H_2 from the organic substrate to dioxygen. This overall reaction represents two half-reactions, involving two-electron oxidation (or reduction) of the substrates:

$$\text{ox} \quad RCH_2OH \rightarrow RCHO + 2\,e^- + 2\,H^+ \quad (2)$$

$$\text{red} \quad O_2 + 2\,e^- + 2\,H^+ \rightarrow H_2O_2 \quad (3)$$

While reaction 1 (alcohol oxidation) has received the most attention by biochemists, the reduction of O_2 to H_2O_2 (reaction 3) is physiologically more important, as the enzyme serves as an extracellular peroxide factory for fungi. In spite of the two-electron character of both half-reactions, the enzyme surprisingly contains a single Cu center in its structure, which would seem to be poorly suited to performing two-electron redox catalysis. The second electron is accounted for by a radical redox site in the protein, which participates directly in the redox chemistry.

XIII.5.2. Active Site Structure

The X-ray crystal structure of galactose oxidase (PDB code: 1GOG) (Fig. XIII.5.1)[6,7] shows that the Cu is bound in roughly square-pyramidal coordination with two His and two Tyr ligands. One of the tyrosines (Tyr495) occupies a pseudo-axial position, while the other (Tyr272) is coordinated in the plane of the histidines and solvent. Closer inspection reveals a unique feature of this second Tyr, a covalent cross-link between an ortho carbon of the Tyr272 phenolic ring and the sulfur atom

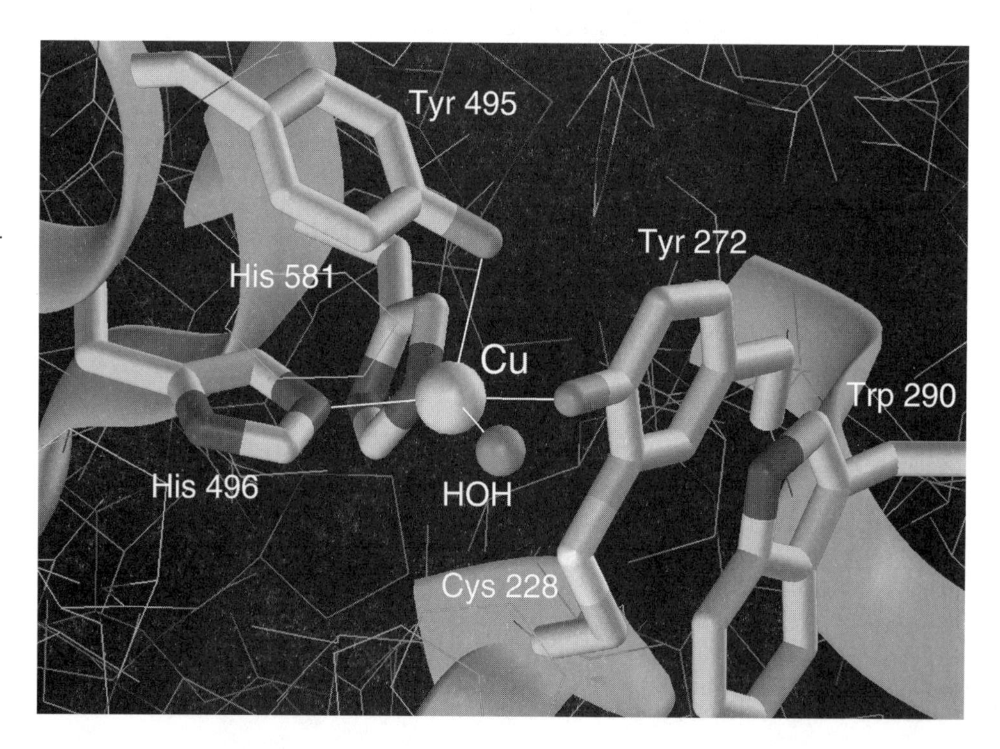

Fig. XIII.5.1. The active site of galactose oxidase. The amino acid side chains coordinated to the active site copper ion together with the stacked tryptophan residue (Trp 290) are represented in stick form. The polypeptide backbone is represented as a ribbon. Based on PDB code: 1GOG and rendered using InsightII.

of a Cys residue (Cys228), forming a cysteinyl-tyrosine dimer in the active site. The indole side chain of a tryptophan residue (Trp290) protects the cysteinyl-tyrosine dimer, lying over it like a shield.

XIII.5.3. Oxidation–Reduction Chemistry

The cysteinyl-tyrosine group is redox active, and the radical formed by oxidation of the metal-free apoenzyme can be detected by EPR spectroscopy. In the copper-containing holoenzyme, the radical is antiferromagnetically coupled to Cu(II), resulting in an EPR silent, singlet ($S = 0$) ground state for the complex. Calculations of the valence electronic structure of a simple model of the cysteinyl-tyrosyl radical (Fig. XIII.5.2) show that the SOMO, or redox orbital, includes significant unpaired electron density on the phenoxyl oxygen and is extensively delocalized over the aromatic side chain, with a large contribution from the ortho sulfur substituent. These features may contribute to the stabilization of the radical in the protein.

The Cu site is also redox active, being converted to Cu(I) by mild reductants. The EXAFS measurements show that the reduced Cu is two- or three-coordinate in the protein, and is most likely bound by the equatorial set of protein ligands (His496, His581, and Tyr272). The combination of two redox centers (one protein and one metal center) makes three distinct oxidation levels accessible in the enzyme (Fig. XIII.5.3). The fully oxidized enzyme (Fig. XIII.5.3, oxidation level 1) contains a Cu(II) metal center ligated by a cysteinyl-tyrosyl radical (YC·). One-electron reduction of this complex eliminates the radical, leaving a simple Cu(II) center (Fig. XIII.5.3, oxidation level 2), which may be further reduced by one electron to a Cu(I) complex (Fig. XIII.5.3, oxidation level 3). The fully oxidized and fully reduced complexes are catalytically active and are interconverted by two-electron redox steps for substrate oxidation (Eq. 2) and O_2 reduction (Eq. 3), while the one-electron reduced complex is catalytically inactive.

Although the description of the Cu-binding site as square pyramidal is geometrically accurate, the Cu–OH_2 bond is actually the longest [corresponding to the weakest ligand interaction for the Cu(II) ion]. The Cu center binds small anions, which replace the coordinated solvent in the equatorial plane of the complex, displacing the pseudo-axial Tyr (Tyr495) from the metal ion. The anion complexes (e.g., acetate, PDB code: 1GOF) are roughly planar, with relatively strong metal–anion

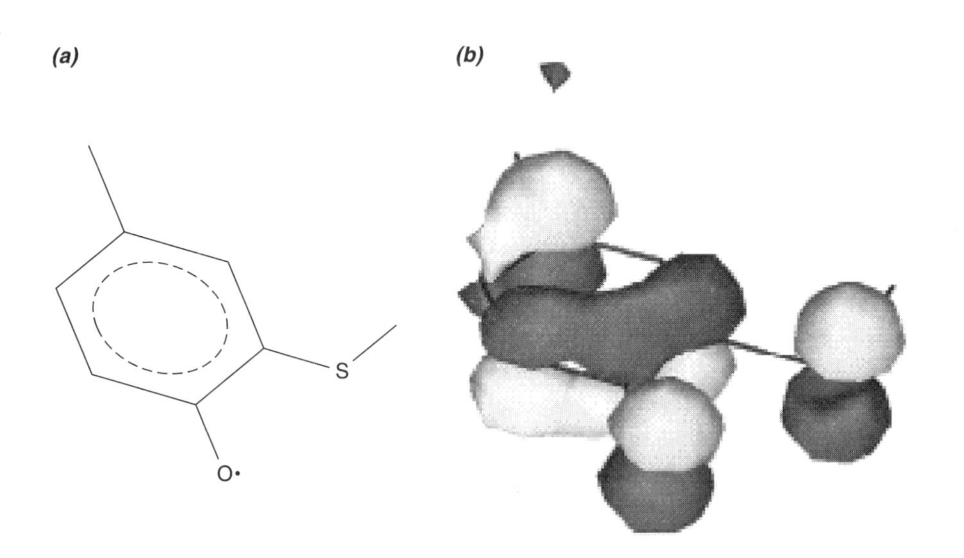

Fig. XIII.5.2. Cysteinyl-tyrosine free radical. (*a*) Molecular structure showing the ortho thioether substitution in the phenoxyl free radical. (*b*) Electronic structure calculated for 4-methyl-2-methythio phenoxyl free radical, a model for cysteinyl-tyrosine. The isosurface of the SOMO is shown contoured at $0.5\,e\,Å^{-3}$. Note the delocalization over the oxygen and sulfur atoms as well as the aromatic π-system.

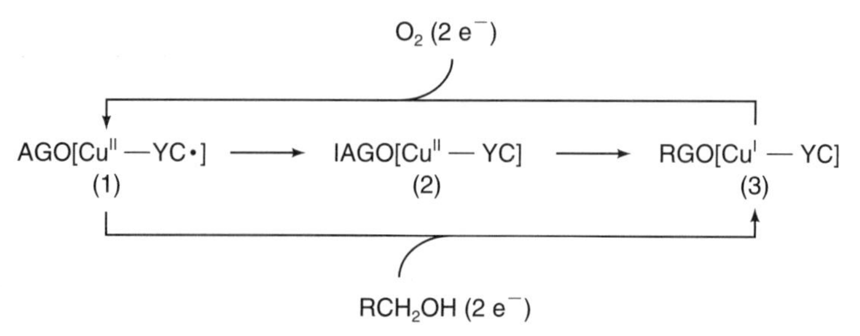

Fig. XIII.5.3.
Oxidation–reduction reactions of the galactose oxidase active site. Catalytic turnover involves two-electron transformations between fully oxidized (metalloradical, 1) and fully reduced [Cu(I) nonradical, 3] complexes. The one-electron reduced form (2) is catalytically inactive. YC = cysteinyl-tyrosine cofactor.

interactions replacing the weak aqua interaction of the resting enzyme. The displaced tyrosinate ligand becomes protonated in these complexes, coupling protonation with ligand binding. Proton transfer from the coordinated solvent to the pseudo-axial Tyr can also be accomplished by lowering the temperature, which stabilizes a hydroxide complex. These transformations of the active site define a proton-transfer path between coordinated hydroxyl groups and the Tyr495 phenolate, allowing the axial Tyr to serve as a general base in catalysis.

XIII.5.4. Catalytic Turnover Mechanism

The catalytic mechanism of galactose oxidase is outlined in Fig. XIII.5.4. Primary alcohols binding to Cu through the hydroxyl group will be deprotonated by tyrosine Tyr495 in an acid–base reaction that leaves the substrate coordinated as an alkoxide anion. This anion is much easier to oxidize than the starting alcohol, so the proton transfer activates the substrate for subsequent redox steps. Isotopic labeling experiments have shown that the enzyme stereospecifically removes the *pro-S* hydrogen of the substrate during turnover, implying that the hydroxymethylene group is oriented as shown in Fig. XIII.5.4. The C–H bond cleavage of this type is most favorable energetically when a group capable of forming a new bond to hydrogen is present. In galactose oxidase, the cysteinyl-tyrosine phenoxyl radical serves this role as a hydrogen atom acceptor site. Coupled proton and electron transfer to the phenoxyl radical will convert it to a phenol, containing a newly formed O–H bond that compensates for the loss of the C–H bond in the substrate. The other electron removed from the substrate is delivered to the Cu(II) center, reducing it to Cu(I). The overall reaction involves transfer of two electrons and two protons (the equivalent of dihydrogen) from the alcohol substrate to the enzyme active site, resolved into proton transfer, electron transfer (ET), and hydrogen atom transfer steps (Fig. XIII.5.5).

As often occurs in a detailed analysis of an enzyme reaction mechanism, the precise ordering of these steps is still unclear. Two possible schemes for ET and hydrogen-transfer steps in substrate oxidation by galactose oxidase are shown in Section XIII.5.5. If the coordinated alkoxide is sufficiently activated for C–H bond cleavage, the hydrogen atom abstraction may occur first, producing a ketyl radical that subsequently reduces Cu in an inner-sphere ET reaction (Fig. XIII.5.5*a*). Alternatively, a redox equilibrium may be established in the alkoxide complex, forming an alkoxyl radical–Cu^{I} intermediate that undergoes hydrogen transfer to the active site phenoxyl (Fig. XIII.5.5*b*). Deuterium kinetic isotope measurements confirm that C–H bond cleavage is fully rate limiting for substrate oxidation, but leave the ordering of the redox events somewhat ambiguous.

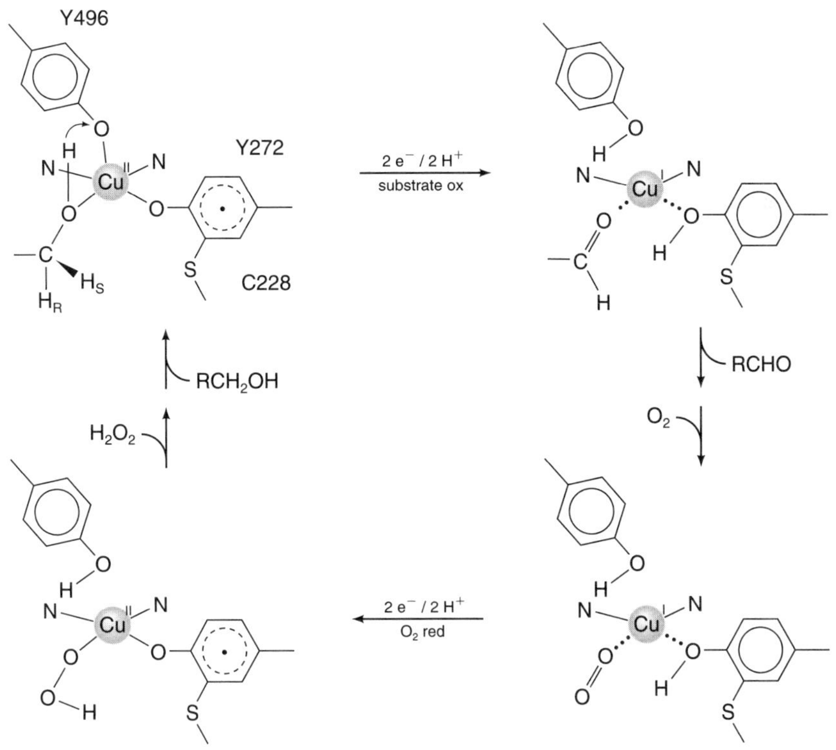

Fig. XIII.5.4.
Galactose oxidase catalytic cycle. The metalloradical complex in the active form of the enzyme activates bound substrate for two-electron oxidation, forming the corresponding aldehyde. In a second stage of the reaction, the reduced, Cu(I)-containing enzyme is reoxidized by O_2, forming hydrogen peroxide product.

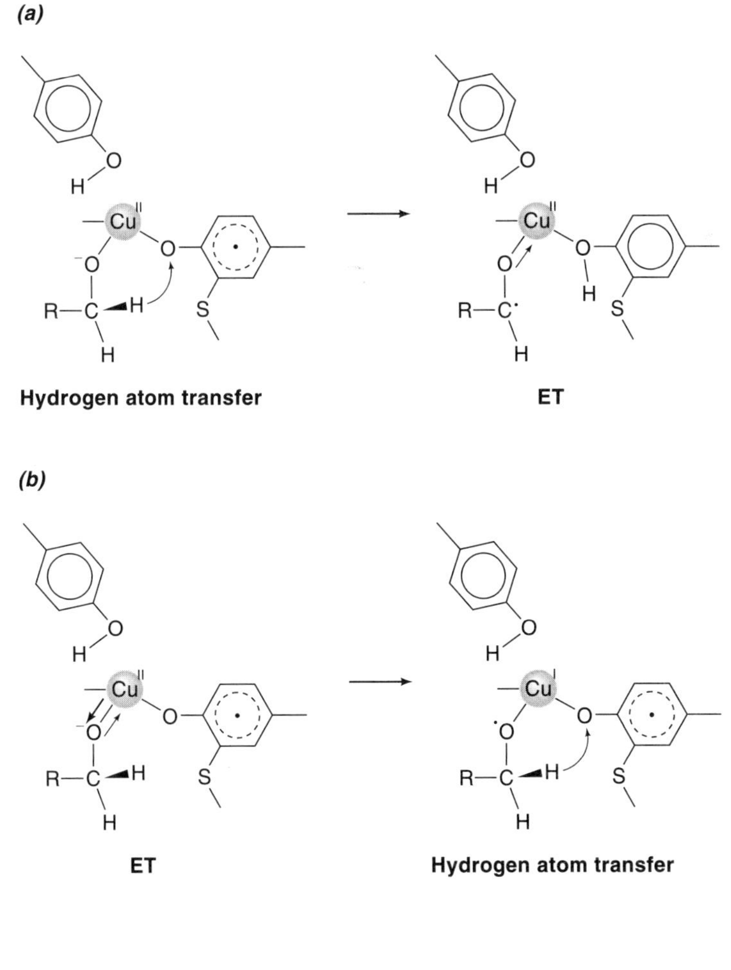

Fig. XIII.5.5.
Substrate oxidation mechanisms. Two possible alternative mechanisms for substrate oxidation are shown. (*a*) Initial hydrogen atom transfer forms a ketyl free radical that undergoes inner-sphere electron transfer to reduce the metal ion. (*b*) An initial, unfavorable inner-sphere oxidation of the coordinated alcohol forms an alkoxyl free radical species that undergoes atom transfer to form the aldehyde product. (See Tutorial I for one letter residue abbreviations.)

These redox events complete only one of the half reactions (Eq. 1) required for turnover. Following dissociation of the aldehyde product, the reduced enzyme complex reacts extremely rapidly with O_2 in what is likely to be a reversal of the steps involved in alcohol oxidation. A weak Cu(I)–O_2 complex could thus be reduced by inner-sphere ET from the Cu(I) to form a superoxide adduct that would abstract the hydrogen atom from the Tyr272 phenol to produce hydroperoxide bound to the Cu^{II}–radical complex. Transfer of the Tyr495 proton to the coordinated oxygen of the hydroperoxide would assist in displacement of the H_2O_2 product as Tyr495 rebinds at the end of the turnover cycle (Eq. 3).

Galactose oxidase illustrates how ligand reactivity can expand the chemistry of a simple metal complex, a principle that has been applied in the successful synthesis of functional models for the enzyme active site. These models incorporate a redox-active ligand as an essential element of their structure, allowing a mononuclear Cu complex to serve as an efficient alcohol oxidation catalyst.

XIII.5.5. Mechanism of Cofactor Biogenesis

Galactose oxidase requires a specific post-translational modification, the covalent cross-link between Tyr and Cys side chains, for enzymatic activity. The protein is expressed as a precursor that can spontaneously form the Tyr-Cys redox cofactor in the presence of Cu and O_2, in the absence of any other proteins. This self-processing reaction demonstrates that the active site has an additional essential function beyond catalytic turnover, directing the biogenesis of the redox cofactor.

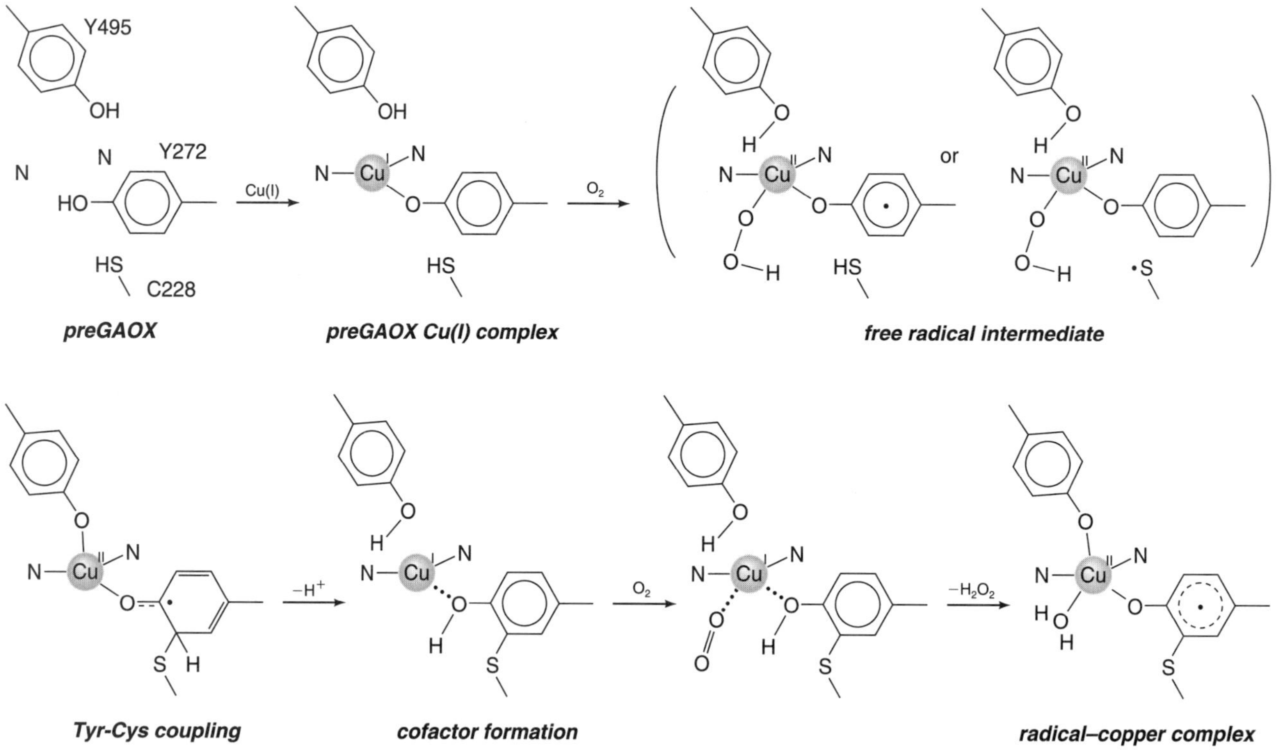

Y495 OH N N Y272 HO HS C228 — Cu(I) →

preGAOX

preGAOX Cu(I) complex

free radical intermediate

Tyr-Cys coupling

cofactor formation

radical–copper complex

Fig. XIII.5.6.
Mechanism for galactose oxidase cofactor biogenesis. The formation of the covalent cross-link between tyrosine and cysteine in the active site depends on Cu(I) and two molecules of O_2. The molecular mechanism for cross-linking involves the formation of either tyrosyl phenoxyl or cysteinyl thiyl free radicals. Four electrons are removed from the active site in converting the preGAOX Cu(I) complex to the mature metalloradical.

A simple electron-counting analysis of the overall transformation is a useful starting point for thinking about the biogenesis mechanism (Fig. XIII.5.6). This analysis predicts that the Cu(I) oxidation state for the metal ion will be most suitable to drive the reaction, since it yields an overall four-electron process for the conversion to the active enzyme, allowing more favorable two-electron steps for O_2 reduction. In contrast, starting with Cu(II) would require three-electron reduction of O_2, which is unfavorable. Experimentally, reaction of pregalactose oxidase with Cu(I) in the presence of O_2 results in rapid (seconds time scale) formation of fully oxidized, cofactor-containing active enzyme. The reaction with Cu(I) is nearly 4000× faster than the reaction with Cu(II) (hours time scale), which may actually need to be reduced to support cofactor biogenesis. Thiyl or phenoxyl free radical intermediates are predicted to be involved in the biogenesis reaction, for which a detailed mechanism has been proposed (Fig. XIII.5.6).[8]

References

General References

1. Whittaker, J. W., "Galactose oxidase", *Adv. Prot. Chem.*, **60,** 1–49 (2002).
2. Whittaker, J. W., "Free radical catalysis by galactose oxidase", *Chem. Rev.*, **103,** 2347–2363 (2003).

Specific References

3. Wang, Y., DuBoise, J. L., Hedman, B., Hodgeson, K. O., and Stack, T. D. P., "Catalytic Galactose Oxidase Models: Biomimetic Cu(II)-Phenoxyl-Radical Reactivity", *Science*, **279,** 537–540 (1998).
4. Itoh, S., Taki, M., and Fukuzumi, S., "Active site models for galactose oxidase and related enzymes", *Coord. Chem. Rev.*, **198,** 3–20 (2000).
5. Jazdzewski, B. A. and Tolman, W. B., "Understanding the copper-phenoxyl radical array in galactose oxidase: contribution from synthetic modeling studies", *Coord. Chem. Rev.*, **200–202,** 633–685 (2000).
6. Ito, N., Phillips, S. E. V., Stevens, C., Ogel, Z. B., McPherson, M. J., Keen, J. N., Yadav, K. D. S., and Knowles, P. F., "A novel thioether bond revealed by a 1.7 Å crystal structure of galactose oxidase", *Nature (London)*, **350,** 87–90 (1991).
7. Ito, N., Phillips, S. E. V., Yadav, K. D. S., and Knowles, P. F., "Crystal structure of a free radical enzyme, galactose oxidase", *J. Mol. Biol.*, **238,** 794–814 (1994).
8. Whittaker, M. and Whittaker, J. W., "Cu(I)-dependent biogenesis of the galactose oxidase redox cofactor", *J. Biol. Chem.*, **278,** 22090–22101 (2003).

XIII.6. Amine Oxidases

Contents

David M. Dooley
Department of Chemistry
and Biochemistry
Montana State University
Bozeman, MT 59717

XIII.6.1. Introduction

For most of the 70 years following their discovery, copper-containing amine oxidases remained an enigma for two reasons. First, although Cu was recognized as an essential cofactor, it proved extremely difficult to determine the mechanistic role of the spectroscopically detectable Cu(II) in the oxidation of primary amines catalyzed by

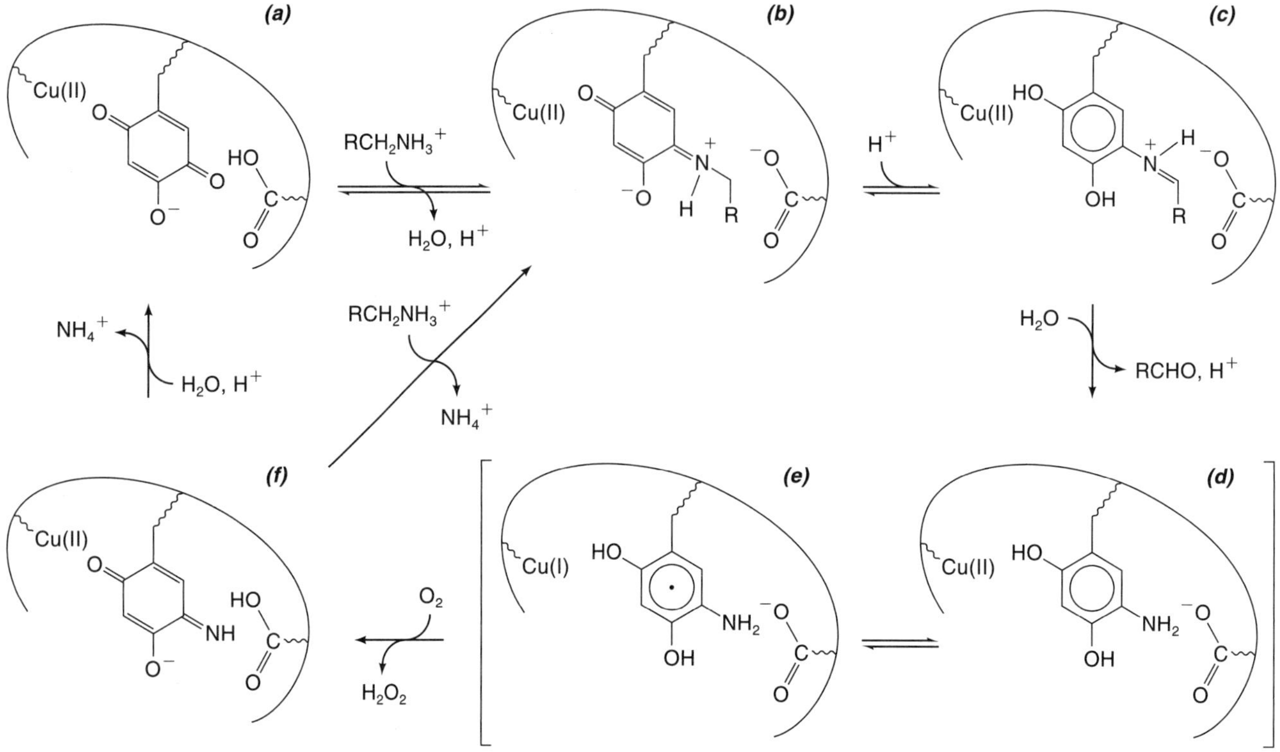

Scheme XIII.6.1.

amine oxidases (Eq. 1). Second, it also proved extremely difficult to identify conclusively the essential "organic cofactor", which had been recognized as an intrinsic component of copper-containing amine oxidases. Both of these difficulties were directly related to uncertainties concerning the molecular similarities of amine

$$RCH_2NH_2 + O_2 + H_2O \rightarrow RCHO + NH_3 + H_2O_2 \quad (1)$$

oxidases isolated from different sources.

Over the past several years,[1–5] the critical first steps have been taken toward achieving a comprehensive understanding of the structure and function of amine oxidases. Mechanistic studies of several representative amine oxidases have established that a common mechanism is operative, which is outlined in Scheme XIII.6.1.

Note the central role of the quinone group in substrate binding and oxidation. Essentially the quinone oxidizes the amine substrate by two electrons, and the reduced quinone is subsequently reoxidized by molecular oxygen. The quinone (2,4,5-trihydroxyphenylalanine quinone, TPQ) is derived from a Tyr residue in the active site, as discussed below.

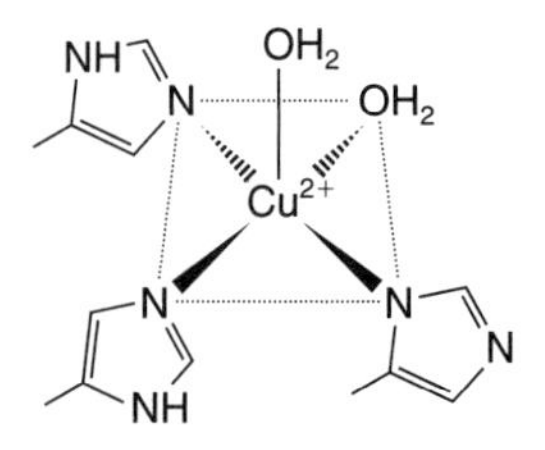

Fig. XIII.6.1.
Model for the Cu(II) site in amine oxidases (AO).

XIII.6.2. Structural Characterization

Extensive spectroscopic studies employing EPR, X-ray absorption spectroscopy (XAS), paramagnetic NMR, and circular dichroism/magnetic CD (CD/MCD) demonstrated that the structure of the Cu(II) site was highly conserved among amine oxidases, and led to the five-coordinate, approximately square-pyramidal model for the Cu(II) sites shown in Fig. XIII.6.1.[6–9]

Both equatorial and axial water ligands are included in the model; investigation of the ligand-substitution chemistry of the Cu(II) centers in amine oxidases was important in reaching the conclusion that two water molecules are coordinated. Because the displacement of water by exogenous ligands (e.g., entering groups like N_3^- or CN^-) invariably inhibited the enzymes,[10,11] a role for Cu in the oxidation of amines was implicated. In addition, reactions at TPQ influenced the Cu(II) coordination chemistry, thus suggesting that these two cofactors were in proximity.

All aspects of the model (Fig. XIII.6.1) were confirmed by the crystal structures of several amine oxidases,[12–15] summarized in Fig. XIII.6.2. The remarkable success of spectroscopic experiments, especially in conjunction with investigation of the coordination chemistry in accurately defining the geometry and ligands of metalloprotein metal ion sites provides confidence for the continued application of physical methods in metallobiochemistry.

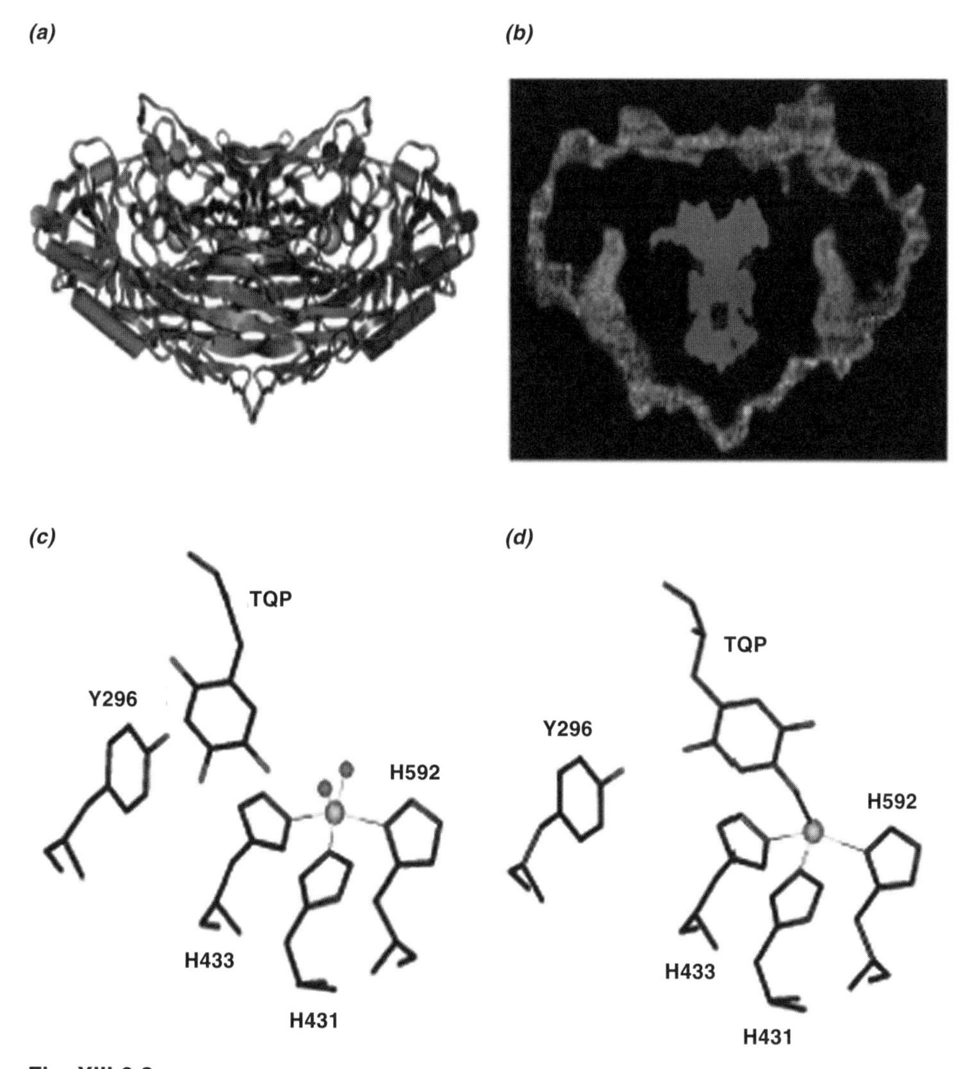

Fig. XIII.6.2.
(*a*) Overall AO structure. (*b*) Cavity–channel section of *Arthrobacter globiformis* amine oxidase (AGAO). (*c*) Active site of AGAO in the *active* conformation (PDB code: 1AV4). (*d*) Active site of AGAO in the *inactive* conformation (PDB code: 1AVL).

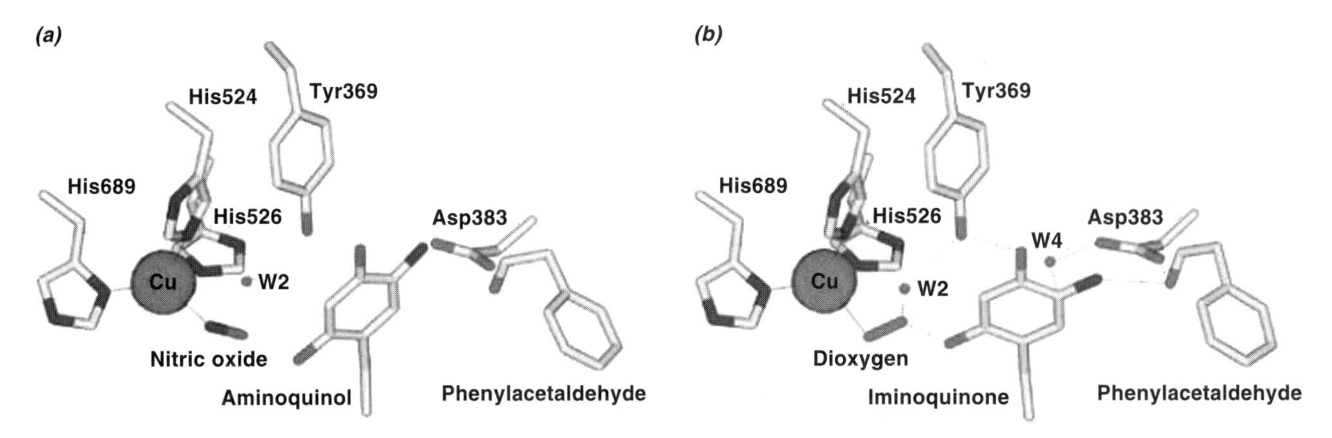

Fig. XIII.6.3.
(*a*) Active site features of the anaerobic substrate-reduced complex of *E. coli* amine oxidase (ECAO) and nitric oxide. (*b*) Crystal structure of the equilibrium turnover species in ECAO with substrate.[19]

XIII.6.3. Structure–Function Relationship

Examination of Fig. XIII.6.2 suggests numerous correlations between the structures of amine oxidases and their mechanisms and biological roles. All structurally characterized amine oxidases are dimers (Fig. XIII.6.2*a*). A remarkable feature of the dimer is the large, internal, solvent-filled cavity (Fig. XIII.6.2*b*). This cavity may provide an alternative path to or from the active site for small molecules such as the substrate O_2 or the product H_2O_2. Given the presence of the solvent cavity, the major interactions that hold the dimer together are those associated with the remarkable "arms" that extend from one subunit to the other in the dimer (Fig. XIII.6.2*a*). These arms may provide a link from one active site to the other. Within the active site, TPQ and Cu(II) are in close proximity and can, in fact, coordinate Cu(II) in some states of the enzyme (Fig. XIII.6.2*c* and *d*). A conserved Asp residue (Fig. XIII.6.3*b*) functions as the base for H^+ abstraction (Scheme XIII.6.1). A channel, whose location is shown in Fig. XIII.6.2*b*, provides substrates with access to the active site. The side chain of an aromatic residue (Fig. XIII.6.2*c*) occupies the bottom of the channel; the ring must move aside to permit substrates to reach the TPQ moiety. Because the electrophilic carbonyl group of TPQ should react with all primary amines, the substrate specificity of amine oxidases must be controlled by the electrostatic and steric properties of the channel and the residues surrounding TPQ.

XIII.6.4. Mechanistic Considerations

As illustrated in Scheme XIII.6.1, the oxidation of primary amines by amine oxidases can be divided into two sequential phases: (1) the oxidation of the amine to an aldehyde with the reduction of TPQ to a catechol amine; and (2) the subsequent oxidation of the catechol amine back to TPQ by O_2 with the release of NH_3 and the production of H_2O_2. Considerable structural and spectroscopic information supports the general mechanism in Scheme XIII.6.1. The intramolecular electron-transfer equilibrium (**D** and **E,** Scheme XIII.6.1) is temperature dependent[16,17] and the Cu^{I}–TPQ• state can be observed at physiological temperatures by EPR and electronic spectroscopy (see Scheme XIII.6.1), but not at low temperature.[18] Consequently, the Cu^{I}-semiquinone state escaped detection for many years (of low-temperature spectroscopic studies). It is not yet known whether O_2 reacts with the Cu^{I}–semiquinone form (**E**) or the Cu^{II}–aminoquinol (**D**) during the catalytic oxidation of amines. The crystal structures of a likely intermediate in turnover, and the structure of the NO adduct (a possible analogue for the O_2 adduct) are shown in Fig. XIII.6.3.[19]

In the NO adduct (Fig. XIII.6.3*a*; PDB code: 1D6Y) the NO is bound between Cu and TPQ, and the oxygen atom of NO appears to be hydrogen bonded to O_2 of the aminoquinol (reduced) form of TPQ. The nonlinear geometry suggests that the predominant species is Cu^{II}–NO^- (although some Cu^I–$NO^{\bullet}$ may also be present). Analysis of the hydrogen-bonding pattern and the electronic absorption spectrum of the trapped dioxygen-containing intermediate (Fig. XIII.6.3*b*; PDB code: 1D6Z) indicates that the TPQ is oxidized and that peroxide is the oxygen species bound close to the copper. Interestingly, the TPQ adopts slightly different conformations in the NO and peroxide complexes; in the NO complex TPQ has rotated ~20° about the C1–C4 axis of the ring to accommodate the NO (cf. Fig. XIII.6.3*a* and *b*). These structures provide strong evidence that the O_2-binding and reaction site is located in a pocket between Cu and TPQ. The results also support solution studies that suggest both protons needed to form H_2O_2 are derived from the TPQ. Collectively, the mechanistic and structural data indicate that copper and TPQ function as a unit to oxidize amines and transfer the electrons to O_2.

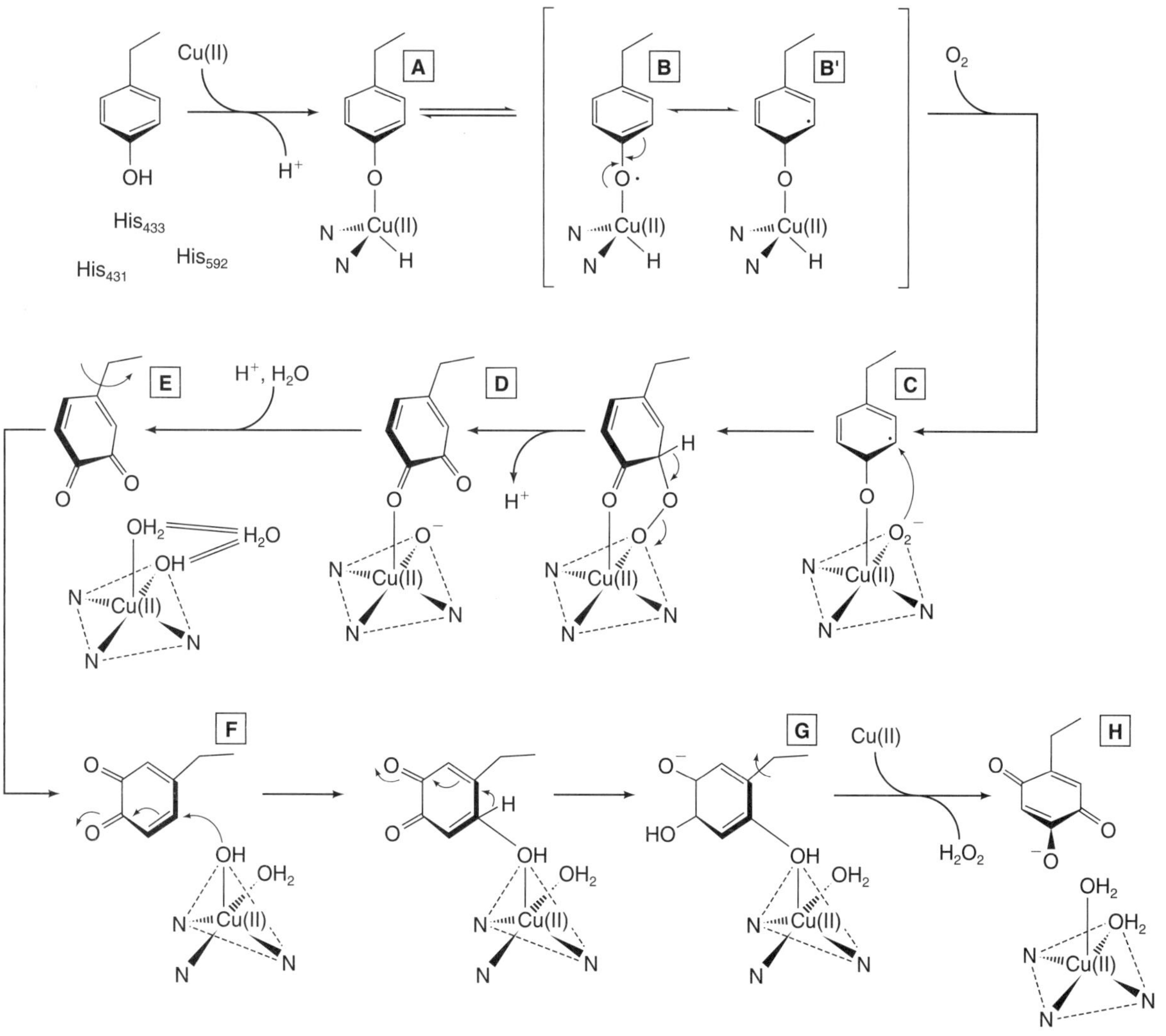

Scheme XIII.6.2.

XIII.6.5. Biogenesis of Amine Oxidases

One of the most fascinating questions concerning these enzymes is the mechanism of oxidation of the active-site Tyr residue to form TPQ. This six-electron oxidation requires only the apoprotein, Cu(II), and O_2 (Eq. 2).[20,21]

$$2\ O_2 + \text{Tyr} \rightarrow \text{TPQ} + 2\ H_2O_2 \qquad (2)$$

The generation of the catalytically competent enzyme from the inactive precursor form is a post-translational, self-processing reaction (as in galactose oxidase, see Section XIII.4). The structure of one unprocessed, apoamine oxidase has been determined. A comparison of the structure of the unprocessed protein to a structure with TPQ coordinated to Cu(II) (Fig. XIII.6.2*d*) established that Tyr can coordinate to Cu(II) when it binds in the Cu site of the unprocessed protein. This observation leads to the hypothesis that Tyr coordination to Cu(II) activates the phenol ring for oxidation by O_2. A possible mechanism for the oxidation of Tyr to TPQ is illustrated in Scheme XIII.6.2.[21]

Species **B** and **B′** are drawn to suggest ring activation via coordination; no evidence for radical intermediates in the Tyr oxidation reaction has been obtained at this time. However, all of the species in Scheme XIII.6.2 are compatible with the structure (Fig. XIII.6.2*d*), and the addition of H_2O to the ring illustrated as **D** → **E** → **F** → **G,** has been demonstrated. Elucidating all the fascinating details of this reaction will require further mechanistic, structural, and model studies.

XIII.6.6. Conclusion

The discovery of self-processing redox enzymes (see Sections XIII.3 and XIII.5) may be relevant to understanding aspects of the evolution of enzymes. Metal ion mediated redox chemistry with oxygen can modify several amino acids, especially Tyr, Trp, Cys, and His, which may have provided a path to generate new redox cofactors prior to the advent of complex biosynthetic pathways.

References

General References

1. Knowles, P. F. and Dooley, D. M., "Amine oxidases", in *Metal Ions in Biological Systems,* Vol. 38, Sigel, H. and Sigel, A., Eds., Marcel Dekker, New York, 1994, p. 361.
2. McPherson, M. J., Parson, M. R., and Wilmot, C. M., "Prokaryotic copper amine oxidases", in *Handbook of Metalloproteins,* Vol. 2, Messerschmidt, A., Hubert, R., Poulos, T., and Weighardt, K., Eds., John Wiley & Sons, Inc., Chichester, 2001, p. 1245.
3. Wertz, D. L. and Klinman, J. P., "Eukaryotic copper amine oxidases", in *Handbook of Metalloproteins,* Vol. 2, Messerschmidt, A., Hubert, R., Poulos, T., and Weighardt, K., Eds., John Wiley & Sons, Inc., Chichester, 2001, p. 1258.
4. Dooley, D. M., "Structure and biogenesis of topaquinone and related cofactors", *J. Biol. Inorg. Chem.,* **4,** 1 (1999).
5. Mure, M., Mills, S. A., and Klinman, J. P., "Catalytic mechanism of the topa quinone containing copper amine oxidases", *Biochemistry,* **41,** 9269 (2003).

Specific References

6. Dooley, D. M., Scott, R. A., Knowles, P. F., Colangelo, C. M., McGuirl, M. A., and Brown, D. E., "Structures of the Cu(I) and Cu(II) forms of amine oxidases from x-ray absorption spectroscopy", *J. Am. Chem. Soc.,* **120,** 2599 (1998).
7. Dooley, D. M., McGuirl, M. A., Cote, C. E., Knowles, P. F., Singh, I., Spiller, M., Brown, R. D., III, and Koenig, S. H., "Coordination chemistry of copper-containing amine oxidases: Nuclear magnetic relaxation dispersion studies of copper binding, solvent-water exchange, substrate and inhibitor binding, and protein aggregation", *J. Am. Chem. Soc.,* **113,** 754 (1991).
8. McCracken, J., Peisach, J., and Dooley, D. M., "Cu(II) Coordination of amine oxidases: Pulsed EPR studies of histidine imidazole, water and exogenous ligand coordination", *J. Am. Chem. Soc.,* **109,** 4064 (1987).
9. Scott, R. A. and Dooley, D. M., "X-ray absorption spectroscopy of the Cu(II) sites in bovine plasma oxidase", *J. Am. Chem. Soc.,* **107,** 4348 (1985).

10. Dooley, D. M. and Cote, C. E., "Copper(II) coordination chemistry in bovine plasma amine oxidase: Azide and thiocyanate binding", *Inorg. Chem.*, **24,** 3996 (1985).
11. McGuirl, M. A., Brown, D. E., and Dooley, D. M., "Cyanide as a copper-directed inhibitor of amine oxidases: Implications for the mechanism of amine oxidation", *J. Biol. Inorg. Chem.*, **2,** 336 (1997).
12. Kumar, V., Dooley, D. M., Freeman, H. C., Guss, J. M., Harvey, I., McGuirl, M. A., Wilce, M. C. J., and Zubak, V. M., "Crystal structure of a eukaryotic (pea seedling) copper-containing amine oxidase at 2.2 Å resolution", *Structure,* **4,** 943 (1996).
13. Wilce, M. C. J., Dooley, D. M., Freeman, H. C., Guss, J. M., Matsunami, H., McIntire, W. S., Ruggiero, C. E., Tanizawa, K., and Yamaguchi, H., "Crystal structures of the copper-containing amine oxidase from *Arthrobacter globiformis* in the holo and apo forms: Implications for the biogenesis of topaquinone", *Biochemistry,* **36,** 16116 (1997).
14. Parsons, M. R., Convery, M. A., Wilmot, C. M., Yadav, K. D. S., Blakeley, V., Corner, A. S., Phillips, S. E. V., McPherson, M. J., and Knowles, P. F., "Crystal structure of a quinoenzyme: copper amine oxidase of *Escherichia coli* at 2 Å resolution", *Structure,* **3,** 1171 (1995).
15. Li, R. B., Klinman, J. P., and Mathews, F. S., "Copper amine oxidase from *Hansenula polymorpha:* the crystal structure determined at 2.4 Å resolution reveals the active conformation", *Structure,* **6,** 293 (1998).
16. Turowski, P. N., McGuirl, M. A., and Dooley, D. M., "Intramolecular electron transfer rate between active-site copper and topa quinone in pea seedling amine oxidase", *J. Biol. Chem.*, **268,** 17680 (1993).
17. Dooley, D. M. and Brown, D. E., "Intramolecular electron transfer in the oxidation of amines by methylamine oxidase from *Arthrobacter P1*", *J. Biol. Inorg. Chem.*, **1,** 205 (1996).
18. Dooley, D. M., McGuirl, M. A., Brown, D. E., Turowski, P. N., McIntire, W. S., and Knowles, P. F., "A Cu(I)-semiquinone state in substrate-reduced amine oxidases", *Nature (London),* **349,** 262 (1991).
19. Wilmot, C. M., Hajdu, J., McPherson, M. J., Knowles, P. F., and Phillips, S. E. V., "Visualization of dioxygen bound to copper during enzyme catalysis", *Science,* **286,** 1724 (1999).
20. Tanizawa, K., "Biogenesis of novel quinone coenzymes", *J. Biochem. (Tokyo),* **118,** 671 (1995).
21. Ruggiero, C. E., Smith, J. A., Tanizawa, K., and Dooley, D. M., "Mechanistic studies of topa quinone biogenesis in phenylethylamine oxidase", *Biochemistry,* **36,** 1953 (1997).

XIII.7. Lipoxygenase

Judith Klinman
Department of Chemistry and Department of Molecular and Cell Biology
University of California, Berkeley
Berkeley, CA 94720

Keith Rickert
Department of Cancer Research WP26-462
Merck & Co.
P. O. Box 4
West Point, PA 19486

Contents

XIII.7.1. Introduction

Lipoxygenases are a family of enzymes that catalyze the peroxidation of unsaturated fatty acids (Fig. XIII.7.1). They are found in a wide variety of organisms, ranging from cyanobacteria to higher plants and animals. The amino acid sequence of these proteins is highly conserved, even between such distantly related organisms as humans and higher plants, including soybeans.[1,2]

Lipoxygenases play a role in a number of important biological processes. In mammalian cells, lipoxygenases synthesize a wide range of compounds that strongly influence the process of inflammation.[3] These compounds, known as leukotrienes and lipoxins, serve to attract white blood cells to sites of infection or injury and can also affect the endothelial cells that line blood vessels to enhance this process. Leukotrienes mediate some of the destructive side effects of inflammation, such as the

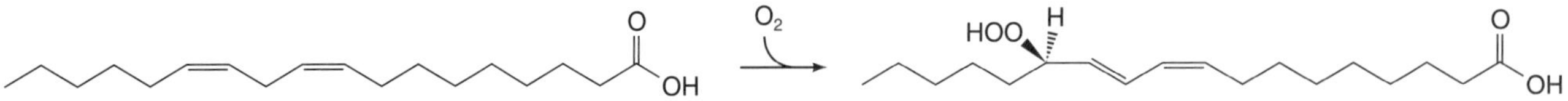

Fig. XIII.7.1.
Reaction catalyzed by lipoxygenases.

bronchial constriction associated with asthma. Additionally, lipoxygenases may have importance in the pathogenesis of atherosclerosis, and in the selective degradation of organelle membranes.[4]

Plant lipoxygenases are thought to play a role in response to injury and infection; in fact, the biological functions are surprisingly similar to those in mammalian cells, although the substrates and final products involved are quite distinct. Historically, the plant enzymes have been more readily available, more robust, and more easily assayed than mammalian enzymes. Thus, they are more commonly used for biochemical and mechanistic studies; in particular, soybean lipoxygenase-1 (SBL-1) has been extensively studied with a variety of kinetic and spectroscopic tools and under a wide range of conditions.

XIII.7.2. Structure

The lipoxygenases are large, monomeric proteins, with a single nonheme Fe center.[5] Perhaps surprisingly for an enzyme with a lipid substrate, they are generally soluble proteins, most typically found in the cytoplasm. Crystal structures for both the soybean and a mammalian enzyme each show a two-domain structure, with significant similarity to each other (Fig. XIII.7.2).[6,7] The N-terminal domain is composed mainly of β-sheet structure. By analogy to the lipases, its function has been suggested to be that of extracting lipid substrates from a membrane, although there is no direct evidence for this. The C-terminal domain, which is principally α-helical, comprises ~65% of the total enzyme and contains the nonheme Fe and putative substrate-binding site. In the crystal, where the enzyme is in its resting, Fe(II) state, the Fe ligands are three His residues, the C-terminal carboxylate, a water molecule, and a single Asp side chain (this is replaced by His in mammalian lipoxygenases) (Fig. XIII.7.3).

The Fe site is immediately adjacent to a long, narrow, hydrophobic cavity, with a single Arg residue at one end, which is thought to be the substrate-binding pocket, with the Arg near the substrate carboxylic acid. Using molecular modeling to dock a substrate molecule into this cavity, the Fe site appears to be in close proximity to the unsaturated regions of the substrate where the reaction takes place. However, there are no structures available of substrate bound to the enzyme, or of the enzyme in its active, Fe(III) state. Nor is it clear how substrate enters this cavity, as the cavity is entirely closed off from bulk solvent.

XIII.7.3. Mechanism

The mechanism by which lipoxygenases are believed to effect the reaction in Fig. XIII.7.1 is shown in Fig. XIII.7.4.[8] Oxidized enzyme binds substrate and irreversibly abstracts a hydrogen atom, reducing the Fe center and resulting in a radical intermediate. This intermediate then reacts with triplet molecular oxygen to give a peroxyl radical, which is subsequently reduced by the Fe center to return the enzyme to its oxidized state. The product hydroperoxyl anion can then be protonated and released into solution.

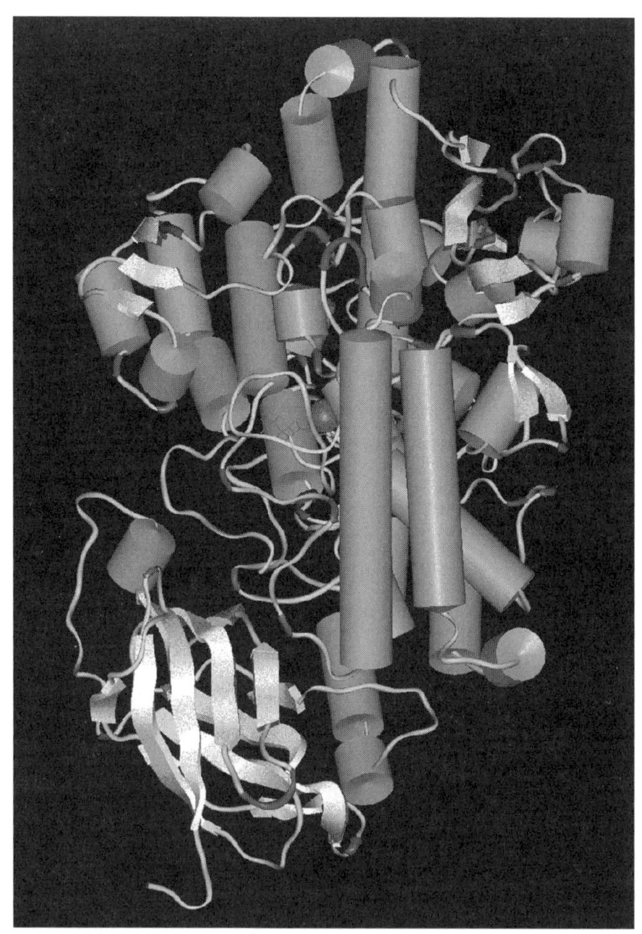

Fig. XIII.7.2.
Crystal structure of SBL-1, based on the structure solved by Minor et al.[6] The N-terminal domain can be seen at the lower left of the structure (PDB code: 1YGE). [See color insert.]

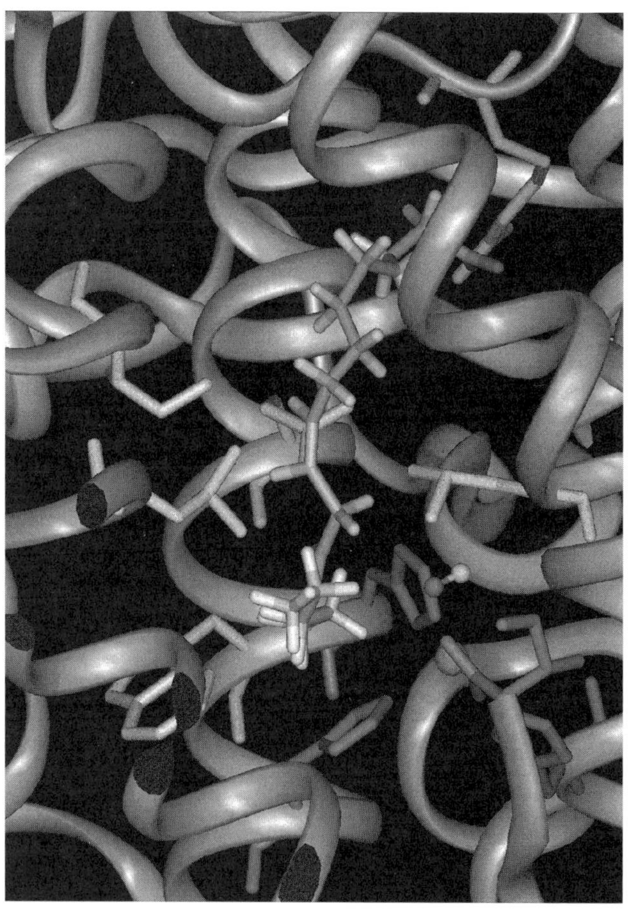

Fig. XIII.7.3.
Crystal structure of SBL-1, based on the structure solved by Minor et al.[6] The Fe and its ligands are shown at the lower right. (PDB code: 1YGE). [See color insert.]

Fig. XIII.7.4.
Postulated mechanism for SBL-1 reaction.

The state of the Fe center throughout the catalytic cycle has been extensively studied by a variety of spectroscopic techniques, including EPR, UV–visible, X-ray, and Mössbauer spectroscopy. The feature most readily studied is the oxidation state of the Fe center, and all methods agree that the Fe center undergoes reduction from Fe(III) to Fe(II) upon anaerobic interaction with substrate.[4,9] Extended X-ray absorption fine structure (EXAFS) and other X-ray techniques are also capable of probing the coordination state and nature of the ligands. This has shown that while the Fe(II) enzyme appears to be coordinated to water, Fe(III) lipoxygenase has a single hydroxide ion bound to the metal, an important feature in the above mechanism.[10] No evidence has been seen for oxygen binding directly to Fe, either in the Fe(II) or Fe(III) states,[11] ruling out a number of possible mechanisms. The reduction potential of the iron center has been estimated at 0.6 V,[12] making it a relatively strong oxidant. The available crystal structures show that there are no other cofactors or transition metals involved, and biochemical evidence is clear that the enzyme acts as a monomer,[6] suggesting strongly that there is a one-electron oxidized intermediate, whether on the protein itself or as part of the substrate.

Further understanding of this issue has come from EPR spectroscopy of lipoxygenase quenched during turnover. These spectra reveal the existence of organic radicals, derived from substrate, in the reaction mixture. However, it has not been conclusively proven that these species represent a catalytic intermediate and not a side reaction of some sort. More detailed analysis of a substrate-derived radical formed upon incubation of enzyme with the product 13-hydroperoxy-9,11-(*Z,E*)-octadecadienoic acid (13-HPOD) indicates that the species formed is a 9,10,11-allylic radical bound to enzyme (Fig. XIII.7.5*b*), which at higher oxygen concentrations is in equilibrium with a 9-peroxyl radical.[13] It is not clear if or how this species might proceed to form the observed 13-HPOD product, or instead if it represents a side product. Formation of an allylic radical rather than a pentadienyl radical (Fig. XIII.7.5*a*) represents a useful mechanism for regiochemical control, if it can be observed along the main reaction pathway. Nelson et al. suggest a mechanism involving the 9,10,11-allylic radical as an intermediate, allowing subsequent oxygen activation by the Fe center and attack of oxygen upon C13 of the substrate.[13] However, this mechanism has not yet been fully tested.

Additional evidence for a radical intermediate is given by analysis of the product mixtures for some lipoxygenases. In contrast to SBL-1, these lipoxygenases generate a product with a significantly lower stereochemical purity, suggesting that oxygen attack on the substrate occurs with lower stereochemical control in these enzymes.[5] This result is most consistent with a radical mechanism, as the radical species might interact with triplet oxygen readily regardless of the precise stereochemistry; it is more difficult to explain if oxygen attack occurs from an oxygen species coordinated to the metal center.

The existence of a substrate-derived radical in the mechanism has also been explored by use of unusual substrate analogues, designed to trap radicals. These species were observed to inactivate the enzyme at low oxygen concentrations and to produce

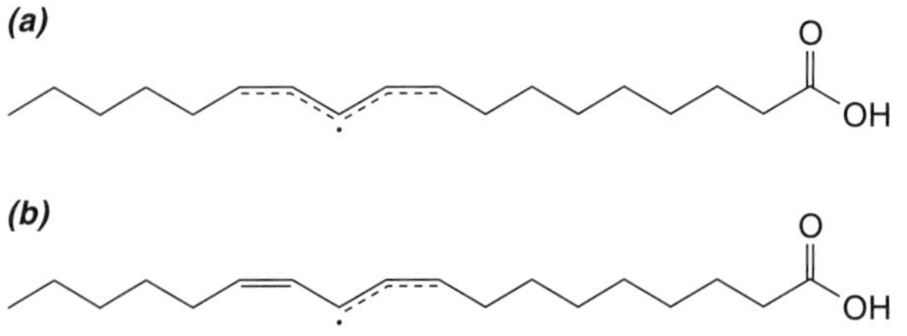

Fig. XIII.7.5.
Possible radical intermediates: (*a*) pentadienyl radical and (*b*) 9,10,11-allylic radical.

(a)

O
OH

(b)

O
OH

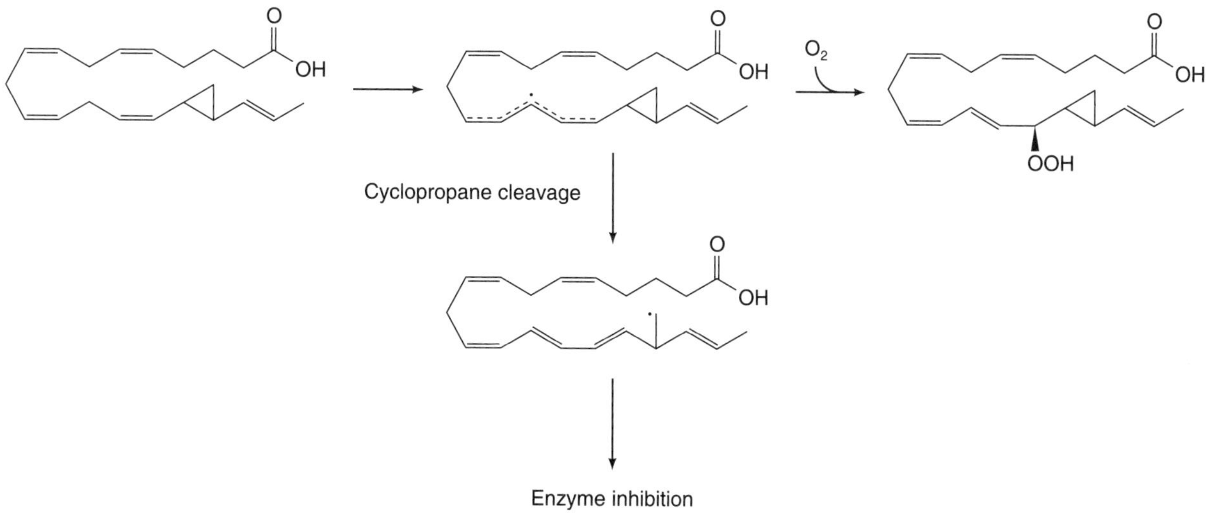

Cyclopropane cleavage

Enzyme inhibition

Fig. XIII.7.6.
Postulated reaction pathways for a radical trap substrate.

unrearranged oxidized products at higher oxygen concentrations.[14] The results were taken as evidence against a radical intermediate, by comparison to peroxidation by a free radical chain reaction in solution, where only the rearranged product was observed. However, both observations are potentially compatible with an enzyme-bound radical intermediate. As already observed, the enzyme may restrict the delocalization of a radical by constraining substrate geometry, and a similar mechanism could allow rearrangement-free oxidation of the radical clock substrate under high oxygen pressure (Fig. XIII.7.6). The radical produced by rearrangement, however, may interact with enzyme and form a covalent adduct, consistent with the ability of this compound to act as an irreversible inhibitor of lipoxygenase at lower oxygen pressures.

XIII.7.4. Kinetics

A number of unusual features in lipoxygenase kinetics have been observed, particularly extended lag phases preceding steady-state kinetics; these are believed to be associated with the slightly unusual mechanism for activation of the enzyme.[15] As isolated, the lipoxygenase iron center is present in the Fe(II) oxidation state and does not oxidize to Fe(III) with exposure to oxygen. However, the mechanism described above requires that enzyme be present as Fe(III) to catalyze the reaction. Since oxygen itself does not oxidize the enzyme, how is enzyme activated in the presence of both substrate and oxygen? The answer seems to be that autoxidation of substrates leads to a small amount of non-enzymatically produced hydroperoxides. These lipid hydroperoxides have been shown to be capable of oxidizing the enzyme, allowing a small portion of the enzyme to achieve the active state and begin catalysis. Since the product of enzymatic catalysis is also a lipid hydroperoxide, and thus capable of oxidizing Fe(II) enzyme, all of the enzyme can reach the active state via this autocatalytic mechanism. This initial period accounts for the observed lag phases, which can be reduced or eliminated when enzyme is oxidized before the kinetic study.

Another notable feature of the lipoxygenase reaction is the extreme size of the isotope effects observed with deuterated substrates, $k_H/k_D \sim 80$.[16] These are far larger than are normally observed with enzymatic systems and greatly exceed the theoretical maximum of 7–10 predicted using a semiclassical transition state theory for a room temperature deuterium isotope effect. Several possible explanations for such large isotope effects have been explored; the one most consistent with the data is that the hydrogen transfer occurs via tunneling.[17] Tunneling is a quantum mechanical process by which the hydrogen travels through the energy barrier, rather than over it, and can potentially account for both the large isotope effects and the very low observed activation energies (1–2 kcal mol^{-1}).

Lipoxygenase represents a completely different strategy for oxygenation of an organic substrate than other oxygenases (see Section XI.5). Most other oxygenases catalyze a reaction where oxygen is first reduced to a more reactive species, such as a metal–oxo complex, by a metal, or other redox-active cofactor. This reactive oxygen can then attack the organic substrate or abstract hydrogen from it (see Section XI.5). In contrast, with lipoxygenase, there is activation of the substrate by the metal center, followed by attack of the activated substrate upon molecular oxygen. In part, this difference may arise from the highly oxidizing nature of the lipoxygenase metal center. It may also reflect the unusual nature of the active site and its ability to catalyze quantum mechanical hydrogen transfer under conditions where a classical transfer would be a very slow process.

References

General References

1. Sigal, E., "The molecular biology of mammalian arachidonic acid metabolism", *Am. J. Physiol.*, **260,** L13–L28 (1991).
2. Siedow, J. N., "Plant lipoxygenases: structure and function", *Annu. Rev. Plant Physiol. Mol. Biol.*, **42,** 145–188 (1991).
3. Samuelsson, B., Dahlén, S.-E., Lindgren, J. Å., Rouzer, C. A., and Serhan, C. N., "Leukotrienes and lipoxins: structures, biosynthesis, and biological effects", *Science,* **237,** 1171–1176 (1987).

Specific References

4. Nelson, M. J. and Seitz, S. P., "The structure and function of lipoxygenase", *Curr. Opin. Struct. Biol.*, **4,** 878–884 (1994).
5. Gardner, H. W., "Recent investigations into the lipoxygenase pathway of plants", *Biochim. Biophys. Acta,* **1084,** 221–239 (1991).
6. Minor, W., Steczko, J., Stec, B., Otwinowski, Z., Bolin, J. T., Walter, R., and Axelrod, B., "Crystal structure of soybean lipoxygenase L-1 at 1.4 Å resolution", *Biochemistry,* **35,** 10867–10701 (1996).
7. Gillmor, S. A., Villaseñor, A., Fletterick, R., Sigal, E., and Browner, M. F., "The structure of mammalian 15-lipoxygenase reveals similarity to the lipases and the determinants of substrate specificity", *Nat. Struct. Biol.*, **4,** 1003–1009 (1997).
8. Glickman, M. H. and Klinman, J. P., "Nature of rate-limiting steps in the soybean lipoxygenase-1 reaction", *Biochemistry,* **34,** 14077–14092 (1995).
9. Feiters, M. C., Boelens, H., Veldink, G. A., Vliegenthart, J. F. G., Navaratnam, S., Allen, J. C., Nolting, H.-F., and Hermes, C., *Recl. Trav. Chim. Pays-Bas,* **109,** 133–146 (1990).
10. Scarrow, R. C., Trimitsis, M. G., Buck, C. P., Grove, G. N., Cowling, R. A., and Nelson, M. J., "X-ray spectroscopy of the iron site in soybean lipoxygenase-1: changes in coordination upon oxidation or addition of methanol", *Biochemistry,* **33,** 15023–15035 (1994).
11. Glickman, M. H. and Klinman, J. P., "Lipoxygenase reaction mechanism: demonstration that hydrogen abstraction from substrate precedes dioxygen binding during catalytic turnover", *Biochemistry,* **35,** 12882–12892 (1996).
12. Nelson, M. J., "Catecholate complexes of ferric soybean lipoxygenase I", *Biochemistry,* **27,** 4273–4278 (1988).
13. Nelson, M. J., Cowling, R. A., and Seitz, S. P., "Structural characterization of alkyl and peroxyl radicals in solutions of purple lipoxygenase", *Biochemistry,* **33,** 4966–4973 (1994).
14. Corey, E. J. and Nagata, R., "Evidence in favor of an organoiron-mediated pathway for lipoxygenation of fatty acids by soybean lipoxygenase", *J. Am. Chem. Soc.*, **109,** 8107–8108 (1987).
15. Schilstra, M. J., Veldink, G. A., and Vliegenthart, J. F. G., "The dioxygenation rate in lipoxygenase catalysis is determined by the amount of iron(III) lipoxygenase in solution", *Biochemistry,* **33,** 3974–3979 (1994).
16. Glickman, M. H., Wiseman, J. S., and Klinman, J. P., "Extremely large isotope effects in the soybean lipoxygenase-linoleic acid reaction", *J. Am. Chem. Soc.*, **116,** 793–794 (1994).
17. Knapp, M. J. and Klinman, J. P., "Environmentally coupled hydrogen tunneling: Linking catalysis to dynamics", *Eur. J. Biochem.*, **269,** 3113–3121 (2002).

CHAPTER XIV

Metal Ion Receptors and Signaling

XIV.1. Metalloregulatory Proteins

Dennis R. Winge
Departments of Medicine
and Biochemistry
University of Utah Health
Sciences Center
Salt Lake City, UT 84132

Contents

XIV.1.1. Introduction: Structural Metal Sites

Metal ions participate in multiple roles in protein structure and function. In addition to their roles in catalysis, electron transfer, and ligand binding discussed in previous chapters, metal ions are important for the structural integrity of many proteins. In proteins containing structural metal ions, the role of the bound metal ion is to contribute to the stabilization energy of the folded conformation. Even in proteins containing catalytic metal ions, the bound metal often contributes to the stabilization energy of the folded conformation. Metal ions usually stabilize domains or subdomains of larger modular proteins. The ligands of structural metal ions are often clustered in buried sites. Metal ion binding reduces the number of accessible conformations of a polypeptide chain, analogous to other interactions that stabilize protein structure including hydrogen bonds, ion pairs, van der Waals contacts, or cross-links. Stabilization of domains by metal ions may be more common in intracellular proteins owing to the reducing environment of the cell, which limits disulfides as common stabilizing cross-links.

A subset of metalloproteins exists in two stable and interconvertible conformational states under physiological conditions. Metal ion binding can stabilize a conformation distinct from that of the metal-free molecule. The dynamics of the reversible conformers allows a function to be switched on or off. Since metal ions can stabilize one conformer of these proteins, regulatory metal ions will be treated as a subclass of structural metal ions. Structural metal sites will be examined for insights that may be applicable to regulatory metal sites.

The class of proteins containing structural metal ions is dominated by domains stabilized by Zn ions. Zinc ions may be the preferred metal ion to stabilize domains for four major reasons.[1–6] First, the borderline hardness of Zn(II) ions permits it to coordinate to a variety of potential side-chain ligands including histidine, cysteine, and carboxylates. Second, Zn(II) can form four, five, and six coordinate complexes and the lack of any ligand-field stabilization effects for Zn permits conversion between these various coordination numbers. Third, Zn(II) is resistant to redox changes and cannot generate radicals. Fourth, Zn(II) is an abundant metal ion with minimal toxic side effects. Although Zn levels in mammalian cells approach millimolar concentrations, most of the cellular Zn is protein bound. The Zn levels are regulated to minimize any deleterious effects.[7] Excess Zn ions inhibit a variety of enzymes with nanomolar stability constants. The combination of these effects makes Zn(II) suitable as a structural metal ion.

Calcium ions are also structurally important in many proteins. Many extracellular proteins require bound Ca(II) ions for function.[8] Calcium binding often stabilizes a conformation appropriate for ligand binding and Ca(II) is an important structural cation extracellularly because of its high millimolar concentration and ability to form a variety of coordinate complexes. The coordination numbers 6, 7, and 8 are the most common.[9]

Copper and iron ions are less commonly used as structural metal ions, probably due to their redox activities. Reduced copper and iron ions can catalyze formation of deleterious oxygen radicals. The reactivity of these metal ions dictates that cells control their concentrations. However, copper and iron ions do still function in structural and regulatory roles. Iron ions present as Fe–S clusters are used as structural elements in proteins such as endonuclease III and MutY.[10] In addition, copper and iron ions modulate the function of a series of metalloregulatory proteins to be discussed.

XIV.1.2. Structural Zn Domains

Multiple distinct Zn structural domains have been identified.[1–3,11] The prototype is the classical Zn finger motif also described as the CCHH motif for the order of ligands (2 cysteines and 2 histidines). This motif is found in ~0.7% of all proteins in the yeast *Saccharomyces cerevisiae* and the metazoan *Caenorhabditis elegans*. Other distinct Zn motifs include the LIM (originally observed in three proteins Lin-11, Isl-1, and Mec-3) domain, RING (really interesting new gene) motif, GATA domain, CRD (cysteine-rich domain) motif, nuclear hormone receptor motif, the Zn ribbon, and the binuclear Zn cluster.[11] These motifs primarily contain cysteinyl Zn ligands, but differ in their folding architecture and the arrangement and spacing of Zn liganding amino acids. Some Zn modules (e.g., Zn fingers, LIM, and CRD) can be repeated in the same structure, whereas a RING motif is usually present only once.

The nuclear hormone receptor motif, classical Zn finger, and RING motifs are three of the 10 most abundant structural domains in the *C. elegans* genome. Curiously, the nuclear hormone receptor motif is not found in the yeast genome, whereas the binuclear Zn cluster is the most abundant Zn motif in yeast. The *C. elegans* genome contains only one candidate binuclear Zn motif. None of these structural Zn motifs have been observed in the *Escherichia coli* genome. The abundance of structural Zn motifs in eukaryotic cells suggests that eukaryotes have evolved mechanisms to permit greater Zn(II) retention relative to prokaryotic cells.

Some general features emerge about structural Zn(II) sites in proteins. Unlike catalytic Zn sites, structural sites commonly have distorted tetrahedral geometries with four protein side-chain ligands. In these cases, metal binding constrains the protein conformation by tethering the protein at four different sites. A limited number of structural Zn sites exhibit trigonal-bipyramidal, square-pyramidal, or octahedral geometries that increase constraints on the protein conformation. Ligand groups tend to be Cys, His, and to a lesser extent aspartate or glutamate. A unique feature of cysteinyl ligands in these Zn motifs is that they often lay within a reverse turn stabilized by a hydrogen bond between a cysteinyl sulfur and a main-chain amide (Fig. XIV.1.1). This hydrogen-bonded loop, designated the rubredoxin knuckle, requires that the backbone be extended before and after the Zn site. When histidine serves as a Zn(II) ligand, two tautomers of the imidazole side chain are available for Zn binding. The ε ring nitrogen is the usual ligand rather than the δ nitrogen. The ratio of the Nε to Nδ tautomer is 7:3. The imidazole is typically neutral in charge. Structural Zn sites involving histidyl ligands frequently have hydrogen bonds from the non-metal bound nitrogen to an oxygen atom of water or a carboxylate. The orientation and ligand strength of the imidazole side chain is influenced by the hydrogen-bonded oxygen group.

The structural Zn domains differ in their physiological function and most appear to be interfaces for macromolecular interactions. Classical Zn finger motifs are important interfaces for protein–nucleic acid interaction and can function as either transcriptional activators or repressors. The classical Zn finger domain consists of two antiparallel β strands followed by an α helix (Fig. XIV.1.2). Zinc(II) binding brings the β strands in juxtaposition with the helix creating a hydrophobic core. Residues forming the hydrophobic core are highly conserved. Zinc ligands come from two cysteines in the β hairpin turn and two histidyl residues in the α helix. Three residues in the helix commonly contact contiguous bases within the major groove of a DNA duplex with additional residues making DNA backbone contacts. In proteins

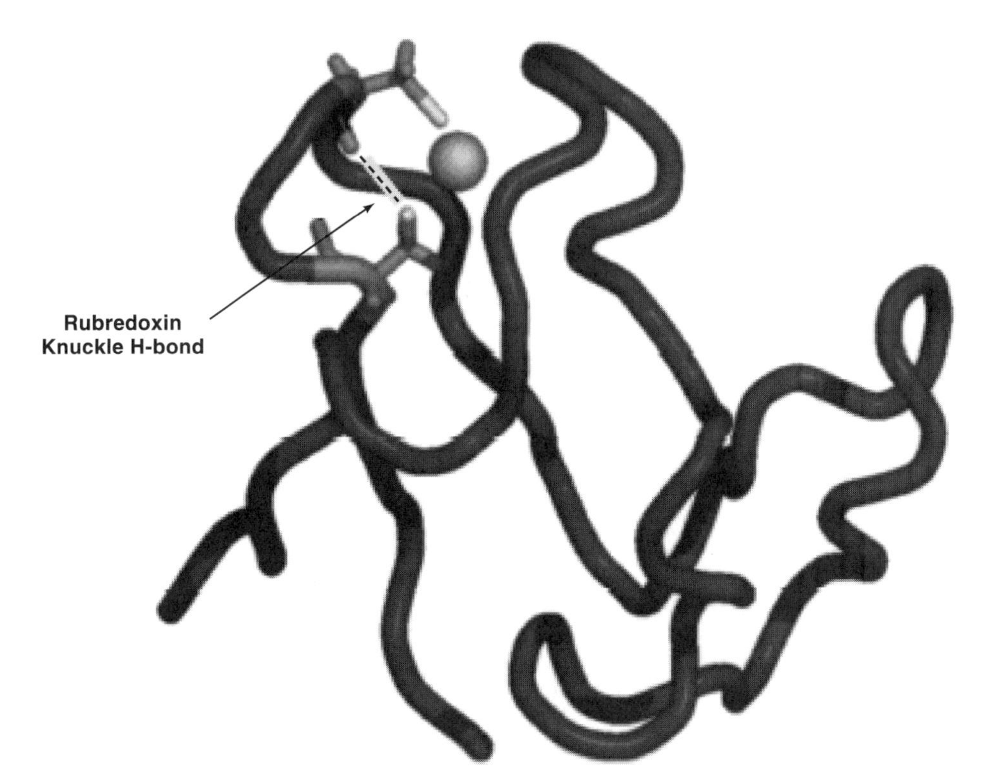

Fig. XIV.1.1. Knuckle motif seen in the metal site in *Clostridium pasteurianum* rubredoxin (PDB code: 1IRO). The dashed line shows the hydrogen bond between the backbone amide hydrogen and the cysteinyl sulfur.

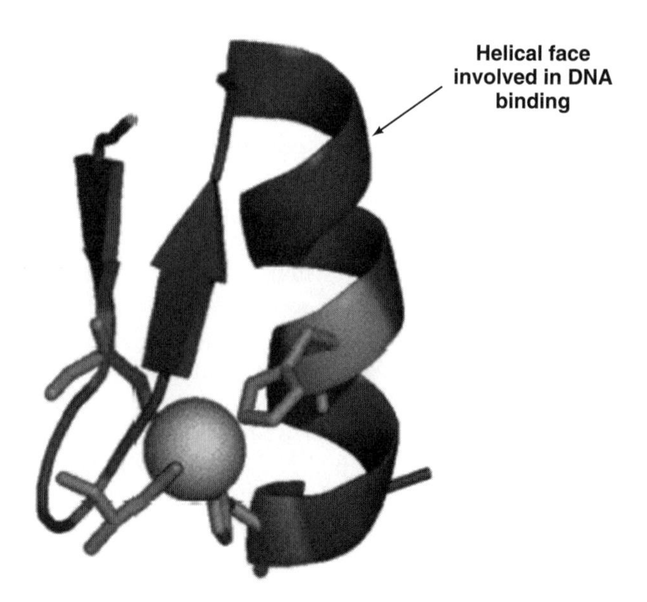

Fig. XIV.1.2.
Classical Zn finger motif in mouse Zif268 transcription factor (PDB code: 1ZAA). One zinc finger motif from Zif268 is shown. Two cysteinyl ligands from the beta hairpin and two histidyl ligands from the helix ligate the Zn(II) in tetrahedral geometry.

Fig. XIV.1.3.
The LIM domain in rat CRIP (cysteine-rich intestinal protein) (PDB code: 1IML). The two finger motifs are packed together with hydrophobic contacts. The Zn(II) ions are separated by 17.6 Å.

Fig. XIV.1.4.
Structure of the zinc domain of the glucocorticoid receptor from rat liver (PDB code: 1GLU). The two Zn ions are separated by 15.4 Å.

with more than three Zn finger motifs, a subset of Zn finger domains make DNA base or backbone contacts, while others are important for protein/protein contacts with other fingers.[11]

The RING, LIM, nuclear hormone receptor, and CRD domains consist of single structural units integrating two Zn sites within the fold. The two Zn(II) ions are independently coordinated within these domains and are separated by 13–18 Å. This contrasts with a >26-Å separation of Zn(II) in adjacent classical Zn finger domains. The number of residues within the bimetallic domains is nearly double the size of classical Zn finger motifs, which is ~30 residues. Despite the commonality of two Zn sites in these four motifs, the overall structural features of each bimetallic domain are unique with limited similarities within subdomains. One difference is that LIM, GATA, and nuclear hormone receptor domains contain contiguous ligands for the two Zn sites creating adjacent subdomains (LIM and glucocorticoid receptor domain shown in Figs. XIV.1.3 and XIV.1.4). In contrast, the RING and CRD motifs have interleaved, noncontiguous ligands for the two bound Zn ions (RING and CRD domains in Figs. XIV.1.5 and XIV.1.6).

The LIM domain is an independent structural module of nearly 55 residues containing two Zn finger-like motifs packed together, forming a hydrophobic core region (Fig. XIV.1.3). The core structure of the LIM domain is predominantly an antiparallel β sheet with the two Zn ions located at the end of the hydrophobic core. The spacer between Zn subdomains is a short two residues, which restricts the separation of the β-strand loops. The LIM domain is present as 1–5 repeated motifs in proteins and also found in combination with other functional domains such as homeodomains or kinase motifs. The observation that a single LIM domain acts as a specific protein/protein interface is consistent with the fact that proteins containing multiple LIM domains may have adapter functions, mediating assembly of multiple components into macromolecular complexes. The LIM domains are found in proteins involved in cell growth and differentiation.

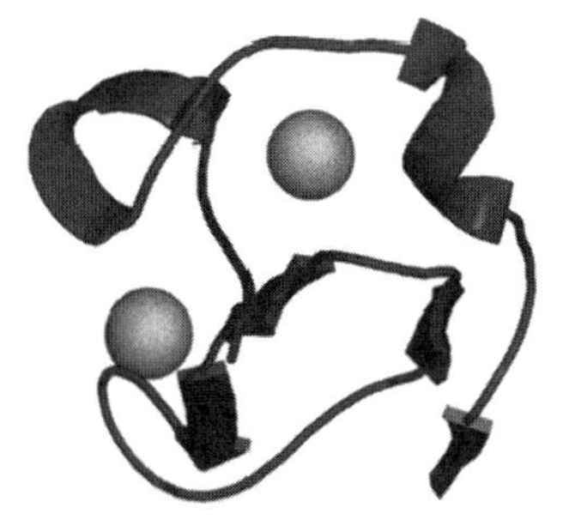

Fig. XIV.1.5.
Structure of the RING domain of the Pml A proto-oncoprotein (PDB code: 1BOR). The Zn–Zn distance is 14.2 Å.

The nuclear hormone receptor proteins are DNA-binding proteins. The domain responsible for DNA binding consists of two similar modules, each nucleated by a Zn ion and folded together into a single unit (Fig. XIV.1.4). The Zn ions are tetrahedrally coordinated by four cysteinyl residues. Within each module, an extended loop exists between pairs of cysteines and an α helix begins after the third cysteine. Residues within the two helices interact, stabilizing the overall globular fold. The interaction between adjacent modules distinguishes this domain from classical Zn fingers in which no interaction exists between adjacent Zn modules. The GATA DNA-binding subdomain structurally resembles the N-terminal motif of the nuclear hormone receptor, but the two Zn-binding domains interact with DNA differently. The prototype is the GATA-1 transcription factor, which binds a core DNA consensus sequence of GATA in the enhancer element of various erythroid genes.

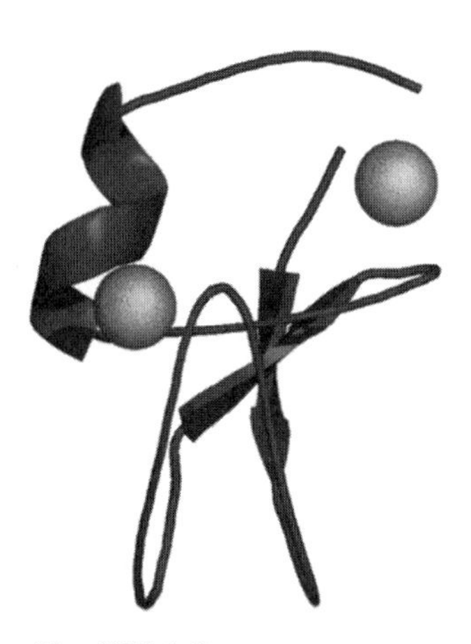

Fig. XIV.1.6.
The CRD domain found in mouse protein kinase C delta (PDB code: 1PTQ). The two Zn(II) sites are coordinated by three thiolates and one imidazole. The Zn ions are separated by 14.9 Å.

The RING motifs (also designated C_3HC_4) are found in proteins with disparate functions ranging from regulation of gene expression, DNA repair, recombination, and peroxisomal assembly. A number of RING proteins are oncogenes such as the BRCA1 breast cancer susceptibility gene. The RING motif consists of a triple-stranded antiparallel β sheet and a short helix creating a conserved hydrophobic core between the two Zn sites (Fig. XIV.1.5). The first and third pairs of ligands in the RING domain bind one Zn(II) and the second and fourth ligand pairs ligate the second Zn(II). The RING motif can adopt subtle changes in Zn ligand position without losing overall conformity to the motif. An unusual feature in the RING motif is the single residue spacer between two ligands (Cys-x-His) in one Zn half-site. This feature may account for the observed Zn(II) ligation by the δN of His, a situation that differs from the more common coordination through the εN of His.

The CRD motif is found in a number of signal transduction proteins including protein and diacylglycerol kinases. The CRD structure consists of one Zn site formed by two loops connecting four β strands with clusters CxxC and HxxC ligand pairs in the two loops. The second Zn site is formed by a CxxC ligand pair from a loop and isolated Cys and His residues at the two chain termini of the domain (Fig. XIV.1.6). The δN of each His serves as the ligand. Both Zn(II) ions lie in buried cavities explaining why Zn(II) ions are integral components of the structure. The CRD in several proteins contains separate binding sites for diacylglycerol and phosphatidylserine. Lipid binding may direct CRD-containing molecules to membrane surfaces. The CRD domain in Raf1, for example, does not bind diacylglycerol, but does bind phosphatidylserine and makes a protein–protein interaction with Ras.

The Zn ribbon motif is formed by a three stranded β structure anchored by a Zn(II) ion bound by four cysteines. This motif, found in the regulatory domain of protein kinase CK2, mediates protein dimerization.

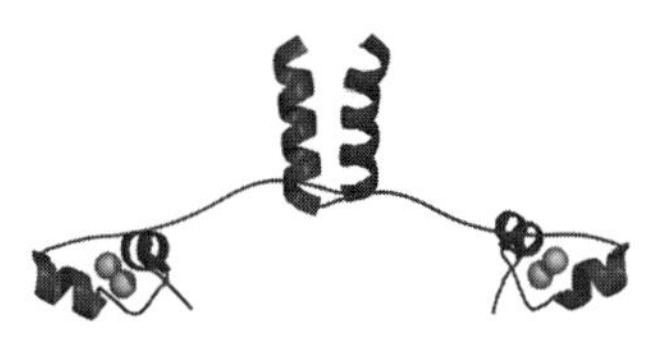

Fig. XIV.1.7.
Binuclear Zn cluster in the yeast Gal4 DNA-binding domain (PDB code: 1D6G). The coiled coil dimerization motif is shown with the two globular domains, each containing a binuclear Zn(II) thiolate cluster. The two Zn(II) ions are separated by 3.1 Å.

The binuclear Zn domain common in fungal transcription factors contains ~32 residues held together by two Zn ions tetrahedrally coordinated by six cysteines. Two of the cysteines coordinate both Zn(II) ions, creating the binuclear cluster with a close 3.1-Å Zn–Zn distance (Fig. XIV.1.7). The prototype of this domain is the Gal4 DNA-binding domain. These metal modules bind DNA as a dimer with dimerization occurring through a separate coiled-coil motif.

XIV.1.3. Metal Ion Signaling

Metal ion signaling requires a change in the cellular concentration, flux, or compartmentalization of a metal ion. Alternatively, signaling may arise from a modification reaction that poises a protein to respond to existing metal ion levels. A modification that poises a protein to respond to calcium levels is γ carboxylation of glutamate residues, commonly seen in extracellular proteins in the blood coagulation cascade. Metal ion signaling is also dependent on proteins that sense the metal ion signal and transduce the signal into a physiological response. The sensory and transduction processes are often metal ion specific to distinguish the signaling metal ion from other biologically important ions. An important question is how biology confers metal ion specificity on the signaling and sensory components.[12]

Metal ion signaling is best understood with respect to calcium ions (see Section XIV.2).[8] Calcium(II) ion signaling arises from physiological effectors that stimulate release of cellular Ca pools from internal stores, such as the endoplasmic reticulum and mitochondria. The intracellular Ca(II) concentration may rise from 10^{-8} to 10^{-5} M, resulting in propagation of the signal. Two major types of regulatory motifs are known that bind Ca(II) ions in the 10^{-6} M concentration range, making them ideally suited for Ca regulation.

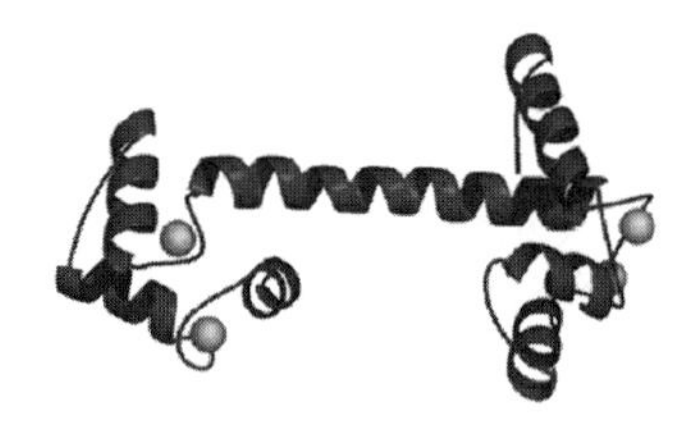

Fig. XIV.1.8.
Calcium-bound EF hand motifs in *Drosophila* calmodulin (PDB code: 4CLN). Each of the two globular domains contains two Ca(II) sites consisting of EF structural motifs.

The classical Ca-regulatory motif, originally defined in the parvalbumin protein, is designated as the EF hand, consisting of a helix–loop–helix substructure that binds a single Ca(II) ion within the 12 residue loop between nearly perpendicular helices.[9] Calcium sites within EF hands exhibit pentagonal bipyramidal geometry with five oxygen ligands coming from the loop segment and the remaining two oxygens from a bidentate carboxylate located outside the loop. A single water molecule is often present in the coordination sphere. Multiple EF hands are often present in Ca-regulatory proteins; the prototype is the Ca-regulatory protein calmodulin (Fig. XIV.1.8). Half-maximal activation of calmodulin occurs at 1-μM Ca. A common mechanism of modulation is through a Ca-induced conformational change exposing hydrophobic interactive surfaces. The conformational change alters the orientation of the two helices within an EF hand from nearly antiparallel to more perpendicular in the Ca-bound conformer. This type of change is confined to regulatory EF hands and is not as apparent in structural EF hands such as those found in calbindin. The EF hand motif serves as a molecular switch enabling the cell to detect a stimulatory influx of Ca(II), thereby transducing the signal into a cellular response. Typically, the Ca-bound form is the conformer that interacts with a specific target protein.

The EF hand motif exhibits selectivity for Ca(II) ions, permitting discrimination between Ca(II), Mg(II), and monovalent cations. Several factors contribute to the Ca selectivity. Coordination by seven oxygen atoms arranged in a pentagonal bipyramid favors Ca(II) binding over Mg(II), which prefers octahedral coordination. In addition, the high negative charge in these sites selects against binding by monovalent cations.

A class of Ca-regulatory EF hand proteins is known in which Ca binding within two EF hand motifs triggers a conformational switch exposing an N-terminal myristate (14 carbon) fatty acid moiety that directs membrane association (Fig. XIV.1.9). In the absence of calcium, the myristate is packed within a deep hydrophobic pocket. The prototype of the calcium–myristoyl switch is recoverin.[13] Recoverin contains four EF hand motifs within two domains; only one of the two EF motifs in each domain binds Ca(II). The major conformational change induced by Ca(II) binding in the C-terminal domain of recoverin occurs curiously in the non-calcium binding EF hands of the N-terminal domain.

The second major class of Ca-regulatory motifs is the C_2 domain (Fig. XIV.1.10). In contrast to Ca signaling through mononuclear sites in EF motifs, Ca signaling in C_2 domains is through formation of a polycalcium cluster.[14] The C_2 domain consists of nearly 130 residues and is found in a variety of proteins involved in generation of lipid second messengers and membrane trafficking. The general function of most C_2 domains is in Ca-regulated phospholipid binding. The C_2 domain consists of two four-stranded β strands forming a β sandwich with multiple Ca(II) ions binding between loops at the top of the sandwich structure. Aspartate carboxylates bridge twc or three Ca(II) ions resulting in a Ca cluster with Ca–Ca distances as short as 3.7 Å. Phospholipid binding requires Ca cluster formation. Not all Ca(II) coordination spheres are complete; the incomplete coordination sites are available for target molecule binding. Although Ca binding does not induce major conformational changes, the loop region is stabilized and a number of side-chain positions are altered, changing the electrostatic surface of the domain. Thus, the C_2 domain appears to be a Ca-regulated electrostatic switch. Both Mg(II) and K(I) fail to stimulate phospholipid binding, although Ba(II) and Sr(II) have limited activity. The observed Ca specificity of C_2 domain function presumably lies in the clustered Ca center.

(a)

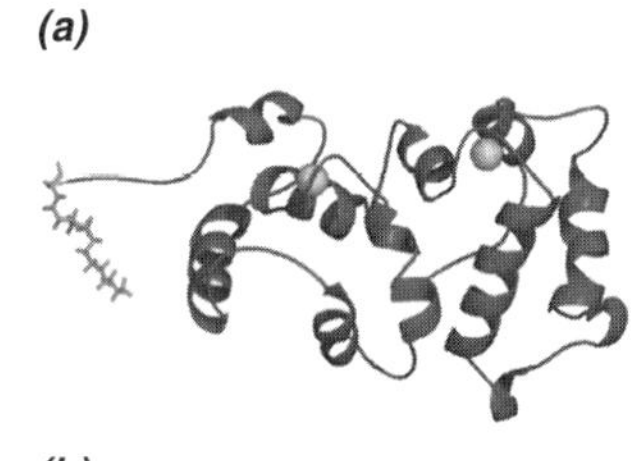

(b)

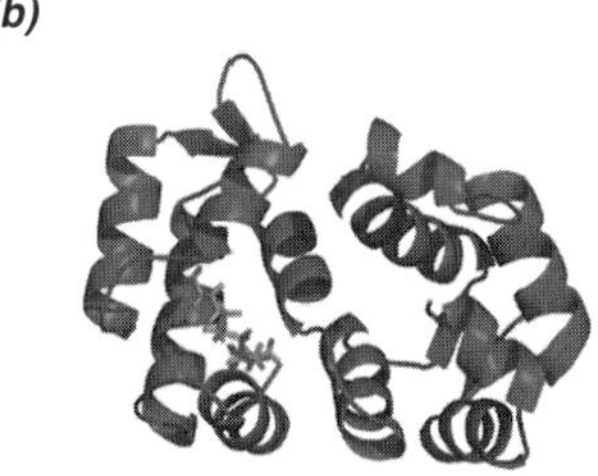

Fig. XIV.1.9.
Calcium-induced myristoyl switch in recoverin (PDB codes: 1JSA and 1IKU). The myristate group shown in stick form, projects out from the protein-fold in the Ca conformer, but is tucked within the protein in the apo conformer.

Analogous to Ca signaling, signaling by other metal ions may depend on modulation of metal ion pools within a cell, organelle, or compartment. Signaling through Zn(II) ions is known. One example of Zn ion signaling is neuromodulation by Zn(II) ions.[15] Neurons communicate to each other through transmissions across a 20-nm wide cleft formed by a junction between neurons, called the synapse. Nerve impulses are communicated across the synapse by diffusible neurotransmitters. Synapses of one class of neurons that use glutamate as the major neurotransmitter contain high levels of both glutamate and Zn(II) ions within synaptic vesicles. Zinc levels in the synapse can reach nearly 300 μM. Excitation of these neurons releases the content of Zn-containing vesicles into the cleft separating the pre- and postsynaptic membranes. The released Zn ions can modulate postsynaptic ion channel receptors. Zinc is known to inhibit one class of glutamate receptor channels and stimulate the activity of a second class. Therefore, depending on the distribution of the various types of glutamate receptors on postsynaptic neurons, Zn(II) ions may modulate the effects of glutamate. Although Zn is abundant in these neurons, synaptic Zn is not critical for neurotransmission since mice, lacking a transporter responsible for Zn loading of the synaptic vesicles, have normal neural activity. Zinc also blocks certain channel receptors for the neurotransmitters γ-aminobutyrate (GABA) and glycine in the 10^{-6} M concentration range. These GABA and glycine receptors mediate inhibitory synaptic transmissions.

Fig. XIV.1.10.
Polycalcium cluster in the C2 domain of rat synaptotagmin (PDB code: 1BYN).

A second example of Zn ion regulation occurs in a subset of Ca-binding EF hand proteins, designated S100 proteins, whose Ca-binding function is modulated by Zn binding at distinct sites. One S100 molecule, S100A3, is solely modulated by Zn and has a micromolar Zn-binding affinity suggestive that the site is regulatory and not structural. Structural Zn sites exhibit 10^{-9}–10^{-12} M Zn affinities.

Zinc ions may also promote protein oligomerization. Zinc-mediates homodimerization of the human growth hormone and the endotoxin superantigen. Zinc is important for the heterodimerization of the Lck tyrosine protein kinase to the CD4 T-cell receptor. This intracellular complex formation is necessary for T-lymphocyte activation. A pair of cysteines on each molecule binds the Zn ion in a structure called a zinc clasp motif.

XIV.1.4. Metalloregulatory Proteins

Proteins that transduce metal ion signals into changes in gene or protein expression are designated metalloregulatory proteins.[12,16] Prokaryotic cells often contain separate sensory and transduction molecules creating a two component signal transduction pathway. Metal ion sensors within the membrane modulate downstream effector molecules through post-translational modification reactions. In contrast, many metalloregulatory proteins in eukaryotic cells have combined metal sensory and transduction activities. The latter class of metalloregulatory proteins are often transcription factors or translational regulators.

Metalloregulatory proteins often function in metal ion homeostasis or detoxification.[16] Cells have mechanisms to control the levels of essential metal ions ensuring adequate, but not toxic, intracellular quantities. Such mechanisms are necessary since cells in the natural world encounter a changing environment with fluctuations in nutrient concentrations. Cell survival requires physiological responses to such changes. Homeostatic controls depend on the presence of sensory systems that detect fluctuations in intracellular metal levels and discriminate among various metal ions. Several metalloregulatory proteins have been identified that function in nutritional regulation of metal ion acquisition and storage. Many cells also have specific mechanisms to minimize cytotoxicity induced by nonessential metal ions. In prokaryotes, genetic systems exist that detoxify a variety of nonessential metal ions including cadmium, mercury, and silver. Bacterial genes functioning in metal ion detoxification are often arranged in operons in which expression of the contiguous gene array is controlled by metalloregulatory factors.

Many metalloregulatory proteins that function in metal homeostasis and detoxification exhibit metal ion specificity in their responses. Discrimination between metal ions may be achieved by three mechanisms. First, metal-specific structural motifs exist. The conformation of such motifs orients metal-liganding amino acids in a coordination sphere favoring a specific stereochemistry at the metal center that creates specific metal–ligand bond distances and angles. The combination of these constraints and the ligand types imposes metal ion selectivity for the protein, which is the basis for metal ion discrimination in a class of bacterial metal ion sensors described below. Second, the arrangement of metal ligands in a protein may favor formation of polymetallic centers in which coordination properties of the clusters impose metal ion selectivity, analogous to the polycalcium C_2 domains. Cluster formation may enhance discrimination between metal ions beyond that possible with mononuclear sites. Third, selectivity in metal binding may arise by specific metal ion uptake and/or intracellular translocation to organelles containing the metalloregulatory molecules. Vectorial transport of Cu ions to sites of copper insertion appears to be specific and protein mediated.

XIV.1.5. Metalloregulation of Transcription

Saccharomyces cerevisiae presents the most complete picture of metalloregulation in a eukaryote. Metalloregulatory factors mediating Cu, Fe, and Zn homeostasis are known.[7,12] These include Mac1 and Ace1 for copper sensing, Aft1 and Aft2 for Fe sensing, and Zap1 for Zn sensing. The response of yeast to nutritional deprivation of

Cu, Fe, or Zn ions is the transcriptional activation of genes encoding components for high-affinity uptake of these metal ions. Metal ion uptake for Cu and Fe ions involves specific permeases and cell surface metalloreductases. The metalloreductases mobilize oxidized and insoluble Fe(III) and Cu(II) ions in the environment through reduction to Fe(II) and Cu(I) ions. The Aft1, Mac1, and Zap1 metalloregulatory factors are transcription factors that control the expression of Fe, Cu, and Zn uptake genes, respectively. Ace1 (for activation of *CUP1* expression) is a transcriptional activator that mediates Cu-induced expression of Cu(I) buffering proteins.

Gene-specific transcriptional activators in yeast typically have two distinct and independent activities, DNA binding and transactivation. The proteins contact specific DNA sequences in the target gene's promoter and recruit other general transcription factors. Transactivation domains are responsible for the assembly of the transcription preinitiation complex, involving the TATA element binding protein, mediator complex, and the RNA polymerase. Metalloregulation can modulate either DNA binding, transactivation, localization, or post-translational modification reactions.

Iron transcription factors have been studied in fungi. Two distinct types of these fungal Fe-responsive transcription factors are known. In the yeast *S. cerevisiae,* the trans-activating factors Aft1 and Aft2 activate gene expression in response to low iron. In other fungi, GATA-type transcription factors repress gene expression under high Fe conditions. The mechanism of Fe inhibition of Aft1 and Aft2 function in *S. cerevisiae* is not known, but the GATA factor Fep1 from the fission yeast *Schizosaccharomyces pombe* appears to bind Fe directly. The fungal Fe-responsive GATA factors contain two adjacent Zn motifs that flank a region containing four conserved cysteine residues that may be the Fe-binding site.

Copper-mediated gene activation was first demonstrated for the *CUP1* metallothionein in *S. cerevisiae*. Yeast Cup1 is a low molecular weight Cu(I) ion buffer. The factor that mediates Cu-induced transcription of *CUP1* is the transcription factor Ace1. A homologous Cu-activated factor, Amt1, mediates the Cu-induced expression of several metallothionein genes in the yeast *Candida glabrata*. The DNA-binding interface of Ace1 and Amt1 consists of two distinct metal modules; a Zn structural domain and an adjacent domain stabilized by a tetracopper thiolate cluster. The Zn domain is a novel single Zn motif consisting of 35 residues. A contiguous domain of 60 residues enfolds the tetracopper cluster. This Cu-regulatory domain contains eight cysteinyl residues present in Cys-x-Cys or Cys-x-x-Cys sequence motifs found commonly in metallothioneins (see Section VIII.4).[12] The conformation of this domain, lacking a hydrophobic core typically found in globular domains, is dictated by Cu(I) binding. The carboxyl-terminal half of Ace1 possesses transactivation activity.

Evidence for the tetracopper cluster comes primarily from Cu K-edge extended X-ray absorption fine structure (EXAFS) analysis. The presence of an outer-shell scatter peak at 2.7 Å, seen also in crystallographically defined $[Cu_4(SPh)_6]^{2-}$ tetracopper thiolate clusters, is diagnostic of the polycopper cluster in Ace1 and Amt1. The all-or-nothing formation of a tetracopper species in Ace1 is consistent with a cluster nuclearity of 4. The cluster structure in proteins is expected to be largely maintained by μ-bridging thiolates and noncovalent interactions. The target of Cu-activated Ace1 transcription, the Cup1 metallothionein, itself contains a heptacopper cluster. Copper(I) binding by Cup1 is largely responsible for Cu(I) buffering in yeast. The Cu–Cup1 solution structure consists of two large parallel polypeptide loops separated by a deep cleft containing a heptacopper cluster (Fig. XIV.1.11). The Cup1 polycopper cluster coordinates Cu(I) ions with thiolate ligands in both digonal and trigonal geometries.

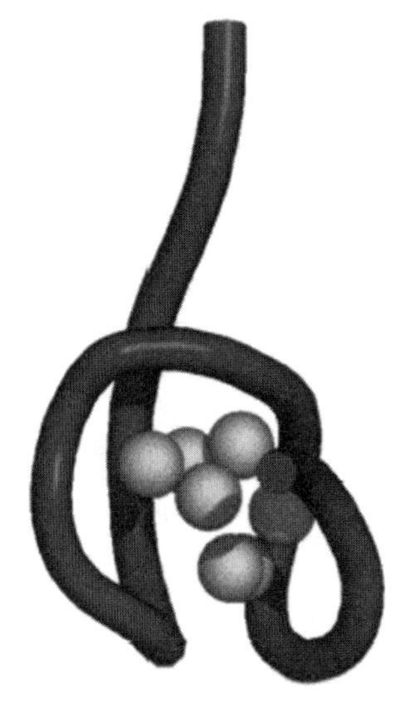

Fig. XIV.1.11.
The polycopper-thiolate cluster in yeast Cup1 metallothionein (PDB code: 1AQS). The cluster consists of seven Cu(I) ions coordinated by cysteinyl thiolates. Bridging thiolates anchor the cluster.

A polycopper cluster is also important in the metalloregulation of the Mac1 transcription factor. Yeast genes transcriptionally expressed under Cu-deficient conditions are regulated by Mac1. Expression of these genes is attenuated when the extracellular Cu(II) levels exceed 1 n*M*. Copper(I) binding within two C-terminal Cys-rich motifs inhibits activation activity and also DNA binding by the N-terminal DNA binding domain. A tetracopper cluster forms in each of the two Cys-rich motifs, triggering a conformational switch resulting in an intramolecular interaction with the N-terminal DNA-binding domain.

The importance of a tetracopper cluster as the structural unit within the activated transcription factors may be twofold. First, a polycopper cluster formed by eight cysteinyl residues, as in Ace1, is able to organize and stabilize a larger structural unit than a single-bound metal ion. Second, a polycopper cluster provides metal ion specificity. A second metal ion that can activate Ace1 is Ag(I). Silver(I) is known to form similar metal–thiolate cage clusters as seen with Cu(I). Thus, the metal ion specificity of Ace1, Amt1, and Mac1 may arise from the ability of a metal ion to form a metal–thiolate cage structure that brings multiple ligands in proper juxtaposition to create a DNA-binding interface in Ace1 or inhibit the activation domain in Mac1.

Polymetallic centers have already been mentioned as important regulatory components in regulatory C_2 domains (polycalcium centers) and iron regulatory proteins (iron–sulfur clusters). Iron–sulfur clusters are also important sensors of oxygen and nitric oxide.[10]

Gene regulation that is tied to the nutritional metal ion status is also seen in other species. *Chlamydomonas reinhardtii* can functionally interchange two proteins for a key electron-transfer step during photosynthesis depending on the copper status of the cell. Under conditions of Cu ion deficiency, cytochrome c_6 replaces the Cu-containing plastocyanin as a key electron-transfer molecule. The heme moiety in cytochrome c_6 becomes the redox active cofactor. Genes for cytochrome c_6 and the heme biosynthetic enzyme coproporphyrinogen oxidase are transcriptionally activated in Cu-deficient algae by Crr1. Unlike Mac1 from *S. cerevisiae*, Crr1 appears to respond to Cu(II) ions.[4]

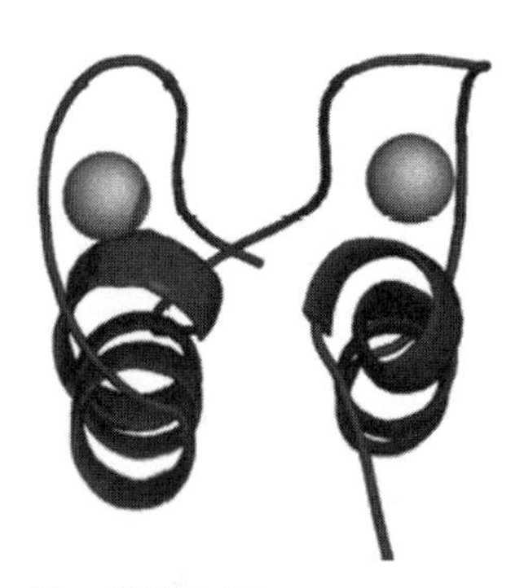

Fig. XIV.1.12.
Gli Zn-finger pair motif from the five finger human Gli–DNA complex (PDB code: 2GLI). Hydrophobic contacts mediate the packing of the two finger motifs together. The two Zn(II) ions in this finger pair are separated by 12.5 Å.

A series of Zn metalloregulatory proteins are known. Two eukaryotic Zn metalloregulatory proteins, Zap1 and MTF-1, contain multiple classical Zn finger motifs and regulation may involve the Zn finger domains. Zinc negatively regulates the Zap1 transcription factor in *S. cerevisiae*. In contrast, Zn activates gene expression in animal cells through MTF-1.Zap1 activates the expression of >40 genes in cells deficient in Zn(II), a number of which encode Zn(II) ion transporters. The Zap1 activation of gene expression in yeast involves binding to a DNA promoter element (ZRE for Zn Responsive Element). A ZRE located within the *ZAP1* promoter allows Zap1 to regulate its own expression; and Zap1 is inhibited in Zn-replete cells through Zn-inactivation of two transactivation domains, designated AD1 and AD2. The Zn inhibition of AD2 involves Zn(II) binding to two labile zinc finger motifs that pack together to form a finger pair structure analogous to a finger pair structure in the mammalian transcription factor Gli (Fig. XIV.1.12). Unlike the Zn(II) binding to most Zn finger motifs, which are structural, Zn(II) binding to the two zinc fingers in AD2 is highly labile. The Zn(II) occupancy of the two zinc finger motifs appears to result in a stable finger pair structure that masks residues essential for transactivation. Thus, AD2 in Zap1 appears to be under kinetic control, responding to changes in the flux of intracellular Zn(II) levels. This is the first example of a regulatory zinc finger motif. Zinc inhibition of AD1 in Zap1 does not involve zinc finger motifs. Rather, zinc metalloregulation may occur through formation of adamantane-type clusters resembling the Zn clusters in mammalian metallothioneins.

The *MTF-1* gene is essential in the mouse, and deletion of *MTF-1* results in *in utero* lethality owing to degeneration of hepatocytes. The MTF-1 protein binds to a DNA enhancer element, designated MRE (for Metal-Regulatory Element) in a Zn-dependent manner, and mediates Zn activation of MRE-containing genes including a family of mammalian metallothionein genes, a Zn-specific efflux pump, and a gene encoding the first enzyme in the glutathione biosynthesis pathway. A subset of the six zinc finger domains in MTF-1 is important for MRE binding. Zinc-activation of MTF-1 involves multiple steps and MTF-1 appears to normally cycle between the nucleus and cytoplasm. The addition of zinc salts to mammalian cells results in the nuclear accumulation of MTF-1. Zinc is also required to induce DNA-binding activity in MTF-1. One model of Zn activation is reversible Zn(II) binding within two regulatory zinc-finger motifs.

A different class of Zn-metalloregulatory proteins exists in bacteria. Two distinct cyanobacteria have related Zn-responsive transcriptional repressors.[17,18] The *Synechococcus* SmtB regulates Zn ion sequestration through regulated expression of a metallothionein. *Synechocystis* ZiaR regulates expression of a Zn-efflux transporter. The SmtB functions as a repressor, binding to the *smt* DNA operator-promoter in Zn-limiting cells and dissociating in Zn-supplemented cells. The SmtB and ZiaR factors are members of a family of bacterial metal-regulated repressors that differ in metal ion selectivity. The SmtB family of proteins consists of dimeric molecules with a classical helix–turn–helix structural motif. Two classes of metal sites exist, with ligands for each coming from residues on both subunits. The regulatory Zn(II) site pair in SmtB lies near the C-terminus of each polypeptide chain at the dimer interface (Fig. XIV.1.13). This site consists of carboxylates and imidazoles that favor binding of harder Lewis acids. In contrast, a pair of Zn(II) sites exists within the DNA-binding motif, which consists of thiolates, and therefore favors binding by thiophilic metal ions. Variants of SmtB exist in which metal-specific regulation of DNA binding occurs through the C-terminal pair of sites. Metal selectivity is due to a combination of metal ion availability in the bacterial cytoplasm and coordination geometry. Thus, the mycobacterium NmtR regulator responds to Ni(II) or Co(II) ions that tend to bind in octahedral coordination geometry, whereas the Zn(II) specificity of SmtB arises from tetrahedral coordination. Studies on the bacterial SmtB family of regulators have shown that metal ion coordination geometry is more significant than metal ion affinity as a determinant of metal ion selectivity.

Two zinc-metalloregulatory transcription factors, Zur and ZntR, are known in *E. coli*. The Zn–Zur complex is a repressor of transcription of the Zn(II) uptake

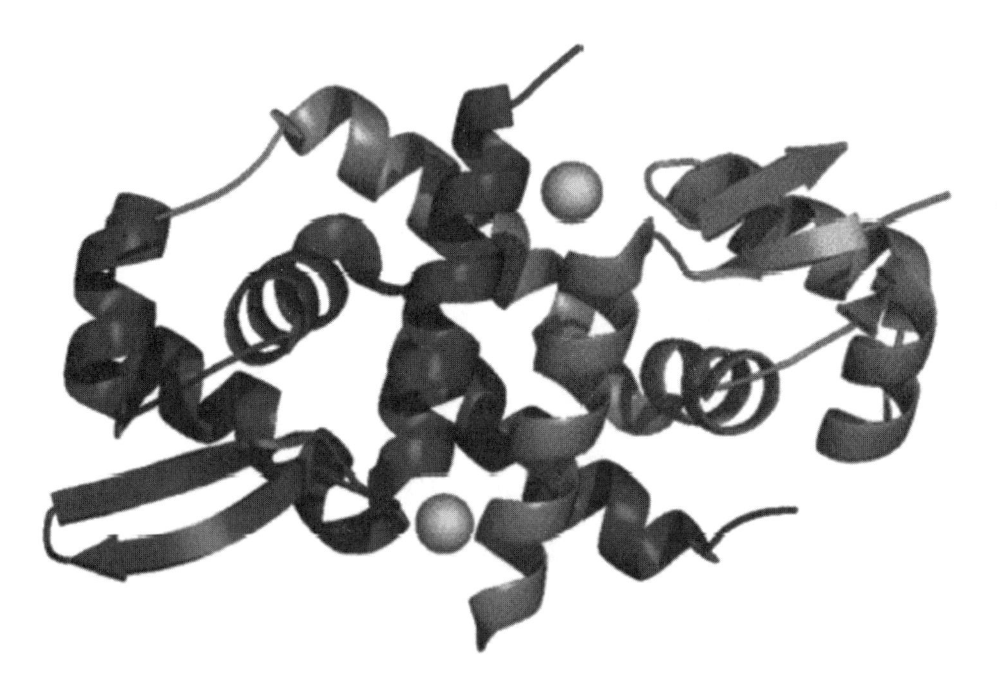

Fig. XIV.1.13. Structure of cyanobacteria SmtB repressor with Zn(II) ions bound in the regulatory alpha5 sites at the dimer interface (15) (PDB code: 1R22). The DNA-binding helix–turn–helix motif lie at the outside edge of this view.

transporter gene(s), whereas the Zn–ZntR complex activates expression of a Zn(II) efflux transporter. A well-studied bacterial metalloregulatory protein, related to ZntR, is MerR. The MerR metalloregulatory protein controls expression of bacterial mercury resistance genes arranged in a genetic operon. One resistance gene encodes an enzyme that reduces ionic Hg(II) to volatile mercury vapor Hg(0).[19] MerR represses transcription of the Mer operon in the absence of Hg(II), but Hg(II) binding converts MerR to a transcriptional activator. The MerR proteins differs from SmtB in that the Hg-bound MerR remains bound to DNA. Although the structure of MerR is unresolved, DNA binding appears to occur through HTH domains in the dimeric protein. Apo-MerR binds DNA between the −10 and −35 promoter elements forming an inactive complex in which the DNA helix is bent. The kinked DNA promoter element is a poor template for RNA polymerase. Mercury(II) binding occurs in a domain adjacent to the DNA-binding domain and relaxes the MerR-induced DNA bend. Thus, MerR regulation is an example of metal-responsive modulation of DNA structure, although the structural basis for the Hg-induced unbending of DNA is unresolved. Other metal ions can activate the Mer operon but at 100-fold higher concentrations. The basis for the selectivity rests in the tricoordinate cysteine binding site consisting of two cysteinyl residues of one subunit and a cysteine from the second subunit.

The MerR-like molecule, CueR, mediates Cu-metalloregulation of expression of the bacterial copper efflux pump.[16] In *E. coli,* CueR induces expression of the CopA effluxer under conditions when the dimeric CueR binds two Cu(I) ions, each in a linear S–Cu^{I}–S geometry by thiolate ligands. The structure of the Cu(I) sites in the CueR dimer is shown in Fig. XIV.1.14. The anionic charge of the buried S–Cu–S center may be partially neutralized through a positive helical dipole. The combination of the linear coordination geometry and the charge neutralization of a bound +1 charged ion are likely important for the observed Cu(I) selectivity of CueR.

Transcriptional regulation by molybdenum oxyanions is known in bacteria. Molybdate-bound ModE binds DNA and represses transcription of an *E. coli* operon encoding a high-affinity molybdate transport system, but activates transcription of genes encoding molybdoenzymes and proteins important for molybdenum cofactor biosynthesis. Molybdate binds to a tandemly repeated, conserved structural motif in the C-terminal dimerization domain of ModE. The DNA-binding function maps to the winged helix–turn–helix motif in the N-terminal domain. The molybdate structural motif is specific, selecting molybdate or tungstate against phosphate

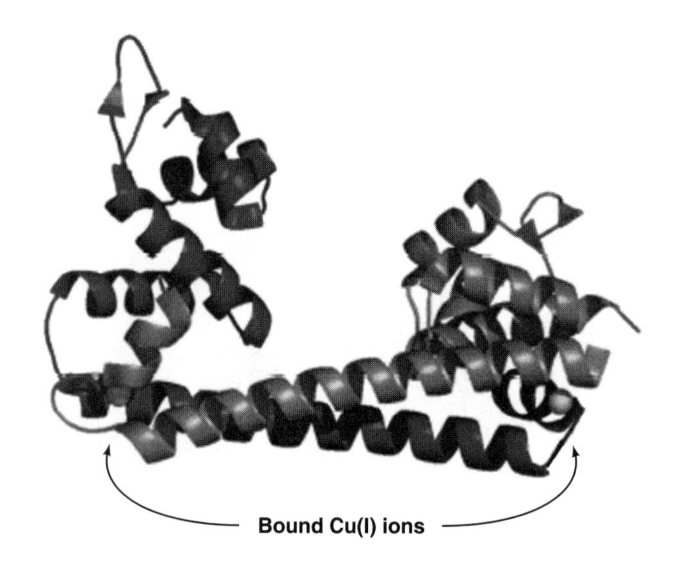

Fig. XIV.1.14.
Structure of the *E. coli* Cu(I) metalloregulatory protein, CueR (PDB code: 1Q05). The two chains are colored in dark gray and light gray. The Cu(I) ions are at each end of the dimerization helix.

or sulfate. Molybdate binds with micromolar affinity, whereas sulfate and phosphate anions exhibit only millimolar affinities. The basis for this selectivity may resemble that in the periplasmic molybdate-binding protein ModA. In ModA, molybdate is tetrahedrally bound in a deep cleft at the junction of four helices through seven hydrogen bonds. Selectivity may arise from size selection, the volume of the binding site destabilizing hydrogen bonding to smaller anions such as sulfate.

XIV.1.6. Metalloregulation of Post-Transcriptional Processes

Iron metabolism is regulated in cells to ensure adequate, but not excessive, levels of cellular iron. Cellular iron homeostasis is regulated in animal cells by post-transcriptional mechanisms. Two cytoplasmic iron regulatory proteins, IRP1 and IRP2, modulate the translation or stability of messenger ribonucleic acids (mRNAs). These proteins bind RNA hairpin structures located in mRNAs encoding proteins involved in iron metabolism. The RNA-hairpin structures, known as iron-responsive elements, are located either in 5′ or 3′ untranslated sequences of mRNAs, where they mediate iron-responsive control of translation (5′ IREs) or mRNA stability (3′ IREs). The IREs are stem–loop RNA structures with a conserved terminal loop and a base-paired stem interrupted by a critical internal bulge. The RNA binding by IRP occurs only in Fe-deprived cells. The IRP binding to IREs located in the 5′ untranslated sequences represses translation. The cytoplasmic iron sequestration molecules H- and L-ferritin mRNAs contain 5′ IREs. In contrast, IRP binding to IREs located in the 3′ untranslated sequences prevents degradation of mRNA and, therefore, translation is increased. Thus, mRNA levels for two proteins containing 3′ IREs, the transferrin receptor, and the divalent metal transporter (DMT1), are increased in Fe-deprived cells. In Fe-replete cells mRNAs with 5′ IREs are translated, whereas mRNAs with 3′ IREs are degraded because of the presence of a separate rapid degradation structural motif (see Sections VIII.1 and VIII.2 for discussions of transferrin and ferritin).

IRE binding by IRPs in Fe-deficient cells results in enhanced transferrin receptor and DMT1 mRNA translation. Upregulation of the transferrin receptor and DMT1 is a cellular mechanism to acquire iron from the circulating iron reservoir, transferrin. The transferrin receptor binds diferric transferrin on the cell surface and the complex is internalized into endosomal vesicles where Fe ions are released. The DMT1 protein is responsible for transport of the released endosomal iron into the cytoplasm. Tight regulation of iron uptake is necessary as excess cellular iron can be toxic.

The IRP1 molecule is a stable, bifunctional protein in which the presence or absence of an iron–sulfur cluster dictates whether the molecule is an RNA-binding protein or a non-RNA-binding molecule with aconitase enzymatic activity. In Fe-replete cells, IRP1 is an aconitase with a functional 4Fe–4S cluster. In contrast in Fe-deprived cells, IRP1 lacks the FeS cluster, and therefore lacks aconitase activity, but is competent to bind the IRE element. The RNA-binding function of IRP1 is, therefore, regulated by reversible assembly of the FeS cluster. The RNA-binding region appears to overlap with the catalytic sites, explaining the mutually exclusive functions of IRP1.

The Fe-dependent loss of IRE binding differs for IRP1 and IRP2 in that IRP2 is degraded by the proteasome in Fe-replete cells in contrast to the formation of the FeS cluster in IRP1. In addition, IRP2 lacks an aconitase function. The mechanism for Fe-induced degradation of IRP2 may involve Fe-mediated oxidation of IRP2 followed by ubiquitin-mediated proteasomal degradation. Additional modulation of IRP function occurs by nitric oxide signaling and hypoxia.

Targeted disruption of IRP2 in the mouse results in cerebellar dysfunction, whereas IRP1 null mice lack any obvious phenotype. One interpretation of the null mouse phenotypes is that IRPs exhibit selectivity for different IREs. The IRE variations and different tissue IRP expression ratios may provide a range of mRNA-specific responses to a change in iron flux.

XIV.1.7. Post-Translational Metalloregulation

Metal ions regulate a number of post-translational processes involving membrane proteins. These include ligand–membrane receptor complex formation, degradation, and membrane trafficking of plasma membrane transporters. Receptors and transporters within the plasma membrane are important in nutrient signaling and acquisition. Homeostatic mechanisms not only regulate the biosynthesis of these receptors and transporters, but also modulate their activity. Post-translational metalloregulation permits rapid responses to changes in metal ion availability.

An example of regulated ligand–receptor complex formation is transferrin. A major function of transferrin (Tf) is the transport of iron from sites of storage and absorption to peripheral tissues. Tissue uptake of transferrin iron is mediated by the transferrin receptor (TfR). At physiological pH, TfR preferentially interacts with the diferric form of transferrin in preference to metal-free transferrin, permitting productive uptake of iron ions. Discrimination between the two forms of Tf occurs through Fe-induced conformational dynamics in transferrin. Iron(III) binding to Tf induces a hinge-bending motion between domains, resulting in closure of the bilobe structure. The cleft separating the two lobes is reduced by 20 Å (Fig. XIV.1.15). The transferrin receptor can discriminate between these conformers at physiological pH (see Section VIII.1).

Post-translational down regulation of transporter function occurs with several metal ion transporters in yeast. The Zrt1 transporter, responsible for high affinity Zn(II) ion uptake, is stably present in the plasma membrane in Zn-deficient cells. Exposure of the cells to elevated extracellular Zn levels results in a Zn-induced endocytosis of the transporter and subsequent proteolytic degradation within the vacuole. Likewise, elevated copper levels result in the internalization and degradation of the yeast Ctr1 Cu(I) transporter.

Metal-induced endocytosis is also observed in the mammalian cell surface prion precursor protein PrP. The normal cellular PrP protein is a precursor to the pathogenic prion molecule responsible for spongiform encephalopathy in cattle and Creutzfeldt–Jacob disease in humans. The PrP protein may have a role in copper

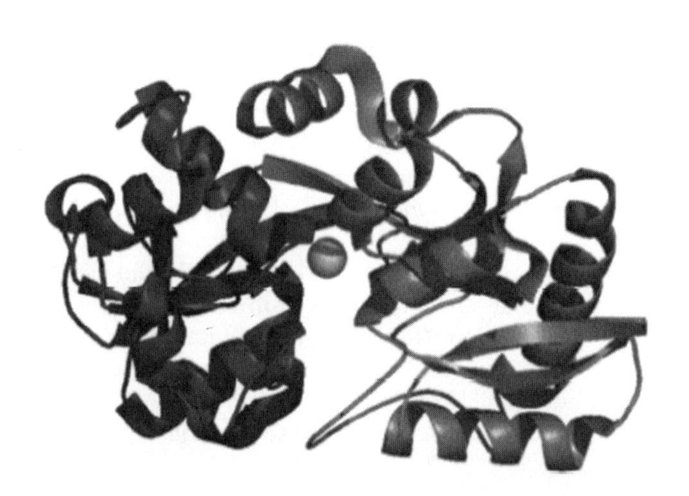

Fig. XIV.1.15. Iron-induced closure of the bilobe structure of porcine serum transferrin (PDB code: 1H76). The two lobes in transferrin are shaded differently and are shown closed down on the ferric ion.

metabolism in the brain. Experiments in cultured cells revealed that copper rapidly stimulates endocytosis of PrP from the cell surface. The N-terminal region of PrP contains octapeptide repeats that bind Cu(II) ions with micromolar affinity.

Metal ion regulated membrane protein trafficking occurs in mammalian cells with two Cu(I) translocating P-type ATPases, designated MNK and WND. Mutations in the MNK and WND genes are responsible for Menkes' and Wilson's diseases, respectively, in humans (see Section VIII.5). Both proteins are localized in the trans-Golgi network where they function in transporting Cu(I) ions into the trans-Golgi lumen for incorporation into cuproenzymes as they progress through the secretory pathway. In copper-loaded cells, MNK translocates to the plasma membrane and functions to efflux copper ions from the cell. The translocation of MNK and Cu efflux function are significant components of mammalian copper ion detoxification. A subset of the six Cu(I)-binding motifs is required for the observed Cu-dependent translocation to the plasma membrane.

The combination of transcriptional, translational, and post-translational regulatory mechanisms permits fine tuning of metal ion homeostasis. The involvement of multiple mechanisms in the homeostasis of a given metal ion permits a graded response of a cell to its metal status. Additional factors in metal homeostasis may likely exist. For example, under conditions of partial metal deficiency, a cell may redistribute its metal quota to ensure that the essential metalloenzymes receive their metal quota at the expense of less important metalloproteins. One unresolved issue is how cells partition their metal quota to achieve optimal metabolism.

References

General References

1. Alberts, I. L., Nadassy, K., and Wodak, S. J., "Analysis of zinc binding sites in protein crystal structures", *Prot. Sci.*, **7,** 1700–1716 (1998).
2. Berg, J. M. and Shi, Y., "The galvanization of biology: A growing appreciation for the roles of zinc", *Science,* **271,** 1081–1085 (1996).
3. Christianson, D. W., "Structural biology of zinc", *Adv. Protein Chem.*, **42,** 281–355 (1991).
4. Glusker, J. P., "Structural aspects of metal liganding to functional groups in proteins", *Adv. Protein Chem.*, **42,** 3–76 (1991).
5. Harding, M. M., "Geometry of metal-ligand interactions in proteins", *Acta Crystallogr.*, **D57,** 401–411 (2001).
6. Holm, R. H., Kennepohl, P., and Solomon, E. I., "Structural and functional aspects of metal sites in biology", *Chem. Rev.*, **96,** 2239–2314 (1996).
7. Rutherford, J. C. and Bird, A. J., "Metal-responsive transcription factors that regulate iron, zinc and copper homeostasis in eukaryotic cells", *Eukaryotic Cell* (in press) (2003).
8. Evenas, J., Malmendal, A., and Forsen, S., "Calcium", *Curr. Opin. Chem. Biol.*, **2,** 293–302 (1998).
9. McPhalen, C. A., Strynadka, N. C. J., and James, M. N. G., "Calcium Binding Sites in Proteins: A Structural Perspective", *Adv. Prot. Chem.*, **42,** 77–144 (1991).
10. Beinert, H., Holm, R. H., and Munck, E., "Iron–Sulfur Clusters: Nature's Modular, Multipurpose Structures", *Science,* **277,** 653–659 (1997).
11. Schwabe, J. W. R. and Klug, A., "Zinc mining for protein domains", *Nat. Struct. Biol.*, **1,** 345–349 (1994).
12. Winge, D. R., "Copper Metalloregulation of Gene Expression", *Adv. Prot. Chem.*, **60,** 51–92 (2002).

Specific References

13. Ames, J. B., Ishima, R., Tanaka, T., Gordon, J. I., Stryer, L., and Ikura, M., "Molecular Mechanics of Calcium-Myristoyl Switches", *Nature (London),* **389,** 198–202 (1997).
14. Rizo, J. and Sudhof, T. C., "C_2-domains, structure and function of a universal Ca^{2+}-binding domain", *J. Biol. Chem.*, **273,** 15879–15882 (1998).
15. Huang, E. P., "Metal ions and synaptic transmission: Think zinc", *Proc. Natl. Acad. Sci. U.S.A.*, **94,** 13386–13387 (1997).
16. Finney, L. A. and O'Halloran, T. V., "Transition metal speciation in the cell: Insights from the chemistry of metal ion receptors", *Science,* **300,** 9331–9936 (2003).
17. Eicken, C., Pennella, M. A., Chen, X., Koshlap, K. M., VanZile, M. L., Sacchettini, J. C., and Giedroc, D. P., "A metal–ligand–mediated intersubunit allosteric switch in related SmtB/ArsR zinc sensor proteins", *J. Mol. Biol.*, **333,** 683–695 (2003).
18. Pennella, M. A., Shokes, J. E., Cosper, N. J., Scott, R. A., and Giedroc, D. P., "Structural elements of metal selectivity in metal sensor proteins", *Proc. Natl. Acad. Sci. U.S.A.*, **100,** 3713–3718 (2003).
19. Ansari, A. Z., Bradner, J. E., and O'Halloran, T. V., "DNA-bend modulation in a repressor-to-activator switching mechanism", *Nature (London),* **374,** 371–375 (1995).

XIV.2. Structural Zinc-Binding Domains

John S. Magyar
Beckman Institute
California Institute of Technology
Pasadena, CA 91125

Paola Turano
CERM and
Department of Chemistry
University of Florence
Sesto Fiorentino,
Italy 50019

Contents

XIV.2.1. Introduction

Section XIV.1 on metalloregulatory proteins introduces the concept of structural zinc sites in proteins and describes several types of structural zinc-binding domains [e.g., GATA domains, LIM domains, the RING finger, and the classical (TFIIIA-type) zinc finger]. Here, we further discuss these proteins, focusing on the macromolecular interactions between structural zinc-binding domains and nucleic acids and on the coordination chemistry of the metal-binding site itself.[1–10]

Properly folded states in proteins are stabilized by a variety of intramolecular contacts, including both covalent and noncovalent interactions. In the case of structural zinc-binding domains, which are relatively small (~18–35 amino acids), tetrahedral metal binding is required for proper protein folding. The native metal in these sites is a zinc(II) ion, tetrahedrally coordinated by histidine and/or cysteine residues. The coordination geometry of Zn^{2+} closely resembles that of iron in rubredoxins (Chapters IV and X.1). The zinc ion can be replaced by a variety of other metals, with differing effects on the structure and activity of the protein; some of these effects will be discussed later in this section.

XIV.2.2. Molecular and Macromolecular Interactions

Structural zinc-binding domains participate in a range of molecular interactions, including protein–DNA, protein–RNA, protein–protein, and protein–lipid interactions. Here, we present a specific case to illustrate some of the chemical features of such intermolecular interactions.

DNA-binding proteins that use zinc finger domains to contact DNA almost invariably contain two or more zinc finger domains, which work in tandem for nucleic acid recognition. Each domain interacts primarily with three base pairs of DNA. The first zinc finger to be characterized by X-ray crystallography was the DNA-binding region of the murine transcription factor Zif268.[11,12] The Zif268 structure contains three CCHH-type zinc fingers. This transcription factor was crystallized as a complex with an oligonucleotide corresponding to a favored DNA-binding site.

To accompany the discussion that follows, we recommend that the reader download and view the Zif268 structure from the Protein Data Bank (PDB code: 1ZAA; see Appendix D; see Figs. XIV.2.1 and XIV.2.2). In this structure, the amino acid residues in positions −1, 3, and 6 of the first and third zinc finger domain (numbered relative to the start of the helix) are positioned to make specific, one-to-one contacts with the 3′-, central, and 5′-bases of the DNA triplet subsite on the primary strand. The first and third zinc finger domains of Zif268 have Arg in position −1, Glu in

Fig. XIV.2.1.
The molecular structure, determined by X-ray crystallography to 2.1-Å resolution, of the three zinc fingers of Zif268 complexed to DNA (PDB: 1ZAA). The three zinc fingers fit neatly along the major groove of the DNA double helix, with the α-helices of the zinc fingers interacting directly with the DNA. Each zinc ion is bound to two Cys sulfurs and two His nitrogens. Figure was prepared from the PDB file (1ZAA) using UCSF Chimera. From the Computer Graphics Laboratory, University of California, San Francisco (supported by NIH P41 RR-01081). [See color insert.]

position 3, and Arg in position 6. The guanidinium groups of these Arg residues participate in a pair of hydrogen bonds with guanine bases in the binding site. The Glu residues do not participate in hydrogen bonds with the DNA, but they sterically favor cytosine bases.

The second zinc finger domain of Zif268 has Arg in position 1, His in position 3, and Thr in position 6. The Arg residue forms two hydrogen bonds with a guanine, while the His side chain binds guanine through a single hydrogen bond to N7 of the base. The Thr does not contact the DNA. In addition to these contacts, each zinc finger domain has an Asp residue in position 2. The carboxylate group of each Asp forms a hydrogen-bonded salt bridge with the Arg residue in position 1 and a hydrogen bond with an adenine or cytosine base on the complementary strand immediately preceding that zinc finger domain's subsite. Given this interaction, adjacent zinc finger domains actually contact partially overlapping four-base-pair subsites. In addition to these base contacts, Zif268 makes a number of base-independent contacts with the DNA phosphodiester backbone. The NH group from the imidazole of the first metal-bound His residue from each domain participates in a hydrogen bond with an oxygen from a backbone phosphate. In addition, several Lys and Arg side chains within each domain may form direct or water-mediated hydrogen bonds with the backbone.

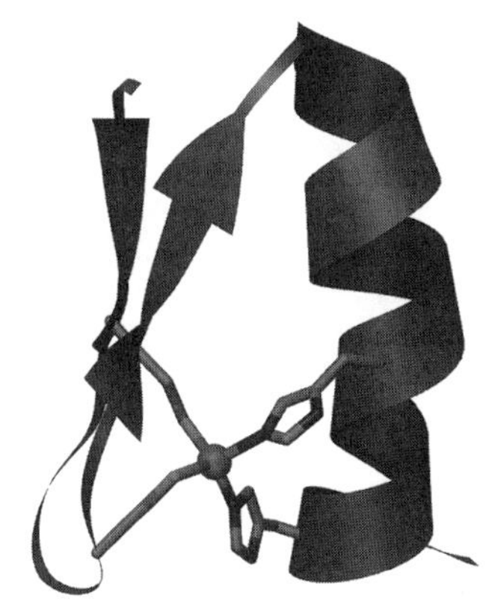

Fig. XIV.2.2.
A closeup view of the N-terminal zinc finger of Zif268 (PDB: 1ZAA) showing the canonical ββα motif and the tetrahedral Cys_2His_2 coordination of the Zn^{2+} ion. Figure was prepared from the PDB file (1ZAA) using UCSF Chimera.

The linkers joining one zinc finger domain to the next are flexible in the absence of DNA but become structured upon DNA binding. While the linker region does not appear to make any direct contacts with the DNA, structural comparisons and mutagenesis experiments suggest that it plays a role in DNA binding by stabilizing the positioning of individual zinc finger domains relative to one another.

The description of the above structure is just one example of how the chemical features of amino acids at the interface with DNA modulate the recognition between the protein and the nucleic acid. The guanidinium groups of arginines recognize guanines, histidines can interact with cytosine, aspartates interact with adenines and cytosines. Moreover, when proteins contain multiple fingers, each finger binds to adjacent subsites within a larger DNA recognition site, thus allowing a relatively simple motif to bind specifically to a wide range of DNA sequences. The combination of several zinc finger domains in the same protein molecule also ensures specificity: each domain recognizes a group of four bases, and the sequences recognized by the different domains must be at a particular position on the DNA molecule. These macromolecular interactions are the key for the regulatory function of zinc finger-containing proteins.

XIV.2.3. Metal Coordination and Substitution

XIV.2.3.1. Zinc and Cobalt

Structural zinc-binding domains gain their structure from the coordination of a single Zn^{2+} ion in the center of the domain. In a sense, using a metal ion to direct folding of this small domain frees up the other amino acids for their roles in intermolecular interactions with nucleic acids. However, the very feature that makes Zn^{2+} ideal for its structural role in zinc fingers and other similar sites in transcription factors (i.e., its lack of redox activity) makes it extremely difficult to study directly. The Zn^{2+} ion, with the electron configuration $[Ar]3d^{10}$, is often described as "spectroscopically silent". Since its *d* orbitals are completely filled, the *d–d* transitions that give the other transition metals their bright colors cannot occur; and solutions of Zn^{2+} are colorless. In addition, zinc-thiolate charge-transfer bands (from zinc-cysteine coordination) appear at high energies and are difficult to observe.

In place of Zn^{2+}, many studies of zinc-binding proteins have substituted Co^{2+}. Unlike Zn^{2+}, Co^{2+} ($[Ar]3d^7$) has strong *d–d* transitions in the visible region of the spectrum, making it easily observable by visible absorption spectroscopy; Co^{2+}-thiolate charge-transfer bands also appear in an accessible region of the ultraviolet (UV) absorption spectrum. In addition, Co^{2+} and Zn^{2+} are nearly the same size (Zn^{2+}, $r = 0.60$ Å; Co^{2+}, $r = 0.58$ Å)[13] and have been shown to adopt the same tetrahedral coordination in these protein sites. In catalytic zinc-binding sites, such as those described in Chapter IX.1, the enzymes are often still active with cobalt bound in place of zinc; and cobalt-substitution studies have been critical to the development of our understanding of these proteins' structures and mechanisms.[14,15]

The first clue that the coordination of Co^{2+} changes upon protein binding comes from visual observation—"free" $[Co(H_2O)_6]^{2+}$ is pink; the solution of the Co^{2+}–peptide complex is blue. On protein binding, the absorption maximum shifts from ~510 nm ($\varepsilon = 5\,M^{-1}\,cm^{-1}$) to ~640 nm ($\varepsilon \sim 500–900\,M^{-1}\,cm^{-1}$), with an associated dramatic increase in intensity. Both the 100-fold increase in ε and the shift in the absorption maximum to lower energy are characteristic of a change from octahedral to tetrahedral coordination; in fact, a similar color change is observed for the classic case in coordination chemistry:[16]

$$\underset{\text{Pink}}{2[Co(H_2O)_6]Cl_2} \rightleftharpoons \underset{\text{Blue}}{Co[CoCl_4]} + 12H_2O$$

The ligand-field spectra of Co^{2+} are now well understood, and a wealth of information about the protein metal-binding site is available from these spectra.

In order to learn more about the metal coordination environment in structural zinc-binding domains, Berg and co-workers[17,18] prepared a series of zinc-binding consensus peptides based on 131 naturally occurring zinc finger sequences. They prepared three different peptides, varying the relative numbers of cysteines and histidines in the metal-binding site, but otherwise retaining the same amino acid sequence. The peptides were given the names CP-CCHH, CP-CCHC, and CP-CCCC, the "CP" for "consensus peptide", and the other letters denoting the metal-binding residues. Berg and co-workers[17,18] then investigated the Co^{2+}-binding properties of the three peptides and were able to observe the effect that changing the ligands has on the spectra.

For all three peptides, absorption bands at $\sim$640 nm ($\varepsilon \sim$ 500–900 M^{-1} cm^{-1}) appear on metal binding, indicating tetrahedral metal coordination. As the number of cysteine thiolate ligands is increased, the absorption bands shift to lower energies; the splitting of the bands also changes with the change in coordination.[18]

The Co^{2+}-peptide spectra also make it possible to study the metal-binding thermodynamics of the system by spectrophotometric titrations. Briefly, a protein sample is titrated with aliquots of a Co^{2+} solution, and the increase in the characteristic Co^{2+}-peptide spectrum is monitored. From these data, a Co^{2+}-peptide binding constant, $K_b{}^{Co}$, can be determined. By titrating the Co^{2+}-peptide solution with Zn^{2+} and monitoring the decrease in the intensity of the absorption bands as Zn^{2+} replaces Co^{2+} in the protein, it is then possible to also obtain $K_b{}^{Zn}$, the zinc-binding constant.[5,18]

Studies of this type have now been performed for a number of different structural zinc sites; and although the absolute-binding constants vary, Zn^{2+} consistently binds more tightly than Co^{2+} does in the same site;[5,18–24] significant specificity for Zn^{2+} over other transition metals (e.g., Mn^{2+}, Fe^{2+}, Ni^{2+}, Cu^{2+}) has also been observed.[25,26]

If Co^{2+} and Zn^{2+} both stabilize the same geometry, why does Zn^{2+} bind so much more tightly than Co^{2+}? The answer lies in ligand-field theory.

As described earlier, Co^{2+} goes from octahedral to tetrahedral coordination geometry on peptide binding. This change in geometry is accompanied by a change in the ligand-field stabilization energy (LFSE) (see Tutorial II).[27] The LFSE for the octahedral complex $[Co(H_2O)_6]^{2+}$ is -21.3 kcal mol^{-1}; that for the Co^{2+}–CP–CCHH complex is -16.8 kcal mol^{-1}.[18,28] From these energies, it is clear why Zn^{2+} binds more tightly than Co^{2+}. Since its *d* orbitals are filled, Zn^{2+} experiences no change in ligand-field stabilization energy when it goes from octahedral to tetrahedral coordination on peptide binding. In contrast, Co^{2+} pays a 4.5-kcal mol^{-1} penalty for the shift to tetrahedral geometry. Since Zn^{2+} and Co^{2+} are so similar in size and polarizability, the other factors that generally affect the specificity of metal binding, the LFSE plays the dominant role in this case.[13,18,29] Thus, for the metal substitution reaction

$$Co^{2+}\text{–peptide} + [Zn(H_2O)_6]^{2+} \rightleftharpoons Zn^{2+}\text{–peptide} + [Co(H_2O)_6]^{2+}$$

the LFSE accounts for much of the energy driving this equilibrium to the right.

Thus the Zn^{2+} ion seems uniquely suited for its structural role in transcription factors. It is redox inactive, so it will not generate reactive species (radicals, etc.) that might degrade the DNA and RNA that bind to the protein. The Zn^{2+} ion binds the protein tightly, and it neatly stabilizes the proper domain structure. Finally, there is significant specificity for Zn^{2+} in these domains owing to the destabilizing LFSE effects on other transition metal ions.

There are several metals, however, that can bind as tightly (or more tightly!) than Zn^{2+} to structural zinc-binding domains; and it is these metal ions that we turn to next.

XIV.2.3.2. Arsenic, Cadmium, and Lead

The As^{3+}, Cd^{2+}, and Pb^{2+} ions have the potential to bind to and interfere with structural zinc-binding domains.[30–33] The ions As^{3+} ([Ar]$3d^{10}4s^2$), Cd^{2+} ([Kr]$4d^{10}$), and

Pb^{2+} ([Xe]$5d^{10}6s^2$) all have closed shells; like Zn^{2+}, they do not have any *d–d* transitions. For Cd^{2+} and Pb^{2+}, however, the metal–thiolate charge-transfer bands in the absorption spectrum appear at a long enough wavelength (>250 nm) to be easily observable.[6,22,32,34] The presence of these bands means that unlike Zn^{2+}, Cd^{2+} and Pb^{2+} ions are not spectroscopically silent when cysteine binding is involved; and increasingly, these bands are being used to monitor metal binding and determine binding affinities directly.[5,6,22]

These studies of metal-binding affinity reveal that both Cd^{2+} and Pb^{2+} bind tightly to structural zinc-binding domains; and both bind more tightly to sites with more cysteines.[6,18,22] This result is consistent with Pearson's theory of hard and soft acids and bases (see Tutorial II); Cd^{2+} and Pb^{2+} are both relatively "soft", so they prefer binding to soft ligands like cysteine sulfurs over harder histidine nitrogens.[35] The Zn^{2+} ion, a "borderline" hard–soft metal, does not show such a dramatic preference.

The metal's effect on the protein, however, differs for the two metals. The Cd^{2+} ion binds tetrahedrally; and at least in some cases, the protein retains specific DNA recognition, indicating proper folding.[18,33] In contrast, Pb^{2+} does not stabilize the proper form of structural zinc-binding domains even though binding is tight; and, for example, when Pb^{2+} binds to the Cys_3 site of the metalloregulatory protein CadC, it causes CadC to dissociate from DNA.[6,32,36,37] The Pb^{2+} ion also binds to classical (CCHH) zinc fingers and inhibits their DNA binding.[38–41] Lead's interference with protein structure may help to account for some of the developmental problems associated with lead poisoning.[37] As research progresses in this area, it is likely that additional potential targets and mechanisms will be identified.

XIV.2.4. Zinc Fingers and Protein Design

The small size, relatively simple structure, and modular nature of zinc fingers and other structural zinc-binding domains have made them a natural target for de novo protein design. As noted above, many critical contributions to our understanding of zinc finger metal binding, structure, and thermodynamics have come from studies of small model peptides, including the designed "consensus peptides".[17,18]

Zinc fingers have played an important role in the development of methods for de novo design. The zinc finger domain is small, but its ββα motif includes the three major elements of protein structure, that is, α-helix, β-sheet, and turn regions. These features have made the zinc finger domain a superb design target. An early success in computer-driven protein design was the development of a ββα peptide, a "zincless" zinc finger, using fully automated sequence selection.[42,43] The designed peptide was then synthesized, and investigations of its structure by nuclear magnetic resonance (NMR) spectroscopy confirmed the structure. In this structure, stacking aromatic residues in the center of the peptide "finger" stabilize the domain in place of the metal ion.

Zinc fingers have also been used in the area of metal ion sensors, taking advantage of the change in protein structure upon metal binding. A zinc finger peptide is labeled with a fluorescent dye molecule or molecules. If zinc is present in the medium, the zinc binds to the peptide, which folds; and a change in the fluorescence emission occurs. Depending on the dyes used, this emission change results either from energy transfer between a pair of dyes, or from a change in the polarity of the local environment of the fluorophore.[44–46] Thus, protein folding can be used to sense trace metals in solution.

By far the largest area of de novo zinc finger design, however, is in the study of DNA binding and gene transcription, the natural role of many structural zinc-binding proteins. A wide variety of artificial zinc fingers have been designed to study specific DNA recognition. It is now possible to make biomimetic transcription factors that will regulate specific targeted genes, and researchers are learning how to increase specificity through series of two-finger units and through incorporation of unnatural amino acids.[7,47–49] Although much work remains to be done, there is real promise for these technologies to be useful in gene therapy and clinical medicine.

Understanding both metal coordination chemistry and larger-scale macromolecular interactions are critical for our understanding of biological processes. In structural zinc-binding domains, metal binding and protein folding are intimately linked. Without the metal, these domains will not fold; with the wrong metal, they may misfold. Since proper structure is required for nucleic acid binding, recognition, and specificity, misfolding or not folding the protein has the potential to lead to major negative consequences for gene recognition. Fundamental studies of proteins such as zinc fingers, including their metal coordination chemistry and macromolecular interactions, move us ever closer to being able to apply our basic knowledge to understanding complex biological processes, improving health, and healing disease.

References

General References

1. Klug, A. and Rhodes, D., "'Zinc fingers': a novel protein motif for nucleic acid recognition", *Trends Biochem. Sci.*, **12,** 464–469 (1987).
2. Wolfe, S. A., Nekludova, L., and Pabo, C. O., "DNA Recognition by Cys_2His_2 Zinc Finger Proteins", *Annu. Rev. Biophys. Biomol. Struct.*, **3,** 183–212 (1999).
3. Berg, J. M. and Shi, Y., "The galvanization of biology: a growing appreciation for the roles of zinc", *Science,* **271,** 1081–1085 (1996).
4. Berg, J. M. and Godwin, H. A., "Lessons from Zinc-binding Peptides", *Annu. Rev. Biophys. Biomol. Struct.,* **26,** 357–371 (1997).
5. Magyar, J. S. and Godwin, H. A., "Spectropotentiometric analysis of metal binding to structural zinc-binding sites: accounting quantitatively for pH and metal ion buffering effects", *Anal. Biochem.*, **320,** 39–54 (2003).
6. Payne, J. C., ter Horst, M. A., and Godwin, H. A., "Lead Fingers: Pb^{2+} Binding to Structural Zinc-Binding Domains Determined Directly by Monitoring Lead-Thiolate Charge-Transfer Bands", *J. Am. Chem. Soc.*, **121,** 6850–6855 (1999).
7. Jantz, D., Amann, B. T., Gatto, Jr., G. J., and Berg, J. M., "The Design of Functional DNA-Binding Proteins Based on Zinc Finger Domains", *Chem. Rev.*, **104,** 789–799 (2004).
8. Regan, L., "The Design of Metal-Binding Sites in Proteins", *Annu. Rev. Biophys. Biomol. Struct.*, **22,** 257–281 (1993).
9. Folkers, G. E., Hanzawa, H., and Boelens, R., "Zinc Finger Proteins", *in* Bertini, I., A. Sigel, and H. Sigel, Eds., Marcel Dekker, New York, pp. 961–1000, 2001.
10. Lippard, S. J. and Berg, J. M., "Principles of Bioinorganic Chemistry", University Science Books: Mill Valley, CA, pp. 178–184, 213–215, 1994.

Cited References

11. Elrod-Erickson, M., Rould, M. A., Nekludova, L., and Pabo, C. O., "Zif268 protein-DNA complex refined at 1.6 Å: a model system for understanding zinc finger-DNA interactions", *Structure,* **4,** 1171–1180 (1996).
12. Pavletich, N. P. and Pabo, C. O., "Zinc finger-DNA recognition: crystal structure of a Zif268-DNA complex at 2.1 Å", *Science,* **252,** 809–817 (1991).
13. Shannon, R. D., "Revised atomic radii and systematic studies of interatomic distances in halides and chalcogenides", *Acta Crystallogr., Sect. A.*, **32,** 751–767 (1976).
14. Bertini, I. and Luchinat, C., in "The Reaction Pathways of Zinc Enzymes and Related Biological Catalysts", I. Bertini, H. B. Gray, S. J. Lippard, and J. S. Valentine, Eds., University Science Books: Sausalito, CA, pp. 37–106, 1994.
15. Bertini, I. and Luchinat, C., "High Spin Cobalt(II) as a Probe for the Investigation of Metalloproteins", *Adv. Inorg. Biochem.*, **6,** 71–111 (1984).
16. Basolo, F. and Johnson, R. C., "Coordination Chemistry", W. A. Benjamin: New York, 1964.
17. Krizek, B. A., Amann, B. T., Kilfoil, V. J., Merkle, D. L., and Berg, J. M., "A Consensus Zinc Finger Peptide—Design, High-Affinity Metal-Binding, a pH-Dependent Structure, and a His to Cys Sequence Variant", *J. Am. Chem. Soc.*, **113,** 4518–4523 (1991).
18. Krizek, B. A., Merkle, D. L., and Berg, J. M., "Ligand variation and metal-ion binding specificity in zinc finger peptides", *Inorg. Chem.*, **32,** 937–940 (1993).
19. Posewitz, M. C. and Wilcox, D. E., "Properties of the Sp1 Zinc Finger 3 Peptide: Coordination Chemistry, Redox Reactions, and Metal Binding Competition with Metallothionein", *Chem. Res. Toxicol.*, **8,** 1020–1028 (1995).
20. Witkowski, R. T., Ratnaswamy, G., Larkin, K., McLendon, G., and Hattman, S., "Equilibrium Metal

Binding of the Translational Activating Protein, COM", *Inorg. Chem.*, **37,** 3326–3330 (1998).

21. McLendon, G., Hull, H., Larkin, K., and Chang, W., "Metal binding to the HIV nucleocapsid peptide", *J. Biol. Inorg. Chem.*, **4,** 171–174 (1999).
22. Chen, X., Chu, M., and Giedroc, D. P., "Spectroscopic characterization of Co(II)-, Ni(II)-, and Cd(II)-substituted wild-type and non-native retroviral-type zinc finger peptides", *J. Biol. Inorg. Chem.*, **5,** 93–101 (2000).
23. Payne, J. C., Rous, B. W., Tenderholt, A. L., and Godwin, H. A., "Spectroscopic Determination of the Binding Affinity of Zinc to the DNA-Binding Domains of Nuclear Hormone Receptors", *Biochemistry,* **42,** 14214–14224 (2003).
24. Ghering, A. B., Shokes, J. E., Scott, R. A., Omichinski, J. G., and Godwin, H. A., "Spectroscopic Determination of the Thermodynamics of Cobalt and Zinc Binding to GATA Proteins", *Biochemistry,* **43,** 8346–8355 (2004).
25. Berg, J. M. and Merkle, D. L., "On the Metal Ion Specificity of "Zinc Finger" Proteins", *J. Am. Chem. Soc.*, **111,** 3759–3761 (1989).
26. Krizek, B. A. and Berg, J. M., "Complexes of Zinc Finger Peptides with Ni(II) and Fe(II)", *Inorg. Chem.*, **31,** 2984–2986 (1992).
27. Orgel, L. E., "An Introduction to Transition-Metal Chemistry: Ligand-Field Theory", Methuen & Co., London, 1960.
28. Holmes, O. G. and McClure, D. S., "Optical Spectra of Hydrated Ions of the Transition Metals", *J. Chem. Phys.*, **26,** 1686–1694 (1957).
29. Irving, H. and Williams, R. J. P., "Order of Stability of Metal Complexes", *Nature (London),* **162,** 746–747 (1948).
30. Kaltreider, R. C., Davis, A. M., Lariviere, J. P., and Hamilton, J. W., "Arsenic Alters the Function of the Glucocorticoid Receptor as a Transcription Factor", *Environ. Health Perspect.*, **109,** 2001 (2001).
31. Godwin, H. A., "The biological chemistry of lead", *Curr. Opin. Chem. Biol.*, **5,** 223–227 (2001).
32. Claudio, E. S., Godwin, H. A., and Magyar, J. S., "Fundamental Coordination Chemistry, Environmental Chemistry, and Biochemistry of Lead(II)", *Prog. Inorg. Chem.*, **51,** 1–144 (2003).
33. Kuwahara, J. and Coleman, J. E., "Role of the Zinc(II) Ions in the Structure of the Three-Finger DNA Binding Domain of the Sp1 Transcription Factor", *Biochemistry,* **29,** 8627–8631 (1990).
34. Fitzgerald, D. W. and Coleman, J. E., "Physicochemical properties of cloned nucleocapsid protein from HIV. Interactions with metal ions", *Biochemistry,* **30,** 5195–5201 (1991).
35. Pearson, R. G., "Hard and Soft Acids and Bases", *J. Am. Chem. Soc.*, **85,** 3533–3539 (1963).
36. Busenlehner, L. S., Cosper, N. J., Scott, R. A., Rosen, B. P., Wong, M. D., and Giedroc, D. P., "Spectroscopic Properties of the Metalloregulatory Cd(II) and Pb(II) Sites of *S. aureus* pI258 CadC", *Biochemistry,* **40,** 4426–4436 (2001).
37. Magyar, J. S., Weng, T.-C., Stern, C. M., Dye, D. F., Rous, B. W., Payne, J. C., Bridgewater, B. M., Mijovilovich, A., Parkin, G., Zaleski, J. M., Penner-Hahn, J. E., Godwin, H. A., "Reexamination of Lead(II) Coordination Preferences in Sulfur-Rich Sites: Implications for a Critical Mechanism of Lead Poisoning", *J. Am. Chem. Soc.*, **127,** 9495–9505 (2005).
38. Hanas, J. S., Rodgers, J. S., Bantle, J. A., and Cheng, Y. G., "Lead inhibition of DNA-binding mechanism of Cys(2)His(2) zinc finger proteins", *Mol. Pharmacol.*, **56,** 982–988 (1999).
39. Razmiafshari, M., Kao, J., d'Avignon, A., and Zawia, N. H., "NMR Identification of Heavy Metal-Binding Sites in a Synthetic Zinc Finger Peptide: Toxicological Implications for the Interactions of Xenobiotic Metals with Zinc Finger Proteins", *Toxicol. Appl. Pharmacol.*, **172,** 1–10 (2001).
40. Quintanilla-Vega, B., Hoover, D. J., Bal, W., Silbergeld, E. K., Waalkes, M. P., and Anderson, L. D., "Lead Interaction with Human Protamine (HP2) as a Mechanism of Male Reproductive Toxicity", *Chem. Res. Toxicol.*, **13,** 594–600 (2000).
41. Huang, M., Krepkiy, D., Hu, W., and Petering, D. H., "Zn-, Cd-, and Pb-transcription factor IIIA: properties, DNA binding, and comparison with TFIIIA-finger 3 metal complexes", *J. Inorg. Biochem.*, **98,** 775–785 (2003).
42. Dahiyat, B. I. and Mayo, S. L., "De Novo Protein Design: Fully Automated Sequence Selection", *Science,* **278,** 82–87 (1997).
43. Sarisky, C. A. and Mayo, S. L., "The ββα Fold: Explorations in Sequence Space", *J. Mol. Biol.*, **307,** 1411–1418 (2001).
44. Walkup, G. K. and Imperiali, B., "Design and Evaluation of a Peptidyl Fluorescent Chemosensor for Divalent Zinc", *J. Am. Chem. Soc.*, **118,** 3053–3054 (1996).
45. Godwin, H. A. and Berg, J. M., "A Fluorescent Zinc Probe Based on Metal-Induced Peptide Folding", *J. Am. Chem. Soc.*, **118,** 6514–6515 (1996).
46. Walkup, G. K. and Imperiali, B., "Fluorescent Chemosensors for Divalent Zinc Based on Zinc Finger Domains. Enhanced Oxidative Stability, Metal Binding Affinity, and Structural and Functional Characterization", *J. Am. Chem. Soc.*, **119,** 3443–3450 (1997).
47. Moore, M., Choo, Y., and Klug, A., "Design of polyzinc finger peptides with structured linkers", *Proc. Natl. Acad. Sci. U.S.A.*, **98,** 1432–1436 (2001).
48. Moore, M., Klug, A., and Choo, Y., "Improved DNA binding specificity from polyzinc finger peptides by using strings of two-finger units", *Proc. Natl. Acad. Sci. U.S.A.*, **98,** 1437–1441 (2001).
49. Jantz, D. and Berg, J. M., "Expanding the DNA-Recognition Repertoire for Zinc Finger Proteins beyond 20 Amino Acids", *J. Am. Chem. Soc.*, **125,** 4960–4961 (2003).

XIV.3. Calcium in Mammalian Cells

Torbjörn Drakenberg
Department of
Biophysical Chemistry
Lund University
SE-22100 Lund, Sweden

Bryan Finn
IT Department
Swedish University of
Agricultural Sciences
SE-23053 Alnarp, Sweden

Sture Forsén
Department of
Biophysical Chemistry
Lund University
SE-22100 Lund, Sweden

Contents

XIV.3.1. Introduction

On one hand, calcium compounds make up a major part of the biomaterials we see around us, for example, the exoskeletons of seashells and the endoskeletons of mammals, our own included. On the other, Ca^{2+} ions are also the prime movers in an intracellular regulatory system that seemingly pervades all eukaryotic systems, from single yeast cells to insects, plants, and mammals.

This section will mainly deal with this latter role of Ca^{2+} and some of the molecular mechanisms by which the regulatory systems operate will try to be outlined. This will be done by comparing the interplay of Ca^{2+} ions with three well studied Ca^{2+}-binding proteins: calmodulin, troponin C, and calbindin D_{9k}. Calmodulin is a pivotal protein in the Ca^{2+}-dependent intracellular regulatory networks, troponin C is important for muscle contraction, while calbindin D_{9k} is involved in the vitamin D dependent uptake of Ca^{2+} from the intestine and the transport of Ca^{2+} through the epithelial cells of the placenta. Calbindin D_{9k} seems to play more of a buffering or storage role in higher organisms. As an introduction, however, some general comments about the concentration levels of Ca^{2+} ions and their regulation in mammals will be made. For more detailed information see Refs. 1–13.

XIV.3.2. Concentration Levels of Ca^{2+} in Higher Organisms

In mammals, we may consider bone, the extracellular, and the intracellular fluids as three compartments among which Ca^{2+} ions may be exchanged. The total Ca^{2+} concentration in our blood is kept constant within a narrow range $\sim 2.45\,\text{m}M$. The mechanism by which this extracellular concentration is regulated has recently been unearthed. The plasma membrane of several specialized cells, like those of the kidneys and certain hormone secreting glands, contains a protein that acts as a sensor of the Ca^{2+} concentration in the blood serum. In this way, the excretion of Ca^{2+} through the urine, the uptake of Ca^{2+} through the intestines, and the release and uptake of Ca^{2+} in the bones work in synchrony to keep the extracellular concentration constant.

Note that the National Institutes of Health (NIH) approved recommendations for the daily intake of calcium[14] range from 400 mg for newborns to 1500 mg for men and women over 65. The danger of an overdose is small—our systems seem to cope with intakes of at least 2000 mg day^{-1} without any ill effects.

The intracellular concentration of Ca^{2+} is a matter of some importance to us. Here we must make a distinction between the total concentration of Ca^{2+} in a compartment, $[Ca^{2+}]_0$, and the "free" Ca^{2+} concentration in the same compartment, $[Ca^{2+}]_i$. The value of $[Ca^{2+}]_0$ represents the sum of all species that contain calcium—be it inorganic matter, macromolecules with tightly bound Ca^{2+}, or "free" hydrated Ca^{2+} ions—while the value of $[Ca^{2+}]_i$ refers to the concentration of the "free", hydrated, Ca^{2+} ions only. Thus $[Ca^{2+}]_0$ is invariably larger than $[Ca^{2+}]_i$. The intracellular concentration of "free" Ca^{2+} is kept much lower than the extracellular by orders of magnitude. The extracellular concentration of free Ca^{2+} in humans is lower than the total blood concentration and generally taken to be of the order of 1.2 m*M*. The intracellular concentration of free Ca^{2+} in resting cells has been determined by the use of Ca^{2+}-specific fluorescent chelators to be in the range 10–100 n*M*. As we will discuss further below, this low resting concentration level makes feasible the role of the Ca^{2+} ion as an intracellular signaling entity—a "second messenger" by analogy to the phrase used to characterize the role of cyclic adenosine monophosphate (cAMP).

The low intracellular value of $[Ca^{2+}]_i$ is maintained in two ways. The Ca^{2+} ions are extruded from cells either by an adenosine triphosphate (ATP) driven pump in the plasma membrane, a Ca^{2+}-adenosine triphosphatase (ATPase), or through an exchange mechanism by which Na^+ ions going inward are traded with Ca^{2+} ions going outward (i.e., through a membrane bound Na^+/Ca^{2+} exchanger). Lest the consequence would be an intracellular overload of Na^+, cells also have an ATP driven Na^+/K^+ pump that extrudes Na^+.

A low intracellular (or rather cytoplasmic) value of $[Ca^{2+}]_i$ can also be maintained, at least temporarily, by pumping Ca^{2+} into specialized membrane-enclosed intracellular compartments, "organelles", like, for example, the endoplasmic reticulum (ER) or its equivalent in muscle cells, the sarcoplasmic reticulum (SR). Also, mitochondria may be used as a storage device. The membranes of ER–SR contain a Ca^{2+}ATPase that in this case pumps cytoplasmic Ca^{2+} into the organelles. In muscle cells, the density of Ca^{2+} ATPase pumps in the SR membranes is impressive (some 25,000 molecules μm^{-2}). In fact these pumps use up a large portion of the metabolic energy of an active mammal!

The combined effect of all these pumping or extrusion devices is that a temporary rise in the cytoplasmic $[Ca^{2+}]_i$ can be very rapidly reduced if needed.

XIV.3.3. The Intracellular Ca^{2+}-Signaling System

There are two characteristic features of the intracellular Ca^{2+}-dependent signaling system. The first is that an extracellular signal will result in a temporary increase of the intracellular Ca^{2+} concentration (local or global). The second is that in the cytosol there are Ca^{2+}-specific target molecules that respond in some way to increased concentrations of Ca^{2+}.

A transient increase in the cytosolic Ca^{2+} level may be accomplished through the entry of Ca^{2+} ions from the extracellular space via the plasma membrane, through the release of Ca^{2+} from intracellular stores, or both. A well-documented mechanism for the release of Ca^{2+} from intracellular stores involves a low molecular weight compound, inositol 1,4,5-trisphosphate (InsP3). This compound is an intracellular "messenger" produced through the enzymatic hydrolysis of a phosphoinositol phospholipid in the plasma membrane. The enzyme involved is a specific phospholipase C, which becomes activated when G-protein-coupled, or tyrosine kinase-coupled, plasma membrane receptors are stimulated by hormones, growth factors, neurotransmitters, or other extracellular signals. The InsP3 will diffuse from its site of production and reach InsP3 receptors on the ER and SR.

The binding of InsP3 to these four-subunit receptors triggers the release of Ca^{2+} from the ER–SR stores. The InsP3 receptor, or InsP3 release channel, has the interesting property that the initial Ca^{2+} ions released will further stimulate the release of additional Ca^{2+} ions. But, when the local concentration of Ca^{2+} approaches a certain level, the release of Ca^{2+} ions will gradually be turned off. The net result is that a transient and initially localized cloud of Ca^{2+} ions is created around the release channel. The Ca^{2+} ions will diffuse away and may help trigger the release of Ca^{2+} from nearby InsP3 receptors–release channels. The final consequence may eventually be a more global transient increase in the Ca^{2+} concentration. Diffusion of InsP3 close to the receptor laden membrane surface may also influence the time course and spatial distribution of the Ca^{2+} concentration.

The SR of cardiac and skeletal muscle cells are abundant in a special Ca^{2+} release channel called the ryanodine receptor (RyR). The name is derived from the plant alkaloid ryanodine that was first observed to trigger the release of Ca^{2+} from the SR at concentrations $< 10\,\mu M$. The RyR, or rather the family of RyRs, is activated by Ca^{2+}, but most likely also by other entities. Thus interestingly Ca^{2+} release through the two major Ca^{2+} release channels is triggered by Ca^{2+} itself—at least at low Ca^{2+} concentrations. As a curiosity for coffee addicts, caffeine has an activating effect on the RyR, but an inhibitory effect on the InsP3 receptor.

As briefly mentioned above, Ca^{2+} may also enter the cytoplasm from the extracellular space. This mechanism is particularly important in cells involved in the transmission of nerve impulses and in the triggering of muscle contraction. Here, specialized voltage-dependent Ca^{2+} channels enable these cells to increase the cytosolic Ca^{2+} levels dramatically. The depolarization of the membrane initiates the opening of the channels and the influx of Ca^{2+} ions will further activate the RyRs on the SR. A schematic picture of the major elements of the Ca^{2+} signaling system is shown in Fig. XIV.3.1. The different means and routes by which Ca^{2+} may enter the cytoplasmic space may result in significantly elevated intracellular Ca^{2+} concentrations that range from 1 to $100\,\mu M$ and, most importantly, can be localized as well as global.

The Ca^{2+}-specific molecules that respond to a transiently increased Ca^{2+} concentration constitute the second leg of the Ca^{2+}-signaling system. These Ca^{2+}-binding molecules should be able to bind Ca^{2+} ions at elevated Ca^{2+} concentrations and undergo some kind of exploitable modification or conformational change with a minimal time delay. The binding site must be able to discriminate between Ca^{2+} and other cations—in particular Mg^{2+}, which is abundant in the cytosol of eukaryotic cells. Well-researched representatives of such protein molecules are calmodulin, a protein present in seemingly all eukaryotic cells and the activator of a wide variety of cellular activities, and troponin C, which is abundant in skeletal muscle cells and involved in the contractile process. Characteristic of these proteins is that in the Ca^{2+} loaded form they become competent to interact with specific regions of other protein molecules—their targets. As a consequence, the biological activity of these target molecules may then become enhanced or suppressed.

As mentioned in the introduction, not all eukaryotic Ca^{2+}-binding proteins so far discovered are directly involved in the signaling system. Some molecules may use Ca^{2+} in a structure-stabilizing role while other molecules seemingly serve as soluble Ca^{2+}-buffering or -transport molecules. The need for such molecules is particularly obvious in placental cells or cells of the intestinal brush border membranes, where Ca^{2+} ions are to be ferried across the cytoplasm without interfering with the Ca^{2+}-signaling system. In these latter cases, the binding of Ca^{2+} need not be accompanied by a conformational change. Calbindin D_{9k} is a representative of the buffering or transport type of protein molecules.

It appears that, at least in eukaryotic systems, Nature has fulfilled the requirements for a versatile intracellular Ca^{2+}-binding site in a unique way—through a type of Ca^{2+}-binding motif that for historical reasons is commonly called the "EF-hand". The EF-hand Ca^{2+}-binding sites are somewhat paradoxically observed not only in the proteins that respond to Ca^{2+} binding with a significant conformational change, but also in proteins that in comparison undergo only insignificant conformational changes. In the following section, the general structural features of EF-hands will be discussed in greater detail, how minor variations in the sequence of amino acids

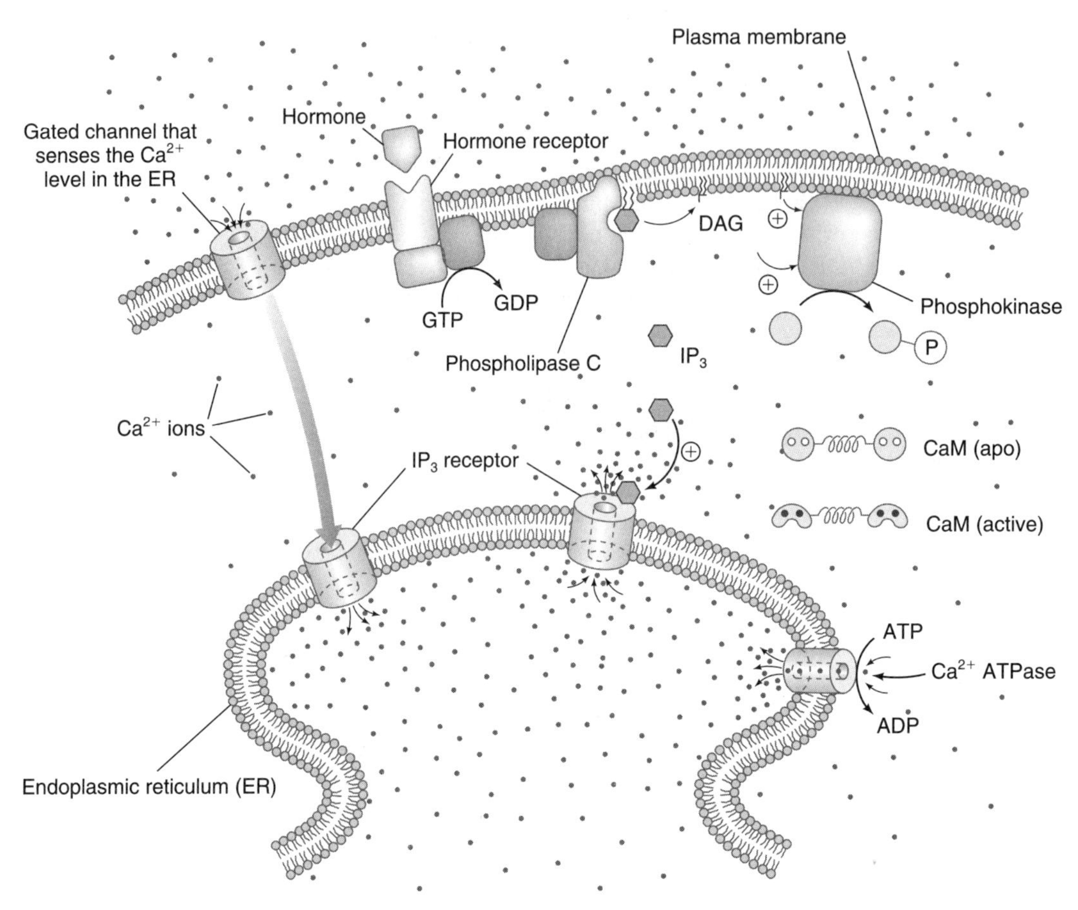

Fig. XIV.3.1.
Schematic picture of some major elements of the intracellular Ca^{2+}-signaling system in mammalian cells. The binding of an agonist, for example, a hormone, to a G-coupled receptor on the plasma membrane will result in the activation of a membrane-bound phospholipase C. This enzyme specifically cleaves off the head group from phosphatidyl inositol. The phosphorylated head group, that is, 1,4,5-triphosphoinositol (IP3), will diffuse away and bind to specific IP3 receptors (or Ca^{2+} release channels) on the ER. The tetrameric IP3 receptors then undergo a conformational change resulting in a temporary release of Ca^{2+} ions from the ER into the cytoplasm. The Ca^{2+} concentration in the cytoplasm and the ER are restored by a Ca^{2+} ATPase that pumps Ca^{2+} ions back into the ER. The Ca^{2+} ions may also be pumped out of the cytoplasm by means of a Na^+/Ca^{2+} ion exchanger (not shown). Furthermore the high Ca^{2+} levels in the ER may be restored by a less well understood mechanism by which Ca^{2+} ions from the extracellular space will enter the ER—possibly involving direct physical contact between an IP3 receptor and a plasma membrane gated channel.

At the transiently increased Ca^{2+} levels in the cytoplasm, following upon the opening of the IP3 receptor channels, certain cytoplasmic Ca^{2+}-binding proteins will become loaded with Ca^{2+}. As a consequence, as in the case of calmodulin (CaM), these proteins may undergo conformation changes that poise them to interact with other cytoplasmic proteins and affect their biological activity. The lipid part of the phosphatidyl inositol cleaved by phospholipase C, that is, diacylglycerol (DAG), will act in a concerted way with Ca^{2+} ions to increase the activity of a phosphokinase at the plasma membrane. This kinase will then phosphorylate other proteins and in this way influence their activity.

provide an explanation for their paradoxical behavior, and for the fact that the Ca^{2+}-affinity EF-hand proteins cover a wide range of Ca^{2+} affinity values—from $K_a < 10^5\ M^{-1}$ to well above $10^{10}\ M^{-1}$.

XIV.3.4. A Widespread Ca^{2+}-Binding Motif: The EF-Hand

An example of a typical EF-hand Ca^{2+}-binding loop is depicted in Fig. XIV.3.2*a* and the corresponding three-dimensional (3D) structure of the protein in which it occurs is shown in Fig. XIV.3.2*b*. The 12 amino acid residue long loop contains all the Ca^{2+}-chelating ligands and is flanked on both sides by an α-helix. A survey of a large number of loops shows that the three most highly conserved amino acids are Asp1, Gly6, and Glu12. Positions 3 and 5 (the numbering system is explained in Fig. XIV.3.2*a*) are most frequently either Asp or Asn. Side chains from residues 1, 3, 5, and 12 are ligands as is the backbone carbonyl of residue 7. The side chain of residue 9 is either a Ca^{2+} ligand or it is hydrogen bonded to a water molecule that constitutes a ligand. The specific nature of residue 9 seems to be one of the major factors that control the rate of Ca^{2+} binding and release, and thus also affinity, of EF-hands.

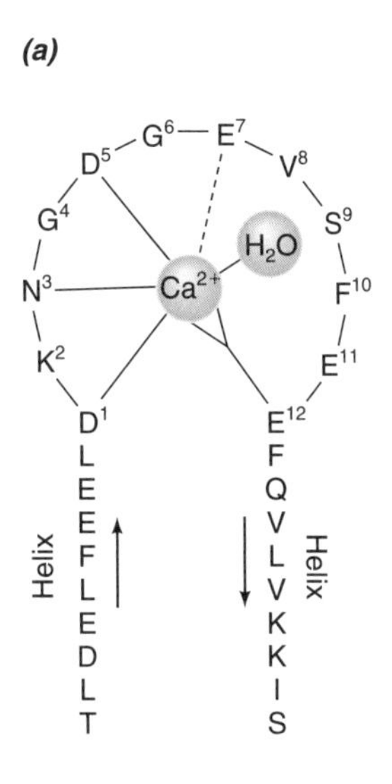

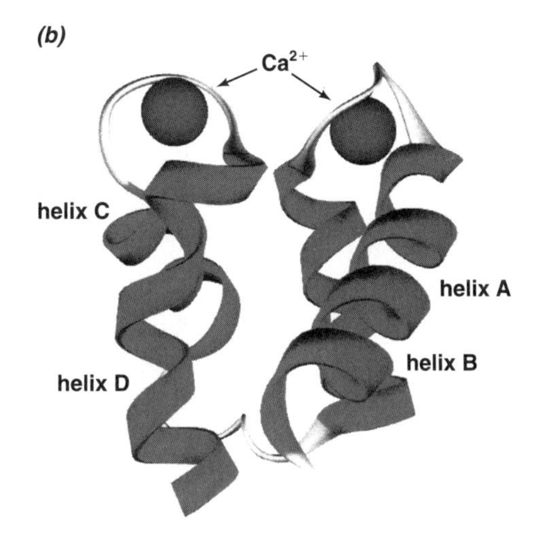

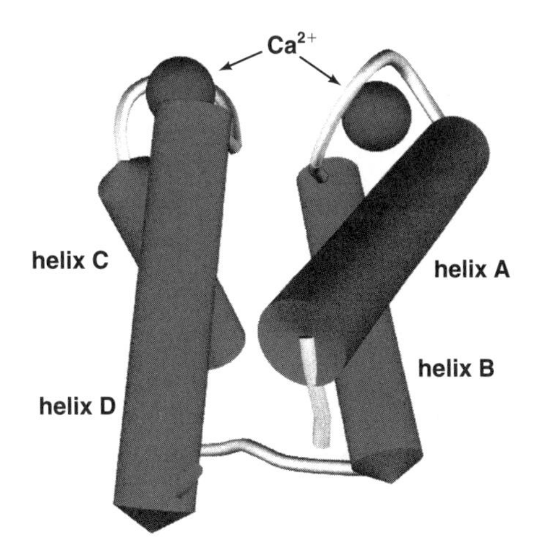

Fig. XIV.3.2.
(*a*) A schematic drawing of the second EF-hand Ca^{2+}-binding subdomain from calbindin D_{9k} (residues 45–74). Ligands from carboxyls are indicated in solid lines; the ligand from a backbone carbonyl is indicated by a dashed line. The final glutamic acid ligand in the loop forms a bidentate ligand with the Ca^{2+} ion, a common feature of EF-hands. (*b*) Two representations of the EF-hand Ca^{2+}-binding protein calbindin D_{9k}. The top view is a ribbon diagram showing the secondary structure of the protein. The Ca^{2+} ions are indicated as spheres. In the lower view, the helices are shown as cylinders in order to emphasize their arrangement in a pairwise fashion in each EF-hand subdomain and as a four-helix bundle in the whole protein. [See color insert.]

The carboxylate group of Glu12 constitutes a bidentate ligand to the Ca^{2+} ion, which will thus be surrounded by a total of 7 ligands. The 3D structure may be described either as a distorted octahedron or as a pentagonal bipyramid.

An isolated EF-hand loop does not bind Ca^{2+} very strongly compared to an intact EF-hand with helices flanking the loop. Furthermore, almost without exception, the functional unit in proteins is not a single EF-hand, but a pair of EF-hands (Fig. XIV.3.2*b*). If both EF-hands in a pair will bind Ca^{2+}, that is, there are no deleterious substitutions among the amino acids in the loops, they most often bind in a cooperative manner. There are, however, examples in Nature of EF-hand pairs in which one of the loops has lost its Ca^{2+}-binding ability because of the loss of a coordinating amino acid side chain.

There are several direct interactions between the two EF-hands in a pair, first through a short β-sheet comprising the loop amino acids at positions 8 and 9, and second through the helices. The first helix in the N-terminal EF-hand, helix A, interacts with the second helix in the C-terminal EF-hand, helix D. Likewise the second helix in the N-terminal EF-hand, helix B, interacts with the first helix in the C-terminal EF-hand, helix C.

Also, conserved amino acids are found outside the Ca^{2+}-binding loop. For example, in an EF-hand pair residues −1 and −4 in helix A of the N-terminal hand normally have aromatic side chains as have residues 10 and 13 in the C-terminal EF-hand. See Fig. XIV.3.3 for the consensus sequences for the two EF-hands (note the two different numbering systems currently in use in Fig. XIV.3.3). These amino acids form an aromatic cluster in the folded protein and are preserved also in most EF-hands that have been modified in the loop in such a way that they no longer bind Ca^{2+}. Occasionally, these aromatic amino acids are replaced by other long hydrophobic amino acids. The presence of these aromatic amino acids is certainly contributing to the fact that isolated EF-hands, prepared by enzymatic cleavage in the linker region between a pair of EF-hands, prefer to form heterodimers instead of homodimers. Even though a homodimer contains the same number of aromatic residues close to the binding loops, they are not in suitable positions to form a tight aromatic cluster unless the structure is rearranged significantly compared to that in an intact two EF-hand domain. This is necessary because the residues are at opposite ends of the β-sheet.

In general, an EF-hand pair can be seen as a four-helix bundle, with a very short β-sheet at the top. This β-sheet holds the two Ca^{2+}-binding loops together and is normally also present in the absence of Ca^{2+}. This short β-sheet and the aromatic cluster seem to keep helices A and D and the β-sheet relatively insensitive to metal ion binding. In contrast, helices B and D in some proteins swing out upon Ca^{2+} binding and in the process expose a large hydrophobic surface, used to bind target proteins as discussed in Section XIV.3.5.

Fig. XIV.3.3.
Consensus sequences for the EF-hands somewhat modified by taking the differences between the two hands in a pair into account. The asterisk (*) indicates no preference, h is for hydrophobic, and (h) is for mostly hydrophobic and a for aromatic. Note that helix A in the C-terminal end has two aromatic residues on the same side of the helix and that the same is true for the N-terminal end of helix D.

EF-hand I

helix A																						helix B						
									1	*2*	*3*	*4*	*5*	*6*	*7*	*8*	*9*	*10*	*11*	*12*								
1	2	3	4	5	6	7	8	9	10	11	12	13	14	15	16	17	18	19	20	21	22	23	24	25	26	27	28	29
E	h	*	*	(h)	a	*	(h)	a	D	*	D	*	D	G	*	I	D	*	*	E	h	*	*	h	h	*	*	(h)

EF-hand II

helix C																						helix D						
									1	*2*	*3*	*4*	*5*	*6*	*7*	*8*	*9*	*10*	*11*	*12*								
1	2	3	4	5	6	7	8	9	10	11	12	13	14	15	16	17	18	19	20	21	22	23	24	25	26	27	28	29
*	(h)	*	*	*	h	*	*	*	D	*	D	*	N	G	*	I	D	a	*	E	a	*	*	h	h	*	*	*

XIV.3.5. Ca^{2+} Induced Structural Changes in Modulator Proteins (Calmodulin, Troponin C)

Here, calmodulin is mainly used (Fig. XIV.3.4) to discuss in more detail the structural changes in a pair of EF-hands as a function of metal ion binding. Since we are most interested in functional aspects, we should consider physiological conditions as much as possible. In the case of Ca^{2+}-binding proteins, we have to consider ionic strength effects and effects attributable to the presence of other bivalent metal ions, especially Mg^{2+}, which has an intracellular concentration in the millimolar range. Even though the affinity of EF-hands for Mg^{2+} is observed to be about a factor of 10^3–10^4 less than for Ca^{2+}, this difference will be at least partially compensated for by the much higher Mg^{2+} concentration. In a resting cell with a low cytoplasmic Ca^{2+} concentration, many EF-hands may in fact be almost fully occupied by Mg^{2+}, which can of course have major implications for the role of Mg^{2+} in the Ca^{2+}-activation process. This fact is, however, rarely explicitly considered. The physiologically important structural change in calmodulin as a result of Ca^{2+} binding thus does not begin from the apo-form, but from the Mg^{2+} form (though the apo-form may be a transient intermediate between Mg^{2+} release and Ca^{2+} binding).

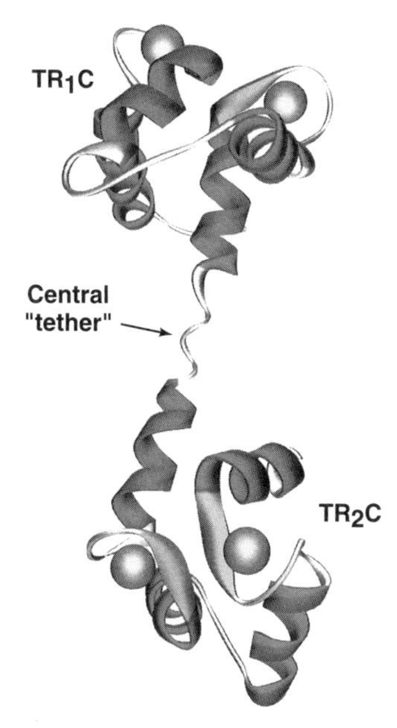

Fig. XIV.3.4.
A ribbon diagram of Ca^{2+}-loaded calmodulin. The two domains, TR_1C and TR_2C are indicated as well as the central tether that connects them.

At present, very limited structural information is available regarding the Mg^{2+}-forms of EF-hand proteins. No 3D structures have been published for the Mg^{2+}-form of either calmodulin or troponin C, even though it is well known that under low Ca^{2+} conditions the C-terminal half of troponin C is saturated with Mg^{2+}. Because of the lack of information regarding the magnesium bound forms of calmodulin and troponin C, we will use the apo-form in the following discussion. Spectroscopic data indicates that this is a reasonable assumption at least for calmodulin and the N-terminal EF-hand pair of troponin C.

It has been shown on various occasions using a number of experimental techniques that the biophysical properties of the two halves of calmodulin, comprising one EF-hand pair each, are essentially identical whether present in the intact protein or occurring as separate molecular entities. Therefore, we can focus on a single two EF-hand domain of calmodulin. Note that this assertion is not valid for the biological activity, which, as will be seen in Section XIV.3.6, requires the concerted action of both halves of the protein. Figure XIV.3.5 shows a superposition of the apo- and $(Ca^{2+})_2$-forms of the C-terminal half of calmodulin, CaM–TR_2C. The structural change resulting from Ca^{2+} binding can be visualized as a hinge motion around amino acids at the end of β-strand I and the beginning of β-strand II. This change will clearly result in the movement of helices F and G away from helices E and H,

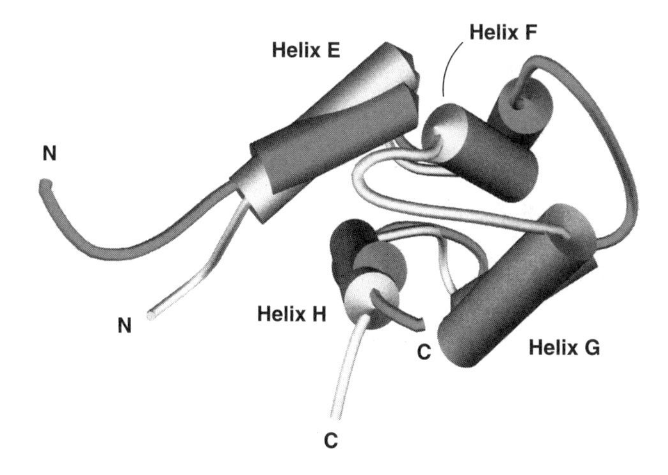

Fig. XIV.3.5.
A diagram showing the conformational change in the TR_2C domain of calmodulin induced upon Ca^{2+} binding. Helices E and H are aligned in the two structures in order to emphasize the relative movement of helices F and G and the loop between them relative to helices E and H. The Ca^{2+}-binding sites are located in the loops between helices E and F and helices G and H.

but preserve the β-sheet as well as the aromatic cluster formed by the four aromatic residues at the end of helix E and the beginning of helix H. As pointed out above, the comparison should preferably be between the Mg^{2+}- and Ca^{2+}-form. There is, however, no compelling evidence that the 3D structure of the Mg^{2+}-bound form of CaM–TR_2C differs significantly from the apo-form. Furthermore, there are so far no reports of Ca^{2+}–calmodulin activated processes that are also activated by Mg^{2+}. It may be worth pointing out that Mg^{2+} binding to the C-terminal half of troponin C has been assumed to induce a structural change similar to Ca^{2+} binding, resulting in a structure at least compatible with the formation of a complex with troponin I—a target protein for troponin C involved in skeletal muscle contraction.

Very similar structural changes upon Ca^{2+} binding have been observed for both halves of calmodulin as well as for skeletal muscle troponin C. On the other hand, the N-terminal half of cardiac muscle troponin C, cNTnC, does not show this dramatic structural change upon Ca^{2+} binding. The cNTnC has only one intact EF-hand; the first loop has been made defunct, in terms of Ca^{2+} binding, by replacing two side-chain ligands with non-liganding residues. The 3D structure of Ca–cNTnC shows no sign of an opening up of the structure. In solution, the structure could still exist as an equilibrium between the closed and open forms with the equilibrium displaced toward the closed form. Very similar observations have been made for mutants of CaM–TR_2C, where one or the other Glu12 in the EF-loops have been replaced by Gln (see Fig. XIV.3.2). This substitution reduces the Ca^{2+} affinity of the mutated site dramatically and the Ca_1-form is found to be in equilibrium between two conformations, presumably an "open" apo-form and a "closed" Ca_1-form, with close to 50:50 population ratios.

Recent results indicate that the rate of exchange between the closed and open conformations in EF-hand pair domains of calmodulin and troponin C is of the order $10^3\ s^{-1}$. The Ca^{2+} on-rates of EF-hand loops have been determined to be of the order of $10^8\ M^{-1}\ s^{-1}$, which probably means that in a typical physiological situation these proteins will respond to a transient increase of the Ca^{2+} concentration within a millisecond or so.

Structural biology has played a central role in furthering our understanding of other facets of calmodulin's structure and function. One of these is the extent of communication between the two domains of calmodulin. The first X-ray crystal structure of Ca^{2+}-bound calmodulin determined in 1985 showed that each EF-hand pair domain has a hydrophobic binding "half-site". These two "half-sites" from each domain would together make up the binding site for target proteins. However, the domains were arranged in a "dumbbell" shape in which an apparently rigid helical linker joins the two globular domains together. This poses the question: Since most of the target peptides known to bind calmodulin were too small to span the gap between the two domains and bind to both simultaneously, how did a single calmodulin bind a single peptide? The answer is provided by other structural techniques such as X-ray and neutron scattering and especially by high-resolution structural and dynamical information from nuclear magnetic resonance (NMR). In solution, as opposed to the crystal, the central "helix" is found to be flexible, and therefore the "rigid" helix was most likely a crystallization artifact.

Structures of calmodulin–target peptide complexes also support this view. Here, however, X-ray crystallography and other studies agree that the two domains must come together to bind the target peptide. The first several structures of calmodulin–target complexes from X-ray and NMR all show the same result, that calmodulin's two domains bind together around a helical target peptide. As shown in Fig. XIV.3.6, the hydrophobic binding sites of the EF-hands interact with the side chains on the surface of the helical peptide. Thus the "central helix" concept has been replaced by the "central-tether".

(a)

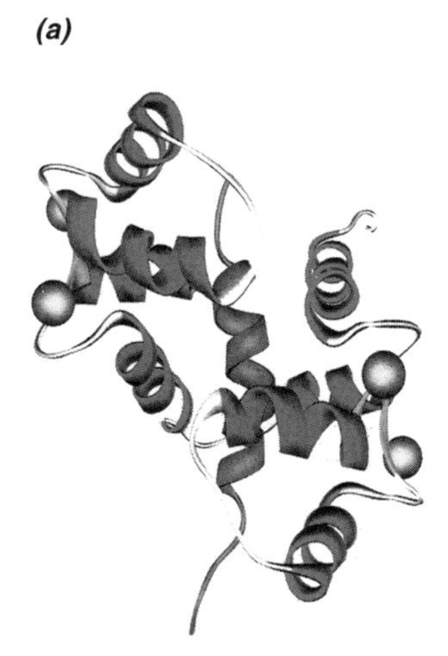

(b)

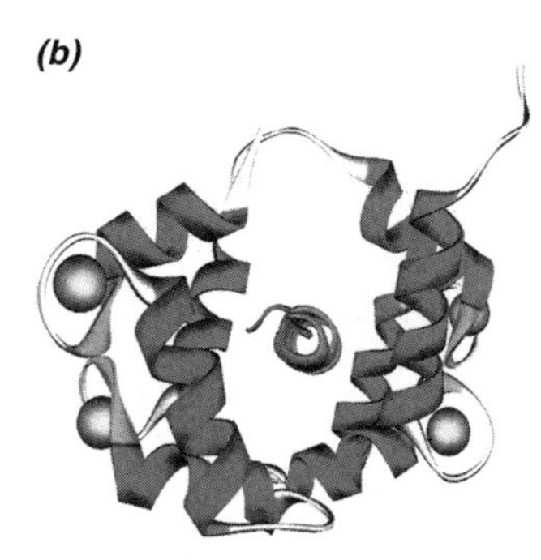

Fig. XIV.3.6.
Ribbon diagrams showing the complex between calmodulin and the calmodulin-binding peptide from myosin light-chain kinase (MLCK). (*a*) and (*b*) are rotated 90° relative to one another. [See color insert.]

Structures of calmodulin and its targets address a second major question as well, namely, how calmodulin can bind to a large number of target peptides with differing sequences while still maintaining a high affinity. Calmodulin is known to have >100 unique protein targets, each with a correspondingly unique calmodulin-binding sequence. This is an unusual situation for target recognition and molecular recognition in general. Most targets, whether they are in proteins or DNA, have evolved well-conserved sequences in order to retain high specificity and affinity for the protein that must recognize them. Any variation in the sequences leads to weaker binding, and therefore reduced activity. However, calmodulin is unique in that it has evolved a structure that can recognize a wide range of target sequences and bind them tightly (though conversely calmodulin's own sequence is very highly conserved).

The structures of the first calmodulin–target complexes shed some light on the mechanism of calmodulin's propensity to recognize a broad range of targets. First, it was found that in comparing several structures, the two EF-hand pair domains could maneuver relative to one another and thereby adapt to many sequences. Second, the overabundance of long flexible hydrophobic side chains such as methionine also allows local adaptation in the binding surface to fit a variety of sequences. However, these properties may prove to be only part of the answer. The peptides used in these initial structures were but a small fraction of the known calmodulin-binding sequences. On top of that, they show a great deal of similarity to one another in the arrangement of hydrophobic and negatively charged residues and therefore probably only represent one type of recognition sequence. Thus, additional levels of complexity in the calmodulin–target binding interactions are needed to explain all calmodulin-binding sequences.

A recent structure of calmodulin complexed with a peptide from the calmodulin-binding domain of rat Ca^{2+}/calmodulin-dependent protein kinase kinase (CaMKK) determined by NMR spectroscopy has demonstrated that this is the case. In this complex, rather than having the usual helical conformation, the peptide adopts a helix-loop conformation in which both types of structure are important for calmodulin binding (Fig. XIV.3.7). However, there is yet another level of complexity added to the mix. Not only does this conformation differ from the originally characterized complexes with respect to secondary structure, it also differs in orientation. For the initially studied, all-helical peptides, the amino terminus of the peptide interacts predominantly with the carboxy-terminal domain of calmodulin; and conversely the carboxy-terminal end of the peptide interacts with the amino-terminal domain of calmodulin. However, the orientation of the helix–loop peptide is opposite to that observed in the original all-helical peptide complexes. The explanation for this behavior is that hydrophobic interactions and electrostatic charges not only specify the target sequences, the electrostatics also specify the peptide orientation. In addition, the central linker plays a more direct role in the binding than its previously proposed "tethering" role had suggested. Fig. XIV.3.8 showns a schematic cartoon for CaM activation.

The observed structural changes may also be discussed in terms of free energy changes. As shown in Fig. XIV.3.9, the difference in thermodynamic terms between skeletal and cardiac TnC may not be as dramatic as indicated by the structures alone. The free energy supplied by the binding of two Ca^{2+} ions to skeletal TnC is sufficient to overcome the energy cost to expose the hydrophobic surface, whereas a single Ca^{2+} ion bound to the N-terminal half of cardiac TnC does not supply sufficient energy. This does not mean that cardiac troponin C cannot function in the same way as skeletal troponin C. If everything else is the same, this means that the affinity of cardiac troponin C for its troponin I peptide will be less than that of skeletal troponin C. The development of this weaker binding in the cardiac troponin complex almost certainly has a functional implication, even though we have not yet figured that out.

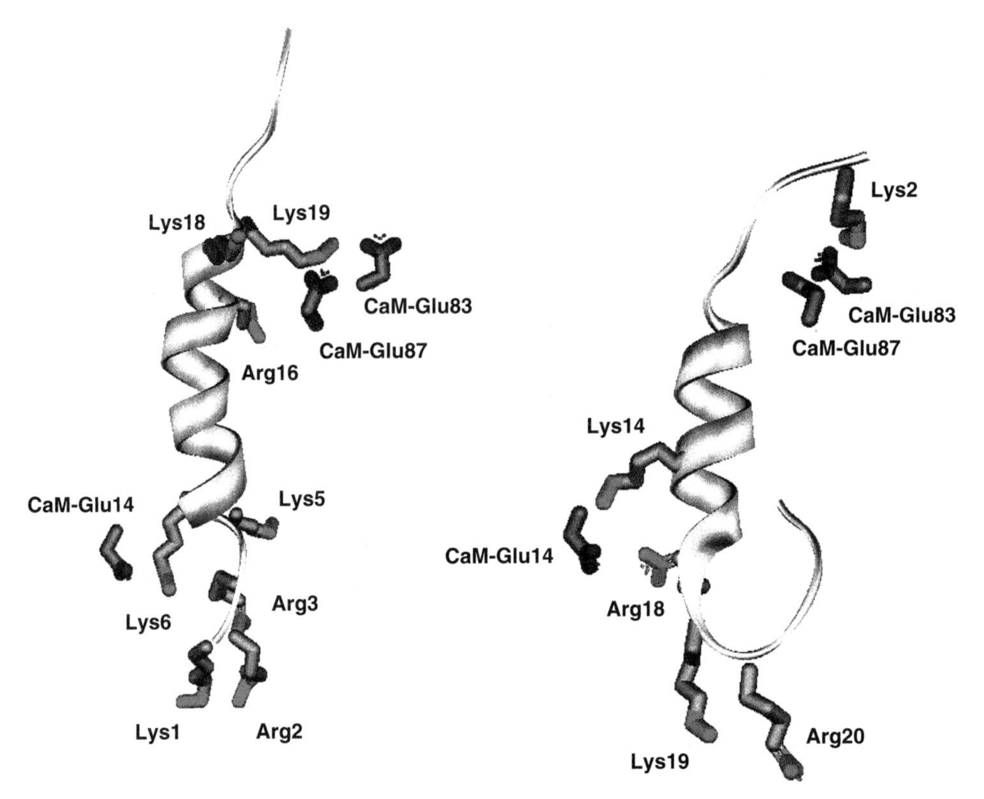

Fig. XIV.3.7.
A comparison of peptides know to bind calmodulin, but with different seconday structures. On the (*a*) is the complex between calmodulin and the calmodulin-binding peptide from MLCK. In this complex, the target is a regular helix. On the (*b*) is the complex between calmodulin and the CaMKK. In this case, the peptide forms a helix–hairpin loop structure. The other major difference between the two is the orientation of the peptides that are opposite to one another. The electrostatic interactions between acidic side chains on calmodulin and basic residues on the peptide, which are believed to control this orientation, are indicated. While the structures differ substantially, the spatial orientation of these charged groups are surprisingly similar, especially the pairs MLCK-Lys19:CaMKK-Lys 2, MLCK-Lys6:CaMKK-Arg18, MLCK-Arg2:CaMKK-Arg20, and MLCK-Lys1:CaMKK-Lys19.

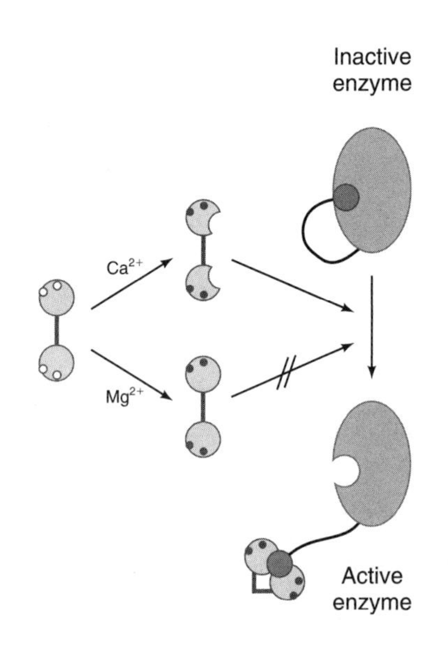

Fig. XIV.3.8.
A cartoon representation of the action of calmodulin. Note that even though Mg^{2+} binds to the Ca^{2+}-binding sites, the EF-hands, this binding does not induce any dramatic conformational change. Therefore the Mg^{2+}-bound form of calmodulin will not induce any activation of the target enzymes.

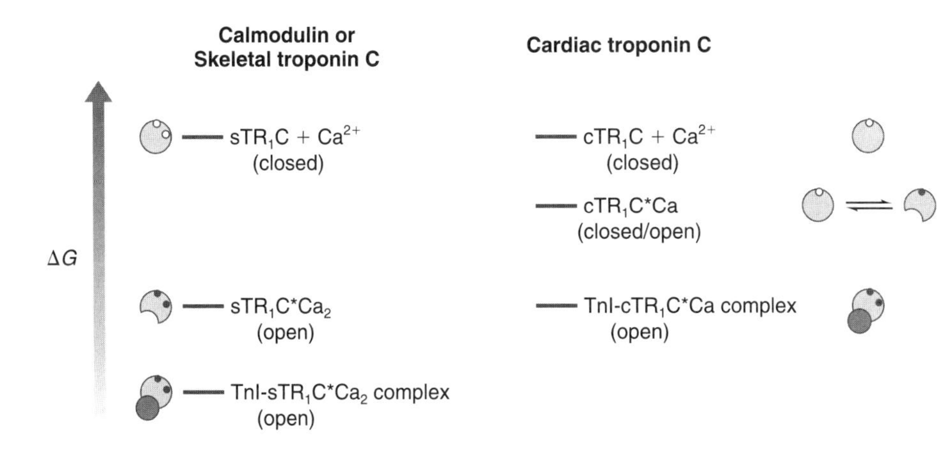

Fig. XIV.3.9.
Schematic free energy diagram visualizing the effect of Ca^{2+} and peptide binding to the N-terminal half of calmodulin, skeletal and cardiac muscle troponin C. The conformational state of the protein is indicated as open or closed.

XIV.3.6. Ca^{2+} Binding in Buffer or Transporter Proteins

A second category of EF-hand proteins has no modulating activities, but these proteins are assumed to work as Ca^{2+} buffers or transporters. Two typical proteins in this group are calbindin D_{9k} and parvalbumin. Both have a pair of EF-hands with strong and cooperative Ca^{2+} binding. In calbindin D_{9k}, the loop of the N-terminal EF-hand deviates from the canonical structure in that it has 14 instead of 12 amino acids. It is frequently referred to as a "pseudo-EF-hand". Calbindin D_{9k} has been extensively investigated using many biophysical methods. The 3D structure of the Ca^{2+} form of calbindin D_{9k} is shown in Fig. XIV.3.10. The coordination pattern in the C-terminal EF-hand is similar to that in calmodulin and troponin C. In the pseudo-EF-hand, most of the ligands are, however, supplied by backbone carbonyls. Only the last amino acid in the loop, Glu14, is coordinated through its side chain and is a bidentate Ca^{2+} ligand as in the archetypical EF-hands. The four aromatic amino acids that form the aromatic cluster in archetypical EF-hand pairs are present in calbindin D_{9k} as well. Note, however, that the Ca^{2+} loaded form of calbindin D_{9k} is compact and very unlike the open forms of Ca^{2+} loaded EF-hand pairs. The NMR studies of the apo-form of calbindin D_{9k} show it to have a 3D structure strikingly similar to that of the Ca^{2+} loaded form. Although Ca^{2+} binding has little effect of the average structure, NMR studies show that the flexibility of the calbindin D_{9k} molecule becomes much reduced.

This section has stressed the need for information about the Mg^{2+}-forms of EF-hand proteins. Fortunately, in the case of the buffer–modulator group of proteins, structural information on the Mg^{2+}-bound forms is available. The X-ray structure of the Mg^{2+}-form of calbindin D_{9k} is shown in Fig. XIV.3.10*b*. It is obvious that the small Mg^{2+} ion cannot accommodate all the available Ca^{2+} ligands in its inner-coordination sphere: a water molecule is inserted between the ion and the important bifurcating ligand at the end of the loop. If this phenomenon occurs also in calmodulin, it would explain the lack of evidence for a conformational change in its Mg^{2+} loaded form.

It is interesting to note that, while Mg^{2+} binding seemingly has little effect on the structure of calmodulin, in the case of calbindin D_{9k} the apo- and Ca^{2+}-forms are quite similar, while the Mg^{2+}-form is significantly different from the other two. The structural differences are, however, far smaller than those caused by Ca^{2+} binding to calmodulin.

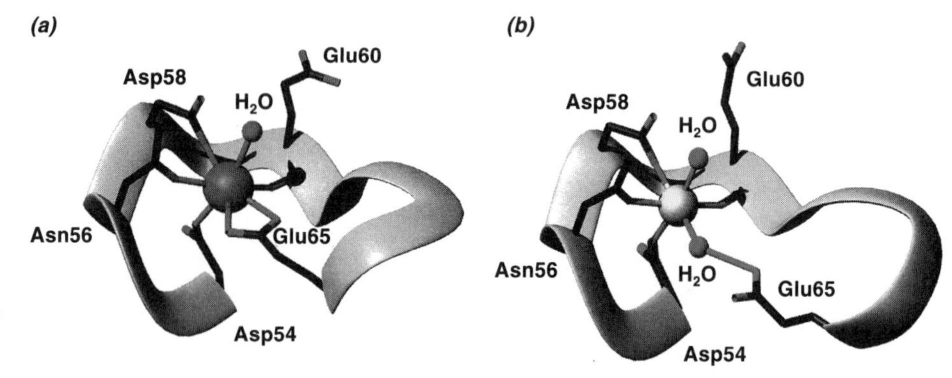

Fig. XIV.3.10. (*a*) A ribbon diagram showing the EF-hand Ca^{2+}-binding loop from calbindin D_{9k} with Ca^{2+} bound. The residues and water that chelate Ca^{2+} are indicated. (*b*) A ribbon diagram showing the EF-hand Ca^{2+}-binding loop from calbindin D_{9k} bound to a Mg^{2+} ion. Due to the difference between Mg^{2+} and Ca^{2+} an extra water molecule is needed to bind to Glu 65.

It is too simplistic and premature to single out one particular cause for the lack of conformational change upon Ca^{2+} binding in calbindin D_{9k}. The fact that the pseudo-EF-hand loop is significantly longer than the canonical loops would appear to play an important role since the approach of the bifurcated Glu14 side chain to an incoming Ca^{2+} ion does not require a concomitant movement of the adjacent α-helix. This appears, however, not to be the whole story. Through site directed mutagenesis, the pseudo-site in calbindin D_{9k} has been turned into a normal EF-hand without any major changes in the overall 3D structure upon Ca^{2+} binding, except locally in the binding loop. This result indicates that the loop structure as such cannot explain the difference been regulators and sensors. One striking difference between calbindin D_{9k} and calmodulin is, however, that calmodulin has an unusually high methionine content, whereas calbindin does not have a single methionine. It has therefore been speculated that the presence of these methionines is somehow responsible for the fact that calmodulin (troponin C) will open up its structure upon Ca^{2+} binding.

References

General References

1. Bertini, I., Gray, H. B., Lippard, S. J., and Valentine, J. S., *Bioinorganic Chemistry,* University Science Books, Mill Valley, CA, 1994.
2. Carafoli, E., The intracellular homeostasis of calcium: an overview, *Ann. N.Y. Acad. Sci.,* **551,** 147 (1988). Carafoli, E. and Klee, C. B., *Calcium as a cellular regulator,* Oxford University Press, New York, 1999.
3. Celio, M. R., "Guidebook to the Calcium Binding Proteins", in *Guidebook Series,* Oxford University Press, New York, 1996.
4. Evenäs, J., Malmendal, A., and Forsén, S., "Calcium", *Curr. Opin. Chem. Biol.,* **2,** 293 (1998).
5. Falke, J. J., Drake, S. K., Hazard, A. L., and Peerse, O. B., "Molecular tuning of ion binding to calcium signaling proteins", *Q. Rev. Biophys.,* **27,** 219 (1994).
6. Finn, B. E. and Forsén, S., "The evolving model of calmodulin structure, function and activation", *Structure,* **3,** 7 (1995).
7. Finn, B. E. and Drakenberg, T. D., "Calcium-binding Proteins", *Adv. Inorg. Chem.,* **46,** 441 (1999).
8. Gagné, S. M., Li, M. X., McKay, R. T., and Sykes, B. D., "The NMR angle on troponin C", *Biochem. Cell Biol.,* **76,** 302 (1998).
9. Muranyi, A. and Finn, B. E., "Calcium and Its Enzymes", in *Handbook on Metalloproteins, 37,* Marcel Dekker, Basel, 2000.
10. Nemeth, E. F. and Carafoli, E., "The role of extracellular calcium in the regulation of intracellular calcium and cell function", *Cell Calcium,* **11,** 319 (1990).
11. Santella, L. and Carafoli, E., "Calcium signaling in the cell nucleus", *FASEB J.,* **11,** 1091 (1997).
12. The EF-Hand Ca^{2+}-Binding Proteins Data Library http://structbio.vanderbilt.edu/cabp_database
13. Cellular Calcium Information Server: http://calcium.oci.utoronto.ca/

Specific Reference

14. NIH Consensus Conference. Optimal calcium intake. NIH Concensus Development Panel on Optimal Calcium Intake., *JAMA,* **272,** 1942–1948 (1994).

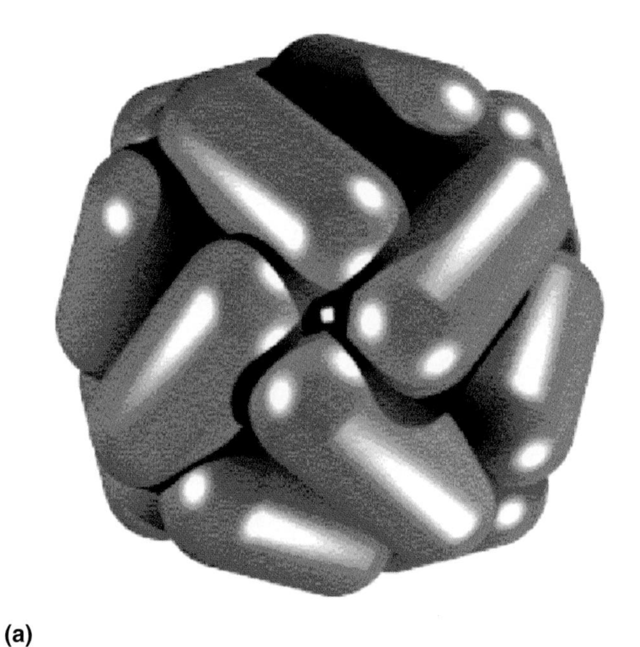

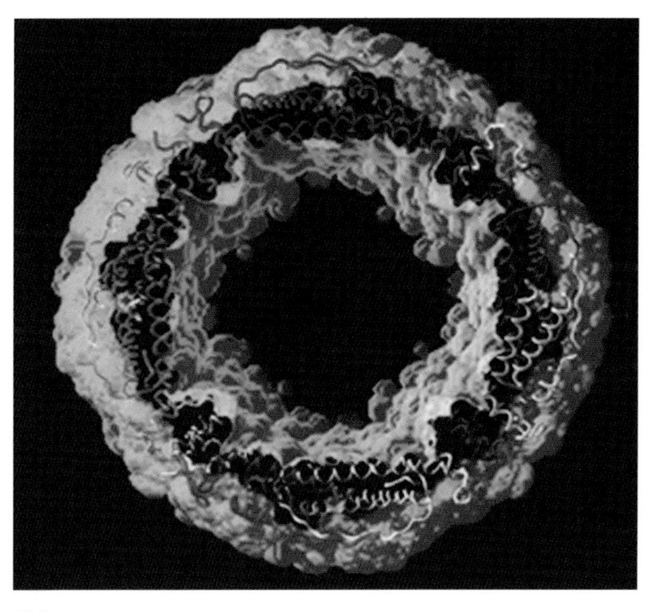

(a) (b)

Fig. VIII.2.1.
Ferritin protein assembled (*a*) and as a cross-section through the center of the protein sphere (*b*). The volume of the protein is ~864 nm^3 and the volume of the nanoreactor cavity, where the ferric iron mineral (hydrated ferric oxide) forms, is ~256 nm^3, and is colored black. Amino acid side chains are shown in orange and are deleted over the subunit polypeptide backbones (multicolored).

Sites Fe-2 and Fe-5 are better studied than Fe-1, Fe-3, and Fe-4. Sites for movement of Fe(II) and Fe(II) mineral precursors have not yet been identified.
Fe-1 = Ferritin F_{ox} Site
Fe-2 = Ferroxidase (F_{ox}) site
Fe-3 = Mineral nucleation site
Fe-4 = Nanomineral
Fe-5 = Fe Exit Pore

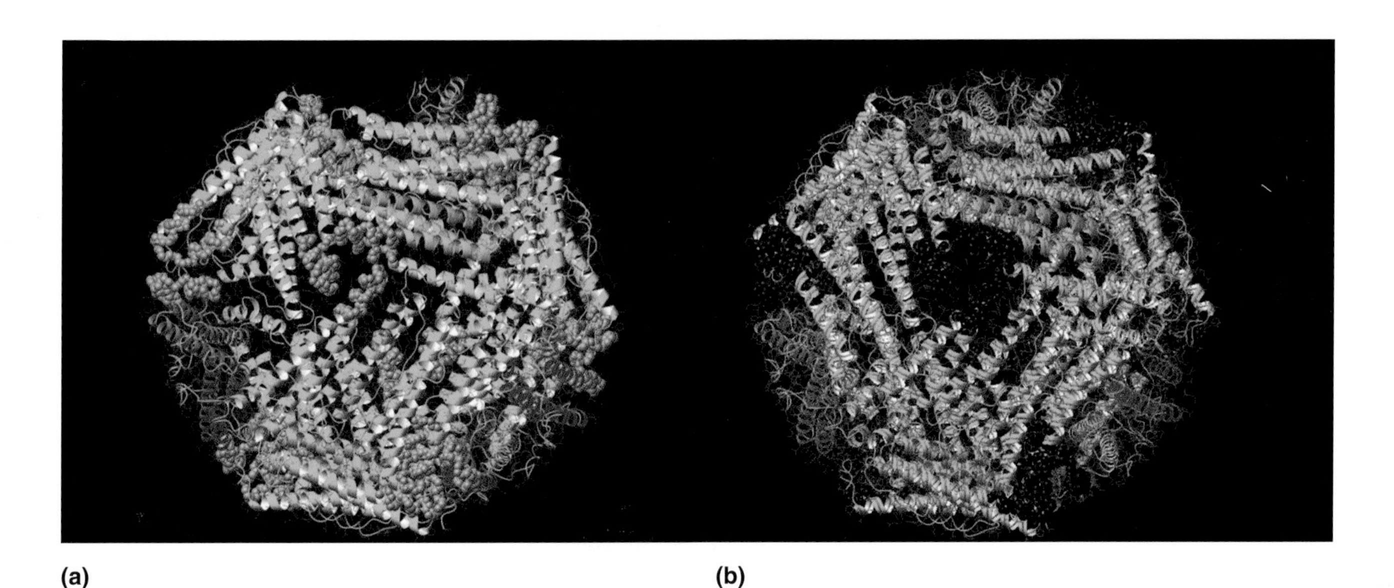

(a) (b)

Fig. VIII.2.2.
Gated ferritin pores. Pore helices, composed of three pairs of helices, as in ferritin (one set from each subunit), assemble around the threefold axis of 24-subunit ferritins, and are exquisitely sensitive to heat, low concentrations (m*M*) of chaotropes, such as urea or guanidine, and mutation of conserved residues (pores). The gates open when the pore helices unfold differentially, with the global protein structure intact, to accelerate iron removal through increased access between the iron mineral, reductants, and chelators. (*a*) Closed gates with pore helices in gold; (*b*) Open gates, where pore helices are unfolded and so disordered that they appear structureless in protein crystals, although the polypeptide chains are intact.

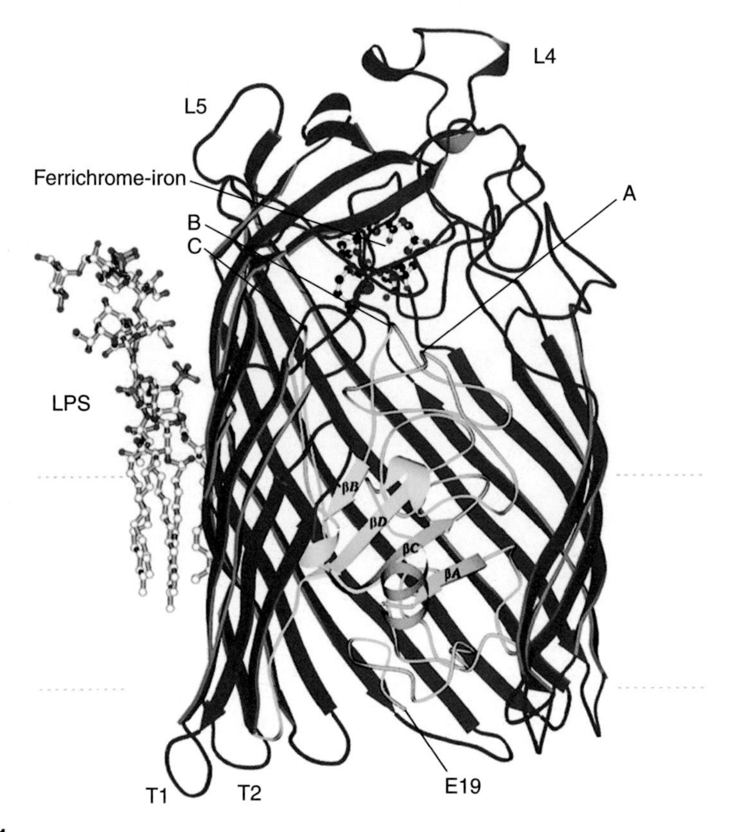

Fig. VIII.3.1.
Crystal structure of the FhuA–ferrichrome-iron complex showing a single molecule of lipopolysaccharides LPS noncovalently associated with the outer membrane (OM) protein complex.

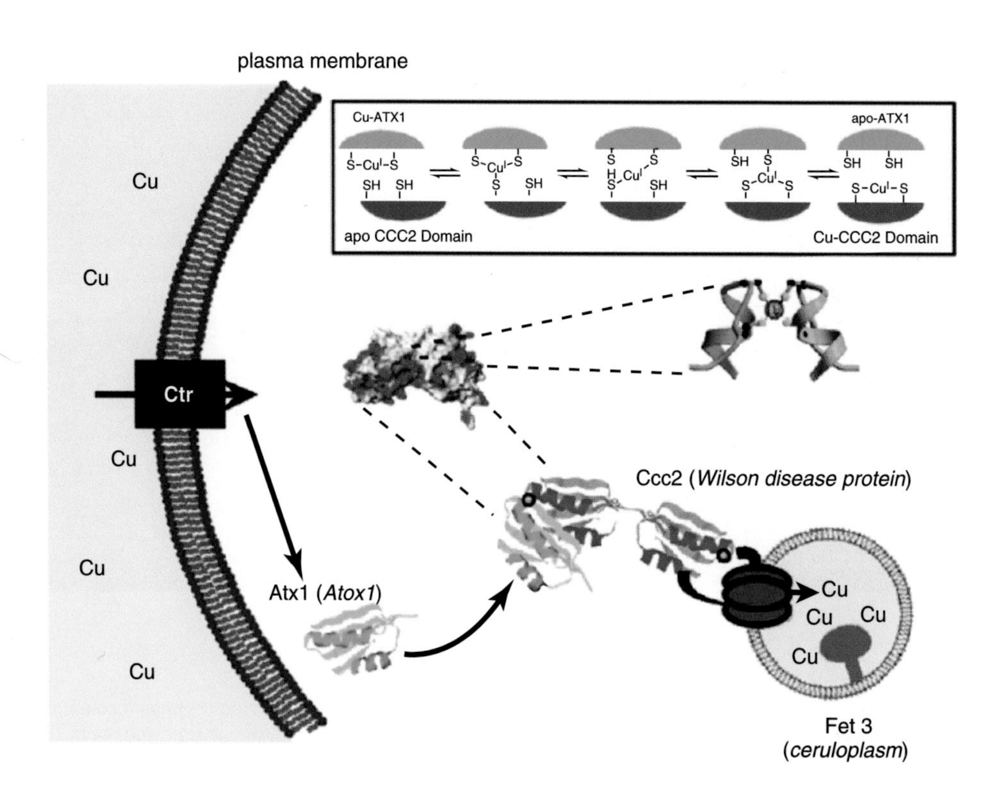

Fig. VIII.6.2.
Inorganic chemistry and structural biology of the ATX1 copper chaperone trafficking pathway (as summarized in Ref. 8). The Cu(I) ions are brought into the cell via unknown mechanisms, some of which involve the Ctr1 family of membrane-bound transporter proteins. Once loaded, the Cu–Atx1 complex (green) can dock with a structurally homologous domain of a partner protein, the P-type ATPase Ccc2 (purple). Electrostatic complementarity between the Atx1 chaperone and its Ccc2 (see docked structures) provides a basis for partner recognition and for correct orientation of donor and acceptor Cys residues. A mechanism of copper transfer involving a series of two- and three-coordinate Cu(I)-thiolate intermediates (inset) is corroborated by spectroscopic, thermodynamic, and structural studies. Similar mechanisms are proposed for the mammalian forms (names in italics) of these yeast proteins. The PDB locators are Cu-Hah1: 1FEE; Atx1: 1FD8; Cu-Ccc2a: 1FVS.

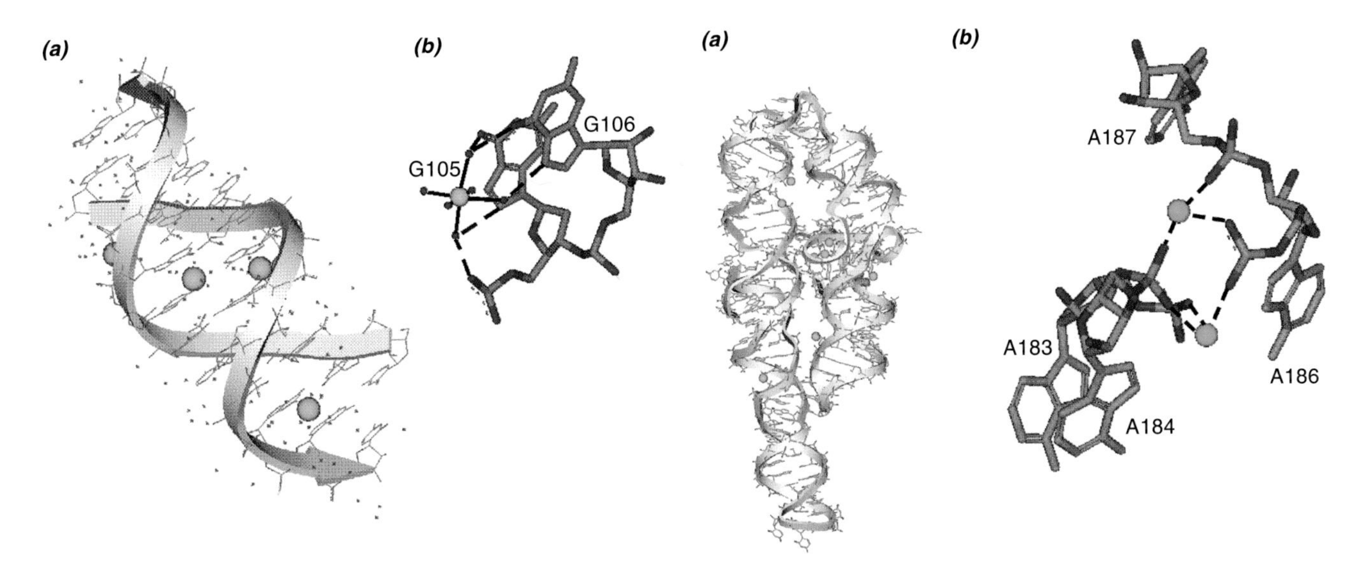

Fig. IX.5.6.
(*a*) The crystal structure of an internal loop E of a 5S rRNA with Mg^{2+} (PDB code: 354D). (*b*) An outer-sphere binding of Mg^{2+} (green) involving six water molecules (red spheres) observed in the crystal structure.

Fig. IX.5.7.
(*a*) Crystal structure of the P4–P6 domain of the group I ribozyme with Mg^{2+} (PDB code: 1GID). (*b*) Structure of the A-rich bulge with two Mg^{2+} ions (green) that coordinate to the phosphate of the A-rich bulge of the P4–P6 domain of the group I ribozyme. Both Mg^{2+} ions bind directly to the phosphates (inner-sphere binding). The average phosphate-oxygen- Mg^{2+} distance is 2.2 Å.

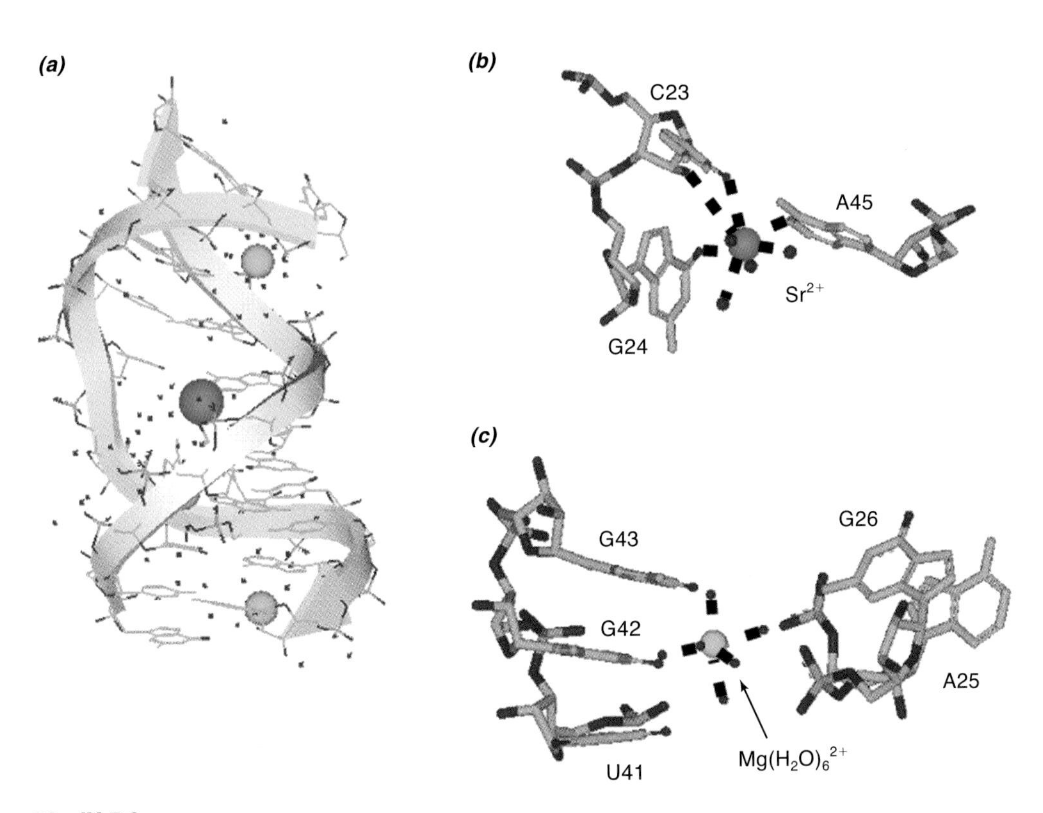

Fig. IX.5.8.
Crystal structure of the leadzyme ribozyme with Mg^{2+} and Sr^{2+} (*a*) (PDB code: 1NUV). Inner-sphere binding of Sr^{2+} (*b*). Outer-sphere binding of Mg^{2+} (*c*).

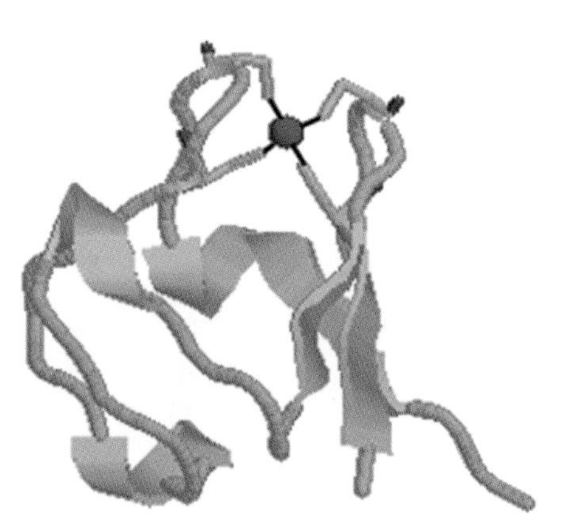

Fig. X.1.4.
The X-ray crystal structure of rubredoxin from *Pyrococcus furiosus* (PDB code: 1BRF).

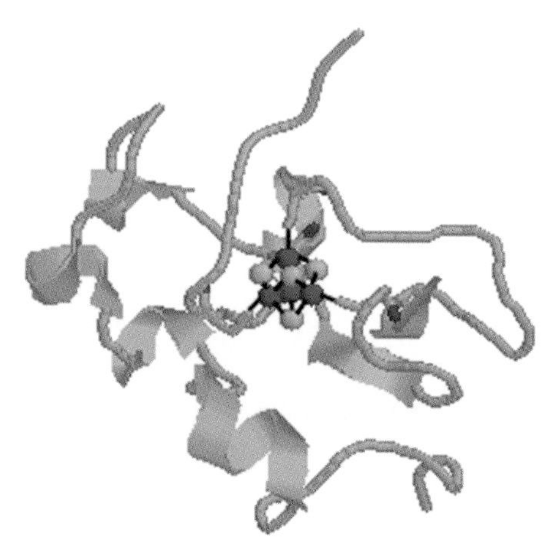

Fig. X.1.5.
The X-ray crystal structure of *Chromatium vinosum* HiPIP (PDB code: 1CKU).

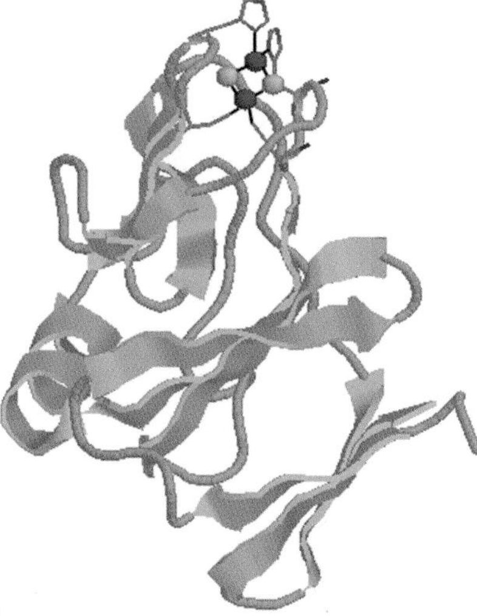

Fig. X.1.6.
The X-ray crystal structure of a water soluble fragment of the Rieske iron–sulfur protein of the bovine heart mitochondrial cytochrome bc_1 complex (PDB code: 1RIE).

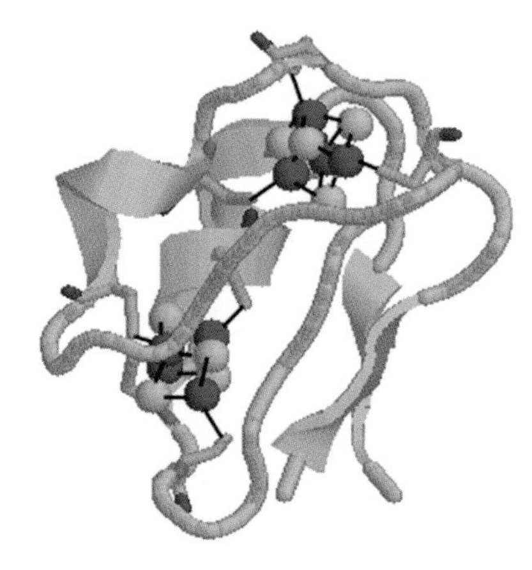

Fig. X.1.7.
The X-ray crystal structure of the oxidized 2Fe–2S ferredoxin from *Haloarcula morismotui* (PDB code: 1DOI).

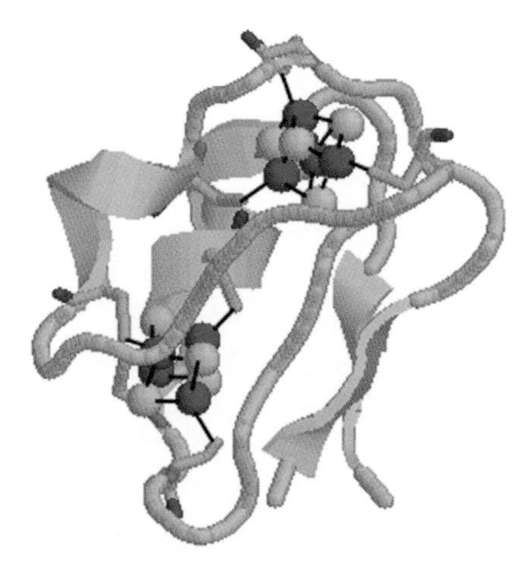

Fig. X.1.8.
The X-ray crystal structure of the 8Fe–8S *Clostridium acidiurici* ferredoxin (PDB code: 2FDN).

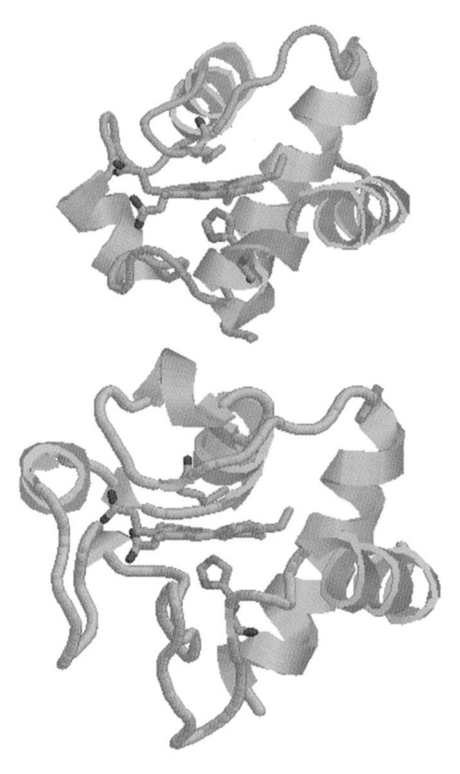

Fig. X.1.10.
The structure of cytochrome c_{553} from *Desulfovibrio vulgaris* (strain Miyazaki; constituted by 79 amino acids; PDB code: 1C53) is compared with the structure of horse heart cytochrome *c* (containing 105 amino acids; PDB code: 1HRC). The presence of an extra loop facing the heme propionates in the larger cytochrome is evident from this protein orientation.

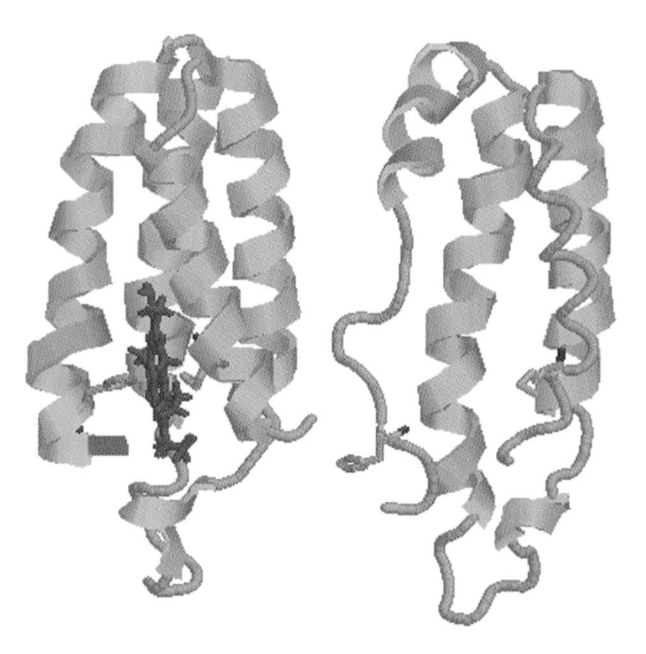

Fig. X.1.11.
The solution structures of the (*a*) holo- and (*b*) apo-form of cytochrome b_{562} from *E. coli* (PDB codes: 1QPU and 1APC, respectively).

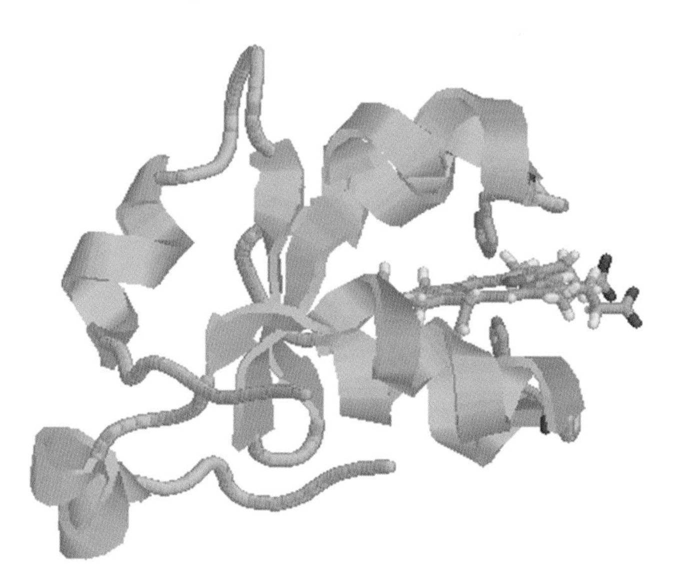

Fig. X.1.12.
The solution structure of the soluble fragment of oxidized rat microsomal cytochrome b_5 (PDB code: 1BFX).

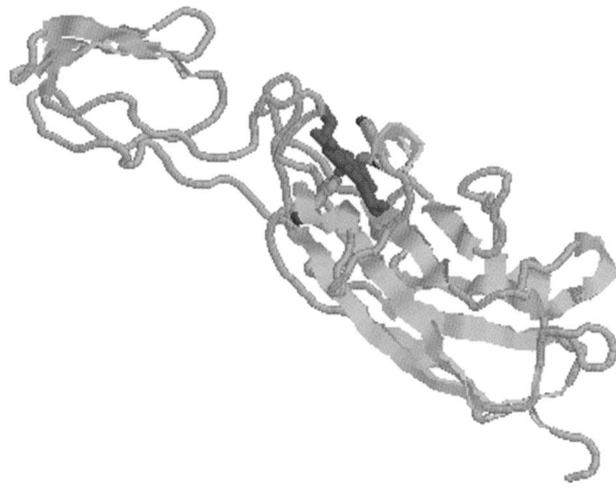

Fig. X.1.13.
The X-ray crystal structure of the lumen-side domain of reduced cytochrome *f* (PDB code: 1HCZ).

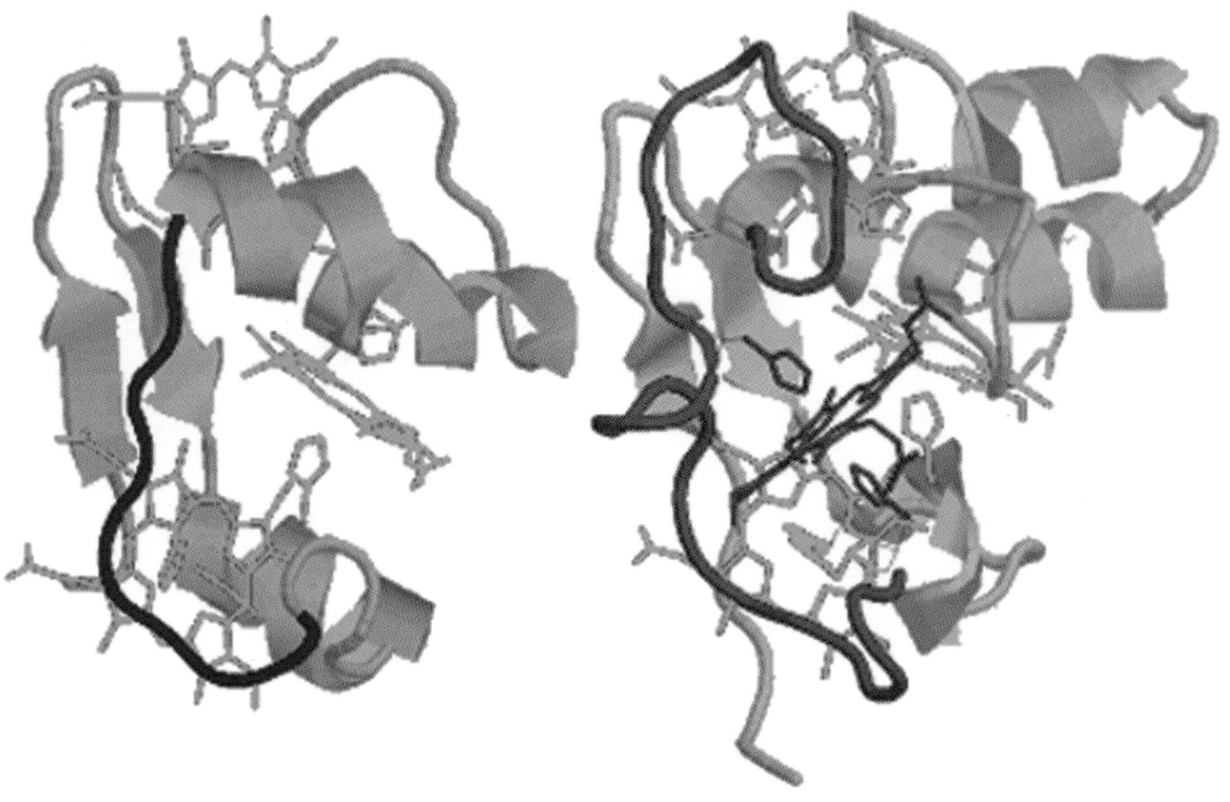

Fig. X.1.14.
Comparison of the structures of the (*a*) three-heme cytochrome c_7 from *Desulfuromonas acetoxidans* and of the (*b*) four-heme *Desulfovibrio desulfuricans* cytochrome c_3. Regions that differ due to the different length of the protein and different number of hemes are shown in black (PDB codes: 1EHJ and 3CYR, respectively).

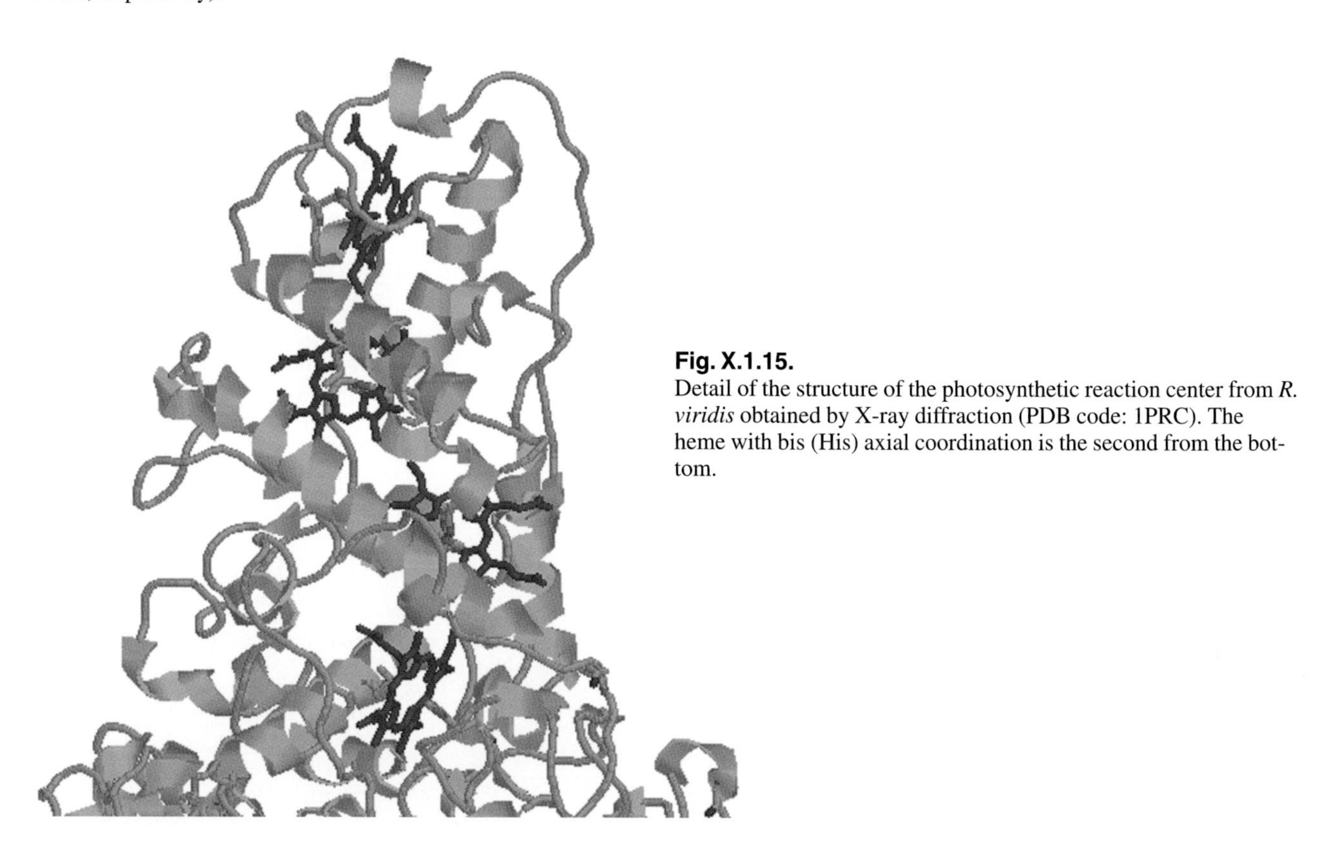

Fig. X.1.15.
Detail of the structure of the photosynthetic reaction center from *R. viridis* obtained by X-ray diffraction (PDB code: 1PRC). The heme with bis (His) axial coordination is the second from the bottom.

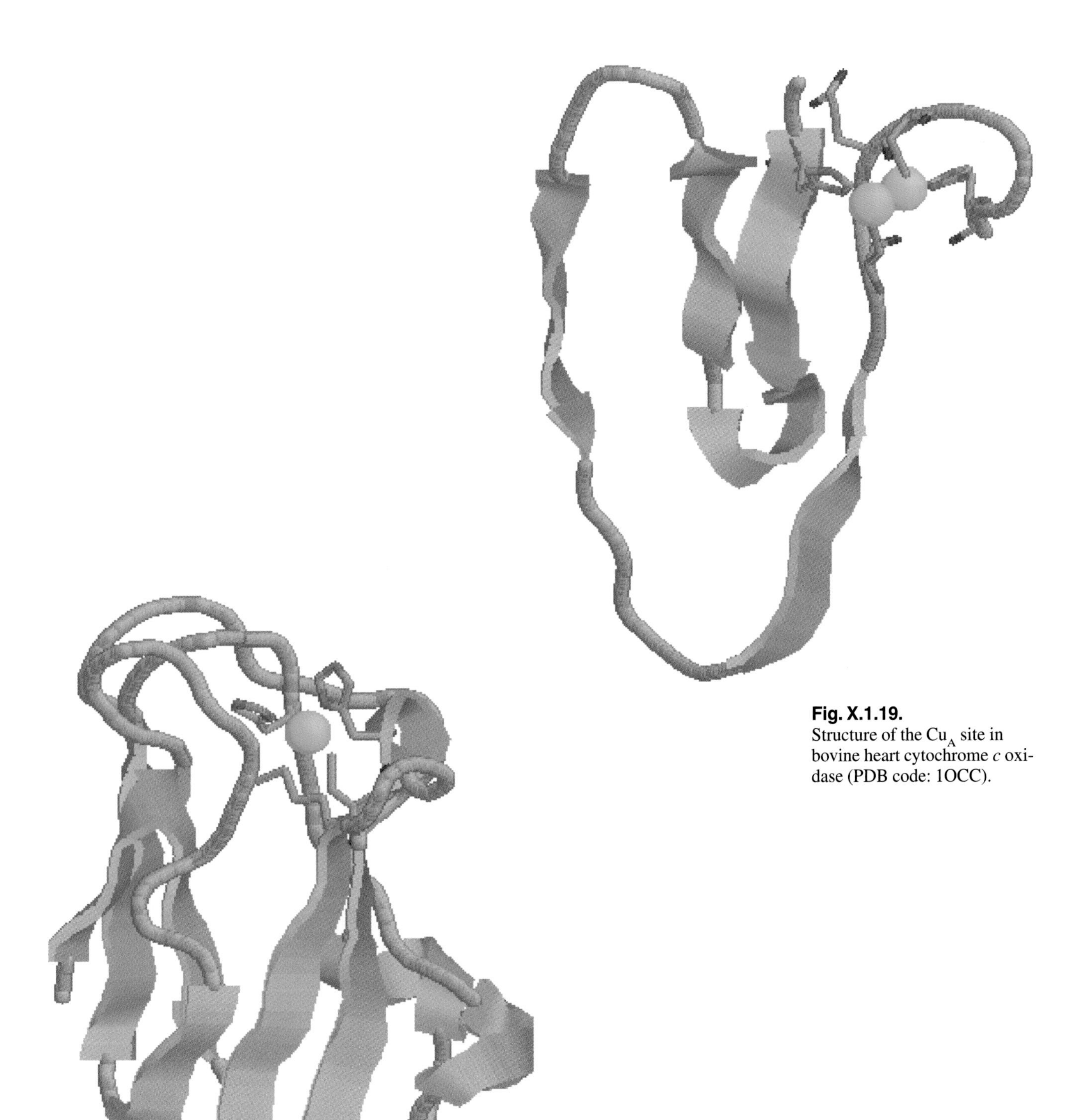

Fig. X.1.19.
Structure of the Cu_A site in bovine heart cytochrome *c* oxidase (PDB code: 1OCC).

Fig. X.1.17.
The X-ray crystal structure of the plastocyanin from spinach (PDB code: 1AG6).

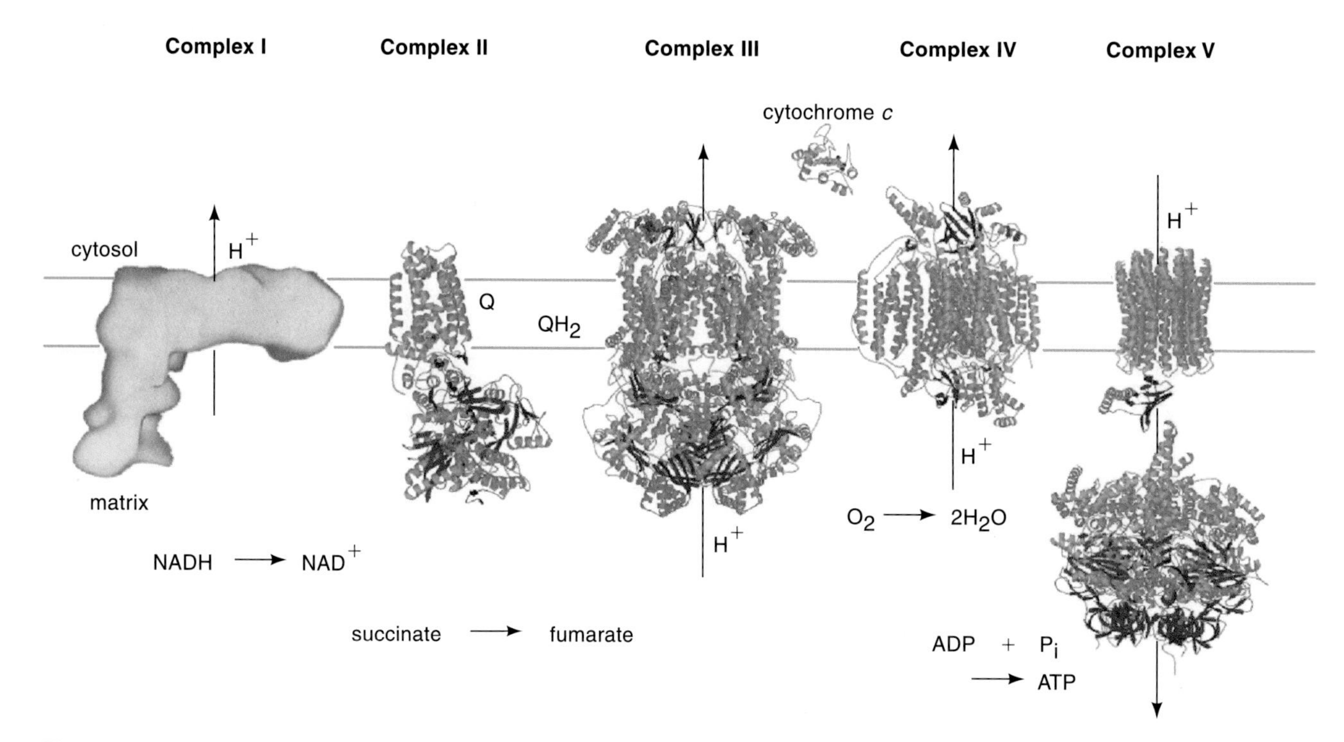

Fig. X.3.5.
Components of the mitochondrial electron-transfer chain. Four of the five complexes shown are represented by their actual crystal structures (see text) as they would sit in the inner-mitochondrial membrane. Complex **I,** or NADH-ubiquinone(Q) oxidoreductase, is represented by the current low-resolution picture of its overall shape (not to scale). Complex **II,** or succinate dehydrogenase, is not a part of the proton-pumping machinery, but is a supplier of electrons and protons in the form of ubiquinol (QH_2). [Adapted from Schultz, B. E. and Chan, S. I. *Annu. Rev. Biophys. Biomolec. Structure* 30, 23–65 (2001) with permission from Annual Reviews, Palo Alto, CA.]

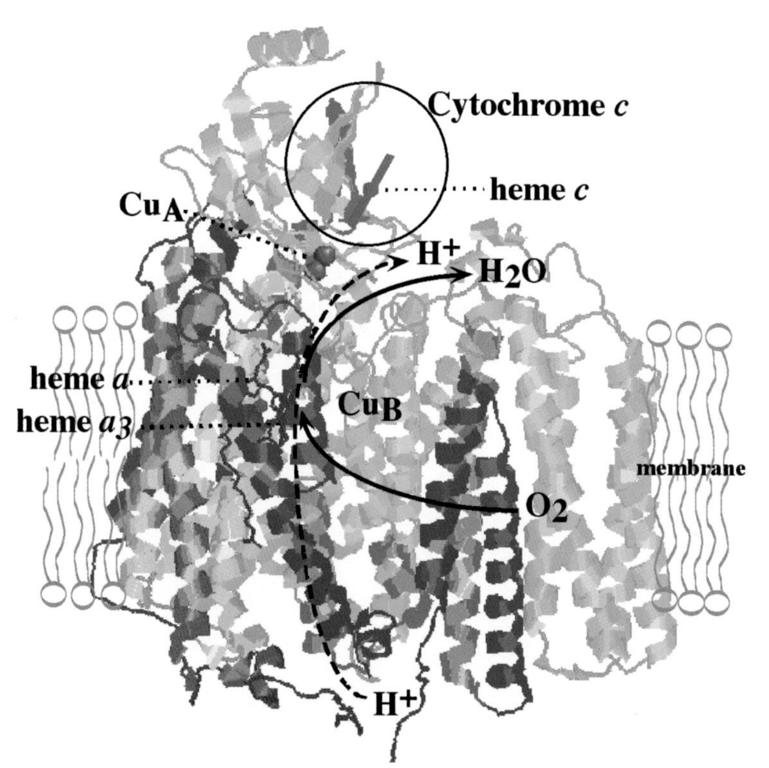

Fig. X.3.7.
Cytochrome *c* oxidase oriented in the membrane. The three largest subunits of the bovine oxidase are shown, with two hemes and three copper ions. Cytochrome *c* is represented at its interaction site on the outside surface, where electrons are donated and protons are released. (Drawing made in Rasmol and Canvas from PDB code: 1OCO.)

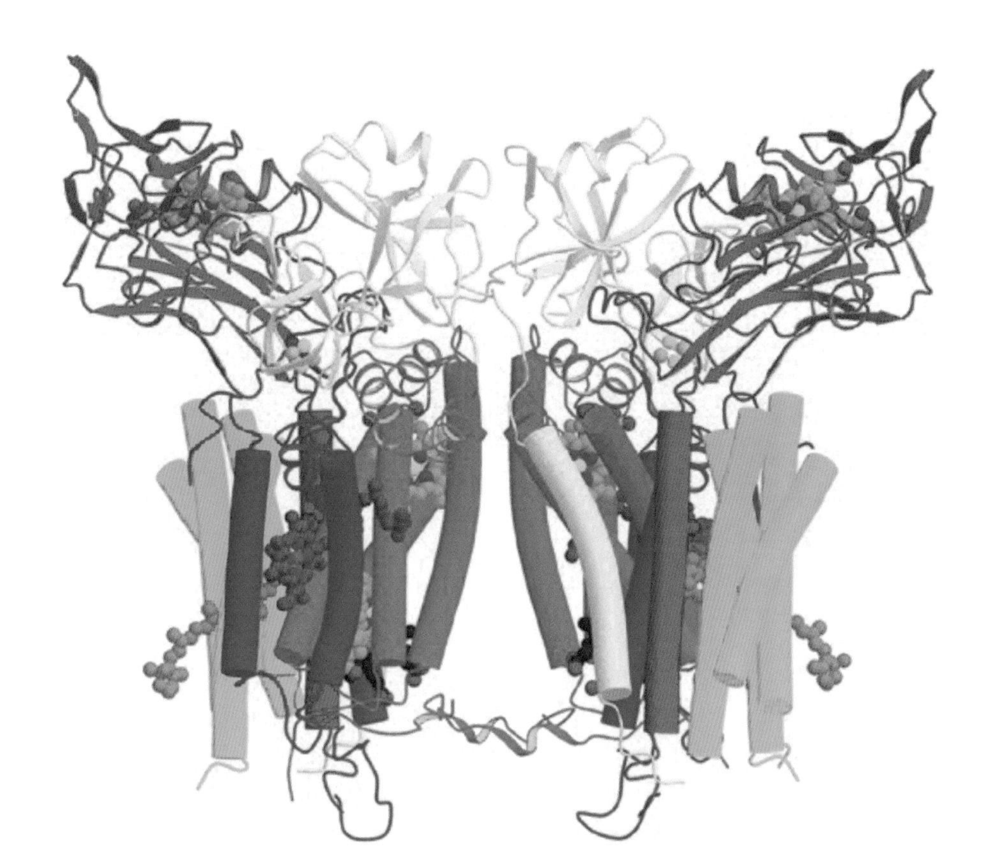

Fig. X.3.12.
Structure of the cytochrome b_6f complex. Shown is an illustration of the cytochrome b_6f protein complex, which is critical for photosynthesis. The eight colors represent the eight protein components of the cytochrome complex; the cylinders are the 26 segments of the complex that cross the photosynthetic membrane; the colored rings made of little balls that are embedded in protein are the groups that actually carry the electrons stimulated by light absorbed in photosynthesis (gray = heme *f,* heme *b;* green = chlorophyll; orange = carotenoid). [Adapted from a model by H. Zhang/W. A. Cramer] (http://news.uns.purdue.edu/UNS/html4ever/031002.Cramer.photo.html) Crystal structures of two photosynthetic b_6f complexes are available (PDB codes: 1UM3 and 1Q90).

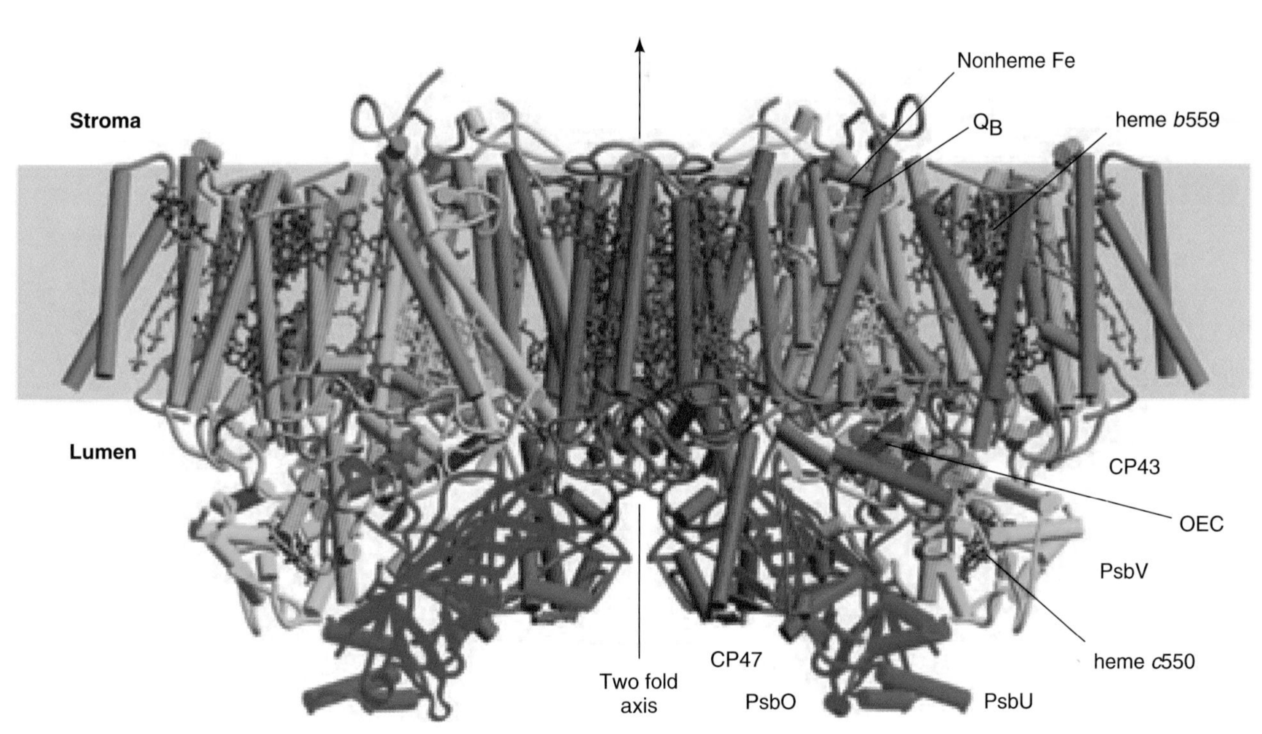

Fig. X.4.2.
The 3.5-Å resolution structure of PSII from the thermophilic cyanobacterium *T. elongatus.* The structure consists of a dimer of reaction centers of the enzyme, split by the twofold axis shown in the figure. The major integral membrane polypeptides shown here are colored as follows: D1; D2; CP47; CP43; and cytochrome b_{559}. The extrinsic polypeptides are manganese stabilizing protein, or PsbO, PsbU, cytochrome c_{550}, or PsbV. Chlorophylls are shown in green, and carotenoids in orange. The position of the Mn cluster is indicated by "OEC" (O_2 evolving complex). Descriptions of the functions of the polypeptides are given in the text (PDB code: 1S5L). The PDB codes for the other structures are 1FE1 and 1IZL.

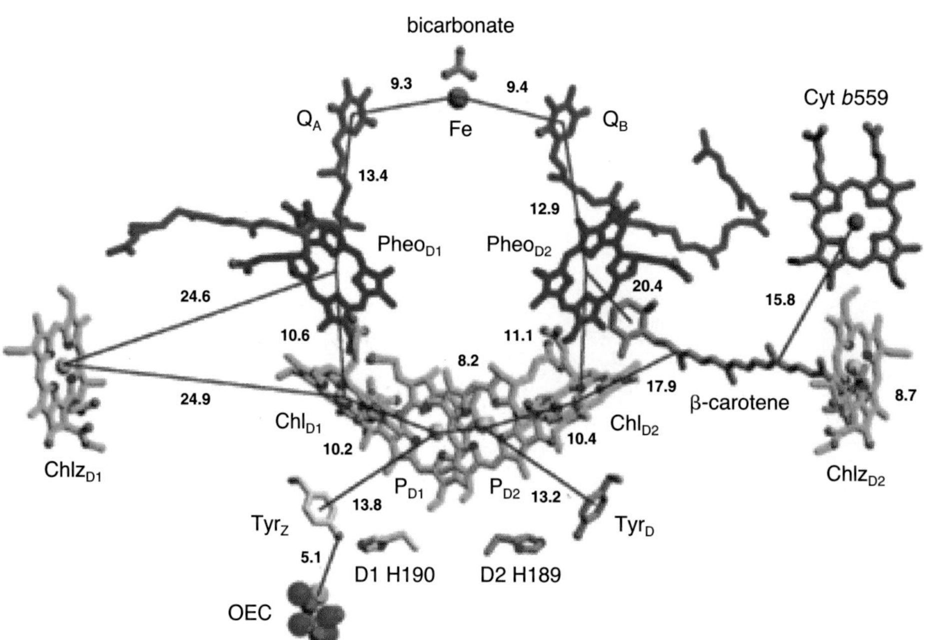

Fig. X.4.3.
Distances between the principal cofactors of the PSII reaction center. The numbers next to the lines connecting molecules are the distances in angstroms. Both Chl_{D1} and $Pheo_{D1}$ are used to identify the pigments ligated by the D1 polypeptide of the reaction center. A similar notation is used for the pigments associated with D2. Likewise, D1 H190 and D2 H189 identify two His residues that that are proposed to hydrogen bond to tyrosine residues Tyr_Z and Tyr_D.

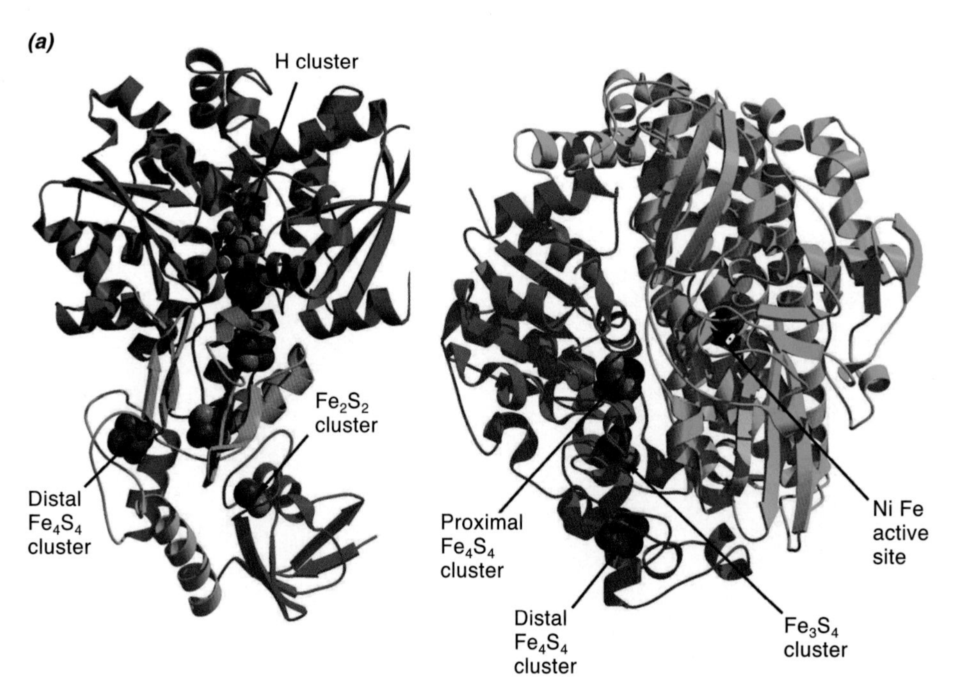

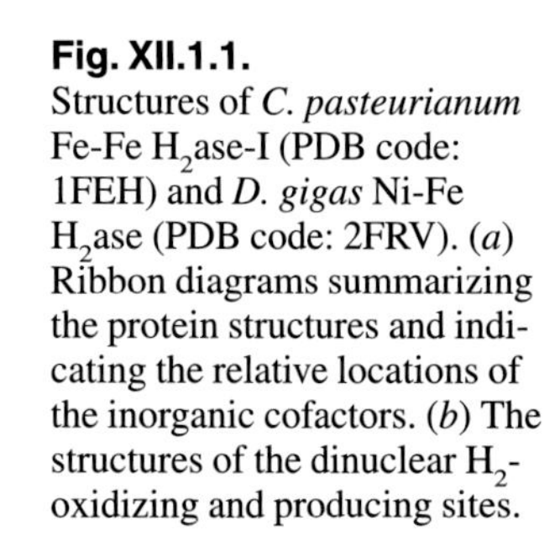

Fig. XII.1.1.
Structures of *C. pasteurianum* Fe-Fe H_2ase-I (PDB code: 1FEH) and *D. gigas* Ni-Fe H_2ase (PDB code: 2FRV). (*a*) Ribbon diagrams summarizing the protein structures and indicating the relative locations of the inorganic cofactors. (*b*) The structures of the dinuclear H_2-oxidizing and producing sites.

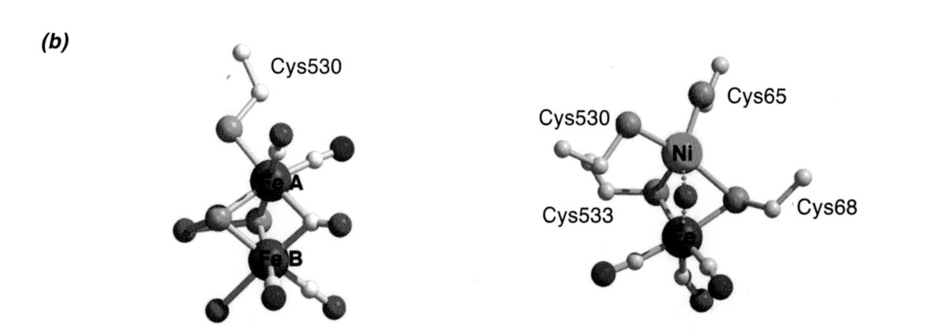

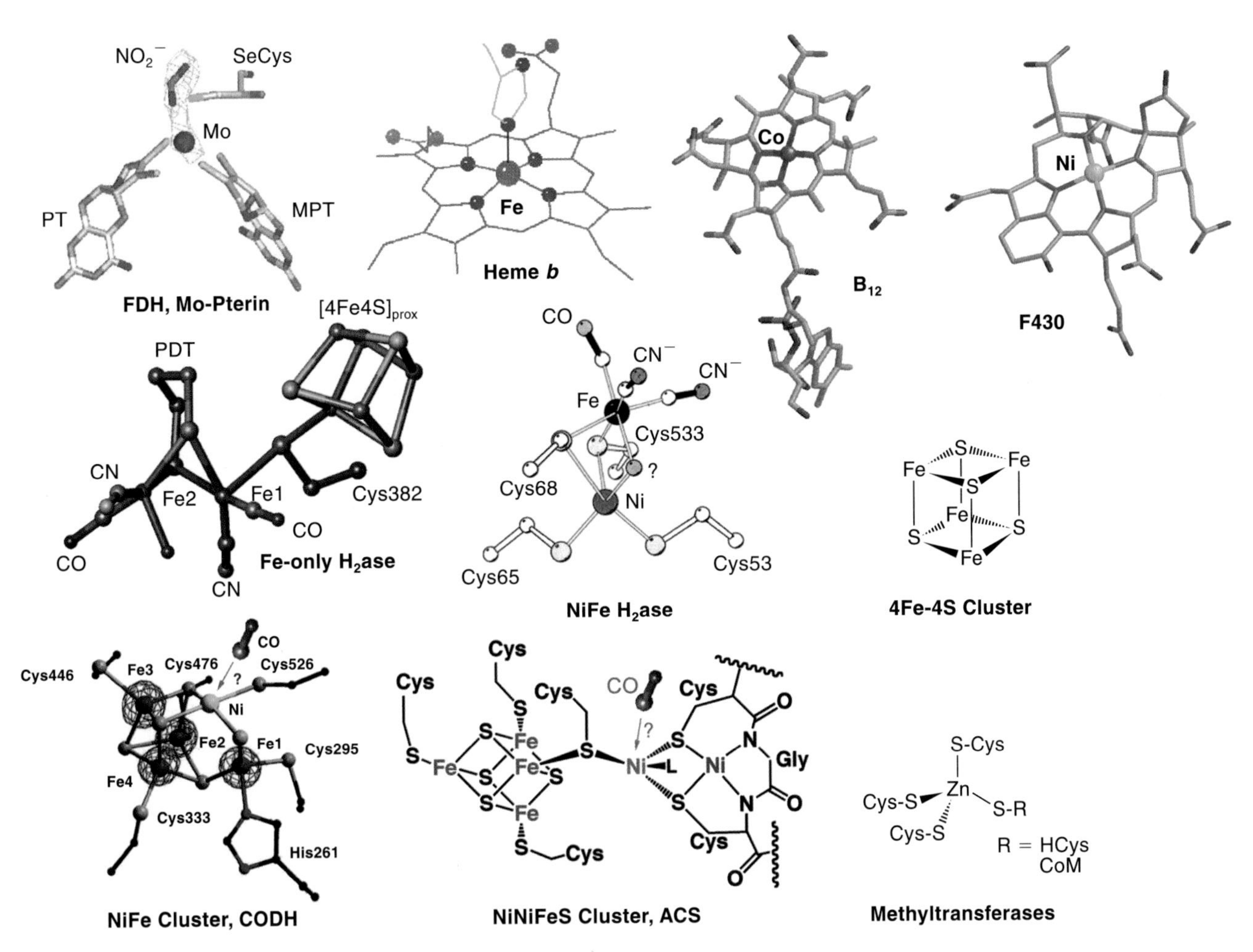

Fig. XII.2.2.
Metal sites in the pathways of one-carbon reduction.

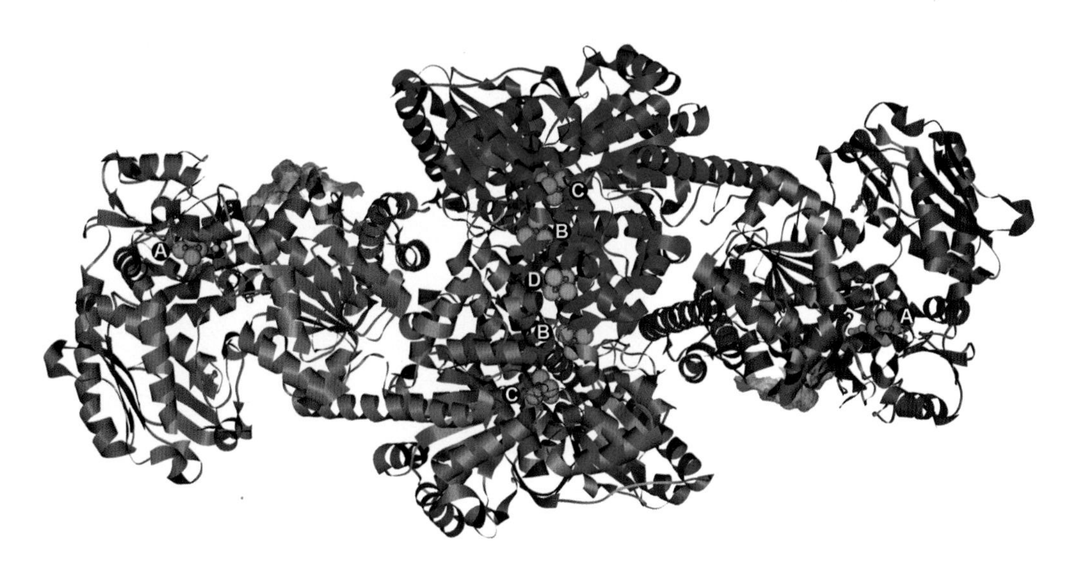

Fig. XII.2.4.
Structure of CODH/ACS. The CODH subunit is composed of the two core subunits, while the ACS subunits are attached to each central CODH subunit.

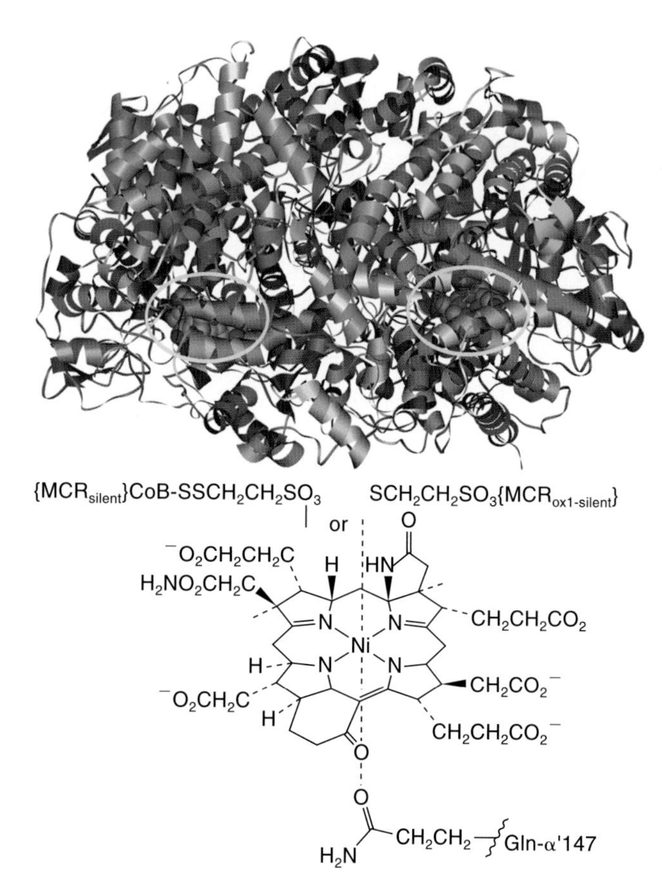

Fig. XII.2.7.
The structure of CH_3-CoM reductase. Each of the subunits are differentially colored and CoF_{430} is circled in yellow as a space-filling model and shown diagrammatically below the structure. The structure of MCR has PDB code: 1HBM.

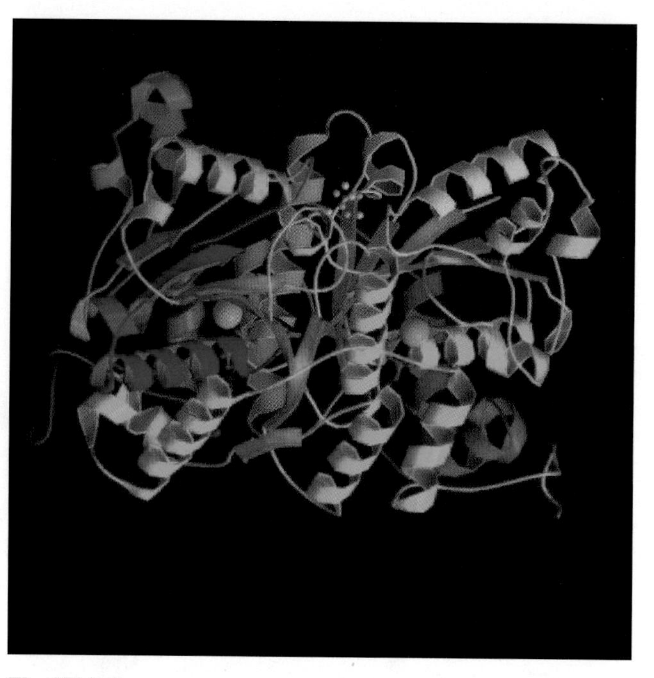

Fig. XII.3.5.
The overall polypeptide fold of the Fe-protein dimer complexed with MgADP, showing the [4Fe–4S] cluster bound between the two subunits and with the MgADP and the switch regions individually indicated (PDB code: 1FP6).

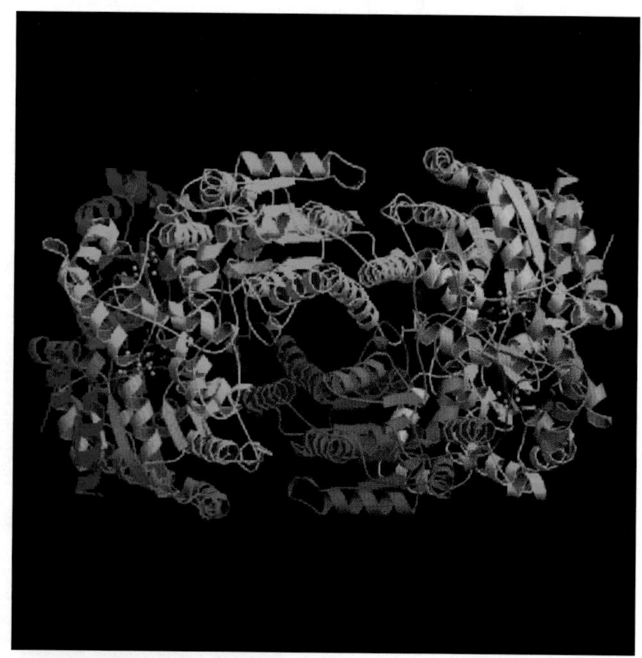

Fig. XII.3.6.
The overall polypeptide fold of the *A. vinelandii* nitrogenase MoFe protein tetramer. (*a*) As viewed along its twofold axis, showing the lack of interactions between the α-subunits. (*b*) A view at approximately right angles to view (*a*). The α- and β-subunits and prosthetic groups are individually colored. Each FeMo-cofactor is situated completely within an α-subunit, whereas each P cluster sits at an interface of the α- and β- subunits (PDB code: 3MIN).

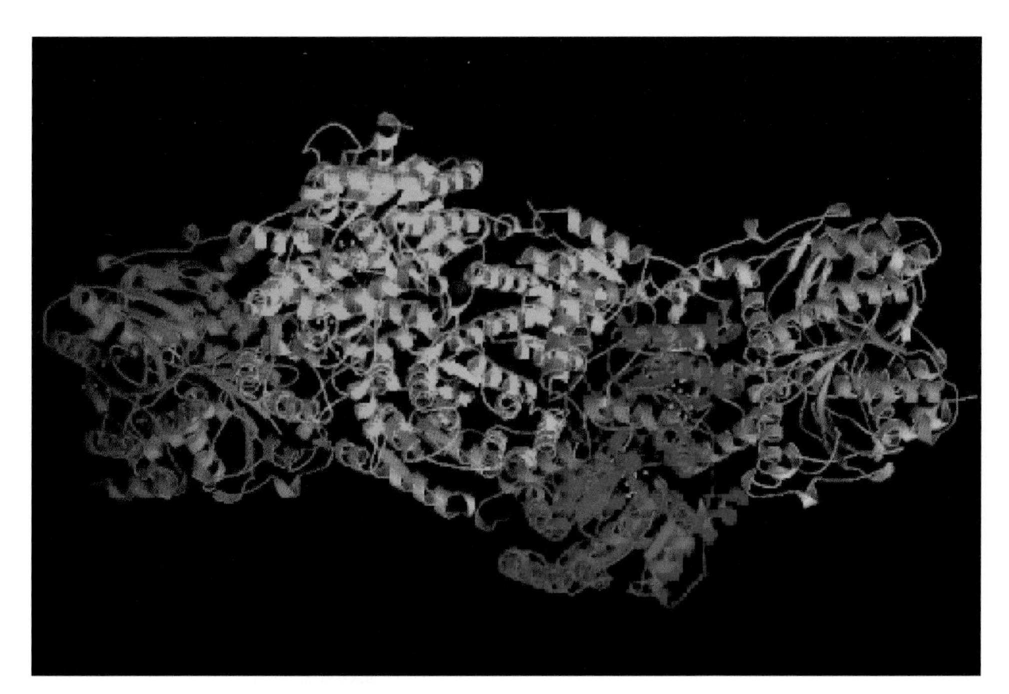

Fig. XII.3.9.
The structure of the 2:1 Fe protein–MoFe protein complex of the *A. vinelandii* nitrogenase stabilized by MgADP plus AlF_4^-. The Fe-protein molecules and the α- and β- subunits of the MoFe protein are individually colored. Each Fe-protein component docks over a MoFe-protein α/β-subunit interface and is juxtaposed with a P cluster (PDB code: 1N2C).

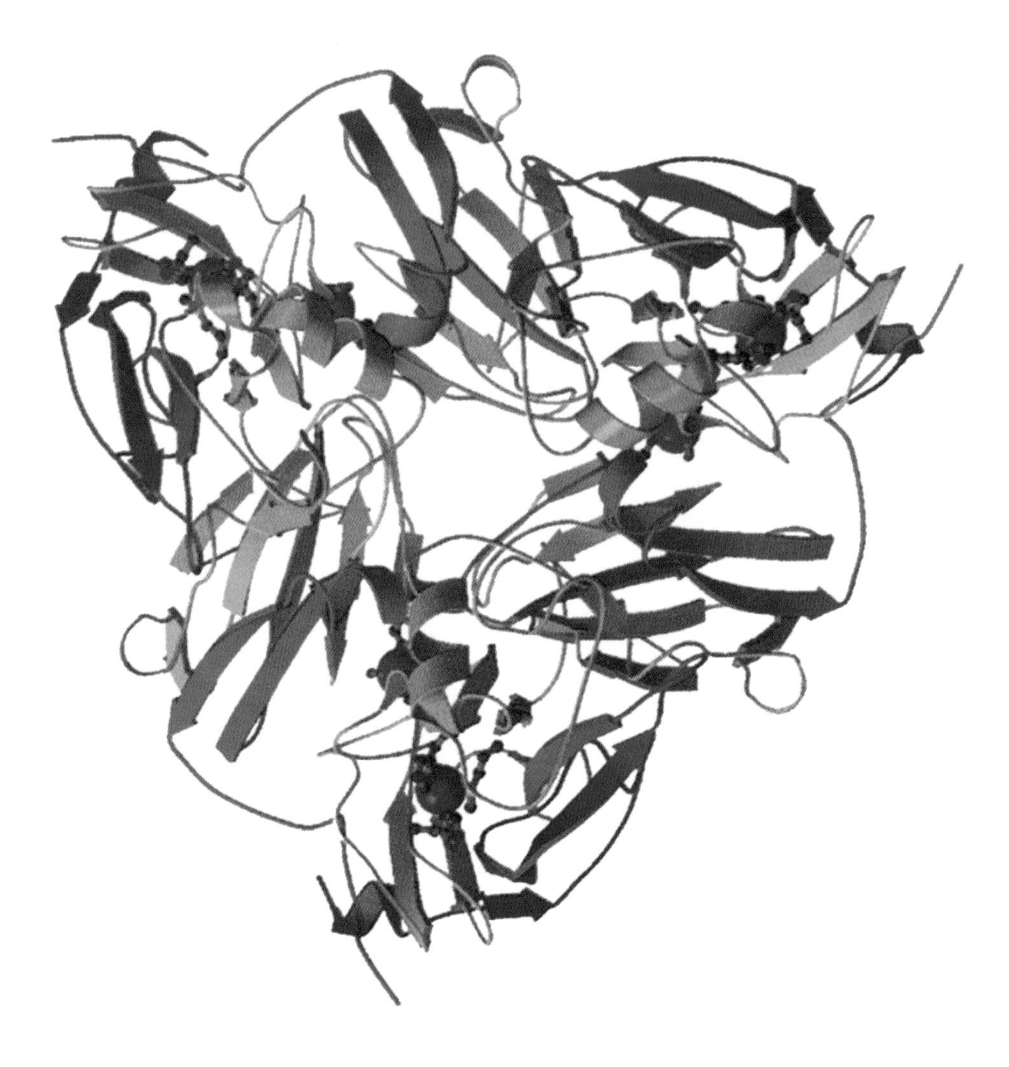

Fig. XII.4.4.
The structure of the dimeric heme cd_1-nitrite reductase from *Pseudomonas pantotrophus*. The cytochrome *c*-like domains are red, while the heme d_1-containing β-propeller catalytic domains are shown in a rainbow of colors (PDB code: 1QKS).

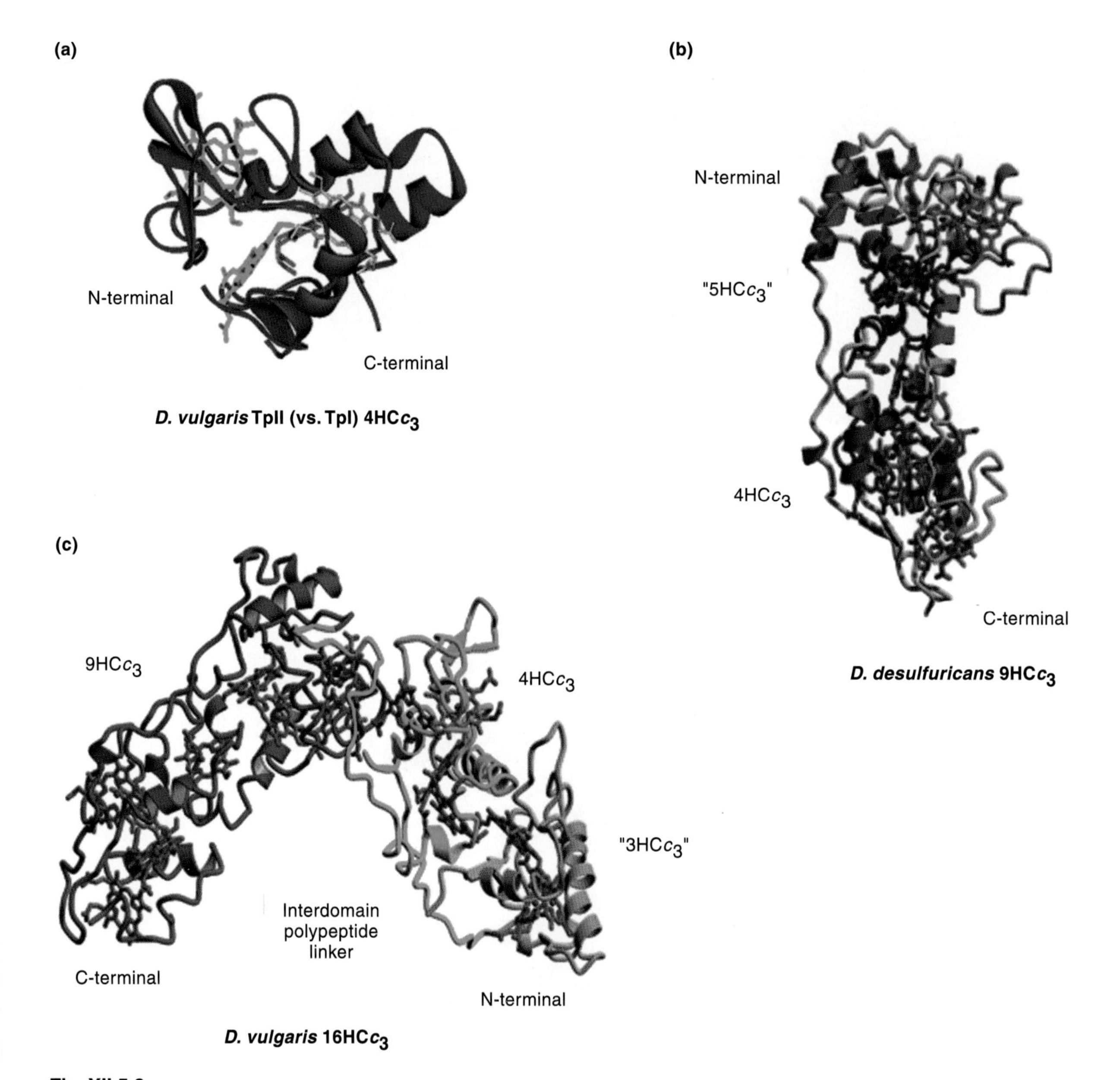

Fig. XII.5.3.
Multiheme cytochromes c_3. (*a*) Comparison of *D. vulgaris* 4HCc_3TpII (red) and 4HCc_3TpI (blue); superimposing the hemes, although only one set of hemes is shown: Heme 1 green, heme 2 pale blue, heme 3 yellow, heme 4 orange. (*b*) *Desulfovibrio desulfuricans* 9HCc_3; the 5HCc_3 region refers to a 4HCc_3-like domain plus an extra heme in an extra loop. (*c*) *Desulfovibrio vulgaris* 16HCc_3; the 3HCc_3 region refers to a 4HCc_3-like domain missing one heme.

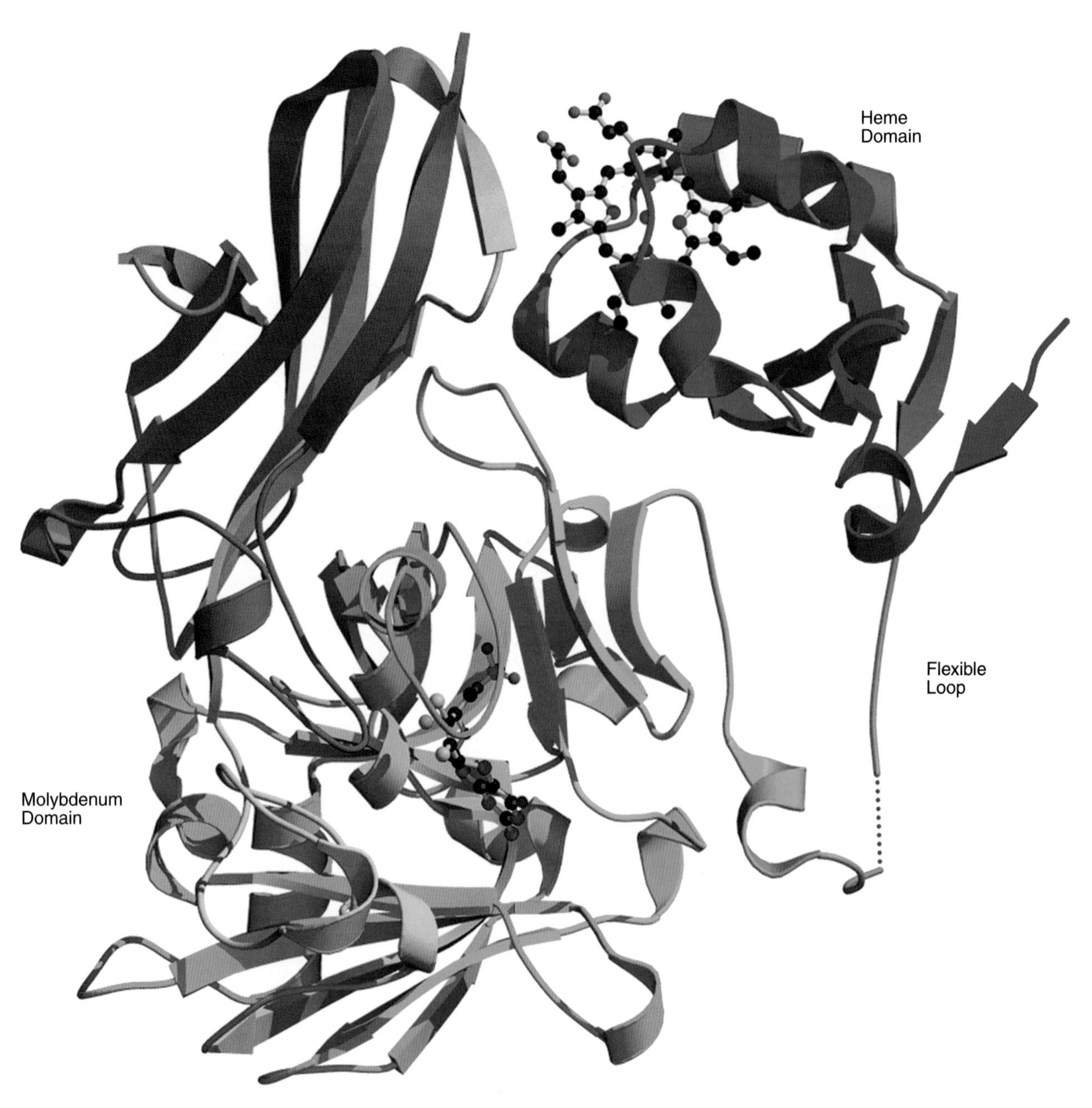

Fig. XII.6.13.
Schematic representation of the structure of monomer of sulfite oxidase from chicken liver, showing the smaller heme domain and the larger Mo domain (PDB code: 1SOX).

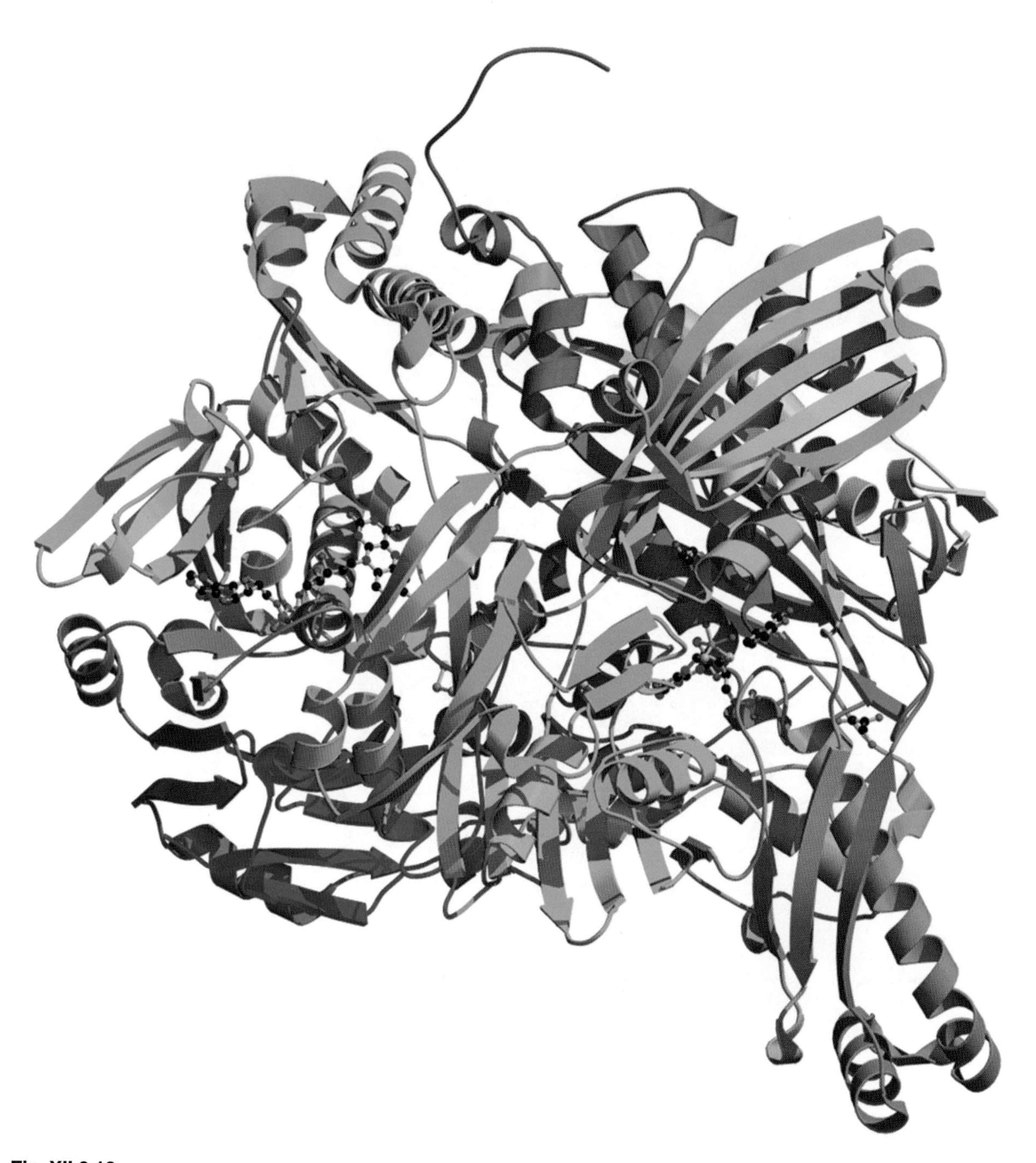

Fig. XII.6.18.
The structure of the monomer of bovine milk XDH (PDB code: 1FO4).

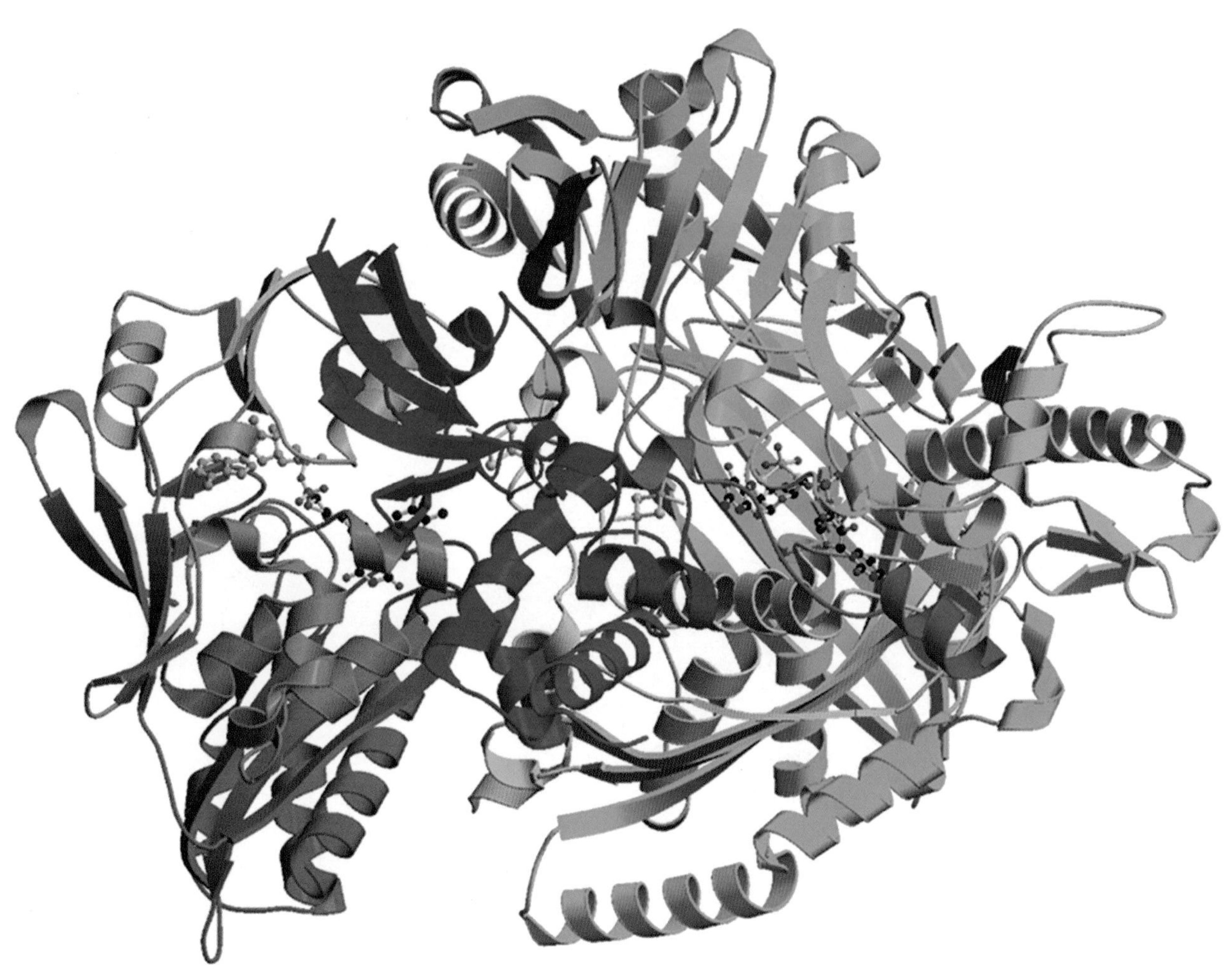

Fig. XII.6.21.
The structure of the CO dehydrogenase from *O. carboxidovarans* (PDB code: 1N5W).

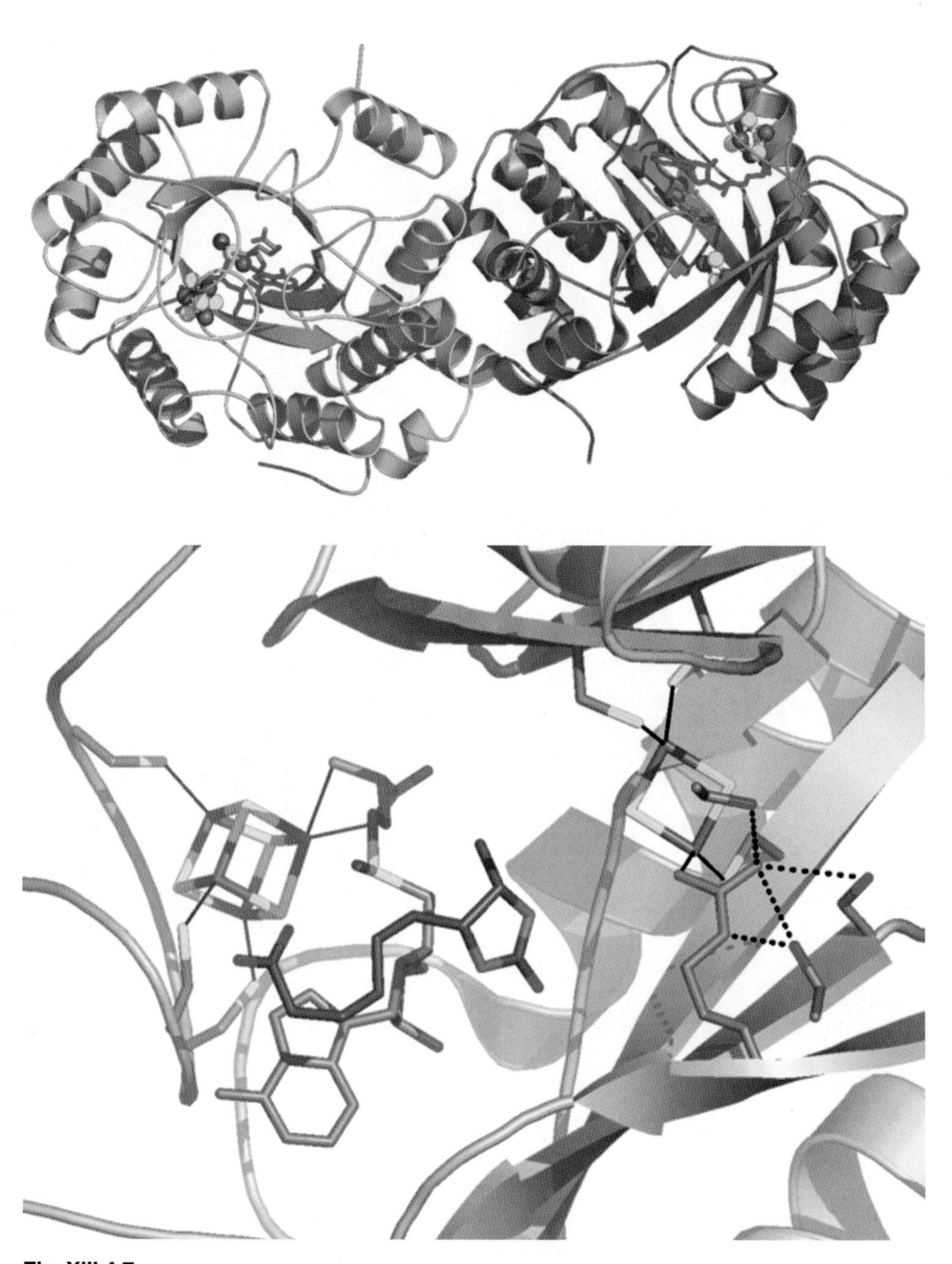

Fig. XIII.4.7.
(*a*) Overall X-ray crystal structure of the biotin synthase dimer, showing the [4Fe–4S] and [2Fe–2S] clusters (brown and yellow spheres), SAM (red), and dethiobiotin (blue). (*b*) The active site of biotin synthase, showing SAM coordinated to an iron of the [4Fe–4S] cluster, dethiobiotin bound between SAM and the [2Fe–2S] cluster, and the unusual arginine ligand to the [2Fe–2S] cluster (PDB code: 1R30). (Figures courtesy of F. Berkovitch and C. Drennan.)

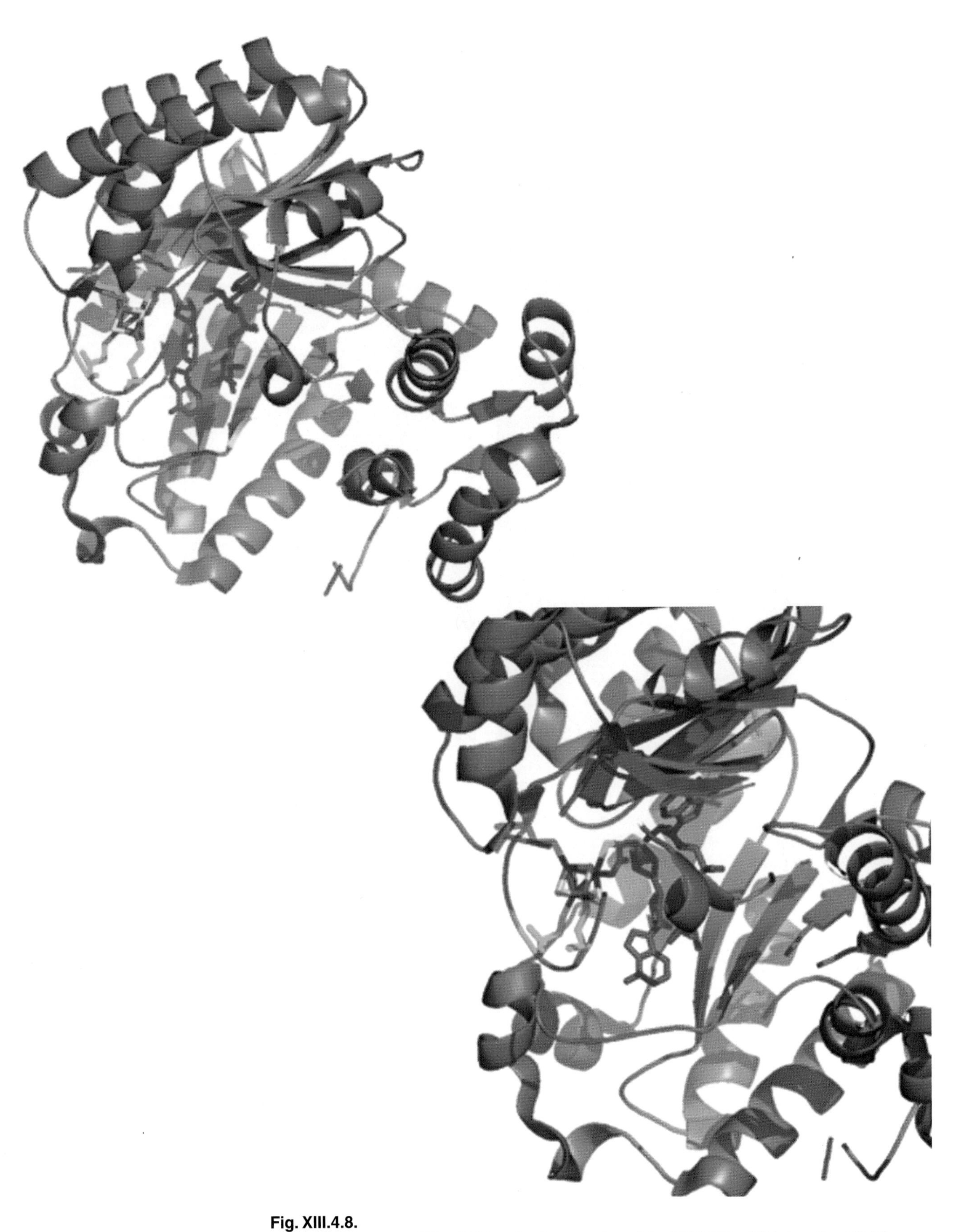

Fig. XIII.4.8.
The overall structure of HemN (*a*) and a close-up view of the SAM-[4Fe–4S] interaction (*b*). The [4Fe–4S] cluster is in red and yellow, the iron ligands are shown in green, and the two SAM molecules are in magenta (PDB code: 1OLT).

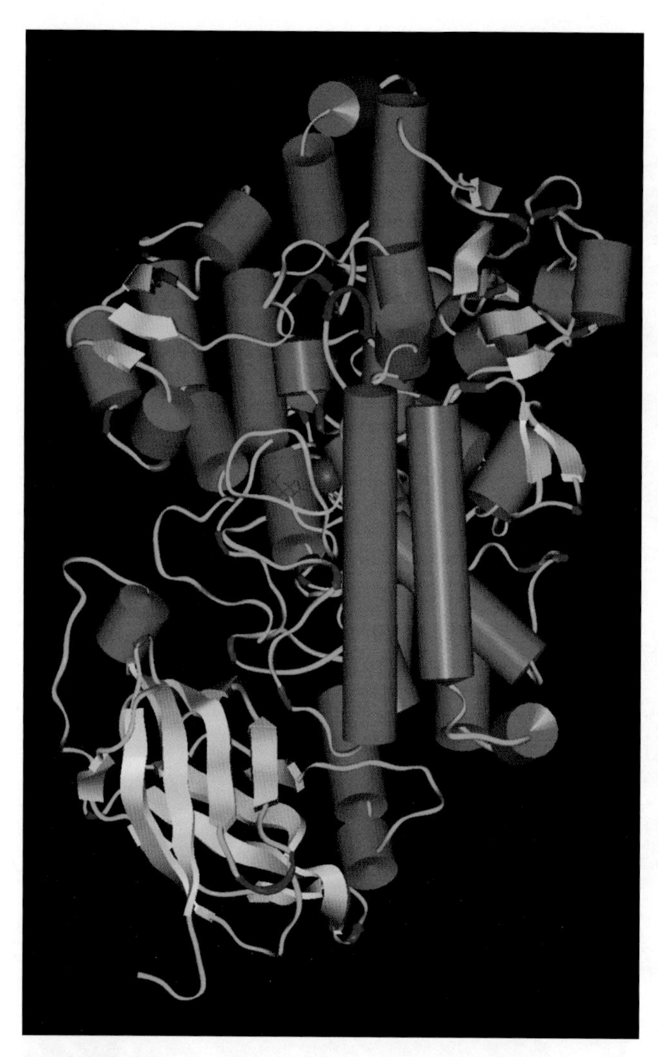

Fig. XIII.7.2.
Crystal structure of SBL-1, based on the structure solved by Minor et al. The N-terminal domain can be seen at the lower left of the structure (PDB code: 1YGE).

Fig. XIII.7.3.
Crystal structure of SBL-1, based on the structure solved by Minor et al. The Fe and its ligands are shown in red at the lower right. The substrate has been modeled in green, and the residues forming the substrate pocket are outlined in yellow (PDB code: 1YGE).

Fig. XIV.2.1.
The molecular structure, determined by X-ray crystallography to 2.1-Å resolution, of the three zinc fingers of Zif268 complexed to DNA (PDB: 1ZAA). The three zinc fingers (purple) fit neatly along the major groove of the DNA double helix (sea green), with the α- helices of the zinc fingers interacting directly with the DNA. Each zinc ion (pink) is bound to two Cys sulfurs (yellow) and two His nitrogens (blue). Figure was prepared from the PDB file (1ZAA) using UCSF Chimera.

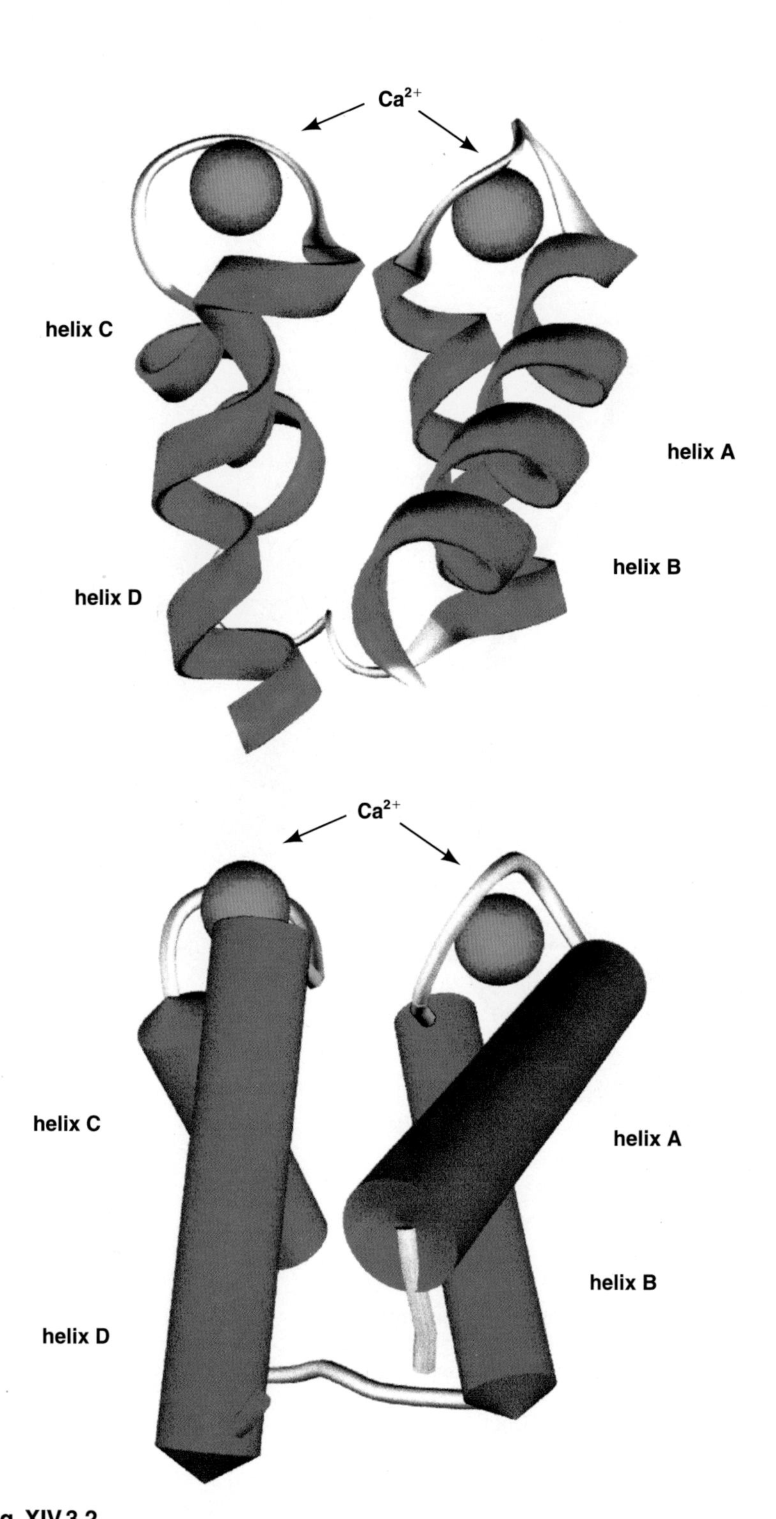

Fig. XIV.3.2.
Two representations of the EF-hand Ca^{2+}-binding protein calbindin D_{9k}. The top view is a ribbon diagram showing the secondary structure of the protein. Helices are colored red, β-sheets blue, and loops are in white. The Ca^{2+} ions are indicated in green. In the lower view, the helices are shown as cylinders in order to emphasize their arrangement in a pairwise fashion in each EF-hand subdomain and as a four-helix bundle in the whole protein.

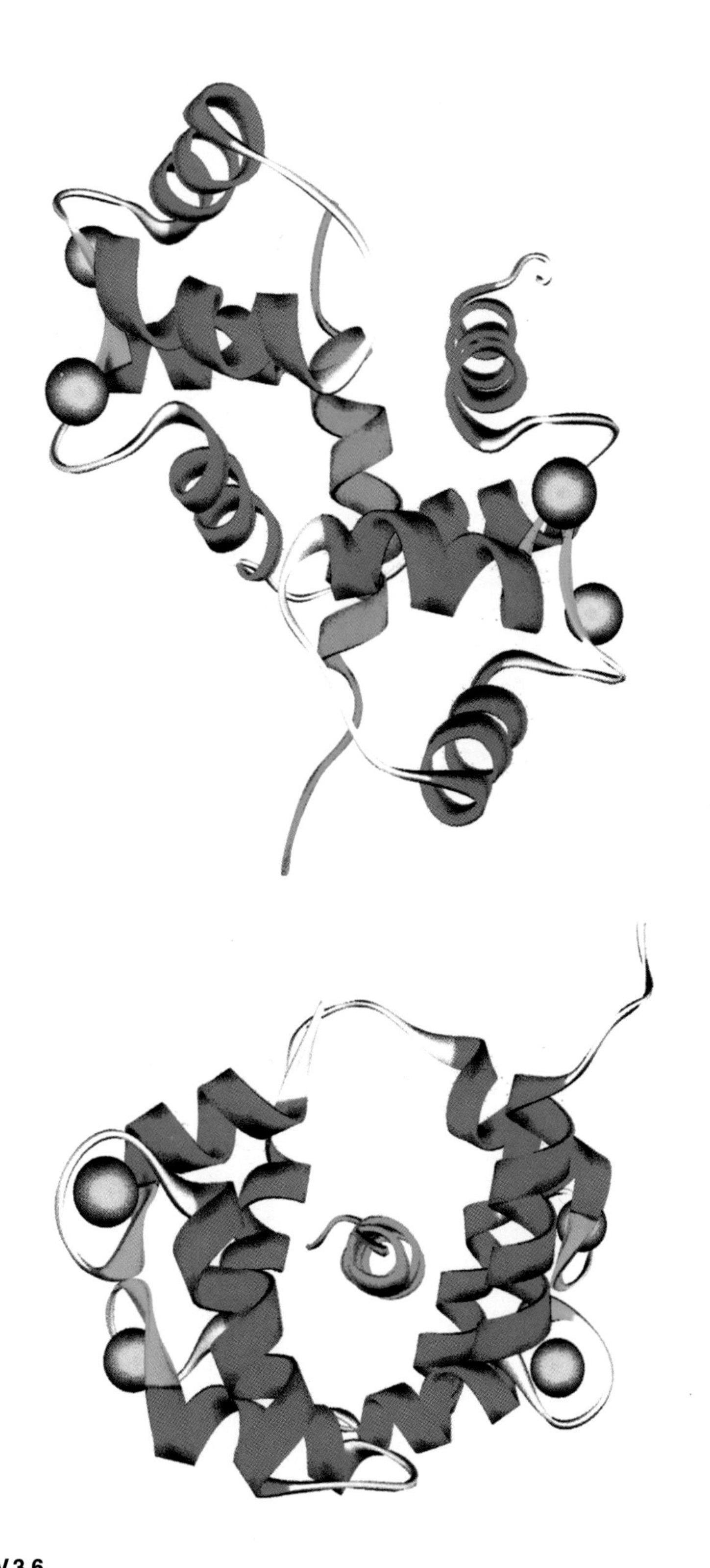

Fig. XIV.3.6.
Ribbon diagrams showing the complex between calmodulin and the calmodulin-binding peptide from myosin light-chain kinase (MLCK). The helices of calmodulin are indicated in red and the peptide is indicated in green. The Ca^{2+} ions are colored yellow and the small β-sheets connecting the pairs of EF-hand subdomains are colored blue. The two figures are rotated 90° relative to one another.

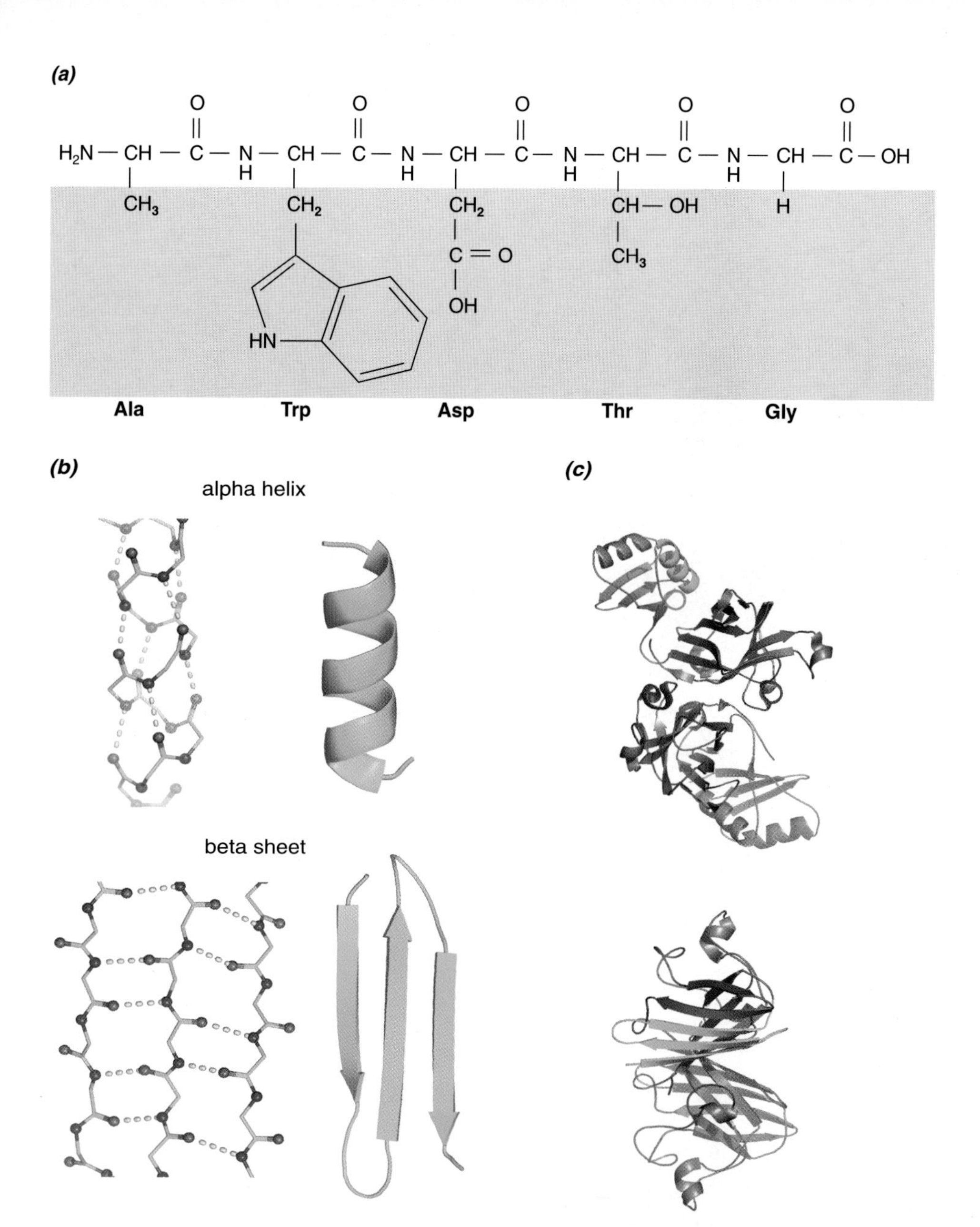

Fig. T.I.10.

Protein Structure. (*a*) Primary Structure. Amino acids are joined together in a linear polymer in a specific order. (*b*) Secondary Structure. Groups of neighboring amino acids form secondary structures when their main-chain atoms hydrogen bond with each other. The two main forms of secondary structures are the alpha helix (top) and beta sheet (bottom). On the left, the backbone structure is shown, with the structure-forming hydrogen bonds indicated. On the right is a typical ribbon diagram of the structure. In both cases, the side chains have been left off for clarity. In an alpha helix, the side chains stick out away from the main axis of the helix (vertical in this illustration) and the hydrogen bonds form between amino acids that are close neighbors in the primary structure. In a beta sheet, the side chains alternate sticking up and down relative to the plane of the sheet. Note that a beta sheet can form between noncontiguous stretches of amino acids. An antiparallel beta sheet is shown, in which the strands run in opposite directions. Beta sheets can also form between parallel strands with a slightly different hydrogen-bonding pattern (not shown). The primary sequence determines the secondary structure that is preferred. (*c*) Tertiary and quaternary structure. The structure formed when a whole polypeptide chain folds is referred to as its tertiary structure, and may includes areas of different kinds of secondary structures, turns, and random coils. If two or more different polypeptide chains interact to form a single protein, it is called quaternary structure. The examples of alpha helix and beta sheet were taken from real proteins whose whole structures are illustrated here. The positions of the secondary structure examples are highlighted (in green). The example of alpha helix was taken from the copper chaperone for superoxide dismutase (CCS). Its tertiary structure is a mixture of alpha helix and beta sheet. The protein is a dimer of identical subunits (oriented one above the other in this illustration) and thus has quaternary structure as well. It is also an excellent example of a protein with more than one domain, which can be seen most clearly in the subunit at the top of the page, where two separately folded units are connected by a single strand of random coil. The beta sheet example is from copper, zinc superoxide dismutase (CuZnSOD, metal ions not shown), a homodimer that is primarily a beta sheet and has no domains.

XIV.4. Nitric Oxide

Thomas L. Poulos

Departments of Molecular Biology and Biochemistry, Chemistry and Physiology and Biophysics, University of California, Irvine, Irvine, CA 92617

Contents

XIV.4.1. Introduction: Physiological Role and Chemistry of Nitric Oxide

Nitric oxide (NO) is a potent signaling molecule with a wide range of physiological functions.[1,2] The importance of NO was first recognized in the 1980s when it was discovered that NO is the endothelial derived relaxing factor (EDRF),[3,4] the molecule responsible for smooth muscle relaxation. This discovery was awarded the Nobel Prize in Physiology and Medicine in 1998. Since the initial discovery, NO has been implicated in a large number of physiological systems. In addition to its role in the cardiovascular system, NO is involved in neural transmission and in the immune defense system where NO can act as a cytotoxic molecule.[1,5] Superoxide (O_2^-) is another molecule generated during oxidative stress (see Section XI.1). Nitric oxide reacts with superoxide in a nearly diffusion controlled reaction[6] to generate peroxynitrite ($ONOO^-$), another highly reactive cytotoxic molecule.[7]

The primary target for NO is a heme-containing enzyme called guanylate cyclase (GC), which converts guanosine-5′-triphosphate (GTP) to cyclic guanosine-3′,5′-monophosphate (cGMP)[2], the signaling molecule leading to the observed physiological effects of NO (Fig. XIV.4.1). Nitric oxide is known to bind tightly to the iron of heme groups and its stereochemistry with various model heme complexes has been worked out.[8] The binding of NO to the GC heme activates the enzyme and hence, stimulates the production of cGMP.[9] An unusual feature of NO as a diatomic ligand in heme systems is that NO has an unpaired electron. Upon formation of the NO–Fe complex, the NO unpaired electron resides in an orbital with substantial metal d_{z^2} character (Fig. XIV.4.2) that serves to weaken the protein ligand–Fe axial bond.[8] This effect is not observed in other diatomic ligands with paired spins like CO. The strong interaction between the NO and heme iron serves to weaken the other axial ligand trans to the NO, usually a His residue. As a result, NO binding breaks the His–Fe bond in GC.[10] In GC, the His–Fe bond is weaker, as judged by the His–Fe stretching frequency,[11] than in other hemoproteins, which further facilitates the breaking of the His–Fe bond when NO binds. The NO binding and rupture of the His–Fe bond results in an ill-defined conformational change in GC that leads to activation of enzyme activity.[12]

Nitric oxide binds very tightly to the heme iron and in the Fe^{2+} globins, $K_{eq} \sim 10^{12}$, leading to a nearly irreversible complex.[13] This very high affinity is due primarily to a diffusion-limited rate of NO binding to the heme iron.[14,15] This high affinity presents an interesting biological challenge in using NO as a regulatory molecule. To operate as an effective messenger molecule, NO must be able to reversibly bind to the heme iron. Since the rate of NO binding to heme is diffusion limited, proteins must modulate K_{eq} by adjusting k_{off} since $K_{eq} = k_{on}/k_{off}$, where k_{off} = the

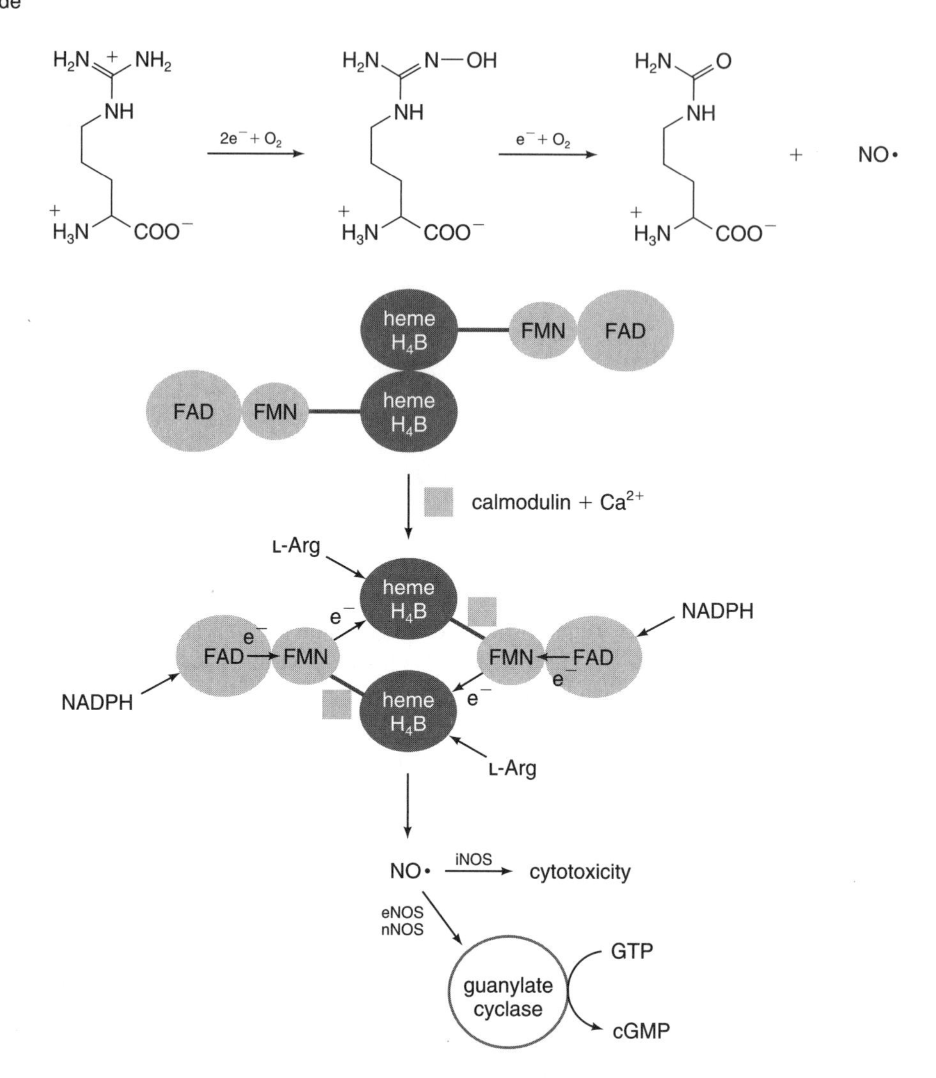

Fig. XIV.4.1.
An overall view of the nitric oxide synthase (NOS) catalytic mechanism. The NOS oxidizes one L-Arg guanidinium N atom to NO in two distinct steps. The NOS first hydroxylates L-Arg followed by a second oxidation step giving NO and L-citrulline as products. Nitric oxide synthase is a complex dimeric enzyme with the heme and cofactor, tetrahydrobiopterin (H_4B), located in one domain and the electron-delivering flavins (flavin adenine dinucleotide, FAD, and flavin mononucleotide, FMN) located in another domain. The C-terminal end of the heme domain is attached to the N-terminal end of the flavin domain by a linker peptide that binds calmodulin. In a complex with Ca^{2+}, calmodulin binds to the linker region and activates electron transfer from the reductase flavin domain to the heme domain leading to the production of NO. The NO, in turn, activates guanylate cyclase to give cGMP, which leads to the various physiological effects of NO. In the immune system, iNOS produces NO as a cytotoxic molecule that helps to battle invading pathogens.

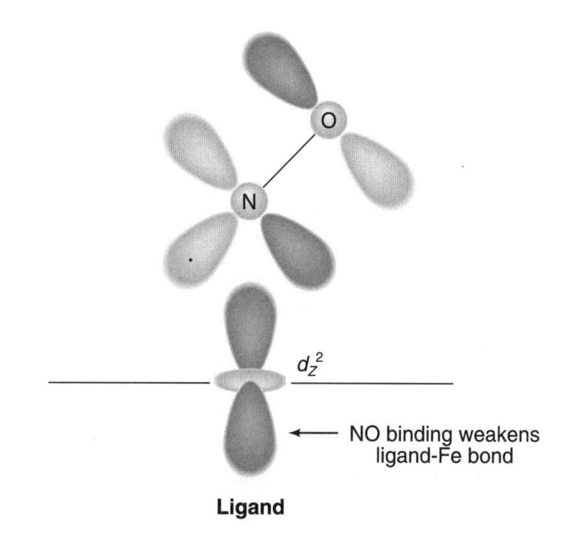

Fig. XIV.4.2.
A schematic molecular orbital (MO) picture of NO complexed with the heme iron. The NO has an unpaired spin that enables an unusually strong bond to form with the iron. This results in a weakening of the ligand–Fe bond trans to the NO.

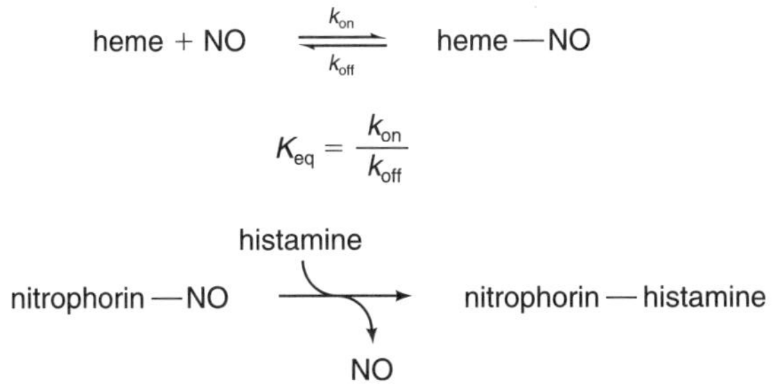

$$\text{heme} + \text{NO} \underset{k_{off}}{\overset{k_{on}}{\rightleftharpoons}} \text{heme—NO}$$

$$K_{eq} = \frac{k_{on}}{k_{off}}$$

$$\text{nitrophorin—NO} \xrightarrow[\text{NO}]{\text{histamine}} \text{nitrophorin—histamine}$$

Fig. XIV.4.3.
The relationship of k_{off} and k_{on} to the equilibrium constant, K_{eq}. In nitrophorin, the NO is displaced by histamine since the affinity for histamine is higher than for NO. This "readjustment" of binding affinities is accomplished by providing a heme-binding site within the protein that preferentially binds histamine but also decreases the affinity for NO.

rate of NO dissociation and k_{on} = rate of NO association (Fig. XIV.4.3). A careful analysis of the rate of dissociation of NO from heme complexes shows that the rate of NO dissociation from GC is unusually fast.[15] It thus appears that the GC protein architecture is such as to decrease the affinity of NO by increasing k_{off}.

A similar problem is faced by an interesting heme protein called nitrophorin (Fig. XIV.4.3). Nitrophorin is found in the saliva of certain insects that feed on mammalian blood. In addition to binding NO to the heme iron, the nitrophorin heme also has a high affinity for histamine. When the insect "bites", the exposure of nitrophorin to the higher pH of blood and histamine results in the release of NO.[16] The NO then exercises its normal physiological role of causing vasodilation and prevention of platelet aggregation, both of which make it easier for the insect to feed on human blood. The K_{eq} for NO binding to nitrophorin is remarkably low, ranging from $\sim 5.3 \times 10^6\ M^{-1}$ at low pH to $\sim 5.9 \times 10^5\ M^{-1}$ at high pH, while the binding of histamine is much tighter, $\sim 5.3 \times 10^7\ M^{-1}$.[17,18] The crystal structure of nitrophorin in a complex with histamine is known,[18] but precisely why nitrophorin exhibits a tighter affinity for histamine over NO remains an unsolved problem.

XIV.4.2. Chemistry of Oxygen Activation

Nitric oxide is produced by an enzyme called nitric oxide synthase or NOS. As shown in Fig. XIV.4.1, NOS catalyzes the formation of NO in two distinct step. In the first step, one O_2-derived O atom is used to oxidize one guanidinium N atom of L-arginine. The overall reaction is very similar to cytochromes P450, which are enzymes designed to hydroxylate aromatic and/or aliphatic molecules (see Section XI.5). The reaction mechanism requires coordination of O_2 to the heme iron followed by cleavage of the O–O bond (Fig. XIV.4.4). In this mechanism, the substrate (SH in Fig. XIV.4.4) enters the active site and displaces a water molecule coordinated to the iron. Next, the iron is reduced from Fe^{3+} to Fe^{2+}, which enables O_2 to coordinate with the iron. A second electron reduces the oxy complex to the peroxide level. Intermediates **4**–**6** in Fig. XIV.4.4 have not been directly observed in either P450 or NOS, but the mechanism of O–O bond cleavage is thought to be similar to peroxidases (Chapter XI.3). As in peroxidases, the leaving O atom must first be protonated before departing as a hydroxide ion or water molecule. The source of protons is a complex issue in these enzymes, but undoubtedly involves key active site groups that help direct the transfer of protons to the oxy complex. After cleavage of

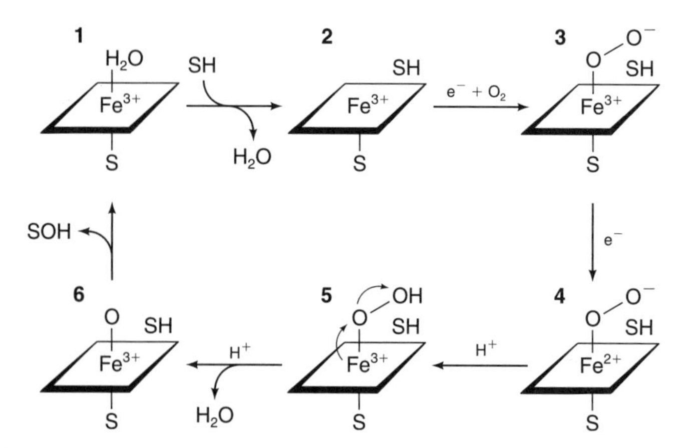

Fig. XIV.4.4.
The catalytic cycle for P450 and, by analogy, NOS. The substrate (SH) binds leading to **2**, the substrate–enzyme complex. In going from **1** to **2**, a water molecule coordinated to the heme iron is displaced leading to a change from low- to high-spin heme and an increase in redox potential, which makes subsequent electron-transfer reactions thermodynamically favored. In other words, the Fe^{3+} in **2** is easier to reduce than in **1.** The oxy complex in **3** is shown as the Fe^{3+}–superoxide complex. A second electron-transfer step gives the hypothetical peroxy complex in **5.** In order to achieve cleavage of the O–O bond, the leaving O atom must be protonated, possibly with the assistance of suitably positioned active-site groups. After cleavage of the O–O bond, the potent oxidizing agent in **6** hydroxylates the substrate that is held by the enzyme to within 4–5 Å of the iron.

the O–O bond, the oxygen atom bound to the iron has only six valence electrons and is a potent oxidizing agent. In peroxidases, the actual distribution of oxidizing equivalents gives Fe^{4+}=O and a porphyrin radical, but the precise electronic structure of this intermediate in both NOS and P450 remains unknown. Nevertheless, the net result is the formation of a potent oxidizing intermediate that is used to hydroxylate the substrate.

XIV.4.3. Overview of Nitric Oxide Synthase Architecture

There are three main isoforms of NOS: endothelial NOS (eNOS, cardiovascular system), neuronal NOS (nNOS, neuronal system), and inducible NOS (iNOS, immune system). All three isoforms share a high level of sequence and structural similarity. NOS is a homodimer ranging in size from 1153–1429 residues per monomer. As shown in Fig. XIV.3.1, NOS consists of a heme-binding domain and an FAD/FMN reductase domain. The reductase domain is very similar in sequence to cytochrome P450 reductase. In the P450s, electrons are delivered from reduced nicotinamide adenine dinucleotide (NADH) or reduced NAD phosphate (NADPH) via the FAD and FMN of the reductase. Since the reductase and P450 are separate polypeptide chains, an intermolecular complex must form prior to electron transfer. In NOS, the reductase and heme domains are on the same polypeptide giving the following domain architecture: heme–FMN–FAD. As in P450s the electron flow is NADPH → FAD → FMN → heme. The linker connecting the heme and reductase domains binds the regulatory protein, calmodulin (see Section XIV.2). When Ca^{2+} binds to calmodulin, the calmodulin is able to interact with the NOS linker between the heme and reductase domains, which leads to an ill-defined conformational switch that enables electrons to flow from the reductase flavins to the heme.[19] Both eNOS

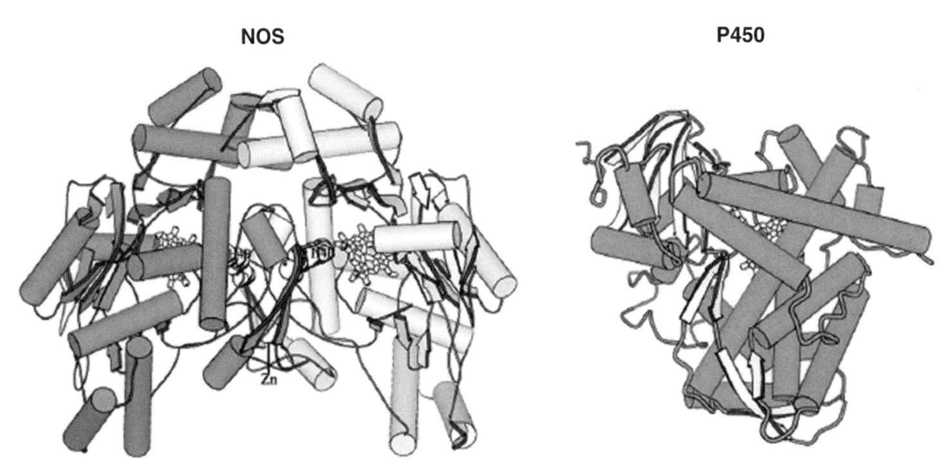

Fig. XIV.4.5.
The structure of the NOS (PDB code: 1FOP) heme domain dimer and cytochrome P450 (PDB: code 1JME). The particular P450 shown, P450BM-3, consists of 455 residues, while each NOS monomer consists of ~480 residues. Hence, the functional NOS heme dimer is about twice as large as P450. The NOS consists of a mixture of sheets forming the inner core of the active site that is surrounded by helices. In sharp contrast, P450s are divided into helical-rich and β-sheet-rich domains. There is essentially no structural homology between the two despite the close similarities in chemistry, spectral properties, and heme ligands. The NOS also contains a structural Zn^{2+} ion tetrahedrally coordinated to symmetry related pairs of Cys residues at the bottom of the dimer interface. That is, two Cys residues from one subunit pair with the symmetry-related Cys residues in the neighboring subunit to provide four Cys ligands.

and nNOS are regulated by the reversible binding of Ca^{2+}–calmodulin. However, iNOS has calmodulin bound as a permanent subunit and hence, is not regulated by the reversible binding of calmodulin. Instead, iNOS is regulated at the level of transcription in response to stimulation of the immune response.[20]

Crystal structures of the eNOS, iNOS, and nNOS heme domains are known.[21–23] Since the chemistry of NOS is similar to that of cytochromes P450, it is of interest to compare the NOS and P450 structures. As shown in Fig. XIV.4.5, the overall architecture of the two enzymes is totally different even though both carry out similar chemistry, both have a Cys residue as an axial heme ligand, and both exhibit very similar spectral properties. Note, too, that NOS is a homodimer while P450 is a monomer, so the functional unit for the NOS heme domain is about twice as large as P450.

XIV.4.4. Nitric Oxide Synthase Mechanism

The active site architecture for NOS is shown in Fig. XIV.4.6. The L-Arg is situated within 4 Å of the heme, which is similar to P450s, where substrates are known to bind 4–5 Å from the iron.[24] Generally, it is assumed that the active hydroxylating species is Fe–O (**6** in Fig. XIV.4.2) so positioning of the substrate atom to be hydroxylated close to the iron-linked O atom ensures that the correct product is formed. A distinct difference between NOS and P450, however, is the requirement for the cofactor, tetrahydrobiopterin (H_4B, Fig. XIV.4.6). As shown in Fig. XIV.4.6, H_4B hydrogen bonds to the same heme propionate that hydrogen bonds to the substrate, so there is an indirect hydrogen bonded link between the cofactor H_4B, and substrate, L-Arg. The role of H_4B in NOS catalysis is to donate an electron to

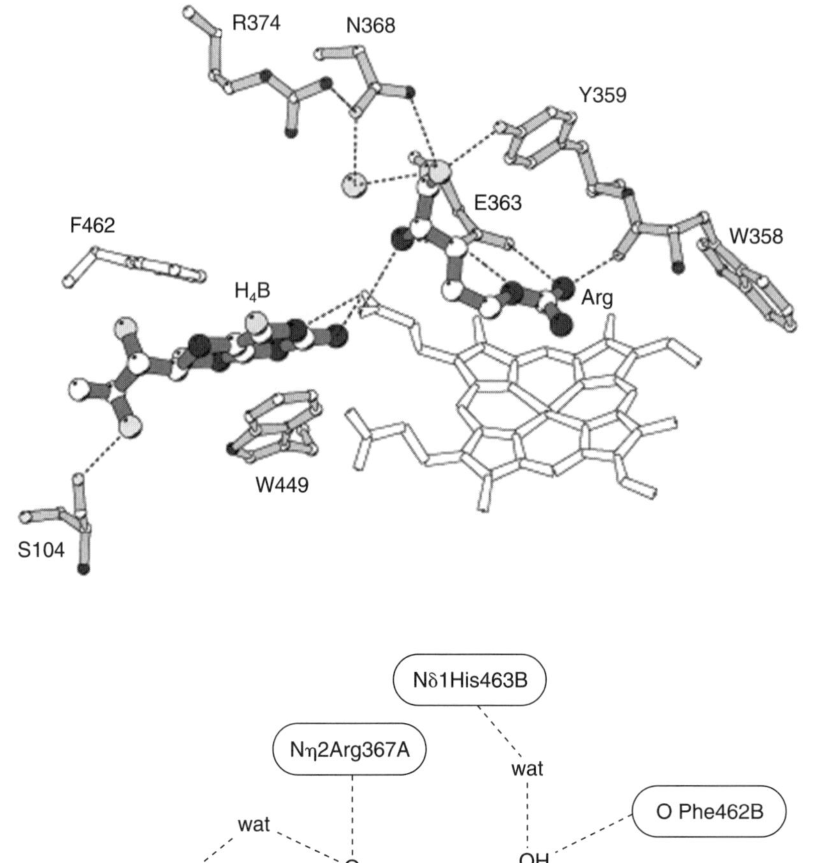

Fig. XIV.4.6.
The active site of NOS. The substrate, L-Arg, is held in place by a series of hydrogen bonds with highly conserved active-site groups. These interactions hold the N atom to be hydroxylated ~4 Å from the iron. The pterin cofactor, H_4B, is situated at the dimer interface and interacts with groups from each subunit. One end of H_4B hydrogen bonds with the same heme propionate that hydrogen bonds with the substrate amino group. This helps to explain why there is a synergy between H_4B and L-Arg binding.

the oxy complex thus giving a pterin radical (Fig. XIV.3.7).[23,25,26] In the scheme shown in Fig. XIV.4.7, H_4B donates an electron to the heme that effectively reduces the oxy complex to peroxide. The H_4B is ultimately reduced by an electron provided by the flavins in the reductase domain. Hence, there is no net oxidation or reduction of H_4B during the catalytic cycle.

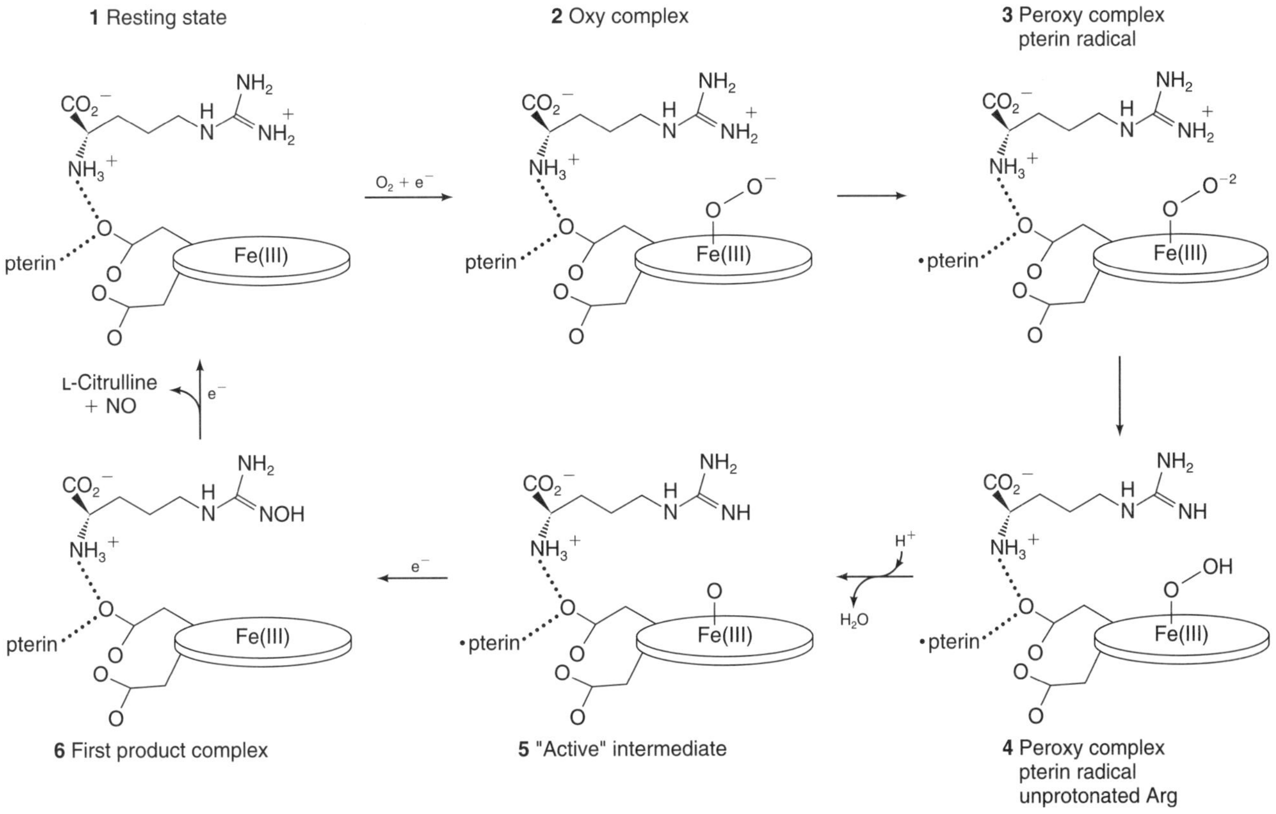

1 Resting state

2 Oxy complex

3 Peroxy complex pterin radical

4 Peroxy complex pterin radical unprotonated Arg

5 "Active" intermediate

6 First product complex

Fig. XIV.4.7.
A hypothetical scheme for NOS catalysis and the involvement of the pterin (H_4B) as a free radical. As in Fig. XIV.3.4, the oxy complex, **1,** is shown as the Fe^{III}–superoxide complex. One possible role of the pterin is to donate an electron to the heme to give the peroxy complex, **3.** While pterins do not form stable radicals under normal physiological conditions, the local protein environment and neighboring aromatic groups could possibly help to stabilize such a radical. The peroxy dianion in **3** is very likely a potent basic group capable of abstracting a proton from L-Arg to give **4.** Therefore, NOS may have solved the problem of acid catalysis by having the proton-rich substrate directly donate a proton to the leaving O atom.[21–23] Once the O–O bond is cleaved, the chemistry is essentially the same as in P450s with the Fe^{III}–O intermediate, **5,** serving as the first hydoxylating agent.

References

General References

1. Bredt, D. S. and Snyder, S. H., "Nitric oxide: a physiologic messenger molecule", *Annu. Rev. Biochem.*, **63,** 175–195 (1994).
2. Hobbs, A. J., "Soluble guanylate cyclase: the forgotten sibling", *Trends Pharmacol. Sci.*, **18,** 484–491 (1997).
3. Furchgott, R. F., "Studies on relaxation of rabbit aorta by sodium nitrite: the basis for the proposal that acid-activatable inhibitory factor from bovine retractor penis is the inorganic nitrite and the endothelium-derived relaxing factor is nitric oxide", in Anhoutte, P. M., Ed., Vasodilatation: Vascular Smooth Muscle, Peptides, Autonomic Nerves, and Endotheleium, Raven Press, New York, 1988, pp. 401–414.
4. Ignarro, L. J., Byrns, R. E., and Wood, K. S., "Biochemical and pharmacological properties of endothelium-derived relaxing factor and its similarity to nitric oxide radical", in Anhoutte, P. M., Ed., Vasodilatation: Vascular Smooth Muscle, Peptides, Autonomic Nerves, and Endotheleium, Raven Press, New York, 1988, pp. 427–435.
5. Griffith, O. W. and Stuehr, D. J., "Nitric oxide synthases: properties and catalytic mechanism", *Annu. Rev. Physiol.*, **57,** 707–736 (1995).

Specific References

6. Huie, R. E. and Padmaja, S., "The reaction of NO with superoxide", *Free Radic. Res. Commun.*, **18,** 195–199 (1993).
7. Beckman, J. S., Beckman, T. W., Chen, J., Marshall, P. A., and Freeman, B. A., "Apparent hydroxyl radical production by peroxynitrite: implications for endothelial injury from nitric oxide and superoxide", *Proc. Natl. Acad. Sci. U.S.A.*, **87,** 1620–1624 (1990).

8. Scheidt, W. R. and Ellison, M. K., "The synthetic and structural chemistry of heme derivatives with nitric oxide ligands", *Acc. Chem. Res.*, **32,** 350–359 (1999).
9. Ignarro, L. J., Degnan, J. N., Baricos, W. H., Kadowitz, P. J., and Wolin, M. S., "Activation of purified guanylate cyclase by nitric oxide requires heme. Comparison of heme-deficient, heme-reconstituted and heme-containing forms of soluble enzyme from bovine lung", *Biochim. Biophys. Acta.*, **718,** 49–59 (1982).
10. Zhao, Y., Hoganson, C., Babcock, G. T., and Marletta, M. A., "Structural changes in the heme proximal pocket induced by nitric oxide binding to soluble guanylate cyclase", *Biochemistry,* **37,** 12458–12464 (1998).
11. Tomita, T., Ogura, T., Tsuyama, S., Imai, Y., and Kitagawa, T., "Effects of GTP on bound nitric oxide of soluble guanylate cyclase probed by resonance Raman spectroscopy", *Biochemistry,* **36,** 10155–10160 (1997).
12. Dierks, E. A., Hu, S. Z., Vogel, K. M., Yu, A. E., Spiro, T. G., and Burstyn, J. N., "Demonstration of the role of scission of the proximal histidine-iron bond in the activation of soluble guanylyl cyclase through metalloporphyrin substitution studies", *J. Am. Chem. Soc.*, **119,** 7316–7323 (1997).
13. Addison, A. W. and Stephanos, J. J., "Nitrosyliron(III) hemoglobin: autoreduction and spectroscopy", *Biochemistry,* **25,** 4104–4113 (1986).
14. Cassoly, R. and Gibson, Q., "Conformation, cooperativity and ligand binding in human hemoglobin", *J. Mol. Biol.*, **91,** 301–313 (1975).
15. Kharitonov, V. G., Sharma, V. S., Magde, D., and Koesling, D., "Kinetics of nitric oxide dissociation from five- and six-coordinate nitrosyl hemes and heme proteins, including soluble guanylate cyclase", *Biochemistry,* **36,** 6814–6818 (1997).
16. Ribeiro, J. M., Hazzard, J. M., Nussenzveig, R. H., Champagne, D. E., and Walker, F. A., "Reversible binding of nitric oxide by a salivary heme protein from a bloodsucking insect", *Science,* **260,** 539–541 (1993).
17. Andersen, J. F., Champagne, D. E., Weichsel, A., Ribeiro, J. M., Balfour, C. A., Dress, V., and Montfort, W. R., "Nitric oxide binding and crystallization of recombinant nitrophorin I, a nitric oxide transport protein from the blood-sucking bug *Rhodnius prolixus*", *Biochemistry,* **36,** 4423–4428 (1997).
18. Weichsel, A., Andersen, J. F., Champagne, D. E., Walker, F. A., and Montfort, W. R., "Crystal structures of a nitric oxide transport protein from a blood-sucking insect", *Nat. Struct. Biol.*, **5,** 304–309 (1998).
19. Abu-Soud, H. M. and Stuehr, D. J., "Nitric oxide synthases reveal a role for calmodulin in controlling electron transfer", *Proc. Natl. Acad. Sci. U.S.A.*, **90,** 10769–10772 (1993).
20. Cho, H. J., Xie, Q. W., Calaycay, J., Mumford, R. A., Swiderek, K. M., Lee, T. D., and Nathan, C., "Calmodulin is a subunit of nitric oxide synthase from macrophages", *J. Exp. Med.*, **176,** 599–604 (1992).
21. Crane, B. R., Arvai, A. S., Ghosh, D. K., Wu, C., Getzoff, E. D., Stuehr, D. J., and Tainer, J. A., "Structure of nitric oxide synthase oxygenase dimer with pterin and substrate", *Science,* **279,** 2121–2126 (1998).
22. Fischmann, T. O., Hruza, A., Niu, X. D., Fossetta, J. D., Lunn, C. A., Dolphin, E., Prongay, A. J., Reichert, P., Lundell, D. J., Narula, S. K., and Weber, P. C., "Structural characterization of nitric oxide synthase isoforms reveals striking active-site conservation", *Nat. Struct. Biol.*, **6,** 233–242 (1999).
23. Raman, C. S., Li, H., Martasek, P., Kral, V., Masters, B. S., and Poulos, T. L., "Crystal structure of constitutive endothelial nitric oxide synthase: a paradigm for pterin function involving a novel metal center", *Cell,* **95,** 939–950 (1998).
24. Poulos, T. L., Cupp-Vickery, J., and Li, H., "Cytochrome P450: Structure, mechanism and biochemistry", in Ortiz De Montellano, P., ed., Structural studies on prokaryotic cytochromes P450. Plenum, New York, 1995, pp. 125–150.
25. Bec, N., Gorren, A. C., Voelker, C., Mayer, B., and Lange, R., "Reaction of neuronal nitric-oxide synthase with oxygen at low temperature. Evidence for reductive activation of the oxy-ferrous complex by tetrahydrobiopterin", *J. Biol. Chem.*, **273,** 13502–13508 (1998).
26. Hurshman, A. R., Krebs, C., Edmondson, D. E., Huynh, B. H., and Marletta, M. A., "Formation of a pterin radical in the reaction of the heme domain of inducible nitric oxide synthase with oxygen", *Biochemistry,* **38,** 15689–15696 (1999).

TUTORIALS

Cell Biology, Biochemistry, and Evolution

Tutorial I

Edith B. Gralla
Department of Chemistry
and Biochemistry
UCLA
Los Angeles, CA 90095

Aram Nersissian
Chemistry Department
Occidental College
Los Angeles, CA 90041

Contents

T.I.1. Life's Diversity

Cells are the basic building blocks of life, the minimal entities that contain all of the information and machinery necessary for self-replication and other activities of living. They vary tremendously in size, shape, and even composition, but certain characteristics are shared by all cells: A lipid bilayer membrane separates the cell from the outside world and virtually all cells contain deoxyribonucleic acid (DNA) as the genetic material. All cells have the ability to convert energy into a usable form, to biosynthesize, to replicate, and to self-protect. In order for cells to live in the world around them, a variety of signal transduction mechanisms are at work to detect changes in the environment and respond accordingly. For a single-celled organism, this environment is the physical world directly; for a multicelled organism, it is neighboring cells in its densely packed interior, as well as signals arising from the outside world (often sensed by specialized cells). Multicellularity allows organisms to get much larger, and perhaps more importantly, to have extensive cellular specialization. Many functions are carried out more efficiently by cells devoted to one or a few tasks and located where the tasks need to occur.

In biology, there are always exceptions, even to the most basic rules, as stated above. For example, mature mammalian red blood cells (erythrocytes) have no DNA, and the mycelia of certain fungi are not divided into individual cells, but have a continuous cytoplasm containing hundreds of nuclei (syncytium). The ability to reproduce, as a criterion in the definition of a living cell, becomes a little blurred

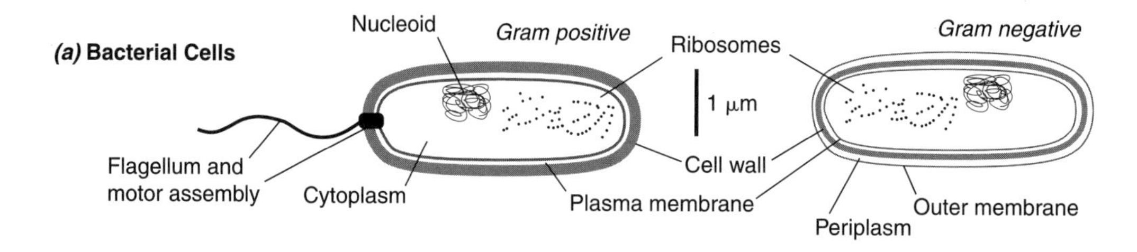

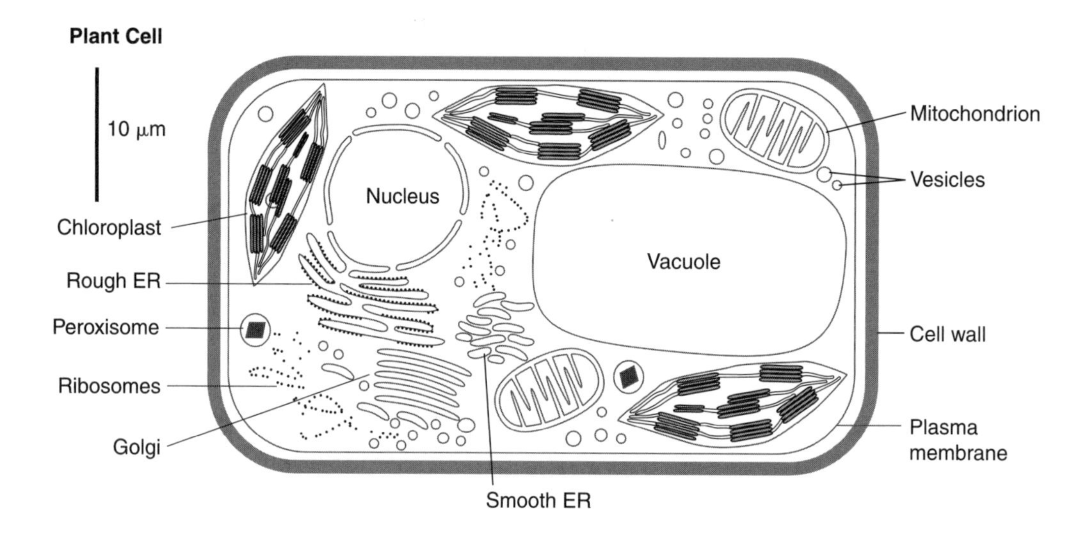

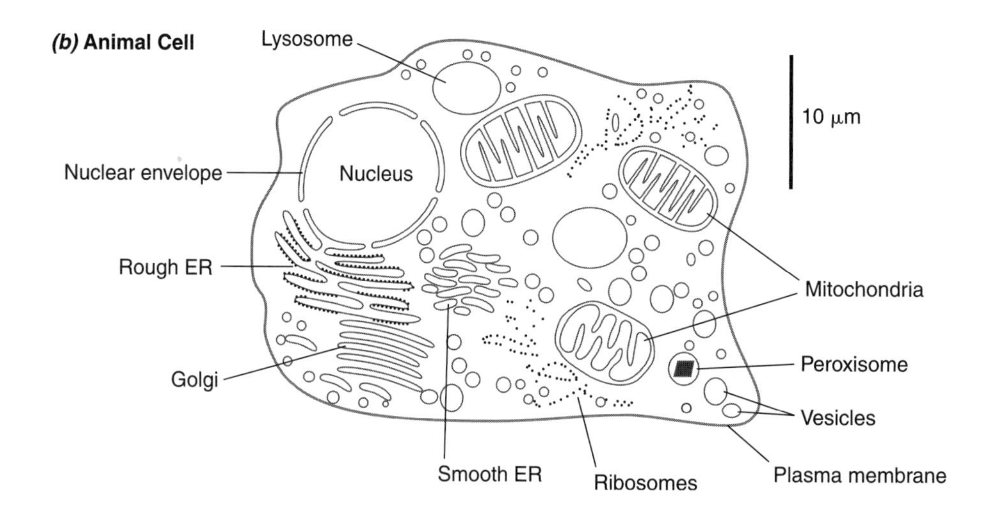

Fig. T.I.1.
Diagrams of typical cells. (*a*) Prokaryotic cells. Note the relatively simple organizational structure, and the size, which is much smaller than a typical eukaryote. Pictured on the left is a Gram-positive bacterium, which has a single cell membrane and a thick peptidoglycan cell wall. Gram-negative bacteria (on the right) have a much thinner cell wall and another membrane outside of the wall, creating a compartment called the periplasm, or the periplasmic space. (*b*) Eukaryotic cells. These cells are much bigger than prokaryotic cells and contain a nucleus as well as multiple organelles. Animal cells are enclosed by a single membrane, the plasma membrane. Plant cells have a thick cell wall made of cellulose outside the plasma membrane, and contain a large vacuole and chloroplasts, in addition to the components that are present in animal cells. Various structures are labeled. For more information about each type of organelle, see Table T.I.3.

when you consider some individual cells of multicellular organisms that do not survive outside the organism except under very controlled conditions. However, by and large, the parameters listed above define the cell, whether it is a whole single-celled organism, such as a bacterium or an amoeba, or a component of a much larger multicellular organism—a tree, a human, or a slug. As you read this tutorial or study biological systems, it pays to remember that for any stated mechanism or property, there is probably an organism somewhere that is different.

Life falls into two broad forms, based on cellular organization: prokaryotes and eukaryotes. Prokaryotes are classified into two domains Archaea (archaebacteria) and Bacteria (eubacteria), and are generally single-celled organisms with a simple internal architecture. Eukaryota comprise the third domain and are more complex in structure (see below). Prokaryotes (Fig. T.I.1*a*) lack intracellular membrane-bounded organelles and their genetic material, DNA, is in the form of a single circular chromosome. Most prokaryotes are surrounded by a strong and rigid cell wall that protects the cells against turgor (osmotic) stress, preventing them from exploding in dilute solution. Such a simple organizational plan limits the possible size of prokaryotes, which are consequently relatively small (Fig. T.I.2). The dimensions of a typical rod-shaped bacterium are $1 \times 3\,\mu m$. Most of the prokaryotes with which we are familiar, such as the well-studied organism *Escherichia coli, Lactobacillus* (used to make yogurt), and human pathogens are eubacteria (members of the domain Bacteria). Archaea were discovered relatively recently, and are often organisms that live in extreme environments. Although they have the prokaryotic body plan, in certain key features Archaea are more similar to eukaryotes than to the eubacteria—a discovery that has spurred reassessment of our beliefs about early evolution. Current thought has it that Archaea and Bacteria are as different from each other as either is from the Eukaryotes. Some important differences and common features among Archaea, Bacteria, and Eukaryota are presented in Table T.I.1.

Eubacteria fall into two major classes based on the thickness of their cell wall, and visualized by their ability to be stained by the Gram stain. Gram-negative bacteria have a relatively thin peptidoglycan layer and an outer lipid membrane (the periplasmic membrane) outside the cell wall in addition to the plasma membrane. Gram-positive bacteria have a much thicker peptidoglycan layer and no second membrane. The bacterial cell wall is made of peptidoglycan—carbohydrate polymers cross-linked with short peptides—and is quite different from the wall of plant cells, which is made of cellulose (see Section T.I.4.4 Carbohydrates). Archaeal cell walls differ from bacterial walls and are not all alike. They do not contain peptidoglycan, but instead are made from either pseudopeptidoglycan, polysaccharide, or protein (the S-layer), depending on the species.

Eukaryotes (Fig. T.I.1*b*) are fundamentally different from prokaryotes in two important ways: the presence of organelles and the organization of their DNA. Other differences are summarized in Table T.I.2. The eukaryotic cell interior is organized into a number of different kinds of membrane-bound compartments or organelles, each carrying out a specific task or tasks. Prime examples are the mitochondria, which carry out respiration, and are present in all eukaryotes (with the exception of a group of anaerobic single-celled eukaryotes that contain hydrogenosomes), and the chloroplasts, found in plants, which carry out photosynthesis and carbon fixation. The genetic material of eukaryotes is located in the nucleus and organized into chromosomes—linear DNA molecules heavily coated with proteins. The number of chromosomes depends on the species; humans, for example, have 46 (23 pairs). In addition, a much smaller amount of DNA is present in mitochondria and chloroplasts, organized in circular chromosomes reminiscent of bacterial DNA. The characteristics and functions of eukaryotic organelles are summarized in Table T.I.3.

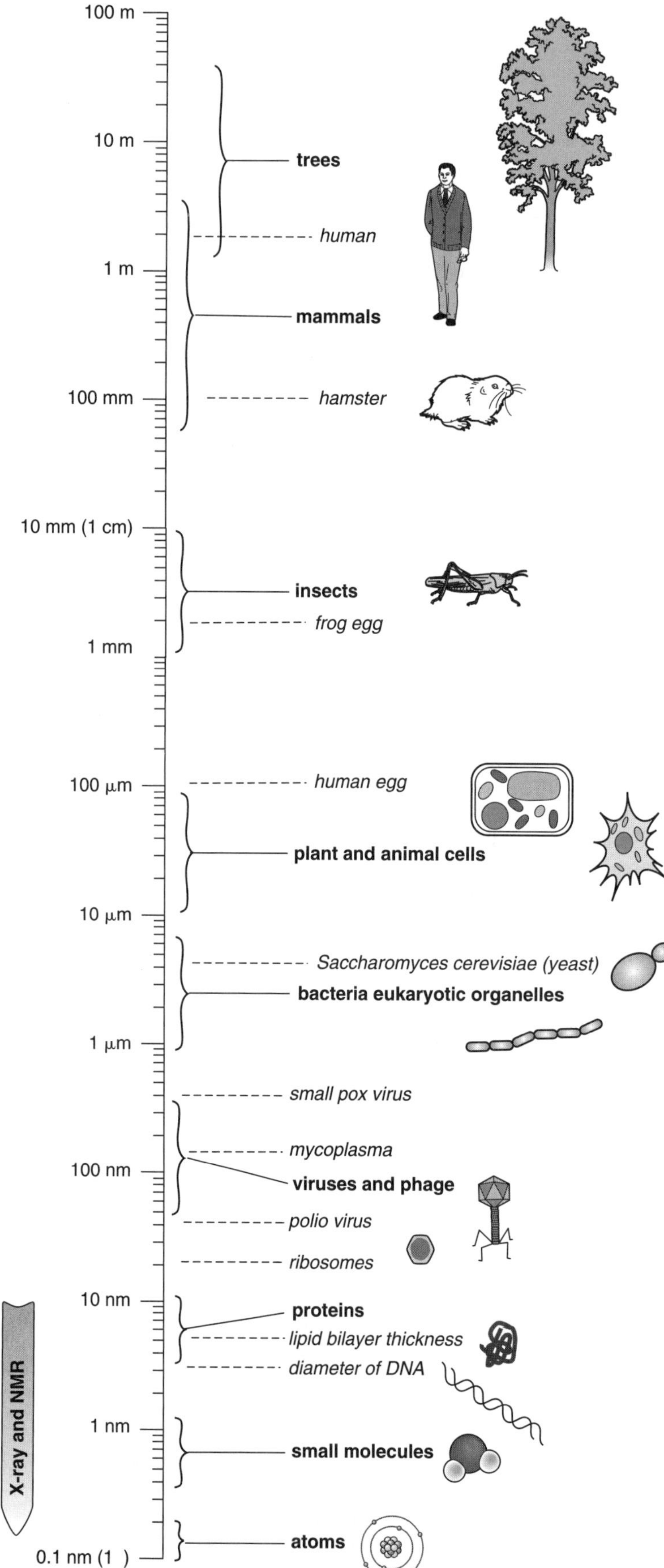

Fig. T.I.2.
Sizes of living things and their components. Approximate sizes are shown for organisms, cells, and their components. More precise sizes are shown for specific organisms or structures. On the left are indications of the tools necessary to visualize items in each size range. "X-ray and NMR" refer to X-ray crystallography and nuclear magnetic resonance spectroscopy, respectively, used to determine molecular (protein) structures.

Table T.I.1.
Some Features that Are Similar or Different among the Three Domains of Life, Archaea, Bacteria, and Eukaryota[a]

Characteristic	Bacteria	Archaea	Eukaryota
Cell wall	Yes, Contains Peptidoglycan	Yes, Contain Glycoprotein. Not Peptidoglycan	None or Cellulose (plants) or Chitin (fungi)
Membrane phospholipids	Ester linked	Ether linked, some tetraethers	Ester linked
Ribosome size[b]	70S	70S	80S
Ribosome sensitive to diphtheria toxin	No	Yes	Yes
Ribosome sensitive to antibiotics chloramphenicol, streptomycin, kanamycin	Yes	No	No
Initiator tRNA	Formylmethionine	Methionine	Methionine
Introns in DNA	No	Yes, limited	Yes, extensive
RNA polymerase	One, 4 subunits	Several, 8–12 subunits each	Three, 12–14 subunits each
DNA packed into nucleosomes	No	Yes (scaffold resembles histones)	Yes (histones)
N-Linked glycosylation	No	Yes	Yes
Polycistronic operons	Yes	Yes	No
Size (linear dimension)	1–10 μm	1–10 μm	5–100 μm

[a] Key differences among the three domains of life. In many of the features that distinguish archaea from bacteria, the archaea more closely resemble eukaryotes. Similarities between archaea and eukaryotes prompted the three-domain classification originally proposed by Carl Woese. As we can see, archaea and bacteria are as different from each other as each is different from eukaryota.

[b] The unit S stands for Svedburg, a measure of sedimentation rate, which is related to the size and shape of a molecule.

Eukaryotes are members of the third domain, Eukaryota, which contains four of the five kingdoms of the traditional divisions: Animalia, Plantae, Fungi, and Protista. (The fifth kingdom, Monera, consists of the prokaryotes: Archaea and Eubacteria.) Eukaryotes can be either unicellular; such as protists and some algae and fungi; or multicellular, such as plants (viridiplantae) and animals (metazoa), and the rest of the algae and fungi. The typical eukaryotic cell is 5–200 μm in diameter, but some unicellular algae can be as big as 5 mm long. Multicellularity allows living things to reach much larger sizes (see Fig. T.I.2 for examples) and confers other important advantages related to the ability of cells to specialize in function.

The essential feature that allows millions of cells to work together in harmony in a multicellular organism is their ability to communicate by sending, receiving, and responding to signals. Multicellular organisms are almost invariably eukaryotes, but it is interesting to note that single-celled organisms are sometimes found organized into simple structures, such as filaments, colonies and biofilms, and, when they are in such forms, they too have mechanisms of intercellular communication.

The major groups of eukaryotes (plants, animals, fungi, and protists) differ tremendously in size, shape, and appearance. However, in basic metabolism they are relatively uniform (although there are some interesting variants among the protists).

Table T.I.2.
Prokaryotes and Eukaryotes

A. SOME IMPORTANT DIFFERENCES BETWEEN PROKARYOTES AND EUKARYOTES

Prokaryotes	Eukaryotes
Eubacteria, Archaea	Plants, animals, fungi, protists
1–10 μm (linear dimension)	5–100 μm (linear dimension)
Mainly *unicellular,* can grow in colonies or mixed populations (e.g., biofilms)	Often *multicellular;* many differentiated cell types present in the same organism
No membrane-bound organelles, few or no visible internal features, can have inclusions for storage of metabolites or nutrients	Many visible structures: *membrane-bound organelles* that carry out specific functions: nucleus, mitochondrion, chloroplast, endoplasmic reticulum, golgi, vacuole, and so on (see Table T.I.3)
Single circular chromosome in cytoplasm (rare cases have >1), located as a tangled mass in the nucleoid region. Small circular independently replicating DNA *plasmids* are also important.	DNA packaged in *long linear chromosomes with extensive protein scaffolding* (histones), which are located in the nucleus; number of chromosomes varies by species. In addition some organelles (mitochondria and chloroplasts) have their own small genomes that resemble bacterial ones
RNA and protein synthesized in same compartment (cytoplasm) and closely coupled	RNA made in nucleus; protein synthesis in cytoplasm; Extensive RNA processing in the nucleus, including addition of "cap" and polyA tail, and removal of introns (intervening sequences) by splicing.
Limited cytoskeletal features; external cell wall provides shape and support	*Elaborate cytoskeleton* composed of several kinds of protein filaments (actin, myosin, microtubules, intermediate filaments) used for structural support and internal movement of components and materials.
No bulk (vesicular) transport in and out of cell. Transport of nutrients and dissolved substances across plasma membrane by simple diffusion, facilitated diffusion, or active transport	Bulk transport of materials via *endocytosis, exocytosis, and internal vesicular trafficking.* (In endocytosis plasma membrane in-folds, pinches off and forms interior vesicles. Exocytosis is the opposite—interior vesicles fuse with the plasma membrane, releasing contents to the exterior.) As in prokaryotes, diffusion, facilitated diffusion and active transport are present
In cell division, replicated DNA molecules pulled apart by attachment to plasma membrane	In cell division, replicated chromosomes pulled apart by spindle apparatus
Reproduce by fission; sexual reproduction is unusual and almost always consists of one-way transfer of partial genetic information	Reproduce by mitosis; sexual reproduction with complete mixing of genomes is usual. Meiosis to produce haploid germ cells (egg and sperm) is a common mechanism

Table T.I.2. (Continued)

A. SOME IMPORTANT DIFFERENCES BETWEEN PROKARYOTES AND EUKARYOTES	
Prokaryotes	Eukaryotes
Peptidoglycan cell walls in eubacteria, in archebacteria walls are pseudopeptidoglycan or may be absent.	Many (plants, fungi) have cell walls, but composition is simpler: plants have cellulose, fungi have chitin, yeast have glucan and mannan; cells without walls often have polysaccharide coat (glycocalyx), particularly if they are exposed directly to the environment as in single celled organisms
Cell (plasma) membrane has no sterols and no exterior polysaccharides	Cell (plasma) membrane contains *sterols as well as glycolipids, glycoproteins, and proteoglycans* that form an exterior polysaccharide coating
Flagella (if present) made of two protein components, movement is a rotation of the whole thing from a motor at the base	Flagella and cilia (if present) much larger, made of microtubules surrounded by plasma membrane, move by wave propagation
Energy metabolism is *broadly diverse:* Glycolysis and citric acid cycle in cytoplasm, electron-transport chain in plasma membrane. Anaerobic respiration using alternate final electron acceptors such as nitrate or sulfate is present in many members. A variety of fermentation mechanisms as well	Energy metabolism is more uniform: Glycolysis in cytoplasm, citric acid cycle in mitochondrial matrix, electron transport chain in inner-mitochondrial membrane; oxygen is generally the terminal electron acceptor in respiration
Photosynthesis, when it occurs, is anoxygenic (electrons derived from reduced organic or inorganic compounds) except in cyanobacteria where it is oxygenic	Photosynthesis, when it occurs, is oxygenic (electrons derived from water) and is carried out in chloroplasts.
Small genome size: 0.6–9 megabase (Mb) pairs; 500–8000 genes (potential protein coding regions)	Larger genome size: 12–3000 or more megabase pairs; 5000–50,000 potential protein coding regions

B. FEATURES COMMON TO BOTH PROKARYOTES AND EUKARYOTES

Plasma membrane

DNA → RNA → protein

Self-replicate

Basic unit is the cell

In general, eukaryotes use organic carbon as an energy source and utilize dioxygen (molecular oxygen, O_2) as the final electron acceptor for respiratory metabolism. Plants, unlike animals, are also able to grow *photoautotrophically,* that is, they use sunlight as the energy source for synthesis of organic materials from CO_2. The photosynthetic apparatus of all plants is very similar.

Prokaryotes as a group are much more flexible than eukaryotes in their metabolic capabilities. They have the ability to utilize a wide variety of both organic and inorganic energy sources, as well as light, and they can use a great diversity of oxidants, in addition to dioxygen, as the final electron acceptor for respiration (see Chapter II). They also display extraordinary variation in their metabolic properties,

Table T.I.3.
Eukaryotic Organelles, Their Functions and Key Characteristics

Organelle	Function
Nucleus	Contains DNA, site of DNA replication, transcription and processing of RNA, splicing; double membrane bound, with fairly large pores for cytoplasmic communication
Mitochondrion	Cellular powerhouse; site of respiration and other functions; relatively permeable outer membrane, folded inner membrane containing proteins for oxidative phosphorylation; site of Fe–S cluster synthesis, heme synthesis, and parts of amino acid synthesis pathways; has own genetic material—small circular genome encodes 6–15 proteins (depending on species) and transcription and translation machinery
Chloroplast	Site of photosynthesis and synthesis of glucose from CO_2; three membrane layers—innermost stacked thylakoid membranes where photosynthetic machinery is located and inner and outer surrounding membranes; own genome encoding ~80 proteins in higher plant chloroplasts and between 60 and 200 proteins in plastids of algae
Rough endoplasmic reticulum (ER)	In-folded layers of membrane, continuous with nuclear envelope; "rough" because of presence of bound ribosomes; site of synthesis and processing of membrane and secreted proteins
Smooth ER	More tubular elaboration of membrane; site of synthesis of lipids and other biomembrane components
Golgi apparatus	Layers of membrane flattened sacks in "stacks", involved in processing proteins for secretion and delivery to other organelles
Vacuole	In plants and fungi; large organelle involved in storage and degradation
Lysosome	In animal cells; degradation of cellular wastes
Peroxisome	Contain oxidative enzymes; fatty acid oxidation; hydrogen peroxide generation and disposal
Centriole	Cell division and formation of cilia and flagella
Vesicles	Various kinds of small membrane sacs that transport components around the cell, between compartments, and between the interior and exterior of the cell
Cytoplasm	The parts of the interior of the cell are not enclosed by another membrane; the site of glycolysis, the first pathway in glucose utilization and energy generation; contains the ribosomes and thus is the site of protein synthesis; contains all or parts of other synthetic (anabolic) pathways for amino acids and other small cellular components; contains degradative (catabolic) pathways for small cellular components

cellular structures, and habitats (see Fig. T.I.3 for a display of the variety of possible metabolic schemes and examples of organisms that use them).

Only ~6000 prokaryotic species have been identified to date, but evidence points to hundreds of thousands, perhaps even millions, of others yet to be discovered. In a 100-cm^3 sample of a forest soil sample, it was estimated that there are ~6000 prokaryotic species, and 10 L of a nutrient-rich freshwater sample was shown to contain ~160 different microbes! Moreover, more and more evidence is now accumulating that life extends well into the Earth's crust and that there is a large biota in the oceanic and terrestrial subsurfaces. Eukaryotic diversity may well extend to 500,000 different species or more.

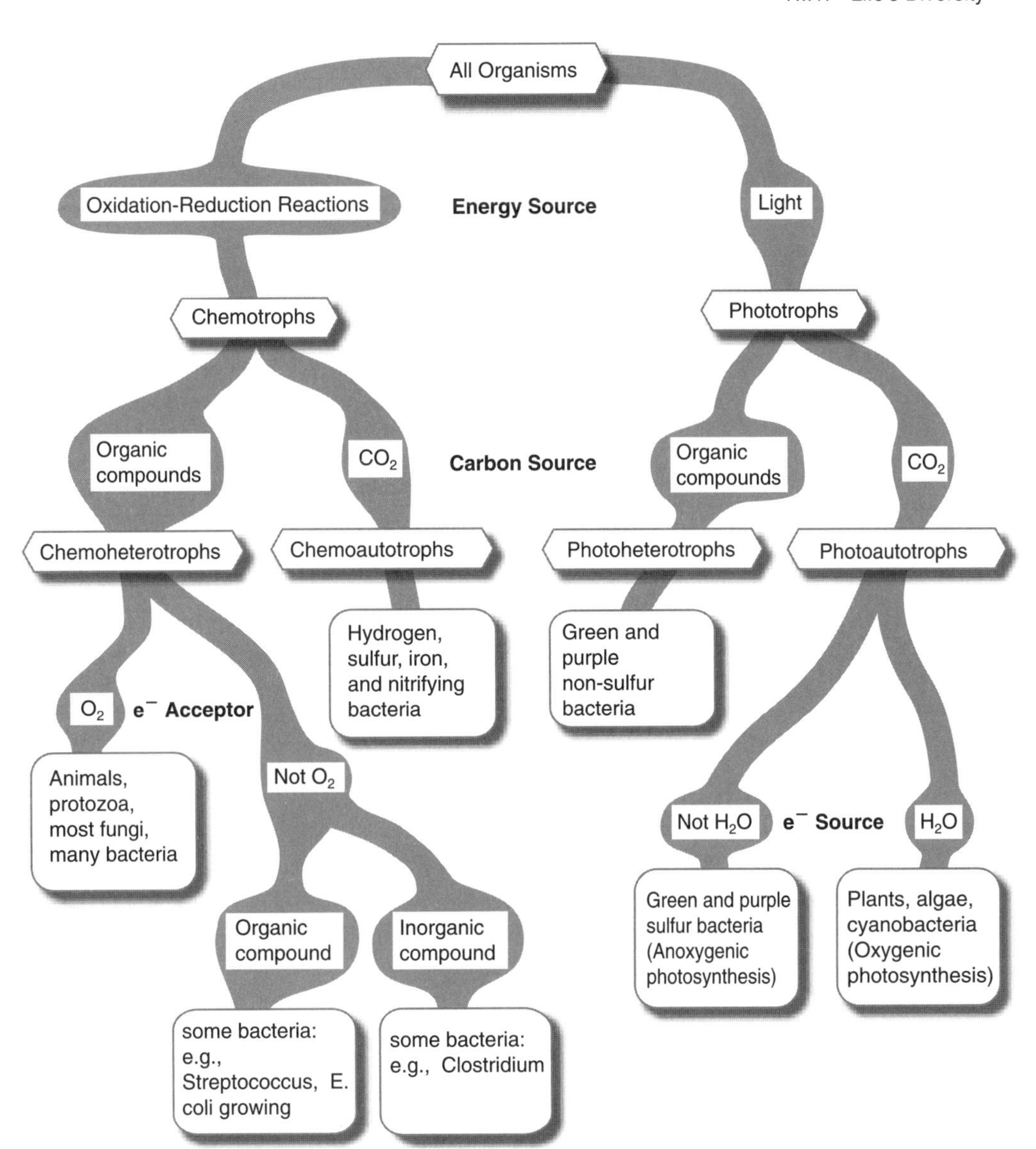

Fig. T.I.3.
A classification of living things. Organisms can be classified according to the ultimate source of their energy and where their carbon comes from. In this diagram the shadowed boxes represent groups of organisms, which are divided and subdivided according to metabolic criteria that are detailed on the gray areas. All organisms get their energy from either sunlight or oxidation–reduction reactions (energy source), giving rise to the first major branch: Phototrophs derive their energy from light, and chemotrophs derive energy from oxidation–reduction reactions. These two groups can each be further subdivided by whether they can reduce CO_2 and use it as a source of carbon to synthesize biomolecules, or whether they require organic compounds that are already reduced (carbon source). Heterotrophs require organic carbon while autotrophs can reduce and use CO_2. The terms can be combined so that, for example, photoautotrophs get energy from light and can use CO_2 as the carbon source. In this diagram, the organisms on the two outermost pathways are the most familiar to most of us—plants on the far right, and animals on the left. Most of the organisms in the interior pathways are bacteria or Archaea that live lives far less obvious to the casual observer, which are nevertheless extremely important because of the ways in which they shape the biosphere.

T.I.2. Evolutionary History

Astrophysicists and geologists have estimated that our solar system was formed 4.6 billion years ago (Bya). The period of heavy bombardment ended ~4 Bya and it was around that time that the planet Earth cooled to temperatures permissive for the assembly of cellular life. Although the results have been questioned, the analysis of stromatolites (microfossils formed from natural colonies of photosynthetic cyanobacteria) lead to the conclusion that the first life appeared surprisingly soon after that cooling, only 3.5 Bya (perhaps even 3.8 Bya). The composition of the atmosphere at that time is not certain, but it is believed to have been quite reducing. The activities of living organisms using and changing the environment have played a profound role in changes that have resulted in the oxidizing atmosphere we know today (N_2, O_2, and a small amount CO_2). Cyanobacteria, the first organisms to develop oxygenic photosynthesis, were responsible for the initial production of dioxygen, which at first was quickly sopped up by reactions with reduced iron and other species in the Earth's crust, oceans, and atmosphere, and resulted in an atmosphere of ~1% oxygen. Around 2.3 Bya, dioxygen began accumulating to higher levels in the air, although the penetration of oxygen into the depths of the oceans took much longer. These atmospheric changes had dramatic effects on the availability of certain metallic elements, and thus implications for the way organisms evolved to use them. Iron, for example, is practically insoluble in an oxidizing atmosphere, while copper and molybdenum are more soluble. (These issues are considered in much more detail in Chapter II, Biogeochemical Cycles, and Chapter VI, Biomineralization.)

The first evidence of the existence of single-celled eukaryotes dates to ~2.2 Bya, coincident with an increase in atmospheric oxygen (to 5–18% O_2). Shortly thereafter, the progenitors of modern eukaryotic cells were generated by addition of mitochondria through endosymbiosis. It is now well established that mitochondria were originally free-living bacteria that invaded, or were engulfed by, a unicellular protist. Chloroplasts arose in a similar manner. Comparison of ribosomal RNA (rRNA) from cyanobacteria with that in the genome of the chloroplasts of contemporary eukaryotes, including plants and algae, shows that oxygenic photosynthesis evolved only once and that the closest organism to the chloroplasts is the contemporary cyanobacterium *Prochlorococcus* spp. The contemporary organisms (bacteria) closest to the mitochondria are species of *Rickettsia*. A scheme displaying the proposed evolutionary relationships between hosts and endosymbionts is shown in Fig. T.I.4.

Once the formerly free-living mitochondrial and chloroplast precursors were established inside the new host, the sizes of their genomes were reduced in size by extensive transfer of genes from the organelles into the nucleus of the host cell. In modern organisms, most, but not all, mitochondrial and chloroplast genes are located in the nucleus, synthesized in the cytosol as precursor proteins, and imported into the organelle across the lipid membrane(s) using specific signal sequences in their precursors. Each organelle still maintains a small genome, which produces a few essential proteins and RNAs, the precise number of which varies from species to species.

Around 540 million years ago (Mya), a period of tremendous change occurred, popularly known as Cambrian explosion, which led to a great proliferation of multicellular organisms. Within 5 million years (a very short time span from a geological standpoint), all of the evolutionary and ecological frameworks of modern ecosystems were formed. An important part of this change was the evolution of sophisticated methods of cell-to-cell communication and cellular specialization and differentiation. Some highlights of 4 billion years of evolution are shown in a time line in Fig. T.I.5.

Evolution occurs through changes in the DNA that sometimes result in altered, more fit organisms. Such changes range from small point mutations to larger

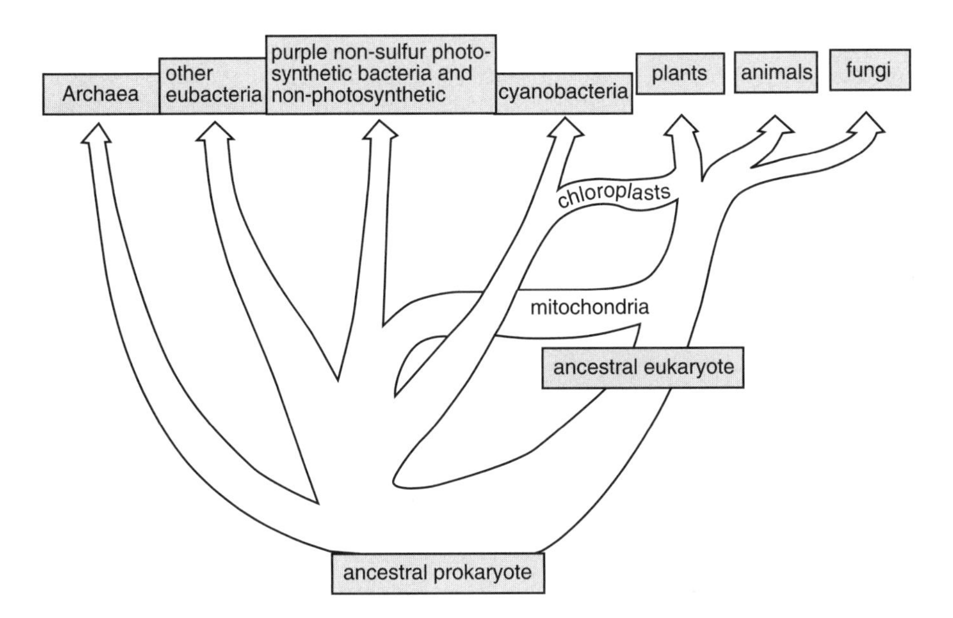

Fig. T.I.4.
A diagram of one current understanding of the evolutionary relationships among living groups of organisms. The ancestral prokaryote gave rise to the two groups of prokaryotes (eubacteria and Archaea) as well as an ancestral eukaryote that is not well defined except that it is thought to have had a nucleus. Symbiotic bacteria became incorporated into these cells as mitochondria, and, in a later event the ancestors of plants incorporated cyanobacteria, which became chloroplasts.

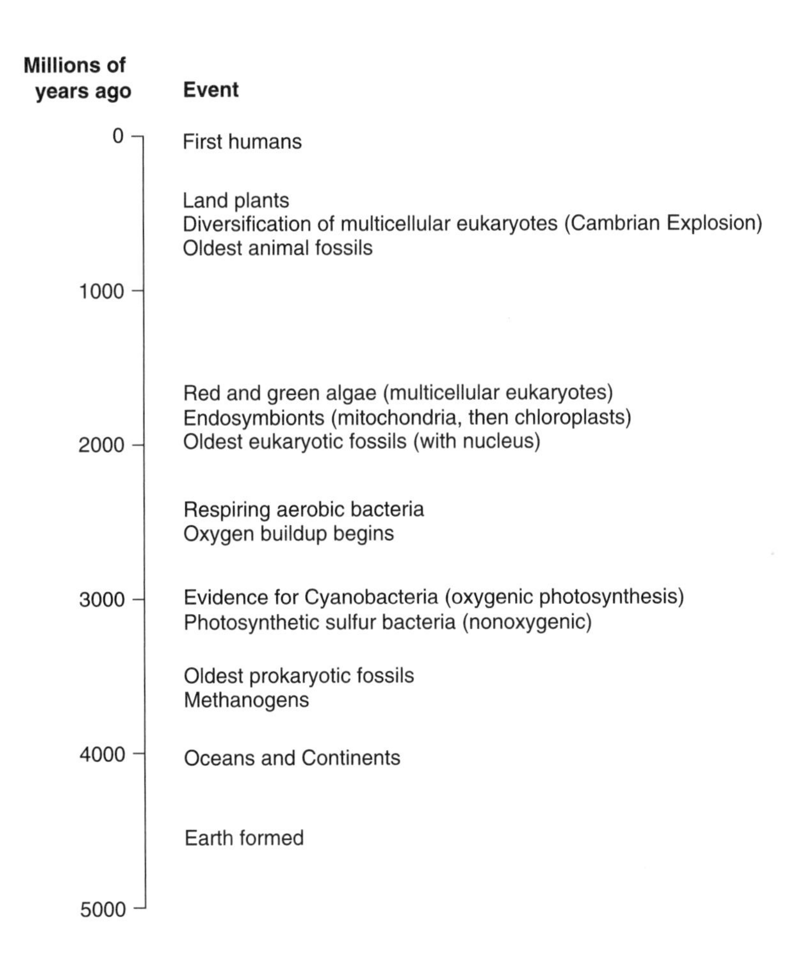

Fig. T.I.5.
A time line of some important evolutionary events. Modern oxygen levels were reached ~600–550 Mya, and the availability of energy that this provided may have been an important factor in the Cambrian Explosion, which occurred 540 Mya and established the major groups of eukaryotes. In the billion years or so before this, oxygen concentrations were lower and varied; estimates range between 5 and 18%.

deletions, replacements, or additions. Eukaryotes evolved mostly through the modification of their own existing genetic information, which frequently included events of large-scale, genome-wide duplications, and chromosomal rearrangements as well as smaller base changes, insertions, and deletions. In contrast, some bacteria have obtained a significant portion of their genetic diversity through the acquisition of sequences from distantly related organisms, through extensive lateral transfer of genetic materials between prokaryotes via bacterial viruses, bacteriophages, conjugation, and/or plasmids. The latter are autonomously replicating small circular DNA elements.

T.I.3. Genomes and Proteomes

The nucleotide sequences of >300 genomes of different organisms, including humans, have been deciphered and many more sequencing projects are currently underway. The number is increasing daily, and up-to-date information can be found at various public web sites (see links in additional reading and reference material). This growing body of information has opened new approaches to solving problems that have mystified scientists for centuries.

Although the DNA sequences of many genomes are complete, there is still an enormous amount of work to be done to determine the proteomes—the catalog of proteins (or more correctly, gene products) that are actually produced. These gene products are mainly proteins (structural elements or enzymes). However, a small but essential portion of any genome codes for cellular RNA structures that are critical components of the protein synthetic apparatus [ribosomal RNA (rRNA), transfer RNA (tRNA), and parts of the RNA-splicing machinery]. In both prokaryotes and eukaryotes, a surprisingly high number of predicted gene products, up to 30%, are unrelated to known proteins and are of unknown function.

Many of the bacterial genome sequences completed to date are for species that cause human diseases (e.g., *Rickettsia prowazekii,* which causes typhus). But, in many cases, the clinical importance couples with other academic interests. For example, the close relationship of the genomic sequences of *Rickettsia* with those of mitochondria provides clues as to how this organelle evolved. A number of archaeal and eukaryotic genomes have also been sequenced.

Genome and predicted proteome sizes for some representative organisms are listed in Table T.I.4. Eubacterial genomes vary widely in size, ranging from 0.58 Mb in *Mycoplasma genitalium,* a parasite of the human urogenital tract, to 8.67 Mb in the soil-dwelling filamentous bacterium *Streptomyces coelicolor* A3. The former encodes 484 and the latter ~7700 proteins. The reduced gene numbers in many parasites can be explained by their dependence on the host to provide key metabolites. On the other hand, the second domain of prokaryotes, Archaea, does not display such diversity in genome size; the 25 species sequenced as of 2005 have genomes ranging between 1.5 and 5.5 Mb.

The genome sizes of eukaryotes are much larger and vary by three orders of magnitude from 11.7 Mb in baker's yeast *Saccharomyces cerevisiae* to ~16,000 Mb in wheat. An important factor contributing to the large genome size in many eukaryotes is the presence of considerable amounts of DNA that does not code for proteins, or "noncoding" DNA, both between genes (mostly repetitive elements) and within genes, in the form of introns. Only 10% of the genome of a typical bacterium is noncoding DNA sequences, while in eukaryotes up to 98.5% of the DNA may be noncoding. In humans, protein coding regions (exons) are $<1.5\%$ of total DNA.

Typical multicellular eukaryotes have on the order of 20,000–40,000 real or potential coding regions (see Table T.I.4). The larger size of the proteome in these or-

Table T.I.4.
Genome and Proteome Sizes of Some Selected Members of Different Domains of Life[a]

	Genome size (Mb)	Predicted protein coding genes[b]
EUKARYOTA		
Animals		
Homo sapiens (human)	3,152	20,000–25,000
Mus musculus (house mouse)	2.500	~30,000
Fugu rubripes (tiger pufferfish)	365	~35,000
Caenorhabditis elegans (nematode, worm)	97	19,000
Drosophila melanogaster (fruit fly)	123	13,500
Anopheles gambiae (malaria mosquito)	278	13,700
Ciona intestinalis (sea squirt)	155	16,000
Plants		
Oryza sativa (monocot, rice)	430	~50,000
Arabidopsis thaliana (dicot)	125	25,500
Protists		
Plasmodium falciparum (human malaria parasite)	23	5,300
Plasmodium yoelii yoelii (rodent malaria parasite)	23	5,800
Fungi		
Neurospora crassa (bread mold)	40	10,000
Saccharomyces cerevisiae (baker's yeast)	11.7	6,300
Schizosaccharomyces pombe (fission yeast)	13.8	4,800
PROKARYOTA		
Eubacteria		
Bradyrhizobium japonicum	9.10	8,300
Mycoplasma genitalium	0.58	484
Synechocystis sp. PCC 6803	3.57	3,167
Rickettsia conorii	1.27	1,374
Escherichia coli K12 (laboratory strain)	4.64	4,288
Escherichia coli O157:H7 (pathogenic strain)	5.50	5,361
Salmonella typhimurium LT2	4.86	4,451
Bacillus subtilis	4.21	4,112
Archaea		
Thermoplasma acidophilum	1.56	1,482
Sulfolobus solfataricus	2.99	2,977

[a] Based on GenBank annotations of full-genome sequence.

[b] Note that only a small fraction of the genes are for RNAs (50–300 depending on species).

ganisms is probably due to the elaboration of gene families that encode from a few to dozens of similar proteins carrying out related function(s) in different organelles, in different cell types, or at different developmental stages of a multicellular organism. However, only a fraction of genes, perhaps 10%, are actively expressed in a particular cell type at a particular developmental stage. Another source of protein variety in eukaryotes is alternative RNA splicing, where the choice of splice junction can result in proteins of different function or location being expressed from the same gene (see Section T.I.4.1 and Fig. T.I.8).

It is interesting to consider how (and whether) genome and proteome size are related to an organism's complexity. Clearly, unicellular organisms (bacteria, Archaea, and protists) have smaller genome sizes than multicellular eukaryotes, but within multicellular eukaryotes the number of apparent genes does not correlate particularly well with our current perceptions of the relative complexity of organisms (see Table T.I.4). For example, the small weedy plant *Arabidopsis thaliana* has nearly twice as many genes as the fruit fly *Drosophila melanogaster,* and humans have 5–10% more genes than the tiny nematode worm *Caenorhabditis elegans*. Despite large difference in their apparent gene numbers, it has been suggested that ~10,000 different types of proteins may be the minimal genome complexity required by multicellular eukaryotes to execute development and respond to their environment.

T.I.4. Cellular Components

The interior of a cell is a very crowded place, as Fig. T.I.6, a picture of bacterial cytoplasm drawn by David Goodsell using known concentrations of various cellular components, dramatically demonstrates. The macromolecular, organic components that occupy the space in all cells are principally nucleic acids, proteins, lipids, and carbohydrates. Each of these components is built from basic building blocks according to specific rules, and each plays specific and essential roles in physiology. The following sections give a very basic overview of the structures and functions of the organic components.

T.I.4.1. Nucleic Acids: DNA and RNA

The closely related nucleic acid polymers DNA and RNA are key players in heredity and the use of genetic information. The role of DNA is to carry all the genetic information that is necessary for survival of the organism and to transfer this hereditary endowment to the next generation. The RNA is used to relay the genetic information to the protein synthetic machinery, as well as playing a number of essential structural, enzymatic, and regulatory roles in this machinery: Ribosomal RNAs form the active core of ribosomes (see below), tRNAs are involved in bringing the correct amino acids to the growing protein chain, and messenger RNAs (mRNA) carry the code for protein sequences and are "read" by the ribosomes to synthesize the particular protein.

The basic building blocks for DNA and RNA are very similar: DNA consists of 2′-deoxyribose coupled at the 1′ position to one of four bases—guanine, adenine, thymine, or cytosine (G, A, T, or C)—and at the 5′ position to phosphate (Fig. T.I.7*a*). Ribonucleic acid differs from DNA in two ways: Ribose is used instead of deoxyribose and uracil (U) takes the place of thymine. Each strand of DNA or RNA is a linear polymer of nucleotides, where the phosphate links the 5′ position of one nucleotide to the 3′ position of the next (Fig. T.I.7*b*). A typical DNA molecule consists of two such polynucleotide strands that form a double helix held together by hydrogen bonds between pairs of bases; A pairs with T and G pairs with C (Fig. T.I.7*c*). Deoxyribonucleic acid is relatively straight and stiff, and long strands are packaged by wrapping them around protein scaffolds, while RNA is generally single stranded, has a tendency to fold into more globular shapes, and is chemically less stable than DNA.

DNA has two important properties that make it suitable for its role in inheritance—it is chemically stable, and its structure can be exactly duplicated, facilitating the passage of genetic information from one generation to the next. In DNA replication, the two strands of DNA unzip, and new second strands are synthesized using the templates provided by each of the first strands. This process results in exact duplication of the DNA molecule, and is referred to as "semiconservative" because each new double helix consists of one old strand and one newly synthesized strand. The precise sequence of the four nucleotides carries the genetic information. The DNA sequence determines the sequences of all the proteins, according to the genetic code, and of all the structural and functional RNA molecules. In addition, DNA sequences specify binding sites for DNA-binding regulatory proteins (transcription factors, etc.) and signals for DNA and RNA processing enzymes.

In order to utilize the information stored in the genetic material, cells must express the genes, that is, use the genetic code to make the correct proteins. Gene expression requires transcription and translation (Fig. T.I.8). Transcription is the copying of one

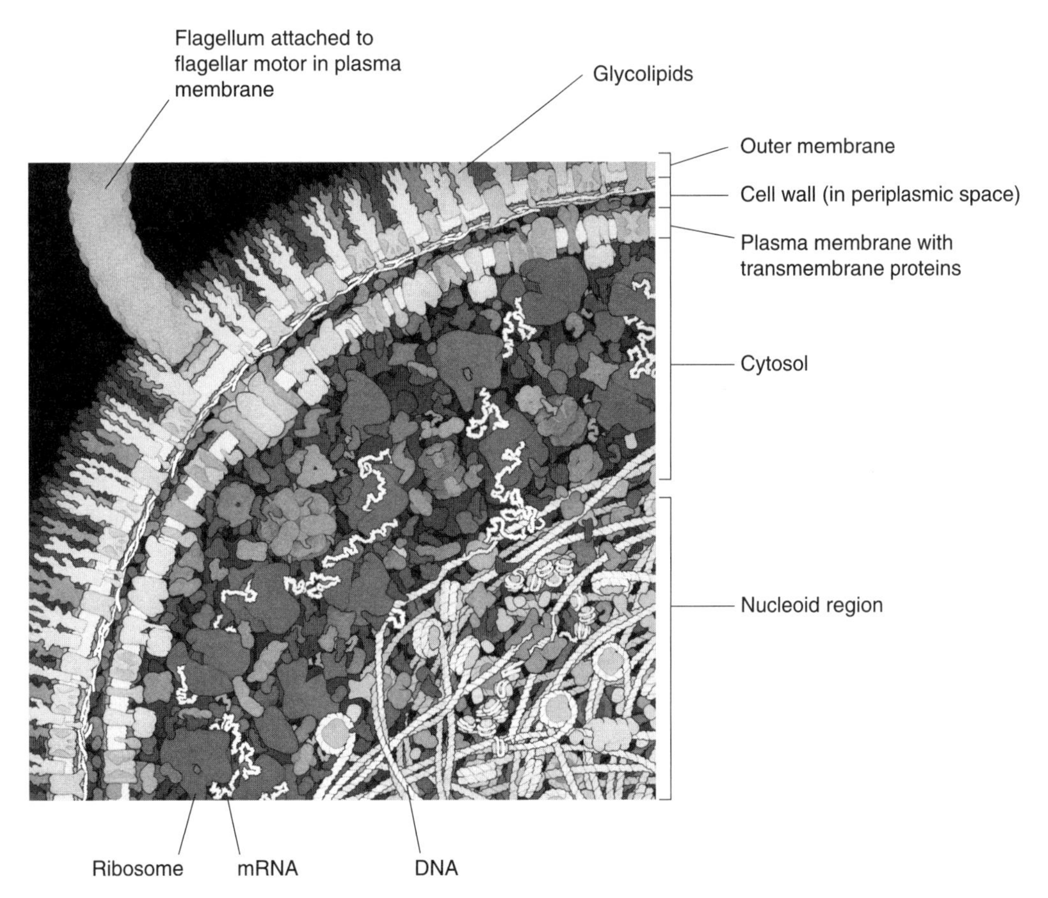

Fig. T.I.6.
A typical cell interior with components drawn to scale and taking account of known concentrations of various components. It is clear that the interior of a cell is a very crowded place. This illustration shows a cross-section of a small portion of an *E. coli* (bacterial) cell, but a eukaryotic cell would be equally crowded. The cell wall and the concentric inner and outer membranes studded with transmembrane proteins are indicated near the top. A large flagellar motor crosses the entire wall, turning the flagellum that extends upward from the surface. The cytoplasmic area is darker gray. The largest molecules are ribosomes and the wiggly white strands are mRNA. Other enzymes and cellular components are shown in varying shades of gray. The nucleoid region is shown in lighter shades. The DNA is long and ropelike and in places is wrapped around HU protein (bacterial nucleosomes). In the center of the nucleoid region shown here, you might find a replication fork, with DNA polymerase replicating DNA. (Copyrighted © David S. Goodsell 1999, used by permission.)

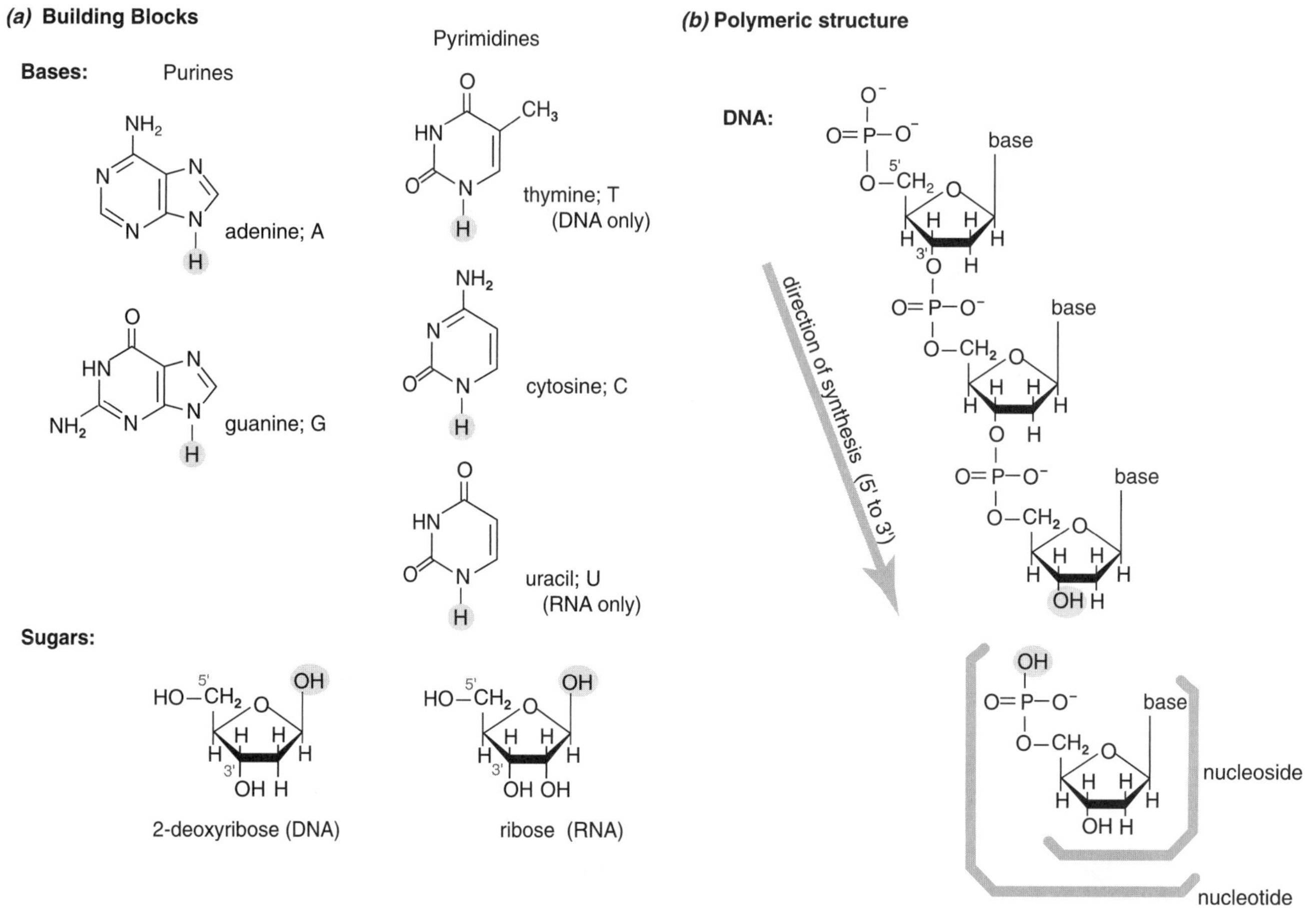

(a) Building Blocks

Bases: Purines

adenine; A

guanine; G

Pyrimidines

thymine; T (DNA only)

cytosine; C

uracil; U (RNA only)

Sugars:

2-deoxyribose (DNA)

ribose (RNA)

(b) Polymeric structure

DNA:

direction of synthesis (5′ to 3′)

nucleoside

nucleotide

Fig. T.I.7.
Nucleic acid structures. DNA and RNA are polymers of nucleotides, which are in turn made from more basic building blocks (bases, sugars, and phosphate). Each nucleotide has one base, one sugar, and one phosphate. (*a*) The bases and sugars that are used in DNA and RNA are shown. Note that three of the bases are the same in each, but DNA contains thymine while RNA contains uracil. The bases are important because they carry the genetic sequence, and their ability to form specific hydrogen bonds with only one other base allows precise replication of DNA and RNA. The sugar used by DNA and RNA also differs, DNA uses 2-deoxyribose, and RNA uses ribose. (*b*) The polymeric structure of a strand of DNA is shown. (RNA is the same except ribose is used in place of 2-deoxyribose.) The backbone chain consists of alternating deoxyribose (or ribose) and phosphate units and is formed by phosphodiester bonds between the 3′ hydroxyl of one nucleotide with the 5′ hydroxyl of the next one. (*c*) The bases form base pairs in a very specific manner: A only pairs with T, and G with C, as shown in the figure. A double-stranded DNA molecule consists of two DNA strands held together by hydrogen bonds between the bases, with the phosphate-sugar backbones winding around the outside. Note that the two strands run in opposite directions. The base pairs are planar, and the plane is oriented at approximately a right angle to the helical axis. By convention, when DNA sequences are written out, the 5′ end is written at the left, or first, and corresponds to the nucleotides N-terminus of the protein for which it codes. (*d*) Small molecule nucleic acids are also extremely important in life. Here two essential ones are shown. ATP is the cellular energy currency, and is also a good example of a nucleotide triphosphate. Both NAD^+ and $NADP^+$ are electron carriers for respiration, photosynthesis, and many anabolic and catabolic pathways.

strand of DNA sequence into single-stranded mRNA molecules, which carry the coded information to the protein synthetic apparatus. Translation is the process by which the information stored in the mRNA is decoded to produce a protein. Translation is carried out by subcellular particles called ribosomes, very large enzymes with RNA and protein components. The mRNAs are generally temporary molecules. In prokaryotes, they are degraded rapidly after synthesis of the protein. In eukaryotes, mRNA stability can vary widely depending on the specific message, the stage of growth, or other signals.

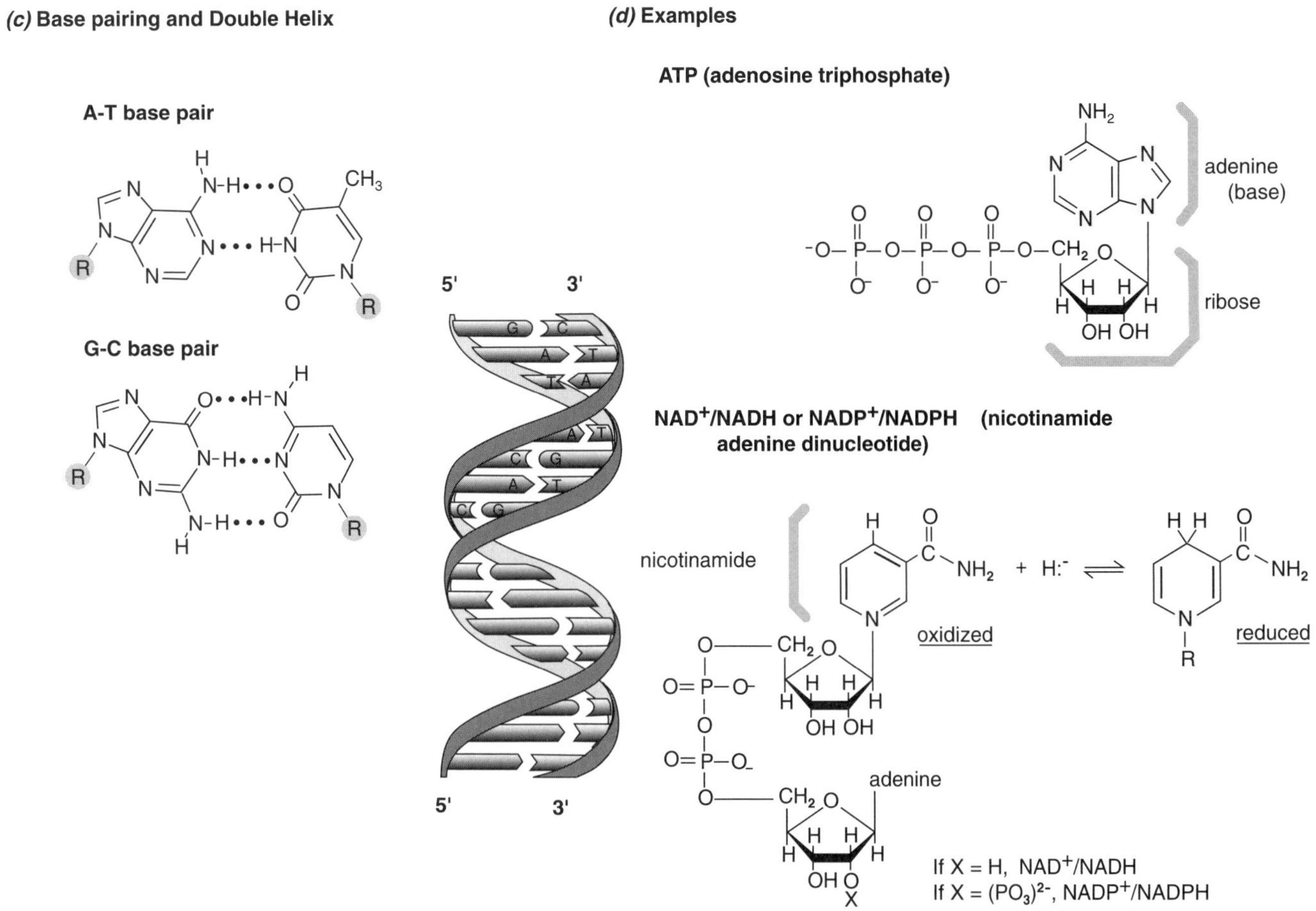
(c) Base pairing and Double Helix
A-T base pair
G-C base pair
5'
3'
5'
3'
(d) Examples
ATP (adenosine triphosphate)
adenine (base)
ribose
NAD+/NADH or NADP+/NADPH (nicotinamide adenine dinucleotide)
nicotinamide
oxidized
reduced
adenine
If X = H, NAD+/NADH
If X = (PO3)2-, NADP+/NADPH

Fig. T.I.7.
(Continued)

Regulation of gene expression plays a pivotal role in all types of cells. Different patterns of gene expression control responses to the environment and to stress, as well as development and morphology in multicellular organisms. All cells in a multicellular organism carry the same genetic information. However, different types of cells synthesize different sets of proteins, some of which are unique to that particular type of the cell. For example, insulin is synthesized exclusively in certain pancreatic cells and immunoglobulins are synthesized exclusively in B lymphocytes.

Prokaryotic Gene Expression. In prokaryotes, the DNA is not separated from the cytosol by a nuclear membrane, and transcription and translation are closely associated, occurring in the same cellular compartment, the cytosol. Ribosomes begin translating mRNA at one end while the other end of the mRNA is still being transcribed from the DNA (Fig. T.I.8*a*). Because of this close association, prokaryotes can change their gene expression patterns extremely quickly, initiating the synthesis of a needed protein within a few minutes. Prokaryotic genes are most often organized into operons, groups of genes whose proteins function in the same pathway and that are transcribed together in one continuous mRNA. With this arrangement, one regulatory region can control the expression of a number of proteins. Most prokaryotic genes are located on the single circular chromosome. However, genes can be present on plasmids, much smaller circular DNA elements that replicate autonomously. Plasmids are important for two reasons: genetic engineering in the laboratory relies

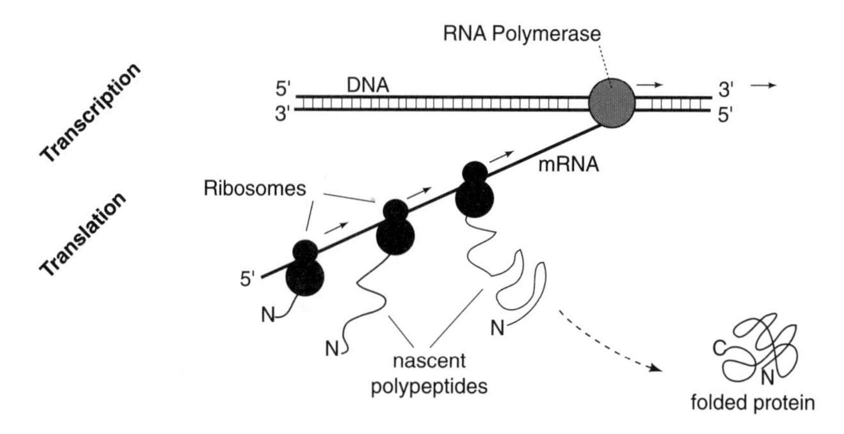

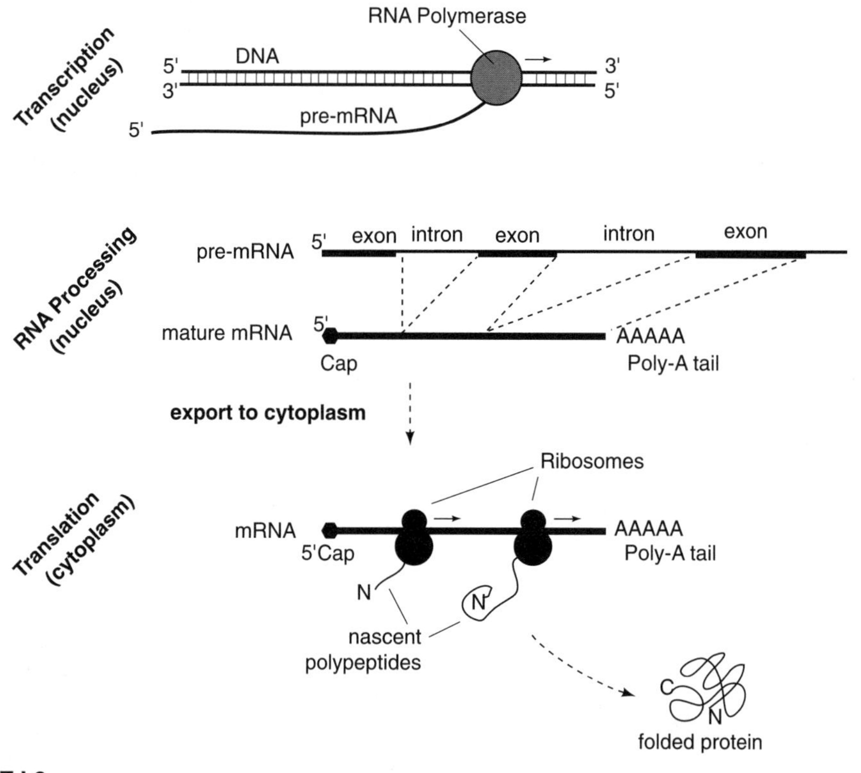

Fig. T.I.8.
Gene Expression. Gene expression in prokaryotes and eukaryotes is similar in that DNA is copied (transcribed) into mRNA, and the mRNA is translated into protein. Differences arise from the facts that (1) in eukaryotes transcription is in the nucleus, while translation takes place in the cytoplasm and (2) eukaryotes do far more processing of the mRNA. (*a*) Prokaryotic transcription and translation are coupled. While RNA polymerase, the enzyme that transcribes the DNA into mRNA, is still working, ribosomes find the mRNA chain and start translation. As shown, transcription proceeds in a 5′ to 3′ direction, and the ribosomes also carry out translation in this direction, reading 5′ to 3′ and synthesizing from

heavily on plasmids to move genes easily between organisms, and antibiotic resistance genes are also often transferred from species to species on plasmids, contributing to current problems with the spread of antibiotic resistant pathogens.

Eukaryotic Gene Expression. In eukaryotes, transcription occurs in the nucleus and translation in the cytoplasm, necessitating the addition of extra steps to detach RNA transcripts from the DNA and move them to the cytoplasm. Before export from the nucleus, eukaryotic mRNAs are marked with a special nucleotide-based structure, called the cap, at the 5′ end. A string of adenosines, the poly-A tail, is often, but not always, added to the 3′ end. In addition, many eukaryotic genes are interrupted by one or more "introns", segments of noncoding sequence that appear between the "exons", or fragments of coding sequence. In a process called splicing, the introns are clipped out and the exons are joined together prior to export from the nucleus. A mature mRNA that is exported from the nucleus codes for only one protein, is capped on its 5′ end, has had its introns removed, and usually has a poly-A tail (Fig. T.I.8*b*). While this elaborate system adds complexity, it introduces greater opportunity for precise control. Much regulation is accomplished at the level of transcription initiation, but alternative splicing is another possible route. If splice junctions are skipped, or added, proteins with different molecular weights, functionalities, or locations can be formed from the same full-length gene. The mRNA molecules are translated into proteins on ribosomes in the cytoplasm. For some genes, regulation of their expression occurs at the translational level: the mRNA is always present, but is only translated when certain conditions are met.

T.I.4.2. Proteins

Proteins are linear polymers of amino acids (see Fig. T.I.9) linked from carboxyl group to amino group by peptide bonds, a type of amide linkage. Proteins are synthesized by ribosomes following the information provided on mRNA templates, beginning with the 5′ end of the mRNA and the N-terminus of the protein. Upon release from the ribosome, or even before, a newly synthesized protein folds into a compact form whose structure is usually determined by the sequence of the protein polymer itself. The process of protein folding is very complex and is currently under intense study. Some proteins need the help of "chaperones", or helper proteins, to fold correctly, but others fold unassisted.

Fig. T.I.8.
(Continued)
N-terminus toward the C-terminus. The polypeptide often begins to fold up before it is released from the ribosome. In many cases, several genes are transcribed on the same RNA and thus are expressed together. In general, the correspondence between DNA, RNA, and protein is strictly linear, and most of the control over gene expression is exerted at the transcriptional level. (*b*) In eukaryotes, transcription is similar, although the polymerase is much bigger and involves more factors. The pre-mRNA is subject to processing at both ends and in the middle before it reaches its mature form and is sent to the cytoplasm. A poly-A tail is added to the 3′ end and a special nucleotide called the cap is added to the 5′ end. Eukaryotic gene coding sequences are interrupted with intervals of noncoding DNA, called introns or intervening sequences. The introns are removed and the exons, or protein-coding regions, are spliced together to make an RNA with a 1:1 correspondence with the protein it codes for. Once these three modifications have occurred, the mRNAs are sent to the cytoplasm where they are translated by ribosomes. Eukaryotic ribosomes are bigger that prokaryotic ones, but work in a similar way. In general, one eukaryotic mRNA codes for one protein. In eukaryotes, gene expression can be regulated at the transcriptional level or at the translational level.

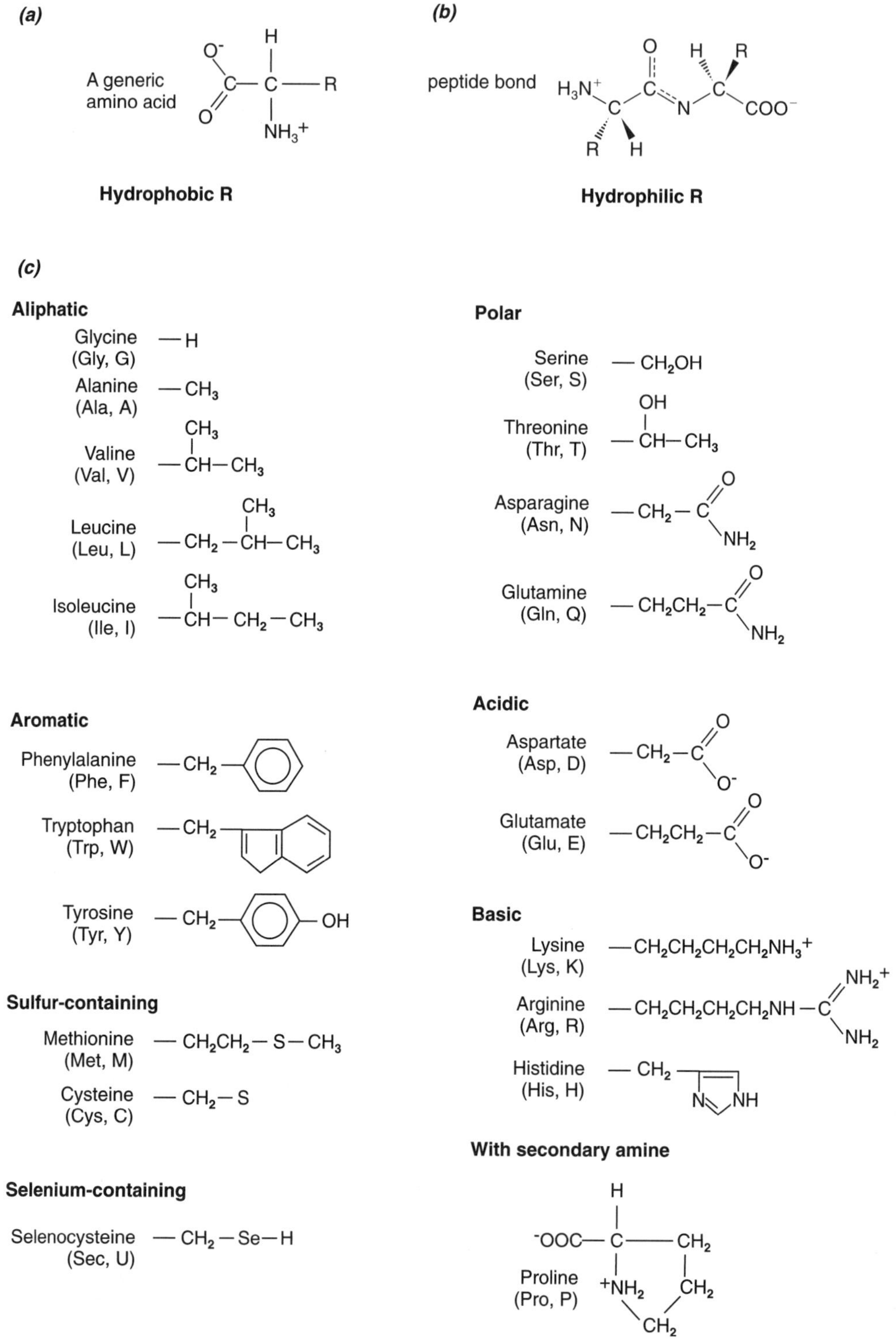

(a)

A generic amino acid

O^-—C(=O)—C(H)(NH_3^+)—R

Hydrophobic R

(b)

peptide bond H_3N^+—C(R)(H)—C(=O)—N—C(H)(R)—COO^-

Hydrophilic R

(c)

Aliphatic

Glycine (Gly, G) —H

Alanine (Ala, A) —CH_3

Valine (Val, V) —CH(CH_3)—CH_3

Leucine (Leu, L) —CH_2—CH(CH_3)—CH_3

Isoleucine (Ile, I) —CH(CH_3)—CH_2—CH_3

Aromatic

Phenylalanine (Phe, F) —CH_2—(phenyl)

Tryptophan (Trp, W) —CH_2—(indole)

Tyrosine (Tyr, Y) —CH_2—(phenyl)—OH

Sulfur-containing

Methionine (Met, M) —CH_2CH_2—S—CH_3

Cysteine (Cys, C) —CH_2—S

Selenium-containing

Selenocysteine (Sec, U) —CH_2—Se—H

Polar

Serine (Ser, S) —CH_2OH

Threonine (Thr, T) —CH(OH)—CH_3

Asparagine (Asn, N) —CH_2—C(=O)—NH_2

Glutamine (Gln, Q) —CH_2CH_2—C(=O)—NH_2

Acidic

Aspartate (Asp, D) —CH_2—C(=O)—O^-

Glutamate (Glu, E) —CH_2CH_2—C(=O)—O^-

Basic

Lysine (Lys, K) —$CH_2CH_2CH_2CH_2NH_3^+$

Arginine (Arg, R) —$CH_2CH_2CH_2CH_2NH$—C(=NH_2^+)—NH_2

Histidine (His, H) —CH_2—(imidazole, N, NH)

With secondary amine

Proline (Pro, P) ^-OOC—C(H)—CH_2—CH_2—CH_2—$^+NH_2$ (ring)

Fig. T.I.9.
Amino Acids. Amino acids are the building blocks for proteins. (*a*) A generic amino acid is shown. (*b*) The polypeptide chain (protein) is formed by links between the amino group of one amino acid with the carboxyl group of the previous one, resulting in the net loss of water and formation of a peptide bond. Proteins are synthesized from left to right as shown, such that the amino group of the first amino acid is free and the carboxyl group of the last amino acid is free. The ends are referred to as the N- and C-termini, respectively.

Naturally occurring proteins are constructed from 21 different amino acids (AAs). These AAs are classified according to the properties of their side chains and are summarized in Fig. T.I.9. All except glycine have a chiral center at the α-carbon atom and, consequently, can exist in two enantiomeric forms, L and D. However, in normal proteins, cells use only the L enantiomers of amino acids.

Proteins have primary, secondary, tertiary, and quaternary structures, which are diagrammed in Fig. T.I.10. Primary structure is the amino acid sequence (Fig. T.I.10*a*). Secondary structure involves the local formation of particular structural units such as alpha helices, beta sheets, and random coils (Fig. T.I.10*b*). Preference for a specific structural form is inherent in the primary sequence. These areas of local structure are folded further to form the tertiary structure, in a manner that once again is very dependent on the precise amino acid sequence (Fig. T.I.10*c*). At this point, a monomeric protein has reached its final folded shape. The noncovalent binding of several subunits (monomers) together to form a multimeric protein (dimer, trimer, tetramer, etc.) is called the quaternary structure. Multimeric proteins can be composed of the same or different types of monomers. A dimer of identical subunits is called a homodimer; a dimer of different subunits is called a heterodimer. Many proteins are divided into domains, that is, areas of the same polypeptide chain that fold separately. Often, two domains in one protein will carry out related but different functions, and their enforced proximity improves the efficiency of the reaction. A common example of domains occurs in transcription factors, where one domain binds the DNA and another domain binds parts of the transcription apparatus, thus coupling one to the other and promoting expression of the gene.

Various forces hold a folded protein in its correct shape. The weaker interactions, such as hydrogen bonds, van der Waals forces, electrostatic forces, and hydrophobic interactions, are very important in proteins because they stabilize interatomic interactions in molecules. These forces are often key determinants not only of protein folding but also of the catalytic function that a protein may be "designed" (have evolved) to carry out. Alpha helix and beta sheet structures are formed by specific patterns of hydrogen bonding of backbone atoms. Interactions that stabilize tertiary structure are more varied and often involve atoms in the amino acid side chains as well as the backbone. *In vivo,* proteins normally fold up correctly, either spontaneously or with the help of chaperones. However, sometimes small errors or mutations cause dramatic changes that can lead to misfolding and subsequent aggregation of the protein into nonfunctional forms, which in several instances are associated with disease (examples include sickle cell anemia and many neurological diseases).

A typical protein is 100–1000 amino acids long. However, as is usual in biology, extreme examples can be found. In the human genome, for example, the dystrophin

Fig. T.I.9.
(Continued)
(*c*) The 20 standard amino acids that appear in proteins are shown, grouped by the salient characteristic of the R group, or side chain. These amino acids are represented in the DNA by the standard genetic code. Note that proline, unlike any of the others, has a secondary amine that participates in the peptide bond. This property has profound effects on the flexibility of the polypeptide chain, and proline generally causes a kink in protein structures, disrupts secondary structure and is often found at turns. Also shown is a twenty-first amino acid, selenocysteine, which is present at the active site of a limited, but growing, number of proteins. Selenocysteine is inserted as the protein is synthesized by the ribosome, although the process is somewhat more complicated than that used for the other amino acids. (Other unusual amino acids, e.g., hydroxyproline found in collagen, are created by post-translational modification of the standard amino acids.)

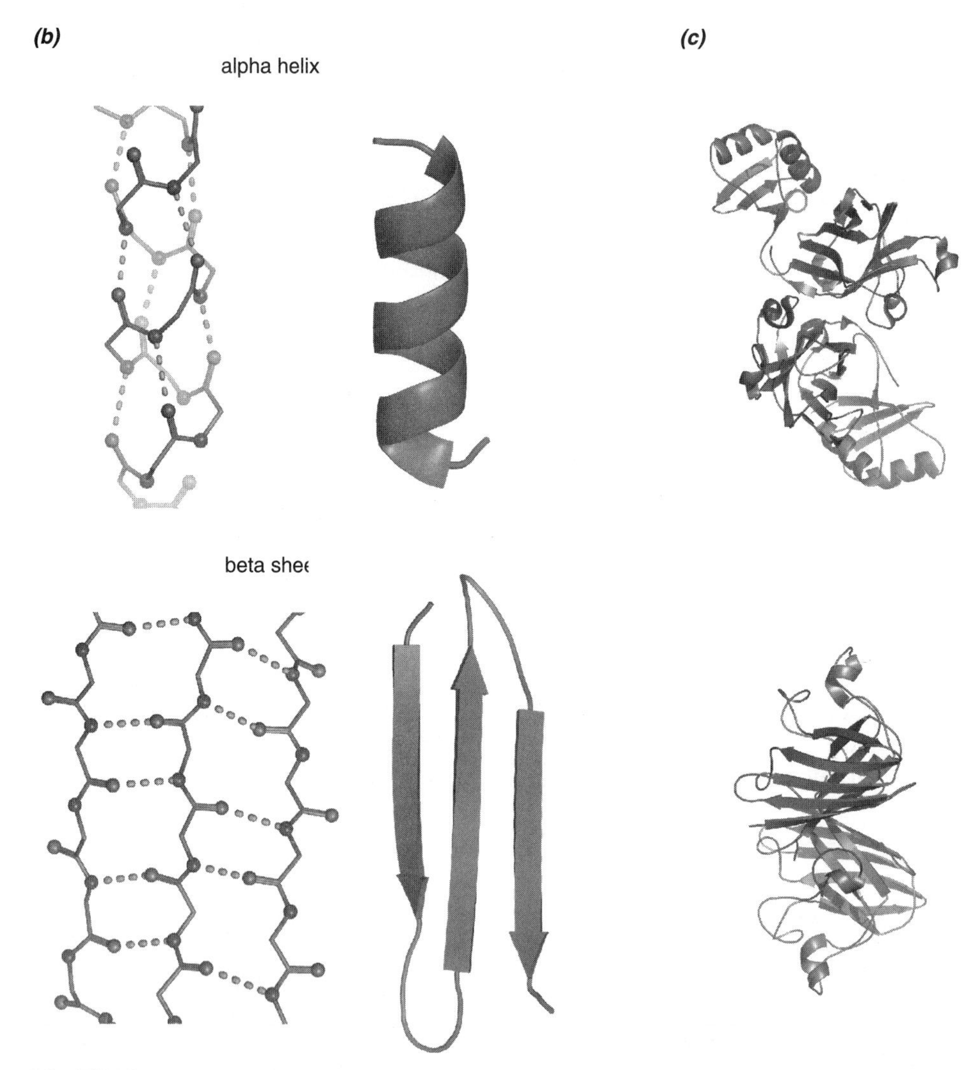

Fig. T.I.10.
Protein Structure. (*a*) Primary Structure. Amino acids are joined together in a linear polymer in a specific order. (*b*) Secondary Structure. Groups of neighboring amino acids form secondary structures when their main-chain atoms hydrogen bond with each other. The two main forms of secondary structures are the alpha helix (top) and beta sheet (bottom). On the left, the backbone structure is shown, with the structure-forming hydrogen bonds indicated. On the right is a typical ribbon diagram of the structure. In both cases, the side chains have

gene is 2.3 Mb long, nearly as long as some entire bacterial chromosomes. The gene is 99% noncoding intron sequences; the rest consists of 79 exons and produces a polypeptide of 3685 amino acids. There are also noteworthy examples of gigantic proteins, such as 3814-kD titin. It is encoded by a 300-kb gene, with nearly 300 exons, which produces a single-chain polypeptide that is 35,000 amino acids long. At the opposite end of the spectrum are such important proteins as insulin (51 amino acids), the small metal-binding protein, metallothionein (60 amino acids), and many ferredoxins and metallochaperones (80–100 amino acids).

Longer proteins are often divided up into multiple sequence domains that fold separately. During the course of evolution, genes have duplicated or fused to produce these more complex structures. Such proteins often carry out functions radically different from those of their ancestral proteins. Thus, Nature makes efficient use of existing protein folding topologies to craft new proteins that carry out important new functions. Approximately 1000 different protein-folding topologies have been described to date.

Proteins whose final location is not the cytosol are usually synthesized as precursors that carry information for targeting them to specific compartments, to specific membranes, or to the exterior of the cell. This information often resides at the beginning of the protein sequence (the N-terminus), in which case it is called the signal or transit peptide. This signal peptide binds to a translocation complex in the membrane, usually before translation is complete, such that translation of the protein sequence and its translocation across the membrane can be coordinated. Generally, the signal sequence is removed by proteolytic cleavage during or after translocation of the protein to its final destination. Such sequences are found at the N-termini of the precursors of many chloroplast- and mitochondria-targeted proteins, as well as those secreted to the outside of the cell or to the periplasm (in bacteria). In eukaryotes, proteins destined for secretion to the cell exterior are sent across the membrane of the endoplasmic reticulum, and undergo further processing including proteolytic

Fig. T.I.10.
(Continued)
been left off for clarity. In an alpha helix, the side chains stick out away from the main axis of the helix (vertical in this illustration) and the hydrogen bonds form between amino acids that are close neighbors in the primary structure. In a beta sheet, the side chains alternate sticking up and down relative to the plane of the sheet. Note that a beta sheet can form between noncontiguous stretches of amino acids. An antiparallel beta sheet is shown, in which the strands run in opposite directions. Beta sheets can also form between parallel strands with a slightly different hydrogen-bonding pattern (not shown). The primary sequence determines the secondary structure that is preferred. (*c*) Tertiary and quaternary structure. The structure formed when a whole polypeptide chain folds is referred to as its tertiary structure, and may includes areas of different kinds of secondary structures, turns, and random coils. If two or more different polypeptide chains interact to form a single protein, it is called quaternary structure. The examples of alpha helix and beta sheet were taken from real proteins whose whole structures are illustrated here. The positions of the secondary structure examples are highlighted. The example of alpha helix was taken from the copper chaperone for superoxide dismutase (CCS). Its tertiary structure is a mixture of alpha helix and beta sheet. The protein is a dimer of identical subunits (oriented one above the other in this illustration) and thus has quaternary structure as well. It is also an excellent example of a protein with more than one domain, which can be seen most clearly in the subunit at the top of the page, where two separately folded units are connected by a single strand of random coil. The beta sheet example is from copper, zinc superoxide dismutase (CuZnSOD, metal ions not shown), a homodimer that is primarily a beta sheet and has no domains. [See color insert.]

cleavage and addition of carbohydrates before secretion via the Golgi apparatus. Signal sequences for different locations have somewhat different characteristics, and efforts to predict the final location of a protein from its sequence have been only modestly successful. Proteins destined to be anchored to the cell surface via a lipid component contain in addition a C-terminal lipid attachment signal. Other localization signals are embedded in the protein sequence, and are less well understood. Proteins with such internal signals include those destined for the peroxisomes and some mitochondrial membrane proteins.

Many enzymes use small molecules as cofactors or prosthetic groups to help in their activities. These can be permanently bound or diffusible, and include many molecules that we know as vitamins (thiamine, riboflavin, etc.) as well as metal ions or metal ion complexes, electron carriers such as nicotinamide adenine dinucleotide (NAD^+) and NAD phosphate (NADP), and the universal energy carrier, adenosine triphosphate (ATP) (Fig. T.1.7.d). The metal-containing cofactors are discussed in detail in Chapters III and IV.

In many cases, proteins undergo additional processing, termed post-translational modification, in which new chemical groups are introduced by modification of the amino acid side chains. The various kinds of modifications serve different purposes, including signal transduction, modulation of enzymatic activity, and protection against, or signaling for, degradation. Phosphorylation, attachment of phosphate groups to the protein (catalyzed by kinases), and the reverse process, dephosphorylation (catalyzed by phosphatases), are two of the most important intracellular modifications, modulating cell-to-cell communications, cellular responses to signals, and key aspects of development. The most common phosphorylation sites are the hydroxyl groups of Thr, Ser, and Tyr. It has been estimated that nearly 10% of eukaryotic cytosolic proteins are subject to phosphorylation.

Glycosylation, the covalent attachment of branched chains of sugars and amino sugars to the protein via the side-chain nitrogen atom of asparagine (N-linked glycosylation) or hydroxyl groups of Thr or Ser (O-linked glycosylation), is another common post-translational modification that occurs in all types of cells. Glycosylated proteins are typically located on the outside of the cell, either attached to the outer surface or floating in the extracellular fluid, and it is believed that carbohydrate components increase resistance to protease degradation, resistance to heat denaturation, and solubility. In eukaryotes, most glycoproteins acquire their carbohydrate chains in the lumen of the rough ER and the Golgi apparatus, prior to being exported, although a few are glycosylated in the cytosol. Bacteria, lacking organelles, add carbohydrates to proteins at the cell surface. In many instances, the carbohydrate components of surface glycoproteins play roles in pathogenicity and/or immune recognition.

Many proteins are enzymes: catalysts that speed up important biological reactions and enable energetically favorable reactions to occur readily. During billions of years of evolution some enzymes such as carbonic anhydrase and superoxide dismutase have achieved a state close to catalytic perfection, functioning at nearly the diffusion controlled limit. For multistep reactions, a strategy commonly used in living things is to gather the necessary enzymes in multienzyme complexes, in which the product of one enzyme is a substrate for the neighboring enzyme in a cascade of reactions. The set of reactions in such a cascade can go much faster than would be the case if the enzymes were separated because diffusion is minimal and a high fraction of reactant reaches the next enzyme. There is some evidence that certain multienzyme complexes may even direct the substrates from one active site to the next, controlling or eliminating free diffusion and leaving little to chance. Multienzyme complexes can be soluble or membrane bound and vary widely in size and number of components.

T.I.4.3. Lipids and Membranes

One of the key features of life is the separation of the living entity from the surrounding environment, self from non-self. For cells, this separation is accomplished using a lipid bilayer membrane. The water-based (hydrophilic) environment is divided into two compartments (inner and outer) by a continuous lipid (hydrophobic) membrane, and ionic and polar molecules do not pass freely between the two compartments. Thus, cells are able to maintain an internal environment that is quite different from the outside.

The basic building block of biomembranes is an amphipathic molecule with a charged or polar head group and a highly lipophilic tail (Fig. T.I.11). Such molecules spontaneously aggregate in water to form bilayers with the hydrophobic tails pointed inward and the head group facing the water, precisely the form of a biomembrane. A typical eukaryotic or eubacterial membrane is made of phospholipids—glycerol coupled through phosphate to a polar head group at one hydroxyl and to two fatty acids through ester linkages to the other two –OH groups (Fig. T.I.11*a*). Interestingly, Archaea use isoprenoid alcohols instead of fatty acids, and these alcohols are linked to the glycerol via ether linkages. Archaeal membranes also contain diglycerol tetraethers, which span the membrane (Fig. T.I.11*b*). These features likely help increase membrane stability in the extreme environments in which many of these organisms live.

There is biological variety at both ends of the phospholipid (or phospho-glycerol-ether) molecule. The fatty acids can be saturated, monounsaturated, or polyunsaturated to varying degrees, and longer or shorter (typically 16–20 carbons atoms). The predominant head groups vary among the different taxonomic groups, among different cell types in the same multicellular organism, among different membranes in the same cell, and even between the two sides of the same membrane. Typical head groups include choline, ethanolamine, serine, glycerol, and inositol, but many less common ones also exist.

Biological membranes are "fluid", that is, there is substantial lateral mobility of membrane components, which is often called the "fluid mosaic model" of membrane structure. In many eukaryotes, sterols are important structural components of membranes: cholesterol in animal cells, ergosterol in yeast and other fungi. Plants and prokaryotes generally lack sterols. Plants tend to have more polyunsaturated fatty acids than animals.

Biomembranes also contain many other components (Fig. T.I.11*c*). Prominent among these are proteins, which can be fully embedded in the membrane (integral membrane proteins) or more loosely attached to the membrane via short hydrophobic peptide or lipid anchors (peripheral membrane proteins). The cell's exterior surface in particular often acquires a carbohydrate coat through the presence of glycoproteins and glycolipids, carbohydrate moieties anchored to the membrane by a protein or lipid component, respectively.

The plasma membrane separates a cell from its environment, but the cell must live in its surroundings. At the very least, it needs to obtain nutrients, excrete toxic byproducts, and sense and respond to the environment. Therefore, the existence of the membrane barrier dictates the necessity for cross-membrane transport and signal transduction across the membrane, and a significant fraction of the cellular apparatus is devoted to these functions. Proteins embedded in the membrane act as specific transporters and sensors to carry out these functions. Chapters V and VIII cover these processes where metal ions are involved.

The embedding of proteins also allows the membrane to function as a matrix or scaffold on which to mount other components. Membrane location restricts motion

***(a)* Typical phospholipid structure**

O
‖
O–C
H_2C O
| ‖
O HC–O–C
‖ |
R–O–P–O–CH_2
|
O^-

(saturated fatty acid, e.g., stearic acid, 18:0)

(polyunsaturated fatty acid, e.g., linoleic acid, 18:2)

Representative R groups:

ethanolamine	$-CH_2CH_2NH_3^+$
choline	$-CH_2CH_2N(CH_3)_3^+$
serine	$-CH_2CHNH_3^+$ (with COO^-)
glycerol	$-CH_2CHOHCH_2OH$

Cartoon representation

***(b)* Samples of Archaeal phospholipid structures**

diether

OH
|
R–O–P–O–
‖
O

tetraether

OH
|
R–O–P–O–
‖
O

–OH

***(c)* Typical biological lipid bilayer membrane, with example proteins**

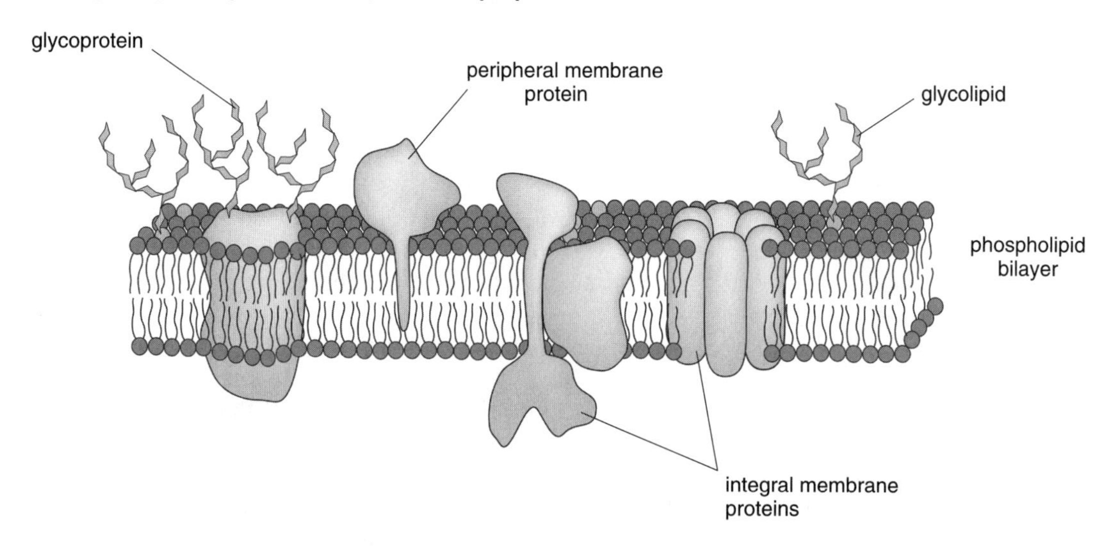

Fig. T.I.11.
Lipid structures. A phospholipid is an amphipathic molecule with a phosphate and polar head group at one end and a hydrophobic tail at the other end. Phospholipids are the basic building blocks for all biological membranes. Part (*a*) shows a typical eukaryotic or eubacterial phospholipid. The hydrophobic portions are fatty acids that are attached to the glycerol core by ester linkages. Some typical head groups are shown. Phospholipids come in a variety of types, depending

to two dimensions (2D), rather than three dimensions (3D) as with soluble proteins. Proteins that form part of a metabolic pathway, such as the electron-transport chain of respiration, can be held in close proximity so that substrates can move easily from one reaction to the next, greatly speeding the process.

T.I.4.4. Carbohydrates

Carbohydrates have multiple important roles in biological systems. Monosaccharides are the fuel for the cell (e.g., glucose, fructose, glycerol), some polysaccharides are used for energy storage (e.g., glycogen, starch), other polysaccharides play important structural roles either alone (e.g., cellulose) or in combination with polypeptide components (e.g., peptidoglycan of the bacterial cell wall). Polysaccharides also play important roles in cell–cell communication and immune recognition. (The ABO blood group antigens, e.g., are carbohydrates attached to the surface of cells.) Other important biomolecules have carbohydrate components. Notably, DNA and RNA are phosphodiester polymers of amino derivatives of the five-carbon sugars D-deoxyribose and D-ribose, respectively (Fig. T.I.7), and the core on which phospholipids and triglycerides (fat) are assembled is the C_3 sugar glycerol (Fig. T.I.11*a*).

The variety of possible carbohydrate structures is enormous, and only a few are discussed here in a general way (Fig. T.I.12*a*). The smallest carbohydrates, and the ones that serve as building blocks for the rest, are the monosaccharides. These compounds have aldehyde or ketone functional groups and multiple hydroxyl groups with an empirical formula of $(CH_2O)_n$. In biologically important monosaccharides, n ranges from 3 to 7. In solution, most carbohydrates with five or more carbons exist primarily as hemiacetal (or hemiketal) rings, formed by the reaction between the aldehyde (or ketone) group and a hydroxyl group (alcohol) (Fig. T.I.12*b*). This reaction results in the creation of a chiral center on the aldehyde or ketone carbon; the new stereoisomers are designated α (alpha) and β (beta). Polysaccharides with α and β linkages (Fig. T.I.12*c*) can have very different biological properties.

With all their functional groups, monosaccharides in aqueous solution undergo quite a number of biologically important chemical reactions, including phosphorylation, formation of amino derivatives, oxidation, reduction, and polymerization. Phosphorylations of glucose and other glycolytic intermediates are key steps in energy metabolism. Replacement of a hydroxyl group with an amine results in an amino sugar, and some of these, along with their N-acetylated derivatives, are important components of bacterial cell walls (e.g., glucosamine, *N*-acetyl glucosamine). The open-chain forms of aldehyde-containing monosaccharides are capable of reducing metal ions, such as Ag^+ and Cu^{2+}, leading to their designation as "reducing sugars".

Polysaccharides are used in energy storage and for structural purposes in virtually all living things. Variations in the sugar subunit(s) used, the type of bond formed, the polymer length, and the degree of branching generate a tremendous variety of

Fig. T.I.11.
(Continued)
on the identity of the head group and the length and degree of unsaturation of the fatty acid tails. (*b*) Archaeal phospholipids are different in that their hydrophobic tails are formed from isoprenyl groups and are bound to the glycerol through ether linkages. Interestingly, some Archaea also have what are known as tetraethers, which have glycerol and polar groups at both ends and a double-length middle. They span the whole membrane. Part (*c*) is a diagram of a typical biological membrane. Several kinds of membrane proteins are shown, including integral membrane proteins, a peripheral membrane protein, a glycoprotein, and some glycolipids. In the fluid mosaic model, the membrane proteins have considerable lateral mobility. Typically, the carbohydrate component of glycolipids and glycoproteins is located on the exterior of the cell. Depending on the membrane, up to 50% or more of its weight may be protein.

***(a)* Six-carbon monosaccharides**

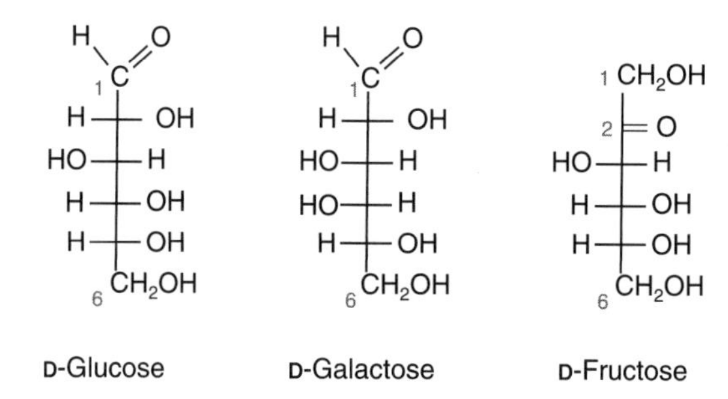

***(b)* The cyclized forms of glucose**

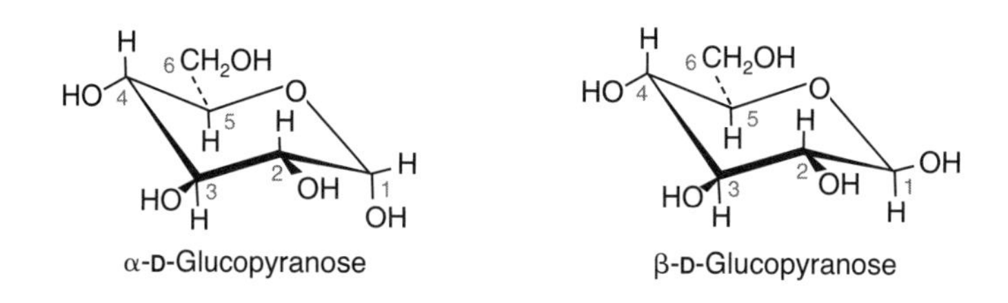

***(c)* Sample bonds that link monosaccharides**

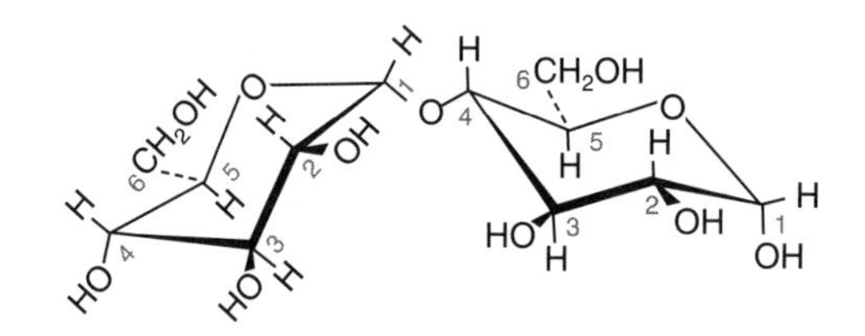

alpha(1-->4) linkage, found in glycogen, starch, etc.

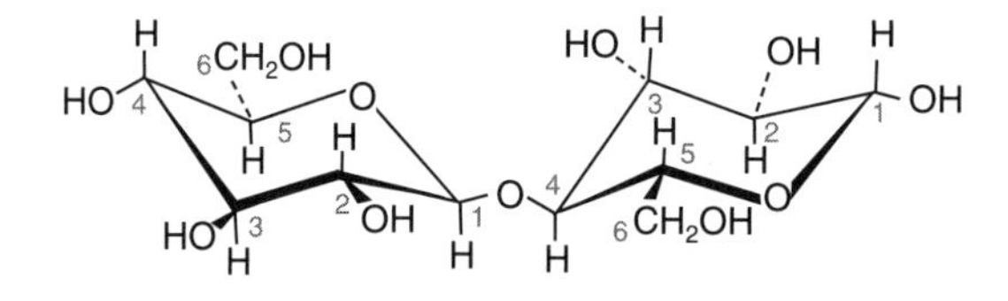

beta(1-->4) linkage, found in cellulose

Fig. T.I.12.
Carbohydrates. (*a*) Six-carbon monosaccharides are shown in their linear form. (*b*) The cyclized forms of glucose, most correctly refered to as glucopyranoses, are most common biologically. (*c*) Alpha- and beta-glycosidic bonds are shown. As described in the text, the end product is very different in alpha- and beta-linked glucose polymers.

substances with widely differing characteristics. As an example of this variety, we consider the difference between polysaccharides formed with the α and β stereoisomers of glucose. When the C1 of glucose is in the α configuration, the polymer produced has bond angles that allow it to fold up into a compact helical structure, ideal for a storage function. Thus, the storage carbohydrates glycogen and starch are branching or linear polymers of glucose with α linkages. On the other hand, if the C1 atoms of the glucose units are in the β configuration, the resulting polymer is cellulose—the main component of the plant cell wall and of wood, and the most abundant single substance in biomass. The β configuration favors a straight fiber held in place by extensive intrachain hydrogen bonding; using interchain hydrogen bonds, the fibers bundle together, resulting in strong and rigid structures. An additional property of the β linkages in cellulose is that they are indigestible by most organisms. Higher organisms that are able to utilize cellulose as food (e.g., ruminants and termites) can do so only because they harbor specific bacteria in their guts that contain the enzyme cellulase and are able to break down cellulose.

Other examples of structures in which carbohydrate polymers participate include: prokaryotic cell walls, which are cross-linked networks of carbohydrates and amino acids, and plant cell walls, which are mainly a polymer of glucose (cellulose). Animals lack cell walls, but the extracellular matrix, which connects and supports cells, is made up of glycoproteins (e.g., collagen, fibronectin, elastin, laminin), proteoglycans (molecules that are mainly carbohydrate, but have a small protein component), and complex carbohydrates. Another biologically important structural polysaccharide, chitin, is the main component of the exoskeleton of arthropods (insects, spiders, crustaceans, etc.) and is also made by many fungi.

Oligosaccharides also act in signaling processes. In plants, some oligosaccharides regulate the expression of genes that control growth and development as well as playing roles in two major physiological processes in plants—defense and nodulation. These signaling oligosaccharides have extremely complex structures. In animals, carbohydrate components serve as destination labels for some proteins, and play roles as mediators for specific cell–cell and cell–matrix interactions. These signaling components act in processes as varied as cell migration during development, blood clotting, and self–non-self recognition.

T.I.5. Metabolism

Throughout this tutorial, we have emphasized the great diversity of life and the underlying similarities that link all living things together in common ancestry. The pathways by which cells extract and store energy are good examples of this juxtaposition of unity and diversity. In broad outline, all known living things have similar bioenergetic pathways. Cells use a combination of processes that occur in the soluble fraction (glycolysis, fermentation, the citric acid cycle) and in the membrane fraction (respiration, photosynthesis) to convert energy available in the environment to forms that can be used by the cell. See Fig. T.I.13 for a schematic overview of these processes. The basic short-term storage forms for energy in all cells are the membrane potential (see below for more) and "high-energy" molecules, such as ATP. Reducing power is generally stored and transmitted in NADH (reduced NAD^+) and NADPH (reduced $NADP^+$) (Fig. T.I.7*a*).

The great variety of life becomes evident when one looks at the sources of energy used by various organisms (see Fig. T.I.3 and Chapter II). The nutritional requirements of an organism reflect its source of metabolic free energy. Thus, autotrophs can synthesize everything they need from small inorganic compounds (H_2O, CO_2,

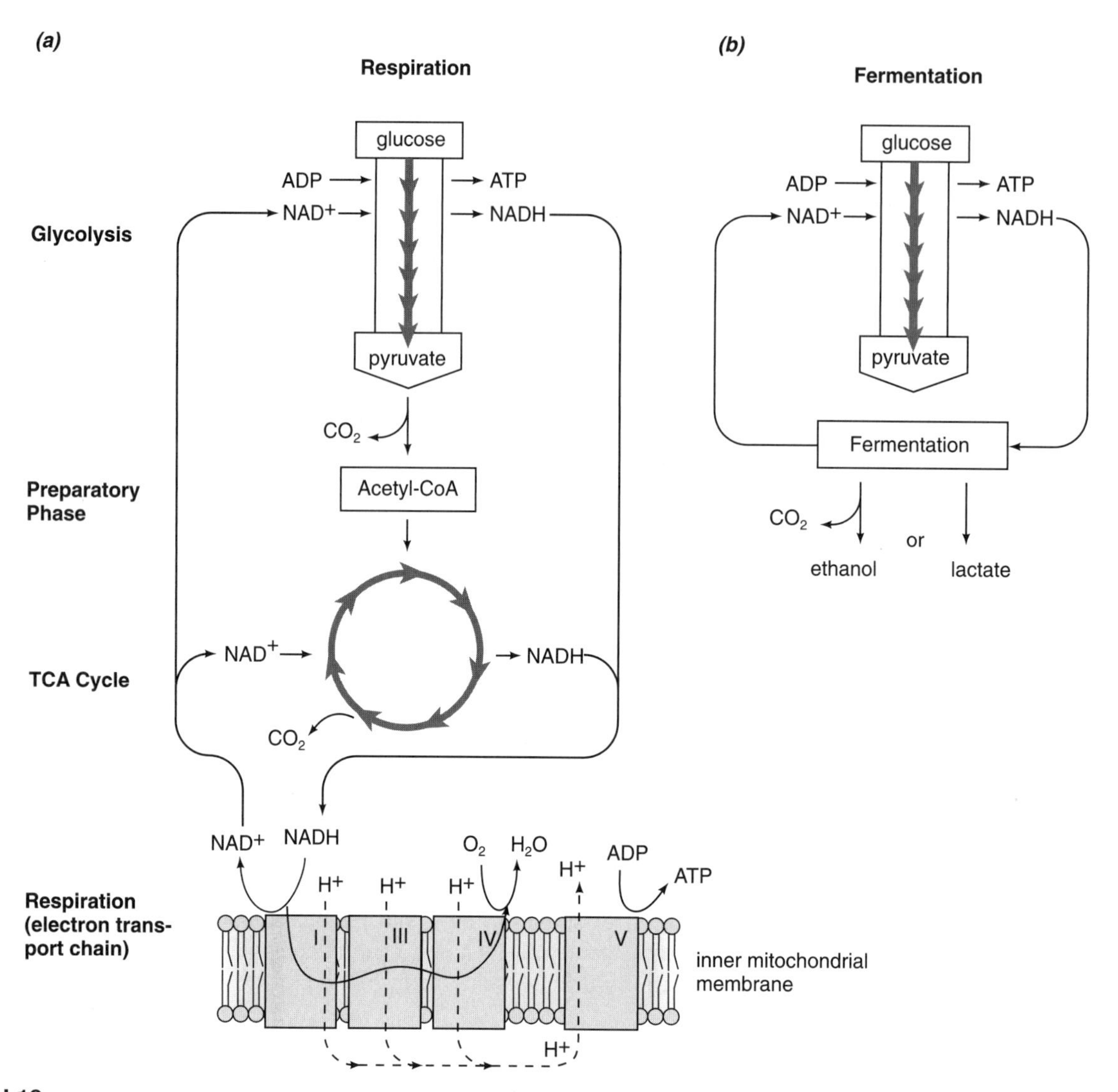

Fig. T.I.13.
Basic energy metabolism and carbon flow. (*a*) Respiration. The flow of carbon is shown in bold arrows and the flow of electrons is shown in thin lines. Glycolysis generates a limited amount of ATP and of reduced NADH as it converts glucose to the three-carbon compound pyruvate. The TCA cycle (tricarboxylic acid cycle, also called the Krebs cycle or the citric acid cycle) generates large amounts of reduced NADH. The NADH donates electrons to the electron-transport chain, where the energy is harvested by transfer through a series of redox centers to a terminal electron acceptor. In the diagram of the electron-transport chain, the thin solid line represents the flow of electrons, and dashed lines show the flow of protons. The latter are actively transported across the membrane by complexes I, III, and IV, forming a pH and charge gradient. Protons flow back through complex V, driving ATP synthesis. The bulk of cellular ATP in respiring organisms comes from this enzyme. Diagrammed here is respiration using oxygen as the terminal electron acceptor. Organisms that use other final acceptors may express a different kind of complex IV, but the overall electron-transport pathway is similar. (*b*) Fermentation. In organisms that ferment, energy is derived from glycolysis. Small carbon compounds are reduced to oxidize the NADH so that it can be reused. This results in the buildup of compounds such as ethanol or lactate. The specific end product depends on the organism and the conditions. (*c*) Carbon Flow. This figure shows the same pathways as in (*a*), but focused on the flow of carbon through the central metabolic pathways and TCA cycle. In a typical oxygen-respiring, non-photosynthesizing organism, nutrients in the form of proteins, fats, and carbohydrates are broken down and their component parts feed into the pathways at different points (catabolic pathways). Small building blocks are also taken out of the pathways to make the larger organic molecules in the cell (amino acids, nucleotides, fatty acids, and sugars), which are used to build the cell (anabolic pathways). This metabolic diagram is vastly simplified, but it gives an indication of how the anabolic and catabolic pathways are coupled and interrelated.

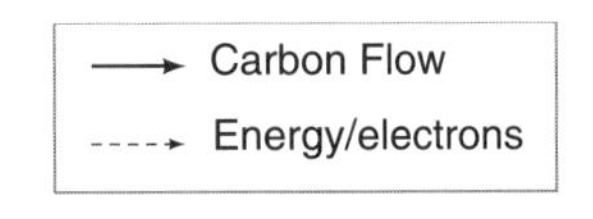

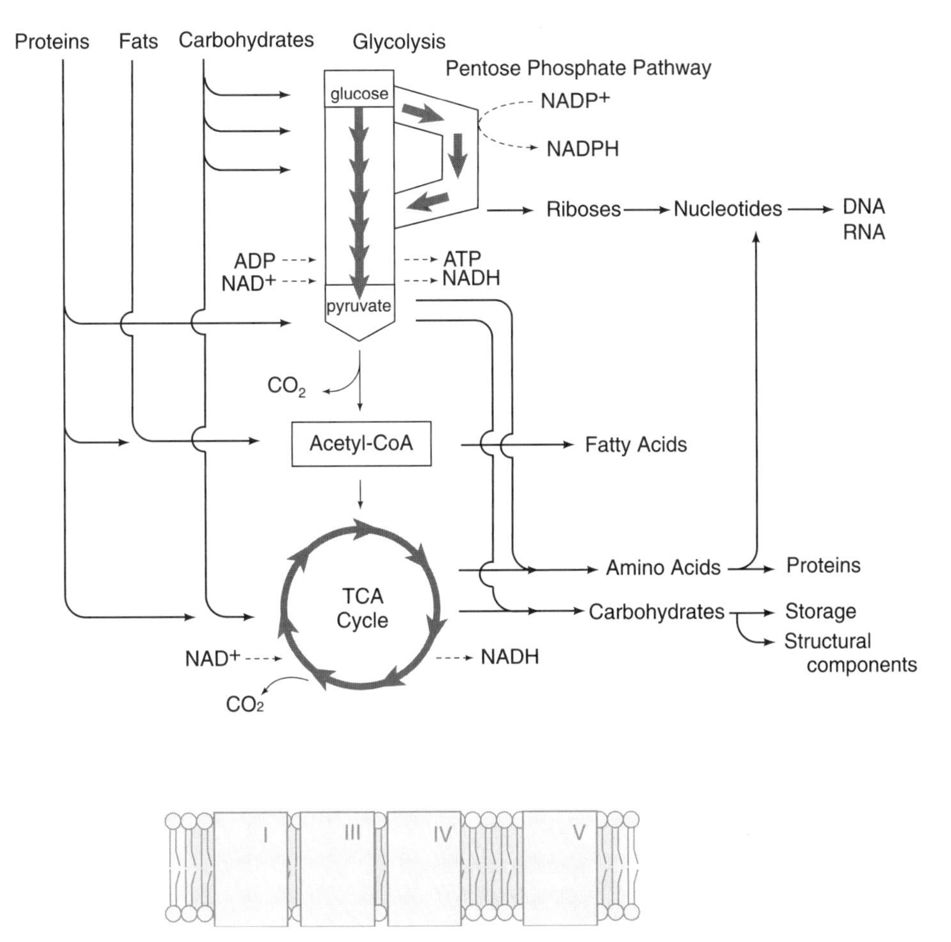

Fig. T.I.13.
(Continued)

H_2S, NO_3^-, NH_3). This group is subdivided into chemotrophs, which oxidize reduced inorganic compounds such as NH_3, H_2S, or Fe^{2+} to obtain energy, and photoautotrophs, photosynthetic organisms, which harness light energy to reduce CO_2 and make carbohydrates. Heterotrophs, on the other hand, require nutrients in the form of organic compounds (hydrocarbons, carbohydrates, lipids, proteins), which are oxidized or sometimes disproportionated to generate energy. Thus, the very existence of heterotrophs depends on the autotrophs (or some other source of organic compounds).

In the following sections, a brief overview is presented of the central components of pathways that living things (autotrophs and heterotrophs) use to derive energy from organic precursors (mainly glucose): glycolysis, the citric acid cycle, respiration, fermentation. Figure T.I.13 shows the relationships among these different processes. In addition, a brief overview of photosynthesis is presented, as an example of energy-harvesting pathways of autotrophs (Fig. T.I.14).

(a)

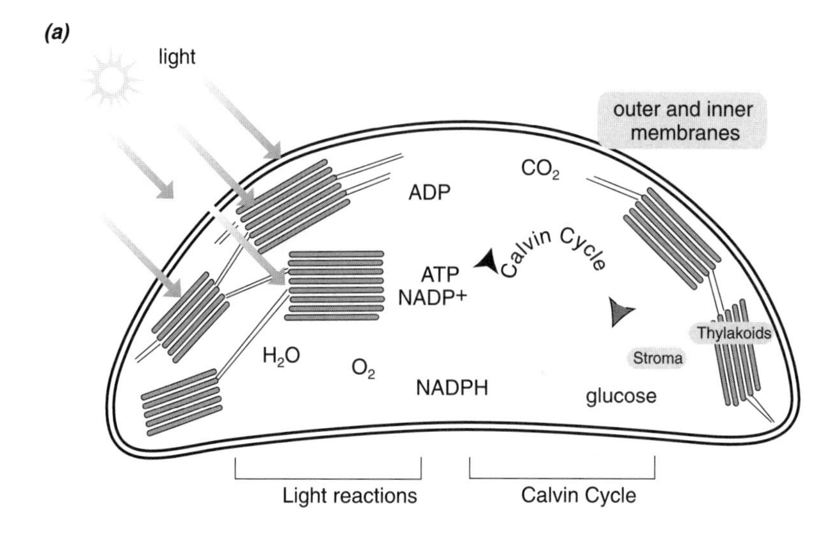

(b)

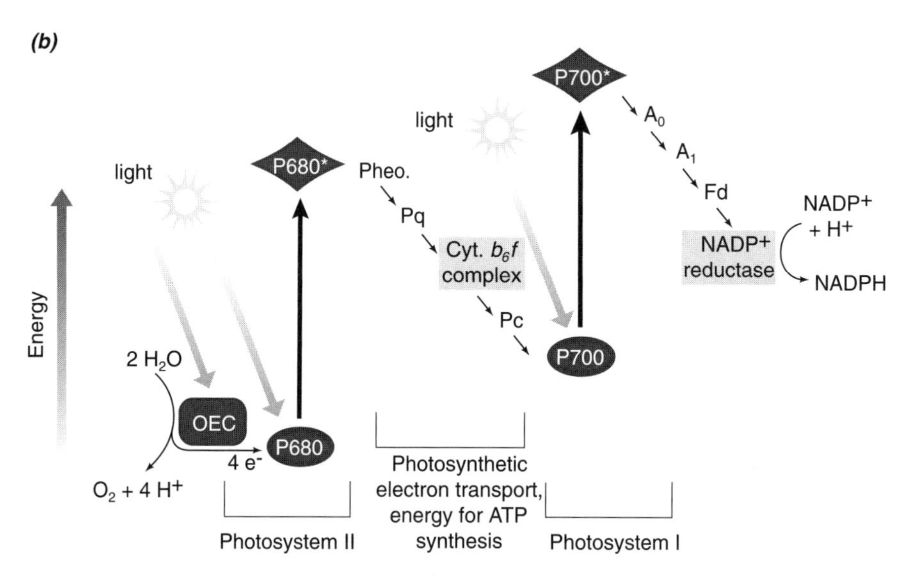

(c)

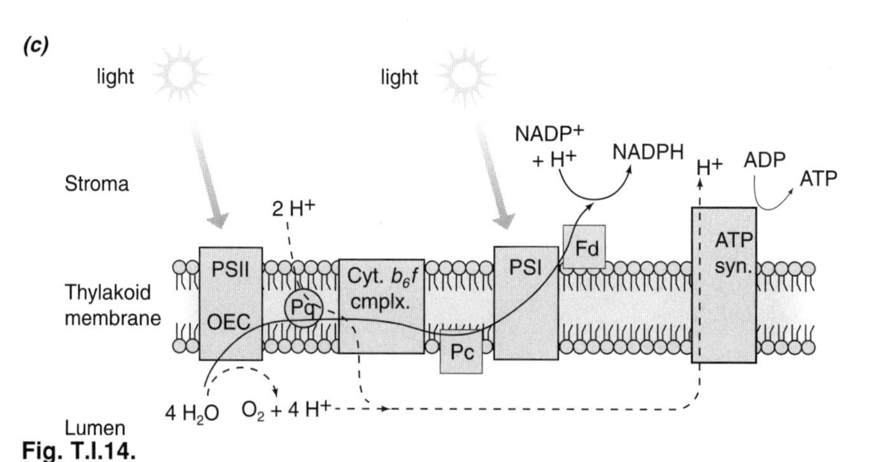

Fig. T.I.14.
Part (*a*) is a diagram of the chloroplast showing the major reactions of photosynthesis in green plants and their locations. Protein complexes located in the thylakoid membranes absorb photons and use their energy to generate ATP and NADPH. Since light is required, these processes are known as the light reactions. The ATP and NADPH thus generated are used in turn by the Calvin cycle, located in the stroma, to reduce CO_2 for synthesis of glucose.

T.I.5.1. Energy Storage

Energy-harnessing pathways result in the formation/regeneration of ATP from ADP or adenosine 5′-monophosphate (AMP). Energy stored in ATP is then used to drive a multitude of energy-requiring processes. Because of this critical role, the small diffusible ATP molecule is often called the common currency of cellular energy. The enzymatic coupling of ATP hydrolysis (to ADP or AMP) with thermodynamically unfavorable, but biologically necessary, reactions is what makes life possible. Cells also keep their internal environments more reduced than their surroundings. In the energy-harnessing pathways, two-electron reduction of NAD^+ to NADH and of $NADP^+$ ($NADP^+$ oxidized form) to NADPH occurs at several points. The NADH is used mainly to deliver electrons to the respiratory electron-transport chain, while NADPH is used in many biosynthetic reactions (see Fig. T.I.7*a* for their structures).

The generation of energized membranes is one of the basic processes in living systems. The most widespread example of a membrane potential is the electrochemical proton gradient generated by respiration. At the inner membrane of mitochondria and the plasma membrane of prokaryotes, the unidirectional transport of protons across the membrane in an outward direction results in an "energized"

Fig. T.I.14.
(Continued)
This process is also known as "fixing carbon" or the "dark reactions". (*b*) The "Z-Scheme" of photosynthesis, which operates in green plants and cyanobacteria. This figure is a schematic diagram of the energy of electrons as they flow from water through Photosystem II, the photosynthetic electron-transport chain, PhotosystemII, and ferredoxin to the final acceptor, $NADP^+$. Electron flow is represented by bold black arrows. These arrows represent the excitation of an electron when a photon is absorbed by the reaction center chlorophyll (P680 or P700). (*c*) A simplified schematic of the organization in the membrane of the protein complexes involved in photosynthesis. Dashed arrows show the path of protons that produce an electrochemical gradient across the membrane, which drives ATP synthesis by the chloroplast ATP synthase. While some protons are actually translocated across the membrane, much of the gradient is created because protons are released from water on one side and removed by reaction with $NADP^+$ on the other side of the thylakoid membrane. Solid arrows show the path of electrons from H_2O to $NADP^+$. More detail is given in Section X.3.

Abbreviations

PSII, Photosystem II

OEC, oxygen evolving complex, which contains Mn and is part of PSII

P680 and P680* the reaction center of PSII in the normal and excited states, respectively

Pheo, Pheophytin, the primary acceptor for the electron from P680*

Pq, plastoquinone

Cyt, cytochrome

Pc, Plastocyanin

PSI, Photosystem I

P700, the reaction center of PSI in the normal and excited states, respectively

A_0, the primary acceptor for the electron from P700*, probably a special chlorophyll

A_1, phylloquinone

Fd, ferredoxin

$NADP^+$ reductase, Fd

$NADP^+$ oxidoreductase

ATP syn, the chloroplast ATP synthase (analogous to the mitochondrial one)

membrane—more positively charged outside and a higher concentration of hydrogen ions (lower pH) outside. This energy can be quantified and is usually called the proton motive force, or Δp. It has a chemical and an electrical component. The chemical (or concentration) gradient exists because there is more H^+ on the outside, and the electrical gradient exists because there is more positive charge on the outside. The proclivity of the H^+ ions to return across the membrane (and collapse the gradient) is harnessed to do work. Much of the stored energy generated by respiration is used to drive ATP synthesis by the F_0F_1ATPase, but other reactions—for example, transport of ions, metabolites, or polypeptides, and the turning of the flagellar motor in bacteria—are also driven by an energized membrane. In some phototrophic bacteria, light energy is used to generate the membrane potential.

The ability of membranes to store energy as chemical and/or electrical gradients has led to specialized uses, particularly in eukaryotes. Assymmetric transport of other ions across membranes can generate other forms of membrane potentials; a good example is the sodium/potassium gradient in the plasma membrane of animal cells. The hydrolysis of ATP is used to drive a pump that transports Na^+ out of the cell and K^+ in, resulting in concentration gradients (in opposite directions) for Na^+ and K^+. A negative-inside electrical gradient is also formed because (1) K^+ leaks out faster than Na^+ leaks in, and (2) the interior counterions are large and cannot easily follow the K^+ out. In most cells, the membrane potential drives transport of various solutes, nutrients, and so on. In nerve cells, the Na^+/K^+ gradient is harnessed to carry nerve impulses. Rapid alteration of this membrane potential (depolarization followed by repolarization) carries the signal along the length of the nerve cell axon.

T.I.5.2. Glycolysis

For almost all organisms, the central metabolite is glucose, and, for most organisms the starting point for its use (or central metabolic pathway) is glycolysis. In this pathway, a set of enzymes works sequentially to convert glucose through a series of intermediates finally to form pyruvate. The intermediates serve as precursor metabolites to the other pathways, and the pathway generates ATP and NADH (Fig. T.I.13*a* and *c*). Glycolysis is truly central, as it is practically ubiquitous in living things. Two other related pathways exist. The pentose phosphate pathway shares some steps with glycolysis, is present in many of the same organisms as glycolysis, and provides precursors for nucleotide synthesis as well as generating NADPH for other biosynthetic reactions. A third central pathway, the Entner–Doudoroff pathway, is widespread among prokaryotes but practically nonexistent in eukaryotes and includes steps that enable these organisms efficiently to metabolize gluconate as well as glucose. These pathways are located in the cytoplasm.

T.I.5.3. Citric Acid Cycle

In respiring organisms, the pyruvate produced by glycolysis is oxidized to acetyl-coenzyme A (acetyl-CoA) and CO_2. Acetyl-CoA is fed into the citric acid cycle, also known as the Krebs cycle or the tricarboxylic acid (TCA) cycle, and the NADH formed in this process is used in respiration (Fig. T.I.13*b*). Catabolism of other metabolic fuels such as fats, proteins, and other carbohydrates also produces acetyl-CoA, so this cycle is important for recovering energy from a variety of sources. In nonrespiring organisms such as yeast and fermenting anaerobes, pyruvate oxidation is handled a little differently, to avoid the buildup of reduced NADH while still harvesting some energy (see later).

The citric acid cycle is present in most organisms that respire, whether aerobically or anaerobically. The first step is the transfer of the acetyl group of acetyl-CoA to oxaloacetate to form citrate. The subsequent steps are a series of oxidations resulting in regeneration of the oxaloacetate, release of two CO_2, and reduction of three NAD^+ to NADH per turn of the cycle. Citric acid cycle intermediates also function as starting materials for the biosynthesis of other cellular constituents. One ATP is produced, but the major bioenergetic product is NADH, which feeds electrons into the electron-transport chain for respiration. Most of the citric acid cycle enzymes are soluble. In prokaryotes, the enzymes are cytoplasmic, and in eukaryotes they are located in the matrix of mitochondria. One important TCA cycle enzyme, aconitase, is discussed in detail in Section VIII.4.

T.I.5.4. Respiration

The NADH formed by the citric acid cycle delivers electrons to the electron-transport chain, a group of membrane-bound protein complexes that carries out a series of electron-transfer reactions, passing the electrons through redox centers of progressively greater standard reduction potentials (affinity for electrons) (Fig. T.I.13*a*, bottom). The final step is reduction of a terminal substrate, generally oxygen, but other electron acceptors (NO_3^-, SO_4^{2-}, etc.) are used by some organisms. This process is called respiration, and the work that is done as the electrons move in the energetically favorable direction is the movement of protons across the membrane to the outside (in the energetically nonfavorable direction), resulting in gradients of both pH and electrical charge. In other words, the energy harvested from electron transfer is converted to and stored as a membrane potential. The tendency of the protons to flow back across the membrane is then used to power the F_0F_1ATPase, which is responsible for the synthesis of a large fraction of the cellular ATP in respiring organisms. In mammals, one glucose molecule yields two ATP molecules via glycolysis, and 36 via respiration. In other organisms, particularly if they are not using oxygen as the terminal electron acceptor, the number of ATP molecules derived from respiration may be lower.

From a human viewpoint, we generally think of respiration as consuming oxygen, and many organisms, including all higher organisms, do use molecular oxygen as the terminal electron acceptor, reducing it to water. However, many prokaryotic organisms, and a few eukaryotes, respire anaerobically using other molecules as the terminal electron acceptor—nitrate, sulfate, selenate (SeO_4^{2-}), Fe(III), and Mn(IV) are used by various organisms. As an example, organisms that use sulfate are responsible for the odorous accumulation of H_2S around some hot springs. Usually, these organisms are obligate anaerobes, meaning they only grow in the absence of oxygen. However, there are also facultative aerobes (or facultative anaerobes) such as the old standby experimental organism, *E. coli,* that respire using oxygen if it is available, but can switch to other terminal electron acceptors (nitrate or fumarate) if oxygen is absent. Different terminal oxidases are expressed, depending on the growth conditions.

Enzymes of the electron-transport chain complexes contain multiple metal ions at their active sites. A smooth flow of electrons requires centers that can be easily and reversibly reduced and oxidized, which is a characteristic of redox-active metal ions. The enzyme that carries out the four-electron reduction of O_2 to H_2O, cytochrome *c* oxidase, or Complex IV, contains two heme iron centers and two copper centers (one dinuclear); without these metal components the cytochrome oxidase would be unable to collect, store, and then use the four electrons required for the concerted reduction of O_2 to take place. Complexes I, II, and III contain heme iron and iron-sulfur clusters. These enzymes are located in the plasma membrane of prokaryotes and the mitochondrial inner membrane in eukaryotes. Respiration and the enzymes of the electron-transport chain are discussed in detail in Section X.3 (Photosynthesis and Respiration) and Sections X.3 and XI.6 (Cytochrome Oxidase).

T.I.5.5. Fermentation

If, for some reason, respiration is not possible, then some organisms and cell types are able to carry out fermentation (Fig. T.I.13*b*). In fermentative growth, energy is derived from glycolysis but not from the subsequent cycles, which are absent. Since respiration is not present, the NADH must be reoxidized by metabolites produced by the pathway. Although not nearly as much ATP per glucose is produced by glycolysis as by respiration, glycolysis can proceed very quickly, and organisms such as bread yeast (*S. cerevisiae*) can grow rapidly by glycolysis alone when sufficient glucose is present. Oxidation of NADH is accomplished by reducing the product of glycolysis (pyruvate) to produce fermentative end products, often, acctate, lactate, or, in yeast, ethanol, plus CO_2. Fermenting bacteria produce a variety of reduced organic acids and alcohols as fermentative end products, as well as CO_2 and H_2. Which end products are produced depends on the species of bacteria and the substrates available to it. Fermentation by brewers' yeast is responsible for the presence of alcohol in beer and wine, while fermentation by the related baker's yeast causes bread to rise.

Fermentation also occurs in active muscle cells to allow rapid regeneration of ATP when insufficient oxygen is available. In this case, lactic acid (lactate) is the product. The interconversion of pyruvate and lactate is reversible, so when oxygen becomes available again, pyruvate can be regenerated and respiration continued. In muscle, much of the lactate produced during high activity is carried by the blood to the liver where it is used to make glucose.

T.I.5.6. Photosynthesis

All living organisms can carry out glycolysis or a related pathway and either respiration or fermentation or both. Autotrophs have additional pathways that enable them to use the energy in simple reduced inorganic compounds or to capture the energy of light to synthesize glucose from CO_2. Photosynthesis is the best known and most widespread of these pathways. It is present in several genera of bacteria and in all algae and green plants. Photosynthesis can be oxygenic, where water is used as the source of electrons and molecular oxygen is produced, or anoxygenic, where other compounds (organic compounds, inorganic sulfur compounds, or hydrogen) are used as the source of electrons and oxygen is not produced. Anoxygenic photosynthesis is limited to bacteria, and only occurs when the organisms are growing anaerobically. Oxygenic photosynthesis is used by plants, green algae, and cyanobacteria (formerly known as blue-green algae). Fig. T.I.14 shows an overview of the major steps of photosynthesis in a chloroplast.

The basis of photosynthesis is the use of light to form an excited state in chlorophyll. The high-energy electron thus produced is passed down a series of carrier molecules, called the photosynthetic electron-transport chain, which is somewhat analogous to the respiratory electron-transport chain run in reverse. As in respiration, the process generates energy that is stored in a proton gradient across the membrane and this energy is used to synthesize ATP from ADP and inorganic phosphate (P_i). The electron carrier $NADP^+$/NADPH is used in photosynthesis instead of the NAD^+/NADH used in respiration—and, the reaction goes in the opposite direction. In photosynthesis, the oxidized version of the electron carrier ($NADP^+$) is the final electron acceptor and is reduced to NADPH. (In respiration, the NADH is the electron donor and is oxidized to NAD^+.) The electrons given to NADPH are replaced by electrons derived from water by the action of the oxygen evolving complex (OEC), which contains four Mn and one Ca atoms (see Section X.4); dioxygen and protons are released, and electrons move into the photosynthetic electron-transport chain. The NADPH provides reducing equivalents that are necessary to synthesize glucose from CO_2 (i.e., to fix carbon), in a process called the Calvin cycle. Although details of the Calvin cycle are beyond the scope of this chapter, it is important to remember that the Calvin cycle is critical to human life, as it provides the organic nutrients we all need to live.

All eukaryotic photosynthesizers and cyanobacteria use a system consisting of two photosystems working together. Photosystem II primarily oxidizes water to O_2, and Photosystem I primarily reduces $NADP^+$ to NADPH. These two systems are coupled via the cytochrome b_6f complex, a counterpart of the mitochondrial complex III (cytochrome bc_1), and must work in a concerted fashion. Non-cyanobacterial prokaryotic phototrophs use systems resembling one or the other of the eukaryotic photosystem, but not both. The eukaryotic–cyanobacterial system likely evolved from the simpler single photosystems.

In eukaryotes, photosynthesis takes place in chloroplasts, organelles with three membranes that most likely evolved through the endosymbiosis of cyanobacteria. The innermost membrane is organized in stacks of disks, called thylakoids. The thylakoid membranes hold Photosystems I and II and the rest of the photosynthetic electron-transport chain. The inner compartment is called the lumen, and is the location for the chlorophyll and other light absorbing pigments. The aqueous compartment outside the thylakoids but inside the two outer membranes is called the stroma. The Calvin cycle is located in the stroma, where mostly soluble enzymes synthesize glucose from carbon dioxide, using the NADPH produced by photosynthesis to supply the necessary reducing power and ATP for energy. Simplified schemes for eukaryotic photosynthesis and the two kinds of prokaryotic photosynthesis are shown in Fig. T.I.14. The inorganic biochemistry of photosynthesis is considered in detail in Section X.3, Photosynthesis and Respiration, and Section X.4, Dioxygen Production: Photosystem II.

Bibliography

Additional Reading and Reference Material

For basic biochemistry any college level textbook will be useful. Most are oriented toward eukaryotic systems in general and the human in particular. The one listed here is a good one. Most publishers now have supplementary material accessible on the web that can be particularly useful as they usually include animations and movies.

D. L. Nelson and M. M. Cox, *Lehninger Principles of Biochemistry,* 3rd ed., Worth Publishers, New York, 2000. A fourth edition is in preparation, and supplemental material is available on the web at http://www.worthpublishers.com/lehninger/ (for the 3rd ed.).

For the biochemistry of prokaryotic organisms, there are many microbiology textbooks available, but we find the following two particularly useful for chemists and biochemists. The first one includes a lot of molecular detail and is organized by topic, rather than organism; the second, a new edition of a classic in the field, is organized by organism and includes treatment of eukaryotic microorganisms.

D. White, *The Physiology and Biochemistry of Prokaryotes,* 2nd ed., Oxford University Press, New York, 2000.

M. T. Madigan, J. Martinko, and J. Parker, *Brock Biology of Microorganisms,* 10th ed., Prentice Hall, 2002.

For further reading on eukaryotic cell biology and organelles, we particularly like

H. F. Lodish, A. Berk, P. Matsudaira, C. A. Kaiser, M. Krieger, M. P. Scott, S. Lawrence Zipursky, and J. Darnell, *Molecular Cell Biology,* 5th ed., W.H. Freeman & Company, New York, 2003.

And for general biology, we like the following, although there are many good alternatives.

N. A. Campbell and J. B. Reece, *Biology,* 6th ed., Pearson Higher Education, 2001.

A good introduction to protein structure is found in either of the two following books.

C. Branden and J. Tooze, *Introduction to Protein Structure,* 2nd ed., Garland Publishing, New York and London, 1999.

G. A. Petsko and D. Ringe, *Protein Structure and Function,* Sinauer Associates, Sunderland MA, 2003.

For further reading on bioinformatics, genomics, and sequence analysis the following books will probably prove useful.

R. Durbin, S. R. Eddy, A. Krogh, and G. Mitchison, *Biological Sequence Analysis: Probabilistic Models of Proteins and Nucleic Acids,* Cambridge University Press, Cambridge and New York, 1999.

D. W. Mount, *Bioinformatics: Sequence and Genome Analysis,* Cold Spring Harbor Laboratory Press, Cold Spring Harbor, New York, 2001. The latter has an associated web site http://www.bioinformaticsonline.org/

Web-Only Resources

There are three major depositories of DNA and protein sequences. They maintain extensive websites that are freely accessible to the public for data retrieval. They also provide free software tools for sequence analysis.

GenBank database at the National Center for Biotechnology Information (NCBI), USA:
http://www.ncbi.nlm.nih.gov/

European Molecular Biology Laboratory Nucleotide Sequence Database (known as EMBL-Bank) maintained by European Bioinformatics Institute:
http://www.ebi.ac.uk/

DNA Data Bank of Japan (DDBJ):
http://www.ddbj.nig.ac.jp/Welcome.html

And finally, for beginners interested in an introduction to genetics and molecular biology we recommend

The Dolan DNA Learning Center (Cold Spring Harbor Laboratory) which can be accessed at
http://www.dnalc.org/ and http://www.dnaftb.org/dnaftb/

Fundamentals of Coordination Chemistry

Tutorial II

James A. Roe
Department of Chemistry and Biochemistry
Loyola Marymount University
Los Angeles, CA 90045

Bryan F. Shaw and Joan Selverstone Valentine
Department of Chemistry and Biochemistry
UCLA
Los Angeles, CA 90095

Contents

T.II.1. Introduction

The study of "coordination complexes", molecules or ions formed from metal ions bound to water or other ligands, is called coordination chemistry. In a coordination complex, the properties of both the metal and the ligand are altered by the complexation (ligation) process. The study of metalloenzymes, metallodrugs, and many other aspects of biological inorganic chemistry relies on a knowledge of coordination complexes.

T.II.2. Complexation Equilibria in Water

A cation dissolved in aqueous solution is bound by water molecules. The dipolar water molecules orient themselves so as to point the negative end (and a lone pair of electrons) toward the center of the positive charge. Cations derived from metallic elements, such as Na^+, Ca^{2+}, or Fe^{2+}, are surrounded in aqueous solution by a shell of water molecules. The result is called a *coordination complex,* in which water molecules are the *ligands,* which are coordinated directly to the metal ion, forming its first *coordination sphere.* In addition to or in place of water molecules, other ligands can also coordinate to metal ions. Many large and small molecules found in living systems have high affinities for metal ions. So-called "free metal ions" are generally considered those bound only to water molecules (or water molecules plus hydroxide or oxide, depending on the pH). Complexed metal ions are bound to other ligands, including small biological ligands such as amino acids or macromolecules such as proteins, nucleic acids, and so on. The concentrations of virtually all metal ions are very tightly controlled and regulated in living organisms in a process called metal ion *homeostasis.*

For convenience, water ligands are often left out of chemical formulas or equations, but they should never be forgotten. For example, the equilibrium expression for complexation of an ammonia molecule to a metal ion is frequently written as

$$M^{x+} + NH_3 \rightleftarrows M(NH_3)^{x+}$$

$$K_f = [M(NH_3)^{x+}]/[M^{x+}][NH_3]$$

where K_f is the equilibrium constant for the formation of the monoammonia complex of the metal. However, a fuller description of what is actually happening includes the water ligands and is given by

$$M(H_2O)_y{}^{x+} + NH_3 \rightleftarrows M(H_2O)_{y-1}(NH_3)^{x+} + H_2O$$

$$K_f = [M(H_2O)_{y-1}(NH_3)^{x+}][H_2O]/[M(H_2O)_y{}^{x+}][NH_3]$$

Recent studies suggest that concentrations of "free" metal ions are extremely low in normal living systems, which allows us to conclude that the other ligands present have much greater affinities than water and clearly outcompete water molecules as ligands, despite the fact that the water molecules are in great excess.

The *formation constant* K_f (also known as a stability constant) measures the ability of a ligand to bind a metal ion in place of water. Thus, a small or large K_f does not necessarily mean that the ligand is a weak or strong metal ion binder per se, but that it binds weakly or strongly compared to water.

When metals can undergo multiple ligand substitutions, the *overall formation constant* (β) is the product of the stepwise formation constants:

$$M + L \rightleftarrows ML \qquad K_1 = [ML]/[M][L]$$

$$ML + L \rightleftarrows ML_2 \qquad K_2 = [ML_2]/[ML][L]$$

$$ML_2 + L \rightleftarrows ML_3 \qquad K_3 = [ML_3]/[ML_2][L]$$

$\beta_3 = [ML_3]/[M][L]^3$ and it can be shown arithmetically that

$$\beta_3 = K_1K_2K_3$$

The numerical value of a formation constant for metal complexes involving biological ligands sometimes can help explain why living systems employ certain metals for one function and other metal ions for others.

One way that metal complexes are stabilized is through the *chelate effect.* A *chelating ligand* is a molecule that simultaneously binds a metal with more than one donor atom. This simultaneous binding produces a complex that is more stable than an analogous compound that binds independent ligands. For example, ethylenediamine (en) is a bidentate ligand with two amine donor groups (it is called *bidentate,* two teeth). When Cu^{2+} binds en, a more stable complex is produced than when Cu^{2+} binds two individual ammonia groups (Fig. T.II.1):

$$[Cu(H_2O)_6]^{2+} + 2\,NH_3 \rightleftarrows [Cu(H_2O)_4(NH_3)_2]^{2+} + 2\,H_2O$$

$$\beta_2 = K_1K_2 = 5.0 \times 10^7$$

$$\Delta H = -46\,\text{kJ mol}^{-1}$$

$$\Delta S = -8.4\,\text{J K}^{-1}\,\text{mol}^{-1}$$

$$[Cu(H_2O)_6]^{2+} + \text{en} \rightleftarrows [Cu(H_2O)_4(\text{en})]^{2+} + 2\,H_2O$$

$$K = 3.9 \times 10^{10}$$

$$\Delta H = -54\,\text{kJ mol}^{-1}$$

$$\Delta S = +23\,\text{J K}^{-1}\,\text{mol}^{-1}$$

An explanation for the chelate effect is found in the thermodynamic parameters for these two reactions. The reaction enthalpies, ΔH, are fairly similar, because two Cu–N bonds are formed in each case, but the reaction entropies (ΔS) differ greatly. The difference is understood in terms of the change in the total number of molecules in each reaction. In the ammonia reaction, there is no net change in the total number of molecules: two ammonia molecules become coordinated to the copper ion releasing two water molecules. In contrast, for the reaction with en, the net number of molecules increases: one en molecule becomes coordinated to the copper, causing the release of two water molecules. Consequently, the disorder, or entropy, of the system increases more in the case of the en reaction. This entropy difference makes the en reaction thermodynamically more favorable, accounting for much of the chelate effect.

Many metalloproteins and other biomolecules are capable of functioning as chelating ligands. Polypeptide chains can twist and fold around metal ions to act as large multidentate ligands (see Chapter III). Other metals are chelated by porphyrins

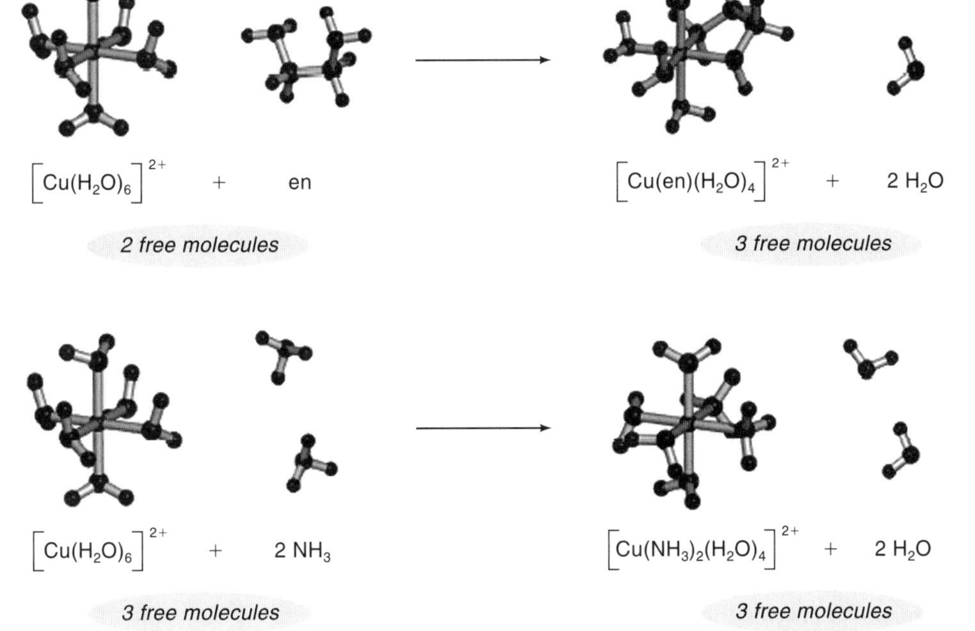

Fig. T.II.1. The "chelate effect": Binding of the bidentate ligand ethylenediamine (en or $NH_2CH_2CH_2NH_2$) to $[Cu(H_2O)_6]^{2+}$ is entropically favored over the binding of two ammonia ligands, because the binding results in an increase in the total number of molecules.

or corrins, cyclic tetradentate macrocyclic ligands or dithiolenes (see Chapter IV), as well as by protein ligands. The chelate effect clearly has a great impact on biological coordination chemistry.

T.II.3. The Effect of Metal Ions on the pK_a of Ligands

When a protic ligand is bound to a metal ion, the ligand generally becomes more acidic, because the positively charged metal ion stabilizes the anionic conjugate base of the ligand. For example, the enzyme carbonic anhydrase contains a Zn^{2+} ion at its active site with a 'water' molecule bound to it. A free water molecule in bulk water has a pK_a of 15.7. Binding of the water molecule to the zinc ion in carbonic anhydrase lowers that pK_a to ~7. The Zn^{2+} bound aqua ligand is therefore deprotonated to a significant extent at physiological pH, giving a Zn^{2+}–hydroxo ligand complex. The hydroxo group acts as a nucleophile and attacks CO_2 to form HCO_3^- in the enzymatic mechanism.

The alkali and alkaline earth metal ions (e.g., Na^+, K^+, Mg^{2+}, and Ca^{2+}) have fixed oxidation states, but those of the transition metals can vary. The highest oxidation state metal ions tend to decrease ligand pK_a values the most because this higher positive charge stabilizes the conjugate base to a greater degree. Sometimes both protons dissociate from the aqua ligand to form mononuclear oxo, O^{2-}, complexes. In the case of molybdenum and tungsten, for example, stable Mo=O and W=O complexes are quite common in the oxidation state VI (see Chapters XII.6 and XII.7). For other metal ions, the oxo ligands may form bridges between two or more metal ions. The ions Fe^{3+} and Al^{3+}, for example, form hydroxo- and oxo- bridges even at physiological pH. Oxo- and hydroxo-bridged Fe^{3+} clusters are a frequent motif in iron-containing proteins (see Chapters VIII and XI).

T.II.4. Ligand Specificity: Hard versus Soft

The Brønsted–Lowry theory of acid–base equilibria recognizes one acid, the proton (H^+), and defines a base as a species that bonds with, or accepts an H^+. In the more general Lewis theory, any species that forms a bond by accepting an electron pair is a *Lewis acid* and the species that donates the electron pair is a *Lewis base*. Lewis acids and bases need not be charged species. The lone electron pairs on water, nitrogen, and ammonia render these species Lewis bases and empty *p* atomic orbitals in trivalent boron compounds can accept two electrons, making them Lewis acids. The Brønsted–Lowry model is a specific example of the more general Lewis model, since the Brønsted acid, H^+ is also a Lewis electron-pair acceptor.

Trends in the thermodynamic stability of complexes formed between Lewis acids and Lewis bases (by determining the equilibrium formation constant, K_f) have led to the classification scheme commonly called "hard and soft" acids and bases (HSAB). The halides are a convenient set of Lewis bases (ligands) that can be used to gauge the relative hardness or softness of a particular Lewis acid. Fluoride is the hardest base while iodide is the softest. Hard acids, for example, form halide complexes with increasing K_f in the order:

$$I^- < Br^- < Cl^- < F^-$$

whereas soft acids demonstrate the opposite trend:

$$F^- < Cl^- < Br^- < I^-$$

In this scheme, a *hard acid* reacts with a *hard base* to form a more stable complex than that formed with a soft base.

Soft acids and bases are usually larger and more polarizable than hard acids and bases, with hard acids and bases being smaller, and less polarizable. For species that can have multiple oxidation states, the lower oxidation state is softer than the higher oxidation state. For example, cuprous ion (Cu^+) having 10 valence electrons ($3d^{10}$) has a larger ionic radius and a lower charge than cupric ion (Cu^{2+}, $3d^9$), and cuprous is therefore softer than cupric ion. Relative to cupric, cuprous ion therefore prefers to bond to softer Lewis bases.

Hard acids and bases bond differently to each other when compared to soft acids and bases. Since hard acids are more compact, with smaller ionic radii, and are less polarizable, their bonds have greater electrostatic character, with substantially greater electron density from the bond remaining on the donor atom. Soft acids are larger and more polarizable. Their bonding is more covalent in nature, and there is more sharing of the electron density of the bond between the acid and the base.

Many biological acid–base pairs can be recognized from Fig. T.II.2, and the HSAB scheme partly explains their presence in living organisms. For example the hard acid Mg^{2+} commonly stabilizes hard bases, including adenosine diphosphate (ADP) and transfer ribonucleic acid (tRNA) via its strong interaction with phosphate groups. The Ca^{2+} ion forms such stable complexes with carbonate and phosphate that these are often incorporated in to the solid-state structures of bones, teeth, and shells (see Chapter VI).

An important example of the application of HSAB theory in biology is the complexation of different metal ions with amino acids and peptides. Amino acids, either monomeric or linked to other amino acids, such as di-, tri-, or polypeptides (or proteins), can bind metals by a variety of modes (see Chapter III). Monomeric amino acids have amine (RNH_3^+) and carboxylate ($RCOO^-$) functional groups, hard bases that therefore bind hard acids such as Na^+, Mn^{2+}, Mg^{2+}, and Ca^{2+} in preference to softer acids such as Cu^+, Cu^{2+}, Fe^{2+}, and Zn^{2+}. When amino acids are condensed or polymerized to form polypeptides or proteins, however, the amine and carboxylate groups are converted into amides (RNHCOR′). Amide functional groups are not particularly good ligands for metal ions and, moreover, they tend to be tied up in the hydrogen bonds that maintain the α-helical, β-sheet, and loop secondary structures commonly observed in proteins. As a result, metal binding to proteins often (but not always) occurs with the side chains of the amino acid residues. Nitrogenous residues (e.g., histidine) and sulfur ligands (e.g., cysteine and methionine) provide softer ligands that are more suitable for bonding with Cu^+, Co^{2+}, Ni^{2+}, Fe^{2+}, and Zn^{2+}. Amino acids with harder side chains such as aspartate and glutamate ($RCOO^-$), serine and tyrosine (ROH), have oxygen donor atoms that preferably bind to harder metal ions, such as Mg^{2+} and Fe^{3+}.

Hard acids

H^+, Li^+, Na^+, K^+
Mg^{2+}, Ca^{2+}, Mn^{2+}
Al^{3+}, Cr^{3+}, Co^{3+}, Fe^{3+}

Hard bases

F^-, Cl^-, H_2O, OH^-, O^{2-}
$RCOO^-$, ROH, RO^-, R_2O, phenolate
NO_3^-, ClO_4^-, CO_3^{2-}, SO_4^{2-}, PO_4^{3-}
NH_3, RNH_2

Borderline acids

Fe^{2+}, Co^{2+}, Ni^{2+}, Cu^{2+}, Zn^{2+}
Rh^{3+}, Ir^{3+}, Ru^{3+}
Sn^{2+}, Pb^{2+}

Borderline bases

Br^-, NO_2^-, N_3^-
SO_3^{2-}
pyridine, imidazole

Soft acids

Cu^+, Ag^+, Au^+, Cd^{2+},
Hg^{2+}, Pd^{2+}, Pt^{3+}

Soft bases

I^-, H_2S, HS^-, S^{2-}, RSH, RS^-, R_2S
CN^-, CO, R_3P

Fig. T.II.2. The HSAB classification of some common Lewis acids and bases.

T.II.5. Coordination Chemistry and Ligand-Field Theory

Coordination complexes of transition metal ions are often found to be six or four coordinate (i.e., bound to six or four donor ligands), but other coordination numbers are observed as well. The properties of coordination complexes are dependent on the identities of the ligands and on the coordination numbers and geometries of the complexes. Ligand-field theory (originally an extension of the crystal-field theory) provides a useful way to describe the effect of ligand binding on the properties of transition metal ions. In the simplest form of this theory, the ligand electron pairs that point toward the metal ion are modeled as point negative charges residing on the binding axes. As the ligands are brought progressively closer to the metal ions along the binding axes, they begin to interact with the five *d* orbitals of the metal center (Fig. T.II.3). The ligands raise the energy of each of the *d* orbitals because of repulsive interactions. However, the relative destabilization of each of the five *d* orbitals depends on the coordination number and geometry of the coordination complex.

T.II.5.1. The Octahedral Field

We first consider the case of an ML_6^{n+} complex, as depicted in Fig. T.II.4*b*, where the ligands are arranged in a six-coordinate octahedral geometry around the metal ion. Initially, before the ligands move close to the metal ion, all five *d* orbitals are equivalent in energy, as is the case of a gas-phase transition metal ion with no ligands

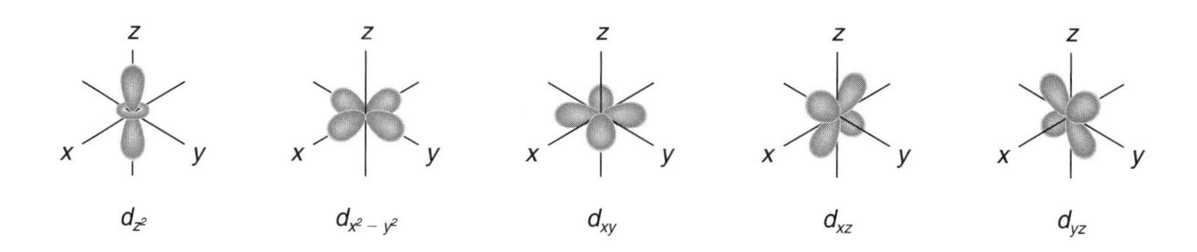

Fig. T.II.3.
Atomic *d* orbitals.

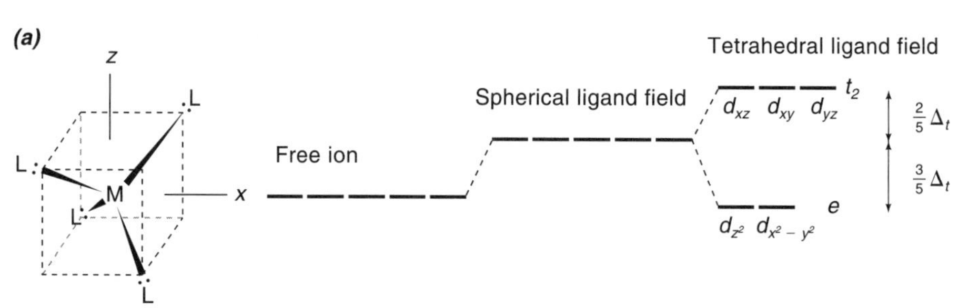

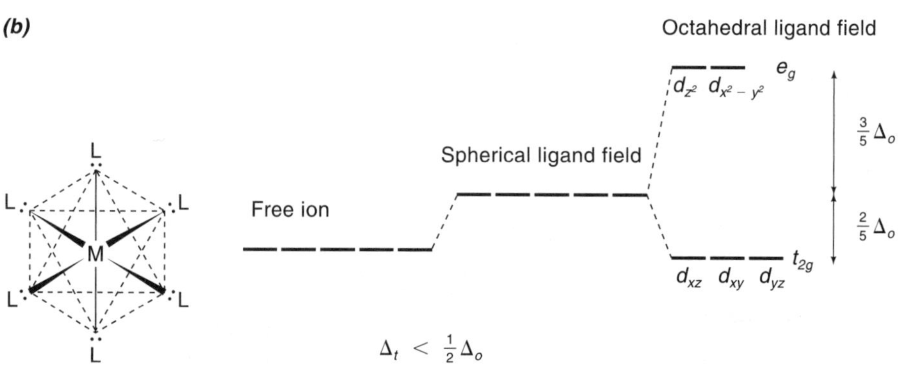

Fig. T.II.4.
Energy diagram showing the splitting of metal *d* orbital energies in (*a*) a tetrahedral ligand field and (*b*) an octahedral ligand field.

around it. As the six negative charges representing the six ligands approach the metal ion, the energies of the *d* orbitals are raised. If the charges were uniformly spread over the surface of a sphere instead of being localized along the axes, the effect would be to raise the energies of each of the *d* orbitals by the same amount. Instead, the charges are localized on the *x, y,* and *z* axes, and, consequently, they have more effect on the d_{z^2} and the $d_{x^2-y^2}$ orbitals than on the d_{xy}, d_{xz}, and d_{yz} orbitals because they are closer to the former than to the latter (see Fig. T.II.4).

The d_{z^2} and the $d_{x^2-y^2}$ comprise the upper pair of orbitals and are referred to as the e_g orbitals. The lower set of three orbitals (d_{xy}, d_{xz}, and d_{yz}) is referred to as the t_{2g} orbitals. The g stands for *gerade* (meaning even) and indicates that the orbitals have inversion symmetry. The two levels, t_{2g} and e_g, are separated by an amount, Δ_o, which is known as the ligand-field splitting parameter. The "o" indicates that this parameter belongs to an octahedral field. The magnitude of Δ_o depends on the identity of the metal ion, its charge, and the nature of the ligands. The splitting energy often corresponds to the visible region of the spectrum, and consequently, transition metal complexes with partly filled *d* orbital shells are often colored, absorbing visible light due to *d–d* transitions.

T.II.5.2. The Tetrahedral Field

We now consider the case of an $ML_4{}^{n+}$ complex where the ligands are arranged in a tetrahedral geometry around the metal ion. A useful way to visualize this geometry is depicted in Fig. T.II.4*a,* in which the ligands are placed at the opposite corners of a cube that is defined by the *x, y,* and *z* coordinate system. The tetrahedral field splits the *d* orbital energy levels into two sets of equivalent energy. In this case, however, the *d* orbitals that lie closer to the *x, y,* and *z* axes, namely, d_{z^2} and $d_{x^2-y^2}$, are lower in energy. These are referred to as the *e* orbitals. The d_{xy}, d_{xz}, and d_{yz} orbitals are higher in energy because they interact more strongly with the ligands, which lie off-axes in the tetrahedral field. They are referred to as the t_2 orbitals.

Notice that for the tetrahedral field there is no "g" associated with either level (i.e., t_2 instead of t_{2g} and e instead of e_g), because there is no center of inversion in a tetrahedral metal ion complex. This difference between the symmetry of the octahedral field and that of the tetrahedral is significant in the spectroscopy of metal ions with partially filled *d* orbitals. For the tetrahedral field we can again identify a crystal-field splitting parameter, Δ_t, whose magnitude is a function of the identity of the metal ion, as well as its charge and the nature of the ligand.

It is valuable to compare the magnitude of Δ_o with that of Δ_t for two complexes, one six coordinate and one four coordinate, in which the metal ions, the ligands, and the M–L bond lengths are the same. Intuition suggests that Δ_t should be $<\Delta_o$ simply because it is caused by interaction with four rather than six ligands. This finding is indeed the case, and it can be shown that in the electrostatic model Δ_t is, in fact, four-ninths the value of Δ_o, when all else is equal.

T.II.5.3. Two Other Cases: Axially Distorted Octahedral and Square-Planar Fields

Two other commonly encountered crystal (ligand) fields are the axially distorted octahedron and the square plane, which are related. In the first case, the octahedral complex is distorted either by "compression" in which the axial ligands along the *z* axis are closer to the metal ion, or by "elongation" in which the axial ligands are farther from the metal than those that lie in the *xy* plane. Either distortion affects most strongly the energy of those *d* orbitals that have a *z* component, namely, d_{z^2}, d_{xz}, and d_{yz}. Compression raises both the d_{xz} and d_{yz} equally, and the d_{z^2} even more. Elongation has the opposite effect, lowering the d_{z^2}, d_{xz}, and d_{yz} and leaving the $d_{x^2-y^2}$ as

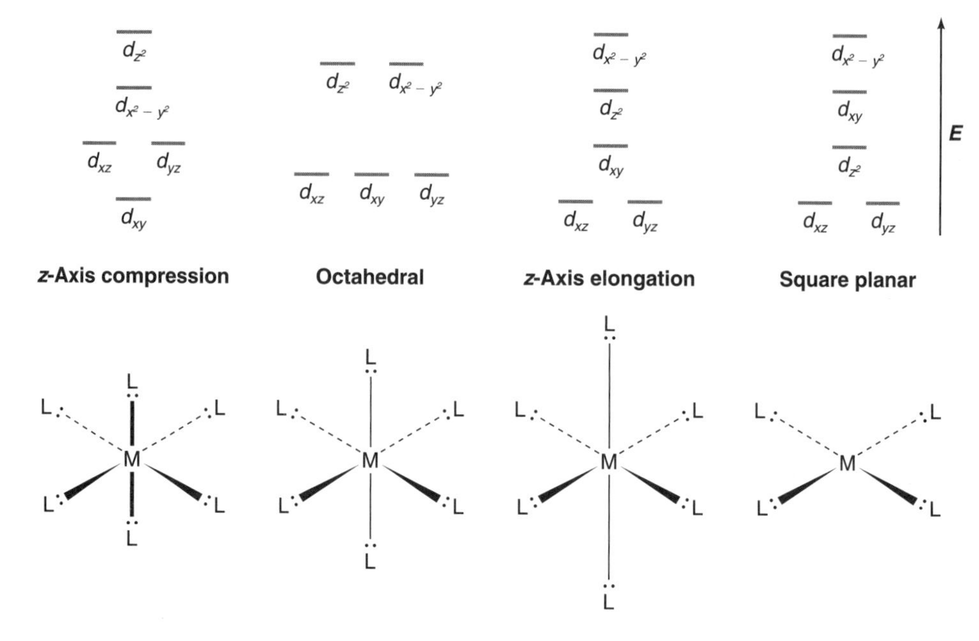

Fig. T.II.5. Energy diagrams depicting the effect of axial bond compression and elongation on *d* orbital energies in ML_6 octahedral complexes.

the highest orbital. The ordering of the energy levels for both cases is depicted in Fig. T.II.5. This type of distortion is also known as tetragonal.

As the axial ligands are pulled farther and farther away from the central ion, the orbitals containing a *z* component become lower in energy. When the axial ligands are removed altogether, the resulting field is called square planar, where the four ligands are placed equidistantly from the central metal ion at the corners of a square in a plane that includes the central ion. The arrangement of the ligands around the central metal ion, as well as the ordering of the energy levels, is illustrated in Fig. T.II.5.

It is important to note that regular geometries of ligation around transition metal ions are rarely observed in metalloenzymes and other metalloproteins. In the tug of war to optimize folded protein structure and the metal ion binding to the folded protein, the tendency of the metal ion to adopt a regular geometry rarely wins out. The resultant irregular coordination geometries often have interesting spectroscopic, magnetic, and reactivity properties.

T.II.5.4. Ligand-Field Splitting: The Spectrochemical Series

For a given ligand geometry and specific metal ion, the crystal-field parameter, Δ, is dependent on the nature of the ligand. For example, coordination by the ligands CN^- or CO always leads to a large Δ relative to coordination by halide ligands such as I^- or Br^-. Measurements on many complexes reveal an ordering of common ligands in terms of their relative splittings. This ordering is known as the spectrochemical series and includes the following:

$$CN^- \sim CO > NO^{2-} > \text{bpy} > \text{en} > NH_3 > \text{edta}^{4-}$$
$$> NCS^- > H_2O > OH^- > F^- > Cl^- > Br^- > I^-$$

(bpy = 2,2′-bipyridine, and edta = ethylenediaminetetraacetate). The ligands toward the high end of the series are known as *strong-field ligands,* and those on the low end as *weak-field ligands*. Strong-field ligands typically result in larger orbital splittings than do weak-field ligands.

T.II.6. Consequences of Ligand-Field Theory

The coordination geometry and the nature of the ligands determine the magnitude and complexity of the *d* orbital splitting in a transition metal ion. The splitting in turn contributes to the geometric, spectroscopic, and magnetic properties of the complex, as well as to its stability.

T.II.6.1. Electronic Absorption Spectroscopy

The splitting of the energy levels of the *d* orbitals has major effects on the ultraviolet–visible–near-infrared (UV–vis–NIR) electronic absorption spectra of transition metal ion complexes. The energy of spectral bands due to transitions between *d* orbitals is strongly influenced by the splitting constant Δ, while the number of bands is determined (generally) by the number of levels that are formed. An undistorted regular octahedral complex that has one electron residing in the t_{2g} energy level (an ion in a $3d^1$ configuration) would have only a single *d–d* electronic absorption transition available, that is, the promotion of that electron to the e_g manifold. This electronic transition would absorb electromagnetic radiation with energy equal to Δ_o. For an undistorted regular tetrahedral complex with a d^1 configuration, there is again a single transition from the *e* to the t_2 level. If there is no electron in the lower set of *d* orbitals, there can be no *d–d* absorption, as in the case of Ca^{2+}. Another case where no *d–d* absorption is observed is with Zn^{2+} or Cu^{+}. For these ions, which have $3d^{10}$ configurations, there is no "hole" in the filled upper level to accommodate an electron excited from the lower energy orbital. Hence, no *d–d* electronic absorption is possible. Thus electronic absorption spectroscopy, which is of great use in studying many other biologically important metal ions, is of limited use in studying Zn^{2+} or Ca^{2+} complexes.

In general, tetrahedral complexes absorb with more intensity than octahedral ones, albeit at a lower energy (longer wavelength) due to the lower energy of the *d–d* transitions ($\Delta_t < \Delta_o$) for a given ligand. The greater intensity of *d–d* transitions in tetrahedral versus octahedral complexes is due to differences in their symmetries.

In a strictly octahedral complex, the *d* orbitals retain their centers of symmetry because the ligand arrangement has inversion symmetry. Therefore, the ligands perturb the energies of the orbitals without perturbing their symmetries (i.e., they are still *gerade*). Electronic transitions between levels of the same symmetry are referred to as "Laporte forbidden". So why do we observe these low intensity *d–d* transitions in octahedral complexes at all? One reason is that the symmetry of the *d* orbitals is instantaneously distorted from pure octahedral when the ligands vibrate asymmetrically.

In contrast, the ligand arrangement in tetrahedral complexes does not have inversion symmetry, and the *d* orbitals consequently lose their inversion symmetries because of perturbation by the ligands. Thus *d–d* transitions in tetrahedral complexes are not Laporte forbidden, and they are therefore significantly more intense.

A good example of this difference in intensity can be seen in comparing solutions of $[CoCl_4]^{2-}$, which is tetrahedral, with those of $[Co(H_2O)_6]^{2+}$, which is octahedral. At the same concentrations of cobalt, the solution of the former complex is intensely blue, while that of the latter is a pale pink. The blue color is due to absorption in the orange part of the visible spectrum, and the pink from absorption in the green, a shorter wavelength. The difference in intensity, however, has nothing to do with the energies of the *d* orbitals, but instead is attributed to the presence of inversion symmetry in the pale octahedral complex and the absence of inversion symmetry in the deep blue tetrahedral complex.

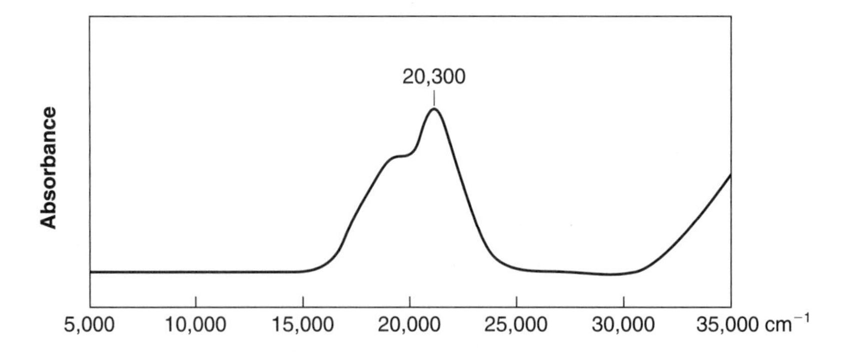

Fig. T.II.6.
Optical absorption spectrum of $[Ti(H_2O)_6]^{3+}$.

Multiple bands are observed when the symmetry of a coordination complex is distorted from regular tetrahedral or octahedral. A simple example is the complex $[Ti(H_2O)_6]^{3+}$, which has a $3d^1$ configuration. This complex is an axially compressed octahedron, and two electronic transitions are observed in the visible, one from d_{xy} to $d_{x^2-y^2}$ and one from d_{xy} to $d_z{}^2$ (Fig. T.II.6).

In addition to electronic transitions due to metal ion centered *d–d* transitions, transition metal complexes also can have electronic transitions in which electrons transfer from ligand to metal or vice versa. For example, a ligand-to-metal charge-transfer (LMCT) excitation arises from the excitation of an electron in a ligand-centered orbital into a *d* orbital on the metal ion. Such transitions typically give very intense absorptions, since there are often no symmetry restrictions involved in the process. The LMCT bands can vary from UV excitations for hard ligands (O,N) to visible excitations for soft ligands (e.g., S). In general, the lower the electronegativity and the greater the polarizability of the ligand, the less energy it takes to remove that electron from an orbital on the ligand. For example, $[Ti(H_2O)_6]^{3+}$ has an LMCT oxygen to titanium(III) transition in the UV, but complexes of Cu(II) or Fe(III) with sulfur ligands have intense LMCT transitions in the lower energy visible region of the spectrum. The colors of proteins containing iron–sulfur clusters are due to such LMCT absorption bands (see Chapter IV).

T.II.6.2. Paramagnetism

The magnetic properties of transition metal complexes depend on the number of unpaired electrons that reside in *d* orbitals, which in turn depends on the strength of the field created by the surrounding ligands (i.e., the magnitude of the *d*-orbital splitting). If the splitting Δ is greater than the energy required to pair electrons in a single orbital (i.e., the pairing energy), then the metal ion exists in what is called a "low-spin" state. On the other hand, if the pairing energy is greater than the splitting, then a "high-spin" state will occur. A good example compares two different octahedral complexes of Fe(III), both of which have $3d^5$ configurations. When complexed by six water molecules, the splitting constant for Fe(III) is smaller than the pairing energy, and each of the five *d* orbitals (t_{2g} and e_g) is populated with one electron yielding five unpaired electrons and a total spin of $\frac{5}{2}$ (each electron having spin $S = \frac{1}{2}$). In contrast, when Fe^{3+} is complexed by six cyanide ligands (strong field ligands), the splitting is larger than the pairing energy and all five electrons go into the t_{2g} orbitals. The complex has a spin of $\frac{1}{2}$, which is depicted in Fig. T.II.7. The general rule of thumb, then, for octahedral complexes, is that strong-field ligands lead to low-spin states, while weak-field ligands lead to high-spin states.

In the case of tetrahedral complexes, the small Δ_t splitting is always less than the pairing energy, and tetrahedral complexes are therefore all high spin.

d_{z^2} $d_{x^2-y^2}$

d_{xy} d_{xz} d_{yz}

Weak field $3d^5$
$S = \frac{5}{2}$
$[Fe(H_2O)_6]^{3+}$

d_{z^2} $d_{x^2-y^2}$

d_{xy} d_{xz} d_{yz}

Strong field $3d^5$
$S = \frac{1}{2}$
$[Fe(CN)_6]^{3-}$

E

Fig. T.II.7.
The effect of a strong-field ligand (CN^-) and weak-field ligand (H_2O) on *d* orbital splittings in octahedral Fe^{3+}. Note that the weak-field ligand produces a high spin while the strong-field ligand produces a low-spin configuration.

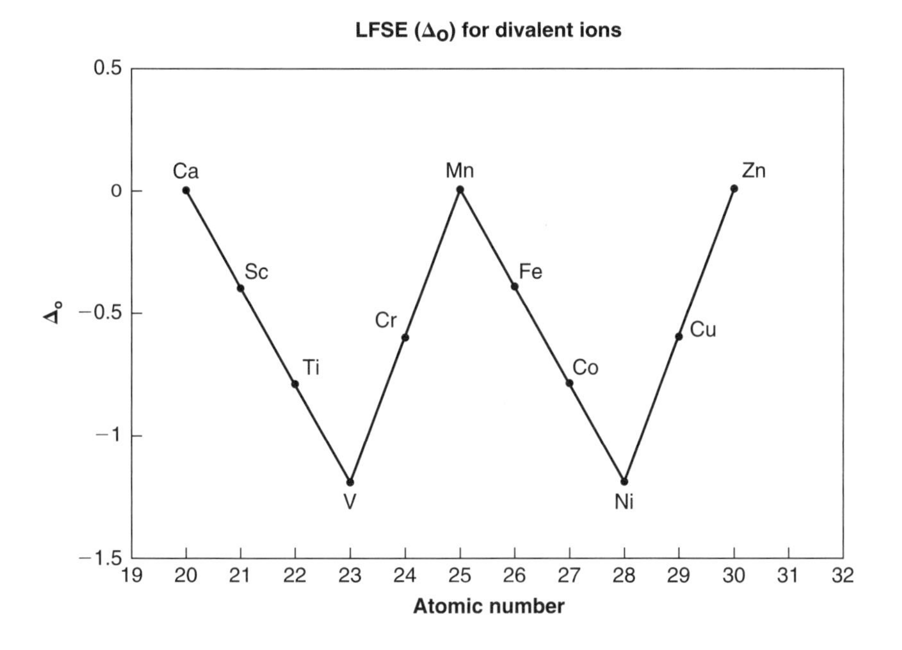

Fig. T.II.8.
Ligand-field stabilization energies for octahedral M^{2+} ions. The Δ_o values for transition elements Sc through Zn are for high-spin ions.

T.II.6.3. Ligand-Field Stabilization Energies and Periodic Properties

Ligand-field stabilization energy (LFSE) is a stabilization due to preferred population of the lower energy set of *d* orbitals. The magnitude of this stabilization is illustrated in Fig. T.II.8 for a series of high-spin octahedral complexes of divalent transition metals. The splitting is a periodic function of *d*-orbital occupancy. The LFSE is zero for Ca(II), because it has no *d* electrons and reaches a maximum for V(II), which has three electrons in the t_{2g} level. From Sc(II) to V(II), each added electron contributes to the stability of the complex. After this point, the next two electrons populate the e_g level, destabilizing the complexes. The Mn(II) ion has no LFSE, because the stabilization contributed by the three electrons in the t_{2g} is canceled by the two in the e_g. The periodic nature becomes apparent when additional electrons are added, again first stabilizing the complex by increasing the population in t_{2g} until another maximum LFSE is obtained for Ni(II). Further addition of electrons to the e_g again diminishes the LFSE until we reach the $3d^{10}$ configuration of Zn(II), which displays no LFSE at all.

A good example of LFSE is shown in Fig. T.II.9. The figure graphically depicts the heats of hydration for the first-row divalent and high-spin transition metal ions. These values represent the heat that is released when the metal ion is taken from the gas phase and is solvated by six water molecules in the presence of an excess of water.

$$M^{2+}(g) + H_2O(excess) \rightarrow [M(H_2O)_6]^{2+}$$

The general trend can be clearly seen as an increase in the amount of heat released from Ca(II) through Zn(II). The trend shows periodicity, as can be observed by the double humps. In fact, there are two contributions to these heats of hydration: one is due to LFSE, as indicated by the humps, and the second is a monotonic change that is the result of decreasing ionic radii with increasing atomic number (see Fig. T.II.10).

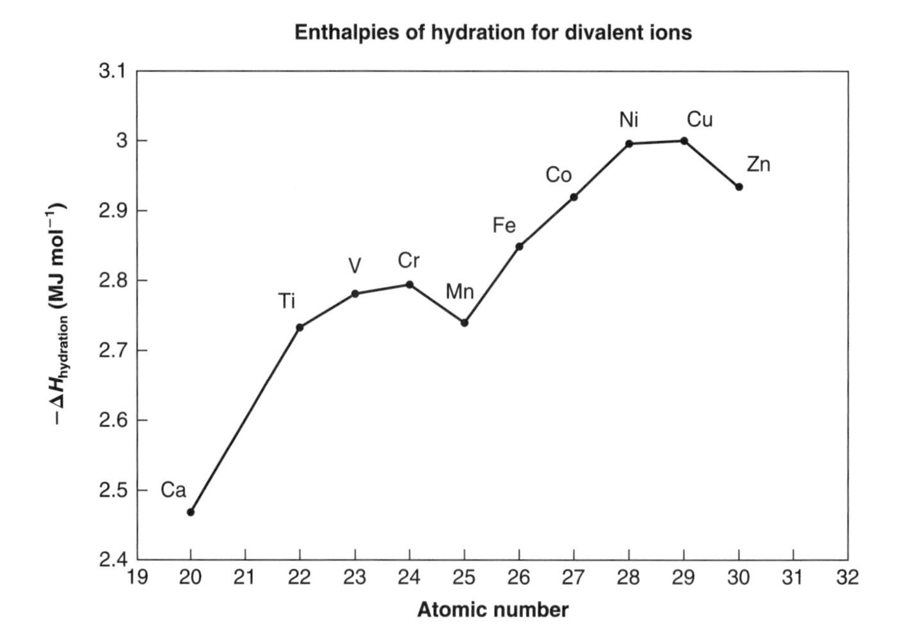

Fig. T.II.9. Enthalpies of hydration for divalent metal ions without correcting for ligand-field stabilization values.

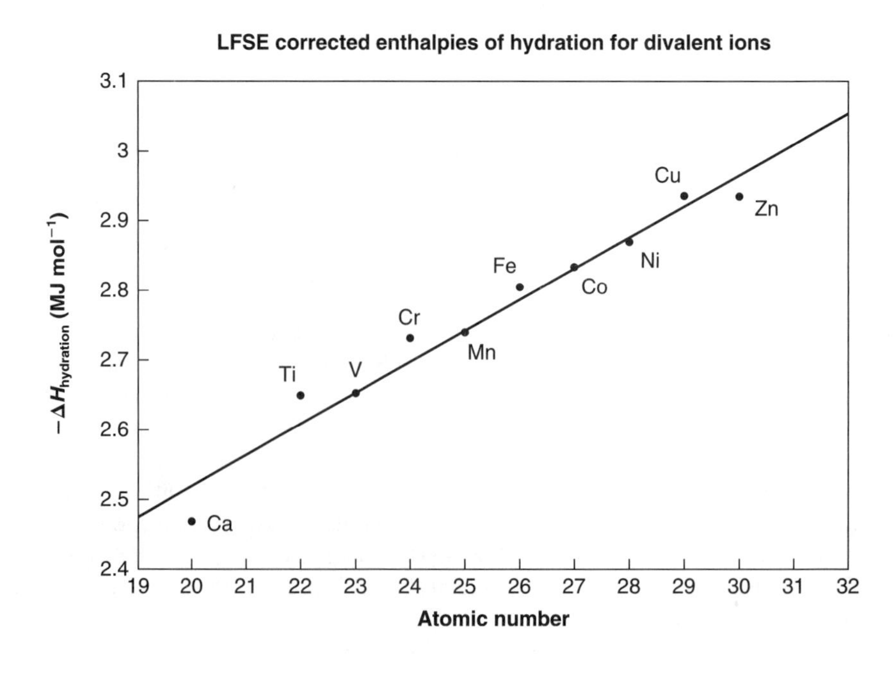

Fig. T.II.10. These enthalpies of hydration for divalent ions have been corrected by subtracting ligand field stabilization energies in Fig. T.II.8 from the enthalpies of hydration in Fig. T.II.9 (after the proper conversion of units).

This second phenomenon is almost linear and is the result of the decrease of the distance between ligand and the charged metal ion, which creates a stronger ineraction.

T.II.6.4. The Jahn–Teller Effect and Distorted Coordination Complexes

Some of the complexes that one might expect to be very symmetric in structure have been shown by physical methods (e.g., X-ray crystallography) to be distorted. Consider the following structures: $[NH_4]_2[Ni(OH_2)_6][SO_4]_2$ and $[Cu(OH_2)_6][ClO_4]_2$. In both cases, there are six identical H_2O ligands coordinated to the central metal ion, and an octahedral coordination might be expected. For the nickel(II) complex, this expectation is met. The six O–Ni bonds are identical in length and arranged symmetrically. However, this is not true for the copper complex, in which the axial bonds are elongated relative to the equatorial bonds, that is, the axial ligands have been pushed farther way from the metal ion. These distortions in this (and many other) Cu(II) complexes are the result of the Jahn–Teller effect.

Nickel(II) is a $3d^8$ transition metal ion with a $3d$ configuration of $(t_{2g})^6(e_g)^2$ depicted in Fig. T.II.11. Each of t_{2g} orbitals is fully populated, and there is one electron each in the d_{z^2} and the $d_{x^2-y^2}$ orbitals, as per Hund's rule. The configuration indicates that the electron cloud about Ni(II) is very symmetrical and suggests that all bond lengths should be equal.

In the case of Cu(II), another electron has been added to the $3d^8$ configuration to make it $3d^9$. The question now arises, however, into which orbital the electron should be placed: the d_{z^2} or the $d_{x^2-y^2}$? Both cases are depicted in Fig. T.II.11. By placing the additional electron into the $d_{x^2-y^2}$, the electron density along the x and y axes increases. Consequently, the axial bond lengths (along the z axis) are shorter than those in the xy plane. This compression also affects the ordering of the d orbitals, by lowering those having an x or y component and raising those having a z component. This finding is illustrated on the right side of Fig. T.II.11.

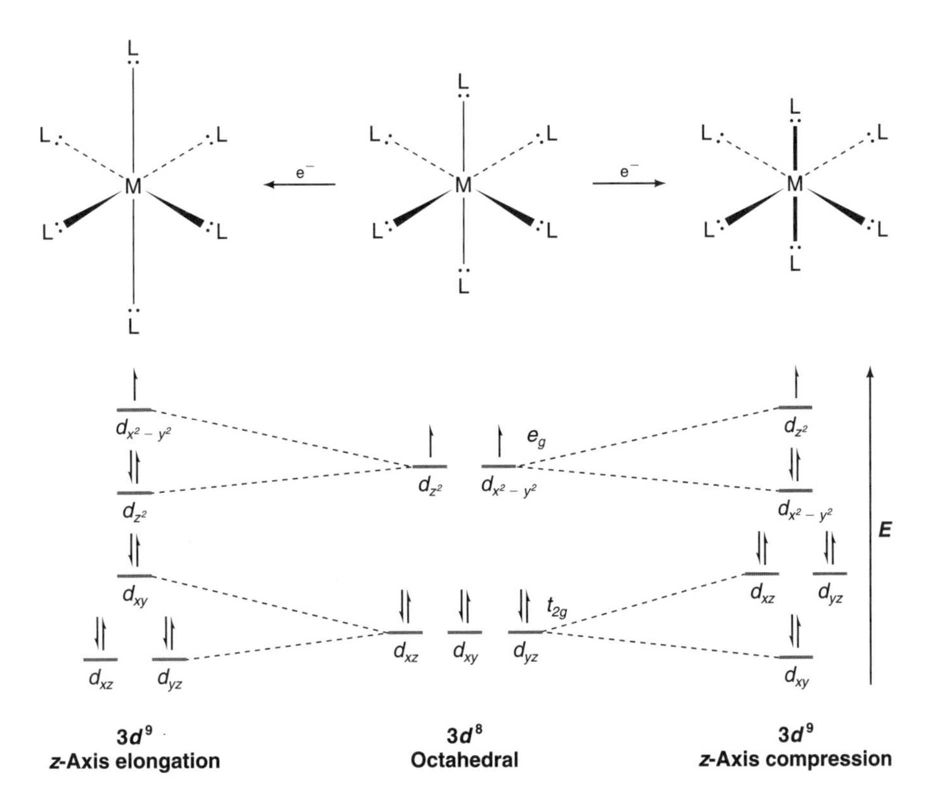

Fig. T.II.11. The addition of 1 electron to an octahedral d^8 ML_6 complex may result in axial elongation or axial compression of the d^9 ML_6 complex.

In the same way, the addition of an electron to the d_{z^2} orbital increases the electron density along the z axis, pushing the axial ligands away from the metal center. The final consequence is the lowering of the energy of orbitals having a z component, and the lifting of the others as depicted in on the left side of Fig. T.II.11. The Cu(II) complexes typically distort by elongation, as in the case $[Cu(OH_2)_6][ClO_4]_2$.

T.II.7. Kinetic Aspects of Metal Ion Binding

T.II.7.1. Ligand Exchange Rates

Table T.II.1 lists the water exchange rates for some metal ion aqua complexes. The lability of the $M{-}OH_2$ bond clearly varies greatly among the metal ions: from over 1 billion times per second for Ba^{2+} to the very low rate of $<10^{-8}\ s^{-1}$ for Rh^{3+}. In general, the magnitude of metal–ligand exchange rates follows the same trends as the aquo complexes listed in Table T.II.1.

The charge density on the metal ion is a significant contributor to the exchange rate. Typically, ligand exchange rates are greater for less highly charged M^{2+} ions than for M^{3+} ions. For example, the water exchange rate for the Fe^{2+} ion is more than 2×10^4 times greater than that for the Fe^{3+} ion.

The size of the ion is another contributor to its relative lability. Smaller diameter ions have high charge density, which tends to decrease exchange rates. Consider the trend of exchange rates for first-row divalent transition metal ions from Cr^{2+}, which has an exchange rate of $10^9\ s^{-1}$, to that of Ni^{2+} having a rate of $4 \times 10^4\ s^{-1}$. However, the trend is not smooth due to the effect of Jahn–Teller distortions that may lead to higher exchange rates. Compare the rate for Ni^{2+}, a $3d^8$ ion with that of Cu^{2+}, a $3d^9$ configuration. The exchange rate for the aqua complex of Cu^{2+} is more than four orders of magnitude greater than that for the Ni^{2+} complex. The explanation lies in the distortion of the aqua complex of Cu^{2+} by elongation along the z axis, which imparts greater lability to the elongated axial ligands, as compared with the equatorial. This distortion is attributable to the Jahn–Teller effect in the Cu^{2+} ion (see above). The Ni^{2+} ion undergoes no such distortion.

In the case where a metal ion is bound to a chelating or macrocyclic ligand such as porphyrin, the metal ion may have additional coordination positions not occupied by the chelating or macrocyclic ligand. The additional ligands at these "open" positions

Table. T.II.1.
Exchange Rates for Water Molecules in the First Coordination Sphere of Selected Metal Ions at 25°C

Ion	k_1 (s^{-1})	Ion	k_1 (s^{-1})	Ion	k_1 (s^{-1})
Li^+	4×10^8	V^{2+}	8×10^1	Sn^{2+}	$>10^4$
Na^+	7×10^8	Cr^{2+}	1×10^9	Hg^{2+}	4×10^8
K^+	1×10^9	Mn^{2+}	2×10^7	Al^{3+}	1
Be^{2+}	8×10^2	Fe^{2+}	4×10^6	Fe^{3+}	2×10^2
Mg^{2+}	6×10^5	Co^{2+}	3×10^6	Ga^{3+}	2×10^9
Ca^{2+}	3×10^8	Ni^{2+}	4×10^4	Gd^{3+}	2×10^9
Ba^{2+}	2×10^9	Cu^{2+}	1×10^9	Bi^{3+}	$>10^4$
		Zn^{2+}	2×10^7	Cr^{3+}	2×10^{-6}
				Co^{3+}	$<10^{-6}$
				Rh^{3+}	6×10^{-9}

frequently have substantial exchange rates, for example, axial ligands in metalloporphyrin complexes may exchange rapidly, while the chelating or macrocyclic ligands themselves exchange much more slowly, if at all.

T.II.7.2. Mechanisms of Exchange

There are two limiting mechanisms involved in ligand exchange: dissociative and associative. A dissociative mechanism is a first-order process in which the initial step is the loss of a ligand from the complex; it is analogous to an S_N1 reaction in organic chemistry. An associative mechanism is a second-order process in which the initial step is addition of a ligand; it can be compared to an S_N2 reaction. In simple complexes, the operating mechanism is often dictated by steric effects. Metal ion complexes having low coordination numbers ($n \leq 4$) are more likely to have room to accommodate an expanded first coordination sphere, and they typically exchange by an associative pathway. In contrast, those having high coordination numbers ($n \geq 6$) often exchange dissociatively.

T.II.8. Redox Potentials and Electron-Transfer Reactions

T.II.8.1. Redox Reactions

Oxidation is loss of one or more electrons and a species that loses one or more electrons has been *oxidized*. The species that accepts this electron(s) is reduced or has undergone *reduction*. A reaction in which one reactant is reduced while another is oxidized is referred to as a *redox* reaction.

Individual redox reactions can be split into two half-reactions, one for the oxidation and one for the reduction that together represent the overall reaction. The reduction of H^+ by metallic iron, for example, can be split into two half-reactions:

Overall redox reaction: $Fe^0 + 2\,H^+ \rightarrow Fe^{2+} + H_2(g)$

Oxidation half-reaction: $Fe^0 \rightarrow Fe^{2+} + 2\,e^-$

Reduction half-reaction: $2\,H^+ + 2\,e^- \rightarrow H_2(g)$

The potential for a redox reaction at standard conditions ($P = 1$ bar, $T = 25°C$, pH 0, and solids and solutes at unit activity) is $E°$. By convention, half-reactions are usually written as reduction reactions, with their potentials listed as reduction potentials, that is, with the electrons to the left of the arrow. When the reduction half-reaction is written as an oxidation (i.e., with the electrons on the right), the sign of $E°$ is reversed.

In order to be able to tabulate reduction potentials for half-reactions, the $E°$ for the standard hydrogen electrode or SHE, $2\,H^+ + 2\,e^- \rightarrow H_2(g)$, is arbitrarily set to zero, and $E°$ values for all other half-reactions are therefore measures of their ability to reduce hydrogen. Tables of standard reduction potentials report the $E°$ values of hypothetical galvanic cells that combine the SHE with the listed reduction half-reduction of interest.

The $E°$ of a redox reaction is related to the Gibbs standard free energy ($\Delta G°$) by the following equation:

$$\Delta G° = -n\mathscr{F}E°$$

where

$$\Delta G° = \Delta H° - T\Delta S°$$

$$\mathscr{F} = \text{Faraday constant} = 96{,}500\,\text{kJ V}^{-1}$$

$$n = \text{number electrons transferred in reaction}$$

A spontaneous or thermodynamically favorable reaction has a negative $\Delta G°$ and a positive $E°$ value for a spontaneous reaction at standard conditions. The potential for the reduction of H^+ by Fe, at standard conditions, is the difference of the two standard reduction potentials for the half reaction:

$$Fe^{2+} + 2\,e^- \rightarrow Fe^0 \qquad E^{\circ}_{Fe} = -0.44\,V$$

$$2\,H^+ + 2\,e^- \rightarrow H_2(g) \qquad E^{\circ}_{H} = 0\,V$$

$$Fe^0 + 2\,H^+ \rightarrow Fe^{2+} + H_2(g) \qquad E^{\circ}_{overall} = E^{\circ}_{H} - E^{\circ}_{Fe} = +0.44\,V$$

However, the positive $E^{\circ}_{overall}$ tells us that the reaction is thermodynamically favorable, but it is important to note that $E^{\circ}_{overall}$, like $\Delta G°$, contains no kinetic information. The rate of the reaction depends on the activation energy rather than the overall thermodynamic driving force, and so a highly favorable reaction with a high $E^{\circ}_{overall}$ (equivalent to negative $\Delta G°$) may nevertheless be very slow if the activation energy is high.

The Nernst equation is used to determine the redox potential of a reaction that occurs under all of the standard conditions except with different concentrations.

$$aA + bB \rightarrow cC + dD$$

$$E = E° - \frac{0.05915}{n} \log Q$$

where n is the number of electrons and Q is the reaction quotient:

$$Q = [C]^c[D]^d/[A]^a[B]^b$$

Tables of reduction potentials for biological molecules typically list values of $E^{\circ\prime}$, which is defined at pH 7 (rather than of $E°$, which is defined at pH 0). To convert one to another, the Nernst equation is used. One particularly important oxidizing agent in biology is dioxygen, O_2. The reduction potential $E°$ for dioxygen is 1.229 V. To convert $E°$ to $E^{\circ\prime}$, we take into account the difference in the concentration of H^+.

$$O_2 + 4\,H^+ + 4\,e^- \rightarrow 2\,H_2O \qquad E° = 1.229\,V$$

$$E^{\circ\prime} = E° - \frac{0.05915}{n} \log Q$$

$$E^{\circ\prime} = 1.229 - \frac{0.05915}{4} \log \frac{1}{[H^+]^4}$$

$$E^{\circ\prime} = 1.229 - \frac{0.05915}{4} \log 10^{28} = 0.815\,V$$

T.II.8.2. Electron-Transfer Reaction Mechanisms

For electron transfer between two coordination complexes in solution, two very different types of mechanisms have been demonstrated experimentally: *outer sphere* and *inner sphere*. The "sphere" refers to the first coordination sphere of ligands directly attached to the metal ion center. Outer-sphere electron-transfer reactions occur by the transfer of electrons from the reducing metal ion to the oxidizing metal ion through both intact first coordination spheres. In contrast, inner-sphere electron transfer occurs through a temporarily bridging ligand that is simultaneously bonded to both metal centers.

T.II.8.3. Outer-Sphere Mechanisms

Consider the following outer-sphere electron transfer (ET) reaction:

$$[Fe(H_2O)_6]^{2+} + [Fe^*(H_2O)_6]^{3+} \rightarrow [Fe(H_2O)_6]^{3+} + [Fe^*(H_2O)_6]^{2+}$$

This is the simplest type of electron-transfer reaction, often called a self-exchange reaction because the reactants and products are chemically identical. The outer-sphere electron-transfer mechanism is further discussed in Chapter X, Section 2.

T.II.8.4. Inner-Sphere Mechanisms

Unlike outer-sphere electron transfer, where no making or breaking of chemical bonds occurs, inner-sphere electron transfer involves the breaking and reforming of metal–ligand bonds. The first experimental demonstration of an inner-sphere electron-transfer mechanism was made by Henry Taube and co-workers using the exchange inert complex $[Co^{III}(NH_3)_5Cl]^{2+}$ reacting with the exchange labile complex $[Cr^{II}(H_2O)_6]^{2+}$.

$$[Co^{III}(NH_3)_5Cl]^{2+} + [Cr^{II}(H_2O)_6]^{2+} \rightarrow \underset{\mathbf{A}}{[(NH_3)_5Co^{III}Cl\cdots Cr^{II}(H_2O)_5]^{4+}} \rightarrow$$

$$\underset{\mathbf{B}}{\{[(NH_3)_5Co\cdots Cl\cdots Cr(H_2O)_5]^{4+}\}} \rightarrow \underset{\mathbf{C}}{[(NH_3)_5Co^{II}\cdots ClCr^{III}(H_2O)_5]^{4+}} \rightarrow$$

$$[(NH_3)_5Co^{II}]^{2+} + [ClCr^{III}(H_2O)_5]^{2+} \xrightarrow{H^+}$$

$$\underset{\mathbf{D}}{[Co^{II}(H_2O)_6]^{2+}} + 5\,NH_4^{+} + [Cr^{III}Cl(H_2O)_5]^{2+}$$

The electron-transfer reaction can be broken down into four steps: the formation of a precursor complex **A** upon collision; the formation of the bridged dinuclear activated complex **B,** where chloride ion is simultaneously bonded to both metal ions; formation of the successor complex **C** in which the electron has transferred from the Cr(II) center to the Co(III) center to give Cr(III) and Co(II); and finally, the dissociation of the product complexes (**D**). The Co(II) complexes are labile, that is, they exchange their ligands rapidly. The labile cobalt–amine complex therefore immediately hydrolyzes to form the hexaaqua cobalt(II) complex. The Cr(III) complexes, on the other hand, are inert and thus the $[Cr^{III}Cl(H_2O)_5]^{2+}$ that is formed remains unchanged. The transfer of the chloride ion from the inert Co(III) center to the inert Cr(III) center indicates that chloride was simultaneously bonded to both metal ions when the electron transfer took place.

When this reaction is run in the presence of free $^{36}Cl^-$, very little $[Cr^{III\,36}Cl(H_2O)_5]^{2+}$ is formed, further illustrating the inertness of the Co(III) starting complex and the Cr(III) product and also supporting the existence of an intermediate $[Cr–Cl–Co]^{4+}$ species that does not exchange with free chloride in solution.

Chloride, iodide, fluoride, and hydroxide are good bridging ligands and the electron-transfer rate constants are between 10^5 and $10^6\,M^{-1}s^{-1}$ for reactions of $[Cr^{II}(H_2O)_6]^{2+}$ with $[Co^{III}(NH_3)_5X]^{2+}$ for X = Cl, I, F, OH. Water is not a very good bridging ligand, and the reaction proceeds 10 million times slower, with a rate constant of $0.1\,M^{-1}s^{-1}$, between $[Cr^{II}(H_2O)_6]^{2+}$ and $[Co^{III}(NH_3)_5(H_2O)]^{3+}$. When $[Co^{III}(NH_3)_6]^{3+}$ is used in place of $[Co^{III}(NH_3)_5X]^{2+}$, electron transfer is very slow because no ligand bridging can occur at all. The ammonia has only one lone electron pair, which it uses for bonding to Co(III) and thus cannot act as a bridging ligand. The resulting electron-transfer reaction therefore occurs by an outer-sphere mechanism and is sluggish, happening 100 billion times more slowly, with a rate constant on the order of $10^{-5}\,M^{-1}s^{-1}$.

Bibliography

Basolo, F. and Johnson, R., *Coordination Chemistry,* Benjamin, Menlo Park, CA, 1964.

DeKock, R. L. and Gray, H. B., *Chemical Structure and Bonding,* University Science Books, Mill Valley, CA, 1989.

Lippard, S. J. and Berg, J. M., *Principles of Bioinorganic Chemistry,* University Science Books, Mill Valley, CA, 1994.

Riehens, D. T., *The Chemistry of Aqua Ions: Synthesis, Structure, and Reactivity: A Tour through the Periodic Table of the Elements,* John Wiley & Sons, New York, 1997.

Roat-Malone, R. M., *Bioinorganic Chemistry:* A Short Course, Wiley-Interscience, Hoboken, NJ, 2002.

Wilkins, P. C. and Wilkins, R. G., *Inorganic Chemistry in Biochemistry,* Oxford University Press, New York, 1997.

Winter, M. J., *d-Block Chemistry,* Oxford University Press, New York, 1994 (reprinted 1996 with corrections).

Abbreviations

ABC	ATP binding cassette
ABTS	2,2′-Azinobis-(3-ethylbenzothiazoline-6-sulfonic acid)
Ac	Acetyl
ACCO	1-Aminocyclopropane-1-carboxylate oxidase
Ace1	Activation of CUP1 expression
Acetyl-CoA	Acetyl coenzyme A
ACS	Acetyl-coenzyme A synthase
Acn	Aconitase
ACP	Amorphous calcium phosphate
ACV	δ-(L-α-aminoadipoyl)-L-cysteinyl-D-valine
Ad	Adenosine
ADH	Aldehyde dehydrogenase
AH	Acetylene hydratase
AdoCbl	Adenosylcobalamin
ADP	Adenosine diphosphate
AdoMet	S-Adenosylmethionine (also SAM)
AGAO	*Arthrobacter globiformis* amine oxidase
AHA	Acetohydroxamic acid
AIDS	Acquired immune deficiency syndrome
AMO	Ammonia monooxygenase
AO	Amine oxidase
AO	Ascorbate oxidase
AO	Atomic orbital
AOR	Aldehyde ferredoxin oxidoreductase
APS	Adenosine phosphosulfate
APX	Ascorbate peroxidase
ATP	Adenosine 5′-triphosphate
ATPase	Adenosinetriphosphatase
ATP syn	ATP synthase
AV	Areolar vesicles
bdt	Benzene-1,2-dithiolate
BLM	Bleomycin
BMA	Bis(methylamide)
BME	β-Mercaptoethanol
BMOV	Bis(maltolato)oxovanadium(IV)
Bn	Benzyl
bopta	(9*R*,*S*)-2,5,8-Tris(carboxymethyl)-12-phenyl-11-oxa-2,5.8-triazadodecane-1,9-dicarboxylate
BphC	Biphenyl 2,3-diol catechol dioxygenase
BPU	*Bacillus pasteurii*
bpy	2,2′-Bipyridine
4,4′-bpy	4,4′-Bipyridine
BSS	Bismuth subsalicylate
Bya	Billion years ago
CA	Carbonic anhydrase
CAcn	Cytosolic aconitase
CaMKK	Calmodulin-dependent kinase kinase
CAM-Tr_2C	C-terminal half of calmodulin
CaMKK	Ca^{2+}/calmodulin-dependent protein kinase
cAMP	Cyclic adenosine monophosphate
CAR	Carboxylic acid reductase
Cbl	Cobalamin
CBS	Colloidal bismuth subcitrate
Ccc2	Copper transport protein
CcP	Cytochrome *c* peroxidase (also CCP)
CCS	Copper chaperone
CD	Circular dichroism
cdta	1,2-Cyclohexanediaminetetraacetate
CEPT	Coupled electron and proton transfer
CHNOPS	C,H,N,O,P,S
cit	Citrate
CGMP	Cyclic guanosine-5′-monophosphate
CMP	Cytidine 5′-monophosphate
CoA	Coenzyme A
CODH	Carbon monoxide dehydrogenase
CoM	Coenzyme M
CoQ	Ubiquinone (also UQ)
COX	Cyclooxygenase
CP	Ceruloplasmin
Cp	Cyclopentadienyl
CPA	Carboxypeptidase A
CRD	Cysteine-rich domain
CRIP	Cysteine-rich intestinal protein

CTD	Catechol 1,2-dioxygenase
Cyt	Cytochrome
1,5-DClm	1,5-Dicyclohexylimidazole
2D	Two dimensional
3D	Three dimensional
5′-dA	5′-Deoxyadenosine
5′-dA•	5′-Deoxyadenosine radical
5′-dAdo	5′-Deoxyadenosine
dach	1,2-Diaminocyclohexane
DAG	Diacylglycerol
DAOCS	Deacetoxycephalosporin
DBH	Dopamine β-hydroxylase
DD	Diol dehydratase
DFP	Diferric peroxide
DFT	Density functional theory
DgAOX	*Desulfovibrio gigas* aldehyde oxidoreductase
DMB	Dimethylbenzimidazole
DMPS	2,3-Dimercapto-1-propanesulfonic acid
DMS	Dimethyl sulfide
DMSA	Dimercaptosuccinate
dmso (ligand)	Dimethyl sulfoxide
DMSO (solvent)	Dimethyl sulfoxide
DMSP	Dimethylsulfoniopropionate
DMT	Divalent metal transporter
DNA	Deoxyribonucleic acid
DO3A-butrol	1,4,7-Tris(carboxymethyl)-10-(1-(hydroxymethyl)-2,3-dihydroxypropyl)1,4,7,10-tetraazacyclododecane
DOE	Department of Energy
Dopa(ligand)	3-(3,4-Dihydroxyphenyl)alanine
DOTA	1,4,7,10-Tetraazacyclododecane-1,4,7,10-tetraacetate
dppe	1,2-Diphenylphosphinoethane
dps	DNA protection during starvation
DTPA	Diethylenetriaminepentaacetate
equiv	Equivalent
EDTA	Ethylenediaminetetraacetic acid
edta(ligand)	Ethylenediaminetetraacetato
EDTMP	Ethylenediaminetetramethylenephosphonate
ER	Endoplasmic reticulum
Er	Erythrocruorin
en	Ethylenediamine (ligand)
EXAFS	Extended X-ray absorption fine structure
EPR	Electron paramagnetic resonance
ENDOR	Electron–nuclear double resonance
ET	Electron transfer
ES	Enzyme substrate complex
ECAO	*Escherichia coli* amine oxidase
EAL	Ethanolamine ammonia lyase
FAD	Flavin adenine dinucleotide (oxidized form)
FADH•	Flavin adenine dinucleotide (radical form)
Fd	Ferredoxin
Fd_{ox}	Ferredoxin (oxidized)
Fd_{red}	Ferredoxin (reduced)
FDA	U S Food and Drug Administration
FDH	Formate dehydrogenase
Fdx	Ferredoxin
FeMoco	Iron–molybdenum cofactor
FhuA	Ferric siderophore receptor
Fld	Flavodoxin
F(M)DH	*N*-Formylmethanofuran dehydrogenase
FUR	Ferric uptake receptor
FMN	Flavin mononucleotide
FNR	Ferredoxin:NADP reductase
FOR	Formate ferredoxin oxidoreductase
FTIR	Fourier transform infrared
GABA	γ-Aminobutyrate
GAP	GTP activating protein
GAPOR	Glyceraldehyde-3-phosphate ferredoxin oxidoreductase
GC	Guanylate cyclase
GC	Gas chromatography
GdmCl	Guanidinium chloride
GI	Gastrointestinal
GIF	Growth inhibiting factor
GMP	Guanosine 5′-monophosphate
GSH	Glutathione
GTP	Guanosine 5′-triphosphate
^1H NMR	Proton (hydrogen) nuclear magnetic resonance
H_2ase	Hydrogenase
$H_2folate$	Dihydrofolate
HAO	Hydroxylamine oxidoreductase
HAP	Hydroxyapatite
Hb	Hemoglobin
Hc	Hemocyanin
Hcit	Citric acid
HEDP	Hydroxyethylidene-1,1-diphosphonate
HFE	Hemochromatosis iron overload protein (or gene)
HiPIP	High potential iron–sulfur protein
HIV	Human immunodeficiency virus
HLH	Helix–Loop–Helix
HMG	High mobility group
HNE	*trans*-4-Hydroxy-2-nonenal
HOMO	Highest occupied molecular orbital
HP-DO3A	Hydroxypropyl derivative of DOTA
13-HPOD	13-Hydroperoxy-9,11-(*Z,E*)-octadienoic acid
Hr	Hemerythyrin
HRP	Horseradish peroxidase
HXT	Hexose
IHP	Inositol hexaphosphate
Im	Imidazole
IMP	Inosine 5′-monophosphate
IPNS	Isopenicillin *N*-synthase
IR	Infrared
IRE	Iron-responsive element
IRPI	Iron regulatory protein
IRP2	Iron regulatory protein

JBU Jack bean urease

KAU *Klebsiella aerogenes urease*

LAC Laccase
LFSE Ligand-field stabilization energy
LHCP Light-harvesting chlorophyll protein
LIM Three proteins: Lin-II, Isl-1, and Mec-3
L-N_3 Hydrotris(3,5-dimethyl-1-pyrazolyl borate)
LUMO Lowest unoccupied molecular orbital

mAb Monoclonal antibody
madd 1,12-bis(2-hydroxy-5-methoxybenzyl)-1,5,8,12-tetraazadodecane
MAG Mercaptoacetylglycine
Mb Megabase
Mb Myoglobin
MCD Magnetic circular dichroism
MDA Malondialdehyde
MeCbl Methylcobalamin
metHb Met-hemoglobin
Met Methionine
MGD MPT guanine dinucleotide
MHC Major histocompatibility complex
MLCK Myosin light-chain kinase
MLCT Metal-to-ligand charge transfer
MMCoA Methylmalonyl-CoA
MMO Methane monooxygenase
MMOA Methane monooxygenase protein A
MMOH Methane monooxygenase hydroxylase
MMP Matrix metalloproteinase
mnt Maleonitrile-1,2-dithiolate
MO Molecular orbital
Moco Molybdenum cofactor
MP Metalloprotein
MPT Metal-binding pyranopterin ene-1,2-dithiolate (also molybdopterin)
MPTppG MPT guanine dinucleotide
MRE Metal-regulatory element
MRI Magnetic resonance imaging
mRNA Messenger ribonucleic acid
MS Methionine synthase
MT Microtubules
MT Metallothionein
MWC Monod–Wyman–Changeux model
Mya Million years ago
MyoHr Monomeric form of Hr

NAD^+ Nicotinamide adenine dinucleotide (oxidized form)
NADH Nicotinamide adenine dinucleotide (reduced form)
NAD(P) Nicotinamide adenine dinucleotide phosphate
NADPH Nicotinamide adenosine dinucleotide phosphate (reduced form)
NaR Nitrate reductase
NDO Naphthalene-1,2-dioxygenase
NHE Normal hydrogen electrode
NIH National Institutes of Health (US)
NIO Nitrite oxidoreductase
NIR Near-infrared
NiR Nitrite reductase
NMDA *N*-Methyl-D-aspartate
NMR Nuclear magnetic resonance
NMRD Nuclear magnetic relaxation dispersion
nNOS Neuronal NOS
NO Nitric oxide
NoR Nitric oxide reductase
NoS Nitrous oxide reductase
NOS Nitric oxide synthase
NTP Nucleotide triphosphate
Nuc Nucleoside

OCP Octacalcium phosphate
OEC Oxygen evolving complex
OM Outer membrane

P450 Cytochrome P450
P_i Inorganic phosphate
P-loop Walker A motif
PAL Peptidyl-α-hydroxyglycine-α-amidating lyase
PAPS Phosphoadenosine phosphosulfate
Pc Plastocyanin
PC Phytochelatin
PCD Protocatechuate 3,4-dioxygenase
pCdta 1,2-Propylenediaminetetraacetate
PDB Protein Data Bank
PET Positron emission tomography
PheH Phenylalanine hydroxylase
Phen 1,10-Phenanthroline
Pheo Pheophytin
PHM α-hydroxylating monooxygenase
PLP Pyridoxal 5′-phosphate
pMMO Particulate methane monooxygenase
PNP Peanut peroxidase
ppb Parts per billion
PPD Phenylphosphorodiamidate
ppmV Parts per million by volume
PQ Plastoquinone
PRC Photosynthetic reaction center
PSI Photosystem I
PSII Photosystem II
PUFA Polyunsaturated fatty acid
Py Pyridine

QH• Ubisemiquinone

RBC Ranitidine bismuth citrate
RC Reaction center
Red Reduced
RFQ Rapid freeze quench
RING Really interesting new gene
rms Root mean square
RNA Ribonucleic acid

RNR	Ribonucleotide reductase
ROS	Reactive oxygen species
rRNA	Ribosomal ribonucleic acid
RTPR	Ribonucleoside triphosphate reductase
RyR	Rianodine receptor
SA	Streptavidin
salen	Bis(salicylidene)ethylenediamine
SAM	S-Adenosylmethionine (also AdoMet)
SBL-1	Soybean lipoxygenase-1
SDV	Silica deposition vesicles
SHE	Standard hydrogen electrode
SIR	Sulfite reductase
sMMO	Soluble MMO
SOD	Superoxide dismutase
SOMO	Spin occupied molecular orbital (or singly occupied molecular orbital)
SOR	Superoxide reductase
SP	Spore photoproduct
SPECT	Single-photon emission computed tomography
SR	Sarcoplasmic reticulum
SR	Sulfite reductase
SRB	Sulfate reducing bacterium
TACN	1,4,7-Triazacyclononane
TauD	Taurine:α-ketoglutarate dioxygenase
TCA	Tricarboxylic acid
TCO	Terminal cytochrome oxidase
Tex	Texaphyrin
TFIIIA	Transcription factor IIIA
Tf	Transferrin
TfR	Transferrin receptor
THF	Tetrahydrofuran (solvent)
TIM	Triosephosphate isomerase
TMAO	Trimethylamine-*N*-oxide
TMC	1,4,8,11-Tetramethyl-1,4,8,11-tetraazacyclotetradecane
Tp	Hydridotris(pyrazolyl)borate monoanion
TP	Tunneling pathway
TPA	Tris(2-pyridylmethyl)amine
TPP	Tetraphenyl porphyrin
TPQ	2,4,5-Trihydroxyphenylalanine quinone
tpy	2,2′:6,2″-Terpyridine
UB	Uniform barrier
UQ	Ubiquinone
UV	Ultraviolet
WT	Wild type
XANES	X-ray absorption near-edge spectroscopy
XAS	X-ray absorption spectroscopy
XDH	Xanthine dehydrogenase
XO	Xanthine oxidase
ZRE	Zn responsive element

Glossary

A

Acetogens: Bacteria that obtain energy for cell growth by producing acetate as an end product, often using hydrogen and carbon dioxide as growth substrates.

Aconitase: The iron–sulfur enzyme that catalyzes the interconversion of citrate and isocitrate through the unsaturated intermediate aconitic acid.

Active site(s): The site(s) at which biological function is expressed; for example, in myoglobin, the active site for dioxygen binding is at the iron atom of the heme *b* group embedded in the protein.

Adenosylcobalamin (AdoCbl): The cofactor involved in B_{12}-catalyzed rearrangement reactions.

Adiabatic electron transfer: An electron-transfer reaction, characterized by strong electronic coupling between reactants and products, in which there is unit probability of an electronic transition from reactants to products at the transition state nuclear configuration.

Aerobe: A microorganism that can grow only when oxygen gas is present (also called an obligate aerobe).

Aerobic respiration: Respiration in which molecular oxygen is the terminal acceptor of electrons.

Allostery: A phenomenon whereby the conformation of an enzyme (or other protein) is altered by binding at a site other than the substrate-binding site, usually with a small molecule, referred to as an effector, which results in either decreased or increased activity of the enzyme.

Amine oxidase: An enzyme that catalyzes the two-electron oxidation of an amine by molecular oxygen to an aldehyde with concomitant production of hydrogen peroxide and ammonia.

Amorphous: Without shape; when applied to biominerals refers to noncrystalline phases, such as amorphous silica.

Anemia: A deficiency in the number of red blood cells or in their hemoglobin content.

Anaerobe: An organism that can only grow in the complete absence of O_2.

Angiogenesis: The process of vascularization of a tissue involving the development of new blood vessels.

Anoxic: A descriptor indicating the absence of oxygen.

Anoxygenic photosynthesis: Photosynthesis without water splitting and oxygen production.

Antimetastatic: Combating the spread of cancer from one part of the body to another by way of the lymphatic system or bloodstream.

Antioxidant: A low molecular weight molecule that serves to protect the cell from oxidative damage usually caused by dioxygen and reactive oxygen species (ROS) derived therefrom.

Antiporter: A cotransporter, where the cosubstrates are transported in opposite directions across a membrane.

Apoprotein: A protein from which the metal center or centers or other cofactors are missing or have been removed, generally by chemical means.

Apoptosis: Programmed cell death; an organized, controlled, cell dismantlement, often called cellular suicide, which is a critical part of development and is crucial to the pathologies of many diseases.

Aquation: Incorporation of water into the inner-coordination sphere of a metal complex.

Archaea: Microorganisms that constitute one of the three domains of life (along with bacteria and eukaryotes); like bacteria, they lack a nucleus and are classified as prokaryotes (singular, archaeon).

Astrobiology: The science of the study of planets and life, which includes the search for extraterrestrial life and the study of the origins of life on Earth.

ATP: Adenosine 5′-triphosphate; a nucleotide that is the principal energy carrier of cells.

ATPase: Adenosinetriphosphatase, an enzyme that hydrolyses ATP (in the presence of Na^+, K^+, or Mg^{2+}).

Autopoiesis: The ability of organisms to regenerate parts of themselves on a continuing basis.

Autosomal recessive gene: A eukaryotic gene wherein both copies must be at the same locus to express the full observable phenotype.

Autotroph: An organism capable of growth solely on inorganic molecules, for example, one using CO_2 as the sole source of carbon.

B

B-DNA: Right-handed double-helical DNA, containing 10 residues per turn, and a helix pitch of 34 Å.

Bacteroid: A modified bacterium that is no longer actively growing and that has become specialized for inhabiting the root nodules of leguminous plants as a symbiont, where it performs nitrogen fixation.

Bioavailability: The amount of an element in a particular environment that is usable by a living organism.

Biogenesis: The process of forming new structures in biology.

Biomineral: A hard structure that generally contains a particular phase of an inorganic solid as well as macromolecular components.

Biomineralization: The process by which living systems produce solid components, that is, biominerals.

Bipolar affective disorder: A mood disorder characterized by swings from mania (exaggerated feeling of well-being) to depression.

Blue copper protein: An electron-transfer protein containing a mononuclear copper center, where copper shuttles between Cu(I) and Cu(II); characterized by an intense absorption in the visible region ($\varepsilon > 3000\,M^{-1}\,cm^{-1}$ at 600 nm) arising from S(cysteine) $\rightarrow$ Cu(II) charge transfer.

Bohr effect: A heterotropic allosteric effect; for example, in hemoglobin, where dioxygen binding is affected by protons (pH).

C

Carcinoma: A form of cancer that develops in tissues covering or lining organs of the body, such as skin, uterus, lung, or breast.

Carriers (permeases): Transporters that bind their substrate on one side of the membrane, undergo a conformational change, and then release it on the other side.

Catalase: An enzyme that catalyzes the conversion of two molecules of H_2O_2 to O_2 and two H_2O molecules, with H_2O_2 acting as both oxidant and reductant.

Catalysis: The acceleration of a chemical transformation by entities (called catalysts, which in this book are mostly enzymes) that are not consumed in the reaction.

Catalytic nucleic acid: Nucleic acids (DNA or RNA) that perform catalytic functions.

Ccm genes: A pool of genes that, in Gram-negative bacteria and plant/protozoa mitochondria, encode for the proteins involved in cytochrome *c* biosynthesis.

Chagas' disease: An insect-transmitted parasitic disease.

Chaotropes: Small molecules or ions that disrupt protein structure.

Chemiosmotic hypothesis: The explanation, first offered by Peter Mitchell, for the generation of proton gradients across a cell membrane by coupling the pumping of protons to the flow of electrons from reductant to oxidant; the chemiosmotic hypothesis is a cornerstone of bioenergetics.

Chemolithotropy: Growth by organisms on inorganic components with no input from organic components or light (as opposed to heterotrophy or phototrophy).

Channel: A pore-forming membrane protein that allows movement of substrate along a concentration gradient.

Chlorin: A heterocyclic ring consisting of three pyrroles and one reduced pyrrole coupled through four methine linkages; a chlorin is largely aromatic, although not conjugated through the entire circumference of the ring.

Chlorophyll: A magnesium-containing chlorin.

Chloroplast: An organelle found in green plants wherein the light and dark reactions of photosynthesis take place.

Classical zinc finger: A structural zinc-binding domain with a Cys_2His_2 metal-binding site, as in transcription factor IIIA (TFIIIA).

Clathrin: The protein that coats cellular membrane pits and vesicles.

Cofactor: A non-protein group that along with the protein is essential to the function of a particular enzyme; cofactors may be covalently or noncovalently bound, and extrinsic (e.g., flavin nucleotides, NAD derivatives) or intrinsic (e.g., the Tyr-Cys redox cofactor in galactose oxidase).

Cofactor biosynthesis: The protein-dependent process leading to the *in vivo* assembly of a cofactor.

Compound I: The reaction product formed during peroxidase or catalase turnover with peroxides; both oxidizing equivalents of the peroxide are stored in the enzyme—one as Fe^{4+} and one as a free radical.

Compound II: The one-electron reduction of Compound I gives Compound II.

Conserved motif: An amino acid sequence in a protein that has remained essentially unchanged during evolution.

Coordination: Binding interactions, for example, between a metal ion and its ligands.

Cooperativity: An allosteric effect in which the binding of a ligand to one subunit influences the binding of that ligand to other subunits; for example, in hemoglobins and hemocyanins, cooperativity is positive, that is, affinity for ligand increases as more is bound.

Copper A (Cu_A): The dinuclear copper center of the electron-transfer domain of cytochrome *c* oxidase.

Copper nitrite reductase: The non-heme subset of dissimilatory nitrite reductases (NiRs), which use copper centers in catalysis.

Corrin: A tetrapyrrole ring in which three pyrrole moieties are joined on opposite sides by a C-CH_3 methylene link, while two of the pyrroles are joined directly; the ligand in the cobalt-containing vitamin B_{12}.

Crystal structure: The three-dimensional structure of a crystalline protein determined by X-ray diffraction analysis, wherein the electron density is revealed; the strength of the X-ray scattering is proportional to the number of electrons around an atom, and hydrogen atoms are usually not observed in protein X-ray structures.

Consensus sequence: An amino acid sequence common to a class of sites that may be identified in a particular family or in several families of proteins; genome searching for consensus sequences for metal-binding sites has become a powerful tool in metallobiochemistry.

Cu_Z: The tetranuclear copper center that constitutes the site at which nitrous oxide is reduced to dinitrogen by nitrous oxide reductase.

Cu-MT: The copper derivative of the cysteine-rich metallothionein (MT) protein.

Cubane: The structural motif of the 4Fe-4S cluster, constituted by four iron and four sulfide ions.

Cupredoxin: A common fold for electron-transfer copper domains, typically with seven strands organized in two sheets according to a Greek-key motif.

Cyanobacteria: Prokaryotic organisms that carry out oxygenic photosynthesis using photosystems that are very similar to those found in higher plants and green algae. (Formerly known as blue-green algae or blue-green bacteria.)

Cytochromes: Heme proteins that act as electron carriers by shuttling between ferrous iron(II) and ferric iron(III) in their active sites, with irons usually bound to two axial ligands.

Cytokine: Intercellular chemical messenger protein released by white blood cells.

Cytosolic domain: An independently folded unit within a protein located in the cytoplasm of the cell (in comparison, e.g., to a transmembrane unit, which is embedded in a cellular membrane).

D

Dcytb: The Fe^{3+} reductase of duodenal iron-absorbing cells.

Dehydrogenase: An enzyme that catalyzes a reaction involving oxidation and reduction that typically uses a nicotinamide cofactor as the electron carrier.

Demetalated: A protein or other biological construct from which the metal ions have been removed (see, Apoprotein).

Denitrification: The anaerobic use by bacteria of nitrogen oxide molecules and ions, such as nitrate and nitrite, as terminal electron acceptors instead of oxygen, leading ultimately to the production of nitrous oxide (N_2O) or dinitrogen (N_2).

Deoxy: Prefix denoting unligated functional state of an oxygen-binding or activating protein; for example, deoxyhemerythrin.

Deoxyribozyme: DNA that can perform enzymatic functions; also called β-catalytic DNA, DNA enzymes, or DNAzymes.

Dephosphorylation: Removal of a phosphate group from a functional unit of a protein nucleic acid, sugar, or polysaccharide.

Diamond: Trivial name for the structural motif of the 2Fe-2S cluster, constituted of two iron and two sulfide ions.

Diazotroph: An organism that can fix (reduce) atmospheric N_2 as the sole source of nitrogen for growth.

Dielectric constant (ε): The attractive force between opposite charges is reduced ε-times with respect to vacuum (where $\varepsilon = 1$); the dielectric constant is 80 for H_2O, and varies greatly within the nonuniform medium of a protein, although the average protein dielectric constant is estimated to be very low compared to H_2O.

Diferric peroxo (DFP): A reaction intermediate formed in proteins when two ferrous ions in proximal protein-binding sites react with dioxygen to form two ferric ions bridged by peroxo; proposed as an intermediate in ferritin mineralization, methane monooxygenase (MMO), and other di-iron metalloproteins.

Dismutation: See, Disproportionation.

Disproportionation: A reaction in which a reagent reacts with itself to produce two new species, one in higher oxidation state and one in lower oxidation state [e.g., $2\ Cu^{(I)} \rightarrow Cu^{(II)} + Cu^{(0)}$]; also called dismutation.

Dissimilatory nitrate reduction: The energy-yielding reduction of nitrate to nitrite and gaseous nitrogen-containing products.

Distal histidine: In heme proteins, the histidine that is in close proximity to the ligand-binding site.

Distance decay constant (β): In electron-transfer reactions, the exponential decay constant describing the decrease of rates with increasing donor-acceptor separation.

DNAzymes: Enzymes that use DNA (instead of protein or RNA) as their macromolecular component. (See Deoxyribozyme.)

Domain: In proteins, a recognizable part of a polypeptide chain that has a distinct tertiary stucture and, often, a distinct function.

E

EF Hand: A small protein domain (usually ~40 residues long, although multiple domains are found in some proteins) that binds calcium for functional and/or structural purposes.

Electron carrier: A protein or molecule that can accept and donate electrons upon interaction with other proteins or enzymes.

Electronic coupling: In electron-transfer reactions, the interaction between the electronic states of the reactants and products at the transition state nuclear configuration; characterized by the matrix element, H_{AB}, which is equal to half the energy splitting at the avoided crossing between reactant and product potential energy surfaces.

Electron tunneling: Nonadiabatic electron transfer, in which the electron goes 'through' rather than over the barrier.

Electron paramagnetic resonance (EPR): A spectroscopic technique based on resonance absorption of radiation by molecules undergoing transitions between electronic spin states split in a magnetic field; used to detect and characterize paramagnetic species, such as metal ions and free radicals, by their g-values and hyperfine splittings.

Encephalopathy: Disease of the brain.

Endonuclease: A nuclease that hydrolyzes the nucleic acid backbone within the chain, but not from the 3′ or 5′ ends.

Endopeptidase: An enzyme that mediates hydrolytic cleavage of the peptide bonds in a protein polymer from within the polymer, but *not* the C- or N-terminal peptide bonds.

ENDOR: Electron–nuclear double resonance; a magnetic resonance technique in which the electron spin is monitored as the nuclear spin states are perturbed at radio frequencies to detect those nuclei that are coupled to the electron spin.

EPR: See, Electron paramagnetic resonance.

Eukaryote: A cell that contains distinct subcellular particulate bodies called organelles, including a true organized nucleus in which its DNA resides; compare with prokaryotes, which do not have their DNA in such visible organized structures.

EXAFS: See, Extended X-ray absorption fine structure.

Exogenous ligand: A non-protein ligand of a metal ion in a metalloprotein.

Exonuclease: A nuclease that hydrolyzes the nucleic acid backbone from either the 3′- or 5′-end, removing the terminal nucleotides consecutively.

Exopeptidase: An enzyme that mediates hydrolytic cleavage of the C- or N-terminal peptide bond in a protein polymer.

Extended X-ray absorption fine structure (EXAFS): The region of the X-ray absorption spectrum beyond the near edge in which the amplitudes and frequencies of sinusoidal oscillations yield information about: the type of atoms surrounding the absorbing atom; the number of atoms bound; their distances from the absorber; and their disorder or temperature factor.

Extremophiles: Microorganisms that grow under extreme conditions of temperature, pressure, pH, and so on.

Extrinsic proteins: Water-soluble proteins that bind to intrinsic membrane proteins.

F

Facultative anaerobe: An organism that is capable of growth both in the presence or absence of O_2; for example, *Klebsiella pneumoniae* is a facultative anaerobe that can only fix N_2 in the absence of O_2.

Fermentation: Anaerobic biodegradation (catabolism) of organic substrates, such as carbohydrates, for the purpose of supplying energy for microbial growth.

Ferredoxin: A low molecular weight protein (often <10 kDa) that contains an iron–sulfur cluster (one or more of: [2Fe–2S], [4Fe–4S], or [3Fe–4S], or combinations thereof) and functions in transferring electrons, usually one at a time, between redox-active proteins and enzymes.

Ferric reductase: An enzyme that reduces Fe^{3+} to Fe^{2+} making iron more bioavailable.

Ferrochelatase: A protein that catalyzes the insertion of ferrous iron into the porphyrin macrocycle.

Ferroportin1: A membrane iron (Fe^{2+}) exporter (alternative names, IREG1, MTP1).

Ferrous oxidase: An enzyme that oxidizes Fe^{2+} and is coupled with other iron-transport proteins.

Flavodoxin: A flavin-containing one- or two-electron carrier that can function as an electron donor, for example, to the Fe protein of nitrogenase.

Flavin mononucleotide (FMN): A common organic redox cofactor in biological systems

FMN: See, Flavin mononucleotide.

Free radical: Any molecule containing an unpaired electron in its valence shell.

G

Genome sequence: The complete linear nucleotide sequence of all of the DNA contained within an organism.

Greenhouse effect: The effect that atmospheric gases including CO_2, CH_4, N_2O, and others have on warming the Earth by their transparancy to visible and UV light and absorption of IR radiated from the surface.

Green-sulfur bacteria: Anoxygenic phototrophic bacteria that oxidize sulfide.

Guanylate cyclase (GC): A heme-containing enzyme that, upon NO binding to heme, is activated to convert guanosine triphosphate (GTP) to cyclic guanosine monophosphate (GMP); GC is one of the primary targets for NO.

H

Hb: See, Hemoglobin.

Hc: See, Hemocyanin.

Hem-: A prefix, denoting blood, as in *hem*oglobin, *hem*ocyanin, or *hem*erythrin.

Heme: An iron-containing porphyrin prosthetic group; specific forms (named a, b, c, etc.) differ in the chemical nature of the peripheral substituents on the porphyrin skeleton.

Heme a: A noncovalently bound heme found in cytochrome *c* oxidase.

Heme b: The most common type of heme, iron protoporphyrin IX, which is *not* covalently bound to the protein.

Heme c: The most common type of heme that is covalently bound to the protein.

Heme d_1: The heme in the active site of iron nitrite reductases, which are nitrite reductases (NiRs) that use heme centers in catalysis.

Hemerythrin (Hr): A dioxygen-binding protein, whose active site has an asymmetric triply bridged di-iron motif wherein dioxygen binds to one iron.

Hemocyanin (Hc): A dioxygen-binding protein whose active site has an approximately symmetrical dicopper motif.

Hemoglobin (Hb): Heme-containing dioxygen carrier, which binds dioxygen cooperatively at the iron of the embedded heme b group.

Hemoglobin family: The entire family of multimeric and monomeric heme-containing dioxygen-binding proteins that share the globin fold.

Heteroatom: Any atom other than carbon or hydrogen in a molecular structure; in protein crystallography, the term "heteroatom" refers to an atom that does not normally occur in a polypeptide.

Heterocyst: A cell, found in certain filamentous cyanobacteria, that is specialized for nitrogen fixation.

Heterolytic bond cleavage (bond heterolysis): Referring to a bond cleavage process in which an electron pair is divided unequally between products with one product getting both electrons and the other product none.

Heterotroph: An organism (sometimes called an organotroph) that obtains energy and carbon for growth from complex organic compounds.

Heterotropic allosteric effector: A molecule that binds to a protein and alters the binding of another unlike molecule ligand at a different site; for example, in hemoglobin, 2,3-

biphosphoglycerate binds and reduces the affinity of hemoglobin for dioxygen.

Histocompatability: Immunological tolerance to transplanted tissue.

HIV: Human immunodeficiency virus, the retrovirus that causes AIDS.

Hodgkin's disease: Malignancy (cancer) of lymphoid tissue found in lymph nodes, the spleen, the liver, and bone marrow.

HOMO: The _h_ighest _o_ccupied _m_olecular _o_rbital in a molecule.

Homolytic bond cleavage (bond homolysis): Referring to cleavage of the A–B bond into separated A and B, with each atom retaining one of the two electrons of the original bond.

Homotropic allosteric effector: A ligand which, when bound to one or more subunits, alters the binding (kinetics and thermodynamics) of the same ligand to other subunits: For example, in hemoglobin, binding of dioxygen to a subunit increases the affinity of remaining unligated subunits of the tetramer for dioxygen.

Hydrogen tunneling: A process whereby hydrogen travels through rather than over an energy barrier.

Hydrogenase: Any one of a number of metalloproteins (either iron or nickel-iron containing) that are either bidirectional (both evolve H_2 and take up H_2) or unidirectional (take up H_2 to form protons and electrons).

Hydrogen atom abstraction: A reaction in which a hydrogen atom (one proton + one electron) is removed from a molecule.

Hydrogen cycle: The biogeochemical cycle, carried out mainly by microorganisms, in which H_2 is produced by fermentation and/or phototrophy and consumed by a variety of organisms as a source of reducing equivalents to supply energy for growth.

Hydratase: Enzyme that inserts a water molecule into its substrate.

Hydrolase: An enzyme that mediates hydrolysis (splitting by water) of a chemical bond.

Hydrosphere: The part of the Earth that consists of bodies of water including oceans, lakes, rivers, aquifers, and a variety of smaller compartments.

Hypercalcemia: Abnormally high levels of calcium in the blood.

Hyperthermophiles: Microorganisms with an optimal growth temperature $> 80°C$.

Hypotensive: Having abnormally low blood pressure.

I

Inorganic sulfide: The S^{2-} ligand found in Fe-S proteins, which is acid labile, that is, easily liberated and detected as H_2S gas upon treatment of the protein with acid.

Intrinsic proteins: Proteins that are associated with the lipid bilayers of membranes; usually only soluble in detergents.

Inverted region: In electron-transfer reactions, the regime in which rates decrease with increasing reaction driving force; in semiclassical theory, the inverted region occurs when the driving force ($-\Delta G°$) is greater than the reorganization energy (λ).

Iron–sulfur cluster: A structural motif found in iron metalloproteins, where the coordination of two or more metal ions is accomplished through bridging inorganic sulfide ions.

Iron–sulfur proteins: Electron-transfer proteins and enzymes in which one or more of the metal cofactors consists of clusters of iron and inorganic sulfide ions (see, iron–sulfur cluster).

Isomerase: An enzyme that mediates conversion of a substance into an isomeric form.

K

Kinetic isotope effect: The effect on the rate of a reaction by substituting one isotope of an element for another.

L

Labile ligand: Undergoing facile exchange with solvent or other ligands.

Laue diffraction: Use of multiple wavelength radiation for X-ray diffraction studies; especially using polychromatic sources of X-rays, available at synchrotrons, wherein time-resolved crystallography from milli- to picosecond time scales is possible.

Legume: The collective name for a large family of dicotyledonous plants that form nitrogen-fixing root (or infrequently stem) nodules in symbiosis with rhizobia bacteria.

Ligand: A donor that binds to an acceptor, which is often a metal ion.

Ligase: An enzyme that mediates synthesis of a large molecule from two smaller ones by forming a covalent bond.

LUMO: The _l_owest _u_noccupied _m_olecular _o_rbital in a molecule.

Lyase: An enzyme that mediates nonhydrolytic addition or removal of functional groups; for example, cytochrome *c* heme lyase, found in fungi, invertebrates, and mitochondria, catalyzes the attachment of heme c to the protein chain.

Lysosome: A subcellular compartment, delineated by a lipid bilayer, that maintains a low pH ($\sim$4.5) and proteolytic enzymes to effect the destruction and recycling of old proteins and mitochondria in the cell.

M

Magnetic resonance imaging (MRI): A magnetic resonance technique in which high spatial resolution of proton NMR yields detailed images that provide information on morphology, pathology, and functional localization.

Magnetotactic bacteria: Anaerobic or microaerophilic microorganisms that sense the Earth's magnetic field, usually by using magnetite biomineral constructs, and move along the lines of force of that field.

Met: A prefix denoting an oxidized, nonfunctional state of an oxygen-binding protein; for example, in *met*hemocyanin the copper ions are in the Cu^{II} state, which does not bind dioxygen. (Met is also the three-letter abbreviation of the amino acid methionine.)

Metalloregulatory proteins: Proteins that transduce metal ion signals into changes in gene or protein expression.

Methane monooxygenase (MMO): The enzyme that converts methane to methanol, which is the first step in the oxidative metabolism of methane by methanotrophs.

Methylcobalamin (MeCbl): The cofactor involved in methyl-transfer reactions.

Metallochaperone: A protein whose role is to deliver metal ions to specific metal-requiring enzymes.

Metallothionein (MT): A sulfur (cysteine)-rich protein that can bind Zn, Cu, and/or Cd.

Methanogens: Archaea that obtain energy for cell growth by producing methane gas, often using hydrogen and carbon dioxide as growth substrates.

Mitochondria: Organelles found in aerobic eukaryotic organisms responsible for most of the synthesis of ATP by making use of redox (respiratory) chains (singular, mitochondrion).

Mitochondrial complex I: NADH-quinone oxidoreductase; one of three energy-transducing enzyme complexes of the mitochondrial respiratory chain, which catalyzes the reaction:

$$NADH + Q + H^+ + nH_{in}^+ \rightarrow NAD^+ + QH_2 + nH_{out}^+$$

$$(Q = \text{quinone};\ n = 3\text{–}5)$$

Motility: The ability of organisms or parts of organisms to move.

Morphogenesis: The formation of shapes and structures; in biomineralization, the process that leads to the specific shape of biomineralized structures such as shells, spikes, bones, and teeth.

Multiplicity: The number of substates in a given electronic configuration; for example, the degeneracy of the electron spin system with spin quantum number S is $(2S + 1)$.

Mutagenesis: The alteration of a gene either by natural causes or intentionally (e.g., as in site-directed mutagenesis) to produce a new gene that (often) codes for a new protein that differs from the original due to the gene alteration.

MWC Model: A model for cooperativity first proposed by Monod, Wyman, and Changeux in which an oligomeric protein has two (or more) distinct quaternary structures, each with its own affinity for ligand.

Myoglobin (Mb): Monomeric heme-containing dioxygen-binding protein, which in vertebrates stores dioxygen and regulates nitric oxide levels.

N

NADH: Reduced nicotinamide adenine dinucleotide.

Nanoreactor: A protein structure with physical dimensions in the nanometer range, where biochemistry and/or inorganic chemistry occur.

Neurotransmitter: A small molecule, such as acetyl choline or dopamine, responsible for the transmission of a nerve impulse across a synapse.

Nicotinamide adenine dinucleotide: An organic cofactor that uses a pyridine derivative as a two-electron, one-proton carrier (hydride equivalent) [abbreviated as NAD(H) or NADP(H)].

Nitrate reductase (NaR): An enzyme that catalyzes the reduction of nitrate (NO_3^-) to nitrite (NO_2^-); different types of nitrate reductase initiate the nitrate assimilation pathway and the denitrification pathway, but both are molybdenum cofactor enzymes.

Nitrification: The oxidative process that microorganisms use to convert ammonia (NH_3) into nitrite (NO_2^-) and/or nitrate (NO_3^-).

Nitrite reductase (NiR): A respiratory enzyme that catalyzes the reduction of nitrite (NO_2^-) to nitric oxide (NO) in the dissimilatory pathway; in contrast, one-electron assimilatory nitrite reductase catalyzes the six-electron reduction of nitrite to ammonia/ammonium.

Nitric oxide (NO): A short-lived diatomic gas that is an intermediate in the denitrification pathway and serves as a potent signaling molecule in the cardiovascular, neuronal, and immune systems.

Nitric oxide synthase (NOS): The enzyme responsible for the production of nitric oxide.

Nitric oxide reductase (NoR): An enzyme that catalyzes the reduction of nitric oxide (NO) to nitrous oxide (N_2O).

Nitrogen mustard: A toxic bifunctional alkylating agent, resembling mustard gas in structure, which is important in cancer treatment.

Nitrogenase: A two component MgATP-utilizing metalloenzyme system that catalyzes the reduction of atmospheric N_2 to ammonia.

Nitrogen cycle: The natural cycle by which nitrogen passes from the atmosphere to organisms (soils and waters) and back to the atmosphere, driven by a series of redox processes catalyzed by metalloenzymes in a variety of organisms. The main inorganic pathways are nitrogen fixation, nitrification, nitrate assimilation, and denitrification.

Nitrogen fixation: The conversion of atmospheric N_2 through natural or synthetic processes to form ammonia and/or ammonium ion, which are readily utilizable by living things.

Nitrophorin: A heme protein in blood-sucking insects that binds NO and releases it to dilate victim blood vessels to facilitate blood flow.

Nitrous oxide reductase (NoS): The respiratory enzyme that catalyzes the reduction of nitrous oxide (N_2O) to dinitrogen (N_2) in the dinitrification pathway.

Nodule: A swelling on a legume plant root that is induced through colonization by nitrogen-fixing bacteria, which form a symbiosis with the plant.

Nodulation: The process by which certain bacteria induce the formation of nodules on the roots of their specific host plants.

Nonadiabatic electron transfer: An electron-transfer reaction in which there is extremely low probability of an electronic transition from reactants to products at the transition state nuclear configuration; electron transfer is facilitated by quantum mechanical tunneling.

Nuclease: An enzyme that mediates the hydrolysis of the phosphodiester backbone of a nucleic acid.

Nuclear magnetic resonance (NMR): The spectroscopic technique in which strong magnetic fields are used to split nuclear spin states that can then be interrogated with radio frequency radiation to yield information about the chemical surroundings of the nucleus in question.

Nucleotide: A molecule that consists of a nitrogenous base (adenine, cytosine, guanine, thymine, or uracil), a sugar (ribose or 2′-deoxyribose), and phosphates; and forms the monomeric unit of DNA and RNA.

Nuclease: An enzyme that mediates the hydrolysis of the phosphodiester backbone of a nucleic acid.

Nucleosynthesis: The process by which the chemical elements of life are synthesized in the Big Bang, by nuclear fusion in stars, and in supernova explosions.

O

Organometallic: Possessing a metal–carbon bond.

Ortho- and para-hydrogen: Nuclear spin isomers of diatomic hydrogen molecules; ortho-hydrogen has parallel nuclear spins of ($I = \frac{1}{2} + \frac{1}{2} = 1$) and para-hydrogen has antiparallel spins ($I = \frac{1}{2} - \frac{1}{2} = 0$) and the ortho- to para-hydrogen conversion involves a forbidden triplet-to-singlet transition.

Osteoblast: A bone-forming cell.

Osteoclast: A cell that breaks down and resorbs bone material.

Oxidases: Enzymes that use molecular oxygen as their oxidant, but do not incorporate either of the O_2 atoms in the products (the oxygen in the products coming from water).

Oxidative phosphorylation: Redox-linked synthesis of ATP from ADP and P_i.

Oxidoreductase: An enzyme that catalyzes a reaction involving oxidation and reduction. Many enzymes in the oxidoreductase family have more specialized names.

Oxy: A prefix denoting a dioxygen-bound state; for example, oxyhemoglobin.

Oxygenases: Enzymes that use molecular oxygen as their oxidant and incorporate one (monooxygenase) or both (dioxygenase) atoms of oxygen into their products.

P

Paramagnetic: Containing unpaired electrons.

Peptidase: An enzyme that mediates hydrolytic cleavage of a peptide bond, usually in a protein polymer.

Periplasm: The contents of the periplasmic space, the space between the inner- and outer-cell membranes in Gram-negative bacteria (cells that are not stained by the Gram technique), which contains proteins secreted by the cell.

Permeases (Carriers): Transporters that bind their substrate on one side of the membrane, undergo a conformational change, and then release it on the other side.

Peroxidases: Enzymes that use the two oxidizing equivalents in H_2O_2 to oxidize a variety of biological molecules.

Peroxynitrite ($ONOO^-$): A potent cytotoxic anion resulting from the reaction of nitric oxide (NO) and superoxide (O_2^-).

Phosphorylation: Attachment of a phosphate group to a functional unit of a protein, nucleic acid, sugar, or polysaccharide.

Photodynamic therapy (PDT): Treatments using drugs that become active when exposed to light.

Phototroph: An organism that obtains energy from light.

Phototrophic: A descriptor for organisms that use light as an energy source.

Photosynthesis: The use of light energy to drive the conversion of CO_2 into carbon compounds utilized for growth.

Photosystem: A membrane protein complex containing chlorophyll and other redox cofactors that effects the conversion of light energy into oxidation–reduction reactions.

Plastoquinone: Lipid-soluble *p*-benzoquinone carrier of electrons that facilitates electron transfer in oxygenic photosynthesis.

Pi-acceptor ligand (π-acid ligand): A ligand that can form a π bond to a metal by accepting electrons donated by the metal.

Pi-donor ligand: A ligand that can form a π bond by donating electrons to a metal.

Polymerase: An enzyme that catalyzes the formation of a nucleic acid polymer by condensing nucleotide building blocks to form a series of phosphate diester links.

Polymorph: A particular phase of a solid-state material; for example, the polymorphs of calcium carbonate include calcite, aragonite, and vaterite.

Polymorphonuclear leukocyte: A type of white blood cell.

Porphyrin: A heterocyclic aromatic ring made from four pyrrole units joined on opposite sides through four methine links, which coordinates metal ions by the four pyrrole nitrogen atoms in the central "hole" of the molecule.

Primary structure: The amino-acid sequence of a protein.

Prokaryote: Cells that are distinguished by the absence of any membrane-bounded subcellular compartments (i.e., organelles), most notably a true nucleus, and usually has its DNA present as a single molecule.

Prosthetic group: A non-protein, low molecular weight "helper" group (such as heme in hemoglobin) that is not part of the primary product of gene expression, but is recruited into proteins to play a major functional role (i.e., the protein is generally inactive without the prosthetic group).

Proton pump: A membrane-based system that transports protons from one side of a membrane to the other against a proton concentration gradient with energy supplied by electron-transfer chains associated with the membrane.

Protoplasm: Ensemble of cellular contents.

Protoporphyrin IX: The porphyrin moiety of heme b.

Proximal histidine: In the heme protein family, the absolutely conserved histidine coordinated to the iron atom of the heme group on the opposite side of the substrate binding site.

Psoriasis: Chronic skin condition characterized by inflamed, red, raised areas that develop silvery scales.

Pump: A transporter that uses the direct input of energy, usually in the form of ATP, to transport an ion across a membrane against its concentration gradient.

Purple sulfur bacteria: Anoxygenic phototrophic bacteria that oxidize reduced sulfur compounds and store sulfur.

Pyrrole: A five-membered aromatic ring, four elements of which are carbon atoms and one is a nitrogen atom; molecular formula C_4H_5N.

Q

Quaternary structure: The noncovalent assembly of subunits.

R

Radioimaging: Diagnostic image produced by monitoring the nuclear decay of radioactive isotopes in specific chemical compounds that are administered to the patient; Tc^{99m} is commonly used in radioimaging.

Reaction center: Smallest unit of a photosystem that is capable of catalyzing a charge-separation reaction upon light absorption.

Reduction potential: A quantity that measures (in volts) the tendency of an element, a compound, or an ion to acquire electrons.

Reorganization energy: In electron-transfer reactions, the extent of nuclear reorientation that accompanies the movement of charge is characterized by a reorganization energy parameter, λ, defined as the energy required to rearrange the nuclei of the reactants into the equilibrium configuration of the products *without* transferring the electron.

Respiration: The process by which energy is generated by the oxidation of organic substrates using membrane associated electron-transfer chains; dioxygen is the terminal electron acceptor (oxidant) in aerobes, but anaerobes or facultative anaerobes can use nitrate, nitrite, sulfate, Fe(III), and other electron acceptors.

Rhizobia: The name for several genera of bacteria that have the ability to colonize and invade the roots of legumes and produce root nodules.

Ribonucleotide reductase: Metalloenzymes found in all living organisms that catalyze the reduction of ribonucleotides to deoxyribonucleotides, the precursors to DNA.

Ribosome: The functional unit of protein synthesis, composed of a large complex of RNA and protein molecules within the cell.

Ribozyme: An RNA molecule that performs enzymatic functions; also called: catalytic RNA, RNA enzyme, or RNA-zyme.

Reactive oxygen species (ROS): Biologically damaging oxygen-derived compounds, such as hydrogen peroxide, superoxide, and hydroxyl radical.

Root effect: In hemoglobins, a very strong heterotropic allosteric effect caused by protons (especially large in fish).

R-state: The quaternary structural form of hemoglobin that has high relative affinity for dioxygen and other ligands that bind at the heme.

S

Schiff base: The reaction product between a carbonyl group and an amine, which contains the –C=N– functional group.

Secondary structure: Structural motifs such as β sheets and α helices formed from primary sequences.

Septic shock: Potentially lethal drop in blood pressure due to the presence of bacteria in the blood (a common complication of severe burns and abdominal wounds).

Siderophore: A low molecular weight iron(III)-binding compound excreted by microorganisms to facilitate iron acquisition.

Singlet: A nondegenerate spin system having the electron spin quantum number $S = 0$.

Site-bound outer-sphere binding: An interaction between a metal ion and a nucleic acid wherein metal ions bind to specific sites of nucleic acids without forming direct bonds with nucleic acid functional groups (i.e., by forming an outer-sphere complex.)

Site-bound inner-sphere binding: An interaction between a metal ion and a nucleic acid wherein at least one metal-bound water is replaced by a ligand from the nucleic acid.

Solvent accessibility: In a given protein conformation, that part of the van der Waals surface of the amino acid residues and/or prosthetic groups that is exposed to the solvent surrounding the protein.

Solution structure: The 3D structure of a protein in solution, generally determined by NMR.

SOMO: The singly occupied molecular orbital in the electronic structure of a free radical or transition metal ion.

Soret band: The large spectroscopic feature of the absorption spectra of hemes, occurring in the region of 400–450 nm.

Sulfate reducing bacteria (SRBs): Bacteria that perform sulfate respiration under anaerobic conditions producing hydrogen sulfide as the reduced sulfur product.

Sulfate respiration: Respiration in which sulfate is used as the acceptor of electrons and converted to sulfide.

Superoxide: The ion O_2^-.

Superoxide dismutase (SOD): The enzyme that catalyzes the reaction of two equivalents of the ion O_2^- to form O_2 and H_2O_2; the four types of SOD currently known contain, respectively: Cu and Zn; Mn; Fe; and Ni as their active-site metal ions.

Superoxide reductase: The enzyme catalyzing the reduction of O_2^- to H_2O_2, which has so far been found in certain anaerobic bacteria and only known to contain Fe as its active-site metal.

Symbiosis: An intimate association of two dissimilar organisms that is mutually beneficial.

Symporter: A co-transporter where both substrates are transported in the same direction across a membrane.

Synapse: The gap wherein a nerve impulse transfers from one neuron to another, or from a neuron to another cell, often mediated by neurotransmitter molecules.

T

Tertiary structure: Three-dimensional conformation of a polypeptide.

Tetrapyrroles: A class of molecules containing four rings of the pyrrole type, generally linked by single-atom bridges between positions adjacent to the nitrogen atom of the five-membered pyrrole rings.

Thalassemia: An inherited form of anemia caused by faulty synthesis of hemoglobin.

Transcriptional activator: A protein that activates the transcription of DNA to RNA by binding to a specific regulatory sequence in a eukaryotic cell.

Transcriptional repressor: A protein that shuts down the transcription of DNA to RNA by binding to a specific regulatory sequence in a eukaryotic cell.

Trans effect: The effect of a coordinated ligand upon the *rate of substitution* of ligands opposite ($\sim 180°$) to it.

Transferase: An enzyme that mediates the transfer of a group from one molecule to another.

Transmembrane segment: A region in a protein sequence that traverses a membrane.

2,4,5-Trihydroxyphenylalanine quinone (TPQ): The protein-derived cofactor of copper-containing amine oxidases.

Triplet: A triply degenerate spin system, $S = 1$, having three-fold multiplicity.

T-state: The quaternary structural form of hemoglobin that has relatively low affinity for dioxygen and other ligands that bind at the heme.

Tunneling pathway: In nonadiabatic electron-transfer reactions, the series of interatomic contacts (that may include covalent and nonbonded interactions) mediating the electronic interaction between the electron donor and acceptor.

Tunneling timetable: In electron-transfer reactions, a plot of electronic coupling limited electron-transfer time versus donor–acceptor distance.

U

Uniporter: A carrier that simply facilitates diffusion of a substrate across a membrane along its concentration gradient.

Urea: A common metabolite and environmental component of formula NH_2CONH_2.

Urease: The nickel enzyme that catalyzes the hydrolysis of urea to form an ammonium ion and carbamic acid, the latter subsequently hydrolyzing to a bicarbonate and ammonium ion.

V

Vasodilator: A drug that widens blood vessels through the relaxation of smooth muscle, the process being called vasodilation.

X

X-ray Absorption Near-Edge Structure (XANES): That part of the X-ray absorption spectrum beyond the absorption edge, but before the onset of EXAFS (extended X-ray absorption fine structure).

X-linked: Designation of a genetic variation caused by mutations in the X chromosome.

Y

Y_Z, Y_D: Redox-active tyrosine residues.

Z

Zinc finger (classical): A structural zinc-binding domain with a Cys_2His_2 metal-binding site, as in transcription factor IIIA (TFIIIA).

APPENDIX III

The Literature of Biological Inorganic Chemistry

Research in biological inorganic chemistry now permeates numerous areas of science. The growing body of "bioinorganic" work is now published in a variety of specialized journals as well as in classic chemistry, biology, and (bio)physics journals of all kinds.

The brief survey and compilation below is meant to highlight where new results, perspectives, overviews, and reviews of biological inorganic chemistry can be found. Papers with bioinorganic themes are published in increasingly broad areas, ranging from genomics, proteomics, and neuroscience, to astrobiology, environmental science, and biotechnology. The journals and books below present an introduction to the scope of the literature of inorganic biochemistry.

- **Hot items in bioinorganic chemistry are often published in the top general science journals including**

  Nature
  Science
  Cell
  Proceedings of the National Academy of Sciences, USA (PNAS)

- **Journals that predominantly publish papers focusing on bioinorganic chemistry include**

  Journal of Biological Inorganic Chemistry (JBIC)
  Journal of Inorganic Biochemistry (JIB)
  BioMetals

- **Major chemistry journals that publish important bioinorganic articles include**

  Journal of the American Chemical Society (JACS)
  Angewandte Chemie International Edition (English)
  Chemical Communications
  Inorganic Chemistry
  Dalton Transactions
  European Journal of Inorganic Chemistry

- **Major review/perspective journals that publish bioinorganic articles include**

  Accounts of Chemical Research
  Chemical Reviews
  Trends in Biochemical Sciences
  Current Opinion in Chemical Biology
  Current Opinion in Structural Biology
  Chemical Society Reviews

- **Most biology/biochemistry journals publish bioinorganic chemistry articles (representative of the pervasiveness of this area in modern biology). Examples include**

  Journal of Biological Chemistry
  Biochemistry
  Nature Structural Biology
  Journal of Molecular Biology
  Biochimica et Biophysica Acta
  Biochemical and Biophysical Research Communications (BBRC)
  Archives of Biochemistry and Biophysics
  FEBS Letters
  FEBS Journal

- **Bioinorganic-related microbiology is often found in**

  Applied and Environmental Microbiology
  Archives of Microbiology
  FEMS Letters
  Geomicrobiology
  Molecular Microbiology

- **More specialized material with a bioinorganic flavor can be found in many journals, a few of which are**

  Origins of Life and Evolution of Biospheres
  Astrobiology
  Global Biogeochemical Cycles
  Environmental Science and Technology
  Bioconjugate Chemistry
  Journal of Medicinal Chemistry

- **Review series in book form include**

Metal Ions in Biological Systems, H. Sigel et al. Eds., Vols. 1–42, Marcel Dekker, NY.

Advances in Inorganic Biochemistry, Elsevier Biomedical, [New volumes not published since 1995: about 10 volumes in print.]

Progress in Inorganic Chemistry, Vols. 1–52, John Wiley and Sons, New York. [Published regularly with key articles covering bioinorganic chemistry.]

Advances in Inorganic Chemistry (continues *Advances in Inorganic and Nuclear Chemistry*), Vols. 1–59, Academic Press, New York [Published regularly with key articles covering bioinorganic chemistry.]

- **Texts**

Stephen J. Lippard and Jeremy M. Berg, *Principles of Bioinorganic Chemistry,* University Science Books, Mill Valley, CA (1994).

Wolfgang Kaim and Brigitte Schwederski, *Bioinorganic chemistry: inorganic elements in the chemistry of life: an introduction and guide,* John Wiley & Sons, New York (1994).

R. J. P. Williams and J. J. R. Fraústo da Silva, *The Natural Selection of the Chemical Elements: The Environment and Life's Chemistry,* Oxford University Press, Oxford, UK (1995).

James A. Cowan, *Inorganic Biochemistry: An Introduction,* 2nd ed., Wiley-VCH, New York (1997).

R. J. P. Williams and J. J. R. Fraústo da Silva, *Bringing Chemistry to Life: From Matter to Man,* Oxford University Press, Oxford, UK (1999).

J. J. R. Fraústo da Silva and R. J. P. Williams, *The Biological Chemistry of the Elements: The Inorganic Chemistry of Life,* 2nd ed. Oxford University Press, Oxford, UK (2001).

Rosette M. Roat-Malone, *Bioinorganic Chemistry: A Short Course,* John Wiley & Sons, Inc., Hoboken, NJ (2002).

- **Review volumes with collections of articles on biological inorganic chemistry include**

A. Messerschmidt, T. Poulos, and K. Wieghardt, Eds., *Handbook of Metalloproteins,* Vols. I, II, III, John Wiley & Sons, Inc., New York (2001).

I. Bertini, A. Sigel, and H. Sigel, Eds., *Handbook on Metalloproteins,* Marcel Dekker, New York (2001).

L. Que, Jr., Ed., *Physical Methods in Bioinorganic Chemistry: Spectroscopy and Magnetism,* University Science Books, Mill Valley, CA (2000).

J. Reedijk and E. Bouwman, Eds., *Bioinorganic Catalysis,* 2nd Ed., Marcel Dekker, NY (1999).

I. Bertini, H. B. Gray, S. Lippard, and J. Valentine, Eds., *Bioinorganic Chemistry,* University Science Books, Mill Valley, CA (1994).

- **Special issues of journals specifically covering topics in biological inorganic chemistry include**

Bioinorganic Enzymology, *Chemical Reviews,* Vol. 96(7), 2239–3042 (1996). (Assembled by Richard H. Holm, Pierre Kennepohl, and Edward I. Solomon).

Special Section on Bioinorganic Chemistry, *J. Chem. Soc. Dalton Trans.* 3903–4126 (1997). (Assembled by C. David Garner).

Forum on Metalloprotein Folding, *Inorganic Chemistry* 43(25), 7893–7960 (2004). (Assembled by Kara L. Bren, Vincent L. Pecoraro, and Harry B. Gray; Editor, Richard Eisenberg).

Special Issue on Bioinorganic Chemistry, *Proceedings of the National Academy of Sciences, USA (PNAS)* Volume 100(7), 3562–3840 (2003). (Assembled by Kenneth N. Raymond and Jack Halpern).

Biomimetic Bioinorganic Chemistry, *Chem. Rev.* 104(2), 347–1200 (2004). (Richard H. Holm and Edward I. Solomon, Guest Editors).

APPENDIX IV

Introduction to the Protein Data Bank (PDB)

Many of the illustrations in this text display protein structures. Published protein structures are archived in the Protein Data Bank (PDB), where they are accessible to anyone. The PDB site has the URL: http://www.rcsb.org/pdb, with several mirror sites around the world. This site allows access to all crystallographic- and NMR-derived structural information about proteins. At the site, you can download (for free) a variety of programs (e.g., Rasmol, Swiss Protein Database Viewer) that allow you to view and manipulate protein structures on your own computer. You can choose the mode of display (e.g., ball and stick, space filling, ribbons), the color scheme, and so on. Once you download a structure and have the software working, you can rotate images and zoom in so you can get a good idea of the three-dimensional structure of the protein and its active site.

For every protein structure displayed in this text, the PDB ID (four-character code) for accessing that structure is given in the figure caption. Each time you encounter a new structure in this volume you should visit the PDB, enter the four-character code, download the coordinates, and manipulate the image to create renditions that bring out its key points. Viewing complex structures in an interactive mode on your computer is far superior to staring at one, or even, several views in a text. You will find the PDB to be an extremely powerful resource for understanding structural aspects of metallobiochemistry.

The PDB is also a searchable database. A query at the PDB site returns a list of all entries matching the search parameters. Query options allow retrieval of a single structure or multiple structures (e.g., a protein class) and you may search using author names. Advanced queries allow you to customize the search and look for specific ligands and/or prosthetic groups, for structures obtained only by a specific technique (e.g., NMR), or for structures with a given resolution. If only one structure matches the query (for example, when a specific PDB ID is input), a summary Information page is displayed. The same summary page is accessed when clicking on one of the entries from the results list in the case of broader searches.

The summary page provides links to sequence and geometrical data, information on the presence of unusual amino acids, metal cofactors, ligand molecules, and other experimental details. From the summary page, you can access the structural classification of the protein at the bottom of the page. The "Download" link allows you to download PDB files in a variety of formats, in compressed or uncompressed form. The "Display Molecule" link displays the primary sequence, while the "View Structure" page links to various modes of visualization as well as to some still images.

PDB files contain additional information associated with each structure in addition to coordinates. In the header of the file (i.e., the part that precedes the coordinates) experimental details, relevant literature references, primary, and secondary structure information are provided together with the details of the non-polypeptide part of the molecule (ligands, metal ions, special cofactors, etc.). Metal ions and metal-containing cofactors are usually labeled HETATM (heteroatoms) rather than ATOM.

The best way to learn the PDB is simply to go to the site and start looking at structures. The tremendous amount of information available in a protein structure cannot fit in a research paper or textbook. Therefore, routine use of the PDB and other computer tools for visualizing and analyzing structures is now the norm. Learning to use the PDB in conjunction with this text is a prerequisite to getting the most out of the book and to learning how modern biological inorganic chemistry is pursued.

Useful online tutorials that facilitate search and structural analysis are provided directly at the PDB website.

Index

C

D

E

F